Microscopy of Semiconducting Materials 1991

Microscopy of Semiconducting Materials 1991

Proceedings of the Institute of Physics Conference
held at Oxford University, 25–28 March 1991

Edited by A G Cullis and N J Long

S
M M
VII

Institute of Physics Conference Series Number 117
Institute of Physics, Bristol, Philadelphia and New York

CODEN IPHSAC 117 1–801 (1991)

British Library Cataloguing in Publication Data

Institute of Physics Conference (1991 Oxford)
Microscopy of semiconducting materials 1991.
— (Institute of Physics conference series)
I. Title   II. Cullis, A.G.   III. Long, N.J.
IV. Series
537.6

ISBN 0-85498-406-2

Library of Congress Cataloging-in-Publication Data are available

Conference Co-Chairmen
A G Cullis and N J Long

Honorary Editors
A G Cullis and N J Long

Scientific Sponsors
The Institute of Physics
The Royal Microscopical Society
The Materials Research Society

This work relates to Department of the Navy Grant N00014-91-J-9008 issued by the Office of Naval Research European Office. The United States has a royalty-free license throughout the world in all copyrightable material contained herein.

Published under The Institute of Physics imprint by IOP Publishing Ltd
Techno House, Redcliffe Way, Bristol BS1 6NX, England
335 East 45th Street, New York, NY 10017-3483, USA
US Editorial Office: 1411 Walnut Street, Philadelphia, PA 19102, USA

Printed in Great Britain by J W Arrowsmith Ltd, Bristol

Preface

This volume contains the invited and contributed papers presented at the conference on 'Microscopy of Semiconducting Materials' which took place at Oxford University on 25–28 March 1991. The event was organised with scientific sponsorship by the Electron Microscopy and Analysis Group of the Institute of Physics, the Royal Microscopical Society and the Materials Research Society. This conference was the seventh in the series which focuses on the most recent developments in the microscopical studies of semiconductors and delegates from 20 countries gave a comprehensive account of work progressing at the current state-of-the-art across the whole field.

Advanced electronic materials incorporate ever more complex, high-tolerance structures to yield required electrical and/or optical performance. The microscopy techniques arrayed at this conference represent, in many cases, the only methods available to determine materials characteristics with the necessary precision. For example, high resolution microscopy using any of the transmission, scanning transmission, scanning tunnelling and atom probe techniques now routinely yields atomic-scale structural and compositional information. Small-scale optical and electrical properties can be probed with a variety of advanced scanning techniques and these, together with the very wide range of standard characterisation approaches, provide a highly detailed understanding of semiconductor behaviour. The 160 papers presented at this conference cover the broad spectrum of in-depth basic and applied research which relies directly upon the integrated application of microscopy techniques.

Each camera-ready paper submitted for publication in this volume was reviewed by two referees and modified accordingly. The editors are most grateful to the following scientific referees for their rapid and meticulous work:

P D Augustus, L J Balk, U Bangert, S J Barnett, P E Batson, G R Booker, K Bowen, C W T Bulle-Lieuwma, A Cerezo, H Cerva, D Cherns, J-P Chevalier, K Durose, D J Eaglesham, F Glas, P J Goodhew, J P Gowers, M Hockley, D B Holt, R Hull, C J Humphreys, J L Hutchison, P Kightley, J E Macdonald, S Mahajan, S McKernan, T E Mitchell, A G Norman, S J Pennycook, D D Perovic, P Pirouz, P Pongratz, F W Schapink, R Sinclair, D J Smith, A E Staton-Bevan, D J Stirland, J Thibault, A De Veirman, C A Warwick, G C Weatherly, B Wessels, R H Williams, P R Wilshaw and A C Wright.

The conference organisers are especially pleased to acknowledge the financial sponsorship of the event provided by:

British Telecom Research Laboratories
Office of US Naval Research, European Office
Sharp Laboratories of Europe Ltd
Siemens AG Research Laboratories.

Special thanks are due to A Petford-Long (Oxford) and O D Dosser (RSRE) for correcting the proof copies of all individual manuscripts. Furthermore, the work of R Briant (Oxford) was of vital importance to overall conference arrangements, while Mrs D M Handley provided constant, efficient secretarial input.

July 1991   **A G Cullis**

N J Long

Contents

Section 3: Dislocations and grain boundaries

†Invited

Section 4: Processed silicon

†Invited

Section 5: Metal semiconductor contacts and silicides

†Invited

†Invited

†Invited

†Invited

Section 8: Quantum wells and superlattices

†Invited

Section 9: X-ray studies

Section 10: Advanced scanning microscopy techniques

†Invited

Inst. Phys. Conf. Ser. No 117: Section 1
Paper presented at Microsc. Semicond. Mater. Conf., Oxford, 25–28 March 1991

Characterisation of compound semiconductors by high resolution electron microscopy

David J Smith and Ping Lu*

Center for Solid State Science and Department of Physics, Arizona State University, Tempe, Arizona 85287, USA.
*Present Address: Department of Mechanics and Materials Science, Rutgers University, Piscataway, New Jersey 08855, USA.

ABSTRACT: Recent advances in high-resolution electron microscopy have made it possible to extract atomic-level details about defects, interfaces and surfaces in compound semiconductors. The chemical composition of dislocation cores can be deduced by careful matching of experimental micrographs with image simulations, provided that the crystal polarity has first been independently determined. Interfacial quality in multilayer materials can be determined by imaging at optimal values of objective lens defocus and specimen thickness. The nature of surface reconstructions induced by cleaning under ultrahigh vacuum conditions can be established with assistance from image simulations and prior knowledge of the crystal polarity. High-resolution studies of compound semiconductors in our laboratory and elsewhere are reviewed.

1. INTRODUCTION

The electronic properties of devices based on compound semiconductors depend to a large extent on the local atomic structure at defects and interfaces. Moreover, the surface geometry has a critical influence on surface reactions, in particular the initiation of epitaxial growth processes. The advent of high performance, intermediate voltage electron microscopes has greatly increased the level of detail which is visible in high-resolution electron micrographs. The application of electron microscopy to the characterization of compound semiconductors has recently led to a great deal of novel structural information. Our purpose here is to provide a brief but comprehensive overview of these recent developments.

Historically, the study of elemental and compound semiconductors by high-resolution electron microscopy (HREM) was limited by microscope performance to the <110> projection. (Gibson 1984, Hutchison 1984). Resolution of individual atomic columns could not be achieved but imaging in this projection was nevertheless successfully used for investigating dislocation core structures in elemental materials (Anstis et al 1981, Bourret et al 1981, Olsen and Spence 1981), and for studying the epitaxial growth of metal silicides on silicon (Cherns et al 1982, Tung et al 1983). The improvement of interpretable resolution in the intermediate voltage HREM makes it possible, for example, to resolve the Si (220) lattice spacing of 1.92Å. Atomic structure imaging can at least then be achieved in the <100> and <111> zone axis projections, though still not in <110> (Ourmazd et al 1985). This ability to image an interface in more than one projection enables an invaluable three-dimensional view of the interface structure to be assembled.

2. IMAGE SIMULATIONS AND ANALYSIS

It is well-known that the appearance of a high-resolution electron micrograph depends very sensitively upon the objective lens defocus and the specimen thickness, as well as being dependent upon the microscope resolution. In the particular case of small-unit-cell materials, such as the compound semiconductors, the situation is exacerbated in that Fourier or self-images of the crystal lattice recur periodically with defocus (Smith and O'Keefe 1983). Recognition of correct defocus must then rely upon some other image feature such as the characteristic appearance of any amorphous material, or possibly the Fresnel fringe, along the edge of the sample. The sphalerite (zincblende) and wurtzite compound semiconductors are non-centrosymmetric materials. Important relationships between structure factors can be utilized for chemical imaging of interfaces or identifying crystal polarity – refer to Glaisher, Barry and Smith (1989) for detailed discussion.

As a first step towards quantitative interpretation of defect images, it is usually helpful to calculate a series of so-called weak-phase-object (WPO) images of the material of interest. The WPO series entails calculation of the (kinematic) image appearance as a function of resolution alone, but no account is taken of the defocus (effect of transfer function) or thickness (effect of dynamical scattering). As an example, Fig. 1 shows a WPO series for [110] InP from Glaisher, Barry and Smith (1989) – the number of contributing beams and the approximate resolution are indicated. Note the progressive changes in image appearance. The pairs of InP atomic columns are delineated with only 5 beams whereas, with further inclusion of the two chemically-sensitive {200} beams, the image reflects the crystal asymmetry. A strong dumbell-shaped motif is visible with 13 beams and individual atoms are eventually resolved with the inclusion of 35 beams. Similar trends can be identified in WPO series for other compound semiconductors.

In practice, if anything more than the simplest detection or identification of a defect is required, then a systematic study of the image appearance as a function of defocus and specimen thickness must be made. For example, a through-thickness series of image simulations of <110> sphalerite materials typically reveals that, even for a five-beam image, the intensity minimum is initially displaced towards the higher atomic material but, with increasing thickness, there is a shift in the minimum towards the material of lower atomic number (Glaisher, Spargo and Smith 1989a). This variation in asymmetric spot contrast can be best seen in the [110] CdTe simulations shown in Fig. 2 by viewing across the page at a glancing angle. Similar contrast shifts also occur in <1120>-oriented wurtzite materials (Glaisher, Spargo and Smith 1989b). Detailed analysis has shown that the spot positions can be related to the phase differences between the Bijvoet-related {111} pairs (eg. (hkl), (hkl)) of reflections (Glaisher and Spargo 1985)). In thick crystals, however, the effects of inelastic scattering will need to be taken into account (Boothroyd and Stobbs 1989).

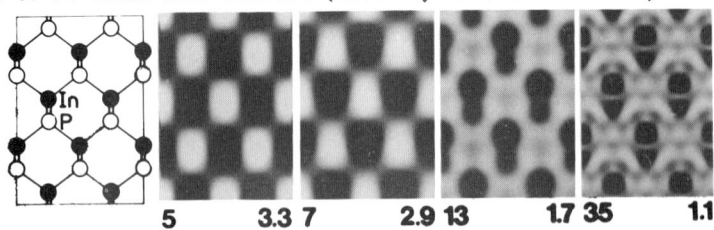

Fig. 1   Weak-Phase-Object series of images for [110] InP. Effective resolution and number of contributing beams are indicated.

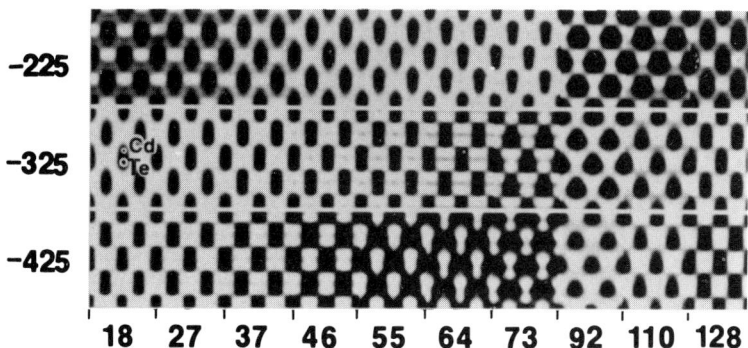

Fig. 2   Image simulations for [110] CdTe (400kV, C_S = 1.00mm) as a function of defocus (top to bottom) and thickness (left to right).   Both scales are in Ångstroms.

Polarity determinations on the basis of asymmetries in multi-beam lattice images have been previously discussed for ZnSe by Shiojiri et al (1982), for CdTe by Yamashita et al (1982) and for GaSb by Wright et al (1989). However, the sensitivity of image appearance to beam and crystal tilt means that these parameters must be carefully adjusted, particularly in the case of wurtzite materials when anomalous image features can be anticipated in certain sensitive projections (Smith et al 1984).   The characteristic shifts in contrast for <110>-oriented sphalerite as a function of crystal thickness facilitates the correct identification of crystal polarity. Polarity determination in this manner has confirmed the chemical nature of the (110) surface reconstruction observed in CdTe (Smith et al 1989a).

Detailed analysis of dynamical scattering behavior for <100>-oriented sphalerite materials again confirms the periodic occurrence of characteristic image contrast over certain defocus and specimen thickness ranges (Glaisher, Barry and Smith 1989, Glaisher and Smith 1989).   The differences between atomic species can be emphasized under certain conditions (Ourmazd et al 1986, Bourret et al 1988).   In the special case of lattice-matched materials, such as GaAs/AlAs or InP/InGaAs, it is usually possible to find a defocus/thickness combination which accentuates the contrast differences between the two materials (Ourmazd et al, 1987), in particular enabling the abruptness of the interface between the two materials to be assessed (McKernan et al 1987).   Experimental results confirming these expectations are summarized in a later section.

3.   LATTICE DEFECTS

The basic morphology of dislocations in (elemental and) compound semiconductors is well-known.   For example, high-resolution images of defects in CdS (Echigoya et al 1982, Suzuki et al 1983), CdSe (Suzuki et al 1983) and CdTe (Sinclair et al 1983) have been reported although the dislocation core structures were not analyzed.   Defects in ZnSe, ZnS and CdTe introduced by routine ion milling could be eliminated by reactive In ion milling (Cullis et al 1985a).   In sphalerite materials, the dislocations generally lie in {111} planes and are typically dissociated into partials separated by an intrinsic stacking fault.   Identification of the partials by imaging in the [110] projection is usually straightforward provided that attention is paid to the image symmetry as a function of defocus (Olsen and Spence 1991).   However, the currently available microscope resolution is insufficient to resolve individual atomic columns in this projection, thereby complicating analysis of the defect structure.

Fig. 3   High-resolution micrograph of CdTe recorded at 400kV showing the dissociated β–form of the 60° dislocation with an intrinsic stacking fault. Atomic positions and identities shown overlaid on the partials.

Several studies of GaAs dislocations have been reported (e.g. Kuesters et al 1986, Ponce et al 1986), and the core structures of 60° dissociated dislocations (Tanaka and Jouffrey 1984) and 30° partial dislocations (Gerthsen et al 1989) have been analyzed in detail.   The misfit dislocations associated with GaAs island growth on Si(111) have also been characterized (Gerthsen et al 1990, Carter et al 1991). With assistance from multislice image simulations, we found that it was possible to differentiate between the glide and shuffle sets of the 30° and 90° partial dislocations induced in CdTe by ion-milling (Lu and Smith 1990).   An example of a dissociated β–form is shown in Fig. 3. Note that unambiguous identification of the chemical nature of the defect core relied on prior knowledge of the crystal polarity, which was established by considering the image behaviour of the perfect crystal adjacent to the defect.

4.   INTERFACES

The examination of interfaces in compound semiconductors by high-resolution electron microscopy requires that both materials be aligned in low index zones, and the interfaces should also be parallel with the incident beam direction.   For the particular cases of epitaxial thin films or multilayer samples, this cross-sectional configuration has the added advantage that the microstructure of the material can be characterized as a function of distance from the interfaces(s) as well as laterally.   However, it is usually assumed that the crystal composition and structure is maintained in the beam direction.   In the case of compound semiconductors, any interface roughness or interpenetration of materials can usually be made apparent by choosing thickness and defocus values such that the contrast differences between materials are highlighted.   Any composition gradient across the interface can then be assessed by the abruptness with which the two contrast motifs terminate at the interface.   Indeed, quantification of interfacial chemistry has been demonstrated for GaAs/AlAs multilayers using pattern recognition algorithms derived from nearly perfect crystal images (de Jong et al 1987, Ourmazd et al 1990).

When there is a large lattice mismatch between materials across an interface, then the significant image features can usually be interpreted without resorting to special defocus/thickness combinations.   A number of such interfaces involving compound semiconductors have been reported, such as GaAs grown by molecular beam epitaxy on Si(100)(Biegelsen et al 1988),

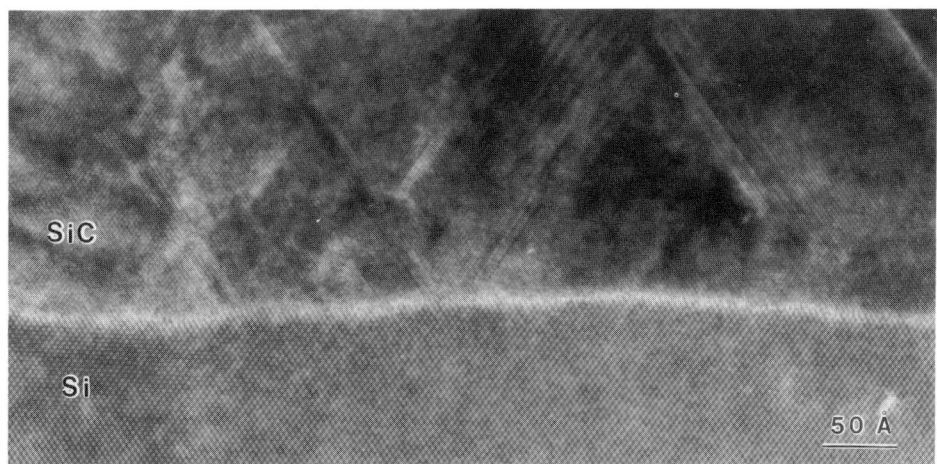

Fig. 4   Interface region between thin film of β–SiC grown on (100) Si
    substrate.   Note microtwins and stacking faults on {111} lattice planes
    and local roughness of interface.

GaAs grown on Al[100] (Kendrick et al 1990) and vicinal GaAs/Si (Hull et al
1987).   The effect of substrate cleaning on GaAs/Si(100) epigrowth has also
been studied (Vanhellmont et al 1989).   As an example, Fig. 4 shows the
interfacial region of a thin film of β–silicon carbide grown by
carburization on a (100) silicon substrate (Nutt et al 1987).   The stacking
faults and microtwins on {111} planes that accommodate the lattice mismatch
(~4:5 ratio of lattice constants) are clearly visible and easily
identified, and the rough and undulating morphology of the interface itself
is also apparent.   A similar study of β–GaN grown epitaxially on GaAs has
just been completed (Strite et al 1991).

As the lattice mismatch is reduced, as for example in III–V or II–
VI/GaAs(001) epitaxial heterojunctions (Cullis et al 1985b, Feuillet 1989,
Mallard et al 1989, Schwartzmann 1990), it is more common to find an array
of interfacial misfit dislocations.   Their nature can be established
directly from high-resolution images (e.g., Di Cioccio et al 1987), and any
subsequent changes as a result of annealing to relieve stress, usually to a
periodic array of perfect Lomer misfit dislocations, can also be followed
(Schwartzmann 1990).   Finally, strained layer superlattices involving
$CdTe/Cd_xZn_yTe/ZnTe$ have recently received increased attention (e.g.
Feuillet 1989 and Mallard et al – these proceedings).   Relaxation
mechanisms for layer thicknesses beyond the critical value will surely
involve similar considerations.

The study of multiple quantum well (MQW) devices based on ultrathin
multilayers of lattice-matched III–V or II–VI compound semiconductors
represents a challenge to the electron microscopist.   In thin crystals, it
is difficult enough at the optimum defocus even to locate the interfaces,
let alone start to assess their flatness.   Initial studies utilized [110]
orientations because of resolution limitations but, following the work of
Hetherington et al (1985), recent attention has been directed towards
imaging in [100] orientations.   This orientation features four {200} beams
which have amplitudes that depend on differences in structure factors.
These beams remain weak in GaAs up to substantial crystal thicknesses
(>300Å) unlike those of AlAs.   Cross-sectional images of GaAs/AlAs MQWs
therefore display a thickness regime where the GaAs layers are dominated by

Fig. 5   Cross-sectional view of GaAs/AlAs MQW imaged in [001] projection
     showing layers of GaAs (dark) and AlAs (light) dominated by crossed 2.0Å
     {220} and 2.8Å {200} lattice fringes respectively.

crossed {220} lattice fringes of 2.0Å, whereas the AlAs layers are
dominated by crossed {200} lattice fringes of 2.8Å rotated by 45° with
respect to those of GaAs (Tanaka et al 1987).   As clearly seen in Fig. 5,
the interfaces between successive layers of GaAs and AlAs can then be
readily identified and interfacial steps can be located.   In the case of
InP/InGaAs the highest contrast occurs in the thinnest area (Buffat et al
1989).   Image simulations suggest that it would, in fact, be preferable to
restrict the image formation process to only five beams using an
appropriate objective aperture (McKernan et al 1987, Glaisher and Smith
1989).   Despite the loss of image resolution, the differences in image
contrast between successive layers should be enhanced in both [100] and
[110] orientations.   The thickness/defocus regime for "chemical imaging" in
[110] is much more limited than in [100], but imaging of atomic steps along
<110> directions may only be possible in the former.   Indeed, interfacial
steps at a GaAs/AlAs interface have been successfully imaged in this
orientation (Ikarashi et al 1990) and a study of InP/InGaAs interfaces also
confirms that interfacial sharpness can be assessed (Petford-Long et al
1989).

5. SURFACES

Application of the technique of surface profile imaging (Smith et al 1989b)
to compound semiconductors is not straightforward because of the problems
associated with removal of the persistent surface oxide/contamination
layer.   Initial high-resolution observations have been made under poorly
controlled conditions.   For example, dynamic events on CdTe surfaces
during electron irradiation have been reported (Sinclair et al 1981), and
reconstruction of {100}, {110} and {111} CdTe surfaces was induced
following sublimation of surface layers due to beam heating (Lu and Smith
1987).   The (110) surface had a (1x1) reconstruction with a characteristic

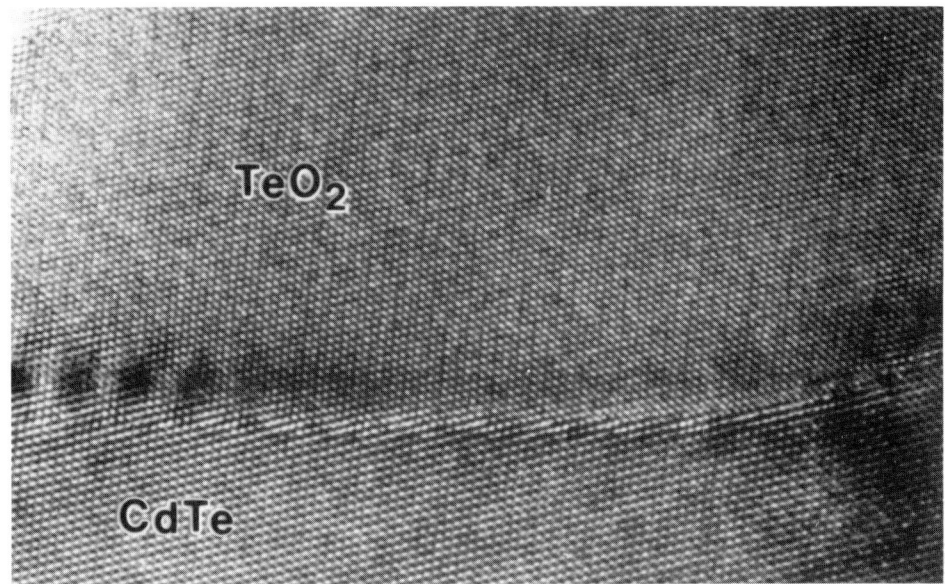

Fig. 6 High-resolution electron micrograph of CdTe crystal pre-heated in air at 260°C for 200 minutes.   Note well-defined epitaxial relationship between surface oxide and bulk semiconductor.

chevron appearance corresponding to relaxation of Te atoms away from the bulk and inwards contraction of Cd, as identified by the polarity determination methods described above (Smith et al 1989a). Similar images have also been reported by Hutchison et al (1989). Intense irradiation of In-compound semiconductors revealed the gradual development of small, randomly oriented, crystallites of In_2O_3, in a process that has been interpreted as being due to the oxidation of In following the electron-beam-stimulated desorption of the anion species (Petford-Long and Smith 1987). Similar in situ observations of ZnTe revealed the development of ZnO crystals whereas ex situ oxidation gave rise to the layered sequence ZnTe/Te/ZnO with an epitaxial relationship between ZnTe and Te (Lu and Smith 1988). An epitaxial relationship giving rise to a coherent interface was observed in the case of $CdTe/TeO_2$ (Ponce 1982) - an example from our studies of the same system is shown in Fig. 6 (Lu and Smith 1990c).

Subsequent observations of reconstructions and dynamic phenomena on CdTe surfaces have been made (Lu and Smith 1991) under ultrahigh vacuum conditions ($\sim3 \times 10^{-9}$ torr). The sample was heated in situ to ~200°C to obtain completely clean surfaces and the (110) surface then reconstructed to the previously observed (1x1) chevron appearance, thereby facilitating identification of the overall crystal polarity. Annealing to ~500°C led to the formation of long and flat (001) surfaces which, upon cooling, displayed a (2x1) reconstruction which transformed reversibly at a temperature of about 200°C to a (3x1) reconstruction. Typical regions with these two reconstructions are shown in Figs. 7 (a) and (b). Confirmation of proposed structural models was achieved by extensive image simulations and careful matching with experimental micrographs. The temperature dependence of surface atomic motion was also followed using an image pickup system attached to the UHV microscope (Lu and Smith 1990b). The (001) surface was found to be much more active than the (110) and (111) surfaces at all temperatures. It was also interesting that (111) surface layers

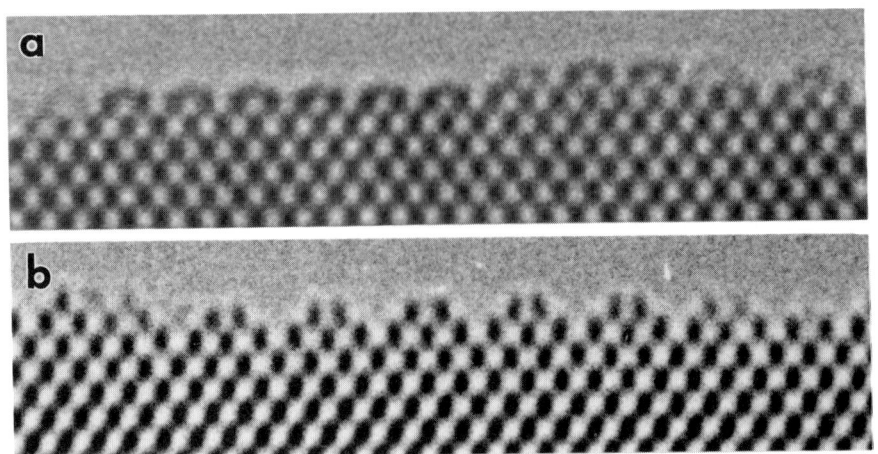

Fig. 7   Profile images of CdTe(001) surface as viewed in [110]:
(a) (2x1) reconstruction at 140°C; (b) (3x1) reconstruction at 240°C.

were usually removed in a layer-by-layer process involving a ledge
mechanism except that, when terminated by the reconstructed (001) surface,
the (111) surface layers were instead removed in a double layer process.

6. DISCUSSION

A great deal of highly useful qualitative information about the typical
microstructure of compound semiconductors is readily extracted from high-
resolution   electron   micrographs   recorded   under   appropriate   imaging
conditions.   For example, it is straightforward to identify the types of
misfit dislocations associated with epitaxial growth and strained-layer
superlattices. Moreover, using image simulations and image analysis, it is
usually possible to establish the optimum crystal thickness and objective
lens defocus so that the separate atomic species can, in principle, be
differentiated.   Not   only   does   this   enable   the   position(s)   of   the
interface(s) to be established but useful chemical details about the
interface(s), such as the nature of any facetting and chemical diffuseness,
can also be determined.   The desirability of observations in higher order
projections and the feasibility of determining interface roughness have
been emphasized (Bourret 1989).   These possibilities have now been realized
experimentally for a number of interfaces (Ourmazd et al 1990), and
extension to other compound semiconductor systems can safely be predicted.
Nevertheless, it should be appreciated that low-resolution imaging can
sometimes more easily provide information about composition profiles in MQW
systems (Stobbs et al 1989, Spycher et al 1989).   The difficulties
associated both with obtaining clean surfaces and then imaging them under
UHV conditions will continue to represent a serious limitation with surface
imaging of compound semiconductors.   Removal of the noise customarily
associated with an amorphous overlayer can substantially enhance the image
contrast.   With similar attention to sample cleanliness, it might also be
anticipated that, as well as chemical identity, atomic positions in the
vicinity of defects and interfaces could then be measured with greater
accuracy.   For example, in our efforts to refine the CdTe(001) surface
reconstructions, we estimate that the atomic column positions could be
located with an accuracy of better than 0.2Å (Lu and Smith 1991).   It can
therefore be concluded that high-resolution electron microscopy, combined
with quantitative image simulations and analysis, should continue to be an
invaluable method for detailed characterization of compound semiconductors.

ACKNOWLEDGEMENT

We are grateful for support from NSF Grant DMR-85-14583.

REFERENCES

Anstis G R, Hirsch P B, Humphreys C J, Hutchison J L and Ourmazd A 1981 Inst. Phys. Conf. Ser. 60 23
Biegelsen D K, Ponce F A, Kursor B S, Tramontana J C and Yingling R D 1988 Appl. Phys. Letts. 52 1779
Boothroyd C B and Stobbs W M 1989 Ultramicroscopy 31 259
Bourret A, Desseaux J and D'Anterroches C 1981 Inst. Phys. Conf. Ser. 60 9
Bourret A, Rouviere J L and Spendeler J 1988 phys. stat. sol.(a) 107 481
Bourret A 1989 Mat. Res. Soc. Symp. Proc. 139 1
Buffat P A, Ganiere J D and Stadelmann P 1989 Mat. Res. Soc. Symp. Proc. 139 111
Carter C B, Andersen G B and Ponce F A 1991 Phil. Mag. A63 279
Cherns D J, Spence J C H, Anstis G R and Hutchison J L 1982 Phil. Mag. A26 849
Cullis A G, Chew N G and Hutchison J L 1985a Ultramicroscopy 17 203
Cullis A C, Chew N G, Hutchison J L, Irvine S J C and Giess J 1985b Inst. Phys. Conf. Ser. 76 22
de Jong A F, Bender H and Coene W 1987 Ultramicroscopy 21 373
di Coccio L, Hewat E A, Million A, Gailliard J P and Dupuy M 1987 Inst. Phys. Conf. Ser. 87 243
Echigoya J, Pirouz P and Edington J W 1982 Phil. Mag. A45 455
Feuillet G 1989 Evaluation of Advanced Semiconductor Materials by Electron Microscopy ed D J Cherns (New York: Plenum) pp. 33-45
Gerthsen D, Ponce F A and Anderson G B 1989 Phil. Mag. A59 1045
Gerthsen D, Biegelsen D K, Ponce F A and Tramontana J C 1990 J. Cryst. Gr. 106 157
Gibson J M 1984 Ultramicroscopy 14 1
Glaisher R W, Barry J C and Smith D J 1989 Evaluation of Advanced Semiconductor Materials by Electron Microscopy ed D J Cherns (New York: Plenum) pp. 1-17
Glaisher R W and Smith D J 1989 Inst. Phys. Conf. Ser. 100 17
Glaisher R W and Spargo A E C 1985 Ultramicroscopy 18 323
Glaisher R W, Spargo A E C and Smith D J 1989a Ultramicroscopy 27 131
Glaisher R W, Spargo A E C and Smith D J 1989b Ultramicroscopy 27 117
Hetherington C J D, Barry J C, Bi J M, Humphreys C J, Grange J and Wood C 1985 Mat. Res. Soc. Symp. Proc. 37 41
Hull R, Fischer-Colbrie A, Rosner S J, Kock S M and Harris J S 1987 Appl. Phys. Lett. 51 1723
Hutchison J L 1984 Ultramicroscopy 15 51
Hutchison J L, Lyster M and Booker G R 1989 Inst. Phys. Conf. Ser. 100 29
Ikarashi N, Sakai A, Baba T and Ishida K 1989 Appl. Phys. Letts. 55 2509
Kendrick A B, Hutchison J L and Cherns D 1990 Inst. Phys. Conf. Ser. 100
Kuesters K-M, De Cooman B C and Carter C B 1986 Phil. Mag. A50 141
Lu P and Smith D J 1987 Phys. Rev. Letts. 59 2177
Lu P and Smith D J 1988 phys. stat. sol. (a) 107 681
Lu P and Smith D J 1990 Phil. Mag. B62 435
Lu P and Smith D J 1990b Electron Microscopy 1990, Vol. 1, pp. 110-111
Lu P and Smith D J 1990c Microscopy of Oxidation ed G Lorimer and M Bennett (Inst. of Metals: London)
Lu P and Smith D J 1991 Surf. Sci. in press
Mallard R E, Wilshaw P R, Mason N J, Walker P J and Booker G R 1989 Inst. Phys. Conf. Ser. 100 331
McKernan S B, de Cooman B C, Conner J R, Summerfelt S R and Carter C B 1987 Inst. Phys. Conf. Ser. 87 201

Nutt S R, Smith D J, Kim H J and Davis R F 1987 Appl. Phys. Letts. 50 203
Olsen A and Spence J C H 1981 Phil. Mag. A43 945
Ourmazd A, Ahlborn K, Ibeh K and Honda T 1985 Appl. Phys. Letts. 47 685
Ourmazd A, Rentschler J A and Taylor D W 1986 Phys. Rev. Lett. 57 3073
Ourmazd A, Tsang W T, Rentschler J A and Taylor D W 1987 Appl. Phys. Lett.
  50 1417
Ourmazd A, Baumann F H, Bode M and Kim Y 1990 Ultramicroscopy 34 237
Petford-Long A K, Booker G R and Hockly M 1989 Ultramicroscopy 31 385
Petford-Long A K and Smith D J 1986 Phil. Mag. A54 837
Ponce F A 1982 Ultramicroscopy 9 215
Ponce F A, Andersen G B, Ballingall J M 1986 Surf. Sci. 168 564
Ponce F A, Andersen G B, Haasen P and Brion H G 1986 Mat. Sci. Forum 11 775
Schwartzmann A F 1990 Mat. Res. Soc. Symp. Proc. 183 161
Shiojiri M, Kaito C, Sekimoto S and Nakamura N 1982 Phil. Mag. A46 495
Sinclair R, Ponce F A, Yamashita T and Smith D J 1983 Inst. Phys. Conf.
  Ser. 67 103
Sinclair R, Yamashita T, Parker M A, Kim K B, Holloway K and Schwartzmann
  1988 Acta. Cryst. A44 965
Sinclair R, Yamashita T and Ponce F A 1981 Nature 290 386
Smith D J, Bursill L A and Wood G J 1984 Ultramicroscopy 16 19
Smith D J, Glaisher R W and Lu P 1989a Phil. Mag. Letts. 59 69
Smith D J, Glaisher R W, Lu P and McCartney M R 1989b Ultramicroscopy 29
  123
Smith D J and O'Keefe M A 1983 Acta. Cryst. A39 139
Spycher R, Buffat P A, Stadelmann P A, Roentgen P, Henberger W and Graf V
  1989 Inst. Phys. Conf. Ser. 100 299
Strite S, Ruan J, Li Z, Manning N, Salvador A, Chen H, Smith D J, Choyke W
  J and Morkoc H 1991 J. Vac. Sci. Tech. B in press
Stobbs W M 1987 J. Physique Colloque C5-33
Stobbs W M, Baxter C S, Bithell E G, Boothroyd C B, Broom R F, Ross F M and
  Williams E J 1989 Inst. Phys. conf. Ser. 100 271
Suzuki K, Takeuchi S, Shino M, Kanaya K and Iwanaga H 1983 Trans. Japan.
  Inst. Mets. 24 435
Tanaka M and Jouffrey B 1984 Phil. Mag. A50 733
Tanaka M, Ichinose H, Furuta T, Ishida Y and Sakaki H 1987 J. Physique
  Colloque C5-101
Tung R T, Gibson J M and Poate J M 1983 Mat. Res. Soc. Symp. Proc. 14 435
Vanhellemont J, De Boeck J, Borghs G and Mertens R 1989 Inst. Phys. Conf.
  Ser. 100 109
Wright A C, Ng T L and Williams J O 1988 Phil. Mag. Letts. 57 107
Yamashita T, Ponce F A, Pirouz P and Sinclair R 1982 Phil. Mag. A45 693

Inst. Phys. Conf. Ser. No 117: Section 1
Paper presented at Microsc. Semicond. Mater. Conf., Oxford, 25–28 March 1991

Effects of atomic number and ionicity on the (110) lattice images of compound semiconductors

N Hashikawa, K Watanabe*, K Hiratsuka, C Tsuruta**, I Hashimoto and H Yamaguchi

Science University of Tokyo, 1-3 Kagurazaka, Shinjuku-ku, Tokyo 162 Japan;
* Tokyo Metropolitan Technical College, 1-10-40 Higashiohi, Shinagawa-ku,
Tokyo 140 Japan; ** Hitachi Keisoku Engineering Co.,Ltd., 882 Ichige,
Katsuta-shi, Ibaraki 312 Japan

ABSTRACT: Lattice images of compound semiconductors, InP, InAs and InSb were taken by High resolution transmission electron microscope (HRTEM) operated at 300kV. By fitting of through-focus and through-thickness images with simulated ones, the effect of the difference in the atomic number and the screening of valence charge electrons due to ionicity on lattice images are examined. As the result, the screening effect is essential to discuss the detailed atomic structure if the difference in the atomic number of these compounds is small.

1. INTRODUCTION

Applications of HRTEM to a wide range of material problems have recently received considerable attention. In order to interpret experimental HRTEM images, it is necessary to carry out n-beam dynamical calculation. In general, the crystal potential used in n-beam dynamical calculations has been constructed from the superposition of neutral atoms. Therefore, the effect of valence charge electrons due to ionicity is ignored. Then, we examined the effect on the lattice images using GaAs, ZnSe and CdTe, because these materials are expected to have a large discrepancy from the neutral crystal potential calculation due to their large ionicity and their small difference in the atomic number of constituent atoms. As a result, the screening is indispensable to derive the accurate atomic information from the (110) lattice image rather than the (100) lattice image (Hiratsuka 1991, Watanabe et al. 1991, Watanabe et al. 1991). However, it is not certain whether this fact for the (110) compound semiconductors usually exists independently of the difference in the atomic number.

In this paper, we will choose compound semiconductors InX(X=P, As and Sb) in which the variation in the atomic number occurs under almost the same ionicity, and will discuss how the difference between the neutral potential and screening one affects the (110) lattice images by fitting between through-focus and through-thickness images with simulations.

2. EXPERIMENTAL PROCEDURE

The (110) slices of compound semiconductors InP, InAs and InSb crystals were first mechanically thinned to a few micrometers and finally milled with a neutral Ar atom beam accelerated at 5kV with a shallow angle of 12 degrees. The specimens were observed by a Hitachi H-9000 UHREM (Cs=0.9mm, Cc=1.5mm) operated at 300kV. Forty-three beams were admitted by objective aperture. A through-focus series of images was taken at 5nm steps with a direct magnification of 300,000. The defocus and thickness were determined by fitting of experimental images with simulated images.

3. IMAGE SIMULATION

Lattice images were simulated with the Bethe's eigenvalue method (Fujiwara 1959) using two kinds of crystal potentials which are the neutral potential and screening potential. The former can be derived from the atomic scattering factor (Doyle and Turner 1968), and the latter is constructed from valence charge density of each ion. The valence charge density is evaluated by non-local pseudopotential (Chelikowsky and Cohen 1976) and ionic atom potential is derived from Hartree-Fock-Slater equation. Fourier coefficients of two crystal potentials (neutral and screening one) are listed in Table 1.

4. RESULTS

Sets of through-focus micrographs for InP, InAs and InSb with a defocus step of 5nm and through-thickness micrographs for InAs and InSb are shown in Figs.1,2 and 3, respectively. In the case of InP, lattice images can be obtained in wide area of specimen, but a variation with thickness cannot be observed. Simulated images are superimposed in each micrograph in which the upper and lower inserted images are simulated by using neutral and screening potentials, respectively. Simulated defocus and thickness are also denoted in Fig.1 for InP and tables 3 and 4 for

Table 1. Fourier coefficients (in Ry) of crystal potentials. The origin is at the bond center.

InP hkl	Neutral real	imag.	Screened real	imag.
111	0.518	-0.160	0.582	-0.027
200	0.0	-0.211	0.022	-0.141
220	0.497	0.0	0.556	0.0
311	-0.300	-0.117	-0.330	-0.120
222	0.0	-0.161	-0.009	-0.168
400	-0.345	0.0	-0.367	-0.141

InAs				
111	0.545	-0.083	0.604	-0.035
200	0.0	-0.108	0.017	-0.049
220	0.539	0.0	0.600	0.0
311	-0.330	-0.006	-0.363	-0.061
222	0.0	-0.082	-0.007	-0.084
400	-0.387	0.0	-0.413	0.002

InSb				
111	0.557	0.023	0.602	0.117
200	0.0	0.030	0.013	0.073
220	0.561	0.0	0.619	0.0
311	-0.347	0.008	-0.379	0.008
222	0.0	0.011	-0.007	0.009
400	-0.411	0.0	-0.438	0.002

InAs and InSb. From InP micrographs as shown in Figs.1(b) and (c), the intensity difference in white spots corresponding to constituent atoms are found, while these white spots are no longer separated in Fig.1(a). The spacing between the closest spots is elongated by about 12% over the true atomic separation along the <100> direction. Both simulations display similar change of intensity with defocus and about 8% elongation. White spots sit on tunneling sites and brighter spots correspond to In sites. There is little difference in spot intensity between images simulated by using the screening and neutral potentials.

A set of through-focus for InAs in Figs.2(a), (b) and (c) shows that the white spots corresponding to constituent atoms are resolved over these defocuses. The large variation of contrast does not occur with defocus like InP, and about 30% elongation is seen. According to the simulations, white spots sit on tunneling sites and brighter spots correspond to In sites. The distance of closed atom pair is elongated by about 20% and 25% for images simulated by using the neutral potential and the screening one, respectively. Both simulations reproduce the difference in spot intensity, but images using the screening potential seem to be clearly separated.

The through-thickness micrographs for InAs are shown in Figs.2(b), (d) and (e). Resolved white spots can be seen in Figs.2(b) and (d). Then they become single at the thin region (Z=1.73nm) in Fig.2(e). In Figs.2 (d) and (e), both simulated through-thickness images make little difference and agree well with experimental images in relation to contrast and shape.

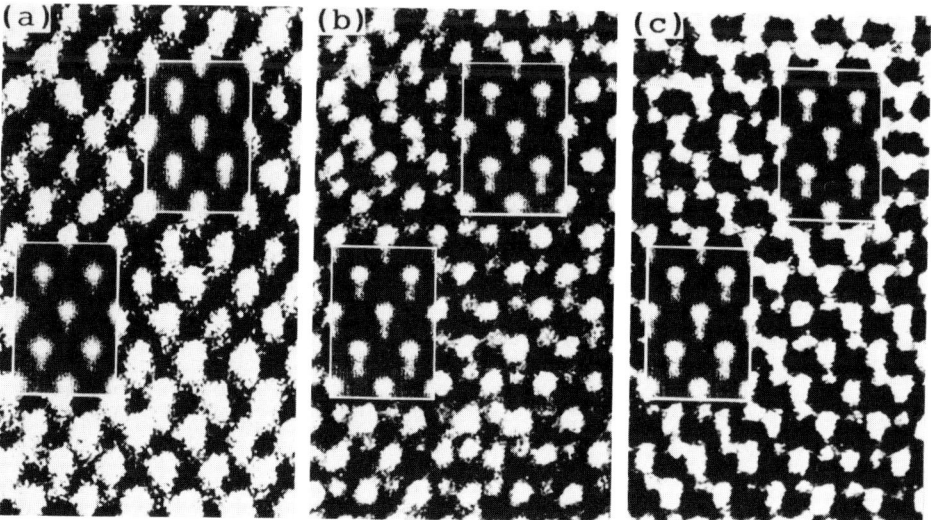

Fig. 1. A through-focus series of (110) InP. Simulated images using the neutral potential and screening potential are respectively located on upper and lower left hand side of each micrograph. The thickness is 10.36nm and 9.53nm for the neutral and screening. (a) Δf_N=148nm, Δf_S=144nm; (b) Δf_N=153nm, Δf_S=149nm; (c) Δf_N=158nm, Δf_S=154nm. Δf_N:defocus for neutral, Δf_S:defocus for the screening.

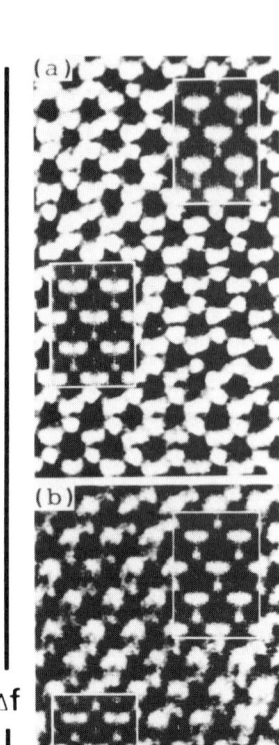

Table 2. Defocus and thickness determined by each simulation for InAs Δf :defocus(nm), Z:thickness(nm)

	neutral		screened	
	Δf	Z	Δf	Z
(a)	144		142	
(b)	149	8.56	147	8.13
(c)	154		152	
(d)	149	6.13	149	6.13
(e)		1.73		1.73

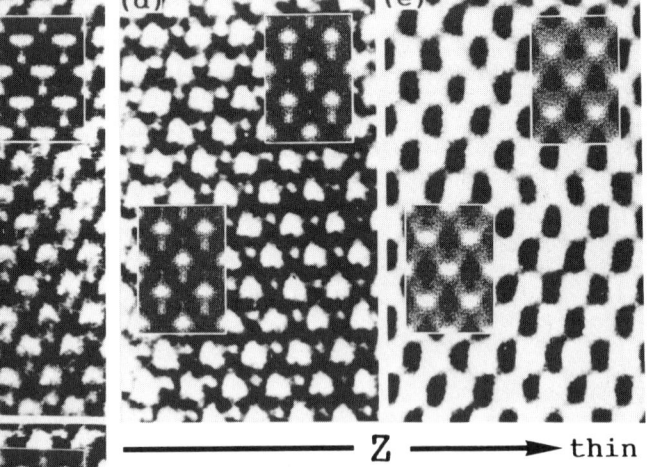

Fig. 2. A through-focus series of (110) InAs is (a),(b) and (c). A through-thickness one is (b),(d) and (e). Simulated images using the neutral potential and the screening potential are respectively located on upper and lower left hand side of each micrograph.

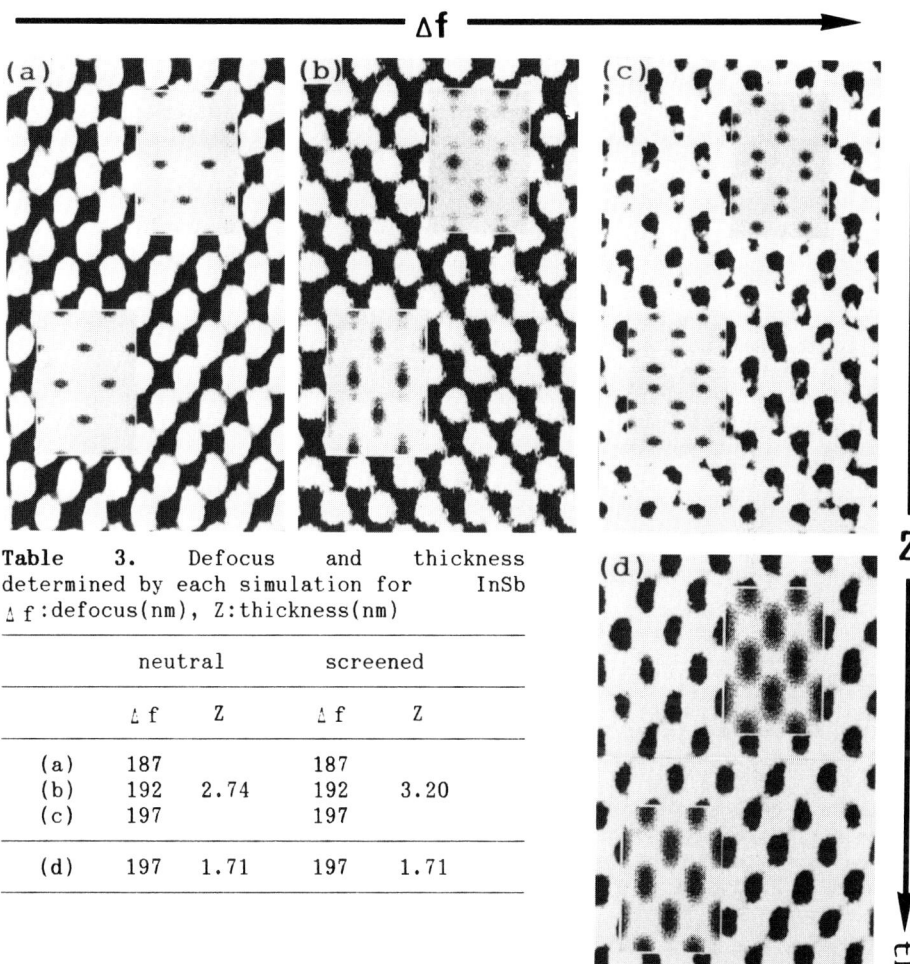

Table 3. Defocus and thickness determined by each simulation for InSb
$\triangle$ f:defocus(nm), Z:thickness(nm)

	neutral		screened	
	$\triangle$ f	Z	$\triangle$ f	Z
(a)	187		187	
(b)	192	2.74	192	3.20
(c)	197		197	
(d)	197	1.71	197	1.71

Fig. 3. A through-focus series of (110) InSb is (a), (b) and (c). A through-thickness one is (c) and (d). Simulated images using the neutral potential and screening potential are respectively located on upper and lower left hand side of each micrograph.

Figure 3 displays the micrographs for InSb. Through-focus images in Figs.3(a), (b) and (c) show black spots. In Fig.3(c), these black spots are clearly resolved and the distance of two nearest black spots are elongated by about 18%. According to simulations, the resolved black spots are elongated by about 9% for both simulations. In

Fig. 3(c), using the neutral potential, the difference in spot contrast cannot be reproduced. On the other hand, the screening potential makes it possible. As for through-thickness images in Figs. 3(c) and (d), the black spots are also seen. Both simulations display approximately similar through-thickness images, but the black spots match with experimental images in size using the screening potential. The black spots sit on atomic columns in all simulated images.

5. CONCLUSION

Our results have shown that the screening effect of valence charge electrons due to ionicity of compound semiconductors is indispensable to describe the detailed atomic structure such as polarity and identification of constituent atoms if the difference in the atomic number of these atoms is small. For a large difference, the neutral potential as well as the screening potential can reproduce the contrast change with defocus and thickness.

ACKNOWLEDGEMENT

The authors would like to thank the Computer Center of Tokyo University for use of HITAC S-810.

REFERENCES

Chelikowsky J R and Cohen M L 1976 Phys. Rev. B14 556
Doyle J M and Turner P S 1968 Acta Cryst. A24 380
Fujiwara K 1959 J. Phys. Soc. Japan 14 1513
Hiratsuka K Philos. Mag. (in press)
Watanabe K, Hashikawa N, Hiratsuka K, Tsuruta C, Nakamura K and Yamaguchi H to be submitted in Philos. Mag.
Watanabe K, Hiratsuka K and Yamaguchi H Philos. Mag. (in press)

Inst. Phys. Conf. Ser. No 117: Section 1
Paper presented at Microsc. Semicond. Mater. Conf., Oxford, 25–28 March 1991

17

The role of specimen relaxation in the high resolution electron microscopy of strained semiconductor heterojunctions

R E Mallard[+], G Feuillet and P -H Jouneau

DRF/SPh/PSC, Centre d'Etudes Nucléaires de Grenoble, 38041, 85X Grenoble CEDEX, France

+ present address: Department of Materials, Parks Road, Oxford, OX1 3PH, United Kingdom

ABSTRACT: It is widely appreciated that the relaxation of coherency strains at the free surfaces of electron microscope specimens can lead to anomalous contrast in images of compositionally modulated semiconductor materials under two-beam analysis conditions. In this paper, we discuss the influence of the elastic relaxation of thin foils for electron microscopy on the accuracy of the assessment of the structure of CdTe/ZnTe multilayer structures by high resolution electron microscopy. Experimental images are interpreted with reference to image simulations which describe idealised "perfect" and diffuse interfaces, and take into account the structural distortion in the microscope specimen which is induced by the relaxation of multilayer coherency strains at the specimen surface. For the very thin foils required in the semi-quantitative analysis of the present interfacial structures, the phenomenon of surface relaxation does not appear to play a significant role in changing the local image intensity. Measurement of the local lattice fringe spacings in both simulated and experimental relaxed thin foils however reveal significant deviations from those expected in the bulk, especially when the specimen thickness is less than the superlattice period.

1. INTRODUCTION

Cross sectional transmission electron microscopy (TEM) has been an important technique for the analysis of the perfection of semiconductor multilayer structures, notably with regard to the interfacial abruptness and morphology. The performance of high resolution electron microscopes is now such that chemical and morphological information from semiconductor interfaces is, in favourable cases, available with atomic resolution. However, the fabrication of the specimens required for TEM analysis potentially alters the stress state of the material, since elastic relaxation of the coherency strains may take place at the free surfaces of the specimen, the occurrence of which may lead to the distortion of the local lattice structure of the material (Treacy and Gibson 1986, Perovic et al 1989). This phenomenon has a profound effect on contrast in conventional 2-beam TEM images of strained heterostructures in cross section, but has not been fully investigated from the viewpoint of high resolution electron microscopy (HREM). The ZnTe/CdTe multilayers which we analyse in the present study are representative of systems composed of regions of differing lattice parameter as well as composition, information on which may be recovered in HREM images and used to describe the material structure on a local scale through a comparison of experimental with simulated images. Using the criteria of local image intensities and lattice fringe spacings we discuss the extent to which this analysis is influenced by specimen elastic relaxation.

2. EXPERIMENTAL

A multilayer structure consisting of 5 periods of nominally 34 monolayers of CdTe and 4.5 monolayers of ZnTe (6% misfit) was deposited by molecular beam epitaxy onto 2.5μm thick $Cd_{0.83}Zn_{0.17}Te$ buffer layer grown on a [001] oriented GaAs substrate. The structure was configured in such a way as to obtain approximately equal total strain partitioning between layers. Electron microscope specimens in the (100) orientation were fabricated by argon ion milling techniques and examined in the JEOL 4000EX HREM. The images were subsequently electronically digitised and examined in detail with regards to relative contrast levels and local lattice spacings using the SEMPER system (image processing software commercially available through Synoptics Ltd.). The HREM contrast of interfacial structures is in general interpreted only with reference to image simulations, which take into account specimen thickness and defocus dependent changes in the diffracted wavefront, as well as Fresnel scattering due to the discontinuity in specimen potential which occurs at the interface. The accuracy of the simulations is of course limited by the reliability of the model used to describe the structure, which for a strained layer superlattice must take into account both potential and bond length variations. In the case of a "bulk" or unrelaxed specimen, the development of such a model is relatively straightforward, but the occurrence of elastic relaxation at the free surfaces of a thin foil introduces structural changes, namely, plane bending and a change in local lattice spacing, which should be considered to fully describe the actual specimen configuration. Treacy and Gibson (1986) have derived expressions describing the axial and shear strains contributing to specimen relaxation for single sinusoidal compositional modulations in a direction perpendicular to the foil normal. Furthermore, they describe how this treatment may be extended to encompass the case of a strained epitaxial layer or superlattice, which is represented by a Fourier series, the components of which relax according to the same set of equations. From their analysis a number of general points in relation to the modified lattice structure of the relaxed thin foils of strained layer superlattices may be made. In the case of a thick TEM foil, that is, where the specimen thickness is much larger than the superlattice period, the material behaves essentially as in the specimen bulk, and as such, the shear strain acting on elemental cells within the material and the relaxation strain of the material from the buffer layer lattice parameter equal 0. The tetragonal strain equals $\varepsilon_0(1+\nu)/(1-\nu)$ where ν is the Poisson ratio of the material and ε_0 is the strain modulation amplitude for the relaxed bulk constituent materials. In the thin foil limit, uniform lattice relaxation occurs throughout the specimen thickness such that the tetragonal strain in the growth direction relaxes to $(1+\nu)\varepsilon_0$. Intermediate cases are however typical of electron microscopy problems, implying that the correct lattice spacings cannot be inferred from an HREM micrograph without knowing the precise strain state of the system. In order to qualitatively assess the HREM contrast within this regime, we perform image simulation experiments based on an extension of the Treacy and Gibson calculation. The actual atomic displacements throughout the foil thickness are obtained by integrating over the strain equations for each Fourier component comprising the superlattice. Having obtained the atomic displacements of the atoms in the relaxed material, the atomic coordinates are calculated, sorted into supercells, and input directly into a multislice image simulation program (Stadelmann 1987). Individual phase gratings are calculated for each supercell. The final specimen wavefront is calculated by passing an incident wavefunction successively through each.

3. RESULTS AND DISCUSSION

HREM at the (100) pole exploits the same pronounced differences in sphalerite {200} structure factor for varying compositions, where regions of high {200} diffracted intensity often give rise to strong lattice fringes with {200} periodicities. Figs 1a and b are experimental HREM micrographs under imaging conditions which were calculated to give rise to a high degree of chemical contrast, as judged by the differing characteristic fringe "motifs" in regions of differing composition. Such a condition exists at an underfocus between -600 to -800Å, the value of which for the present micrographs was confirmed by optical diffractogram analysis. Fig 1a corresponds to a very thin region of the specimen, perhaps 50Å, where the ratio of the specimen thickness to the superlattice period, t/Λ, is approximately 0.4, and fig 1b, to a somewhat thicker region where t/Λ equals 1.0. Background contrast differences between the two materials, that is to say, contrast due to variations in the differential inelastic contrast of the incident beam outside of the objective aperture, are negligible for such thin regions of the specimen. The ZnTe layer is recognised as a region of contrast where the vertical and horizontal {200} fringes are relatively intense. The diagonal {220} fringes are approximately of the same intensity in all regions, as the {220} structure factors are not particularly sensitive to compositional changes. For these diffraction conditions, all of the atoms in the (100) projection are fully resolved and appear white.

One potential means of assessing the interfacial chemical abruptness in these images is to measure the distance across the interface over which this characteristic change in contrast occurs. This may conveniently be done by projecting the total image intensity onto a plane perpendicular to the interface, as shown in figs 1c and d. The ZnTe layer is characterised by peaks with a periodicity corresponding to the (002) fringes, of an overall high intensity, and the CdTe, by oscillations of double the frequency, corresponding to the [010] projected separation of the {220} fringes. These curves were obtained by projecting the image intensity over a length of approximately 200Å along the [010] direction, and therefore represent the specimen composition averaged over this scale length, as well as through the thickness of the specimen. From the standpoint of the characteristic periodicity of the lattice fringes, as well as the overall image intensity, the ZnTe layers would appear to be

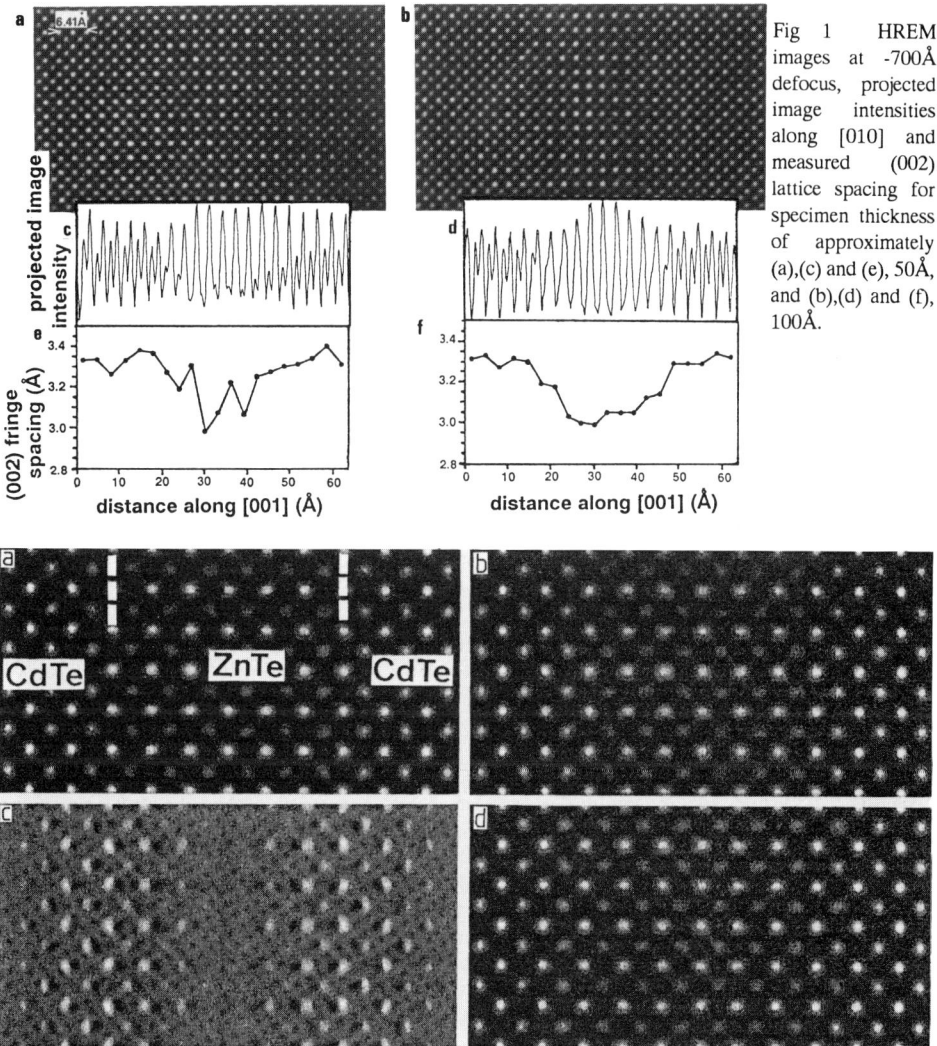

Fig 1 HREM images at -700Å defocus, projected image intensities along [010] and measured (002) lattice spacing for specimen thickness of approximately (a),(c) and (e), 50Å, and (b),(d) and (f), 100Å.

Fig 2 Image simulations for CdTe/ZnTe multilayers at a specimen thickness of 75Å and defocus of -700Å. (a) "bulk" material, (b) including surface relaxation effect, (c) difference between (a) and (b), and (d) interfaces with a compositional grade over three monolayers.

diffuse over several monolayers on each interface. (The degree of this gradual contrast change may however be due to interfacial roughness on a short scale rather than actual chemical diffuseness, as suggested by the direct observation of interfacial roughness in (110) HREM images.) From these same intensity projections, we are able to measure the lattice fringe spacing in the direction of the compositional modulation, simply by measuring the distance between the peaks, as shown in figs 1e and f, representing the variation in the (200) fringe spacing across the interfaces of lattice images in figs 1a and b respectively. The bulk multilayer should have a (002) interplanar spacing of 3.32Å in the CdTe regions and 2.86Å in the ZnTe. For the thinner of the two specimens (fig 1e), the variations in spacing are very irregular, in part due to the difficulty with which precise measurements of this sort may be made over small fields of view, but also perhaps to the high sensitivity of fringe positions to non-uniform phase shifts induced by spherical aberration, and therefore, defocus (Cockayne and Gronsky, 1981). In thicker regions as in fig 1f, the transition in fringe spacings across the interfaces more accurately reflects the interfacial structure implied by the image intensity projection in fig 1d, and suggests that the interplanar spacings have indeed relaxed from the specimen bulk values.

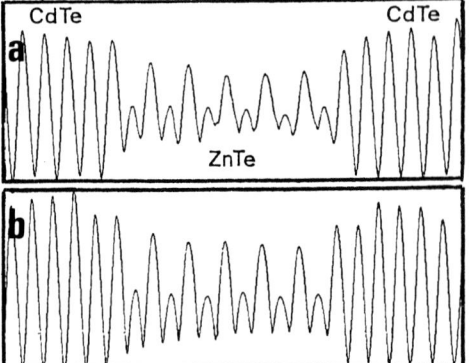

Fig 3 Projection of simulated image intensity parallel to the interfaces in figs 2(a) and (b) respectively.

Image simulations for a six monolayer by six monolayer ZnTe/CdTe multilayer with an interfacial lattice parameter of 6.29Å, roughly agreeing with the observed contrast in fig 1 are shown in fig 2, and correspond to "bulk", elastically relaxed, and chemically diffuse cases. Of interest is the apparent lack of a distinct difference in character between the bulk and relaxed simulations, (a) and (b), compared to the diffuse simulation (d), which clearly exhibits a more gradual change from {200} to {220} dominated contrast across the interface. Fig 2c is the calculated difference between figs 2a and b, and shows that the main change in image contrast due to the surface relaxation effect is in the displacement of the lattice fringes due to the structural distortion through the specimen bulk, rather than having a profound effect on local image intensity. We observe similar effects for a range of analysis conditions, though the observed difference between bulk and relaxed cases subsides as the specimen thickness increases beyond approximately twice the superlattice period. From this viewpoint, it would appear that surface relaxation phenomena are not likely to severely compromise our ability to distinguish between diffuse and abrupt interfaces by HREM.

Projections of the simulated image intensities corresponding to figs 2a and b, bulk and relaxed specimens, are shown in fig 3. In both cases the transition between periodicities characteristic of the two materials occurs over one monolayer. Overall fringe intensities are strongly influenced by the choice of Debye-Waller and absorption parameters and may be responsible for the poor agreement between experimental and simulated images with regards to overall intensity differences between the layers. Close inspection of the local intensities however reveals that some differences between the two simulated cases do exist: the CdTe intensity in the vicinity of the interface in the relaxed simulation appears to have an anomalously low intensity, and does not vary as smoothly across the interface as it does in the bulk case. While this effect may be real for this analysis condition, we consider that the likelihood of observing such a subtle change in image intensity, especially given the noise levels commonly encountered in experimental images (as in fig 1), is very low.

ACKNOWLEDGEMENT

Thanks to E Hewat for help with the image processing, and to J Cibert for the provision of the multilayer sample used in the study.

REFERENCES

Cockayne D J H and Gronsky R, 1981, Phil Mag A, 44, 159
Perovic D D, Weatherly G C, and Houghton D C, 1989, in *Evaluation of Advanced Semiconductor Materials by Electron Microscopy*, ed D Cherns, NATO ASI series, 203 (London:Plenum), 335
Stadelmann P A, 1987, Ultramicroscopy, 21, 131
Treacy M M J and Gibson J M, 1986, J Vac Sci Tech, B4, 1458

Inst. Phys. Conf. Ser. No 117: Section 1
Paper presented at Microsc. Semicond. Mater. Conf., Oxford, 25–28 March 1991

21

High resolution imaging of twin intersections in Si/Ge superlattices on Ge(001) substrates

W Wegscheider, K Eberl and G Abstreiter

Walter Schottky Institut, TU München, Am Coulombwall, 8046 Garching, FRG

H Cerva and H Oppolzer

Siemens AG, Research Laboratories, Otto Hahn Ring 6, 8000 München 83, FRG

ABSTRACT: The formation of twin bands represents the only relaxation mechanism we observed in Si/Ge superlattices grown on Ge(001) substrates. High-resolution electron microscopy was applied to study the dislocation reactions which determine the defect structure of microtwin intersections. Based on these results a general dislocation model for multiple twinning in diamond structure semiconductors leading to polytypic transformations and particularily to the diamond hexagonal phase is proposed.

1. INTRODUCTION

We have recently demonstrated that strain relaxation of Si/Ge heterostructures occurs through the formation of thin twin lamellae on {111} planes provided that the biaxial stress field is tensile (Wegscheider *et al* 1990a). In this case the geometrical arrangement of the atoms on the close-packed {111} planes requires that for the given misfit of about 1% a 90° 1/6⟨112⟩ Shockley partial dislocation has to nucleate in order to resolve part of the mismatch induced strain (Wegscheider *et al* 1990b). Although a 30° Shockley partial dislocation could in principle follow the 90° partial dislocation, thus, forming a dissociated 60° 1/2⟨110⟩ dislocation, it is energetically favourable to introduce further 90° partial dislocations on adjacent {111} glide planes (Wegscheider and Cerva 1991). In this way microtwins as shown in Fig. 1 are formed which could be formally described as an array of dissociated 60° dislocations with the trailing 30° partial dislocations pinned at the surface. This mechanism is further supported by the different mobilities reported for the two types of partial dislocations at typical growth temperatures (Pirouz 1987). In this paper we concentrate on the examination of twin intersections which have recently been subject to renewed attention (Pirouz *et al* 1990a, 1990b, Dahmen *et al* 1989, 1990). Because of the limited number of deformation twins and the high degree of crystal perfection, this superlattice system is well suited to investigate the dislocation reactions associated with the propagation of one microtwin into another by means of high-resolution electron microscopy.

2. EXPERIMENTAL DETAILS

Samples with superlattices having a unit cell consisting of 3 monolayers (ML) Si and 9 ML Ge were prepared by molecular beam epitaxy at a growth temperature of 310°C on Ge(001) substrates. The superlattice thickness was chosen to be well above the critical thickness for pseudomorphic growth ($h_c \leq 70$ nm) to produce a sufficiently high density of microtwins. The [110] oriented cross-sectional specimens

Fig. 1: TEM cross section of a 120–period Si$_3$Ge$_9$ superlattice (SL) using the (004) bright–field condition near the [110] zone axis.

were prepared by mechanical polishing and Ar ion milling. Microscopy was performed in a JEOL 200 CX and a JEOL 4000 EX microscope both operated at 200 kV to avoid electron irradiation damage which would occur at higher accelerating voltages.

3. RESULTS AND DISCUSSION

The simplest case of a twin intersection — the penetration of an intrinsic stacking fault (STF) by one single 90° Shockley partial dislocation — is shown in Fig. 2.

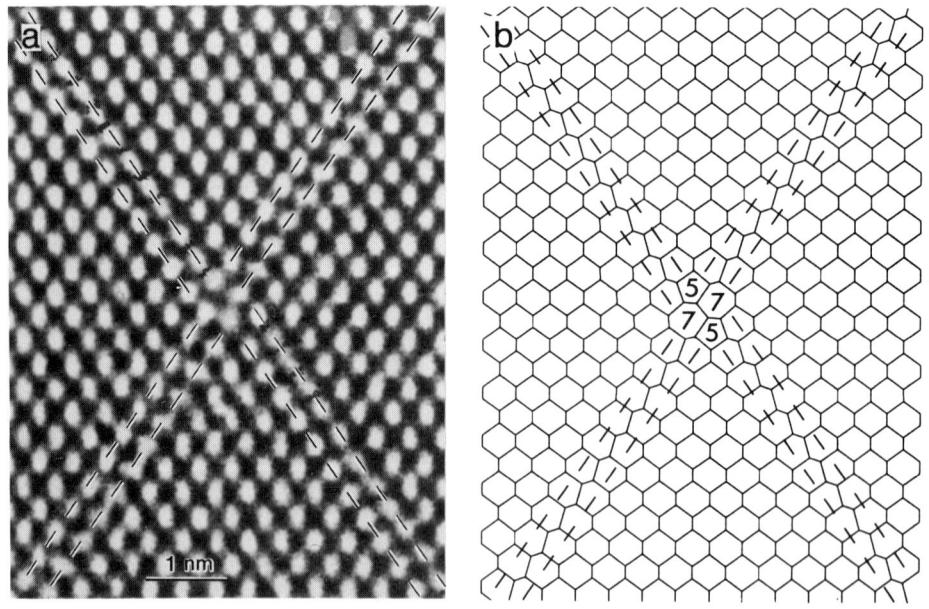

Fig. 2: Black atom high–resolution lattice image in [110] projection (a) and corresponding atom positions (b) of an intrinsic STF which has been crossed by another intrisic STF. At the intersection five- and seven–membered rings are formed.

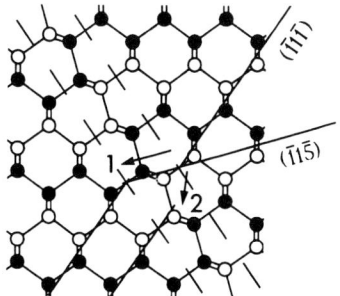

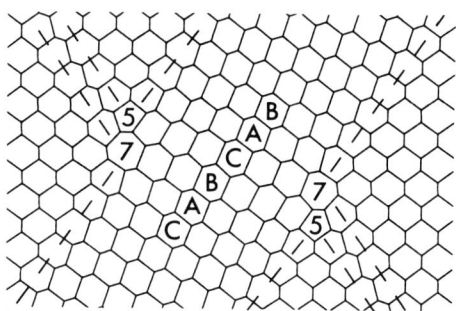

Fig. 3: Alternatives for the continuation of the $(\bar{1}1\bar{1})$ planes in the matrix on both sides of an intrinsic STF.

Fig. 4: Atomic configuration of the intersection of an intrinsic STF with a seven layer microtwin. The diamond cubic stacking sequence ABC is preserved.

As indicated in the schematic drawing of Fig. 3 the $(\bar{1}1\bar{1})$ glide planes in the matrix on both sides of the STF do not coincide. As a consequence the arriving 90° Shockley partial dislocation which forms the second stacking fault has to continue on either of the $(\bar{1}1\bar{1})$ planes denoted with "1" or "2". The first possibility may be understood by the dissociation of the partial dislocation at the first "twin" boundary of the STF leading to a dislocation which is glissile on the $(\bar{1}1\bar{5})$ plane and by the reverse dislocation reaction at the second "twin" boundary. Similar dislocation reactions will be discussed in detail later. However, comparison with the experimentally obtained high-resolution lattice image (atom pairs appear as black dots in Fig. 2(a)) reveals that the second case occurs, i.e. the continuation of the twinning transformation on plane 2, which leads to the atomic configuration depicted in Fig. 2(b). If further 90° partial dislocations cross the STF in the same way on adjacent $(\bar{1}1\bar{1})$ planes, a thick microtwin with a perfect diamond cubic stacking sequence is formed as sketched in Fig. 4. This is in full agreement with our experimental observations.

The more interesting case of an intersection of two thicker microtwins is shown in the high-resolution lattice image of Fig. 5 where atom pairs again appear as black dots. Secondary twinning has occured within the thick microtwin lying parallel to the $(\bar{1}11)$ plane. Furthermore, a STF is located within the secondary twin which finally leads to an ABCBCAC stacking sequence as indicated. According to the notation of Ramsdell (1947) this can be envisaged as two thirds of the stacking sequence of a 9R polytype (complete unit cell ABCBCACAB) with the orientation relationship $(0001)_{9R} \parallel (1\bar{1}0)_{T2} \equiv (\bar{1}1\bar{1})_{T1} \equiv (\bar{1}1\bar{5})_M$ and $[\bar{1}2\bar{1}0]_{9R} \parallel \pm[110]_{M,T}$ to the twinned regions (labeled with "T1" and "T2") and the matrix (labeled with "M") all having the diamond cubic crystal structure. This is identical to the orientation relationship of twin intersections and ribbons with diamond hexagonal structure in hot-indented Si (Dahmen *et al* 1989, Pirouz *et al* 1990a) and of hexagonal Si inclusions in chemical vapour deposited polycrystalline Si (Cerva 1991). Since T1 and T2 are formed by successive glide of 90° partial dislocations on adjacent {111} planes the identification of the dislocation reactions which have to occur at the boundaries of T1 with T2 is straight forward in this material system. Namely, the 90° $1/6[1\bar{1}\bar{2}]$ dislocations glide on the $(\bar{1}1\bar{1})$ planes and hence form T2, the former have to dissociate at the $(\bar{1}11)$ boundary thus producing secondary twinning on the $(\bar{1}1\bar{5})$ plane within T1. This case has been already discussed by Mahajan and Chin (1973). However, all the dislocation reactions suggested by these authors for secondary twinning in T1 produce high energy stackings AA. If strain compatibility is taken into account, the only twinning dislocations which do not form an AA stacking sequence are of the 30° Shockley type i.e. $1/18[\bar{2}\bar{7}\bar{1}]_M \equiv 1/6[121]_{T1}$ and $1/18[72\bar{1}]_M \equiv 1/6[\bar{2}\bar{1}1]_{T1}$. Figure 6 visualizes the

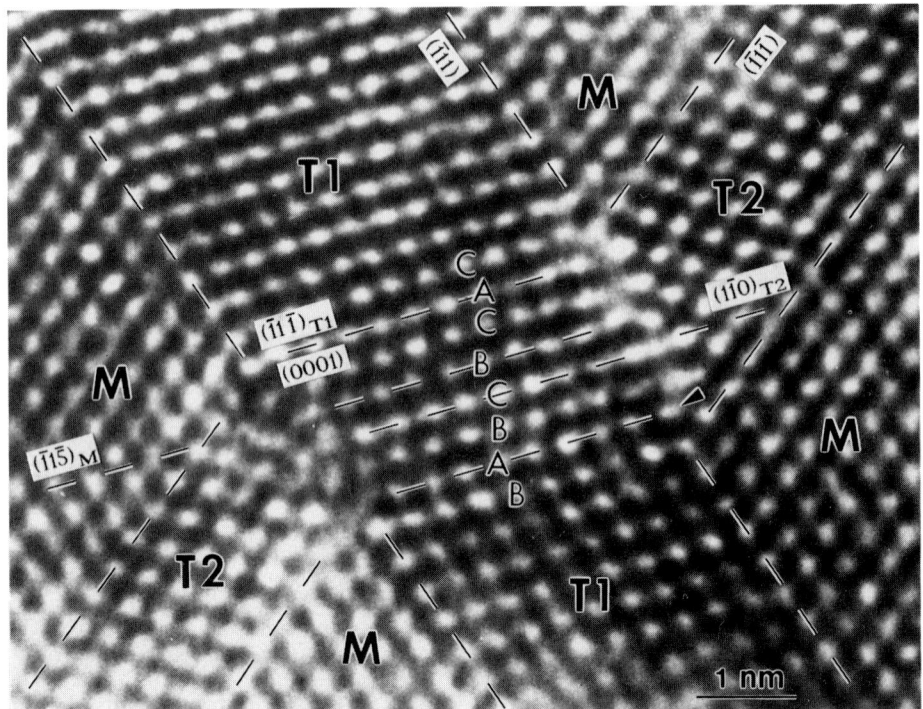

Fig. 5: Black atom high-resolution lattice image in [110] projection of an intersection of two microtwins (labeled "T1" and "T2"). The "9R polytype" stacking sequence within the intersection is indicated.

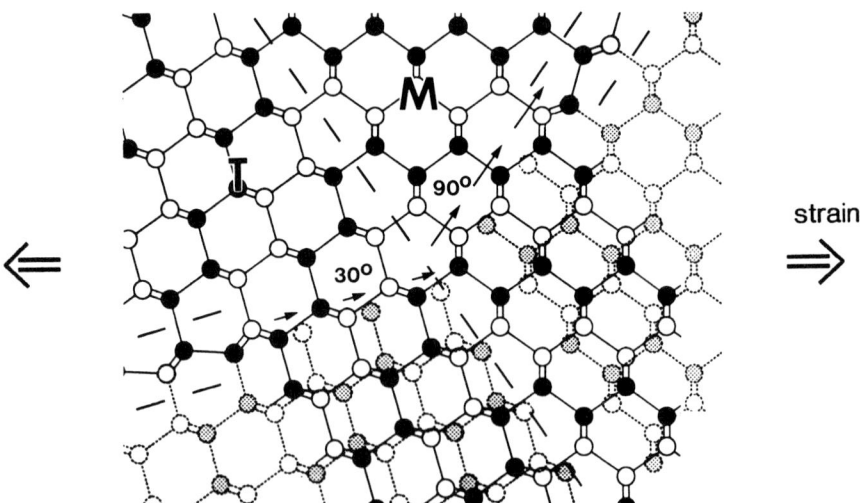

Fig. 6: Schematic drawing of the atomic displacements produced by a 90° partial dislocation in the matrix and by a 30° partial dislocation in the twin. Under the applied stress field the shearing operation of a 90° partial dislocation in the twinned region would involve a violation of close packing on the stacking fault plane.

atomic displacements induced by a 90° partial dislocation on one side of a twin plane. To produce a secondary twin on the other side of the twin plane under the applied stress field the shearing operation has to be performed by a 30° partial dislocation (Fig. 6). In this reaction, however, a dislocation with an associated Burgers vector of $\mathbf{b} = 1/18[54\bar{5}]$ or $\mathbf{b} = 1/18[\bar{4}5\bar{5}]$ remains in the twin plane. These considerations lead to the following dislocation reactions which take place at the right and left twin boundary of T1:

$$
\begin{array}{ccccccc}
90° & 30° & 30° & 60° & 60° & 90° & 60° \\
\downarrow & \downarrow & \downarrow & \downarrow & \downarrow & \downarrow & \downarrow \\
\end{array}
$$

$$6 \times \tfrac{1}{6}[1\bar{1}2] \rightarrow 2 \times \tfrac{1}{18}[\bar{2}\bar{7}\bar{1}] + 2 \times \tfrac{1}{18}[72\bar{1}] + \tfrac{1}{6}[4\bar{1}\bar{1}] + \tfrac{1}{6}[14\bar{1}] + 5 \times \tfrac{1}{6}[\bar{1}1\bar{2}] + \tfrac{1}{2}[0\bar{1}1]$$
$$+ \tfrac{1}{18}[81\bar{5}] \qquad (1)$$

$$
\begin{array}{cccccc}
30° & 30° & 60° & 60° & 90° & 90° \\
\downarrow & \downarrow & \downarrow & \downarrow & \downarrow & \downarrow \\
\end{array}
$$

$$2 \times \tfrac{1}{18}[\bar{2}\bar{7}\bar{1}] + 2 \times \tfrac{1}{18}[72\bar{1}] + \tfrac{1}{6}[4\bar{1}\bar{1}] + \tfrac{1}{6}[14\bar{1}] \rightarrow 5 \times \tfrac{1}{6}[1\bar{1}2] + 3 \times \tfrac{1}{6}[1\bar{1}2] + \tfrac{1}{18}[1\bar{1}2] \qquad (2)$$

For clarification, the projections of the Burgers vectors of these dislocations onto the (110) plane are shown in Fig. 7. The six $1/6[1\bar{1}2]$ partial dislocations, when piled up

reaction (1):　　　　　　　reaction (2):　　　　　　　reaction (3):

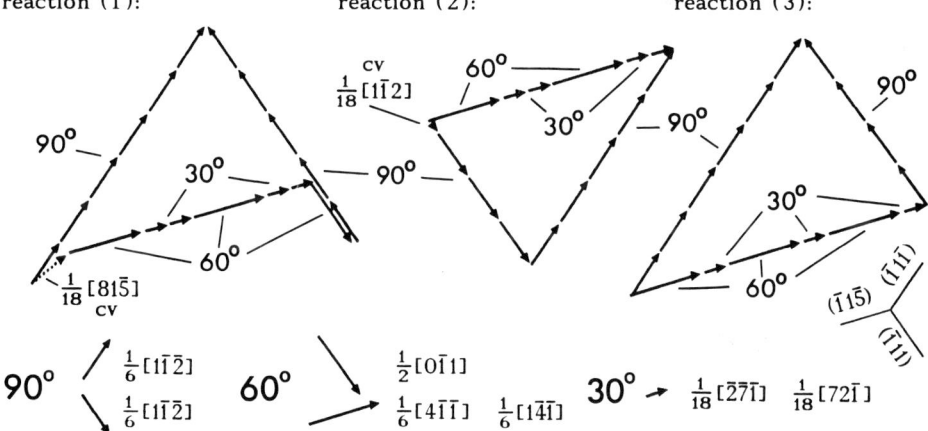

Fig. 7: Projection of the Burgers vectors involved in the dislocation rections (1), (2) and (3) onto the (110) plane. The closure vectors of Burgers circuits around the reaction zones in Fig. 5 correspond to the vectors labeled "cv".

against T1, coalesce to form a $[1\bar{1}2]$ dislocation which subsequently dissociates into dislocations which are glissile either on the $(\bar{1}15)$ or the $(\bar{1}11)$ planes, and a dislocation with an associated Burgers vector $\mathbf{b} = 1/18[81\bar{5}]$. Under the imposed shear stress all but the last dislocation glide away from the reaction zone. The five 90° $1/6[\bar{1}1\bar{2}]$ partial dislocations lead to a thickening of T1 above the intersection whereas the four 30° partial dislocations produce a twinning operation on the $(\bar{1}15)$ planes within T1 which is interrupted on every third plane by the glide of one 60° perfect dislocation ($\mathbf{b} = 1/6[4\bar{1}\bar{1}]_M \equiv 1/2[\bar{1}01]_{T1}$, $1/6[14\bar{1}]_M \equiv 1/2[011]_{T1}$). Before leaving T1, the 30° and 60° dislocations dissociate similarily according to reaction (2), thereby producing a five layer microtwin on the $(\bar{1}1\bar{1})$ planes and a thickening of T1 below the intersection. Under the assumption that T1 had a thickness of six $(\bar{1}11)$ planes before it has been crossed by T2, the glide of the five 90° partial dislocations on the $(\bar{1}11)$ planes according to reaction (1) leads to an eleven layer microtwin above the intersection as observed in' the high-resolution lattice image of Fig. 5. The thickening of T1 below the intersection by three $(\bar{1}11)$ planes (Fig. 5) is then produced by the glide of the three 90° partial dislocations resulting from reaction (2) at the other twin boundary. Thus, these dislocation reactions explain very well the ABCBCAC stacking sequence, the tapering of both microtwins T1 and T2 and the incoherency of the $(\bar{1}1\bar{1})$ planes of T2 with the (0001) planes of the "9R polytype" as indicated by an arrow and which is caused by the 60° $1/2[0\bar{1}1]$ perfect dislocation on the $(\bar{1}11)$ plane. In addition, inspection of the reaction zones by drawing Burgers circuits in the high-resolution lattice image of Fig. 3 results into closure vectors of the circuits which correspond exactly to the projections of the $1/18[81\bar{5}]$ and $1/18[1\bar{1}2]$ dislocations onto the (110) plane. These dislocations were formed according to reactions (1) and (2) and could not glide away. It is obvious that the continuation of the described phase transformation mechanism (diamond cubic → 9R) is energetically unfavorable since the component of the Burgers vector $\mathbf{b} = 1/18[81\bar{5}]$ perpendicular to the $(\bar{1}11)$ glide plane increases with each $1/6[1\bar{1}2]$ twinning dislocation of T2. However, if the 30° partial alternates with the 60° perfect dislocation on every $(\bar{1}1\bar{5})$ plane within the intersection, reaction (1) simplifies to

$$
\overset{90°}{\underset{\downarrow}{}} \quad \overset{30°}{\underset{\downarrow}{}} \quad \overset{30°}{\underset{\downarrow}{}} \quad \overset{60°}{\underset{\downarrow}{}} \quad \overset{60°}{\underset{\downarrow}{}} \quad \overset{90°}{\underset{\downarrow}{}}
$$

$$
6 \times \tfrac{1}{6}[1\bar{1}\bar{2}] \rightarrow 2 \times \tfrac{1}{18}[\bar{2}7\bar{1}] + \tfrac{1}{18}[72\bar{1}] + 2 \times \tfrac{1}{6}[4\bar{1}\bar{1}] + \tfrac{1}{6}[1\bar{4}\bar{1}] + 4 \times \tfrac{1}{6}[\bar{1}1\bar{2}] \qquad (3)
$$

and all dislocations can glide away from the reaction zone as can be seen in the corresponding drawing of the projected Burgers vectors in Fig. 7(c). Due to the twinning operation, which occurs on every other $(\bar{1}1\bar{5})$ plane in this case, a diamond hexagonal crystal structure with an ABAB stacking sequence is formed. Assuming the appropriate dislocation reaction to occur also at the second twin boundary of T1, the interfaces between the diamond hexagonal and the diamond cubic phase are coherent and lie on $(\bar{1}11)$ planes. In a simple picture one would expect a displacement of the two twin halves of T1 below and above the intersection against each other caused by the glide of the dislocations on the $(\bar{1}1\bar{5})$ planes. However, due to the thickening of T1 on the right twin boundary above the intersection and on the left twin boundary below this displacement is exactly compensated.

Thus, a phase transformation mechanism has been formulated which leads to the experimentally observed orientation relationship of diamond cubic Si to diamond hexagonal Si (Dahmen *et al* 1989, Pirouz *et al* 1990a, Cerva 1991) and involves only $1/6\langle112\rangle$ Shockley partial and 60° $1/2\langle110\rangle$ perfect dislocations.

REFERENCES

Cerva H 1991 submitted to J. Mat. Res. for publication
Dahmen U, Hetherington C J, Pirouz P and Westmacott K H 1989 Scripta metall. 23 269
Dahmen U, Westmacott K H, Pirouz P and Chaim R 1990 Acta metall. mater. 38 323
Mahajan S and Chin G Y 1973 Acta metall. 21 173
Pirouz P 1987 Scripta metall. 21 1463
Pirouz P, Chaim R, Dahmen U and Westmacott K H 1990a Acta metall. mater. 38 313
Pirouz P, Dahmen U, Westmacott K H and Chaim R 1990b Acta metall. mater. 38 329
Ramsdell L S 1947 American Mineral. 32 64
Wegscheider W, Eberl K, Abstreiter G, Cerva H and Oppolzer H 1990a Appl. Phys. Lett. 57 1496
Wegscheider W, Eberl K, Abstreiter G, Cerva H and Oppolzer H 1990b Mat. Res. Soc. Symp. Proc. 183 155
Wegscheider W and Cerva H 1991 to be published

Inst. Phys. Conf. Ser. No 117: Section 1
Paper presented at Microsc. Semicond. Mater. Conf., Oxford, 25–28 March 1991

High-resolution Z-contrast imaging of superlattices and heterostructures

S J Pennycook, D E Jesson and M F Chisholm

Oak Ridge National Laboratory, P.O. Box 2008, Oak Ridge, TN 37831-6030

ABSTRACT: The Z-contrast technique for high-resolution imaging provides incoherent images with column-by-column compositional sensitivity, therefore allowing a direct interpretation of interfacial structures to be made to first order. This capability is illustrated through studies of interfaces in epitaxial silicides and ultrathin $(Si_mGe_n)_p$ superlattices. Interface defects observed at the (100) $Si/CoSi_2$ interface are linked to Si surface defects preserved during growth of the thin silicide template. At Si/Ge interfaces grown by MBE, new ordered arrangements have been observed and attributed to a Si/Ge atom exchange mechanism occurring during growth of Si over a 2×1 reconstructed Ge surface. Anticipated performance of a 300 kV high-resolution STEM is briefly discussed.

1. THE Z-CONTRAST IMAGING PROCESS

The high-angle annular detector introduced by Howie (1979) is responsible for the two key advantages of Z-contrast imaging, compositional sensitivity and incoherent characteristics. Compositional sensitivity is built into the image through the almost Z^2 dependence of the high-angle scattering cross section, while the incoherent characteristics arise because the high-angle signal consists predominantly of thermal diffuse scattering (TDS). As shown by Hall (1965), at high-angles the TDS is approximated very well by the Einstein model of independently vibrating atoms, with the result that we can consider the scattering to be generated incoherently by each atomic site. Dynamical effects on the outgoing electrons, which give rise to Kikuchi lines, are averaged out by the wide angular range of the Howie detector, with the result that the image may be calculated as a simple rate of loss from the incident electron wavefunction $\psi(\mathbf{r})$ due to a suitable potential $V^{HA}(\mathbf{R})$ (Pennycook and Boatner, 1988, and Pennycook and Jesson, 1990, 1991), giving

$$I^{TDS} = \frac{2}{\hbar v} \int |\psi(\mathbf{r})|^2 \, V^{HA}(\mathbf{R}) \, d\mathbf{r} \ . \tag{1}$$

This is the standard approach used to calculate total absorption (Hall and Hirsch, 1965), or the generation of secondary effects such as x-ray fluorescence (Cherns, Howie, and Jacobs, 1973) or cathodoluminescence emission (Pennycook and Howie 1980). The analysis below implies that any such signal, if sufficiently localized, could be used to form an incoherent image, although it is the relatively high flux of the high-angle electron scattering which makes such images a practical possibility. The potential for high-angle scattering is sharply peaked around the atom sites, while $\psi(\mathbf{r})$ varies slowly by comparison, so that a sharp potential approximation may be made to simplify calculation giving

$$I^{TDS} = \frac{2}{\hbar v} \sum_i |\psi(\mathbf{r}_i)|^2 \int V^{HA}(\mathbf{R}) d\mathbf{R} = \sum_i |\psi(\mathbf{r}_i)|^2 \sigma_i \ , \tag{2}$$

a sum of the local electron intensity at each atom site i weighted by an atomic cross section σ_i for high-angle TDS. The σ_i are given by

$$\sigma_i = \left(\frac{4\pi\gamma}{\chi}\right)^2 \int_{detector} f_i^2(s)\left[1 - e^{-2M_i s^2}\right]d^2s ,$$ (3)

where the symbols have their usual meanings. Note that in a crystal, the σ_i are reduced from their full isolated atom values by the term in square brackets, the difference between the elastic scattering of a stationary atom and a vibrating atom as shown in Fig. 1. Accurate screened single atom scattering cross sections are required here (see for example Pennycook, Berger, and Culbertson, 1986). Born approximation scattering factors (e.g., Doyle and Turner, 1968) while being appropriate for dynamical calculations of the elastic wavefield in a crystal, give significant errors for the single atom scattering factor at high angles.

Fig. 1. Plots of the scattered intensity $f^2(s)$ from an isolated Si atom at room temperature integrated over a narrow annulus at s (from Doyle and Turner, 1968) and its decomposition into elastic and TDS components. Our detector covers s ~ 1–2 Å-1 (75–150 mrad).

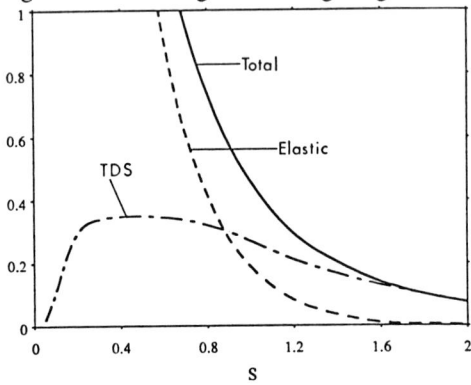

The problem of calculating the image now reduces to the problem of calculating the incident electron intensity at each atomic site. Clearly in the case of a phase object, $|\psi(r)|^2$ is just the incident probe intensity profile $P(R)$ and an incoherent image results. In terms of standard incoherent imaging theory, the image can be expressed as a convolution

$$I_{HA}(R) = P(R) * O(R) ,$$ (4)

where the object function $O(R) = \sum_i \sigma_i \delta(R - R_i)$. Since $P(R)$ is the intensity spread function for the objective lens, we see the expected resolution improvement over bright field phase contrast imaging controlled by the amplitude transfer functions.

To ensure only TDS reaches the detector, Fig. 1 would imply that a very large inner angle is required, reducing image efficiency and statistics. However, lower angles do not necessarily result in a large coherent component since it may be redistributed substantially by the crystal. HOLZ reflections are intrinsically weak and represent a small fraction of the total intensity reaching the Howie detector. Due to the curvature of the Ewald sphere, the zero layer reflections are also generally weak, except in crystals a few atomic layers thick when the shape factors for high-angle reflections become sufficiently broad to intersect the Ewald sphere. In this case, a coherent component must be added to the image. Interestingly, such images are calculated to show similar form and contrast to an incoherent Z-contrast image (Loane et al., 1988, Shin et al., 1989) which may be understood as the result of an effective transverse incoherence introduced by the Howie detector (Jesson, Pennycook, and Chisholm, 1989, Jesson and Pennycook, 1991). Coherent interference effects between neighboring columns cause intensity to be redistributed on the detector plane, but at a scale very fine compared to the size of the detector. The total intensity reaching the detector is therefore unaffected, and

Eq. (4) may still be used to describe the image. However, coherent interference effects along an individual column can remove intensity from the detector, and the object function becomes an oscillatory function of thickness representing the result of the coherent integration of amplitudes along the column. Initially, therefore, the object function is proportional to n^2, where n is the number of atoms along the string, but Fresnel defocussing effects rapidly become important, and the strength of the coherent object function never increases above that of a very thin crystal. The coherent component rapidly becomes swamped by the TDS component which increases approximately as n. For example, in Si at room temperature the coherent image is reduced to only 2% of the TDS image by a thickness of 50 Å (with detector angles 75–150 mrad corresponding to s ~ 1–2 Å^{-1} on Fig. 1). Note that estimates of the coherent fraction based on a single phonon model for TDS (Wang and Cowley, 1990) will be significantly higher since most high-angle TDS is multiphonon scattering (Hall, 1965).

At greater thicknesses, we might anticipate dynamical diffraction effects to complicate significantly the calculation of the effective illumination intensity at each atomic site. However, one of the most remarkable characteristics of Z-contrast imaging, responsible in large part for its practical utility, is how little dynamical diffraction affects the form of the image. The intensity close to the atom sites is almost entirely due to s-type Bloch states tightly bound by the columnar potential (see Fig. 2). Such states are sharply peaked at the atomic sites, and therefore strongly and coherently excited by all incident angles comprising the STEM probe. The Howie detector can therefore be viewed as an efficient Bloch state filter preventing less localized states from contributing to the image. Depth dependent interference effects are reduced to second order allowing intuitive interpretation of relative columnar scattering powers to first order (see Fig. 3). In conventional phase contrast imaging, the filtering effect is entirely reversed through the action of the objective aperture. Only the lowest order Fourier components of the s-states are transmitted to the image, and these can be of comparable amplitude to the corresponding components of the less localized states (Fig. 2). Strong depth dependent interference effects and a substantially less localized image result so that phase contrast images from defects and interfaces must be calculated explicitly. In the case of Z-contrast imaging, the highly localized nature of the s-states means that they are relatively insensitive to the strength and distribution of surrounding strings so that intuitive interpretability holds even at interfaces. Object functions for interfaces, superlattices, or complex unit cells may be constructed by assembling an array of appropriate isolated string strengths calculated using only axial Bloch states and the dispersion surface, with a vast saving in computer time.

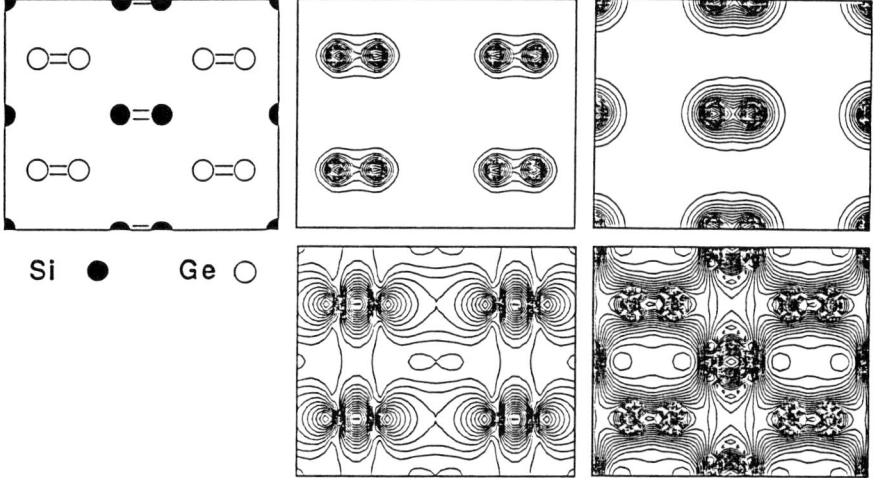

Fig. 2. Bloch state calculations for a $(Si_2Ge_2)_p$ superlattice in the $\langle 110 \rangle$ projection. Bonding s-state contributions are localized about individual Si or Ge dumbbells whereas more dispersive states are not.

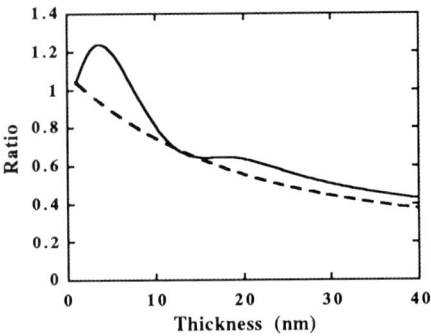

Fig. 3. Ratio of Ge to Si image intensity as a function of thickness (normalized by their respective cross sections). Solid line is the full dynamical calculation, dashed line is the s-state contribution. Although the Ge intensity is enhanced dynamically in very thin crystals, such effects are second order compared to the almost four fold increase in scattering cross section.

For closely spaced identical strings, the s-states themselves may overlap, in which case appropriate molecular orbital states may be used to assemble object function, as shown in Fig. 2 for a Si_2Ge_2 superlattice in $\langle 110 \rangle$ projection (Jesson, Pennycook, and Baribeau, 1990). The major contributor to the relative string strength is in fact the large variation in the scattering cross sections σ_i. The dynamical effects tend to cancel; for example, a deeper potential well will have a more sharply peaked s-state, but its excitation will be correspondingly reduced.

To a good approximation, all the materials dependent terms may be included in the object function and the image is then given by a convolution with the incident probe intensity profile. A simulated defocus sequence for Si$\langle 110 \rangle$ is shown in Fig. 4, in which it is clear that there is only one optimum focus condition. This condition corresponds to the Scherzer (1949) incoherent conditions $\Delta f = -(C_s\lambda)^{1/2}$, $\alpha_{opt} = (4\,\lambda/C_s)^{1/4}$ giving a resolution of $d_{min} = 0.43\,C_s^{1/4}\lambda^{3/4}$ (2.2 Å on our VG Microscopes HB501UX with $C_s = 1.3$ mm). A larger objective aperture can be used to give higher image resolution (1.9 Å by Xu et al., 1990, with $C_s = 0.7$ mm) although the narrowing of the central maximum is accompanied by increased intensity in the tails of the probe, which reduces the compositional resolution (Pennycook and Jesson, 1991). A more promising approach to higher resolution is to realize that with Z-contrast imaging, the high spatial frequency information is directly transmitted from the sample to the detector, the probe size acting as a simple incoherent instrumental broadening function. Thus, intuitive interpretation may be extended significantly below the resolution set by the probe as shown clearly by the elongation of the Si dumbbell along the $\langle 100 \rangle$ direction in Fig. 4. It would seem a relatively simple matter to extend the image resolution to around one half the incident probe size through a suitable computer algorithm. With the advent of a 300 kV STEM having anticipated probe size of 1.3 Å, such a procedure would extend the Z-contrast image resolution well below the 1 Å level.

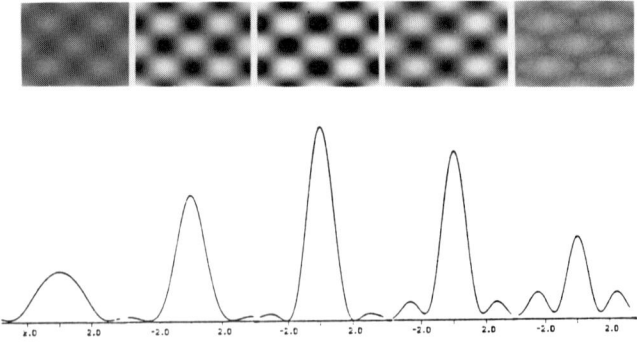

Fig. 4. Simulated defocus series for $\langle 110 \rangle$ Si at 100 kV, $C_s = 1.3$ mm, with corresponding probe intensity profiles.

2. APPLICATION TO SUPERLATTICES AND HETEROSTRUCTURES

The most valuable aspect of the Z-contrast technique is that the optimum imaging conditions are independent of the material being imaged; no pretuning of the microscope conditions is required so that unanticipated interface structures will be immediately apparent. We have found several cases where the actual interface structures were more complicated than previously imagined, which has provided significant insight into not only interfacial properties but the growth process itself at the atomistic level.

Figure 5 shows interfacial defects at an epitaxial $CoSi_2/Si(100)$ interface taking the form of $CoSi_2$ protrusions one monolayer deep into the Si, with a width of one, two, or three atomic spacings. They show strongest contrast in the thinnest regions of the sample and must therefore be ≤ 5 nm in length along the $\langle 110 \rangle$ beam direction. No long-range periodicity is seen in the image, and these defects are presumably unrelated to the long-range 2×1 reconstruction observed by Loretto et al. (1989), which involves Si atoms at the interface. Such a reconstruction would be difficult to detect at our current resolution. Most likely the Co protrusions originate from defects present on the original cleaned substrate surface which are preserved during formation of the film. The film was grown by a template technique (Yalisove et al., 1989) in which two to three monolayers of Co are deposited onto a cleaned Si surface, reacted to form a thin $CoSi_2$ template layer, and the film grown to the desired thickness by coevaporation of stoichiometric $CoSi_2$. STM studies of Si(100) surfaces have shown that a significant density of dimer vacancies may occur which show a substantial ordering even at low temperatures (Hamers et al., 1990). Figure 6 shows a likely mechanism for the formation of $CoSi_2$ protrusions from a surface vacancy channel assumed to be a typical three dimers wide. Since the depth of the template layer is comparable to the depth of the surface defect, this procedure acts as a template not only for the epitaxial orientation, but for the surface morphology itself, leading to protrusions one, two, or three spacings wide as observed.

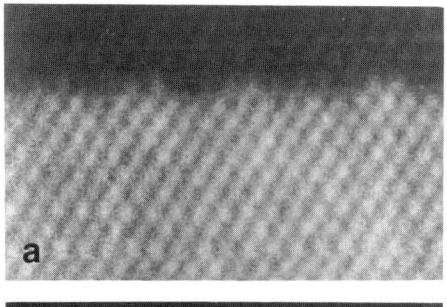

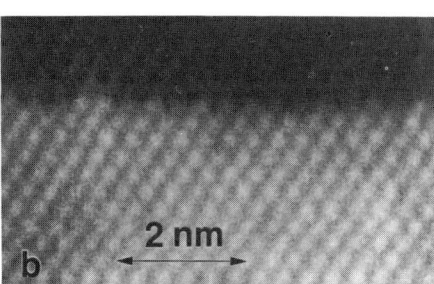

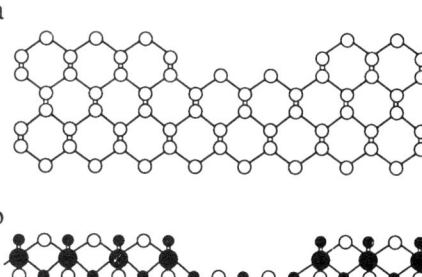

a

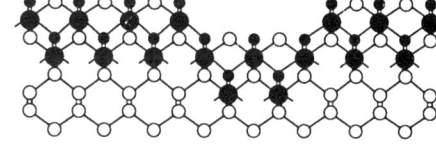

b

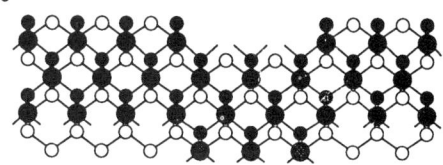

c

2 nm

Fig. 5. Interfacial defects at an $CoSi_2/Si(100)$ interface. Co columns image bright.

Fig. 6. Proposed mechanism for formation of interface defects from a Si surface vacancy channel (a), during reaction to form the template two (b) or three (c) monolayers thick. Open circles represent initial Si, solid circles displaced Si, large solid circles Co.

It may be that such a process is critical to the entire template action since thicker template layers in which such protrusions would be smoothed out no longer act as templates for the epitaxial orientation. In view of the recent studies of faceted NiSi$_2$ interfaces (Tung et al., 1991), the microscopic facets which the protrusions of Fig. 5 represent may even be important in determining the observed Schottky Barrier height.

Another system which has revealed a whole new level of complexity on the atomic scale is that of Si/Ge interfaces. The first indications of unexpected complexity came from diffraction observations of ordering in Si$_x$Ge$_{1-x}$ alloys by Ourmazd and Bean (1985). Long thought to be a perfectly miscible system, much theoretical and experimental work followed, the primary goal being essentially to determine which of the two ordered phases proposed by Ourmazd and Bean was actually present (e.g., LeGoues et al., 1990). Figure 7 shows how Z-contrast imaging would clearly distinguish these phases without averaging over the large crystal volume need for selected area diffraction studies. Conventional phase contrast imaging has not so far been able to image an ordered structure in this system. However, Z-contrast images of an ultrathin superlattice shown in Fig. 8 immediately show the situation to be far more complicated than previously imagined. A different ordered arrangement is seen at each interface, 2 × n interfacial ordering at the top Si on Ge interface, a {111} planar structure in the central Si layer, with Ge threading right through to the next Ge layer, and cross-like structures are visible in the lowest Si layer. Note the asymmetries apparent here. Much Ge is present in the Si layers, but little Si in the Ge layers, and the Si on Ge interfaces are generally much broader than the Ge on Si interfaces. This is inconsistent with the simple notion of strain-enhanced interdiffusion, and suggests that these phases are produced as a result of the growth process itself. In addition, clear images of distinct phases are only observed in the thinnest regions of crystal, consistent with the small island sizes observed by STM for low-temperature growth (Hamers et al., 1990). The STM observations clearly demonstrate that growth at low temperatures occurs via the successive nucleation of 1 × 2 and 2 × 1 monolayer height islands which grow predominantly by the motion of one particular type of step, normally referred to as a type S$_B$ step in the notation of Chadi (1981).

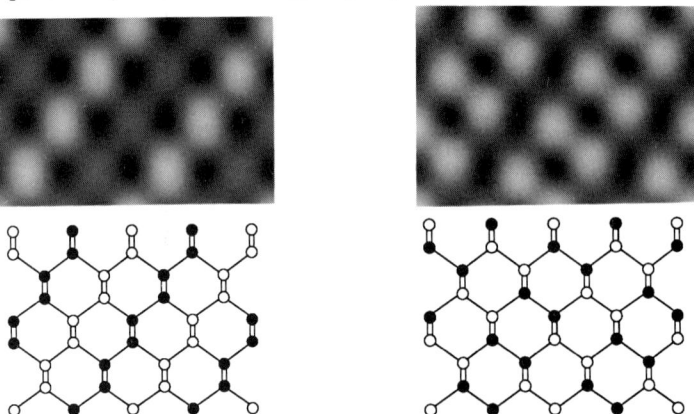

Fig. 7. The two structures proposed for ordered Si$_{.5}$Ge$_{.5}$ with their corresponding Z-contrast image simulation.

Consider the advance of a Si type S$_B$ step over a Ge substrate. Due to the 2 × 1 reconstruction of the Ge surface, the step is forced consecutively through two different configurations as shown in Fig. 9, conventionally referred to as the rebonded and nonrebonded structures. At the rebonded configuration, interchanging the loosely bound Si and Ge atoms at the step edge will bury the Si dangling bond and replace it with a Ge bond. Since Si dangling bonds have a significantly higher energy than Ge dangling bonds, an appreciable driving force exists for this "atom pump" mechanism, whereas at the other step configuration no dangling bonds can be saved by such an exchange. Thus, we obtain a 2 × 1 compositional ordering along the direction of growth. The next monolayer will similarly give ordering along the perpendicular direction. With the third monolayer, there are four

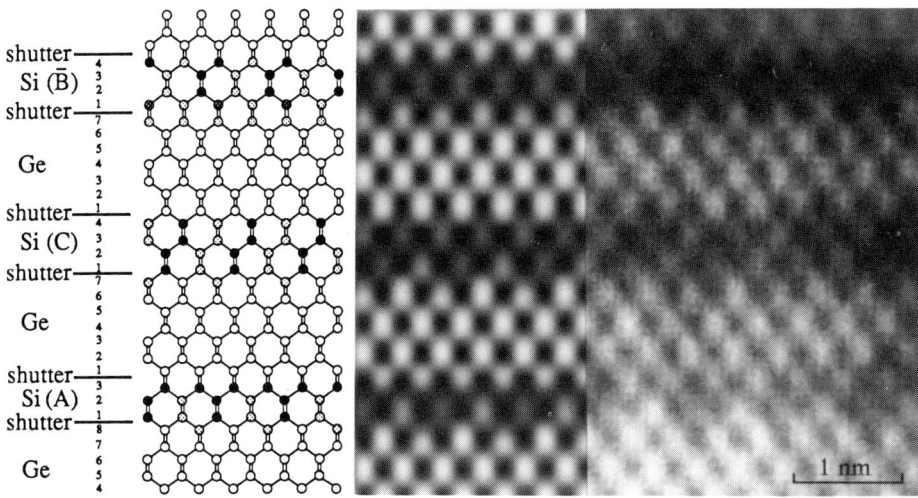

Fig. 8. ⟨110⟩ experimental (right) and simulated (left) images of interfacial ordering in a (Si₄Ge₈)₂₄ superlattice grown on Ge⟨001⟩ at 350°C, with interpretation of the structure based on an atom pump model. Shaded circles represent alloy columns, solid circles Si, and open circles Ge.

possibilities depending on whether growth occurs in the same or opposite direction to the first monolayer, and with the same or opposite phase of dimerization. The ordering may either propagate, terminate, or reverse giving the three ordered phases we observe experimentally. The simulations shown in Fig. 8 are an excellent match with the actual images and allow us to see monolayer by monolayer exactly how the superlattice has grown (Jesson, Pennycook, and Baribeau, 1991).

○ Ge   ● Si

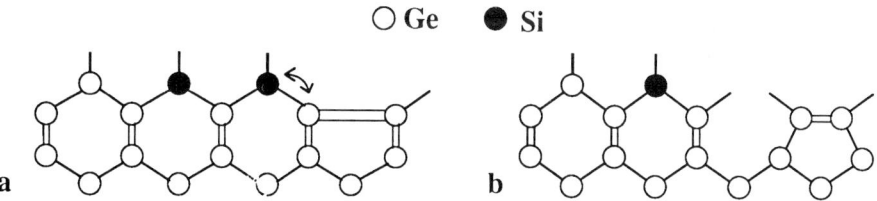

Fig. 9. Rebonded (a) and nonrebonded (b) edge configurations for the type S_B step.

Interfacial dislocations can also be imaged by Z-contrast as demonstrated in Fig. 10. Here, a 7-8 nm thick film of $Si_{.18}Ge_{.82}$ has been grown on Si by the method of implantation and oxidation which constrains the film to grow layer by layer instead of the island growth mode normally seen in such high misfit films. Pseudomorphic growth continues well beyond the predicted critical thickness just as observed in the low misfit regime, and relaxation is clearly seen to occur via the surface nucleation of 30° partial dislocations which glide to the interface leaving stacking faults. Subsequently, 90° partials follow, annihilating the faults and leaving 60° dislocations near the interface (Fig. 10a). If oxidation continues, the 60° dislocations will sweep the moving interface and react to form the more favorable 90° edge dislocations in Fig. 10b. These observations strongly suggest that at lower misfits, internal dislocation sources must be operating (Chisholm et al., 1990 and Chisholm and Pennycook, 1991).

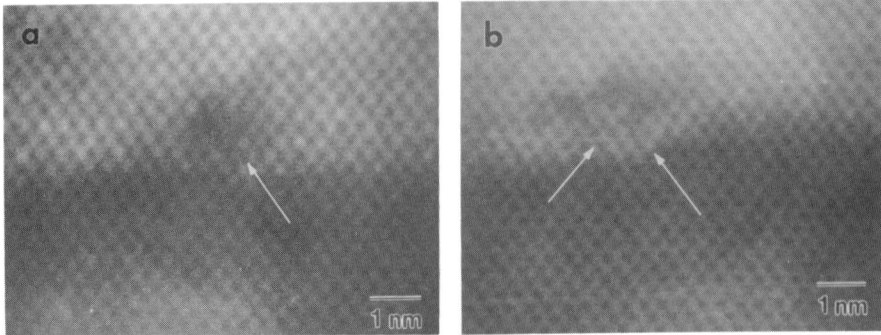

Fig. 10.   60° (a) and edge (b) dislocations lying inside the Ge layer at a Ge/Si interface.
Arrows indicate extra half planes.

Although the dislocation lines are parallel to the beam direction, surface relaxation around the
dislocation cores reduces the image intensity. Even so, since the strain field is long range
compared to the column separation, the abrupt interface can be clearly located, and the dislocation cores clearly seen to be within the elastically softer Ge layers. Since strain can be
considered as frozen in thermal vibrations (Hall et al., 1966) this is an indication of the temperature dependence of our images. Even though (from Fig. 1) we are not collecting all the
TDS from the crystal, increasing the atomic displacements **reduces** the total scattering
reaching the detector. The reduction in the dynamical wavefield at the atom sites is more
significant than the increased high-angle cross section (see Eqs. 3 and 4) and implies that
raising the sample temperature will give less image intensity.

3. CONCLUSIONS

The most important aspect of the Z-contrast technique is the intuitive nature of the images.
Since the optimum imaging conditions do not depend on the anticipated structure, and all
structures contribute to the image, unanticipated effects are immediately apparent and provide
a new depth of insight into the structure of materials at the atomic level.

The advent of a 300 kV STEM will greatly increase the range of applicability of the direct
imaging approach, allowing for example the resolution of the dumbbells in the Si$\langle 110 \rangle$ projection as shown in Fig. 11, and providing compositional sensitivity for the individual strings
comprising the dumbbell. Together with computer enhancement, images showing real
information below the 1Å level seem entirely feasible without the need for any preconceived
ideas concerning the interface structure.

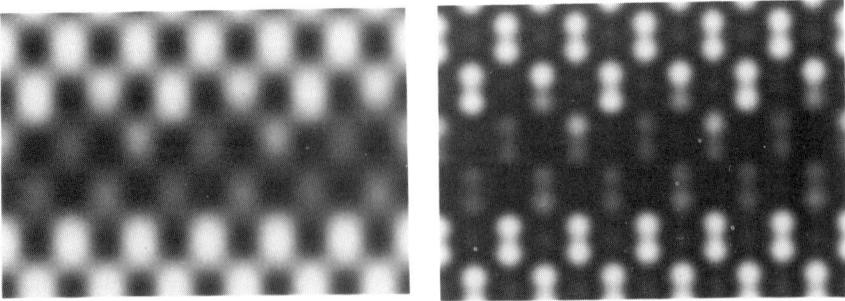

Fig. 11. Simulated images of a particular interfacial arrangement in a $(Si_4Ge_8)_p$ superlattice
at 100 kV (left) and 300 kV (right) accelerating voltage, with optimum probe
conditions and $C_S = 1.3$ mm.

Another exciting avenue to pursue is column-by-column spectroscopy using localized analytical signals. The high-intensity Z-contrast signal could be used to locate the probe over a single column while x-ray or EELS data were collected. Although in the EELS case care would be needed to average sufficiently over dynamical diffraction effects in the outgoing electron flux, such atomic resolution analytical microscopy may soon be achievable with recent advances in x-ray and EELS detector efficiencies and noise performance.

ACKNOWLEDGEMENTS

We should like to thank S. M. Yalisove for provision of the CoSi$_2$ film, J.-M. Baribeau and D. C. Houghton for provision of the ultrathin SiGe superlattices, and S. Carney, T. C. Estes, and J. T. Luck for technical assistance. This research was sponsored by the Division of Materials Sciences, U.S. Department of Energy, under contract DE-AC05-84OR21400 with Martin Marietta Energy Systems, Inc., and in part by an appointment to the U.S. Department of Energy Postgraduate Research Program at ORNL administered by Oak Ridge Associated Universities.

REFERENCES

Chadi D J 1981 Phys. Rev. Lett. 59 1691
Cherns D, Howie A, Jacobs M H 1973 Z. Naturforsch 28a 565
Chisholm M F, Pennycook S J, and Jesson D E 1990 Atomic Scale Structure of Interfaces, eds R D Bringans, R M Feenstra, and J M Gibson (Materials Research Society) p 447
Chisholm M F and Pennycook S J 1991 submitted to Phys. Rev. Lett.
Doyle P A and Turner P S 1968 Acta Cryst. A43 390
Hall C R 1965 Philos. Mag. 12 815
Hall C R and Hirsch P B 1965 Proc. Roy. Soc. A286 158
Hall C R, Hirsch P B, and Booker G R 1966 Phil. Mag. 14 979
Hamers R J, Kohler U K, and Demuth J E 1990 J. Vac. Sci. Technol. A8 195
Howie A 1979 J. Microsc. 117 11
Jesson D E and Pennycook S J 1991 submitted to Ultramicroscopy
Jesson D E, Pennycook S J, and Chisholm MF 1990 Atomic Scale Structure of Interfaces, eds R D Bringans, R M Feenstra, and J M Gibson (Materials Research Society) p 439
Jesson D E, Pennycook S J, and Baribeau J-M 1990 High Resolution Electron Microscopy of Defects in Materials, eds R Sinclair, D J Smith, and U Dahmen (Materials Research Society) p 223
Jesson D E, Pennycook S J, and Baribeau J-M 1991 Phys. Rev. Lett, 66 750
LeGoues F K, Kesan V P, and Iyer S S 1990 Phys. Rev. Lett 64 40
Loane R F, Kirkland E J, and Silcox J 1988 Acta Cryst. A44 912
Loretto D, Gibson J M, and Yalisove S M 1989 Phys. Rev. Lett. 63 298
Ourmazd A and Bean J C 1985 Phys. Rev. Lett. 55 765
Pennycook S J and Boatner L A 1988 Nature 336 565
Pennycook S J and Howie 1980 Philos. Mag. 41 809
Pennycook S J, Berger S D, and Culbertson R J 1986 J. Microsc. 144 229
Pennycook S J and Jesson D E 1990 Phys. Rev. Lett 64 938
Pennycook S J and Jesson D E 1991 Ultramicroscopy (in press)
Shin D H, Kirkland E J, and Silcox J 1989 Appl. Phys. Lett. 55 2456
Tung R T, Levi A F S, Sullivan J P, and Schrey F 1991 Phys. Rev. Lett 66 72
Wang Z L and Cowley J M 1990 Ultramicroscopy 32 275
Xu P, Kirkland E J, Silcox J, and Keyse R 1990 Ultramicroscopy 32 93
Yalisove S M, Tung R T, and Loretto D 1989 J. Vac. Sci. Technol. A7 1472

Inst. Phys. Conf. Ser. No 117: Section 1
Paper presented at Microsc. Semicond. Mater. Conf., Oxford, 25–28 March 1991

Qualitative and quantitative analysis of the GaAs/AlAs interface by high-resolution electron microscopy

S Thoma and H Cerva

Siemens AG, Research Laboratories, Otto-Hahn-Ring 6, 8000 München 83, F.R.Germany

ABSTRACT: In this paper new methods for the qualitative and quantitative analysis of the $GaAs/AlAs$ interface in $\langle 100 \rangle$ projection by high-resolution electron microscopy (HREM) at $400\,kV$ are presented. Accurate determination of the contrast-dominating parameters (defocus Δf and thickness t) and minimizing beam and crystal tilt is achieved with the help of non-linear contrast details. The proposed image-processing algorithm for composition determination is applied to HREM images of cleaved $90°$ wedge-shaped specimens recorded at "minimum contrast" (i.e. $\Delta f = -20\,nm$). Profiles of the Al content x across the interface were determined with an accuracy of ± 0.1 per atomic layer.

1. INTRODUCTION

Since interpretation of the HREM contrast in experimental images is not trivial, all specimen and microscope parameters, which essentially affect the intensity distribution, must be determined as accurately as possible in order to avoid misinterpretations. Only then a qualitative analysis of the interface between $GaAs$ and $AlAs$ by image simulations becomes possible. For quantitative analysis of the Al composition it is additionally necessary to find the optimal imaging conditions under which the contrast difference between $GaAs$ and $AlAs$ is especially emphasized, and variations of the Al content in the interface region may be directly related to contrast variations. Because of the increased contribution from the chemically sensitive $\{200\}$ reflections to the image contrast, imaging in $\langle 100 \rangle$ projection is preferred to $\langle 110 \rangle$. The aim of the work presented here has been to show both theoretically and experimentally that all quantities dominating the image intensity can be sufficiently well controlled to provide reliable qualitative and quantitative characterization of the $GaAs/AlAs$ interface. In this paper cleaved specimens consisting of molecular beam epitaxy (MBE) grown $GaAs/AlAs$ multi quantum wells (MQW) are investigated in $\langle 100 \rangle$ projection at $400\,kV$ in a JEOL 4000EX microscope having a point-resolution of $0.17\,nm$.

2. PARAMETERS DOMINATING THE IMAGE CONTRAST

The non-linear imaging theory yields a quantitative understanding of the dependence of image contrast on the microscope defocus Δf and specimen thickness t (Glaisher et al 1989). This was achieved both theoretically and experimentally for $Al_x Ga_{1-x} As$ ($0 \leq x \leq 1$) in the $\langle 100 \rangle$ projection (Thoma and Cerva 1991a). The intensity distribution results from different interference processes between the important beams, i.e. the transmitted beam, and the $\{200\}$ and $\{220\}$ reflections. It can be written as

$$I(\vec{r}) = 2 \sum_{\vec{g}} I(\vec{g}) \left[\cos(2\pi \vec{g}\,\vec{r}) + \cos(2\pi \vec{g}^R\,\vec{r}) \right], \tag{1}$$

with the image spatial frequencies $\vec{g} \epsilon \{(000), (200), (220), (400), (240), (420), (440)\}$. The vector $\vec{g}^R$ originates from $\vec{g}$ by a counterclockwise rotation of $90°$ about the zone axis. The Fourier components $I(\vec{g})$ which depend on the amplitudes and phases of the beams, the wave aberration function and the envelopes of the spatial coherence, are functions of thickness and defocus (Thoma and Cerva 1991b). The contribution of the (200) contrast to the image intensity distribution is manifested by white spots at the positions of the cations and black spots at the positions of the anions or vice versa. It can be defined by the quantity $s_1 = 4(I(200) + I(240) + I(420))$ as the difference of the intensities at the positions of the cations (III) and anions (V) is given by $I_{III} - I_V = 2 s_1$. The contribution of the (220) contrast which is manifested by white spots at the atom positions and black spots at the tunnel positions or vice versa, is defined by the quantity $s_2 = 4 I(220)$ because the difference of the intensities at the atom (A) and tunnel (T) positions is given by $I_A - I_T = 2 s_2$. Contrast reversal of the (200) and (220) spacings which result from sign changes of s_1 and s_2, can be seen in the $AlAs$ and $GaAs$ layers in the experimental image of Fig.1 which was recorded near Scherzer focus ($\Delta f = -50\,nm$). Cleaved $90°$ wedge-shaped samples have the advantage that the specimen thickness is known everywhere in the image. For an accurate determination of the defocus Δf, images of both $GaAs$ and $AlAs$ layers simulated by the EMS programs of Stadelmann (1987) for different thicknesses were matched with the experimental image (insets in Fig.1). Defocus changes strongly influence the positions of the zeros of s_1 and s_2. Hence, specimen thicknesses at which a contrast reversal of the (200) or (220) spacings occurs, are a very suitable measure for the defocus Δf of an experimental HREM image. The contrast reversal of the (220) spacings in $GaAs$ can be easily recognized by the occurence of (400) spacings, the so-called half-spacing contrast (Thoma and Cerva 1991a). This allows defocus determination with an accuracy of at least $\pm 5\,nm$. For $\Delta f = -50\,nm$ the (400) contrast shows up in $GaAs$ as half-spacings at $t \approx 10\,nm$, and in $AlAs$ as a square pattern at $t \approx 10\,nm$. Both contrasts react sensitively on beam and crystal tilt. This can be demonstrated by image simulations (Thoma and Cerva 1991b). Therefore, the appearance of clear (400) contrast in $GaAs$ and $AlAs$ represents a means to control proper microscope and crystal alignment, i.e. to minimize beam tilt below $1.4\,mrad$ and crystal tilt below $2.8\,mrad$.

3. QUALITATIVE AND QUANTITATIVE ANALYSIS OF INTERFACES

For defoci near minimum contrast at $\Delta f = -20\,nm$, the (220) contrast is dominant in $GaAs$ and the (200) contrast in $AlAs$ for thicknesses $t \leq 15\,nm$ (Fig.2). Additionally, $s_1(x)$ increases monotonically with increasing Al content x in $Al_xGa_{1-x}As$ for $5 < t < 15\,nm$, whereas $s_2(x)$ is almost constant for $t \leq 11\,nm$ (Thoma and Cerva 1991b). Hence, only the Fourier components $I(\vec{g})$ in s_1 should be used to determine the interesting chemical information from digitized experimental images (see below). The optimal imaging conditions for interface analysis, therefore, are $-30 < \Delta f < -10\,nm$ and $7 < t < 13\,nm$. Image simulations for $\Delta f = -20\,nm$ and $7 \leq t \leq 20\,nm$ show that for diffuse interfaces the virtual interface which is defined as the clearly visible boundary between the (200) and (220) contrast, is shifted more and more into the $AlAs$ layer over almost the whole diffuse interface region, when the specimen thickness increases from 7 to $20\,nm$ (Thoma and Cerva 1991b). For a sharp interface, however, the position of the virtual interface remains unchanged with increasing thickness. This enables an estimation of the width of the diffuse interface region. Furthermore, simulated images of diffuse interfaces show that the white spots at the positions of the cations lose intensity when moving from the $AlAs$ to the $GaAs$ layer. In the case of a sharp interface the white spots abruptly vanish. The dependences of the intensity contributions $s_1(x)$ and $s_2(x)$ on the Al content for the optimal imaging conditions is the physical basis of the following algorithm for composition determination. Since only the $\{200\}$ beams contain the chemical information, the spatial resolution of composition determination algorithms is given by $g_{200}^{-1} = 0.28\,nm$. Therefore, the image area is divided into cells of $0.28 \times 0.28\,nm^2$ with the cations in the center and the anions at the corners. As we are only interested in the local composition of the columns of

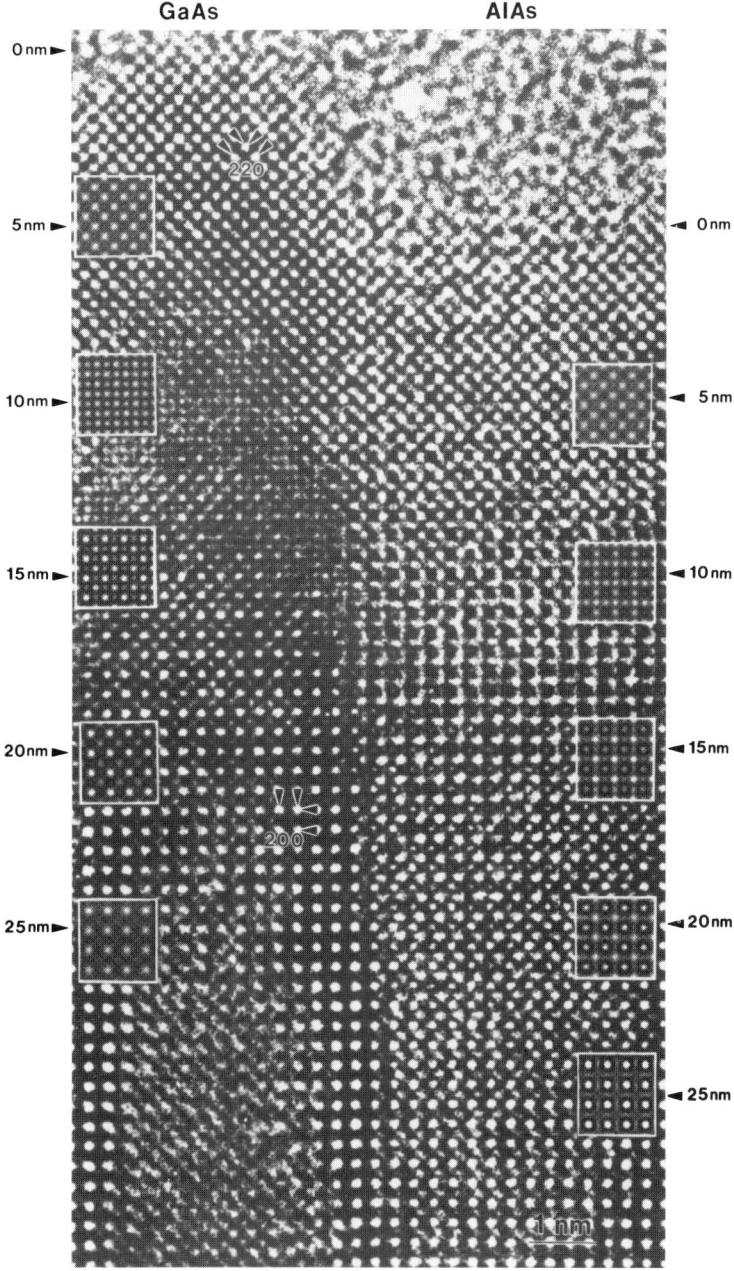

Fig.1. Experimental HREM image of a $90°$ wedge–shaped specimen consisting of $GaAs$ and $AlAs$ layers for the defocus $\Delta f = -50\,nm$. Simulated images for the thicknesses $t = 5, 10, 15, 20, 25\,nm$ are shown as insets.

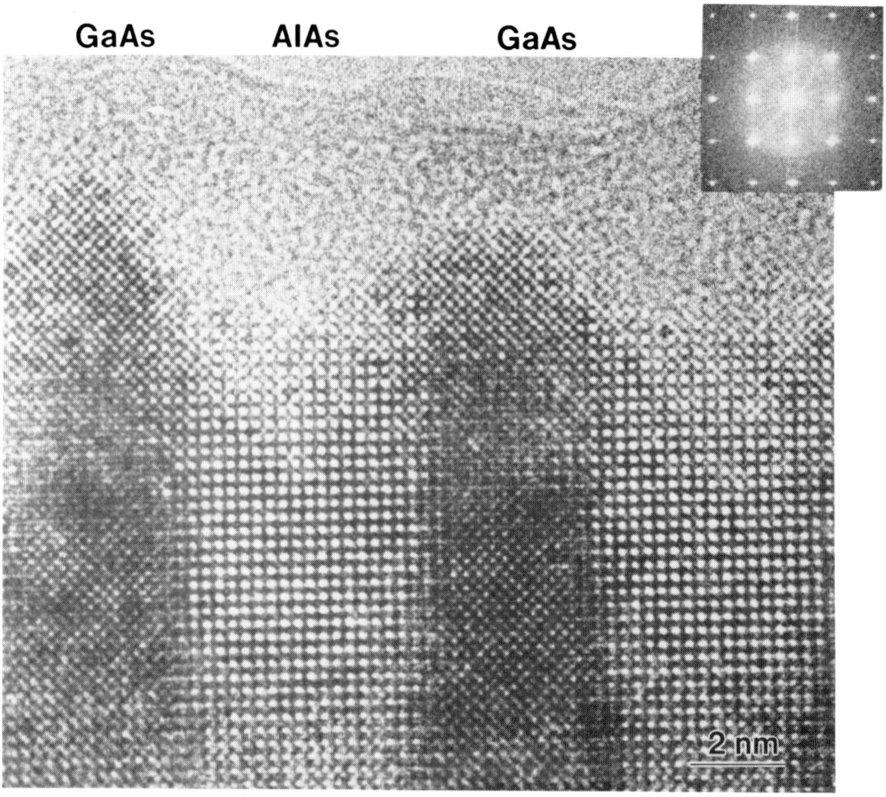

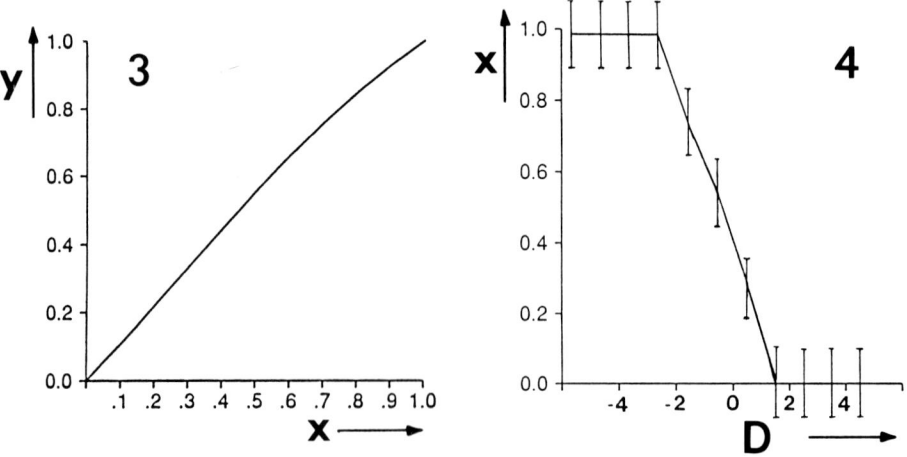

Fig.2. Experimental HREM images of a 90° wedge-shaped specimen consisting of $GaAs$ and $AlAs$ layers for $\Delta f = -25\,nm$

Fig.3. $y(x)$ vs. composition x for $\Delta f = -25\,nm$ and $t = 11.3\,nm$

Fig.4. Composition profile determined from the experimental image of Fig.2 at thickness $t \approx 10\,nm$ by using the new algorithm

cations, we calculate the chemically sensitive Fourier components $I(200)$, $I(240)$ and $I(420)$ which are directly proportional to the amplitude of the $\{200\}$ beams, from the image intensity distributions $I(\vec{r})$ within the individual cells. The quantity s_1 (for each cell) is obtained by $s_1 = 4(I(200) + I(240) + I(420))$. By the normalization

$$y(x) = \frac{s_1(x) - s_1(x = 0)}{s_1(x = 1) - s_1(x = 0)} \tag{2}$$

the chemical signal $y(x)$ becomes 0 for $GaAs$ and 1 for $AlAs$. This function calculated by the non-linear imaging theory for the optimal imaging conditions can be excellently fitted by an arcus tangens function (Fig.3). For application of the suggested algorithm on experimental images it is very important that $y(x)$ is almost insensitive to defocus variations of about $\pm 10\,nm$ and thickness changes of about $\pm 4\,nm$ (Thoma and Cerva 1991b). Testing our algorithm on simulated images of sharp and diffuse interfaces at minimum contrast shows that the composition profiles across the interfaces obtained by the algorithm deviate from the given profiles by only $\Delta x < 0.05$. Furthermore, minimized beam and crystal tilt still cause small errors in calculating the composition profile, as can be shown by image simulations. In experimental images where noise is present, it is certainly most favourable to consider only the significant Fourier components which strongly depend on x (i.e. those in the definition of s_1), for the determination of the composition. The algorithms suggested by Ourmazd et al. (1990) and de Jong et al. (1990) take all components into account. Hence, we prefer our algorithm where only the chemically sensitive Fourier components are selected from the image intensity distribution, for the quantitative analysis of experimental images. Assuming that the defocus is determined with an accuracy $\pm 5\,nm$ and the misalignments are sufficiently well controlled by the behaviour of the half-spacing contrast, the resulting systematic error for determining the composition x of an atomic layer is smaller than 0.1.

The experimental HREM image of Fig.2 recorded at minimum contrast clearly shows the shift of the virtual interface over about three atomic layers of cations from the $GaAs$ to the $AlAs$ layer with increasing specimen thickness. This means that the diffuse interface region extends over about three atomic layers of cations. Furthermore, the gradual reduction of the (200) contrast across the interface is also a proof for a diffuse interface. For the application of our algorithm the negatives were digitized by a charge-coupled-device TV camera. Because of the resolution of the camera each cell contains 16×16 pixels, from which the moduli of the chemically sensitive intensity components $I(200)$, $I(020)$, $I(240)$, $I(4\bar{2}0)$, $I(420)$ and $I(2\bar{4}0)$ are calculated by Fourier transformation. A larger number of pixels within the defined cells does not yield any further interesting information due to the granular contrast appearing in the negative which is mainly caused by the electron noise. The calculation of the moduli stabilizes the algorithm against slight shifts of the cells. The quantity $s_1(x)$ is obtained by $s_1(x) = 2\sum_{\vec{g}} |I(\vec{g})|$. For quantitative analysis of the interface it is indispensable to determine the random error of our algorithm which is caused by the quantum noise, amplified by the photographic process, and the inevitable surface roughness of the specimen. For this purpose, s_1 is calculated for several cells in the $GaAs$ layer at the same thickness, and these results are averaged yielding $\bar{s}_1(x = 0)$. In the same way $\bar{s}_1(x = 1)$ is obtained from several cells in $AlAs$. For noise reduction these quantities are used for the normalization of $s_1(x)$ of a cell with an unknown composition x (eq.(2)). In $GaAs$ and $AlAs$ 80 % of the values s_1 deviate from $\bar{s}_1(x = 0)$ and $\bar{s}_1(x = 1)$ by less than $\sigma(x = 0)$ and $\sigma(x = 1)$. With $\sigma(x = 0)$ and $\sigma(x = 1)$ the random error is given by

$$\Delta x = \frac{\max(\sigma(x = 0), \sigma(x = 1))}{\bar{s}_1(x = 1) - \bar{s}_1(x = 0)} \tag{3}$$

The inverse ratio gives the signal-to-noise ratio of our algorithm. For the experimental images of Fig.2 the error is about 0.1 and the signal-to-noise ratio, therefore, about 10. Neglecting the systematic error of the algorithm this means that the Al content x of an individual cell can be determined with an accuracy of ± 0.1. For the determination of the composition profile the quantity $s_1(x)$ was averaged over four cells along the interface in order to get a higher accuracy which is $\Delta x \approx \pm 0.1$ due to the random and systematic error. This was performed at various thicknesses between 7 and 13 nm. The diffuse interface region extends over three atomic layers with average compositions of $x = 0.28, 0.54$ and 0.76 (Fig.4). The compositional fluctuations within the individual atomic layers which are generated by the random error of the algorithm and the actual fluctuations, are about ± 0.1.

4. DISCUSSIONS AND CONCLUSION

In this paper we investigated interfaces between $GaAs$ and $AlAs$ MBE layers, but the proposed qualitative and quantitative procedure can be also used for the analysis of $GaAs/AlGaAs$ interfaces. Provided that chemically sensitive reflections exist, it should be straight forward to modify our algorithm for interfaces of other heteroepitaxial structures of semiconductors, metallic and superconducting materials.

For thorough examinations of the $GaAs/AlAs$ interface it is necassary to know and control the parameters dominating the contrast. The defocus at which the HREM image is recorded, can be accurately determined with the help of the thicknesses where the half-spacing contrast appears. Beam and crystal tilt can be minimized by the behaviour of fine contrast details. The width of the diffuse interface region is estimated by the shift of the virtual interface with increasing thickness in experimental images of cleaved wedge-shaped samples at $\Delta f \approx -20\,nm$. The optimal imaging conditions for quantitative analysis of the interface between $GaAs$ and $AlAs$ were found to be $-30 < \Delta f < -10\,nm$ and $7 < t < 13\,nm$. We propose a new algorithm for composition determination with a spatial resolution of $0.28\,nm$. Since our algorithm takes into account only the chemically sensitive Fourier components of the intensity distribution in experimental images, we obtain noise reduction compared to earlier proposed algorithms. We applied our algorithm to experimental images of cleaved samples in order to quantify the composition profile across the interface. The achieved accuracy in determining the composition per atomic layer is $\Delta x \approx \pm 0.1$.

REFERENCES

De Jong A F and Van Dyck D 1990 Ultramicroscopy 33 269
Glaisher R W, Spargo A E C and Smith D J 1989 Ultramicroscopy 27 19
Ourmazd A, Baumann F H, Bode M and Kim Y 1990 Ultramicroscopy 34 237
Stadelmann P 1987 Ultramicroscopy 21 131
Thoma S and Cerva H 1991a accepted for publication in Ultramicroscopy
Thoma S and Cerva H 1991b submitted to Ultramicroscopy

Inst. Phys. Conf. Ser. No 117: Section 1
Paper presented at Microsc. Semicond. Mater. Conf., Oxford, 25–28 March 1991

Analysis of diffuse and abrupt HgTe/CdTe interfaces under chemically sensitive imaging conditions

R E Mallard[+], G Feuillet, N Magnea, H Mariette and E Ligeon

Equipe CNRS-CEA "Microstructures à Semiconducteurs II-VI", Centre d'Etudes Nucléaires de Grenoble, 38041, 85X, Grenoble CEDEX, France

+ present address: Department of Material, Parks Road, Oxford, OX1 3PH, United Kingdom

ABSTRACT: Under appropriate analysis conditions of specimen thickness and defocus, HREM images of semiconductor heterointerfaces imaged in the <100> orientation may be obtained in which the differences in lattice fringe motif between the constituent layers are maximized. With the aid of accompanying analysis by image simulation, the interfacial chemical structure may often be illuminated at monolayer resolution using relatively simple image processing techniques by measuring comparative lattice fringe intensities. We present an HREM analysis of the structure of MBE grown HgTe/CdTe interfaces. This system is particularly interesting due to the interfacial contrast available because of the relative phase differences in the {200} fringes occurring in each layer. We demonstrate that the as-grown interfaces have an interfacial abruptness of one monolayer, and furthermore, that small degrees of interfacial roughness and diffusivity induced by ion implantation may quantitatively be measured using HREM.

1. INTRODUCTION

In the analysis of the morphological and chemical character of compound semiconductor/semiconductor interfaces, an important analytical technique is the (002) dark field (DF) transmission electron microscopy (TEM) technique. The (002) structure factor and hence (002) diffracted amplitude, unlike that for other low index sphalerite reflections, is particularly sensitive to compositional variations because it depends on the difference, rather than the sum, of the electronic scattering factors for each individual face centred cubic sublattice comprising the lattice. The assignment of which group of chemical species occupies which FCC sublattice is entirely arbitrary, but determines the sign of the (200) structure factor, $\mathcal{F}_{200}$. As the image intensity itself is proportional to the square of the diffracted amplitude, it is however insensitive to this arbitrary choice. The general trend in the values of the elemental electronic scattering factors is to increase with increasing atomic number, and compound semiconductors with large differences in atomic weight are generally thus imaged as bright regions in (002) DF micrographs.

The same property has been exploited in high resolution electron microscopy (HREM) of multilayer interfaces, especially at the (100) pole, where four of the "chemically sensitive" {200} reflections occur (Heatherington et al 1985, Ourmazd et al 1989). HREM contrast however differs from conventional two-beam microscopy in that the contrast is strongly dependent on the phase of the diffracted information because the image is formed by the interference of a number of scattered beams. Under certain analytical conditions, these intensity/phase relationships may be such that these interference processes are different for regions of differing composition, the occurrence of which is more pronounced when forming the image with a combination of chemically sensitive reflections. In the present analysis, the phase of the {200} diffracted beams is of special significance, as it not only influences the extent to which the corresponding {200} lattice fringes are visible, but also the position of the intensity maxima and minima in the image, which may itself be exploited to yield chemical contrast.

In the HREM image formation process, the electron wavefront, having passed through a single thin slice of specimen, or phase grating (corresponding to the hypothetical weak phase object), is composed of an intense forward scattered beam and a number of relatively weak diffracted beams, the phase difference of which relative to the forward scattered beam is $\pi/2$. Importantly for the {200} reflections, the sense of this phase difference is reversed when the atomic species occupying the sublattices are reversed, and indeed is the sign of the (200) structure factor as we have previously described. Reconstruction of images from diffracted wavefronts at a resolution limited to {200} periodicities, that is, by allowing only the innermost five beams at the (100) pole to pass through the objective aperture, means that only one of the sublattice is resolved. Bright spots occur on either the cation or anion sites depending on the sign of the generalised transfer function for the (200) reflection, T_{200}, which is equal to $\cos(\phi_{200} - \phi_o + \chi_{200})$, where ϕ_{200} is the phase of the (200) diffracted beam, ϕ_o is the phase of the forward scattered beam, and χ_{200} is the phase shift of the (200) diffracted beam due to the spherical aberration of the objective lens (Bourret et al 1975, Glaisher 1989). For a wide range of analysis conditions in the CdTe/HgTe system, the site upon which the bright image intensity occurs is different for the two materials because the sign of the (200) structure factor, chosing a consistent origin, is different for the two materials; {200} lattice fringes traversing the interface thus experience a phase shift of π at the interface, and this situation may be exploited as a means of recognising the interface position and abruptness. The actual image formation mechanism becomes much more complicated with the inclusion of nine beams in the objective aperture (including the four {220} reflections present at the (100) pole, and thus allowing the full resolution of both the cation and anion sites). A similar image characteristic may nevertheless be obtained when the {200} beams from each material make a significant contribution to the overall contrast and remain "out of phase" across the interface.

2. EXPERIMENTAL

A multilayer structure consisting of thick CdTe and HgTe layers followed by a single HgTe well bordered by $Cd_{0.68}Hg_{0.32}Te$ was grown on a [001] oriented $Cd_{0.96}Zn_{0.04}Te$ substrate by molecular beam epitaxy. Cross section specimens were prepared in the [100] orientation by conventional Ar^+ ion milling, and examined in the JEOL 4000EX with a point to point resolution of 0.16nm at extended Scherzer defocus. Multislice image simulation experiments were performed, allowing the determination of the location of the atomic species with respect to the observed intensity variations. Interpretation of the experimental images was simplified using the SEMPER processing system in the following manner: following the high pass filtering of the images to filter out any "background" intensity variations, a simple measurement of the image intensity along specific atomic planes is obtained by projecting the image intensity along <100> directions. Specifically, projection along the [010] direction (parallel to the interfacial plane), enables us to monitor the relative cation/anion intensities for each material, and to measure the interfacial abruptness, averaged over the width of the projected area, by assessing the distance over which the character of the contrast changes in the intensity projection. For comparison to the interfacial structure of the abrupt as-grown material, a section of the same slice was ion implanted with Zn^{++} to a dose of $10^{15}cm^{-2}$ at 180keV, producing an expected interfacial intermixing, and re-examined in the HREM. The implantation was performed at room temperature and at low current in order to minimise local heating effects. Under these conditions, we expect the peak in the Zn concentration to appear at a depth of approximately 100nm from the specimen surface.

3. RESULTS AND DISCUSSION

Figs 1a and b are (200) DF micrographs of the single well structure used in the study, before and after ion implantation. We are primarily interested in the structure of the CdTe/HgTe interface occurring at a depth of 250 nm from the specimen surface. For the alloy compositions contained in the structure, fig 1a exhibits an increasingly bright contrast with increasing x in $Cd_xHg_{1-x}Te$, although the value of $\mathcal{F}^2_{200}$ does not necessarily monatomically increase over the entire range of x from zero to one. Prior to ion implantation all interfaces appear to be abrupt and planar to within the approximately 0.5nm resolution of the technique. In fig 1b, we observe that a substantial degree of interfacial mixing occurs as a result of the ion implantation, particularly for layers within 200 nm of the surface. This effect is especially pronounced at the surface, where the thin HgTe well has been almost completely smeared out. The degradation of the abruptness of the bottom interface of the deeper HgTe layer is much less apparent, and occurs over a distance not greatly exceeding the resolution of the technique.

Fig 2a is an HREM micrograph of the as-grown CdTe/HgTe interface in the [100] orientation, corresponding to a thin region of the foil and at a microscope overfocus of 40nm, as confirmed by optical diffractogram analysis.

Fig 1. (002) DF micrographs of (a) as grown and (b) ion implanted CdHgTe multilayer structure. The implanted material has undergone extensive intermixing in the vicinity of the surface, and contains bands of dislocations at depths from approximately 120 to 200nm.

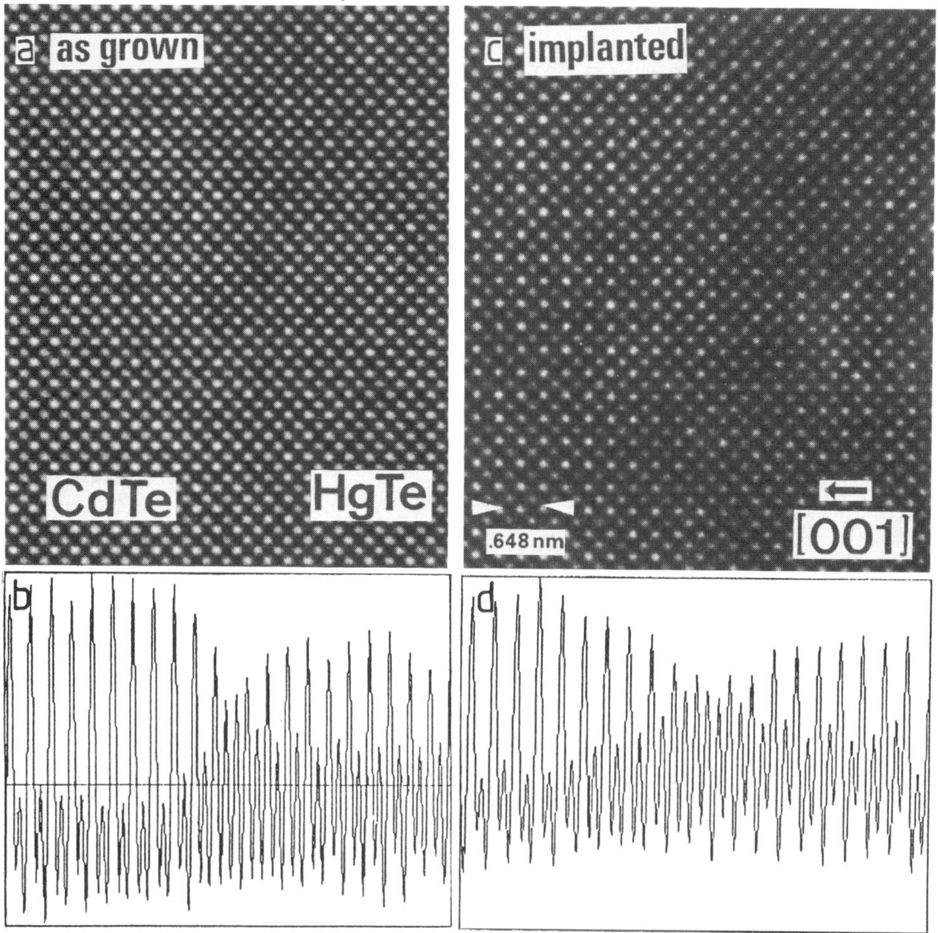

Fig 2. HREM micrographs and total image intensity projections of the HgTe/CdTe interface at a depth of 250nm from the specimen surface. (a) as grown material, (b) intensity projection along [010] corresponding to the image in (a), (c) ion implanted sample, from the same region as in (a), (d) intensity projection along [010] from the image in (c). Note the relative strength of the {200} fringes in CdTe and the inversion in cation/anion relative intensity when traversing the interface.

The objective aperture size was 7nm^{-1}, allowing the transmission of 13 beams and the full resolution of all atomic positions in the projection. A projection of the digitised image intensity along the [010] direction (that is, parallel to the interface) is shown in fig 2b. Image simulation experiments show that for the chosen microscope conditions, we obtain an "atoms appear white" condition, and the position of an atomic column will therefore appear as a bright peak in the image projection. The projected intensity trace illustrates clearly the dominance of the {200} periodicities in the formation of the CdTe image, which is predominantly of anion-type contrast. There is a large corresponding difference in intensity between the peaks representing columns of Te atoms in CdTe (strong) and Cd atoms (relatively weak). Reflecting a change in T_{200} across the interface, a cation-type contrast is indicated on the HgTe side of the interface, but is of reduced importance, as indeed would be expected from the trend in (200) DF intensities revealed in fig 1a. In the region of the interface itself we observe a band of intermediate contrast, 2 atom layers in width, which cannot unambiguously be ascribed to either CdTe or HgTe without the aid of interfacial image simulations. Comparison to simulations indicates that the observed contrast across the interface is consistent with an interfacial abruptness of one to two monolayers over the length of interface from which the intensity projection was obtained (approximately 10nm).

The somewhat diffuse nature of the same interface subsequent to ion implantation is apparent in the HREM micrograph and intensity traces of fig 2c and d. Analysis conditions are similar to those in fig 2a. Although we again observe the characteristically strong {200} periodicities on either side of the interface, the width over which the cation/anion inversion takes place increases to approximately 6 atom layers, or 1nm, accompanied by a gradual change in the overall contrast across this region. It would therefore appear that in this case, the ion implantation induces some intermixing even at a depth of 250 nm from the surface, though the extent of spreading occurs over a width just exceeding the resolution of conventional TEM techniques.

4. SUMMARY

We describe a simple HREM technique for the analysis of interfacial abruptness of compound semiconductor interfaces when a change in sign of the generalised transfer function for the compositionally sensitive {200} reflections, T_{200}, occurs across the interface. In spite of generally low interlayer contrast levels to the naked eye, simple image processing methods may be used to clearly delineate the position and structure of the interfaces. Intermixing on a scale which is unobservable using conventional TEM methods has been quantitatively measured with atom scale precision.

ACKNOWLEDGEMENT

We gratefully acknowledge J Thibault-Desseaux, P Gentile, L Di Cioccio, and E Hewat for invaluable assistance with the work, and to CEA/CENG and the Université Joseph Fourier for the provision of a post-doctoral research fellowship for REM.

REFERENCES

Bourret A, Desseaux J and Renault A, 1975, Acta Cryst, A31, 746
Glaisher R W, 1989, in Proc NATO ASI on Evaluation of Advanced Semiconductors by Electron Microscopy, ed D Cherns, (London:Plenum), 1
Heatherington C J D, Barry J C, Bi J M, Humphreys C J, Grange J, and Wood C, 1985, MRS Proc, 37, 41
Ourmazd A, Taylor D W, Cunningham J, and Tu C W, 1989, Phys Rev Lett, 62, 933

Inst. Phys. Conf. Ser. No 117: Section 1
Paper presented at Microsc. Semicond. Mater. Conf., Oxford, 25–28 March 1991

The use of the Fresnel Method for the measurement of the compositional profile of the interfaces of isolated sub-unit-cell-thick layers of Si in Ge

R E Dunin-Borkowski, W C Shih and W M Stobbs

Department of Materials Science and Metallurgy, Cambridge University, Cambridge, UK

ABSTRACT: The Fresnel Method is used here to determine the compositional abruptness of the interfaces in a multilayer which was grown to have, within a period of the structure, a two monolayer thick Si layer and a much thicker Ge layer. The method has to be applied in different ways for the compositional profiling of fine period layer structures and isolated layers. Data obtained for short wavelength multilayers has indicated that such interfaces can be very sharp and here we examine the potentially simpler application of the method which is possible for well separated layers and discuss the implications of the results in relation both to the accuracy of the method and to the form which the interfaces take.

1. INTRODUCTION

The compositional abruptness of the interfaces in a semiconductor heterostructure can have profound effects upon its electrical and optical properties; the importance of an accurate assessment of this characteristic of the structure is, for example, demonstrated in relation to the photoluminescence behaviour of an $Al_xGa_{1-x}As/GaAs$ tunnelling structure elsewhere in these proceedings (Rimmer et al. 1991). Monolayer accuracy is often needed, particularly for fine structures, if the data is to be really useful in making the behavioural modelling uniquely testable and there are few techniques which can be applied to this accuracy. Compositional mapping approaches using lattice fringe imaging, as normally applied, neglect interface Fresnel effects (Stobbs 1990). High Angle Dark Field (HADF) methods have the spatial accuracy required but the contrast need not be simply Z dependent and can be strongly affected by the thermal diffuse scattering due to lattice distortions (Perovic et al. 1991). Using the Fresnel Method (as recently reviewed by Ross et al. 1991a) it is also probable that, until field emission sources are used and energy filtering becomes available, ambiguity will remain at the atomic level. However we have successfully applied the approach to the characterisation of compositional profiles in layered AlGaAs system heterostructures (Ross et al. 1988, 1991b) to about the level required but these systems have fairly diffuse interfaces. There are few interfaces that really are abrupt at a level which can test the potential problems in the method due to inelastic scattering. It has thus been exciting that a fine Si/Ge multilayer, as appraised using a developed form of the method taking advantage of the effects of scattering from the periodic structure, was found to have abrupt Si/Ge interfaces to near monolayer (a/4) accuracy (Shih et al. 1991). We have been concerned that there could be problems in the method due to the neglect of contributions to the contrast from inelastic/elastic effects when it is applied for such sharp interfaces. We have thus undertaken an appraisal of both the interfaces and the method as applied to relatively isolated Si layers in Ge for which the contrast changes at low defoci are more easy to interpret than are those at higher defoci or for finer multilayers. The initial results of this appraisal are described here.

2. RESULTS OF FRESNEL METHOD APPRAISAL OF Si/Ge INTERFACES

The Si/Ge multilayer examined had a wavelength, (λ), of 3.36 nm, nominally comprising 2 monolayers of Si and 22 of Ge, and we are grateful to British Telecom for providing it. The long wavelength of this specimen, by comparison with that previously assessed (λ=1.34 nm), allowed the examination of the relatively lower contrast individual Fresnel effects for each Si layer, as are seen before strong overlap and multilayer effects dominate the contrast. For the specimen described here overlap effects started to dominate the behaviour only by a defocus of about 600nm so the Fresnel effects (which increase in strength with both foil thickness and defocus for the fairly low ΔV (1.7V) for this system) could be examined for the individual layers at thicknesses for which the effects of layer undulation and/or inelastic scattering should not have been dominant. Nevertheless care had to be taken experimentally both to maintain a low contamination level (causing background phase contrast noise) and to ensure that the edge-on specimens examined had the layering (which was of very low vicinality to the [001] growth direction) vertical for the Fresnel series taken. The experimental methods used, and the general methods needed for the appraisal of the contrast are described elsewhere (Ross et al. 1991a). The Fresnel series examined was taken with a JEOL2000FX at 200 kV using a fairly large objective aperture (Airy disc radius 0.2nm) and a beam convergence of 0.42mr at about 12° from (010). The conventional atomic (rather than mean potential) multislice simulations computed for the assessment of the experimental image series used consistent parameters to the above, Cs=2.5mm and a $\Delta f/f$ value of 1/3 Scherzer defocus. The structure model simulated assumed Si to maintain a constant volume and neglected surface relaxation.

Examples of images from the series taken, at defocus steps of about 37nm, are shown in fig.1 for a range of Δf values above and below the range (around -450nm) where a double peaked feature of the contrast seen between the Si layers proved to be sensitive to the form of the interface composition profile. The high quality of the layering examined is demonstrated by the examples of the contrast profiles shown in fig.2 for averages over distances along the layering of 3.8 to 25 nm (the averaging direction was optimised to a sensitivity of 7.5 mr). The contrast features are defined in fig.3. A set of digitised profiles for coarse Δf differences, and for a foil thickness up to about 100nm, is shown in fig.4. Amongst a wealth of effects we can see the "drift" of contrast features seen at lower defocus levels in thinner foil regions to higher defoci in thicker regions. The region examined in more detail as a function of defocus around -400nm in fig.5, at a thickness of about 80nm, is marked on fig.4. We can see for this area the changes from a triple to a double to a single (strongly overlap affected) peaked character of the fringes between those centred at the marked Si layers as the defocus is increased, noting that the profiles were obtained as averaged over 13nm along the layering.

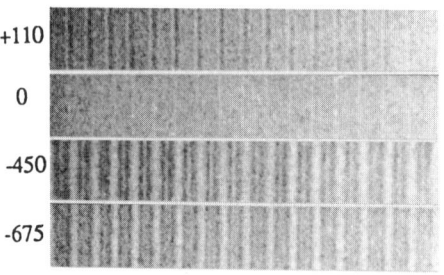

+110

0

-450

-675

Fig1 : Through-focal images of a 3.36 nm wavelength Ge/Si superlattice at the defoci shown (nm).

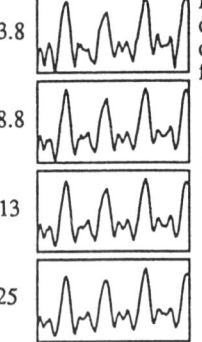

3.8

8.8

13

25

Fig 2 : Digitised Fresnel contrast profiles averaged over the lengths shown (nm) for a defocus of -450nm.

Fig 3: Defns of I_c, I_f, I_s,I_o.

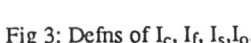

I_c

I_o I_s

I_f

Simulated profiles (at a foil thickness of 80nm) are shown in fig.6 for the interesting defocus values and for the interfaces as atomically abrupt, (0), and as graded over one monolayer, (1), as well as over 1.5 monolayers. The main feature which we should be able to use fairly easily in assessing the experimental profiles (since it changes rapidly with the degree of grading) is the contrast of the central double peaks relative to that of the Si peak. In order to do this we can compare the behaviour of $(I_c-I_f)/I_o$ and $(I_c-I_s)/I_o$ as a function of defocus for the simulated and experimental profiles. We can see immediately from fig.6 that an abrupt interface should show changes of sign of $(I_c-I_s)/I_0$ and this is not seen experimentally. A quantitative appraisal is given by the graphs of $(I_c-I_f)/I_0$ and $(I_c-I_s)/I_0$ as a function of defocus in fig.7 for the experimental data and in fig.8a and b, with a coarser contrast scale, for the three models (exemplary profiles for which are shown in fig.6) as now compared with the experimental values (now as averaged for three thicknesses). The ranges of defocus over which the profiles exhibit triple, double and single intermediate peaks experimentally are marked as 3,2 and 1 respectively in fig.7. For "2" we see an interesting dip in the values of both contrast measures (fig.7) which can be associated with the effects of "overlap" even for the low defoci used. A Fourier approach suggests that this behaviour would indicate a fair degree of squareness of the Si potential profile and yet we have the clear contradictory evidence of the negative values of $(I_c-I_s)/I_0$ for the square model which are not found experimentally (fig.8b). On the other hand the defocus values for the dips (fig.7) alluded to above relate to those at which this feature is seen for the abrupt but not for the more diffuse models (fig.8a) though again the contrast values do not relate well. The experimental contrast is low, and its changes with defocus are less than predicted. This indicates the effects of multiple inelastic/elastic scattering. Further spreading the potential will not yield better fits with the data in fig.8a. Equally, if more qualitatively, comparison of the relative behaviours of the central and second peaks with that of the central and third (at lower defoci) for the three models (fig.6) shows how, while increasing the spread might give better local (in defocus) values of $(I_c-I_s)/I_0$ this would make the third peak, centrally positioned between the Si layers, stronger than that at these layers. Such behaviour is not observed. At this point we should note that experimentally we can see a variety of features of the contrast which indicate rather clearly irregularities in the spacings of the layering. The point of the exercise here was to examine the Fresnel effects for "isolated" layers but the strengthening of the third inter-Si peak relative to other features is a clear indication in the modelling of the regularity of the model layer spacing and related effects must occur for the second peak which could be making the apparent relative agreement of the Fresnel data to the spread relative to the abrupt model in fig.8b illusory. Surprising though this may seem, and remembering that it is contrast form changes with defocus that are more reliably interpreted than their values, there is no good evidence here which is necessarily indicative of the layer interfaces being spread by even a monolayer, particularly when it is noted that the use of the multilayer effects in the Fresnel contrast for finer spaced layers was well interpretable for an interface form intermediate between abrupt and as spread by a monolayer (Shih and Stobbs 1991). This is also the form of the interface which would be expected from the ordered roughness seen by HADF techniques (Jesson et al. 1991).

We may conclude that the real problem here is that we have been forced by low experimental contrast levels to thicknesses at which the effects of layer undulations are bound to smear the contrast changes which we want to examine as will contributions to the contrast from multiple inelastic/elastic scattering as discussed by Ross et al.(1991a). We are now assessing further ways of averaging data along the layers for thinner foils which do not lose the local contrast while comparing these data with models for more isolated layers to reduce further the effects of regular multilayer periodicity. If the agreement we have found to date between theory and experiment seems poor it should be noted that we are attempting to assess the compositional form of a layer only half a unit cell thick. We thank the SERC and GEC for financial support.

REFERENCES

Jesson D E, Pennycook S J and Baribeau J-M in "Evolution of thin film and surface microstructure" eds: C V Thompson, J Y Tsao and D R Srolovitz 1991 MRS in press
Perovic D D, Weatherley G C and Houghton D C 1991 Philos. Mag. in press
Rimmer N, Syme R T, Frost J E F, Ritchie D A, Jones G A C, Kelly M J and Stobbs W M 1991 these Proceedings
Ross F M, Bithell E G and Stobbs W M 1988 Proc.AEM ed.G.Lorimer (IOM) pp 205-208
Ross F M and Stobbs W M 1991a Phil. Mag. A63 37
Ross F M and Stobbs W M 1991b Ultramicrosc. in press
Shih W C and Stobbs W M 1991 Ultramicrosc. in press
Stobbs W M 1990 AMP 2 ed: H.E.Exner and V.Schumacher (DGM Verlag) pp 1007-1017.

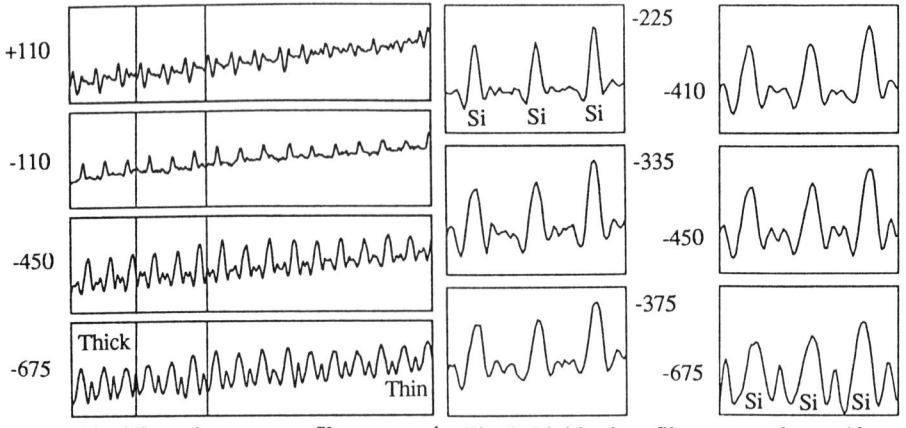

Fig 4 : Digitised Fresnel contrast profiles averaged over 13nm for the defoci shown (nm) and for a thickness range of between 20nm and 100nm.

Fig 5: Digitised profiles averaged over 13nm for the defoci shown (nm).

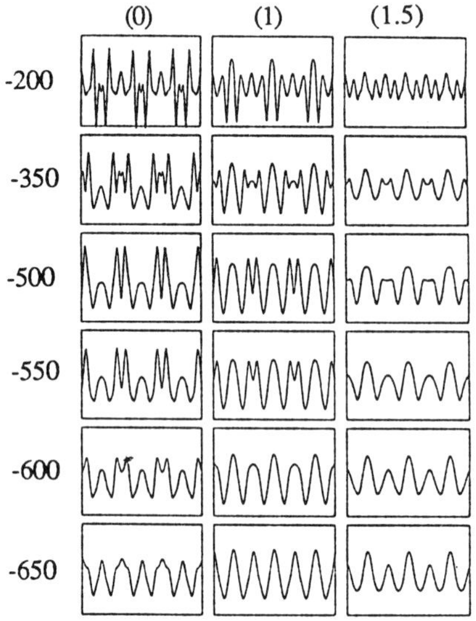

Fig 6 : Abrupt (0), a/4 (1) and 3a/8 (1.5) simulations for the defoci shown (nm) and at a thickness of 80nm.

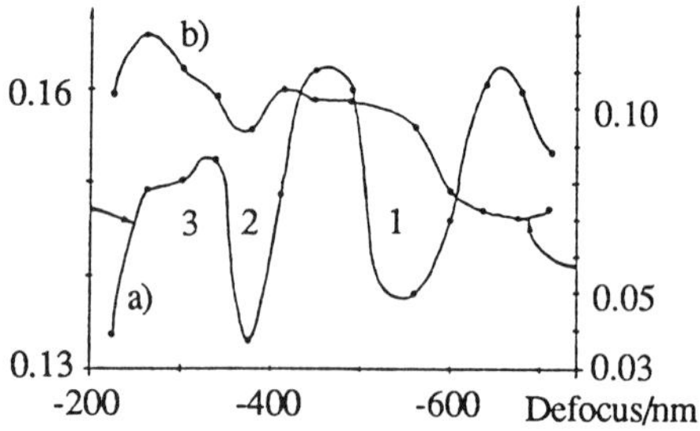

Fig 7 : Experiment :
a) $(I_c-I_f)/I_o$
b) $(I_c-I_s)/I_o$

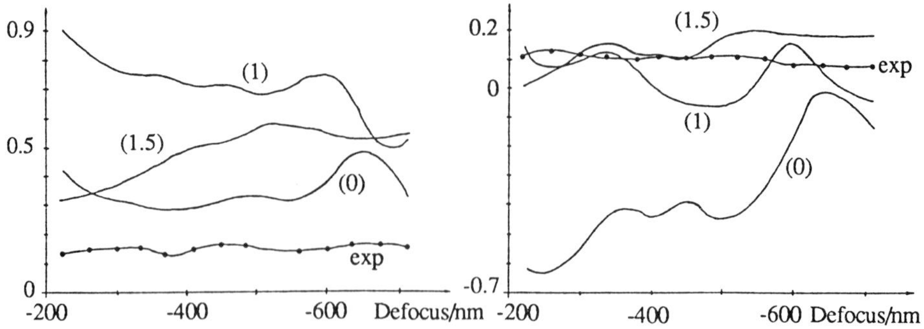

Fig 8 : A comparison of experimental and theoretical for abrupt (0), a/4 (1), 3a/8 (1.5) for
a) $(I_c-I_f)/I_o$ and for b) $(I_c-I_s)/I_o$.

Inst. Phys. Conf. Ser. No 117: Section 1
Paper presented at Microsc. Semicond. Mater. Conf., Oxford, 25–28 March 1991

Electron microscopic observation of five-fold symmetry in germanium particles crystallised in their amorphous matrix

V Bykov[1], H Hofmeister[2], T Junghanns[1] and S Nepijko[1]

[1]Institute of Physics, Academy of Sciences of the UkSSR,
Prospekt Nauki 46, 252650 Kiev, Ukraine
[2]Institute of Solid State Physics and Electron Microscopy,
Weinberg 2, 0 - 4050 Halle (Saale), Germany

ABSTRACT: Crystalline particles grown in amorphous films during vapour deposition of Ge on NaCL substrates have been studied by transmission electron microscopy. Frequently they have a fivefold twinned structure and a facetted outline. In some cases they exhibit several interlinked cyclic twins and parallel twins with segments of subgrain boundaries in between. Based on cross twinning owing to elastic strains a mechanism is proposed for the formation of such twin networks observed for the first time.

1. INTRODUCTION

The formation of multiply twinned structures during solid phase crystallization of amorphous semiconductors (Nachodkin et al. 1987, Sinclair 1990) may be regarded in close relationship to the formation of multiply twinned particles (MTPs) of Si and Ge (Saito et al. 1978, Iijima 1987). Such particles and structures cannot be attributed to one common mechanism of formation. Besides the nucleation of clusters with non-crystallographic stacking of atoms (Gillet 1977, Phillips 1979) growth twinning (Hofmeister 1984, Sinclair 1990) has to be taken into consideration as well as deformation twinning (Hall & Fawzi 1986, Hofmeister et al. 1991). We present a characterization by electron microscopy of multiply twinned structures formed in the amorphous matrix during deposition of Ge thin films and discuss mechanisms involved in the formation of the observed structures. Thereby main attention is directed on the crossing of microtwins because of internal elastic strains.

2. EXPERIMENTAL

Amorphous films of Ge containing small crystalline particles (5-25 nm in size) have been prepared by vapour deposition on air-cleaved NaCl crystals under a vacuum of 5 x 10^{-5} Pa. The vapour deposition was done at substrate temperatures between 200 and 350 °C up to layer thicknesses between 10 and 30 nm at a rate of about 0.9 nm min^{-1}. For the evapo-

ration of highly pure Ge (less than $2 \times 10^{14} cm^{-3}$ impurities) a resis-
tance-heated Ta coil was used. Electron microscopy examination of the Ge
films detached from the substrates and mounted on copper grids was done
at a JEM 100 C microscope operating at 100 kV and a JEM 4000 EX microscope operating at 400 kV, respectively. High resolution electron microscopy (HREM) images of particles in <110> zone axis orientation were
taken with the objective lens slightly defocused to allow imaging of the
channels characteristic of this projection as oblong bright dots.

3. RESULTS AND DISCUSSION

Crystalline particles are formed within thin films of amorphous Ge during vapour deposition if the substrate temperature is nearly 200 °C
or more. Fig. 1 (a) shows the bright field image of a Ge film of 15 nm
thickness obtained at 225 °C. The crystalline nature of the particles is
clearly revealed by the electron diffraction pattern of this area shown
in Fig. 1 (b). Furthermore a slight preference of the (110) orientation
of the particles may be deduced from this.

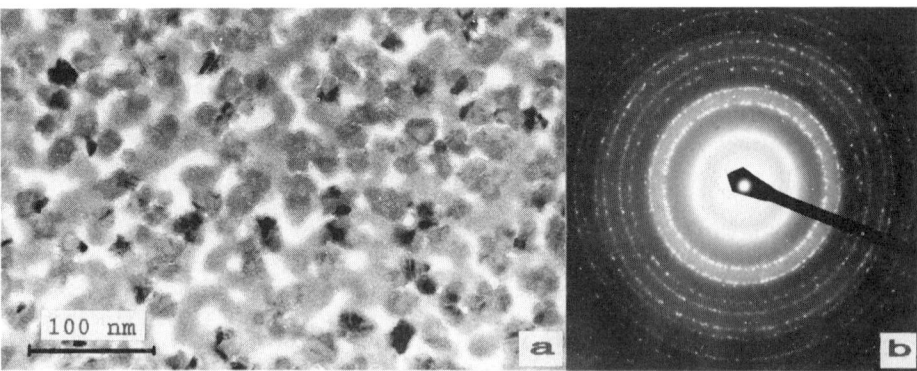

Fig. 1: (a) Amorphous Ge film with crystalline particles grown at 225 °C
(b) electron diffraction pattern of the area imaged in (a).

A more thorough inspection of the particles by HREM reveals clear deviations from a spherical shape. Their internal structure is characterized
by the presence of numerous microtwins. The growth of the particles may
proceed in <112> directions along existing twin boundaries (Nakamura et
al. 1989) or by faulty stacking of atoms so to form new twin boundaries
(Sinclair 1990). Typically the particles exhibit a multiply twinned
structure with the fivefold axis parallel to the overall growth direction of the film. The fivefold cyclic twins may originate from nucleation or from successive twinning during growth. Segments of subgrain
boundaries occur if there is no twin relation between neighboured regions of the particle. Re-entrant faces and grooves at the outer termination of twin boundaries and subgrain boundaries, respectively, lead to
a facetted outline of the particles.
The structural peculiarities described above can be found at the particle shown in Fig. 2. Furthermore this particle exhibits a network of

twins consisting of several interlinked cyclic twins and parallel twins with segments of subgrain boundaries in between. Twin boundaries and subgrain boundaries are marked by dashed lines and thin lines, respectively, in the figure. The four configurations where cyclic twins meet in a fivefold axis are denoted by the letters A to D.

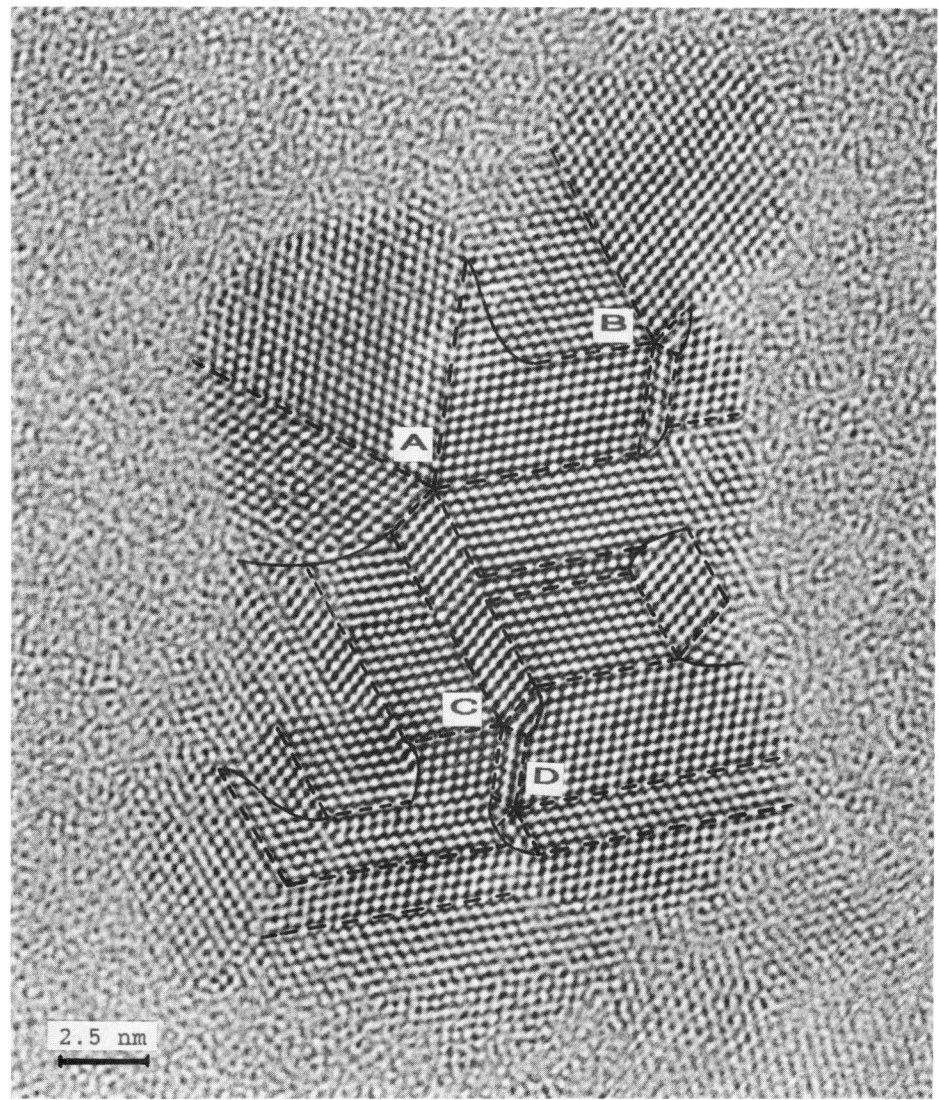

Fig. 2: HREM image of a crystalline particle of Ge with a network of several cyclic twins (A,B,C,D) and additional parallel twins.

Since such structures cannot be attributed to the formation by nucleation or by growth twinning a mechanism is proposed which is based on the crossing of microtwins under the influence of elastic strains

originating at the amorphous-crystalline interface. The corresponding
model is shown in Fig. 3. After a first twin (T1) has passed a matrix

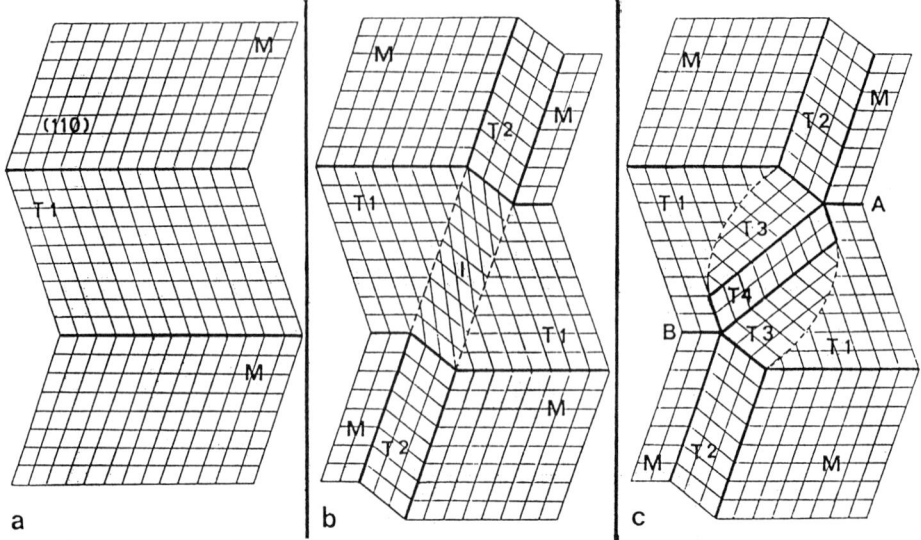

a   b   c

Fig. 3: Formation of multiply twinned structures by cross twinning
(schematic drawing for a f.c.c. lattice in (110) orientation).

crystal (M) a second twin (T2) is crossing the first one which leads to
a distorted lattice structure in the region of intersection (I) (Dahmen
et al. 1990). A transformation of this region preserving the regular
lattice structure and minimizing the length of subgrain boundary segments (dashed lines) may lead to interlinked cyclic twins (A,B) accompanied by additional parallel twins (T3,T4).

REFERENCES

Dahmen U, Westmacott K H, Pirouz P and Chaim R 1990
    Acta metall. mater. 38 323
Gillet M 1977 Surface Science 67 139
Hall C R and Fawzi S A H 1986 Phil. Mag. 54 805
Hofmeister H 1984 Thin Solid Films 116 151
Hofmeister H, Bardamid A F, Junghanns T and Nepijko S A 1991
    Thin Solid Films to be published
Iijima S 1987 Jap. J. Appl. Phys. 26 346,941
Nachodkin N G, Bardamid A F, Novoselskaja A J and Yakimov K I 1987
    Fizika Tverdogo Tela 29 715
Nakamura A, Emoto F, Fujii E, Yamamoto A, Uemoto Y and Senda K 1989
    J. Appl. Phys. 66 4248
Phillips J C 1979 Phys. Rev. Lett. 42 1151
Saito Y, Yatsua S, Mihama K and Uyeda R 1978 Jap. J. Appl. Phys. 17 1149
Sinclair R 1990 Proc. XIIth Int. Cong. Electr. Microsc., eds L D Peachy
    and D B Williams (San Francisco: San Francisco Press) pp 512-513

Inst. Phys. Conf. Ser. No 117: Section 2
Paper presented at Microsc. Semicond. Mater. Conf., Oxford, 25–28 March 1991

Studies of EELS L$_{2,3}$ absorption fine structure in thin silicon

PE Batson

IBM Thomas J. Watson Research Center, Yorktown Heights, New York 10598

ABSTRACT: Electron energy loss fine structure within 5eV of the Si L$_{2,3}$ absorption edge is obtained with 0.2eV resolution from nm-sized regions using the scanning transmission electron microscope. The edge structure is examined and compared with available theory, with the aim of relating the observations to local electronic structure. At the 0.2eV level, there are observable differences between the EELS and x-ray absorption results even in the dipole limit.

1. INTRODUCTION

The detailed shape of the Si L$_{2,3}$ absorption edge should be sensitive to the bonding and electronic structure near the excited atom. Many questions arise, however, as to the practical use of this sensitivity, given the small scattering signals, core excitonic distortions, and a general lack of theoretical calculations for partial cross sections. This discussion centers on work aimed at overcoming some of these issues. It will be shown that sufficient signal is available for sub-nm analysis, and that excitonic distortions are tractable.

2. STANDARD SCATTERING THEORY

Inokuti (1971) provides a thorough review of the inelastic scattering for swift charged particles. The notation here will follow Leapman, et al. (1980). In the limit of a sudden, short interaction of the charged particle with an atom, and assuming the Born limit of weak scattering, the differential scattering cross section can be written:

$$\frac{\partial \sigma}{\partial E} = \frac{4\pi e^4}{\hbar^2 v^2} \int_{q_{min}}^{q_{max}} \frac{dq}{q^3} \mid <f| e^{i\vec{q} \cdot \vec{r}} |i> \mid^2 \rho_{f,i}(E) \tag{1}$$

where v is the incident electron velocity, $|i>$ and $|f>$ are the initial and final one electron states, E is the energy difference between $|f>$ and $|i>$, and $\rho_{f,i}$ is the energy dependent joint density of states (DOS) appropriate for that transition. The limits of integration for the scattering wavevector, q, are determined by the scattering kinematics and physical collection apertures.

For scattering angles in the 10mR range, $q_{max} \approx 2.0 \text{Å}^{-1}$ and $\vec{q} \cdot \vec{r}$ is normally much less than one for typical core excitations. Therefore, in the small q limit, we relate the matrix element above to the optical dipole matrix element:

$$<f| e^{i\vec{q} \cdot \vec{r}} |i> \approx iq <f|\vec{\varepsilon} \cdot \vec{r} |i>, \equiv iq \, M_{f,i}. \tag{2}$$

In this case, the q dependence of the matrix element drops out, leaving only the direction $\vec{\varepsilon}$ of q, and the cross section becomes proportional to the optical absorption of photons having polarization parallel to $\vec{\varepsilon}$,

$$\frac{\partial \sigma}{\partial E} = \frac{4\pi^2 e^2}{\hbar c} E \mid M_{f,i} \mid^2 \rho_{f,i}(E). \tag{3}$$

Evaluation of equation (1) gives typical values for partial cross sections of order $10^{-20} - 10^{-24}$ cm^2/eVatom for collection apertures subtending about 10mR semi-angles at 100KeV ($q_{max} \approx 1.8$Å^{-1}). Weng (1988) gives a partial cross section of 9×10^{-22} cm^2/eVatom for scattering into an 8mR semi-angle for the silicon L$_{2,3}$ edge. Assuming a 1nm diameter probe on a 50nm thick specimen of silicon, (n $= 2 \times 10^{22}$atoms/cm^3), we expect a scattering probability, $\partial P/\partial E$, of about 9×10^{-5}eV^{-1}. The modern scanning transmission electron microscope (STEM) can put 1 nanoampere or about 6×10^9e$^-$/ sec into this probe, producing a signal count rate of 5×10^5 e$^-$/eVsec. To obtain 1% statistics in 0.1eV resolution spectra, we thus require about 2 sec of integration time to acquire $\approx 10^4$ counts in each spectral point.

3. THE INSTRUMENT

The numbers given above clearly appear promising, but they require a high energy loss spectral resolution with reasonably large collection angles. Even then, some sort of array detection is required to make the required 2sec/point integration per point attainable in reasonable total times. During the past few years a system capable of this performance has been developed. (Batson 1986,1988) It consists of a Wien filter spectrometer coupled to a VG Microscopes HB501 STEM. The spectrometer is mounted within a high voltage electrode to allow deceleration prior to energy analysis. A resolution of better than 100 meV using a 10mR collection semi-angle has been demonstrated. The accuracy and stability of the energy axis is ± 20 meV when care is taken. A hybrid detector system using serial scanning across a photo-diode array is used, as first suggested by Shuman and Kruit (1985). In the present application, this allows use of the multiple slit capability of the diode array, while retaining an absolute energy calibration inherent in the scanning system. It also allows averaging over channel to channel gain and background variations. After acquisition, the data can be sharpened by deconvolution of the field emission source distribution to produce a spectral energy resolution of 0.2 eV or better. (Batson, 1991)

4.   TYPICAL L$_{2,3}$ EDGE RESULTS FOR SILICON

Experimental results for the silicon L$_{2,3}$ edge are shown in Figure 1.

Figure 1.   Summary of the background stripped Si L$_{2,3}$ edge, the sharpened edge, and the L$_3$ part extracted by a Fourier technique. The inset shows the estimated improvement in spectral resolution from the 0.4eV wide field emission profile to a symmetric, 0.22eV wide peak.

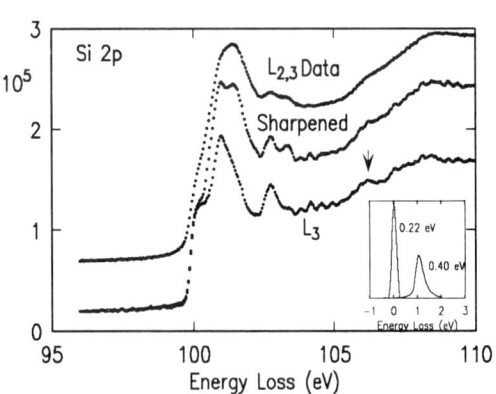

These were obtained in a 60nm thick area, determined by thickness fringe contrast. After resolution sharpening to 0.22 eV, the L$_3$ (spin 3/2) edge intensity is extracted by a Fourier technique assuming a spin orbit splitting of 0.608 and a 2p core occupation ratio of 2:1. (Brown and Rustgi, 1972) The inset shows the a best guess for the resolution function after deconvolution, compared with the field emission incident beam distribution (shifted by 1eV to the right). The deconvolution is possible because the field emission distribution has a sharp onset on the left.

The sharpened result must be treated with some care, because the apparent noise content is now different from the raw data. The numerical algorithm amplifies resolution frequencies out to some value where noise begins to dominate the measurement. The very high frequencies, which are dominated by noise, are not amplified. Thus, apparently smooth structure at the 0.22eV resolution limit may actually have a poor signal/noise ratio. However, it should be possible to gain some feeling for this by examining featureless sections in the original data, for instance below 98 eV. The structure in the final result in that region is most likely to be noise, and should be usable as an estimate of the reliability of the structure at the 101eV peak. Thus the small structure in the 104-105eV range is probably noise related. But the broad peak at 106 eV (arrow) is likely real, and due to a thin oxide layer on the top and bottom of the sample.

This data can now be compared with calculations using equation (1) and with the optical absorption results. Figure 2 summarizes a range of these. The optical absorption results are from x-ray partial photo-yield data obtained by Eberhardt, et al. (1988). (I will abbreviate these results as XAS for convenience, and to emphasize that the x-ray absorption results should be identical.) The L_3 part of their results has been extracted as explained above. The energy loss data have been replotted from above on an expanded scale. The energy loss axis has been carefully checked. Intensities are normalized to a theoretical model described below. It is apparent that the shapes of the two experimental results are not the same at the 0.2eV resolution level, in spite of expectations engendered by equations (1) and (2).

Figure 2. Comparison of EELS silicon $L_{2,3}$ edge with photo-yield data (XAS), (a) the calculated total DOS, (b) the symmetry projected DOS, and (c) a partial cross section which includes an estimate for the matrix elements. The relative normalization of the XAS and EELS results is governed by fits to an exciton theory discussed below. The bandstructure critical points are located relative to the total DOS (a). The calculation (c) is not significantly different when quadrupole terms are included.

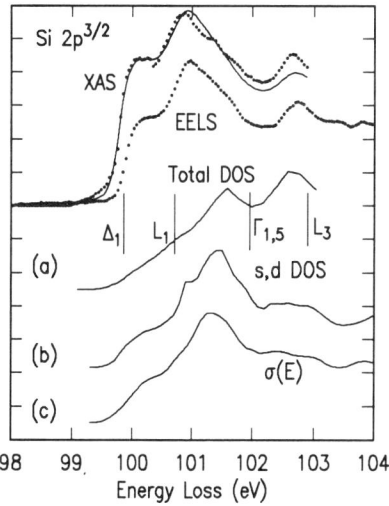

The experimental results are compared with three calculations. As summarized in Equation (1), we must consider both the matrix element and the DOS to predict an expected shape for the edge. The first calculation, due to Chelikowski and Cohen (1974) summarizes the total conduction band DOS. This does not compare very well with the measured results because it is dominated by states having p-symmetry. Since the core state also has p-symmetry, the dipole selection rules require the final states to have either s- or d-symmetry. The s,d-symmetry DOS was calculated by Weng, et al. (1990). It has also a x2 weighting of the s-states to approximately account for expected differences in the matrix elements. The agreement with the data is much better. Finally, Ma, et al. (1990) recently completed a calculation of the partial cross section. Thus, the calculations which include the selection rules properly give very good agreement with the measured EELS data. The vertical lines give the positions of several critical points, band minima (Δ_1 and L_1) and saddle points ($\Gamma_{1,5}$ and L_3). These are defined relative to the total DOS. Transitions to $\Gamma_{1,5}$ are dipole forbidden. The interesting correspondence of Δ_1 and L_1 with positions of edges in the data and in the projected DOS suggests that we may be able to identify these spectral features with the onset of the high density of states at the band minima.

How are we to understand the differences between the photo-yield and EELS data? Ma, et al. (1990) also calculated the quadrupole corrections to equation (1). They did not find that the result was significantly different from the dipole limit. Therefore, we are confronted with a breakdown of the equivalence of EELS and optical absorption even within the dipole limit.

5. EXCITONIC FORMULATION

Discrepancies between the optical data and single electron calculations are well known, and have largely been ascribed to the formation of core excitons. (Altarelli and Dexter 1972, Morar et al. 1985) Briefly, when a core electron is lifted to the conduction band, it can become loosely bound to the positive core hole in a Wannier exciton. Binding energies for typical semi-conductors can range from 0.5 to 0.05 eV. Within an approximate treatment suggested by Elliott (1957), equation (3) must be modified to include a set of bound, discrete states below the absorption onset, and a multiplicative factor (the Sommerfeld factor) which distorts the results of equation (3), enhancing intensities close to the absorption onset. We have

$$\frac{\partial \sigma}{\partial E} = \frac{4\pi^2 e^2}{\hbar c} E \left[\sum_n |\psi_n(0)|^2 \delta(E_n - E) + |\psi_c(0)|^2 |M_{f,i}|^2 \rho_{f,i}(E) \right] \qquad (4)$$

where $|\psi_n(0)|$ are the magnitudes of the bound state envelope functions at the origin, and $|\psi_c(0)|$ describes the magnitude of the continuum solutions. Within the approximation introduced by Elliott, these envelope functions multiply the crystal Bloch functions, essentially modifying the excitation probability to Bloch states as a function of their energy. Equation 4 is quite successful at predicting the shape of the XAS results, given an appropriate symmetry projected DOS. This result is shown in Figure 2 as the line overlying the XAS data. Apparently, then, the differences between the EELS and XAS results may be related to the core exciton. Similar behavior, to be reported elsewhere, is very apparent in diamond. (Batson and Bruley 1991)

In the time dependent perturbation theory, electron energy loss scattering is normally described within the sudden limit. That is, the coulomb interaction between the swift electron and the core is applied instantaneously, and is turned off a short time later. The atomic system is assumed to evolve into a new configuration of electron states which are not influenced by the swift electron. This limit is thought to be good for the soft x-ray region. It becomes questionable for more tightly bound core levels, where the core excitation Bohr period is very short. If the interaction is not sudden, then we must include the swift electron charge in an evaluation of the final state configuration.

The semi-conductors pose an interesting situation in this respect, because the final state configuration is dominated by a very weak core hole. This is the basis for the existence of Wannier excitons. In the case of silicon, the effective hole charge of $0.09e^-$ significantly distorts the XAS edge shape. Thus, even a very weak influence of the swift electron might change the final state configuration. To see whether this possibility could explain the differences apparent in Figure 2, we must extend the Elliott analysis to include the swift electron. This has been done in an approximate way as discussed elsewhere. (Batson and Bruley, 1991) The results are summarized in Figure 3. There, I show the XAS and EELS data together with theoretical fits which use identical materials parameters, (Altarelli and Dexter 1972) but which include the swift electron in the EELS case and not for the XAS case. In the XAS case the standard Elliott theory is recovered. The DOS which is required to fit both curves is also shown.

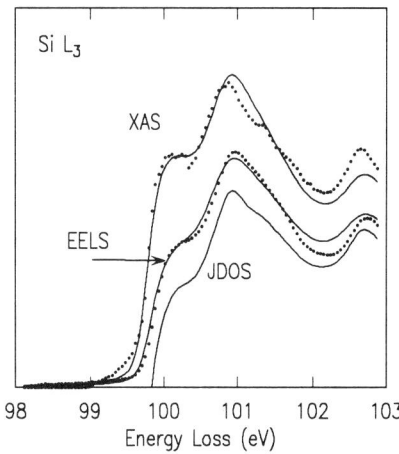

Figure 3. EELS and XAS results compared to theoretical fit using equation 4, with exciton envelope functions calculated on the presence of the swift electron in the EELS case, and without the swift electron in the XAS case. Without the swift electron, the Elliott theory is recovered. The DOS is obtained iteratively, to fit both sets of data. The good correspondence of the EELS and DOS results suggests that the swift electron suppresses the excitonic distortion in the Si case.

An important result for the practical use of this absorption edge, is that the EELS data are not very different from the projected DOS. Thus the swift electron almost entirely suppresses the excitonic distortion for silicon. This explains why the one-electron calculations summarized above can explain the EELS data so well, in spite of the widely held view that the optical absorption is not explainable in one-electron terms. It should be emphasized strongly that this result is unique to materials having a weakly bound exciton. A rule of thumb is that the exciton binding energy should be less than about 0.2 eV.

Interestingly, the EELS results cannot be matched unless a small correlation hole, trailing the swift electron, is included in the analysis. A direct result of this is that the theoretical framework for the Elliott analysis is anisotropic, showing cylindrical symmetry about the path of the swift electron.

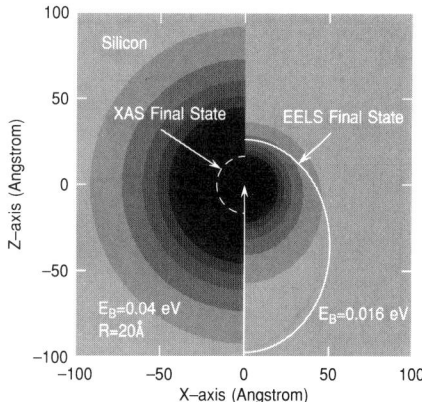

Figure 4. Plots of the exciton well potential and first exciton orbital width for the XAS case (left) and the EELS case (right). In the EELS case, the orbital is elongated along the path of the swift electron. The values E_B are the binding energies for the exciton in the two cases.

The excitonic orbitals which result are therefore elongated along the path of the swift electron. Figure 4 shows a comparison of the exciton well potentials for silicon with and without the swift electron and correlation charge. The small orbits denote the widths at half height of the first bound state of the excitons. Without the swift electron the first bound state energy is 40 meV, and with the swift electron it drops to about 16 meV. If this final state elongation is correct, some measurable anisotropy should be apparent with respect to surfaces in a thin sample. This is, in fact observable. In Figure 5, I show several results for the Si $L_{2,3}$ edge in a thin wedge as a function of thickness, determined from a multiple scattering analysis of the low loss region.

Figure 5. Silicon $L_{2,3}$ edge data acquired as a function of thickness in a wedge shaped specimen. The fine structure is sensitive to the top and bottom surface at thicknesses of order 20nm.

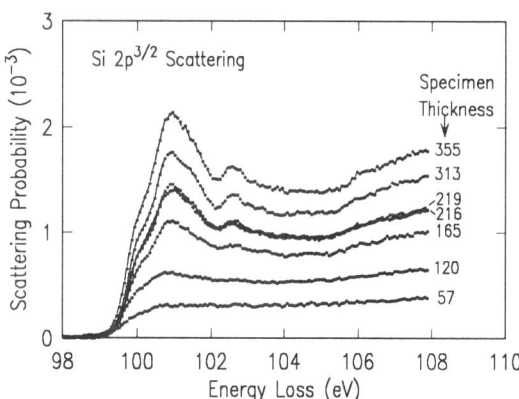

These spectra show two effects. First, the fine structure obviously is sensitive to the proximity of the surfaces on a 20nm length scale. This is much larger than the lengths on which we expect quantum confinement of the bandstructure to occur. However, it is precisely what we expect if the swift electron correlation hole determines the extent of the excited state. This is the case because the correlation hole is simply the first minimum in the plasmon wake, which has a wavelength of the order of 30-50 nm. Second, the scattering cross section, determined from single scattering calculations, is found to be weakly thickness dependent.

In spite of all this, when we perform an experiment in which the probe approaches an interface in the lateral direction, no effect of the surface is found until the probe physically links the surface. This behavior is summarized in Figure 6. In the experiment shown there, the probe is incident on a piece

Figure 6. Variation of the shape of the silicon L_3 edge as a function of lateral distance from a cleaved Si surface. The shape does not change until the 1nm probe physically hits the surface.

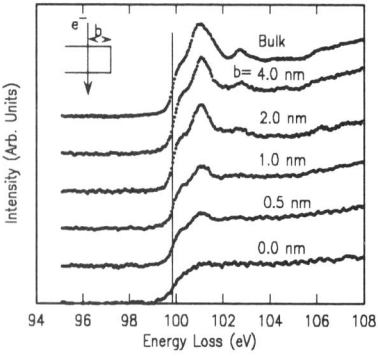

of silicon which was thick enough to yield the bulk edge shape. The probe was moved close to a broken edge. The thickness did not change very much on moving up to the edge. We can see that the shape of the L_3 edge remains constant until the impact parameter is of the order of 1nm, the nominal size of the probe. Thus the volume which is probed is very localized laterally, but is highly elongated along the swift electron path. In silicon, the limiting bulk fine structure shape is obtained only if the thickness is greater than about 40nm.

6. APPLICATION TO SUB-STOICHIOMETRIC SiO_x

Figure 7 shows several results for silicon in various local environments. (Dori, et al, 1990) These are not sharpened or resolved into the L_3 part, but show only the background subtracted $L_{2,3}$ data. Beginning at the bottom, I show thick Si, amorphous Si, thin Si, quartz and SiO. The crystalline Si results show an onset of $99.85\pm0.02eV$, while the amorphous Si result is about 0.1eV lower. The oxides have strongly peaked structure at 106, 108, and 115 eV. These are the result of final state resonances within the SiO_4 tetrahedron. Since the conduction band edge would appear to fall near 107.5 eV from optical and XPS measurements, (Nithianandam and Schnatterly 1988) the lowest peak may be a tightly bound core exciton.

Figure 7. Silicon $L_{2,3}$ absorption spectra obtained for several known environments and for two sub-stoichiometric oxides. The bulk silicon results are different by small amounts due to crystallinity and confinement of the scattering geometry. The Particle result was obtained from fairly large (> 5nm) particle embedded in SiO_x. The Matrix results were obtained from areas which showed no contrast at the 1-1.5nm resolution limit. The difference in the two matrix results can be attributed to the local distribution of excess Si -- uniformly spread in (b) and clustered as a crystal in (a).

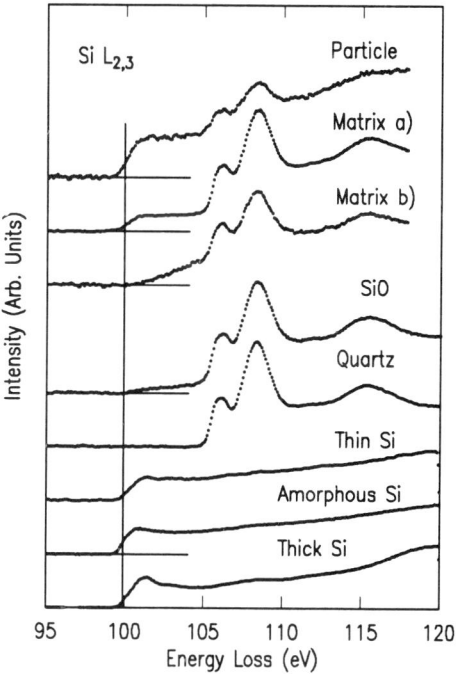

In the crystalline oxide, no intensity is present below the 106eV peak. In the sub-stoichiometric material, there is a smooth ramp of intensity extending down to about the same onset as is observed for the bulk Si. This ramp is a signature of silicon in lower oxidation states. XAS work has resolved structure within this ramp, and shows a correlation of the structure with oxidation state. (Harp, et al, 1988) The top three curves were obtained from sub-stoichiometric oxides grown with the intention of creating small Si islands within the matrix. The top curve is obtained from an obvious small particle. Notice that the onset is featureless, but resides at the crystalline Si onset position. It was confirmed from diffraction that this particle was indeed crystalline. A careful examination of the EELS results confirmed that the shape was distinguishable from the amorphous result due to the small shift in energy of the amorphous Si edge. Within the matrix around the particle, two types of results were obtained. Matrix (a) shows results which are readily decomposed into a thin Si part and an SiO_2 part. Matrix (b) could not be decomposed that way. It appears then, that matrix (a) consists of a matrix of SiO_2 with embedded sub-nm crystalline Si particles. Matrix (b) consisted of a uniform sub-stoichiometric oxide. In other samples, spectra of type (b) transform to the shape of (a) on annealing. Therefore we see that the bonding environment of Si can be deduced from the energy loss spectra, but due care must be employed when dealing with the extremely confined volumes.

7. CONCLUSIONS

In summary, spatially resolved electron energy loss spectroscopy promises to deliver very accurate bandstructure characterizations of sub-nm size regions of silicon based systems. The $L_{2,3}$ edge is readily obtainable now, and interpretation is becoming possible as more and more situations are investigated. At the same time, the detailed theoretical basis for interpretation is still weak. The presence of the core exciton produces some uncertainty, and forces the consideration of many subtle scattering processes which have been thought to be irrelevant in the past. Still, the results of Figure 7 are very encouraging.

A difficult problem for the future is to extend the present performance to the sub-nm probe size. A 0.2nm probe size is possible in the present instrument equipped with a pole piece having a 1.3mm coefficient of spherical aberration. (Shin, et al. 1989, and Pennycook and Jesson, 1990) However, the current available at this resolution is only about 1% of that available at 1nm. Thus, more sensitive detection is necessary than that provided by the diode array which is currently used in this system. An upgrade to a the more sensitive, CCD-based detector is currently in progress to try to attain this sensitivity.

REFERENCES

Altarelli M and Dexter D L 1972 Phys. Rev. Lett. 29 1100
Batson P E 1986 Rev. Sci. Inst. 57 43
Batson P E 1988 Rev. Sci. Inst. 59 1132
Batson P E and Bruley J 1991 submitted to Phys. Rev. Lett.
Brown F C and Rustgi O P 1972 Phys. Rev. Lett. 28 497
Chelikowski J R and Cohen M L 1974 Phys. Rev. B10 5095
Dori L, Bruley J, DiMaria D, Batson P E, Tornello J, and Arienzo M 1990 J. Appl. Phys., in press.
Eberhardt W, Kalkoffen G, Kunz C, Aspnes D, and Cardona M 1978 Phys. stat. sol. (b)88, 135
Elliott R J 1957 Phys. Rev. 108 1384
Harp G R, Zhi L H, Tonner B P 1990 Phys. Scr. T 31 23
Inokuti M 1971 Rev. Mod. Phys. 43 297
Leapman R D, Rez P and Mayers D F 1980 J. Chem. Phys. 72 1232
Ma H, Lin S H, Carpenter R W, and Sankey O F, 1990 J. Appl. Phys. 68 288
Morar J F, Himpsel F J, Hollinger G, Hughes G, and Jordan J L 1985 Phys. Rev. Lett. 54, 1960
Nithianandam V J and Schnatterly S E 1988 Phys. Rev. B 38 5547
Pennycook S J and Jesson D E 1990 Phys. Rev. Lett. 64 938
Shin D H, Kirkland E J and Silcox J 1989 Appl. Phys. Lett. 55 2456
Shuman H and Kruit P 1985 Rev. Sci. Inst. 56 231
Weng X 1988 private communication
Weng X, Rez P, and Batson P 1990 Sol. Stat. Comm. 74 1013

Inst. Phys. Conf. Ser. No 117: Section 2
Paper presented at Microsc. Semicond. Mater. Conf., Oxford, 25–28 March 1991

Elemental mapping in AlGaAs/GaAs heterostructures using parallel EELS

H Lakner, M Maywald, L J Balk and E Kubalek

Universität Duisburg, Werkstoffe der Elektrotechnik, Sonderforschungsbereich 254, Kommandantenstraße 60, D 4100 Duisburg 1, F.R.Germany

ABSTRACT: Parallel detection of EEL spectra in a field emission STEM allows the determination of the Ga–concentration in $Al_xGa_{1-x}As/GaAs$ heterostructures with a spatial resolution of 1 nm and an accuracy of $\pm$ 0.05 in (1–x). The weak signal to noise ratio and problems in background estimation limit the direct measurement of the Al–concentration by means of the Al–K edge. Automation of EELS data acquisition enables the detection of up to 4096 spectra in one experiment. The spectra either represent a linescan or a 2 D map. In practice, the technique is very useful to record elemental compositions along line profiles. The detection of 2 D maps with associated long recording times is in principle possible but still limited by specimen contamination and system instabilities.

1. INTRODUCTION

The quality of semiconducting heterostructures with "tailored" electronic properties for applications in electronic devices depends strongly on the nature of the interfaces and on the chemical composition of the layers. For example, in a heterostructure field–effect transistor, only flat and chemically abrupt interfaces yield superior device performance like high cut–off frequency and low noise behavior. In optoelectronic devices the bandgap and refractive index of AlGaAs layers depend strongly on the relative concentration of the group III elements. Typical layer thicknesses are in the nm–range and less, therefore requiring analytical techniques of very high spatial resolution.

Electron energy loss spectroscopy (EELS) of thin specimens performed in a dedicated field–emission scanning transmission electron microscope (STEM) can give chemical information with high spatial resolution. The localization of inelastic scattering for 100 keV electrons at high energy losses (more than 1 keV) is in the range of 0.2 nm (Kohl and Rose 1985). The element specific Ga–L, As–L, and Al–K energy loss edges start at 1115 eV, 1323 eV, and 1560 eV, respectively. Therefore these losses are well localized. With decreasing energy losses the localization gets worse (e.g. for the Al–L edge at 72 eV). The field–emission electron source can provide probe diameters down to the range of 0.2 nm (Pennycook 1988). Such probes can contain enough current to perform EELS with sufficient signal to noise ratio (Lakner et al 1991). The spatial resolution of EELS is not limited by elastic scattering as compared to x–ray microanalysis (Joy 1986). These considerations should in principle allow us to perform microanalysis of AlGaAs heterostructures via EELS with subnanometer resolution.

However, serial detection of high loss spectra and associated long exposure times (causing specimen damage and decreased spatial resolution due to system instabilities) limited the applicability of EELS to the point analysis mode only. Using parallel detection techniques EEL spectra can be recorded in much shorter recording times overcoming the drawbacks of serial detection.

Up to now, very little work has been published on the application of EELS on AlGaAs/GaAs heterostructures using small electron probes and small probe currents. This paper will give examples for point analysis and chemical line profiles by parallel recorded EELS and demonstrates the efficiency of EELS as a microanalytical tool for the characterization of AlGaAs/GaAs heterostructures. The question of whether parallel detection extends the capability of EELS to record two–dimensional maps within acceptable recording times will be discussed.

2. EXPERIMENTAL PROCEDURE

All spectra and micrographs have been recorded in a field—emission STEM (VG: HB 501) operated at 100 keV which has been modified in the following manner: An ultra high resolution pole piece allows performance of microanalytical work with superior beam parameters. Parallel detection of the spectra (see fig. 1) is achieved by means of an 80^0 magnetic sector field corrected to second order aberrations, three magnetic quadrupole lenses and a thin single crystal scintillator (YAG) optically coupled with a lens to an intensified linear photodiode array (Lakner et al 1989). Beam positioning and data acquisition of the parallel recorded spectra and images (high angle annular dark field and/or bright field image) are controlled by a microprocessor. In detail, the microprocessor acts as master controller of a data acquisition and control system (Hewlett Packard HP 3852 S) which is equipped with two D/A converters, a multiplexer, a digital voltmeter, and a memory (Kaufmann et al 1991). The two D/A converter supply suitable voltages to the scan coils of the STEM and therefore control the beam position. By means of the multiplexer and the digital voltmeter the analog signals as arriving from the annular dark field detector and the bright field detector of the microscope are digitized and stored in the memory. This allows simultanous detection of bright and annular dark field images with 13 bit accuracy. Additionally the microprocessor masters an optical multichannel analyzer (OMA) interface (EG&G PAR 1461) controlling the intensified photodiode array. The digitally recorded images allow selection of specimen locations of interest. A menu controlled program enables the user to select either the point analysis mode, the line profile mode (consisting of n point measurements), or the two dimensional (2 D) mapping mode (n × m pixel²). The timing between beam positioning and spectral data acquisition is achieved by triggers and interrupts sent from the microprocessor to the data acquisition and control system and the OMA interface, respectively. Up to 4096 spectra (with 1024 channels each) can be stored within one experiment either representing a line profile or a two—dimensional map. The diodes of the detector array were cooled down to -30^0C to reduce dark currents. Additionally all spectra are corrected for dark counts and inhomogenous response. Background subtraction in the spectra is performed according to Trebbia 1988. For quantification the stripped edges are compared to standard spectra recorded under identical conditions in specimens of the same thickness. For the case that bright field imaging is requested the light emitted from the YAG is deflected by a mirror and detected by a photomultiplier.

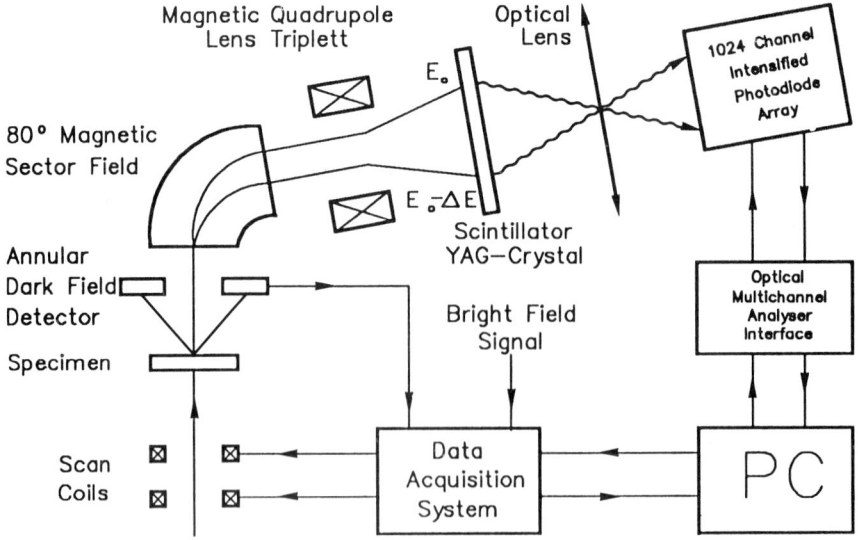

Fig.1:   Digitally controlled data acquisition for high angle annular dark field imaging and parallel EELS in the STEM.

The investigated AlGaAs/GaAs heterostructures have been grown by MOVPE. For cross–sectional preparation standard Ar–ion milling was used. In order to minimize the effect of electron channeling on the EELS data (Long 1989) the orientation of both the standard specimen and those of unknown concentrations have been such that the [110] direction has been parallel to the electron beam.

3. RESULTS

Fig. 2a) shows a high angle annular dark field image (atomic number contrast) of a heterostructure consisting of a GaAs–, an AlAs–, and an $Al_xGa_{1-x}As$–layer. The GaAs(AlAs) appears bright (dark) due to the high (low) mean atomic number. The composition of the $Al_xGa_{1-x}As$–layer is unknown. Fig. 2b) shows typical EEL spectra each recorded within 20 s recording time (1 nA probe current, 1 nm probe diameter). In all spectra the background is subtracted for the Ga edge. The energy resolution of the spectra is 3 eV. The Ga–$L_{2,3}$ edge in the GaAs and the AlGaAs spectra starts at 1115 eV, whereas in the AlAs spectrum no Ga edge is visible. The shape of this spectrum in the region of the Ga edge indicates a good fit of the background. All three spectra exhibit the As–$L_{2,3}$ edge at 1323 eV. The AlAs and AlGaAs spectra show clearly the Al–K edge at 1560 eV which is overlapping the weak As–L_1 edge at 1520 eV being clearly visible in the AlAs spectrum.

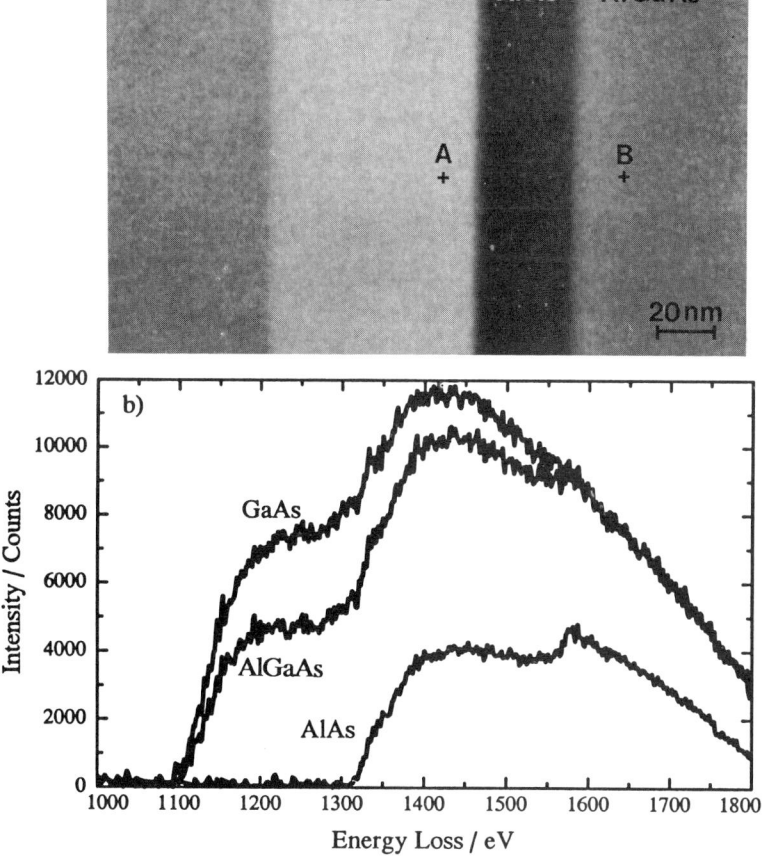

Fig.2:   a)   High angle annular dark field image of an AlGaAs/GaAs heterostructure.
          b)   EEL spectra of GaAs, $Al_{0.3}Ga_{0.7}As$, and AlAs after background subtraction (1 nA probe current, 20 s recording time each).

Comparison of the intensities (integration of the counts from the beginning of the edge over an energy region of a width $\Delta=100$ eV) in the Ga edges of the GaAs and the $Al_xGa_{1-x}As$ spectra indicates $x=0.28 \pm 0.05$ for the $Al_xGa_{1-x}As$ layer. Background substraction under the Al–K edge in the AlAs spectrum was also performed in order to compare the intensities of the Ga–$L_{2,3}$ edge in GaAs and the Al–K edge in AlAs for $\Delta=100$ eV. As a result the Ga edge shows to be more intense than the Al–K edge by a factor of 7 ± 1.5.

An example of an EELS linescan across the heterostructure of fig. 2a) along the line A–B is given in fig. 3. The linescan is extracted from 75 spectra (1 nA probe current, 10 s recording time each). For all 75 spectra the background has been subtracted under the Ga–$L_{2,3}$ and the Al–K edges and the intensities in these edges have been compared to standard spectra. The solid line represents the Ga–concentration $(1-x)$, the dashed line the Al–concentration x. The Ga–concentration is in good agreement with the data obtained from fig. 2b) and it is determined with an uncertainty of ± 0.05 along the line A–B. The Al–concentration along A–B shows an uncertainty of ± 0.2 in x and therefore is more qualitative (compare the discussion below). The spatial resolution in the linescan is estimated to be 5 nm as caused by probe parameters not being optimized for high spatial resolution microanalysis.

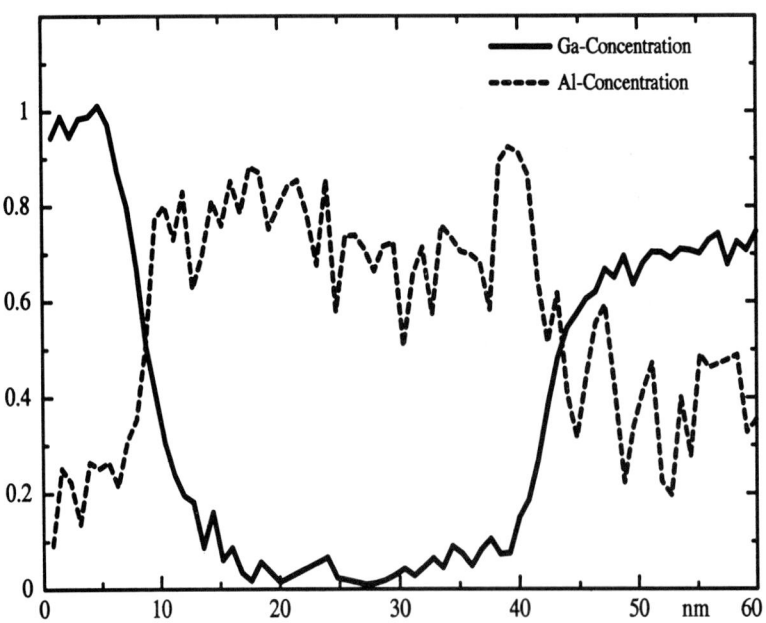

Fig.3:   Ga– and Al– (dashed) concentration along the line A–B as indicated in fig. 2a. The linescan was formed from 75 spectra (1 nA probe current, 10 s recording time each).

For the following example of an EELS linescan emphasis has been put on spatial resolution. Fig. 4a) shows a high angle annular dark field image of an AlGaAs/GaAs superlattice with 2 nm thick AlGaAs layers. Fig. 4b) shows an EELS linescan across the line C–D as indicated in the micrograph. The linescan representing the Ga–concentration ($\Delta=100$ eV) is extracted from 80 spectra (0.1 nA probe current, 4 s recording time each) and it resolves the four AlGaAs layers clearly.

The spatial resolution is 1 nm (edge resolution for a straight boundary: between 20% and 80% of the maximum intensity), but the smaller probe current associated with a smaller probe diameter causes decreased signal to noise ratios in the spectra and therefore increased error (± 0.1) in the determination of the Ga—concentration.

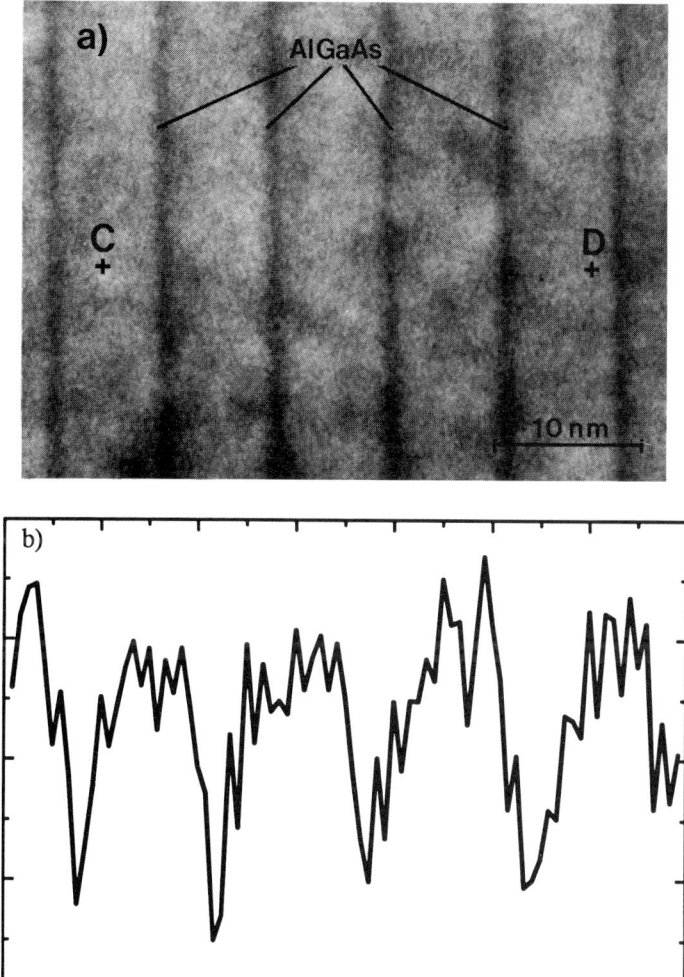

Fig. 4: a) High angle annular dark field image of an AlGaAs/GaAs superlattice.
        b) Ga—concentration across the superlattice (along C—D). The linescan was formed from 80 spectra (0.1 nA probe current and 4 s recording time each).

4. DISCUSSION

The results demonstrate that parallel EELS allows us to determine the chemical composition of AlGaAs/GaAs heterostructures with a spatial resolution of 1 nm. The intensity of the Ga—$L_{2,3}$ edge can be monitored with an accuracy of ± 0.05 whereas getting control of the intensity of the Al—K edge is more complicated. The intensity of the Al—K edge is a factor of 7 weaker than the Ga intensity for $\Delta=100$ eV. This is in good agreement with values for these edges given by Hofer 1991. Additionally the As edges in front of the Al—K edge cause problems in exact background estimation and quantification. A detection of the Al—L edge at 72 eV was not taken into account due to the following reasons: Firstly, such a low loss is not as well localized as high energy losses. Secondly, once again minor As edges in front of the Al—L edge make background estimation troublesome. The automation of the EELS data acquisition allows collection of up to 4096 spectra either representing a line profile or a 2 D map. In practice only the acquisition of up to $\simeq$ 200 spectra of high energy losses (1 keV to 2 keV) is useful. For longer overall recording times specimen contamination and/or system instabilities limit the applied technique. This allows the recording of line profiles even of high energy losses with good accuracy but 2 D maps will have only a small pattern size (15 × 15 pixel^2). The acquisition of more spectra and therefore of 2 D maps with increasing pattern size is possible for the low loss region where recording times are much shorter. The storage of the raw spectra for a linescan or a 2 D map which completes a microanalytical experiment needs a lot of memory capacity. However, the user is able to process the raw spectra in many ways afterwards. The energy windows for background estimation, extrapolation, and edge integration may be varied, or multible element distributions may be monitored. Further improvements in the parallel detection technique will yield increased signal to noise ratios of the spectra and/or shorter recording times. The first item can improve the accuracy of the technique, the second one may decrease the limitations of the technique due to specimen contamination and/or system instabilities. E.g. readout of groups of pixels in the photodiode array can improve the statistics of spectra. The maximum tolerable group size depends on the required energy resolution within the spectra. Such experiments will be performed in the near future. The apparent spatial resolution obtained in these first results is 1 nm. Further experiments will be necessary to check the limit in spatial resolution of the technique described here.

ACKNOWLEDGEMENTS

The authors like to thank the following persons for their help:
K Kaufmann assisted in the development of the digital image acquisition technique with many helpful discussions. Specimen preparation was carried out by Miss U Spitzner. F Scheffer from the department of Solid State Electronics of Duisburg University supplied the MOVPE grown AlGaAs/GaAs specimens.
This work was financially supported by the Deutsche Forschungsgemeinschaft within the Special Collaborative Program 254: Very High Frequency and Very High Speed Circuits Based on III—V Compound Semiconductors.

REFERENCES

Hofer F 1991 Microscopy, Microanalysis, Microstructures 2 (to be published)
Joy D C, Romig A D, Goldstein J I 1986 Principles of Analytical Electron Microscopy (New York and London: Plenum Press) 270
Kaufmann K, Koschinski P, Zinke U, and Balk L J 1991 Proc. 19th Int. Symposium on Acoustical Imaging April 3—5 Bochum F R Germany (to be published by Plenum Press)
Kohl H, Rose H 1985 Adv. Electron. Electron Phys. 65 175
Lakner H, Balk L J, Kubalek E 1989 Beitr. Elektronenmikroskop. Direktabb. Oberfl.22 241
Lakner H, Balk L J, Kubalek E 1991 Microscopy, Microanalysis, Microstructures 2 (to be published)
Long N J 1989 Inst. Phys. Conf. Ser. No. 100 55
Pennycook S J 1988 Ultramicroscopy 26 239
Trebbia P 1988 Ultramicroscopy 24 399

Inst. Phys. Conf. Ser. No 117: Section 2
Paper presented at Microsc. Semicond. Mater. Conf., Oxford, 25–28 March 1991

STEM EDS X-ray mapping, TEM and HREM studies of the effect of gas switching procedures on the chemical abruptness of interfaces in MOCVD GaInAs/InP MQW structures

N J Long, A G Norman*, A K Petford-Long, B R Butler[1+], C G Cureton[1], G R Booker and E J Thrush[1]

Department of Materials, University of Oxford, Parks Road, Oxford OX1 3PH, UK
[1]BNR Europe Ltd., London Road, Harlow, Essex CM17 9NA, UK

ABSTRACT: STEM EDS X-ray mapping, TEM and HREM studies have been performed to investigate the chemical compositional abruptness and planarity of interfaces in MOCVD GaInAs/InP MQW structures grown using different gas switching procedures. The results obtained demonstrate that STEM EDS X-ray mapping is a powerful technique for revealing directly the abruptness of interfaces in MQW structures and that the use of different gas switching procedures can have a major effect on both the abruptness and planarity of interfaces in MOCVD GaInAs/InP MQW structures.

1. INTRODUCTION

There is considerable interest in the growth of GaInAs/InP multiple quantum well (MQW) structures by metal organic chemical vapour deposition (MOCVD) for applications in a wide range of advanced, improved performance opto-electronic devices e.g. MQW lasers. The enhanced performance of these devices, based on the quantum confinement of charge carriers in the wells, depends critically on the planarity and chemical abruptness of the quantum well interfaces. Conventional transmission electron microscopy (TEM) studies of MOCVD GaInAs/InP MQW structures (Spurdens and Hockly 1986, Chew et al 1987, Norman et al 1989, Taylor et al 1989) have revealed a marked asymmetry in the planarity of the well interfaces. The lower (InP-to-GaInAs) interfaces were closely planar whilst the upper (GaInAs-to-InP) interfaces often exhibited undulations, typically 1-2nm in amplitude and of lateral extent 20-100nm. Norman et al (1989) demonstrated that the origin of these undulations was associated with the gas switching at the end of growth of the GaInAs wells and this was later supported by the work of Spurdens et al (1991). A considerable improvement in the planarity of the upper interfaces was achieved by using growth pauses under flowing H_2 (Norman et al 1989).

The chemical compositional abruptness of interfaces in MOCVD GaInAs/InP MQW structures has been studied by a variety of techniques. TEM **g**[002] dark field (DF) studies revealed 1-2nm wide faint bands of contrast in the InP barrier layers immediately above the upper interfaces of the GaInAs wells and it was suggested that these could be associated with an As "tail" in the InP (Chew et al 1987, Norman et al 1989). High resolution electron microscopy (HREM) studies combined with extensive image simulations have enabled abrupt and diffuse interfaces to be distinguished in this system using a microscope defocus technique (Petford-Long et al 1989a,b). These imaging techniques, however, are unable to reveal the chemical composition of the interfaces directly. Scanning transmission electron microscopy (STEM) combined with energy dispersive (EDS) X-ray microanalysis (Chapman et al 1987, McGibbon et al 1988, 1989, Long 1989) and pulsed laser atom probe (PLAP) and position sensitive atom probe (POSAP) (Liddle et al 1989) studies have directly revealed the composition of interfaces in this system. The results revealed an asymmetry in the abruptness of the quantum well interfaces in most samples studied. The lower interfaces were closely abrupt whilst the upper interfaces were diffuse, with significant levels of Ga and especially As being incorporated in the InP barrier layers. Correlations between STEM, HREM and POSAP/PLAP data obtained from the same structures have recently been reported (Liddle et al 1990). The STEM EDS studies mentioned above were performed by taking a series

---

*Present address: IRC for Semiconductor Materials, Blackett Laboratory, Imperial College, Prince Consort Road, London SW7 2BZ, UK
+Present address: BT and D Technologies, Whitehouse Road, Ipswich, Suffolk IP1 5PB, UK

of point-by-point analyses across the quantum well interfaces, moving the beam between analysis points manually. Obtaining such profiles is tedious and time consuming and problems may be encountered with specimen drift and sample beam damage due to the long count times required at each point for good analysis statistics. Many such profiles may also be required to obtain statistically relevant data. A simple mapping scheme has therefore been developed for STEM EDS microanalysis (Long and Glaisher 1990) which enables averaged line profiles to be extrapolated by the use of image processing techniques. This method allows correction for small amounts of linear drift and the shorter analysis times used at each point reduce the effects of specimen beam damage. The averaged line profiles obtained can also lead to improved analysis statistics.

In this paper we report the application of this new high spatial resolution STEM EDS mapping technique to the study of the compositional abruptness of interfaces in GaInAs/InP MQW structures grown by both atmospheric pressure (AP) and low pressure (LP) MOCVD using different gas switching techniques. The results obtained revealed major differences in the compositional abruptness of some of the interfaces and these results correlated well with TEM and HREM results obtained from the same samples.

2. EXPERIMENTAL

[110] cross-sectional specimens for TEM, HREM and STEM EDS microanalysis were prepared by mechanical pre-thinning followed by final thinning using low angle, 2-4kV, Ar^+ ion milling at liquid N_2 temperatures in a turbomolecular-pumped ion mill. The X-ray maps were obtained using a VG HB501 UHV field emission gun STEM, operated at 100kV using a probe size of 1.5nm, fitted with a LINK systems LZ5 windowless EDS detector. The X-ray maps were formed by scanning the beam via 12 bit DACs and counting the LINK AN 10,000 output pulses using a counter/timer board on an IBM compatible PC. The counts were displayed on a Synapse frame-store during acquisition and stored in SEMPER6 data format for future processing. SEMPER6 was used to image the data and to perform any corrections for specimen drift. The data could be summed as an average linescan and the relative profiles superimposed. The approach could be extended to give a fully background-subtracted signal and to yield actual composition profiles with appropriate software. TEM and HREM studies were performed in a JEOL 200CX operated at 200kV and a JEOL JEM 4000EX operated at 400kV (information limit at 0.14nm).

3. RESULTS AND DISCUSSION

The first sample studied consisted of three sets of nominally 10nm GaInAs quantum wells and 26nm InP barrier layers, with each set separated by a $0.13\mu m$ InP spacer layer grown by AP-MOCVD using different gas switching procedures at the interfaces. The growth apparatus, described in detail elsewhere (Briggs and Butler 1987, Butler et al 1988), consisted of a vent/run horizontal reactor, utilising pressure balancing between the vent and run lines, fitted with a linear, constant volume, gas switching manifold. The growth conditions used were: AsH_3, PH_3, trimethylindium (TMI) and trimethylgallium (TMG) as sources, H_2 carrier gas at 6 litres per minute, growth temperature 650°C, V:III ratio 45:1 and growth rates of GaInAs $6\mu m$/hour and InP $4\mu m$/hour.

In Fig. 1a,b,c are shown superimposed the relative Ga and As averaged X-ray count profiles obtained from 16 x 64 point X-ray maps across quantum wells of each of the three sets. An analysis time of 100ms was used at each point and the specimen was tilted slightly off the <110> pole into a [002] symmetry position. The growth direction for all three sets is from left to right. Fig. 1a,b,c shows the results obtained from the first set of wells which were grown with no pauses in growth at the interfaces, the second set with 6s pauses under PH_3 at the lower interfaces and 6s pauses under AsH_3 at the upper interfaces, and the third set with 6s pauses under PH_3 at the lower interfaces and 12s pauses under flowing H_2 at the upper interfaces respectively. The profiles show that the lower interfaces of all three sets of wells are relatively abrupt as assessed from the 20%/80% interface widths, the distance over which the Ga and As count profiles increase from 20 to 80% of their maximum values, which were measured to be ~2.5nm for both the Ga and As profiles for all three sets of wells. The measured 20%/80% widths depend on several factors including the finite size (1.5nm) of the electron probe used, beam spreading effects in the sample, broadening due to projection of the data along a possible non-planar interface and broadening due to non-linear specimen drift, as well as the actual width of the interface. The Ga and As profiles at these interfaces closely follow each other indicating that the Ga and As arrive and are incorporated into the growing layer simultaneously and also that the use of 6s pauses under PH_3 had no significant effect. However there are major differences between the upper interfaces of the first two sets and the third set of quantum wells. For the first two sets, Fig 1a and b, the upper interfaces are more diffuse than the lower interfaces. The As profile (20%/80% width ~4.0nm) lies outside the Ga profile (20%/80% width ~2.5nm),

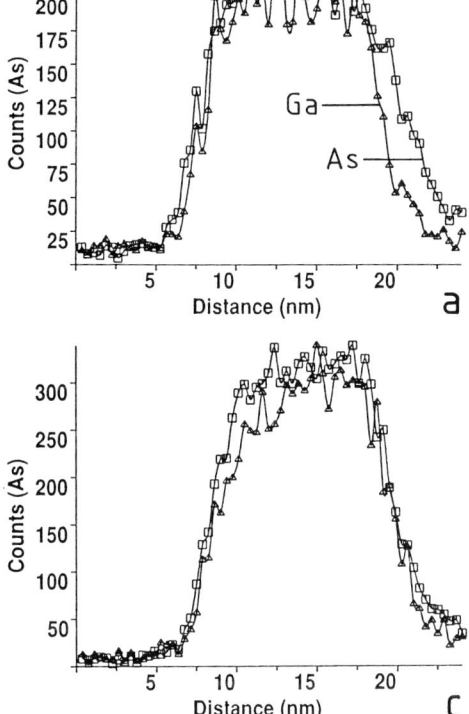

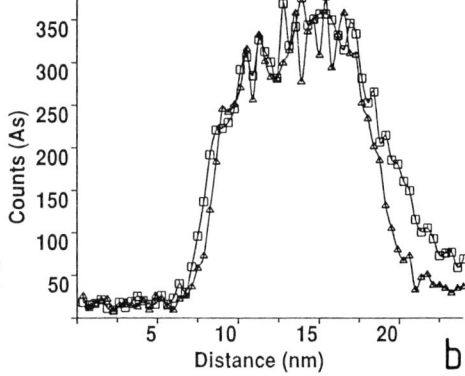

Figure 1. Relative Ga (triangles) and As (squares) average X-ray count profiles obtained from 16 x 64 point X-ray maps (analysis time per point 100ms) across GaInAs quantum wells in AP-MOCVD GaInAs MQW structures grown using different gas switching procedures. Growth direction is from left to right.

a. No pauses at interfaces.

b. 6s pauses under PH_3 at lower (InP-to-GaInAs) interfaces, 6s pauses under AsH_3 at upper (GaInAs-to-InP) interfaces.

c. 6s pauses under PH_3 at lower interfaces, 12s pauses under H_2 at upper interfaces.

indicating the presence of a 2-3nm wide As-enriched layer in the InP barrier layer directly above these interfaces. The profiles also show slight Ga tails in the InP above these interfaces. In comparison the upper interfaces of the third set , Fig. 1c, are more abrupt than the first two sets with both the As and Ga profiles having measured 20%/80% widths of ~2.5nm. The As-enriched layer, present in the InP of the first two sets, has been virtually eliminated and there are now only small As and Ga tails in the InP. Due to the very different pyrolysis characteristics of AsH_3 and PH_3, As is more readily incorporated during growth than P. The flushing of residual AsH_3 from the reactor during the 12s pause under H_2 is thus mainly responsible for the improved abruptness of these interfaces. The abruptness of the upper interfaces in this set of wells is almost as good as that of the lower interfaces.

In Fig. 2a,b and c are shown g[002] DF TEM images of quantum wells from sets 1,2 and 3 described above. The growth direction is upwards for all the micrographs. The lower interfaces of the wells in all three sets are planar and reasonably abrupt, correlating well with the X-ray mapping results. The upper interfaces of the first and second sets, grown with no pauses and 6s pauses under AsH_3 respectively, possess undulations (Norman et al 1989). There is also a faint band of contrast, 1-2nm wide, visible in the InP barrier layers of these two sets just above the upper interfaces. In comparison the upper interfaces of the quantum wells in the third set, grown with 12s pauses under H_2 , are much more planar and there is little evidence of a faint band of contrast in the InP barrier layers above these interfaces (Norman et al 1989). From these results it is concluded that the faint bands of contrast visible in the InP barrier layers, just above the upper interfaces of the first two sets in the g[002] DF micrographs of Fig. 2a and b, are associated with the As-enriched regions revealed in the X-ray profiles obtained from the two sets of wells shown in Fig. 1a and b.

In Fig. 3a,b and c are shown HREM micrographs of quantum wells from sets 1,2 and 3 of the AP-MOCVD GaInAs wells. For all three sets of wells the lower interfaces are distinct and planar, with a planar band of different contrast present extending over two or three monolayers suggesting reasonable abruptness. In the first and second sets of wells the upper interfaces are much less distinct and are rough and diffuse, being spread over

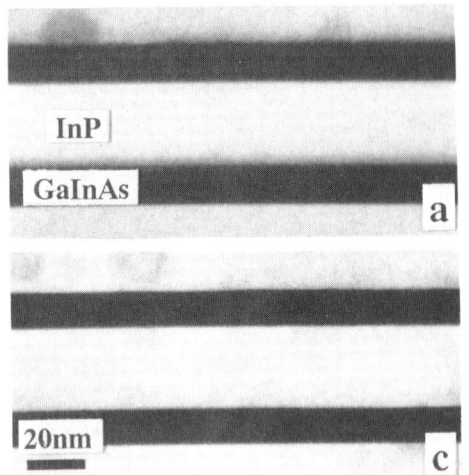

Figure 2. **g**[002] DF micrographs of AP-MOCVD GaInAs/InP MQW structures grown using different gas switching procedures. Growth direction up page.
a. No pauses at interfaces.
b. 6s pauses under PH_3 at lower (InP-to-GaInAs) interfaces, 6s pauses under AsH_3 at upper (GaInAs-to-InP) interfaces.
c. 6s pauses under PH_3 at lower interfaces, 12s pauses under H_2 at upper interfaces.

up to 8 monolayers, but in the third set of wells the upper interfaces are almost as abrupt as the lower ones. The diffuseness of the upper interfaces visible in the HREM micrographs of the first two sets of wells is probably a result of the presence of the As-enriched regions in the InP buffer layers at these interfaces revealed by the X-ray mapping results shown earlier.

Similar studies were performed on a LP-MOCVD 10nm GaInAs/10nm InP MQW structure. This structure was grown in a low pressure reactor, described in detail by Thrush et al (1987), which incorporated a linear switching manifold close-coupled to a simple, horizontal, RF heated, reactor tube of rectangular cross section. Dynamic pressure balancing was included between the vent and run lines using N_2 flows. The switching manifold also possessed two mass flow controlled, H_2 "sub-ballast" lines which were inputted via their own vent/run valves. These lines enabled H_2 flows to be added or subtracted to the reactor stream at the instant of reagent gas switching in order to minimise pressure transients which can degrade interface quality. The growth conditions used were: temperature 650°C, pressure 150 torr, H_2 ballast flow rate 15 litres per minute and a growth rate of $3\mu m$ per hour. "Sub-ballast" switching was used at the interfaces and a 10s pause under H_2 was included at the upper interfaces of the quantum wells to flush residual AsH_3 and TMG out of the reactor before InP growth. Double crystal X-ray diffraction and 4K photoluminescence results indicated that this was a very high quality MQW structure (Cureton et al 1991).

In Fig. 4a are shown superimposed the average relative Ga and As X-ray count profiles obtained from a 16 x 128 point X-ray map of one of the quantum wells of this LP-MOCVD structure. It can be seen that in this sample there is no evidence of any As-enriched layer in the InP barrier layers above the upper interfaces of the quantum wells. The As and Ga profiles closely follow each other at both the lower and upper interfaces indicating good compositional control. The upper interfaces in this sample are as abrupt as the lower interfaces with measured 20%/80% widths of ~2.0nm. Most of this 2.0nm width is due to instrumental factors, and the "true" abruptness of the interface is probably ≤ 0.5nm. TEM and HREM studies also indicated good quality interfaces in this structure. In **g**[002] DF micrographs both the lower and upper interfaces are planar and abrupt with no bands of contrast visible in the InP barrier layers. HREM, Fig. 4b, also indicated that both the lower and upper interfaces were abrupt. The HREM defocus technique (Petford-Long et al 1989a,b) indicated that the lower interfaces were spread over 1-2 monolayers with the upper interfaces slightly more diffuse. The quality of the interfaces in the LP-MOCVD structure was superior to that of the interfaces for all three sets of wells examined in the AP-MOCVD structure.

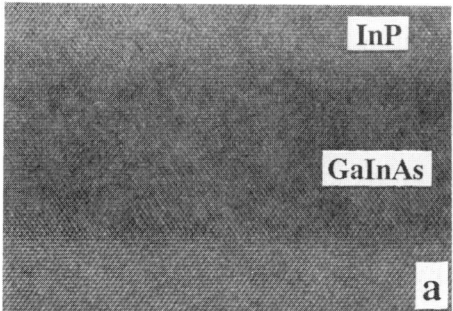

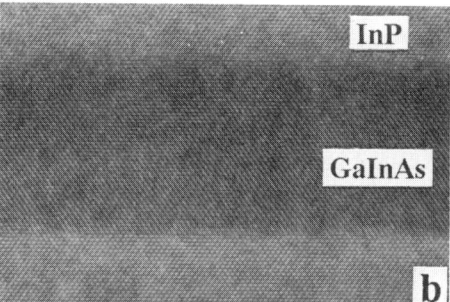

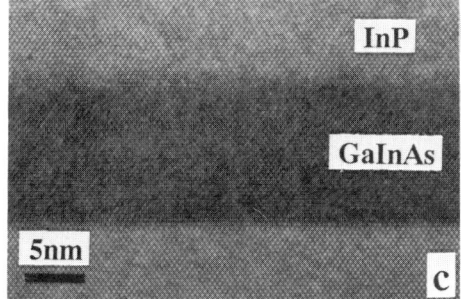

Figure 3. [110] HREM images of AP-MOCVD GaInAs quantum wells grown using different gas switching procedures. Growth direction is up page.

a. No pauses at interfaces.

b. 6s pauses under PH_3 at lower (InP-to-GaInAs) interfaces, 6s pauses under AsH_3 at upper (GaInAs-to-InP interfaces.

c. 6s pauses under PH_3 at lower interfaces, 12s pauses under H_2 at upper interfaces.

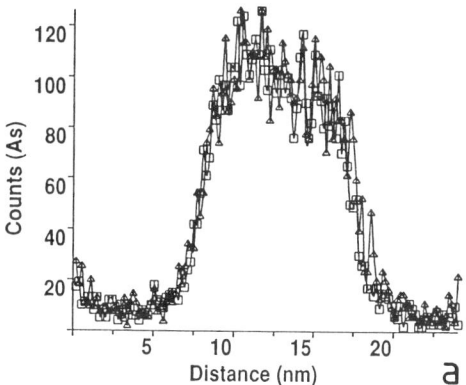

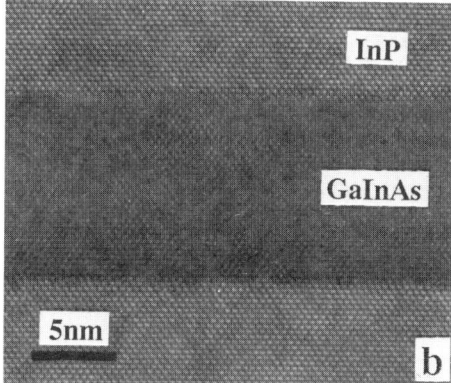

Figure 4. a. Averaged relative Ga (triangles) and As (squares) X-ray count profiles obtained from a 16 x 128 point X-ray map of a GaInAs quantum well from a GaInAs/InP MQW structure grown by LP-MOCVD using "sub-ballast" switching and a 10s pause under H_2 at the upper (GaInAs-to-InP) interfaces. Growth direction is from left to right.

b. [110] cross-section HREM image of GaInAs quantum well from same sample as Fig. 4a. Growth direction is up page.

4. CONCLUSIONS

In conclusion we have demonstrated that high spatial resolution X-ray mapping in the VG HB501 STEM is a powerful technique for investigating the compositional abruptness of interfaces in MQW structures and that correlation of the results obtained with TEM and HREM images from the same samples can lead to improvements in the interpretation of the latter. We have also shown that the use of different gas switching techniques can have a major effect on the abruptness of interfaces in MOCVD GaInAs/InP MQW structures.

ACKNOWLEDGEMENTS

The authors wish to thank Professor Sir Peter Hirsch FRS for provision of laboratory facilities and the SERC, STC plc and the UK Department of Trade and Industry for jointly funding this work under the JOERS and the LINK initiative.

REFERENCES

Briggs A T R and Butler B R 1987 J. Crystal Growth 85 31
Butler B R, Briggs A T R, Thrush E J, Garrett B and Stagg J P 1988 Chemtronics 3 31
Chapman J N, McGibbon A J, Cullis A G, Chew N G, Bass S J and Taylor L L 1987 Proc. Microsc. Semicond. Mater. Conf. Oxford 1987, Inst. Phys. Conf. Ser. No. 87 649
Chew N G, Cullis A G, Bass S J, Taylor L L, Skolnick M S and Pitt A D 1987 Proc. Microsc. Semicond. Mater. Conf. Oxford 1987, Inst. Phys. Conf. Ser. No. 87 231
Cureton C G, Thrush E J and Briggs A T R 1991 J. Crystal Growth 107 549
Liddle J A, Norman A G, Cerezo A and Grovenor C R M 1989 Appl. Phys. Lett. 54 1555
Liddle J A, Long N J and Petford-Long A K 1990 Materials Characterisation 25 157
Long N J and Glaisher R 1990 Proc. EMAG-MICRO89 London 1989, Inst. Phys. Conf. Ser. No. 98 277
Long N J 1989 Proc. Microsc. Semicond. Mater. Conf. Oxford 1989, Inst. Phys. Conf. Ser. No. 100 59
McGibbon A J, Chapman J N, Cullis A G and Chew N G 1988 Proc. Analytical Electron Microscopy Workshop Manchester 1987 219
McGibbon A J, Chapman J N, Cullis A G, Chew N G, Bass S J and Taylor L L 1989 J. Appl. Phys. 65 2293
Norman A G, Butler B R, Booker G R and Thrush E J 1989 Proc. Microsc. Semicond. Mater. Conf. Oxford 1989, Inst. Phys. Conf. Ser. No. 100 311
Petford-Long A K, Booker G R, Hockly M and Taylor M R 1989a Proc. Microsc. Semicond. Mater. Conf. Oxford 1989, Inst. Phys. Conf. Ser. No. 100 281
Petford-Long A K, Booker G R and Hockly M 1989b Ultramicoscopy 31 385
Spurdens P C and Hockly M 1986 Mater. Lett. 4 353
Spurdens P C, Taylor M R, Hockly M and Yates M J 1991 J. Crystal Growth 107 215
Taylor M R, Hockly M, Petford-Long A K, Lyons M H and Spurdens P C 1989 Proc. Microsc. Semicond. Mater. Conf. Oxford 1989, Inst. Phys. Conf. Ser. No. 100 305
Thrush E J, Cureton C G, Trigg J M, Stagg J P and Butler B R 1987 Chemtronics 2 62

Inst. Phys. Conf. Ser. No 117: Section 2
Paper presented at Microsc. Semicond. Mater. Conf., Oxford, 25–28 March 1991

Point-to-point resolution in X-ray microanalysis of thin coatings in the energy range 20–100keV

A G Nassiopoulos and E Valamontes

Microelectronics Institute, NCSR "Demokritos", PO Box 60228, 153 10 Aghia
Paraskevi Attikis, Athens, Greece.

ABSTRACT: The lateral extent of the X-ray signal in X-ray microanalysis of thin overlayers is calculated systematically, by using Monte-
Carlo simulations. Different primary beam energies in the range
20–100keV and different film thicknesses are considered. A Point-
Spread-Function composed of two Gaussian curves is determined for each
point and the Rayleigh criterion is used in order to calculate the
corresponding point-to-point resolution.

1. INTRODUCTION

It is well known that lateral resolution in X-ray microanalysis of bulk
materials is limited by the diffusion of electrons within the specimen.
This diffusion produces an additional signal along the electron trajectory due to backscattering and to characteristic or continuous X-rays
created within the sample by incident electrons, which ionize the atoms
of interest on their way out of the sample. This additional signal, added
to the signal created by the incident beam, limits the lateral resolution
in the micron range. On the other hand, when the sample of interest is a
thin coating on a bulk material, the additional signal due to backscattering and X-rays from the substrate, although it contributes substantially to the total signal [Cazaux et al. 1989, Nassiopoulos et al.
1990], does not significantly affect the spatial resolution when the film
thickness is small [Cazaux et al. 1990, Nassiopoulos et al. 1990,1991] .
The extreme case is the case of a very thin film (maximum thickness: some
hundreds of A), where the lateral resolution is exclusively governed by
the incident probe size. This case is equivalent to Auger Electron Spectroscopy where the analysed depth is imposed by the technique to a maximum of some tens of Angstroms. It is now well established that in Scanning Auger Microscopy the point-to-point resolution is principally influenced by the probe size [El Gomati et al. 1978, Tholomier et al. 1988,
Cazaux et al. 1988, Valamontes et al. 1990, Glezos et al. 1991] and it is
not seriously affected by backscattering or by the X-ray induced signal
from X-rays created by the incident beam.

In this paper, the point-to-point resolution is calcutated by using Monte-Carlo simulations . Different film thicknesses are considered for the
whole energy range 20–100keV. As coating and substrate materials, the
case of a Cu film on Au is chosen (analysis of the CuKa line), because
previous calculations [Nassiopoulos et al. 1990, Valamontes et al. 1990]

proved that in this case, the enhancement signal from the substrate is very important and greater than the primary beam induced signal at high primary beam energies. So it is a very unfavorable case for the point-to-point resolution.

2. RESULTS AND DISCUSSION

The Monte-Carlo programme has been described in detail elsewhere [Nassiopoulos et al. 1990]. The total X-ray signal from the film is calculated by considering all contributions to this signal: the primary beam induced signal, the signal due to backscattered electrons from the substrate and the signal due to both characteristic and continuous X-rays created within the substrate and ionizing the atoms of the film on their way out of the sample. The incident beam is considered to be normally distributed (gaussian curve) around the point of incidence. The primary beam is at normal incidence and the detection angle is 25.5° from the surface of the sample. The radial distribution of the analysed characteristic X-ray line is calculated by assuming isotropic emission of the X-ray signal over 2π steradians. Details about the calculation of the different contributions to the signal are given in the reference cited above (Nassiopoulos et al. 1990).

The calculated total signal is then divided into two parts:

$$I_T = I_P + I_E$$

where I_P is the primary beam induced signal and I_E includes three separate contributions:
-the signal from the film due to backscattered electrons.
-the signal induced by characteristic X-rays from the substrate.
-the signal induced by continuous X-rays also from the substrate.

In the Monte-Carlo programme different signals of interest are calculated as follows:
a)The electron induced signal, including both the signal from incident and backscattered electrons. The electron trajectory is first determined and the X-ray signal is calculated for the part of the electron trajectory which lies within the film.
b)The fluorescence signal from characteristic X-rays from the substrate. All possible characteristic X-rays created in the substrate along the electron trajectory are considered. Absorption within the substrate and within the film is taken into account.
c)The fluorescence signal from continuous X-rays coming from the substrate.
The continuous X-rays created in the substrate by incident electrons are calculated by considering Dyson's formula at the interface film-bulk (Nassiopoulos et al. 1990). In order to separate the primary beam induced signal from the total signal induced by electrons (incident and backscattered) we consider that the primary beam induced signal is equal to the signal from a thin unsupported film, which is calculated separately.

When the incident spot size is small, as in our case ($\sigma = 40\overset{\bullet}{A}$), the lateral extent of the incident beam induced signal is much smaller than that of the three other contributions. So the three last contributions, which have approximately the same lateral extent, are assimilated by one gaussian curve. Another gaussian curve describes the signal from the primary beam.

$$I^r=(I_P/\pi\sigma_P{}^2).exp(-r^2/\sigma_P{}^2)+(T.I_P/\pi\sigma_T{}^2).exp(-r^2/\sigma_T{}^2) \quad (1)$$

where I_P is the collected X-ray signal due to the primary beam and σ_P is the standard deviation of the gaussian curve attributed to this signal. $T.I_P$ is the total enhancement signal from the substrate and σ_T the standard deviation of the gaussian curve describing this total signal. An example of the radial distribution of the total signal is given in fig. 1.

For each primary energy and film thickness the coefficients I_P, σ_P, $T.I_P$, σ_T are calculated by fitting on Monte-Carlo curves. So the Point-Spread-Function, given by expression (1), is determined. The point-to-point resolution is then calculated by using the Rayleigh criterion, according to which two points are resolved if there is at least 26.5% dip in signal intensity at the mid-point between them. This minimum distance, d_m, has been calculated by using expression (1). Results are given in figs. 2,3 and 4. Fig. 2 indicates the point-to-point resolution as a function of energy for a Cu film on an Au substrate and for different film thicknesses. When the film thickness is small, the point-to-point resolution is not affected by backscattering or fluorescence from the substrate, so it is small and approximately independent of energy, in the range 20-100keV. But as the film thickness is increased, the signal from the substrate begins to deteriorate the spatial resolution. For a given film thickness, this deterioration is more important at low primary beam energies. This result is easily expected because at low energies the electron range approaches more the film thickness, so the film is seen by the incident beam as a bulk material. Indeed, by examining the results from a bulk material (fig. 3) we see that a film of 5000Å has, at 20keV approximately the same point-to-point resolution as the bulk.

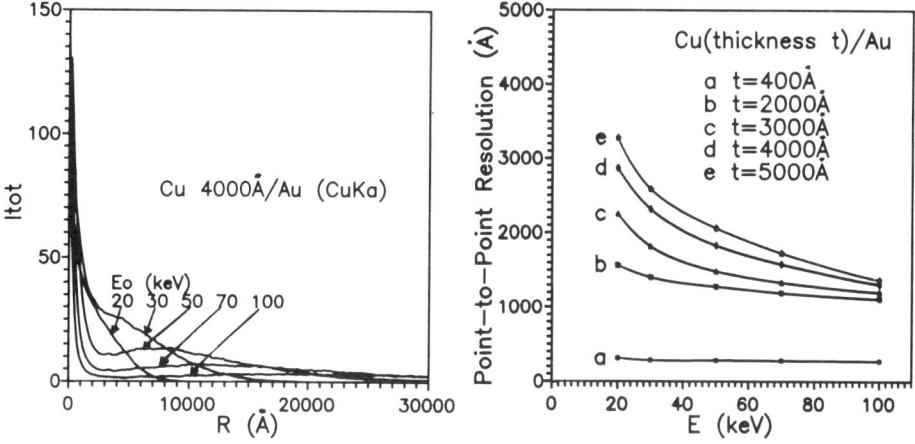

Fig. 1   Fig. 2

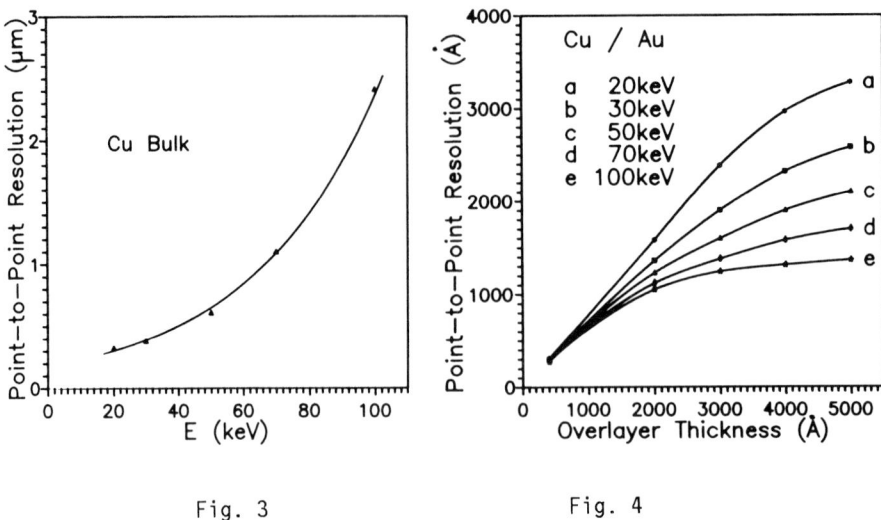

Fig. 3   Fig. 4

Fig. 4 indicates the point-to-point resolution as a function of film thickness. For a given energy, curves are saturated at great thicknesses. The variation of the lateral resolution is greater at low primary energies.

3. CONCLUSION

The point-to-point resolution in X-ray Microanalysis of thin coatings has been calculated as a function of energy and film thickness for the case of a Cu layer on Au. It is demonstrated that for small film thicknesses, the resolution is only governed by the incident spot size, for great thicknesses, approaching the electron range at the corresponding energy, it reaches that of a bulk material, totally governed by the diffusion of the incident beam within the substrate and for intermediate thicknesses it is energy and thickness dependent.

REFERENCES

Cazaux J., 1983 Surf. Sci 125 335
Cazaux J. , Chazelas J. , Charasse M. M. and Hirtz J. P. , 1988 Ultramicroscopy 25 31
Cazaux J., Jbara O., Nassiopoulos A.G., Valamontes E., 1989 Proceed 12th ICXOM 201
Cazaux J., Jbara O. and Thomas X. 1990 Surf. Interf. Anal. 15 567
El Gomati M.M. and Prutton M., 1978 Surf. Sci 72 485
Glezos N. and Nassiopoulos A.G., 1991 Surf. Sci (to appear)
Nassiopoulos A.G. and Valamontes E., 1990 Surf. Interface Anal. 15 405
Nassiopoulos A.G. and Valamontes E. 1991, Microbeam Anal. (to appear)
Tholomier M., Dogmaine D. and Vicario E., 1988 J. Micros. Spectr. Elect. 13 119
Valamontes E., Nassiopoulos A.G., Glezos N., 1990 Surf. Interf. Anal. 16(1) 203

Inst. Phys. Conf. Ser. No 117: Section 2
Paper presented at Microsc. Semicond. Mater. Conf., Oxford, 25–28 March 1991

The use of TEM contrast effects for the assessment of the composition of strained $In_xGa_{1-x}As$ layers

A S Dobson, C S Baxter and W M Stobbs

Department of Materials Science and Metallurgy, Cambridge University, Cambridge, UK

ABSTRACT: We re-examine here the contrast exhibited by thin strained layers of $In_xGa_{1-x}As$ in GaAs and assess the possibility of determining the composition of these layers by a variety of new approaches. Experimentally a number of features of the contrast characteristics which are described here would appear to have the sensitivity and potential reliability to be used quantitatively. However, as is also discussed, some of these contrast features are sensitive to modelling parameters which are difficult to quantify.

1. INTRODUCTION

Finding better methods than are now available for the determination of the composition of strained layers of III-V compounds would be useful for the assessment of the properties of such systems as well as to clarify how the strain level can affect the growth rate and the composition levels for both MOCVD (Monserrat et al. 1988) and MBE. There are invariably a number of TEM techniques which can be applied for a given III-V system but, while attempts have been made to judge which method should have the highest potential accuracy for a given alloy comparison (Bithell et al. 1991), strains reduce the viability of most approaches. Advances in the understanding of the effects of thin foil relaxations on the contrast shown by strained and layered systems (Treacy et al. 1985, 1986) have led to their inclusion for III-V wedges (Bangert et al. 1989) as well as generally (Perovic et al. 1991), and relaxations for wedges can now apparently be accurately measured (Harvey et al. 1991). Ion beam thinned specimens often have to be examined however and the effects of the usually unknown compositional abruptness of the interface make it difficult to assess the degree to which higher Fourier components of the strain relaxation should be included. This can make it difficult to use modifications of the contrast approaches which work well for unstrained systems (Bithell et al. 1989) quantitatively. It might be better to evade the effects of relaxations by avoiding contrast conditions which they affect, and this is one of the approaches considered below.

Here we examine a range of contrast conditions which might be used for the characterisation of strained layers of $In_xGa_{1-x}As$ in GaAs for a specimen with isolated groups of five alloy layers of varying thicknesses (that examined had layers about 5 nm thick with x~0.35). This system was chosen because it is one of the more challenging. For example the 002 structure factor is less than that of GaAs for x less than about 0.45, asymmetric contributions from non systematic row reflections can make the alloy intensity in 002 and 00$\bar{2}$ very different (Baxter et al. 1987), the strains for higher x can be more than 10% , the Debye Waller factor for In is high and the anomalous absorption is difficult to quantify accurately for the system.

2. THE CONTRAST OF $In_xGa_{1-x}As$ FOR 00n AND 0n0 REFLECTIONS

We may be reminded of the gross effects of thin foil relaxation for (001) strained layers, while noting that it is not necessarily the best approach to use the 002 reflection when assessing composition, by examining the 004 dark field image series in Fig.1 taken as a function of deviation parameter through the Bragg condition at a beam normal near to (100). The strong contrast at the edges of the group of five $In_xGa_{1-x}As$ layers is the signature of the foil's relaxation which is greater at its top and bottom surfaces than nearer to its centre. Fig. 1d with s very negative for GaAs and less so for $In_xGa_{1-x}As$ immediately suggests a method of determining the chemistry. The effects of the non uniformity of the relaxation are hence negligible and the deviation parameter for the GaAs could be measured with a convergent probe away from the layers so that the relative contrast of the alloy and matrix could then be compared with Bloch wave simulations for the weak beam conditions of each. The advantage of using 004 rather than 002 in this approach is both that Δg can be larger and that the absolute intensity at moderate deviation parameters is not too weak. The same technique could be used for the 008 reflection, as we can see from the images in Fig.2, the use of $00\bar{8}$ reflection emphasising that it is at a negative deviation parameter for the matrix that the alloy can be markedly brighter than the GaAs. The effects seen in the 006 and $00\bar{6}$ reflections (Fig.3) are interesting and confusing: at first sight it appears that here is a condition at which the alloy, for x in the range specified, exhibits an anomalously higher intensity than GaAs. This should not happen with 006 at the Bragg condition, as was checked using Bloch wave simulations. We have refined neither the absorption values nor the Debye Waller factors for these simulations since our intention has been qualitative rather than quantitative. Nor is this the effect due to non systematic contributions as described by Baxter et al (1987). This was checked by taking such images at a variety of beam normals. Since, as we will see below, the effect is not associated with thermal diffuse scattering either, its probable origin can be inferred from the dp in Fig.3a (for Fig.3b). Such images are weak and are taken after small tilts which tend to be made to maximise the contrast rather than by examining the local dp. We can see that $00\bar{4}$ is strongly excited and that 00-6 for $In_xGa_{1-x}As$ almost certainly has s negative when the anomalous contrast is observed, so that the effect seen for Fig.1d is now enhanced by dynamical contributions from the strong inner reflections. For x in the given range, I_{004} for the alloy can be higher than for GaAs, the extinction distance now being reduced. That thermal diffuse scattering is relatively unimportant is demonstrated by Fig.4a where we see the alloy layers darker than the matrix. The image was taken away from any reflection, at about 0.5° from the central beam towards [010], for the symmetric foil orientation on the 00n row seen in bright field in Fig.4b.

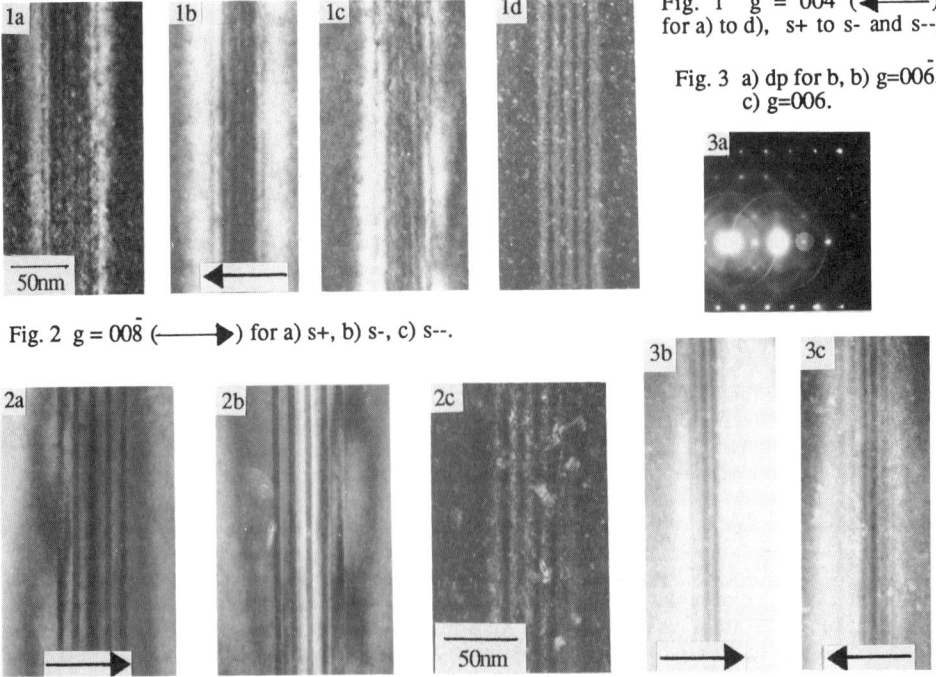

Fig. 1 g = 004 ($\longleftarrow$) for a) to d), s+ to s- and s--.

Fig. 3 a) dp for b, b) g=$00\bar{6}$, c) g=006.

Fig. 2 g = $00\bar{8}$ ($\longrightarrow$) for a) s+, b) s-, c) s--.

50nm

Fig. 4 a) Diffuse Dark Field at 0.5° towards [010], b) Symmetric Bright Field on 00n, c) 060, d) Intensity Profile of 060, e) First thickness fringe for GaAs = 1, and for InGaAs = 2.

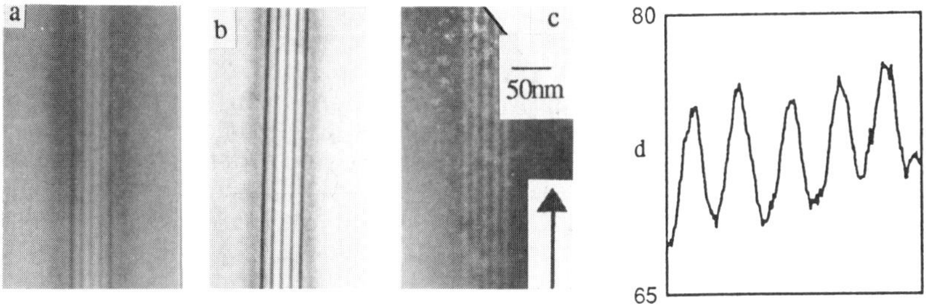

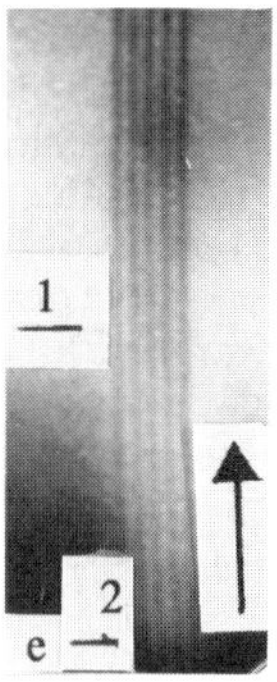

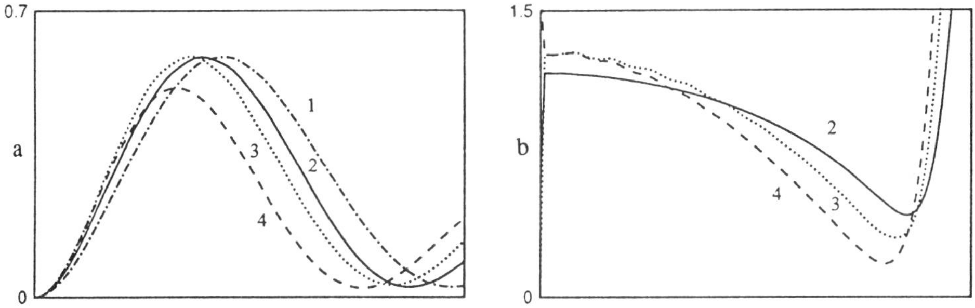

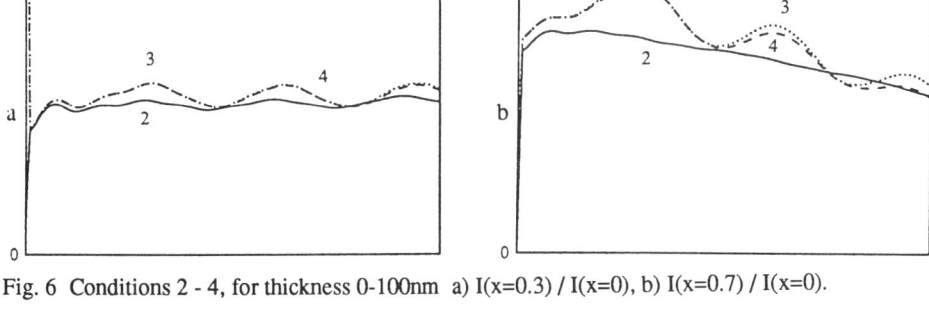

Fig. 5 In$_x$Ga$_{1-x}$As, 2 : 040, 3 : 004, 4 : 004 for GaAs at s=0
a) Thickness fringe profiles for x=0.7 cf for GaAs 040 (1)
b) I(x=0.45) / I(x=0), Conditions 2 - 4.

Fig. 6 Conditions 2 - 4, for thickness 0-100nm a) I(x=0.3) / I(x=0), b) I(x=0.7) / I(x=0).

Fig. 7

a) g = 00$\bar{2}$
b) g = 002
c) g = 020
d) Intensity profile of c.

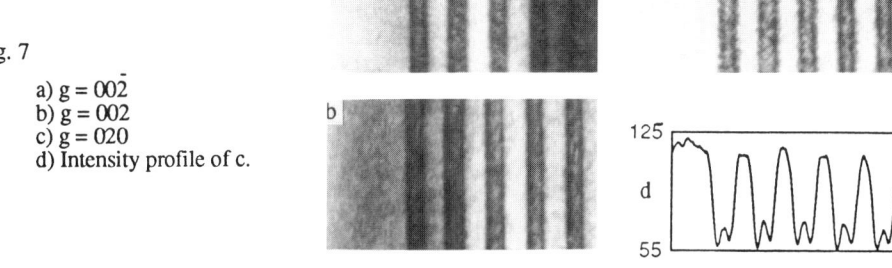

We can of course rid ourselves of the effects of the strains by using 0n0 reflections and this can be done at only about 2° or so from the (100) normal (so that the layers if moderately thick and in a thin foil can still be examined near to edge-on) without seeing the effects of the non systematics on the strains. 060 and 080 images are shown in Figs.4c and e respectively with an intensity profile of the anomalous layer contrast for the former image (again taken with a negative deviation parameter) in Fig.4d. It is clear from Fig.4e that the 008 $In_xGa_{1-x}As$ first bright fringe occurs at a greater thickness than it does for GaAs. This suggests a second method for measuring x based on this effect but the sensitivity of the approach depends on x. It is interesting to examine the extent to which changes in structure factor with angle can be useful and the effects for {004} reflections are exemplified for $In_xGa_{1-x}As$ by the simulated thickness fringes shown in Fig.5a. Here (as for other simulated data below for other x) (1) is 040 for GaAs, (2) is 040, x=0.7, (3) is 004 for x=0.7 (simulated here for no foil relaxation) at a lesser angle for the coherent structure and (4) is 004 x=0.7 again but now at the orientation at which GaAs 004 would be at s=0. The effects are quite strong at fairly low x and a high value is used simply for clarity. Useful potentially quantitative approaches would include comparing thickness fringe positions for 040 and comparing intensities at each first thickness fringe. The intensity ratios for 040 as a function of thickness are shown in Fig.5b. for x=0.45 when the 020 reflection intensities would differ little for the alloy and for GaAs. Since Fig.4e demonstrates that the approach could be readily applied for 080, thinner layers could be analysed using 040, 080 having the greater extinction distance.

Nevertheless the 020 condition remains useful for 0.2<x<0.4 simply because the relative contrast of the reflection to that for GaAs for the alloys changes very rapidly up to and beyond x=0.45 as may be seen from Figs.6a and 6b for, respectively, x=0.3 and 0.7. Examination of the images shown in Figs.7a and b for 00$\bar{2}$ and 002 reminds us of the near impossibility of avoiding the effects of surface relaxations for this **g** as well as the confusions caused by the alloy spread at the edges of the layering leading to a reduction in the layer intensity before it increases again at its centre. This is another reason why the use of 040 can be preferable. The image in Fig.7c was however taken for 020 and an intensity profile of the 5nm (necessarily slightly inclined) layers is shown in Fig.7d. The ratio of the alloy intensity to that of the adjacent GaAs proved to be ~0.64 and on a very coarse interpolation of predicted values this would indicate that x is ~0.38. The value could undoubtedly be improved both by using improved simulations and by using ratio evaluations for a finer set of values of x. It can also be noted from Fig.6 that another advantage of this simplistic method is that values of the ratio remain well behaved as the foil thickness is reduced, particularly for lower x values, so that still thinner layers could be examined (if that is, they were not too heavily graded).

In this rather preliminary appraisal of conditions, other than the well worn 002, that might be used for x value determinations our aim has been simply to demonstrate that there are many other intensity ratio based approaches that can be taken. We can further note that, of the four or five methods we have briefly described here, each would appear to have its own advantages and disadvantages but that each too would appear to have its own range of x over which it would be preferred to the others. We thank the SERC for financial support.

REFERENCES

Bangert U and Charsley P 1989 Philos. Mag. 59A 629

Baxter C S, Stobbs W M, Monserrat K J and Tothill J N 1988 Proc. Analytical Electron Microscopy ed:.G.Lorimer (Manchester,1987) IOM pp 209-212

Bithell E G and Stobbs W M 1989 Philos. Mag. A60 39

Bithell E G and Stobbs W M 1991 J. appl. Phys. 4 2149

Harvey A J, Faux D A, Bangert U and Charsley P 1991 Philos. Mag. Lett. in press

Monserrat K J, Tothill J N,.Haigh J, Moss R A, Baxter C S and Stobbs W M 1988 J. Crystal Growth 93 466

Perovic D D, Weatherley G C and Houghton D C 1991 Philos. Mag. in press

Treacy M M J and Gibson J M 1986 J. Vac. Sci. Technol. B 4 1458

Treacy M M J, Gibson J M and Howie A 1985 Philos. Mag. 51 389

Inst. Phys. Conf. Ser. No 117: Section 2
Paper presented at Microsc. Semicond. Mater. Conf., Oxford, 25–28 March 1991

The effects of the inclination of an interface in an $Al_xGa_{1-x}As$/GaAs heterostructure on its contrast

E J Williams, E G Bithell and W M Stobbs

University of Cambridge, Department of Materials Science and Metallurgy, Cambridge, UK.

ABSTRACT: The use of the compositionally sensitive 002 reflection for the measurement of the abruptness of the composition change at the leading and trailing edges of a layer of $Al_xGa_{1-x}As$ grown in GaAs is examined. It is demonstrated that the inclination of the layer to the beam direction can lead to contrast effects at the edges of the layers which are strongly dependent upon the foil thickness and the deviation parameter.

1. INTRODUCTION

Our aim here is to determine the primary origin of the dark and bright lines which are often seen in 002 and 020 dark field images at the edges of layers of differing $Al_xGa_{1-x}As$ chemistries grown on substrates of near [001] normal. The motivation for undertaking this study is that the band structure of a III–V layer heterostructure is dependent not only upon the thicknesses and compositions of the layers but also upon the form of the concentration changes at the interfaces. For example, it is demonstrated in an accompanying paper (Rimmer *et al.* 1991) that the photoluminescence spectrum of such an $Al_xGa_{1-x}As$/GaAs heterostructure is dependent upon the few atomic layers (approximately two unit cells) over which the composition is usually graded at the interfaces. There are problems with the use of conventional high resolution chemical mapping approaches because of the way the Fresnel effects due to the interface are normally disregarded (Stobbs 1990) but the analysis of the Fresnel contrast itself can be used to provide data to the monolayer level for $Al_xGa_{1-x}As$/GaAs layer structures (Ross *et al.* 1987, Ross *et al.* 1991). 002 dark field images can be analysed to provide x for moderately thick (>3nm) $Al_xGa_{1-x}As$ layers to reasonable accuracy (Bithell *et al.* 1989): it is thus relevant to examine whether the interface contrast can be used in the same simple way. This is useful too in understanding the relative importance of the form of the composition profile and of the interlayer strain in determining the contrast of more complicated III-V layer systems of greater misfit.

2. COMPOSITION FRINGE CONTRAST

Some 002 dark field images of $Al_xGa_{1-x}As$ heterostructures would invariably be discarded as being 'misleading'. An example of such an image for a relatively thick GaAs layer between two $Al_xGa_{1-x}As$ layers is shown in fig. 1. There are gross local increases, as well as decreases, of the interface contrast relative to the adjacent contrast levels, obviously associated with the waviness of the boundaries and these would normally lead to the image being set aside. Yet images like this, showing interfaces with grossly variable inclination to the beam direction, point to the origins of the effects seen and also indicate the dangers of interpreting the interface contrast for more nearly vertical layers (which must exhibit lesser effects of the same type) as a simple representation of the local composition profile. The interest of the MBE grower in the abruptness of the composition change at the initiation and completion of $Al_xGa_{1-x}As$ layers has however led to the use of 002 dark field images for the qualitative assessment of the form of interface composition changes. [010] bright field thickness fringe images of cleaved wedges are similarly interpreted. Even when effects of the type seen in fig. 1 are not obvious, the

Fig. 1. 002 dark field image of a layer of
$Al_xGa_{1-x}As$ with very undulating interfaces.

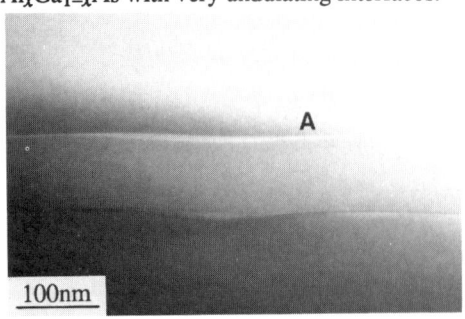

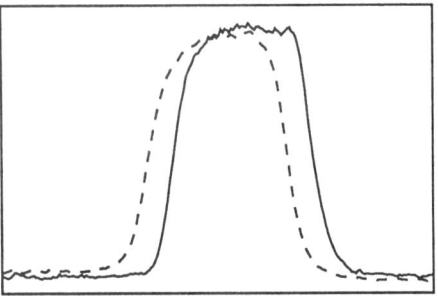

A

100nm

Fig. 2. Intensity profiles across 002 (solid) and
00$\bar{2}$ (broken line) images of an AlAs layer.

dangers of such approaches are demonstrated by the intensity traces in fig. 2. These are 002 and 00$\bar{2}$ images of the same region of an AlAs layer (~4nm thick) in GaAs, at a foil thickness of only ~30nm. The growth direction is to the right: while the shapes of the contrast changes at the two interfaces are 'as would be expected' for 002 (i.e. more diffuse at the top surface of the AlAs: note especially the shapes of the lower corners at the left of the profile), they are essentially symmetric for 00$\bar{2}$. Although the inclination of a high vicinality layer can be measured reasonably accurately (Boothroyd *et al.* 1987), this is more difficult for layers which are less than 1° from (001). In the light of the interface contrast effects seen in fig. 1 it would be unwise to assert which of the above images had the layer the nearer to vertical.

We will now consider in more detail interface contrast effects which are intermediate between those described for fig. 1 and fig. 2 and which at first sight suggest an increase in Al concentration at the interface. This latter interpretation has been used to justify the form of δ-fringe contrast found for highly inclined interfaces (Bangert *et al.* 1987), though it seems to be more generally considered that such edge contrast has its primary origin in the effects of the strains even for the small misfit exhibited by this system. Since the analyses of the contrast effects due to periodic strains in thin foils made by Treacy *et al.* (1985, 1986), the contrast seen for a variety of heavily strained layer structures has been successfully assessed on this basis (e.g. Gibson *et al.* 1985). Strains, as well as the inclination of the foil surfaces to the beam, can undoubtedly affect the contrast but here we will demonstrate that the primary cause of enhanced edge contrast for low misfit layers lies simply in the foil thickness dependence of the intensity scattered into a 002 beam for a given composition. This interpretation is actually clear from fig. 1, when it is noted that in this image the high intensity edge contrast is clearest at A; here the thickness of the foil and the deviation parameter are such that the GaAs intensity is high in comparison with that for $Al_xGa_{1-x}As$, which is falling rapidly. A composition intermediate between GaAs and the $Al_xGa_{1-x}As$ in the bulk of the layer (such as occurs in the projected composition profile at the interface) could thus scatter more strongly than either of these, at thicknesses where the individual alloy intensities are behaving in this way.

A high quality (001) $Al_{0.3}Ga_{0.7}As$/GaAs multilayer with relatively flat interfaces is shown in fig. 3a. Figs. 3b to 3e are enlargements of 002 and 00$\bar{2}$ images of this sample using beam directions about 5° on either side of the [100] normal. In fig. 3b (region A of fig. 3a) there is no edge contrast and the intensities in the layers are increasing together. The remaining images are for thicknesses and deviation parameters such that the layer intensities are changing locally (as at A in fig. 1). Now we see approximately symmetric edge contrast for fig. 3e and more asymmetric interface contrast for figs. 3c and 3d. That this asymmetry could not be reversed on reversing g is presumably because the angle change was insufficient to change the sense of the layer inclination; nevertheless the interfaces appear to be less steep for figs. 3c and 3d than for fig. 3e. If symmetric contrast can be found, and in particular bright edge fringes, this must suggest the influence of the compositional diffuseness which often occurs in such layers (Ross *et al.* 1987). In order to calculate approximately the effects which might be expected for such interfaces, conventional column approximation Bloch wave image simulations were carried out

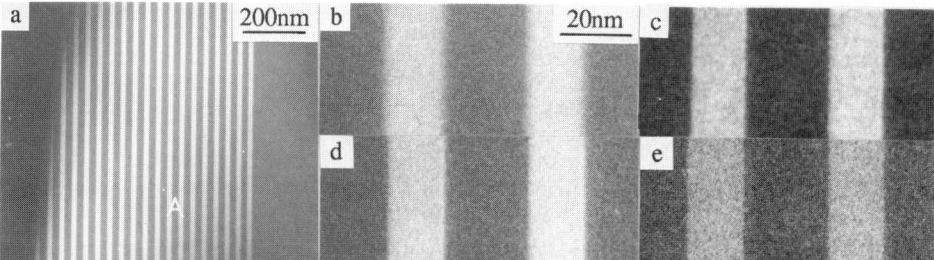

Fig. 3. (a) Low magnification image of a GaAs/Al$_x$Ga$_{1-x}$As multilayer. (b,e) 002 dark field; (c,d) 00$\bar{2}$ dark field. (b,d) are to one side of the [100] zone axis and (c,e) to the opposite side.

Fig. 4. 9 beam Bloch wave simulations of Al$_{0.3}$Ga$_{0.7}$As layers tilted by 0.75° in a wedged foil (t = 0 to 300nm) at 200keV. (a) 002 Bragg condition, low absorption; (b) g (1.5g) on 002, low absorption; (c) 002 Bragg condition, high absorption.

for layers tilted by 0.75° (fig. 4). Although column approximation calculations are of dubious validity for interfaces at such high tilts it is nonetheless significant that these simulations predict edge fringes which are qualitatively very similar in intensity to those seen experimentally, even though the lattice parameter of the alloy was taken to be identical to that for GaAs for these and all other simulations. Non–column Bloch wave or multislice simulations would be needed for accurate simulation of such near vertical interfaces, but our purpose here is more qualitative. For the simulations shown in figs. 4a and 4b a low level of absorption was included (V'_g = (0.04+0.04g)V_g), and for both s = 0 and s > 0 symmetric bright fringes are seen. An increased level of absorption (V'_g = (0.1+0.1g)V_g: fig. 4c) fails to introduce significant asymmetry or top/bottom effects, and it is this aspect of the experimental contrast which our simplistic calculations do not reproduce. Asymmetric edge fringes are predicted by multislice simulations of much narrower, slightly inclined layers (a few unit cells thick) and are thus likely to be reproduced by non–column Bloch wave calculations. It remains probable that such asymmetry can be further affected by strain and foil relaxation.

The effect was exaggerated experimentally by examining the multilayer using the 020 reflection at 5.5° to [100] for the s = 0, (g,2g) and (g,–g) conditions as shown in figs. 5a, 5b and 5c respectively. Simulations for these situations are shown in figs. 6a, 6b and 6c, and for this less steeply inclined interface the calculated and experimental images are closely similar. It can also be demonstrated that we are now simply seeing 'projected composition fringes', by comparing a full simulation for layers at the Bragg condition with one which was obtained by assuming that the contrast is given by that for the average composition of a given column. The difference between the two models, due to the stepped phase change in the column rather than continuous addition of amplitude, is shown in fig.7. The maximum thickness is 200nm; the intensity scale runs from –0.021 (black) to +0.0085 (white), this range being 6.6% of the maximum intensity in the Bloch wave calculation.

We have already noted that the effects of strain and foil inclination can increase the asymmetry of 'composition fringe' contrast, even for Al$_x$Ga$_{1-x}$As, as can Fresnel effects (though the convergence used is rarely low enough to show this last effect, except at large defoci). Having demonstrated that the primary origin of enhanced edge contrast for slightly inclined, or compositionally spread, $Al_xGa_{1-x}As$ layers is due to the effects of the changing projected composition we should note that there are additional secondary effects. For example, 'half spacing' lines are commonly seen between vertical layers (fig. 8) which are probably due to the 'superlattice periodicity' imposed by the regular multilayer. Such fringes would only be reproduced in simulations which included the periodic potential due to the presence of the multilayer. Similar but very weak fringes are common in images of fine $Al_xGa_{1-x}As/GaAs$ superlattices. It is however difficult to explain why these lines tend to be so thin unless this is related to the differences in the Fourier transforms of the shape functions of the two layers in each period.

REFERENCES

Bangert U and Charsley P,   MSM 87 eds. A Cullis and P Augustus (IoP, Bristol) pp 89–94.
Bithell EG and Stobbs WM 1989. Philos. Mag. A60 39.
Boothroyd CB, Britton EG, Ross FM, Baxter CS, Alexander KB and Stobbs WM,   MSM 87 eds. A Cullis and P Augustus (IoP, Bristol) pp 195-200.
Gibson JM, Hull R, Bean JC and Treacy MMJ, 1985. App. Phys. Lett. 46 649.
Rimmer N, Syme RT, Frost JEF, Ritchie DA, Jones GAC, Kelly MJ and Stobbs WM, 1991, these Proceedings.
Ross FM, Bithell EG and Stobbs WM, AEM 88 ed. G Lorimer (IoM, London) pp 205-208.
Ross FM and Stobbs WM, 1991. Ultramicroscopy, in press.
Stobbs WM, 1990. AMP 2, eds. HE Exner and V Schumacher (DGM Verlag) pp 1007-1017.
Treacy MMJ and Gibson JM, 1986. J. Vac. Sci. Technol. B 4 1458.
Treacy MMJ, Gibson JM and Howie A, 1985. Philos. Mag. 51 389.

Fig. 5.   020 dark field images of inclined $GaAs/Al_xGa_{1-x}As$ layers, with diffraction conditions as follows: (a) g (g); (b) g ($2g$); (c) g ($-g$). (d) $-g$ ($2g$) (inset: profile across two layers. Note the fine detail.).

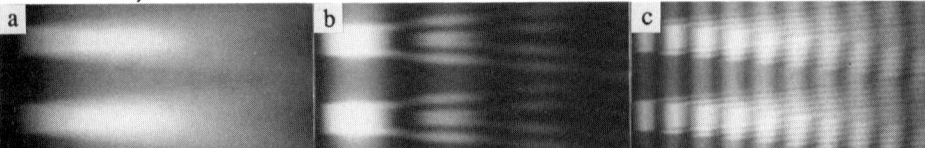

Fig. 6.   (a,b,c) 9 beam Bloch wave simulations corresponding to figs. 5a, 5b and 5c respectively. The foil thickness increases from zero (left) to 300nm (right).

Fig. 7. Difference between a 9 beam Bloch wave calculation for inclined interfaces and the same calculation using the projected composition instead of the abrupt interface.

Fig. 8.   002 dark field image of layers of $In_xGa_{1-x}As$ in GaAs.

Inst. Phys. Conf. Ser. No 117: Section 2
Paper presented at Microsc. Semicond. Mater. Conf., Oxford, 25–28 March 1991

Three dimensional visualisation of semiconductor multi-quantum well interfaces

R A D Mackenzie, A Cerezo and C R M Grovenor

Oxford University, Department of Materials, Parks Road, Oxford OX1 3PH

ABSTRACT: Individual interfaces in multiquantum well materials have been imaged in three dimensions using the position sensitive atom probe. The location and chemical identity of single atoms within a small volume have been determined and a constant composition contour constructed in three dimensions. This form of observation makes it possible to directly assess the interface morphology.

1. INTRODUCTION

Semiconductor devices based on quantum wells are now used in an increasingly large range of applications. The two major factors which influence the properties of a quantum well based device are (i) the compositions of the well and barrier layers and (ii) the abruptness of the interfaces between the wells and the barriers. The assessment techniques in routine use today fall into two basic categories, those which permit indirect measurement of composition and of interface morphology, and those which can be used to directly generate two dimensional maps or images. It has been shown that pulsed laser atom probe techniques can be used to generate two dimensional composition maps for semiconductor materials (comparable to those generated using high resolution x-ray microanalysis methods), Liddle et al, 1990. The next stage in this development is to extend the analysis techniques into three dimensions (3D). The potential of the position sensitive atom probe (POSAP) for generating three dimensional images in metallic systems has already been demonstrated, Cerezo et al 1989 and 1990. This paper reports the latest progress in applying 3D visualisation methods to semiconductor quantum well materials.

2. THE POSITION SENSITIVE ATOM PROBE

The technique of POSAP microanalysis has been described in detail elsewhere (Cerezo et al 1988). This permits a small volume of material (typically a cylinder 20 nm in diameter, and from 5 to 30 nm deep) to be analysed at near atomic resolution. For each atom recorded from this volume, its position in three dimensions is determined in addition to its chemical identity. Once this information has been obtained it is necessary to display the data in a meaningful way. The most straight-forward method of visualisation is to simply represent each atom by a dot (perhaps colour-coded) for each species present. This method, whilst clearly of some value in representing two dimension data, is not appropriate for detailed three dimensional information: layers close to the surface tend to obscure other layers and much of the three dimensional information is effectively lost.

3. VISUALISATION OF THREE DIMENSIONAL DATA

The method which has been found to be most satisfactory for representing compositional information obtained from the POSAP is based on the technique of generating a solid surface at a particular composition (Cerezo et al, 1990). This isosurface gives a clear impression of the morphology of interfaces present in the volume of analysis. In order to move from a series of isolated points to a constant composition surface requires several stages. Firstly, the entire block of data is divided into a series of small three dimensional cells, and the composition of each cell calculated. The actual size of these cells will differ from sample to sample, but in each case the cell width is $1/_{64}$ of the diameter of the image. In practice this means that each cell will be between 0.25 and 0.50 nm square. The cell dimension in the third direction can be scaled independently, in most cases this will be kept the same as the other two dimensions, but in cases where higher resolution information is required it may be possible to reduce this size. Using the composition of these cells it is then possible to construct an isosurface which joins up all the points within the three dimensional data-set which have the same composition. The resulting surface can then be rendered using standard visualisation techniques. In the Oxford Field Ion Microscope group, visualisation of atom probe data is carried out using a Stardent ST1000 graphics mini-supercomputer running AVS™ (Application Visualisation System).

These visualisation techniques have been applied to a variety of multiple quantum well (MQW) samples observed using the POSAP. The images in Fig. 1 show a single interface in a GaInAs-InP quantum well stack. Fig. 1(a) is a two dimension composition map from the data showing an interface running down the centre of the field of view. The area to the left of the interface is indium-rich (a barrier layer), and the area to the right gallium-rich (a quantum well). It is also clear that in this sample the barrier layer contains very little gallium, suggesting a well controlled growth process. Using the visualisation techniques described above, we can now rotate the block of data around to permit the interface to be seen in three dimensions, Fig. 1(b). In this case the isosurface has been constructed around the gallium content so that the area enclosed represents the inter-well barrier layer. The threshold layer has been chosen to be approximately half-way between the gallium content of the barrier (23 at%) and that of the well (0 at%). The entire image area is approximately 27 nm in diameter and it appears from this image that the interface is rough on the scale of 3-4 nm. In the image shown in Fig. 1(b) the depth scale magnification is 5 times the width scale magnification. The techniques required for the quantification of interface morphology from POSAP data are still under development. Initial results on the morphological characterisation of complex three dimensional interfaces in spinodally decomposed alloys have recently been presented, Cerezo et al, 1991. In the case where two interfaces are present in the same volume of data it is possible to make a direct visual comparison of the two. An example (from another GaInAs-InP stack) of this is shown in Fig. 2, where the top and bottom interfaces from a single well can be seen. The 2D composition maps in Fig. 2(a) show the orientation of the specimen; the two dark bands in the left hand (indium) image correspond to the bright bands in the right hand (gallium) image - these are the barrier layers enclosing the single well. Fig. 2(b) is an isosurface generated from the indium content in the sample (the threshold level has been chosen to be intermediate to the indium content of the barrier (50 at%) and of the well (26 at%)). The viewpoint chosen (using a perspective visualisation method) permits both interfaces to be seen simultaneously and shows that the left hand interface is smoother than the right hand interface. Similar observations have been made in high-resolution electron microscopy studies of these structures, however the POSAP provides direct 3D chemical mapping of the interface. The result is a more quantitative, as well as a more graphic characterisation of the MQW interface.

4. CONCLUSIONS

The position sensitive atom probe produces three dimensional compositional information. Using modern data visualisation techniques it is possible to reconstruct the data recorded to produce a three dimensional object which accurately indicates the morphology of very fine scale structures. This makes it possible to directly visualise a single interface in three dimensions within a quantum well stack, and will ultimately make it possible to directly relate the interface morphology to the growth conditions used.

ACKNOWLEDGEMENTS

The authors are grateful to Professor Sir Peter Hirsch FRS and Dr G D W Smith for the provision of laboratory space. The semiconductor research program within the Oxford Field Ion Microscopy Group is funded by the Science and Engineering Research Council under the Low Dimensional Structures and Devices Initiative. AC is grateful to The Royal Society and Wolfson College, Oxford for support during the course of this work.

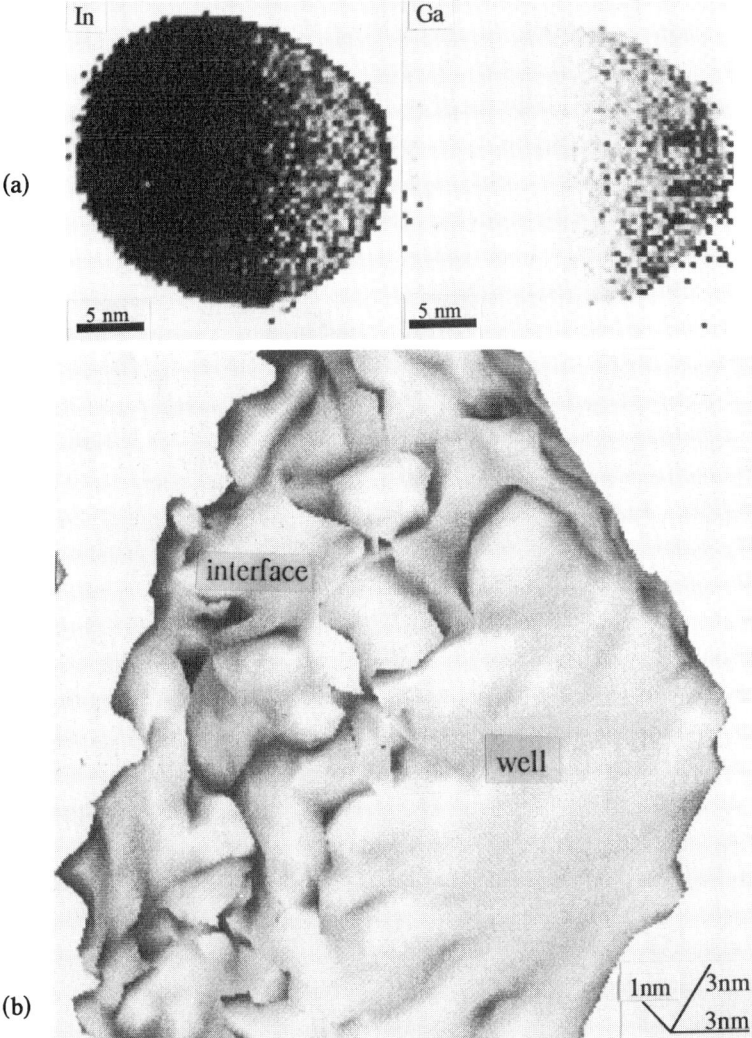

Fig. 1. A single interface in a GaInAs-InP multiquantum well stack. (a) Two dimensional composition maps showing the distribution of In and Ga near the interface(the dark regions are rich in the element indicated). The barrier layer (left hand side) can be seen to indium-rich but almost completely free of gallium. (b) A chemical isosurface constructed from the gallium content. The interface extends over approximately 3 nm.

REFERENCES

Cerezo A, Godfrey T J and Smith G D W, Rev. Sci. Instrum. **59** (1988) 862
Cerezo A, Godfrey T J, Grovenor C R M, Hetherington M G, Hoyle R M, Jakubovics J P,
    Liddle J A and Smith G D W, J. Microsc. **154** (1989) 215;
Cerezo A, Godfrey T J, Grovenor C R M, Hetherington M G, Hyde J M, Liddle J A,
    Mackenzie R A D and Smith G D W, EMSA Bulletin **20** (1990) 77
Cerezo A , Hetherington M G and Hyde J M, submitted to Scripta Materiala (1991)
Liddle J A, Long N J and Petford-Long A K, Mater. Charact. **25** (1990) 157

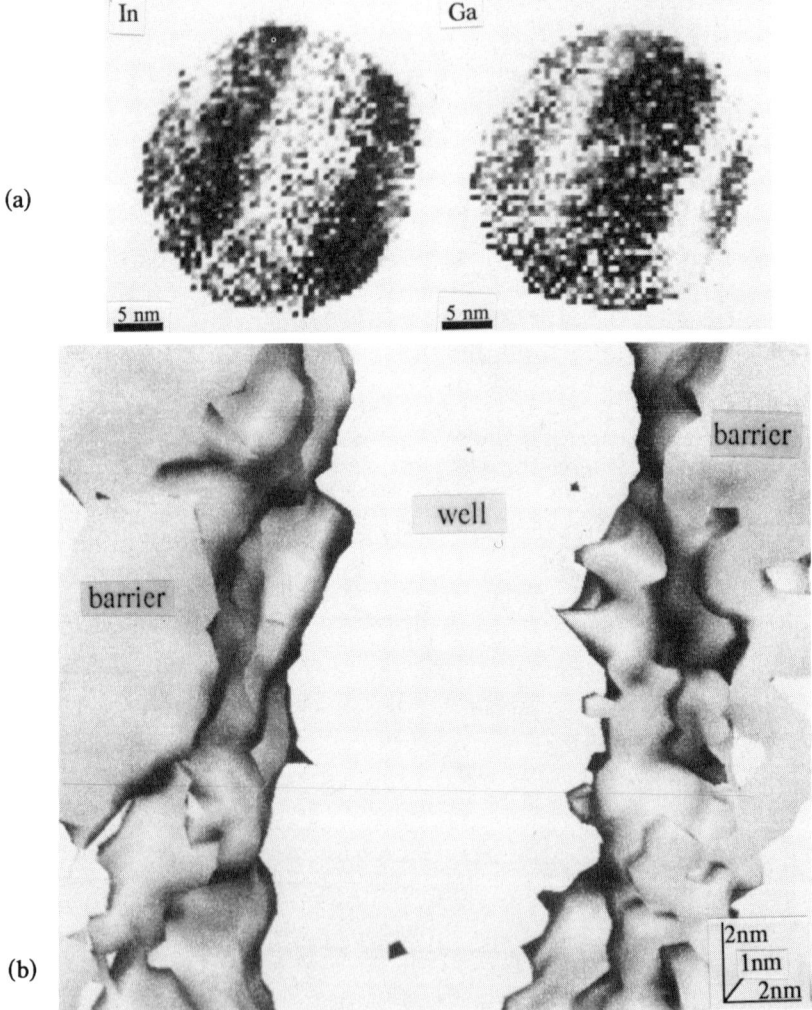

(a)

(b)

Fig. 2. GaInAs quantum well surrounded by InP barrier layers. (a) Two dimensional maps showing the relative position of the well and barrier regions. The barriers appear as the dark stripes in the indium image, and as the bright stripes on the gallium image. (b) An indium chemical isosurface showing the top and bottom interfaces of a single quantum well. The left hand interface appears to be smoother in three dimension than the right hand one.

Inst. Phys. Conf. Ser. No 117: Section 2
Paper presented at Microsc. Semicond. Mater. Conf., Oxford, 25–28 March 1991

Atom probe analysis of Au/GaAs and Ag/GaAs interfaces

Qiu-Hong Hu, Anders Kvist and Hans-Olof Andrén

Department of Physics, Chalmers University of Technology,
S-412 96 Göteborg, Sweden

ABSTRACT: Au/GaAs and Ag/GaAs interfaces were studied using transmission electron microscopy and atom probe microanalysis. The needle shaped GaAs atom probe specimens were field evaporated to obtain a clean surface before metal deposition in UHV. Some of the clean specimens were exposed to air before deposition to obtain an oxide layer. Silver showed no interaction with clean GaAs but single silver atoms were found to have penetrated into the substrate when there was oxygen present. For gold on a clean surface intermixing occurred, but this was blocked by the presence of an oxide layer.

1. INTRODUCTION

One of the basic problems in electronic materials is the understanding of the Schottky barrier formation. Research has been going on for many years in order to solve this problem and several models have been proposed (Mönch 1990). The contacts used for model study of the barrier are usually formed by deposition of metals on semiconductors cleaved in ultrahigh vacuum (Liliental-Weber 1990), while in conventional integrated circuit fabrication the semiconductor substrates are exposed to air prior to the deposition of a metal (Rhoderick and Williams 1988). This exposure allows a thin contamination layer to form on the substrates by which the height of the Schottky barrrier and the stability of the contact are affected (Miret et al 1988, Newman 1988).

In our paper we present a method to study the interface between metal and semiconductor where we use transmission electron microscopy (TEM) and atom probe field ion microscopy which has a very high spatial resolution. Our aim was to characterise in detail the composition of the interfaces formed by gold and silver both on atomically clean GaAs substrates and those formed on air exposed surfaces of GaAs. Some of the results have been reported earlier (Andrén 1987, Hu et al 1991).

2. EXPERIMENTAL

Needle-shaped specimens for atom probe field ion microscopy were prepared from GaAs wafers, Si-doped to 3.9×10^{18} cm^{-3} (1.3 mΩcm), by cleaving rods with a <110> axis and chemically etching to form a sharp tip (Andrén 1987). Each specimen was first inspected by TEM and then inserted into a field ion microscope (FIM) and cooled to 77 K. The tip region of the specimens were cleaned by means of field evaporation in order to get a smooth, clean and approximately hemispherical surface at the tip with a radius of 60 to 100 nm. Metal/GaAs interfaces were produced by thermally evaporating Au or Ag onto the specimens from a spherically shaped source, supported by a molybdenum filament, in the field ion microscope. The thickness of the deposited metal overlayer was typically between 0.5 and 3 nm; it was controlled with a quartz crystal oscillator of a resolution better than 0.05 nm. The metal deposition was usually done immediately after field evaporation in order for the interface of Au or Ag to form on the clean GaAs substrate. During the metal deposition the pressure in the FIM did not exceed 5×10^{-10} torr, and the specimen temperature probably raised only slightly above 77 K. The specimen was then allowed to warm up to room temperature, taken out in air and examined by TEM to determine the morphology of the overlayer. Atom probe analysis (Miller and Smith 1989, Tsong 1990) was then performed using electrical pulsing in an energy compensated instrument described elsewhere (Andrén and Nordén 1979,

Andrén 1986). Some of the analyses were made at 90 K and with a pulse factor of 15%. It was later found, however, that these conditions give a systematic error due to preferential field evaporation of Ga (Andrén 1987). Most of the analyses were therefore made using a specimen temperature of 40 K, which prevents preferential evaporation of Ga. In order to make certain that no Au or Ag was lost during analysis, the pulse fraction was increased to 25% during some analyses, although this may introduce systematic errors because of the limited energy focusing capability of the energy compensator. The area of analysis varied between 6 and 10 nm in diameter, and the depth of analysis was typically at most a few nm.

In addition to the study of clean interfaces, the interaction of Au and Ag with an oxidised substrate was also studied. For these studies, the field evaporated specimen was taken out from the FIM and kept in dry air for a few days before insertion back into the FIM, cooling, and deposition of Au or Ag. A total of 18 specimens have been analysed during the course of the work. A large number of analyses were attempted but terminated prematurely due to specimen fracture under the intense electric field in the FIM or the atom probe; the failure rate was approximately 50 %.

3 . RESULTS

3.1 Ag/GaAs interface

By TEM we observed, after silver deposition, island formation on the shank of the specimen, where the contamination layer from the chemical etching remained after the field evaporation. At the apex of the GaAs atom probe needle one large island was formed with a denuded zone close to the apex. These islands were formed on both atomically clean and on air exposed surfaces, Fig. 1a. Atom probe analysis of the island on the clean surface showed that it was a pure silver island on top of the GaAs substrate with a few oxygen atoms in between. Neither oxygen nor silver was observed below this interface in the GaAs substrate. In Fig. 2 atom probe data are presented in the form of a ladder diagram, i.e. the cumulative number of atoms detected of a certain element vs. the total number of atoms detected (a constant composition will in such a diagram give a line with a constant slope, a pure layer results in a line 45° against the axes, etc.). The analysis of the layer on an air exposed surface showed similar features, but in this case we observed a small number (<1at%) of arsenic, gallium and oxygen atoms spread out in the silver layer. We also observed about one percent of oxygen and silver in the GaAs substrate close to the interface. This oxygen and silver was found in a 1-2 nm thick layer only; further away from the interface, pure GaAs was registered. The interface was not as abrupt as on the clean surface, but had a thickness of about 0.2 nm, i.e. approximately one monolayer of intermixing, Fig. 3.

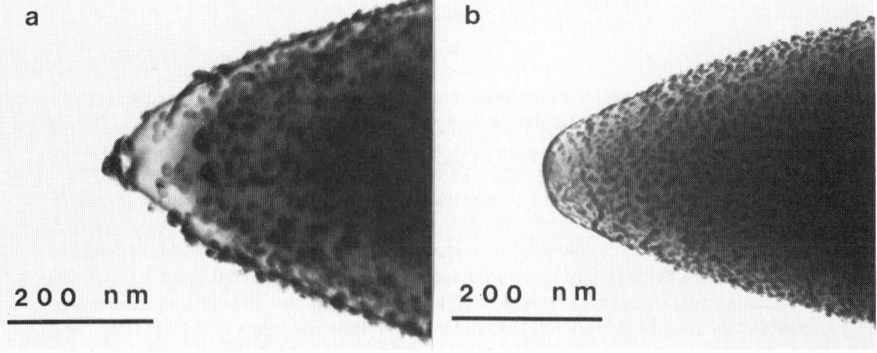

Fig. 1. TEM micrographs of GaAs atom probe specimens with a) silver and b) gold on the surface. The denuded zone close to the apex in a) can clearly be seen as well as the smooth apex in b).

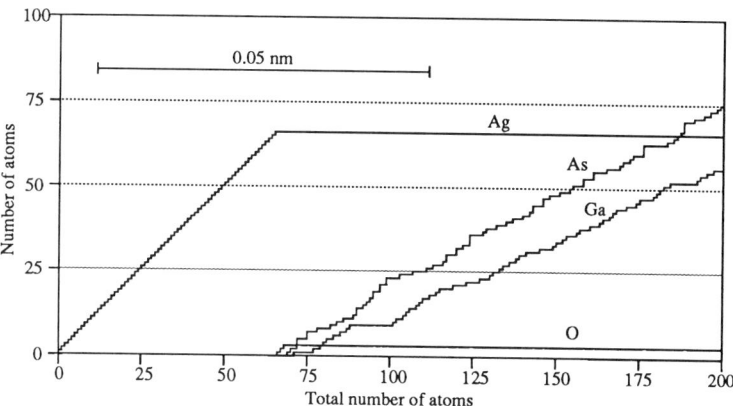

Fig. 2. Ladder diagram of an atom probe analysis of silver on a clean GaAs surface after removing most of the layer during field-ion imaging. The atomically abrupt interface is evident.

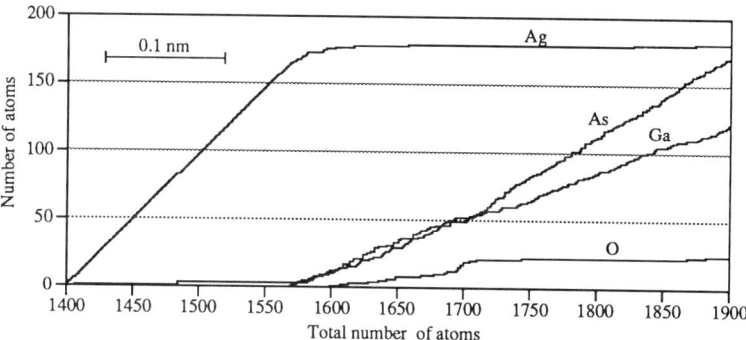

Fig. 3. Portion of a ladder diagram close to the interface of an atom probe analysis of silver on an air exposed GaAs surface. As can be seen the interface is not as abrupt as in Fig. 2.

Fig. 4. Ladder diagram of an atom probe analysis of gold on an air exposed gallium arsenide surface. Here we can see that oxygen effectively blocks intermixing.

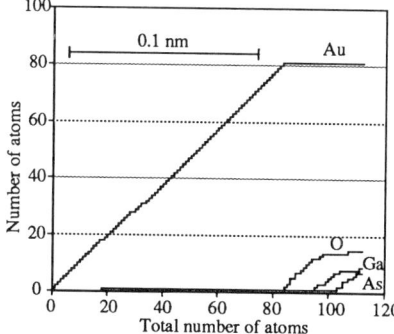

3.2 Au/GaAs interface

After deposition of gold, similar island formation as found for silver was observed in TEM, both on the shank of the specimen and at the tip apex of the air exposed substrate but with a smaller island size. On the other hand, on the clean substrate we could not see anything but a very smooth surface at the tip apex, Fig. 1b. Atom probe analysis of an island on the air exposed GaAs showed an almost pure layer of gold, with only single atoms of arsenic, gallium and oxygen in the layer, on top of a thin oxide layer on the semiconductor surface, Fig. 4. In contrast to this, in the case of the clean surface, we observed an intermixed layer of gallium, arsenic and gold with less than 1at% of oxygen in the layer, Fig. 5. The gold concentration in the layer was approximately 10at%. On top of this layer some small amounts of oxygen, arsenic and gallium was found. No oxygen or gold was detected below this mixed layer in the gallium arsenide substrate.

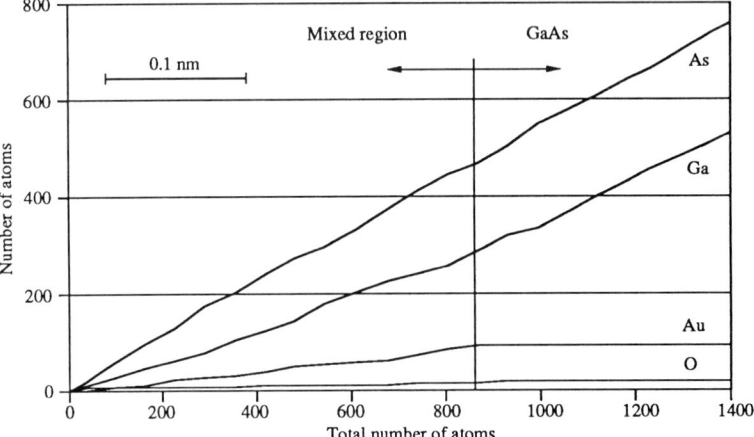

Fig. 5. Ladder diagram of an atom probe analysis of gold on a clean gallium arsenide surface. The intermixed layer here has a depth of 0.3 nm.

4. DISCUSSION

4.1 Interfaces formed on clean substrates

In the atom probe analyses, the number of silver atoms detected correlated well with the amount of silver deposited on the GaAs specimen. However, in the case of gold considerably less gold atoms were recorded than expected. This is probably due to a difference in surface diffusion resulting in very thin gold containing layers on the clean GaAs surface (Fig. 1b).

The atomically sharp interface of Ag formed on the clean substrate indicates that Ag does not react with either Ga or As at and below room temperature. This observation agrees very well with the room temperature photoemission yield, LEED and AES data of Bolmont et al 1982, and also with the UPS and XPS measurements by Ludeke et al 1983 and Chin et al 1985. Recent work with HREM also shows a flat interface (Liliental-Weber 1990). Evidently Ag on GaAs is a good example of an almost ideally abrupt interface.

In the case of Au on a clean GaAs substrate we have clear evidence of an intermixed layer of gallium and arsenic with a gold concentration varying between 10 and 40 at% and less than one percent oxygen. The thickness of this layer may vary; we have observed thicknesses between 0.1 nm and 0.6 nm. These data are in agreement with AES, XPS and UPS measurements made by Chye et al 1978. It would seem that our data contradict those of Liliental-Weber 1990 where no intermixing was observed. However, after annealing at 405°C a 0.5-1 nm distorted layer, somewhat richer in arsenic than gallium was reported by Liliental-Weber. This observation after annealing could very well be similar to what we observe already at room temperature, possibly because of a different specimen geometry. It is interesting to observe that we also observe more arsenic than gallium in the layer (Fig. 5), although we cannot rule out the possibility of a systematic loss of gallium due to preferential field evaporation.

4.2 Interfaces formed on air exposed substrates

It can be concluded from atom probe analysis of Au on air exposed substrates that oxygen prevents the intermixing and makes the interface sharp. Some little gallium and arsenic was found at the layer surface but not inside the layer. Similar results have been obtained by Lü et al 1983 with UPS and XPS.

For silver the oxide does not seem to completely prevent diffusion since we observed a few silver atoms below the interface together with some oxygen. Furthermore the interface was not as abrupt as for the clean surface and also a few arsenic, gallium and oxygen atoms were detected in the silver layer. The diffusion zone into the GaAs is only about 1-2 nm wide and contains less than 1 at% of silver and slightly more of oxygen. Thus there is a large difference between this silver diffusion and the much more gold rich intermixed layers. Another important difference seems to be that whereas oxygen effectively blocks intermixing for gold, oxygen seems necessary for silver diffusion to occur. A completely sharp silver interface was formed when no oxygen was present. The small amount of gallium and arsenic observed in the silver layer (a total of 1 at%) probably stem from the oxide layer. This gallium and arsenic was distributed in the silver layer and not, as in the case of gold, concentrated to the surface.

In summary our results clearly suggest that oxygen has a strong effect on the exact structure of the Au/GaAs and Ag/GaAs interfaces. It is therefore important to know the oxygen content when studying metal/semiconductor interfaces.

5. SUMMARY

• The interface between Ag and a clean GaAs substrate was found to be atomically abrupt.
• On an air exposed GaAs substrate, Ag deposition resulted in an Ag layer with a total of 1 at% Ga and As and less oxygen. Beneath the oxide, a diffusion zone of oxygen and silver was found with a depth of 1–2 nm. The Ag content was low, less than 1 at%, and the O content only slightly higher.
• On a clean GaAs substrate, Au forms an intermixed layer 0.1 to 0.6 nm wide and with 10 to 40 at% of Au.
• An oxide layer formed in air on GaAs effectively blocks Au intermixing. A small amount of As and Ga was seen at the surface of the almost pure Au layer.

ACKNOWLEDGEMENTS

This work was supported by the Natural Science Research Council (NFR) and the National Board for Technical Development (STU).

REFERENCES

Andrén H-O 1986 J. de Physique 47 C7-483
Andrén H-O 1987 J. de Physique 48 C6-463
Andrén and Nordén 1979 Scand. J. Metall. 8 147
Bolmont D, Chen P, Proix F and Sebenne CA 1982 J. Phys. C: Solid State Phys. 15 3639
Chin K K, Pan S H, Mo D, Mahowald P, Newman N, Lindau I and Spicer W E 1985 Phys Rev. B 32 918
Chye P W, Lindau I, Pianetta P, Garner C M, Su C Y and Spicer W E 1978 Phys. Rev. B 18 5545
Hu Q-H, Kvist A and Andrén H-O 1991 Surf. Sci. (in press)
Liliental-Weber Z 1990 EMSA Bulletin 20 54
Ludeke R, Chiang T C and Miller T 1983 J. Vac. Sci. Technol. B 1 581
Lü Z M, Petro W G, Mahowald P H, Oshima M, Lindau I and Spicer W E 1983 J. Vac. Sci. Technol. B 1 598

Miller M K and Smith G D W 1989 Atom probe microanalysis: Principles and applications of materials problem (Pittsburgh: MRS)

Miret A, Newman N, Weber E R, Liliental-Webber Z, Washburn J and Spicer W E 1988 J. Appl. Phys. 63 2006

Mönch W 1990 Rep. Prog. Phys. 53 221

Newman N, Liliental-Weber Z, Weber E R, Washburn J and Spicer W E 1988 Appl. Phys. Lett. 53 145

Rhoderick E H and Williams R H 1988 Metal-Semiconductor Contacts (Oxford: Clarendon)

Tsong T T 1990 Atom-probe Field Ion Microscopy (Cambridge: Cambridge University Press)

Inst. Phys. Conf. Ser. No 117: Section 2
Paper presented at Microsc. Semicond. Mater. Conf., Oxford, 25–28 March 1991

Atomic scale chemistry of Co- and Ni-Si(100) interfaces

A Cerezo, C R M Grovenor and R Ozsanlav

Department of Metallurgy and Science of Materials, University of Oxford, Parks Road, Oxford OX1 3PH U.K.

ABSTRACT: This paper presents results on the atomic scale chemistry at the Co-Si(100) and Ni-Si(100) interfaces, obtained using atom probe techniques. Room temperature deposition of Co on Si(100) oriented specimens is found to form a reacted layer of composition $CoSi_2$. A layer of this composition appears to persist at the interface during annealing at 250°C as the metal reacts to form the CoSi silicide. The position–sensitive atom probe has been applied to the analysis of metal–semiconductor interfaces for the first time, showing the potential of this technique for mapping the variation in chemistry over an area of the interface.

1.   INTRODUCTION

It has long been realised that the detailed atomic-scale chemistry and structure of metal-semiconductor interfaces are crucial in determining Schottky barrier heights. However no clear overall picture has emerged of the exact correlation, with crystal structure, defect formation and interface phases all being cited as controlling factors in various systems (see, for example, Rhoderick and Williams, 1989). In the case of metal contacts on silicon, the measured Schottky barrier height is found to be remarkably independent of either the metal chosen (in the case of reactive metals) or the annealing temperature (and thus nominal silicide composition).

The ultra–high spatial resolution of the atom probe field–ion microscope, and more particularly the pulsed laser atom probe (Kellogg and Tsong, 1980), provides a powerful method by which information can be gained on the chemistry and structure of metal-semiconductor interfaces. (For a review of atom probe microanalysis, see Miller and Smith, 1990.) In this technique, field evaporation is used to remove atoms from the surface of a specimen in the form of a sharp needle point, with end–radius about 100nm. Those atoms from a region about 2nm in diameter on the specimen surface pass through a selection aperture into a time–of–flight mass spectrometer. The high field sensitivity of field evaporation makes this technique highly surface sensitive, resulting in a composition-depth profile with 2nm lateral resolution, and atomic layer depth resolution. In the more recently developed position-sensitive atom probe (POSAP) (Cerezo *et al.*, 1988), both the chemical identity and source position is obtained for atoms field evaporated from a region 20nm in diameter on the specimen surface. Although the technique has lower mass resolution than the conventional atom probe, it is capable of reconstructing, in three dimensions, the chemical variations originally present in the sample, and so permit analysis of the interface structure with sub-nanometre (but not atomic) resolution. In this paper these two techniques have been used to study the chemical nature of the Co–Si(100) and Ni–Si(100) interfaces after low temperature annealing.

2.   EXPERIMENTAL TECHNIQUE

For the atom probe analysis of metal–semiconductor interfaces, the specimens must be in the form of sharp needles of end radius about 100nm. Wafer material is sliced using a diamond saw into 0.5mm square cross-section blanks, which are subsequently chemically polished using nitric/hydrofluoric acid mixtures to form a field-ion 'tip'. The blanks are cut along the (100) direction so that the surface of interest is at the centre of the imaged area. The tip is imaged in the microscope of the VG FIM100 atom probe, and field evaporated to give an atomically clean surface. A filament evaporator and heating stage in a second, independently pumped chamber on the instrument allows metallisation and annealing of the specimen. The base pressure in this chamber is below 10^{-9}mbar and evaporation takes place at a pressure of about 10^{-7}mbar (rate of about 1 monolayer per second) with the specimen at room temperature. After annealing, the specimen is returned to the main chamber for analysis, either by pulsed laser atom probe or position sensitive atom probe techniques.

3.   RESULTS AND DISCUSSION

The analysis of a series of Co–Si(100) interfaces as deposited and annealed at 250°C for 1 hour and 15 hours is shown in Fig. 1. These composition–depth profiles are shown in the form of 'ladder diagrams' in which the detection of individual ions is represented in a vector plot. The local composition at some point on the diagram can be obtained from the slope of the line; thus a horizontal line represents pure Si, and a line of gradient 1 gives a local stoichiometry of CoSi. Of course, the statistics involved in detection of individual atoms means that even a region of homogeneous composition will produce an irregular line. The scale bars on these diagrams should be taken as representing the number of atoms detected for a given depth of material removed (which varies with end radius of the specimen) and is measured along the line of the diagram.

The analysis of an unannealed contact (Fig. 1a) shows that even without heating, a distinct reacted layer about 1nm thick is formed. Gibson et al (1987) have found that after deposition of 2nm of Co onto Si(111) at room temperature, a substantial fraction of the layer reacted to form Co_2Si. In our case, we find a layer of composition $CoSi_2$ is formed. (The straight line provides a guide to the slope that would be expected for this composition). This is surprising, since this is normally considered to be the final equilibrium phase in this system, and would only be expected after annealing at several hundred degrees. The single atomic layer of Co at the surface suggests that this result is not merely due to the limited supply of metal. In any case, it is clear that atomic movement over the 1nm layer thickness has occurred during deposition at room temperature. Once a layer of this composition is formed, it would be expected to remain as the thickness of the metal layer increases. Part of the analysis of a thicker metal layer, this time annealed at 250°C for 1 hour after deposition, is shown in Fig. 1(b). Here two regions of approximately constant slope appear between the metal overlayer (which contains some dissolved silicon) and the substrate. The first has a slope which corresponds to composition Co_2Si, which is a recognised low temperature interface phase. This is separated from the silicon by a region approximately 2nm thick of composition $CoSi_2$ once again. It is interesting that no region of the intermediate composition CoSi is seen. After annealing for 15 hours at 250°C, most of the overlayer has reached the CoSi stoichiometry, but the interface between this silicide and the silicon is not abrupt. Instead, there is a intermediate layer, approximately 1nm thick, which is much larger than could be expected from an atom probe analysis of a sharp interface. While this layer is not clearly delineated, the overall slope of the region is consistent with a composition of $CoSi_2$.

Fig. 2 shows the POSAP analysis of a thin Ni layer deposited on Si(100) and annealed at 300°C for 15mins. In this diagram, individual atoms from a region 15nm in diameter are shown as dots, the data being shown in cross-section, that is to say at right angles to the direction of analysis. A reacted layer can clearly be seen above the pure silicon of the underlying substrate, with a sharp interface (apparent intermixing on the scale of a single

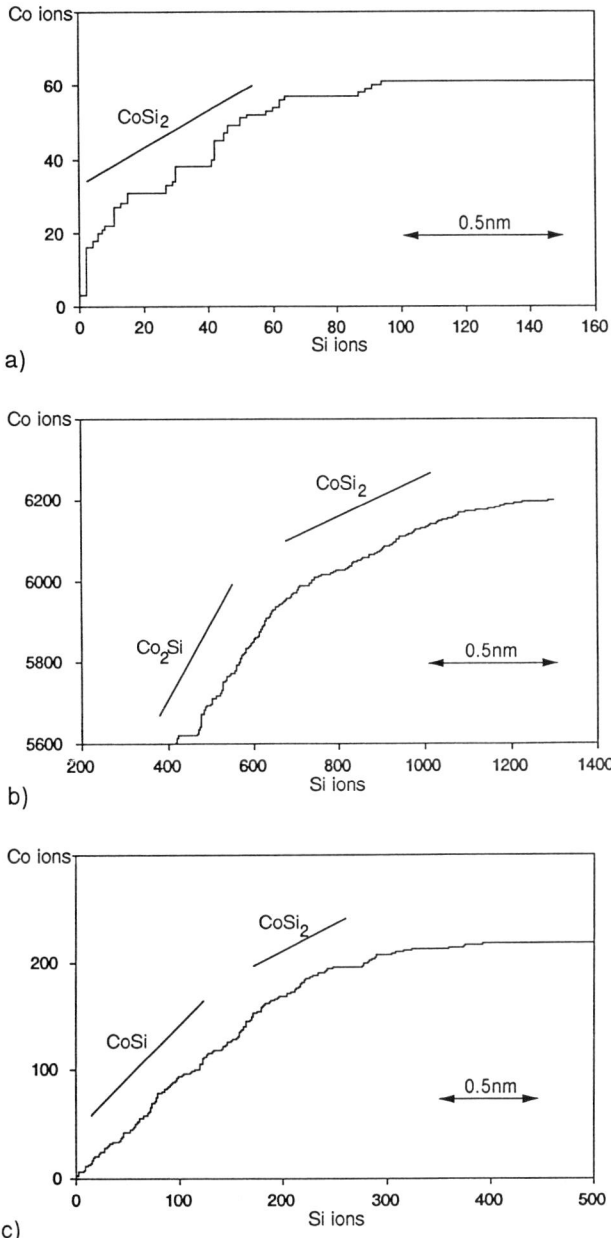

Fig. 1. Analysis of the cobalt-silicon(100) interface as-deposited (a), and after annealing at 250°C for 1 hour (b) and 15 hours (c), showing the development of the layer composition. Distance scales are approximate and relate to the number of atoms, so should be measured along the plotted line, while lines of slope corresponding to specific silicide compositions are provided for reference. It is interesting to note the formation of a layer (about 1nm thick) of composition $CoSi_2$ in the unannealed sample. As the reaction progresses (b-c), a layer of this composition appears always to be present at the interface.

atomic layer) between the two. The measured composition of the reacted layer corresponds to a stoichiometry of $NiSi_4$, a measurement which has been reported by Nishikawa *et al.* (1984) for a very thin Ni layer deposited on Si(111) and annealed at 525°C for 4 mins. Further work is required to determine whether this composition is retained when the deposited metal layer is of a greater depth. It should be noted that while figure 2 represents a view through 15nm of material, similar in this respect to a transmission electron micrograph, this is just one view from many which could be constructed from the POSAP data. In general is is possible to reconstruct the 3–dimensional nature of the interface as well as deriving compositional data.

4.   CONCLUSION

Atom probe analysis of thin (1-30nm) Co layers on Si(100) has indicated the formation of a $CoSi_2$ layer after room temperature deposition, and a layer of this composition appears to remain at the interface during annealing at 250°C while a silicide of composition CoSi is formed in the overlayer. Since any Schottky barrier heights will be dependent on the environment within 1nm of the metal–semiconductor interface, the presence of such a layer would be expected to control the actual barrier height observed. Indeed, if an interface layer of the high temperature phase was present even after low temperature annealing, as indicated here, this might provide an explanation for the constancy of barrier heights for different notional silicide compositions. Initial experiments have shown the applicability of the POSAP to the analysis of metal–silicon interfaces, and this will provide further information in the studies of Co–Si and Ni–Si systems.

ACKNOWLEDGEMENTS

We would like to thank Professor Sir Peter Hirsch FRS for the provision of laboratory space. AC is grateful to to The Royal Society and Wolfson College, Oxford for support during the course of this work. The Oxford Atom Probe facility is funded by the Science and Engineering Research Council.

REFERENCES

Cerezo A, Godfrey T J and Smith G D W 1988 Rev. Sci. Instrum. <u>59</u> 862
Gibson J M, Batstone J L and Tung R T 1987 Appl. Phys. Lett. <u>51</u> 45
Kellogg G L and Tsong T T 1980 J. Appl. Phys. <u>51</u> 1184
Miller M K M and Smith G D W 1989 Atom Probe Microanalysis: Principles and
    Applications to Materials Problems (Pittsburgh: MRS)
Nishikawa O, Shibata M, Yoshimura T and Nomura E 1984 J. Vac. Sci. Technol. B <u>2</u> 21
Rhoderick E H and Williams R H 1988 Metal–Semiconductor Contacts (Oxford: OUP)

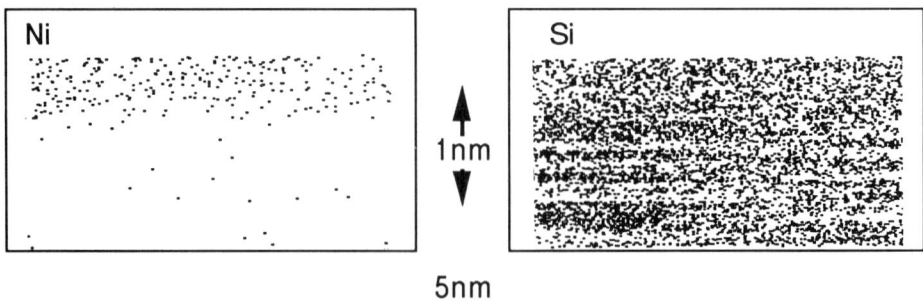

Fig. 2. POSAP analysis through a thin Ni layer deposited on Si(100) and annealed at 300°C for 15mins. In these plots, each marker represents the position of an individual Ni or Si atom. A reacted region (composition $NiSi_2$) and the underlying substrate are clearly seen, separated by an abrupt interface. Note the different vertical and horizontal scales.

Inst. Phys. Conf. Ser. No 117: Section 2
Paper presented at Microsc. Semicond. Mater. Conf., Oxford, 25–28 March 1991

Proton microscopy of a microcircuit layer structure

M B H Breese, F Watt, G W Grime and P J C King

University of Oxford, Department of Nuclear Physics, Keble Rd., Oxford, OX1 3RH, U.K.

ABSTRACT: The composition, thickness and distribution of buried microcircuit layers are difficult to determine without etching techniques because of a limited penetration depth of optical and electron beam methods. MeV protons typically have a range of tens of microns through a microcircuit which allows the full thickness of the layer structure to be studied. The Oxford Scanning Proton Microprobe (SPM) can focus a 3 MeV proton beam to sub–micron spot sizes, and its ability to identify defects in buried microcircuit layers using three analytical techniques is shown here.

1. INTRODUCTION

Proton Induced X–ray Emission (PIXE) (Johansson *et al* 1970) causes K or L shell electrons of the specimen atoms to be ejected by the incident beam. The vacancy can be filled by electrons falling from outer shells, with emission of a characteristic X–ray. Because the incident MeV proton beam causes very little bremsstrahlung radiation, the best sensitivity of this technique using a Si(Li) detector is at the ppm levels for atomic weights greater than 11. Rutherford Backscattering Spectrometry (RBS) (Chu *et al* 1978) measures the energy of incident MeV protons recoiling from the specimen, giving quantitative compositional information and depth profiles through several microns of the specimen. Since backscattering cross–sections of 3 MeV protons on many light elements show non–Rutherford resonance effects, this must be taken into account for accurate compositional simulation. The application of these two techniques to microcircuit analysis has until recently been hampered by insufficient spatial resolution, but the advent of high resolution proton microprobes (Watt and Grime 1987), (Jamieson *et al* 1989), capable of focusing 100 pA of 3 MeV protons to 0.3 μm allows these MeV ion analytical techniques to study the smallest microcircuit features with no sample preparation.

The third analytical technique presented here is Scanning Transmission Ion Microscopy (STIM) (Overley *et al* 1983). This is used here as a rapid method of determining the uniformity and distribution of metallisation layers, even if they are buried under other metal layers. The microcircuit substrate is mechanically polished away until the remaining layer thickness is about 20 μm. A 3 MeV proton beam has sufficient energy to travel through the remaining layers, and the energy of the transmitted beam depends on the composition and thickness of the layers traversed. Since metal layers cause the MeV beam to undergo a much larger energy loss than the Si layers, the image contrast is mainly due to the various metallisations present. The transmitted energy spectrum from the beam scanning over the surface is divided into individual energy windows, and these are used to generate images showing different energy loss regions of the microcircuit. The energy loss information in these individual STIM images can also be combined to generate an image showing the average energy loss through each point of the scan. Because every particle is used to form the STIM images, a beam current of only 2000 protons per second is needed. This is achieved by reducing the size of the beam defining apertures, so a spatial resolution of less than 0.2 μm is attainable. This average energy loss STIM image is thus analogous to a high resolution X–ray microradiograph.

The test microcircuit area studied with these three techniques is shown in figure 1(a), and the distribution of the metallisation layers present in the central area is shown in figure 1(b). There is a TiSi$_2$ layer under the first metallisation to assist adhesion to the underlying BPSG (borophosphosilicate glass). Overlaid on parts of the first metallisation layer, and separated by a layer of BPSG is the second metallisation of aluminium. There is a thick SiO$_2$ passivation layer over the surface except on the contact pads which can be seen on the far right.

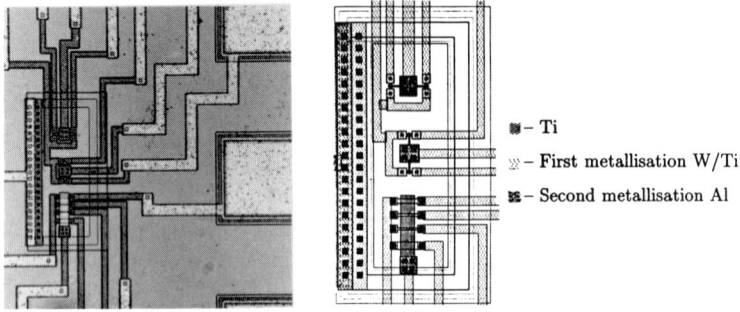

- Ti
- First metallisation W/Ti
- Second metallisation Al

Figure 1. (a) 400μm^2 optical image.   (b) metallisation scheme.

2. RBS ANALYSIS

RBS is used to give the depth distribution of the microcircuit layer structure. Areas as small as 0.5 μm^2 can be studied, and STIM is used to position these areas as shown in section 3. Figure 2 shows three RBS spectra from different areas of the microcircuit, measured with a surface barrier detector at an angle of 135° to the incident beam. The simulation to determine the composition and depth distribution is shown as a smoothly varying line on each spectrum.

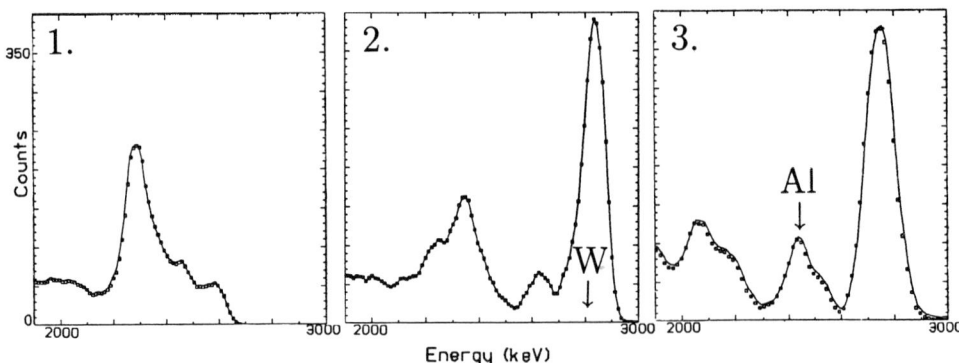

Figure 2. RBS spectra.

Spectrum 1 is taken from an area with no metallisation, so only a 4.1 μm SiO$_2$ layer and the Si substrate are detected. Spectrum 2 is taken from inside the first metallisation contact hole shown in the last STIM image of figure 4. The large tungsten peak from the first metallisation is clearly seen, and simulation shows this layer to be 0.7 μm thick. Spectrum 3 is taken inside a via hole from the second to the first metallisation layers. The position of aluminium in the spectrum is indicated and the tungsten peak is slightly shifted down in energy compared to spectrum 2 because of this overlying layer. Simulation shows that the second metallisation aluminium layer is 0.9 μm thick. RBS has thus determined the thickness of the metallisation layers present.

3. STIM AND PIXE ANALYSIS

PIXE and STIM are used to examine the spatial distribution and uniformity of the metallisation layers. The PIXE spectrum from a 400 μm^2 area slightly shifted to the left of that shown in figure 1(a) showed the presence of aluminium, silicon, titanium and tungsten.

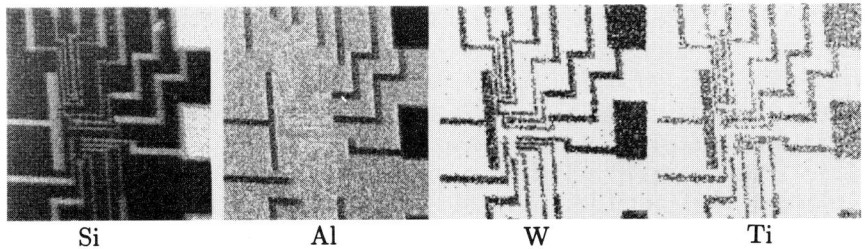

| Si | Al | W | Ti |

Figure 3. 400×400μm PIXE images.

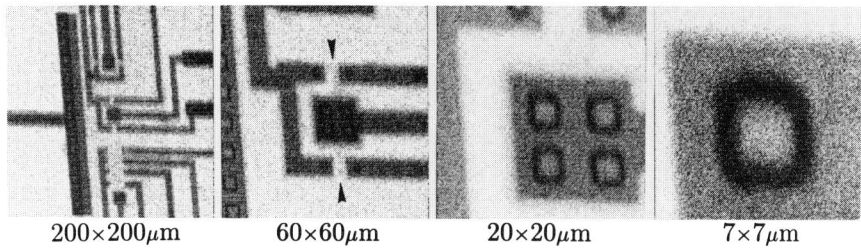

| 200×200μm | 60×60μm | 20×20μm | 7×7μm |

Figure 4. Average energy loss STIM images.

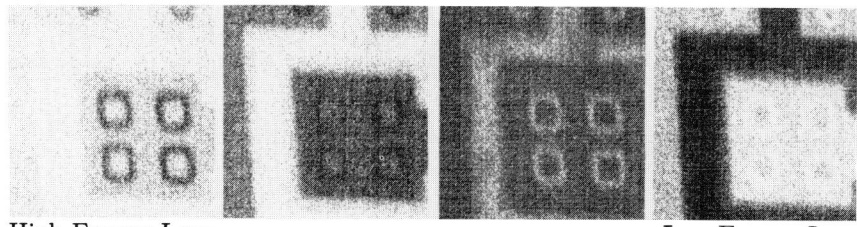

High Energy Loss   Low Energy Loss

Figure 5. 20×20μm individual energy loss STIM images.

Figure 6. 20×20μm PIXE images

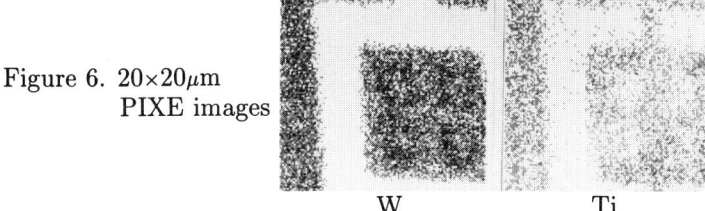

W   Ti

Energy windows were set on the Al K, Si K, Ti K and W L X–ray lines, and the distributions of these elements in this area is shown in figure 3. The Si X–ray image shows low intensity from the metallised regions, because Si X–rays generated in deeper levels are absorbed by the overlying metal. There is even lower Si intensity from the contact pads because there is no passivation layer over these. The Al X–ray image shows that it occupies only the second metallisation regions. There is a stronger Al signal from the large contact pads than from the Al tracks because there is no absorption from the passivation layer. The Ti and W X–ray images show similar distributions corresponding to the first and second metallisation stages.

The average energy loss STIM images of the same region area as the PIXE images of figure 3 are shown in figure 4, with decreasing scan sizes of 200, 60, 20 and 7 μm^2, homing in on a small contact hole through the first metallisation to the Si substrate. The dark regions of the images indicates high energy loss, and light areas indicates low energy loss. The thick tracks of tungsten produce the greatest energy loss and thus the strongest contrast in these STIM images. The additional second metallisation over the tungsten tracks is discernible on the 200 μm^2 image as the darker still regions, indicating even higher energy loss. The contact holes visible on the 60, 20 and 7 μm^2 images have higher energy loss around their edges because the thickness of W traversed by the beam increases over this region. It can be seen from these images that the first metallisation is a highly uniform layer, even inside and around the edges of the contact holes. On the 60 and 20 μm^2 images however there is indication that the metal distribution is not exactly as shown in figure 1(b). The upper of the two faint lines indicated by the arrows on the 60 μm^2 image is slightly displaced from the centre of the gap in the first metallisation layer. This defect is more obvious on the individual STIM images of the 20 μm^2 area shown in figure 5. The image on the left has the highest energy loss and corresponds to   thickest regions of tungsten whereas the right hand image has the least energy loss and corresponds to regions with no overlying tungsten. On the third image there is clear evidence that this track is misplaced.

This defect has been detected with STIM, but PIXE is needed to determine the nature of fault giving rise to the contrast variation in the STIM images. Figure 6 shows the distribution of W and Ti in the same 20 μm^2 area. The distribution of W is the same as the layout of figure 1 (b), but the Ti distribution is indeed displaced in the upper 1 μm wide track, possibly because of a small defect in the mask used to lay down the Ti.

CONCLUSIONS

A combination of PIXE, RBS and STIM is able to determine the depth, distribution and uniformity of deep metal layers without the need to use etching or sectioning techniques. This paper is the first to demonstrate the ability of the Scanning Proton Microprobe to image the smallest microcircuit features using STIM, and to determine the depth distribution through areas as small as 0.5 μm^2 using RBS. A combination of these three techniques has determined the likely failure mechanism of a test microcircuit.

MHS is gratefully acknowledged for donating the test microcircuit.

REFERENCES

Johansson T B, Akselsson R and Johansson S A E 1970 *Nucl. Instrum. Meth.* **84** 141

Chu W-K, Mayer J W, Nicolet M-A 1978 *Backscattering Spectrometry* (Academic Press)

Watt F and Grime G W eds 1987 *Principles and Applications of High Energy Ion Microbeams* (Bristol: Adam Hilger)

Jamieson D N, Grime G W and Watt F 1989 *Nucl. Instrum. Meth.* **B40/41** 669

Overley J C, Connolly R C, Sieger G E, MacDonald J D and Lefevre W H, 1983 *Nucl Instrum. Meth.* **218** 43

Inst. Phys. Conf. Ser. No 117: Section 3
Paper presented at Microsc. Semicond. Mater. Conf., Oxford, 25–28 March 1991

An HREM study of the entrance and the decomposition of dislocations into deformed $\Sigma = 19$ and $\Sigma = 9$ grain boundaries

J Thibault, J L Putaux, H M Michaud*, X Baillin*, A Jacques,** A George**.

DRFMC/SP2M, Centre d'Etudes Nucléaires, 85X-38041 Grenoble-France.
*CEA/DTA/CEREM, Centre d'Etudes Nucléaires, 85X-38041 Grenoble-France.
**Lab. Physique des Solides, Ecole des Mines, 54402 Nancy-France.

ABSTRACT: The deformation in compression of a $\Sigma=19$ symmetrical tilt grain boundary (GB) in germanium will be described and compared to the results obtained with the $\Sigma=9$ silicon tilt GB deformed in tension. It will be shown that in both cases, special GB defects similar in structure but having almost opposite Burgers vector are found to be the intermediate step of the entrance and the decomposition of the incoming 60°dislocations.

1. INTRODUCTION

The evolution under deformation of the structure of the $(\bar{1}2\bar{2})_I[011]$ $\Sigma=9$ silicon tilt grain boundary (GB) in silicon has been studied either in compression or in tension (the subscript refers to grain I). The results have been described in detail in previous papers by Elkajbaji and Thibault (1988), George, Jacques, Baillin, Thibault-Desseaux and Putaux (1989), and Thibault-Desseaux and Putaux (1990). The structure of the defects induced by the entrance and the decomposition of the dislocations has been determined. By the use of the combination of 1 Mev in-situ, transmission and high resolution electron microscopies, the authors were able to achieve a comprehensive view of the phenomenon. The results obtained on $\Sigma= 9$ are summarized on table 1.

To a large extent, the results on the structural evolution are compatible with the theory of minority and majority structural units (SU's), developed by Sutton and Vitek (1983) and Balluffi and Bristowe (1984). The minority units, i.e. the so called C and T units, correspond to elementary $\Sigma=9$ DSC dislocations whose Burgers vectors accommodate a small deviation from the exact Σ position. However, it has been shown that, depending on the temperature, the deformation can lead to two different structures for the same final state, i.e. the same final GB: for instance, the $\Sigma=11$ GB (resulting from the compression of $\Sigma=9$) could have two structures. The first one contained T units embedded in the SUs of $\Sigma=9$. These T units, related to the $\Sigma=3$ GB, are "favoured SUs" in accordance with the theory. On the other hand, the second structure contained a new SU which in fact was related to a non favoured structure of $\Sigma=3$. The two $\Sigma=11$ GBs (same misorientation angle) differed by the translation state between both grains. Moreover, as shown by Putaux and Thibault-Desseaux (1990), computer simulations led to only slightly different GB energies. Despite a simple description, the deformation can lead to unpredictable configurations, as recently emphasized by Sutton (1990).

This paper deals with the evolution of tilt GBs whose misorientation angle is included between 26.53° ($\Sigma=19$) and 38.94° ($\Sigma=9$) around [011]. We will concentrate on the deformation of a germanium $\Sigma=19$ GB in compression and compare the resulting configurations with the ones obtained by the deformation in tension of the silicon $\Sigma=9$ GB. The $\Sigma=19$ Ge bicrystal was deformed at 440°C, under a 20 MPa stress applied along the <321> axis parallel to the GB plane. The strain was about 1%. Due to the limitation of the tensile machine, the silicon $\Sigma=9$ GB was deformed under 15MPa along the <411> axis parallel to the GB plane at 750°C. The strain was about 1.3%. The bicrystals were cooled down under load in order to freeze the configurations. The HREM observations were performed on a JEOL 200CX. electron microscope along the [011] common axis of the bicrystals.

$\Sigma=1$ ($\theta=0°$)	adding C units in tension	two descriptions of $\Sigma=9$ ($\theta=38.94°$)		adding T units in compression	$\Sigma=3$ ($\theta=70.5°$)
			M⁺	M⁻	T
C	L⁻	L⁻			
			M⁺	M⁺	T
C	L⁺	L⁺			
			M⁻	T	T
C	C	L⁻			
			M⁺	M⁻	T
C	L⁻	L⁺			
			M⁻	M⁺	T
C	L⁺	L⁻			

Table 1. Schematic evolution of the $\Sigma=9$ GB under deformation. According to Vaudin, et al.(1983), $\Sigma=9$ has two possible descriptions, one with a 7-5 atom ring unit called L, corresponding to the Lomer dislocation core and the other one, with a different arrangement of the 7 and 5 atom rings, called M (see fig.1).

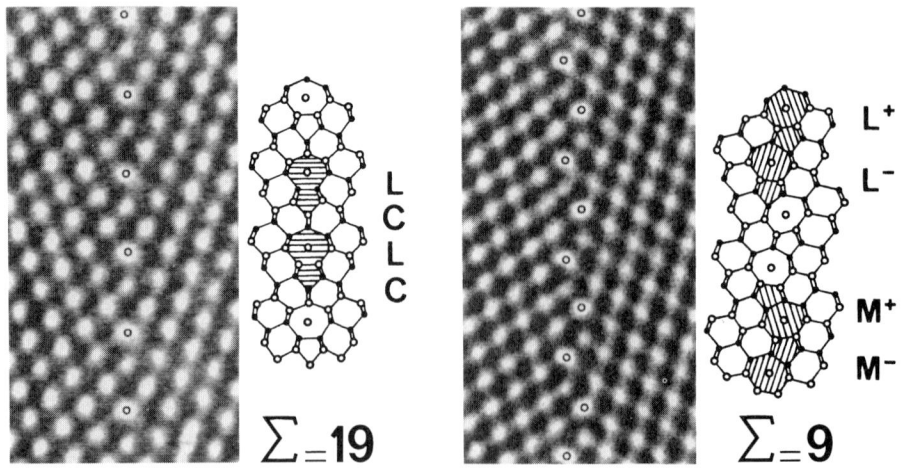

Fig.1: HREM images of the perfect $\Sigma=19$ and $\Sigma=9$ symmetrical tilt GBs (black atoms). The misorientation angles around [011] are 26.53° and 38.94°, the GB planes are $(1\bar{3}3)$ and $(1\bar{2}\bar{2})$ respectively. The structures deduced from these images are shown in the frames.

2. STRUCTURE OF THE Σ=19 AND THE Σ=9 TILT GBs BEFORE DEFORMATION

The HREM images obtained with the undeformed bicrystals are shown on fig.1. The structure deduced from these images is shown in the frame.
The structure of the perfect Σ=9 is described by the repeating sequence {L$^+$L$^-$}. The + and - superscripts refer to the zig-zag connection between the L units. With the convention used in table 1, the period structure of Σ=19 can be described by the SU sequence : {LCLC}. The GB structure is a pure mirror. As a matter of fact, this structure can be imagined starting from the Σ=9 structure and adding C units, basic SUs of the perfect crystal (Σ=1), in between each Σ=9 L units, following the scheme presented table 1. The period is doubled to take the atom level along the tilt axis into account.

3. DEFORMATION OF Σ=19 IN COMPRESSION AND Σ=9 IN TENSION

3.1. Geometry of the deformation

The defects resulting from the entrance and the decomposition of the deformation induced dislocations will be referred to the Σ=19 or the Σ=9 DSC lattices. Fig.2 shows the coincidence site lattice (CSL) and the DSC lattice of both bicrystals. Σ=9 DSC lattice is a body centered one. Σ=19 DSC lattice is an orthorhombic lattice with one centered face parallel to the $(\bar{1}3\bar{3})$ plane. The sign convention and the Burgers vector determination have been described in detail in the paper by Elkajbaji and Thibault-Desseaux (1988). The dislocations are assumed to arrive from grain I. However, in a particular case, in order to facilitate the comparison between structures, the dislocations will be referred to grain II. The consequences are simple : the structure of the induced defects are mirror related, the component of the Burgers vector normal to the GB remains unchanged whereas the one parallel to the GB is reversed. In the following, the two elementary DSC vectors normal to the tilt axis will keep the same name whatever GB they are referred to :

	$(\bar{1}3\bar{3})$ Σ=19	$(\bar{1}2\bar{2})$ Σ=9
b$_c$	a/19$[\bar{1}3\bar{3}]$I	a/9$[\bar{1}2\bar{2}]$I
b$_g$	a/38$[\bar{6}11]$I	a/18$[\bar{4}11]$I

The b$_c$ GBDs move along the GB plane by pure climb whereas the b$_g$ GBDs move by pure glide. The b$_c$ GBDs will contribute to the misorientation angle variation, whereas the movement of the b$_g$ GBDs will introduce a GB sliding.
In compression, the strain of the Σ=19 GB results in an increase of the misorientation angle by the accumulation of GBDs with bc type component whereas the strain in tension of Σ=9 results in a decrease of the misorientation angle by the accumulation of GBDs with -bc type component.
It has to be noticed that the incoming dissociated 60° dislocations have a reversed Burgers vector in compression and in tension. That implies that the partial dislocations are also reversed. In compression, the 90° partial is the leading one; in tension, it is the 30° partial.

3.2. The "2b$_c$" GBD in Σ=19

The defect corresponding to this Burgers vector is shown on fig.3a. This is not an elementary DSC vector. It is often found in a compact configuration. It has no GB step associated with it. The main point is that the structure of 2b$_c$ is equivalent to the one of a Σ=9 period. In fact, the removal the two C units of a Σ=19 period leads to a Σ=9 period. The presence of this defect can be described by the sequence :
..LCLC(LL)LCLC...

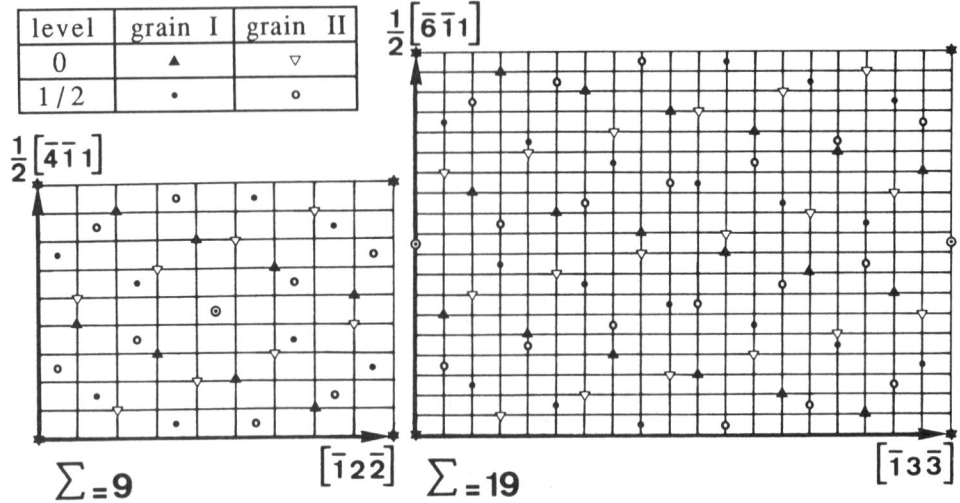

level	grain I	grain II
0	▲	▽
1/2	•	o

$\frac{1}{2}\left[\overline{6}\,\overline{1}1\right]$

$\frac{1}{2}\left[\overline{4}\,\overline{1}\,1\right]$

$\Sigma = 9$ $[\overline{1}2\overline{2}]$ $\Sigma = 19$ $[\overline{1}3\overline{3}]$

Fig.2: Coincidence site lattice and DSC lattice for the $\Sigma=9$ and $\Sigma=19$ GBs.

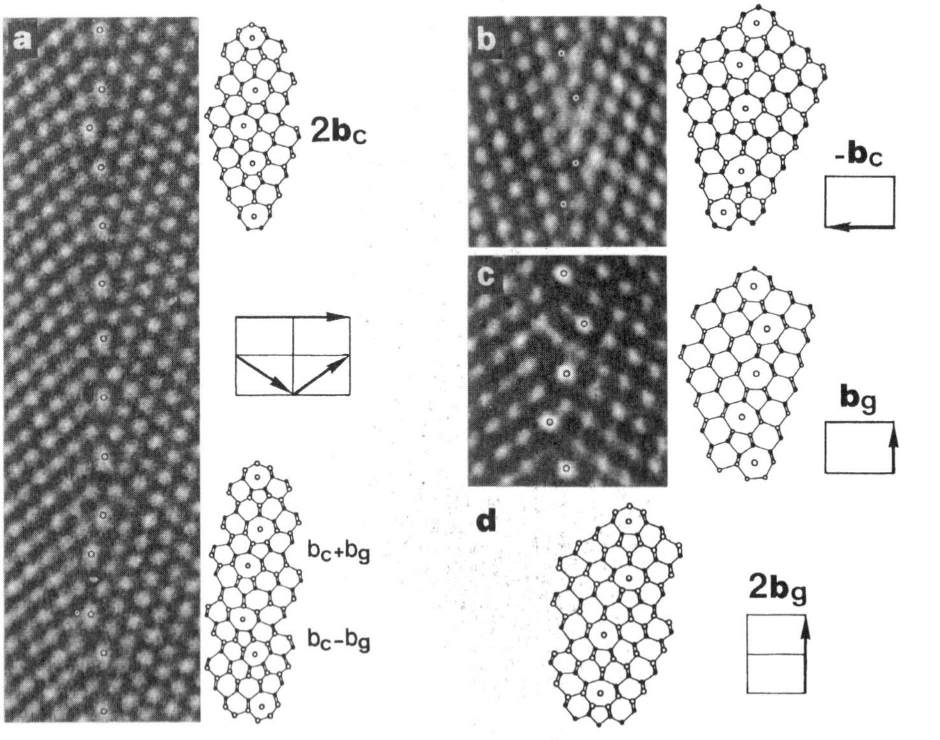

$2b_C$

b_c+b_g

b_c-b_g

$-b_C$

b_g

$2b_g$

Fig.3: HREM images of $\Sigma=19$ and $\Sigma=9$ DSC dislocations a) ($\Sigma=19$) $2b_c$ in a compact configuration, and b_c+b_g and b_c-b_g which in fact correspond to the decomposition of $2b_c$, b) $-b_c$ ($\Sigma=9$), c) b_g ($\Sigma=9$). (a, c black atoms, b white atoms),d) scheme of $2b_g$ ($\Sigma=19$).

Due to the two possible ways of connecting the L units in $\Sigma=9$, the same Burgers vector is associated with

$$...LCLC(L^+L^-)LCLC...\quad or\quad ...LCLC(L^-L^+)LCLC...$$

3.2. The "-b_c" GBD in $\Sigma=9$

This defect does not introduce a GB step. Its core corresponds to the introduction of a C unit between the L^+ and L^- SUs in $\Sigma=9$ (fig. 3b). C is the basic unit of the $\Sigma=1$ "perfect bicrystal" with the period {CC}(table1). -bc can have two structures :

$$...L^+L^-L^+(C)L^-L^+L^-...\quad or\quad ...L^+L^-(C)L^+L^-L^+L^-...$$

In both cases, the C unit undergoes equal and opposite shears parallel to the GB plane, although it remains associated to the same Burgers vector. This defect can be viewed as the insertion of half a period of $\Sigma=19$ in $\Sigma=9$ such as :

$$...L^+L^-(LC)L^-L^+L^-...$$

3.3. The "b_g" and "-b_g" GBDs in $\Sigma=9$

The core of the defect corresponds to an additional L unit in the $\Sigma=9$ period (fig.3c). The sequence related to bg is $...L^+L^-(L^-)L^+L^-...$, associated with a GB step whose height is $h=-h_0=-a/3$. -b_g and b_g GBD's have opposite Burgers vectors and mirror related cores. Thus, the -bg core is associated with one L^+ unit and a $+h_0$ GB step.

3.4 The "$2b_g$" GBD in $\Sigma=19$

The elementary b_g GBD in $\Sigma=19$ is geometrically associated with a GB step whose height would be $h=+8h_0$ ($h_0=a/19[\bar{1}3\bar{3}]$). On the other hand, the step associated with the $2b_g$ DSC dislocation would be $h=+3h_0$ which makes it more likely observable. Unfortunately, due to the symmetrical deformation, this defect has not been detected along the interface. As a matter of fact the entrance of two 60° dislocations coming from both grains leads to $4b_c$ and the glissile components (like the screw ones) annihilate, thus, the glissile $2b_g$ GBD is unlikely to be seen. Nevertheless the structure could be easily imagined from the study of the geometrical conditions to obtain a glissile edge dislocation with a small step along an interface of this type. The structure we proposed, is shown on fig.3d.

3.5. The "b_c+b_g", and "b_c-b_g" GBDs in $\Sigma=19$

As a matter of fact, the $2b_c$ defect is also found decomposed into two defects whose Burgers vector are b_c+b_g and bc-bg respectively. On fig.3b the two defects are separated by half a period of $\Sigma=19$. bc+bg corresponds to ..LCLC(L$^-$)LCLC.. and bc-bg to ..LCLC(L$^+$)LCLC.. Each defect is equivalent to the insertion of half a period of $\Sigma=9$ into $\Sigma=19$ GB. One may notice that the configuration shown fig.3b can also be seen as the structure of the $\Sigma=27$ GB, whose period is {L$^-$L$^+$CL$^+$L$^-$C} with a mirror glide symmetry and whose misorientation angle is 31.57° just in between $\Sigma=19$ and $\Sigma=9$.

3.6. The $b_r{}^2$ GBD in $\Sigma=19$ and the $b_r{}^4$ GBD in $\Sigma=9$

Unlike the previous GBDs, these defects have a screw component of $\pm a/4[011]$. The components of the Burgers vector normal to the [011] axis are

$$b_r{}^2 = (b_c+b_g)+(b_c-b_g)\ -1/2b_g \qquad \text{in the case of } \Sigma=19,$$
$$b_r{}^4 = -3/2b_c + 1/2b_g \qquad \text{in the case of } \Sigma=9.$$

The atomic structure of br^2 seems rather complex. However it is exactly the same as the one of the so called $b_r{}^4$ GBD found in $\Sigma=9$ deformed in tension and resulting

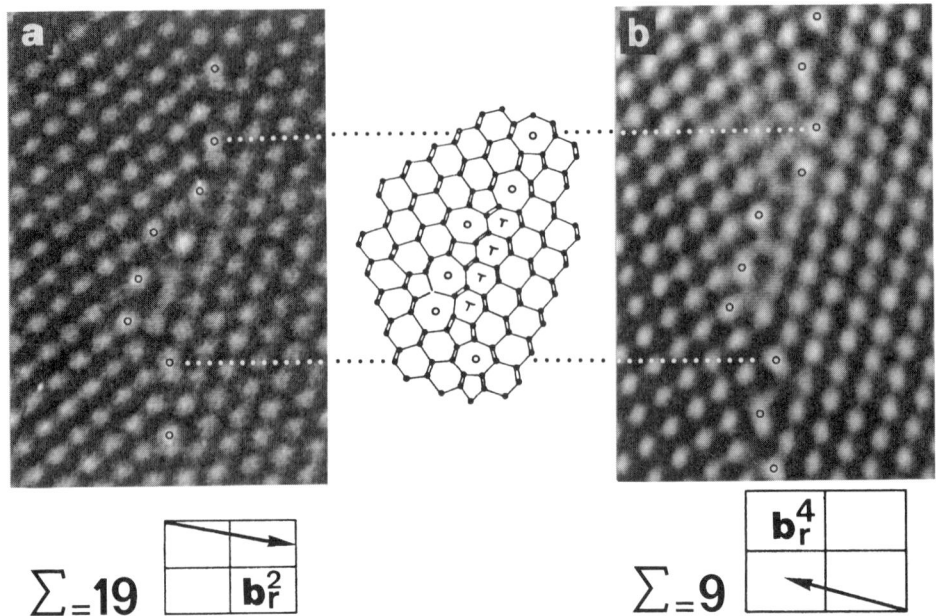

Fig.4: a) br^2 GBD in $\Sigma=19$ GB. Its structure is the same as the br^4 one in the $\Sigma=9$ GB. (a and b in black atoms).

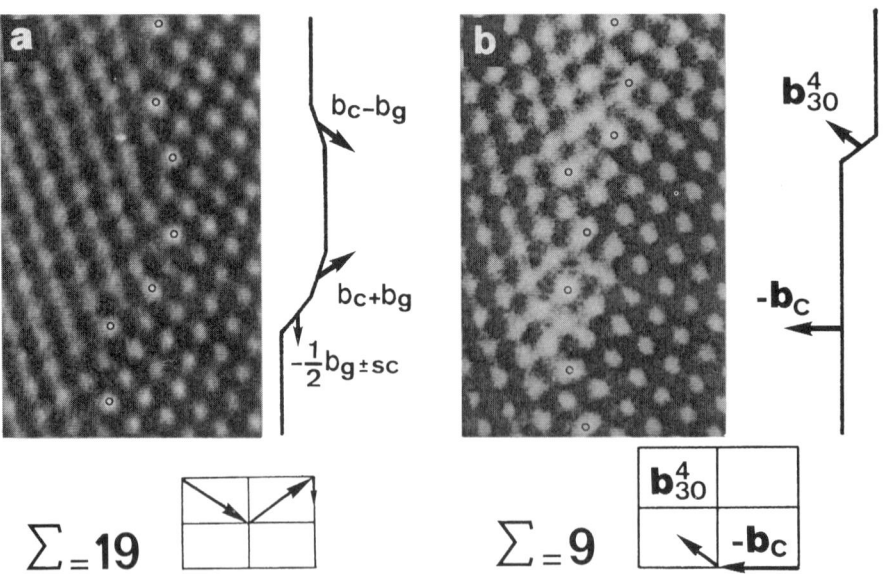

Fig.5 : Configuration resulting from the decomposition into elementary DSC dislocations of a) br^2 GBD in $\Sigma=19$ GB and b) br^4 GBD in $\Sigma=9$ GB (black atoms).

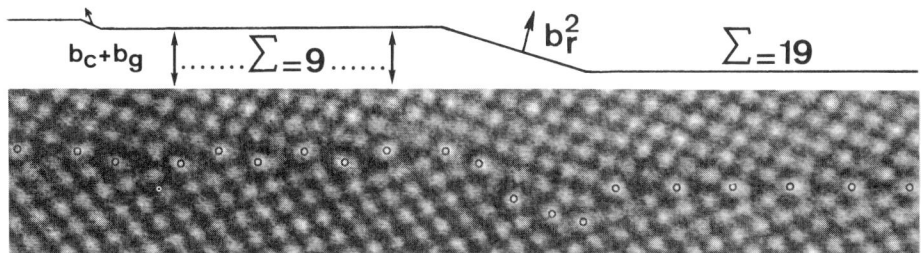

Fig.6) b_r^2 GB defect at the junction between $\Sigma=19$ and a few periods of $\Sigma=9$.

from the entrance of a matrix dislocation in grain I and its decomposition into b_r^4 and two b_g's (referred to $\Sigma=9$ GB). The fig 4 presents the b_r^2 defect in $\Sigma=19$. This defect results from the entrance of a dislocation in grain II. A dislocation entering from grain I would have left a b_r^1 residue structurally the mirror image of b_r^2. as already mentioned in §3.1 The point here, is that the component normal to the GB plane is reversed if referred to $\Sigma=19$ or $\Sigma=9$. To describe the core of b_r^4, we proposed a model mixing the usual 7 and 5 atom rings with T units (boat shaped 6 atom rings of the first order twin $\Sigma=3$). One particular reconstruction along [011] had to be introduced to account for the screw component of the defect.

An interesting configuration is presented on fig.6 where the b_r^2 defect is found at the junction between $\Sigma=19$ and a small area of $\Sigma=9$ which may result from the accumulation of $2b_c$'s produced by the entrance and the decomposition of matrix 60° dislocations coming from both grains.

3.7   Decomposition of the "b_r" GBDs

As in the $\Sigma=9$ case, the complex b_r^2 GBD is often found in a decomposed configuration into two smaller DSC defects.

b_r^2 leads to $2b_c$ - $(1/2b_g + sc)$ referred to $\Sigma=19$ whereas
b_r^4 leads to b_c + $(b30)$ referred to $\Sigma=9$.
$(1/2b_g + sc)$ is one defect and sc means a a/4[011] screw component. $b30$ contains the same screw component and can be written $(1/2b_c+1/2b_g+sc)$ referred to $\Sigma=9$. It has to be kept in mind that HREM gives the projection of the structure, thus the screw components are not detectable. The decomposed configurations of b_r^1 and b_r^4 are shown on fig 5.

4.   ENTRANCE OF A DISLOCATION INTO $\Sigma=19$

As explained in the case of $\Sigma=9$ by Thibault-Desseaux, Putaux, Bourret, Kirchner, 1989.and as confirmed by George et al. (1989) and Baillin et al. (1990) by in situ and conventional TEM experiments the entrance of a matrix dislocation and its decomposition result into glissile and non glissile residual dislocations.
The main point is that thermal activation makes slip into the GB possible and blocks the 60°D at the GB. The direct crossing of the in-coming dislocation is unlikely. However, it has been clearly shown that the successive emission of partial dislocations in the adjacent grain is favoured at relatively low temperature by a dislocation pile-up at the GB

As in the $\Sigma=9$ case, we could imagine two ways for the entrance of the 60° dislocation in $\Sigma=19$. On one hand, the emission of $2b_g$ (associated with a small GB step) is not an a-priori favourable path, but will make the entrance of the trailing 30° partial easier. The presence of the b_r^1 or b_r^2 GBD, sustains this first assumption and b_r might be considered as an intermediate stage of the entrance. On the other hand, the emission of one $(b_c + b_g)$ GBD is energetically favorable. The climb decomposition of the head partial in the GB could be driven only by the repulsive interaction between the $(bc + bg)$ and the residual GBD left at the junction of the stacking fault and the GB plane; this GBD has a partial DSC vector. In fact the applied stress does not provide any climb force. The observation of the decomposition of the $2b_c$ GBD sustains the fact that climb could not be excluded.

An important point mentioned in § 3.6, is that the b_r component normal to the GB plane is reversed as referred to $\Sigma=19$ or $\Sigma=9$. Thus the same structural defect will contribute to decrease the angle if located in $\Sigma=9$ and to increase it if located in $\Sigma=19$. As a matter of fact, the major part of the br type GBDs (see fig.6) is similar to the structure used by Bourret et al. (1987) to describe particular steps in as-grown $\Sigma=27$ bicrystals. These steps locally split, following the reaction:

$$(\overline{2}55)_I \ \Sigma=27 \ \Rightarrow \ (11\overline{1})_I \ \Sigma=3 + (\overline{1}2\overline{2})_I \ \Sigma=9$$
$$\theta=+31.57° \ \Rightarrow \ \theta=+70.51° + \theta=-38.94°$$

When $\Sigma=19$ is deformed in compression, the misorientation angle increases towards the one of $\Sigma=27$. Similarly, when $\Sigma=9$ is deformed in tension, the angle decreases towards the one of $\Sigma=27$. So it seems that the br type defect, even though it has a rather complex extended core, is favoured (because not so much distorted) within the range of misorientation between $\Sigma=19$ and $\Sigma=9$ and contributes to the relaxation of the incompatibility between the two delimiting GBs.

5. CONCLUSION

As in the case of $\Sigma=9$ GB, the complete decomposition of a 60°D in $\Sigma=19$ leads to residual GBDs with a small Burgers vector and   associated with a small GB step. These residues move along the interface by glide and climb resulting in a more or less homogeneous GBDs distribution depending on the temperature. The key point is that these Burgers vectors are not necessarily basic DSC vectors. (see for instance bc+bg in $\Sigma=19$). This might be explained by the fact that the basic DSC vectors would be associated with a high but geometrically necessary   step and this seems to be energetically unfavorable.

Another important point to notice is that, in both cases of deformation ($\Sigma=9$ in tension and $\Sigma=19$ in compression) and despite the reversed position of the in-coming partials, the intermediate step of entrance leads to GBDs with a $(b_{60}-(nb_g))$ type Burgers vector. This argues in favour of a mechanism of dislocation entrance facilitated by the emission of glissile GBs as proposed by Thibault, Putaux, Bourret, Kirchner (1989). This intermediate stage is observed at a medium temperature. At higher temperature, the climb and glide decomposition would result rapidly in primitive defects, the intermediate step would be unlikely to be observable.

REFERENCES

Baillin X, Pelissier J., Jacques A., George A., 1990, Phil. Mag., A61, 329.
Balluffi R.W. , Bristowe P.D., 1984, Surf. Sci., 144, 28.
Bourret A, Bacmann J.J., (1987), Rev. Phys. Appl., 22, 563.
Elkajbaji M. , Thibault-Desseaux J., 1988, Phil. Mag. A58, 325.
George A, Jacques A., Baillin X., Thibault-Desseaux J., Putaux J.L., 1989, Inst. Phys. Conf ., 104, 349.
Putaux J.L., Thibault J., 1990, Jour. Phys. C1-51, 323.
Sutton A., Vitek V., 1983, Trans. Roy. Soc; London, A309, 1, 37, 55.
Thibault-Desseaux J., Putaux J.L., 1989, Inst. Phys. Conf., 104, 1
Thibault-Desseaux J.,Putaux J.L., Bourret A., Kirchner H.,1989, Jour. Phys. , 50, 2525.
Vaudin M., Cunningham B, Ast D, 1983, Scripta. Met., 17, 191.

Inst. Phys. Conf. Ser. No 117: Section 3
Paper presented at Microsc. Semicond. Mater. Conf., Oxford, 25–28 March 1991

The structure of twist units in a Ge $\Sigma = 27$ $<110>$ tilt grain boundary

Stuart McKernan, Zvi Elgat*, and C Barry Carter

Department of Materials Science and Engineering, Cornell University, Ithaca, NY 14853.
*Present Address: LUZ Industries Israel, Jerusalem 19073, Israel.

ABSTRACT: General high-angle tilt grain boundaries may be described by an arrangement of repeating structural units. A particular type of defect can occur in this repeating structural-unit arrangement in a $\Sigma = 27$ related <110> tilt grain-boundary in Ge. This defect accommodates a slight misorientation of the boundary away from the perfect tilt configuration; it introduces a twist component into the tilt grain-boundary. The defect, which can be described as a twist unit, is characterised by a short segment of (111) facet at the interface, and is associated with a screw dislocation in the boundary plane.

1. INTRODUCTION

High-resolution electron microscope (HREM) images of different tilt grain-boundaries in many materials display a qualitative similarity of the atomic configuration at the grain boundary. Low-angle tilt grain-boundaries are frequently described in terms of a particular grain-boundary structural unit separated by 'perfect' crystal. Higher-angle tilt grain boundaries may be described in terms of an array of different structural units along the grain boundary. In general there will be different structural units (or different spacings between the structural units) along the boundary, according to the local configuration of the grain-boundary plane, so that any arbitrary tilt grain-boundary may be modelled. In general, images from these boundaries are obtained from regions of the boundary where the structure is particularly clear, and the the boundary is parallel to the electron beam, and in the exact tilt configuration. Deviations of these boundaries away from the pure tilt orientation result in dislocations being present in the interface, which causes defects to appear in the repeating structural unit configuration of the boundary. The HREM images are determined only by the projected structure of the boundary and not by the full three-dimensional structure. The interpretation of these images is made more complicated by the presence of defects (such as dislocations) which are inclined to the beam direction, or steps in the boundary plane which occur within the thickness of the foil. In general, therefore, it is not possible to determine from a high-resolution image alone the complete structure of a defect observed in the grain boundary; particularly whether or not it is associated with a dislocation which has a Burgers vector parallel to the electron beam. If the presence of these defects can be determined independently, a more reliable interpretation of the high-resolution images can be obtained.

As an example of this extended defect analysis, a particular defect in the grain-boundary structure of a Ge bicrystal has been re-examined (Elgat and Carter 1985, Smith et al. 1989, Krakow et al. 1990). The boundary plane of this grain-boundary changes in orientation along the length of the boundary. An analysis of the lattice planes on either side of the boundary shows that the misorientation also varies locally by a few degrees. For the exact $\Sigma = 27$ condition the misorientation across the boundary is 32°. This local variation has been shown to be accommodated by different arrangements of 5-, 6- and 7-fold rings forming discrete structural units at the boundary plane. It has also been noted that short steps occur along the boundary. Although most of these papers describe the effects of a single step, this defect is actually fairly common. In the segment of the boundary discussed here the steps form a regularly spaced array. In order for the structure of the boundary to be the same on either side of the step, the presence of a dislocation at the step with an $a/4[\bar{1}10]$ screw component must be inferred (Elgat and Carter 1985, Smith et al. 1989, Krakow et al. 1990). Alternatively Krakow et al. (1990) have suggested that a sheet of 5-fold coordinated Ge atoms may exist on one side of the step. In this case alternating bands of 4-fold and 5-fold coordinated Ge would have to exist at the boundary plane. The purpose of this paper is to examine the structure of this grain-boundary step more detail.

2.   EXPERIMENTAL

Germanium bicrystals were grown from oriented seeds using a modified Czochralski technique (Skrotski et al 1984). The seeds were oriented to produce an incoherent $\Sigma = 3$ boundary, and "bicrystals" ~12 cm long and ~5 cm in diameter were grown. After cutting the bicrystals normal to the boundary plane it was found that, within a short distance from the seed crystals, the lateral $\{112\}$ $\Sigma = 3$ boundary had dissociated into several coherent $\Sigma = 3$ twin boundaries. The residual boundary after this dissociation is a 30.5° tilt-boundary which is very close to the $\Sigma = 27$ third-order twin configuration. Thin TEM samples were prepared from this boundary by cutting 3mm discs from the transverse slices of bicrystal with the boundary lying along a diameter of the disc. The discs were mechanically polished and ion-milled to perforation. Specimens with the boundary plane contained in the plane of the disc were also prepared in the same manner. The specimens were examined in a JEOL 4000EX and a Siemens 102 electron microscope.

3.   RESULTS

Low magnification, bright-field images and selected area diffraction patterns of the tilt grain-

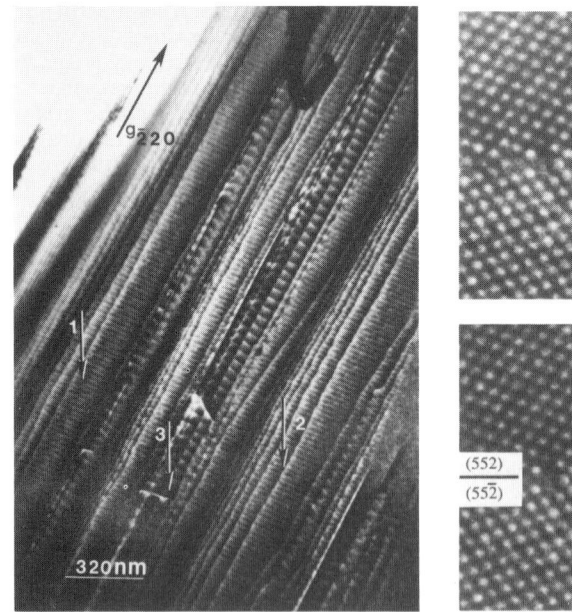

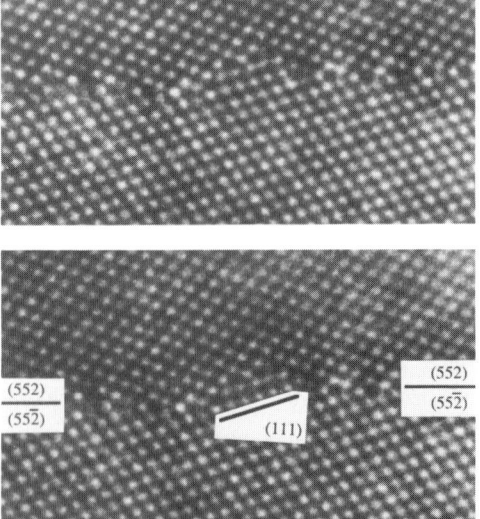

Figure 1   Low magnification bright-field image of the flat-on grain-boundary showing three different types of dislocation in the boundary.

Figure 2   HREM image of the $\Sigma = 27$ <110> tilt grain-boundary. Various facet planes are marked in the copy of this image shown beneath it for clarity.

boundary viewed edge-on, show that the average grain-boundary plane is close to {552}. It is evident that the boundary is not flat and that local changes in orientation of the boundary plane frequently occur. Periodic strain contrast is observed along the interface, similar to that expected from dislocations emerging from the foil. When the boundary is viewed flat-on (i.e. normal to the boundary plane) under weak-beam conditions, many dislocations are found to be present in the interface (figure 1). The dislocations appear to occur as three different arrays; an array of edge dislocations with a separation of ~ 3.5 nm, and two dislocation arrays which are close to screw orientation. These screw dislocations impart a twist component to the boundary of ~0.35°, and have a spacing which varies between 18 nm and 100 nm. The varying separation of the dislocations indicates that the misorientation of the grain-boundary may vary along the boundary. This separation is also consistent with the separation of the step defects observed in the HREM images. This set of dislocations is viewed end-on in the HREM images, the other set lie obliquely in the plane of the boundary. The unprocessed HREM image in figure 2 shows the structure of the grain boundary close to a step in the boundary plane. Structural units consisting of 5-, 6-, and 7-membered rings can be identified on either side of the step. The step is bounded in one grain by a {111} facet which is 5 or 6 planes long, depending on how the facet is terminated. The other side of the interface is more distorted, but lies approximately parallel to a {771} or a {551} plane. A schematic diagram of the structure of part of this boundary is shown in figure 3. This has been constructed to preserve the 4-fold coordination of the Ge atoms everywhere except at the step. The open and closed circles represent atoms at different heights above the plane of the diagram. The grey circles represent the effect of a dislocation with a screw component of $^a/_4$ [110] at the interface. The bonding at the step is not shown in order to emphasize the crystal planes on either side of the boundary. It is possible to fully reconstruct this step using the 5-, 6-, and 7-membered rings and the boat-shaped 6-membered ring characteristic of the $\Sigma = 3$ twin boundary (see e.g. Bourret and Bacmann (1987) and Thibault-Desseaux and Putaux (1989) for reconstructions of similar defects in $\Sigma = 9$ and $\Sigma = 19$ tilt grain-boundaries).

4.   DISCUSSION

The structure of the step consists of a close-packed (111) surface adjacent to a much less

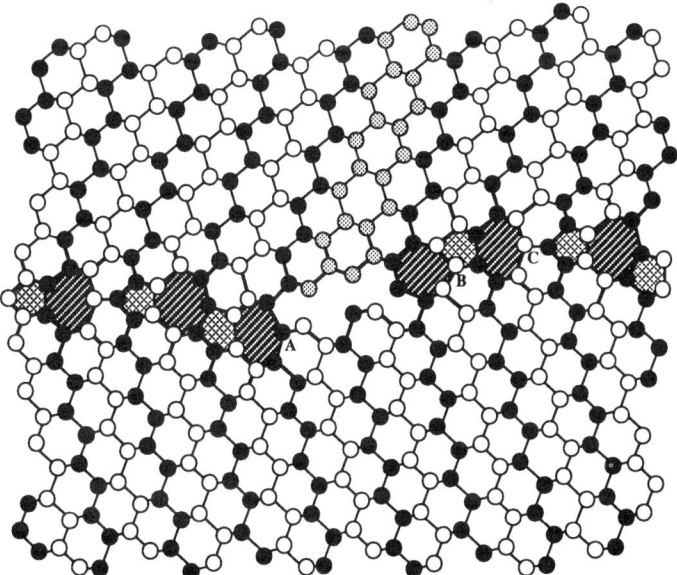

Figure 3   Schematic diagram of the high-resolution electron micrograph shown in figure 4. See text for details. No attempt has been made to indicate the bonding arrangement at the step.

dense (771) or (551) surface of the second grain (these facets are described by AC or AB respectively in figure 3). The matching of the two facets across the boundary is much better for the (771) facet in terms of boundary misorientation, but is better for the (551) facet in terms of plane matching. The angle between the (111) face and the (552) boundary normal is 19.5°. This is closer to the 21.6° angle between the (771) face and the normal. In both these cases the grain boundary structure is close to periodic with a centered unit cell which is approximately twice the length of the observed facet (the misfit is very small in the case of the (551) facet and somewhat larger in the case of the (771) facet). The effect of an $a/4[\bar{1}10]$ screw dislocation on the repeat structure of the facet is to distort the centered grain boundary unit cell into a primitive unit cell of half the length. This observed step length is uniform in all steps analyzed, indicating a uniformity of structure. This difference between the two different possible facets in the second grain arises mainly as a result of the inherent disorder of the $\Sigma = 27$ grain-boundary, which contains both the connected 7-fold and 5-fold rings of the $\Sigma = 9$ tilt-boundary, and the separated 7-fold and 5-fold rings of the $\Sigma = 19$ tilt-boundary, which makes it difficult to determine accurately the ends of the different facets. By defining the facet to include only those elements not present in either the $\Sigma = 9$ or $\Sigma = 19$ end members, the structure can best be described as a (111) / (551) facet combined with a grain-boundary dislocation containing an $a/4[\bar{1}10]$ screw component. It would be expected from energetic considerations that most of the distortion introduced by the dislocation would occur on the less dense, higher-energy side of the step. This is also borne out by the relative clarity of the (111) face compared with the (551) face. The strain energy associated with this arrangement is quite large and may provide a sufficient driving force to reconstruct the boundary. The grey circles in figure 5 which represent the $a/4[\bar{1}10]$ vertical displacement of the atoms are shown localised in a small region. If this region is extended then the distortions associated with any single atom become extremely small. As these steps occur in an array, then almost all of the atoms away from the boundary will be associated with some vertical displacement, providing a continuous shear in the $[\bar{1}10]$ direction. In combination with the second array of screw dislocations in the grain-boundary plane a twist component is introduced into the grain-boundary.

5.   CONCLUSIONS

A repeating structural unit has been recognised in a $\Sigma = 27$ related <110> tilt grain-boundary in Ge. This consists of a (111) facet of length $a[11\bar{2}]$ in one grain in contact with (551) facet in the second grain. The lengths of these two facets are very closely matched. The (551) facet has suffered a shear caused by the presence of a grain-boundary dislocation with an $a/4[\bar{1}10]$ screw component. This particular configuration accomplishes a lateral step of the boundary plane and, in combination with a second such array, incorporates a twist component into the boundary. This configuration, consisting of a high-density, low-energy facet on one side of the bicrystal, and a low-density high-index facet approximately commensurate with it on the other bicrystal, could be considered to be a general mechanism for the incorporation of a screw component into a tilt grain-boundary. In the case of a symmetric or nearly symmetric boundary a lateral step in the boundary plane would therefore necessarily be produced.

ACKNOWLEDGMENTS
The authors would like to thank Mr R. Coles and Ms M. Fabrizio for technical support. The electron microscopes are part of a central facility provided by the Materials Science Center at Cornell and are supported, in part, by National Science Foundation. This research is supported by the the DoE under grant No DE-FG02-89ER45381.

REFERENCES
A. Bourret and J.J. Bacmann; Rev. Phys. App., **22**, 563 (1987)
Z. Elgat, and C.B. Carter; Ultramicroscopy, **18**, 313 (1985)
W. Krakow; A.A. Levi, and S.T. Pantelides; J. Mat. Res., **5**, 587 (1990)
W. Skrotzki, H. Wendt, C.B.Carter; and D.L. Kohlstedt ; Mat. Res.Soc. Symp. Proc.
    25, 341 (1984)
D.A. Smith, Z. Elgat, W. Krakow, A.A. Levi and C.B. Carter; Ultramicrosc., **30**, 8 (1989)
J. Thibault-Desseaux, J.L. Putaux; Inst. Phys. Conf. Ser .**104**, 1, (1989)

Inst. Phys. Conf. Ser. No 117: Section 3
Paper presented at Microsc. Semicond. Mater. Conf., Oxford, 25–28 March 1991

117

Observation by *in situ* HVEM and X-ray topography of mobile grain boundary dislocations in a $\Sigma=9$ ($1\bar{2}2$) bicrystal of silicon

H A Benhorma*, A Jacques*, A George*, X Baillin** and J Pelissier**

* Laboratoire de Metallurgie Physique - Sciences des Materiaux, URA CNRS Nr 155, Ecole des Mines, parc de Saurupt, 54042 Nancy cedex, France
** CEA-IRDI, DMECN, Département de Métallurgie, Centre d'Etudes Nucléaires de Grenoble, 85 X, 38041 Grenoble cedex, France

ABSTRACT: The velocity of glissile grain boundary dislocations has been measured in a $\Sigma = 9$ bicrystal of silicon. Screw segments are much faster than edge ones parallel to the [011] tilt axis of the bicrystal. The apparent activation energy for the glide of fast segments is 1.2 ± 0.2 eV, at τ = 20 MPa. Climb in the interface of non-glissile grain boundary dislocations was also observed.

1. INTRODUCTION

The velocities of usual lattice dislocations were accurately measured long ago in silicon but nothing has been reported so far about the mobility of grain boundary dislocations (GBDs), except in the special case of Shockley partials moving in a $\Sigma = 3$ twin boundary (Yasutake et al 1986). The mobility of GBDs is expected to be very different from that of lattice dislocations owing to the different atomic arrangement and should play an important part in processes like GB migration or dislocation transmission across GBs. We report here the observations and measurements made in a $\Sigma = 9$ bicrystal of silicon. The $\Sigma = 9$ symmetric tilt GB (angle: 38.94°, axis: [011], GB plane: $(\bar{1}22)_I/(12\bar{2})_{II}$) and its GBDs were studied by HREM by El Kajbaji and Thibault-Desseaux (1988). GB atoms are arranged in 7 and 5 atom rings with 1/2[011] and $1/2[41\bar{1}]_I$ periods in the interface. Elemental DSC GBDs of concern here have either a $b_g = 1/18[41\bar{1}]_I$, or $b_c = 1/9[1\bar{2}2]_I$ Burgers vector. The former is glissile, the latter can only move by climb.

2. VELOCITY OF GLISSILE b_g DISLOCATIONS

2.1 X-Ray Topography Experiments

b_g dislocations were produced in initially dislocation-free bicrystalline tensile samples by decomposition of lattice dislocations trapped at the GB plane (Baillin et al 1990). A $[6\ 5\ \bar{5}]_I/[\bar{8}2\ 11\ \bar{1}1]_{II}$ axis was chosen, with the [011] tilt axis parallel to the intersection of the GB plane with large sample free surfaces. Dislocation half-loops gliding on $(111)_I$ and $(\bar{1}11)_I$ planes were created from microhardness indentations and pushed towards the GB plane at T = 1023 K, under nominal stress $\sigma \approx 40$ MPa. b_g dislocations emitted from the impact lines with the GB plane could be imaged by X-ray topography, using the 400_{II} selective reflection (Fig. 1). With this diffraction vector g.b is large <u>and</u> lattice dislocations of grain I are out-of-contrast, which makes them visible only along the cross made by the intersection lines between their glide planes and the GB.

The emitted GBD loops move under the applied shear stress between or outside the branches of the cross. Outside the cross, they are formed of long straight segments

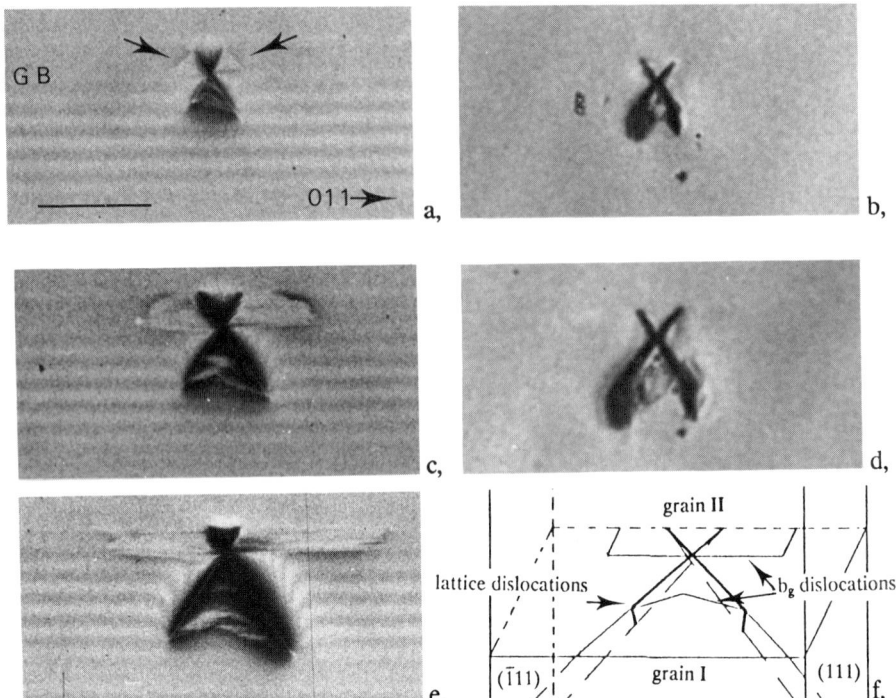

Fig. 1 Formation and movement of glissile b_g dislocations. t = 10mn (a,b). t = 18 mn (c,d). t = 26 mn (e). X-ray topographs. g = 400_{II} (a,c,e) or g = 022 (b,d). marker = 500 μm. See text for details.

parallel to the surface i.e. to [011], which are linked to the surface by shorter lines with an average screw character. "Screw" segments are in fact heavily kinked, trailing edge dipoles. Only the velocity of these faster screw segments could be determined; edge dislocations were so slow that no motion could be detected before lattice dislocations totally invaded the field of view. Fig. 2 is an Arrhenius plot of the velocity of screw b_g

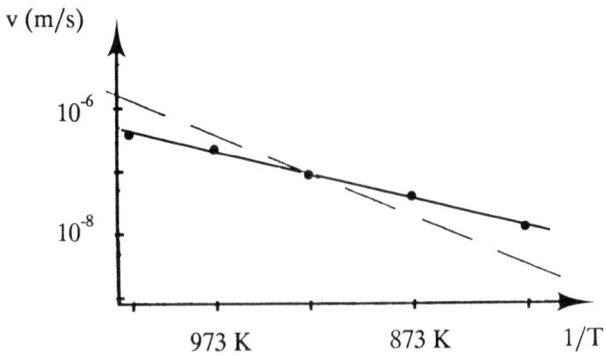

Fig. 2 Velocity of screw b_g dislocations as a function of temperature at τ = 20 MPa (full dots) . Screw lattice dislocations (broken line) are shown for comparison.

GBDs at the resolved shear stress, τ, of 20 MPa between 823 and 1023 K. The activation energy is 1.2 ± 0.2 eV, significantly lower than the 2.2 eV value characterizing the glide of lattice dislocations (George and Champier 1979). In the investigated range, edge b_g dislocations are at least fifty times slower than screws. Lattice dislocations are about as fast as the latter at T = 923 K, τ = 20 MPa.

2.2 In Situ HVEM Observations

Slow edge-type b_g GBDs could be observed moving in the 1 MeV electron microscope of CENG working at 400 kV, at T $\simeq$ 973 k, under undetermined stress conditions (probably $\tau > 50$ MPa). The recorded sequence of Fig. 3 proved that moving [011] segments are very rigid, except when they reach the obstacle of another GBD. Here,

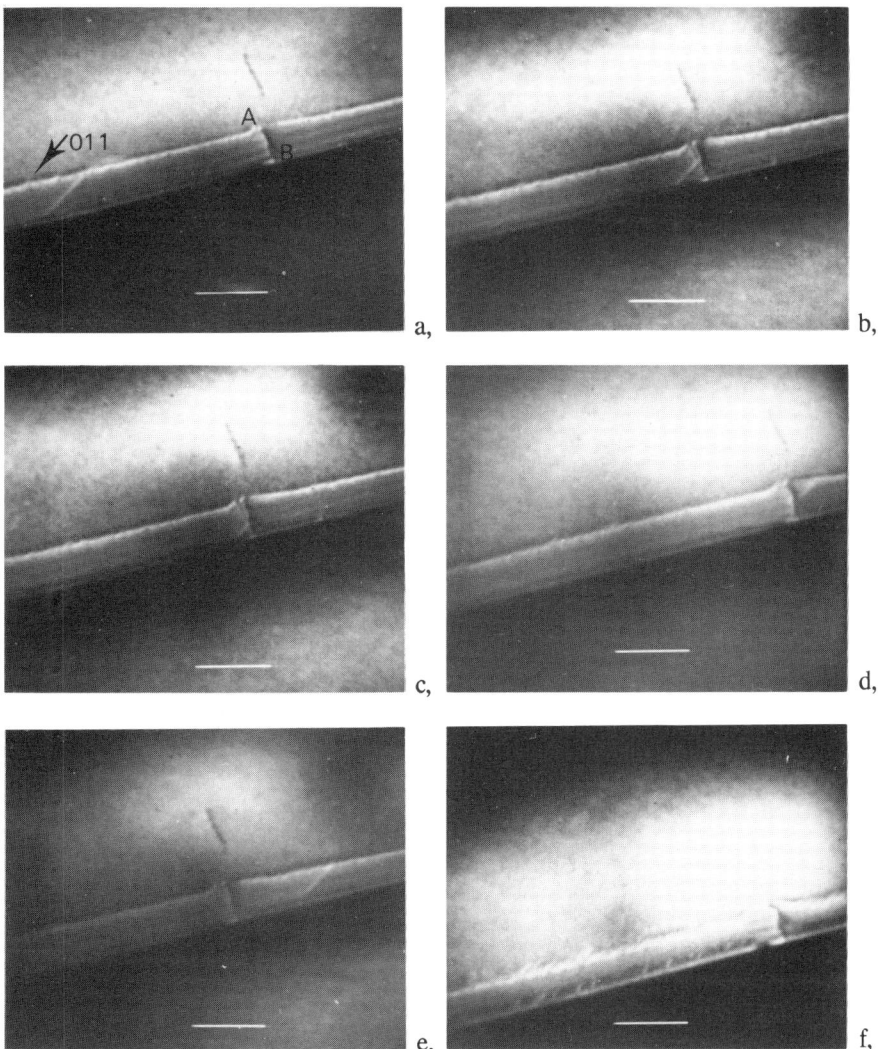

Fig. 3 b_g dislocation gliding in a Σ = 9 GB.T = 973 K.t = 0 s (a), t = 98 s (b), t = 118s (c), t = 162 s (d), t = 199 s (e), t = 705 s (f). HVEM. g = 2 0 2$_I$. marker 1 μm

the obstacle AB was a lattice dislocation probably with the $1/2[101]_I$ Burgers vector trapped in the GB plane which had already emitted a b_g GBD, and was further dissociating by climb. The incoming b_g dislocation aligned itself progressively along the obstacle. About 70 seconds after the first extremity was stopped at A a b_g [011] segment reappeared at the same point and got free by unzipping from the trapped dislocation. Such an event was observed several times. Far from the obstacle, the measured velocity was 22 nm/s.

3.   CLIMB OF b_c DISLOCATIONS IN THE GB

In the course of *in situ* HVEM observations, lines of contrast appeared at the foil surfaces and moved slowly towards the centre of the foil (Fig. 3e to f). Fig. 4 is a conventional electron micrograph of a neighbouring area of the same foil, taken after the interruption of in situ straining. The near-surface lines appeared to result from the coalescence of half-loops elongated in the [011] direction. Some similar closed loops are also seen. These loops are not b_g dislocations. Their extinction conditions are consistent with the b_c Burgers vector. Such lines did not appear in samples which had been submitted to a short chemical polishing in order to remove the defects left by ion milling, but closed loops remained. The length of half-loops i.e. the distance travelled in the [011] direction, was maximum in those parts of samples which received the highest doses of electron irradiation , suggesting a mobility enhanced by the 400 KeV electron beam. This maximum mobility was 0.16 nm/s. From the width of non coalesced half loops, we can deduce for [011] segments a velocity lower than a fiftieth of that of fast segments.

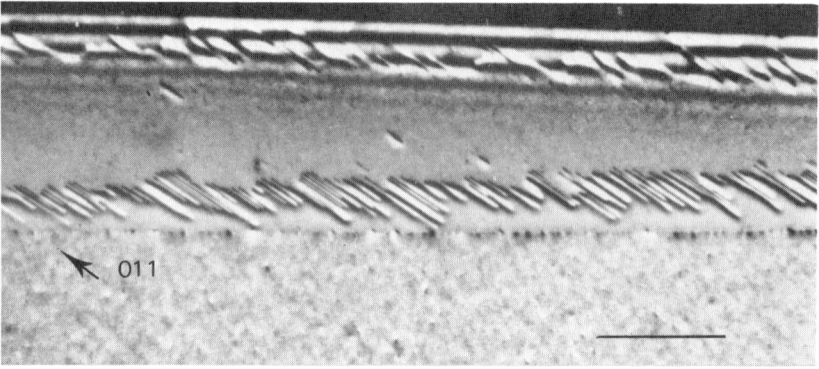

Fig. 4 b_c dislocations climbing in the GB from the foil surfaces. Conventional TEM. g = 2 0 2$_I$. marker 0.5 μm.

REFERENCES

Baillin X, Pélissier J, Jacques A and George A 1990 Phil. Mag. A 61 329
El Kajbaji M and Thibault-Desseaux J 1988 Phil. Mag. A 58 325
George A and Champier G 1979 Phys. Stat. Sol. (a) 53 529
Yasutake K, Shimizu S,Umeno M and Kawabe H 1986 J. Appl. Phys. 61 940

Inst. Phys. Conf. Ser. No 117: Section 3
Paper presented at Microsc. Semicond. Mater. Conf., Oxford, 25–28 March 1991

In situ electron microscopy study of interstitial cluster formation in silicon and germanium

A L Aseev and L I Fedina

Institute of Semiconductor Physics, Academy of Sciences of the USSR, Siberian Branch, 630090 Novosibirsk, USSR

ABSTRACT: The processes of formation, growth and shrinkage of interstitial clusters have been investigated by in situ electron irradiation of Si and Ge crystals in electron microscopes JEM-1000 and EM-200 at energies of 1 MeV and 200 keV, respectively. The characteristics of the formation of interstitial atom clusters have been determined.

1. INTRODUCTION

The experiments on the irradiation of Si and Ge crystals in the electron microscope (EM) were carried out over a wide range of electron irradiation intensity ($1\cdot10^{18}$– $5\cdot10^{20}$ electrons$\cdot$ cm$^{-2}\cdot$s^{-1}), crystal temperature (-100 to 700°C) and impurity concentration of the crystals. Various coatings on the crystal surface irradiated in the EM were used.

2. INTERSTITIAL ATOM CLUSTERS AS ROD-LIKE AND {113}-DEFECTS

According to the models for the atomic structure of rod-like and {113}-defects presented by Salisbury and Loretto (1979), Tan (1981), Pasemann et al (1983) they should be considered as interstitial atom clusters in metastable configurations. A typical image of a {113}-defect in the irradiated Si crystal is presented in Fig.1.

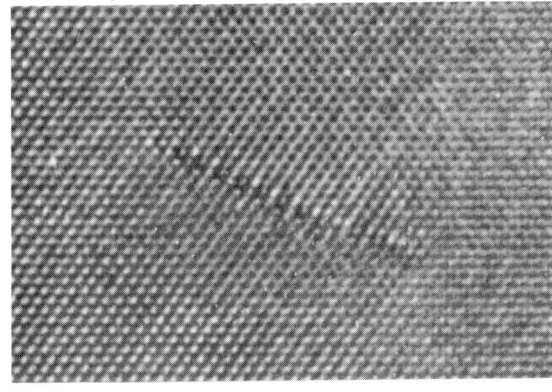

Fig.1. Many – beam image of {113}-defect in Si crystal produced by irradiation in EM.

The main features of interstitial atom cluster formation are: the existence of an incubation period; strong dependence on the concentration of electrically inactive impurities; not uniform spatial distribution (with the maximum of defect concentration for Si crystals in the vicinity of the oxidized surface) and nonlinear dependence of the rate of RLD and {113}-defect formation on the irradiation intensity (I) - (fig.2). According to the theory of quasichemical reactions between point defects and impurity atoms, the rate of defect formation is given by $\dot{C}_d = k_{id} \cdot C_i^n \cdot C_p^m$, where C_d is the concentration of interstitial atom clusters; k_{id} is the coefficient for interaction between the interstitial atoms and defects; $C_i \sim I^q$ is the concentration of interstitial atoms in the irradiated crystal; $0.5 \le q \le 1.0$ for Si and $q \approx 1$ for Ge; C_p is the concentration of impurity atoms; n and m are the numbers of interstitial atoms and impurity atoms in the nucleus corresponding to the critical size for the formation of {113} defects . It follows from the experimental data for $\dot{C}_d$ dependence on I and C_p that n=1-4 and m=1-2 for various values of irradiation intensity (fig.2). The dependence of incubation time on irradiation intensity is found to be $\tau_d \sim I^{-1}$ for the silicon crystals. Such a dependence can be associated with the activation of impurity atoms and their radiation induced diffusion upon direct collision with the high energy electrons of the electron beam during the formation of nuclei of the interstitial atom clusters.

Under constant irradiation conditions, a nonlinear dependence of defect growth rate on irradiation time has been found for the <110> direction (fig.3).This result can be explained by pipe diffusion of interstitial atoms along the axis of the growing defects producing their initial accelerated growth. From the temperature dependence $v_{<110>} = L_{<110>}/t^*$ (fig.3), it follows that the activation energy of the pipe diffusion is 0.8 eV for Ge and 0.6 eV for Si. The occurrence of a stationary growth stage $L_{<332>} \sim t$ (fig.3) is evidence for a quasi equilibrium between the generation rate of point defects and the rate of their migration to the surface and to the {113}-defects. From a comparison of the observed generation rates it follows that the impact separation coefficient for vacancy-interstitial (v - i) pairs is about 10^{-3} both for Si and Ge. The high rate of v - i pair annihilation is caused by the strong crystal ionization arising from the intensive irradiation in the EM. A weak dependence of the growth rate in the <332> direction on the temperature was observed in the experiments providing evidence for the high mobility of vacancies and interstitials generated in Si and Ge by electron irradiation.

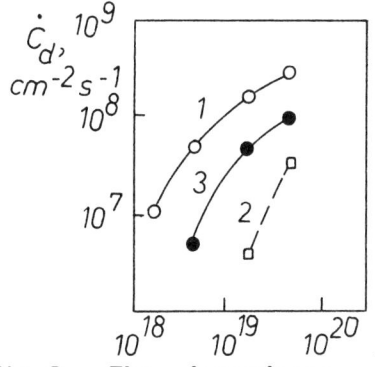

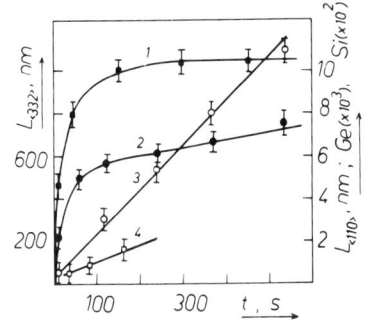

Fig.2. The dependence of formation rate of RLD on intensity of irradiation for Cz-Si (1), Fz-Si (2) and Ge (3). T=500°C for Si and -115°C for Ge. The data for Ge crystals given after treatment same as experimental results by Kiritani and Hirata (1981).

Fig.3. The dependence of sizes of {113}-defects in directions <110> (1,2) and <332> (3,4) for Si (1,4) and Ge (2,3) on irradiation time. T=500°C (Ge), 400°C (Si), I=2·10¹⁹ el/cm²·s.

An analysis of the conditions for size decrease and shrinkage of interstitial atom clusters and {113}-defects during irradiation in the EM is carried out. Defect shrinkage is observed in all cases where the condition $C_i D_i = C_v D_v$ is fulfilled, with $C_{i,v}$ and $D_{i,v}$ being the concentrations and migration coefficients of i and v, respectively. The above mentioned condition is fulfilled, for example during irradiation of the crystal with the surface being a neutral sink for interstitials and vacancies.

3. FORMATION OF DISLOCATION LOOPS

It is difficult to transform an interstitial atom cluster into a dislocation loop or dipole because of the presence of the energetic barrier for interstitial transition from the metastable position to the regular lattice position. The analysis of the dislocation climb rate temperature dependence given by Aseev et.al. (1987) and Fedina and Aseev (1986) shows the value of this barrier to be about 0.5 ev for Ge and 1.3 eV for Si. Therefore, the formation of dislocation loops of extrinsic type occurs during interstitial interaction with vacancy complexes. The intermediate stage of loop formation comprises the interstitial accumulation in metastable configurations with the quantity exceeding the number of vacancies in the complex. The existence of such a process is assumed because of the occurrence of dislocation loops in Si and Ge crystals doped with phosphorus and antimony, respectively. In this case the vacancy-impurity complexes were formed at the first step of the

irradiation, so the interstitial interaction with these complexes leads to the formation of dislocation loops of extrinsic
type (fig.4).

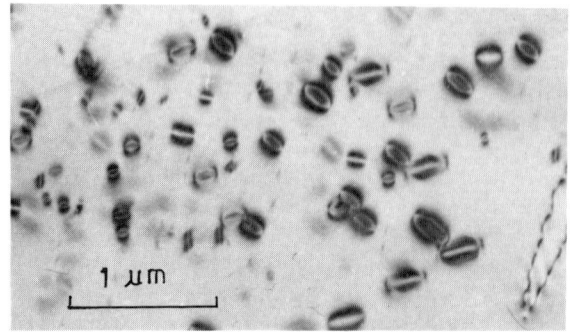

Fig.4.   Dislocation
loops in irradiated Si
doped by P with con-
centration 10^{19} cm^{-3}.
T=650°C, I= $6 \cdot 10^{18}$ el/
cm^2·s, time of irradi-
ation is 300 s.

REFERENCES:

Aseev A L, Ivakhnishin V M, Hoehl D and Bartsch H 1987 Phys.
    Stat. Sol. (a) 100 431
Fedina L I and Aseev A L 1986 Phys. Stat. Sol. (a) 95 517
Kiritani M and Hirata M 1981 Defects and Radiation Effects in
    Semiconductors, ed R R Hasiguti (Bristol and London: The
    Institute of Physics), pp 449-454
Pasemann M, Hoehl D, Aseev A L and Pchelyakov O P 1983 Phys.
    Stat. Sol. (a) 80 135
Salisbury I G and Loretto M H 1979 Philosophical Magazine A
    39 317
Tan T Y 1981 Philosophical Magazine A 44 101

Inst. Phys. Conf. Ser. No 117: Section 3
Paper presented at Microsc. Semicond. Mater. Conf., Oxford, 25–28 March 1991

125

The pinning effect of phosphorus on dislocation cores in silicon

M Heggie, R Jones and A Umerski

Department of Physics, University of Exeter, Stocker Road, Exeter EX44QL.

ABSTRACT: The interaction of phosphorus with 90 degree partial dislocations in silicon is examined using a cluster method with local density functional pseudopotential theory. We describe several states of phosphorus at dislocation cores which are (i) normally reconstructed and (ii) which contain solitonic reconstructed bonding patterns. Our overall conclusion is that there is a clear tendency for phosphorus to migrate towards the dislocation core, and assume three fold coordination, thus (i) breaking reconstructed bonds across the core and (ii) passivating the solitonic dangling bonds. In each case the P loses its donor characteristic.

1. INTRODUCTION

The 60 degree dislocation in Si may be obtained by inserting/removing an extra half plane such that : (a) it terminates on closely spaced {111} planes (also known as the glide plane), (b) its terminating edge lies parallel to <1$\bar{1}$0> (the dislocation line direction) and, (c) gives rise to a Burgers vector b_{60}=a/2<0$\bar{1}$1> (a = 5.43Å - cubic unit cell length). The enormous elastic energy required to create such a dislocation (proportional to the square of the Burgers vector length) makes it energetically favourable for it to dissociate into two dislocations - the 30 and the 90 degree partials (b_{30}=a/6<1$\bar{2}$1>, b_{90}=a/6<1$\bar{1}$2>), which lie in the same plane separated by a stacking fault, parallel to the original, and are able to move in this glide plane by the destruction and reconstruction of local bonds. Theoretical work [Heggie and Jones 1987, Jones 1979, Lodge *et al* 1989, Moller 1978] shows that the bonds across the dislocation core (which are broken in the 60 degree case) are able to reconstruct making the Si atoms four fold coordinated, thereby reducing the electrical activity of the core, and furthermore making these partials even more stable (our calculations also support this view - see §4b below). In this paper we investigate several complications to this scenario for 90° partials, including reconstruction defects, the interaction of P atoms with reconstructed and solitonic cores, as well as the interaction of P atom pairs with reconstructed cores.

2. DISLOCATION MOBILITY

The glide of dislocations occurs by the generation of double kinks and their subsequent migration along the line until they meet an oppositely moving one or reach a pinning point or the end of the line. The reconstruction of dislocations and kinks implies a large activation barrier to kink motion, and a theory in this case, for pure materials, is given by Hirth and Lothe [Hirth and Lothe 1982]. The dislocation velocity is

$$v = v_o \tau \exp\left[\frac{-E}{kT}\right]$$

Where v_0 is constant depending on the material, τ the applied stress, T is the temperature, and E the kink activation energy. Experimentally [Imai and Sumino 1983] the kink activation energy is found to be 2.2eV for 60° dislocations and 2.35eV for screw in pure Si.

2.1 THE INFLUENCE OF IMPURITIES

For impure specimens, which have been aged at elevated temperatures (typically 600 - 900 °C and 0 - 5000 secs) dislocations effectively getter impurities, and they are observed to be pinned at specific points along their length (with interaction energy ~3eV), so that a release stress (which is both temperature and age duration dependent) is required to free the dislocation and begin motion [Sumino and Imai 1983]. P (unlike C) has an extremely strong pinning effect on dislocations within a very short time period. This combined with the fact that the diffusion rate of P (like C) is very small and consequently the mean distance between pinning particles along the core very large implies substantial binding energies. Since the typical (elastic) interaction energy of a size misfit atom is ~0.5eV, P pinning particles have previously been thought to occur as clusters of impurity atoms - we show here that, clusters of only one or two P atoms are capable of producing such energies.

Once the dislocation has been freed, and under high stress, the dislocation velocity is generally enhanced if the impurity is of donor character, whereas light element impurities (O,N,C) have no effect [Imai and Sumino 1983]. In Si:P the activation energy (for dislocation motion) - E drops to 1.8eV and 1.7eV for 60° and screw dislocations respectively [Erofeev and Nikitenko 1971, Patel *et al* 1976], and the dislocation mobility is enhanced [Hirsch 1979, Jones 1980]. A theory for dislocation velocity which includes pinning effects is given by Celli *et al* [1963], and this has been generalised to take dissociation into account by Moller [1978].

SUMMARY

A fundamental problem is to explain why P for example should have such a dramatic locking effect on dislocations. Here we summarise the results of our calculations (previously described in Heggie *et al* [1989, 1990, 1991]) showing that (i) P is strongly attracted (has binding energy ~1.5eV) to a soliton or reconstruction defect on a 90 degree partial, (ii) that P actually breaks the reconstruction and that pairs of P defects strongly bind to the line (with binding energy ~2eV). In addition we show that there is little or no tendency for these P pair pinning particles to aggregate. In all these cases the valence state of P changes at the dislocation core from 4-fold coordination into 3-fold (which is the preferred state of P in a-Si [Street 1982]). Thus as the dislocation advances, either the 3-fold structure reverts to a 4-fold one, or the P is necessarily dragged along with it. We thus explain the pronounced locking effect of P in terms of pinning particles of only one or two atoms.

3. METHOD

We use a local density functional scheme which has shown itself reliable in calculating structures and dynamical properties of semiconductors and their defects [Briddon and Jones 1989, Jones and Sayash 1986, Jones 1988a]. The technique employed here was to initially construct a large cluster of 500 Si atoms containing the 90 degree partial dislocation and the particular defect we wished to investigate. A valence force routine [Tores and Stoneham 1985] was used to relax all the atoms to an equilibrium state, and then a cluster of up to 50 atoms surrounding the dislocation core and containing the defect of interest was cut. The dangling bonds of the surface atoms were then saturated with H atoms, and this molecule provided the starting point for the pseudo potential scheme. The total energy of the cluster is found and the force on each atom is evaluated semi-analytically. All the atoms (including H's) which lie in the glide plane passing through the dislocation core were allowed to relax in response to these forces, whilst all others were kept fixed - in this way we allowed a reasonable degree of freedom to atoms undergoing complex interactions, whilst (to some extent) simulating the embedding of this small cluster in a larger one. Unfortunately the terminating H atoms have bonding and anti-bonding states which couple strongly with states at the top of the valence band and at the bottom of the conduction band and hence artificially enlarge the gap. Consequently the band gap for a 56 atom cluster $Si_{26}H_{30}$ is 4.5eV. Nevertheless as we shall see it is possible to distinguish deep from shallow levels. The method is described in detail in [Jones 1988b].

4. SUMMARY OF EARLIER RESULTS

The results of our calculations may be summarised as follows:

(a)   P sits on a substitutional site in crystalline Si, with a P-Si bond length of 2.41Å, compared to 2.35Å for Si-Si. It contributes a half filled donor state lying 0.6eV below the conduction band edge (recall the gap is enlarged to ~4.5eV in our calculations) (Fig. 1a).

(b)   The 90° reconstructed partial in Si (Fig. 2a) is a stable structure, the reconstructed bonds across the core (6-13, 8-15, 10-17, ...) being only 2.36Å (~1% stretch), and giving a clear gap in the eigenvalue spectrum (Fig. 1b).

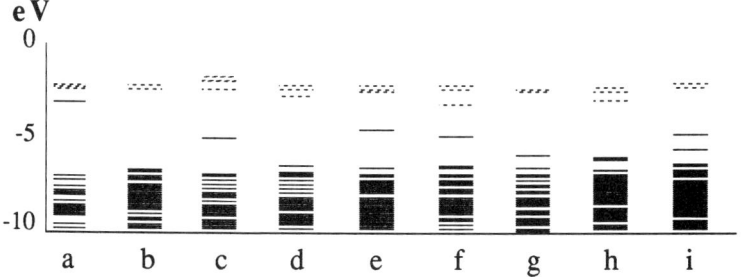

Figure 1 : Eigenvalue spectrum near the gap for relaxed clusters (----- denotes empty levels).

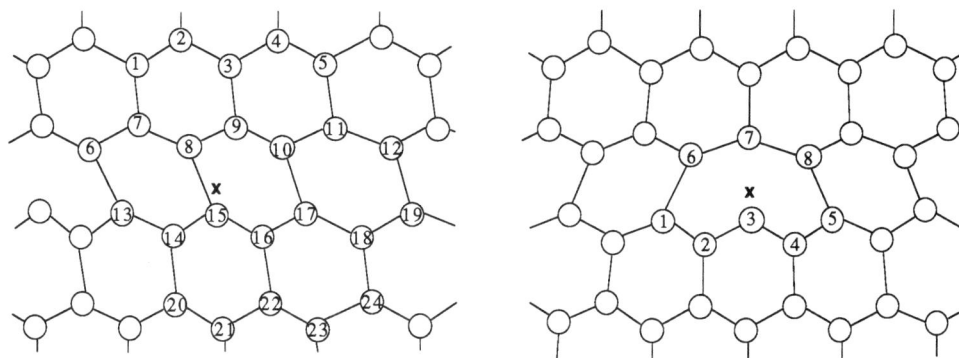

Figure 2a : Reconstructed 90° partial. Figure 2b : 90° partial with solitonic defect.

(c) The 90° partial with a solitonic or reconstruction defect core (Fig. 2b) is a stable structure with a half filled state at mid gap - E_v+1.5 eV (Fig. 1c). The wave function for this level is concentrated in the core, 0.8Å off the solitonic Si atom (3 in Fig. 2b) and so is associated with a dangling bond (indicated by x in Fig. 2b). The Si-Si bond lengths of the solitonic atom (3-2, 3-4) are about 5% shorter than normal (because of the increased sp^2 hybridization) whilst those on the opposite side of the core (6-7, 7-8) are about 3% longer.

(d) P has an affinity for solitonic sites (migrates to position 3 in Fig. 2b with a binding energy of ~1.5eV), where it becomes 3 fold coordinated (the three P-Si bonds reducing by 0.08Å to about 2.24Å), allowing the mid gap (highest) filled level to drop down to the valence band edge.at E_v+0.1eV (Fig. 1d) and thus clearing the gap of states. This state is also associated with a dangling bond pair in the core, concentrated 0.8Å off the P atom at x in Fig. 2b.

(e) A single P atom has no particular affinity for the core of a reconstructed partial (eg. position 15 in Fig. 2a), however once there it prefers to break the reconstructed bond (8-15 lengthens to ~2.8Å), leaving both itself and the Si (in position 6) 3 fold coordinated. Its lone pair lies in some state in the valence band, whilst the partially filled donor level drops by 0.5eV to E_v+2eV in the middle of the gap (Fig. 1e). This state is now associated with a high charge density about 0.8Å off the P (not the Si!!) in the core as indicated in Fig. 2a at x.

(f) Pairs of P's will migrate towards reconstructed cores assuming three fold bonding patterns in adjacent sites (eg 8 and 15 in Fig. 2a), with binding energies of ~2.3eV. The first stage of this process is similar to that described in (e) except that the presence of the other P yields a single filled mid gap state at E_v+1.6eV (Fig. 1f), whose wave function corresponds to a dangling bond 0.8Å off the P atom at the core (at x in Fig. 2b). The remaining P now migrates to the adjacent core site becoming three fold coordinated, and the P lone pairs have weak bonding and antibonding interactions with energies at the valence band edge (E_v+0eV and E_v+0.6eV in Fig. 1g). This configuration leads to almost no electrical activity of the core.

(g) An alternative configuration in which P atoms were placed at positions 10 and 15 (Fig. 2a), was tried. It was found that the two P-Si bonds 8-15 and 10-17 across the core broke, and the two Si atoms (8 and 17) formed a bond of length 2.33Å. This lead to two (filled) states at the valence band edge - E_v+0.4eV and E_v+0.55eV (Fig. 1h).which are identified with very weak bonding and antibonding wave functions between the P atoms. The equilibrium energy of this configuration is 0.5eV greater than that described in (f), so that under annealing this situation is far less likely to occur.

(h)   When four P atoms are placed at a dislocation core (eg. at sites 8, 10, 15, 17 in Fig. 2a) representing two defects of the type described in (f), we find that both P-P bonds (8-15,10-17) across the core break (lengthen to ~3.0Å) leaving two weak bonding and two weak antibonding filled states at E_v+0eV, $E_v+0.1eV$, $E_v+0.9eV$, $E_v+1.5eV$ (Fig. 1i), between the two bonds. This system has a binding energy of ~4.0eV, which is (approximately) twice that of the P pair pinning points described in (f), so that there must be little or no tendency for these to aggregate.

5. CONCLUSIONS

The calculations have shown that there are at least two mechanisms through which P binds to dislocations. Individual P atoms have binding energies of 1.5eV with reconstruction defects (solitons), whilst P atom pairs bind with interaction energies of 2.3eV through the breaking of reconstructed bonds. We have determined that there is little or no tendency for the P pair pinning particles to aggregate, and claim that as few as two P atoms are required to reproduce the experimentally observed values for the pinning point interaction energy. The key to the success of P in locking dislocations is then, that it is able to take up a three fold coordinated structure at the core. We would expect similar behaviour for B, and believe these results are also applicable to 30° partials and other dislocations, as well as to a-Si and grain boundaries [Werner 1989] where reconstruction occurs. Another consequence would be that the concentration of chemical donors (i.e. four fold coordinated P atoms) would be reduced i.e. the dislocation would act as having acceptor states with no levels in the gap. Because P at the dislocation is electrically neutral it should lead to an increase in the mobility of the remaining electrons. Of course, jogs and other defects would introduce gap states which would tend to mask these effects. We suggest that N similarly forms such defects. It is surprising that B does not seem to behave in the same way.

Calculations are at present underway to determine the O pinning mechanism.

Preliminary results indicate an analogous effect to that of P - in that single O atoms seem to have little or no affinity for the core of a reconstructed 30° partial in Si, whilst O pairs do seem to have a strong interaction energy. These results will be the subject of a forthcoming paper.

REFERENCES

Briddon P, Jones R, 1989, *J. Phys. : Condensed Matter*, 1, 10361.
Celli V, Kabler M, Ninomiya T, Thomson R, 1963, *Phys. Rev.*, 131, 58.
Erofeev VN, Nikitenko VI, 1971, *Soviet Phys. Solid State*, 13, 116.
Heggie M, Jones R, 1987, Institute of Phys. Conf. Series 87, p 367.
Heggie M, Jones R, Lister GMS, Umerski A, 1989, Institute of Phys. Conf. Series 104, p 43.
Heggie M, Jones R, Umerski A, 1990, Mat. Sci. Forum, Vols 65-66 (1990), Trans Tech Pub., pp265-270
Heggie M, Jones R, Umerski A, 1991, *Phil. Mag. A*, 63, 571.
Hirsch PB, 1979, *J. de Physique*, 40, C6, 117.
Hirth JP, Lothe J, 1982, Theory of Dislocations, John Wiley & Sons, p 531-545.
Imai M, Sumino K, 1983, *Phil. Mag. A*, 47, 599.
Jones R, 1979, *J. de Physique*, 40, C6, 33.
Jones R, 1980, *Phil. Mag. A*, 42, 213.
Jones R, Sayyash A, 1986, *J. Phys. C: Solid State Phys.*, 19, L653.
Jones R, 1988a, *J. Phys. C: Solid St. Phys.*, 21, 5735.
Jones R, 1988b, *J. Phys. C: Solid St. Phys.*, 20, L271.
Lodge KW *et al*, 1989, *Phil. Mag. A*, 60, 643.
Moller HJ, 1978, *Acta. Metall.*, 26, 963.
Patel JR, Testardi LR, Freeland PE, 1976, *Phys. Rev. B*, 13, 3548.
Street RA, 1982, *Phys. Rev. Lett.*, 49, 1187.
Sumino k, Imai M, 1983, *Phil. Mag. A*, 47, 753.
Torres VJ, Stoneham AM, 1985, Handbook of Interatomic Potentials, UKAEA (Harwell).
Werner JH, 1989, Inst. Phys. Conf. Ser. N° 104, 1989. p 63.

Inst. Phys. Conf. Ser. No 117: Section 3
Paper presented at Microsc. Semicond. Mater. Conf., Oxford, 25–28 March 1991

Impurity effects on the morphology of fresh dislocations in GaAs

I Yonenaga, K Minowa and K Sumino

Institute for Materials Research, Tohoku University, Sendai 980, Japan

ABSTRACT : The morphology of dislocations freshly introduced in impurity-doped GaAs under applied stress is investigated by means of X-ray topography. α dislocations are preferentially generated in Si-doped GaAs from surface irregularities while β dislocations are generated in Zn-doped GaAs, both trailing long screw dislocations behind them. The morphology of dislocations generated in In-doped GaAs depends on the magnitudes of the applied stress and temperature. The characteristics are interpreted as being related to the strong interaction between α dislocations and In impurity.

1. INTRODUCTION

Doping of certain kinds of impurities into compound semiconductors is known to be effective in reducing the density of grown-in dislocations, which is attributed to the suppression of dislocation generation related to dislocation-impurity interaction. Clarifying the characteristics in such dislocation-impurity interaction is important from both basic and practical points of view. Impurity effects on the dynamic activities of dislocations may be divided into two categories : one is the effect on the velocity of dislocations in motion and the other the immobilization of dislocations when they are at rest. The present authors (Yonenaga and Sumino 1987, 1989) have shown that a variety of impurities in GaAs immobilize dislocations on segregating along the latter. Electrically active impurities affect the velocity of dislocations in motion.

The above kinds of observations are usually conducted using the etch pit technique in III-V compounds. However, the etch pit technique does not reveal the morphology of dislocations in motion, which directly reflects the dynamic properties of various types of dislocations in the crystal. The mobility of a dislocation depends very strongly on the type of dislocation. In III-V compounds the mobility of the fast type of dislocation is usually higher than that of the slow type of dislocation by orders of magnitude. X-ray topography applied to a rather thick sample makes it possible to reveal the characteristics of the morphology of dislocations in motion. Up to now, a very limited number of X-ray topographic studies of the morphology of dislocations in III−V compound semiconductors are available (George, Jacque and Coquille 1985, Burle-Durbec, Pichaud and Minari 1987, Di Persio and Abbas 1989). However, they all reported the morphology of the slow type of dislocations. Observations of the fast type of dislocations have not yet succeeded except for the work of Di Persio and Kesteloot (1983) for InSb.

In this paper, we report the characteristics of the morphology of dislocations, especially concentrating on the fast type of dislocations in GaAs doped with In, Si or Zn, which are freshly introduced under applied stress and observed with X-ray topography.

2. EXPERIMENTAL

Observations have been made on three types of GaAs crystals with low densities of grown-in dislocations : Those doped with Si ([Si] = 4 $\times 10^{18}$ cm^{-3}, grown with the boat-technique), doped with Zn ([Zn] = 3 $\times 10^{19}$ cm^{-3}, grown with the boat-technique), and doped with In ([In] = 2 $\times 10^{20}$ cm^{-3}, grown with the LEC technique).

Specimens with a rectangular shape of dimensions approximately 5 $\times$ 6 $\times$ 15 mm^3, having the long axis parallel to the [011] direction and the side surfaces parallel to the (100) and (01$\overline{1}$) planes, were prepared from the above crystals and were finished by chemical polishing of the surfaces. Then, a scratch was drawn with a diamond stylus in a diagonal direction on the (100) and ($\overline{1}$00) surfaces of a specimen as a preferential generation center for fresh dislocations.

The specimen was stressed by compression along the long axis in a vacuum. The duration of stressing was chosen in such a way that the characteristics of the morphology of dislocations travelling from the scratch were most clearly seen in X-ray micrographs. Thin plates were cut from the stressed specimen parallel to one of active primary slip planes and thinned to a thickness of 0.4 mm with chemical polishing. The morphology of dislocations on the primary slip plane was observed by means of X-ray topography using the Borrmann effect with 220 diffraction of Mo K α_1 radiation from an X-ray generator RIGAKU RU-500. Topographs were taken in such a geometry that the incident X-ray beam fell on the As ($\overline{1}\overline{1}\overline{1}$) surface and the diffracted beam came out from the Ga (111) surface.

3. RESULTS

3.1. Dislocations in GaAs doped with Si or Zn

Figure 1 shows a topographic image of dislocations generated and moved under a stress of 5 MPa at 450 °C in Si-doped GaAs. The arrows indicate the locations of scratches drawn on the side surfaces. Two sets of groups of dislocations are seen to be generated and have travelled toward the inside of specimen only from the scratch on the (100) surface, though some dislocations are also seen to be generated from the edge of the specimen. Dislocations marked with Da1 and Da2 are on the primary slip plane which is parallel to the plate plane, while those marked with da1 and da2 are on another primary slip plane which intersects the plate plane. Dislocations marked with GI are grown-in dislocations. Dislocations Da2 are half loops with a shape of a smoothly curved spearhead. From topographs taken under different 220 and 422 diffraction conditions, the Burgers vectors of the half loops were determined to be parallel to the elongated portion of the loops, indicating that those portions are screw dislocations. Moreover, a geometrical analysis of the nature of the half loops expanding towards the inside of specimen under the uniaxial compressive stress identifies the head parts of the loops to be dislocations of α type (i.e., the extra half-plane ends with a row of As atoms in the glide set model). Dislocations Da1 are elongated screw dislocations which remained after the leading parts (α dislocations) of the half-loops generated from the scratch had slipped out of the specimen.

Figure 2 shows a topograph of dislocations in Zn-doped GaAs formed under a stress of 5 MPa at 450 °C, in which the same notations as those used in Fig. 1 are adopted. Contrary to the case of Si-doped GaAs, two sets of dislocation groups are seen to be generated and have travelled toward the inside of specimen only from the scratch on the ($\overline{1}$00) surface. Dislocations marked with Db1 and Db2 are on the primary slip

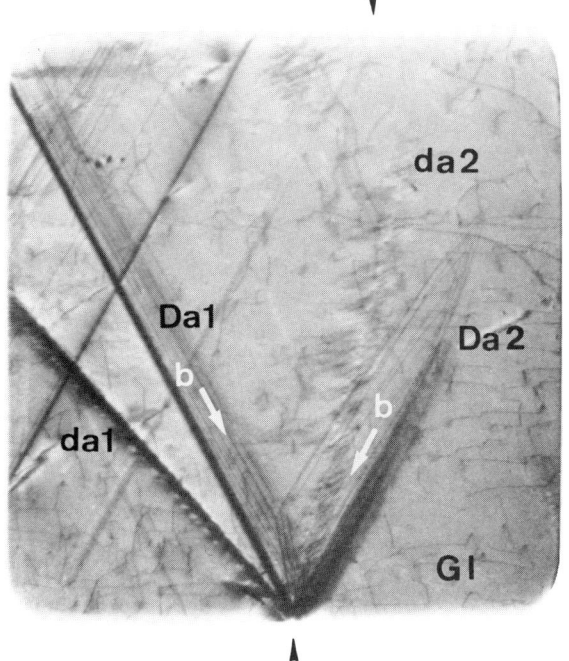

Fig. 1. X-ray topograph of dislocations in Si - doped GaAs stressed at 5 MPa for 20 min at 450 ℃. The front side is Ga (111) surface. b shows the Burgers vector of dislocations.

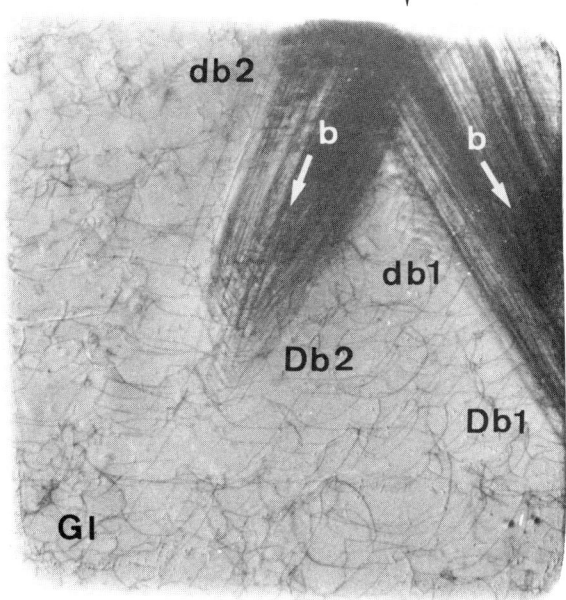

Fig. 2. X-ray topograph of dislocations in Zn - doped GaAs stressed at 5 MPa for 5 min at 450 ℃. b shows the Burgers vector of dislocations.

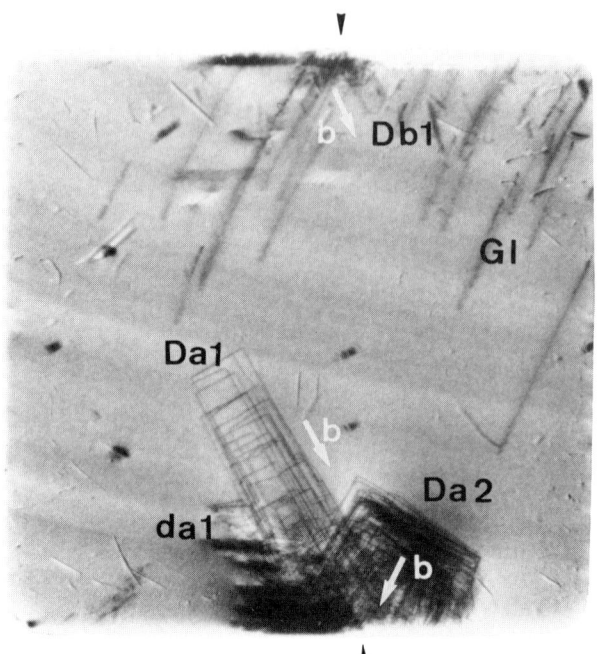

1 mm

Fig. 3. X-ray topograph of dislocations in In-doped GaAs stressed at 5 MPa for 20 min at 450 ℃. b shows the Burgers vector of dislocations.

plane which is parallel to the plate plane. Those marked with db1 and db2 are on another primary slip plane. Dislocations Db2 are half loops with the shape of a smoothly curved spearhead. The diffraction analysis again shows that the Burgers vectors of the dislocation loops are parallel to the elongated portions of the loops. The geometrical analysis in this case identifies the head part of the half loops as of β type (i.e., the extra half plane ends with a row of Ga atoms in the glide set model). Dislocations Db1 are screw dislocations remaining after the leading parts of loops of β type had slipped out of the crystal.

The morphology of fresh dislocations observed above reflects the difference in the velocities of relevant dislocations. Topographs in Figs. 1 and 2 suggest that α dislocations move faster than screw and β dislocations in Si-doped GaAs while β dislocations move faster than screw and α dislocations in Zn-doped GaAs. These characteristics in dislocation velocities are in good agreement with the results obtained with direct measurements of dislocation velocities in GaAs doped with these impurities (Yonenaga and Sumino 1989). It must be noted that α or β portions of leading parts of half loops in GaAs doped with Si or Zn are not in the exact 60° orientation.

3.2. Dislocations in GaAs doped with In

The characteristics of the morphology of dislocations formed in In-doped GaAs stressed under a high stress of 20 MPa at a low temperature of 300 ℃ is rather similar to those in Figs. 1 and 2 except that the head parts of half loops assume the shape of a half hexagon. Two sets of groups of dislocation half loops are generated and travel toward the inside of specimen only from the scratch on the (100) surface.
    According to geometrical considerations, the morphology is achieved by the generation and motion of dislocations of α type, in preference to other types of dislocations, trailing screw dislocations.

In contrast, a different type of dislocation morphology was produced in In-doped GaAs under a low stress but at a high temperature. Figure 3 shows a topograph of fresh dislocations in In-doped GaAs stressed under 5 MPa at 450 °C. Dislocation half loops were generated from the scratch on both the (100) and ($\bar{1}$00) surfaces of the specimen. Half loops generated from the scratch on the ($\bar{1}$00) surface denoted with Db1 assume the rather regular shape of a half hexagon and indicate that the β and screw portions have moved at comparable velocities. It is noticeable that the α portions of half loops generated from the scratch on the (100) surface are not in the 60° orientation but nearly in the edge orientation.

Under a stress of 3 MPa at 450 °C, dislocation loops are generated and travel from the scratch on the ($\bar{1}$00) surface by means of the motion of β and screw dislocations, while α dislocations do not penetrate into the interior of the specimen. This observation is consistent with the findings in earlier papers (Yonenaga and Sumino 1987, 1989) which showed that α dislocations were selectively immobilized by locking due to In solute atoms.

4. DISCUSSION

It is generally accepted that the stable dislocations in a semiconductor crystal with diamond or sphalerite structure are those with the line orientation parallel to < 110 > directions. Namely, 60° dislocations and screw dislocations are energetically stable and two types of 60° dislocations, α and β types, are to be distinguished according to the core structure in a sphalerite-type crystal. It is usually assumed that a dislocation in a semiconductor crystal glides by the mechanism of double kink formation and, as a consequence, a loop moving on the slip plane consists of straight segments parallel to < 110 > directions and some number of kinks. A number of papers reported a similar hexagonal shape for moving dislocations in silicon. Such a dislocation shape was also observed for portions of the slow type of dislocations in some III-V compounds. However, the morphology of dislocations of the fast type as well as that of dislocations immobilized by impurity observed in the present work is remarkably different from that observed for the slow type of dislocations.

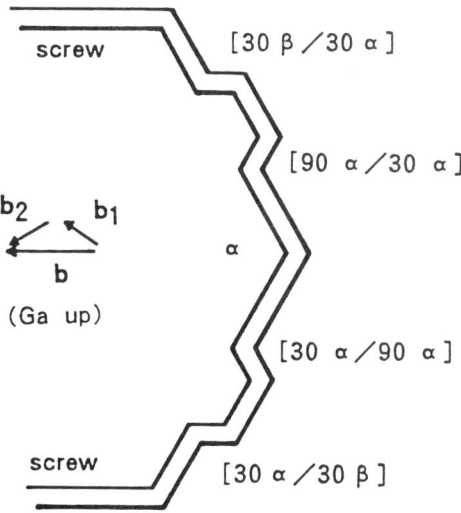

Fig. 4. Geometry of kinks on an α dislocation in GaAs.

First, we discuss how the shape of a smoothly curved spearhead is realized for the leading parts of half loops in GaAs doped with Si or Zn. Though a kink on a dislocation has a length of the order of one interatomic distance, it may have some extended structure since dislocations which are connected by the kink are extended. Such a model has been pictured by Hirsch (1979) for Si and by De Cooman and Carter (1989) for GaAs. Figure 4 schematically shows kinks on an α segment of an hexagonal dislocation loop in GaAs, where, for convenience, the kinks as well as the dislocation segments are drawn along the < 110 > directions. Kinks are denoted with the notations such as [90 α / 30 α] which means a 60° kink consisting of a leading 90° partial kink of α type and a trailing 30° partial kink of α type. If [90 α / 30 α] and [30 α / 90 α] kinks move faster than [30 β / 30 α] and [30 α / 30 β] kinks, the morphology of the leading parts of half loops characteristic of Si-doped GaAs is reasonably explained. In a similar way, the morphology of leading parts of loops of Zn-doped GaAs can be explained if [90 β / 30 β] and [30 β / 90 β] kinks move faster than [30 α / 30 β] and [30 β / 30 α] kinks. It has experimentally been established that an α dislocation (consisting of a 90° α partial and a 30° α partial) has a much higher mobility than a screw dislocation (consisting of a 30° α partial and a 30° β partial) in Si-doped GaAs, while a β dislocation (consisting of a 90° β partial and a 30° β partial) has a much higher mobility than a screw dislocation in Zn-doped GaAs. The above differences among the mobilities of various kinds of kinks are highly conceivable if the elementary process of kink pair formation and that of kink migration involve some common atomic and/or electronic processes.

On the other hand, α dislocations in In-doped GaAs formed at high temperature under low stress are observed to be in the edge orientation. Such a morphology may be realized during glide of the dislocations if [90 α / 30 α] and [30 α / 90 α] kinks move slower than [30 β / 30 α] and [30 α / 30 β] kinks. However, an α dislocation has been observed to move faster than a screw dislocation in In-doped GaAs (Yonenaga and Sumino 1989). Thus, the morphology seems not to be explained in terms of the difference of the mobilities among various kinds of kinks. In atoms in a GaAs lattice are known to segregate preferentially along the α portion of dislocations and immobilize them (Yonenaga and Sumino 1987, 1989). We think that the α dislocations were not in the edge orientation when they were in motion as supposed from the observations under high stress. An α dislocation which has ceased moving is thought to have turned into the edge orientation since it reduces the free energy of the system by increasing the interaction energy of the dislocation and In impurity and also by shortening the length of the α dislocation.

ACKNOWLEDGEMENTS

Crystals used in the present work were offered by Mitsubishi Chemical Company and Sumitomo Electric Industry Ltd. to which the authors express their gratitude.

REFERENCES

Burle-Durbec N, Pichaud B and Minari F 1987 Phil. Mag. Lett. 56 173
De Cooman B C and Carter C B 1989 Phil. Mag. A60 245
Di Persio J and Abbas M 1989 Inst. Phys. Conf. Ser. No.100 pp 391 − 6
Di Persio J and Kesteloot R 1983 J. Physique 44 C4 − 283
George A, Jacque A and Coquille R 1985 Inst. Phys. Conf. Ser. No.76 pp 439 − 44
Hirsch P B 1979 J. Physique 40 C6-117
Yonenaga I and Sumino K 1987 J. Appl. Phys. 62 1212
Yonenaga I and Sumino K 1989 J. Appl. Phys. 65 85

Inst. Phys. Conf. Ser. No 117: Section 3
Paper presented at Microsc. Semicond. Mater. Conf., Oxford, 25–28 March 1991

135

The indentation of GaAlAs and GaInAs at room temperature

R Haswell, P Charsley and U Bangert

Department of Physics, University of Surrey, Guildford, Surrey, GU2 5XH, UK.

ABSTRACT: The occurrence of microtwinning in the <110> rosette arms is compared, after indentation of the (001) surface, for $Ga_{0.47}In_{0.53}As$ and $Ga_{0.7}Al_{0.3}As$. The effect of doping in the GaAlAs alloy has been studied and the orientation of the {111} twinning planes determined. It is concluded that p-doping and/or the substitution of In for Al changes the twin plane orientations compared with n-doped GaAlAs.

1. INTRODUCTION

The plastic deformation due to the indentation of an {001} surface of GaAs is predominantly in the form of rosettes along the <110> directions. These rosettes do not show 4-fold symmetry (Warren, Pirouz and Roberts, 1984, and Lefebvre, Androussi and Vanderschaeve, 1987). The TEM studies (Höche and Schreiber, 1984, and Lefebvre et al, 1987) have revealed microtwins along one of the two <110> rosette directions. We have previously shown that when GaAs is alloyed with AlAs there is a tendency to form single stacking faults in the other rosette arm for n-doped material (Haswell and Charsley, 1989) and more recently (Haswell, Bangert and Charsley, 1991) that p-doped GaAlAs alloys as well as n-doped GaInAs alloys form microtwins along both <110> rosette directions.

In this paper we report on the nature of the stacking faults which form the microtwins and the orientation of the {111} fault planes in relation to the indented surface. Etching has been used to distinguish between (GaAl) and As {111} planes but this technique for differentiating between {111} and {$\bar{1}\,\bar{1}\,\bar{1}$} has not yet been used for the GaInAs alloys.

2. EXPERIMENTAL METHODS

These have been discussed in more detail in Haswell et al (1991). The indented face is given the indices (001), this surface being the exit surface for the electron beam. The (111) and ($\bar{1}\,\bar{1}$ 1) are (GaAl) planes which intersect the (001) surface converging below the indenter along [$\bar{1}$ 10] assuming a right-hand co-ordinate system. The As planes are (1 $\bar{1}$ 1) and ($\bar{1}$ 1 1) which intersect the surface converging below the indenter along the [110] direction. These planes and directions are shown diagrammatically in Haswell et al (1991).

The specimens were grown by MOCVD and an indentation load of 5g was used, with a Vickers diamond indenter, at room temperature. TEM specimens were produced by chemical thinning from the un-indented surface; all microscopy was at 200kV using a JEOL 2000FX microscope.

3. RESULTS

Using the comparison of edge fringes in bright-field and centred dark-field the natures of the stacking faults and the slopes of the habit planes in relation to the indentation have been determined. In the case of $Ga_{0.47}In_{0.53}As$ (n-type) the micrographs in **Fig 1** show sets of overlapping stacking faults in the rosettes parallel to the [110] and [1$\bar{1}$0] directions. The area containing the indentation has not been included; a mosaic at lower magnification of the whole region has recently been published, Haswell et al (1991). It should be noted here that the mosaic shows that the extent of these overlapping sets of faults is approximately equal in the two arms of the rosette and that extended single stacking faults are also visible. We consider that the overlapping faults are in fact microtwins because in other cases, where selected area electron diffraction was easier, the expected additional twin spots have been detected, as described in Haswell et al (1991).

In **Fig 2** bright and dark-field images of a microtwin are shown, in n-doped $Ga_{0.7}Al_{0.3}As$ in the [110] rosette arm. The fringes at positions determined by $(3q + 1)$, where q is an integer which defines the stacking fault within the sequence, are intrinsic in nature so that the microtwins are formed from intrinsic stacking faults on adjacent planes - these planes converge below the indentation. Although single stacking faults are observed in the [1 $\bar{1}$ 0] rosette arm there are no microtwins. In the case of p-doped $Ga_{0.7}Al_{0.3}As$, microtwins are formed in both rosette arms to approximately equal extents; stacking faults are also more frequent. The microtwins are formed, as before, from intrinsic faults but the slope of the twinning planes in relation to the indentation in the [110] arm is changed, compared with the n-doped material so that they diverge below the surface. In the [1 $\bar{1}$ 0] arm the planes converge below the indentation. We have not distinguished, at present, the {111} set of planes from the {$\bar{1}$ $\bar{1}$ $\bar{1}$} set for the GaInAs material, however, we observe the same difference in slope between the rosette arms as seen in p-doped GaAlAs. It would seem probable that the orientations for GaInAs are similar to p-doped GaAlAs but this point remains to be determined.

4. DISCUSSION

The effect of changing the doping from n-type to p-type has a much more dramatic effect on microtwin formation than does the replacement of a fraction of the Ga atoms by Al atoms in GaAs. This has been established for Al compositions up to $Ga_{0.7}Al_{0.3}As$. However, n-doped $Ga_{0.47}In_{0.53}As$ shows a very similar behaviour to the p-doped GaAlAs alloy. The changes in orientation of the twinning planes in relation to the indentation can be considered to result from changes in the core atom type of the leading partial dislocations. This has been confirmed by our analysis of the magnitudes and signs of the Burgers vectors of the dislocations involved, which will be described in detail elsewhere. These Burgers vectors of the 1/6 <112> type are consistent with the formation of microtwins from intrinsic stacking faults on adjacent planes. The model described by Pirouz (1987) would seem to be consistent with our observations in this respect. There is a marked tendency for the production of only one microtwin in any of the <110> rosette arms in which twinning is observed; this is particularly true for the GaAlAs alloys where the origins of the microtwins appear to be close to the corners of the indentations. This supports the idea that high stresses with specific directions are involved in the nucleation of the microtwins and that these nucleation conditions are not present elsewhere. The Pirouz model requires a large stress on both the

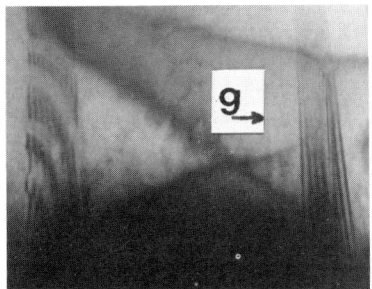 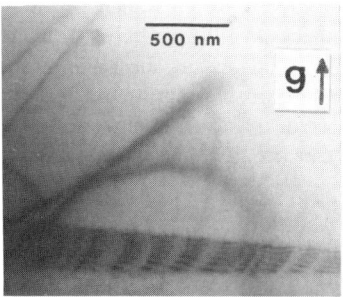

Fig 1 Bright Field $Ga_{0.47}In_{0.53}As$
$g = 2\bar{2}0$ (on left) $g = 220$ (above).

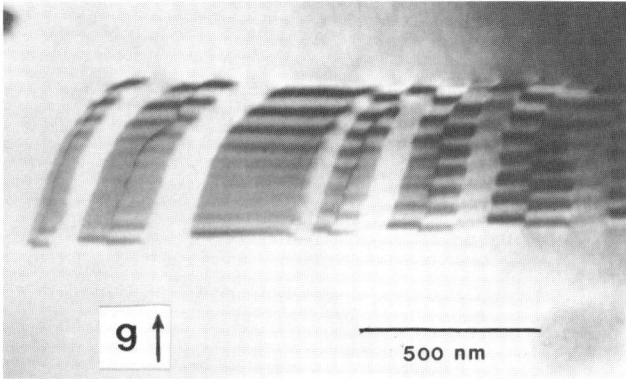

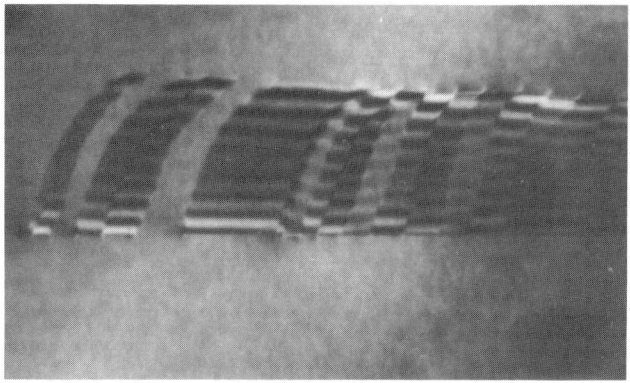

Fig 2 Bright Field (top) and Centred Dark Field (bottom) $Ga_{0.7}Al_{0.3}As$ $g = 220$

primary and cross-slip systems and such a stress system is most easily achieved near the indentation corners. We do not observe microtwins which are continuous along the positive and negative directions of any rosette arms as suggested by Lefebvre et al (1987). This can be explained by supposing that the nucleation sites are at the indentation corners where the appropriate stresses exist. The suggested picture due to Lefebvre et al (1987) implies twin nucleation further out from the indentation.

The changes in twin plane orientation when the doping is changed from n-type to p-type can be understood on the basis that the leading partial dislocations have Burgers vectors with edge components which have the extra half-plane on the same side of the slip plane as the indentation. The slopes of the twin planes along $\pm$ [1 1 0] correspond to the As(g) partials being faster than Ga(g) partials for n-doped material (Alexander and Gottschalk, 1989); this situation is reversed for p-doped material with corresponding changes in the slopes. Similar changes in slope can be accounted for on the same basis for the $\pm$ [1 $\bar{1}$ 0] rosette arms.

REFERENCES

Alexander H and Gottschalk H, 1989, Structure and Properties of Dislocations in
    Semiconductors, Institute of Physics Conference Series No 104, 281.
Haswell R and Charsley P, 1989, Philos Mag Lett, 59, 165.
Haswell R, Bangert U and Charsley P, 1991, Philos Mag Lett, 63, No 1.
Hoche H R and Schrieber J, 1984, Phys Stat Sol (a), 86, 229.
Lefebvre A, Androussi Y and Vanderschaeve G, 1987, Phys Stat Sol (a), 99, 405.
Pirouz P, 1987, Scripta Met, 21, 1463.
Warren P D, Pirouz P and Roberts S G, 1984, Philos Mag A, 50, L23.

Inst. Phys. Conf. Ser. No 117: Section 3
Paper presented at Microsc. Semicond. Mater. Conf., Oxford, 25–28 March 1991

The measurement of the rigid body translation across the {111} APB facet in GaAs

Stuart McKernan, D René Rasmussen and C Barry Carter

Department of Materials Science and Engineering, Cornell University, Ithaca, NY 14853.

ABSTRACT: At planar boundaries a relaxation of the atoms at the boundary plane will generally occur. This relaxation consists of a local re-arrangement of the atoms close to the boundary, and a rigid-body translation which affects the relative position of the grains far from the boundary. The rigid-body translation associated with a {111} APB facet in GaAs, a non-stoichiometric facet consisting entirely of one type of anti-site bonds, has been determined. For the Ga-Ga configuration studied, the relaxation is normal to the facet plane, and corresponds to an expansion of 0.0098 ± 0.002 nm.

1. INTRODUCTION

At a planar boundary a relaxation of the atoms at the boundary plane will generally occur. This relaxation of the boundary consists partly of a local re-arrangement of the atoms close to the boundary plane, and partly of a rigid-body translation which affects the relative position of the grains far from the boundary. An accurate experimental determination of this translation provides an important constraint for subsequent atomistic structure calculations. For antiphase boundaries in GaAs several different classes of boundary are possible depending on the stoichiometry of the boundary plane. Type I boundaries (such as a {110} facet) contain equal numbers of Ga-Ga and As-As anti-site bonds, type II boundaries (such as a {111} facet) consist solely of either Ga-Ga or As-As bonds, and type III boundaries (such as a {112} facet) contain unequal numbers of Ga-Ga and As-As bonds (Holt 1969). The relaxation associated with each of the different boundary types may be expected to be different. In particular, the rigid-body translation associated with each type may be expected to show some variation with the changing volume of the grain boundary as a result of the different stoichiometry. The rigid-body translation associated with the type I {110} APB facets has recently been determined by Rasmussen et al. (1991) by matching experimental intensity profiles from {110} APB facets to simulated image profiles. The rigid-body translation in this case was found to lie in the APB plane, with a magnitude resulting in the Ga-Ga and As-As separations at the interface being very close to that predicted purely on the basis of atomic radii.

This method is applied to determine the magnitude of the rigid-body translation across other antiphase boundary facets in GaAs. In particular, to the type II non-stoichiometric {111} facets in GaAs. The {111} facets occur much more infrequently than the {110} facets, but because of the extreme difference in stoichiometry of the interface plane, a substantially different relaxation may be expected at this facet.

2. EXPERIMENTAL

Epilayers of GaAs grown on (001) Ge substrates by MOCVD were provided by Dr D.K. Wagner (Cho et al. 1985). TEM specimens were prepared by cutting 3 mm-diameter discs. The discs were then mechanically polished from the Ge side and thinned from the same side by ion-milling with Ar+ at 4 kV until perforation. Bright-field images and high-resolution images were formed using a JEOL 4000EX at 200 kV. The thickness of the region containing the APBs was determined either geometrically or by CBED technique. The absolute polarity of the GaAs on either side of the APB was determined by the "bright/dark cross" technique described by Taftø and Spence (1982). A "test negative", where regions of the image plate were exposed to the electron beam for various times, was processed along with the actual micrograph, thus allowing a calibration of the film density in terms of the actual transmitted electron-beam intensities. This procedure is essential since the response of the plate emulsion to the electron beam is not linear. The calibration curve was used to modify the experimental profile before matching to the calculated image. The experimental profiles were averaged over a small length of the boundary to reduce the noise using the SEMPER image processing program. Simulated intensity profiles were calculated for the experimental conditions used with the COMIS program (Rasmussen et al. 1991). The calculation geometry used a four-beam, column approximation, which was shown in earlier studies to produce acceptable matches between simulated and experimental profiles.

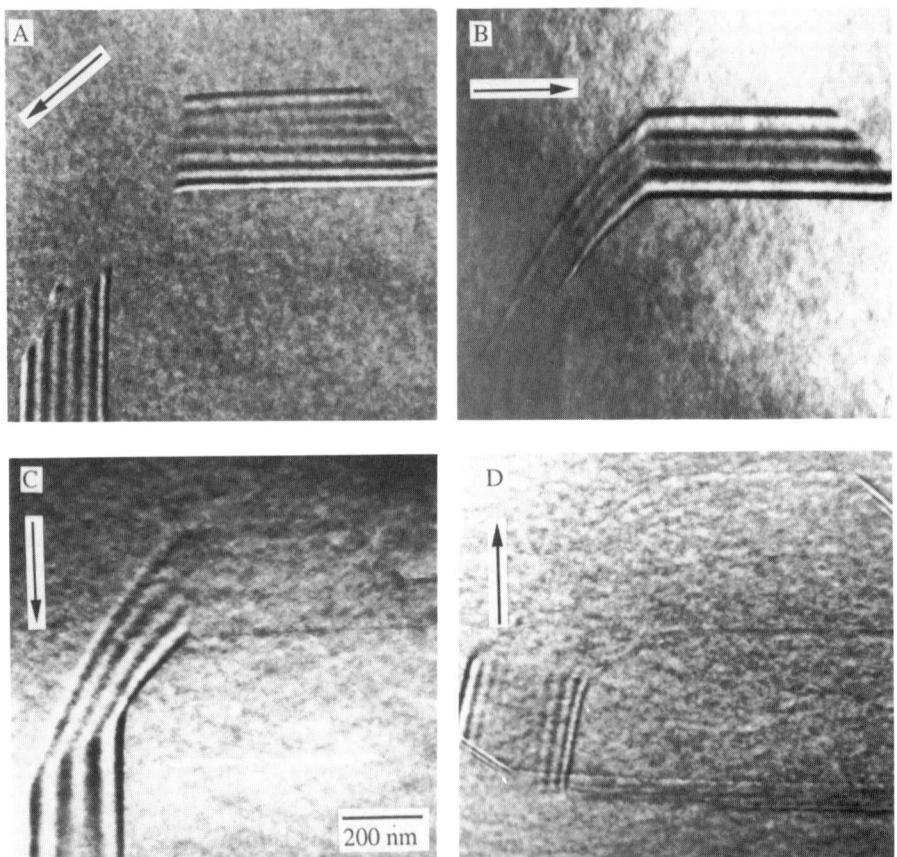

Figure 1 Electron micrographs of a (111) APB facet in GaAs recorded under four different 2-beam conditions: A) $\bar{2}20$ dark-field, B) $0\bar{4}0$ bright-field, and C) $\bar{4}00$ dark-field images all recorded close to the (001) pole and D) $\bar{2}02$ bright-field image recorded close to the (101) pole. The (111) facet is out of contrast in images A and D, whereas the neighbouring {110} facets are visible.

3. RESULTS

Several images of a {111} APB are shown in fig.1. The crystallographic orientation of the boundary was determined from the comparison between the image and the corresponding selected-area diffraction pattern including tilting the interface to an edge-on orientation. With the facet tilted edge-on, it can be seen that the interface is not exactly flat, but curves away from the (111) plane slightly. It is closest to the (111) plane where it meets the horizontal (101) facet in figure 1. All the comparisons were therefore performed on this area. The 2-beam fringes were found to vanish for several different **g**-vectors (e.g. figure 1 A and D). The direction of the relaxation was thus determined, from the **g.R** =0 criterion, to be normal to the APB plane. The relaxation is therefore in the form of a pure expansion or contraction with no shear component present. To determine the magnitude, R, of the translation across the APB, the experimental fringe profiles were compared with image simulations. In general, a phase change is experienced by the electrons on crossing the APB; it is produced by a combination of the rigid-body translation (an effect similar to that produced by a stacking fault) and the reversed polarity of the crystals. The effect of the inversion has been described by Holt (1969) for sphalerite structures, and was incorporated in the simulations for the reflection used. The simulation of the images requires a knowledge of the anomalous absorption coefficient and the extinction distance for the reflection used, and the appropriate specimen thickness; all these quantities must be determined experimentally.

For comparison between the simulated and experimental micrograph, the experimental micrographs were digitized, and intensity profiles were obtained by averaging along the fringes over a length of ~30 nm. The averaged profiles were then normalized using the intensity calibration data. For each diffraction condition, bright-field and dark-field images were recorded under identical conditions. The deviation parameter was then determined for the exact location of the fringes by matching the ratio of the contrast of the outermost fringes at the top and bottom exit surfaces of the APB in the dark-field image, Δ_g. The contrast of the outermost fringes at the top and bottom exit surfaces of the APB in the bright-field image, Δ_0 is also evaluated. To minimize the effect of absorption on the result, the quantity ρ is used to match the simulated and experimental images, where ρ is given by:

$$\rho = \Delta_0 + \Delta_g / 2\Sigma$$

where Σ is the sum of the background intensities in the bright-field and dark-field images.

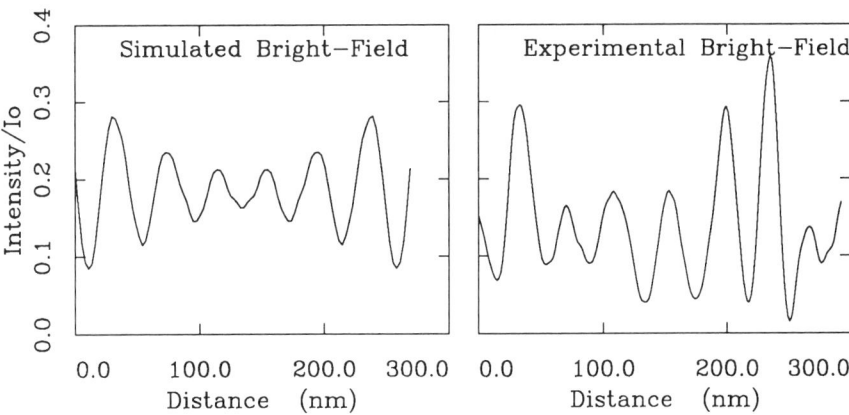

Figure 2 Comparison of experimental and simulated profiles for the $\bar{2}\bar{2}0$ bright-field fringes. A uniform background ramp has been removed from the experimental image.

A value is estimated for R and simulations are performed to obtain a close match with the experiment by varying the anomalous absorption coefficient. The best value of R is then obtained by scaling to produce an exact match of the quantity ρ. This analysis can be performed independently on the outermost fringes at top and bottom exit surfaces of the APB, giving a check on the result. The best value for R obtained using this analysis is 0.0098 ± 0.002 nm. However, since this value depends on so many independent quantities, the precise matching of simulated and experimental profiles is still in progress. A comparison of the experimental and simulated profiles is shown in figure 2. The experimental profile appears rather rough, as the averaging was only performed over a short segment of the facet, since the facet was curving away from the (111) plane, giving rise to a geometrical variation in the projected width of the facet. This curvature may also be responsible for the slight variation of the fringe appearance along the facet.

4.   DISCUSSION

Despite the fact that the facet examined curves away from the (111) plane, and the specimen is slightly bent so that the background intensity varies across the facet, a reasonably consistent value for the rigid-body translation across the APB has been obtained. The rigid-body translation corresponds to an expansion across the APB plane of 0.0098 ± 0.002 nm. The size of this expansion is very close to that which would be predicted by the hard sphere model. The Ga-As bond length (0.245 nm) is almost identical to the sum of the Ga and As tetrahedral radii in their elemental form (0.126 and 0.118 nm respectively). The Ga-Ga bond would therefore be 0.252 nm long; an expansion of 0.007 nm. Since there is no shear component to the relaxation, the bond angles of the atoms at the boundary are all preserved. This is similar to the results for the rigid-body translation across the {110} facets, where the bond lengths across the boundary were close to the predicted hard-sphere values. In this case however, since the translation involved was a pure shear, the bond angles at the boundary suffered some distortion.

5.   CONCLUSIONS

The rigid-body translation across a Ga-Ga (111) APB facet has been determined. This translation is an expansion at the boundary of magnitude 0.0098 ± 0.002 nm. This result is consistent with the value obtained on the basis of a hard-sphere model.

ACKNOWLEDGMENTS

The authors would like to thank Mr R. Coles and Ms M. Fabrizio for technical support. The electron microscopes are part of a central facility provided by the Materials Science Center at Cornell and are supported, in part, by the National Science Foundation. This research is supported by the DoE under grant No DE-FG02-89ER45381.

REFERENCES

Cho N-H, De Cooman B C and Carter C B, Appl. Phys. Lett. 1985 **47** 879
Head A K, Humble P, Clarebrough L M, Morton A J, Forwood C T, Computed Electron
    Micrographs and Defect Identification, 1973, Amsterdam: North Holland.
Holt D B, J. Phys. Chem. Solids 1969 **30** 1297
Rasmussen D R, McKernan S, and Carter C B, Phil. Mag. 1991 in press
Rasmussen D R, and Carter C B, J. Elect. Micr. Tech. 1991 in press
Taftø J and Spence J C H, J. Appl. Cryst. 1982 **15** 60

Inst. Phys. Conf. Ser. No 117: Section 3
Paper presented at Microsc. Semicond. Mater. Conf., Oxford, 25–28 March 1991

Excitation-enhanced mobility of dislocations in ZnS: a TEM *in situ* study

C Levade, G Vanderschaeve, J J Couderc, A Faress and D Caillard[*]

Laboratoire de Physique des Solides, INSA, Avenue de Rangueil, 31077 Toulouse Cedex, France.

(*) CEMES-LOE , B.P. 4347, 31055 Toulouse, France.

ABSTRACT : The mobility of dislocations under stress in ZnS has been determined by TEM in situ straining experiments in the temperature range 300-520 K. At constant stress (τ) , temperature (T) and electron beam intensity (I), the velocity (v) of a dislocation segment is proportional to its length (L). The dislocation mobility is strongly enhanced by electron excitation. For $I \leq 1000$ A/m^2, the dislocation velocity obeys a general law $v = v_0(\tau,T)L\ I$, but at higher intensities, this law is no longer valid. The activation energy for dislocation glide under irradiation is smaller than for non-irradiation (I = 0).

1. INTRODUCTION

An increase in dislocation mobility under the electron beam has been observed in a variety of large band gap semiconductors by scanning electron microscopy in the cathodoluminescence mode (for a recent review see Maeda et al. 1990). As pointed out by these authors, the radiation enhanced dislocation glide exhibits features expected from the recombination enhanced defect motion (Weeks et al. 1975, Sumi 1984) : i.e., the dislocation motion is assisted by the energy released upon non-radiative recombination of electron-hole pairs at the defect.

Dislocation glide in semiconductors is governed by nucleation and migration of kinks. According to Hirth and Lothe (1968), two regimes of dislocation velocity are expected when dislocations are submitted to lattice friction, depending on whether the length of moving segments (L) is much smaller, or much larger than the mean free path of kinks (X) along the dislocation line. In the first case, the dislocation velocity is proportional to the dislocation length, whereas in the second one (i.e. when kink-kink collision occurs) it is length independent.

In their recent analysis of the excitation-enhanced dislocation mobility in semiconductors, Maeda and Takeuchi (1989) considered that lattice vibrations induced by non-radiative capture of a carrier by a deep-level defect could greatly enhance either the nucleation or migration of kinks, through the reduction of the corresponding activation energies : to account for the experimental results, and particularly for the linear relationship between the dislocation velocity (v) and the electron beam intensity (I), they concluded that the enhancement of kink migration, even if it occurs, does not contribute to the total dislocation mobility enhancement. (Notice that recent intermittent loading experiments by Maeda et al (1990) provide evidence that kink migration is not affected by electron excitation). Maeda and Takeuchi (1989) suggested that, experimentally, the condition X >> L was actually realized (i.e. kink-kink collision does not occur), but no definite proof was given. Indeed in most experiments dislocation velocities are averaged over travel distances of a few 10 μm, and the dislocation behaviour in the early stages of source working (short dislocation segments) cannot be analyzed. That is why we have performed in situ straining experiments in an electron microscope on II-VI ZnS microsamples at different electron beam intensities and the results of this work are reported in the present paper.

2. EXPERIMENTAL PROCEDURES

The material used in this study was natural single crystals of sphalerite ZnS. Fe (content $\leq$ 0.07 at.%) and Cd (content $\leq$ 0.03 at.%) were the main impurities, as detected by electron microprobe analysis ; EPR analysis revealed also the presence of Mn^{2+}ions. In order to introduce a sufficiently high density of fresh dislocations, the crystals were first 1% prestrained at 620 K by uniaxial compression. Tensile microsamples (foil surface (011); tensile axis [21$\bar{1}$]) suitable for in situ straining experiments were made in the following way : slices were cut using a wire saw, mechanically polished with an abrasive powder and finally chemically thinned until perforation. The experiments were performed in a JEOL 200 CX electron microscope operated at 200 kV in the temperature range 300 - 520 K using a special tensile holder (Kubin and Veyssière 1982) and video recording.

As the dislocation behaviour was observed to be strongly influenced by electron irradiation, the intensity of the electron beam was varied in our experiments by changing the excitation of the condenser lens of the electron microscope. The beam intensity was measured using the photosensitive cells located near the observation screen.

Finally, the value of the local shear stress acting on the moving dislocations was determined by measuring the radii of curvature of the curved parts of the dislocations (Hirth and Lothe 1968).

3. EXPERIMENTAL RESULTS

In general, the dislocation glide was observed to be smooth and continuous over all the studied temperature and intensity ranges. Quantitative dislocation velocity measurements have been performed at 450 K and 390 K. As the results of the 450 K experiments have already been published (Levade et al. 1990), we shall focus here on the new experimental results obtained at 390K.

3.1. Dislocation sources

Fig.1 shows a single-ended dislocation source which rotates counterclockwise. It emits dislocations on the secondary slip system (111) 1/2 [01$\bar{1}$], the screw dislocations being almost parallel to the slip traces. The local shear stress, evaluated as described above, is found to be equal to 40 ± 12 MPa.

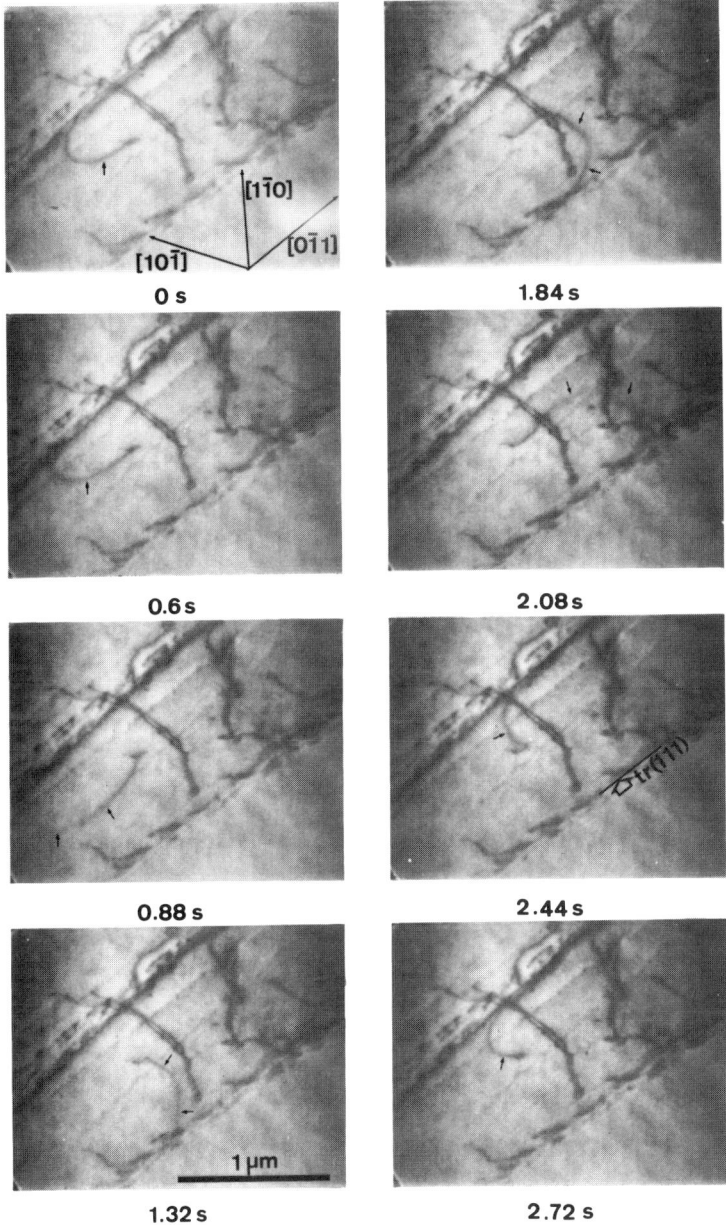

Fig. 1 Dislocation source in ZnS T = 390 K ; τ = 40 ± 12 MPa ; I = 5600 A/m^2.

3.2. Velocity measurements

At a given beam intensity I, measurements of the dislocation velocity were made for each straight segment. Care was taken to consider only segments that did not emerge at the sample surfaces.

Contrary to what is observed in the III-V compounds (Caillard et al 1987, 1989), no significant difference between the velocity of screw and 60° dislocations was observed (see also Levade et al. 1990).

It is noteworthy that the velocity v of the straight moving segment shows a linear dependence with its length L. This is illustrated in fig.2 for the screw parts of the source displayed in fig. 1 (I = 5600 A/m^2). Note that the V(L) line does not intercept the L axis at the origin but at some length L_o ($L_o \cong 0.12$ μm at 390 K). Finally, no saturation in dislocation velocity was observed in the studied length range.

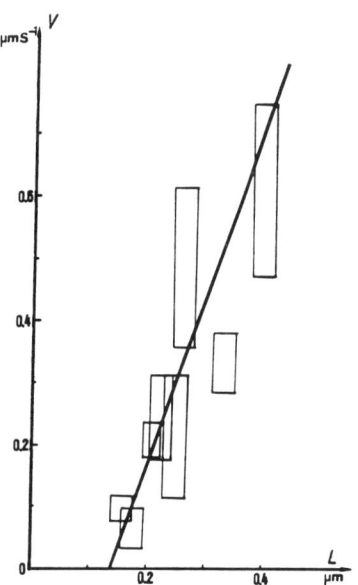

Fig. 2 Velocity of screw dislocations versus their length.
T = 390 K
$\tau = 40 \pm 12$ MPa
I = 5600 A/m^2

3.3. Cathodoplastic effect

In all the studied temperature range (300 - 520 K) and beam intensity range (35-5600 A/m^2) the dislocation velocity varies linearly with the length of the moving segment. Moreover, the dislocation velocity is strongly enhanced by electron excitation, as can be seen from fig.3 which reports the variation of the slope of the v(L) curve versus I at 390 K and 450 K. For I ≤ 1000 A/m^2, v/L varies linearly with I at both temperatures. For higher beam intensities, v/L (I) at 390 K deviates from a linear relationship and a better fit of the experimental results is obtained by considering a logarithmic dependence of v/L with I.

4. DISCUSSION

The viscous motion of dislocation segments indicates that lattice friction is acting on them during glide (Peierls regime). These friction forces are higher in darkness (i.e. I = 0) than under electron excitation, since dislocations are considerably slowed down in darkness.

Following the Hirth and Lothe (1968) analysis, the linear relationship between

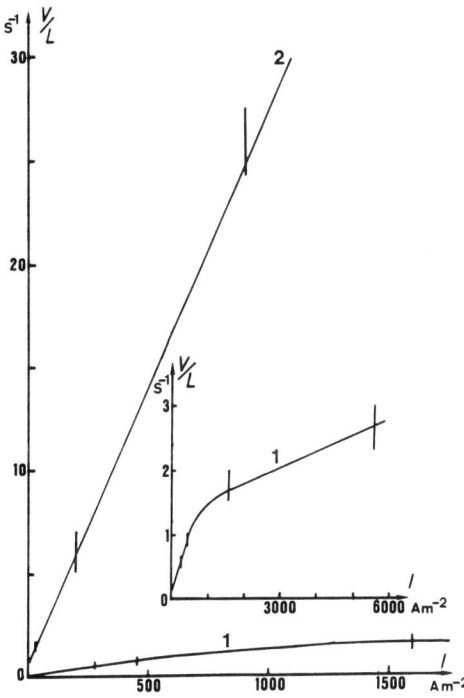

Fig. 3 Slope of the v(L) curve versus electron beam intensity. $\tau = 40 \pm 12$ MPa
Curve 1 : T = 390 K
Curve 2 : T = 450 K

v and $(L - L_o)$, as well as the lack of saturation in dislocation velocity, are consistent with a mean free path of kinks much larger than the length of the mobile segments; that is the kink velocity is high, either in darkness or under illumination. It is suggested that this is related to the pronounced ionic character of ZnS bonding which prevents from bond reconstruction.

As pointed out above, the v(L) straight line intercepts the L axis at some length L_o. L_o is probably related to the gradual transition between 60° and screw parts, which results in an accumulation of kinks at both ends of the moving segment, so that the effective length of the rectilinear segments is $(L - L_o)$. This again is consistent with a high mobility of kinks along the dislocation line.

Our results give the first experimental evidence for an enhancement of the dislocation glide by an electron excitation during the early stages of source working (i.e. when the dislocation motion occurs without kink-kink collisions). For $I \leq 1000A/m^2$, they quantitatively agree with Maeda and Takeuchi's analysis (1989). The dislocation velocity for non-irradiation ($I = 0$) being negligible, the dislocation velocity under excitation can be written as $v_I = v_0 (\tau,T)L\ I$. This is accounted for by considering that the enhancement of dislocation motion is promoted by the recombination of injected carriers at dislocations and that the recombination rate is proportional to the electron-hole pair generation rate, which is itself proportional to the electron beam intensity. At higher intensity levels, our results do not agree with this statement, since the slope of the v/L (I) curve strongly decreases with increasing I (fig. 3). This could be discussed in terms of the general theory of recombination enhanced defect motion : indeed Sumi (1984) pointed out the possibility of a saturation of the radiation enhanced defect reaction rate at high levels of injection currents (i.e. high electron beam intensity in the present case).

From the experiments performed at 390 K and 450 K, the activation energy ΔG_I for dislocation glide under electron excitation in ZnS can be determined ; indeed, $v_0(\tau, T)$ is proportional to $\exp(- \Delta G_I/kT)$. Of course this is done for $I \leq 1000A/m^2$ when the curves v/L (I) are linear. At $\tau = 40 \pm 12$ MPa , we found $\Delta G_I = 0.7 \pm 0.1$ eV for screw dislocations, which is of the same order of magnitude than the values reported by Maeda and Takeuchi (1983) for the radiation enhanced motion of dislocations in III-V compounds GaAs (β and screw dislocations) and InP (β dislocations). Due to the very slow motion of dislocations for non-irradiation ($I = 0$), a precise evaluation of the activation energy in this condition is rather difficult. A crude estimation gives $\Delta G_d \cong 1.2$ eV. According to Maeda and Takeuchi (1989) the reduction in activation energy corresponds to the energy released upon non-radiative capture of excited carriers at the straight dislocation sites (i.e, it assists the initial unit kink pair formation). It is correlated to the dislocation electronic energy level in the band gap.

REFERENCES

Caillard D, Clément N, Couret A, Androussi Y, Lefebvre A and Vanderschaeve G 1987 *Inst. Phys. Conf. Ser.* <u>87</u> 361 ; 1989 *Inst. Phys. Conf. Ser.* <u>100</u> 403
Hirth J P and Lothe J 1968 *Theory of Dislocations (New-York , McGraw-Hill)*
Kubin L P and Veyssière P 1982 *Proc. 10th Int. Cong. Electr. Microscopy* <u>2</u> 531
Levade C, Couderc J J, Caillard D and Couret A 1990 *Electron Microscopy in Plasticity and Fracture Research of Materials, Eds U Messerschmidt F Appel J Heydenreich and V Schmidt (Berlin Akad. Verlag)* p. 199
Maeda K and Takeuchi S 1983 *J.Phys.* <u>44</u> C4-375
Maeda K, Yamashita Y, Maeda N and Takeuchi S 1990 *Mat. Res. Soc. Symp. Proc.* <u>184</u> 69
Maeda N and Takeuchi S 1989 *Inst. Phys. Conf. Ser.* <u>104</u> 303
Sumi H 1984 *Phys.Rev.* B <u>29</u> 4616
Weeks J D, Tully J C and Kimmerling L C 1975 *Phys.Rev.* B <u>12</u> 3286

Inst. Phys. Conf. Ser. No 117: Section 3
Paper presented at Microsc. Semicond. Mater. Conf., Oxford, 25–28 March 1991

The role of dislocations in the 3C $<->$ 6H SiC polytypic transformation

P Pirouz[*], J W Yang[*], J A Powell[**], and F Ernst[***]
[*]Department of Materials Science and Engineering,
 Case Western Reserve University, Cleveland, OH 44106, U.S.A.
[**]NASA Lewis Research Center, Cleveland, OH 44135, U.S.A.
[***]Max-Planck-Institut für Metallforschung, Stuttgart, Germany.

ABSTRACT: Experimental evidence is presented for a dislocation mechanism of polytypic transformation in SiC.

1. INTRODUCTION

Polytypism is the occurrence of different crystal structures which are variants of a one-dimensional stacking sequence. It occurs readily in some compound semiconductors such as SiC or ZnS. The model of polytypic transformation considered in this paper was proposed recently (Pirouz, 1989a) and was a generalization of a twinning mechanism in semiconductors based on the different mobilities of partials in these materials (Pirouz, 1987,1989b). The present paper discusses some experimental evidence for this model. It should be emphasized that the model is kinetic and discusses a dislocation mechanism for transformation from one polytypic phase to another. It is not a thermodynamic model and does not consider the stability, or metastability, of different polytypes at different phase conditions (e.g. pressure and temperature). Hence, it assumes *a priori* that a phase transformation is thermodynamically possible and just discusses the micromechanism of transformation. However, recently, there has been considerable progress in theoretical considerations of the stability of different phases (see e.g. Heine, 1990).

Although the model under consideration is quite general and can explain any of the polytypic transformations, the evidence presented in this paper pertains to a particular transformation, namely 3C→6H or, possibly, 6H→3C in SiC. In the following, the model is briefly reviewed before the experimental work is presented. First, it is recalled that the stacking sequence of hexagonal 6H-SiC is ...*ABCACBABCACB*... while that of cubic 3C-SiC is ...*ABCABC*... where each letter represents two neighboring (0001) or {111} layers: a layer of Si and a layer of C separated by a distance $\sqrt{3}a/4$ where a is the lattice parameter of 3C-SiC.

2. BRIEF REVIEW OF THE MODEL

It is convenient to classify the polytypic transformations into two different types: (i) Non-cubic→cubic, and (ii) cubic→non-cubic. In this section we shall just review the models for the particular cases of 6H→3C [i.e. type (i)] and 3C→6H [i.e. type (ii)] in SiC which is the subject of the experimental observations discussed in section 3.

2.1. Dislocations in SiC: Glide dislocations in non-cubic (either the rhombohedral, R, or the hexagonal structures, H) SiC lie on the basal (0001) planes and have a Burgers vectors $\frac{1}{3}<11\bar{2}0>$. In the cubic phase they have a Burgers vector $\frac{1}{2}<1\bar{1}0>$ and lie on the {111} planes.

SiC is known to have a very low stacking fault energy (~2.5 mJ/m^2 in 6H-SiC; Maeda *et al.*, 1988), i.e. the dislocations in this material are dissociated into two widely separated partials bounding a ribbon of stacking fault. A perfect screw dislocation **AB** (Hirth & Lothe, 1982), with a $\frac{1}{3}<11\bar{2}0>$ Burgers vector (in 6H), or a $\frac{1}{2}<1\bar{1}0>$ (in 3C), is dissociated into two 30° partials with $\frac{1}{3}<10\bar{1}0>$ and $\frac{1}{3}<01\bar{1}0>$ (in 6H), or $\frac{1}{6}<2\bar{1}\bar{1}>$ and $\frac{1}{6}<1\bar{2}1>$ (in 3C), Burgers vectors, respectively. It is important to study the the core structure of these two 30° partials. Two points are noteworthy about these core structures: (i) the atoms along one partial are all Si [the Si(g) partial] while the atoms along the other partial are all C [the C(g) partial] and (ii) there are rows of broken bonds along each of the partials. Because of the high energy of a dangling bond, an unreconstructed configuration, which contains a row of them, would not be stable. As a result, and in analogy with other semiconductors, reconstruction of the broken bonds along a dislocation line is likely, specially since both C and Si are tetravalent. The reconstructed cores of the 30° partials in SiC are shown on the primary glide plane in Fig. 1.

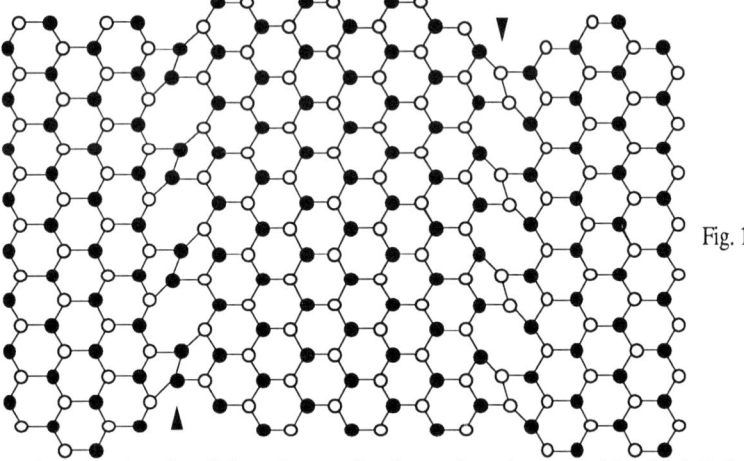

Fig. 1

The accepted mechanism for dislocation motion in semiconductors with a high Peierls barrier is kink nucleation and kink migration. Both of these processes involve the breakage of bonds along the dislocation line and re-formation of bonds in the next Peierls valley (primary Peierls valley for kink pair nucleation and secondary Peierls valley for kink migration). Note that in Fig. 1, motion of one of the partials involves the sequential breakage (and re-formation) of Si-Si bonds, while the motion of the other partial involves the breakage (and re-formation) of C-C bonds. Using a crude analogy of the comparative bond strength of bulk diamond versus bulk silicon, it may be reasonably expected that the formation and migration of kinks along the Si(g) partial is much easier than that along the C(g) partial. It follows that the thermal energy required for the motion of Si(g) partial is much less than that required for the motion of the C(g) partial. In other words, the activation energy, E, for motion of these two partial dislocations is very different and $E_{Si} \ll E_C$. Consequently, it may be expected that the Si(g) partial would become mobile at much lower temperatures than the C(g) partial.

The high stacking fault densities in plastically deformed SiC may, in fact, just reflect the large difference in the mobilities of Si(g) and C(g) partials. Note that macroscopic plastic deformation of SiC by slip requires the motion of both partials because the motion of only one partial on the slip plane prevents the motion of subsequent partials and, thus, processes such as dislocation multiplication will not be possible (see section 3). Since the mobility of a dissociated dislocation is controlled by the mobility of the slow partial, the high temperatures (>1400°C) required for the plastic deformation of SiC may imply the high temperatures required for the onset of motion of the slow C(g) partial dislocation [see also Maeda *et al.* (1988) and Rabier and Boivin (1990)]. On the other hand, formation of stacking faults requires the motion of only a single partial, and since the movement of Si(g) partial has a low activation energy, it would be very simple to form wide stacking faults in SiC even at relatively low temperatures.

2.2. Multiplication of Partial Dislocations: In Fig. 2a a screw dislocation is shown pinned at points X and X' on the primary glide plane (pgp) which is a (0001) plane in 6H-SiC and a {111} plane in 3C-SiC. The dislocation is dissociated into two partial dislocations: the leading partial δA and the trailing partial $B\delta$ (Hirth & Lothe, 1982). It is assumed that the leading partial, δA, is the fast Si(g) 30° dislocation, while the trailing partial, $B\delta$, is the slow C(g) 30° dislocation. Under a sufficiently high resolved shear stress, τ_{pgp}, on the screw dislocation, the leading partial breaks away from the trailing partial and forms a faulted loop. After the formation of the faulted loop (Fig. 2b), the nature of partials is reversed, i.e. the leading partial is the slow C(g) and the trailing partial is the fast Si(g) dislocation. Hence, the screw dislocation in Fig. 2(a) is Si(g)/C(g) while the screw dislocation in Fig. 2(b) is C(g)/Si(g). It has been recently shown that, on the basis of the Escaig's mechanism (Escaig, 1968; Rabier and Boivin, 1990), such a situation is very favorable for the double cross-slip of the screw dislocation onto a neighboring pgp (Pirouz & Hazzledine, 1991). Hence after one faulted loop forms on a primary plane, the screw dislocation cross-slips on the cross slip plane (csp) until it finds an opportunity to double cross-slip back on a parallel pgp. Subsequent cycles of faulted loop formation and cross-slip to a next pgp will change the stacking sequence in a systematic manner and thus transform a given polytype into another. Different patterns of available primary glide planes will, accordingly, lead to different polytypes.

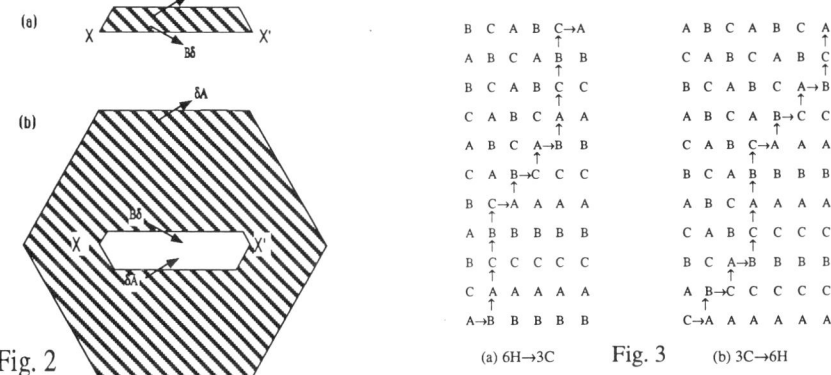

Fig. 2

(a) 6H→3C   Fig. 3   (b) 3C→6H

2.3. The 6H→3C polytypic transformation: Consider the stacking of double layers parallel to the basal plane in 6H-SiC as shown in the first column of Fig. 3a. In the second column, a faulted loop has formed on a basal plane by the mechanism of Fig. 2 thus shearing the atoms above the pgp from A sites to B sites. In Fig. 3, this is shown by a horizontal arrow. As mentioned before, after the formation of the faulted loop, because of compressive stresses on the two partials, the screw dislocation cross-slips onto the csp [the {1$\bar{1}$00} prism plane]. Cross-slip is shown by vertical arrows in Fig. 3. Because of a reversal in the nature of the screw dislocation from C(g)/Si(g) to Si(g)/C(g) (Pirouz & Hazzledine, 1991), and also because $\tau_{pgp} \gg \tau_{csp}$, there is a driving force on the screw dislocation to immediately cross-slip back on the basal plane, dissociate, and form another faulted loop. This is only possible after the screw dislocation cross-slips on the csp past three basal planes; a dissociation on any of these basal planes would result in a high energy *AA* type stacking. The first possible double-cross-slip is then shown by a horizontal arrow in column 3 shearing the atoms from the C-sites to the A sites. The sequence of dissociation, formation of a faulted loop, and the subsequent cross-slip of the screw dislocation to the next allowed basal plane, is shown in Fig.3a and, in a schematic fashion, in Fig. 4a.

2.4. The 3C→6H Transformation: The stacking sequence in a 3C-SiC crystal is shown in the first column of Fig. 3(b). In a 3C crystal, following the formation of a faulted loop on a certain (111) pgp, double cross-slip can take place on every neighboring (111) plane. In such a case, the crystal is transformed into a twinned variant (Pirouz, 1987). This is possible if the mobility of the leading partial is large enough on the pgp so that the faulted loops expand rapidly enough. If, however, the loops do not expand sufficiently rapidly, their back stress stops the operation of the source (Pirouz, 1989b). Once faulted loops form on *m* neighbouring (111) planes, the operation of the source stops until the loops expand far enough or, alternatively, the cross slip of the C(g)/Si(g) screw dislocation continues for *n* planes before it can double cross-slip back on a parallel pgp. In a 3C→6H transformation, *m=n*=3. These changes in the stacking sequence are shown in Figs. 3(b) and 4(b).

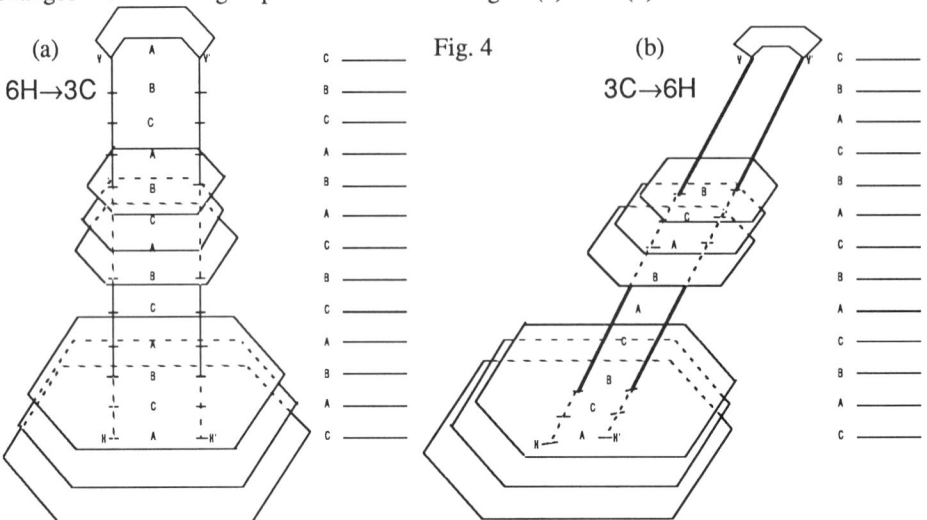

(a) 6H→3C   Fig. 4   (b) 3C→6H

2.5. The Residual Dislocation: The cross-slip of the screw dislocation on the csp leaves a residual dislocation on this plane which is shown in Figs. 4(a) and 4(b) (XYY'X'). The top part of the diagram illustrates the most recent stage in the formation of a faulted loop on the pgp. The Burgers vector of the residual dislocation is of course the same as the original screw dislocation, i.e. **AB**. Hence, the two segments XY and X'Y' are near-edge dislocations lying

on the csp, i.e. the $(1\bar{1}00)$ prism plane in 6H and the inclined $(\bar{1}11)$ plane in 3C crystal.

According to this model, the formation of a regular polytype requires a very uniform stress

distribution in the crystal such that τ_{pgp} and τ_{csp} are roughly constant throughout the region in which the transforming dislocation is active. If there are local non-uniformities in the resolved

shear stress on the pgp or the csp, for instance if, at a certain stage, τ_{csp} becomes equal to or

greater than τ_{pgp}, then cross-slip on the prism plane may continue without the screw dislocation switching back to the pgp even when it is allowed. Thus, the regularity of faulted loop formation on the appropriate pgp can be easily perturbed by local stress non-uniformities.

3. EXPERIMENTAL RESULTS AND DISCUSSION

In order to verify the mechanism of polytypic transformation outlined in §2, occurrence of residual dislocations lying on the csp with the appropriate Burgers vector should be sought.

High quality single crystal, single polytype, 1" wafers of 6H-SiC have recently become available through Cree Research Inc., U.S.A. Powell *et al.* (1991) have used such wafers as substrates for the growth of 3C-SiC or 6H-SiC by CVD. The growth temperature is typically 1450°C and the inlet gases are silane and propane. The 3C/6H interface in such systems is usually smooth and coherent with hardly any defects (Pirouz and Yang, 1990).

During a XHREM study of defects in the 3C-SiC films grown on Cree 6H-SiC substrates, we have been looking at dozens of specimens. In one of the samples (#1140), grown under a high nitrogen concentration, interweaved bands of 3C- and 6H-SiC were observed near the 3C/6H interface. An example of these bands is shown in Fig. 5 and an XHREM micrograph of a small part of this region is shown in Fig. 6. Unfortunately, it has not been possible to identify the initial location of the interface in these micrographs and, consequently, it is not known whether transformation had taken place in the 3C epilayer or the 6H substrate. In Fig. 5 a number of dislocations may be observed that cross the 3C/6H bands. Many of these dislocations were in the form of semi-loops. An example is shown in Fig. 7 where most of the region is 6H and the dislocation crosses a narrow band of 3C-SiC. Since the direction of

Fig. 5   Fig. 6

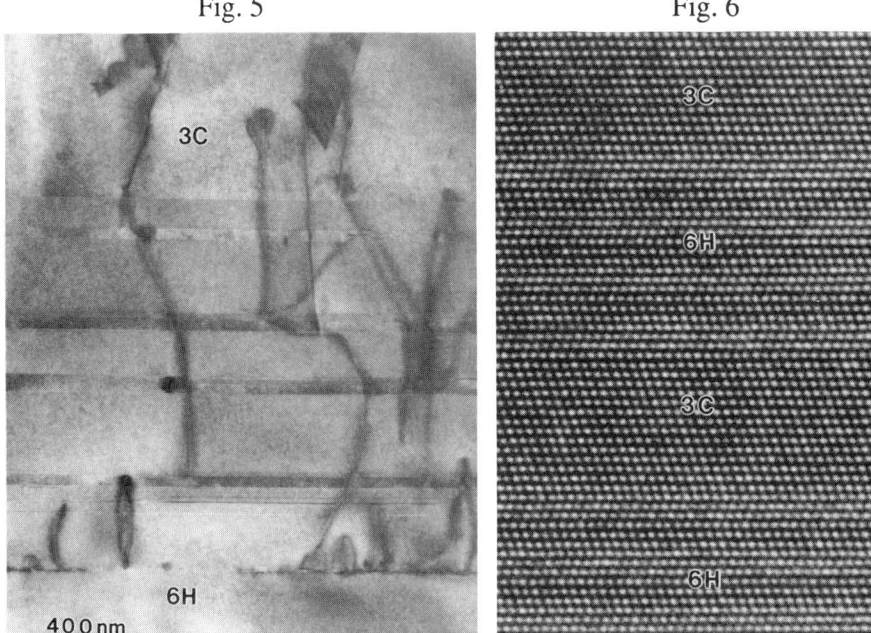

the incident electron beam is [11$\bar{2}$0], the basal planes are normal to the plane of the paper; in Fig. 7, traces of these planes are along the horizontal direction. Hence the dislocations are lying on planes normal, or slightly inclined, to the basal planes. Tilting experiments have shown that all the dislocations examined lie on planes normal to the basal plane and that, indeed, they lie on the {1$\bar{1}$00} prism planes. The Burgers vector of these dislocations has been determined by the **g.b** criterion; they are perfect dislocations with **b**=$\frac{1}{3}$[11$\bar{2}$0].

A different dislocation semi-loop in this region is shown in Fig. 8. The incident beam direction is [11$\bar{2}$0] and the dislocation crosses a number of narrow 3C bands. Note that the top segment of the semi-loop, i.e. the segment lying on the basal plane is dissociated, i.e. the top part is in the form of a faulted loop. One boundary of this faulted loop is shown by an

Fig. 7

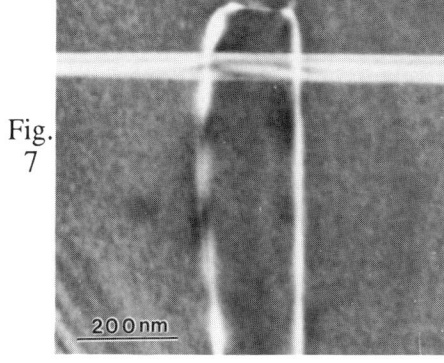

Fig. 8

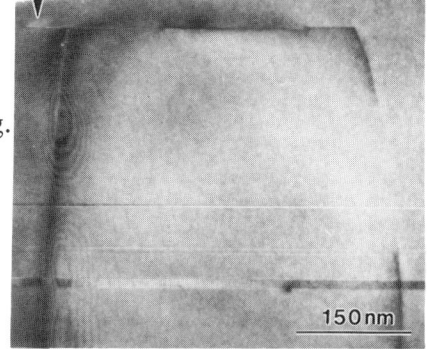

arrow in Fig. 8. Unfortunately, lack of sufficient tilt in the TEM used for taking this micrograph (JEOL 4000EX high resolution electron microscope) made it impossible to tilt the semi-loop around the [1$\bar{1}$00] axis in order to see the dissociation of the screw dislocation more clearly. Another specimen from a different sample (#1180), however was studied in a different microscope (JEM 200CX) which had ±30° double tilt capability. Although mixtures of 3C/6H bands were not observed in this sample, when the specimen was tilted about the

[1$\bar{1}$00] axis, a number of perfect dislocation semi-loops were observed which were surrounded by faulted loops. An example is shown Fig. 9 where the specimen is tilted about

the [1$\bar{1}$00] axis to the [4$\bar{4}$01] pole. Two faulted loops surround the lower part of the semi-loop while a third faulted loop is in the process of formation. The front and back portion of the faulted loop have been cut by the thin foil. This corresponds very closely to the configuration shown in Fig. 2 (also see the uppermost loop in Fig. 4). It should be pointed out that observation of a faulted loop within a regular low-order polytype would be very difficult: the stacking faults would be very close to each other and the resulting contrast would then be the usual one from overlapping faults that are often observed in TEM micrographs of twin bands.

Not being able to identify the initial location of the 3C/6H interface, it is difficult to conclude as to which direction the 3C↔6H transformation is proceeding. However, the 6H→3C polytypic transformation is rather unusual since the α-SiC polytypes are generally thought to be more stable at temperatures up to ~2000°C. We have discussed both the 6H→3C and the 3C→6H transformations because the experimental evidence can be interpreted in terms of both of these transformations. Although the latter appears more likely, the possibility cannot be excluded that Fig. 5 was originally in the 6H substrate, with the resulting structures arising from limited 6H→3C transformation.

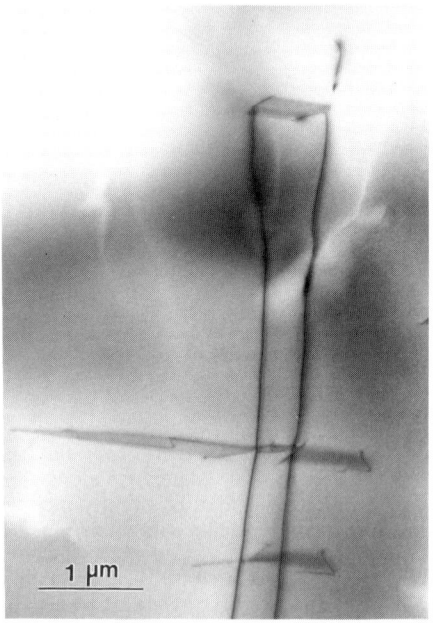

1 μm

Fig. 9

ACKNOWLEDGEMENT

This work was supported by grant number NAG3-758 from NASA. P.P. would also like to acknowledge the support of Max-Planck Foundation and Professor M. Rühle during his stay in the Institut für Metallforschung, Stuttgart, in the summer of 1990 where part of this work was carried out. Useful discussions with Professor A. H. Heuer are acknowledged.

REFERENCES

B. Escaig, J. Physique **29**, 225 (1968).
V. Heine and C. Cheng, in *From Geometry to Thermodynamics*, Ed. J. C. Tolidano, Plenum Press, New York (1990).
J. P. Hirth and J. Lothe, *Theory of Dislocations*, John Wiley & Sons Inc., New York (1982).
K. Maeda *et al.*, Phil. Mag. A**57**, 573 (1988).
P. Pirouz, Scripta Met. **21**, 1463 (1987); **23**, 401 (1989b)
P. Pirouz, Inst. Phys. Conf. Ser. **104**, 49 (1989a).
P. Pirouz and J. W. Yang, Mat. Res. Soc. Symp. Proc. **183**, 173 (1990).
P. Pirouz and P. M. Hazzledine, Scripta Met. (1991). In press.
J. A. Powell *et al.*, Submitted to Appl. Phys. Lett. (1991).
J. Rabier and P. Boivin, Phil. Mag. A **61**, 673 (1990).

Inst. Phys. Conf. Ser. No 117: Section 4
Paper presented at Microsc. Semicond. Mater. Conf., Oxford, 25–28 March 1991

TEM materials characterisation in Si technology

H Cerva

Siemens AG, Research Laboratories, Otto Hahn Ring 6, 8000 München 83, Federal Republic of Germany

ABSTRACT: The superior resolution of TEM is required to characterise critical features in Mbit DRAMs with device dimensions below 0.8 μm. Weak spots occurring in thin oxide-nitride-oxide (ONO) dielectrics are identified. Future DRAM generations need thinner ONO films or novel dielectrics like Ta_2O_5 or TiO_2. Inspection of metallization film homogeneity in narrow contact holes becomes important to ensure low sheet resistances. The interaction of implantation defects and stress inducing films is shown to generate dislocations in trench structures. Unintentional copper contamination during CMOS technology processing leads to small isolated precipitates showing a high strain field contrast.

1. INTRODUCTION

The continuous reduction of lateral dimensions in Si devices (currently at 0.6-0.8 μm) which most obviously is observed in the various DRAM generations, requires precise control of processing conditions. The high spatial resolution and analytical capabilities of TEM in the investigation of geometries, interfaces and material reactions, and crystal defects in cross-sectional and plan view specimens makes this technique an indispensable tool for device characterisation (e.g. Oppolzer (1987,1988)). A few recent examples describing problems encountered during DRAM and other CMOS device technology development will be discussed.

2. ISOLATED STACKED TRENCH CAPACITOR (ISTTC)

Figure 1a shows a cross section through a trench capacitor of a 16Mbit DRAM. Due to the higher packing density in 16Mbit memories trench capacitors have to be more closely spaced than in 4Mbit memories. To avoid electrical trench-to-trench leakage and harmful influences from the substrate the capacitor is stacked into a trench covered with an isolating oxide (Küsters et al (1989), Dietl et al (1990)). The capacitor dielectric consists of a thin oxide-nitride-oxide (ONO) triple layer sandwiched between two poly-Si electrodes. TEM analysis concentrates mainly on roughness and thickness determination of the dielectric layers between the two poly-Si layers (Fig. 1b,c) and on the trench contact which connects the storage electrode to the source/drain region of the transfer gate (A in Figs. 1a,b). For the interfacial oxide between the n-doped poly-Si storage node and the n-doped Si substrate at the trench contact, an optimized process has to be established which has similarities with that of poly-Si diffusion sources used in bipolar transistors (Oppolzer (1988)).

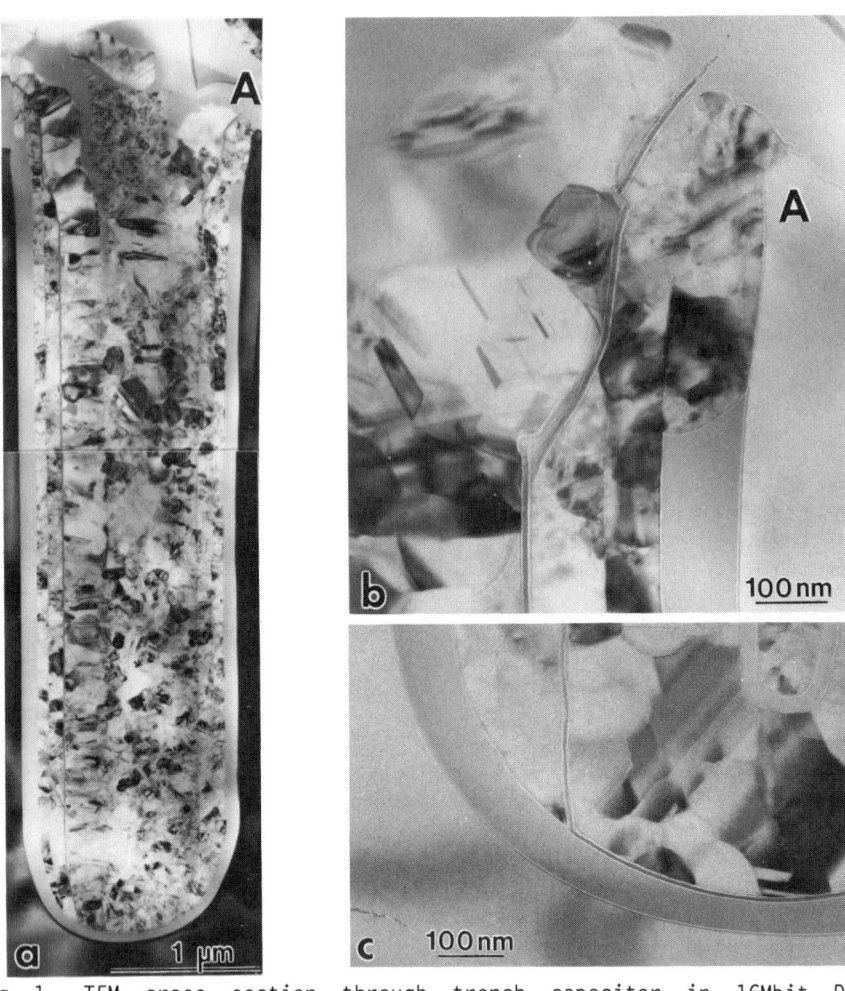

Fig. 1. TEM cross section through trench capacitor in 16Mbit DRAM.
Capacitor stacked into a trench isolated with oxide from the substrate (a).
Trench contact of capacitor to substrate (b). Bottom edge of capacitor (c).
Oxide-nitride-oxide dielectric lies between two poly-Si electrodes.

3. DIELECTRIC LAYERS

Triple layer dielectrics consisting of an Si_3N_4 layer sandwiched between
two SiO_2 layers (ONO) have been introduced with 4Mbit DRAMs (Oppolzer et al
(1987)). The bottom oxide in contact to the Si substrate is thermally grown
whereas the nitride is deposited in a low-pressure chemical vapour
deposition (LPCVD) reactor, and the top oxide is obtained by oxidation of
the nitride. An ONO dielectric surpasses a thin single thermal oxide by
having higher breakdown fields and lower electrical defect densities
(Hönlein and Reisinger (1989)). The most striking advantage is the
conformal step coverage of the LPCVD nitride over trench capacitor edges as
was shown in TEM images (Oppolzer et al (1987)) whereas thermal oxidation
of these edges leads to a great thickness reduction of the oxide film.
Further, ONO layers benefit from the higher dielectric constant of Si_3N_4

($\varepsilon \approx 7$) compared to SiO_2 ($\varepsilon \approx 4$). The LPCVD process provides electrically stable, dense, and homogeneous nitride films in the thickness range 5-10 nm. However, if a thinner nitride is subjected to thermal oxidation to produce the top oxide, the nitride turns out to contain leaky spots for oxygen diffusion (Hönlein and Reisinger (1989)). Since the oxidation rate of Si is much higher than that of Si_3N_4, the bottom oxide will form lens-shaped areas (swellings) at these spots (Fig. 2a). Generally the nitride layer shows a surface roughness of about 1 nm (Fig. 2b). The bottom oxide is flat whereas the top oxide surface follows the nitride surface roughness conformally. Inspection by TEM of the surface roughness of a thicker single Si_3N_4 layer reveals that this roughness comes from the nitride deposition process (Figs. 3a-c). Only in defocused images the roughness may be assessed easily. Figure 4 shows a nominally 2 nm thick nitride layer with a CVD oxide on top. Since the nitride film is broken up into small balls, 1 to 2 nm in size, without oxidation, it is concluded that the surface roughness of thicker layers results from the nucleation of isolated Si_3N_4 islands at the beginning of the deposition.

For the capacitors in 16 Mbit and 64 Mbit DRAMs progressively lower thicknesses of the dielectric will be required (Spitzer et al (1990)). In a 64 Mbit DRAM an effective SiO_2 thickness of 5 nm is needed, which can be realized with two 1.5 nm thick oxide layers sandwiching a 4 nm thick nitride film. Figure 5a shows such a thin SiO_2-Si_3N_4 film on Si substrate. Since such dielectrics will lie between two poly-Si electrodes the interfaces will not be planar, and TEM imaging in projection is only possible in limited regions (Fig. 5b). This problem, however, is already encountered in 16 Mbit DRAMs as can be seen in Fig. 1a-c. Another problem with ONO layer thickness uniformity arises if the thermal bottom oxide is

Fig. 2. Oxide-nitride-oxide (ONO) dielectric. At pores of the nitride, islands of increased oxide thickness grow during thermal oxidation of the nitride (a). ONO dielectric with slight surface roughness of the nitride (b).

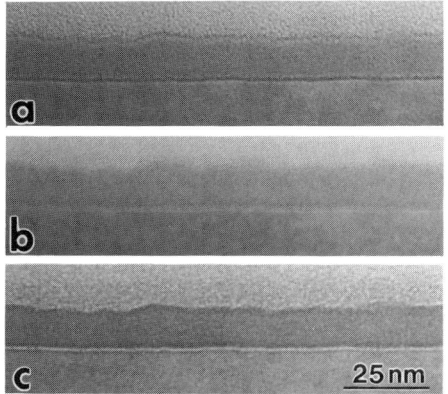

Fig. 4. Thin 2 nm thick inhomogeneous nitride with a CVD deposited oxide on top. Nitride islands were formed during the initial LPCVD process.

Fig. 3. Focus series of single nitride layer. Overfocus (a), in focus (b), and underfocus (c) images.

grown on n⁺-doped poly-Si deposited in the crystalline state: due to dopant segregation at the grain boundaries and preferential oxidation of these sites, dielectrics with poor breakdown characteristics are grown (Oppolzer (1988)). This problem is overcome with amorphously deposited and subsequently crystallized poly-Si layers (Fig. 1a-c)(Hirschler et al (1989)). Only the randomly oriented Si grains cause thickness differences in the bottom oxide film because of the dependence of the oxidation rates on crystal orientation (Fig. 6). But this effect may be minimized by optimizing the oxidation conditions.

The development of future DRAM generations requires alternative dielectric materials to reach charge densities that are larger than those of a 5 nm thick SiO_2 film. Tantalum pentoxide exhibits a higher dielectric constant ($\varepsilon \approx 20$) and may be considered as an alternative if the film is in the amorphous state (Shinriki et al (1988)). The Ta_2O_5 layer in Fig. 7a was sputtered and rapid thermally annealed at 900°C. The film is amorphous and a thin SiO_2 layer was formed on the interface to the Si substrate. The sputtered Ta_2O_5 layer in Fig. 7b crystallized because of the higher annealing temperature of 1200°C. Grooves resulting in thickness reductions at the grain boundaries are responsible for high leakage currents and low breakdown fields of this layer. Another candidate is TiO_2 in the rutile phase ($\varepsilon \approx 100$)(Spitzer et al (1991)). A crystalline film can be formed by reactive sputtering from a pure Ti target in Ar/O_2 atmosphere and subsequent annealing. Figure 8a shows a film which was annealed in Ar/O_2 ambient and reveals grooves at the grain boundaries, large thickness variations (up to 50%) as well as a thick interfacial oxide layer to the

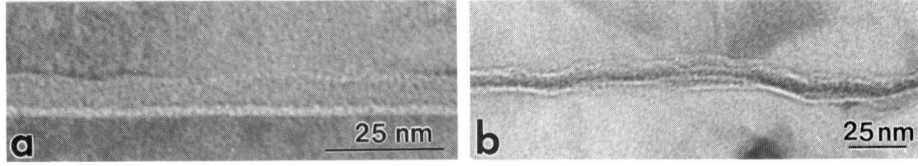

Fig. 5. Thin ONO dielectrics on Si substrate (a) and on poly-Si (b).

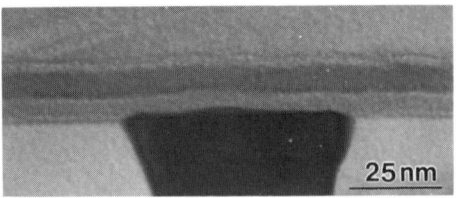

Fig. 6. Varying thickness of thermal bottom oxide in an ONO dielectric due to different orientations of poly-Si grains.

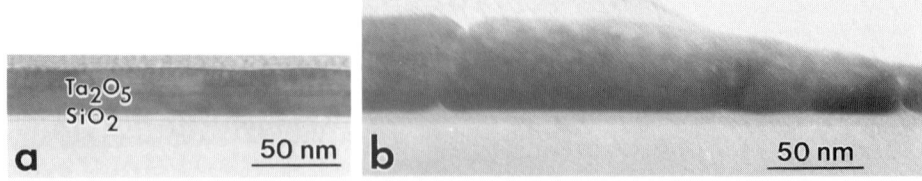

Fig. 7. Sputtered Ta_2O_5 dielectric in the amorphous (a) and the crystalline state (b) after annealing.

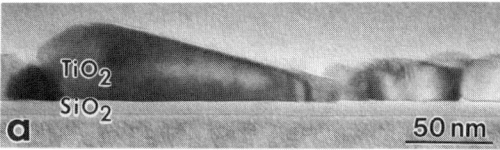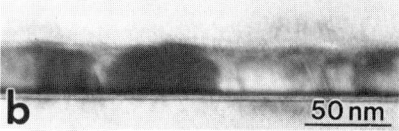

Fig. 8. Sputtered TiO_2 dielectric annealed in Ar/O_2 (a) and in Ar (b).

substrate. A homogeneous TiO_2 layer with a thin interfacial oxide film can be obtained by annealing in Ar ambient (Fig. 8b). With the advent of such dielectrics TEM of thin cross sections will not only be challenged to analyse thicknesses of thin interfacial layers and tiny grooves, but also crystal orientation, structure and material reactions of small grains.

4. CONTACT HOLES

Interconnects in CMOS circuits increasingly use silicides of refractory metals such as molybdenum on top of highly doped poly-Si films. Such double layers benefit from very low sheet resistances and thus reduce delay times. With decreasing lateral device dimensions it is essential to produce homogeneous films also in narrow and steep contact holes. During sputtering of molybdenum silicide a fine columnar structure may be formed in the amorphous material only at the steep side walls of contact holes (Fig. 9a). The bright lines represent low-density material forming channels with a honeycomb structure. If annealing is performed in an oxidizing ambient, oxidation proceeds fast along the fine channels between the columns (Fig. 9b). Then, the $MoSi_2$ no longer forms a continuous film, since the individual grains are surrounded by SiO_2. It is obvious that this effect, which does not occur in the planar film regions, increases the sheet resistance of such interconnects crossing contact holes or surface steps drastically. By changing the annealing conditions such that initial annealing is performed in nitrogen, $MoSi_2$ films with a continuous microstructure and low sheet resistances are obtained.

5. PROCESS-INDUCED DEFECTS

5.1 Defects below Implantation Mask Edges

Source/Drain(S/D) regions of modern MOS-transistors are doped according to the LDD-concept (lightly doped drain) (Fig. 10a). During processing the poly-Si gate acts as a mask for a low dose ($\sim 10^{13} cm^{-2}$) phosphorus implantation (region 2 in Fig. 10a) whereas the spacer oxide on both sides of the gate serves as a mask for a high dose ($\geq 10^{15} cm^{-2}$) arsenic implantation. The phosphorus dose is well below the critical dose ($\sim 10^{14} cm^{-2}$) for amorphization of the substrate and produces no defects upon annealing. The arsenic dose, however, exceeds the critical dose of $\sim 2 \cdot 10^{13} cm^{-2}$ by far and, hence, a 50 keV/$5 \cdot 10^{15}$ As cm^{-2} implantation produces an approx. 63 nm deep amorphous zone at the substrate surface (A in Fig. 10b). The amorphous/crystalline (a/c) interface is sharply curved below the edge of the spacer oxide mask. An annealing step at 900°C for implantation damage recovery and activation of the dopants leads to the configuration shown in Fig. 10d. The amorphous zone is completely crystallized and a crystal defect is formed just below the oxide mask edge. TEM images of plan view specimens revealed that these defects run continuously along the mask edge. High resolution imaging of several defects performed on <110> cross sections identified most of the defects

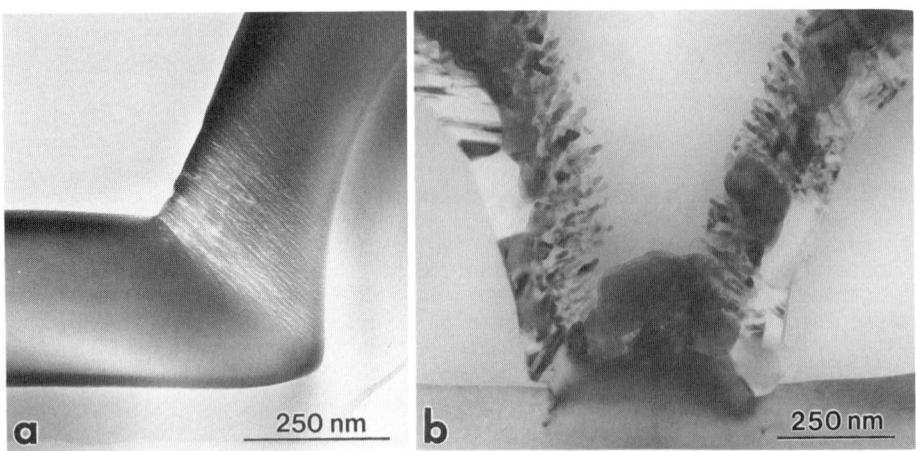

Fig. 9. MoSi$_2$-poly-Si interconnect in contact hole after sputtering (a) and annealing in oxidizing ambient (b).

either as vacancy-type half-loop dislocations lying on a {111} plane or as vacancy-type V-shaped defects (Cerva and Küsters (1990), Cerva and Oppolzer (1990)). The size of the defects clearly reveals that they were formed within the former amorphous region. To study the defect formation, an intermediate stage of the crystallization process was investigated by annealing an implanted sample only at 500°C for 30 min. Figure 10c shows that the amorphous Si is not completely crystallized in this case, but a notch is created in the crystallization front below the mask edge. When crystallization proceeds, the two sides of the notch merge and produce a defect on the {111} plane which points towards the mask window (Figs.

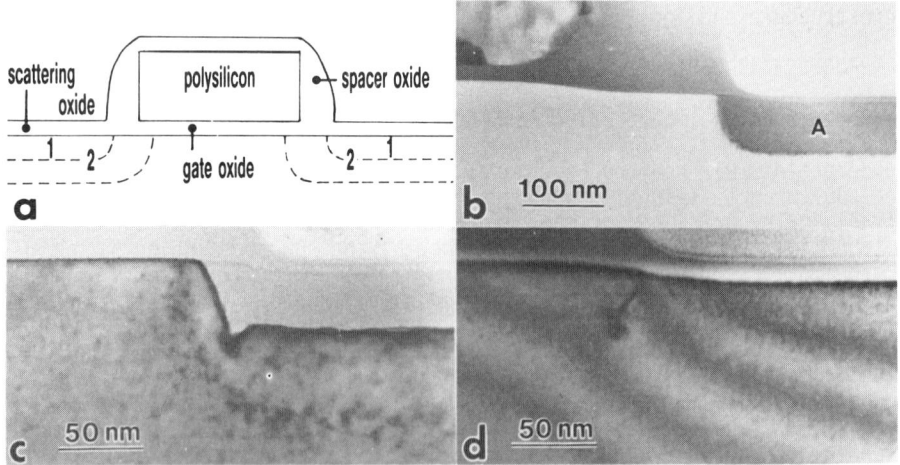

Fig. 10. Schematic cross section through a lightly doped drain (LDD) transistor (a). TEM images after successive process steps: (b) amorphous Si layer formed by As implantation (A); (c) partial regrowth (500°C/30 min), (d) complete regrowth after 900°C/60 min. (b) Bright-field g=±400, (c,d) <110> aligned bright-field images.

10c,d). The formation of the notch may be understood by considering the different crystallization rates on the various Si lattice planes present on the originally bent a/c interface (Fig. 10b). Such a model indeed leads to a notch in a sharply bent a/c crystallization front (Cerva and Küsters (1990)). These defects are possible causes of failure in a transistor structure leading to leakage currents or they can act as nucleation sites for dislocations during further processing. The defects could be avoided either by changing the shape of the spacer oxide mask, which changes the curvature of the a/c-interface, or by applying rapid annealing.

5.2 Dislocations in Trench Structures

Trench structures were found to be potential sites of dislocation generation (Kolbesen et al (1989), Stiffler et al (1990)). However, not all dislocations in the vicinity of the trench are directly induced by the trench. The following example demonstrates that process optimization during technology development cannot be restricted to optimizing only a few process steps of a single functional unit, but all the process steps of the whole structure have to be optimized in conjunction such that no detrimental configurations may occur. In a DRAM the transfer gate is connected to the storage electrode of the capacitor (i.e. the Si substrate in case of a 4Mbit memory) via a source/drain (S/D) region which is usually formed as described in section 5.1. This area of a 4Mbit DRAM is displayed in Fig. 11a. The S/D region lies between the two spacer oxide layers which isolate the poly-Si gate and the upper capacitor electrode, and define the window for the high dose arsenic implantation. As can be seen in Fig. 11a several dislocations were generated close to the poly-Si gate and to the upper trench corner. By tilting the <110> orientated cross section along the {220} Kikuchi lines the Si substrate surface and the nucleation points of the dislocations can be seen more clearly (Fig. 11b). Obviously, the source of the dislocations lies just below the edge of the spacer oxide where the small mask edge defects (induced by the arsenic implantation) as described above were produced. Defect etching could not clarify whether the defects originated from the trench corners or elsewhere because of the large extension of the etch pits. Burgers vector analysis showed that the dislocations are all of the type $b=a/2<110>$ gliding on {111} planes. When analyzing the various process steps it turned out that the 50 nm thick nitride layer between the two spacers in Fig. 11a which is necessary for

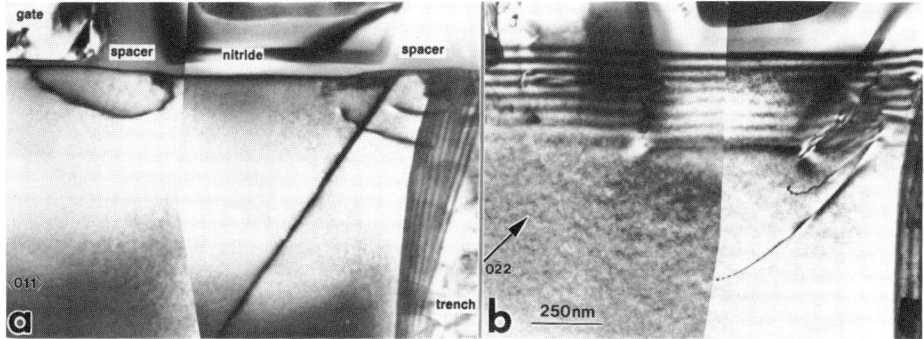

Fig. 11. Source/drain area between transfer gate and trench capacitor. Dislocations nucleated just below the spacer oxide and nitride edge at implantation defects. (a) <110> aligned bright-field image, (b) g=022.

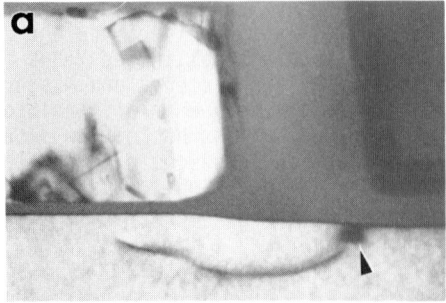

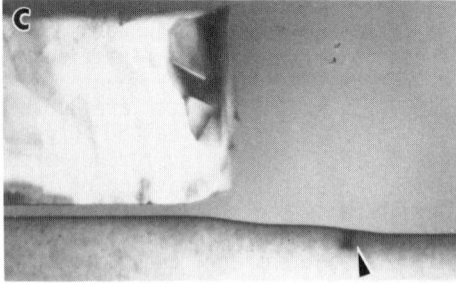

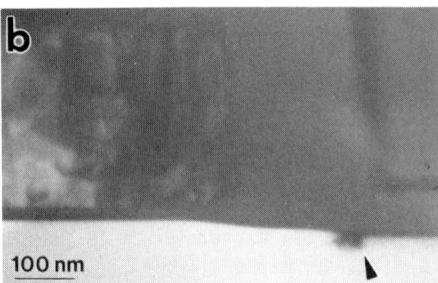

Fig. 12. Nucleation of dislocation loop at a residual implantation defect induced by stress of a thick nitride film in (a). No nucleation with thin nitride (b) and no nitride (c) film.

pro~essing the bitline contact (not visible in Fig. 11a), is responsible for dislocation nucleation. This was demonstrated with test samples the results of which are shown in Figs. 12a-c. The arsenic implantation leads to the mask edge defects in the substrate below the spacer edge (arrows in Figs. 12a-c). Subsequent deposition of a 50 nm thick nitride layer exerts stress due to the different thermal expansion coefficients of the various materials. Thus, glide-dislocations are nucleated at the former implantation defects. A thinner nitride (Fig. 12b) or no nitride layer (Fig. 12c) reduces the stress and no glide dislocations are generated. The implantation defects certainly act as nucleation sites for the dislocations which may glide deep into the substrate. The vicinity of the trench corners leads to an additional stress source which is superimposed and may - if the stresses add positively - even further lower the activation barrier to dislocation nucleation. These defects result in leakage currents during transistor performance (Küsters et al (1987)) and cross the depletion zone of the capacitor. Therefore, optimizing the nitride layer thickness and the arsenic implantation conditions helped to avoid these particular dislocations.

5.3 Copper precipitates

Metal contamination is a crucial problem in device fabrication since it may degrade any kind of silicon device (e.g. Kolbesen et al (1989)). In order to understand the formation of metal precipitates during various technology processes like implantation, oxidation or high temperature annealing, intentionally contaminated samples were studied by TEM in the past (e.g. Honda et al (1987), Wendt et al (1989), Cerva and Wendt (1989), Seibt (1990)). Copper contamination at high concentrations $\geq 10^{18} cm^{-3}$ usually forms small (20-40 nm) silicide particles which are arranged in colonies lying on {110} planes that are surrounded by edge-dislocation loops of extrinsic type. Since the Cu silicides are metal-rich and lead to a volume expansion, Si-interstitials (Si_I) are emitted and are accomodated either in extrinsic dislocation loops or at the Si surface to release the strain.

Such high contamination levels, however, are not realistic in modern device production and, therefore, the appearance and structure of Cu precipitates in a real CMOS technology process can be different. During failure analysis a high density of 10-40 nm precipitates, 1-5 µm apart, was detected in a zone 3-15 µm below the Si surface. The precipitates lie on {111} planes and exhibit a strong strain contrast if a component of the excited reflection is perpendicular to their habit plane (Fig. 13a,e). Over- and underfocused images (Fig. 13b,c) reveal Fresnel fringes along the defect which indicate that the interior of the defect has a larger mean atomic potential (i.e. a higher density and/or a higher atomic number) than the surrounding silicon. A <110> lattice image of the same precipitate shows two additional {111} planes inserted into the Si matrix (Fig. 13d). Evidence for a copper precipitate was finally given by energy dispersive X-ray analysis. Another precipitate lying on a {111} plane which is inclined to the TEM cross section reveals the disk-shape of the particle. Crossed Moiré fringes are visible parallel to {200} and {220} planes and remind one of the Moiré fringes observed for the precipitates in the colonies. The precipitates were not attached to dislocations or stacking faults. Obviously these Cu precipitates correspond to very early stages of precipitation as described by Seibt (1990). Copper atoms agglomerate on {111} planes to form a silicide and adopt a disk-shape to release strain. The volume of the silicide has not yet exceeded the critical value for Si_I emission because no extrinsic dislocation loops for Si_I accomodation were associated with it. The strain field contrast in the bright-field images of these precipitates (Fig. 13a,e), and the absence of this contrast in images of the precipitates in the colonies which represent a later stage of precipitation, corroborate the above assumption. For successful defect characterisation and failure analysis of a metal contaminated device by TEM the prior knowledge of possible specific defect morphologies is extremely

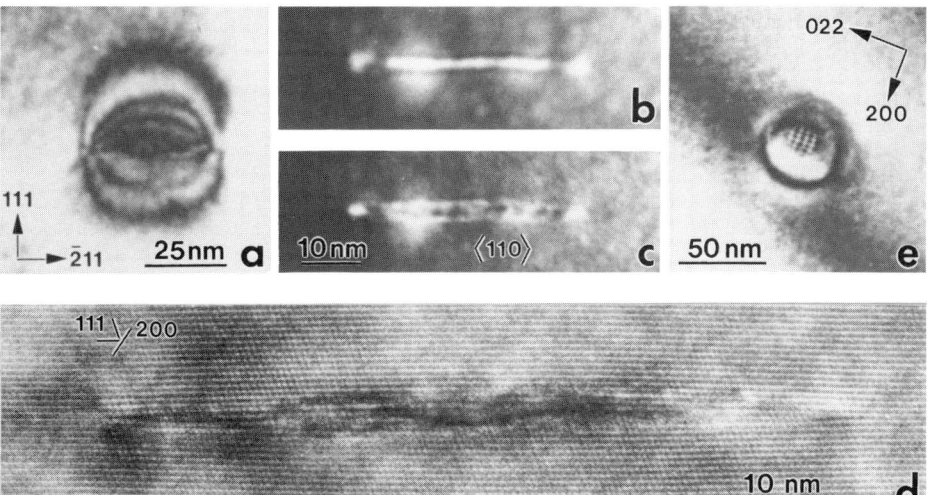

Fig. 13. Single isolated Cu precipitates on {111} planes in Si. <110> aligned bright-field images of precipitate in projection (a), and on inclined {111} plane (e). Projected precipitate in overfocused (b) and underfocused (c) <110> aligned bright-field image. Lattice image of the same defect (d).

helpful in identifying the precipitates. The single Cu precipitates are very small and may easily be overlooked when searching in a device structure. Only their strong strain contrast can help to find them.

ACKNOWLEDGEMENT

The author is indebted to V. Huber and C. Walz for TEM work and technical assistance, and to K.H. Küsters , A. Spitzer, and H. Oppolzer for valuable discussions.

This work has been supported by the Federal Department of Research and Technology of the Federal Republic of Germany (sig. NT 2788 0). The author alone is responsible for the contents.

REFERENCES

Cerva H and Wendt H 1989 Mat. Res. Soc. Symp. Proc. 138 533
Cerva H and Küsters K H 1990 J. Appl. Phys. 66 4723
Cerva H and Oppolzer H 1990 Mat. Res. Soc. Symp. Proc. 183 67
Dietl J, Do Thanh L, Küsters K H, Kusztelan L, Mühlhoff H M, Müller W and
    Stelz F X 1990 Proc. ESSDERC 90, eds W Ecclestone and P J Rosser
    (Bristol: Adam Hilger) pp 465-8
Hirschler J, Do Thanh L, Küsters K H and v Sichart K 1989 Proc. ESSDERC 89,
    eds A Heuberger, H Ryssel and P Lange (Berlin: Springer) pp 357-60
Hönlein W and Reisinger H 1989 Appl. Surf. Sci. 39 178
Honda K, Nakanishi T, Ohsawa A and Toyokura N 1987 Inst. Phys. Conf. Ser.
    87 463
Kolbesen B O, Bergholz W, Cerva H, Gelsdorf F, Wendt H and Zoth G 1989
    Inst. Phys. Conf. Ser. 104 421
Küsters K H, Do Thanh L, Stelz F X, Kellner W U, Mühlhoff H M and Müller W
    1989 Proc. ESSDERC 89, eds A Heuberger, H Ryssel and P Lange (Berlin:
    Springer) pp 907-10
Oppolzer H, Cerva H, Fruth C, Huber V and Schild S 1987 Inst. Phys. Conf.
    Ser. 87 433
Oppolzer H 1988 Inst. Phys. Conf. Ser. 93 73
Seibt M 1990 Proc. Semiconductor Silicon 1990, eds H R Huff, K G
    Barraclough and J Chikawa (Pennington: ECS) pp 663-74
Shinriki H, Kisu T, Kimura S, Nishioka Y, Kawamoto Y and Mukai K 1988 Symp.
    VLSI Technology, Digest of Technical Papers, San Diego, CA, pp 29-30
Spitzer A, Reisinger H and Hönlein W 1990 Proc. ESSDERC 90, eds W Eccleston
    and P J Rosser (Bristol: Adam Hilger) pp 307-10
Spitzer A, Reisinger H, Willer J, Hönlein W, Cerva H and Zorn G 1991 Proc.
    INFOS'91 (Liverpool, UK) to be published
Stiffler S R, Lasky J B, Koburger Ch and Bevy W S 1990 IEEE Trans. Electron
    Devices 37 1253
Wendt H, Cerva H, Lehmann V and Pamler W 1989 J. Appl. Phys. 65 2402

Inst. Phys. Conf. Ser. No 117: Section 4
Paper presented at Microsc. Semicond. Mater. Conf., Oxford, 25–28 March 1991

165

TEM study of Lopos structures for submicron isolation

A Romano-Rodríguez[1], J Vanhellemont, I De Wolf, H Norström[2] and H E Maes

Interuniversity Microelectronics Center (IMEC), Kapeldreef 75, B-3001 Leuven, Belgium
[1]Permanent address: Càtedra d'Electrònica, University of Barcelona, Diagonal 647, E-08028 Barcelona, Spain
[2]Present address: Swedish Institute of Microelectronics, P.O. Box 1084, S-16421 Kista, Sweden

ABSTRACT: A TEM study of a LOPOS isolation process is presented. Advantages of this process are the reduction of the bird's beak compared to conventional LOCOS, the negligible asymmetry of the bird's beak for highly packed structures and the much flatter silicon/pad oxide interface. The stresses inherent to LOPOS increase strongly and can give rise to different stress relaxation mechanisms, which are harmful for the device fabrication. Correlation with micro-Raman spectroscopy is presented. More severe field oxide thinning happens compared with LOCOS for small oxidation window size.

1. INTRODUCTION

LOCOS (LOCal Oxidation of Silicon), first proposed by Appels et al (1970), is the major isolation method employed in integrated circuit fabrication. The advantages of this process are its simplicity and the good electrical characteristics of the fabricated devices. A drawback of this isolation scheme is the extended lateral oxidation below the oxidation mask into the active silicon area, giving rise to the so called bird's beak (BB). This strongly reduces the area available for processing. With the scaling down of the geometries LOCOS can no longer be used for submicron dimensions and adequate field oxide (FOX) thicknesses. Extensive research is going on to modify the LOCOS process in order to reduce the BB length while still keeping its simplicity. A promising approach is LOPOS (Local Oxidation of Polysilicon Over Silicon), also called poly-buffer LOCOS (Han and Ma 1984), which consists in the introduction of an additional polysilicon layer between the pad oxide and the nitride mask. In our experiment this polysilicon layer has been replaced by a microcrystalline one, which transforms into polysilicon during the field oxidation step.

2. EXPERIMENTAL PROCEDURE

5 inch (001) oriented p-type Czochralski pulled silicon wafers, with a medium interstitial oxygen content, $5 \cdot 10^{17}$ cm^{-3}, have been used. A dedicated periodic test structure consisting of parallel lines along [110] directions with a variable width and spacing is used to study the influence not only of the oxidation and oxidation mask parameters, but also of the width and spacing of the lines, on the isolation edge topography.

The mask parameters employed in this study were: pad oxide ranging between 10 and 15 nm, the microcrystalline silicon was 50 nm thick, the nitride mask thickness varied between 10 and 15 nm and the field oxidation has been performed in wet oxygen at temperatures ranging from 900 to 1050°C. Plan view and cross-section specimens have been prepared using a thinning procedure for prespecified regions described in detail elsewhere (Romano et al

1990). For the plan view study all the layers on top of the substrate were removed by wet etching.

Micro-Raman investigations have been performed on the same isolation structures. A detailed micro-Raman study of stresses on LOCOS structures will be published elsewhere (De Wolf et al 1991). The spectra have been obtained in the backscattering configuration, with the light incidence normal to the (001) surface and using an Ar laser line of 457.9 nm. The spot size was less than 1 µm in diameter and the steps between two recorded points was 0.2 µm, achieved using a X-Y table. The accuracy of the measurements was 0.05 cm⁻¹, obtained by fitting the Raman peak to a Lorentz function.

3. RESULTS AND DISCUSSION

3.1 TEM results

Fig. 1 shows a typical cross-section image of a LOPOS structure oxidized at 950°C. The BB is about 250 nm long for a FOX thickness of 0.6 µm. This corresponds to a 50% reduction when compared with an optimized LOCOS structure with the same FOX thickness. In this figure a second small BB can be observed between the nitride and the polysilicon layer. From the oxygen that is flowing through the oxidation window one part is used to oxidize the substrate, a second part is used to form the BB. Finally, a third part is used to oxidize the polysilicon edge and gives rise to the formation of the second BB.

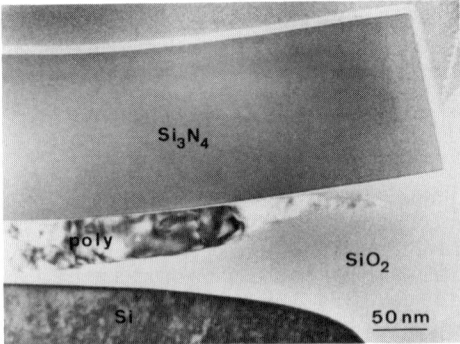

Fig. 1: Typical cross-section image of a LOPOS structures. The mask parameters are: pad oxide 10 nm, microcrystalline silicon 50 nm, nitride mask 150 nm. The oxidation is performed at 950°C. Observe the presence of a second BB between the polysilicon and the nitride.

The reduction of the BB length leads however to a strong increase of the stresses at the nitride edge, which can give rise to the generation of substrate defects (mainly dislocations) in the substrate. This is shown in the plan view image of fig. 2a, corresponding to oxidation at 1050°C. Dislocations, running parallel to the nitride mask and having alternating Burgers vectors, which gives rise to a more efficient stress relief a the film edges (Vanhellemont et al 1987), are clearly seen. In addition surface stacking faults appear below the FOX window, as a result of the high injection rate of silicon self-interstitials during the oxidation. This higher supersaturation of self-interstitials leads to a lower yield stress for defect nucleation (Vanhellemont and Claeys 1988). For the thicker pad oxide the generation of defects is accompanied by plastic deformation of the substrate below the BB. This stress relief mechanism is the dominant at high temperatures. However other relaxation mechanisms have been observed:
- Breaking of the nitride and separation from the polysilicon layer (fig. 2b), which only happens at oxidation temperatures below 950°C..
- Breaking of the pad oxide (fig. 2c) occurs in the whole temperature range and is the dominant stress relief mechanism at temperatures up to 1000°C.

3.2 Stresses Measured By Micro-Raman Spectroscopy

Fig 3. shows micro-Raman spectra obtained from a LOPOS sample, oxidized at 1000°C, for line widths ranging from 2 to 15 µm. The spectra have been plotted in such a way that the vertical axis indicates the shift of the Raman frequency from the stress free value (520 cm-1) while the horizontal axis gives the position along the width of the line, with the origin taken at

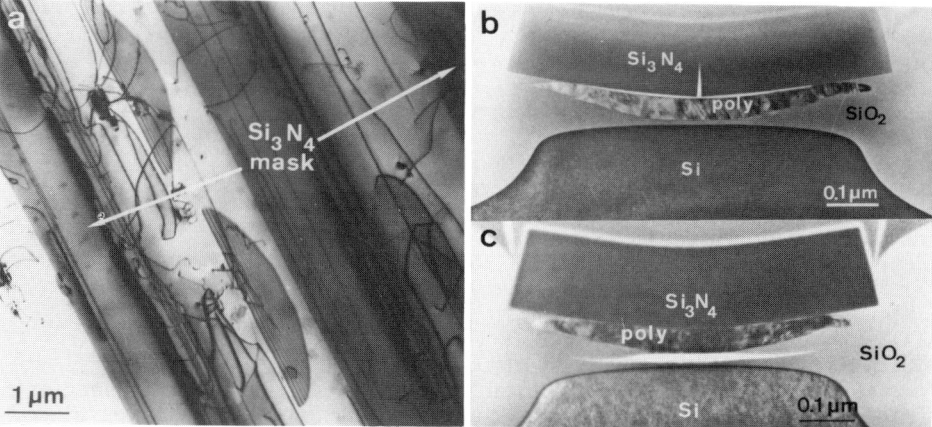

Fig. 2: The three stress relaxation mechanisms observed in this study: a) Generation of dislocations running parallel to the nitride lines and surface stacking faults in the field oxide areas (1050°C); b) Breaking of the nitride and separation from the polysilicon layer (950°C); c) Breaking of the pad oxide (950°C).

Fig. 3 (right): Raman shifts for different line widths obtained from a LOPOS sample (pad oxide: 10 nm, nitride: 150 nm, 1000°C).

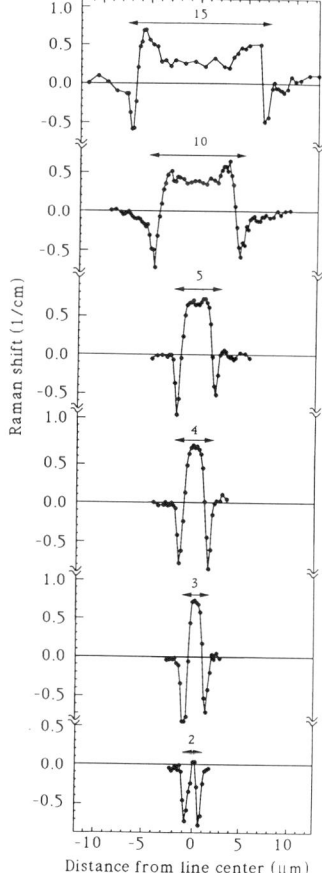

its centre. The positive or negative Raman shift can be associated, respectively, to a compressive or tensile stress in the silicon substrate (Ganesan et al 1970). The following features can be deduced from the figure:
- Below the BB a strong tensile stress is observed. The shift is independent of the line width, indicating that it is only consequence of the BB geometry. The measured shift is double the value of that corresponding to an optimized LOCOS structure.
- In the centre of the lines a large compressive stress is observed. Its value decreases with increasing line width.
- For lines smaller than 4 μm the compressive stress at the centre of the line decreases and becomes even tensile for lines smaller than 2 μm. This is due to the influence of the tensile stress at the BB tip.

3.3 Field Oxide Thinning

With the reduction of the oxidation window size a thinning of the field oxide occurs, as reported by Hui et al (1985). This thinning is more severe the smaller the oxidation windows. Intuitively this can be explained because the amount of oxygen that penetrates through a window is proportional to the window's size. The enhanced thinning in LOPOS structures can be understood by the fact that a larger silicon surface is available for oxidation below the mask for the same window size.

Fig. 4 shows a plot of the FOX thinning as a function of the oxidation window for LOPOS and for an optimized LOCOS structures. The plotted lines correspond to the fitting using the model by Lutze et al (1990). It can be seen that the thinning is more severe for LOPOS than for LOCOS structures.

3.4 Asymmetry In Highly Packed Structures

In the highly packed LOCOS structures of fig. 5 an asymmetry of the different BB can be observed. The FOXs which are at the edges of the figure correspond to large oxidation windows for which no oxygen supply problems arise, i.e., no FOX thinning. Starting from the left, the first BB is longer than the left one of FOX 2; at the same time the right BB of FOX 2 is shorter than the left of FOX 3. The right BB of FOX 3 is equally as long as the left of FOX 4 because the test structure is symmetrical between FOX 3 and 4. Processing in these areas could produce devices with different electrical characteristics. This effect is negligible for LOPOS isolation, where the asymmetry is much smaller.

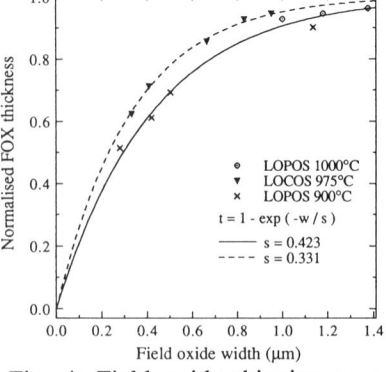

LOPOS 1000°C
LOCOS 975°C
LOPOS 900°C

$t = 1 - \exp(-w/s)$

—— $s = 0.423$
- - - - $s = 0.331$

Field oxide width (µm)

Fig. 4: Field oxide thinning as a function of the oxidation window size. The thinning is more severe for LOPOS than for LOCOS.

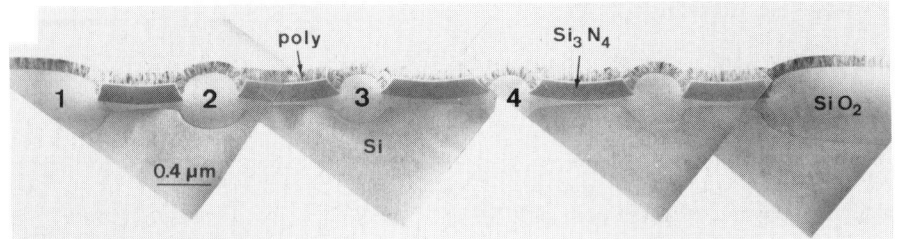

Fig. 5: Asymmetry in highly packed structures obtained using a LOCOS process.

ACKNOWLEDGEMENTS

The TEM work has been performed using the equipment of the University of Antwerpen (RUCA). A.R.-R. and H.N. were supported by the Spanish Ministry of Science and Education and the Swedish Board of Technical Development, respectively.

REFERENCES

Appels J A, Kooi E, Paffen M M, Schatorje J J H and Verkuylen W H G G 1970 Philips Res. Repts. 25 118
De Wolf I, Vanhellemont J, Romano-Rodríguez A, Norström H and Maes H E 1991 (to be published in the Proc. of the "Third International Symposium on Ultra Large Scale Integration Science and Technology", 179th Meeting of The Electrochemical Society, Washington DC (USA))
Ganesan S, Maradudin A A and Oitmaa I 1970 Annals of Phys. 56 556
Han Y and Ma B 1984 Electrochem. Soc. Ext. Abst. 84-1 98
Hui J, Vande Voorde P and Moll J 1985 IEDM Tech. Dig. 392
Lutze J W, Perera A H and Krusius J P 1990 J. Electrochem. Soc. 137 1867
Mizuno T, Sawada S, Maeda S and Shinozaki S 1987 IEEE Trans. Electron Dev. ED-34 2255
Romano A, Vanhellemont J and Bender H 1990 Mater. Res. Soc. Symp. Proc. 199 167
Vanhellemont J, Amelinckx S and Claeys C 1987 J. Appl. Phys. 61 2176
Vanhellemont J and Claeys C 1988 J. Electrochem. Soc. 135 1509

Inst. Phys. Conf. Ser. No 117: Section 4
Paper presented at Microsc. Semicond. Mater. Conf., Oxford, 25–28 March 1991

169

TEM/etching studies of fabrication-induced defects in 4M DRAMs

M Dellith[1], F Gelsdorf[2], G R Booker[1], W Bergholz[2] and B O Kolbesen[2]

[1] University of Oxford, Department of Materials, Parks Road, Oxford OX1 3PH, UK
[2] Siemens AG, Otto-Hahn Ring 6, D-8000 Munich 80, FRG

ABSTRACT: Dislocations in 4M (megabit) DRAMs (dynamic random access memories) with trench capacitor cells can decrease the device performance if they occur in electrically active regions. Optical microscope (OM) and scanning electron microscope (SEM) examinations were made of etched specimens and transmission electron microscope (TEM) examinations of etched and unetched specimens. Two different types of fabrication-induced dislocations were distinguished and characterised.

1. INTRODUCTION

Reduction in costs is the driving force in the development of VLSI (very large scale integration) technology. This can be achieved by further miniaturisation as well as optimisation of the processing procedure. A DRAM (dynamic random access memory) is one of the simplest silicon devices used in VLSI, and so its investigation is likely to provide results of importance for the development of more complicated ones.

In a DRAM (Fig.1) the information is stored by the charging of capacitors cells. To reduce the chip-size in the present 4M (megabit) DRAM, a reduction of cell size is achieved by using three-dimensional trenches as capacitors, which are fabricated in the bulk silicon by dry etching. They are 5µm deep and 1 µm wide, contain a thin dielectric and are filled with polysilicon. The region in the vicinity of the trench capacitor is heavily n-doped by in-diffusion of As atoms to establish one plate of the capacitor; the bulk silicon is p-doped. Furthermore, the DRAMs contain MOS (metal oxide semiconductor) transfer gates and so-called FOBICs (fully overlapping bit-line contacts) which provide the electrical contact between the polycrystalline silicide bit-line and the transfer gate. Under the FOBIC a heavily n-doped region is formed by ion implantation.

During operation of the DRAM the stored charge decreases due to leakage currents (Chatterjee et al 1979) and needs to be refreshed every few milliseconds. Dislocations can play a role in increasing these leakage currents if they cross the p-n junction located around the trench (Kolbesen et al 1989a). Consequently they are harmful to the device and their formation should be avoided.

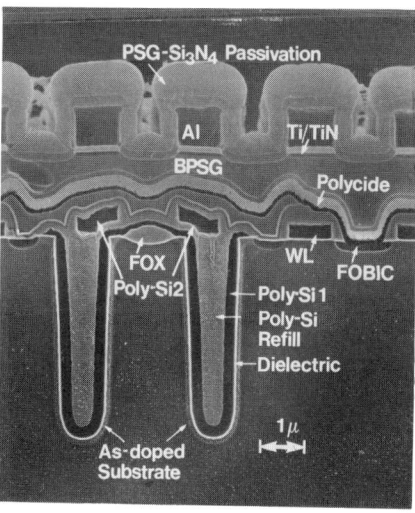

Fig. 1: SEM cross-section of a 4M DRAM showing the trench memory cells and the FOBIC. PSG: phosphor silicate glass; BPSG: boron phosphor silicate glass; FOX: field oxide; WL: word line

2. PREPARATION TECHNIQUES

Before carrying out these studies of dislocations in the bulk of the processed silicon wafers, the surface technology layers, consisting of metal silicides, SiO_2, polysilicon etc., were removed using HF. After this treatment polysilicon residues remained in some of the trenches. To reveal the dislocations that intersected the surface of the bulk Si wafers, the chips were Secco etched (Secco 1972). TEM examinations were performed on (001) plan-view foils prepared by cleaving portions from the wafers followed by mechanical polishing and ion beam thinning from the back side. For the TEM studies of etched specimens, the bulk specimens were etched before the thin foils were prepared. The TEM images were obtained using two-beam conditions, the labelled arrows in the figures indicating the diffraction vectors. The accelerating voltage was 120 kV.

3. TRENCH INDUCED DISLOCATIONS

The first type of dislocation investigated was the 'trench induced dislocation' (TID). OM examination of bulk specimens etched for 30 s revealed etch pits with densities of up to 10^5 cm^{-2}. SEM examination of the same specimens additionally showed inclined etch channels running from the pit at the surface of the specimen down to meet the edge of a memory cell, eg PP' in Fig.2. (Kolbesen et al 1989b)

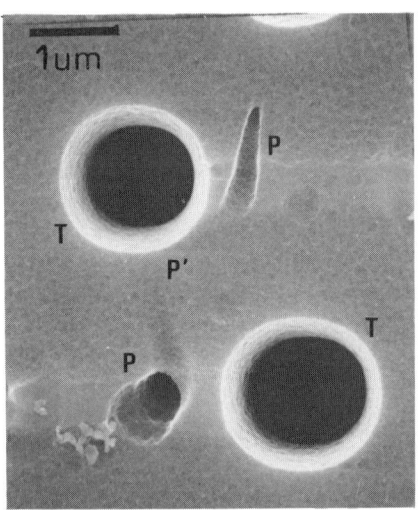
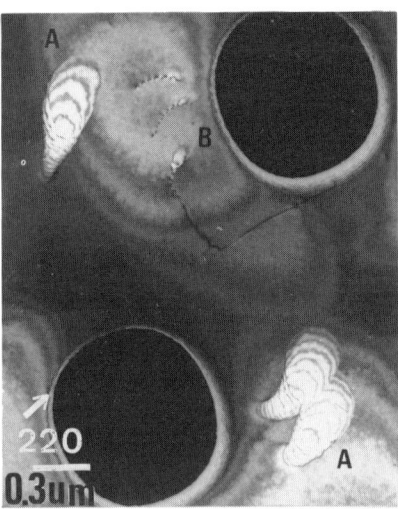

Fig.2: SEM plan-view of a bulk specimen after 30 s Secco etching. Trench holes (T) and dislocation etch pits (P) can be seen.

Fig. 3: TEM plan-view after 10 s Secco etching. Trench holes and fully (A) as well as slightly (B) etched dislocations can be seen.

TEM examinations of specimens etched for 10 s showed either large etch pits (Fig.3, A) where the associated dislocations had been substantially removed, or small etch pits (Fig.3, B) where the associated dislocations were still present, ie the dislocations were attacked at different rates . TEM examinations of similar specimens etched for 30 s showed only large etch pits with the dislocations fully removed. These etching/OM/SEM/TEM studies demonstrated that no new dislocations were introduced during the preparation of the thin foils, and that all of the TIDs intersecting the initial wafer surface gave etch pits.

They also indicated that the dislocation densities determined from OM examinations of specimens etched for 30 s gave slightly low values because the closely spaced dislocations would not have been resolved. Most of the subsequent TEM studies were performed on specimens that had not been etched.

TEM examinations showed that the most common form of the TIDs was as shown in Fig.4 at A. Tilting experiments in the TEM and the observation of stereo-pairs of micrographs showed that the dislocations ran from the initial wafer surface down towards the edge of a cell wall. Moreover, all the dislocations analysed were on inclined {111} planes, had a/2 <110> Burgers vectors inclined to the (001) surface and were close to being 60° type dislocations. These TIDs were not distributed evenly around the trench holes, but occurred more frequently in the regions between the closely spaced pairs of holes (Fig.1).

Investigation of DRAMs fabricated with different materials and thicknesses (5-15 nm) for the dielectric layer within the trench memory cell showed different densities of the TIDs. The TID density was smaller for a thick SiO_2 layer than for a thin one and very few TIDs were observed for a $SiO_2/Si_3N_4/SiO_2$ layer. TEM investigation showed the presence of small dislocation loops (Fig.5, A) in all these wafer types. They occurred in the bulk Si near the trench wall and close to the wafer surface.

From the geometry of the TIDs it is clear that many of them cross the p-n junction located around the trenches and it is likely that they would cause an increase in the leakage current of the memory cell. Indeed, electrical measurements revealed a strong correlation between the presence of TIDs and the occurrence of refresh failures (Kolbesen et al 1989a).

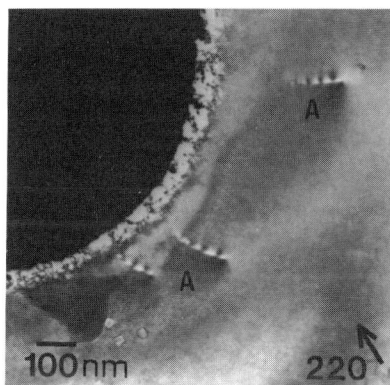

Fig. 4: TEM plan-view specimen showing the distribution of TIDs (A) around the trench hole.

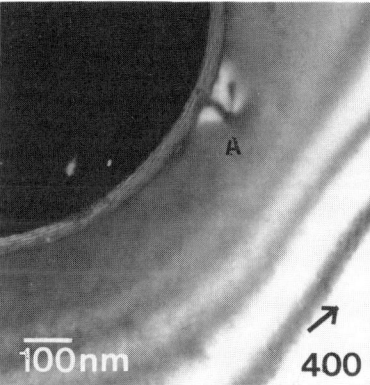

Fig. 5: TEM plan-view specimen with Si_3N_4/SiO_2 as a dielectric inside the trenches, showing a small dislocation half loop (A).

4. FOBIC DISLOCATIONS

The FOBIC lies in between two memory cells (Fig.1). When the technology layers were removed, the HF also dissolves most of the heavily n-doped region below the contact, so that this region occurred as a hole with uneven, inclined edges in the TEM foils.

TEM showed dislocations associated with many of the FOBICs (Fig.6, A), which ran along the <110> direction parallel to the longer edge of this region. These dislocations were parallel to and just below, the wafer surface, with their ends running up to the surface. They had a/2 <110> Burgers vectors lying perpendicular to the dislocation line, and so the main segments were 90° Lomer type dislocations.

For those dislocations, whose ends ran up to the FOBIC region itself (Fig.7, A), TEM showed similar, fine granular contrast for the ends as for the edges of the FOBIC regions. This suggests that pipe-diffusion of this dopant may have taken place down these dislocations, eg during the high temperature processing steps after the fabrication of the FOBICs. The geometry of these dislocations indicatesthat most of them do not cross the p-n junction associated with the FOBIC. There is no electrical evidence that they adversely affect the device performance.

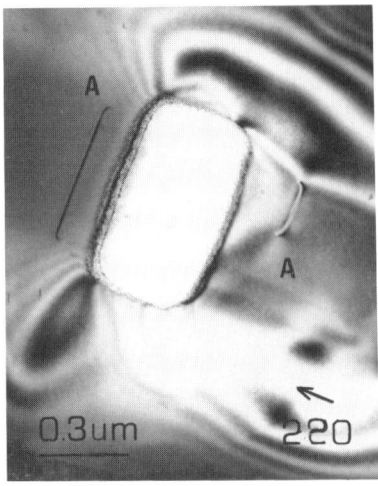

Fig. 6: TEM plan-view specimen showing dislocations (A) close to the FOBIC.

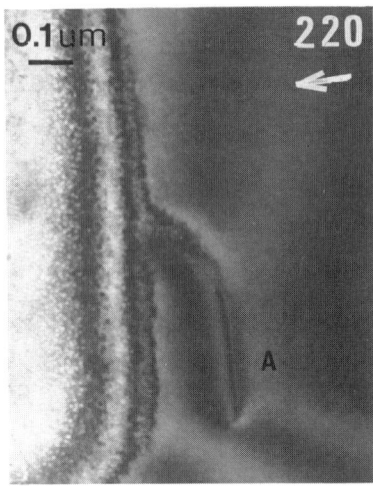

Fig.7: TEM plan-view specimen showing a dislocation (A) close to the FOBIC. One end of the dislocation is connected with the n-doped region of the FOBIC.

5. DISCUSSION AND CONCLUSIONS

The results of the investigation suggest that the TIDs are initiated by small loops that form at the edge of the trench memory cells during the fabrication of the DRAMs. Depending on the particular design of the memory cell, eg the thickness and type of the dielectric layer within the cell, stresses arise during the fabrication. These can cause the loops to glide away from the cells on inclined {111} slip planes. Such dislocations can intersect the p-n junction located around the trenches, and this can cause leakage of charge during operation of the DRAMs.

The FOBIC edge dislocations have not been previously reported and the mechanism whereby they occur has not yet been established. They do not adversely affect the device performance.

REFERENCES

Chatterjee P K, Taylor G W, Tasch A F and Fu H S 1979 IEEE Trans. Electron Devices ED-26 564-575
Kolbesen B O, Bergholz W, Cerva H, Gelsdorf F, Wendt H and Zoth G 1989a, Structure and Properties of Dislocations 1989, eds S G Roberts, D B Holt and P R Wilshaw (IOP Publishing Ltd., Bristol) 421-430
Kolbesen B O, Bergholz W and Wendt H 1989b Mat. Sci. Forum 38-41 1-12
Secco F 1972 J. Electrochem. Soc.: Solid-state Science and Technology 119 948-951

Inst. Phys. Conf. Ser. No 117: Section 4
Paper presented at Microsc. Semicond. Mater. Conf., Oxford, 25–28 March 1991

173

Scanning tunneling and atomic force microscopy imaging of VLSI Si test structures

U Memmert, H E Hessel, J Wiechers, R Houbertz, A Feltz, H Cerva[1] and R J Behm

Institut für Kristallographie und Mineralogie, Universität München,
Theresienstrasse 41, W-8000 München 2, Fed. Rep. Germany

[1] Siemens AG, Research Laboratories,
Otto-Hahn-Ring 6, W-8000 München 83, Fed. Rep. Germany

ABSTRACT: VLSI Si test-structures were investigated by scanning tunneling microscopy (STM) and atomic force microscopy (AFM). AFM measurements were used for topographic imaging in air. For STM measurements, the samples were imaged either under a protective layer of $0.1m$ H_2SO_4 or in vacuum to prevent re-oxidation of the Si surface. Spectroscopic measurements, in vacuum, reveal the local doping type. The reliability, resolution and information content of combined topographic and spectroscopic measurements are demonstrated in a study of the lateral dopant migration during processing of the structures.

1. INTRODUCTION

The sub-micron size structures of modern silicon devices require new microscopic and analytical methods with the ability to characterize their geometrical and electronic structures on a nanometer scale. Atomic force microscopy (AFM) and scanning tunneling microscopy (STM), both of which have sufficient lateral resolution, appear to be promising methods to contribute to the solution of these problems. Their potential is demonstrated in this study of the topography and the local doping character of the $Si-SiO_2$ interface in VLSI Si test-structures. In these measurements surface structures created during device processing are clearly resolved. They can be correlated with the local doping characteristics giving information on the effects of lateral dopant scattering during implantation and on the (lateral) migration of dopant atoms during recrystallization of the sample or in later annealing steps. Measurements following different annealing conditions can give access to the diffusion characteristics of the respective doping species. The reliability of the STM/AFM data was confirmed by transmission electron microscopy (TEM) and secondary ion mass spectroscopy (SIMS) measurements on identical test-structures.

2. SAMPLE MATERIAL

In this study two types of Si test-structures were investigated. The first kind of structure was prepared from a p-doped Si(100) substrate (B, $2\text{-}3 \cdot 10^{15}$ cm^{-3}). It consisted of periodic arrays, ca. 1 mm x 2 mm in size, of long polycrystalline Si stripes (poly-Si) on top of a thin gate oxide layer. The poly-Si stripes were pairwise arranged, with a pair-to-pair distance of 3.8 μm and the stripes within a pair 1.5 μm apart. They were then overlaid with a thick (ca. 300 nm) oxide ('spacer oxide') which acts as a mask during As^+ implantation. A 20 nm thick oxide layer ('scattering oxide') covered the areas between the poly-Si stripes. Subsequent ion implantation

(As$^+$, $5 \cdot 10^{15}$ cm^{-2} , 50 keV) amorphized the Si surface between the poly-Si stripes changing its doping characteristics to n$^+$ type. It resulted in a dopant concentration at the Si/SiO$_2$ interface of ca. 10^{21} cm^{-3} . The amorphous Si was then recrystallized by annealing. A detailed description of this test-structure and the recrystallization process was published earlier by Cerva and Küsters (1990). The processing scheme is typical for the fabrication of MOS transistors, in which the poly-Si stripes would correspond to the gates in a MOS structure.

The second type of structure was produced in a similar way, except for two important differences; a) the poly-Si stripes were significantly wider (ca. 1μm), b) the structure was exposed to an earlier implantation process (P$^+$, 10^{13} cm^{-2} , 80 keV) between patterning of the poly-Si and formation of the spacer oxide. This changed the area below the spacer oxide layer to n-type material. The subsequent As$^+$ implantation process, following patterning of the spacer oxide, was similar to that in the above samples. After implantation this sample was annealed at 900°C for 60 min.

On all samples for the AFM/STM measurements, the oxide cover was removed by etching in 40% HF solution (12 h). By under-etching the poly-Si gates were also lifted off from the surface. Because of negligible etch rates of Si in this solution, the resulting Si surface is identical to the previous Si/SiO$_2$ interface, therefore STM/AFM images reproduce the topography of the previous interface. STM measurements on these surfaces were performed in constant current mode, typically at 2-3 V tunnel voltage, while AFM data were recorded in repulsive force mode, in air. The large area of equivalent structures on these samples made their positioning in the STM/AFM uncritical.

3. RESULTS AND DISCUSSION

The structure of the first type samples is resolved in the TEM image in fig.1, which shows a cross-section perpendicular to the stripes. The poly-Si stripes and the surrounding thick oxide are clearly identified as brighter and darker areas, respectively. In the area between the gates, the scattering oxide is shown as a thin layer. Closer inspection of the Si/SiO$_2$ interface reveals 20nm high steps at the edges of the spacer oxide and a gentle ridge just below the poly-Si gate, which originated from the patterning of the spacer oxide and the under-oxidation of the poly-Si gates. These features are marked by arrows in fig. 1.

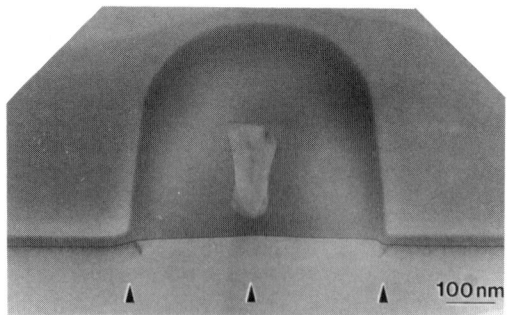

Figure 1: TEM cross-section of a gate on the 900°C/60 min sample.

The large scale image in fig. 2 (15 μm x 15 μm), recorded by AFM, gives an overview of the interface topography. The characteristic stripes - 20 nm high and 600 nm wide - correspond to the elevated areas underneath the spacer oxide stripes. Since the poly-Si gates were very narrow on this sample, the corresponding steps are very close to each other and the elevated area in between is visible only as a smooth ridge in the center of the stripes.

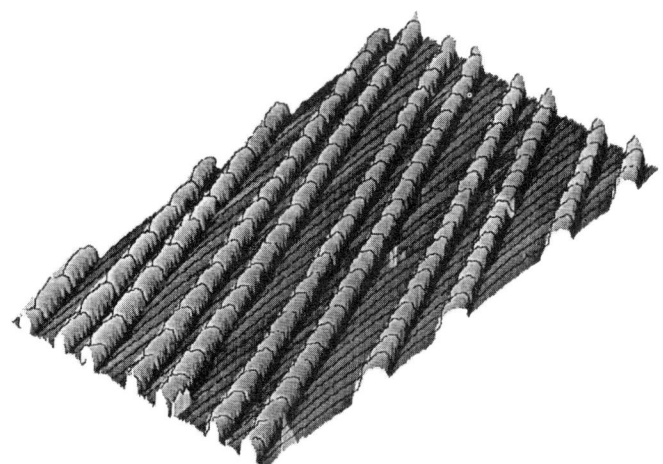

Figure 2: Overview AFM image (15μm x 15μm) of the Si test-structure.

STM imaging of these samples led to similar results. Topographic measurements were performed either under a protective layer of H_2SO_4 or under UHV conditions in order to prevent re-oxidation of the sample. In the first case, after removal of the spacer oxide and the poly-Si stripes by the HF etch, the sample was immediately covered with 0.1 m H_2SO_4 and kept at a cathodic potential with respect to the electrolyte (Tomita et al. 1990, Houbertz et al. 1991). In the second case, the samples were transferred into a UHV system, after the HF etch, via a fast sample load lock mechanism. The sample reached UHV conditions within 10 minutes.

STM measurements offer the possibility of imaging also the electronic structure of the surface by spectroscopic methods. This can be used to identify the local doping character simultaneously with its topography (Muralt et al. 1987, Hosaka et al. 1988, Kordic et al. 1990, Hessel et al. 1991). In our measurements a sinusoidal AC tunnel voltage with a peak-to-peak amplitude between 2 V and 3 V was applied instead of the commonly used DC tunnel voltage. For topographic imaging, the tip-surface distance was kept constant by regulating the AC tunnel current to a constant amplitude, by use of a lock-in amplifier. For determination of the local doping type the AC tunnel current was averaged using a low pass filter. Due to the different shapes of the current-voltage curves ('I-V curves') (Bell et al. 1988, Jahanmir et al. 1989) on differently doped regions of the sample, polarity and quantity of the averaged current depend on the doping type, even if the amplitudes are identical (Hessel et al. 1991). Hence they can be used as a finger-print of the doping type in the imaged region.

Combined topographic/spectroscopic STM images (1.5μm x 1.5μm) of three samples of the first type structure are reproduced in fig. 3. The topographic information is displayed in a three-dimensional line scan representation, while the averaged current signal is overlaid in a gray scale, with dark areas corresponding to n-type and light areas to n^+-type material. In all of these images, the samples exhibit two distinct regions with different values of the average tunnel current. I-V curves recorded on the different regions of the sample reveal that the change in the averaged current corresponds to an n^+/n transition at a dopant concentration of ~10^{19} cm^{-3} (Hessel et al. 1991). In order to assess the effects of dopant migration during sample annealing, which is commonly performed during processing, we investigated a series of differently annealed samples of this structure. The first and second sample were annealed in a furnace for 480 minutes at 500°C (500°C/480min) and for 60 minutes at 900°C (900°C/60min), respectively. The third sample was treated in a "rapid thermal annealer" at 1100°C for 30 seconds (1100°C/30s). The three

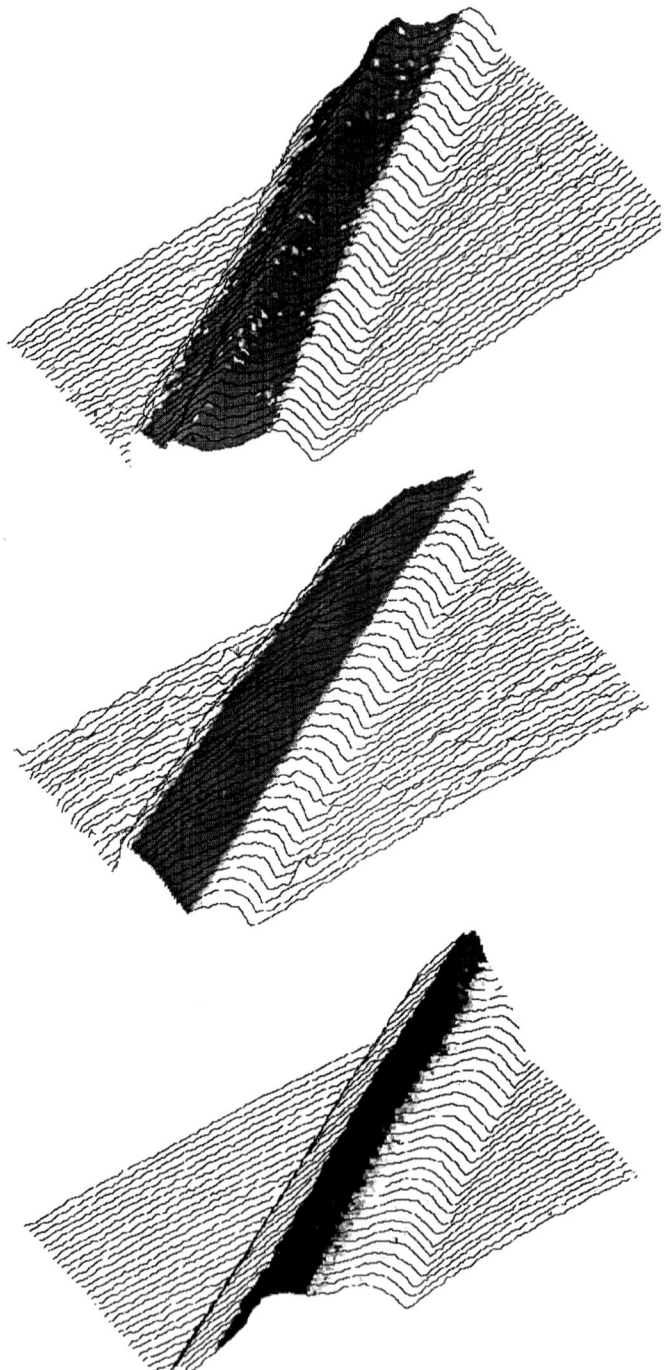

Figure 3: Combined topographic/spectroscopic STM images (1 μm x 1 μm) of differently annealed Si test-structures (dark: n-type material, light: n$^+$-type material). a) 500°C/480min sample, b) 900°C/60min sample and c) 1100°C/30s sample.

Annealing condition	Width of n-area (STM)	lat. displace-ment (STM)	lat. displace-ment (TEM)	vert. displace-ment (SIMS)
500°C/480min	505 ± 10 nm	-	-	0 nm
900°C/60min	390 ± 10 nm	60 nm	56 nm	55 nm
1100°C/30s	285 ± 10 nm	110 nm	104 nm	140 nm

Table 1: Lateral and vertical displacements, respectively, of the 10^{19} cm^{-3} concentration line in Si test-structures by thermal treatment for three different annealing conditions, as determined by STM (lateral), TEM (lateral) and SIMS (vertical).

images clearly demonstrate the strong influence of the annealing procedure on the width of the remaining n-type area, which is reduced with increasing annealing time/temperature. Assuming that for the 500°C/480min sample the diffusion process was negligible (see below), one can evaluate the lateral diplacement of the n+/n transition due to the annealing process from these images. These data are compiled in table 1.

For comparison, the vertical diplacement of the 10^{19} cm^{-3} concentration line upon annealing was determined by SIMS measurements. These showed no difference between an unannealed sample ('as-implanted') and the 500°C/480min sample, confirming our above assumption that there is no significant dopant diffusion under these annealing conditions. In contrast, the two other samples exhibited distinct dopant migration due to annealing. The vertical displacements of the position of the 10^{19} cm^{-3} concentration line in these samples are included in table 1. In the second last column the lateral dopant displacement determined from TEM/selective etch measurements is listed, where the n-type material had been thinned by a doping selective etch procedure (Cerva 1991). The data in table 1 show quantitative agreement between STM measurements and SIMS/TEM results.

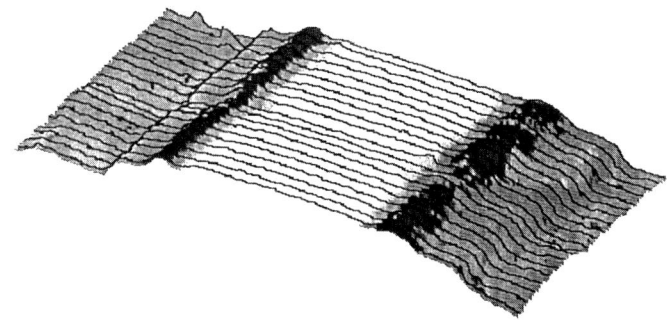

Figure 4: Combined topographic/spectroscopic STM image (2 μm x 1 μm) on a VLSI test-structure with n+/n and p/n transitions (light: p-type material, dark: n-type material, medium: n+-type material).

In STM measurements on these samples, no p-type areas were detected in the center region of the poly-Si stripes. This is different in the second type of test-structure, which had significantly broader poly-Si stripes and where the p-type areas protected against As+ implantation were wider. In these areas p/n transitions were resolved in STM measurements. This is documented in the image in fig. 4, which shows an area of 2 μm x 1 μm recorded on this sample. As before, the doping type is included in a gray scale representation where light colors correspond to p-type, medium colors to n^+-type and dark colors to n-type material. The two surface steps on this sample mark the position of the edges of the poly-Si stripes (inner steps) and the spacer oxide (outer steps), which correspond to the mask edges for P^+ and As^+ implantation, respectively. The regions modified by As implantation contain n^+-type material. The areas below the spacer oxide, where the smaller dose of P^+ implantation had taken place, consist of n-type material, and finally, well below the poly-Si gate, p-type material can be identified. Similar to the images shown in fig. 3 the respective transitions have moved below the former masks due to scattering and diffusion processes.

4. CONCLUSION

We have successfully demonstrated the application of AFM/STM measurements for the characterization of topographic and electronic surface properties of VLSI device structures, on a nanometer scale. Topographic imaging, by AFM or STM, yields well-resolved three-dimensional topographic information consistent with TEM cross-sectional images. Spectroscopic imaging, by STM, gains information on the local electronic properties, e.g. the local doping type. The use of combined topographic/spectroscopic measurements was demonstrated in a systematic study on (lateral) dopant migration in the surface region during processing of the test-structure, where the lateral displacement of a n^+/n junction was determined for different annealing conditions. The very good agreement of the STM results with TEM/selective etch and SIMS measurements lends credibility to the STM data. Further STM measurements and their correlation to device simulation calculations will allow detailed studies of dopant mobility during processing of Si devices.

Acknowledgement

We gratefully acknowledge financial support by the Bundesministerium für Forschung und Technologie (13N5571) and the Deutsche Forschungsgemeinschaft (SFB 338), and fellowships from the Volkswagen foundation (AF) and the Ludwig-Maximilians-Universität München (RH, AF).

REFERENCES

Bell L D, Kaiser W J , Hecht M H and Grunthaner F J 1988 Appl. Phys. Lett. <u>52</u> 451
Cerva H and Küsters K-H 1990 J. Appl. Phys. <u>66</u> 4723
Cerva H 1991 J. Vac. Sci. Technol. <u>B</u> submitted
Hessel H E, Memmert U, Cerva H and Behm R J 1991 J. Vac. Sci. Technol. <u>B</u> in press
Hosaka S, Hosoki S, Takata K, Horiuchi K and Natsuaki N 1988 Appl. Phys. Lett. <u>53</u> 487
Houbertz R, Memmert U and Behm R J 1991 Appl. Phys. Lett. <u>58</u> 1027
Jahanmir J, West P E, Young A and Rhodin T N 1989 J. Vac. Sci. Technol. <u>A7</u> 2741
Kordic S, van Loenen E J, Dijkamp D, Hoeven A J and Moraal H K 1990 J. Vac. Sci. Technol. <u>A8</u> 549
Muralt P, Meier H, Pohl D W and Salemink H W M 1987 Appl. Phys. Lett. <u>50</u> 1352
Tomita E, Matsuda M and Itaya K 1990 J. Vac. Sci. Technol. <u>A8</u> 534

Inst. Phys. Conf. Ser. No 117: Section 4
Paper presented at Microsc. Semicond. Mater. Conf., Oxford, 25–28 March 1991

179

Transmission electron microscopy characterisation of permeable base transistor structures

P Perret, P A Badoz, C Morin, J L Regolini and B Vuillermoz

Centre National d'Etudes des Télécommunications - CNS, B.P. 98, 38243 Meylan Cedex, France.

ABSTRACT: Transmission electron microscopy has been used to characterize permeable base transistor structures. Two kinds of technology have been investigated using silicon epitaxial growth over two different materials used as base. The purpose of these observations has been to check the geometry of the device, the behaviour of the metallic film and the quality of the silicon epitaxy. Depending on the conditions of the process, the epitaxial silicon can show different defects : stacking faults, dislocations, microtwins and a polycrystalline structure. Specific examinations have been made of the interface between the Si substrate and epitaxial Si.

1. INTRODUCTION

The permeable base transistor (PBT) is a device of great interest for its expected high frequency performance. The PBT, which resembles a vertical field-effect transistor, consists of a metal grid embedded in a semiconductor. The electrons flow in the semiconducting channel being controlled by the bias applied to the metal grid. The first PBT's fabricated on a GaAs substrate and using a tungsten grid achieved f_{max} values of 200 GHz (Hollis 1985). Several developments have since been made in order to fabricate this device on silicon. In the overgrown Si PBT version, described in this communication, the metallic base is embedded in epitaxial silicon. In this paper we report the characterisation by TEM of the Si PBT for the following two cases : WSi_2 base (Badoz 1990a) and W base (Badoz 1990b). WSi_2 was chosen for its stability and high resistance to the thermal treatments used in the process. In the second fabrication, in order to decrease the layer resistivity, W was used instead of WSi_2. In this case, a diffusion barrier is formed around the W fingers using plasma nitridation in order to prevent the formation of WSi_2 during thermal treatments.

2. TECHNOLOGICAL PROCESS

The starting material is 1 Ωcm, n type Si (100) in both cases. The epitaxial growth of silicon is obtained in a rapid thermal processing, low pressure chemical vapor deposition (RTP/LPCVD) reactor using SiH_4 diluted in H_2 (Regolini 1989). One of the problems encountered with this buried grid structure is the reactivity of the grid material to the surrounding silicon during epitaxy. This consideration has led to the epitaxial process mentioned above, and in particular much attention has been paid to growth temperature.

2.1 First case with WSi_2 base

After an RCA type clean, a WSi_2 layer is deposited by W and Si sputter codeposition followed by a rapid thermal annealing at 900 °C for 20 s to form the silicide. After a lithography step defining a grating, the pattern is transferred by SF_6 anisotropic reactive ion etching to the WSi_2 layer. Then the resist is removed and a second RCA type clean is performed before loading the sample into the CVD reactor where, prior to the silicon epitaxy, the wafer is heated to 900 °C for 30 s in hydrogen. After the epitaxial growth at 900 °C for 90 s, in order to achieve a good ohmic contact, a low energy (40 keV) selective arsenic implantation (1.5 10^{15} As/cm^2) is performed, associated with a lithography step and followed by a thermal anneal at 650 °C for 30 min under N_2. Finally, an Al metallisation layer is deposited.

2.2 Second case with W base

First a very thin nitride layer is grown at 950 °C on top of the Si substrate using a nitridation plasma treatment (URANOS system : Unit of Reactive Anodization for Nitridation and Oxidation of Semiconductors, French

Patent No 83, 1983). Then a tungsten layer is deposited in a DC magnetron sputtering system. After a lithography step defining the grating, the pattern is transferred by anisotropic reactive ion etching to the W and nitride layers, followed by removal of the resist. A second plasma nitridation is then performed in the URANOS system. The nitride layer created over the silicon substrate is removed using a spin clean procedure. The wafer is loaded into the epitaxy reactor where, before the epitaxial silicon growth, it is heated, firstly at 150 °C for 60 s and secondly at the temperature of the epitaxy for 60 s. Subsequently the same arsenic implantation as previously is performed (40 keV, $1.5 \, 10^{15}$ As/cm^2). Finally a W metallisation layer is deposited.

3. TEM RESULTS

The TEM observations have been made on a cross-sectional specimen in the [110] silicon direction.

3.1 First technology : WSi$_2$ base

In Fig 1 we can see the geometry of the structure. No modification of the silicide fingers occurs during the silicon epitaxy, due to the stability of the WSi$_2$ which can be seen fully covered by the epitaxial silicon. The Si channel between the metallic fingers is defect free and the original surface of the Si substrate is undetectable except near the metallic finger where sometimes there is a short line of dotted defects (Fig 1e). The vertical silicon epitaxy extends by a lateral epitaxial growth over the WSi$_2$ film which can be explained by a diffusion of Si species on the surface of the silicide. This lateral epitaxy is accompanied by the formation of microtwins and stacking faults probably caused by the roughness of the silicide. The density of these defects increases in the middle of the fingers in the area where the two lateral growths join each other. In this area the silicon thickness is smaller leading to a roughness of the epitaxial silicon surface. Figs 1b,1c,1d show the SA diffractions of σ3 twins with two diffferent <111> twin axes depending on the side of the lateral epitaxy (spots of double diffraction appear). Fig 1a shows a magnification of these twins. Over the silicide pad (Fig 2a) the lateral epitaxy can extend for up to 6 μm, the silicon becoming polycrystalline, with preferential orientations. At the same time, there is under the silicide a vertical epitaxial growth of silicon diffusing through the WSi$_2$. The original surface of the Si substrate is observed as a line of dotted defects. HREM observations do not reveal specific precipates or silicide grains. However microtwins seed from this defect line (Fig 2b). This could be a

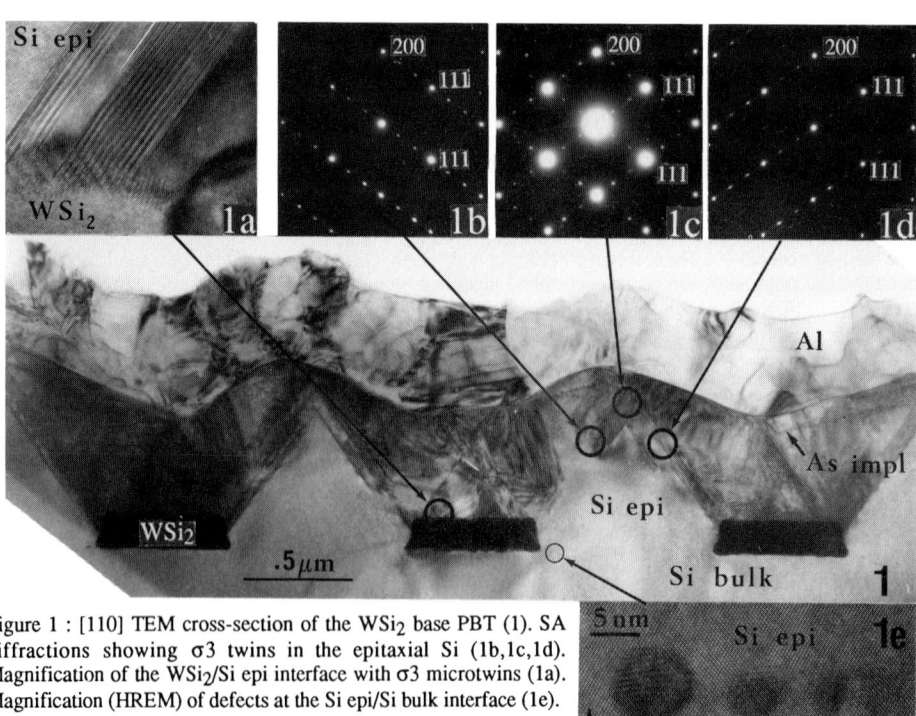

Figure 1 : [110] TEM cross-section of the WSi$_2$ base PBT (1). SA diffractions showing σ3 twins in the epitaxial Si (1b,1c,1d). Magnification of the WSi$_2$/Si epi interface with σ3 microtwins (1a). Magnification (HREM) of defects at the Si epi/Si bulk interface (1e).

problem of insufficient cleaning of the Si surface at the beginning of the process before the silicide formation. Regarding As implantation, a defect line about 70 nm below the upper surface can be seen (Fig 1).

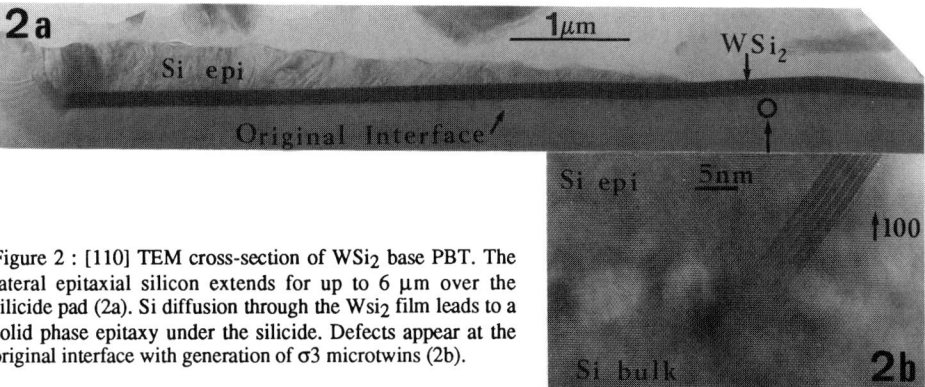

Figure 2 : [110] TEM cross-section of WSi$_2$ base PBT. The lateral epitaxial silicon extends for up to 6 μm over the silicide pad (2a). Si diffusion through the Wsi$_2$ film leads to a solid phase epitaxy under the silicide. Defects appear at the original interface with generation of σ3 microtwins (2b).

3.2 Second technology : W base

Three combinations of two different temperatures for W finger nitridation and Si epitaxy have been tested. A first approach with nitridation and Si epitaxy at 850 °C and a second with nitridation at 850 °C and Si epitaxy at 800 °C show a similar result (Fig 3a). A third approach with nitridation at 700 °C and Si epitaxy at 800 °C led to a reaction between the tungsten and silicon during epitaxy (Fig 4) : Energy Dispersive X-Ray analysis has revealed the presence of W in the Si between the grid and upper metallisation. These results indicate that the temperature of 700 °C for the nitridation is not sufficient to form a diffusion barrier. The Si$_3$N$_4$ nitride grown at 950 °C on the Si substrate shows a thickness of 40 Å measured using the lattice parameter. Contrast from stresses between the silicon substrate and the nitride film can be observed in Fig 3b. The nitridation forming the diffusion barrier is not really visible although an amorphous contrast appears in some areas around the W fingers. The N incorporation into the W has been shown to be mainly localised at the metal surface by Auger spectroscopy characterisation. Fig 5 shows a single crystal growth of silicon over the silicon seeds and a polycrystalline growth over the W lines. A defect line can be seen at the substrate/epitaxial silicon interface (Fig 3c), a contrast similar to that presented in Fig 1e for the WSi$_2$ PBT being observed. This defect contrast can be explained by residual impurities left by incomplete cleaning of the Si surface prior to epitaxy. At the edge of the W grid where the silicon becomes polycrystalline, dislocations, stacking faults in the (111) plane and twins are observed that could be caused by stresses. In particular in Fig 4 where the W has reacted with Si, these defects can run as far as 1μm into the substrate. Fabrication of the structure also has been carried out using the same

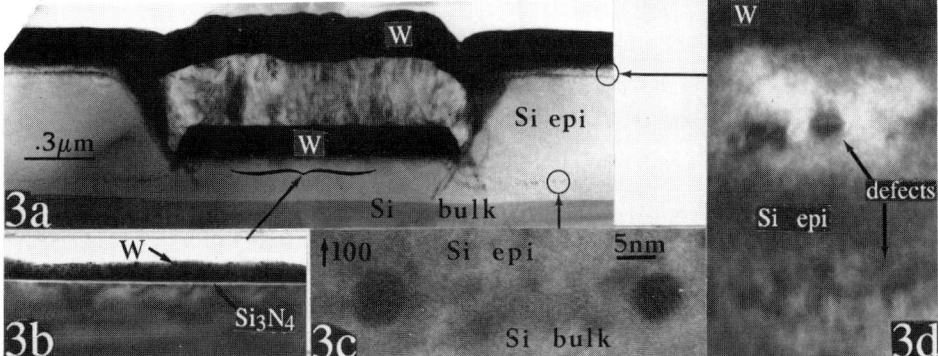

Figure 3 : [110] TEM cross-section of W base PBT with nitridation at 800 °C and 850 °C. Dislocations and stacking faults in the (111) plane at the W finger edge (3a). Stresses at the Si$_3$N$_4$/Si interface (3b). Defects at the Si epi/Si substrate interface (3c). Two defect lines (250 Å and 600Å below the epitaxial Si surface) caused by As implantation : dislocation loops in the first defect line (3d).

process (nitridation at 850 °C) but on a (111)Si wafer. In this case stacking faults and twins in the (111) plane run parallel to the wafer surface (Fig 6). On this sample the Si epitaxy/Si substrate interface is undetectable, and no defect line is visible. The As implantation step has generated two defect lines visible under the top surface of the epitaxial silicon respectively at 250 Å and 600 Å (according to the conditions of the implantation the R_p is localized at 269 Å). The types of defects are mainly dotted defects and dislocation loops (Fig 3d). In the WSi_2 PBT only one defect line is seen and this difference has not been explained. Other experiments will be carried out in order to investigate this problem.

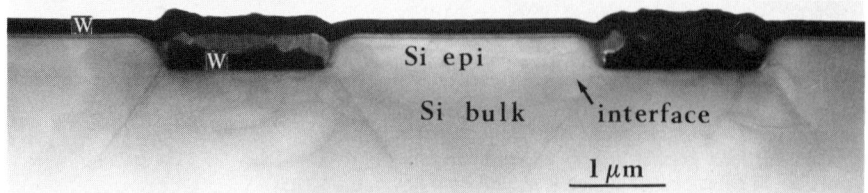

Figure 4 : [110] TEM cross-section of W base PBT with nitridation at 700 °C. W and epitaxial Si have reacted. The nitridation barrier is not efficient. Dislocations and stacking faults in the (111) plane run deep into the substrate.

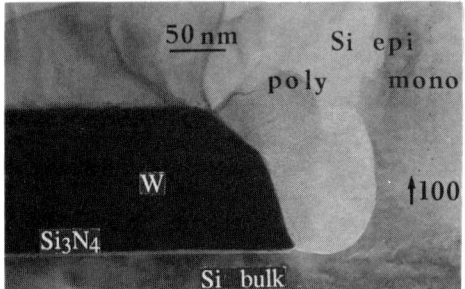

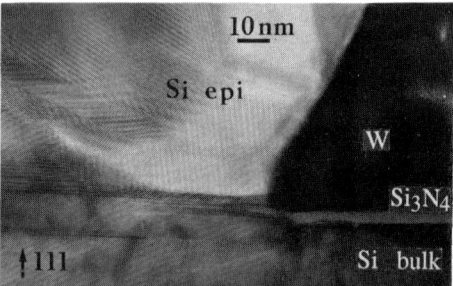

Figure 5 : [110] TEM cross-section of W base PBT on a 100 wafer with nitridation at 850 °C. Transition between monocrystalline and polycrystalline epitaxial silicon covering the metallic finger.

Figure 6 : [110] TEM cross-section of W base PBT on a 111 wafer with nitridation at 850 °C. Stacking faults in the (111) plane at the edge of the metallic finger.

4. CONCLUSION

The interface between the epitaxial silicon and silicon substrate often shows a black dotted defect line, similar contrast being observed in all the samples but with different defect densities. These defects could be explained by remaining species concentrated in some areas due to gettering effects. No specific grains or precipitates have been observed. A thermal cleaning procedure has not been optimised specifically for nitride on Si, and more work should be undertaken in this direction. The silicon epitaxy is not selective, occurring on the W where there is a polycrystalline silicon deposit. On the WSi_2 there is a diffusion throughout the film indicating that there is probably an adsorption of Si species on the WSi_2 and perhaps reaction. Moreover the lateral extension of Si, over the silicide pad, does not correspond to that observed in the case of real selective epitaxy as on SiO_2 (Jastrzebski, 1983). The PBT with a W base exhibits more defects than the PBT with a WSi_2 base, in particular dislocations, stacking faults and a polycrystalline structure at the edge of the metallic finger. These defects could be more critical with the reduction in stuctural dimensions. These developments have shown the feasibility of such devices with good electrical results in spite of the defects (g_m = 50 mS/mm for both technologies) and the TEM observations have allowed these technologies to be checked and better understood.

REFERENCES

Badoz P A, Bensahel D, Guérin L, Perret P, Puissant C and Regolini J L 1990a J. of Electr. Mat., 19 1123
Badoz P A 1990b proceedings of the 2nd intern. conf. on electronic materials (ICEM)
Hollis M A 1985 IEDM Tech Digest, 102
Jastrzebski L 1983 J. Cryst. Growth 63 493
Regolini J L, Bensahel D, Scheid E and Mercier J 1989 Appl. Phys. Lett. 54 658

Inst. Phys. Conf. Ser. No 117: Section 4
Paper presented at Microsc. Semicond. Mater. Conf., Oxford, 25–28 March 1991

183

Scanning ion microscopy investigation of semiconductor devices

JP Gonchond, G Mascarin, E Jourde and R Pantel

Centre National d'Etudes des Télécommunications
BP 98, 38243 MEYLAN, France

ABSTRACT : Submicron CMOS devices have been investigated by Scanning Ion Microscopy using a 30 keV Gallium Focused Ion Beam operated in a SEIKO SMI8300 FIB microscope. Application examples in the fields of Process Monitoring and Failure Analysis of VLSI are described.

1. INTRODUCTION

Scanning Ion Microscopy (SIM) using a Focused Ion Beam (FIB) is becoming popular in both laboratory and industrial environments as a powerful tool for investigating semiconducting materials and IC technologies. The effectiveness of the technique in microsectioning and in situ imaging the sample structure has been demonstrated (Nikawa et al 1989).

After a brief description of the SEIKO SMI8300 FIB Microscope we present some application examples in the fields of Process Monitoring and Failure Analysis of submicron CMOS devices.

2. DESCRIPTION OF THE F.I.B. SYSTEM

THE SEIKO SMI8300 F.I.B Microscope is a dedicated System for making precisely located cross-sections and for their in situ observation. The Focused Ion Beam is obtained from a Ga+ Liquid Metal Ion Source and can be accelerated up to 50 kV (30 kV for normal operation). Beam current values varying from 2 pA to 4 nA can be achieved on the sample surface, depending on the condenser lens excitation and objective aperture selection. The smallest available spot size is about 500 Å. Basically, the SMI8300 System has three main functions :

      i) <u>Micro-deposition</u>

Tungsten layers with a resistivity of 100 - 200 $\mu\Omega$-cm can be locally deposited from a Tungsten hexacarbonyl source via the FIB assisted CVD (Chemical Vapor Deposition) surface reaction :

$$W(CO)_6 \xrightarrow{\text{FIB (Ga+)}} W + 6\,CO$$

In this condition, deposition rate in the range of 0.5 to 1 μm^3/s at 2 nA ion beam current can be achieved.

      ii) <u>Micro-etching</u>

As the ion beam scans the sample surface, atoms and molecules are sputtered out, resulting in a clean cut through the structure.

      iii) <u>in situ SIM imaging</u>

The cross-section is displayed as a secondary electron image after the sample is tilted. In order to minimize sputtering of the sample surface during observation and to get the maximum resolution, an ion beam current in the pA range should be used .

3. PROCESS MONITORING

Test patterns consisting of transistors with different lengths are used to check the electrical parameters of our CMOS process. In order to control the device morphology and to make a comparison between electrical and physical parameters, it is useful to investigate one transistor structure designed at a given length. Microsectioning of this MOS transistor has been accomplished using the FIB techniques as shown in the optical micrograph of **Fig.1a**. One can see the area where tungsten has been first deposited and the small cut obtained through the structure after the local etching. We deposited FIB-CVD tungsten in this case, because :

  - it planarizes the top surface of the device and therefore eliminates etching artefacts due to topography;

  - it can be easily grounded to substrate by making a hole with the beam spot in order to minimize charging effects.

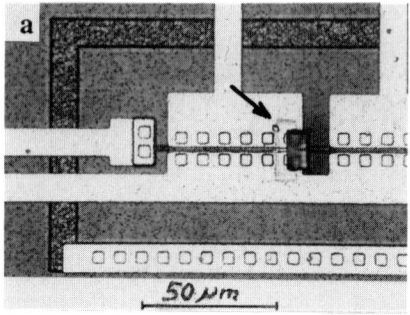

Figure 1a : Optical view of the FIB microsection of the MOS transistor

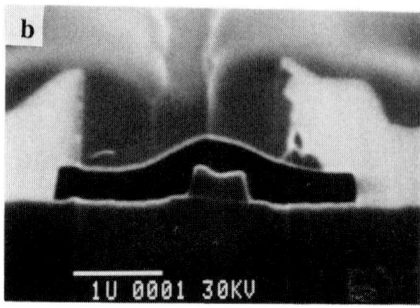

Figure 1b : in situ SIM image of the microsection shown in Fig.1a, after tilting the sample by 60°.

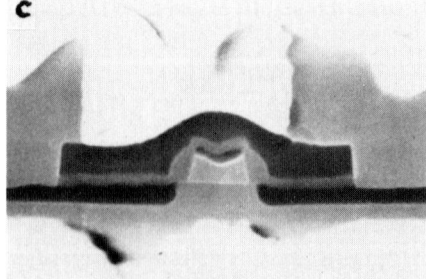

Figure 1c : SEM micrograph of the MOS transistor after chemical delineation of the structure.

The MOS structure obtained by in situ SIM imaging after 60° tilt is shown in **Fig.1b**. In this cross-sectional view, thick oxide layers appear dark whereas the thin gate oxide (15 nm thick) is imaged as a bright line. This is explained as follows : under the ion beam irradiation, thick oxide layers charge-up, resulting in a shift of a few eV in the energy of the secondary electron signal which is then not detected. On the other hand, charge pile-up cannot occur in the case of the gate oxide because its thickness is less than the smallest available spot size : charges are eliminated in the surrounding grounded material. As a result, even if the SIM resolution is limited to about 50 nm, very thin oxide layers (about 15 nm) will be imaged as a bright line in a cross-sectional SIM observation.

Then, we subjected this FIB cross-section to a chemical etching (HF/HNO3/Acetic acid 1:3:20 for 20 s.) for sample structure and junction delineation, and observed them in a conventional SEM, as illustrated in **Fig.1c**. Comparing Figs.1b and 1c, it is shown that SIM and SEM images contain complementary information : contact structure is clearly shown by SIM, whereas the different oxide layers, gate spacers and junctions are only visible in the SEM micrograph.

SIM imaging of metallic interconnexions presents another interesting particularity: channeling of the Ga+ ions inside the crystal lattice gives rise to grain contrast in polycrystalline and metallic layers, as shown in **Fig. 2**. Two experimental requirements should be met to get good channeling contrast in a SIM image : i) overlaying native oxide must be etched before observation; ii) sample chamber vacuum must be less than 1×10^{-6} torr.

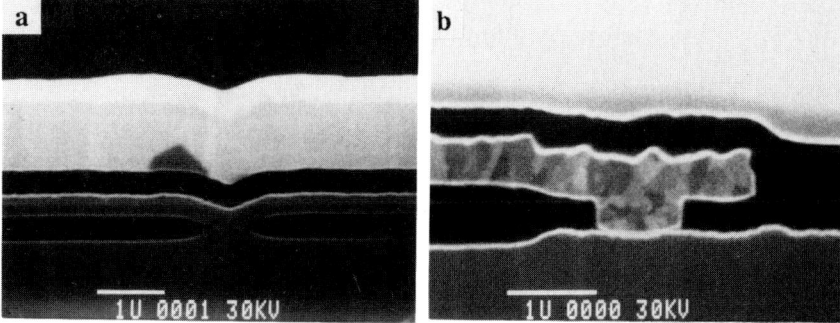

Figures 2 : SIM Micrographs showing a silicon grain in an Al-Si alloy (a), and the bulk grain structure of a CVD tungsten interconnect (b).

4. FAILURE ANALYSIS

We will give an example of failure analysis, where the FIB technique has been successfully applied.Some contacts between first (W) and second (Al) metal of the "WAL CMOS" Process developed at CNET were found defective. In this process, the planarization scheme of the interlevel dielectric layer uses SOG (Spin On Glass). To get a cross-sectional view of this structure, the commonly used sample preparation procedures such as mechanical polishing or cleaving, were found to be not efficient : Mechanical polishing is a time consuming and difficult operation due to the small feature size, whereas cleaving is hazardous and not easy due to the different hardness between tungsten and aluminum.

Using the FIB techniques in this case, we first deposit some tungsten to protect the structure. Then, microsectioning with ions is performed step by step, with a beam positioning accuracy of about 100 nm, in order to get to the center of the submicron contact area. The SIM image of the cross-section at 60° tilt is shown in **Fig 3a**. One can see some voids in the Aluminum layer in the vicinity of the contact. After chemical delineation of the structure, the SEM micrograph of **Fig.3b** shows that some SOG is emerging at the edge of the "via" holes. These observations give support to the well-known "via poisoning" effect (Ting et al 1987).

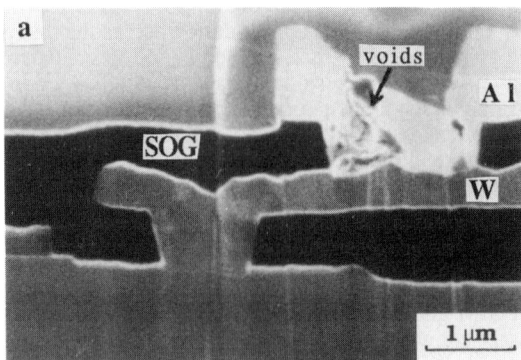

Figure 3a : SIM image of the contact structure.

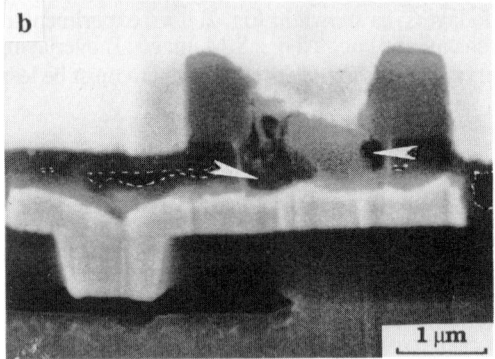

Figure 3b : SEM image after chemical delineation of the structure.

5. CONCLUSION

We applied successfully the FIB techniques to monitor some critical steps of our CMOS process and to identify some failure mechanisms in submicron devices. These techniques make process monitoring and failure analysis easier and less time consuming.

SIM imaging gives rise to specific contrasts which must be discussed in terms of charge pile-up in non conductive materials and ion channeling effects in polycrystalline layers, such as Al-Si alloy and tungsten. In particular, very thin oxide down to 15 nm have been imaged as a bright line, although the SIM resolution is limited to about 50 nm.

Combining in situ SIM imaging with SEM examination after chemical delineation of the sample structure, we found that both methods provide complementary information, which must be taken into account to get an overall understanding of device morphology.

REFERENCES

Nikawa K., Nasu K., Murase M., Kaito T., Adachi T. and Inoue S. 1989 27th Annual Proc.of International Reliability Physics Symposium, Phoenix, USA (IEEE Publish. : New-York), pp 43-52
Ting C.H., Lin H.Y., Pai P.L., and Oldham W.G. 1987 Proc. of the 4th VLSI Multilayer Interconnect. Conf., Santa Clara, (IEEE Publish.: New-York), pp 61-77

Inst. Phys. Conf. Ser. No 117: Section 4
Paper presented at Microsc. Semicond. Mater. Conf., Oxford, 25–28 March 1991

187

The oxidation of silicon observed *in situ* by imaging and diffraction of surface monolayers

Frances M Ross and J Murray Gibson

AT&T Bell Laboratories, 600 Mountain Avenue, Murray Hill, NJ 07974, USA.

ABSTRACT: Using a modified electron microscope we have observed the progress of the initial stages of the oxidation reaction on the Si (111) 7x7 surface. By analysis of the intensities of forbidden reflections we can observe the reaction proceeding with sub-monolayer resolution. Our results suggest that the reaction occurs one monolayer ($1/6$ unit cell) at a time with bilayer steps not moved during the process.

1. INTRODUCTION

Our understanding of the initial stages of the important silicon oxidation reaction has been hampered in the past by a lack of detailed, sub-monolayer resolution information about the progress of the reaction. By allowing the oxidation reaction to proceed *in situ* in an electron microscope we have been able to extract such information by imaging and diffraction of the surface layers, observing the reaction as it proceeds with sub-monolayer resolution. In this paper we present our results and consider the exciting consequences of having this sort of information, which will enable us to examine the oxidation kinetics in detail and to describe the surface structures formed and the behaviour of steps with atomic precision.

We have examined the oxidation reaction both by diffraction, which is sensitive to the reconstruction of the surface monolayers, and by imaging the monatomic surface steps directly using "forbidden" surface reflections. Exposure to O_2 or H_2O is carried out *in situ* in the microscope, with fast reactions being captured on video and slower ones by conventional imaging. Computer modelling is used to relate the diffracted intensities seen to the atomic positions near the silicon surface. This combination of imaging and diffraction techniques allows us to characterise both highly ordered, flat surfaces and those on which a reaction is taking place unevenly. Using these techniques, we have studied the etching of the silicon surface at elevated temperatures and lower pressures of oxygen as well as its roughening and oxidation at higher O_2 and H_2O pressures. In this paper we concentrate on the oxidation of the Si (111) 7x7 surface both by water and by oxygen. We will firstly describe the experimental procedure and the theory of imaging using forbidden reflections, and give an overview of the etching behaviour of the silicon surface. We will then describe the oxidation process in the initial stages, considering in particular the behaviour of steps and the structure of the oxidising surface.

2. EXPERIMENTAL DETAILS

The silicon specimens used in this study were cut from (111) oriented p-doped 10 Ωcm silicon wafers mechanically dimpled and chemically thinned to perforation. Once inserted into the microscope, a JEOL 200CX modified to achieve a base pressure in the region of the specimen in the 10^{-9} Torr range (McDonald et al. 1989), the specimens were cleaned by direct resistive heating to about 1200°C. This produces clean surfaces showing the 7x7 reconstruction, with suitable areas for analysis less than about 200nm thick, as we will demonstrate below, and having terrace widths which can be larger than 1μm with careful preparation. O_2 or H_2O gas was introduced into the specimen chamber through a leak valve with the pressure measured by a quadrupole mass analyzer. Most oxidation experiments were carried out at room temperature but elevated temperatures were achieved when desired by resistive heating, with the actual specimen temperature calculated from the heat dissipated using an equation calibrated by pyrometry taking radiative losses into account.

To monitor the progress of surface reactions both diffraction and imaging techniques provide useful and complementary information. The superlattice spots arising from the reconstruction of the two surfaces are clearly visible even from specimens as thick as 200nm and have intensities typically 10^{-6} of the incident intensity. Matching intensities with kinematical simulations can yield both the initial configuration of the surface atoms (Takayanagi et al. 1985) and the changes occurring after dosing with O_2 or H_2O. These superlattice spots fade and disappear after doses as low as a few L of oxygen (1L=10^{-6} Torr sec.) when irradiated by the electron beam (Gibson 1990). However even after much higher doses the "forbidden" $^1/_3$ 422 reflections remain visible, and we have found that images formed using these reflections can yield a great deal of interesting information: the positions of surface steps are clearly visible and the intensity levels on the terraces between steps are sensitive to the surface structure at the silicon/oxide interface. Furthermore the intensity in this spot is great enough to allow video recording of such images using a Gatan image intensifier, and thus dynamical records of the experiments can be obtained.

The behaviour of the $^1/_3$ 422 reflection is important in understanding the results presented here and it is easiest to appreciate if we consider the hexagonal unit cell (figure 1a). Kinematical theory leads to the amplitude-phase diagram shown in figure 1b for a crystal at a two-beam condition. The triangle is traversed many times within the thickness of the specimen but it is the relative positions of the start and end points which determine the final amplitude of the $^1/_3$ 422 reflection (Cherns 1974; Ourmazd et al. 1983). For example, if there is an integral number of unit cells in the beam direction the intensity is zero but if a monolayer step ($^1/_6$ unit cell) is present the intensity jumps to a value of 1. Four different amplitudes (0, 1, √3, 2) are possible assuming that the top and bottom of the specimen can terminate at any of the six layers. However if we only permit one type of surface layer (A, B, C or a, b, c) the number of intensities is reduced to two, i.e. 0, 2, 2 in the first case and 0, 1, 1 in the second. The importance of this becomes clear if we consider that the A- and a-type surface layers are physically different, with the A-type leaving three dangling bonds per atom but the a-type only one. We will refer to these surface terminations as "shuffle" and "glide" respectively by analogy with dislocation theory.

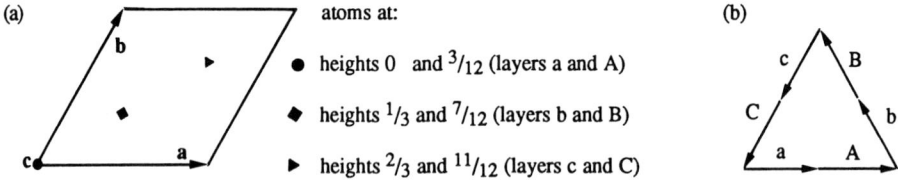

(a)

atoms at:

● heights 0 and $^3/_{12}$ (layers a and A)

◆ heights $^1/_3$ and $^7/_{12}$ (layers b and B)

▶ heights $^2/_3$ and $^{11}/_{12}$ (layers c and C)

(b)

Figure 1: (a) The hexagonal unit cell of silicon. $\mathbf{a}=^1/_2[1\bar{1}0]$ (0.38nm), $\mathbf{b}=^1/_2[10\bar{1}]$ (0.38nm), $\mathbf{c}=[111]$ (0.94nm). The six atoms lie in six different layers (z coordinates are given above) where atoms in layer A lie directly above those of layer a, etc. (b) A simple amplitude-phase diagram showing the contributions to the $^1/_3$ 422 reflection from the six layers at a two-beam condition (for a 220 beam).

This general behaviour is confirmed by multislice computer simulations which include HOLZ effects by treating each layer as a separate grating. In simulations at the exact (111) zone axis the black-white-white contrast is in fact replaced by dark-pale-paler contrast, but the factor of 2 in amplitude (4 in intensity) between glide and shuffle terminations is preserved. Tilting the crystal modulates the intensities further by an envelope function whose period depends on the tilt. At tilts of only 5° this period becomes comparable to the 0.9nm repeat of the $^1/_3$ 422 modulation itself leading to complicated contrast levels with long repeat periods. We have in fact observed these effects experimentally for tilted specimens. Finally we briefly mention the effect of including the the 7x7 reconstruction in simple calculations of $^1/_3$ 422 intensity levels. Expected intensity levels are midway between the two extremes of the shuffle and glide terminations. The registration of the 7x7 unit cells between the top and bottom surfaces of the specimen has a complex effect on the contrast, enabling domain boundaries to be imaged (figure 2), but the effect appears to be only second order.

3.   RESULTS

3.1 Etching

Having discussed the expected appearance of images obtained in the $^1/_3$ 422 reflection we can now describe the importance of such images in studies of both etching and oxidation. It is well known (Lander and Morrison 1962; Smith and Ghidini 1982) that the silicon surface can be either etched or oxidised by O_2, depending on the pressure and temperature, and we have observed both reactions *in situ* using both H_2O and O_2 gases. We have already described the nature of the "phase diagram" of etching and oxidation (Ross and Gibson 1991a) so here we just note that oxidation takes place at higher pressures and lower temperatures with etching requiring high temperatures and lower pressures. Etching is a result of the reaction of Si and O_2 to form the volatile oxide SiO, and video recordings show that the process occurs by step flow, limited we believe by the nucleation of the steps rather than by the speed at which they flow once nucleated (Ross and Gibson, 1991b). The simple expedient of counting steps until the specimen perforates shows that a typical foil thickness used experimentally is about 120nm. Furthermore analysis of the intensity levels of several hundred successive terraces (Ross and Gibson 1991b) allows us to conclude that the etching removes *double layers*, leaving the surface with either a shuffle or a glide termination (or the 7x7 reconstruction if the temperature is below 860°C).

3.2 Oxidation

The first process which occurs at the Si (111) 7x7 surface upon exposure to oxygen or water at room temperature is the destruction of the 7x7 structure. This is confirmed both by the fading of the superlattice spots and by the disappearance of the triangular regions which we believe to be domain boundaries in the 7x7 structure (figure 2). This takes very little time - only a few seconds at pressures of 10^{-6}Torr. Analysis of 7x7 spot intensities (Gibson 1990) shows that oxygen initially attacks the backbond of the adatoms in the 7x7 structure, and after a dose of as little as 1L the spots start to fade indicating that the surface is becoming disordered. Note that this effect is beam induced, as doses of even 10^5L do not destroy the surface structure of unirradiated areas.

In oxygen, higher doses (for example 75 minutes at 10^{-4}Torr O_2) cause a fading of the step contrast in irradiated areas. Much higher doses of oxygen eventually cause a *reversal* of contrast in the images - this was observed for example after 20 minutes oxidation at 1atm O_2 (Ross and Gibson, 1991a), an experiment done without electron beam irradiation. We interpret such contrast reversals as due to the consumption of one or more layers of crystalline silicon to form amorphous oxide. Oxidation occurs much more readily with H_2O, as is well known, and contrast reversals, again beam induced, can be seen readily at doses about 400 times lower (figure 3). We have concentrated on this H_2O oxidation at pressures of typically 10^{-7}Torr because "true" oxidation (without electron beam irradiation) takes too long in our UHV system at pressures below 10^{-3}Torr O_2.

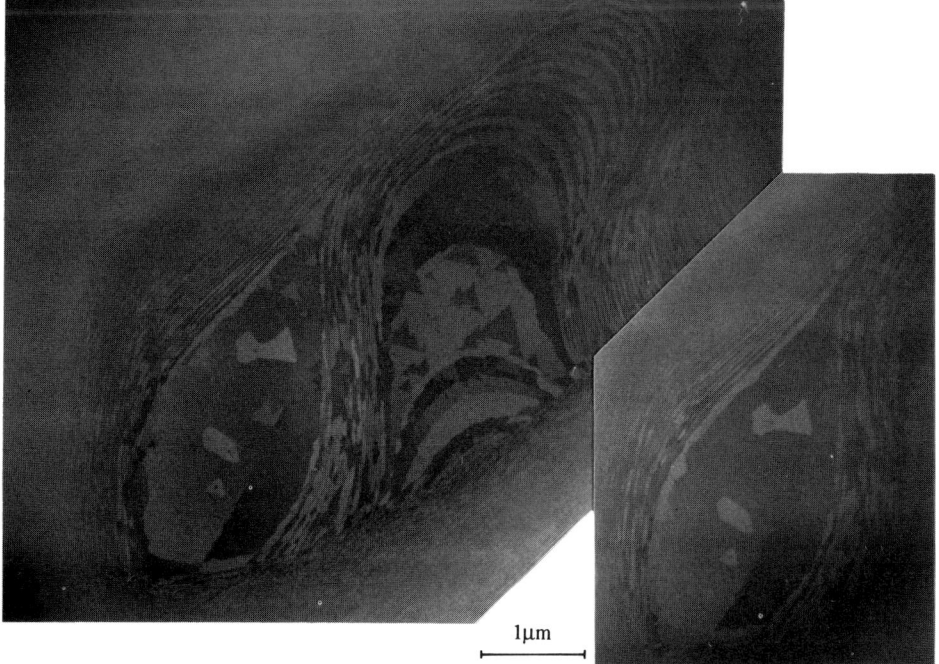

1µm

Figure 2 A clean (1.3L O_2 dose) silicon surface at room temperature, and (inset) after oxidation for 2.40 minutes in 9.0×10^{-7}Torr O_2 (250L dose). The triangles have disappeared but there is no other contrast change. Note how three repeating contrast levels are seen, strongly suggesting that the steps are all double (i.e. $1/3$ unit cell).

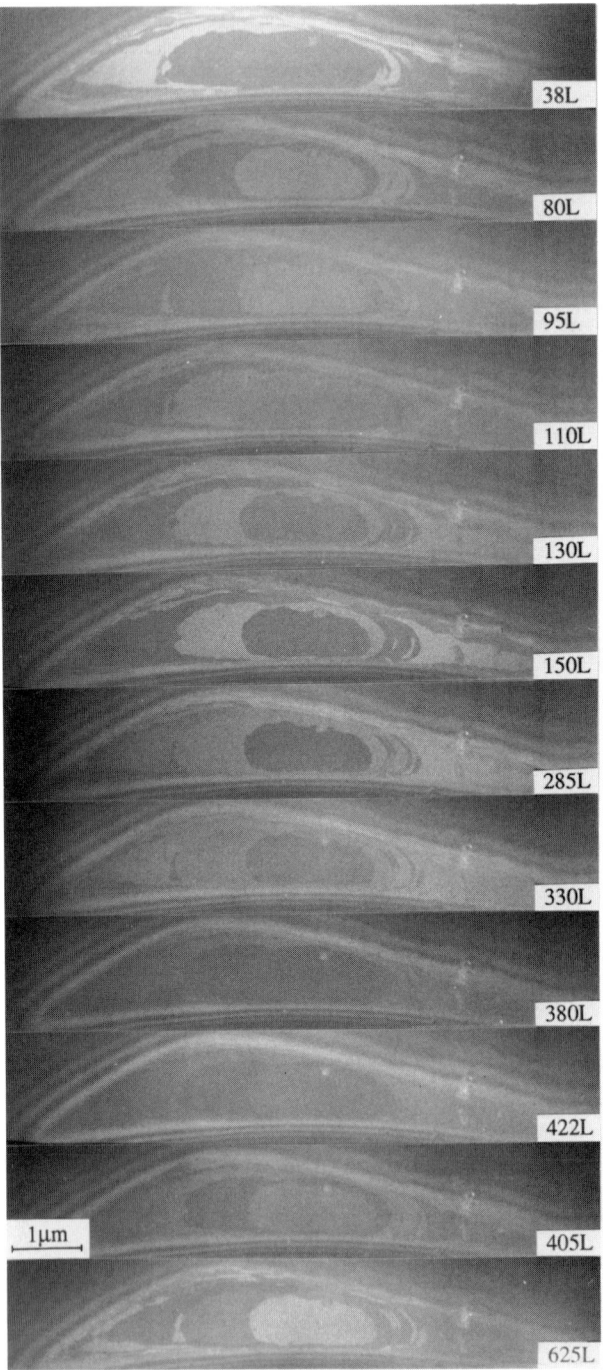

Figure 3 Part of a series of images taken over 40 minutes under RT exposure to H_2O at 2.5×10^{-7} Torr. Doses are given in L. Several contrast reversals can be seen but the steps do not move.

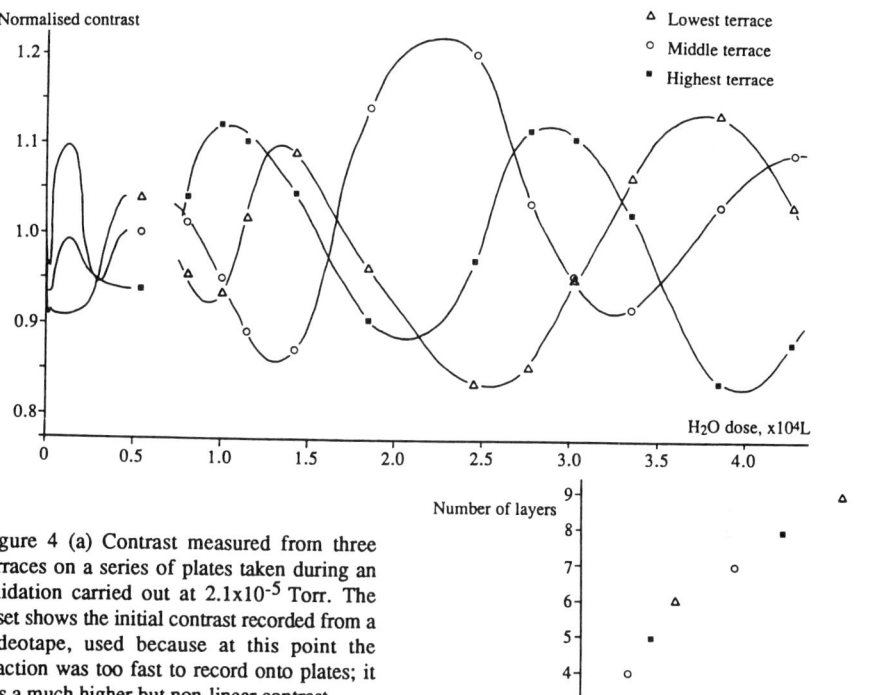

Figure 4 (a) Contrast measured from three terraces on a series of plates taken during an oxidation carried out at 2.1×10^{-5} Torr. The inset shows the initial contrast recorded from a videotape, used because at this point the reaction was too fast to record onto plates; it has a much higher but non-linear contrast. (b) The number of layers consumed against dose, measured from the position of the maxima on the previous graph.

It is important to emphasise that we see no evidence for step movement during this process, and hence the oxidation mechanism involves attack at the terraces. Examination of contrast data, an example of which is given in figure 4, gives valuable information about the oxidation process. In particular we feel that the following points are significant:

1) The general form of figure 4 is as expected, with the contrast repeating after three maxima. However it is not obvious why the contrast levels should return to a high value after each layer is oxidised. Random attack of unit cells at the silicon/oxide interface would lead, after a few layers, to a gradual roughening of the terraces and a decrease in contrast. Each layer is thus oxidised as a whole, with a lower layer not attacked until the upper one is fully (or mostly) reacted. TEM evidence has existed for a while (Krivanek et al. 1978; d'Anterroches 1984) that the Si/SiO_2 interface is very sharp even after hundreds of nanometers of thermal oxidation. It seems to us that models including a strained interface layer (Stoneham et al. 1985) have the best chance of explaining this.

2) From the *order* in which the curves reach their maxima, we deduce that if the three terraces are separated by double steps then the oxidation must be occurring *one* layer ($1/6$ unit cell) at a time, rather than two. The reasoning used to reach this important conclusion is detailed in figure 5, and relies on knowing the sense of the steps on the specimen. The suggestion of bilayer rather than monolayer steps appears justified as it is consistent with both the etching data mentioned in 2.1 and the repeating contrast we see generally at series of terraces, as in figure 2. We suspect that the termination preferred is the glide, both because it has fewer dangling bonds so is energetically favoured, and also because it should show lower contrast than the shuffle (section 1): we have found that contrast levels generally decrease in going both from the 7x7 to the oxidised surface at room temperature and from the 7x7 to the 1x1 reconstruction at 860°C. The suggestion of bilayer steps but monolayer oxidation can also explain why the steps do not move, as a mechanism for this would be hard to visualise if the steps were only a monolayer deep.

3) From the *shape* of these contrast curves we hope to obtain information about how each terrace oxidises.

However this information can not be extracted directly, since the intensity of the terrace as a whole depends non-linearly on the fraction of the unit cells at the silicon/oxide interface which has been oxidised.

4) The dose of H_2O needed to oxidise each layer is shown in figure 4b. This is a very accurate graph of oxide thickness against time and should be useful in distinguishing kinetic models of the initial oxide growth.

5) Finally, the lack of any extra diffraction spots apart from the $1/3$ 422 shows that if any crystalline oxide is present it either has a very low coverage (<10%) or one of a restricted range of structures.

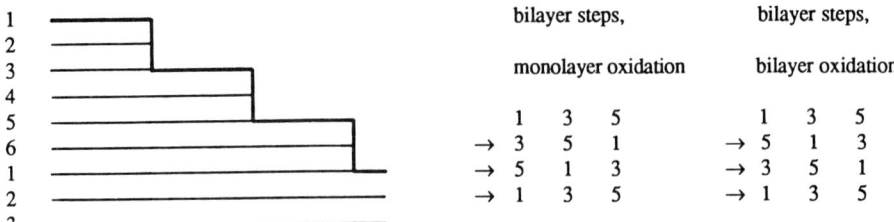

Figure 5 The specimen shown schematically, assuming that each step is double. Six different contrast levels are possible with initial levels seen being 1, 3 and 5. As the oxidation proceeds remember that both surfaces are reacting so 2n monatomic layers must be consumed between one contrast maximum and the next. If n=1 the contrast seen after one change will be 3, 5, 1 (see first table), but if n=2 the second table will apply. Several different experiments were consistent with n=1 and not n=2. Note that this relies on knowing the direction of the steps on a given specimen: for example in the situation shown in figure 3 we assume that the central terrace is the deepest. If instead we suppose that the steps are single, to agree with experiment the oxidation would have to occur in bilayers and furthermore we would have to assume that contrast levels 1=4, 2=5 and 3=6.

4.   CONCLUSIONS AND FURTHER WORK

Observations of the initial stages of the oxidation of silicon by a combination of diffraction and imaging techniques have demonstrated that the double steps present on the clean surface do not take part in the oxidation reaction, which instead proceeds by an attack on the terraces. We believe that we can see the points at which each monolayer ($1/6$ of the hexagonal unit cell) has reacted and before the next monolayer is attacked.

The possibilities of an oxidation study where sub-monolayer accuracy can be achieved in measuring the progress of the reaction are very exciting. We hope to be able to distinguish between models of the initial stages and the structure of the oxide adjacent to the interface using these detailed observations of oxide thickness, interface structure and the behaviour of steps. We also hope to resolve the ambiguity in our discussion of interface structure by isolating the *single* silicon surface in an oxidising environment, possibly by coating one side of the specimen with a silicide. Finally, since interface steps can be seen even under much thicker oxides we would like to look at higher pressure oxidations which are not dependent on the electron beam.

REFERENCES
D. Cherns (1974), Phil. Mag. **30**, 549
C. d'Anterroches (1984), J. Microsc. Spectrosc. Electron. **9**, 147
J. M. Gibson (1990), in *Atomic Scale Structure of Interfaces*, edited by R. D. Bringans, R. M. Feenstra and J. M. Gibson, Mat. Res. Soc. Proc. **159**, 179
O. L. Krivanek, T. T. Sheng and D. C. Tsui (1978), Appl. Phys. Lett. **32**, 437
J. Lander and J. Morrison (1962), J. Appl. Phys. **33**, 2089
M. L. McDonald, J. M. Gibson and F. C. Unterwald (1989), Rev. Sci. Instrum. **60**, 700
A. Ourmazd, G. R. Anstis and P. B. Hirsch (1983), Phil. Mag. **48**, 139
F. M. Ross and J. M. Gibson (1991a), in *Advances in Surface and Thin Film Diffraction*, edited by P. I. Cohen, D. J. Eaglesham and T. C. Huang, Mat. Res. Soc. Proc. **208**, in press
F. M. Ross and J. M. Gibson (1991b), in preparation
F. Smith and G. Ghidini (1982), J. Electrochem. Soc. **129**, 1300
A. M. Stoneham, C. R. M. Grovenor and A. Cerezo (1987), Phil. Mag. B **55**, 201
K. Takayanagi, Y. Tanishiro, S. Takahashi and M. Takahashi (1985), Surf. Sci. **164**, 367

Inst. Phys. Conf. Ser. No 117: Section 4
Paper presented at Microsc. Semicond. Mater. Conf., Oxford, 25–28 March 1991

Influence of residual defects on impurity diffusion in preimplanted Si

Y Kikuchi, M Kase, M Kimura and M Yoshida

Electronic Devices Group, Fujitsu Limited, 1015, Kamikodanaka, Nakahara-ku, Kawasaki 211, Japan

ABSTRACT : In preimplanted Si, the influence of residual defects on impurity diffusion is investigated using TEM and SIMS. The preimplantation by Si^+ produces the crystalline/amorphous(c/a) interface with a wider transition layer than that by Ge^+. TEM clarifies that this result leads to a higher density of end-of-range defects on the Si^+ case. These defects can eliminate the enhanced boron diffusion under the condition of Si^+ preimplantation by 40keV at doses ranging from 4.0×10^{14} to 5.0×10^{14} cm^{-2}.

1. INTRODUCTION

Using a conventional ion implantation technique, it is difficult to form a shallow p^+n junction required for ultralarge scale integration (ULSI) fabrication because of μ-channeling. The preimplantation technique, involving Si^+ or Ge^+ implantation prior to that of B^+ or BF_2^+ has been successful in eliminating a channeling tail[Ajmera et al 1986, Tsai et al 1979] even with a dose lower than the threshold for amorphization[Kase et al 1990]. However, a subsequent annealing process is required for the removal of implanted damage and the activation of dopant. Furthermore, excess interstitials originating in the preimplanted damage cause both enhancement of dopant diffusion and evolution of residual defects during the annealing process. Therefore, many reseachers have optimized this technique to make shallow p^+n layers and eliminate end-of-range defects.

The objectives of this work are to determine the conditions under which shallow p^+n layers are made by Si^+ or Ge^+ preimplantation with an energy as low as 40keV and to clarify the relation between residual defects and boron diffusion taking note of end-of-range defects at the vicinity of the c/a interface.

2. EXPERIMENTAL

The substrates were Czochralski grown <100> oriented n-type Si wafers with resistivity of 10Ωcm. Si^+ and Ge^+ were preimplanted with 40keV at doses ranging from 2.0×10^{13} to 1.0×10^{15} cm^{-2} at room temperature. The samples were implanted with 3×10^{13} cm^{-2} BF$_2^+$ at 10 keV to avoid damage during implanting. Subsequent annealing was performed at 800°C for 30 min in N$_2$.

Transmission electron microscopy(TEM) studies were on (011) cross-sections using the JEOL 4000EX. TEM specimens were produced by sandwiching, dimpling, and ion-milling at about 4kV. Impurity profiles were measured by secondary ion mass spectroscopy(SIMS), HITACHI IMA-3 and ATOMICA 6500.

3. RESULTS AND DISCUSSIONS

3.1 Ge$^+$ Preimplantation

The Ge$^+$ preimplantation has the advantage that amorphization occurs with lower doses than for Si$^+$ bombardment. This advantage leads to eliminatation of the channeling tail without serious damage[Kase et al 1990]. However, as shown in the SIMS boron profile of fig.1, the effect of the retardation of boron diffusion is not apparent below a dose of 2.0×10^{14} cm^{-2}, the threshold dose for amorphization [Kase et al 1990] and the diffusivity of boron increases slightly depending on the doses of preimplanted ions. This result shows that the excess interstitials originating in the implantation damage extend in the region of boron diffusion as the doses of preimplanted ions increase.

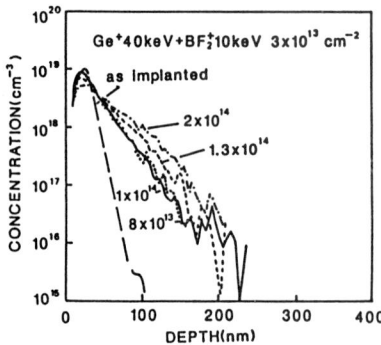

Fig.1   SIMS profiles of BF$_2^+$ implanted samples at 10keV with a dose of 3×10^{13} cm^{-2}, with preimplantation of Ge$^+$ at 40keV for doses of 8.0×10^{13}, 1.0×10^{14}, 1.3×10^{14} and 2.0×10^{14} cm^{-2}.

In order to clarify the mechanism of this diffusion, TEM observation was performed on the specimen with a Ge$^+$ ion dose of 2.0×10^{14} cm^{-2}. Fig.2(a) and (b) show TEM photographs of the sample before and after a 30 min anneal at 800°C, respectively. Fig.2(a) shows that the Ge$^+$ preimplantation produces a crystalline/amorphous(c/a) interface with a shallow transition layer between these phases. Fig.2(b) shows that a few end-of-range defects are formed in the vicinity of the c/a interface.

From these results, it is found that Ge$^+$ preimplantation with energy as low as 40keV is inappropriate for making a shallow

p^+n junction and the increase of preimplanted ions leads to the slight enhancement of boron diffusion.

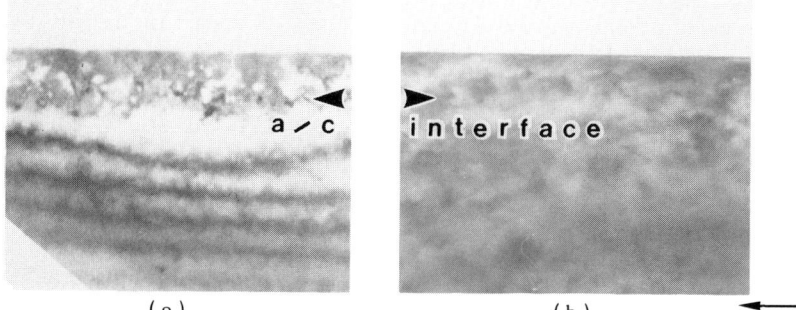

(a)   (b)

50 nm

Fig.2 TEM photographs taken before (a) and after a 30 min anneal at 800°C(b), respectively, with Ge^+ preimplantation by 40keV at a dose of 2.0×10^{14} cm^{-2}.

3.2 Si^+ Preimplantation

The influence of preimplantation by Si^+ as well as by Ge^+ was investigated. Fig.3(a) and (b) show SIMS boron profiles with doses ranging from 2.0×10^{14} to 4.0×10^{14} and from 5.0×10^{14} to 1.0×10^{15} cm^{-2}, respectively. Fig.3(a) shows that the boron diffusion retards as the dose of Si^+ preimplantation is increased. In contrast, a further increase of Si^+ dose leads to enhancement of boron diffusion as shown in fig.3(b). These results show that Si^+ preimplantation can eliminate the enhancement of boron diffusion under 40keV ion bombardment at doses ranging from 4.0×10^{14} to 5.0×10^{14} cm^{-2}, as suggested by Sedgwick[Sedgwick T O, 1989].

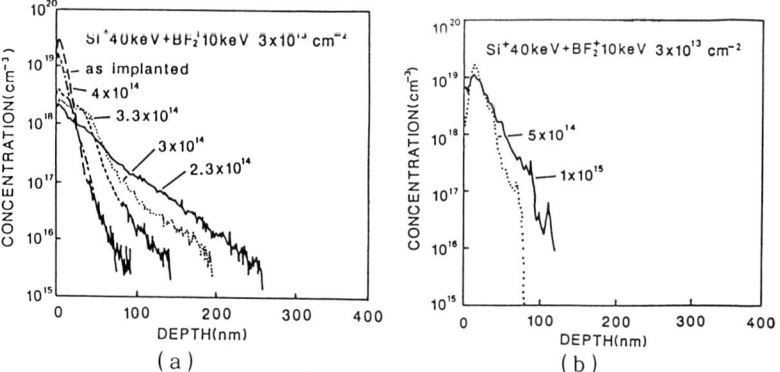

(a)   (b)

Fig.3 SIMS profiles of BF_2^+ implanted samples at 10keV with a dose of 3×10^{13} cm^{-2}, with preimplantation of Si^+ at 40keV with doses ranging from 2.0×10^{14} to 4.0×10^{14} cm^{-2} (a) and from 5.0×10^{14} to 1.0×10^{15} cm^{-2} (b).

TEM observation was performed for a specimen with retarded boron diffusion. Fig.4(a) and (b) show TEM photographs taken before and after annealing under the condition of the dose of 5.0×10^{14} cm^{-2}. In fig.4(a) a wide transition layer was

observed in the vicinity of c/a interface. Fig.4(b) shows that
the depth of this undulation corresponds to that of end-of-
range defects having a high density.
From these results, it is found that preimplantation by Si^+ is
useful for making a shallow p^+n layer but it is difficult to
avoid the formation of end-of-range defects with a high
density caused by the wide transition layer at the c/a
interface, when furnace annealing at around 800°C is employed.

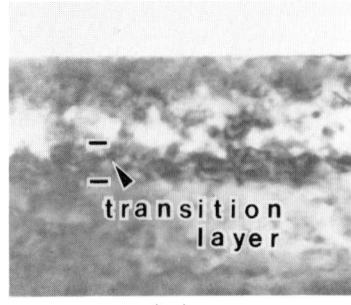

| (a) | (b) | 50 nm |

Fig.4 TEM photographs taken before (a) and after a 30 min
anneal at 800°C(b), respectively, with Si^+ preimplantation by
40keV at a dose of 5.0×10^{14} cm^{-2}.

4. CONCLUSION

We investigated the influence of residual defects on boron
diffusion in preimplanted Si. The following conclusions were
obtained by taking note of preimplantation species.
   1. Si^+ preimplantation with an energy as low as 40keV can
eliminate enhanced boron diffusion at doses ranging from
4.0×10^{14} to 5.0×10^{14} cm^{-2}, but Ge^+ preimplantation can not.
   2. High density end-of-range defects are due to a wider
transition layer of c/a interface.
   3. End-of-range defects play an important part in the
reduction of excess interstitials.

From the facts mentioned above, we conclude that it is
possible to make a shallow p^+n junction using a
preimplantation technique by controlling the depth of end-of-
range defects which cause junction leakage.

ACKNOWLEDGMENT
We would like to thank Mr.T.Ogawa for helpful discussions.

REFERENCES
Ajmera A C and Rozgonyi G A 1986 Appl. Phys. Lett.49 1269
Kase M, Kimura M, Mori H and Ogawa T 1990 Appl. Phys.
   Lett.55 1231
Sedgwick T O 1989 Nucl. Instr. and Meth. B37/38 760
Tsai M Y and Streetman B G 1979 J. Appl. Phys.50 183

Modelling and measurement of dislocation loop shrinkage in germanium pre-amorphised silicon during subsequent rapid thermal annealing

C D Meekison[1], C Hill[2], C D Marsh[1], D P Gold[1], D R Boys[2] and G R Booker[1]

[1]Department of Materials, University of Oxford, Parks Road, Oxford, OX1 3PH, UK.
[2]GEC - Plessey Semiconductors, Caswell, Towcester, Northants., NN12 8EQ, UK.

ABSTRACT: Silicon specimens were implanted with 1×10^{15} cm^{-2} of 400 keV ^{70}Ge$^+$ ions. Controlled etching was used to reduce the thickness of the resulting amorphous layer by various amounts. Rapid thermal annealing at 1100°C in nitrogen produced dislocation loops. TEM showed that the number of interstitial atoms per unit area in the loops decreased with increasing annealing time until the loops disappeared; this occurred more rapidly as the amorphous layer was made thinner. A model based on diffusion of self-interstitials predicts the timescale of this process to within a factor of about 2.

1. INTRODUCTION

The pre-amorphisation process, in which silicon is amorphised by implantation of a non-doping ion prior to boron implantation in order to avoid channelling, and is subsequently recrystallised by annealing, tends to leave a layer of interstitial dislocation loops just below the original amorphous-crystalline interface (Brotherton et al. 1986). Avoidance of these loops is found to become easier as the amorphous layer is made thinner (Ajmera and Rozgonyi 1986). This effect has been attributed either to the reduced distance over which interstitials have to diffuse to the surface (Thornton and Hill 1989) or glide of loops to the surface due to the image force (Narayan and Jagannadham 1987). Meekison et al. (1989) varied the amorphous layer thickness by controlled etching, with fixed implantation conditions, but no definite effect of thickness variation was revealed. However the interpretation of the latter results was complicated by interstitial injection due to the oxidising annealing ambient. In the present work an etching technique followed by rapid thermal annealing in nitrogen has been used to study the effect of amorphous layer thickness on the loop annealing process.

2. EXPERIMENTAL

(001) Czochralski silicon slices (p type, 25 ohm cm) were implanted, at room temperature, through an 18 nm thermal oxide layer, with a dose of 1×10^{15} cm^{-2} of 400 keV ^{70}Ge$^+$ ions, to give a surface amorphous layer ~390 nm thick. The slices were annealed at 500°C for 180 s in nitrogen to stabilise the optical properties of the amorphous layer. Pieces of these slices were etched with dilute (1 : 280) HF/HNO$_3$ so that amorphous layer thicknesses (x) of ~200 and 80 nm remained. These amorphous layer thicknesses were accurately measured by ellipsometry (Table 1). Samples were susceptor-annealed in a Heatpulse 610 rapid thermal annealer in a nitrogen ambient. The annealing temperature profiles consisted of a fast ramp to 1000°C (4 s from 800 to 1000°C), a slow ramp from 1000 to 1100°C (10 to 26 s), a steady state temperature of 1100°C for times (t_a) of either 0, 3, 10, 20 or 30s, and a fast cool down (4.5 s from 1100 to 800°C). The temperature was monitored on the silicon susceptor: the specimen temperature is considered to be up to 10°C higher. Specimens were examined by plan-view TEM with a Philips CM12 microscope operating at 120 kV.

3. RESULTS

A layer of dislocation loops was present in some of the resulting specimens, examples being shown in Fig. 1. As in our previous work (Meekison et al. 1989), most of these loops are of Frank type ($\underline{b}$ = 1/3 a<111>) while a small proportion are unfaulted ($\underline{b}$ = 1/2 a<110>). The loop radii r_L and areal densities of loops n were measured. The numbers of interstitial atoms per unit area contained within the loops (c) were determined according to our previous procedure (Meekison et al. 1989). Fig. 2 displays c as a function of amorphous layer thickness and steady state anneal time. The approximate boundary between the areas in which loops were and were not present is indicated. The overall trends of a decrease in the number of interstitial atoms with either an increase of annealing time at fixed amorphous layer thickness, or a decrease of amorphous layer thickness at fixed time, are apparent, although Specimen 1 (amorphous layer 389.5 nm, 10 s) is slightly anomalous in that it shows an increase in c relative to Specimens D and 7.

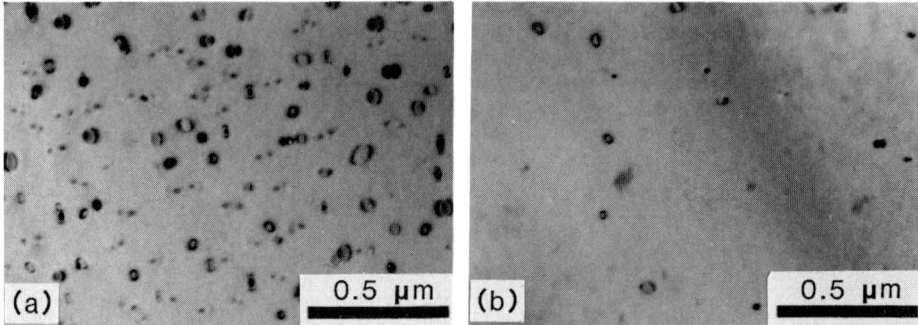

Fig. 1. Plan-view electron micrographs showing dislocation loops in specimens annealed at 1100°C for 0 s (see text), for amorphous layer thicknesses before annealing of (a) 389.5 nm (specimen D), (b) 89.6 nm (specimen G).

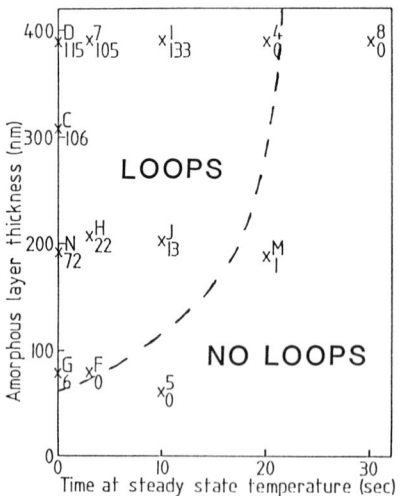

Fig. 2. Map of specimens examined. The crosses indicate the time at the steady state temperature and the amorphous layer thickness before annealing for each specimen. To the right of each cross is the specimen number (above) and the number of interstitial atoms per unit area, c, in units of 10^{12} cm^{-2} (below).

4. DISCUSSION

During the initial stages of annealing, the amorphous layer recrystallises leaving a layer of fine damage, containing excess interstitials, immediately below the original amorphous/crystalline interface (Brotherton et al. 1986, Thornton et al. 1988). The number of excess interstitials per unit area should be unaffected by the removal by etching of part of the amorphous layer before annealing. When interstitials become mobile during the anneal, interstitial-vacancy recombination occurs and the remaining ("excess") interstitials cluster into dislocation loops. Since however thermal emission of interstitials from loops and subsequent diffusion are possible, we expect coarsening of loops and loss of interstitials to the surface to occur. The decrease of the measured interstitial atom density c with increasing time indicates such a loss of interstitials, and the dependence on amorphous layer thickness shows that this loss is faster at smaller thicknesses. This indicates a diffusion-controlled process, for if interstitial emission at loops or absorption at the surface were the rate-controlling process, no effect of amorphous layer thickness would be expected. Loss of interstitial atoms by the glide of loops to the surface is not considered to be likely in the present experiments since (a) most of the loops observed here are sessile, and (b) in similar experiments using similar anneal times in an oxidising ambient (Meekison et al. 1989) loops were found to coarsen without any decrease in c, and moreover when the amorphous layer thickness was reduced to 49.4 nm, the loops were stable even when they intersected the surface.

It is possible to model the shrinkage of loops by transport of interstitials to the surface if simultaneous coarsening of the loops is neglected, although a limitation is imposed by our incomplete understanding of diffusion mechanisms in silicon and uncertainties in the properties of point defects (Van Vechten 1988, Ravi 1988). We assume here that the dominant process is diffusion of uncharged single self-interstitials and neglect any possible coupling to dopant diffusion. In our model all loops have the same radius r_L and are at a depth x from the surface. It is assumed that equilibrium is established at the surface (interstitial concentration C_0), so this model is only applicable to anneals in a non-reactive ambient. A loop is modelled as a toroidal surface of radius ρ (approximately equal to the Burgers vector b) around a circle of radius r_L, on which the interstitial concentration C is equal to the concentration in equilibrium with the loop, C_L, calculated as in our previous work (Meekison et al. 1989). The steady state diffusion condition $\nabla^2 C = 0$ (Laplace's equation) is assumed. By solving this equation with the appropriate boundary conditions and making some approximations we find, after some algebra, that the loop shrinkage rate is

$$-\frac{dr_L}{dt} = \frac{2\pi L D_{Si} (C_L/C_0 - 1)}{b (1 + S - U)} \tag{1}$$

where

$$L = 1/\ln(8r_L/\rho), \quad S = 4\pi^2 Lnxr_L \quad \text{and} \quad U = \pi Lr_L/2x \tag{2}$$

and D_{Si} is the interstitial component of the self-diffusion coefficient and is taken to be equal to $D_i C_0/N$, where D_i is the diffusion coefficient of self-interstitials in silicon. S can be taken as a measure of the effect on the shrinkage rate of interaction between the loops, and U as a measure of the interaction of an individual loop with the surface. Integration of equation (1) allows us to predict the time taken for loops to disappear, given x, n and the initial value of r_L. A reduction in x at fixed n and r_L increases the shrinkage rate through its effects on S and U.

We have experimental data for three amorphous layer thicknesses, approximately 390, 200 and 80 nm, as a function of annealing time. The experimental values of n and r_L have been used with equation (1) (taking $\rho = b$) to predict the additional annealing times (t_p) required to remove the loops (Table 1). The loop depth x is taken to be the same as the amorphous layer thickness. The value of D_{Si} is taken from Gosele (1986) and the temperature is taken as

1100°C. The experimental additional annealing times (t_e) required for the same amorphous layer thickness are also given. The predicted times agree within a factor of 2.5, except for specimen 1 which as pointed out earlier appears anomalous. While our model is not capable of predicting the variations of n with time observed, due to the neglect of coarsening, it nevertheless evidently gives a reasonable prediction of the overall rate of loss of interstitials in the later stages of the annealing process.

The physical behaviour occurring can be qualitatively explained as follows. For the 390 nm amorphous layer, $t_a = 0$, specimen the mean distance between the loops $m = n^{-1/2}$ is much smaller than the distance x from the loops to the surface (Table 1). Consequently, as t_a increases, one expects significant diffusion of interstitials between loops in addition to diffusion to the surface. For the 200 nm specimen, at $t_a = 0$, m and x are similar and some diffusion to the surface has already occurred. As t_a increases, there is less diffusion between loops and more to the surface. For the 80 nm specimen, at $t_a = 0$, m is much greater than x, considerable diffusion to the surface has already occurred, and the loops are subsequently rapidly removed. Consideration of the initial values of S (equation 2) leads to similar conclusions to this comparison of m and x.

Table 1. Experimental measurements of dislocation loops and comparison with theoretical predictions: x = amorphous layer thickness, t_a = time at steady state annealing temperature, r_L = mean loop radius, n = areal density of loops, m = mean loop spacing, c = areal density of interstitial atoms in loops, t_e = additional anneal time required to remove loops deduced from experimental data, and t_p = predicted additional time required.

Specimen number	x (nm)	t_a (s)	r_L (nm)	n (10^9cm^{-2})	m (nm)	c (10^{12}cm^{-2})	t_e (s)	t_p (s)
D	389.5	0	20	5.7	132	115	10 - 20	41
7	389.5	3	20.5	5.1	140	105	7 - 17	40
1	389.5	10	23.5	4.9	143	133	0 - 10	60
N	192.7	0	18	4.5	149	72	20	17
H	206.8	3	20	1.1	302	22	17	15
J	203.2	10	16.5	1.0	316	13	10	8.9
G	89.6	0	12.5	0.83	347	6	0 - 3	3.7

ACKNOWLEDGEMENTS

This work was supported by the SERC, the Alvey Directorate through MOD Proc. Exec., and Plessey Semiconductors Ltd. and thanks are due for funding and permission to publish. John Thornton of Surrey University provided the germanium implantations.

REFERENCES

Ajmera A C and Rozgonyi G A 1986 Appl. Phys. Lett. 49 19
Brotherton S D, Gowers J P, Young N D, Clegg J B and Ayres J R 1986 J. Appl. Phys. 60 3567
Gosele U 1986 Semiconductor Silicon 1986, Electrochemical Society Proc. Vol. 86-4, eds H R Huff and T Abe p 541
Meekison C D, Gold D P, Booker G R, Hill C and Boys D R 1989 Inst. Phys. Conf. Ser. No. 100 p 507
Narayan J and Jagannadham K 1987 J. Appl. Phys. 62 1694
Ravi K V 1988 Properties of Silicon, EMIS Dataviews Series No. 4 (London: INSPEC) p 256
Thornton J, Webb R P, Wilson I H and Paus K C 1988 Semicond. Sci. Technol. 3 281
Thornton J and Hill C 1989 Semicond. Sci. Technol. 4 53
Van Vechten J A 1988 Properties of Silicon, EMIS Dataviews No. 4 (London: INSPEC) p 47

Inst. Phys. Conf. Ser. No 117: Section 4
Paper presented at Microsc. Semicond. Mater. Conf., Oxford, 25–28 March 1991

TEM *in situ* investigations of the crystallisation of a-Si thin films

M Reiche

Institut für Festkörperphysik und Elektronenmikroskopie, Weinberg 2,
Halle/S., 4050, Fed. Rep. of Germany

ABSTRACT: The crystallization of a-Si layers deposited on SiO_2-
or Si_3N_4 films has been investigated by in situ annealing in an
HVEM. The different formation mechanisms of crystallites by a
surface-induced crystallization are presented. Interfacial stresses were deduced to be the main reason for nucleation.

1. INTRODUCTION

Thin films of polycrystalline silicon (PSi) have been widely used in
integrated circuits. Herein they are incorporated into a system of
other layers such as, e.g. insulators (SiO_2), interconnects etc.
Therefore the silicon films have to fulfil a number of requirements
with respect to their 3-dimensional arrangement, their internal
morphology (grain orientation and size) and their flatness. The latter two strongly depend on the preparation conditions. Amorphous
deposited layers, for instance, become more homogeneous PSi films
after subsequent annealing with less rough interfaces compared to
the deposition of PSi-material (Reiche et al 1990). Up to now, however, there is no explanation that satisfactorily describes this
fact. To gain more information about the mechanism of the crystallization of amorphous films, in situ heating experiments were carried out in a TEM. The present paper is mainly concerned with the
nucleation of crystallites in the amorphous layer, whereas their
growth especially on SiO_2 films was described elsewhere (Reiche and
Hopfe 1990).

2. EXPERIMENTAL

Amorphous silicon (a-Si) films were deposited by LPCVD on (100) silicon substrates covered with SiO_2- or Si_3N_4 layers of different
thicknesses. After this, some of the silicon films were implanted
with phosphorus and arsenic, resp., using doses of $5 \cdot 10^{15}$ or
$1 \cdot 10^{16} cm^{-2}$. Further details of the preparation were described elsewhere (Reiche and Hopfe 1990).

Plan-view and cross-sectional samples were prepared for investigation in an HVEM. A heating stage was attached to the microscope allowing annealing up to 800°C. The dimensions of the grains
and their distances were measured by computer-aided interpretation
of TEM images.

3.　RESULTS

3.1　As Deposited Layers

The structure of the Si layers characteristically depends on the deposition temperature and on the substrate. For instance, the deposition on SiO_2 films yields entirely amorphous layers at $T \leq$ 560°C. Increasing the temperature increases the crystalline fraction. Entirely polycrystalline layers are formed at $T \geq$ 625°C. It is important to note that crystallization begins at the a-Si/SiO_2 interface. Furthermore, it was also proved that doping by ion implantation without any subsequent annealing causes the formation of small crystallites having diameters of 5 to 10nm (Fig. 1a).

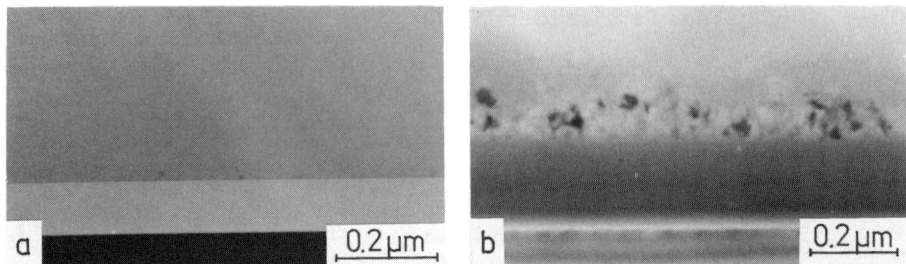

Fig. 1: TEM cross-sectional images of silicon layers deposited on SiO_2 (a) and Si_3N_4 films (b), resp. The layers were implanted with phosphorus ($D = 5 \cdot 10^{15} cm^{-2}$) after deposition at 560°C.

Results similar to those of the deposition on SiO_2 layers were obtained by direct deposition on (100) substrates. On the other hand, the crystalline fraction in the silicon layers drastically increases by deposition on Si_3N_4 films. As shown in Fig. 1b, even the deposition at 560°C induces a high density of crystallites having dimensions of about 50 nm. HREM investigations reveal <111> growth (Fig. 2). In addition, the density of the small crystallites varies with the thickness of the underlying Si_3N_4 film.

3.2　Annealing

In situ annealing experiments on plan-view and cross-sectional samples were used for the 3-dimensional recording of the crystallization of the films. Moreover, the nucleation and further growth of individual grains as well as coalescence stages were analyzed. Fig. 3 shows one example of annealing at 650°C.

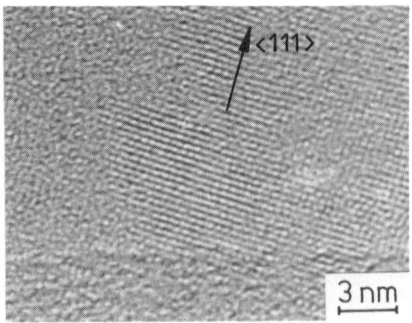

Fig.2: HREM image of a small crystallite at the interface.

Statistical measurements of the grain size have shown that the crystallization of the films is characterized by a $t^{1/2}$ dependence which is independent of the doping and of the substrate, respectively. The same is true also of the morphology of the grains. A needle-like

shape is found, which is in contrast to other investigations suggesting predominantly hexagonal forms of the grains (cf. Koleshko et
al 1988). The ratio of the mean axes of a grain, D1/D2, is determined to be nearly 2.

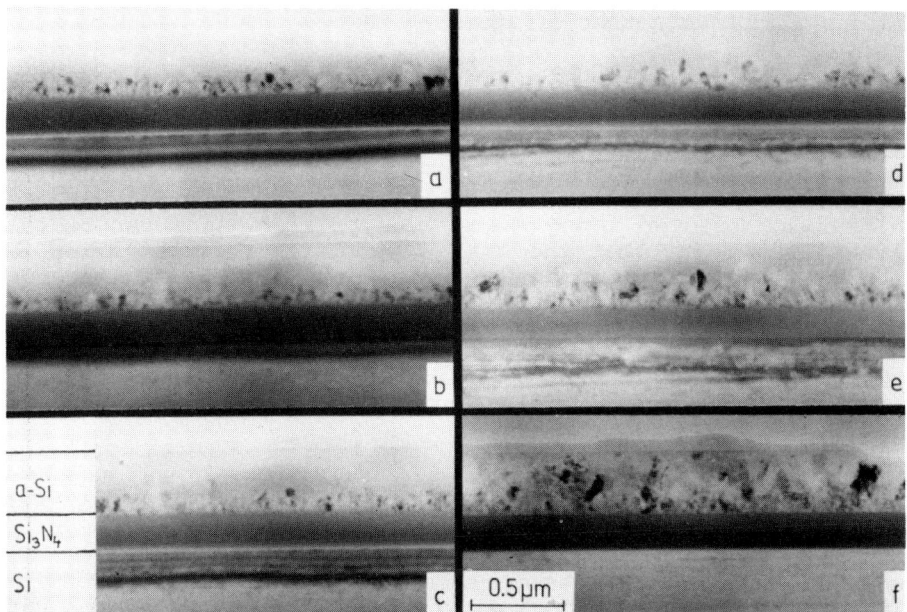

Fig. 3: In situ crystallization of an amorphous, undoped Si layer
deposited on Si_3N_4. As deposited at 560°C (a) and after annealing
at 600°C (b). Annealing at 650°C for 5 (c), 15 (d), 30 (e), and 60 minutes (f), respectively. Note that the micrographs show the same
object region.

Furthermore, it was proved that preexisting crystallites, detected
in the as implanted layers, do not act as nuclei for further
growth. Instead, they are reconstructed or dissolved during the
first stages of annealing (incubation time). On the other hand, the
annealing experiments show the continuous growth of crystallites if
the Si films were deposited on Si_3N_4. Their {111} morphology is preserved. This is the main reason for the origin of {111}-twins, frequently observed in the grains. But twins are still clearly shown to
be generated during the coalescence stage.

4. DISCUSSION

The reason for the formation of small crystallites, identified at
the a-Si/SiO_2 interface after ion implantation, is not clearly
understood. But heterogeneous nucleation at impurity centres is
shown to be almost impossible. The formation of the crystallites
having densities of only $5 \cdot 10^4$ crystallites per 1cm interface length,
is reproducible and not comparable to the high crystallite density
in impurity-contaminated samples. In general, impurities reduce the

nucleation barrier and favour the bulk-induced crystallization in addition to the dominating surface-induced process. Moreover, the formation of crystallites can also not be explained by direct interactions between the ion beam and the sample (radiation damage).

Assuming that a crystalline nucleus formed in an amorphous matrix contains a minimum number of atoms m_C that is stable

$$m_C = [\frac{2 \Delta G_C}{g}]^{2/3} , \qquad (1)$$

Zellama et al (1979) have determined $m_C \approx 100$ atoms at 600°C. In eq. (1) ΔG_C is the Gibbs energy for the formation of a crystallite with the critical radius r^* and Δg is the Gibbs energy for crystallization. Moreover, using m_C, a value of $r^* \approx 0.6$nm can be calculated. This means that the small crystallites detected after ion implantation are only 5 times larger than r^*; they should therefore be metastable. During annealing at higher temperatures r^* increases with the crystallites becoming unstable, i.e. they are dissolved. On the other hand, crystallites identified in the layers after deposition on Si_3N_4 have radii of more than 50 times larger than r^*. The crystallites are stable and continuously grow during subsequent annealing.

The different observations of crystallites in the as deposited layers with SiO_2 and Si_3N_4 underlayers, on the one hand, and their varying density by varying the thickness of the Si_3N_4 layer, on the other, refer to the strong influence of stresses on the nucleation process. Increasing stresses result in deviations of ΔG_C and therefore reduce r^*. At the same time the nucleation rate increases, too (cf. Hirth and Pound 1963). Amorphous silicon films deposited by LPCVD at 560°C have compressive stresses of about 350 MPa (Koleshko et al 1988), which are nearly equivalent to the intrinsic stresses in SiO_2 layers (Fitch and Lucovsky 1988). On the other hand, stresses in Si_3N_4 layers reach values one order of magnitude larger than in SiO_2 films (Groothuis and Schroen 1987) and increase with increasing thickness (in the range considered here). This favours the formation of crystallites even during the deposition of a-Si.

REFERENCES

Fitch J T and Lucovsky G 1988 Mat. Res. Soc. Symp. Proc. 105 151
Groothuis A K and Schroen W H 1987 Proc. 25th Conf. on Reliability
    Physics (New York: IEEE) pp 1-8
Hirth J P and Pound G M 1963 Condensation and evaporation, Pro-
    gress in Materials Science, Vol. 11, ed B Chalmers (Oxford:
    Pergamon Press) pp 15-40
Koleshko V M, Belitsky V F and Kiryushin I V 1988 Thin Solid Films
    165 181
Reiche M and Hopfe S 1990 Ultramicrosc. 33 41
Reiche M, Hopfe S Pappe B and Schaarschmidt A 1990 Proc. 35.
    Intern. Kolloqu. (Ilmenau: Techn. Hochsch.) 4 pp 29-32
Zellama K, Germain P, Squelard S, Bourgoin J C and Thomas P A 1979
    J. Appl. Phys. 50 6995

Inst. Phys. Conf. Ser. No 117: Section 4
Paper presented at Microsc. Semicond. Mater. Conf., Oxford, 25–28 March 1991

205

Dislocations in heavily boron-doped silicon

X J Ning and P Pirouz
Department of Materials Science and Engineering, Case Western Reserve University,
Cleveland, OH 44106, U.S.A.

ABSTRACT: A plan-view and cross-sectional TEM study of dislocations induced by the heavy boron doping of silicon is presented.

1. INTRODUCTION

Diffusion-induced dislocation generation in silicon has long been of interest in semi-conducting devices. The first investigations concerning dislocations induced by diffusion were carried out in the early sixties (Queisser, 1961; Prussin, 1961). Schwuttke and Queisser (1962), using transmission x-ray diffraction microscopy, investigated the dislocations in both boron- and phosphorus- diffused silicon and, in the latter case, characterized them as $\frac{1}{2}<1\bar{1}0>$ edge type. Later, a TEM study of dislocations induced by diffusion of phosphorous in silicon was carried out by Washburn *et al.* (1964). To the authors' knowledge, a TEM investigation of dislocations in boron-diffused silicon has not yet been made. The dislocation configuration of boron-diffused Si is believed to be quite different from that of phosphorus-diffused Si.

The p$^+$ etch-stop technique, which is widely used in the fabrication of microsensors and actuators, is based on the heavy doping of Si by boron to concentrations greater than 5×10^{19} cm^{-3}. The heavy boron doping is usually carried out by diffusion and, as a result of the size mismatch between silicon and boron atoms, the lattice of the diffused layer undergoes a contraction. To relieve the mismatch strains, dislocations are induced in the diffused layer which, depending on the dopant concentration, can be of high density. During the diffusion process, as the diffusion front moves deeper down into the wafer, more dislocations are generated and, in addition, the pre-existing dislocations glide down to greater depths. Consequently, a large number of dislocation interactions may take place within the diffused layer. These dislocations play a critical role on the mechanical properties of the p$^+$ layer. In fact, it has been recently reported that as-diffused and thermally oxidized heavily boron-doped silicon exhibit different mechanical properties (Maseeh and Senturia 1990, Ding 1990). Hence, a study of boron-diffused dislocations in silicon wafers is of interest in determining the mechanical properties of devices produced by the p$^+$ etch-stop technique.

2. EXPERIMENTAL AND RESULTS

(001) silicon wafers with a thickness of ~250 μm were diffused with boron at 1125°C for 3 hours in an atmosphere of 98% nitrogen and 2% oxygen. During the boron diffusion, a ~250 nm thick layer of borosilicate glass (BSG) formed on the wafer surface. To prepare TEM specimens, the BSG layer was first removed. Subsequently, plan view TEM specimens were made by chemical etch-stop thinning followed by ion beam milling from both sides to electron transparency. Cross-sectional TEM specimens with the foil surface parallel to {110} and {111} planes were made by the conventional sandwich technique.

Fig. 1 is a plan view TEM micrograph with the electron beam parallel to the [001] direction. A high density of dislocation reactions may be observed. The arrows in this figure show dislocation dipoles which have probably formed by the mechanism proposed by Tetelman (1962) during the glide of dislocations from the surface to the bulk.

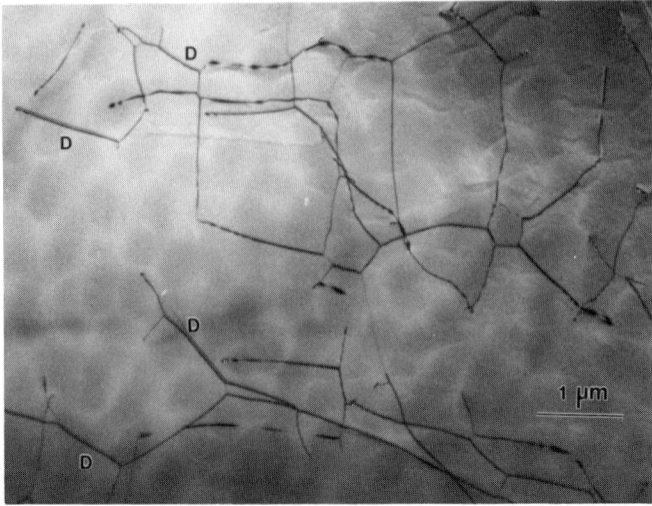

Fig. 1. A [001] plan-view TEM micrograph showing a high density of dislocation interactions. The dipoles are labelled as "D".

The dislocations in this figure lie at different depths within the diffused region of the wafer. In fact, no dislocations could be observed from foils close to the top surface of the wafers or those deeper than, say, 10 µm from the surface. This is verified by Fig. 2 which is a low magnification (110) cross-sectional TEM micrograph. It shows a dislocation-free zone in the vicinity of the wafer surface; this was found in all the samples studied. Past this zone, there is a distribution of dislocations up to a depth of ~5 μm from the surface. The thickness of the dislocation-free zone, as well as that of the dislocated layer, is a function of diffusion time (Ning *et al.*, 1991). In the present case, the thickness of the dislocation-free zone is ~1.7 μm and that of the dislocated layer is ~3.3 μm. The dislocation distribution within the dislocated layer is relatively uniform with a density of the order of 10^9 cm^{-2}.

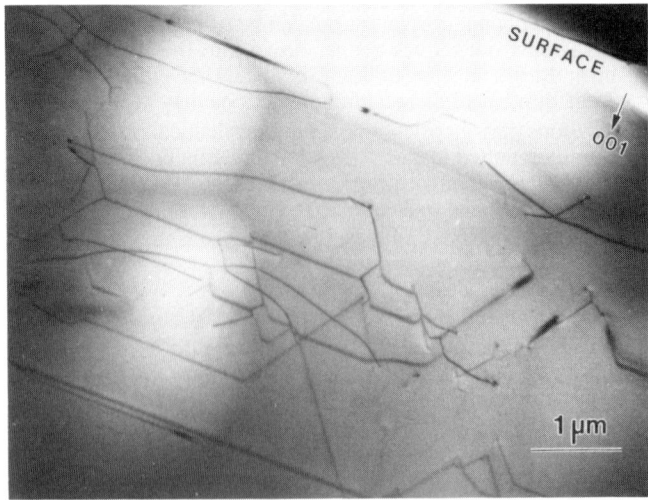

Fig. 2. A <110> cross-sectional TEM micrograph of the B-doped Si layer

The Burgers vector, **b**, of dislocations were determined by the **g·b** criterion and was invariably $\frac{1}{2}$<110>; the dislocations were mostly of the 60° type, with a smaller fraction of 0° (screw) dislocations. Most of the dislocations were dissociated and had reacted with dislocations on other {111} planes to form different types of dislocation products. For example, the long dark lines parallel to the surface in Fig. 2 were Hirth locks (Hirth and Lothe, 1982): composite dislocations resulting from the reaction of two dissociated 60° dislocations on two intersecting obtuse {111} planes. In particular, the leading partials of the two 60° dislocations react to form a $\frac{1}{3}$[001] edge dislocation at the intersection of the {111} planes, leaving a stacking fault on each of these planes bounded by $\frac{1}{6}$<11$\bar{2}$> partial dislocations. In the (001) plan-view specimens and also in the {110} cross-sectional specimens, the observation of dislocation dissociation was difficult because of the small projected width of the stacking fault ribbon. For this purpose, and to confirm the dissociation

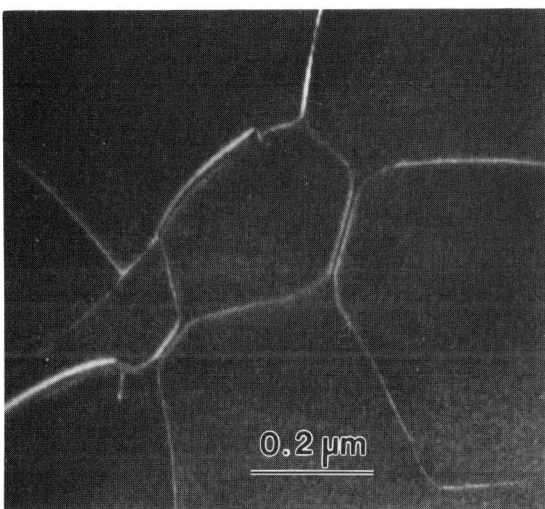

of dislocations, cross-sectional TEM specimens were prepared with their foil surfaces parallel to the {111} planes on which the dislocations lie. This enabled the observation of maximum dissociation width along the beam direction. Fig. 3 is a weak beam image of a {111} cross-sectional specimen. Dissociated dislocation segments and extended nodes can be observed quite clearly in this projection. It should be noted that since there is a symmetrical biaxial stress in the (001) plane of the wafer, the same shear stress exists on all four {111} glide planes. Hence, similar dislocation configurations are expected on all the latter planes. This was confirmed by TEM investigation of different {111} cross-sectional specimens.

Fig. 3. A <111> cross-sectional TEM micrograph of the B-doped Si layer.

3. DISCUSSION

Since boron has a smaller covalent radius than silicon, its incorporation into the silicon lattice leads to lattice contraction. This was verified by Pearson and Bardeen (1949) and Horn (1955) by X-ray diffraction and also by the epitaxial growth experiments of Sugia *et al.* (1969). According to Prussin (1961) and Czaja (1966), the strain, ε, due to boron diffusion in silicon is proportional to the boron concentration, C:

$$\varepsilon = -\beta C \quad ....(1)$$

where $\beta=(r_{Si}-r_B)/r_{Si}$ is the lattice contraction coefficient. From the covalent radii of Si and B (r_{Si}=0.117 *nm*, r_B=0.088 *nm*), β=0.25. This value is in agreement with that determined by Horn (1955) and nearly twice the value reported by Pearson and Bardeen (1949) for polycrystalline material. The stress distribution is $\sigma_{xx}=\sigma_{zz}=\sigma$, $\sigma_{yy}=0$ with the wafer surface parallel to the x-z plane and normal to the y-axis. According to Timoshenko and Goodier (1970) and Prussin (1961), for non-symmetrical diffusion, the stress $\sigma(y,t)$ at depth y and time t is given by:

$$\sigma(y,t) = \frac{\beta E}{(1-\nu)}[C(y,t) - \frac{1}{h}\int_0^h C(y',t)dy' - \frac{12(y-0.5h)}{h^3}\int_0^h (y'-0.5h)dy'] \quad ....(2)$$

where E is the elastic modulus, v is the Poisson's ratio, and h is the wafer thickness. From this equation, provided that the boron concentration distribution in the wafer is known, the stress distribution can be calculated. The boron concentration in the wafer can be described by a complimentary error function:

$$C(y,t) = C_s erfc \sqrt{\frac{y^2}{4Dt}} \quad ....(3)$$

with $C_s = C(0)$ being the boron concentration at the wafer surface. The solubility of boron in Si at 1125°C is ~1% (i.e. $C_s \sim 5 \times 10^{20}$ cm^{-3}) (Olesinski and Abbaschian, 1984). The diffusion coefficient of boron in silicon, D, is given by (Wolf and Taube, 1986):

$$D = 0.76 \exp(-\frac{3.42}{kT}) \quad cm^2 s^{-1} \quad ....(4)$$

However, from resistivity experiments on wafers that had undergone similar treatments as the present wafers, Mehregany (1986) found that D is higher than the value expected from eq. (4). In order to fit eq. (3) to his measurements, D should be nearly an order of magnitude larger than the value predicted by equation (4). The profile for a 3 hour diffusion anneal using eq. (3) with $D = 2.55 \times 10^{-12}$ $cm^2 s^{-1}$ is shown in Fig. 4. It should be mentioned that the experimental concentration profile is not given exactly by the complementary error function; in fact the boron concentration is nearly constant over the first few microns from the surface (Mehregany, 1986). This is consistent with the fact that boron diffusivity in Si, D, is not a constant but is enhanced with increasing concentrations of boron.

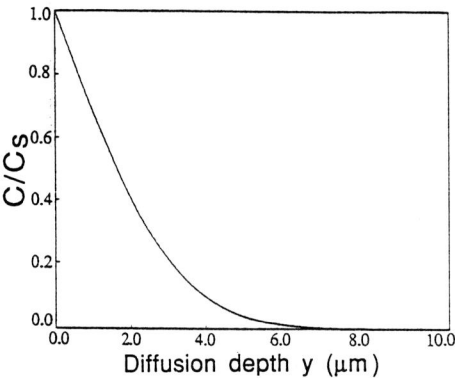

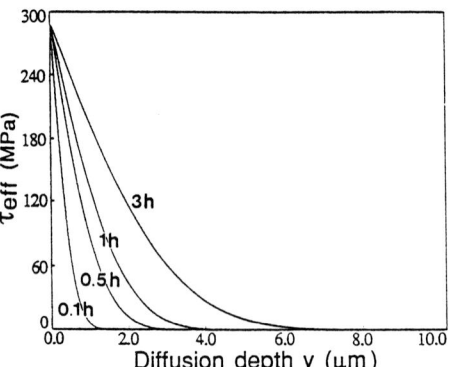

Fig. 4. The boron concentration normalized to C_s vs. diffusion depth.

Fig. 5. The effective shear stress τ vs. y at different diffusion times.

As boron diffusion proceeds, the region on top of the Si wafer will have a higher boron concentration with a lower lattice parameter than that of virgin Si. The situation in the present case is in some ways analogous to a heteroepitaxial layer deposited on a Si substrate with the difference that the interface is not sharp but diffuse with a graded compositional change. This implies that in the diffused region of the wafer with a graded composition, there is a gradual change in the lattice parameter from a_s at the surface to a_o (=0.543 nm) deep down in the virgin Si. Since the boron concentration $C(y,t)$ is a function of depth and time, the lattice parameter in the diffused region, $a(y,t)$, will also vary with depth and time. Consequently there are varying strains with depth in the diffused layer which changes as the composition profile changes with time. Despite this difference, one may use some of the concepts that are used in heteroepitaxial growth. In the following, the term substrate refers to the region of the wafer below which there is a negligible content of B (i.e. its lattice parameter is ~a_o(=0.543 nm), and the term epilayer refers to the diffused region of the wafer with a smaller, but variable, lattice parameter $a(y,t)$. Although there is no sharp interface, at any instant of time, t, there is a certain depth, $y_i(t)$, below which the boron concentration is very small and

$a(t) \sim a_o$. As diffusion proceeds, the epilayer thickness, $y_i(t)$, grows and the strain energy builds up. Up to a certain thickness, $y_i(t)=h_c$, the strain is accommodated elastically and the epilayer is coherent with the substrate; this corresponds to a critical diffusion time, t_c. Above h_c, or beyond t_c, however, the strain energy of the epilayer is sufficient to generate misfit dislocations which then accommodate the lattice mismatches within the diffused layer and thus release the elastic strain energy partially (or totally). Unlike the heteroepitaxial case, however, where the interface is usually sharp and the misfit dislocations lie at the interface, in the present case there is a distribution of dislocations lying within the diffused layer. As the boron concentration starts saturating in the upper layers of the diffused region, and the composition difference decreases, so the need for misfit accommodating dislocations also decreases. Simultaneously, as diffusion proceeds to greater depths, there is need for misfit dislocations deeper down in the wafer and, thus, the misfit dislocations from the upper part of the wafer glide down to lower depths where they are needed.

As mentioned before, Mehregany's measurements (1986) show a relatively constant boron concentration in the uppermost part of the diffused region which may be due to the enhancement of diffusivity with boron concentration. We ascribe the dislocation-free zone to this: all the misfit dislocations in the region of constant composition will be redundant and they will be repelled from this region leaving a dislocation-free zone in the vicinity of the wafer surface. The dislocation loops in this region may either shrink and collapse, or they may glide down to deeper regions where they will be needed.

Thus, as boron diffusion starts, the diffused layer on top of the Si wafer is initially deformed elastically but stays coherent with the underlying material for the first t_c minutes. After this, misfit dislocations are generated and part of the elastic strain is released by plastic deformation. A plausible way for the generation of misfit dislocations is by the heterogeneous nucleation of dislocation semi-loops from the surface and their glide on {111} slip planes until they reach the interface. However, as mentioned before, since the interface is not sharp in the present case, a distribution of dislocations will result throughout the diffuse, graded, interface which moves down with the interface during further diffusion.

It may be seen that a proper treatment of this problem requires that the process be considered in a dynamic fashion. Estimation of h_c, or t_c, is rather difficult because the calculation of strain energy requires taking into account the varying strains in the diffused layer. However, as a first approximation, the situation will be considered in a static manner, i.e. it will be assumed that the diffusion cycle takes place according to the profile shown in Fig. 4 while, during the whole diffusion time, the boron-rich layer is accommodated elastically. Subsequent to this, the resulting elastic stress is relaxed by the generation and movement of dislocations. In the following, it is assumed that $v=1/3$, $E=190\ GPa$, $\beta=0.25$ and $h=250\ \mu m$. The highest biaxial stress at the surface is then $\sim714\ MPa$ which is of the same order of stress as estimated by Prussin (1961) for phosphorus diffusion in Si at 1225°C ($\sim600\ MPa$).

The resolved shear stress on 60° dislocations gliding on {111} planes is $\tau=\sigma\cos60°\cos35.5°=0.408\sigma$. The energy, E_s, required by a unit length of 60° dislocation to glide down on the slip plane to a depth y is given by (Czaja 1966, Hirth and Lothe 1982):

$$E_s(y) = b\int_0^s \tau \cos30°\, ds' \;=\; b\,\cos30° \int_0^y \tau \frac{dy'}{\cos35.3°} \quad ....(5)$$

where $s\ (=y\ sec35.3°)$ is the length in the direction of dislocation motion (it is assumed that the dislocations move in the $<11\bar{2}>$ directions perpendicular to their $<1\bar{1}0>$ line directions). This energy is, of course, provided by the release of the elastic strain energy. The energy involved in the movement of a dislocation segment through the diffused layer is associated with the kinetic energy of dislocation motion and also the line energy of the extra lengths of dislocation created (Freund, 1987; Nix, 1989). According to Nix (1989), the energy required to form a unit length of 60° dislocation is:

$$E_d(y) = 0.95 \frac{b^2\mu}{4\pi(1-v)} \ln(\frac{\varphi y}{b}) ....(6)$$

where $\varphi=0.755$. The effective shear stress for dislocation motion, τ_{eff}, is then (Nix, 1989):

$$E_s(y) - E_d(y) = b\frac{\cos30°}{\cos35.3°}\int_0^y \tau_{eff}\, dy' = 1.06b\int_0^y \tau_{eff}\, dy' ....(7)$$

Hence:

$$\tau_{eff} = \frac{0.94}{b} \frac{d}{dy}(E_s - E_d)....(8)$$

According to this equation, at film thicknesses greater than the critical thickness h_c, the slope of the $E_s(y)$-$E_d(y)$ curve is proportional to the net driving force on a 60° dislocation. Fig. 5 shows the evolution of the effective shear stress τ_{eff} vs. the diffusion depth with time. From this figure, it can be seen that at earlier times of diffusion τ_{eff} decreases linearly with depth from the wafer surface in the first ~2 μm. After this distance, the stress falls much less rapidly. Hence, the glide of dislocations is specially pronounced in the first ~2 μm from the surface because of the rapid fall-off of the stress, i.e. there is a large driving force on the dislocations to glide down deeper into the crystal where the τ_{eff} on them is lower. This depth roughly corresponds to the thickness of the dislocation-free zone in Fig. 2. As the dislocations move down, the stress in the diffused layer is progressively relaxed. Since the dislocation velocity is proportional to τ_{eff}, the dislocations move into the boron diffused Si layer at relatively high velocities in the beginning, and slow down in the deeper parts of the wafer until they stop at ~5.0 μm from the surface. Also, the first dislocations move faster than those generated at a later time. Fig. 2 shows that there are numerous dislocations parallel to the surface which are located in a layer between the dislocation-free zone (~1.7 μm from the surface) and the interface (~5 μm from the surface). Characterization of a number of these dislocations has shown them to be mostly sessile dislocation locks. It is possible that, because of the high density of generated dislocations, there is a high probability that a number of them gliding down on a certain set of {111} planes meet other dislocations gliding down on intersecting {111} planes and interact with them to form various dislocation locks; examples of these are the Hirth locks shown in Fig. 1. As a result of these intersections, the dislocations become sessile and would not be able to move down any further. It is interesting to note that the Hirth locks, with $\mathbf{b} = \frac{1}{3}[001]$, do not make any contribution to the accommodation of lattice mismatch. It is possible that the formation of these particular locks by the interaction of 60° dislocations is favorable during the saturation of the top part of the wafer: as the upper layers acquire a more uniform composition, there is a lesser need for misfit accommodating dislocations.

ACKNOWLEDGEMENTS
The authors thank Dr. X. Ding for providing the diffused samples and Professors M. Mehregany and A. H. Heuer for useful discussions.

REFERENCES
W. Czaja, J. Appl. Phys. **37**, 3441 (1966).
X. Ding, Ph.D Thesis, Case Western Reserve University (1990).
L. B. Freund, J. Appl. Mech. **54**, 553 (1987).
J. P. Hirth and J. Lothe, *Theory of Dislocations*, J. Wiley & Sons, New York (1982).
F. H. Horn, Phys. Rev. **97**, 1521 (1955).
F. Maseeh and S. D. Senturia, Sensors and Actuators A21-A23, 861 (1990).
M. Mehregany, M.S. Thesis, MIT (1986).
X. J. Ning, P. Pirouz, M. Mehregany, and W. Chu, in *Transducers '91*, (1991). In press.
W. D. Nix, Met. Trans. **20A**, 2217 (1989).
R. W. Olesinski and G. J. Abbaschian, Bull. Alloy Phase Diagrams, 5(5), (1984).
G. L. Pearson and J. Bardeen, Phys. Rev. **75**, 865 (1949).
S. Prussin, J. Appl. Phys. **32**, 1876 (1961).
H. J. Queisser, J. Appl. Phys. 32,1776 (1961).
C. H. Schwuttke and H. J. Queisser, J. Appl. Phys. **33**, 1540 (1962).
Y. Sugia, M. Tamura and K. Sugawara, J. Appl. Phys. **40**, 3089 (1969).
A. S. Tetelman, Acta Met. **10**, 813 (1962).
S. P. Timoshenko and J. N. Goodier, *Theory of Elasticity*, McGraw-Hill, New York (1970).
J. Washburn, G. Thomas and H. J. Queisser, J. Appl. Phys. **35**, 1909 (1964).
S. Wolf and R. N. Taube, *Silicon Processing for the VLSI Era*, Lattice, California (1986).

Inst. Phys. Conf. Ser. No 117: Section 4
Paper presented at Microsc. Semicond. Mater. Conf., Oxford, 25–28 March 1991

TEM study of nitrogen enhanced oxygen precipitation in nitrogen-doped Czochralski-grown silicon

X Zhou, A R Preston and C J Humphreys

Department of Materials Science and Metallurgy, University of Cambridge,
Pembroke Street, Cambridge CB2 3QZ, England

ABSTRACT: Thermal induced oxygen precipitates in nitrogen doped Czochralski-grown silicon have been examined by TEM and IR spectroscopy. A concentration of nitrogen about 100 ppba was found to have a strong effect on increasing precipitate nucleation rate and decreasing the threshold of initial oxygen concentration for precipitation between 750 °C and 1000 °C because of nitrogen-related microdefects in as-grown silicon. After annealing at 750 °C or 900 °C, much decrease in interstitial oxygen concentration and a higher defect density were observed in specimens with high nitrogen concentration ($\geq$ 100 ppba) and low oxygen concentration ($\leq$21.5 ppma) compared to those with low nitrogen concentration (<20 ppba) and high oxygen concentration (32.5 ppma).

1. INTRODUCTION

Interest in the behaviour of nitrogen in silicon has increased over the past ten years for two main reasons: first, nitrogen has different properties from those of other group V impurities in silicon (e.g. Abe et al 1981; Itoh et al 1989); second, silicon may be grown by the Czochralski method in a nitrogen rather than in an argon atmosphere (Que et al 1986). Nitrogen will be incorporated into a silicon crystal during crystal growth, so silicon grown by this method will contain a certain level of nitrogen.

It has been discovered that most of the nitrogen in silicon exists as N-N pairs (Stein 1986), which are electrically inactive but can interact with other impurities and lattice defects (Suezawa et al 1986; Itoh et al 1989). Nitrogen in silicon has a strong effect on pinning dislocations and strengthening silicon wafers in device manufacturing processes, this is particularly helpful for float-zone grown silicon (Abe et al 1981). In Czochralski-grown silicon (CZ-Si) nitrogen can combine with oxygen to form oxygen-nitrogen complexes (Suezawa et al 1986) and can accelerate oxygen precipitation (Shimura and Hockett 1986).

Oxygen precipitation in CZ-Si has a key effect on defect generation and on intrinsic gettering. For nitrogen-doped (N-doped) CZ-Si it is necessary to understand the influence of nitrogen on oxygen precipitation. Shimura reported that in N-doped CZ-Si oxygen precipitation was enhanced even at high temperature (1100 °C) and precipitates were formed in the near surface region of the wafers. So no intrinsic gettering effect was observed (Shimura and Hockett 1986). But so far the detailed role of nitrogen in the process of oxygen precipitation is not clear. In this article we report our preliminary results of a study of oxygen precipitation in N-doped CZ-Si by means of transmission electron microscopy (TEM) and infrared spectroscopy in order to investigate the behaviour of nitrogen in the process of oxygen precipitate nucleation and growth. We confirm that nitrogen plays an important part in precipitate nucleation by generating grown-in microdefects as heterogeneous nucleation centres. The possibility that oxygen-nitrogen complexes are these nucleation centres is discussed.

2. EXPERIMENT

A 2-inch diameter dislocation-free single crystal silicon ingot was grown in a <111> direction by the Czochralski method with N_2 as the protective atmosphere and was doped with phosphorus (n-type, 6-10 Ω cm). In order to obtain specimens with different oxygen and nitrogen content, wafers were cut from near the seed and tail end of the ingot respectively. The wafer thickness was about 2 mm; 4 mm wafers were also used to measure the initial nitrogen content. Specimens were double side mirror-like polished. None of the specimens had been subjected to any prior thermal treatment, such as thermal donor killing annealing. A single step heat treatment at a constant temperature between 750 °C and 1100 °C for 20 or 48 hours was used and was carried out in an Ar ambient.

Interstitial oxygen, substitutional carbon and N-N pair content ([Oi], [Cs] and [N]) before and after each heat treatment were measured by Fourier transform infrared spectroscopy (FTIR) at room temperature using the conversion factors for determining [Oi], [Cs] and [N] given by ASTM F121-1979, Newman (1965) and Itoh (1985) respectively. TEM work was carried out with a Philips EM 400 electron microscope, using 120 kev electrons. TEM specimens were prepared from the inner region of the wafers, about 150 μm from the surface. Defect density was determined by optical microscopy after Sirtl etching for densities below 10^9 /cm^3 or by TEM for higher densities.

3.   RESULTS

Initial [N] ([N]$_0$) and [Oi] before and after the heat treatment in each specimen are summarized in Table 1. Specimens are divided into two groups, A and B, according to their initial nitrogen content. Because nitrogen has a very small equilibrium segregation coefficient (7 x 10^{-4}) it is always concentrated in the tail end of an ingot, while oxygen segregates in the opposite direction. Hence group A specimens have a high initial oxygen content ([Oi]$_0$) with [N]$_0$ being below the detectable limit (< 20 ppba), but group B specimens have lower [Oi]$_0$ and higher [N]$_0$. Initial [C$_s$] in each specimen was below the detectable limit (<0.1 ppma).

Several as-grown specimens from both group A and group B were analyzed by TEM and no microdefects were observed even by the weak-beam technique, which indicated that the size of microdefects, if they exist, was smaller than about 60 Å (e.g. Shimura 1982, for weak-beam microscopy, the size limit depending on the nature of the strain field).

Table 1:   Summary of the heat treatment condition and FTIR and defect density
measurement results

Specimen	[N]$_0$ (ppba)	Temp. (°C)	Time (hrs)	[Oi]$_0$ (ppma)	[Oi]$_f$ (ppma)	Defect Density (/cm^3)
Group A	≤ 20	750	48	32.4	30.8	~ 1 x 10^6
	≤ 20	900	20	32.5	31.4	~ 1 x 10^6
	≤ 20	1000	20	32.5	16.7	~ 2 x 10^9
	≤ 20	1100	20	32.4	31.9	< 10^6
Group B	132	750	48	19.4	8.2	~ 5 x 10^{12}
	128	900	20	21.5	6.7	~ 2 x 10^{11}
	100	1000	20	20.2	8.6	~ 3 x 10^{10}
	128	1100	20	21.5	21.2	< 10^6

We describe below the results in three temperature ranges that relate to different oxygen precipitation behaviour.

1) 750 °C--900 °C: Heat treatment in this temperature range gives rise to precipitate nucleation, but it can also cause precipitate growth depending on annealing time and temperature.

At 750 °C, a long annealing time (48 hours) was used to reveal the precipitates. [Oi] change (Δ[Oi]), defect density and defect morphology are different between the two groups of specimens. In group A no distinct [Oi] decrease was observed, whilst in group B [Oi] dropped dramatically from 19.4 ppma to 8.2 ppma (Table 1). In accord with the larger Δ[Oi], the defect density in group B specimens was about 10^6 times higher than that in group A specimens. The dominant defect morphologies in group B specimens were small precipitates and dislocation loops with sizes in the range 100 Å--1000 Å (Fig. 1). A low density of rod-like defects was also observed (Fig. 1c). In group A specimens defects were difficult to detect due to their low density. A typical precipitate is shown in Fig. 2.

At 900 °C, in group A specimens there was still no significant [Oi] change but in group B specimens there was a larger [Oi] decrease than that at 750 °C. Fig. 3b shows that a precipitate in a group B specimen has grown to a large size and punched out a dislocation loop. Extrinsic stacking faults were also observed, which were mostly decorated with precipitates. The defect density was about 10^{11}/cm^3, lower than that at 750 °C. For group A specimens, the defect density was still low (~10^6 /cm^3) and the precipitate morphology was similar to that at 750 °C (Fig. 3a).

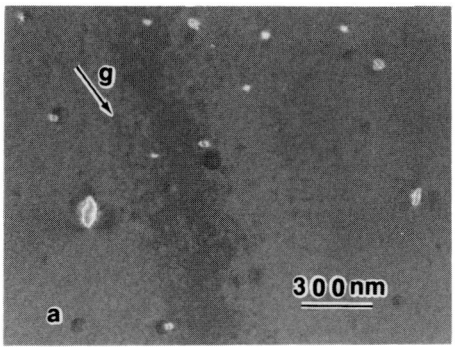

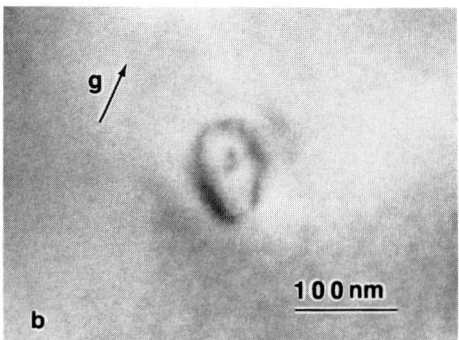

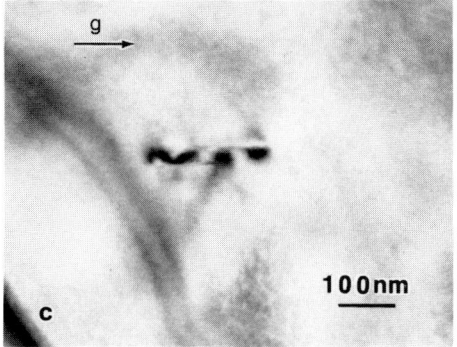

Fig. 1 TEM micrographs of defects in a group B specimen annealed at 750 °C for 48 hours. a) Weak-beam image, showing the defect distribution, g(2g) g= 220. b) Bright-field image of a typical small dislocation loop g= 202. c) Bright-field image of a rod-like dislocation g= 220.

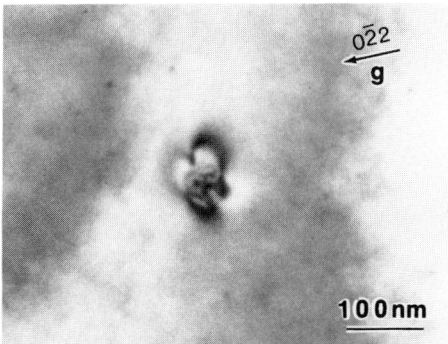

Fig. 2 A precipitate in a group A specimen annealed at 750 °C for 48 hours, Bright-field, g=022

2) 1000 °C: A large $[O_i]$ decrease was observed in both group A and group B specimens. Final $[O_i]$ ($[O_i]_f$) in the group B specimens had nearly saturated after only 20 hours annealing (solubility at this temperature is 6.74 ppma, Shimura and Tsuta 1982). Defect morphologies in both the group A and B specimens are large plate-like {100} precipitates, punched-out dislocation loops and extrinsic stacking faults. Fig. 4 shows typical defects at this temperature. From the micrographs it was found that precipitates in group B specimens had very few related dislocations or dislocation loops, whereas most precipitates in group A specimens were surrounded by punched-out dislocation loops.

3) 1100 °C: No distinct drop in $[O_i]$ was observed and no defects of any kind were visible using the weak-beam technique in both groups of specimens after 20 hours annealing. Sirtl etching revealed that there were some defects in the specimens with a density of about 10^5 /cm^3. A TEM specimen from group B which had no visible defects by the weak-beam analysis was subjected to Sirtl etching for 10 seconds and then re-examined by TEM. Triangular etch pits were observed this time (Fig. 5) and the etch pit density was estimated to be about 10^6 /cm^2. This indicates that there are small defects in the specimen that are too small to be detected by TEM.

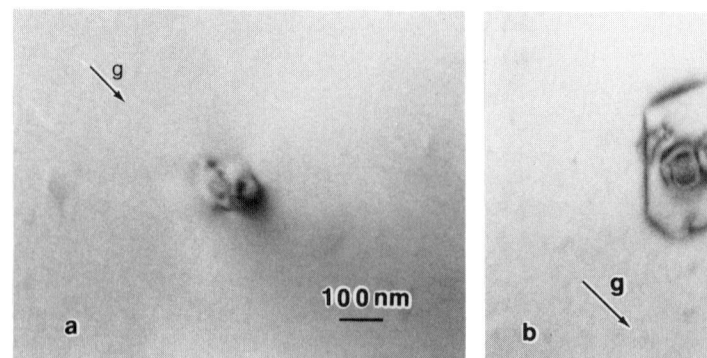
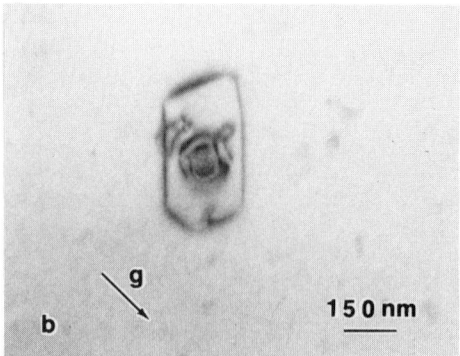

Fig. 3 Bright-field images of the defects in the specimens after 20 hours annealing at 900 °C.
a) group A specimen g= 0$\bar{2}$2 b) group B specimen g= $\bar{2}$20, the precipitate lies on (010)plane and the loop lies on (011) with Burgers vector b=a/2[011]

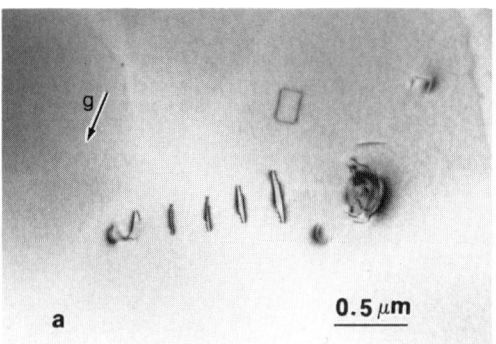
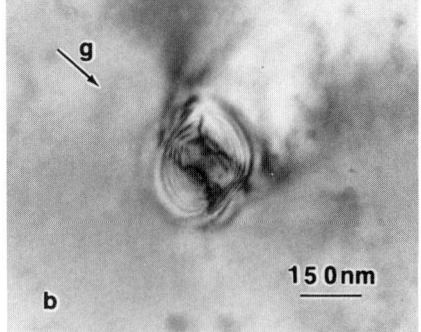

Fig. 4 Bright-field images of the defects in the specimens after 20 hours annealing at 1000 °C.
a) group A specimen g= $\bar{2}$02. b) group B specimen g= $\bar{2}$20

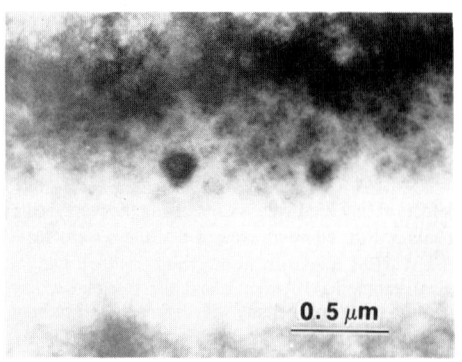

Fig. 5
Triangular etch pits in a group B specimen annealed at 1100 °C for 20 hours, the TEM specimen has been Sirtl etched for 10 seconds, indicating there are small defects that can not be revealed by TEM

4. DISCUSSION AND CONCLUSIONS

Although the growth process of oxygen precipitates has been extensively investigated, it is still controversial whether the precipitate nucleation process is homogeneous or heterogeneous (e.g. Inoue 1981, Tan 1986) But it has been recognised that some kind of impurities or defects, such as carbon and swirl defects, can act as heterogeneous nucleation sites and enhance oxygen precipitation (e.g. Shimura 1989, Tan 1986). According to homogeneous nucleation theory nucleation rate depends on interstitial oxygen supersaturation. Therefore at a given temperature, e.g. 750 °C, high $[Oi]_0$ should have high nucleation rate and vice versa. Furthermore nucleation will not occur if $[Oi]_0$ is below a certain level. Shimura et al (1982) showed that in N-free CZ-Si nucleation was very difficult for $[Oi]_0$ less than 28 ppma. On the contrary, our results revealed that precipitate nucleation became easier in group B specimens than that in group A specimens and the threshold initial oxygen concentration for precipitation ($[O_i]_T$) was decreased to at least 19.4 ppma. This implies that the nucleation process in N-doped CZ-Si is heterogeneous. Since the specimens were not been subjected to any pre-annealing, we propose that there must be a large number of microdefects in as-grown specimens, whose radii are larger than the critical radius at the annealing temperature. It is these microdefects that act as the nuclei of oxygen precipitate nucleation.

Now we will discuss the possible identity of the grown-in microdefects. The thermal history of a specimen is a sensitive factor for oxygen precipitation because it affects microdefect formation. To simplify the experimental conditions specimens from a single ingot were used. Although the cooling process during the crystal growth in different portions of the ingot was not identical, compared with specimens from different ingots or from different vendors the thermal history of our specimens is more comparable. On the other hand due to a low cooling rate at the seed end it should be more favourable for forming microdefects homogeneously in this portion (Shimura 1982). If only oxygen atoms are involved in the microdefects, which means that the microdefects are formed homogeneously, microdefect density should be higher at the seed end rather than at the tail end. But experimental results showed the opposite. Carbon, like nitrogen, is concentrated at the the tail end of the ingot and is too low in content to have a significant effect on oxygen precipitation in our specimens. So we assume that the microdefects consist of oxygen and nitrogen atoms, i.e. oxygen-nitrogen complexes (ONC).

ONC were discovered by FTIR (Wagner 1988, Suezawa et al 1986), photoluminescence (Steel 1990) and electron paramagnetic resonance analysis (Hara 1989). They were found both in as-grown and in annealed N-doped CZ-Si.The nature and types of ONC are still unknown. Some have been found to be shallow donors which are more stable than oxygen thermal donors (Suezawa et al 1986). ONC infrared absorption lines were observed in our high [N] specimens and they were still detectable after 5 hours annealing at 600 °C. An ONC probably comprises one oxygen atom and two or three pairs of nitrogen atoms (Suezawa et al 1988). So its size should be larger than the critical size for nucleus formation (e. g. for temperature below 1100 °C and $[O_i]_0=20$ppma the critical radius was calculated less than 2 Å, Craven, 1981).

Shimura reported that, at 1000 °C, in N-free CZ-Si, a large number of precipitates occurred in the high $[O_i]_0$ specimen, but few in the low $[O_i]_0$ specimen ($[O_i]_0<28$ ppma) (Shimura 1982). The nucleation centres at this temperature were believed to be mainly the lattice defects introduced at early stage of the heat treatment, the so-called secondary heterogeneous nuclei. In N-doped CZ-Si, precipitates occurred in both high $[O_i]_0$ and low $[O_i]_0$ specimen. We assume that secondary heterogeneous nuclei and nitrogen related nuclei exist simultaneously in N-doped CZ-Si and the latter contributes primarily to precipitation in low $[O_i]_0$ specimens. In group B specimens few punched-out dislocation loops were observed whilst in group A specimens most of the precipitates were surrounded by the punched-out dislocation loops. This may be another evidence that N-N pairs can aggregate at defect sites to reduce the lattice strain field (Itoh et al 1989).

Since the high $[O_i]_0$ with high [N] specimens have not been studied in this experiment, the effect of nitrogen on the high $[O_i]_0$ specimen is unknown. In this work most heat treatment times were short (20 hours). Further work is being conducted on high resolution electron microscopy analysis and on studying the precipitates after a longer annealing time so that they are close to their steady state.

In summary, TEM and infrared absorption measurement have been used to study the effect of nitrogen on oxygen precipitation in N-doped CZ-Si. It is clear that nitrogen in CZ-Si increases the oxygen precipitate nucleation rate and decreases the threshold initial oxygen content for precipitation. After heat treatment between 750 °C and 1000 °C, a large number of oxygen precipitates were observed in high [N] specimens and the $[O_i]_T$ was not higher than 19.4 ppma. With the same annealing conditions, defect morphologies in low [N] specimens and high [N] specimens are different. The effective initial nitrogen content on oxygen precipitation appears to be above 20 ppba. Nitrogen effect on oxygen precipitation is attributed to the nitrogen related microdefects, maybe oxygen-nitrogen complexes, which exist in as-grown N-doped CZ-Si and act as heterogeneous nuclei during the subsequent heat treatment.

ACKNOWLEDGEMENT

We wish to express our thanks to the Research Institute for Semiconducting Materials of Zhejiang University, China, for supplying the specimens and Department of Electrical and Electronic Engineering, University of Liverpool for providing the annealing facilities. This work is financially supported by SERC.

REFERENCES

Abe T, Kikuchi K, Shirai S, Muraoka S   1981   in *Semiconductor Silicon 1981* eds Huff H, R, Kriegle R J and Takeishi Y (Electrochem. Soc. Pennington, 1981) 54
Craven R A   1981   in *Semiconductor Silicon 1981* eds Huff H R, Kriegle R J and Takeishi Y (Electrochem. Soc. Pennington, 1981)   254
Freeland P E, Jackson K A, Lowe C W and Patel J R   1977   Appl. Phys. Lett. 30 31
Hara A, Fukuda T, Miyabo T and Hirai I   1989   Jpn. J. Appl. Phys. 28 142
Inoue N, Wada K and Osaka J   1981 in *Semiconductor Silicon 1981* eds Huff H R, Kriegle R J and Takeishi Y (Electrochem. Soc. Pennington, 1981)   282
Itoh T, Nozaki T, Masui T and Abe T   1985   Appl. Phys. Lett. 47 488
Itoh T, Hayamizu Y, and Abe T   1989   Materials Science and Engineering B4 309
Newman R C and Willis J B   1965   J . Phys. Chem. Solids 26 373
Que D, Li L and Lin Y   1986   Chinese Patent CN 85100295 (1986)
Shimura F and Tsuya H   1982   J. Electrochem. Soc. 129 1062
Shimura F and Hockett R S   1986   Appl. Phys. Lett. 48 224
Shimura F   1989   *Semiconductor Silicon Crystal Technology* (Academic Press Inc, 1989) Chapt. 7
Steel A G, Lenchyshyn L C and Thewalt M L W   1990   Appl. Phys. Lett 56 148
Stein H J   1986   in *Oxygen, Carbon, Hydrogen and Nitrogen in Crystalline Silicon* eds Mikkelsen J C Jr, Pearton S J, Corbett J W and Penneycook S J (Materials Research Society , Pittsburgh, 1986)   523
Suezawa M, Sumino K, Harada H and Abe T   1986   Jpn. J. Appl. Phys. 25 L859
Suezawa M, Sumino K, Harada H and Abe T   1988   Jpn. J. Appl. Phys. 27 62
Tan T Y and Kung C Y   1986   in *Semiconductor Silicon 1986* eds Hull H R, Abe T and Kolbesen B (Electrochem. Soc. Inc. Pennington NJ 1986)   864
Wagner P, Oeder R and Zulehner W   1988   Appl. Phys. A   46 73

Inst. Phys. Conf. Ser. No 117: Section 4
Paper presented at Microsc. Semicond. Mater. Conf., Oxford, 25–28 March 1991

217

Climb of dislocations induced by oxygen precipitation in silicon

K Minowa, I Yonenaga and K Sumino

Institute for Materials Research, Tohoku University, Sendai 980, Japan

ABSTRACT : The characteristics of the climb process of glide dislocations caused by excess interstitials produced by oxygen precipitation in Czochralski-grown silicon were investigated by means of the weak beam TEM technique. Dislocations were observed to be extended in all stages of climbing. The climb process was interpreted with a model based on the idea originally proposed for FCC crystals by Thomson and Balluffi.

1. INTRODUCTION

Climb motion of a dislocation in a crystal reflects the process by which the dislocation incorporates point defects. It was first thought that the constriction of a ribbon of stacking fault had to take place when an extended dislocation underwent climb. However, electron microscopical observations showed that dislocations in irradiated or quenched Cu-Al, Si and GaAs were extended after climbing without any constriction of stacking faults (Cherns et al. 1980, Ourmazd et al. 1981, Decamps et al. 1983, Cherns and Feuillet 1985). Since such dislocations were observed to be tangled with prismatic dislocation loops, the climb of an extended dislocation was thought to be facilitated by the condensation of point defects into prismatic dislocation loops along both of two partial dislocations of the extended dislocation. The climb process of extended dislocations observed in this work is thought to be the typical one when the concentration of excess point defects is very high. It is probable that some other kind of climb process of an extended dislocation may operate when the concentration of excess point defects is moderate.

Czochralski-grown (CZ) silicon crystals are supersaturated with oxygen impurity at temperatures below about 1200 °C. Silicon self-interstitials are known to be emitted from SiO_2 particles during precipitation of oxygen at relevant temperatures. In a dislocated crystal such excess interstitials are mostly absorbed by dislocations and promote climbing of the dislocations. The characteristic of climb motion in this case is that the process proceeds at some moderate rate since interstitials are generated one by one at the interfaces of precipitate particles, the emission rate of interstitials being controlled by diffusion of oxygen atoms in silicon.

In this paper, we report the climb process of extended dislocations in CZ-Si crystals caused by excess interstitials emitted from SiO_2 particles observed with the weak beam TEM technique.

2. EXPERIMENTAL

Specimens were prepared from a n-type CZ-Si crystal (P-doped, $5-10$ Ω cm, $[O_i] = 9 \times 10^{17}$ cm^{-3}). They were first deformed at 750 °C to introduce dislocations and, then, were annealed at 1300 °C to dissolve oxide precipitates and also to eliminate defects other than dislocations which were formed during the deformation. Such specimens were annealed at 900 °C to develop SiO$_2$ precipitates. Foils with the surface parallel to the (111) primary slip plane were cut from the specimens and were thinned by chemical polishing. Weak beam TEM observations were made with a JEM 2000EX at an accelerating voltage of 200 kV.

3. RESULTS and DISCUSSION

Dislocations observed in a specimen annealed at 1300 °C are approximately straight and smooth in shape. Upon subsequent annealing at 900 °C, precipitate particles are observed to be nucleated preferentially on dislocations. After annealing at 900 °C for some duration precipitates are observed to align discretely along the lines on which the dislocations are thought to be located before annealing. Depending on the orientation of the dislocation, the change in the location as well as that in the shape of the dislocation are induced due to annealing, and are thought to be caused by absorption of self-interstitials emitted from the precipitates. A 60° dislocation shifts from its original location by climbing uniformly along the dislocation line and retains the straightness in shape. A screw dislocation becomes helical winding around the line connecting the precipitates.

Figure 1 (a) shows a weak-beam image of a 30° dislocation that has undergone climb in the specimen annealed at 900 °C for 48 h. The dislocation is dissociated into two Shockley partials over the whole length with a width of about 10 nm. Small dislocation loops seen around oxygen precipitates, the Burgers vectors of which are confirmed to

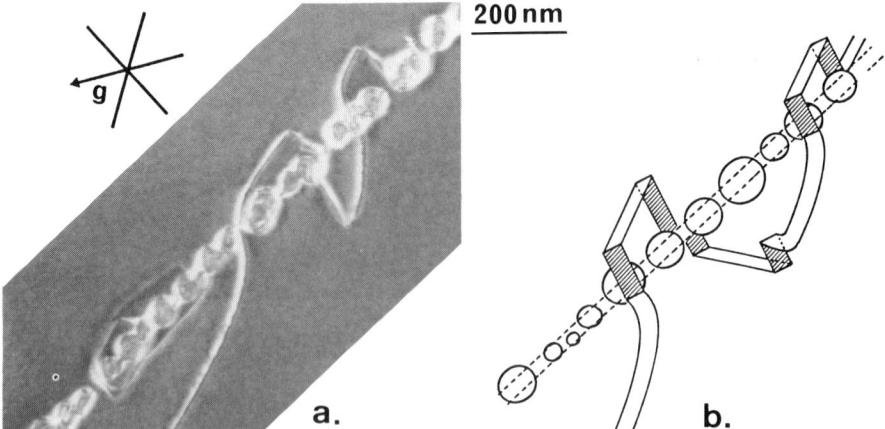

Fig.1. (a) Weak beam image of a 30° dislocation which underwent climbing during annealing at 900 °C for 48 h. (b) Schematic picture for the dislocation configuration.

be different from that of the dislocation in question, are identified as being emitted from oxygen precipitates and are not related to the climbing of the dislocation. Figure 1 (b) shows the three dimensional configuration of the dislocation in (a) determined with a stereographic observation. Dislocation segments with unhatched stacking faults lie on the primary slip planes of different elevations, while dislocation segements with hatched stacking faults lie on the {111} planes inclined to the primary slip plane. It is seen that a 30° dislocation in which the screw component dominates becomes helical during the process of climbing. The above observations show that dislocations which are undergoing climbing in silicon are extended like gliding or stationary dislocations.

Since dislocations undergoing climb are observed to be extended over their whole lengths, it is unlikely that an extended dislocation repeats recombination and dissociation at every step of the climb process. Furthermore, the partial dislocations are observed not to be tangled with any other dislocations after climbing. This means that the climbing of an extended dislocation by means of condensation of self-interstitials into prismatic loops along the two constituent partials is unlikely. The climbing of extended dislocations observed in this work may be described with the model originally proposed by Thomson and Balluffi (1962) for FCC crystals with the idea of an extended jog pair.

Figure 2 shows schematically a series of climb steps of an extended 60° dislocation based on the model applied to a diamond-type crystal. (a) shows the projection of an extended 60° dislocation consisting of a 90° partial P and a 30° partial Q. (b) shows that the first interstitial atom I is absorbed by the 90° partial P that has a larger edge component than the 30° partial Q. The absorbed atom I has two dangling bonds. Figure 2 (c) shows that the second interstitial atom II is absorbed by the

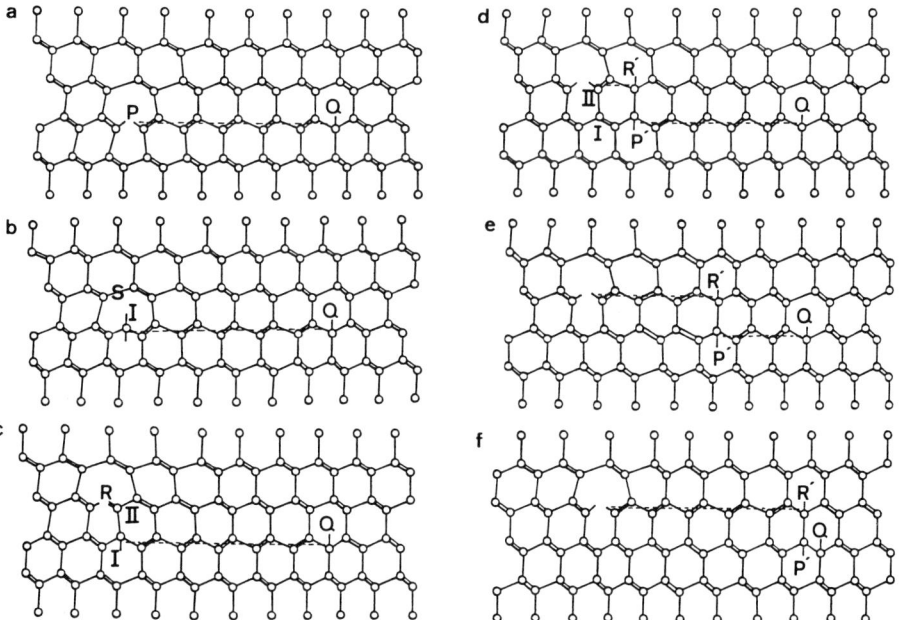

Fig. 2. Successive steps in climb process of an extended 60° dislocation in a diamond-type crystal.

90° partial P and attached to the atom I by breaking the bond S shown in (b). When such an incorporation process of interstitial atoms takes place in every atomic plane along the 90° partial, a perfect 60° dislocation is formed at the position R which is coupled with a 30° dislocation at I. The perfect 60° dislocation starts to be dissociated into two partials on the plane one inter-atomic-plane above the initial stacking fault. Now, the 90° partial has undergone climb upwards by one inter-atomic-plane distance and a dipole of Shockley partials $P' - R'$ has been formed as shown in (d). The dipole glides across the stacking fault toward the 30° partial Q. Figure 2 (e) shows the dipole halfway to Q, resulting in a step on the stacking fault. In the final stage the partial dislocation P' of the dipole meets the partial Q of the original dislocation, leading to pair annihilation, and only the partial R' remains as shown in (f). The dislocation PQ has now undergone climb upwards by one inter-atomic-plane distance and the original configuration of an extended 60° dislocation is restored.

An extended screw dislocation consists of two partial dislocations for which dangling bonds extend in opposite directions to each other with respect to the slip plane. In principle, absorption of point defects cannot cause the climb of a screw dislocation. However, if a dislocation slightly deviates from the screw orientation, it can be regarded to be many segments in the exact screw orientation each connected with some lengths of dislocations having edge components. Such segments with the edge component can undergo climb in a manner similar to a 60° dislocation. The shape of such a screw dislocation becomes helical during the climbing motion.

4. CONCLUSION

The characteristics of the climb process of dislocations caused by excess interstitials associated with oxygen precipitation in CZ silicon were investigated using the weak beam TEM technique. Dislocations introduced by plastic deformation were extended in all stages of climbing. This seems to imply that some mechanism is available by which an extended dislocation climbs without constriction of the stacking fault ribbon. A model based on one originally proposed by Thomson and Balluffi for FCC crystals seems to account for the present observation.

REFERENCES

Cherns D, Hirsch P B and Saka H 1980 Proc. Roy. Soc. A371 213
Cherns D and Feuillet G 1985 Phil. Mag. A51 661
Decamps B, Cherns D and Condat M 1983 Phil. Mag. A48 123
Ourmazd A, Cherns D and Hirsch P B 1981 Inst. Phys. Conf. Ser. No.60 39 − 44
Thomson R M and Balluffi R W 1962 J. Appl. Phys. 33 803

Inst. Phys. Conf. Ser. No 117: Section 4
Paper presented at Microsc. Semicond. Mater. Conf., Oxford, 25–28 March 1991

Rod-like defects in Czochralski silicon crystal and their transformation

L M Sorokin, K L Malyshev, A A Sitnikova

Ioffe Physical-Technical Institute of the USSR Academy of
Sciences, Leningrad 194021, USSR.

ABSTRACT: The TEM diffraction strain contrast of inclined
RLD is studied for silicon annealed at 750°C for 1h and
the comparison of experimental patterns with computer simulated images is performed. Three elastic models of the
RLD with the coesite structure were proposed. Transformation of the RLD was investigated in Si subjected to additional heating at high temperature (~1000°C). At these
treatments dislocation dipoles and loops are generated
around the RLD divided on the train of the particles.

1. INTRODUCTION

It is well known that the heat treatment of Cz-grown silicon
crystals, containing a high interstitial oxygen concentration
at low temperature (450-650°C) results in rod-like defects
(RLD) (Freeland 1980). Yamamoto et al (1983) studied the diffraction contrast of RLD and showed that they are of interstitial type. Bourret et al (1984) showed by means of HREM
that the image of RLD perpendicular to the projection plane
(110) is consistent with the projected structure of coesite,
which is the monoclinic high-pressure phase of SiO_2. Although
many investigations have been reported on the characterization of different defects by means of TEM (Bender 1984, Reiche
et al 1988) disagreement still exists concerning the nature
of RLD, their transformation after a subsequent high temperature treatment.

2. EXPERIMENTAL PROCEDURE

The experiments were performed on Cz-grown silicon wafer with
a high interstitial oxygen content of $1.3 \cdot 10^{18}$ atoms/cm^3.
The crystal was grown in the <111> direction with a diameter
of 60 mm. Low temperature heating for the formation of rod-
like defects was performed at 750°C for 1h. The high temperature annealing of bulk material was done in a wet oxygen ambient at 1020°C for 1h and 5h.

3. RESULTS AND DISCUSSION

3.1. Elastic models

All elastic models of the RLD reported here are choosen on

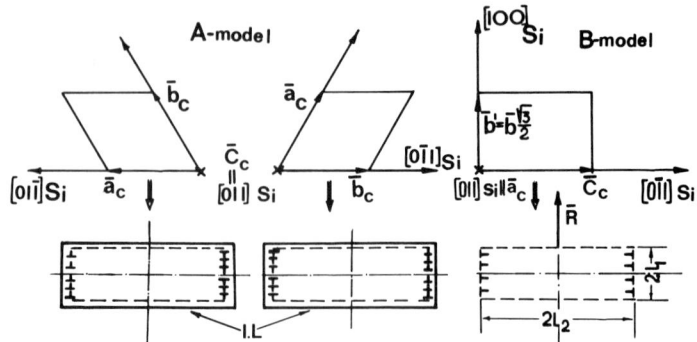

Fig.1 Orientation of coesite (C) unit cell (a=b=7.17 A°, c=12.38 A°, γ=120°) with respect to silicon matrix and corresponding cross-sections of RLD (elastic models). I.L-interstitial dislocation loops.

the basis that the coesite phase of SiO_2 with rectangular cross-section is inserted into the silicon matrix. Three possible orientations of the coesite phase with respect to the silicon matrix and the corresponding elastic models are represented in Fig.1. Since the dimensions of the RLD cross-section are several orders of magnitude smaller than the average length of the RLD, construction of the stress field around the RLD reduces to a planar problem. For the above possible mutual orientations of the silicon and coesite lattices the following misfit parameters arise.

$$\omega_{hkl} = \Delta d/d \qquad \omega_{011} = (12.38-7.68)/7.68 = 0.61$$

$$\omega_{0\bar{1}1} = (7.17-7.68)/7.68 = -0.07$$

$$\omega_{100} = (6.21-5.43)/5.43 = 0.14$$

$$\text{angle misfit} \qquad \omega_{\gamma} = 7.17 \cos60°/5.43 = 0.66$$

From analysis of the misfit parameters two elastic models (A, B) follow. Two possible orientations of the coesite phase underlie a A-model. In this case the elastic stress field around the RLD is described by superposition of solutions for axial rectangular prismatic dislocation loops and for two finite dislocation walls with opposite signs. A B-model takes into account the misfit parameter $\omega_{100}=0.14$ characterizing distortion in the direction normal to the habit plane of the RLD. In this case the stress field around the RLD is described by two continual dislocation rows on lateral sides of the RLD cross-section. The deformation sign corresponds to a defect of interstitial type and the effective displacement vector $\bar{R}$ is perpendicular to the $(100)_{Si}$ habit plane. The components of the proper distortion tensor β_{ij} involving the ω parameter were calculated for both models (Malyshev et al 1990). It should be noted that exactly this same tensor is used in diffraction contrast simulations on the basis of the Howie-Whelan equations. For both models distortion tensors are written as:

$$\text{A-model} \quad \beta = \beta_l + \beta_w$$

$$\text{B-model} \quad \beta = \beta_r$$

where l, w, r stand for "loop", "wall", "row" respectively. Relations obtained for β_{ij} depend on variable (l_1, l_2-dimensions of the RLD cross-section) and fixed (ν , ω) parameters. It is known that the coesite phase (SiO_2) will be the stable phase if the hydrostatic pressure ($P^{A,B}$-one third of stress tensor spur) is higher than a critical value Pc =26.5 kbar, i.e. $|P^{A,B}| > Pc$. Analysis of the behaviour of the P^A function for the A-model of RLD showed that regions of stability of the coesite phase ($|P^A| \geqslant Pc$) are extremly small as compared to the area of cross-section at any ratio l_1/l_2. That means that for the A-elastic model there are no conditions for formation of the stable coesite phase. It will be transformed into another crystalline or amorphous SiO_2 phase . For the B-model an analysis of the behaviour of the P^B function gives rise to the following result. The hydrostatic pressure was calculated to be a negative value. This result indicates that compressive strain occurs around the RLD. The corresponding condition for existence of the stable coesite

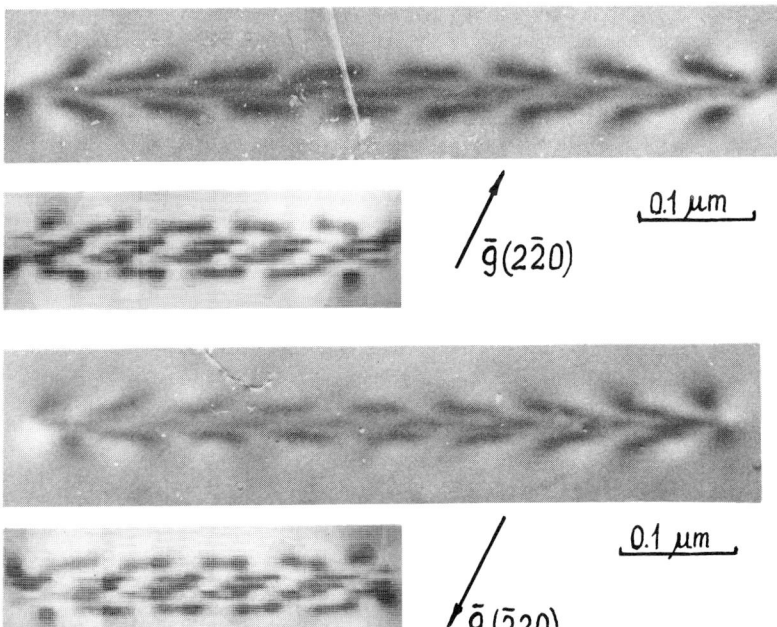

Fig.2 Comparison of the experimental and computer simulated images for an inclined RLD elongated along $[01\bar{1}]$ (B-model), w=s·ξ_g=0. To save simulation time the theoretical images were calculated for a thickness of 5 ξ_g.

phase is given by the inequality (Malyshev et al 1990).

$$-P_{max} (\pm l_1, 0) = \left[2 \mu \cdot \omega (1+\gamma)/3\pi (1-\gamma)\right] \cdot arctg(2l_1/2l_2) \geqslant 26.5.$$

After substitution of μ = 6.41·10^{11} din/cm^2, γ =0.215 and ω=0.14 we can rewrite this relation in the form:

$$(\omega \cdot arctg(2l_1/2l_2))\Big|_{\omega =0.14} > 0.126 \text{ or } l_1/l_2 > 0.63$$

I.e. the coesite phase will not be stable for any dimensions (l_1, l_2) of the RLD cross-section. To establish whether the above elastic models of the RLD are correct or not computer simulated images for both were calculated using the Howie-Whelan equations and comparison of experimental patterns with theoretical ones was performed. Both the experimental investigation and the computer simulation were realized for a (001) surface of Si specimen with an inclined RLD. In this case image contrast is not uniform along the length and is characterized by black-white oscillations of intensity. For an inclined RLD its image is more informative as compared with the image of a RLD parallel to the surface since it contains more detail. To save time during computer simulation of the image, the thickness parameter was taken as to be 5 ξ_g where ξ_g is the extinction distance of the corresponding diffraction reflection whilst the whole specimen thickness under study, as judged by total number of oscillations on the RLD image, is as much as 8 ξ_g. Nevertheless calculated images are quite suitable for establishing the similarity or difference between experimental and simulated images and on the base of this fact to make a judgement about the adequacy of suggested elastic models of RLD to the real situation in the bulk of the specimen. The theoretical images for corresponding elastic models of RLD are represented in Fig.2. The study showed that the best fitting of the experimental images with computer simulated ones are achieved for B-elastic model of RLD at the dimensions of the cross-section of the RLD such as $2l_1:2l_2=10A°$:120A° =1:12. But from the above estimations it was clear that the existence of a stable coesite phase is possible when the following inquality is valid: $l_1/l_2>0.63$. To adjust both results it should be suggested that the RLD contains another phase in addition to the coesite phase (for example, amorphous SiO_2).

3.2. Transformation of RLD

For the study of RLD transformation two temperature regimes were used: 750°C (1h)+1020°C (1h) and 750°C (1h)+1020°C (5h). It is known that at a high heating temperature for Cz-silicon the formation of a stable modification of SiO_2 (β -quartz) occurs (Sorokin et al 1975). Therefore it was expected that after two step heating a new phase of SiO_2 will arise. It was noted that additional heating at 1020°C for 1h gives rise to substantial changes in the external view of the RLD. After this heating a dislocation dipole is generated along almost the whole length of the RLD (Fig.3). In addition the RLD itself has undergone certain changes. Some regions of the RLD

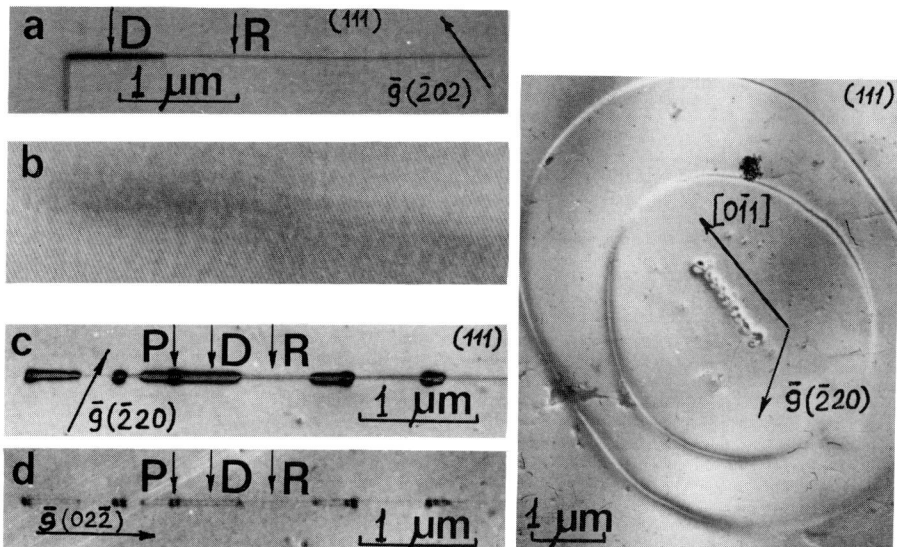

Fig.3 a-initial RLD,   b-lattice image of RLD parallel to the
      surface (both after low temperature treatment: 750°C-
      1h),   c,d-two step   treatment: 750°C-1h, 1020°C-1h.
      R,D,P stand for RLD, dislocation dipole, partical res-
      pectively.   Note change diffraction contrast in depen-
      dence upon g̅-vector.

Fig.4 Two step treatment: 750°C-1h, 1020°C-5h. RLD is divi-
      ded on the train of hexagonal shape precipitates of
      β-quartz (SiO₂). Note anomalous wide image of dislo-
      cation loop (aureol contrast).

are transformed into isodimensional particles of a new phase.
The   ununiform width of the dipole along its length is a direct indication of the presence of small particles. In particular it is clearly seen on the image taken for the g̅-vector
parallel to the RLD length and for s≠0 . At these diffraction conditions the contrast of the RLD disappears absolutely
and the intensity of dipole image decreases. Only small particles having specific contrast with appearence of a couple
of arcs are observed instead of the initial RLD. The study of
the contrast behaviour of the observed defect allows us to
conclude that the R̅-displacement vector for the initial RLD
and the b-Burgers vector of the dislocation dipole equal to
1/2 <110> differ in the direction. Using a standard diffraction technique (inside-outside contrast) and stereomicrographs, the nature of dislocation dipole was determined to be
interstitial in character. The habit planes of the RLD and
the dislocation dipole differ. Because of the small width of
the dislocation dipole it was difficult to establish its habit plane. Perhaps it is (111) or (113). The increase in duration of the heat treatment of silicon gives rise to a further change in the complex: RLD+dislocation dipole. The RLD

is completely divided on the train of separated particles which have strain contrast around them. At diffraction conditions close to a weak beam condition they look like plate-like hexagons. The train of these particles is oriented in the same direction as the initial RLD (Fig.4). They are crystalline particles of a new phase: in some cases microdiffraction patterns taken from similar trains of particles show additional reflections which are identified on the base of a β-quartz lattice. The dislocation dipole is transformed into a dislocation loop lying in the (111) plane. They spread outwards from the particles bit by bit as the particles are growing. It was determined that these dislocation loops are of interstitial type.

4. CONCLUSIONS

On the base of the results obtained and published data, the mechanism of formation of secondary defects (dislocation dipoles and loops) can be proposed. In the early stages of the precipitation process when a particle of new phase did not reach critical size the growth of the precipitate particle is limited by the diffusion of oxygen atoms to the nuclei of the new phase (for example in this case-SiO_2 (coesite)). During this process stresses are introduced into the matrix (silicon) since the formation of one silicon oxide molecule is acompanied by the appearance of the excess of a volume equal to volume of silicon atom. The relaxation of the stresses is accomplished by arising interstitial silicon atoms. Further growth of the precipitate particle is controlled by diffusion of silicon interstitials outward from the interface because the inflow of oxygen atoms to the precipitate particle depends on the concentration of interstitial silicon atoms: the lower the concentration of Is_i the larger the flow of oxygen atoms. Outdiffusion of interstitial silicon atoms from the interface gives rise to the formation of a dislocation dipole with the (111) habit plane differing from the (100) habit plane of the RLD. The climb of the dislocation dipole by joining the interstitial silicon atoms during the precipitation process results in large perfect dislocation loops with the (111) habit plane.

REFERENCES

Bender H 1984 Phys.Stat.Sol.(a) 86 245
Bourret A, Thibault-Desseaux J and Seidman D N 1984 J.Appl. Phys. 55 825
Freeland F E 1980 J.Electrochem. Soc. 20 754
Malyshev K L, Romanov A E, Sitnikova A A and Sorokin L M 1990 Fiz.Tverd.Tela USSR 32 3659
Reiche M, Reichel J and Nitzsche W 1988 Phys.Stat.Sol.(a) 107 851
Sorokin L M and Mosina G N 1975 Zbornik:Legirov.Poluprovodn. (Moskva:Nauka) pp 96-99
Yamamoto N, Petroff R M and Patel J R 1983 J.Appl.Phys. 54 3475

Inst. Phys. Conf. Ser. No 117: Section 4
Paper presented at Microsc. Semicond. Mater. Conf., Oxford, 25–28 March 1991

227

In situ TEM studies of the nature of precipitates in preliminary thermal annealing-induced colonies in Cz-silicon crystal

L M Sorokin and G N Mosina

Ioffe Physical-technical institute of the USSR Academy of Sciences 194021 Leningrad, USSR

ABSTRACT: The behaviour of colonies of precipitate particles has been studied in Czochralski-grown silicon by in-situ annealing (850°C for 10h). The rate of particle growth was estimated to be 0.2 nm/min. Microdiffraction investigations showed that these particles belong to different modifications of SiO_2. Observation of changes to stacking faults during in-situ heating is also reported.

1. INTRODUCTION

Although there are many publications on the structure of thermally annealed silicon single crystals, discussion of the nature of small precipitate particles in colonies observed in silicon after heat treatment (>1000°C) of the bulk material is not yet complete (Ueda et al 1986). In our opinion for the solution of this problem the possibilities of in-situ TEM were not exhausted. It is well known that the colonies of small particles in silicon are produced both in Czochralski-grown material (Sorokin et al 1975) and float-zone crystal (Nes et al 1972). Their origin in Cz-silicon was considered to be the heterogeneous stage of an aging solid solution the decomposition process in the areas enriched by interstitial oxygen atoms which are observed on electron microscopic images as individual black regions (impurity clouds).

2. EXPERIMENTAL PROCEDURE

The experiments were carried out on Cz-grown silicon crystal with a high interstitial oxygen content of $6.9 \cdot 10^{17}$ atoms/cm^3. The initial silicon ingot was grown in the [111] direction with a diameter of 30mm. The silicon wafers were annealed in vacuum at 1100°C for 5h, 10h, 30h and 70h. After that they were thinned by standard techniques to a thickness transparent to electrons accelerated by 100kv. The thin silicon specimens were heated during investigation at 850°C for several hours.

3. RESULTS AND DISCUSSION

For silicon specimens with a (111) surface two type of colonies are observed: perpendicular to and inclined to the sur-

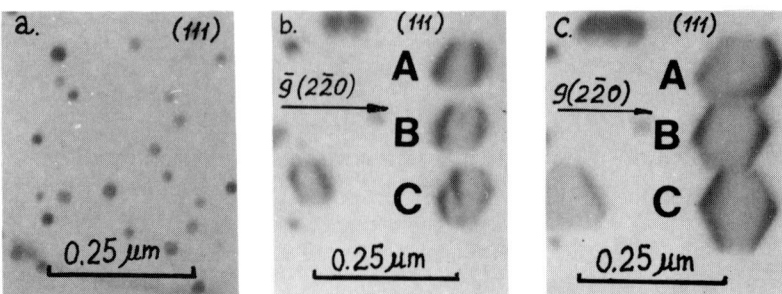

Fig.1 Successive stages of changing in size of selected for observation particles (A,B,C) in colony during in-situ heating: a-initial state of colony, b-after 850°C-1,5h, c-after 850°C-8h.

face. Crystallographical analysis showed that these colonies have the same habit plane of (110) type. As observed both type of colonies in silicon annealed previously at 1100°C for 30-70h underwent substantial changes during in-situ heating. It was noted that several small particles united to form a single large one. Particles increased in size (up to 100 nm) and there was a wide variation in the shape of the precipitate particles (hexagonal, triangle etc.) while in the initial colony no distinct shapes of particles were observed. An initial view of an inclined colony and its successive states after in-situ heating at 850°C for 1.5h and 8h are shown in Fig.1. For studying kinetics of particle size change, detailed observation was carried out for three closely spaced particles marked A,B,C in Fig.1. The rate of particle growth was measured to be ~0.2 nm/min. Agglomeration of closely spaced and enlarged during in-situ heating particles gave rise to additional reflections on the microdiffraction patterns which are indentified as reflections from crystalline silicon oxides (e.g. α, β-quartz, α, β-crystobalite). The results of this calculation are tabulated in Table 1. Also essential changes occured for colonies of particles perpendicular to the (111) surface of the specimen. In the initial colony the particles had no particular shape while after in-situ heating these particles became elongated and hexagonal in shape in most cases

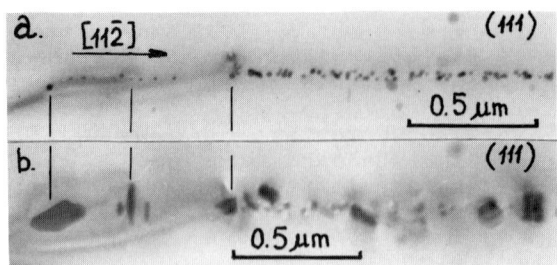

Fig.2 Changes to the perpendicular colony during in-situ heating: a-initial structure of colony, b-after 850°C-6h.

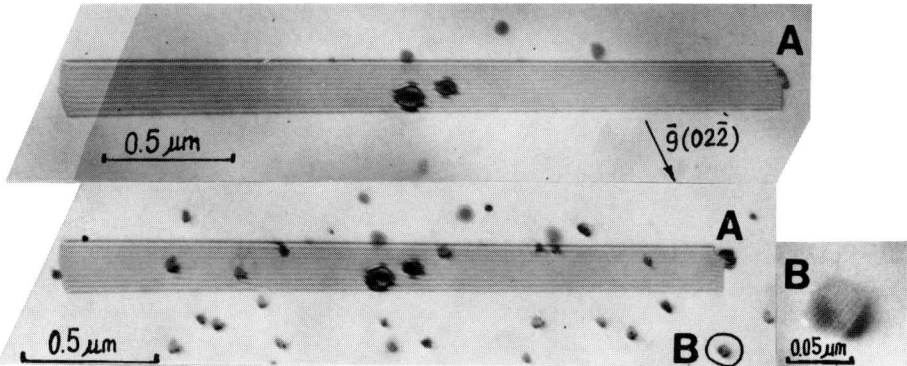

Fig.3 Behaviour of stacking fault of interstitial type during in-situ heating (850°C-8h). Note interaction of partial dislocation with particle (A) and moire contrast on enlarged image of particle (B) (β-quartz).

(Fig.2). Analysis of the particle geometry showed that for this colony, unlike the first type, the habit plane of the hexagonal particles does not coincide with the colony habit plane but lies in another plane of (110) type forming an angle of 35° to the specimen surface. There are also changes in the specimen beyond the colonies. It is known that stacking faults of interstitial type are generated during annealing bulk material at high temperature (~1000°C). In-situ observations of the changes in stacking faults were performed for a silicon wafer which had been preliminary heated at 1100°C for 10h. The stacking fault shown in Fig.3 decreased in size during in-situ heating at 850°C for 8h by 115 nm/min. The rate of shrinkage of the stacking fault was estimated to be ~0.2 nm/min. Remarkable changes of this defect are related to the interaction of the partial dislocation with the particle which increased in size with duration of in-situ heating. It is clearly seen that a partial dislocation "recedes" from the particle (A). The shrinkage of the stacking fault means that this defect is a source of interstitial silicon atoms which possibly compensate the strain field

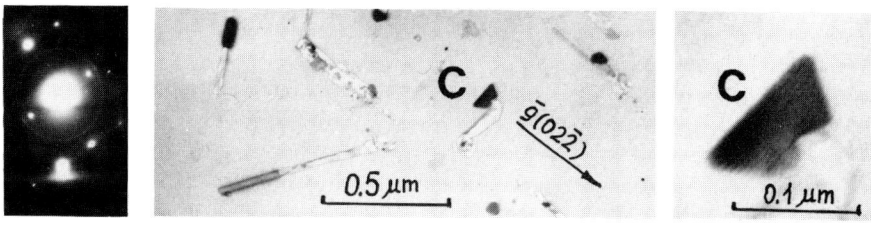

Fig.4 Structure change in thin area of specimen during in-situ heating (850°C-10h). No defects were in this place after preliminary treatment at 1100°C for 10h. Note moire contrast on enlarged image of particle (A) (β-quartz).

Table 1. Lattice spacing distances for precipitate particles in in-situ heated silicon: T=850°C.

experiment	d_{hkl} (Å)					
	α-SiO$_2$ quartz (ASTM)	β-SiO$_2$ quartz (ASTM)	α-SiO$_2$ crystob. (ASTM)	β-SiO$_2$ crystob. (ASTM)	β-SiC exp.	β-SiC (ASTM)
4.35		4.34			2.523	2.51
2.918				2.92	2.175	2.17
2.84			2.84		1.538	1.54
2.018		2.01			1.306	1.31
1.77			1.757			
1.648				1.641	1.085	1.087
1.625		1.624			0.994	0.998
1.45	1.453					
1.293	1.288					

around the growing particle of a new phase. In this case the strain field has to be of vacancy type, unlike that around a particle of SiO$_2$. This fact allows us to suggest that the particle interacting with the dislocation is of β-silicon carbide type. A little later we shall see that this is true. Moreover not far from the stacking fault particles with moire contrast having the shape of truncated triangles are observed. The lattice spacing distance for such particles calculated from moire fringe spacing corresponds to that of β-quartz (SiO$_2$). Further in-situ heating of the specimen gave rise to an increased particle concentration in thin areas on its surface. Debye rings on microdiffraction patterns obtained from these pacticles corresponded to a polycrystalline phase of β-SiC. The corresponding diffraction pattern is shown in Fig.4. Their growth is unoriented on the specimen surface and indicates that these particles are possibly the result of the interaction of the silicon surface layer with hydrocarbons from the ambient of the electron microscope.

4. CONCLUSIONS.

The growth of particles in colonies during in-situ heating of silicon samples and their unambiguous identification as a crystalline phase of SiO$_2$ allow us to conclude that the hete- rogeneous decomposition of the solid solution occurs in the regions enriched by oxygen which are formed as segregation striations during the growth process of the single crystal silicon.

REFERENCES

Nes E and Washburn J 1972 J.Appl.Phys. 43 1835
Sorokin L M and Mosina G N 1975 Zbornik:Legirov.Poluprovodn. (Moskva:Nauka) pp 96-99
Ueda O, Nauka K, Lagowski J and Gatos N C 1986 Proc. 14-th Int.Conf. on Defects in Semicond., ed. H.J von Bardeleben (Switzerland:Trans.Techn.Publ.Ltd.) pp 145-150

Inst. Phys. Conf. Ser. No 117: Section 4
Paper presented at Microsc. Semicond. Mater. Conf., Oxford, 25–28 March 1991

231

Gettering of copper in silicon: precipitation at extended surface defects

M D de Coteau, P R Wilshaw and R Falster[*]

Department of Materials, University of Oxford, Parks Road, Oxford OX1 3PH

[*]MEMC Electronic Materials, Viale Gherzi 31, 28100 Novara, Italy

ABSTRACT: Selected results from a TEM study of copper precipitation at extended surface defects in silicon are reported. The relative gettering effectiveness of surface pits, oxidation induced stacking faults, and their bounding partials is compared. Copper-silicide precipitate colonies on {111} planes are observed for the first time and are found to nucleate at the bounding partials of oxidation induced stacking faults.

1. INTRODUCTION

The precipitation of transition metals (e.g.Cu, Ni, Au and Fe) in device quality silicon has long been associated with catastrophic device failures. The strains introduced into regions of the wafer by complex device structures make the active device regions attractive sites for the precipitation of transition metal contaminants. Gettering procedures, whereby crystal defects are deliberately introduced into predetermined regions of the wafer, are widely practised to control precipitation (see, for example Fair et al. (1985)). The purpose of these structures is to provide stable alternative sites for precipitation which are favoured over those provided by the actual devices.

It has been established that copper will readily precipitate when provided with an appropriate heterogeneous nucleation site--this fact was used by Dash (1957) to decorate dislocations for infrared microscopy. Further, TEM studies by Nes and Washburn (1973), and Seibt and Graff (1988) have characterized the well-known copper silicide precipitate colonies which form upon cooling a Cu-contaminated silicon crystal. In order to develop an optimal gettering system, however, it is important to understand the mechanisms of copper precipitation and the importance of the various defect structures involved in the gettering process. In this investigation, the precipitation behaviour of copper in the presence of extended surface defects used in external gettering systems is studied. Four sets of wafers were used; each set containing a different type and density of surface getter centres. A controlled amount of copper was then diffused into the wafers. The controlled, ordered set of specimens has enabled us to make a direct comparison of the relative effectiveness of different external gettering centres.

2. EXPERIMENTAL

The wafers used in this study were p-type, low oxygen ($5.5 \times 10^{17} \mathrm{cm}^{-3}$), (001) CZ-Si. The initial gettering process employed is known as "soft backside damage" (SBD). In this process, a high velocity silica slurry is directed at the surface; this produces damage in the form of surface pits ≈ 0.1-$0.5 \mu m$ in diameter. SBD is generally reserved for the unpolished side of the wafers; for this study we have performed it on the polished surfaces to facilitate analysis. Four sets of wafers were examined: 1) Damage free. 2) SBD treated; containing $\approx 10^{8} \mathrm{cm}^{-2}$ surface pits. These pits were irregular in shape and weak beam dark field TEM revealed dislocations associated with some of the pits. 3) Type 2 wafers given a dry oxidation at 1000°C for 45 minutes. After oxidation, the pits were smoother and more rounded in appearance than the pits in type 2 wafers. The dry oxidation produced a low density ($\approx 4.5 \times 10^{4}$-$4 \times 10^{5} \mathrm{cm}^{-2}$) of oxidation induced stacking faults (OISF) $\approx 1 \mu m$

long. Each of the OISF's was observed to lie on a pit located near its centre. Only a very small proportion of the pits had served as sites for OISF nucleation; the remaining pits in these wafers had not nucleated OISF's during oxidation. 4) Type 2 wafers steam oxidised at 1100°C for 80 minutes to produce ≈6x10⁵ cm⁻² OISF's 10-15 μm long. OISF's lie on {111} planes and intersect the (001) surface along perpendicular <110> directions. In these specimens the pits and their associated dislocations had been completely removed by the oxidation process. Examples of these defect structures can be found in de Coteau et al. (1990). The oxides grown for wafer types 3 and 4 were stripped by an HF dip prior to copper decoration. Copper was pulse sputtered onto the unpolished wafer surfaces opposite the damage, and diffused in for 10 minutes in nitrogen at either 950°C or 1150°C. Rutherford backscattering spectrometry measurements determined the amount deposited to be 4.1 x 10¹⁵ cm⁻², just below the solid solubility of copper in silicon for the wafer thickness and temperatures studied. The wafers were then quickly pulled from the furnace and cooled in air.

For detailed characterisation of the defect structures and precipitates, (001) plan view and (110) cross-section TEM (XTEM) specimens were prepared from the damaged surface regions. All analysis was carried out at 120kV on a Philips CM12 TEM.

3. RESULTS AND DISCUSSION

3.1 Effect of surface getter centres

In all four wafer types, the majority of the copper precipitation observed was within large colony structures whose intersection with the (001) surface ranged from ≈2-15μm in length. Examples are shown in Fig. 1. A colony is a planar arrangement of copper-silicide precipitates bounded by an interstitial edge dislocation loop b=a<100> or b=a/2<110> lying on {100} or {110} planes, respectively. It was found that the density of colonies intersecting the surface of specimens which contained gettering centres (types 2-4) had increased, on average, by at least an order of magnitude relative to undamaged specimens, from < 5 x 10⁴ cm⁻² to 2-7 x 10⁵cm⁻². The surface defect structures had served as sites for the initial nucleation of precipitates, and precipitation would then proceed with the growth of colony structures. It is important to note that the majority of the surface pits were not decorated. It can be concluded that the colony itself is in effect a gettering centre; that is, a colony, once nucleated, continues to grow, whilst further nucleation and precipitation at the damage centres themselves does not occur. This is illustrated in Fig. 1a, where, despite the presence of a large OISF, all of the Cu has been gettered into the colony structure. The OISF itself does not appear to be decorated. Based on these observations, we believe that the colony is the most effective gettering centre observed in these specimens.

Fig. 1 shows examples of the range of precipitate morphology observed in these specimens. Large surface precipitates .25-.5μm in size were observed in these wafers, as well as the smaller colony precipitates (150-500 Å), which have been previously reported by other workers. Single crystal surface precipitates were observed in plan view TEM (Fig. 1b), and from XTEM (Fig. 1c), it is found that the precipitates are in the form of square-based pyramids, with facets on {111} planes. Kola et al. (1989) have observed similar precipitates in rapid thermal processed silicon; this is the first report of such precipitates in furnace annealed silicon.

3.2 Precipitation at OISF's

In type 3 and 4 wafers, the bounding partial dislocations of the OISF's were found to be preferred sites for precipitation, however the stacking fault plane was not observed to be decorated. In a study of 130 bounding partials in a type 4 specimen, 41% of the partials were "clean"--i.e. no precipitate decoration was observed in TEM (Fig. 1a). Pyramidal precipitates formed at 3% of the bounding partials (Fig. 2a), and 56% were decorated with many small precipitates and had a "ragged-edge" appearance, as shown in Fig. 2b.

As noted above, other workers have reported copper silicide precipitate colonies on both {100} or {110} planes; both types of colonies were observed in this study. However, it was found that in the presence of OISF's , the formation of {100} colonies was suppressed. In type 3 and 4 wafers, nearly all colonies intersected the (001) surface along <110> directions, indicating that these colonies lie on either {110} or {111} planes.

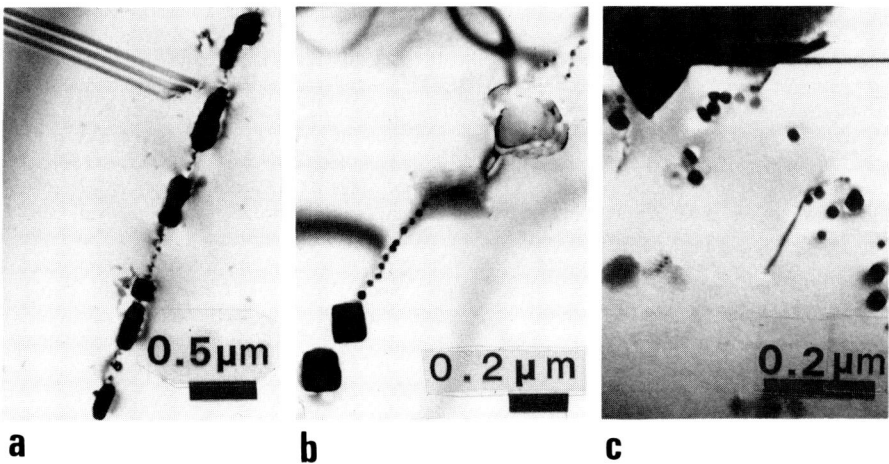

Figure 1: (a) {110} colony in type 4 wafer
(b) {100} colony in type 2 wafer
(c) XTEM image of a colony in type 2 wafer, note pyramidal precipitate at surface.

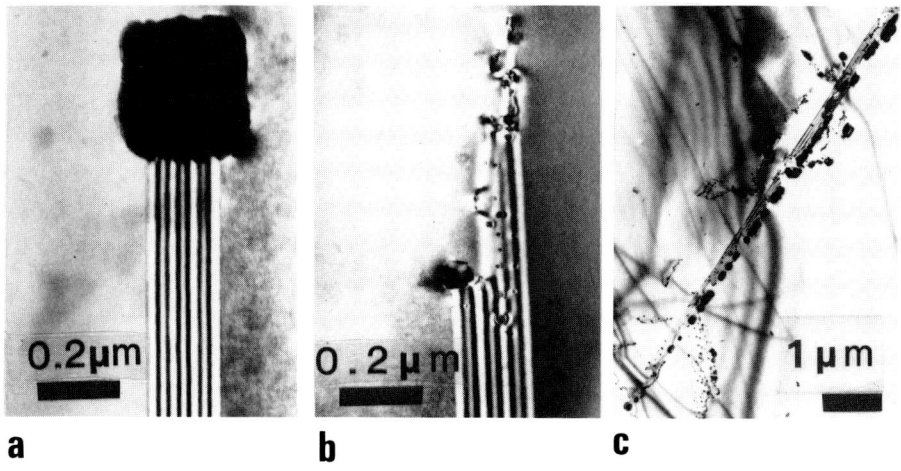

Figure 2: (a) pyramidal precipitate decorating OISF in type 4 wafer.
(b) {111} colony which has nucleated on OISF in type 4 wafer.
(c) {111} colony in type 3 wafer which has caused the growth of a preexisting OISF.

We believe that the suppression of {100} colonies is due to preferential nucleation of colonies at the bounding partials of the OISF's; these colonies are observed to lie on {111} planes. This is the first identification of {111} colonies for the Cu-Si system. The bounding partial of an OISF is a favourable site for segregation of the copper atoms because the matrix neighbouring an OISF is in tension. A copper silicide precipitate which is nucleated on the bounding partial emits silicon self-interstitials as it grows, these interstitials condense on the bounding dislocation, causing it to climb, leaving the precipitates behind. The {111} colony is then generated by repeated precipitation on the climbing dislocation in a manner analogous to that described for {110} and {100} colonies by Nes and Washburn (1973). In Fig. 2c, the growth of a {111} colony in a type 3 wafer has caused the growth of a pre-existing OISF from $\approx 1 \mu$m to over 10 μm in length. Self-interstitials emitted by precipitation in the growing colony have also caused the growth of this extrinsic stacking fault. In some cases the colony continues to grow on the {111} plane with bounding partial $b=1/3<111>$ (hence the "ragged-edge" appearance of some bounding partials). In other cases it has undergone a reaction with dislocations nucleated at other precipitates, thus converting to $b=a/2 <110>$ type dislocation; the colony then continues growth as a {110} colony. The strain fields associated with the bounding partials of the OISF's favour the nucleation of {111} colonies at the bounding partials; after this critical initial step, the growth of some of these colonies on {111} or {110} planes is an important gettering mechanism.

In conclusion, we have found that colonies are the most common form of precipitation in all specimens. The presence of surface damage increases the colony density by nearly an order of magnitude or more relative to that for an undamaged surface. It was found that pits (and their associated dislocations), as well as the bounding partials of OISF's were important nucleation centres for precipitate colony nucleation. The stacking fault plane was not a useful site for gettering Cu. It is important to note that a density of $\approx 10^8$cm^{-2} surface pits resulted in colony densities up to 3 x 10^5cm^{-2} in type 2 and 3 wafers, whereas an OISF density of ≈ 6 x 10^5cm^{-2} produced colony densities up to 7 x 10^5cm^{-2} in type 4 wafers. Therefore the colonies themselves are such efficient getter centres that an increase in surface damage or density of getter centres above the threshold level which activates increased colony nucleation will not improve the gettering efficiency of the system.

REFERENCES

Dash W 1957 Dislocations and Mechanical Properties of Crystals eds. Fisher J, Johnston W, Thomson R, and Vreeland J (New York: Wiley) pp57-68

de Coteau M, Wilshaw P, and Falster R 1990 Phys. Stat. Sol. (a) 117 403

Fair R, Pearce C, and Washburn J, eds.1985 Impurity Diffusion and Gettering in Silicon, Proc. MRS. 36 (Pittsburgh:MRS)

Kola R, Rozgonyi G, Li J, Rogers W, Tan T, Bean K, and Lindberg K 1989 Appl. Phys. Lett. 55 2108

Nes E, and Washburn J 1971 J. Appl. Phys. 42 3562

Seibt M, and Graff K 1988 J. Appl. Phys. 63 4444

Inst. Phys. Conf. Ser. No 117: Section 4
Paper presented at Microsc. Semicond. Mater. Conf., Oxford, 25–28 March 1991

235

High resolution SEM study of porous silicon

JP Gonchond, A Halimaoui and K Ogura *

Centre National d'Etudes des Télécommunications
BP 98, 38243 MEYLAN, France

*) JEOL Ltd, 1-2 Musashino 3-chome, Akishima Tokyo 196, Japan

ABSTRACT : Porous Silicon Layers (PSL) obtained from highly doped (p+) and lightly doped (p) substrates with pore sizes in the 3 to 10 nm range are examined using High Resolution Scanning Electron Microscopy (HRSEM). Both horizontal and vertical structures of the pores have been determined. The comparison with conventional Transmission Electron Microscopy (TEM) demonstrates the advantages of HRSEM over TEM in some cases.

1. INTRODUCTION

During the past few years, motivated by applications in Silicon On Insulator technology, many properties of porous silicon such as surface area and pore size distribution, crystalline structure and optical properties have been extensively studied by a few research groups. Reviews analysing most of these results have been published (Bomchil et al. 1988).

Porous silicon is a material which is obtained by the anodic dissolution of single-crystal silicon in concentrated hydrofluoric acid solutions. Depending upon the substrate resistivity, current density and hydrofluoric acid concentration, the pore size, porosity and the texture of the material can vary over a wide range. Porosities between 10 and 80% and pore sizes of 50 nm down to less than 1 nm are easily obtained by adjusting the electrochemical parameters (Beale et al. 1985 and Herino et al. 1987). Recently, this material has attracted strong interest since photoluminescence phenomena in the visible range at room temperature have been reported for porous silicon layers (Canham 1990).

In this paper we report results on the examination of this material using High Resolution Scanning Electron Microscopy (HRSEM) with spatial resolution in the nanometer range . To the author's knowledge, there has to date been no detailed analysis of PSL using this technique.

2. EXPERIMENTAL PROCEDURE

2.1 Description of the High Resolution SEM

Observations were made in the JEOL 890 Field Emission SEM equipped with a highly excited objective lens. Cold cathode field emission is known to provide a high brightness (1.10^9 A/cm^2-sr) and small virtual source (30 Å) of low energy spread (0.2 eV). It allows low accelerating voltages which in turn avoid electron irradiation damage and electron induced charging of samples. In this context, the JEOL 890 F has been equipped with a F.E. gun using conical anodes which minimize virtual source position for a wide range of low accelerating voltages (Miyokawa et al. 1988).

The samples must be of small size (LxWxH = 23x6x3 mm) and are located between the upper and lower parts of the objective pole piece where the aberration coefficients of the lens are minimum. The achievable resolution at 35 kV is about 7 Å in the secondary electron imaging mode, and 15 Å in the backscattered electron mode.

2. 2 Sample preparation

Porous samples used for these experiments are prepared in p-type (100) silicon wafers with a resistivity of 0.01 Ω-cm (p+), and 1 Ω-cm (p). The anodization is performed at constant current densities of 100 mA/cm^2 and 20 mA/cm^2 for p+ and p samples respectively; the HF concentrations being respectively 25% and 35%. The porosities of the resulting layers are of about 50% on p+, and 60% on p-substrate. The pore radius is in the range of a few nanometers as already determined by gas adsorption experiments (Herino et al. 1987).

Epitaxial growth of silicon is performed onto porous silicon by LRP-CVD (Limited Reaction Processing) at 830 °C. More details on the epitaxial step are given elsewhere (Oules et al.1989).

For the cross-sectional HRSEM observations, the samples are introduced in the SEM chamber via the air-lock system immediately after cleaving. This procedure is intended to reduce contamination and/or oxidation of the freshly cleaved sample surface.

3. RESULTS

3. 1. Cross-sectional views of porous silicon layers

Cross-sectional HRSEM micrographs of the 1.2 µm thick porous silicon layer formed from a p+ substrate are shown in **Figs. 1a** and **1b** at magnifications of 50,000 and 150,000 respectively. Pore diameters are in the range of 10 nm and the typical "tree structure" and anisotropic texture of the pores are clearly shown.

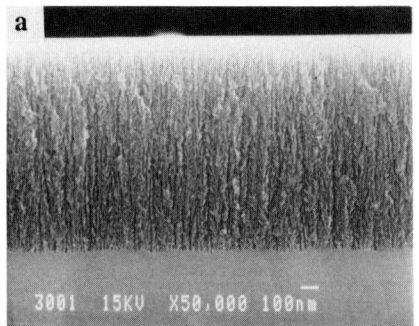

Fig. 1 : HRSEM micrographs of porous silicon layer (p+ substrate) at magnifications of 50,000 (a), and 150,000 (b).

Porous silicon formed from lightly doped substrate (p) exhibits a smaller pore size (**Fig. 2.**). The pore density increases and the "tree structure" changes into an almost random network of interconnected pores.

3. 2. 60° tilt observation of porous silicon layers

60° tilt observations of PSL (p and p+ substrates) are illustrated in Fig.3. These two-dimensional views, which can not be obtained from TEM observations, provide information on both the horizontal (surface) and the vertical (cross-section) structures of the pores.

 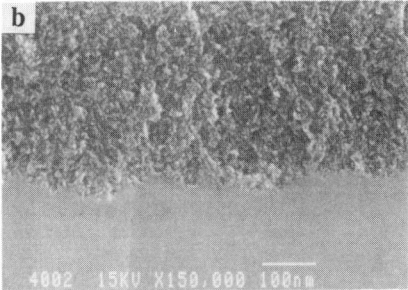

Fig. 2 : HRSEM micrographs of porous silicon layer (p substrate) at
magnifications of 50,000 (a), and 150,000 (b).

 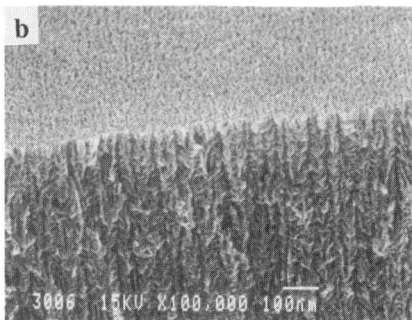

Fig. 3 : 60° tilt HRSEM micrographs of PSL obtained from p-substrate (a), and
p+ substrate(b).

3. 3. Epitaxial Si onto porous silicon

During the epitaxial growth, the thermal treatment leads to pore coarsening, as
shown in **Fig.4a.** More details are visible in **Fig 4b** which is an enlarged view of the
same structure.

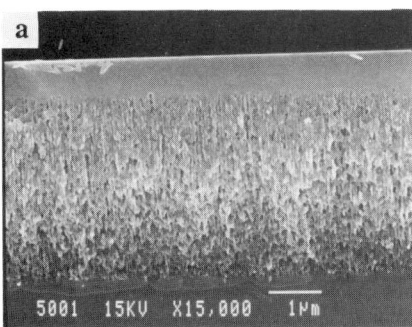 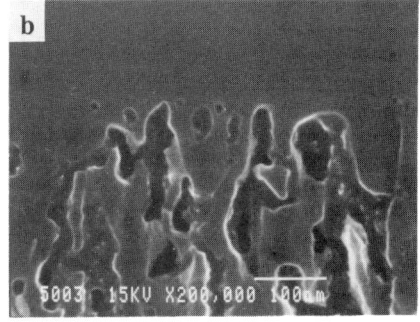

Figs 4 : Low magnification (a), and high magnification (b) HRSEM images of
epitaxial silicon onto PSL (p+ substrate).

For comparison, TEM micrographs of epitaxial silicon onto PSL formed from p+
and p substrates are shown in **Fig.5**. TEM observations reveal structural defects such as

twins and dislocations (Fig.5a) which are not visible from HRSEM micrograhs. However, TEM micrographs do not allow a detailed view of the individual pores due to the overlapping of the random morphology.

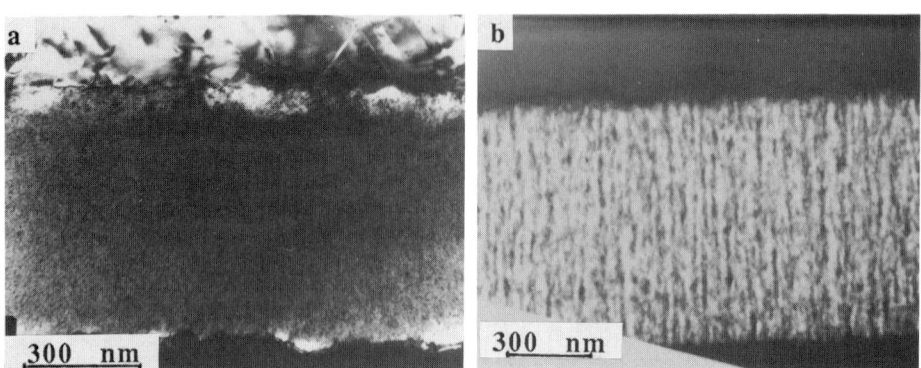

Figs. 5 : TEM pictures of epitaxial silicon onto PSL formed from p substrate (a) and p+ substrate (b)

4. CONCLUSION

The microstructure of Porous Silicon Layers formed from p+ and p silicon substrates was studied in detail by cross-sectional High Resolution Scanning Electron Microscopy. The results are compared to those obtained by Transmission Electron Microscopy. We found that HRSEM is a very attractive technique, since it does not require any sample preparation. Both cross-sectional and 2D imaging are easily obtained on freshly cleaved samples. HRSEM is a useful tool to investigate microstructures as fine as porous silicon. However, Transmission Electron Microscopy, which requires sophisticated sample preparation, is more suited for crystalline defect investigations.

REFERENCES

Beale MIJ, Chew NG, Uren MJ, Cullis AG and Benjamin JD 1985 Appl.
    Phys. Lett. 46 86-8.
Bomchil G, Halimaoui A and Herino R 1988 Microelectron. Eng. 8 293-310
Canham LT 1990 Appl. Phys. Lett. 57 1046-8
Herino R, Bomchil G, Barla K, Bertrand C and Ginoux JL 1987 J. Electrochem.
    Soc. 134 1994-2000.
Miyokawa T, Norioka S and Goto F 1988 Proceed. of the 46th Annual Meeting of the
    Electron Microscopy Society of America (GW Bailey Ed., San Francisco Press), 978-9
Oules C, Halimaoui A, Regolini JL, Herino R, Perio A, Bensahel D and Bomchil G.
    1989 Materials Science and Engineering, B4, 435-9

ACKNOWLEDGEMENTS

The authors wish to thank G Mascarin and A Perio for TEM work.

Inst. Phys. Conf. Ser. No 117: Section 4
Paper presented at Microsc. Semicond. Mater. Conf., Oxford, 25–28 March 1991

239

Optical microscopy of silicon wafer surfaces with subnanometre vertical sensitivity

A J Pidduck, V Nayar and A M Keir

Royal Signals & Radar Establishment, Malvern, Worcs. WR14 3PS, UK

ABSTRACT Scanning optical microscopy with differential phase contrast has been used to characterise the surface roughness of Si wafers before and after surface cleaning, epitaxial growth, and thermal oxidation. Atomic steps are detected on a (100) epitaxial wafer of especially low misorientation, demonstrating the Angstrom-level vertical sensitivity of the technique.

1. INTRODUCTION

Future Si microelectronics will rely on critical layers of just a few nanometres in thickness, for instance 5-10nm MOS gate oxides and 50nm bipolar base-widths. Hence the amplitude of microscopic interface roughness must be controlled within subnanometre tolerances. Differential phase contrast (DPC) microscopy using a laser scanning optical microscope (SOM) has previously been shown to be sensitive to Si surface roughness in this range, and has been used to characterise the polishing texture on as-received silicon wafers by Pidduck and Nayar (1991). We now report a preliminary study showing how the roughness of the starting wafer surface is affected by wafer cleaning, epitaxy and thermal oxidation processes.

2. EXPERIMENTAL

A Bio-Rad Lasersharp SOM150 microscope, operated in DPC mode, was used. The optical principle and geometry has been described in detail by, for instance, Hamilton and Sheppard (1984). SOM-DPC contrast in reflection, like Nomarski microscopy, depends on surface slope. However, electronic contrast enhancement enables higher vertical sensitivity. Lateral resolution is determined by the focussed spot size, which for an HeNe laser source (λ=633nm) with either 0.85NA or 1.3NA (oil immersion) objective lenses is ~0.4μm or 0.3μm respectively.

3. RESULTS & DISCUSSION

3.1 Polished Wafers

As-received Si wafer surfaces have been compared using SOM-DPC, and the observed textures found to depend strongly on wafer supplier and year of origin (Pidduck and Nayar (1991)). The image contrast was assigned to Si surface roughness. Ridge-valley type polishing marks were observed (see for instance fig.3A), although on state-of-the-art polished surfaces these are much reduced, leaving a mesh of faint criss-crossing scratch marks superimposed on an unresolved isotropic submicron-scale roughness.

3.2 Atomic Step Imaging

When the misorientation between the optical wafer surface and the low-index crystal plane is reduced to about 0.01°, the mean atomic-height step-spacing can exceed the optical resolution limit. Fig.1A shows an image from a polished (111) wafer in which the resultant array of surface misorientation steps can be seen running approximately horizontally. Hahn et. al. (1990) reported the misorientation of this wafer to be 0.02°, and demonstrated light diffraction corresponding to the 0.9μm average spacing of the step array. This is consistent with a 0.3nm step height, which is the Si (111) monolayer spacing. In fig.1A, diagonal ridge-valley type polishing marks are evident in the background. We have not observed such steps on polished (100) wafers of similar misorientation.

3.3 Epitaxy

Hahn et. al. (1990) have measured light diffraction peaks from epitaxial (100) wafers of very low misorientation. Fig.1B is a SOM-DPC image from one such wafer (a 10μm-thick epilayer grown at a temperature above 1000°C), showing the steps aligned vertically with a spacing of about 0.5-0.7μm. The wafer misorientation was measured, using a high precision X-ray diffraction method, as 0.021±0.002°, indicating a mean step height near 0.22nm. This is consistent with either monatomic-height (0.136nm) or biatomic-height (0.272nm) steps on the (100) surface. Widely-spaced atomic steps have previously been imaged optically on epitaxial (111) Si by Bauser and Strunk (1985), using Nomarski microscopy. We believe fig.1B to be the first optical image of the smaller (100) Si surface steps, confirming the very high SOM-DPC vertical sensitivity. Despite the additional micron-scale texture in fig.1B, the regularity of the misorientation step spacing and direction is clearly evident. This is consistent with a step-flow epitaxial growth mechanism, which has an inevitable surface smoothing effect. To maintain step-flow, Si adatom surface diffusion lengths must be comparable with the atomic terrace length. This is large due to the low misorientation, and so a relatively high growth temperature is necessary. At the centre of the image is a microscopic pit defect. Around this some bowing of the steps may be discernable to the sympathetic eye, suggesting a left-to-right step motion, which is in agreement with the X-ray measurements.

3.4 Si Wafer Cleaning

Wafer cleaning processes, such as the RCA method (Kern (1984)), generally involve sequences of chemical oxide removal and re-oxidation. The effect on Si surface topography is shown in fig.2, which compares (a) an area of an as-received stepped (111) wafer, with the same area after (b) HF-removal of the thin surface oxide by 60s exposure to a 0.5% aqueous HF mist, followed by H_2O rinse and N_2 spin dry, in an FSI 'Mercury' wafer cleaner, and (c) subsequent reformation of a thin oxide (~2nm-thickness) by UV-ozone exposure. These two treatments appear to have no discernable effect on the SOM-DPC image. Most notably, there is no evidence for any change in form or position of the misorientation steps (for instance in relation to the faint particulate feature just to the right of centre) that might indicate any tendency for a lateral (step-edge) re-oxidation mechanism. This is consistent with the TEM findings of Gibson and Lanzerotti (1987), and indeed selective step-edge/kink site oxidation by the highly reactive radical species generated by UV-irradiation of O_2 and H_2O in air would not be expected.

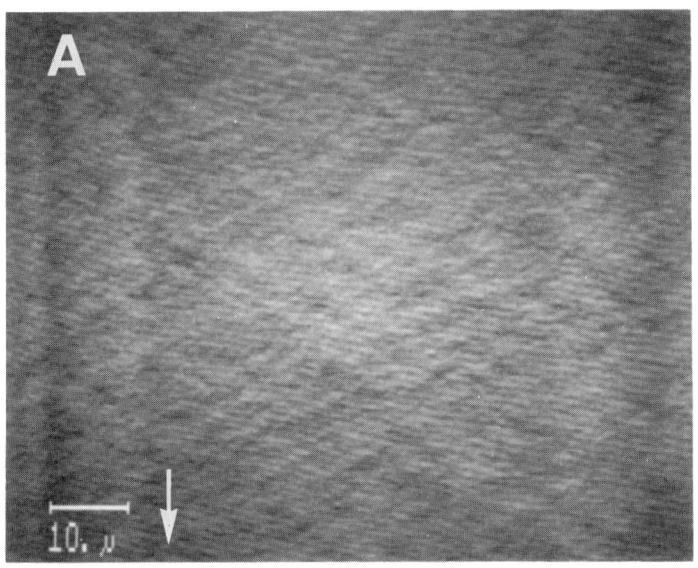

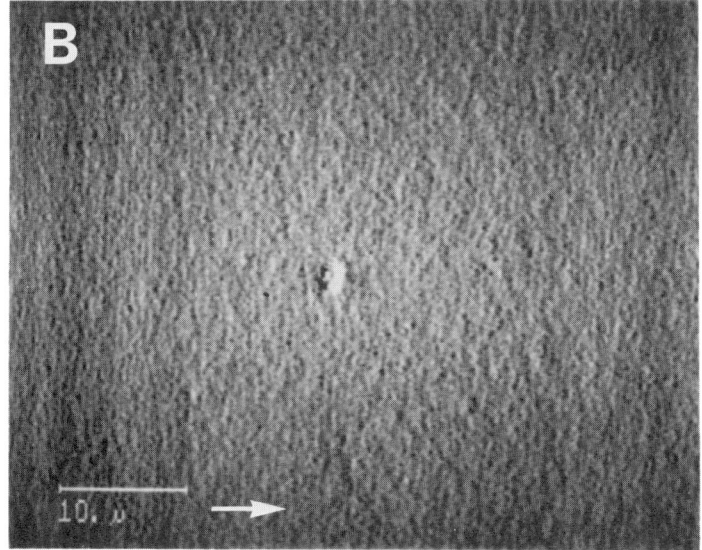

Figure 1. Atomic Step Imaging. SOM-DPC images of very low misorientation Si wafers : A is a polished (111) wafer, and B an epitaxial (100) wafer. B is viewed under oil immersion. Arrows denote the DPC direction (imagine illumination by an oblique light source from this direction).

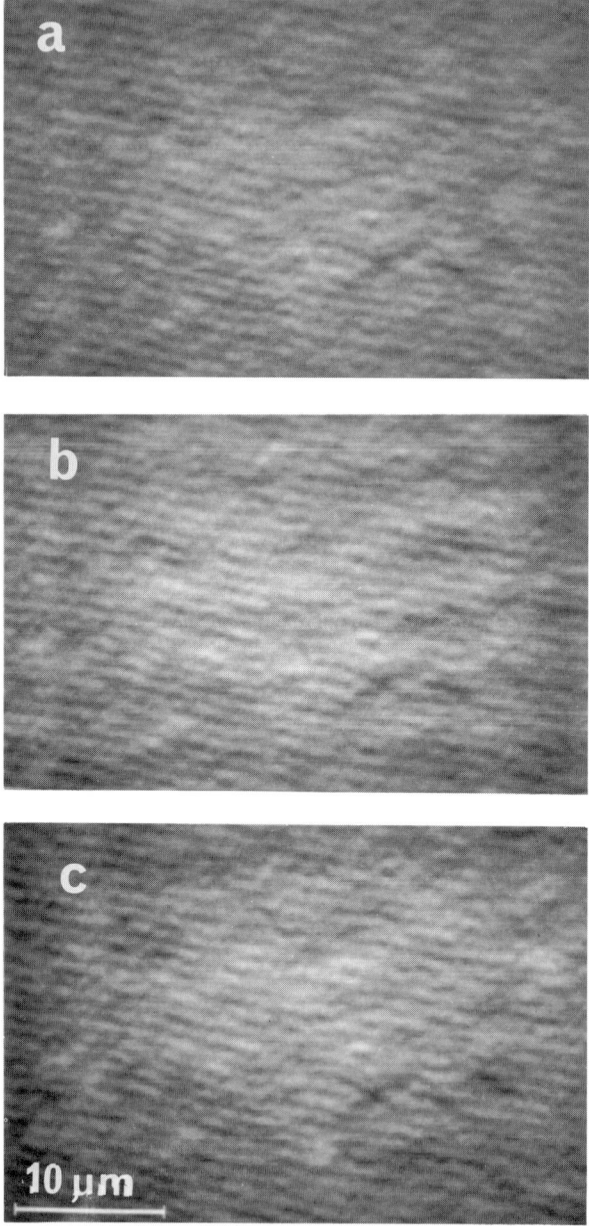

Figure 2. Wafer Cleaning. SOM-DPC study of the effect of simple chemical treatments on the same area of a stepped wafer surface (the same wafer as in fig.1A) : (a) as-received, (b) after HF treatment (60s, 0.5% & H_2O rinse/dry, and (c) after subsequent UV-ozone exposure (10 minutes in air).

3.5 Thermal Oxidation

The Si-(Si oxide) interface dominates the wafer reflectivity, and also (as is evident from fig.2) the SOM-DPC contrast. The technique is thus well-suited to imaging the interface roughness resulting from thermal oxidation. This has previously been studied by light scattering and LEED (Hahn et.al.(1988)), and by TEM and STM (e.g. Niwa et.al.(1990)). Fig.3A shows the nature of a starting surface, prepared by RCA-cleaning (no HF). Polishing texture dominates the surface roughness. Fig.3B shows that after thermal oxidation at 900°C in dry O_2 to produce a 24nm oxide, the starting wafer surface topography is essentially preserved, though the oxidation causes a slight increase in the background submicron-scale roughness. This is very clearly demonstrated by fig.3C which is taken from a wafer oxidised at 1050°C to 100nm. The amplitude of the submicron-scale interface roughness is significantly greater, and now dominates the topography, masking the wider scale polishing features. Fig.3D was taken from a similar wafer which was initially oxidised in a steam ambient at 1150°C to a thickness of 1000nm. Next the oxide was stripped using buffered HF before RCA cleaning and reoxidation to 27nm at 900°C. The wide scale texture is notably reduced, and the submicron-scale texture increased, though not as substantially as during dry oxidation. This trend, of a microroughness increasing with increasing oxide thickness, and more pronounced after dry than after wet oxidation, agrees with the light scattering measurements made by Hahn et.al.(1988). The evolution of roughness depends on a number of interacting mechanisms, such as whether diffusion of oxidising species or interface reaction is rate-determining (Niwa et.al.(1990)), the nature of the interface reaction itself, and the possible annealing effect of higher temperatures (Hahn et.al.(1988)). The present work shows that both roughening and smoothing processes occur on the length scales sampled by the SOM, but is unable to distinguish between the above mechanisms.

4. CONCLUSIONS

This paper demonstrates that optical microscopy with atomic-height vertical sensitivity has a valuable role to play in the study of advanced Si surface processing. Compared with higher resolution microscopies (e.g. TEM, SEM, STM and AFM), SOM-DPC provides information on Si surface roughness on complementary length scales, with the advantage of allowing rapid and non-destructive imaging of relatively large areas.

ACKNOWLEDGEMENTS We are indebted to P.Hahn and M.Grundner (Wacker-Chemitronic GmbH) for the specially-oriented polished and epitaxial wafers, and to R.Jackson for wafer cleaning and oxidation.

REFERENCES
Bauser E and Strunk H P 1985 Mat. Res. Soc. Symp. Proc. 37 109
Gibson J M and Lanzerotti M Y 1989 Nature 340 128
Hahn P O, Grundner M, Schnegg A and Jacob A 1988 The Physics & Chemistry of SiO_2 & The Si-SiO_2 Interface (New York: Plenum) pp 401-412
Hahn P O, Grundner M, Schnegg A and Jacob A 1990 Semiconductor Silicon, Electrochem.Soc.Proc. 90-7, pp 296-312
Hamilton D K and Sheppard C J R 1984 J Microscopy 133 27
Kern W 1984 (April) Semiconductor International 24
Niwa M, Onoda M, Matsumoto M, Iwasaki H and Sinclair R 1990 Japan J. Appl. Phys. 29 2665
Pidduck A J and Nayar V 1991 submitted to Appl. Phys. A

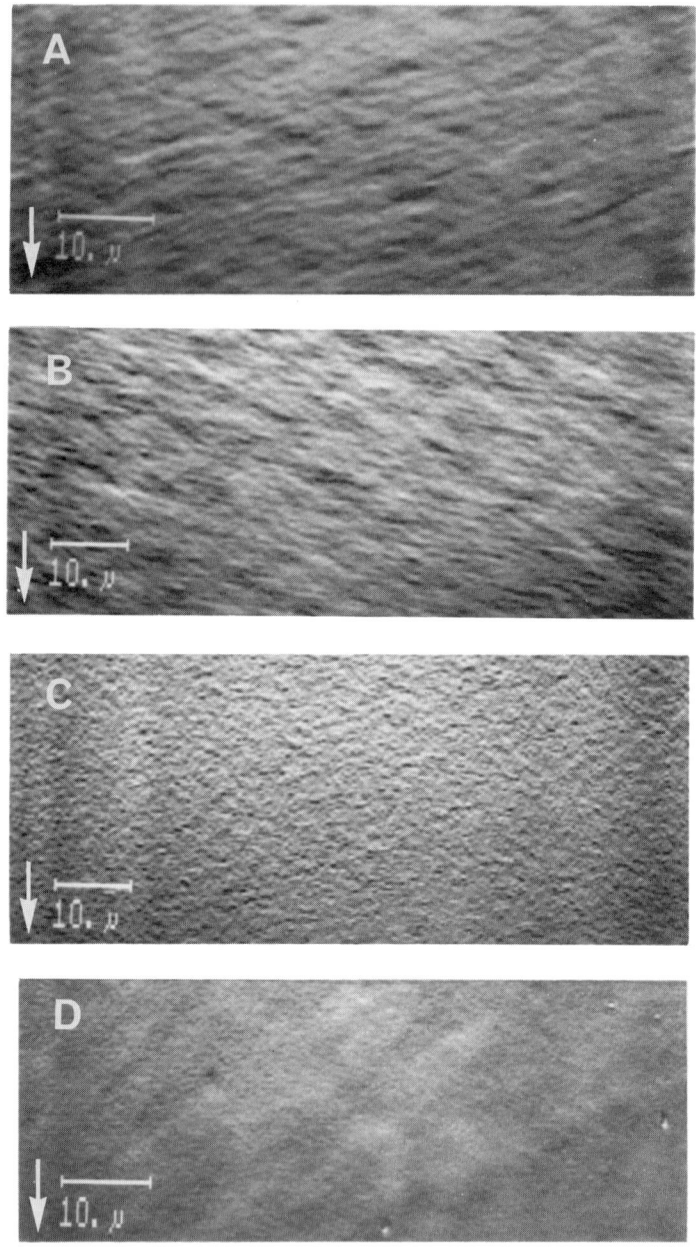

<u>Figure 3.   Thermal Oxidation</u>.   SOM-DPC images showing the effect of
different thermal oxidation conditions on Si surface roughness :   A is an
RCA-cleaned wafer; B and C are similar wafers after oxidation in dry O_2 to
thicknesses of 24nm (at 900°C) and 100nm (at 1050°C) respectively; D is a
wafer with a 27nm oxide grown in dry O_2 after 1000nm sacrificial oxidation
in wet O_2 (at 1150°C) and removal in buffered HF. These images were taken
using the same instrumental contrast settings.

Inst. Phys. Conf. Ser. No 117: Section 5
Paper presented at Microsc. Semicond. Mater. Conf., Oxford, 25–28 March 1991

245

Ballistic electron emission microscopy of semiconductor interfaces

R H Williams

Department of Physics and Astronomy, University of Wales College of
Cardiff, Cardiff CF1 3TH, UK

ABSTRACT: Ballistic Electron Emission Microscopy is a powerful new
technique capable of probing electrical barriers at semiconductor
interfaces on a microscopic scale. We discuss the background of the
method and its application to probe Schottky barriers at metal-
semiconductor interfaces, band discontinuities at heterojunctions, and
conduction band critical points.

1. INTRODUCTION

Internal electrical barriers at metal-semiconductor and heterojunction
interfaces govern the behaviour of a range of solid state electronic
devices. Many experimental techniques are available to probe these
barriers, these include measuring the voltage dependence of current and
capacitance (I-V and C-V) and photoresponse (PR). These methods suffer
from the disadvantage that they probe large contact areas, perhaps tens or
hundreds of microns. However, it is now clear that many aspects
associated with the understanding of electrical barriers at semiconductor
interfaces require techniques which can probe the barriers on a much
smaller scale, preferably close to atomic scale, to enable the influence
of features such as dislocations and interface steps to be investigated.
Ballistic electron emission microscopy (BEEM) promises to make this
requirement a possibility.

2. THE BEEM TECHNIQUE

BEEM was first described by Kaiser and Bell in 1988 and is based on the
scanning tunnelling microscope. Consider first its application to probe
metal-semiconductor contacts as illustrated in Figure 1. A voltage is
placed between the tip of the STM and a thin metal layer on the
semiconductor so that electrons tunnel across the vacuum barrier into the
metal base. Provided the thickness of the base is small enough a
substantial fraction of the electrons reach the metal-semiconductor
interface without energy loss. However, these electrons can only cross
the barrier into the semiconductor provided they have energy in excess of
the Schottky barrier, ϕ_b, and there exists a critical tip-base voltage,
V_b, below which there is no current collected in the semiconductor. The
value of ϕ_b can thus be established from the threshold voltage and its
lateral variation can be probed by moving the STM tip in a plane parallel
to the interface.

Bell and Kaiser (1988) and Ludeke and Prietsch (1991) have considered the elementary aspects of BEEM theory. The former assume that the tip-base tunnel current can be described by the planar tunnelling formalism developed by Simmons (1964). Conservation of energy and of the electron momentum parallel to the metal semiconductor interface have been assumed. For parabolic energy bands in the semiconductor it is deduced that near threshold the collector current should vary as $I \propto (V-V_b)^2$. Ballistic electrons traversing the Au-Si (100) interface are well described by the formalism. Ludeke and Prietsch (1991) have extended the theory to include the energy dependence of the inelastic scattering in the base and of the non-classical transmission across a quantum mechanical barrier. They conclude that near threshold $I \alpha (V-V_b)^{5/2}$, differing slightly from the dependence derived by Bell and Kaiser. A more complete description of the whole process, and in particular the transmission of electrons across an ordered and coherent metal-semiconductor interface, has been given by Stiles and Hamann (1989). The inclusion of phonon scattering has been considered by Lee and Schowalter (1991).

The lateral resolution of BEEM has also been considered by Bell and Kaiser (1988). For a 100Å thick Au layer on the Si(100) surface considerations of parallel momentum conservation suggest that a lateral resolution of around 10Å should be achievable, near the threshold. If correct this offers an enormous leap forward in the development of methods to probe the microscopic nature of electrical barriers at semiconductor interfaces.

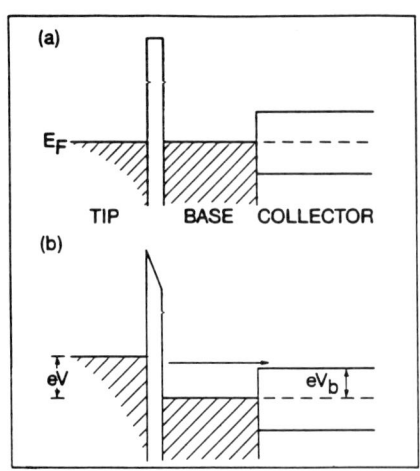

Fig. 1, The BEEM mechanism (a) without bias, and (b) with bias (after Kaiser and Bell).

3.   GOLD CONTACTS TO CADMIUM TELLURIDE

To illustrate the application of BEEM we consider the case of Au contacts on chemically treated n-CdTe surfaces, recently reported by Fowell et al (1990). It is known that such contacts are somewhat difficult to reproduce, sometimes yielding Schottky barrier heights differing by over 0.2 eV for interfaces which appeared to be prepared in an identical fashion. It was suspected that the lower barrier heights of around 0.7 eV were associated with an excess of Te on the surface, following the chemical etching process carried out before the deposition of the gold contacts.

Figure 2 illustrates the BEEM current as a function of the bias for Au-CdTe, for three values of the tunnel current. Close to the threshold the relationship $I\alpha(V-V_b)^2$ provides a good fit to the experimental data, enabling V_b, which we equate to ϕ_b, to be accurately established. By scanning the tip over the gold surface the polycrystalline nature of the layer could be observed. At the same time the BEEM current could be mapped for a fixed tip-base bias and for this particular Au-CdTe combination, it was shown that the BEEM current was not related in a simple way to the polycrystalline nature of the gold overlayer. [However, in some systems the BEEM current is strongly related to the granular nature of the overlayer (Ludeke and Prietsch, 1991)].

In order to probe the lateral homogeniety of the Schottky barrier height I-V relationships were measured for a grid of points 100Å apart. The resulting thresholds are displayed in Figure 3 where a very large variation is observed. The barriers appear to vary from over 1 eV to well below 0.7 eV, a range similar to that observed in normal large contact area I-V measurements.

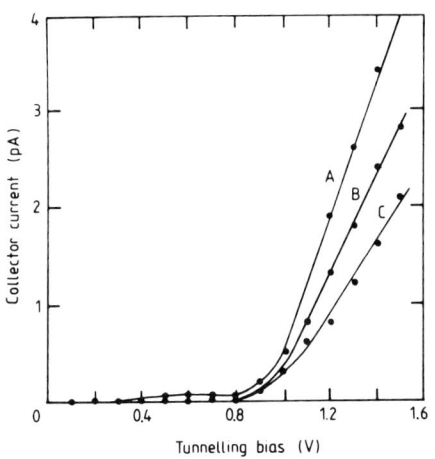

Fig. 2, BEEM current as a function of bias for Au-CdTe for different tunnel currents.

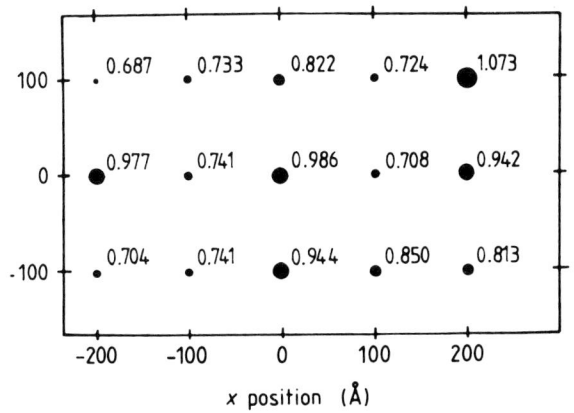

Fig. 3, BEEM thresholds for various positions at the Au-CdTe interface.

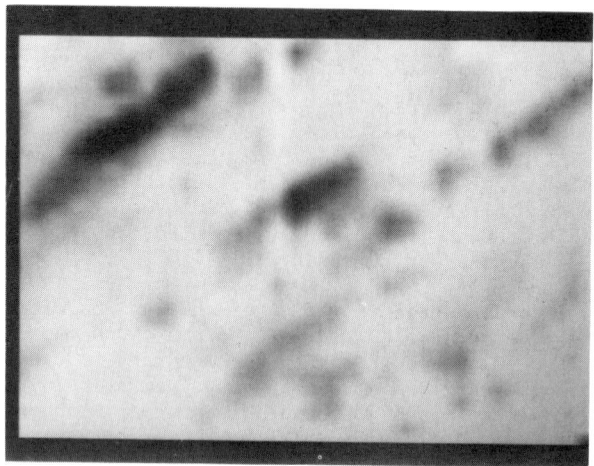

Fig. 4, Variation of BEEM current for a fixed bias for the Au-CdTe
interface. The dark areas correspond to high currents.

In Figure 4 the lateral variation of BEEM current for a fixed tip-base
bias is shown. Here the light and dark regions represent low and high
BEEM currents, respectively. It may be seen that small regions transmit
very efficiently, and details of the order of 20Å are observed, indicating
that the lateral resolution may indeed be of this order. The low current
regions seems to be in linear arrays and it is possible that they are
associated with tellurium precipitates at the interface. It is quite
clear that the interface is highly inhomogeneous on the scale of nm.

Regions where the effective Schottky barrier height ϕ_b are low may well
dominate the I-V relationships for normal large contact area diodes.
Provided the diode current is controlled by thermionic emission the
reverse current J_o is linearly dependent on the contact area but
exponentially dependent on the barrier height ϕ_b. Thus, small areas where
ϕ_b is effectively small can dominate the I-V dependence, as indeed is
observed for Au on chemically treated CdTe.

4.   BEEM OF GOLD ON GaAs

The second example (Fowell et al 1991) deals with a highly perfect
interface, namely Au on atomically clean GaAs (100) surfaces grown by
molecular beam epitaxy. For this interface the BEEM current and threshold
shows remarkable uniformity as a function of position and there are large
areas where the gold layer is smooth and free of grain boundaries.

Figure 5 displays a plot of the square root of the BEEM current as a
function of the tip-base bias for a high quality region of the Au-GaAs
diode. The threshold is at around 0.82 eV and the plot shows linear
behaviour in this region. In addition, there appear to be two further
thresholds displaced from the first by 0.26 eV and 0.41 eV respectively.
The origin of these is believed to be associated with the band structure of
GaAs. The first threshold corresponds to emission of electrons into states
at the Γ point, the bottom of the conduction band, while the second and

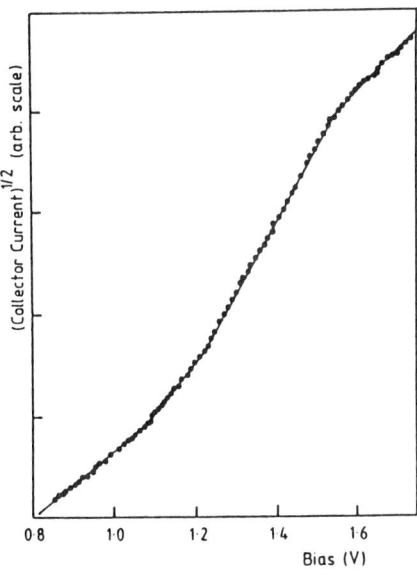

Fig. 5, Square root of
BEEM current as a
function of bias for
the Au-GaAs interface.

third thresholds are associated with the L and X critical points. The ΓL
and ΓX separation predicted theoretically are 0.29 eV and 0.48 eV
respectively. Similar values have also been measured by Kaiser et al
(1989).

The possibility of observing details of the semiconductor band structure by
BEEM is exciting and may offer opportunities to probe the modification of
the band structure on a microscopic level by, for example, local strain
effects. High quality interfaces also offer the possibility of probing the
energy dependence of the hot-electron transmission across the interface.

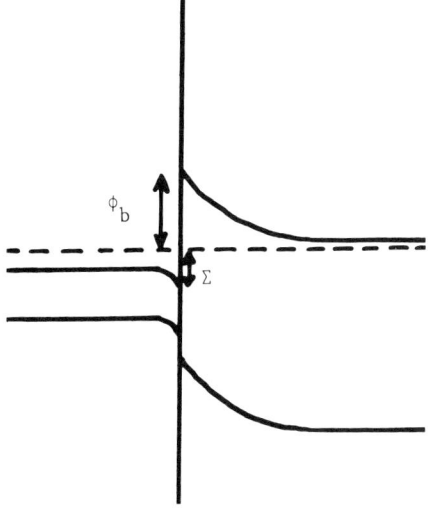

Fig. 6, The form of the
InAs-GaAs heterojunction
used to establish the
conduction band
discontinuity.

5.   PROBING HETEROJUNCTIONS BY BEEM

The semiconducting materials used in most heterojunction device structures are based on lattice matched combinations such as GaAs and AlGaAs. There is, however, increasing interest in systems such as InGaAs-GaAs which are not lattice matched. Beyond the critical thickness misfit dislocations may occur at the interface when the mismatched overlayer relaxes. It is likely that strain is associated with such dislocations and strain, in turn, may influence the electrical barriers at the interface. The possibility exists of probing such effects by BEEM.

Investigations of the InAs-GaAs interface by BEEM have been reported by Shen et al (1991). The structure, sketched in Figure 6 was grown by MBE. The 100Å thick InAs layer was highly n-type ($N_D \sim 4 \times 10^{18} cm^{-3}$) and was coated with 100Å of gold before etching into appropriate mesa form for BEEM.

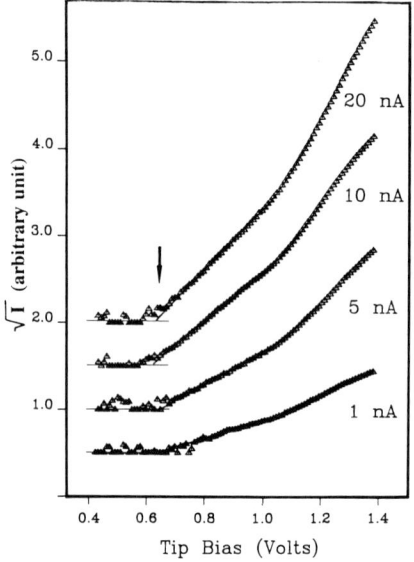

Fig. 7, Square root of BEEM current as a function of bias for InAs-GaAs for various tunnel currents.

Figure 7 shows the relationship between the square root of the current and the tip-base bias. Again a highly reproducible and well defined threshold is observed, at 0.63 eV. This threshold corresponds to ϕ_b in Figure 6 and has the same magnitude as that obtained by conventional I-V investigations. To obtain the magnitude of the conduction band offset, ΔE_c, it is necessary to add the quantity Σ to ϕ_b. The value of Σ was obtained theoretically, by evaluation of Poisson's equation on either side of the interface. The procedure leads to a value of ΔE_c of 0.72 eV.

Figure 8 shows topographic and BEEM current plots for the InAs-GaAs interface. Several features in the BEEM current correlate with the topography, but to date features due to misfit dislocations have not been observed. There are several possible reasons for this and these are now being investigated. Nevertheless, it is clear that band discontinuities at heterojunctions can be probed with a very high lateral resolution by ballistic electron emission microscopy.

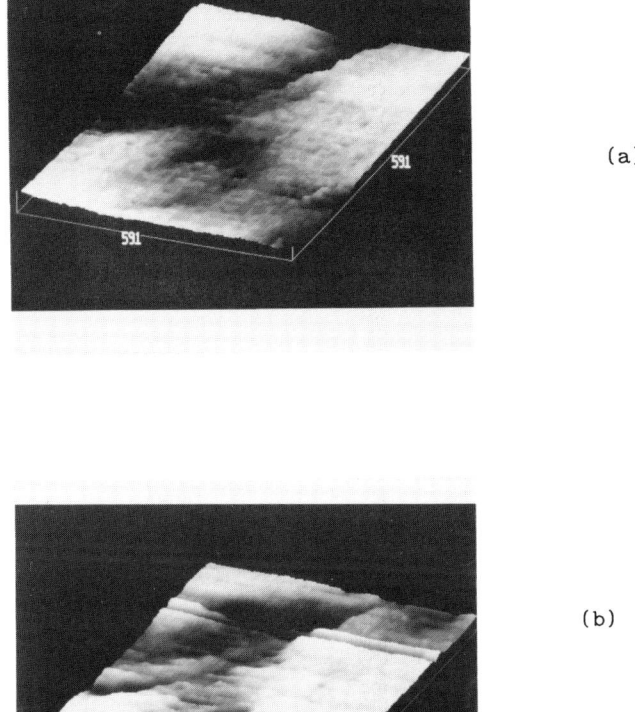

(a)

(b)

Fig. 8, Topographic (a), and BEEM current scans (b) for the InAs-GaAs interface, measured simultaneously.

6. THE FUTURE

It is certain that Ballistic Electron Emission Microscopy will be an exceedingly important method for the investigation of semiconductor interfaces. The technique is in its infancy but is already opening new directions in the investigation of the physics of interfaces. It has been used to probe Schottky barriers to n-type semiconductors and indeed it has been extended by Kaiser et al to yield the magnitudes of Schottky barriers to p-type semiconductors. The technique is clearly applicable to probe heterojunctions and it is feasible that light emission may be observed, enabling another variant to be developed. It has already been applied to investigate carrier-scattering phenomena in the base and is likely to find application in the investigation of hot-electron reflection at interfaces. Extension to barrier structures, such as double-barrier resonance devices is also likely. Clearly, the next few years will see many novel developments of BEEM side by side with an improvement in the understanding of the basic physics of this exciting new technique.

ACKNOWLEDGMENTS

The author wishes to thank Bernard Richardson, Tony Cafolla, Angela Fowell and all colleagues in Cardiff involved with the BEEM technique. He also wishes to thank the SERC, MOD and EEC for financial support for the programme.

REFERENCES

Bell L D and Kaiser W, J Phys. Rev. Letts. (1988) 61 2368
Fowell A E, Williams R H, Richardson B E and Shen T-H, Semicon.
  Sci. Tech. (1990) 5 348
Kaiser W J and Bell L D, Phys. Rev. Letts (1988) 60 1406
Kaiser W J, Bell L D, Hecht M H, Grunthaner F J, J. Vac. Sci.
  Tech. (1989 B7 945
Lee E Y and Schowalter L J, in press
Ludeke R, Prietsch M and Samsavar A, J. Vac. Sci. Tech. 1991 in
  press
Shen T-H, Elliott M, Williams R H and Westwood D, Appl. Phys.
  Letts. (1991)
Shen T-H, Elliott M, Fowell A E, Cafolla A, Richardson B E,
  Westwood D and Williams R H, J. Vac. Sci. Tech. in press
Simmons J G, J. Appl. Phys. (1964) 35 2655
Stiles M D and Hamann D R, Phys. Rev. (1989) B40 1349

Inst. Phys. Conf. Ser. No 117: Section 5
Paper presented at Microsc. Semicond. Mater. Conf., Oxford, 25–28 March 1991

An electron rocking curve determination of the Al(001)/GaAs(001) interface atomic structure

M A Al-Khafaji, D Cherns, C J Rossouw[#] and D A Woolf[*]

H.H.Wills Physics Laboratory, Bristol University, Tyndall Avenue, Bristol BS8 1TL
[#]On leave from:CSIRO Division of Materials Science and Technology, Locked Bag 33,Clayton, Victoria, Australia 3168
[*]Department of Physics, University of Wales, College of Cardiff, P.O. Box 913, Cardiff CF1 3TH

ABSTRACT: A recent generalised Bloch wave treatment for propagating a fast electron wavefunction through an arbitrary number of coherent epitaxial layers is used to describe propagation through a bicrystal. This model is used to interpret electron diffraction data from an Al(001)/GaAs(001) bicrystal to determine the rigid shift ΔR across the interface. Measurements of 220 systematic row rocking curves from regions between misfit dislocations enable the rigid shift to be determined from two possible configurations. Results are consistent with $\Delta R = \frac{1}{4} < 100 >_{GaAs}$, in agreement with previous observations.

1. INTRODUCTION

Two models of the atomic structure across the Al (001)/GaAs(001) interface have been previously considered in several studies (Kiely & Cherns 1989; Eaglesham et al 1989; Kendrick et al 1989): model A assumes Al atoms to be superimposed directly above Ga or As atomic sites, and model B introduces a shift of $\frac{1}{4} < 100 >$ (in terms of the GaAs unit cell). In this paper we will decide which of these models represent the interface structure in this system by calculating the Bloch wave excitation amplitudes in the bottom crystal using an expression derived from a recent model for propagating a fast electron wavefunction through an arbitrary number of coherent layers, each with arbitrary thickness, composition and orientation (Rossouw, et al 1991).

Experimental observations of rocking curve intensities obtained by direct imaging of plan view samples in conventional transmission electron microscopy (TEM) are interpreted in terms of this model. The shape and symmetry of the zeroth order beam rocking curve is sensitive to the interface atomic structure. This effect is attributable to the top Al layer providing a displaced incident wavefunction which preferentially excites different Bloch wave states in GaAs depending on layer thickness, rigid shift and orientation. Sensitivity to rigid shift enables correlation between experimental and theoretical rocking curves to distinguish between models A and B for the interface structure.

2. THEORY

Rossouw et al (1991) show that the eigenstate amplitudes $\alpha^{(j)}_{GaAs}$ in the GaAs crystal on branch (j) of the dispersion surface can be determined from the equation

$$\alpha^{(j)}_{GaAs} = \sum_i C^{(i)*}_{oAl} \sum_g C^{(i)}_{gAl} C^{(j)*}_{gGaAs} \exp[-ig.\Delta R] \exp[ik^{(i)}_{Al} z_{Al}] \qquad (1)$$

and the diffracted beam intensities Ig emerging from the final GaAs layer are

$$I_g = |\sum_j \alpha^{(j)}_{GaAs} C^{(j)}_{gGaAs} \exp[ik^{(j)}_{GaAs} z_{GaAs}]|^2 \qquad (2)$$

where the eigenvalues for Al and GaAs are appropriately indicated. This equation is derived from coupling Bloch waves in GaAs to the total wavefunction at the exit surface of Al, being a coherent superposition of Bloch states at depth z_{Al}. Excitation amplitudes $\alpha^{(i)}$ for isolated Al and GaAs crystals are non-periodic but show similar trends, as in Fig. 1.

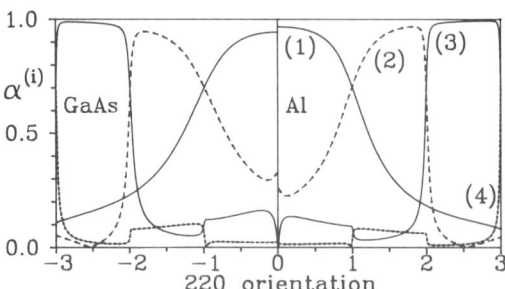

Figure.1 Branch excitation amplitudes $\alpha^{(i)}$ for Al and GaAs.

3. EXPERIMENTAL

The sample examined consists of a (continuous) thin epitaxial Al layer on a clean (001) GaAs substrate grown by molecular beam epitaxy (MBE). In this system, three different epitaxial orientations have been reported [Kiely & Cherns 1989]. In this work we will deal only with Al(001) grains, where the epitaxial relationship is defined Al(001)//GaAs(001) and Al(110)//GaAs(100), i.e the Al unit cell is rotated by 45° about the common <001> zone axis with respect to the GaAs unit cell, resulting in a lattice mismatch of 1.36 %. A continuous layer of Al of thickness 900 Å was evident from examination of cross section and plan view samples.

4. OBSERVATIONS OF Al (001) /GaAs (001)

A zone axis bright field image of 900 Å of Al on 1900 Å of GaAs showing square net of moirè fringes is shown in Fig. 2(a). The area of observation is carefully chosen to lie as far as possible away from the networks of interfacial dislocations which have been located using weak beam images from a small region; see Fig. 2(b). The choice of this area provides a pseudomorphic epitaxial region with relatively uniform strain and rigid shift, which is a difficulty in the case of the Al(001)/GaAs(001) interface because of the existence of a dense network of interfacial dislocations.

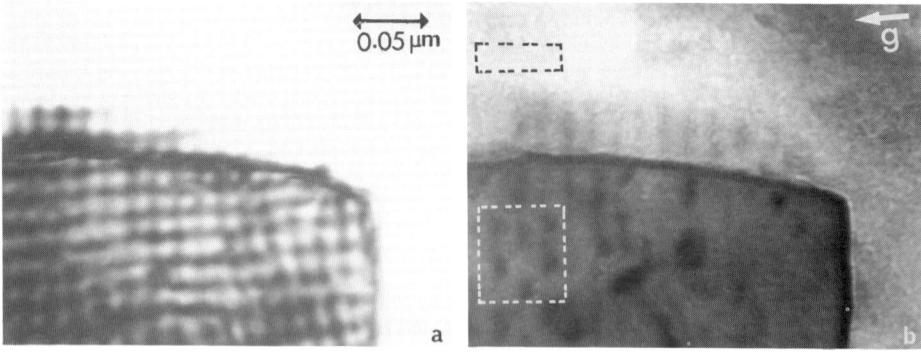

Figure 2. a) zone-axis bright field image of Al(001)/GaAs(001). b) weak beam image showing square net of interfacial dislocation.

Fig. 3(a) shows 4mm symmetry in the zero layer <001> CBED (convergent beam electron diffraction) pattern of the Al/GaAs bicrystal. This symmetry suggests that Al atoms are located on highly symmetric sites with respect to the GaAs lattice (see models A and B in Fig. 3b(i,ii) respectively). For effective one dimensional rocking curve measurements from systematic row diffraction conditions, the 220 row is favourable because the phase factor g.ΔR is different for the two models (i.e. g.ΔR=0 or π for models A and B respectively).

A series of bright field images was recorded at the same exposure time for a range of crystal orientations (the 220 beam being systematically tilted through its exact Bragg condition). Some micrographs from this sequence of orientations are illustrated in Fig. 4. Care was taken to avoid strong excitation of non-systematic reflections. Since the Al deposit is discontinuous in this area (though mostly continuous), it is possible to compare the transmitted intensities from the bicrystal with those from single crystal GaAs

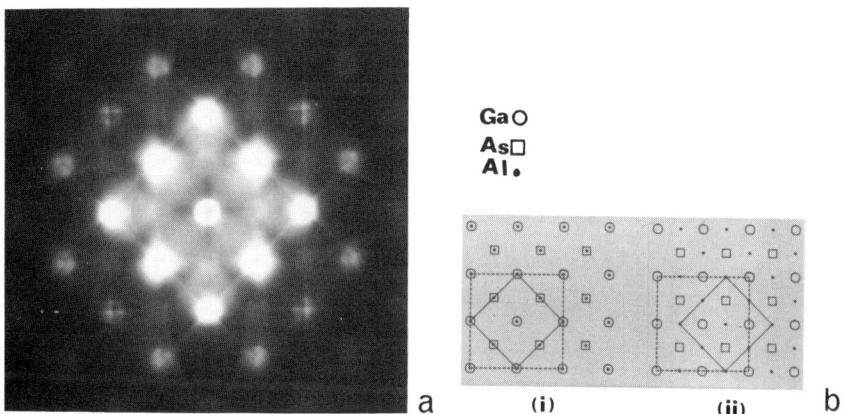

Figure 3. (a) Zone-axis CBED from Al(001)/GaAs(001) grain showing the 4mm projected symmetry. (b) Possible high symmetry projected atomic positions of Al(001) unit cell (solid lines) on GaAs(001) (dashed lines) with (i) ΔR=0 and (ii) ΔR=<$^1/_4$00>.

The two areas highlighted in the region of GaAs and Al/GaAs bicrystal in Fig. 2(b) are those from which intensities were recorded. The centre of the box on the bicrystal represents an undislocated region of interface which was used for comparison with the bright field intensity from box in the GaAs. A bright field rocking curve for 200 keV electrons at room temperature is shown in Fig. 4 for (i) GaAs alone, (ii) the bicrystal with ΔR=<$^1/_4$00> and (iii) the bicrystal with ΔR=0. Orientations at which bright field images were recorded were measured from selected area diffraction patterns and Kikuchi lines. These orientations are indicated as dashed lines.

Note that the overall contrast in the rocking curve for the ΔR=0 model is similar to that for GaAs alone (this is to be expected since similar eigenstates are excited with similar amplitudes in Al and GaAs), Within the 1st BZ (Brillouin zone), branch (1) is preferentially excited in the Al layer. Between the 1st and 2nd BZ boundaries, branch (2) is strongly excited in the Al layer. This leads to strong excitation of branch (2) in GaAs if ΔR=0, leading to good transmission through GaAs. If however ΔR=<$^1/_4$00>, branch (1) in the GaAs layer couples strongly to branch (2) in the Al layer, leading to enhanced overall absorption.

This difference is a result of the rigid shift causing strong coupling of the second Bloch wave branch of the Al to the first branch of the GaAs while also coupling the first branch of the Al strongly to the GaAs branch (2). Superimposed on this effect is an Al diffraction contrast variation which oscillates with thickness. These overall trends are observed in the rocking curves of Fig. 4.

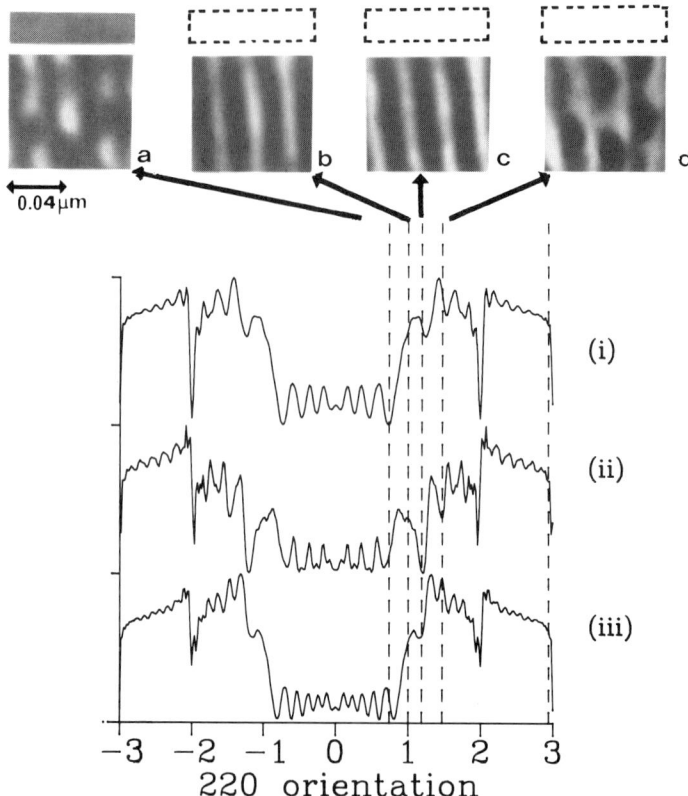

Figure 4 Comparison of simulated 200 keV rocking curves for the bicrystal of 900 Å of Al on 1900 Å of GaAs, for (i) GaAs alone, (ii) Al plus GaAs $\Delta R = \langle 1/_400 \rangle$ and (iii) Al plus GaAs $\Delta R = 0$. (a) orientation 0.75 220, (b) exact 220 Bragg orientation, (c) 1.19 220, and (d) 1.46 220. See fig. (2b) for the position of the chosen areas on top.

5. CONCLUSION

We have used a concise expression to show how a rigid body shift alters the excitation amplitudes of different eigenstates in the bottom layer of the bicrystal. This theory is applicable to a large range of interface analysis problems, with the full dynamical theory of Rossouw et al 1991 enabling the simulation of contrast due to strain and compositional modulations in addition to rigid shifts for an arbitrary multilayer system.

ACKNOWLEDGMENTS

Authors acknowledge helpful discussions with professor J.W. Steeds. CJR acknowledges financial support from the Wolfson Foundation.

REFERENCES

Eaglesham, D.J., Kiely, C.J., Cherns, D. and Missous, M, 1989, Phil. Mag., A60, 161-175.
Kendrick, A.B., Hutchison, J.L., and Cherns, D., 1989, Inst. Phys. Conf. Ser. No. 98, 383-386.
Kiely, C.J., and Cherns, D., 1989, Phil. Mag. A 59, 1-29.
Rossouw, C.J., Al-Khafaji, M.A., Cherns, D., Steeds, J.W., and Touaitia, R., 1991, Ultramicroscopy (in press).

Inst. Phys. Conf. Ser. No 117: Section 5
Paper presented at Microsc. Semicond. Mater. Conf., Oxford, 25–28 March 1991

257

TEM study of NiGa/(001)GaAs and GaAs/NiGa/(001)GaAs heterostructures

B Guenais, A Poudoulec, V Durel, Y Ballini and A Guivarc'h.

Centre National d'Etudes des Télécommunications ; LAB/OCM/MPA, BP 40, F22301, Lannion, Cedex, France.

ABSTRACT : Intermetallic NiGa epitaxial layers were deposited by molecular beam epitaxy, on (001) GaAs substrates at 120°C. The TEM results point out the difficulty in avoiding the nucleation of hexagonal phase(s) (either Ni or Ga rich) together with the NiGa phase. Only the use of a precursor Ni rich surface allows the stabilization of a monocrystalline cubic NiGa layer at the interface, up to a thickness of about 12 nm ; in these conditions, heterostructures with thin (3 to 12 nm) buried good quality NiGa layers were grown. The lower interface is flat at the atomic scale, but the roughness of the upper layer interface leads to a high density of defects in the GaAs overlayer.

1 INTRODUCTION

The NiGa intermetallic compound has been chosen as a candidate material for the achievement of metal-GaAs heterostructures. This choice was based on two main criteria : the thermodynamic equilibrium with GaAs (Guérin and Guivarc'h 1989) and the low lattice mismatch with this substrate. NiGa is a cubic structure (CsCl type) with a lattice parameter close to half that of GaAs : a = 0.2887 nm (Feschotte and Eggiman 1974) ; misfit $\sim$ + 2 %. Its composition homogeneity range extends from 47% to 55% Ni at. (Ni/Ga = 0.88 to 1.22) (Feschotte and Eggiman 1974). Both Ni vacancies and antisite Ni defects occur simultaneously with varying concentrations, in all compositions (Donaldson and Rawlings 1976). The next compounds are, on the Ni rich side, Ni_3Ga_2 and on the Ga rich side, Ni_3Ga_4 and Ni_2Ga_3 ; these two compounds result from an ordering of the constitutional defects present in NiGa (Ellner et al 1969, Bradley and Taylor 1937). Ni_2Ga_3 and Ni_3Ga_2 are both hexagonal pseudo-cubic phases like all the ternary Ni-Ga-As phases (Guivarc'h et al 1989). To avoid agglomeration of the metallic film (Sands et al 1990b) and phase instability in ultra-high vacuum, the substrate temperature must never exceed 350°C during the growth of the heterostructures (Guenais et al 1990, 1989). In this paper, we present the TEM results referring to two types of structure : NiGa / (001) GaAs, and GaAs / NiGa / (001) GaAs.

2 EXPERIMENTAL

Growth of the metallic films on a GaAs buffer layer was carried out by codeposition of Ni and Ga from effusion cells, in a molecular beam epitaxy chamber (MBE) equipped with a rotating holder ; more details are given elsewhere (Durel et al 1990). The substrate temperature was maintained at 120°C and the As background pressure as low as $3 \cdot 10^{-10}$ Torr. The stoichiometry of the layer was monitored by Rutherford backscattering (RBS) on a 20 nm deposit on an oxidized Si substrate ; the Ni/Ga ratio was found to be between 0.9 and 1.1. On the first type of structure, three kinds of starting surface were tested : either a Ga or As stabilized surface, or with the predeposition of about two monolayers of Ni on an As stabilized surface. The terms "Ga surface" or "As surface" or "Ni surface" will be used hereafter. On the second type of structure, a GaAs overlayer was grown at 350°C onto a thin metallic layer (3 or 12 nm) deposited on a starting "Ni surface". TEM observations were performed on (110) cross-sections, at 120 kV for conventional TEM and 400 kV for high resolution TEM (HRTEM).

3 RESULTS

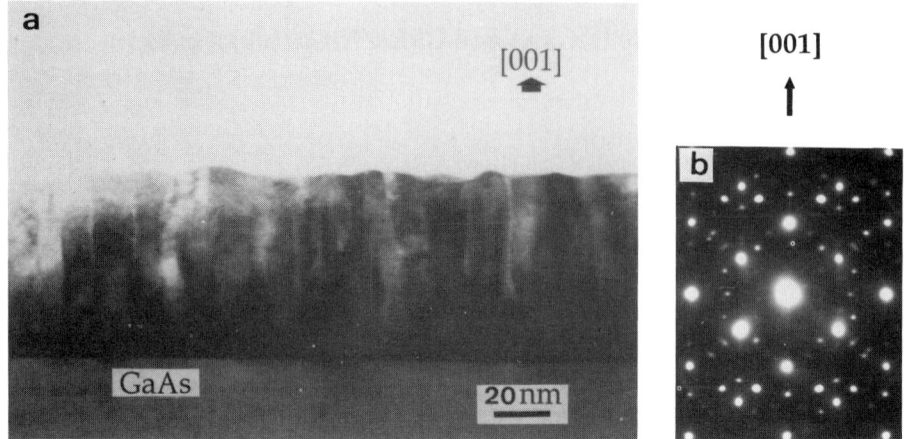

Fig. 1 : (a) bright field cross-section TEM image of 65 nm NiGa / (001) GaAs grown on a "Ga surface". (b) Corresponding diffraction pattern.

NiGa / (001) GaAs structures : Fig. 1(a) is a bright field image of the structure grown on a "Ga surface". The layer appears polycrystalline, with columnar grains of lateral size about 20 nm. More information arises from the diffraction (TED) pattern shown in Fig. 1(b), and two main remarks can be made.

First, the TED pattern reveals at least four orientation relationships of the NiGa grains with the substrate. The first one is the simple "cube on cube" orientation : [100] (001) NiGa // [100] (001) GaAs, but the others correspond to a low match with the substrate : [112] (110) NiGa // [110] (001) GaAs ; [110] (112) NiGa // [110] (001) GaAs ; [110] (110) NiGa // [110] (001) GaAs, and equivalent variants which are presumably present. This latter relationship was also noted on a RHEED pattern for MBE grown NiAl on GaAs (Harbison et al 1988). It is clear that a "Ga surface" has a detrimental effect on the epitaxy relationship of cubic NiGa with a (001) GaAs surface, but the origin of this effect is not so far understood.

Second, the appearance of four symmetrical extra spots around each 111 reflection, is typical of hexagonal pseudo-cubic phases with a ~ 0.4 nm and c ~ 0.5 nm, showing an orientation relationship with the cubic substrate such as [010] (001) hex. // [110] (111) cub.. Because of the fourfold symmetry of the GaAs (001) plane, twin related variants occur, leading to a polycrystalline layer as was previously discussed for Ni_2Ga_3 (Guenais et al 1990). Unlike NiGa, the hexagonal phase is oriented as expected, i.e. according to the best lattice match with the substrate (misfit = 0.6 % for Ni_3Ga_2 and 0.9 % for Ni_2Ga_3). Identification of the corresponding hexagonal phase was not possible because of the very low lattice parameter difference between Ni_2Ga_3 and Ni_3Ga_2. Note that, consequently, the term "hexagonal phase" will be used, whatever the nature and possibly the number of hexagonal phases present. The formation of these compounds could explain the extra diffraction streaks which appear in our RHEED patterns, also observed for MBE grown NiAl films (Sands et al 1990a). This mixture of cubic and hexagonal phases occurs in the majority of our samples. These results suggest that, for these MBE temperature and pressure conditions, in spite of a careful control of the fluxes and a large theoretical homogeneity range of NiGa, the slightest local deviation from stoichiometry results in the nucleation of a hexagonal phase rather than a random distribution of point defects. The ordering of point defects was also observed in bulk non stoichiometric NiAl, but after high temperature heat treatments (Delavignette et al 1972, Reynaud 1976).

For the "As surface" (not shown here), only the "cube on cube " orientation is observed, but contrary to our first assumptions (Guivarc'h et al 1987), the undesirable local nucleation of a hexagonal phase from the interface leads also to a polycrystalline layer.

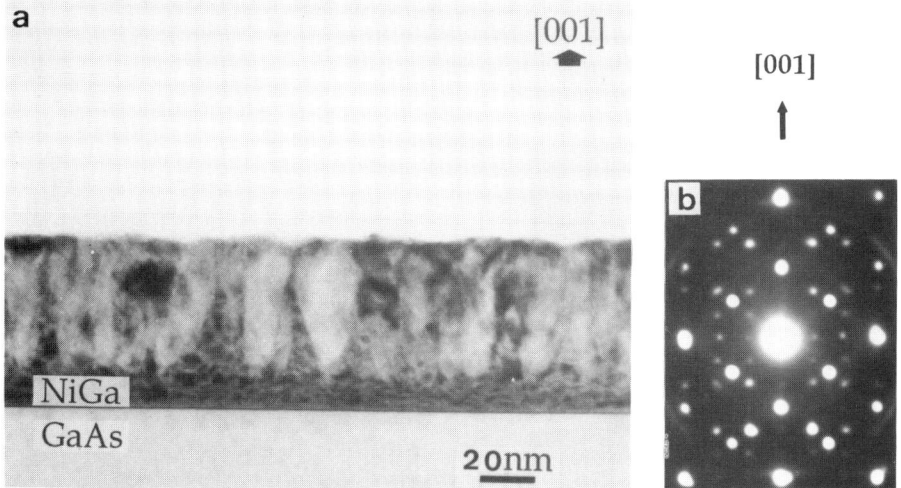

Fig. 2 : (a) bright field cross-section TEM image of 65 nm NiGa / (001) GaAs grown on a "Ni surface". (b) Corresponding diffraction pattern.

Fig. 2(a) is a bright field cross-section of the sample grown on a "Ni surface". The TED pattern (Fig. 2 (b)) reveals a "cube on cube" orientation for NiGa, but again the presence of the four variants of the hexagonal phase. The layer morphology is remarkable : it appears to be composed of two parts, separated by a particularly rough interface. From a dark field contrast analysis, we conclude that the first step of the growth corresponds to monocrystalline NiGa, about 12 nm thick ; beyond this thickness, the monocrystalline epilayer transforms to a polycrystalline structure, containing a mixture of NiGa and hexagonal phase (15 nm grain size). As a conclusion, the deposition of about two monolayers of Ni on the (001) GaAs surface has two favourable effects : first, to promote a single "cube on cube" orientation for the cubic NiGa phase (but an assessment of the nature of this Ni layer influence can be only speculative), and second, to inhibit the nucleation of an hexagonal phase at the interface. However, we note that this latter effect is cancelled beyond a 12 nm thickness, suggesting it could be related to the stress accumulated in the epilayer. Consequently, the Ni rich precursor surface is considered to be the best surface preparation, but the structural transformation, as observed by TEM, limits the thickness of the epilayer to 12 nm on a (001) substrate.

GaAs / NiGa / (001) GaAs structures : in such structures, the metallic layer is grown onto a Ni rich starting surface. Fig. 3 is a bright field cross-section view of a 12 nm thick buried NiGa layer and Fig. 4 shows a HRTEM image of a similar structure with a 3 nm thick NiGa layer ; both 0.2 nm {110} and 0.29 nm {100} fringes can be seen. Note that the metallic layer is monocrystalline, without any extra phase. The lower interface of the metallic layer was shown to be abrupt at the atomic scale, but the roughness of the upper interface (± 2 monolayers) leads to a high density of planar defects and dislocations in the overlayer.

4 CONCLUSION

The structural quality of thick (65 nm) MBE grown NiGa metallic layers on (001) GaAs substrates was assessed by TEM. Among the compared starting surfaces, Ga or As or Ni rich, it is the latter which is the most favourable for a strong epitaxial relationship with the substrate. Indeed the other two surface preparations lead to a polycrystalline layer, containing extra hexagonal phases which are typical of a local deviation from stoichiometry (Ni_2Ga_3 and/or Ni_3Ga_2). Moreover, for the Ga rich surface, several orientation variants are observed for NiGa. Only the Ni rich surface preparation allows the stabilization of a 12 nm thick monocrystalline cubic NiGa layer at the interface. But beyond this thickness, the layer transforms to a polycrystalline mixture of NiGa and extra phases previously quoted. Using the optimized experimental conditions (Ni rich starting surface, thickness ≤ 12 nm), good crystalline quality NiGa layers have been obtained, on which GaAs overlayers have been grown. TEM results show that the quality of the metallic layer top interface has to be improved to lower the defect density in the GaAs overlayer.

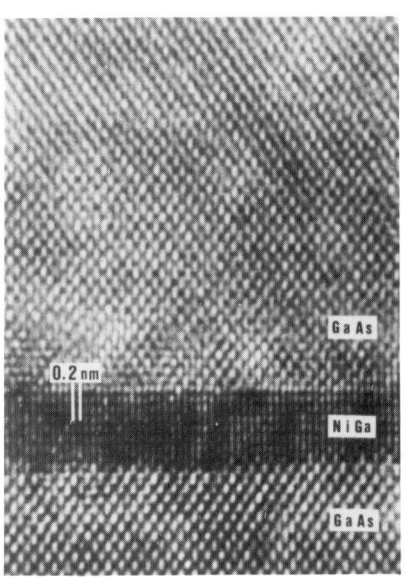

Fig. 3 : dark field image of a 200 nm GaAs / 12 nm NiGa / (001) GaAs structure.

Fig. 4 : HRTEM image (400 kV) of a 750 nm GaAs / 3 nm NiGa / (001) GaAs structure.

ACKNOWLEDGMENTS : the help with HRTEM observations by C. d' Anterroches, and fruitful discussions with R. Guérin are greatfully appreciated.

REFERENCES

Bradley A J and Taylor A 1937 Phil. Mag. A23 p 1049-1067.
Delavignette P, Richel H and Amelincks S 1972 Phys. Stat. Sol. (a) 13 pp 545-55.
Donaldson A T and Rawlings R D 1976 Act. Metall. 24 pp 811-6.
Durel V, Guenais B, Ballini Y, Caulet J, Chomette A, Dupas G, Ropars G, Minier M, Guivarc'h A and Regreny A 1990 Inst. Phys. Conf. Ser. n°112 chapter 3 pp 129-34.
Ellner M, Best K J, Jacobi H and Schubert K 1969 J. Less. Common Metals 19 pp 294-6.
Feschotte P and Eggimann P 1974 J. Less Common Metals 63 p 15-30.
Guenais B, Poudoulec A, Guivarc'h A, Guérin R and Caulet J 1989 Inst. Phys. Conf. Ser. n°100 section 8 pp 665-70.
Guenais B, Poudoulec A, Caulet J and Guivarc'h A 1990 J. Cryst. Growth 102 pp 925-32.
Guérin R and Guivarc'h A 1989 J. Appl. Phys. 66(5) pp 2122-8.
Guivarc'h A, Guérin R and Secoué M 1987 Electron. Lett. 23 pp 1004-5.
Guivarc'h A, Guérin R, Caulet J, Poudoulec A and Fontenille J 1989 J. Appl. Phys. 66(5) pp 2129-36.
Harbison J P, Sands T, Tabatabaie N, Chan W K, Florez L T and Keramidas V G 1988 Appl. Phys. Lett. 53 pp 1717-9.
Reynaud F 1976 J. Appl. Cryst. 9 pp 263-8.
Sands T, Harbison J P, Ramesh R, Palmström C J, Florez L T and Keramidas V G 1990 Mat. Sci. Eng. B6 pp 147-57.
Sands T, Palmström C J, Harbison J P, Keramidas V G, Tabatabaie N, Cheeks T L , Ramesh R and Silberberg Y 1990 Mat. Sci. Reports 5 pp 99-170.

Inst. Phys. Conf. Ser. No 117: Section 5
Paper presented at Microsc. Semicond. Mater. Conf., Oxford, 25–28 March 1991

261

The microstructure of NiInW contacts on GaAs

F Glas, P Hénoc, G Le Roux and C Dubon-Chevallier

Centre National d'Etudes des Télécommunications, Laboratoire de Bagneux, 196 avenue Henri Ravéra, 92220 BAGNEUX, France

ABSTRACT: The microstructure of ohmic and Schottky NiInW contacts on n-type GaAs is investigated by TEM, TED and X-ray microanalysis. The annealing products are identified as NiGa with a little As, and various Ni-Ga-As phases containing In. One of these, close to Ni_4GaAs_3, is reported for the first time. The ohmicity of the contact appears correlated not to the presence of the rarely detected $In_xGa_{1-x}As$, but to that either of NiGa or of the As-rich phases containing In.

1. INTRODUCTION

Low resistivity ohmic contacts on GaAs are needed to fabricate electronic devices based on this material. Several high temperature processing steps require that these contacts also be refractory. For these purposes, Murakami and co-workers (Murakami et al 1988, Shih et al 1989) developed contacts on n-type GaAs formed by Rapid Thermal Annealing (RTA) in an $Ar-H_2$ atmosphere of variously deposited layers of Nickel and Indium capped by Tungsten. By using Transmission Electron Microscopy (TEM), these authors correlated the electrical properties and the microstructure of these contacts. To summarize their findings, for Ni/In/Ni/W deposits (denoting successive and simultaneous deposits by '/' and '-', respectively), $NiAs_2$ and Ni_2Ga_3 were formed at 700°C, and NiAs and $In_xGa_{1-x}As$ at 900°C; for Ni/Ni-In/Ni/W deposits, Ni_2GaAs, $Ni_{11}As_8$ and $In_xGa_{1-x}As$ were formed at 700°C, and NiAs, Ni_3In and $In_xGa_{1-x}As$ (with 0.2<x<0.4) at 900°C. The latter conditions gave the best electrical results. Murakami et al (1988) even reported a direct correlation between the smallness of the contact resistance and the proportion of the GaAs surface covered by $In_xGa_{1-x}As$. They concluded that the ohmicity of the contact is due to the formation of a low barrier height $In_xGa_{1-x}As$ semiconducting layer, and that the other phases simply prevent In diffusion away from GaAs and insure the absence of low melting point metallic Ga and In. In this paper, we show that the microstructure of our own NiInW contacts is indeed very different from that reported by Murakami.

2. SPECIMENS AND EXPERIMENTS

TEM, Transmission Electron Diffraction (TED) and X-ray microanalysis were performed, chiefly on cross-sectionally thinned specimens, in a VG HB5 STEM fitted with a Kevex Si:Li detector and in a Philips CM20 CTEM fitted with a Link HPGe detector; the microscopes were operated respectively at 100 keV and 200 keV. Quantitative analysis was carried out using the following atomic k-factors (for the Kα lines): k_{NiAs}=0.81, k_{GaAs}=0.91, k_{InAs}=6 (for our Si:Li detector) at 100 keV, and k_{NiAs}=0.84, k_{GaAs}=0.92, k_{InAs}=2.1 (for our HPGe detector) at 200 keV. Several types of NiInW contacts on Si-doped ($4x10^{18}$ cm^{-3}) n-type (001)-oriented GaAs were investigated. Type I contacts result from annealing for 7 minutes at 750°C under an As overpressure in a semi-open box of a Ni/Ni-In/Ni/W deposit; the Ni-In layer is obtained by sputtering a Ni target containing 5 at % In (Dubon-Chevallier et al 1990). One of these contacts, which display the lowest contact resistivity so far obtained (10^{-7} Ω.cm^2), was studied. Type II contacts are obtained under conditions close to Murakami's optimum, namely RTA for 10 seconds at 850°C in an $Ar-H_2$ atmosphere; however, the

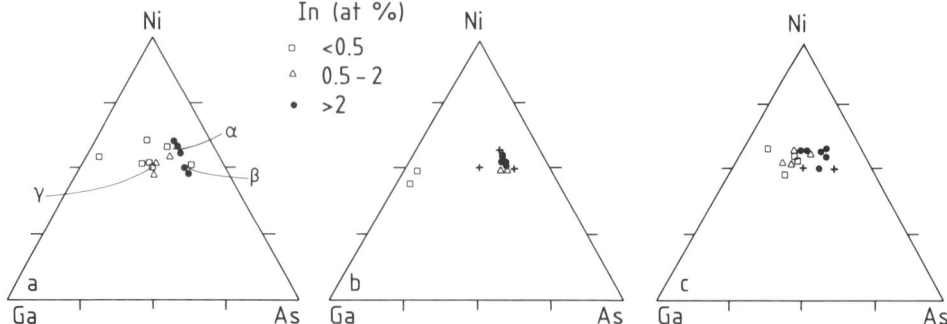

Fig. 1: Results of X-ray microanalysis on various grains of the type I ohmic contact (a), and the type II ohmic (b) and Schottky contacts (c). The relative atomic concentrations of Ni, Ga and As are represented in the usual ternary triangle, and the In content by the symbols. Ni_4GaAs_2, Ni_4GaAs_3 and Ni_2GaAs are indicated by α, β and γ, respectively, in (a) and by crosses in (b,c).

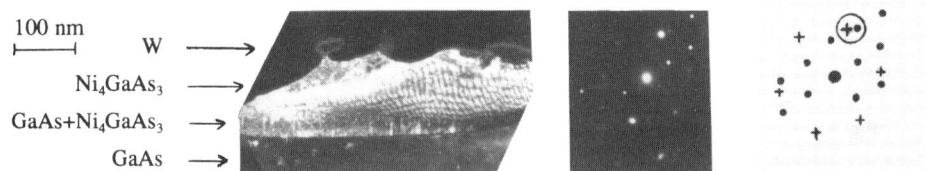

Fig. 2: Dark Field TEM image of the type I ohmic contact, showing a Ni_4GaAs_3 grain, with a TED pattern of the zone where this phase (crosses in the schematic of the pattern) and GaAs (dots) superimpose. The image was formed with the beams circled (311 of GaAs and $11\bar{2}0$ of Ni_4GaAs_3).

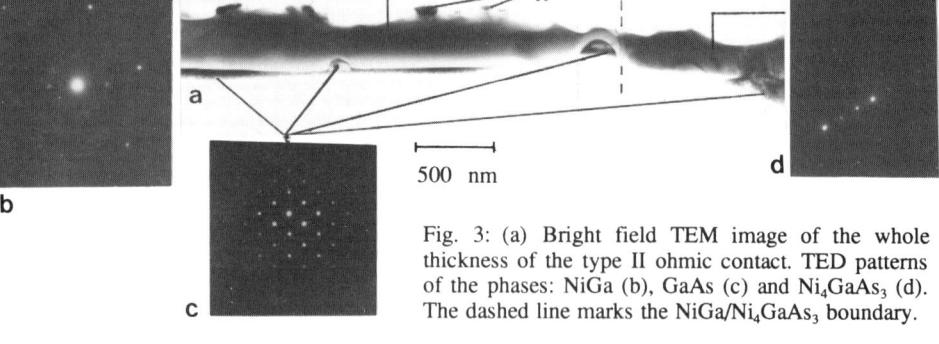

Fig. 3: (a) Bright field TEM image of the whole thickness of the type II ohmic contact. TED patterns of the phases: NiGa (b), GaAs (c) and Ni_4GaAs_3 (d). The dashed line marks the NiGa/Ni_4GaAs_3 boundary.

Fig. 4: High Resolution TEM image of the interface between GaAs and the grain (~Ni_4GaAs_3) seen at the right of Fig. 3. The specimen was slightly tilted from the GaAs <110> orientation to reveal the Ni_4GaAs_3 $(10\bar{1}0)$ lattice planes.

initial deposit consists of a single Ni-In layer capped by W (Hugon et al 1991). Type II contacts potentially contain more In, because of both the richer target used (10 at % In) and the elimination of the Ni layers. We studied two of them, displaying extreme behaviours: an ohmic contact (IIo) having the best contact resistivity then obtained by this technique (10^{-5} Ω.cm^2) and a Schottky contact (IIs). They differ by the thickness of the initial Ni-In deposit, estimated to be respectively 250 nm and 30 nm. In addition, X-ray Diffraction (XRD) was carried out on a series of eight type II contacts.

3. RESULTS

In the three specimens, the NiIn deposit has reacted with GaAs to produce a polycrystalline layer; W has formed roughly spherical grains on top of the structure without reacting with the other elements. Fig. 1 displays, projected in the usual Ni-Ga-As composition triangle, the microanalytical results on various grains. Unless specified otherwise, care was taken to analyse only areas proven by TED to be single-phased. Since In is quantitatively at most a minor constituent of the phases we observe, it will be useful to refer in first approximation to this ternary Ni-Ga-As system, which has already been thoroughly investigated: Ogawa (1980), Sands et al (1986), Lahav et al (1986) and Guérin and Guivarc'h (1989a) studied the interaction of Ni films with GaAs substrates and Guérin and Guivarc'h (1989b) reported recently a comprehensive thermodynamic study.

In specimen I, the reacted layer, about 100 nm thick, exhibits a roughness of several tens of nm (parallel and normal to GaAs). Four main phases were observed (Fig. 1a). One, close to NiGa, with little As and no In, is usually observed just below the capping W. Ignoring In, the other measurements cluster around three stoichiometric compositions: Ni_2GaAs (γ in Fig. 1), Ni_4GaAs_2 (α) and Ni_4GaAs_3 (β). Ni_2GaAs, with no or little In, is usually found just above GaAs, so that it was impossible to ascertain that these analyses detect a single phase. Ni_4GaAs_2 and Ni_4GaAs_3 were isolated; both phases contain In, but always less than 5 at %. The first phase, with a somewhat larger composition range than the others, corresponds to phase A of Guérin and Guivarc'h (1989b). The second has apparently not yet been reported, although an X-ray spectrum of Lahav et al (1986) might correspond to it. The grains of this phase are usually just above GaAs and heavily dislocated (Fig. 2). No $In_xGa_{1-x}As$ was detected in the cross-sectional specimens. This contact was also thinned in two other ways. Firstly, part of it was detached from GaAs with nitric acid; only W and NiGa were found, which confirms that this phase sits just below W. Secondly, thinning from the back with bromine/methanol and analyses through holes in the W cap confirmed the presence and composition of Ni_4GaAs_3; only in this specimen was $In_xGa_{1-x}As$ (with $0.05<x<0.1$) occasionally detected.

In the ohmic IIo contact, grains several tens of nm wide occupy the whole thickness of the layer (~ 300 nm) between GaAs and W (Fig. 3). Only two types were detected: NiGa grains with little As and no In, as in contact I, and grains with compositions between Ni_4GaAs_2 and Ni_4GaAs_3, containing In (Fig. 1b). No $In_xGa_{1-x}As$ was found by either microanalysis or TED. The grain boundaries often correspond to protrusions of GaAs inside the contact layer (Fig. 3). The grains of the second type are semi-coherent. The grain visible in Fig. 3 was studied in detail. It has hexagonal symmetry (as all the ternary phases of Guérin and Guivarc'h 1989b), with a=0.37 nm and c=0.50 nm; at the interface, its $(10\bar{1}0)$ planes are disoriented by 8° from the GaAs (111) planes (Fig. 4), and its $(10\bar{1}2)$ plane is nearly parallel to the (001) GaAs surface. This differs from the orientational relationship between GaAs and various Ni-Ga-As phases observed in previous studies of annealed deposits of Ni on (001) GaAs (Sands et al 1986, Lahav et al 1986, Guérin and Guivarc'h 1989a). High Resolution TEM images of the interface between the grains and GaAs (Fig. 4) confirms the absence of $In_xGa_{1-x}As$.

Finally, in the Schottky IIs contact, no NiGa was found. The analyses performed in the reacted layer (about 50 nm thick) extend widely between Ni_4GaAs_2, Ni_4GaAs_3 and compositions much richer in Ga than in As, which were only observed in this specimen (Fig. 1c). However, it was sometimes impossible to make sure that these are the compositions of the individual grains, here much smaller than previously.

For all the specimens, the In concentration in the Ni-Ga-As phases remains less than 5 at %, but may however amount to 1/3 of the Ga concentration. A substitution of Ga by In is possible, since the concentrations of the two column III elements vary roughly in opposition (Fig. 1). On the other hand, the increases in In and As concentrations appear correlated.

XRD on a series of type II contacts (either Schottky or ohmic with contact resistivities ranging from 10^{-3} to 10^{-5} $\Omega.cm^2$) strengthens the TEM results. Firstly, $In_xGa_{1-x}As$ was searched for near the 004 GaAs reflection: it was very likely detected only once as a definite peak (corresponding to x~0.08 for perfect epitaxy) and possibly several times as a shoulder in the tail of the GaAs peak, always in contacts with intermediate resistivities; it was not found in the best ohmic contact IIo. Secondly, the increase in width of the GaAs double diffraction peak correlates with the decrease of the contact resistivity. Thirdly, we used the Debye-Scherrer method to detect the other phases present in IIs and IIo. NiGa was confirmed to be present in IIo, absent in IIs, and NiAs was not detected. Broad peaks were found, corresponding to spacings around 0.144 nm in IIo and 0.142 nm in IIs. These peaks must be due to the In-containing Ni-Ga-As phases and are indeed in the region of the $20\bar{2}2$ peaks of the ternary phases studied by Guérin and Guivarc'h (1989a).

4. DISCUSSION AND CONCLUSION

The present microstructural study of the NiInW ohmic contacts yields results very different from Murakami's. Bearing in mind Murakami's interpretation of the ohmicity, the main discrepancy is the absence or extreme rarity of $In_xGa_{1-x}As$. The only common findings are the absence of free Ga and In and the occasional presence of Ni_2GaAs. We never found the other phases reported by Murakami, in particular NiAs and Ni_3In. All Murakami's phases are richer in As than in Ga, and the fate of the excess Ga remains unknown. Our detection in the ohmic contacts of both NiGa and In containing As-rich Ni-Ga-As phases does not lead to such a question. In seems to favour the formation of these As-rich phases. Since our ohmic contacts contain both NiGa and these phases, it is not yet clear which determine the ohmicity. Further studies should aim at clarifying this point, and, were the Ni-Ga-As phases found responsible, at deciding if they are semiconductors providing a low barrier height or metals with low work functions.

ACKNOWLEDGEMENT

The authors thank Dr P Blanconnier for annealing the type I contact and Prof B Agius and Dr M-C Hugon from Institut Universitaire de Technologie d'Orsay for depositing the Ni-In/W layers of the type II contacts.

REFERENCES

Dubon-Chevallier C, Glas F, Hénoc P, Hugon M-C, Agius B and Blanconnier P 1990 GaAs and related compounds 1990, Inst. Phys. Conf. Ser. No 112, ed K Singer (Bristol: Hilger) pp 245-250
Guérin R and Guivarc'h A 1989a J. Appl. Phys. 66 2129
Guérin R and Guivarc'h A 1989b J. Appl. Phys. 66 2122
Hugon M-C, Agius B, Dubon-Chevallier C, Blanconnier P, Bresse J-F and Froment M 1991 Accepted for publication in Appl. Phys. Lett.
Lahav A, Eizenberg M and Komen Y 1986 J. Appl. Phys. 60 991
Murakami M, Shih Y-C, Price W H, Wilkie E L, Childs K D and Parks C C 1988 J. Appl. Phys. 64 1974
Ogawa M 1980 Thin Solid Films 70 181
Sands T, Keramidas V G, Washburn J and Gronsky R 1986 Appl. Phys. Lett. 48 402
Shih Y-C, Murakami M and Price W H 1989 J. Appl. Phys. 65 3539

Inst. Phys. Conf. Ser. No 117: Section 5
Paper presented at Microsc. Semicond. Mater. Conf., Oxford, 25–28 March 1991

265

A TEM and SIMS study of Ti/Pd/Au ohmic contacts to p-type GaAs layers for AlGaAs/GaAs heterojunction bipolar transistors

B M Henry[*], A E Staton-Bevan[*], M A Crouch[**] and S S Gill[**].

[*]Department of Materials, Imperial College of Science, Technology and Medicine, London, SW7, 2AZ, U.K.
[**]RSRE, St. Andrews Road, Malvern, Worcs., WR14, 3PS, U.K.

ABSTRACT:Ti/Pd/Au and Pd/Ti/Au multilayer systems were used to form ohmic contacts on thin p^+-GaAs layers. The alloying behaviour of contacts, annealed at 380°C for 4 minutes, was studied using TEM and SIMS. The Ti/Pd/Au (75/75/400nm) contact exhibits the most promising microstructure with depth penetration into the epilayer of the order of a few tens of nms. However the contact resistance (2.88Ωmm) requires improvement. Interchanging the Ti and Pd layers is found to improve the contact resistance (0.24Ωmm) but results in extensive (0.1μm) Pd indiffusion into the epilayer.

1. INTRODUCTION

The n-p-n AlGaAs/GaAs heterojunction bipolar transistor (HJBT) has attracted much interest in recent years since it has many potential advantages over the traditional GaAs MESFET. The adoption of the n-p-n structure for HJBTs has created a demand for shallow low resistance p- and n-type ohmic contacts. This paper reports a microstructural study of the ohmic contact structures to (001) p-type GaAs, based on the Ti/Pd/Au system. Plan view and cross-sectional TEM has been performed together with secondary ion mass spectrometry (SIMS). Phase distributions and interface morphologies have been correlated with contact resistance measurements.

2. EXPERIMENTAL PROCEDURE

Two contact structures were fabricated: Ti/Pd/Au and Pd/Ti/Au, see Fig.1. The metal layers were sequentially electron-beam

Au (400nm)
Pd (75nm)
Ti (75nm)
p-GaAs (500nm) Zn-doped,10^{18}cm^{-3}
SI GaAs

Au (400nm)
Ti (75nm)
Pd (75nm)
p-GaAs (500nm) Zn-doped,10^{18}cm^{-3}
SI GaAs

Fig.1, Schematic cross section of samples studied

evaporated, with the base pressure 10^{-6} Torr, onto MOCVD p$^+$-GaAs epitaxial layers, on SI GaAs (001) wafers. Heat treatments were carried out using a graphite strip heater, at 380°C for 4 minutes. Plan view and cross-section TEM specimens were prepared by Ar$^+$ ion beam milling. TEM studies were performed using a Jeol TEMSCAN 120CX microscope. Analysis by SIMS was performed on an Atomika 6500 instrument using Cs$^+$ sputtering at an impact energy of ≈10KeV. Primary ion currents of ≈30nA, were rastered into an area of ≈400μm x 400μm. The profile data was taken from the central 25% of the crater area. The transmission line method (TLM) (Berger 1972) was used to determine the contact resistance of the ohmic contacts after annealing.

3.   RESULTS AND DISCUSSION

3.1 Transmission Electron Microscopy

Fig.2 shows a cross-section of a Ti/Pd/Au (75/75/400nm) sample annealed at 380°C for 4 minutes. The metallization had remained layered and there was minimal depth penetration (≈2nm) into the semiconductor. SIMS (section 3.2) indicated Ti as the deepest indiffusing species. Previous workers on Ti/GaAs (Kniffin et al 1987, Wada et al 1976) reported that Ti species diffuse into the GaAs during moderate temperature anneals to form an arsenide compound, thought to be AsTi. This reaction also occurs in the presence of an interfacial oxide. TEM and SIMS studies suggest that the light contrast phase (≈15nm thick) on the GaAs substrate in Fig.2 is an oxide. The metallization scheme, above the oxide layer, retained a three-tier structure. EDX analysis and electron diffraction data showed the lower layer (≈65nm thick) to consist of almost pure Ti. A plan view of this region is shown in Fig.3a. The average Ti grain size was ≈35nm. The SADP shown in Fig.3b shows texturing of the Ti grains. A plan view of the non-uniform middle layer (≈70nm) is shown in Fig.3c. EDX analysis of this region confirmed the presence of a Pd(Au) solid solution with an average grain size of 45nm. The uppermost layer (≈400nm thick) is shown in plan view in Fig.3d. This region is composed of heavily twinned α Au(Ga)

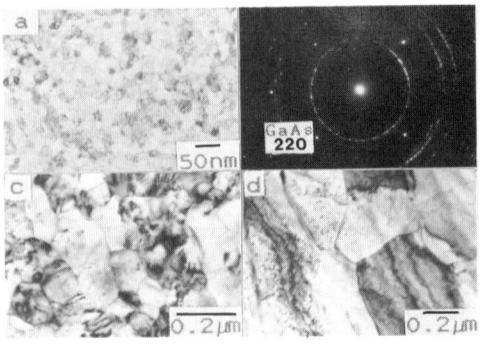

Fig.2, XTEM micrograph of the annealed Ti/Pd/Au contact, 380°C for 4mins.

Fig.3, PVTEM images of TiPdAu annealed at 380°C for 4 mins: a)Ti b)Ti SADP c)Pd d) αAu(Ga) grains of average size ≈550nm. Small (≈6nm), dark second phase particles are visible inside the α Au(Ga) grains: their identity is unknown. Fig.4 shows a cross-sectional bright field image of the Pd/Ti/Au contact heat treated for 4 minutes at 380°C. A layered structure consisting of α Au(Ga), Ti and Pd-rich layers with thicknesses of 450nm, 70nm and 170nm respectively is observed. The Pd/GaAs interface is irregular and Pd phases penetrate deep (≈0.1μm) into the epilayer. The grain size of the gold-rich phases ranges from 0.3μm to 1.3μm. Selected area electron diffraction and x-ray analysis suggest a small scale reaction between the Au and Ti layers: the reaction product is believed to be Au_4Ti. Colgen et al (1987) examined the Au/Ti system and observed a similar result when the sample was annealed at 385°C. Polycrystalline Ti, identified from diffraction and x-ray data, is adjacent to the Pd-rich layer. The average Ti grain size is ≈60nm. SAD and EDX analysis of the interfacial region suggest Pd compound formation. Shown in Fig.5 is a plan view image of the Pd-rich region. The similarity in interplanar spacings between the PdGaAs ternary phases (Sands et al 1987) and the PdGa and $PdAs_2$ compounds makes identification of the reaction products difficult. The ternary phases, $Pd_5(GaAs)_2$ (a=0.673nm, c=0.338nm) and Pd_4GaAs (a=0.92nm, c=0.37nm) provide the best fit for the diffraction data.

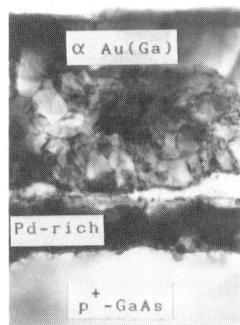

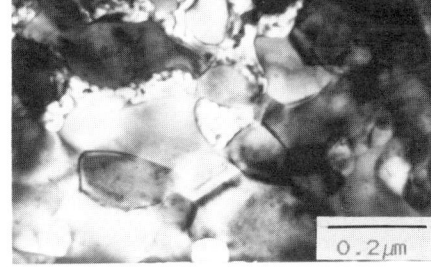

0.2μm

Fig.4 XTEM micrograph of the Pd/Ti/Au contact annealed at 380°C for 4 mins.

Fig.5 BF TEM image of the Pd-rich region in an annealed Pd TiAu contact, 380°C for 4mins

3.2 Secondary Ion Mass Spectrometry

SIMS examination of the Ti/Pd/Au structure established that only Ti penetrates significantly into the GaAs epilayer. Fig.6a shows a SIMS spectrum of the annealed Ti/Pd/Au contact. Au migration into the Pd is seen. The Pd/Ti/Au system (Fig.6b) shows far more mixing between the layers. Pd is seen to diffuse deep into the semiconductor. The Ti remains relatively localised. Broadening of the Au peak into the Pd region is a SIMS artifact due to the "knock-on effect". These findings are in agreement with the TEM observations presented earlier.

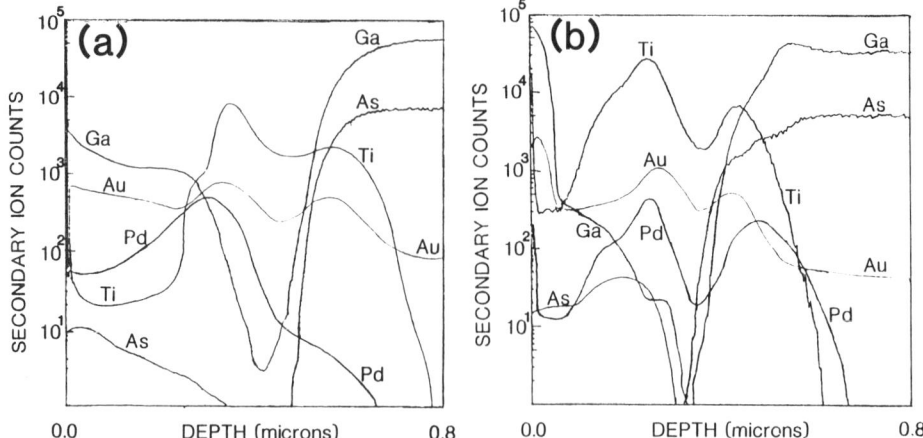

Fig.6, SIMS profiles of annealed (380°C, 4mins) a) TiPdAu and
b) PdTiAu ohmic contacts.

3.3 Electrical Measurements

The contact resistance of the two samples after alloying were
measured using the TLM (Berger 1972). Ohmic behaviour was
observed for both samples. The Ti/Pd/Au structure has a high
contact resistance (2.88Ωmm) which is unacceptable for HJBT
use. The resistance of the Pd/Ti/Au system (0.24Ωmm) is a
factor of ten better. The improvement in contact resistance is
attributed to the doping action of Pd in GaAs.

4. CONCLUSIONS

TEM, SIMS and TLM were used to investigate the microstructual
and electrical behaviour of Ti/Pd/Au structures on p^+-GaAs
epilayers annealed at 380°C for 4 minutes. The Ti/Pd/Au scheme
showed limited reaction between the metallization and the
semiconductor. The contact resistance was found to be high
(r_c=2.88Ωmm). The Pd/Ti/Au structure was found to have better
electrical properties (r_c=0.24Ωmm). However, there is
extensive indiffusion (≥0.1μm) of Pd into the epilayer. Given
the promising microstructures and electrical properties
further development of this system would be worthwhile.

ACKNOWLEDGEMENTS

M.O.D. (UK), is acknowledged for financial support.

REFERENCES

Berger H H 1972 Solid St. Electron. 15 145
Colgan E G and Mayer J W 1987 J. Mater. Res. 2 28
Kniffin M and Helms C R 1987 J. Vac. Sci. Technol. A5 1511
Sands T, Keramidas V G, Gronsky R and Washburn J 1986 Thin
    Solid Films 136 105
Wada O, Yanagisawa S and Takanaashi H 1976 Appl. Phys. Letts.
    29 263

Inst. Phys. Conf. Ser. No 117: Section 5
Paper presented at Microsc. Semicond. Mater. Conf., Oxford, 25–28 March 1991

Electrical and interfacial characteristics of NiAs/n-GaAs, NiAs/Ge/ n-GaAs and Ge/NiAs/n-GaAs structures

RS Rai, S Mahajan, JP Harbison*, T Sands*, M Genut, TL Cheeks* and VG Keramidas*

Department of Materials Science, Carnegie Mellon University, Pittsburgh, PA 15213, USA

*Bellcore, Red Bank, NJ 07701, USA

ABSTRACT :   Multilayer structures consisting of NiAs/n-GaAs, NiAs/Ge/n-GaAs and Ge/NiAs/n-GaAs have been grown by molecular beam epitaxy.   Electrical and interfacial characteristics of these structures have been investigated using current-voltage measurements and high resolution transmission electron microscopy.   Results indicate that in the presence of Ge, the as-deposited multilayer structures exhibit nearly ohmic behavior.   This ohmicity may result from favorable alignment of Fermi levels at the Ge/n-GaAs heterojunction and tunneling at the NiAs/Ge interface.   Implications of these observations to the fabrication of low resistance Au/Ge/Ni contacts to n-GaAs are discussed.

1. INTRODUCTION

Since the introduction of Au/Ge/Ni ohmic contacts to n-GaAs by Braslau et al. (1967), these contacts have formed the backbone of the GaAs technology.   The addition of Ni to the Au-Ge contacts that also have low contact resistance improves the wettability of the (001) GaAs surface by the Au-Ge eutectic film (Braslau 1981).

Recently, attempts have been made to correlate the electrical properties of the above contacts to the atomic structure, chemistry, phase distribution and morphology at the metal-semiconductor interfaces (Kuan et al. 1983 and Kim 1986).   It is observed that in the as-deposited state, the multi-layer metallization consisting of Au/Ge/Ni/Au films forms a Schottky barrier with the underlying (001) n-GaAs.   When the metal-semiconductor composites are annealed at progressively higher temperatures in the range of 350° to 500° C, the current-voltage (I-V) characteristics of the contacts change from the Schottky type to ohmic (Kim 1986).   The lowest contact resistance is observed after annealing at 500° C. The concomitant study of the metal-semiconductor interface by transmission electron microscopy reveals that the majority of the interface is covered with NiAs grains that are epitaxially oriented with respect to the underlying substrate (Kim 1986).   The results of Kuan et al. (1983) are very similar except they suggest Ge is in solid solution in NiAs.   It is emphasized that it is not possible to distinguish between the explanations of Kim (1986) and Kuan et al. (1983) using the characterization techniques which they employ.

An interesting question pertains to the role of Ge in producing low resistance contacts involving the Au/Ge/Ni metallization.   It has been suggested that Ge could heavily dope the underlying semiconductor, and that ohmicity is achieved by carrier tunneling between the NiAs metal and the semiconductor.   To gain a better understanding of the situation, we have grown the following multi-layer structures by molecular beam epitaxy (MBE): NiAs/n-GaAs, NiAs/Ge/n-GaAs, and Ge/NiAs/n-GaAs.   We have evaluated the electrical characteristics of the layered structures and have also investigated the various interfaces using high resolution transmission electron microscopy (HRTEM).   Results of these studies constitute the present paper.

2. EXPERIMENTAL DETAILS

The three types of layered structures listed above were grown in a Riber 1000 unit modified to permit e-beam evaporation of Ni in the main chamber. All other elements were supplied by standard resistively-heated MBE effusion cells. The base pressure was ~ 1 x 10^{-10} torr. The thicknesses of various layers constituting the structures were as follows: .(A) 100nm NiAs/250nm n-GaAs: Si(~1 x 10^{17}cm^{-3})/1μm n$^+$-GaAs:Si (2 x 10^{18}cm^{-3})/(001) n$^+$ - GaAs:Si substrate, (B) 50nm NiAs/30nm Ge/ the sequence of GaAs layers in (A), and (C) 30nm Ge/50nm NiAs/the sequence of GaAs layers in (A). These structures will be, respectively, labeled as A, B and C in the following presentation and discussion. The GaAs layers were grown at 580° C, whereas the temperature for the growth of NiAs and Ge layers was in the range of 280 - 300° C. The respective rates for GaAs, NiAs and Ge were 0.5, 0.2 and 1μm/hr, and the As flux was ~2.5 times of that required to grow stoichiometric GaAs at 0.5μm/hr for a substrate temperature of 580° C. Before the growth of the Ge layers, the As cell was allowed to cool for approximately an hour so that the typical chamber pressure before the deposition of Ge was ~ 3 x 10^{-9} torr. All growths were monitored using reflection high energy electron diffraction (RHEED).

The as-grown structures were fabricated into diodes using standard photolithography and wet chemical etching; the diode area was ~8 x 10^{-15}cm^{-2}. Cross-section samples for HRTEM were prepared by ion milling and were examined in a JEM 4000 EX operating at 400 keV.

3. RESULTS

The RHEED studies showed that all NiAs layers are crystallographically textured having multiple variants as found previously for NiAs/GaAs in Ni/GaAs reacted contacts (Sands et al. 1987). The mosaic spread and surface roughness are greater for the NiAs film in structure B. Shown in Fig. 1 are a sequence of RHEED patterns that are representative of RHEED observations for all three NiAs layers. In addition the Ge layer in structure B is epitaxial with the underlying GaAs, whereas it is fine grained in structure C. As will be shown below, these results are consistent with those of the HRTEM study.

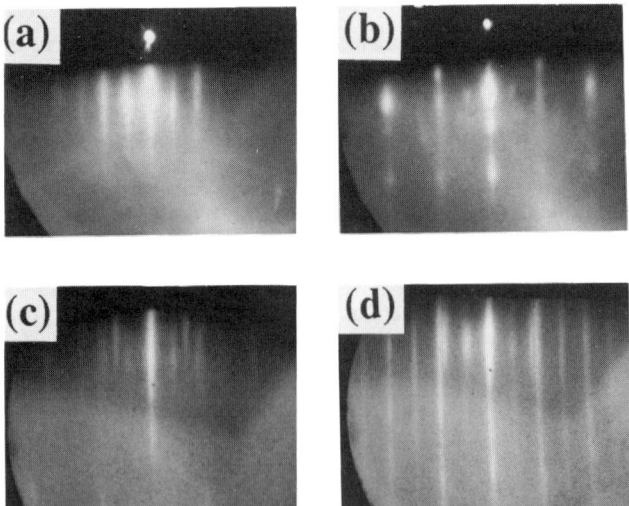

Figure 1. RHEED patterns observed during the growth of NiAs on (001) GaAs: (a) GaAs, [110] azimuth (C-4x4, arsenic stabilized), at 280° C just prior to NiAs deposition, (b) after growth of 5nm NiAs at 280° C, 3.4 x 10^{-7} torr (arsenic), (c) after growth of 100nm NiAs, 280° C, 3 x 10^{-7} torr, and (d) during cooling 3 min. after (c), 250° C, 7 x 10^{-8} torr.

The NiAs layer in structure A forms a Schottky barrier with the underlying semiconductor. The effective Schottky barrier height is 0.73 eV with an ideality factor of 1.1. On the other hand, structures B and C are found to be only weakly rectifying at room temperature as shown in Fig. 2, and thus their barrier heights could not be measured.

Figure 3 shows the HRTEM images of various interfaces in the three structures. The NiAs/GaAs interface, Fig. 3(a), is slightly rough, shows steps and contains misfit dislocations. Furthermore, the orientation of GaAs in Fig. 3(a) is [110], whereas NiAs is oriented along the [01$\bar{1}$1] direction. In other regions of the interface, the following orientation relationship was also observed:

$$[110]_{GaAs} \quad \| \quad [11\bar{2}0]_{NiAs}$$

The above results are consistent with the RHEED observations that the NiAs layer is textured with respect to the underlying GaAs layer.

Shown in Fig. 3(b) are the Ge/GaAs and Ge/NiAs interfaces in structure B. The former is barely discernible, whereas the latter is quite rough on the atomic scale. Also, the NiAs layer is quite defective. The following orientation relationship between the Ge and NiAs layers is observed in Fig. 3(b):

$$[110]_{Ge} \quad \| \quad [11\bar{2}0]_{NiAs}$$

Considering the fact that the Ge and GaAs are in perfect epitaxy, the observed relationship is quite reasonable.

Figure 3(c) shows the HRTEM images of the NiAs/GaAs and Ge/NiAs interfaces in structure C. A

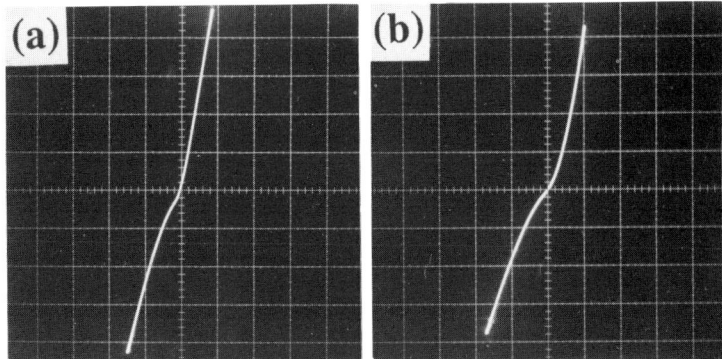

Figure 2. I-V characteristics of as-grown multi-layer structures: (a) 50nm NiAs/30nm Ge/250nm - GaAs:Si (~ 1 x 10^{17}cm^{-3})/1μm n$^+$ - GaAs:Si (2 x 10^{18}cm^{-3}) - structure B, and (b) 30nm Ge/50nm NiAs/250nm-GaAs:Si (~ 1 x 10^{17}cm^{-3})/1μm n$^+$ - GaAs:Si (2 x 10^{18}cm^{-3}) - structure C. The I-V characteristics of structure A (not shown) were rectifying with a Schottky barrier height of .73V as described in text.

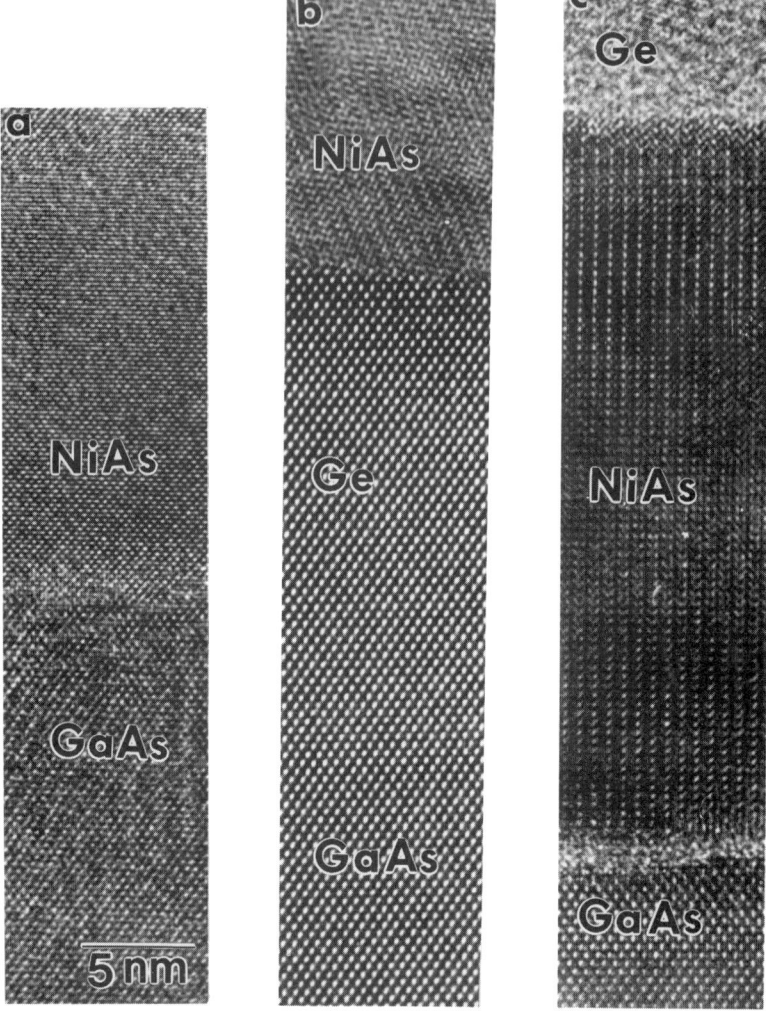

Figure 3. Cross-sectional high resolution transmission electron micrograph showing different interfaces in various structures: (a) The NiAs/GaAs interface in structure A; orientation for GaAs is [110], whereas it is [01$\bar{1}$1] for NiAs. (b) The Ge/GaAs and the NiAs/Ge interfaces; orientation of Ge and GaAs is [110], whereas it is [11$\bar{2}$0] for NiAs. (c) The Ge/NiAs and the NiAs/GaAs interfaces in structure C; orientation of GaAs is [110], whereas it is [11$\bar{2}$0] for NiAs. The Ge layer consists of amorphous and highly defective crystalline regions.

transition region, ~ 1.5nm thick, is observed between the NiAs and GaAs layers that is not discernible in Fig. 3(a). In addition, the Ge layer contains amorphous as well as highly defective crystalline regions. Also, the [110] direction of GaAs is parallel to the [11$\bar{2}$0] direction of NiAs.

4. DISCUSSION

Let us summarize the significant observations which emerge from the preceding study. First, the NiAs layer in structure A forms a Schottky barrier with the underlying n-GaAs and the barrier height is 0.73 eV. Second, the structures B and C in the as-deposited condition exhibit only weakly rectifying behavior at room temperature. Third, the NiAs layer exhibits orientation relationships with the underlying GaAs or Ge layers. Fourth, the Ge layer grown on the NiAs layer is highly defective. Fifth, a transition region, ~1.5nm thick, is observed at the NiAs/GaAs interface in structure C, whereas this is not discernible between the same set of materials in structure A.

One possibility is that the transition region observed between the NiAs and GaAs layers in structure C contains Ge. This suggestion is reasonable in view of the fact that the structure of NiAs is highly open and Ge from the over-layer could diffuse through to the NiAs/GaAs interface. It can be shown that a diffusion coefficient of 6 x 10^{-18}cm^2/sec. for the diffusion of Ge in NiAs would be required to accomplish this, a reasonable value considering the openness of the structure.

Since the GaAs layer immediately underneath the Ge layer or the NiAs layer is only moderately doped and since nearly identical I-V characteristics are observed for structures B and C, the possibility that the low contact resistance is due to tunneling between the degenerately doped GaAs and the metal can be ruled out. A plausible, cogent explanation that is consistent with the results of this study is illustrated schematically in Fig. 4. First, the conduction band edges at the Ge/n-GaAs heterojunction are aligned such that the Fermi level is above the conduction band edge in the Ge layer (Stall et al. 1979). Second, the carrier tunneling occurs at the NiAs/Ge interface. The latter suggestion is reasonable in view of the facts that either Ge reaches the NiAs/GaAs interface by diffusing through the NiAs matrix and could be heavily doped or the surface region of the Ge layer could be doped with As during the growth of NiAs.

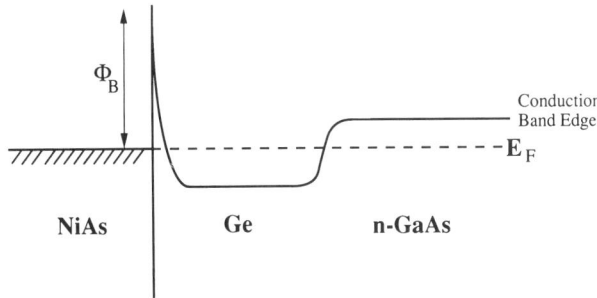

Figure 4. Schematic of the band diagram for the NiAs/n-Ge/n-GaAs structure. Note that the band edges at the Ge/GaAs interface are aligned such that the Fermi level lies above the conduction band edge of Ge. Thus the Ge/GaAs hetero-barrier is transparent to the flow of carriers. Tunneling must occur at the NiAs/n-Ge interface to rationalize the observed nearly ohmic behavior in Fig. 2.

In view of the above results an interesting question arises regarding the role of annealing in producing low resistance Au/Ge/Ni ohmic contacts to n-GaAs. It is plausible that during the contact anneal, the formation of different phases that is caused by thin-film interdiffusion results in a highly doped GaAs surface contiguous to the contact metal; the latter consists of highly textured, epitaxial grains of NiAs and gold-gallium alloy (Kuan et al. 1983 and Kim 1986). As a consequence, the carrier tunneling may now occur at the metal-highly doped GaAs interface, resulting in low resistance ohmic contact.

In summary, by tailoring multi-layer structures consisting of n-GaAs, Ge and NiAs layers we have shown that a nearly ohmic behavior is observed without annealing in NiAs/Ge/n-GaAs and Ge/NiAs/n-GaAs heterostructures. The ohmicity is due to the favorable alignment of the Fermi levels at the Ge/GaAs heterojunction coupled tunneling at the NiAs/Ge interface.

ACKNOWLEDGEMENTS

The work at Carnegie Mellon was supported by the IBM Center for Thin Film Sciences, and RSR, SM and MG gratefully acknowledge the financial support.

REFERENCES

Braslau N 1981 J. Vac. Sci. Tech. 19 803.
Braslau N, Gunn JB and Staples, JL 1967 Solid-State Electron. 10 372.
Kim T 1986, Ph.D. Dissertation, Carnegie Mellon University.
Kuan TS, Batson PE, Rupprecht H and Wilkie EL 1983 J. Appl. Phys. 54 6952.
Sands T, Keramidas VG, Yu KM, Washburn J and Krishnan KM 1987 J. Appl. Phys. 62 2070.
Stall R, Wood CEC, Broad K and Eastman LF 1979 Electron. Lett. 15 800.

Inst. Phys. Conf. Ser. No 117: Section 5
Paper presented at Microsc. Semicond. Mater. Conf., Oxford, 25–28 March 1991

Cross sectional transmission electron microscopic study of Au/A^{111}B^V contacts

B Pécz, G Radnóczi, A Barna, R Veresegyházy and I Mojzes

Research Institute for Technical Physics of the Hungarian Academy of Sciences, H-1325 Budapest, P.O.Box 76, Hungary

ABSTRACT: The application of cross sectional transmission microscopy largely contributed to the understanding of the thermal behaviour of Au/A^{III}B^V contacts. In the Au/GaAs system the formation of pyramidal pits grown into the GaAs was observed. In the Au/GaP system the accumulation of Ga in the contact layer results in the melting of the metallization. The solidified metallization consists of Au_2Ga and Au_7Ga_2 grains. In the Au/InP samples elongated crystals grew into the substrate during annealing. Au_9In_4 grains are situated in these pits in the matrix of the Au_2P_3 monoclinic phase.

1. INTRODUCTION AND SAMPLE PREPARATION

Gold was deposited by thermal evaporation onto the substrates after chemical surface cleaning (see Pécz et al 1990). The samples were annealed in flowing forming gas ($5\%H_2 + 95\%N_2$) for 10 minutes at different temperatures.

After the mechanical grinding to approximately 50 μm thickness we thinned the samples by ion milling with Ar$^+$ ions at 9-10 keV (Barna 1984). The samples were milled at an incident beam of 3-5° to the specimen surface. In the case of compound semiconductors containing In the Ar$^+$ ion bombardment causes the appearance of In dots on the surface during thinning. Therefore in the last stage of ion milling of Au/InP samples we applied only 3 kV Ar$^+$ ions in an iodine ambient to remove these artifacts (Barna and Pécz 1991).

The thinned samples were examined using JEOL 100U, JEOL 100C and Philips CM20 transmission electron microscopes.

2. RESULTS AND DISCUSSION

2.1 Au/GaAs

The first process in this system is the formation of α-Au(Ga) solid solution (Pécz et al 1986) which is followed

by the formation of the hexagonal β-AuGa and the orthorombic
Au_2 Ga phases at higher temperature (Yoshiie et al 1984 and
Pécz et al 1986).

The pyramidal pits grown into the GaAs were detected in
samples annealed at 400°C. In the (110) section these
formations are triangular (Fig.1.a) and in the perpendicular
(1$\bar{1}$0) section they show an elongated form (Fig.1.b). These
results are in good agreement with the findings of Yoshiie
et al (1984). In most cases these pits are single crystals
of α-Au(Ga) solid solution. We identified in this system the
orthorombic Au_2 Ga phase in small amount.

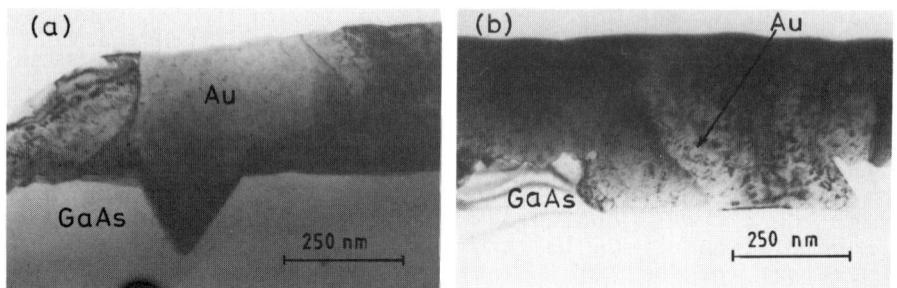

Fig.1. Cross section of annealed (160 nm)Au/GaAs(100)
sample: (110) section (a) and perpendicular (1$\bar{1}$0) section
(b).

2.2 Au/GaP

In the Au/GaP system Piotrowska et al (1985) observed the
formation of α-Au(Ga) in the temperature range of 450-480°C.
In samples annealed at about 500°C the $Au_7 Ga_2$ and the Au_2 Ga
phases were identified using X-ray diffraction method
(Piotrowska et al 1985 and Ginley et al 1984). The whole
interaction process between gold and GaP can be described as
follows.

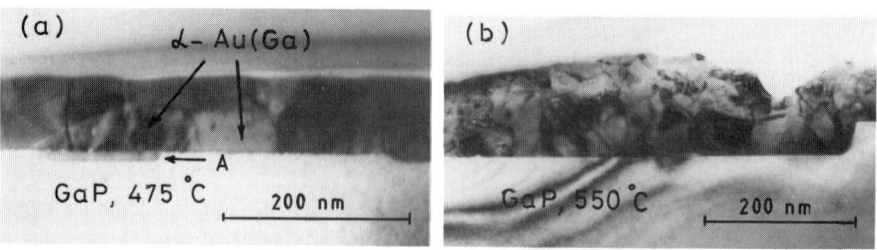

Fig.2. Cross sectional transmission electron micrographs of
Au(80nm)/GaP(111) samples: annealed at 475°C (a) and 550°C
(b).

The interface of the as deposited sample is laterally homogeneous and sharp. The lowest temperature at which very thin, elongated grains grew into the GaP parallel to the interface (grain marked by A in Fig.2.a) was found to be 475°C. These grains are 60-120 nm wide and 10-15 nm thick. Long, flat grains were formed from the thin, elongated ones during annealing at 525°C. These grains consist of α-Au(Ga) solid solution. Up to this temperature no Au-Ga compounds could be identified.

At 550°C (Fig.2.b) the metallization melted due to its high and (increasing) Ga content and broke up into droplets. During cooling the droplets crystallized. By X-ray diffractometry we found orthorombic Au_2Ga and hexagonal β or Au_7Ga_2 phase and the peaks corresponding to gold disappeared. We have also identified the Au_2Ga orthorombic phase by electron diffraction.

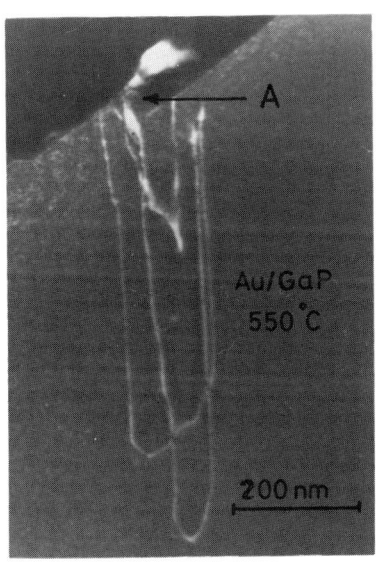

Besides the chemical interaction between the contact layer and the substrate, mechanical stresses are created leading to the formation of dislocation loops in GaP. These loops start from the edge of the steps in the interface (marked by A on Fig.3) and are extending into the GaP. This type of mechanical interaction was observed consequently in this sample.

Fig.3. Weak beam dark field image (g=440) of dislocation loops in GaP created by the interaction between gold and GaP.

2.3 Au-InP

The interface of the as-deposited Au(100nm)/InP(111) sample is sharp and laterally homogeneous. The grain size of the gold layer is about 100nm.

In the first stage of the reaction between gold and InP the formation of α-Au(In) was observed (Piotrowska et al 1981 and Vandenberg et al 1983). At higher temperature the formation of Au_3In and Au_2P_3 were found by Piotrowska et al (1981). In the next stage Au_9In_4 was identified by Vandenberg et al (1983), but the presence of Au_2P_3 and the geometry of the reaction products was not clear.

In the Au(100nm)/InP(111) sample annealed at 375°C the cross sections revealed elongated pits grown into the InP substrate. The pits are 300-600 nm wide and about 150 nm thick. Applying the electron diffraction method and the dark field technique, we have identified the phases constituting these pits. The contact layer consists of cubic Au_9In_4 (Fig.4.). The grains of the Au_9In_4 layer are about 700 nm long. The bright parts of the pits belong to the Au_2P_3 monoclinic phase. Small grains of the Au_9In_4 phase also lie in the pits of Au_2P_3 matrix (Fig.4.). In the sample annealed at 400°C, we have not found a significant change in the interface behaviour compared to samples annealed at 375°C.

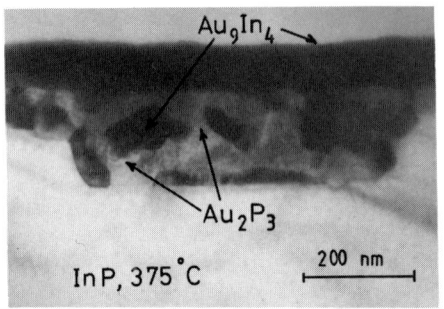

Fig.4. Cross sectional transmission electron micrograph of Au(100 nm)/InP(111) sample annealed at 375°C showing the pits grown into the substrate.

ACKNOWLEDGEMENTS

The authors are grateful to P.B. Barna, B. Pödör and Zs.J. Horváth for helpful discussions. Technical assistance by É. Szabó is also acknowledged.

REFERENCES

Barna A 1984 Proc. 8th Europ. Congr. Electron Microscopy, Budapest eds A Csanády, P Röhlich and D Szabó, 1, 107

Barna A and Pécz B 1991 J. Electron Microsc. Tech. 18, in press

Ginley R A, Chung D D L and Ginley D S 1984 Solid State Electron. 27 137

Pécz B, Jároli E, Radnóczi G, Veresegyházy R and Mojzes I 1986 phys. stat. sol.(a) 94, 507

Pécz B, Veresegyházy R, Mojzes I, Jároli E and Zsoldos É 1990 Vacuum 40, 185

Piotrowska A, Auvray P, Guivarc'h A, Pelous G and Henoc P 1981 J. Appl. Phys. 52 5112

Piotrowska A, Kaminska E, Barcz A, Adamczewska J and Turos A 1985 Thin Solid Films 130 231

Vandenberg J, Temkin H, Hamm R H and Diguiseppe M A 1983 Thin Solid Films 104 419

Yoshiie T, Bauer C L and Milnes A G 1984 Thin Solid Films 111 149

Inst. Phys. Conf. Ser. No 117: Section 5
Paper presented at Microsc. Semicond. Mater. Conf., Oxford, 25–28 March 1991

Comparison of solid state and alloyed AuGeNi contacts to n-In$_{0.53}$Ga$_{0.47}$As for optoelectronic integrated circuits

J S Yu[*], A E Staton-Bevan[*], J Herniman[**] and D A Allan[**]

[*] Department of Materials, Imperial College of Science, Technology and Medicine, London, SW7 2AZ, UK
[**] British Telecom Research Laboratories, Martlesham Heath, Ipswich, IP5 7RE, UK

ABSTRACT: AuGeNi/ZrB$_2$/Au ohmic contacts annealed at 270°C or 445°C were investigated using TEM. Reaction of the 270°C annealed sample took place via solid state diffusion resulting in the formation of AuGeAs and NiGe interfacial phases. In the sample alloyed at 445°C, melting of the contact resulted in a more extensive reaction. The morphology as well as the contact resistivity (0.47Ωmm) was inferior. The phases formed were Au-In compounds, f.c.c. Ni-Ge-As and regrown In$_x$Ga$_{1-x}$As (x<0.3).

1. INTRODUCTION

The current trend in optoelectronic integrated circuits (OEIC's) is towards longer wavelength devices operating at 1.32μm or 1.55μm. InP and compounds such as InGaAs or InGaAsP are normally required for such devices. For these materials, a solid state AuGeNi contact annealed at 270°C provides an adequately low resistivity. Signal processors i.e. transistors, however, can be replaced by more advanced GaAs devices, which require contacts to be alloyed at 445°C. In this case, the contacts to InGaAs also have to undergo the same heat treatment.

In this paper, the microstructures of AuGeNi ohmic contacts to InGaAs annealed at 270°C or 445°C are compared and correlated with their contact resistivities.

2. EXPERIMENTAL PROCEDURE

The contact metals Ni 5nm/Au 45nm/ Ge 20nm, followed by a 50nm ZrB$_2$ diffusion barrier and a 20nm Au overlayer were sequentially deposited on Si-doped (001) n-In$_{0.53}$Ga$_{0.47}$As/InP using electron beam evaporation at <10^{-6}mbar. Solid state contacts were spike-heated at 270°C. Alloyed contacts were initially brought up to 200°C before ramping up to 445°C using RTA. The contact resistivity measurements were carried out by the transmission line method.

Plan view TEM specimens were thinned from the substrate side by grinding, dimple polishing and chemical thinning (0.5 vol% of Bromine in methanol), followed by Ar⁺milling. Cross-sectional TEM specimens were prepared by standard techniques using Ar⁺milling. All Ar⁺milling was carried out with liquid N_2 cooling at 4.5kV. Jeol TEMSCAN 120CX and 2000FX microscopes were used for the TEM examination.

3. RESULTS AND DISCUSSION

Annealing at 270°C caused the Ge layer to diffuse through the Au grain boundaries towards the semiconductor. A small amount of InGaAs was dissociated resulting in the formation of AuGeAs (Fig. 1a). The rest of the Ge either reacted with Ni to form NiGe or remained unreacted at the Au grain boundaries. The NiGe was confined to the metal-semiconductor interface (Fig. 1b). Small traces of In (<10 at%) were observed in the Au grains but little Ga, indicating preferential dissociation of In-As bonds to Ga-As bonds. The diffraction pattern showed no change in the Au lattice parameter confirming that the In dissolution in Au was minimal. The cross-sectional view (Fig. 1b) showed metal penetration into the semiconductor of approximately 80nm. With the presence of a ZrB2 diffusion barrier, this was reduced to 50nm.

When the contact was annealed at 445°C, all the contact metals and the underlying semiconductor melted. Upon solidification, an entirely different microstructure resulted. Fig.2a shows a typical plan view, consisting of dark Au-In compounds, lath shaped Ni-Ge-As and regrown $In_xGa_{1-x}As$ (x<0.3). The cross-sectional TEM demonstrates the extent of the alloying reaction. The Au-In solidified in the semiconductor to a depth of $0.3\mu m$. The In-content in Au varied from 20~50 at% according to EDAX. This suggests a range of Au-In compounds co-existing as observed by

 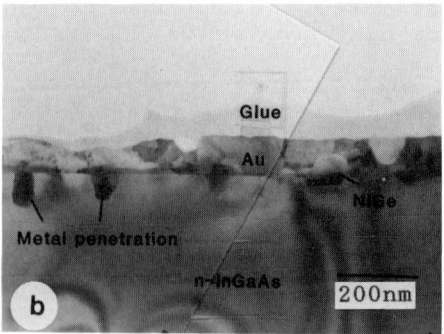

Fig. 1 Solid state AuGeNi contacts to n-InGaAs/InP without ZrB2 diffusion barrier annealed at 270°C a) TEM plan view b) TEM cross-section

Vandenberg et al (1982). There could be a further complication by the addition of Ga. The lath shaped Ni-Ge-As is not the same phase as the hexagonal Ni_2GeAs frequently observed in AuGeNi contacts to GaAs (Kuan et al 1982). It has a f.c.c. structure with a lattice spacing of ~12.0Å (Fig.2c). The average composition from EDAX analysis was $Ni_3Ge_2As_4$. As a consequence of extensive In-dissolution in the Au, In-deficient InGaAs (5~15 at% of In) was regrown from the melt. The regrown layer was discontinuous and highly defective (Fig. 2d).

The contact resistivity of the solid state contact was measured to be 0.13Ωmm while that of the alloyed contact was considerably higher at 0.47Ωmm. The n-type barrier of $In_{0.53}Ga_{0.47}As$ is very low as it is in InP (i.e. exceptions to Mead's rule). This is thought to be responsible for the relative ease of forming n-type ohmic contacts (Kupal 1981). The lower In content in the regrown InGaAs would be a

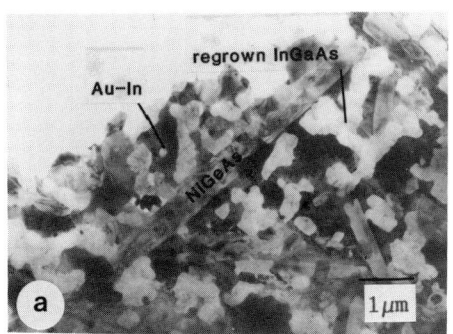
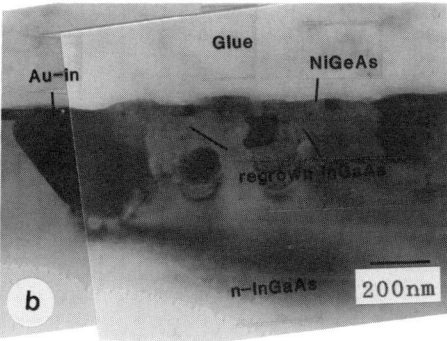

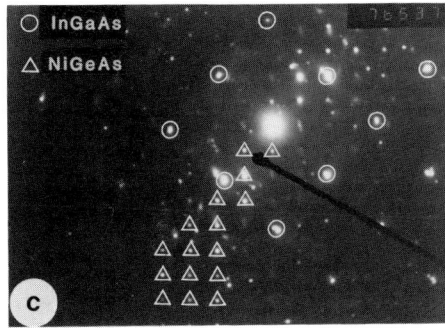
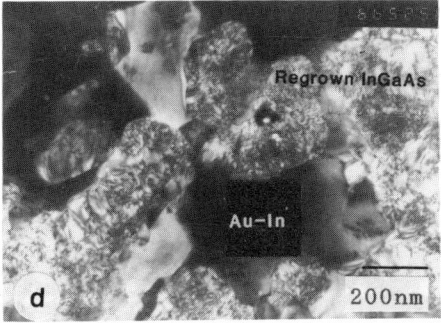

Fig. 2 Alloyed AuGeNi contacts to n-InGaAs/InP annealed at 445°C a) TEM plan view; the ZrB_2 diffusion barrier was removed by brief Ar^+ milling b) TEM cross-section of the barrierless contact c) Diffraction pattern of (001) Ni-Ge-As superimposed on (001) InGaAs from the plan view d) TEM plan view showing dense defects and Moire fringes

departure from this towards GaAs, whose n-type barrier is substantially higher. This may explain the higher contact resistivity of the alloyed contacts.

4. CONCLUSIONS

This AuGeNi contact incorporating a ZrB_2 diffusion barrier was originally developed for n-GaAs (Grimshaw 1989). The same system can be successfully applied to n-InGaAs as a solid state ohmic contact but the alloyed contact requires improvements in both contact resistivity and morphology.

REFERENCES

Grimshaw M P, Staton-Bevan A E, 1989 Mat. Sci. Eng., B5, 21
Kuan T S, Batson P E, Jackson T N, Rupprecht H, Wilkie E L, 1982 J. Appl. Phys.
Kupal E, 1981 Solid-State Electron.,24, 69
Vandenberg J M, Temkin H, Hamm R A, DiGuiseppe M A, 1982 J. Appl. Phys.,53, 7385

Inst. Phys. Conf. Ser. No 117: Section 5
Paper presented at Microsc. Semicond. Mater. Conf., Oxford, 25–28 March 1991

283

Low-temperature reactions at metal-semiconductor interfaces

R Sinclair, T Konno, D H Ko and S Ogawa*.

Department of Materials Science and Engineering, Stanford University, Stanford, CA 94305, USA
*Matsushita Electric Industrial Company, Semiconductor Research Laboratory, Moriguchi, Osaka 570, Japan

ABSTRACT: In this article we consider the interfacial reactions which occur at several metal-semiconductor interfaces. When there is a large free energy of mixing, amorphous phase formation can precede crystal phase nucleation, and this is illustrated by Ti-Si and Pt-GaAs contacts. On the other hand, for a phase separating system (Al-Si), amorphization is not observed, but instead low-temperature (~180°C) silicon crystallization is confirmed.

1. INTRODUCTION

Our work concentrates on understanding the structure and reactions which occur at various semiconductor interfaces. Whenever possible the behavior is correlated with the influence on material properties, and the implications for manufacture and device processing are identified. The primary investigative tool is high-resolution electron microscopy (HREM), with particular emphasis on the application of in situ heating to observe the events occurring directly at the atomic level.

The two most technologically important metal semiconductor interfaces are Ti-Si and Al-Si, because of their preponderance in the metallization of silicon-based integrated circuits. Direct aluminum-silicon contact is normally avoided because of the high solubility of Si in Al, leading to pitting and "spiking" of the semiconductor substrate. Ti-Si is a contact currently employed because of its low resistivity and thermodynamic stability. However they represent interesting, simple scientific systems and we have derived insight into metal-semiconductor behavior by their study. The ideas so developed have been applied to some metal-gallium arsenide interfaces.

2. RESULTS

It was shown definitively by Holloway and Sinclair (1987) that crystalline titanium-amorphous silicon multilayers show a solid-state amorphization reaction on annealing, analogous to that first discovered for Au-La by Schwarz and Johnson (1983). We have recently extended this work in several ways. Firstly differential scanning calorimetry (DSC) has been employed to determine the enthalpy of the amorphization reaction, for a variety of Ti-Si compositions. These data have been used to plot the free energy composition curve of the system, and by using the "law of common tangents", the metastable phase diagram has been derived (Holloway 1989). We tested the validity of the diagram by the following experiment. A crystalline Ti-crystalline Si interface was annealed to "metastable equilibrium" (30 minutes at 430°C). The amorphous Ti-Si alloy was created, as predicted by the thermodynamic analysis and first reported in the last proceedings of this series (Sinclair et al.

1989). Careful energy dispersive spectroscopy across the interface using a 2nm electron probe showed the terminal compositions of the alloy to be about 30% Si in contact with the Ti film, and about 70% Si in contact with the Si substrate (Ogawa et al. 1991). This is in good agreement with the metastable phase diagram for this temperature, showing the usefulness of this fundamental approach for understanding such systems (Sinclair 1990, Ogawa et al. 1991).

Furthermore, the electrical properties of the interface have been measured, particularly the Schottky barrier height as a function of annealing temperature (Ogawa et al. 1991). When the amorphous phase is present, the barrier height is quite large (0.73eV) which is undesirable for contact properties. A reduction occurs (to 0.57eV) on formation of the C49 TiSi$_2$ crystalline phase, by annealing at a higher temperature (above 460°C) (Ogawa et al. 1991).

In the aluminum-silicon system there is a very small (if any) free energy of mixing of the elements and so amorphous phase formation would not be anticipated from a thermodynamic point of view. However there have been reports of metastable phase formation (Hentzell et al. 1987) and possibly amorphization (Ruterana and Buffat 1989) during Al-Si reactions. Accordingly this is a good "test" system to determine whether the thermodynamic guidelines for amorphization, which work so well for Ti-Si, contain the correct predictions for "non-amorphization".

Aluminum-silicon multilayers have been fabricated by sputter-deposition, following the procedures for Ti-Si multilayers (Holloway and Sinclair 1987). As shown in Fig. 1 high quality samples have been produced, with the metal crystalline and the silicon in the amorphous state. Heating, in situ in the microscope and ex situ does not bring about an amorphization reaction, but rather the low temperature crystallization of silicon at about 180°C (this is about 400°C lower than the normal crystallization temperature). The multilayering is rapidly destroyed and the silicon microstructure contains many twins and stacking faults similar to that of crystallized single phase amorphous silicon (Morgiel et al. 1990). This behavior is consistent with earlier reports (Bosnell and Voisey 1970, Herd et al. 1972) of metal-contact induced crystallization of amorphous semiconductors (e.g. a-Si, a-Ge). However ours is the first study of multilayer samples and this is the first time that the reaction has been followed in "real time". We are currently making careful in situ studies of the behavior, combined with DSC measurements and X-ray work, to understand the mechanism of this curious reaction, and to compare it with "pure" silicon in order to establish the role of the metal itself. Our original hypothesis has been established (that Al-Si does not form an amorphous phase).

Since the discovery of amorphous phase formation at metal-silicon contacts, and its occurrence for a variety of systems (see Sinclair 1990 for review), we have been interested as to the possibility of amorphization at metal-GaAs interfaces. We chose the Pt-GaAs system for this study because its phase diagram belongs to category IV of our scheme (Beyers et al. 1987), with the metal thermodynamically unstable in contact with the semiconductor and a large driving force to bring about a chemical reaction. Thus if the equilibrium phase formation is delayed kinetically, amorphization might be a precursor state.

This is a difficult experiment to carry out rigorously. The GaAs substrate must be cleaned as carefully as possible to remove native oxide and contamination, the deposition must be performed in a clean environment and TEM sample preparation should be achieved in a manner that does not induce any interfacial reaction. These precautions have been followed and the results have been successful. The as-deposited Pt-GaAs interface shows a 3nm amorphous phase, analogous in appearance to that at Ti-Si interfaces (Ko and Sinclair 1991). Annealing at low temperature (e.g. 200°C) both in situ and ex situ (Fig. 2) brings about thickening of the amorphous layer prior to the formation of any crystal phases. Prolonged annealing achieves the equilibrium state, with a PtAs$_2$ film in contact with GaAs and PtGa above (Ko and Sinclair 1990). Consequently the original suggestion, that amorphization can

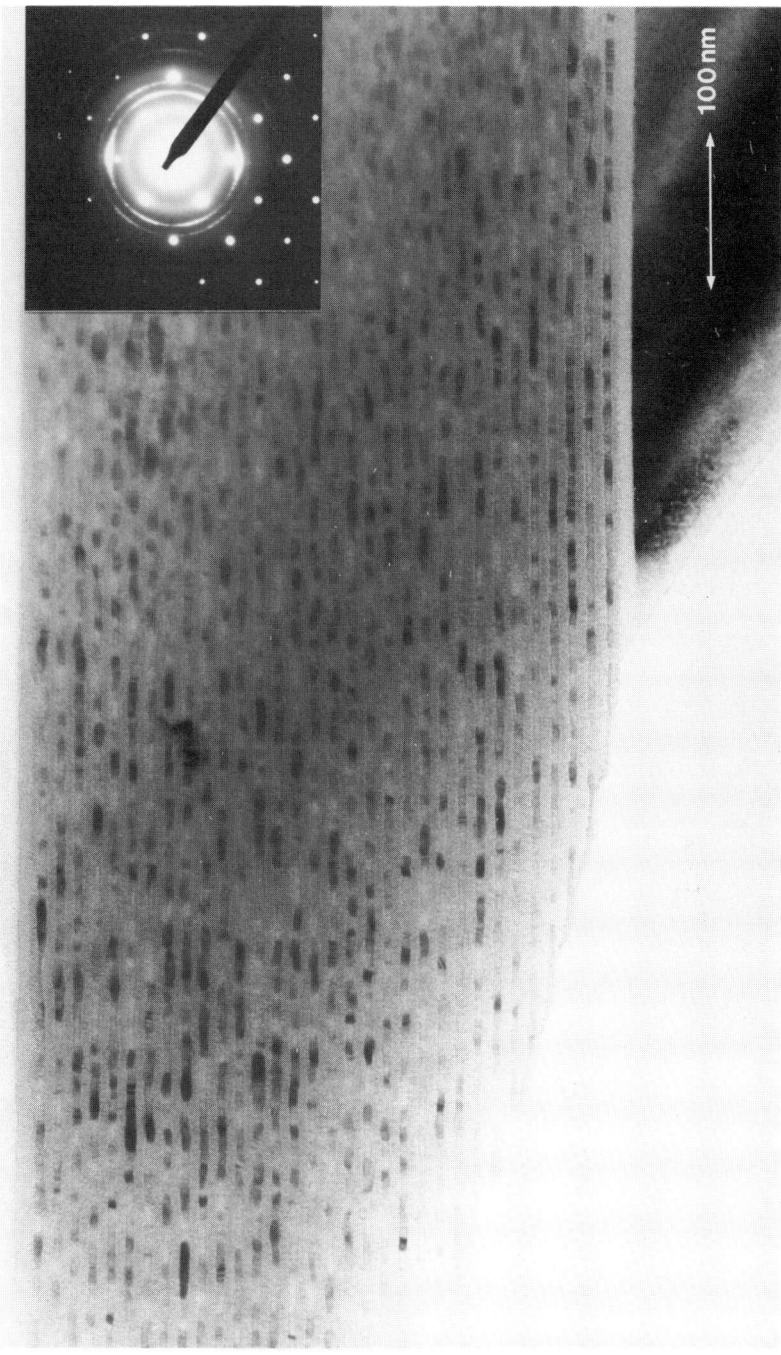

Figure 1. Cross-section TEM micrograph of as-deposited Al/Si multilayer.
BF image and diffraction pattern.

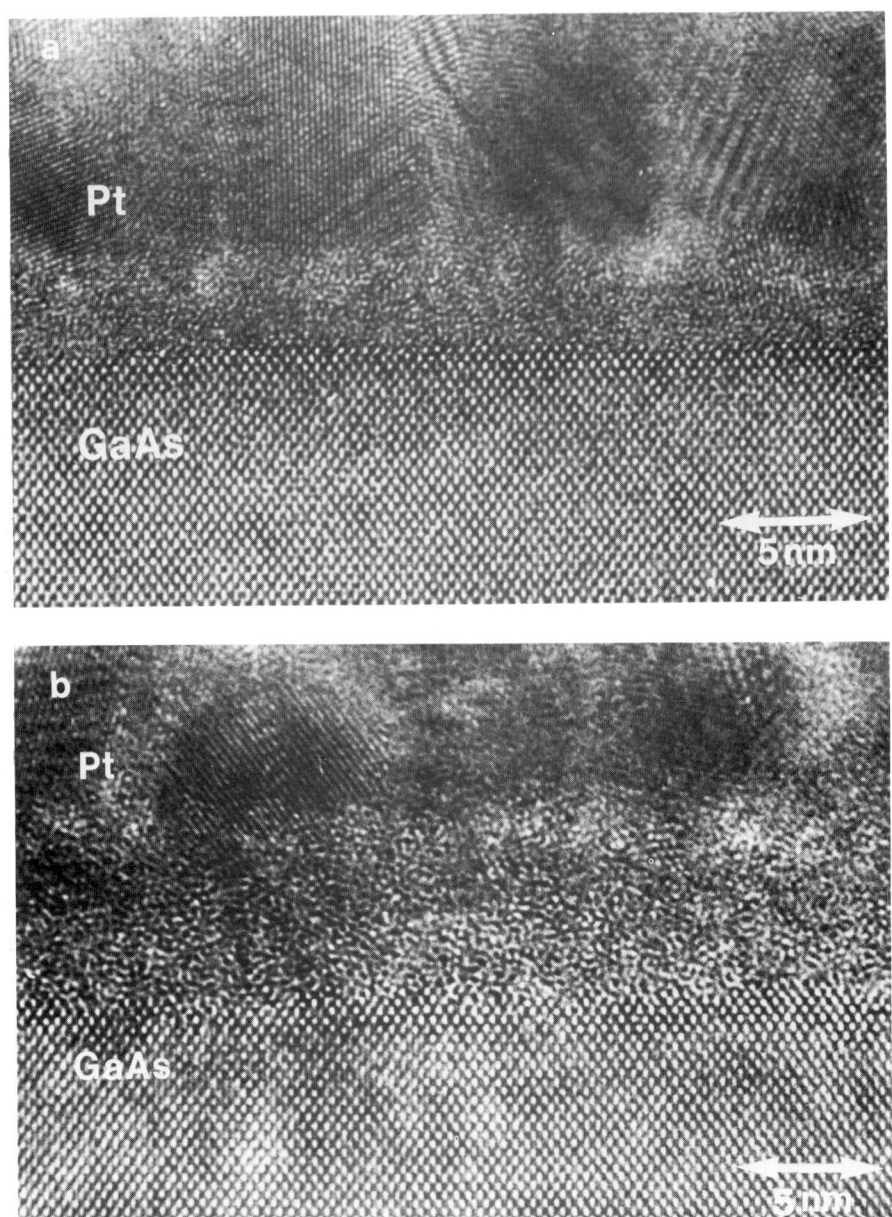

Figure 2. Cross-sectional HRTEM micrographs of the Pt/GaAs interface: (a) as-deposited sample (b) *in-situ* annealed at 200°C for 30 min.

occur at clean metal-GaAs interfaces, even in the as-deposited state, has been fulfilled by this work. Furthermore it has been demonstrated that in situ HREM can reveal the bulk interfacial reaction.

More complex systems such as alloy diffusion barriers for silicon or metal compound contacts to GaAs require the use of higher order phase diagrams for their understanding (e.g. Bhansali et al. 1990), which is beyond the scope of the present article.

3. SUMMARY AND CONCLUSIONS

Basic principles of material science, such as free energy-composition diagrams, can be applied to understand reactions at simple metal-semiconductor interfaces. Amorphous phase formation can occur upon low-temperature annealing when there is a large thermodynamic driving force and crystal phase formation is kinetically retarded (e.g. Ti-Si, Pt-GaAs). For Al-Si, in which there is little or no free energy of mixing, our results confirm the observation of a different low temperature phenomenon, viz. metal-induced crystallization of amorphous silicon.

ACKNOWLEDGEMENTS

The financial support of the National Science Foundation (Grant No. DMR-8902232) is much appreciated.

REFERENCES

Beyers R, Kim K B and Sinclair R 1987 J. Appl. Phys. **61** 2195
Bhansali A S, Ko D H and Sinclair R 1990 J. Electron. Mater. **19** 1171
Bosnell J R and Voisey U C 1970 Thin Solid Films **6** 161
Hentzell H T G, Robertson A, Hultman L, Shaofang G, Hornstrom S E and Psaras P A 1987 Appl. Phys. Lett. **50** 933
Herd S R, Chaudhari P and Brodsky M H 1972 J. Non-Cryst. Solids **7** 309
Holloway K and Sinclair R 1987 J. Appl. Phys. **61** 2195
Holloway K 1989 *Interfacial Reaction in Metal-Silicon Multilayers* Ph.D. Thesis Stanford University 127
Ko D H and Sinclair R 1991 Appl. Phys. Lett. in press
Ko D H and Sinclair R 1990 Mats. Res. Soc. Proc. **181** 333
Morgiel J, Wu I W, Chiang A and Sinclair R 1990 Mats. Res. Soc. Proc. **182** 191
Ogawa S, Yoshida T, Kouzaki T and Sinclair R 1991 J. Appl. Phys. in press
Ruterana P and Buffat P A 1989 Inst. Phys. Conf. Ser. **100** 677
Schwarz R B and Johnson W L 1983 Phys. Rev. Lett. **51** 415
Sinclair R, Holloway K, Kim K B, Ko D H, Bhansali A S, Schwartzman A F and Ogawa S 1989 Inst. Phys. Conf. Ser. **100** 599
Sinclair R 1990 Mater. Trans. Jpn. Inst. Met. **31** 628

Inst. Phys. Conf. Ser. No 117: Section 5
Paper presented at Microsc. Semicond. Mater. Conf., Oxford, 25–28 March 1991

289

Microstructure of Endotaxial CoSi$_2$ layers formed by Co implantation into (100) and (111) Si substrates

C W T Bulle-Lieuwma, A H van Ommen, D E W Vandenhoudt and A F de Jong

Philips Research Labs., P.O. Box 80000, 5600 JA Eindhoven, The Netherlands

ABSTRACT: Hetero-epitaxial Si/CoSi$_2$/Si structures were synthesized by implanting Co ions into (100) and (111) substrates and subsequent annealing. The microstructure of both the as-implanted and annealed structures has been investigated by transmission electron microscopy (TEM), including high resolution TEM (HREM) combined with image simulations. Investigations by HREM revealed that after implantation, Co is present in the form of epitaxial CoSi$_2$ precipitates, which exhibit both the aligned (A-type) and twinned (B-type) orientation. For the highest doses a continuous layer of stoichiometric CoSi$_2$ is already formed during implantation. Annealing of the implants results in the formation of buried CoSi$_2$ layers of aligned orientation. For (100) Si, a domain-like structure is observed consisting of areas with different interface structures. It is found that these layers contain interfacial dislocations of edge-type with Burgers vectors $b=a/4<111>$ and $b=a/2<100>$, which are associated with boundaries separating the domains. For (111) Si it will be shown that the defect structure consists of dislocations of edge-type with Burgers vector $b=a/2<110>$.

1. INTRODUCTION

Hetero-epitaxial structures consisting of Si/CoSi$_2$/Si are of much interest due to their potential applications as metal-base and permeable-base transistors and as buried interconnects in three-dimensional device structures (Hensel et al. 1985). CoSi$_2$ is of special interest because of its low resistivity. Furthermore, CoSi$_2$ grows in a cubic CaF$_2$ structure with a close lattice match with Si ($\Delta a/a = 1.2$ %), and can therefore be grown epitaxially on silicon. A promising technique to form buried CoSi$_2$ layers, first proposed by White et al. 1987, is high-dose Co implantation followed by a high temperature anneal treatment. By this method single-crystalline buried CoSi$_2$ layers of aligned orientation can be formed in both (100) and (111) Si. In this paper, we shall consider the structural features of the CoSi$_2$ precipitates, e.g. its morphology, defect structure, and crystallographic orientation. Finally we shall consider the structural features of the buried layer.

2. EXPERIMENTAL

(100) and (111) oriented Si wafers, 4" in diameter, have been implanted with 170 keV ´Co$^+$ ions at an ion current density of 11 μA/cm^2, and to doses of 1-3x10^{17} Co$^+$ ions/cm^2. The implantation temperature was about 450 °C. Subsequently, the wafers were annealed in a furnace for 30' at a temperature of 1000 °C in a N$_2$/H$_2$ ambient. Characterization was performed using a Philips EM 400T microscope

operated at 120 kV; HREM was carried out at 250 kV with a Philips EM 430ST microscope with a spherical aberration constant C_s = 1.15 mm.

3. RESULTS

3.1 Morphology

Investigation of the as-implanted structures by cross-sectional HREM shows that for the lowest dose of 1×10^{17} ions/cm², the Co is present in the form of isolated $CoSi_2$ precipitates, which exhibit both the aligned (A-type) and twin-oriented (B-type) orientation (van Ommen et al. 1990, Bulle-Lieuwma et al. 1989a, 1991a). A large number of the aligned precipitates is strongly $<111>$-facetted, forming octahedral precipitates. This may be attributed to the different growth kinetics of the facets or to a low interface energy of the $<111>$ facets. The orientation of the B-type precipitates can be described as a rotation over 60° (or 180°) around $<111>$ axes. Most of the B-type precipitates have an elongated shape with the $<111>$ interfaces on which 'twinning' occurs on the long sides. The interface energy is minimized by adopting such an elongated shape. As the dose is increased to 2×10^{17} Co⁺ ions/cm², the density and size of the $CoSi_2$ precipitates increase significantly. In the area of maximum Co concentration, extensive coalescence of the $CoSi_2$ precipitates takes place. For both the (100) and (111) orientation, almost continuous layers of aligned orientation are formed during the implantation. Apparently, with increasing dose, the A-type precipitates become dominant and have coalesced into a buried layer. On both sides of the layer, A- and B-type $CoSi_2$ precipitates are distributed. For the higher dose of 3×10^{17} Co⁺ ions/cm² an approximately 90 nm thick monocrystalline $CoSi_2$ layer is formed during the implantation. The $CoSi_2$ extends almost completely to the surface.

3.2 Interface Structure

The interface structure of both the precipitates and buried layers has been investigated using HREM combined with image simulations. Optimal conditions were established to image the differences between the different interface models (five-eight,- or sevenfold model), referring to the coordination number of the interfacial Co atoms (Bulle et al. 1989b, de Jong et al. 1990).

In fig. 1 an experimental $<110>$ cross-sectional HREM image of an A-type $CoSi_2$ precipitate bounded by $<111>$ facets is shown. The insets are simulated HREM images of the seven-fold model for a sample thickness $t = 4.6$ nm and defocus value of $f = -60$ nm. The discrimination between the eight-fold (or five-fold) model and the seven-fold model can be done most reliably by considering the shifts of the (200) fringes across the $Si/CoSi_2$ (111) interface (de Jong et al 1990). For a defocus value of $f = -60$ nm for all thicknesses investigated (< 12 nm) and for $f = -50$ nm above a sample thickness of 4.6 nm, prominent (200) fringes are formed in the $CoSi_2$ with the low intensity at the position of the Si atom rows. For the seven-fold coordination the (200) fringes shift to the left approximately $1/4\ d_{200}$ when going from the Si to the $CoSi_2$, whereas for the five-fold and eight-fold coordination they shift to the right. In the experimental image of fig. 1 the (200) fringes shift to the left at all four $<111>$ facets, clearly demonstrating seven-fold coordination of the interfacial Co atoms.

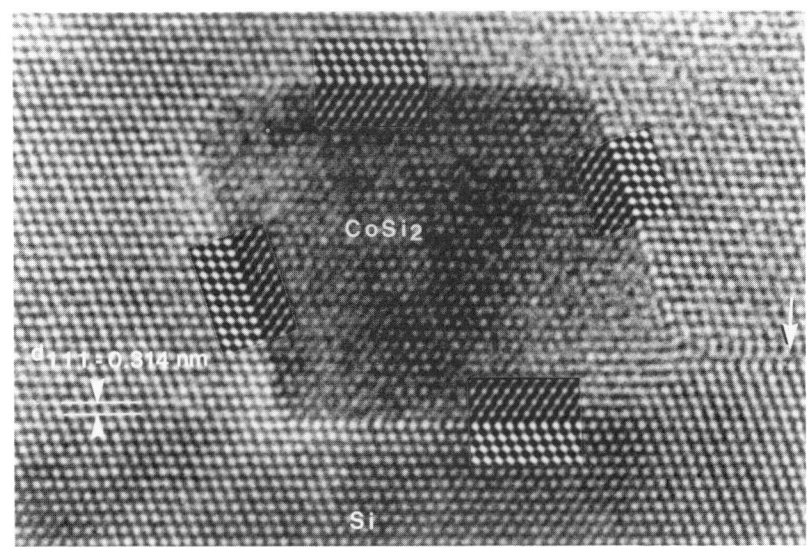

Fig. 1. < 110 > cross-sectional HREM micrograph of an aligned (A-type) CoSi₂ precipitate in Si (111). The insets are simulated images of the seven-fold model for t = 4.6 nm and f = -60 nm.

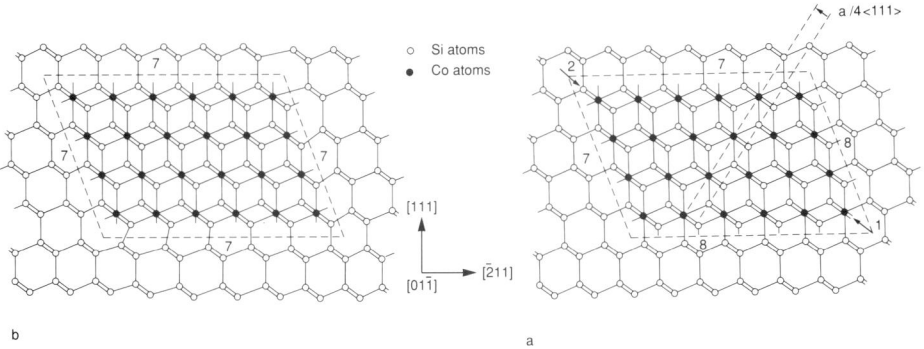

b   a

Fig.2 a) Structure model of the atomic structure of an aligned (A-type) CoSi₂ precipitate bounded by < 111 > facets, showing opposite interfaces with seven-fold coordinated and eight-fold coordinated Co atoms. The numbers 7,8 refer to the number of Si atoms surrounding the interfacial Co atoms.
b) Structure model of a fully seven-fold coordinated CoSi₂ precipitate.

From a geometrical point of view, the most straightforward coordination would be as shown in the structure model of fig. 2a of an aligned CoSi₂ precipitate in a Si matrix projected along < 110 >. A consequence of such a match is that op-posite pairs of (111) interfaces have seven-fold coordinated and eight-fold coordinated Co atoms (see also Hull et al. 1990). However, it is possible to convert the interfacial structure of the CoSi₂ precipitates into fully seven-fold coordinated or fully eight-fold coordinated interfaces. For example, fully seven-fold coordinated interfaces can be obtained by applying a lattice displacement of a/4 < 111 > as

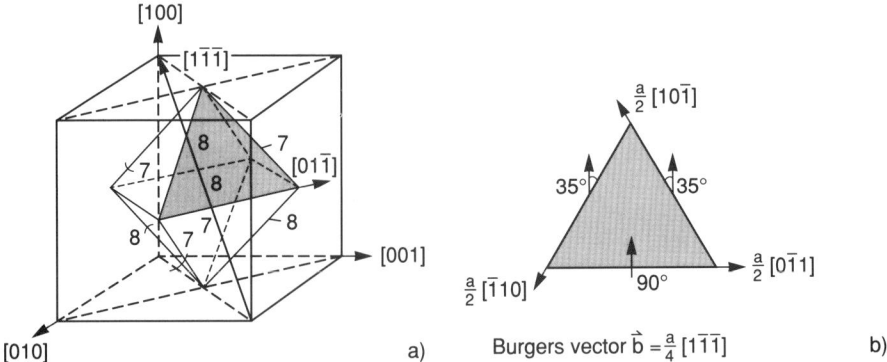

Fig. 3. a) Schematic presentation of an octahedral CoSi$_2$ precipitate b) One of the eight-fold coordinated {111} planes has been converted into a seven-fold coordinated {111} interface. The surrounding dislocation has Burgers vector **b** = a/4 < 111 >.

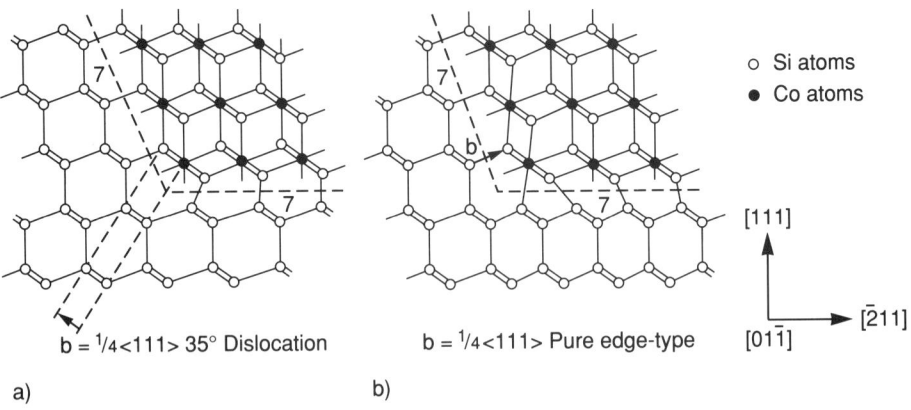

Fig. 4 Structure models projected along [01$\bar{1}$] of an a) a/4 < 111 > 35 ° and a b) 90 ° dislocation.

indicated in fig. 2a (see arrow 1). An extra half-plane of Si atoms is introduced in the CoSi$_2$ precipitate as is shown in the structure model of fig. 2b. A fully eight-fold coordinated interface is obtained by the removal of similar rows of Si atoms. It is envisaged that the addition of layers of atoms is more easily accomplished since the lattice parameter a of CoSi$_2$ is smaller than that of Si. Reality may be more complicated as we have not two-dimensional but three-dimensional CoSi$_2$ precipitates. In fig. 3a, a three-dimensional schematic of an octahedral CoSi$_2'$ precipitate is shown, revealing four equivalent {111} planes with the interfacial Co atoms seven-fold coordinated and four equivalent {111} planes with the Co atoms eight-fold coordinated. One simple way to convert all four eight-fold coordinated interfaces into seven-fold coordinated interfaces requires the introduction of four partial dislocations with Burgers vector b = a/4 < 111 >. This can be made plausible by considering one of the eight-fold coordinated {111} planes of the

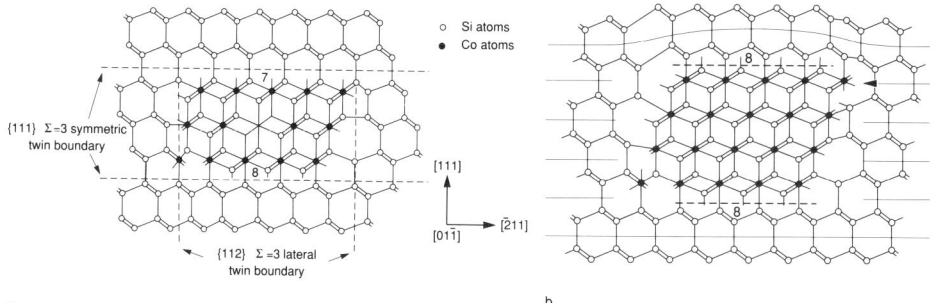

Fig. 5 a) Structure model of the atomic structure of a twinned (B-type) CoSi₂ precipitate bounded by opposite seven-fold and eight-fold coordinated Si/CoSi₂ (111) interfaces. b) B-type CoSi₂ precipitate bounded by eight-fold coordinated interfaces. The numbers 7,8 refer to the number of Si atoms surrounding the interfacial Co atoms. The extra layer in the CoSi₂ precipitate is marked by an arrow.

precipitate (see fig. 3b). By converting this plane into a seven-fold coordinated plane, the (111) plane is surrounded by a dislocation with constant Burgers vector $b = a/4 < 111 >$ in such a way that two sides of the (111) plane are 35° dislocations (right-handed and left-handed) and one side a 90° dislocation. Structure models of both types of dislocations projected along $[01\bar{1}]$ are shown in fig. 4. A net normal displacement of $a/12 < 111 >$ is produced towards the CoSi₂. We would like to point out that whereas the description given above is relatively simple, the actual situation may be much more complicated.

Also in the case of twinned B-type CoSi₂ precipitates, by straightforward incorporation of 3n monolayers of (1$\bar{1}$1) CoSi₂, the perfect stacking sequence is only maintained if the opposite Si/CoSi₂ (111) interfaces have different bonding (seven-fold and eight-fold (or five-fold) coordination of the interfacial Co atoms. The atomic model of such a structure projected on the $[01\bar{1}]$ plane is shown in fig. 5a. In order to have fully seven-fold or fully eight-fold coordinated interfaces at opposite sides, the generation of stacking faults is required in the Si matrix or in the CoSi₂ precipitate. It again gives rise to a large variety of structural options. The case of fig. 5b is still relatively simple. There we obtained eight-fold coordination on both (111) interfaces by the insertion of an extra layer of Si-Co-Si triplets. Removal of such layer would result in seven-fold coordination on both sides. By investigation of the Si/CoSi₂ (111) B interface structure by HREM and image simulations we found evidence that both (111) interfaces of the endotaxial B-type CoSi₂ precipitates in Si are indeed eight-fold coordinated. In fig. 6a, an < 110 > cross-sectional HREM image of a B-type CoSi₂ precipitate is shown. The insets on opposite (111) sides are simulated images of the eight-fold model for sample thickness t = 6.1 nm and f = -90 nm. At this defocus value the bright contrast is at the Co atoms of the Si-Co-Si triplets in the CoSi₂ and at the midpoints of the Si-Si dumbbells in the Si. At f = -90 nm the number of bright {111} fringes in the experimental image of fig. 6a equals the number of monolayers consisting of Si-Co-Si triplets. The B-type precipitate

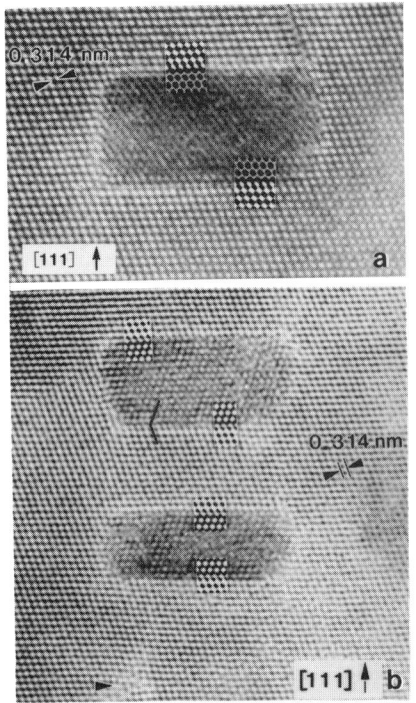

Fig. 6 a) HREM image of a B-type CoSi$_2$ precipitate taken at f = -90 nm and t = 6.1 nm. b) HREM image of B-type CoSi$_2$ precipitates taken at f = -50 nm and t = 3.1 nm.

consists of $17 = (3n + 2)$ monolayers, while the surrounding Si consists of $(3n + 1)$ monolayers of Si-Si dumbbells. The B-type precipitates in the HREM image of fig. 6b are taken at a defocus value of f = -50 nm. At this defocus value and for t < 4.6 nm the bright contrast is at the tunnel positions in both the Si and CoSi$_2$ lattice. The B-type precipitate (upper side) consists of $14 = (3n + 2)$ monolayers. The Si matrix on the left side of the precipitate consists of $(3n + 2)$ monolayers, whereas on the right side the Si matrix consists of $(3n + 1)$ monolayers. The B-type precipitate at the lower side consists of $10 = (3n + 1)$ monolayers, whereas the Si matrix have $(3n + 1)$ monolayers on the right side and 3n monolayers at the left side. It is likely that the extra planes (dislocations or stacking faults) may be introduced in the Si lattice by maintaining both the eight-fold coordination of B-type precipitates and seven-fold coordination of A-type precipitates. It appears that the Si matrix is significantly distorted in this region.

Summarizing, we obtained conclusive evidence for seven-fold coordination of the A-type precipitates and eight-fold coordination for the B-type precipitates. Furthermore, it is worth noting that the extra planes in both the octahedral A-type and elongated B-type CoSi$_2$ precipitates indicates that small precipitates must be highly strained.

3.3 Defect Structure of the Annealed (100) and (111) Layers

(100) Orientation

We have studied the defect structure in plan-view using the weak beam method (Bulle et al. 1991b). The 1.2 % lattice mismatch is accomodated by interfacial dislocations of edge-type with Burgers vector $b = a/4 < 111 >$ running in $< 110 >$ directions and Burgers vector $b = a/2 < 100 >$ running in $< 100 >$ directions. In fig. 7 two micrographs are shown of the same area of an annealed Si (100) wafer implanted with 2×10^{17} Co$^+$ /cm^2 taken with diffraction vectors g = (088) and g = (0$\bar{8}$8) respectively. Interface dislocations appear as lines of contrast lying parallel to the [0$\bar{1}$1] and [011] directions. In fig. 7a, the dislocations with lines of contrast parallel to [0$\bar{1}$1] are out of contrast in fig. 7b for g = (0$\bar{8}$8). In addition, another type of interface dislocations of edge-type is visible with lines of contrast parallel to the [001] and [010] directions having an angle of 45 ° with the $< 110 >$ directions. These dislocations have Burgers vector $b = a/2 < 100 >$.

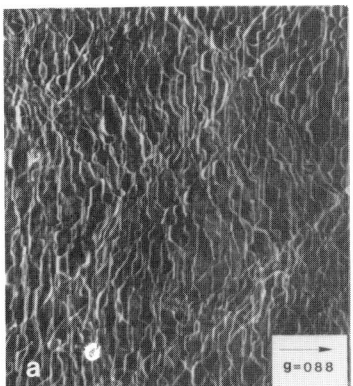
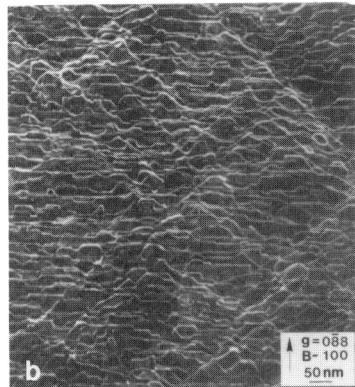

Fig. 7. Plan-view weak beam images of a Si(100) wafer implanted with 2×10^{17} Co$^+$/cm^2 and annealed for 30 minutes at 1000 °C. a) diffraction vector g = (088); b) diffraction vector g = (0$\bar{8}$8)

By HREM in cross-sectional view it is shown that a domain-like structure exists in which the domains have different interface structures; a 2x1 reconstructed interface structure involving a difference in composition and an unreconstructed interface structure, both occurring in two orthogonal domains (Bulle-Lieuwma et al. 1990, 1991b). The boundaries between domains of the same structure are associated with dislocations of edge-type with Burgers vector b = a/4 < 111 >. Domains of different structures are joined by dislocations with Burgers vectors b = a/2 < 100 > and b = a/4 < 111 >.

(111) Orientation

In fig. 8a, a g = (0$\bar{8}$8) weak beam image is shown of a Si (111) substrate implanted with 2×10^{17} Co$^+$ ions/cm^2 and annealed for 30 minutes at 1000 °C. The lattice mismatch is accomodated by interfacial dislocations of edge-type with Burgers vector b = a/2 < 110 >. Interface dislocations appear with lines of contrast parallel to two of the three < 110 > directions lying in the (111) interfacial plane; along [0$\bar{1}$1], [10$\bar{1}$] and [$\bar{1}$10] . The dislocations with lines parallel to [0$\bar{1}$1] are out of contrast.

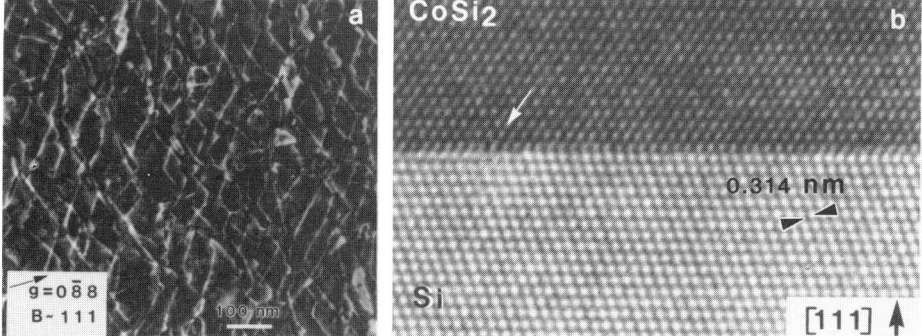

Fig. 8 a) A g = (0$\bar{8}$8) plan-view weak-beam image of a Si (111) wafer implanted with a dose of 2×10^{17} Co$^+$/cm^2 at an energy of 170 keV and annealed for 30 minutes at 1000 °C. b) HREM image of the Si/CoSi$_2$ (111) A interface, revealing a dislocation with Burgers vector b = a/2 < 110 >.

By HREM imaging in cross-sectional view combined with image simulations we found evidence for a seven-fold coordinated Si/CoSi$_2$ (111) A interface. It must be noted that there exists a clear difference between the defect structure of the buried layer and the defect structure of an aligned precipitate, although both exhibit the same seven-fold coordinated structure. The precipitate is associated with $a/4 < 111 >$ dislocations, which are absent in the layer. Only a boundary between sevenfold or eight (or five)-fold coordinated interfaces may be associated with a dislocation with Burgers vector $b = a/4 < 111 >$ inclined to the interface. Such dislocations are associated with interfacial steps. However, we did not found evidence for such a structure. In fig. 8b an $< 110 >$ cross-sectional HREM image of a Si/CoSi$_2$ (111) A interface is shown, revealing an edge-type dislocation with Burgers vector $b = a/2 < 110 >$. The interfacial Co atoms are seven-fold coordinated.

4. CONCLUSIONS

Microstructural TEM analyses of Si implanted with a high dose of Co ions revealed that the Co is present in the form of aligned and twinned CoSi$_2$ precipitates. In small CoSi$_2$ precipitates strain is brought about by the required insertion of additional lattice planes in the precipitates in order to obtain the preferred coordination at the relevant facets. Insertion of lattice planes is more likely than removal, because the lattice parameter of CoSi$_2$ is smaller than that of Si. There is an important difference between the dislocation structure in (100) and (111) Si. In the case of an (100) substrate orientation, the dislocation structure is coupled to the Si/CoSi$_2$ interface structure. There is an unreconstructed and a 2x1 reconstructed interface structure, both occurring in two orthogonal domains. The boundaries between domains of the same structure are associated with edge-type dislocations with Burgers vector $b = a/4 < 111 >$. Domains of different structures are joined by dislocations with Burgers vectors $b = a/2 < 100 >$ and $b = a/4 < 111 >$. In the case of an (111) oriented Si substrate, we found a triangular network of edge-type dislocations with Burgers vector $b = a/2 < 110 >$ lying in the interfacial plane. These dislocations are not associated with domains of the interface structure. Instead, a single interface structure was found with the Co atoms seven-fold coordinated.

REFERENCES

Bulle-Lieuwma C W T, van Ommen A H, van Ijzendoorn 1989a, Appl. Phys. Lett. 54 244
Bulle-Lieuwma C W T, de Jong A F, van Ommen A H, van der Veen J F, Vrijmoeth J 1989b, Appl. Phys. Lett. 55(7) 648
Bulle-Lieuwma C W T, de Jong A F and Vandenhoudt D E W, 1990 Proc. XII Int Conf for Electron Microscopy 1990 Seattle, 4 344
Bulle-Lieuwma C W T, van Ommen A H, Vandenhoudt D E W, Ottenheim J J M and de Jong A F 1991a (submitted to J. Appl. Phys.)
Bulle-Lieuwma C W T, de Jong A F and Vandenhoudt D E W 1991b, Phil. Mag. A, in press
Hensel J C, Levi A F J, Tung R T and Gibson J M 1985 Appl. Phys. Lett. 47 151
Hull R, Hsieh Y F, Short K T, White A E and Cherns D, 1990 Mat. Res. Soc. Symp. Proc. 183 91
De Jong A F and Bulle-Lieuwma C W T 1990, Phil. Mag. A, 62 183
Van Ommen A H, Bulle-Lieuwma C W T, Ottenheim J J M and Theunissen A M T 1990, J. Appl. Phys. 67 1767
White A E, Short K T, Dynes R C, Garno J P and Gibson J M 1987 Appl. Phys. Lett. 50(2) 95

Inst. Phys. Conf. Ser. No 117: Section 5
Paper presented at Microsc. Semicond. Mater. Conf., Oxford, 25–28 March 1991

297

HREM of TiSi$_2$/Si and TiSi$_2$/SiO$_2$ interfaces

A L Vasiliev, N A Kiselev, O I Lebedev, E V Orlova, A G Vasiliev[*] and A A Orlikovsky[*].

Institute of Crystallography, USSR Academy of Sciences
Leninsky pr.59, 117333 Moscow USSR

[*]Institute of Physics and Technology, USSR Academy of Sciences
Krasikova str.25a, 117218 Moscow USSR

ABSTRACT: HREM investigation of TiSi$_2$ on (100) Si, (111)Si and SiO$_2$ interface structure was carried out. Differences in interface structure depending on surface treatment and misfit accommodation ways were defined.

1. INTRODUCTION

Titanium disilicides having the lowest sheet resistance are investigated as promising materials for interconnections in VLSI technology. Sinclair et al (1989) and Catana et al (1990) thoroughly investigated TiSi$_2$/Si interfaces for films obtained by the most common and easy method – annealing of metal film on substrate. Electron-beam coevaporation is promising for obtaining epitaxial films with perfect interfaces (Beyers et al 1984). However, the influence of surface, near-surface defects and impurities has not been fully defined. HREM investigations of interface structure of thin TiSi$_2$ films, grown by electron beam coevaporation in high vacuum on (111) Si, (001) Si and SiO$_2$ are reported.

2. EXPERIMENT

TiSi$_2$ film formation was carried out in a high vacuum chamber with an oil-free pumping system at p=10^{-7} Pa. Films were deposited on (111) Si, (001)Si and SiO$_2$.

Before deposition the substrate surface was cleaned by Ar$^+$ ions with different energies and different incidence angles, with subsequent annealing in the same chamber at temperature T$_{ac}$=750-800°C as follows:

A). E$_{Ar}^+$ < 1 keV normal to the Si surface, T$_{ac}$=750°C;

B). E$_{Ar}^+$ < 1 keV normal to the Si surface, T$_{ac}$=800°C;

C). E$_{Ar}^+$ = 1 keV at grazing angles (5°-10°) to the Si surface, T$_{ac}$=750°C;

D). E$_{Ar}^+$=1.5-2 keV at grazing angles, T$_{ac}$=750°C.

TiSi$_2$ films were formed by Ti and Si coevaporation at substrate temperature T$_s$=300-750°C, deposition rate (V$_{si}$ and V$_{Ti}$) V$_s$/V$_{Ti}$ =(0.04-0.4)/(0.02-0.2) nm/s with further annealing in some cases in the same chamber at T$_{as}$=

=500-800°C. Plan-view specimens and cross-sections were prepared for EM. Some specimens were cooled by LN$_2$ during ion thinning. Investigations were carried out on a Philips EM-400 and a Philips EM-430 ST. Images were processed using band-pass Fourier filtering on an IBAS-2000.

3. RESULTS AND DISCUSSIONS

Plan-view investigations showed that structure, grain dimensions and defect concentrations in the films are highly dependent on film formation conditions. Formation of low-resistance C-54 phase starts at temperature below 600°C.

Cross-section investigations showed that the film-substrate interface structure differs considerably for different substrate cleaning modes.

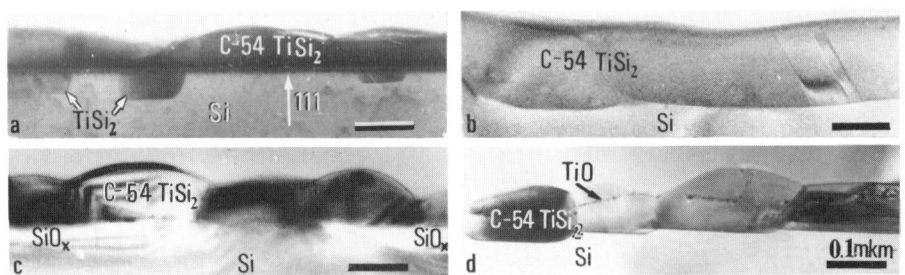

Fig.1 Cross-sectional transmission electron micrographs of TiSi$_2$ films deposited on (111)Si substrates after different treatment: (a) A-type; (b) B-type; (c) C-type; (d) D-type. TiO precipitates are arrowed.

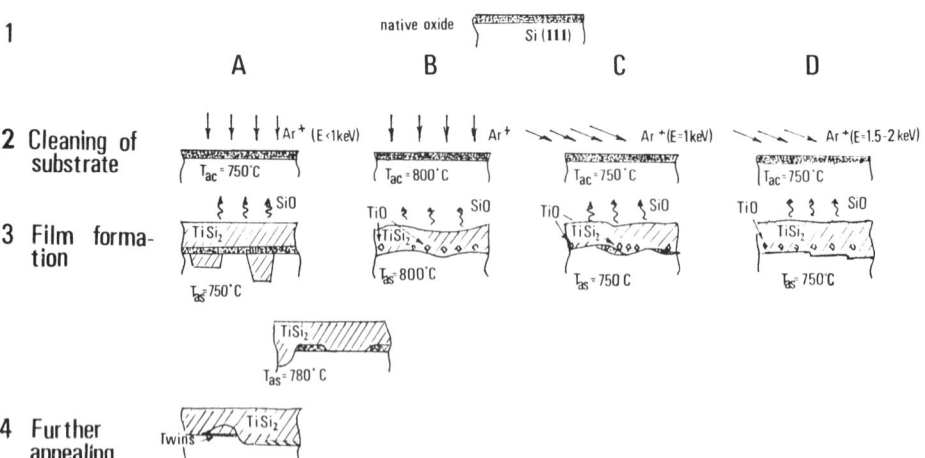

Fig.2 A schematic representation of TiSi$_2$/(111)Si interface formation on substrates after different treatment.

In A-type specimens (Fig.1a) an intermediate amorphous layer 3-6 nm thick was observed on the interface. This amorphous layer is probably an incompletely removed native oxide film. RBS investigations indicated a high concentration of oxygen in that region. Microinclusions, defined by electron and optical diffraction (OD) pattern analysis as crystalline grains of the C-54 and C-49 $TiSi_2$, were observed below the amorphous layer (Vasiliev et al 1988). Their bottom part is flat and parallel to the (111) Si substrate (Fig.2). In specimens annealed at T_{as} =750°C-780°C the amorphous layer looses its continuity, the $TiSi_2$ film forms an interface without an amorphous layer, sometimes penetrating into the silicon along planes parallel to (111) Si (Fig.2). Annealing of specimens with a continuous amorphous layer at T_a=1050°C during 30 minutes resulted in substrate erosion. The amorphous layer on the $TiSi_2$/Si C-54 phase interface disappears while stacking faults and microtwins on the Si substrate as well as Si hillocks are formed. We conclude that if the amorphous layer is not removed Ti atoms diffuse through the amorphous layer along defects into the substrate, forming Ti enriched "pockets", transformed later into $TiSi_2$ grains. Raising of the annealing temperature results in further growth of "pockets"and their coalescence with the $TiSi_2$ film.

In B-type specimens (Fig.1b) (T_{as}=800°C) the interface is sharp, but wavy. The amorphous layer is absent.Microtwins and crystalline precipitates are observed in grains (Fig.2).

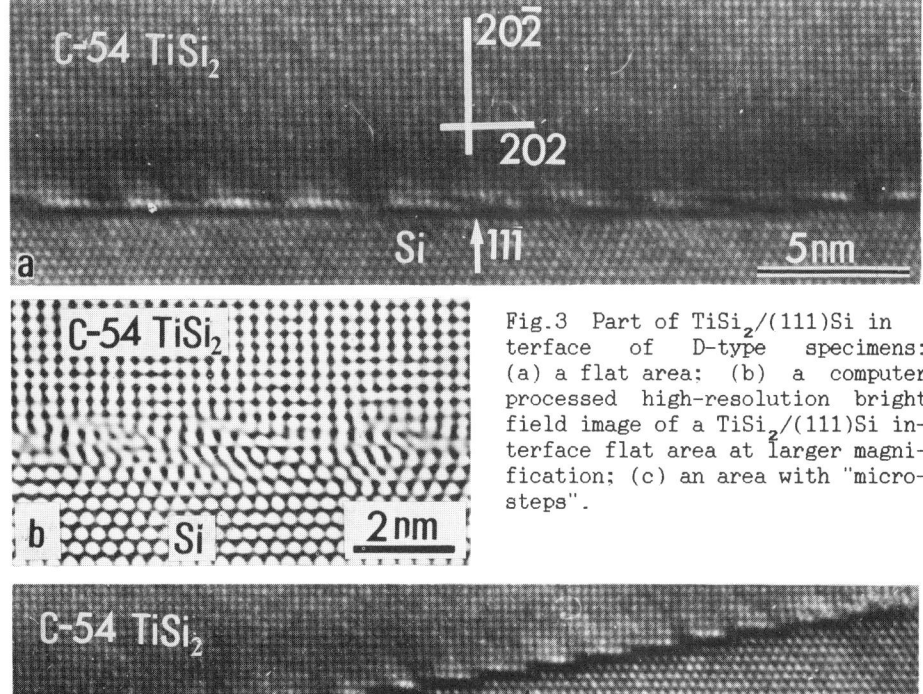

Fig.3 Part of $TiSi_2$/(111)Si interface of D-type specimens: (a) a flat area; (b) a computer processed high-resolution bright field image of a $TiSi_2$/(111)Si interface flat area at larger magnification; (c) an area with "microsteps".

In C-type specimens Fig.1c the interface is uneven. In some areas epitaxial C-54 $TiSi_2$ film growth is observed. The amorphous layer due to the small Ar^+ ion energy is still preserved on broad areas and varies in thickness (Fig.2). Complete absence of the amorphous layer was seen in D-type specimens. Fig.1d shows a $TiSi_2$/(111) Si interface. In some areas steps up to 50 nm high are observed. Their appearance is connected with a 4° misorientation between the Si substrate surface and the (111) plane (technological inclination). Diffraction pattern analysis from flat areas of the interface shows that: (1). (101) C-54 $TiSi_2$// (111) Si, [010] $TiSi_2$ // [110] Si. Investigations revealed in addition the following epitaxial relationships: (2) (515) C-54 $TiSi_2$//(111) Si, [101] $TiSi_2$//[110] Si; (3) (001) C-54 $TiSi_2$//(011) Si. For orientation (1) contrast variations with about 3 nm periodicity are observed (Fig.3a). Inside these areas linkage of the $TiSi_2$ and Si crystalline planes is visible for 3-4 planes only. This correlation is violated in other areas. Thus, lattice mismatch up to 10.7% in this orientation is compensated by defects adjoined to the interface and not penetrating into the $TiSi_2$ lattice (Fig.3b). The area adjacent to a step (Fig.3c) is seen to be formed from monoatomic steps, each one being oriented parallel to {111} Si planes, most stable during Si interaction with $TiSi_2$. If the interface is inclined towards (111) by more than approximately 20°, the interface can be flat and parallel to {551}, not consisting of monoatomic steps.

On some areas of the interface regions of darker contrast are observed in the substrate resembling "pockets" (Fig.4a). Their appearance may be connected with higher Ti atom concentration. HREM image processing revealed microdefects in these "pocket" areas (Fig.4b).

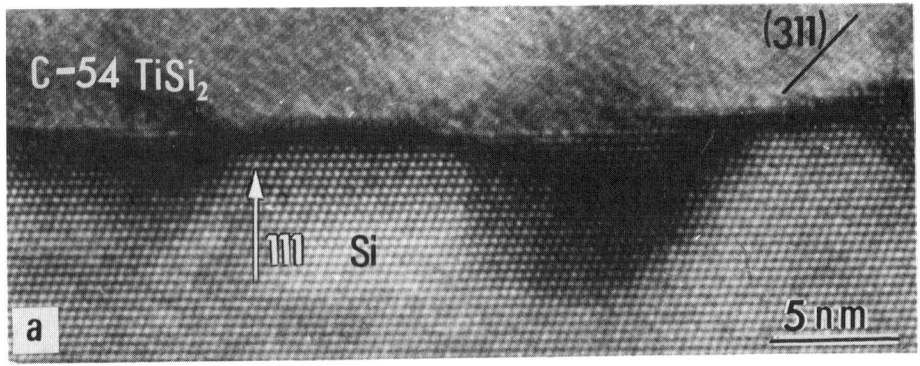

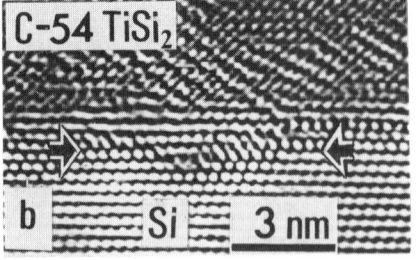

Fig.4 A high-resolution image of the $TiSi_2$/(111)Si interface in the "pockets" area: (a) lattice bright field image; (b) computer processed image, defects in the substrate are arrowed.

Microtwins and crystalline precipitates (Fig.5a) were observed in C-54 TiSi$_2$ films. OD pattern analysis (Fig.5b) and interplane distance measurements indicate that these precipitates are crystalline cubic titanium oxide TiO with lattice parameter: a=0.4177nm ASTM N°8-117.

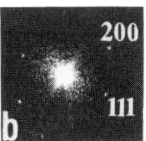

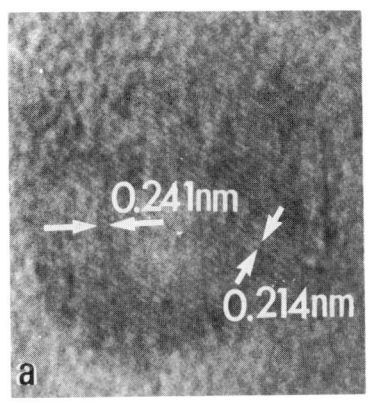

Fig.5 TiO precipitate: (a) high-resolution bright field image; (b) OD pattern.

This effect indicates that the TiSi$_2$/Si interface shifts into the Si substrate. Ti penetrates into the substrate along the near-surface defects. This is confirmed by formation of "pockets" with high Ti concentration and TiSi$_2$ grains. An excess of Ti in the film is thought to be due to partial evaporation of Si from the surface in the form of SiO during film formation and further annealing. This effect was observed by Cohen et al (1987). Early phase TiSi$_2$ film growth studies showed that layer by layer growth of TiSi$_2$ film occurs and results in oxygen diffusion directly towards the film surface. Early phase TiSi$_2$ film growth studies showed that C-54 TiSi$_2$ crystalline grains are formed immediately after deposition of the Ti+Si mixture and the film grows layer by layer. Part of the oxygen from the substrate near the surface interacts with Ti, forming a crystalline titanium oxide TiO. These precipitates in the TiSi$_2$ film mark the initial interface position. Further annealing of films at a temperature of 1050°C results in the interface shifting towards the substrate until a certain amount of oxygen remains (allowing SiO evaporation) in the film and the near-surface region of the Si substrate. After that erosion of the Si surface takes place. Hillocks and microtwins appear at that time.

The TiSi$_2$/(001)Si interface structure is much more wavy, compared with the TiSi$_2$/(111) Si interface, which is attributed to a lower chemical stability of the (001) planes. Nevertheless, flat regions were observed (Fig.6). The interface is strongly faceted. The facet period equals the misfit accommodation period (~3 nm). As in the case with TiSi$_2$/(111)Si the dislocations do not penetrate into the film. Lattice mismatch in this case is ~7%. The TiSi$_2$/SiO$_2$ interface is uneven, and becomes more wavy with increasing substrate temperature. This is explained by film interaction with silicon dioxide. The film and interface structure also depend on the SiO$_2$ layer thickness. The film and substrate begin to interact if the SiO$_2$ layer thickness is below some critical value.

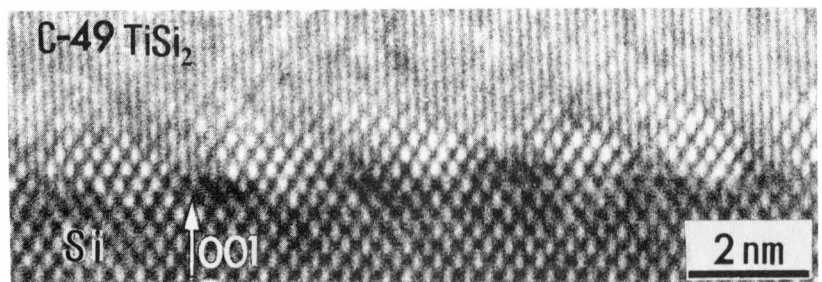

Fig.6 A high-resolution bright field image of the C-49 TiSi$_2$/(001)Si in
terface

4. CONCLUSION

1). HREM of cross-sectional TiSi$_2$/Si specimens obtained by coevaporation
of Ti and Si in high vacuum showed that the interface structure depends on
the Si substrate surface treatment. The best results(epitaxial films with
flat interface) were obtained when Si surface was treated with 2 keV Ar$^+$
ions at grazing angles with subsequent annealing at 750°C.
2). The SiO$_2$ layer works as a buffer only if its thickness exceeds some
critical value.
3). The TiSi$_2$/(111)Si interface moves towards the substrate via "pocket"
formation from excess Ti atoms, transformed later into crystalline TiSi$_2$.
Crystalline TiO precipitate formation is possible in the place of the initial interface.
4). The substrate reacts more slowly along planes parallel to (111)Si. A
4° inclination of the Si wafer surface relative to the (111) plane leads
to step-like structure formation with "micro-" and "mini-" steps.
5). The misfit compensation mechanism is the same for TiSi$_2$/(111) Si and
TiSi$_2$/(001)Si. This mechanism is connected with formation of a dislocation
net in the interface, which does not penetrate into the film.

ACNOWLEDGEMENT

The authors would like to thank V.Globenko for making this paper comprehensible in English.

REFERENCES

Beyers R, Sinclair R, Thomas M E 1984 Mat.Res.Soc.Symp.Proc. 25 pp 601-5
Catana A, Shmidt P E, Heintze M, Levy F, Stadelmann P, Bounet R 1990
    J.Appl. Phys. 67 4 pp 1820-5
Cohen C, Nipoti R, Siejka J, Bentini G G, Berti M, Drigo A V 1987
    J.Appl. Phys. 61 11 pp 5187-9
Sinclair R, Holloway K, Kim K B, Ko D H, Bhansali A S, Schwartzmann A F,
    Ogawa S 1989 Inst.Phys.Conf. Ser.No.100 pp 599-607
Vasiliev A L, Kiselev N A, Vasiliev A G, Orlikovsky A A 1988
    Izv.Acad.Nauk SSSR Ser.Phyzitcheskaya 27 No.7 pp 1288-91 (in Russian)

Inst. Phys. Conf. Ser. No 117: Section 5
Paper presented at Microsc. Semicond. Mater. Conf., Oxford, 25–28 March 1991

303

TEM study of sputter deposited titanium films on silicon and the silicidation reaction during rapid thermal processing

A De Veirman, R Hokke, E Swart and R Wolters

Philips Research Laboratories, P.O.Box 80000, 5600 JA Eindhoven, The Netherlands

ABSTRACT : The microstructure of sputter deposited Ti films on Si is studied. The Ti/Si interface and the texture of the Ti films will be addressed. The C54 $TiSi_2$ layers, formed from C49 after selective etching and subsequent rapid thermal processing, are investigated. The annealing in a nitrogen ambient results in the formation of TiN precipitates in the surface region of the C54 layers.

1. INTRODUCTION

The refractory-metal silicide $TiSi_2$ is used as contact metal in Very Large Scale Integrated (VLSI) circuits. $TiSi_2$ has a low resistivity and can be formed at relatively low temperatures. $TiSi_2$ layers can be prepared by either the deposition of Ti on a Si substrate or the co-sputtering of Ti and Si, followed by an annealing treatment. Annealing at temperatures up to 700°C results in the formation of $TiSi_2$ films with the metastable C49 structure. Annealing at temperatures above $\approx$ 800°C is required to obtain the low resistivity C54 $TiSi_2$ phase (e.g. Beyers and Sinclair 1985).
In the present paper sputter deposited Ti films on Si and the subsequent formation of $TiSi_2$ layers by rapid thermal processing are studied by Transmission Electron Microscopy (TEM).

2. EXPERIMENTAL

A 40 nm thick Ti layer was sputter deposited on a (001) Si substrate. Prior to the Ti deposition, the Si wafers were either cleaned by a 1% HF dip or by an in situ soft Ar sputter etch.
Annealing at 700°C for 30 sec was used to form the C49 $TiSi_2$ phase, selective etching of the Ti(N) top layer and subsequent rapid thermal annealing (RTA) at 850°C for 10 sec yields the C54 $TiSi_2$ phase. The annealing treatments were performed in a N_2 ambient.
The Transmission Electron Microscopy (TEM) investigations are performed in a Philips EM400 microscope operating at 120 kV and a Philips CM30 High Resolution Electron Microscope (HREM) at 300 kV. The latter microscope is equipped with an Energy Dispersive X-ray (EDX) detector.

3. RESULTS AND DISCUSSION

3.1. As-deposited structures

In the as-deposited samples (fig.1,2) a 1.5-2 nm thick amorphous layer is observed at the Ti layer / Si substrate interface. The fact that the formation of this layer is independent of the cleaning procedure indicates that it is due to chemical intermixing of Ti and Si and that it is not related to the native oxide layer. Amorphous Ti/Si interface layers of about the same thickness have been reported by Wang and Chen (1991) and Ogawa et al. (1990). The formation of an amorphous layer due to interdiffusion also occurs for different metal-silicon interfaces (Sinclair et al. 1989).

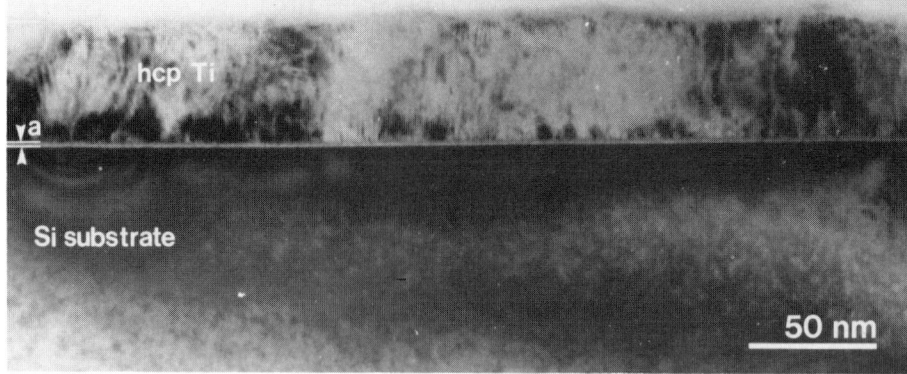

Fig.1 XTEM image of as-deposited sample, which was sputter etched prior to Ti deposition. The amorphous layer (a) at the Ti layer / Si substrate interface is clearly visible.

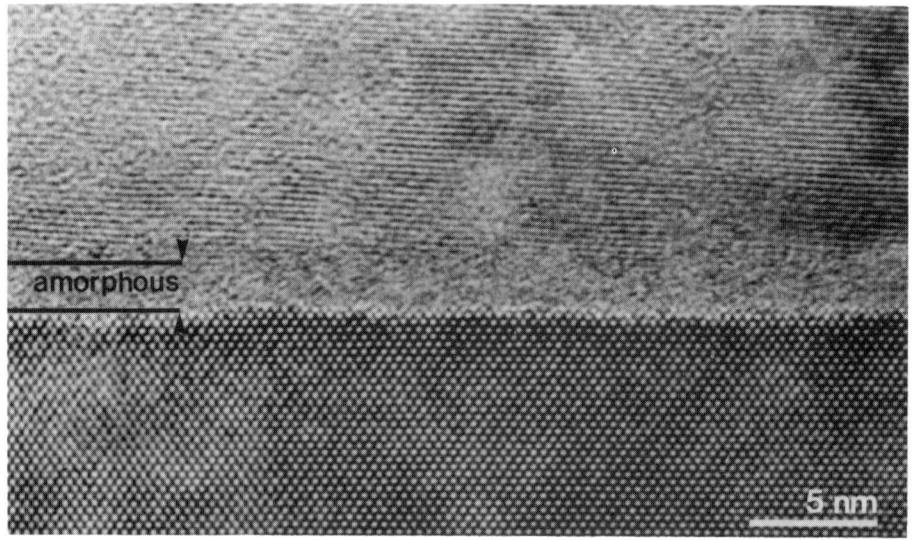

Fig.2 HREM image of Ti/Si interface region, clearly showing the amorphous interface layer. This sample was cleaned by sputter etch prior to Ti deposition.

The Ti layers have a columnar structure with grain sizes between 10-40 nm, as measured from plan-view observations (fig.3). X-Ray Diffraction (XRD) measurements have revealed a difference in texture between the samples which received a different cleaning procedure. The XRD spectrum of the sample HF cleaned before Ti deposition showed an intense 002 peak and a small 010 peak, while in the spectrum of the sputter etched sample, besides the 002 peak, also a considerable 011 peak was present. Plan-view TEM observations, together with the corresponding selected area diffraction (SAD) patterns confirmed the [001] texture, but did not reveal the above mentioned difference (this must be due to the limited sample area analysed by TEM compared to XRD). In both cases the SAD patterns consisted of rings, which corresponded to interplanar distances of 2.56 Å, 1.48 Å and 1.28 Å, being respectively d_{100}, d_{110} and d_{200} of the close packed hexagonal Ti structure (a=2.95Å and c=4.686 Å / ASTM card No. 5-0682). The absence of the intermediate 002, 011 and 012 rings clearly shows a [001] texture for the Ti films for both samples. The [001] texture is very likely for the close packed hexagonal structure, since the close packed (001) planes have the lowest surface energy (Swalin 1972).

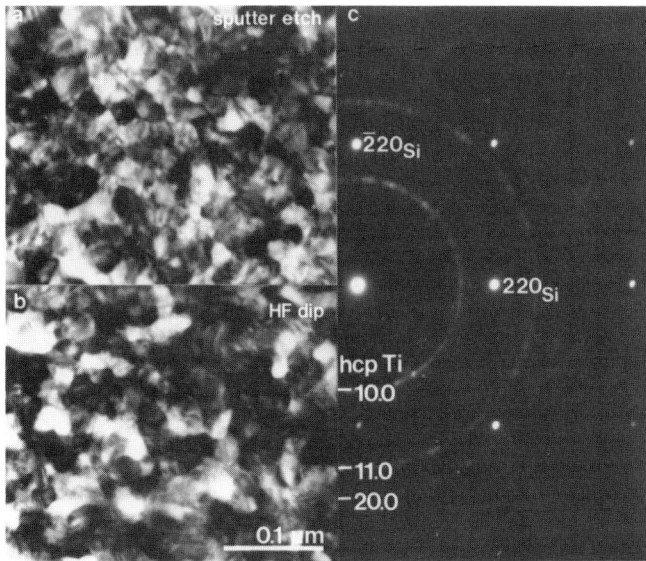

Fig.3 Plan-view image of as-deposited Ti layer.
a. sputter etch
b. HF dip
c. SAD pattern

Only limited information is available in the literature concerning the texture of the Ti film and clear experimental evidence is often not presented. Wang and Chen (1991) mentioned a strong texture with the (010) planes parallel to the surface, but they did not provide experimental arguments. In a study of amorphous Si and crystalline Ti multilayers Holloway and Sinclair (1987) derived the same preferred orientation, but this is not in agreement with the SAD pattern which they show. Brasen et al. (1986) studied similar Ti-Si superlattices and reported a [001] texture. Since no general agreement is obtained, work is continuing to obtain more insight in the lattice orientation of Ti films, both on crystalline and amorphous layers.

3.2. The resulting C54 TiSi$_2$ layer

RTA at 700°C for 30 sec, selective wet etching of the Ti(N) top layer and RTA at 850°C for 10 sec yields a layer of C54 TiSi$_2$. The surface region of the TiSi$_2$ layer contains crystalline precipitates (P) as seen in fig.4.

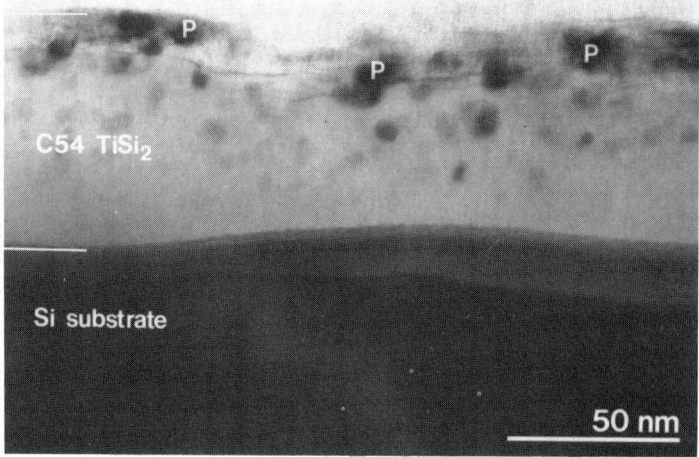

Fig.4 XTEM of C54 TiSi$_2$ layer . Near the surface crystalline precipitates (P) occur.

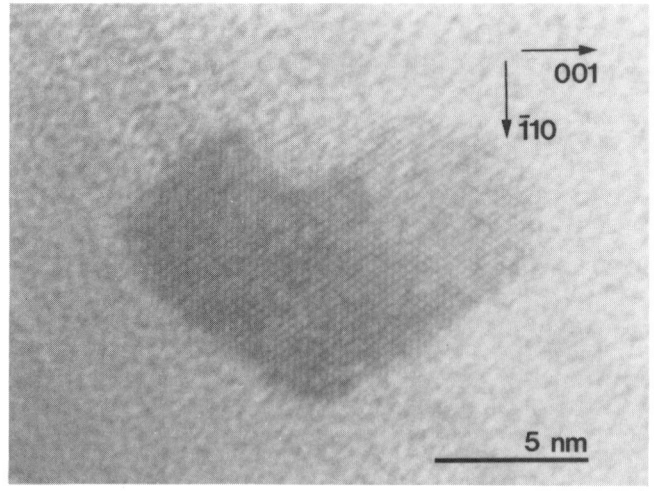

001

$\bar{1}$10

5 nm

Fig.5 HREM image of TiN precipitate, observed along the [110] zone axis.

EDX measurements in nanoprobe mode ($\phi \approx 10$ nm) performed in respectively the bulk of the TiSi$_2$ layer and close to the surface showed an increased Ti/Si concentration ratio in the surface area where the precipitates are present. This EDX result indicates that the precipitates consist of a Ti-rich phase. From HREM images the TiN is deduced to be a likely candidate, which is further shown by comparison with computer simulated images. TiN has the NaCl structure with a=4.24 Å (ASTM card No. 6-0642).

The formation of TiN particles near and at the surface of the TiSi$_2$ layer is plausible since the annealing occurred in a nitrogen ambient. Beyers et al. (1984) showed that TiN is the stable nitride in contact with TiSi$_2$ in the ternary Ti-N-Si system. The formation of surface TiN layers on TiSi$_2$ have been previously reported to occur on annealing TiSi$_2$ layers in N$_2$ (Wittmer 1988) and in ammonia (Kamgar et al. 1989) or also by furnace annealing of Ti in N$_2$ or NH$_3$ (Willemsen et al. 1988). Willemsen et al. (1988) reported that the nitridation reaction, which proceeds downwards from the surface, stops when the TiN layer meets the TiSi$_2$ which grows upwards from the original Ti/Si interface.

The TiSi$_2$ layer is polycrystalline and consists of grains with diameters of a few microns. No particular epitaxial orientation relation exists between the TiSi$_2$ layer and the Si substrate. As seen on fig.4 the TiSi$_2$/Si interface is not flat, but is rather undulating. Detailed studies of the epitaxial relationships of TiSi$_2$ on Si (Nipoti and Armigliato 1985 and Catana et al. 1987) indeed established a rather large number of possible orientation relations.

REFERENCES

Beyers R, Sinclair R and Thomas M E 1984 J. Vac. Sci. Technol. B 2 781
Beyers R and Sinclair R 1985 J. Appl. Phys. 57 5240
Brasen D, Willens R H, Nakahara S and Boone T 1986 J. Appl. Phys. 60 3527
Catana A, Heintze M, Schmid P E and Stadelmann P 1987 Inst. Phys. Conf. Ser. 87 529
Holloway K and Sinclair R 1987 J. Appl. Phys. 61 1359
Kamgar A, Baiocchi F A, Emerson A B, Sheng T T, Vasile M J and Haynes R W 1989 J. Appl. Phys. 66 2395
Nipoti R and Armigliato A 1985 Jap. J. Appl. Phys. 24 1421
Ogawa S, Kouzaki T, Yoshida T and Sinclair R 1990 Mat. Res. Soc. Symp. Proc. 181 139
Sinclair R, Holloway K, Kim K B, Ko D H, Bhansali A S, Schartzman A F and Ogawa S 1989 Inst. Phys. Conf. Ser. 100 599
Wang M H and Chen L J 1991 Appl. Phys. Lett. 58 463
Willemsen M F C, Kuiper A E T, Reader A H, Hokke R and Barbour J C 1988 J. Vac. Sci. Technol. B 6 53
Wittmer M 1988 Appl. Phys. Lett. 52 1573

Inst. Phys. Conf. Ser. No 117: Section 5
Paper presented at Microsc. Semicond. Mater. Conf., Oxford, 25–28 March 1991

307

Antimony doping during formation of self-aligned titanium disilicide by codeposition of titanium and antimony

X-H Li, S F Gong, H T G Hentzell, and J R A Carlsson

Department of Physics and Measurement Technology, Linköping University
S-581 83 Linköping, Sweden

ABSTRACT: Simultaneous Sb doping and self-aligned $TiSi_2$ formation by the reaction between a Ti:Sb film and a Si substrate have been studied by using transmission electron microscopy, X-ray energy-dispersive spectroscopy, Auger electron spectroscopy and secondary ion mass spectroscopy. The results show that, after the sample was subjected to a three-step anneal, shallow n^+-p junctions formed beneath the $TiSi_2$ film.

1. INTRODUCTION

With the further shrinkage of device dimensions in integrated circuits, highly conductive interconnections, low resistivity contacts and shallow junctions are required. Recently, great attention has been paid to dopant diffusion from silicide into silicon, since this method is used to obtain shallow junctions with low contact resistance. It has been reported that by implanting dopant atoms, e.g. As and B (Davari 1982), into a Ti film, shallow junctions and self-aligned $TiSi_2$ could be achieved simultaneously. However, for Sb, a previous study (Gas 1988) has shown that the implanted Sb atoms are immobile in $TiSi_2$ films up to 800 °C and Sb does not diffuse towards the interface between the $TiSi_2$ film and the Si substrate. In this work, Ti and Sb are codeposited on a Si wafer prior to the formation of $TiSi_2$. By studying Sb behavior during silicide formation, a modified self-aligned $TiSi_2$ process with three-step annealing is employed to achieve simultaneous Sb doping and self-aligned $TiSi_2$ formation.

2. EXPERIMENTAL DETAILS

Thin films (150 nm thick) consisting of Ti and Sb with an atomic ratio of Ti:Sb=1:0.045 were deposited on a 10 Ω-cm p-type (100)Si wafer, using an electron-beam evaporation system with a base pressure of 6.0×10^{-7} Torr. For the fabrication of diodes, thermally grown SiO_2 was patterned on the Si wafer before evaporation. In order to prevent Ti from oxidization during later annealing, a codeposited $TiSi_2$ or a SiO_2 cap of 50-nm was deposited on top of the film. The self-aligned $TiSi_2$ process shown in Fig. 1 includes annealing steps at different temperatures and selective etching. A two-step and a three-step annealing sequence were tried respectively in order to obtain simultaneous doping and silicide formation. For the three-step anneal, an additional step in Ar at 500 °C for 40 min was added to the two-step annealing as shown in the dashed frame in Fig. 1. The two-step anneal was annealing in Ar at 650 °C-670 °C for 30 min followed by final annealing in N_2 at 900 °C for 1 h. Selective etching was utilized to remove the unreacted Ti layer before the final annealing, using a mixture of sulphuric acid and hydrogen peroxide (H_2SO_4:H_2O_2=2:1). The $TiSi_2$ or SiO_2 cap on the top of the Ti film was first etched away carefully by a diluted HF acid (HF:H_2O=1:20) without removing the Ti layer.

The structures of the films after annealing were examined through cross-sectional samples in an EM400T transmission electron microscope (TEM). The relative content of elements in the layers was also measured in the TEM using an attached X-ray energy-dispersive spectrometer (EDX). For the EDX measurement, the electron beam was focused to a diameter of 10-20 nm. By measuring a large TiSi$_2$ grain formed by the reaction between a Ti film and a Si substrate at 900 °C, the effective factor K_{TiSi} of X-ray yield from Ti atoms relative to Si was measured to be 0.85. Sb distributions in the film were measured by Auger electron spectroscopy (AES) combined with sputter etching depth profiling which was carried out at a rate around 4 nm/min. A secondary ion mass spectrometer (SIMS) was employed to measure doping profiles of Sb in the Si substrate. A 8.0 keV O_2^+ primary beam was used in the analysis and the qualification of the data was performed with reference to an ion-implanted standard. Characteristics of the n^+-p junction formed were measured using a semiconductor parameter analyzer (hp 4145B).

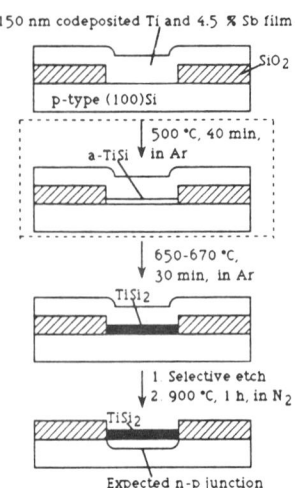

Fig. 1. A self-aligned TiSi$_2$ process flow. The step in the dash frame is an added step for a three-step anneal compared to a two-step anneal.

3. RESULTS

The two-step anneal (no annealing at 500 °C) was first employed in the self-aligned TiSi$_2$ process. After the film (Ti:Sb=1:0.045) was annealed at 650 °C for 30 min, a polycrystalline TiSi$_2$ thin layer was observed between the Ti layer and the Si substrate by a TEM examination. Fig. 2 shows the EDX spectra measured by focusing the electron beam on the TiSi$_2$/Si and the Ti/TiSi$_2$ interface areas. At the TiSi$_2$/Si interface Sb could not be detected. In contrast, a relatively high Sb peak could be observed at the Ti/TiSi$_2$ interface. This implies that there is a Sb depletion region surrounding the TiSi$_2$/Si interface. Diodes were fabricated on the Si wafer with a SiO$_2$ pattern by the process shown in Fig. 1 using the two-step anneal. The reverse leakage current of the diode was measured to be close to 100 μA at a bias voltage of

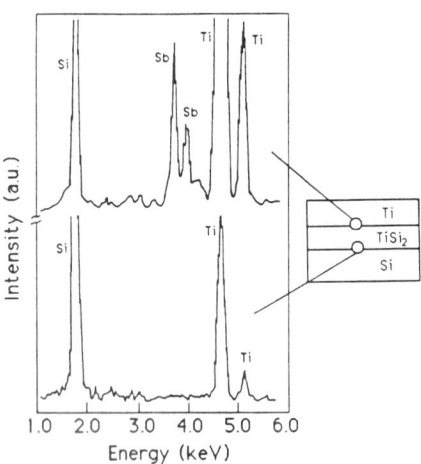

Fig. 2. EDX spectra, showing the relative Sb contents at the TiSi$_2$/Si and the Ti/TiSi$_2$ interface for a sample annealed at 560 °C for 30 min.

-10 V, which is in the order of magnitude for the Schottky diodes fabricated by a pure Ti film using the same self-aligned TiSi$_2$ process. It indicates that no n^+-p junction has formed.

The three-step anneal was then applied instead of the two-step anneal. Fig. 3 (a) is a TEM image showing that an amorphous Ti silicide of about 15 nm thickness had formed after the sample was first annealed in Ar at 500 °C for 40 min. EDX spectra were measured using the cross-section TEM sample by focusing the electron beam on the amorphous layer and the Ti region, as shown in Fig. 3 (b). It is seen that the heights of the Sb peaks measured from the two areas are close with each other. This indicates that the amorphous silicide can dissolve a large quantity of Sb (about 3 at. %) and it seems that the Sb distribution is maintained as that in as-deposited films. By considering the effective factor $k_{TiSi}=0.85$ of Ti relative to Si, the composition of the amorphous silicide can be estimated from EDX spectra as $C_{Ti}/C_{Si}=k_{TiSi}(I_{Ti}/I_{Si})=0.95\approx1$, where $I_{Ti}=49.7$ % and $I_{Si}=47.1$ % are the relative intensities for Ti and Si, respectively. The sample was then annealed in Ar at 670 °C for 30 min. Fig. 4 (a) shows the TEM image in which the amorphous Ti silicide has crystallized forming a 38 nm silicide layer after this annealing step. Fig. 4 (b) shows the Sb depth profile in the film. As seen in Fig. 4 (b), a large amount of Sb accumulates in the unreacted Ti layer and a small amount of Sb piles up at the interface between the $TiSi_2$ layer and the Si substrate. The Sb concentration at the interface (3 %) is roughly as high as that in as-deposited films, while the Sb concentration is lowest in the $TiSi_2$ region, which supports the earlier speculation on the low Sb solubility in $TiSi_2$ (Gas 1988). A final annealing was carried out in N_2 at 900 °C for 1 h. The TEM image shown in Fig. 5 reveals that a 50-nm

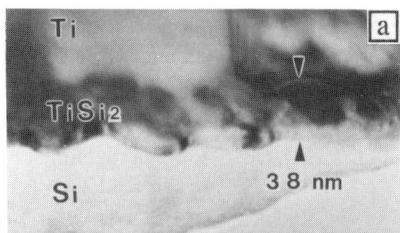

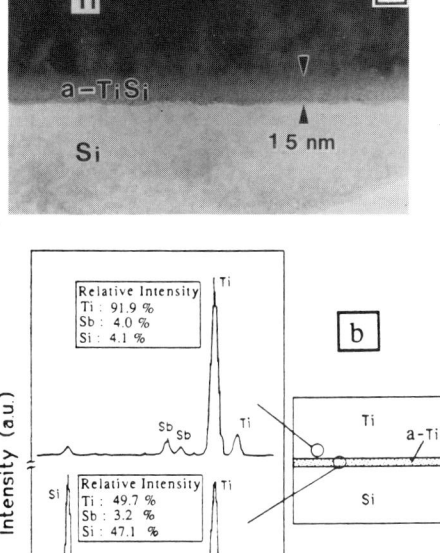

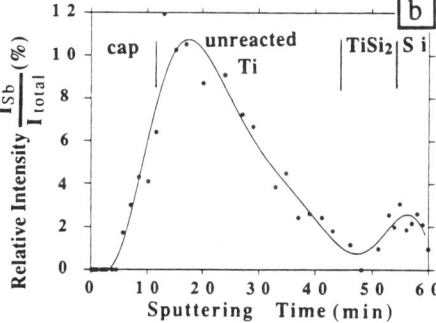

Fig. 4. Ti films with 4.5 at.% Sb annealed subsequently at 670 °C for 30 min. (a) TEM image, showing that a 38-nm crystalline $TiSi_2$ layer has formed. (b) Sb depth profile in the film.

Fig. 3. Ti films with 4.5 at.% Sb annealed at 500 °C for 40 min. (a) TEM image, showing that a 15-nm amorphous silicide layer has formed. (b) EDX spectra, showing Sb content and the relative intensity of each element in the amorphous silicide and the Ti layer.

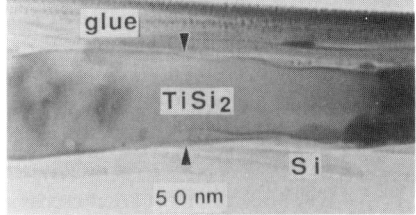

Fig. 5. TEM image, showing a 50-nm $TiSi_2$ film formed after final annealing at 900 °C for 1 h.

TiSi$_2$ film with large grains formed. The doping profile of Sb in the Si substrate measured by SIMS shows that the Sb diffusion depth is 80 nm and the surface concentration of Sb is 10^{19} cm^{-3} which is near the Sb solubility (4×10^{19} cm^{-3}) in crystalline Si at 1000 °C (Einspruch 1983). It is obvious that Sb has diffused into the Si substrate after the three-step anneal. The Sb pile-up at the Si/film interface [see Fig. 4 (b)] served as a diffusion source.

Diodes were fabricated by using the three-step anneal. The reverse leakage current of the diode is 50 nA at a bias voltage of -10 V. Compared to that of Schottky diodes (100 μA) at the same bias, the reverse leakage current of the diodes produced by the three-step anneal is three orders smaller, indicating that a n$^+$-p junction has formed due to Sb diffusion.

4. DISCUSSION

The Sb pile-up in the unreacted Ti layer [see Fig. 4 (b)] indicates that Sb has been rejected by the advancing TiSi$_2$ phase during the reaction. This effect is due to the fact that Si is a fast diffuser (Murarka 1983) during the formation of TiSi$_2$. The Sb rejection may cause Sb depletion in the TiSi$_2$ layer. As a result, when we use an ordinary two-step annealing (i.e., no formation of the amorphous Ti silicide) in which the first silicide phase formed is TiSi$_2$, the Sb rejection by the advancing TiSi$_2$ phase starts from the TiSi$_2$/Si interface, causing Sb depletion at the interface. This is clearly shown by Fig. 2 in which no Sb can be seen at the film/substrate interface in a sample annealed at 650 °C. Consequently, insufficient Sb exists at the interface to serve as a diffusion source. In addition, since the Sb diffusivity in TiSi$_2$ is low (Gas 1988), Sb piled up in the unreacted Ti layer cannot diffuse through the TiSi$_2$ layer into the Si substrate even at 900 °C. Therefore, no n$^+$-p junction can form.

Since we employ a three-step anneal, an amorphous silicide layer first forms at 500 °C [see Fig. 3 (a)]. The amorphous Ti silicide can dissolve a large quantity of Sb [see Fig. 3 (b)]. During the second step of annealing at 670 °C, the amorphous silicide crystallizes and Sb segregates from crystallized amorphous layer towards the TiSi$_2$/Si interface, resulting in a Sb pile-up [see Fig. 4 (b)] since the Sb solubility in TiSi$_2$ is low. Gas (1988) has speculated that the Sb solubility in TiSi$_2$ is equal to or less than 10^{17} cm^{-3} at 800 °C. According to this value, the Sb solubility in TiSi$_2$ is much lower than the Sb concentration ($\sim 10^{21}$ cm^{-3}) in the amorphous silicide film. During the final annealing at 900 °C, the Sb pile-up at the interface serves as a diffusion source to form shallow n+-p junctions.

5. SUMMARY

We have studied the Sb behavior in a codeposited film consisting of Ti and 4.5 at.% Sb during a three-step annealing sequence. An amorphous Ti silicide with a high solid solubility for Sb formed after the first step of annealing at 500 °C. Sb segregated into the interface between the film and the Si substrate after the second annealing step at 670 °C. The segregated Sb served as a diffusion source for Sb doping in the Si substrate at 900 °C. A self-aligned 50-nm TiSi$_2$ film and an 80-nm n$^+$-p junction formed simultaneously during the three-step annealing process.

REFERENCES

Davari B, Taur Y, Moy D, d'Heurle F M and Ting C Y 1982 Proc. 1st Int. Symp. on Ultra Large
        Integration Science and Technology, eds Broydo S, Osburn C M (Pennington) pp 368.
Einspruch N G and Larrabee G B 1983 VLSI Electronics Microstructure Science, (New York: Academic) 6
        49
Gas P, Scilla G, Michel A, LeGoues F K, Thomas O and d'Heurle F M 1988 J. Appl. Phys. 63 5335
Murarka S P 1983 Silicides for VLSI Applications, (New York: Academic), pp 92.

Inst. Phys. Conf. Ser. No 117: Section 5
Paper presented at Microsc. Semicond. Mater. Conf., Oxford, 25–28 March 1991

311

TEM study of ultrathin buried cobalt silicide layers formed by ion beam synthesis

J Vanhellemont, K Maex, M F Wu* and A Romano-Rodríguez **

Inter University Micro-Electronics Center (IMEC), Kapeldreef 75, B-3001 Leuven, Belgium
*Instituut voor Kern- en Stralingsfysika, University of Leuven, B-3001 Leuven, Belgium
Present address: Peking University, Beijing, PR China
**Permanent address: Càtedra d'Electrònica, Facultat de Física, Universitat de Barcelona, Diagonal 647, E-08028 Barcelona, Spain

ABSTRACT: Results are presented of a TEM study of thin buried cobalt silicide layers formed by low energy, low dose cobalt ion implantation. Buried layer formation has been been studied both in (111) and (001) oriented substrates. Special attention is given to the $CoSi_2/Si$ interface quality, defect generation in the silicon substrate and pinhole formation in the ultrathin buried silicide layers.

1. INTRODUCTION

During the last decade ion beam synthesis has become an attractive tool for the formation of buried layers in silicon substrates. Where originally the efforts focussed on the creation of buried isolating layers by the implantation of high doses of oxygen and/or nitrogen ions, recently interest has been growing to create mesotaxial silicide layers by high dose ion implantation of the required metal followed by a high temperature anneal to remove lattice damage and to form the buried silicide layer. Some of the advantages of this approach are that the resistivity of the silicide layers is lower and that their stability during thermal treatments is higher than those films grown by more conventional deposition techniques. Furthermore, it was demonstrated that the buried layers can also be used as seeds for the formation of very thin epitaxial surface layers of $CoSi_2$ (Maex et al 1991). As $CoSi_2$ has the CaF_2 structure with a lattice constant that differs only 1.2 % from that of Si, the buried layers will easily grow epitaxially inside the silicon substrate. Initially most of the research has been devoted to the formation of relatively thick silicide layers by the use of high energy (typically 150-350 keV) and high dose (typically $2\text{-}7\times10^{17}\text{cm}^{-2}$) ion implantation (e.g. Van Ommen et al 1990).

In the present paper results are presented of a TEM study of thin buried cobalt silicide layers formed by low energy (< 100 keV) low dose (< 10^{17} cm^{-2}) cobalt ion implantation. The formation of buried layers with thicknesses ranging from 25 to 40 nm is studied both in (111) and (001) oriented silicon substrates. Special attention is given to the $CoSi_2/Si$ interface quality, defect generation in the silicon substrate and the formation of pinholes in the ultrathin buried silicide layers. The critical film thickness to obtain continuous and stable silicide layers is explored as a function of implantation and anneal conditions.

2. EXPERIMENTAL

Si (111) and (001) substrates are implanted with Co ions with implantation energies between 30 and 100 keV. The substrate temperature is kept between 300 and 400°C and the implanted doses vary between 5×10^{15} cm^{-2} and 1×10^{17} cm^{-2}. Some wafers received isochronal anneals at different temperatures to study the stability of the buried CoSi$_2$ layer and the formation of pinholes in it. On some samples a Co film is deposited after the formation of a buried CoSi$_2$ layer. The resulting structure is then annealed in order to obtain a thin epitaxial silicide layer. Cross-section specimens are prepared following a procedure described in detail elsewhere (Romano et al 1990).

3.   OBSERVATIONS AND DISCUSSION

3.1   (111) substrates

As the interface energy of the {111} planes is minimal, very flat interfaces are observed for buried layers in (111) Si substrates (Wu et al 1990). Very thin films with thicknesses down to 16 nm can be obtained. In one experiment it was demonstrated that double, high quality buried layers of CoSi$_2$ can be formed by sequential implantation and annealing (figure 1).

3.2   (001) substrates

The formation of buried CoSi$_2$ layers with flat interfaces in (001) silicon substrates is found to be more difficult (Hull et al 1990). As the interface energy between CoSi$_2$ and Si is minimal for the (111) planes, buried silicide layers in silicon will have the tendency of forming (111) facets leading to an unwanted roughening of the interface. Another problem, both in (111) and (001) substrates, is the generation of dislocations both in the top silicon layer and in the substrate. The number of defects can strongly be reduced by using higher temperature anneals as illustrated in figure 2. A problem which can arise at these higher temperatures is however the decreasing stability of the buried film, leading to the formation of pinholes. Furthermore the trend towards facetting will enhance pinhole formation in (001) substrates. Until now little attention has been given in the literature to this formation of pinholes in buried layers obtained by ion beam synthesis and its dependence on the implantation and anneal conditions. Most results found in the literature deal with pinhole formation in MBE grown epitaxial CoSi$_2$ layers on (111) (Gibson et al 1987) and (001) (Jimenez et al 1990) silicon.

At low anneal temperatures (≤900°C), the pinholes in our buried layers are so small that it is very difficult to observe them by the normal bright field observation in the TEM. This problem can be circumvented by using dark field imaging. In that case the pinholes appear as dark areas against a bright background and can easily be observed. This is illustrated in figure 3 where the pinholes are nearly invisible in the bright field image while they can be observed quite easily in the dark field image. In (001) substrates the pinholes have a square shape with edges running along the <110> directions of silicon parallel to the (001) surface of the wafer. This is not a surprise as these <110> directions are the intersections of the four {111} planes with the (001) plane. The edges of the pinholes are thus formed by {111} facets having the lowest interface energy.

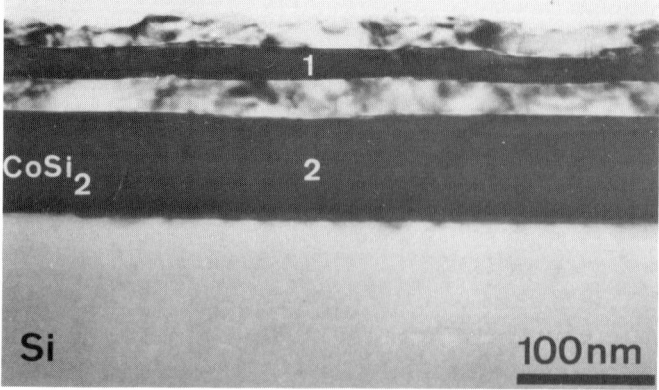

Figure 1: Double buried CoSi$_2$ layer formed by subsequent implantations (200 kEV, 1.5 10^{17} cm^{-2} and 30 kEV, 4.2 10^{16} cm^{-2}) and an anneal at 600°C (1h) and 1100°C (30 min).

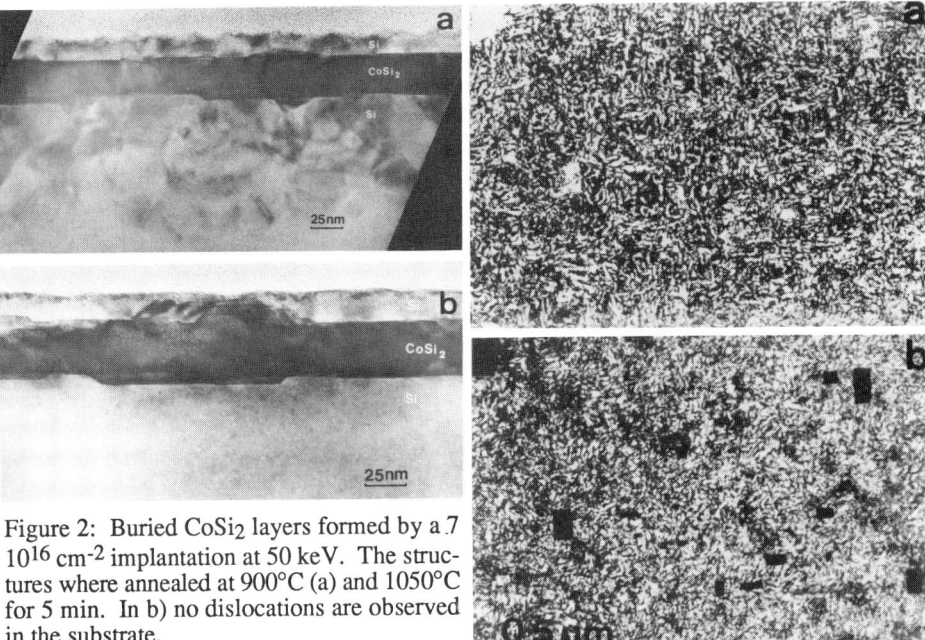

Figure 2: Buried CoSi$_2$ layers formed by a 7 10^{16} cm^{-2} implantation at 50 keV. The structures where annealed at 900°C (a) and 1050°C for 5 min. In b) no dislocations are observed in the substrate.

Figure 3 (right): a) Bright field image of a buried CoSi$_2$ film formed by a 50keV Co ion implantation with a dose of 7 10^{16} cm^{-2} followed by an anneal at 900°C for 5 min. b) The same area observed in dark field. The small pinholes are now clearly visible as dark areas on a bright background.

Increasing the anneal temperature leads to the formation of larger pinholes in the buried silicide layer as illustrated in figure 5. For isochronal 5 min anneals the density of the pinholes remains constant while the size of the pinholes and thus also the percentage of the surface which is covered by them, increases. This suggests a heterogeneous nucleation mechanism whereby the density of pinholes is determined by the implantation conditions rather then by the anneal temperature. Further work is going on to study the growth kinetics of the pinholes as function of anneal temperature and time and to establish which of the process or material parameters are determining the nucleation of pinholes.

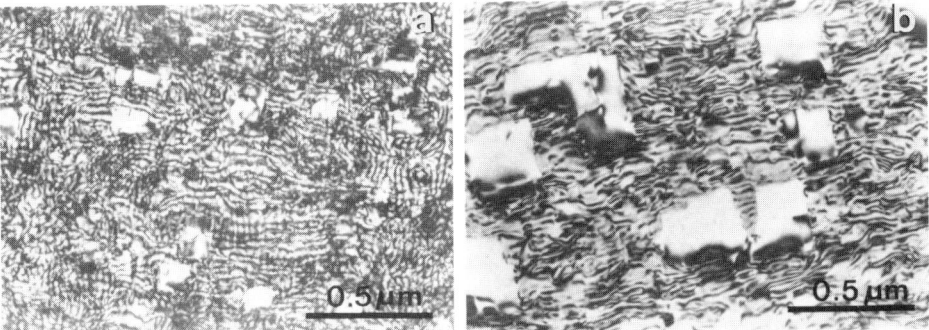

Figure 4: Plan view TEM micrographs showing pinholes formed after a thermal anneal at 900°C (a) and 1050°C (b) for 5 min. In a) a two dimensional Moiré pattern is observed corresponding with a symmetrical orientation of the specimen. In b) two beam imaging leads to only one set of Moiré fringes.

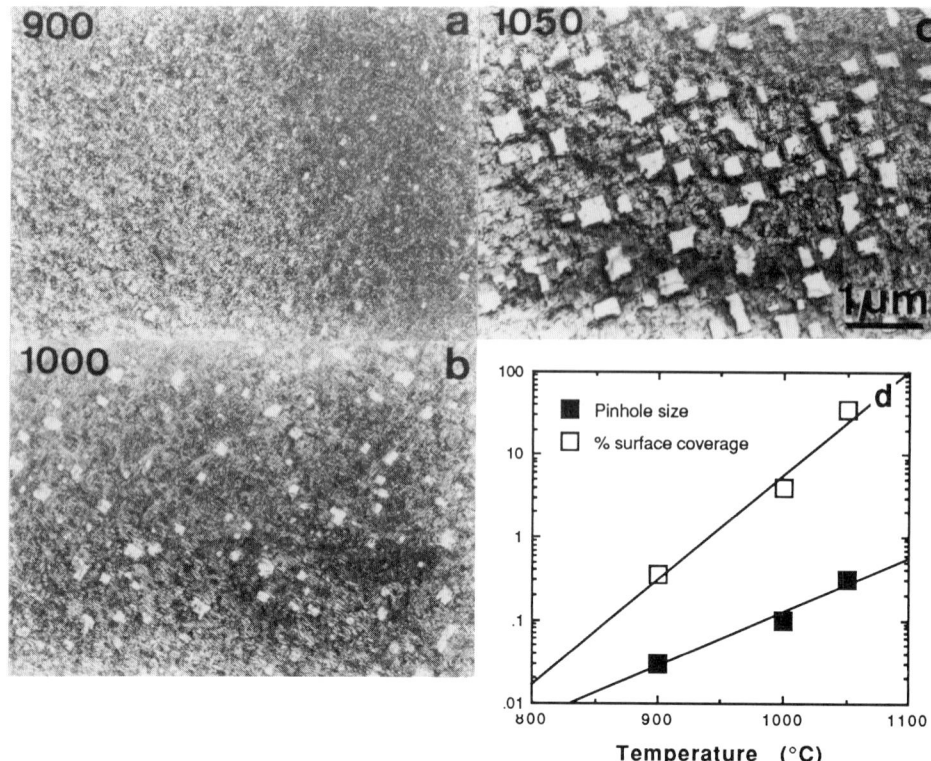

Figure 5: The pinhole size variation as function of the anneal temperature for an isochronal anneal of 5 min. a) 900°C. b) 1000°C. c) 1050°C. In d) the variation of the pinhole size and the percentage of covered surface are represented.

REFERENCES

Gibson J M, Batstone J L and Tung R T 1987 Appl.Phys.Lett. 51 45
Hull R, White A E, Short K T and Bonar J M 1990 J.Appl.Phys. 68 1629
Jimenez J R, Hsiung L M, Rajan K, Schowalter L J, Hashimoto S, Thompson R D and
    Iyer S S 1990 Appl.Phys.Lett. 57 2811
Romano A, Vanhellemont J , Bender H and Morante J R 1990 Ultramicroscopy 31 183
Van Ommen A H, Bulle-Lieuwma C W T, Ottenheim J J M and Theunissen A M L 1990
    J.Appl.Phys. 67 1767
Wu M F, Vantomme A, Pattyn H, Langouche G, Maex K, Vanhellemont J, Vanacken J,
    Vloeberghs H and Bruynseraede Y 1990 Nucl. Instr. and Meth. in Phys. Res. B45 658
Maex K, Brijs B, Vanhellemont J and Vandervorst W 1991 to be published in the proceed-
ings of IBMM 90
Maex K, Vanhellemont J, Petersson S and Lauwers A 1991 to be published in the proc. of
3th workshop on Refractory Metals & Silicides, Stockholm 24-27 March 1991.

Acknowledgement: The TEM work is performed using the equipment of the University of Antwerp (RUCA). Karen Maex is indebted to the National Fund for Scientific Research (NFWO) for her fellowship as a research associate. Albert Romano-Rodríguez is supported by the Spanish Ministry of Science and Education.

Inst. Phys. Conf. Ser. No 117: Section 5
Paper presented at Microsc. Semicond. Mater. Conf., Oxford, 25–28 March 1991

Analysis of Bicrystallographic variants, domains and defects in the CoSi/Si(111) interface

A C Daykin, C J Kiely and R C Pond

Department of Materials Science and Engineering, University of Liverpool, P.O.Box 147, Liverpool L69 3BX, England

ABSTRACT: Bicrystallographic symmetry theory has been used to predict possible variants, domains and defects in the CoSi//Si(111) epitaxial system and group theory applied to determine the conditions under which the domains can be imaged. These predictions have been experimentally verified using Transmission Electron Microscopy (TEM). In addition the planar faults, disclinations and the dislocations present in the CoSi have also been analysed.

1.   INTRODUCTION

CoSi is formed when Co is deposited onto a clean Si(111) surface and annealed to 450°C under UHV conditions. This is an intermediate silicide in the solid state reaction which upon completion at 650°C gives a mixed A and B-type $CoSi_2$ epilayer. Phillips et al (1989) have suggested nucleation mechanisms for A and B-type $CoSi_2$ which are similar to those suggested by Gibson et al. (1989) for the Ni//Si(111) system. However the whole epilayer has been shown to transform to CoSi prior to $CoSi_2$ by Daykin et al (1990) . Consequently, any link between precursor phases and the nucleation mechanisms of A and B-type $CoSi_2$ must therefore be contained with in the CoSi.

The characterisation of bicrystals in which the two component crystals have a large degree of symmetrical disconnexion cannot easily be carried out without a prior knowledge of the possible variants, domains and defects. This paper is a TEM study of the CoSi//Si(111) epitaxial bicrystal using bicrystallographic theory to aid in the interpretation of the results obtained.

2.   EXPERIMENTAL

The CoSi//Si(111) bicrystals were fabricated by depositing 20Å of Co onto a clean Si(111) surface under UHV conditions in an MBE chamber and subsequently annealing to 450°C. The plan view and cross-sectional specimens were prepared by standard methods and were examined in Philips EM 400 T and EM 430 T electron microscopes.

3.   RESULTS AND DISCUSSION

The symmetry theory of bicrystallographic analysis, which can be used to predict the possible variants, domains and defects, has been developed by Pond and his co-workers (1989). The application of this theory to a given bicrystal requires a prior knowledge of the space group of the two component crystals, the interfacial plane and the orientation relationship.

3.1   Variants

Using Transmission Electron Diffraction (TED) the predominant orientation relationship has been found in agreement with Adamski et al.(1989) and Batstone et al. (1989) to be CoSi(111)//Si(111) and CoSi[11$\bar{2}$]//Si[1$\bar{1}$0]. The space groups of CoSi and Si are P2₁3 and Fd3m respectively. Consequently the (111) plane symmetries will be p3 and p3m respectively. If the 3-fold axes of the two component crystals are aligned at the interface then there will be two orientational variants related by a mirror across a (11$\bar{2}$) plane. This is determined by group theoretically decomposing 3m with respect to 3. If the 3-fold axes are misaligned by an in plane translation then there will be two sets of orientational variants each containing three translational variants.The orientational variants can be identified from diffraction information because they will give rise to beams in different positions in reciprocal space. The presence of translational variants can only be confirmed by identifying the dislocations which will delineate them. The [342] zone axis diffraction pattern is shown in Fig.1a. The accompanying schematic, Fig.1b, shows clearly both orientations of CoSi. There has however been no evidence of translational variants in this system.

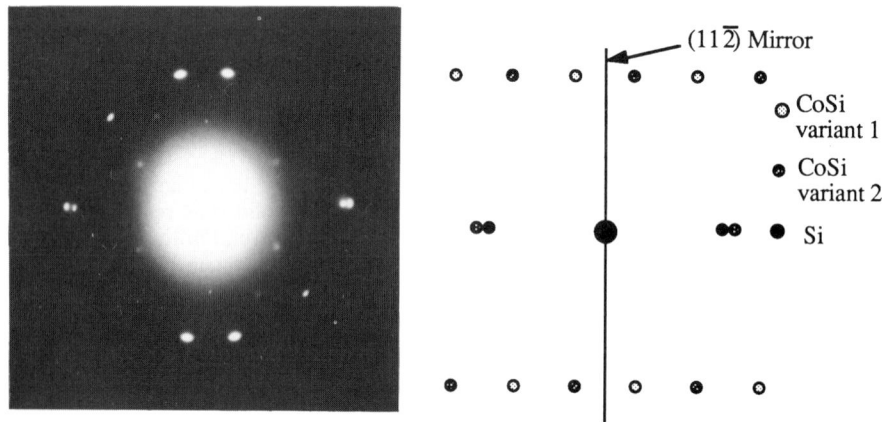

Fig.1a The [342] Zone Axis   Fig.1b Accompanying Schematic for Fig.1a
Diffraction Pattern.

3.2   Domains

CoSi is a non-holosymmetric crystal. This means that it has a lower symmetry than that of it s lattice. Consequently in a homogeneously nucleated solid group theory predicts there will be four domains. If these domains are present in the CoSi//Si(111) bicrystal then they will be non-degenerate. Group theory can also be applied to predict that beams of the type **g**=hkl and **g**=hk0 will give rise to pairs of domains with different kinematic intensities. Fig. 2 shows **g**=210 type dark field micrographs from the same area of the CoSi//Si(111) bicrystal. Figs. 2a and b are different pairs of domains from orientation 1 and Figs.2c and d are different pairs of domains from orientation 2. The relative amounts of the pairs of domains contained with-in each orientation can be seen to be different, confirming that the domains are non-degenerate. Other less well defined epitaxial orientations of the CoSi have also been idenified.

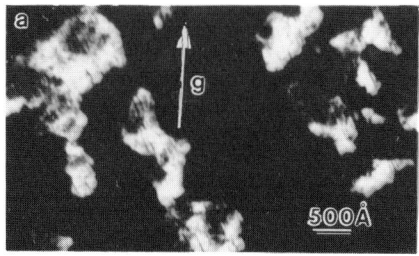 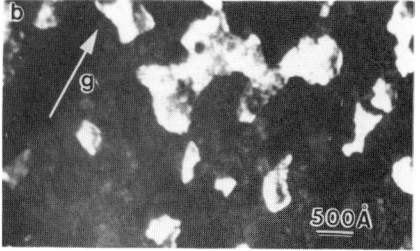

Fig. 2a and b **g**=210 type images showing pairs of domains from orientational variant 1.

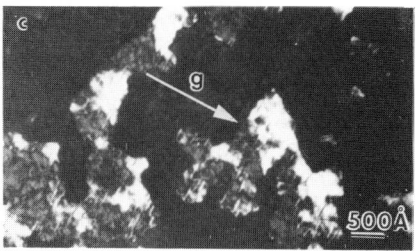 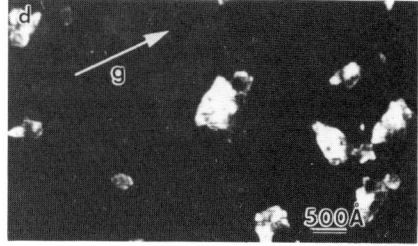

Fig. 2c and d **g**=210 type images showing pairs of domains from orientational variant 2.

3.3 Crystal Defects

CoSi is a primitive cubic crystal with ABCABC stacking of the {111} planes. In all the micrographs shown in Fig.2 there is an abundance of planar faults which intersect the interfacial plane along <1$\bar{1}$0> directions. Fig. 3a is a higher magnification **g**=210 image showing the planar faults in more detail.The measured thickness of these planar faults is consistent with their occurrence on the {111} planes. The **g**=220 moiré image, Fig.3b, shows a moiré fringe shift of a $^1/_3$ across these faults. In **g**=112 dark field micrographs all three types of stacking fault are visible. In **g**=210 darkfield images only one type of stacking fault is visible. And in **g**=321 darkfield images two types of stacking fault are visible. These pieces of evidence are consistent with these planar faults having a shift vector R= $^1/_3$<11$\bar{2}$>. Thus it can be concluded that the planar faults in Fig.2 are stacking faults occurring on {111} planes .

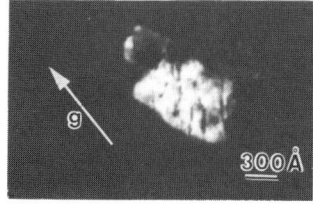  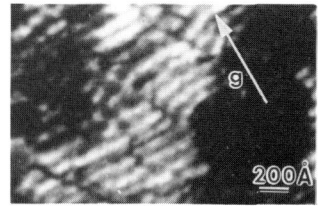

Fig.3a   Planar faults in CoSi.   Fig. 3b   **g**=220 darkfield moiré image.

Through out the pairs of domains imaged in Figs. 2a,b,c and d there can be seen to be large variations in contrast. Plan view tilting experiments show this contrast to reverse uniformly suggesting a rotation of the lattice planes. In cross-section the interface can be seen to be rough and high resolution images suggest disclinations are contained in the CoSi. This is consistent with a non-flat interface because the interfacial steps would give rise to dislocations with Burgers vectors having components perpendicular to the interface.

Weak beam g=220 moiré images show a hexagonal array of dislocations with <11$\bar{2}$> line directions and an approximate spacing of 90Å. These are present at the interface to relieve a misfit of approximately 6%. The possible Burgers vectors of these dislocations can be predicted using broken translation symmetry calculations as detailed by Pond (1989). Obtaining an invisibility condition may prove difficult as the Burgers vector of the dislocations need not be rational. An alternative approach, which we are investigating, could be to calculate the amount of misfit relieved by all the possible dislocation arrays and see which of these is the best fit.

5   CONCLUSIONS

TED has been used to verify experimentally the orientation relationship between the CoSi and the Si. This information has been used to predict the possible variants and domains that can exist in the bicrystal. A combination of darkfield microscopy and group theoretical calculations have been used to confirm these predictions. The fact that more than one domain is present for each orientation suggests that a predominant factor in the mechanism of growth is the periodicity introduced by the epitaxial relationship. Although, the presence of less well defined epitaxial relationships reminds us that the precise interfacial structure is also a governing factor. The defects seen to occur in CoSi are stacking faults, disclinations and dislocations. The stacking faults have been shown to occur on the {111} planes with a characteristic shift of R=$\frac{1}{3}$<11$\bar{2}$>. This is consistent with a discontinuity in the ABCABC stacking of the {111} planes in primitive cubic crystals. The disclinations have been investigated using dark field plan view microscopy and high resolution cross-sectional samples. These have been shown to originate from interfacial roughness. Calculations to determine the possible Burgers vectors of the dislocation arrays observed using weak beam darkfield imaging are in progress. By calculating the misfit relieved by the different arrays it is hoped to determine a suitable candidate.

To date no simple correlation between the crystallographic structure of the CoSi//Si(111) bicrystal and the nucleation of A and B-type CoSi$_2$ has been found.

ACKNOWLEDGEMENTS

We should like to acknowledge J.L. Batstone, Julia M. Phillips and AT&T Bell Laboratories for the supply of samples and SERC for financial support.

REFERENCES

Adamski C, Meiser S, Rahman S H and Baschek G 1989 Mater. Res. Soc. Symp. Proc. EA 18 47

Batstone J L, Daykin A C, Phillips J L and Hensel J C 1989 Inst. Phys. Conf. Ser. No 100: Section 8 641

Daykin A C, Kiely C J, Pond R C and Batstone J L Proc. XII Int. Cong. for Electron Microscopy vol. 4 574

Gibson J M, Batstone J L, Tung R T and Unterwald F C 1989 Phys. Rev. Letts. B(2) vol.60 No. 12 88

Phillips J L, Batstone J L, Hensel J C, Cerullo M and Unterwald F C 1989 J. Mater. Res. 4 144

Pond R C 1989 in Dislocations in solids, Edited by Nabarro J, Vol.8 chapter 38

Inst. Phys. Conf. Ser. No 117: Section 5
Paper presented at Microsc. Semicond. Mater. Conf., Oxford, 25–28 March 1991

319

Thin film manganese silicides

L Zhang and D G Ivey

Department of Mining, Metallurgical and Petroleum Engineering, University of Alberta, Edmonton, Alberta, Canada, T6G 2G6

ABSTRACT: Reactions between manganese thin films and silicon substrates have been studied. Four silicide phases are reported to form. Mn_3Si, $MnSi$ and Mn_5Si_3 form and grow in a layered manner at low annealing temperatures ($<450^{\circ}C$). The three silicides are observed to co-exist and simultaneous growth of Mn_5Si_3 and $MnSi$ takes place. This unusual growth behaviour is explained in terms of changes in diffusion fluxes during the annealing process. $MnSi_{1.73}$ forms from $MnSi$ at higher temperatures and the formation process is nucleation controlled.

1. INTRODUCTION

A small number of transition metal silicides are semiconducting (Bost and Mahan 1987, Bost and Mahan 1985). One of these, $MnSi_{1.73}$, is reported to be a direct band gap semiconductor emitting in the infrared range (Bost and Mahan 1987), making it promising for silicon-based optoelectronic applications, such as optic fibre links and infrared detector arrays. Very little is known, however, about the formation processes and growth kinetics of the Mn/Si system. The only documented investigation of manganese silicide formation in thin film couples has been done by Eizenberg and Tu (1982). Two silicide phases were detected, i.e. $MnSi$ and $MnSi_{1.73}$. $MnSi$ formed first at $400^{\circ}C$ and grew in a layered manner. $MnSi_{1.73}$ formation followed at temperatures as low as $500^{\circ}C$, although the reaction was nucleation controlled and growth was three dimensional.

A systematic study of manganese silicide formation during thermal reaction of Mn/Si thin film couples is reported here. Cross section transmission electron microscopy is utilized as the main characterization technique.

2. EXPERIMENTAL TECHNIQUES

High purity manganese (99.98%) was deposited by thermal evaporation onto <001> oriented, single crystal silicon wafers. Prior to deposition, the silicon wafers were cleaned using buffered hydrofluoric acid (20 parts water to 1 part HF), rinsed with deionized water and dried with high purity Ar. After a Si wafer was loaded into the deposition chamber, the chamber was pumped down to a pressure of 2×10^{-6} Torr. The thickness of the Mn film, measured using cross-sectional TEM of as deposited specimens, was nominally 300nm. After Mn deposition, the Si wafers were cleaved into 1cm × 1cm pieces and loaded individually into a quartz tube furnace. Annealing was done in flowing nitrogen (purified nitrogen, 99.97%) at temperatures ranging from 380-570°C for times up to 256 min.

Silicide formation and growth were monitored using transmission electron microscopy (TEM) techniques. Both plan view and cross-section specimens were prepared for TEM, using standard techniques of dimpling and ion milling. Phase characterization techniques included bright field imaging, selected area diffraction (SAD), convergent beam electron diffraction (CBED) and energy dispersive x-ray spectroscopy (EDX). Reacted layer thicknesses were measured directly from negatives, with 30 randomly selected measurements taken from each negative and 3 negatives measured for each heat treatment. Several electron microscopes were utilized, i.e. a Hitachi H-7000 TEM, a Philips 400 EM and a Hitachi H-600 TEM/STEM equipped with a Be window x-ray detector.

3.   RESULTS AND DISCUSSION

3.1   Low Temperature Silicide Formation (<450°C)

The deposited structure is shown in Fig. 1. A thin
amorphous layer, identified as SiO_2 , is located
between the deposited Mn layer and the Si substrate.
The oxide layer is still present after annealing (Fig.
2d) and therefore can be used as a diffusion marker.

Three phases were identified, i.e. Mn_3Si, MnSi and
Mn_5Si_3, in samples annealed at lower temperatures.
Bright field micrographs depicting the formation
sequence are shown in Fig. 2. The growth process
can be divided into four stages and the kinetics
explained by considering changes in diffusion fluxes.

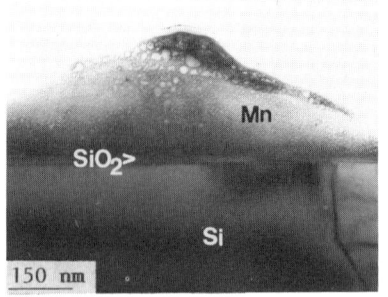

Fig. 1   Bright field micrograph (BF) of cross
section specimen of the as deposited sample.

The growth process, observed here, is unexpected. Most silicide formation processes follow a single layer
growth pattern (Tu 1985, 1986). Generally, only one silicide layer exists and grows until the deposited metal
layer is completely consumed. After the metal is exhausted, a second silicide begins to grow by consuming
the first silicide and so on. However, in this study, three silicide layers co-exist before the Mn layer is
consumed and at least two of the layers grow at the same time. In addition, Mn_5Si_3 formation follows MnSi
formation, which is also unusual. The sequence of formation usually follows a metal-rich to more silicon-
rich pattern (for thin metal films on Si wafers).

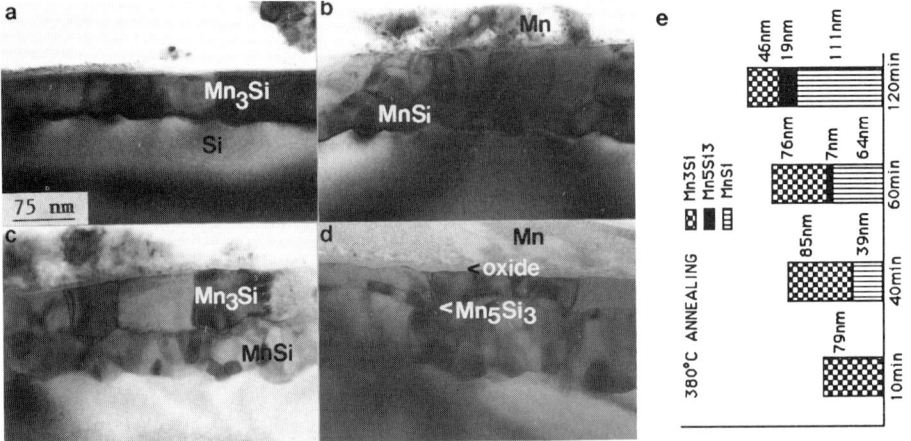

Fig. 2 Bright field micrographs of specimens annealed at 380°C for (a) 10 min, (b) 40 min, (c) 60
min and (d) 120 min. A schematic of the formation sequence is shown in (e).

Average Mn and Si fluxes (J_{Mn} and J_{Si}) in the diffusion region - defined as the region between the oxide layer
and the Si substrate - have been estimated and are plotted in Fig. 3. J_{Mn} is initially larger than J_{Si}. Both
fluxes decrease with annealing time, although J_{Mn} decreases at a faster rate than J_{Si}. This change in diffusion
flux has a decisive effect on silicide formation. The reason for the marked change in Mn flux (almost two
orders of magnitude) may be related to the presence of oxygen, either in the deposited Mn film or from the
annealing environment. Similar phenomena (diffusion flux decrease and multiple layer growth) have been
reported for other thin film systems, i.e. Pt/Si and Ni/Si (Ottaviani 1979) and Mo/Si (Nava *et al* 1981).

Silicide formation is initiated at the oxide/Si interface. Mn_3Si forms according to the following reaction:
$$3 \text{ Mn} + \text{Si} \rightarrow Mn_3Si \qquad (1).$$
Mn diffusion predominates in both the oxide and Mn_3Si and, therefore, subsequent growth takes place at the
Mn_3Si/Si interface with Mn_3Si growing towards the Si substrate.

The rapid decrease in the Mn flux (Fig. 3) brings about the end of Mn_3Si growth and the beginning of the second stage. Silicon diffusion becomes significant and MnSi forms at the Mn_3Si/Si interface and grows towards the surface (Fig. 2b).

$$Mn + Si \rightarrow MnSi \qquad (2).$$

As this reaction proceeds, Si becomes the major diffuser in the growing MnSi layer, so that further growth takes place at the $Mn_3Si/MnSi$ interface. Silicon reacts with Mn that arrives at this interface by diffusing through the Mn_3Si layer. That Si is, indeed, the major diffuser in MnSi is apparent from Fig. 2. Examination of the MnSi layer (Fig. 2b) reveals the presence of many small columnar grains, with an undulating envelope outlining the interface between MnSi and Si. These features are still present after annealing at higher temperatures and for longer times, indicating that the growth of MnSi does not occur at this interface. The shape of the undulation along the MnSi/Si interface follows virtually the same pattern as the $Mn_3Si/MnSi$ interface (Fig. 2b). It is, therefore, reasonable to assert that the position of the MnSi/Si interface is the position of the previous Mn_3Si/Si boundary. Further evidence can be seen in Fig. 4, where many tiny protrusions were found growing at the $MnSi/Mn_3Si$ interface into Mn_3Si.

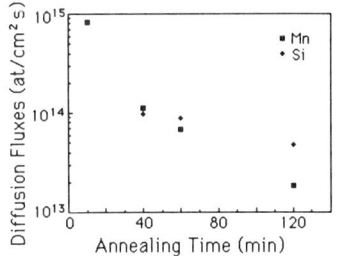

Fig. 3 Plot of average Si and Mn diffusion fluxes vs. time.

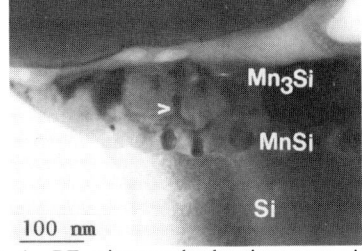

Fig. 4 BF micrograph showing protrusions growing towards Mn_3Si. The specimen was annealed for 40 min at 380°C.

The Mn flux continues to decrease (at a faster rate than the Si flux) so that the flux ratio of J_{Mn} to J_{Si} becomes smaller than one. As less and less Mn arrives at the $Mn_3Si/MnSi$ interface, Si atoms begin to accumulate there. The excess Si then reacts with Mn_3Si to form Mn_5Si_3 according to:

$$Mn_3Si + 0.8\ Si \rightarrow 0.6\ Mn_5Si_3 \qquad (3).$$

This newly formed Mn_5Si_3 layer lies between Mn_3Si and MnSi (Figs. 2c and 2d). Two reactions now occur simultaneously, i.e. MnSi growth at the $Mn_5Si_3/MnSi$ interface and Mn_5Si_3 growth at the Mn_3Si/Mn_5Si_3 interface. Mn_3Si is being consumed by Mn_5Si_3.

With further annealing, the difference between the Mn and Si fluxes continues to widen (Fig. 3). More and more Si accumulates at the $Mn_5Si_3/MnSi$ interface, leading to the reaction of some of this Si with Mn_5Si_3 to form MnSi:

$$0.6\ Mn_5Si_3 + 1.2\ Si \rightarrow 3\ MnSi \qquad (4).$$

Two reactions are taking place at the $Mn_5Si_3/MnSi$ interface. Silicon reacts with Mn diffusing in from the surface layer to form MnSi (reaction (2)) and Mn_5Si_3 decomposes, also forming MnSi (reaction (4)). Mn_5Si_3 is still forming at the Mn_5Si_3/Mn_3Si interface (reaction (3)). Simultaneous growth of the two layers is observed because the Mn_5Si_3/Mn_3Si interface is moving faster than the $Mn_5Si_3/MnSi$ interface. Eventually Mn_3Si and then Mn_5Si_3 will be consumed, leaving only MnSi (Fig.5).

3.2 High Temperature Silicide Formation (>480°C)

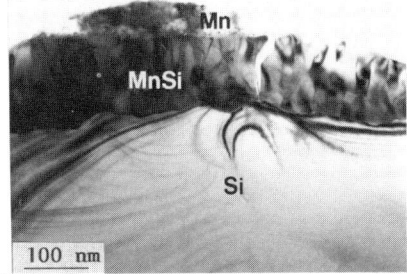

Fig. 5 BF micrograph of specimen annealed for 60 min at 430°C.

At higher annealing temperatures (>480°C), $MnSi_{1.73}$ formation was detected, after a minimum annealing period of 1 hour. $MnSi_{1.73}$ nucleates and grows in islands instead of a layered growth manner (in contrast to the results above), and only begins to form after MnSi growth has finished. Fig. 6 shows evidence of this different growth behaviour, where many islands of large grains ($MnSi_{1.73}$) are surrounded by larger areas with small grains (MnSi). $MnSi_{1.73}$ nucleates preferentially at various positions along the MnSi/Si interface, and then grows in three dimensions. This type of growth process is nucleation controlled.

Nucleation controlled processes for silicide formation have been reviewed by d'Heurle and Gas (1986). They use classical nucleation theory to discuss the phenomena encountered in thin film reactions. According to the classical theory of nucleation, the activation energy ΔG^* for a new phase, AB, forming at an interface between two phases A and B, is given below:

$$\Delta G^* \propto \frac{(\Delta\sigma)^3}{(\Delta G_c - \Delta H_d)^2} \quad (5).$$

$\Delta\sigma$ indicates the increase in surface energy due to the new phase, ΔG_c is the change in "chemical" free energy per unit volume of AB due to the formation of the new phase and ΔH_d is the deformation energy contribution due to the volume change caused by phase transformation.

Fig. 6 BF micrograph of plan view specimen of sample annealed for 60 min at 485°C.

According to d'Heurle and Gas (1986), when metals and silicon react with one another to form silicides, the absolute values of the ΔG's are generally large and the ΔG^*'s are small. As a result, nucleation is so easy and rapid that one can not isolate it and observe it experimentally. However, a nucleation controlled process will occur if the activation energy, ΔG^*, for the new silicide phase to nucleate is large and the reaction temperature is relatively low. This is the case encountered during the annealing reaction of $MnSi_{1.73}$.

$MnSi_{1.73}$ formation begins only after the growth of MnSi has finished. Therefore, the formation process for $MnSi_{1.73}$ has to be:

$$MnSi + 0.73\ Si \rightarrow MnSi_{1.73} \quad (6a) \qquad \text{and not} \qquad Mn + 1.73\ Si \rightarrow MnSi_{1.73} \quad (6b).$$

The standard Gibbs free energy at 485°C for reaction (6a) is fairly small, i.e. -12.2 kJ/mol. In addition, the volume change associated with reaction (6a) is about 40%, which means that ΔH_d will be relatively high. The net result is that the activation energy for reaction (6a) will be quite large, making $MnSi_{1.73}$ formation nucleation controlled. Nucleation controlled kinetics has also been proposed for the following silicides: $NiSi_2$, $ZrSi_2$, PdSi, $HfSi_2$, $OsSi_2$, $IrSi_3$ (d'Heurle and Gas 1986), Rh_4Si_5 and Rh_3Si_4 (Peterson *et al* 1980).

4.   CONCLUSIONS

Three silicides are formed during low temperature annealing (<450°C), i.e. Mn_3Si, MnSi and Mn_5Si_3. These grow in a layered manner, with Mn_3Si forming first followed by MnSi and then Mn_5Si_3. The unusual phenomenon of the coexistence of three silicide phases and simultaneous growth of two silicides has been observed. This has been correlated with changes in Mn and Si diffusion fluxes during annealing. At higher annealing temperatures (>480°C), $MnSi_{1.73}$ forms from MnSi by a nucleation controlled process.

ACKNOWLEDGEMENT

The work was supported by a grant from the Natural Science and Engineering Research Council (NSERC) of Canada.

REFERENCES

Bost M C and Mahan J E 1987 *J. Electr. Mater* 16 389
Bost M C and Mahan J E 1985 *J. Appl. Phys.* 58 2696
d'Heurle F M and Gas P 1986 *J. Mater. Res.* 1 205
Eizenberg M and Tu K N 1982 *J. Appl. Phys.* 53 6885
Nava F, Valeri S, Majni G, Cembali A, Pignatel G and Queirolo G 1981 *J. Appl. Phys.* 52 6641
Ottaviani G 1979 *J. Vac. Sci. Technol.* 16 1112
Peterson C S, Anderson R, Baglin J E, Dempsey J, Hammer W, d'Heurle F M and Laplaca S J 1980 *J. Appl. Phys.,* 51 373
Tu K N 1985 *Ann. Rev. Mater. Sci.,* 15 147
Tu K N 1986 *Advances in Electronic Materials,* eds B W Wessels and G Y Chin (Metals Park, Ohio: ASM) pp 147-180

Inst. Phys. Conf. Ser. No 117: Section 5
Paper presented at Microsc. Semicond. Mater. Conf., Oxford, 25–28 March 1991

323

The use of TEM and Raman spectroscopy to determine the absorption coefficients of titanium nitride and titanium silicide

Louise Weaver, Martine Simard-Normandin and Francois Pagette

Northern Telecom Electronics Limited, P.O.Box 3511, Station C, Ottawa, Ontario, Canada K1Y 4H7.

ABSTRACT: The thickness of TiN and $TiSi_2$ films is usually calculated from four-point probe resistivity measurements on blanket films. This method, however, reveals no information about the thickness variation and stoichiometry of TiN and $TiSi_2$ in sub-micron contact holes on patterned wafers. TEM can provide this information but the technique is destructive and the sample preparation of sub-micron devices is non-trivial. Raman spectroscopy is non-invasive and offers a probe size compatible with current VLSI technology.

The Raman signal from the TiN and $TiSi_2$ films shows little variation across the wafer, but the silicon local mode can be easily detected through the films and is very sensitive to variations in thickness. The intensity of the silicon local mode was measured on a series of 21 samples of TiN on silicon and three samples of $TiSi_2$ on silicon. The precise TiN and $TiSi_2$ film thickness was subsequently determined by cross sectional TEM. From these measurements, the optical constants of TiN and $TiSi_2$ were determined.

1. INTRODUCTION

The traditional interconnect employed in CMOS devices is Al-Si. If junction depths are reduced below 0.3μm, junction spiking can occur (Totta and Sopher 1969). Increasing the Si content of the metal can eliminate spiking, but the formation of silicon precipitates in contact holes leads to unacceptably high contact resistance (Ting and Whittmer 1982). As a solution to these problems, a multilayer contact structure consisting of $TiSi_2$ and TiN can be used (Ho 1989). The silicide layer provides a uniform contact to the Si and the TiN acts as a diffusion barrier between the silicide and the Al-Si metallisation. The thickness of both the $TiSi_2$ and TiN is determined by four-point probe measurements on blanket wafers. Since the film is considerably thicker on the surface of a wafer than at the bottom of a contact hole and also varies across the wafer, this technique is severely limited (Simard-Normandin 1991). Raman microprobe spectroscopy provides a fast, non-destructive measurement technique with a probe size comparable with current VLSI technology. However, Raman microscopy requires calibration against known film thicknesses, measured by TEM, to determine the relevant absorption coefficients.

2 EXPERIMENTAL

The samples in this study consisted of TiN deposited by reactive sputtering and rapid thermally annealed $TiSi_2$. The TiN was deposited on seven p-type, 150mm silicon wafers using an Anelva ILC-1015. The film thicknesses ranged from 8 to 64nm. The $TiSi_2$ films were prepared by sputtering 400Å of titanium on a p-type, 100mm silicon wafer which was then rapid thermally annealed in a N_2 ambient at 650°C for 30s followed by a further 10s at

900°C. The film thickness varied from 32 to 44nm. In both cases the films obtained were of stoichiometric composition as determined by Raman spectroscopy and Auger electron spectroscopy.

Raman spectra were obtained from three positions on each wafer, at the centre, and approximately 15mm from the top and bottom edges along the vertical diagonal. The Raman spectrometer used was an ISA MOLE™ S3000 triple-pass spectrometer with a diode array detector and microscope. The excitation wavelength was the 488nm line of a CW Ar⁺ laser focused to a 1μm spot on the sample. The power was limited to 10mW since excess laser energy can cause TiN to oxidise in air (Begun and Bamberger 1989). An HF cleaned, <100> silicon wafer was used as a reference standard.

TEM samples were prepared from the areas from which the Raman spectra were obtained. The TEM samples were prepared in cross-section by mechanical thinning and ion milling and examined in a JEOL 2000FX STEM.

3. DISCUSSION

Typical Raman spectra from TiN and $TiSi_2$ on silicon are shown in Figs. 1 and 2. The optical phonon of TiN is found at 548rcm^{-1} and this position changes with variations in stoichiometry.   The Raman spectrum of $TiSi_2$ is characterised by four lines at 192, 204, 221, and 224rcm^{-1} and a weaker feature at 402rcm^{-1}. The TiN and $TiSi_2$ signals show little variation with increasing film thickness. If the film is thinner than about 100nm, the 520rcm^{-1} optical phonon from the silicon substrate can be detected through the TiN or $TiSi_2$ films. The thinner the TiN and $TiSi_2$ films, the stronger

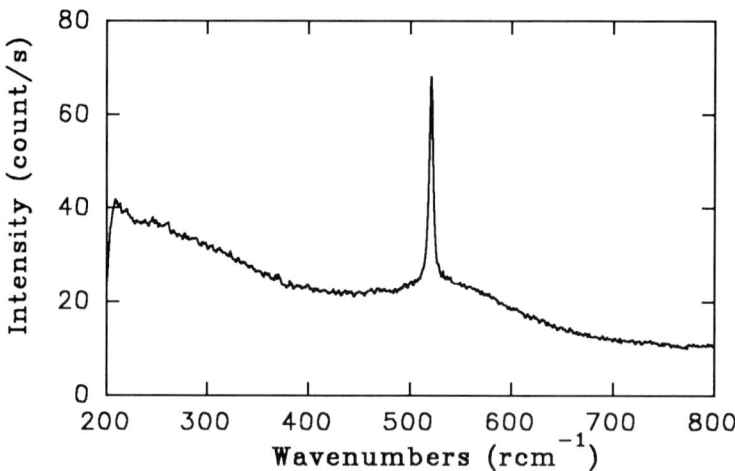

Figure 1: Raman spectrum of TiN on silicon

the silicon signal. The strength of this signal can, therefore, be used to give a precise measure of the TiN and $TiSi_2$ film thicknesses and uniformities. The intensity of the silicon local mode can be expressed as follows (Jenkins and White 1957):

$$I=I_o\exp(-2\alpha t) \qquad (1)$$

where:

I_o is the silicon local mode intensity from a bare silicon layer,
t is the layer thickness,
α is the absorption constant of the film at the wavelength of interest.

Ellipsometry was used to measure α of TiN at three other wavelengths to verify that these values were consistent.

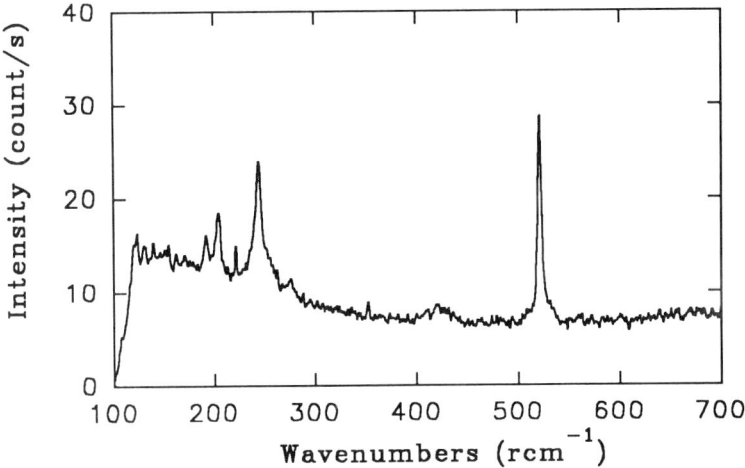

Figure 2: Raman spectrum of TiSi$_2$ on silicon

4. RESULTS

The Raman spectrum from a 51.6nm TiN film is shown in Fig. 1, where the silicon phonon at 520rcm^{-1} is clearly visible. The variation in peak intensity with film thickness is illustrated in Fig. 3. The height of the silicon peak is measured by fitting a Lorentzian profile with a linear baseline to the peak.

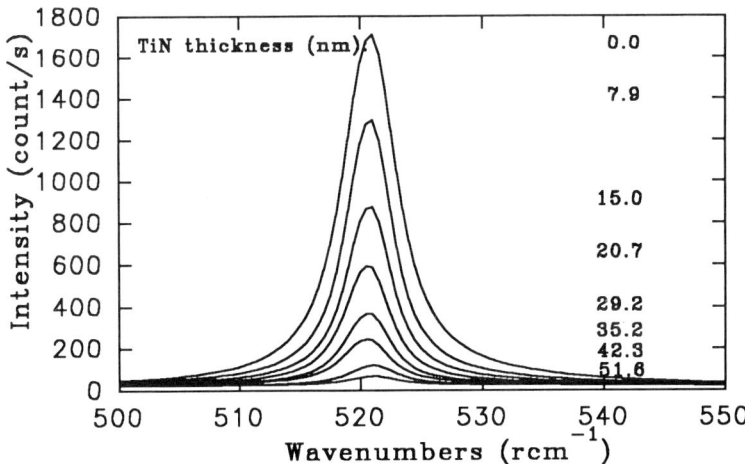

Figure 3: Variation of peak intensity with film thickness for TiN films

The ratio of the intensity of the silicon peak (I/I_0) obtained from Raman spectroscopy is plotted against the thicknesses of the films obtained from

TEM measurements in Fig. 4 for TiN and TiSi$_2$. By fitting the data using equation 1, the absorption coefficient at 500nm (488nm + Raman shift) was found to be 0.030nm^{-1} for TiN and 0.054nm^{-1} for TiSi$_2$.

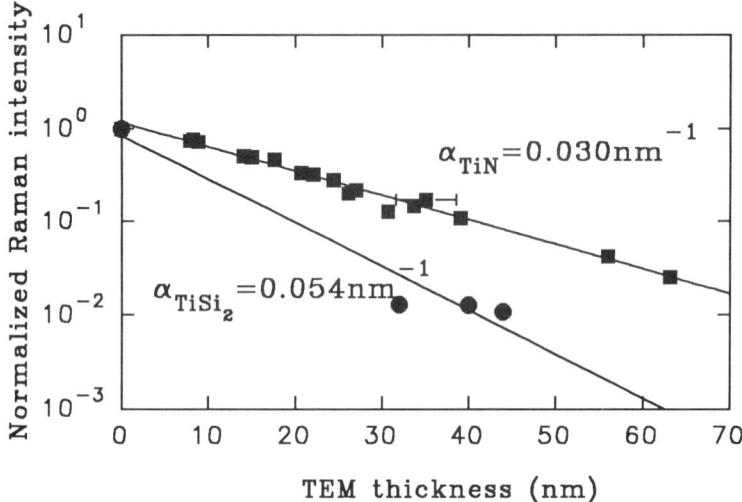

TEM thickness (nm)

Figure 4: Intensity versus thickness for TiN and TiSi$_2$

5. CONCLUSION

The absorption coefficients of TiN and TiSi$_2$ at 500nm have been determined. This permits the use of Raman spectroscopy for the rapid and non-contact measurement of TiN and TiSi$_2$ film thicknesses. Variation in position of the optical phonons for these films also provides a measure of the stoichiometry of the films.

ACKNOWLEDGEMENTS

The authors wish to thank Mr. S. Hambali for the preparation of the titanium films and Dr. V. Q. Ho for useful discussions.

REFERENCES

Totta P.A. and Sopher R.P., 1969, IBM J. Res. Dev. 13, 226
Ting C.Y. and Whittmer M., 1982, Thin Solid Films 96, 327
Ho V. Q., 1989, J. Electronic Mat. 18, 4
Simard-Normandin M., Weaver L., Vacca D., Rogers D., Vitkin A., and Tiedje T., April 1991, Can. J. Phys. 69 (in print)
Begun G.M. and Bamberger C.E., 1989, J. Appl. Spec. 43(1), 134
Jenkins F.A. and White H.E., 1957, Fundamentals of Optics, 3rd edition, McGraw Hill, p200

Inst. Phys. Conf. Ser. No 117: Section 6
Paper presented at Microsc. Semicond. Mater. Conf., Oxford, 25–28 March 1991

Developments in structure studies of bulk GaAs

D J Stirland

Plessey Research Caswell Ltd., Caswell, Towcester, Northants. NN12 8EQ, UK.

ABSTRACT: Experimental observations of macroscopic distributions of dislocations in as-grown, semi-insulating GaAs ingots are described. Possible explanations for the arrangements are considered. Evidence for the existence of micro structure defects, in addition to precipitates associated with dislocations, is presented. Modifications to defect densities and distributions which can be achieved by post-growth heat treatments are described.

1. INTRODUCTION

The characteristics of semi-insulating GaAs, and of MESFETs and ICs fabricated by ion implantation using semi-insulating GaAs substrates, are determined largely by the behaviour of structural defects within the material. Not all defects may be undesirable; indeed, the required electrical properties of semi-insulating GaAs directly result from a compensation process between a deep donor defect, EL2, and residual acceptors. However, inhomogeneities in dislocation and other defect distributions have been linked to lack of uniformity in MESFET threshold voltages, thus impeding the exploitation of large scale integrated circuits (LSI). Considerable improvements in the uniformity of semi-insulating substrates can be achieved by various post-growth heat treatments of as-grown GaAs, although many of the changes which result are not understood.

This contribution describes the principal structural defects which occur in bulk GaAs. It begins with consideration of the macroscopic arrangements of dislocations which are encountered in GaAs, grown by various methods, and continues with a discussion of the explanations which have been proposed to account for the dislocation distributions. Support can be found by microscopic examination of dislocations and associated precipitates. In particular, the microscopic effects of post-growth heat treatments are considered. A microstructural defect, which has been detected by various methods and does not appear to be associated with line dislocations, is described. Attempts to determine the structure of this defect using electron microscopy are reported.

2. DISLOCATION DISTRIBUTIONS

2.1. Cell Structure, Lineage and Slip Dislocations

The general arrangements of dislocations in (001) sections of undoped, semi-insulating, liquid encapsulated Czochralski (LEC) grown GaAs ingots with [001] growth axes have been described extensively (Chen and Holmes 1983, Ono and Matsui 1987, Stirland 1989). The Transmission X-ray Topograph (TXRT) of Fig. 1 illustrates macroscopic dislocation groupings in part of a 2" diam. 300μm thick wafer from the upper (seed) end of an LEC ingot. (Little difference would be evident in a wafer of similar location from a 3" diam. wafer). Dislocation densities are greatest at the periphery, less in the central core, and least

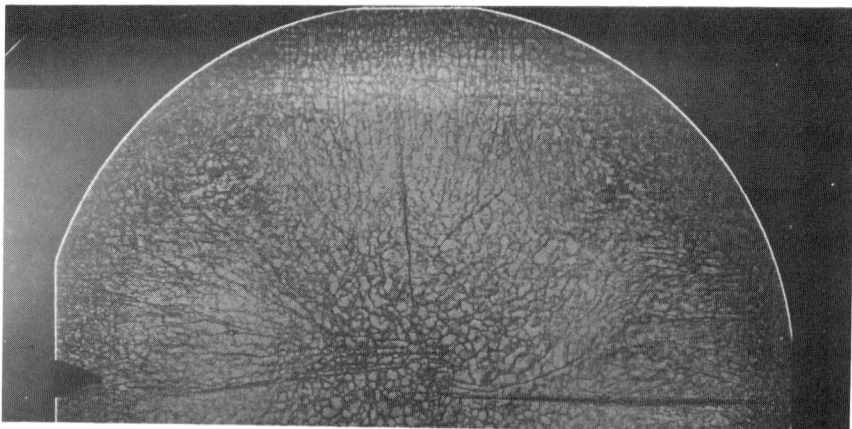

Fig. 1: TXRT of 50 mm diameter undoped, LEC (001) GaAs wafer

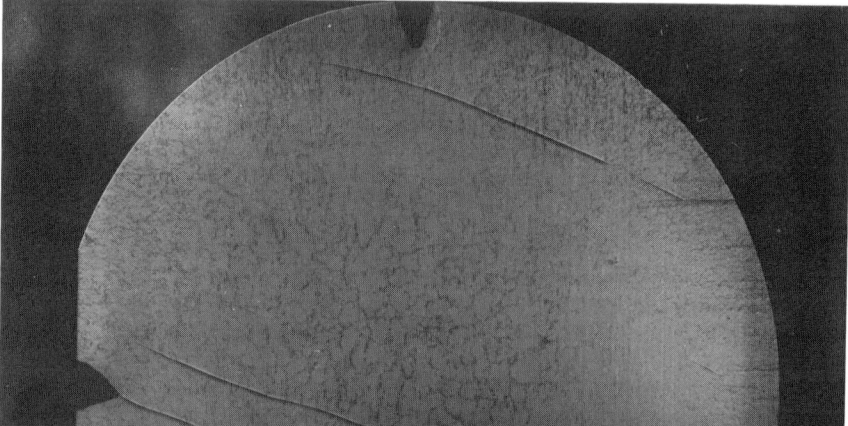

Fig. 2: TXRT of 75 mm diameter undoped VGF (001) wafer. (The dark lines are due to scratches on one surface).

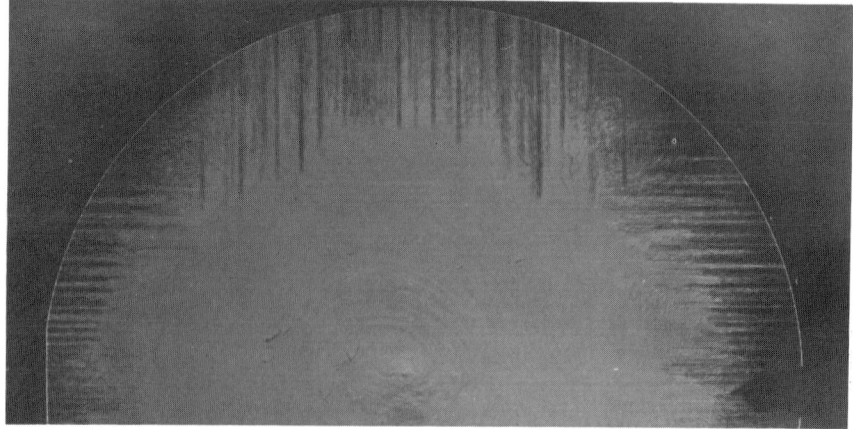

Fig. 3: TXRT of 50 mm diameter indium-doped LEC (001) GaAs wafer

in an annular region between edge and centre. This is the well-known "W-shaped" distribution across a diameter. Microscopically the dislocations are not uniformly distributed, however. Three types of dislocation groupings are observed: cell structure, lineage and slip.

Cell structure is most clearly visible in the central section of the wafer in Fig. 1. The dislocations are clustered into 'cell walls' which enclose nearly dislocation-free regions, 'cell centres'. Cell sizes for 2" and 3" diameter undoped LEC GaAs with dislocation densities $\sim$5-10 x 10^4 cm^{-2} are typically $\sim$500μm, but in general appear to increase as dislocation densities decrease. For example Ikuta et al (1990) noted that cell sizes for liquid encapsulated Vertical Bridgman (VB) grown GaAs of dislocation density 6 x 10^3 cm^{-2} were greater, at 700μm, that those of LEC material, with cell sizes $\sim$300μm at a dislocation density of 5 x 10^4 cm^{-2}. The TXRT of Fig. 2 shows that cell sizes as large as $\sim$2-4 mms can be found in Vertical Gradient Freeze (VGF) material (Gault et al, 1990), with dislocation density $\sim$1-3 x 10^3 cm^{-2}, from a 3" diameter ingot.

Lineage structures consist of closely spaced dislocations extending many millimetres approximately along <110> diameters. Examples can be seen in Fig. 1, but not in Fig. 2. They were shown (Stirland et al 1984) to lie approximately on {110} sheets normal to (00$\bar{1}$), extending several millimetres down the <001> ingot axis. Brown et al (1984) established by double crystal X-ray topography that lattice tilts $\sim\sim$0.01° occurred across regions divided by lineage structures.

The slip dislocations consist of linear arrays along exact orthogonal <110>, forming "tartan" patterns where they overlap. They are particularly well marked on the X-ray topography of an indium-doped LEC GaAs ingot shown in Fig. 3, although they are generally present, less distinctly, in undoped bulk GaAs.

2.2. Proposed Explanations for Dislocation Distributions

Attempts to analyse lineage and cell structures by determining Burgers vectors of individual dislocations, using Transmission Electron Microscopy (TEM) examinations of thinned GaAs foil samples (Brown 1990), have largely been unsuccessful due to the complexity of the dislocation tangles. At best, parts of dislocations at the edges of tangles can be analysed, showing that tangles contain dislocations with at least three different Burgers vectors (Ono and Matsui 1987). Much earlier, Jenkinson and Lang (1962) had encountered a similar problem in attempting analysis of dislocation tangles in float zone silicon. They suggested that the tangles represented regions where slip dislocations were 'trapped' in localised volumes of high dislocation density ($\geq 10^5$ cm^2) and proposed that this might result from Lomer reactions such as:

$$b_1 + b_2 \rightarrow b_3;\ ^1/_2\, a_0\, [\bar{1}10] + ^1/_2\, a_0\, [10\bar{1}] \rightarrow ^1/_2\, a_0\, [01\bar{1}] \qquad \ldots\ldots(1)$$

with dislocation b_1 gliding on (11$\bar{1}$) interacting with dislocation b_2 gliding on (1$\bar{1}$1) to form a sessile edge dislocation b_3 lying along [011]. Such a dislocation reaction was indeed identified.

An alternative approach is to examine specimens with reduced dislocation densities. By alloying GaAs with 10% phosphorus, Ono and Matsui (1987) produced LEC samples with dislocation densities of 1 x 10^4 cm^{-2}. TEM analysis found several dislocation reactions producing sessile Lomer-Cottrell dislocations. By doping with indium ([In]$\sim$10^{20} cm^{-3}), Ono (1988) produced essentially dislocation-free specimens, which were then plastically deformed to simulate stresses present during crystal growth. By analysis of stress distributions (Kitano et al 1986) the observed slip dislocation distributions at the peripheries of wafers could be explained. It was concluded that, in the early stages of dislocation

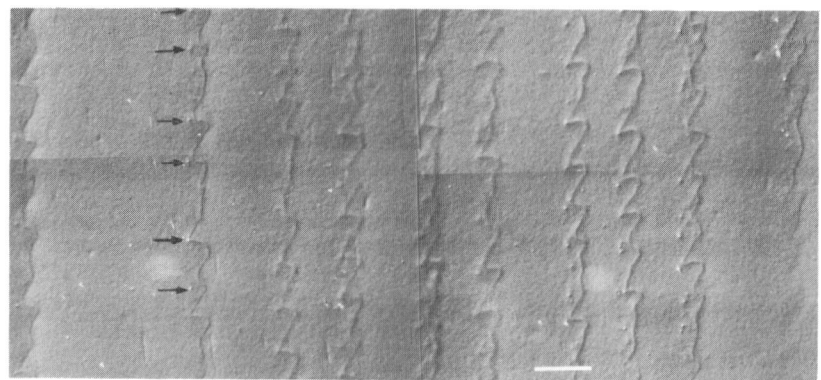

Fig. 4: A/B etched (110) section through centre of In:GaAs ingot. Arrows indicate etch pits from arsenic precipitates. Marker = 20μm.

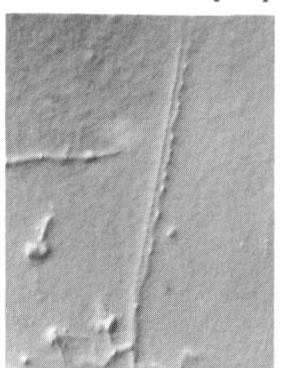 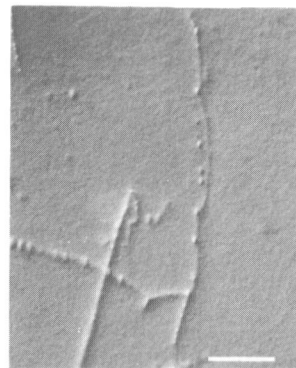

Fig. 5: Dislocation climb in 100 mm diameter undoped LEC (001) GaAs wafer. Marker = 20μm

Fig. 6: Dislocation climb in 50 mm diameter VGF GaAs wafer. Marker = 20μm

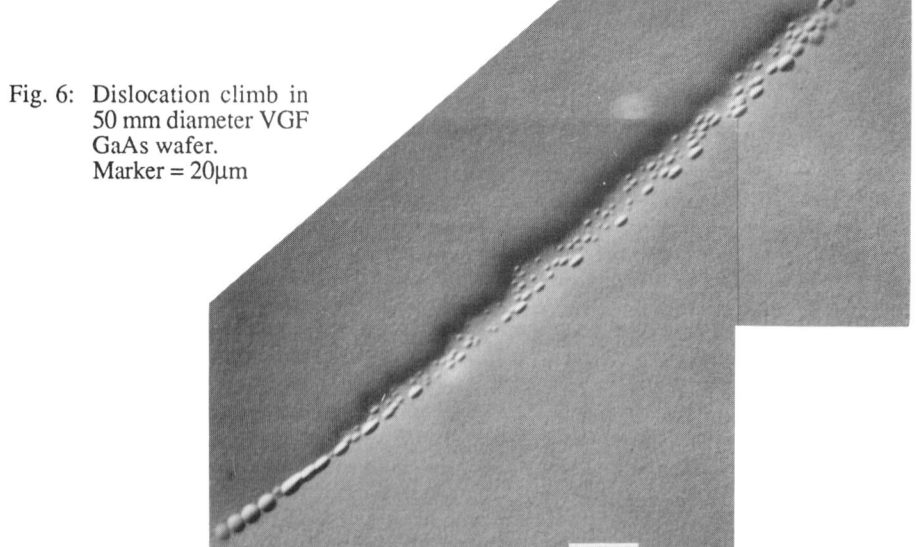

introduction, slip dislocations along <110> directions interact to form sessile <110> dislocations, the precursors of <110> lineage. It is most likely that the cellular distributions of dislocations are due to polygonisation, requiring both glide and climb. Furthermore, point defects are required for climb to occur, which implies that temperatures close to the melting point may be necessary.

Many of these requirements are compatible with results of numerical simulations of thermal stresses in LEC crystal growth (Chabli et al 1990). These indicate thermal stresses with a W-shaped diametric variation across (001) are present in (001) sections near the melt-crystal interface (at ~1200°C) but U-shaped stresses are present near the top of the boric oxide layer (at ~1100°C). Thus the cellular structures can be explained by dislocation interactions involving glide and climb processes under the W-shaped stresses, with climb resulting from the plentiful supply of point defects available at these high temperatures. As the crystal is pulled through the encapsulant the U-shaped stresses take over, at lower temperatures, and these stresses will tend to superimpose a pure glide arrangement of dislocations on the cell structures, limited to the edge regions. In indium-doped LEC GaAs the critical stress for dislocation movement is increased to values not achieved near the melt-solid interface, but only at the edges near the top of the encapsulant. Thus the principal dislocations seen in indium-doped GaAs are the stress-induced glide dislocations at the ingot edge (see Fig. 3). Additional dislocations are generated at the seed-crystal interface and propagate normal to the growth interface (Stirland et al 1987). Alexander and Gottschalk (1989) conclude, in a similar way, that two kinds of dislocations are generated at different temperatures in undoped LEC GaAs: those which form cell walls from vacancy agglomeration followed by climb at high temperature, and slip dislocations which form at lower temperatures by multiplication of dislocations from surface sources. Additional support for many of these ideas has resulted from detailed examination of the microscopic configurations of single dislocations.

3. MICROSCOPIC EXAMINATION OF DISLOCATIONS AND ASSOCIATED DEFECTS

3.1. Evidence for Dislocation Climb

It has been suggested (Stirland et al 1984, 1987; Stirland 1989) following A/B etching of (001) and (110) sections of GaAs, that dislocation bowing between etch pits could be evidence for dislocation climb, rather than pinning of gliding dislocations. The etch pits have been established as the sites of arsenic precipitates on the dislocations (Cullis et al 1980, Stirland et al 1984). An example of complex dislocation bowing between etch pits is shown in Fig. 4, a micrograph of an A/B etched (110) section from the central region of an indium-doped GaAs crystal annealed after growth. The dislocations are edge type with $\underline{b}$ = $\pm \frac{1}{2} a_0 [110]$ or $\pm \frac{1}{2} a_0 [1\bar{1}0]$ (Ono 1990). An important TEM investigation of dislocation configurations in cell walls at the centre, and slip bands at the periphery, of undoped semi-insulating LEC GaAs has been reported recently (Wurzinger et al 1989). Several examples of dislocation climb were analysed, including a clear example of a dislocation bowed between precipitates. In all but one instance the analysis indicated that positive climb (addition of vacancies or emission of interstitials) had occurred. As the authors point out, this contradicts the assumption, based on the fact that semi-insulating GaAs is grown from an arsenic-rich melt, that dislocation climb can act as a mechanism to reduce the concentration of arsenic interstitials [As_i]. It has been suggested that the enhanced concentrations of the deep donor defect EL2 (which is linked with the arsenic antisite defect As_{Ga}) in the vicinity of dislocation cell walls in as-grown, undoped LEC GaAs may result from interstitial climb by such reactions as

$$As_i - CS \rightarrow V_{Ga}; \ As_i + V_{Ga} \rightarrow As_{Ga} \qquad \qquad \ldots\ldots(2)$$

where -CS is the (negative) interstitial climb step. However, if vacancy climb occurs then a different reaction might be:

$$V_{Ga} + CS \rightarrow As_i; \quad V_{Ga} + As_i \rightarrow As_{Ga} \qquad \qquad ......(3)$$

Thus, either mechanism could act to enhance [EL2] in the vicinity of a climbed dislocation.

Figs. 4-6 show typical examples believed to illustrate dislocation climb, in 50 mm diameter indium-doped GaAs, 100 mm diameter LEC GaAs and in 50 mm diameter VGF GaAs. The latter two materials exhibit cell structure, whereas the dislocations in Fig. 4 belong to the central group which originate at the seed-melt interface (Stirland et al 1987, Ono 1990). All three materials thus contain some dislocations which exhibit climb behaviour. Slip dislocations are most pronounced at the periphery of indium-doped GaAs (e.g. Fig. 3). The micrograph of Fig. 7 shows a region near a <100> edge, after an A/B etch. The dislocations essentially lie along orthogonal <110> with few interactions evident. Where precipitates are present there is evidence for climb. Because helical dislocations can be detected in the peripheral regions (Fig. 8), it may be concluded that conditions at the periphery enable some dislocation climb to take place, at dislocations which are essentially formed by slip.

3.2. Effects of Post-Growth Heat Treatments on Defect Distributions

A considerable number of investigations into the effects of annealing on as-grown, semi-insulating, bulk GaAs have been reported (e.g. Rumsby et al 1984, Ogawa 1988, Clark et al 1990). Recently Mori et al (1990) have employed wafer, rather than bulk, annealing procedures. Generally, 'conventional' post-growth heat treatments at temperatures from ~850°C-1000°C do not affect as-grown dislocation densities (Clark et al 1990). However, studies of more extreme treatments such as rapid cooling (quenching) from high ($\geq$1100°C) temperatures have shown that ~10^6 cm^{-2} additional dislocations can be introduced (Stirland et al 1990a). The arrangements that the 'new' dislocations adopt is relevant to the two types of dislocations in as-grown material. TXRT assessment was attempted on a wafer from an undoped LEC GaAs ingot which had been annealed at 1100°C for 5 hours in vacuo, quenched in air, and subsequently reannealed at 950°C for 5 hours. No transmitted 220 diffraction spot could be detected, using a wafer doubly polished to 300μm thickness. However, when the same wafer was examined in the reflection mode a $1\bar{1}5$ diffraction spot of usable intensity was produced. Fig. 9 shows part of the topograph, with dislocations appearing in light contrast. Dislocations at the centre of the wafer are in the form of polygonised cell structure, whereas those at the wafer edges are clearly slip dislocations lying along orthogonal <110>. Figs. 10 and 11 show how these different regions appear after A/B etching. The average cell size at the wafer centre at dislocation density ~1.5 x 10^6 cm^{-2} is 77 ± 30μm, which continues the trend previously noted that cell size decreases with increase in dislocation density. The micrograph of Fig. 10 clearly indicates that the additional dislocations become an integral part of the 'new' cell structure, (Fawcett et al 1989) in the central region of the wafer. However, Fig. 11 clearly indicates that at the periphery of the wafer the additional dislocations do not behave in this way: they form a cross-hatch, or tartan, pattern of long nearly straight dislocations along the orthogonal <110> superimposed on what is presumably the original as-grown cell structure. The size of this cell structure is larger, at ~100-200μm, than that of the central region. This is in contradiction to cell size distributions in as-grown LEC GaAs, which show smaller cells at the wafer edges than in the centre. This lends further support to the view that the cell sizes in the central region of the quenched ingot result from the additional dislocations. Thus the additional dislocations act in two different ways, depending on their location. At the periphery of the wafer their characteristics are those of slip dislocations; at the centre of the wafer they appear to behave as grown-in dislocations which polygonise by climb and glide processes. It can be concluded that temperatures no higher than 1100°C are needed for polygonisation of the new dislocations to take place in the central region of the ingot.

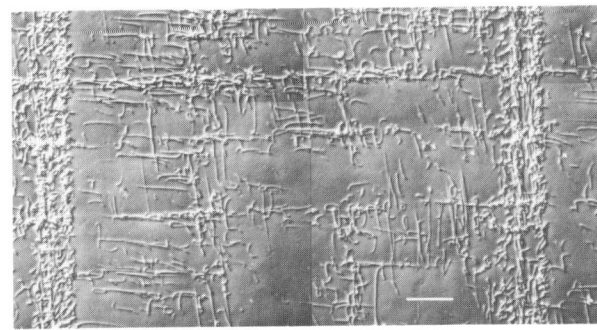

Fig. 7: Slip dislocations near <100> edge of In:GaAs wafer after an A/B etch. Marker = 100μm

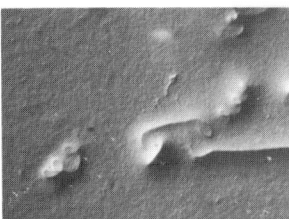
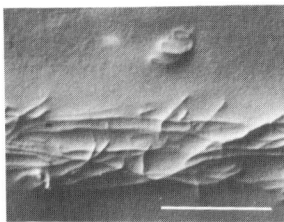

Fig. 8: Helical dislocations near <110> edge in In:GaAs. Marker = 100μm

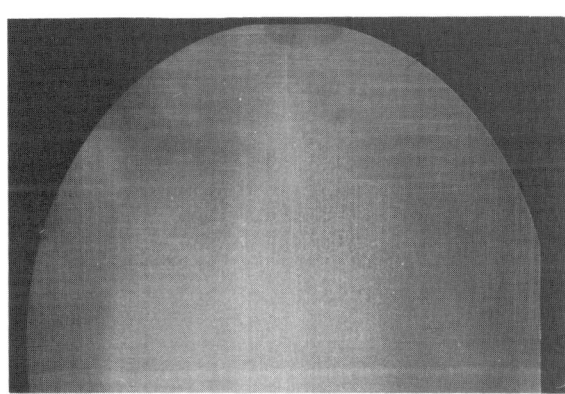

Fig. 9: 115 reflection X-ray topograph from 2" diameter wafer of GaAs ingot quenched from 1100°C

Fig.10: Polygonised dislocations at centre of wafer from quenched ingot

Fig.11: Dislocations at wafer edge, quenched ingot. Marker = 100μm

4. MICROSTRUCTURE DEFECTS IN CELL INTERIORS

Using infra-red Light Scattering Tomography (LST) Inada et al (1989) examined changes in scattering centres, presumably arenic precipitates associated with cell wall dislocations, produced by various annealing schedules. No scatterers were observed in wafers from bulk GaAs quenched after a 10 hour anneal at 1200°C. Similar results were found using the high resolution infra-red Laser Scanning Microscope (LSM) described by Kidd et al (1987). Particles were detected in bulk GaAs at densities of ~5 x 10^7 cm^{-3} after quenches from 1000°C, but none were seen after quenches from 1100°C and 1200°C (Stirland et al 1989). However, Inada et al (1989) showed that a second anneal at 500°C for 30 hours of material quenched from 1200°C resulted in strong LST intensity from dislocation cell wall regions and from the dislocation-free cell centres. They proposed that homogeneous nucleation of supersaturated arsenic at the cell centres was occurring during the 500°C anneal, thereby producing small scattering centres. Similar microstructure has been seen by LST inside dislocation cells in untreated material (Fillard et al 1988, Gall et al 1988, Martin et al 1989). Stirland (1990) has associated roughened areas within dislocation cells, resulting from A/B etching, with the scattering centres. Fig. 12 shows an example of an A/B etched (001) surface which reveals the roughened microstructure in the cell interior. However, in this ingot annealed specimen the cell walls are also surrounded by fine scale microstructure, with a smooth boundary, or denuded zone, between the cell wall and cell interior. The cell wall dislocations and the few isolated dislocations within the cell centre are decorated with arsenic precipitates (etch pits). Stirland et al (1990b) have recently shown that excellent correlation between A/B etched regions, infra-red LST Reverse Contrast images (Mohades-Kassai and Brozel 1990) and low temperature (10K) cathodoluminescence (CL) images is possible. This correlation has allowed the use of CL imaging to locate suitable areas of microstructure for TEM examination. This avoids the use of A/B etched foil samples to locate the microstructure regions, which can show undesirable roughening and surface contamination.

In the initial TEM investigation of microstructure (Stirland et al 1990b) one particle ~50 nm in diameter attached to a possible dislocation loop was found, and although no electron diffraction analysis was possible it was conjectured that this was an arsenic precipitate. Further examinations of two different wafers have recently been made, and in the microstructure regions remote from line dislocations a number of voids have been identified. Fig. 13 shows typical examples. It is believed that this is the first reported observation of such defects in undoped, semi-insulating, LEC GaAs. Further details will be found in Williams et al (1991). It should be noted that the various light scattering results are not in conflict with the view that the microstructure is partially, or entirely, comprised of voids, possibly associated with microprecipitates (of arsenic), since these techniques rely on refractive index changes at the scattering entities.

5. CONCLUSIONS

Dislocation arrangements in undoped, semi-insulating LEC GaAs ingots of 50, 75 and 100 mm diameter consist of cell structure, lineage and slip dislocations. VGF and VB GaAs exhibit cell structure and slip, but not lineage. Lineage probably results from sessile dislocation formations along <110> by interactions between dislocations on inclined slip planes. Cell structure results from polygonisation of dislocations, requiring climb and glide. Cell sizes decrease with increase in average dislocation density. The microscopic examination of dislocations in cell structure regions provides evidence for climb. Examination of a quenched ingot suggests that cell structure can be formed at temperatures as low as 1100°C. Regions of microstructure in cell interiors remote from dislocations contain voids.

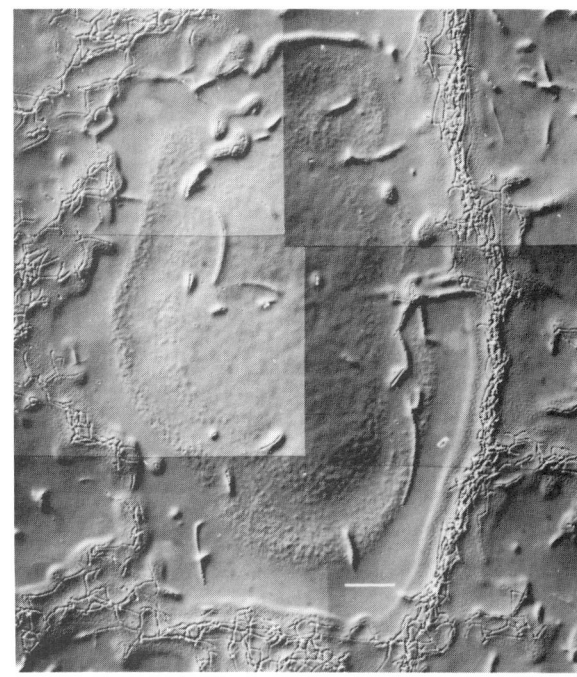

Fig.12: Microstructure region inside cell, revealed by A/B etch. Notice fine structure also surrounds the cell walls.
Marker = 100μm

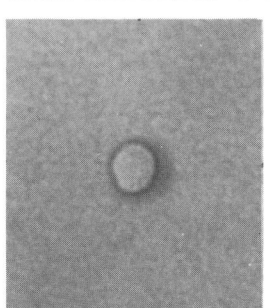 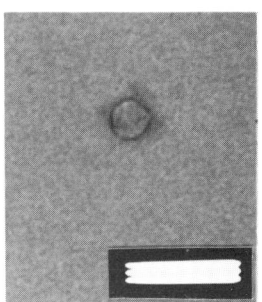

Fig.13: Voids in regions of microstructure in two different GaAs wafer specimens.
Marker = 50 nm

ACKNOWLEGEMENTS

I would like to thank M.R. Brozel,. J.E. Clemans, P. Davies, P. Gall, I. Grant, D.G. Hart, D.T.J. Hurle, R.C. Newman, J.A. Turner, and R.H. Wallis for assistance and encouragement with various aspects of this work. Part of the work was supported and sponsored by the Procurement Executive, Ministry of Defence (Royal Signals and Radar Establishment).

REFERENCES

Alexander H and Gottschalk H 1989 Inst. Phys. Conf. Ser. No. 104 281
Brown G T, Skolnick M S, Jones G R, Tanner B K and Barnett S J 1984 Semi-insulating III-V Materials, Eds D C Look and J S Blakemore (Shiva Publ. Ltd.) pp.76-82.
Brown G T 1990 private communication
Chabli A, Molva E, Le Ludec J P, Bunod P, Perret C and Bertin F 1990 Defect Control in Semiconductors Ed K Sumino (Amsteram: North Holland) pp679-684
Chen R T and Holmes D E, 1983, J. Crystal Growth, 61 111

Clark S, Brozel M R, Stirland D J, Hurle D T J and Grant I 1990 Defect Control in Semiconductors Ed. K. Sumino (Amsterdam: North Holland) pp807-812

Cullis A G, Augustus P D and Stirland D J 1980, J. Appl. Phys. 51 2556

Fawcett T J, Brozel M R and Stirland D J, 1989,. Inst. Phys. Conf. Ser. No. 106 pp.19-24

Fillard J P, Gall P, Weyher J. L, Asgarina M and Montgomery P C, 1988, Semi-insulating III-V Materials Eds G Grossman and L Ledebo (Bristol:Adam Hilger) pp537-542.

Gall P, Fillard J P, Castagne M, Weyher J. L and Bonnafe J, 1988, J. Appl. Phys. 64 5161.

Gault W A, Clemans J E, Conway J H, Dominquez F, Ejim T I, Monberg E M and Simchock F, 1990, Defect Control in Semiconductors, Ed K. Sumino (Amsterdam: North Holland) pp653-660.

Ikuta K, Nakanishi H, Kohda H and Hoshikawa K, 1990 Defect Control in Semiconductors, Ed K Sumino (Amsterdam: North Holland) pp661-666.

Inada T, Otoki Y, Ohata K, Taharasako S and Kuma S, 1989 J. Crystal Growth 96 327.

Jenkinson A E and Lang A R, 1962, Direct Observation of Imperfections in Crystals, Eds J B Newkirk and J H Wernick (New York: Interscience, John Wiley and Sons) pp471-495.

Kidd P, Booker G R and Stirland D J, 1987, Appl. Phys. Lett. 51 1331.

Kitano T, Ishikawa T, Ono H and Matsui J, 1986 Semi-insulating III-V Material, Eds H Kukimoto and S. Miyazawa (Amsterdam: North Holland) pp.91-96.

Martin G M, Suchet P, Deconinck P and Gillardin G 1989, Inst. Phys. Conf. Ser. No. 100 p363-372.

Martin S, Suchet P and Martin G M 1990, Semi-insulating III-V Materials, Eds A G Milnes and C J Miner (Bristol: Adam Hilger) pp1-10.

Mohades-Kassai A and Brozel M R, 1990, J. Crystal Growth 103 303.

Mori M, Kano G, Inoue T, Shimakura H, Yamamoto H and Oda O, 1990, Semi-insulating III-V Materials, Eds A Miles and C J Miner (Bristol: Adam Hilger) pp155-160.

Ogawa O, 1988, Semi-insulating III-V Materials Eds G Grossmann and L Ledebo (Bristol: Adam Hilger, pp477-482/

Ono H and Matsui J, 1987, Appl. Phys. Lett. 51 801.

Ono H, 1988, J. Crystal Growth 89 209

Ono H, 1990, J. Crystal Growth 102 949.

Rumsby D, Grant I, Brozel M R, Foulkes E J and Ware R M, 1984, Semi-Insulating III-V Materials, Eds D C Look and J S Blakemore (Shiva Publ. Ltd.) pp165-170.

Stirland D J, Augustus P D, Brozel M R and Foulkes E J, 1984, Semi-Insulating III-V Materials, Eds D C Look and J S Blakemore (Shiva Publ. Ltd.) pp91-94.

Stirland D J, Hart D G, Grant I, Brozel M R and Clark S 1987, Inst. Phys. Conf. Ser. No. 87 pp269-274.

Stirland D J, Kidd P, Booker G R, Clark S, Hurle D T J, Brozel M R and Grant I 1989, Inst. Phys. Conf. Ser. No. 100 pp373-378.

Stirland D J, 1989 Proc. 6th Int. Symp. Structure and Properties of Dislocation in Semiconductors, Inst. Phys. Conf. Ser. No. 104, 431.

Stirland D J, 1990 Defect Control in Semiconductors Ed. K Sumino (Amsterdam: North Holland) pp783-794.

Stirland D J, Brozel M R, Breivik L, Clark S, Williams G M and Cullis A G, 1990a, Proc. XIIth State of the art program on compound semiconductors, Eds D C D'Avanzo, A T Macrander, R E Enstrom and D. Decoster (Publ. Electrochem. Soc. Inc. New Jersey) Vol 90-15 pp76-88.

Stirland D J, Gall P, Brozel M R, Breivik L, Williams G M, Cullis A G and Fillard J P, 1990b, Inst. Phys. Conf. Ser. No. 112 pp55-60.

Williams G M, Cullis A G and Stirland D J, 1991, submitted to Appl. Phys. Lett.

Wurzinger P, Oppolzer H, Pongratz P and Skalicky P 1989, Inst. Phys. Conf. Ser. No. 100, pp385-390.

Inst. Phys. Conf. Ser. No 117: Section 6
Paper presented at Microsc. Semicond. Mater. Conf., Oxford, 25–28 March 1991   337

TEM investigations of defects in undoped semi-insulating GaAs crystals

P Schloßmacher, H Rüfer* and K Urban

Institut für Festkörperforschung, Forschungszentrum Jülich GmbH
D–W5170 Jülich, Germany
*Wacker Chemitronic GmbH, D–W8263 Burghausen, Germany

ABSTRACT : For the first time a complete analysis of nearly 800 dislocations in Liquid Encapsulation Czochralski (LEC) GaAs wafers was performed. By calculation of Schmid factors for all glide systems using reported values of thermal stresses it could be shown that dislocations building up the cell walls penetrate the crystal from the outer periphery. Moreover, Arsenic (As) precipitates were observed in all wafers as a second defect type. The distributions of their sizes before and after post-growth annealing were studied yielding valuable information on As diffusion in GaAs.

1. INTRODUCTION

The main requirements for GaAs as a substrate material for device applications are a low dislocation density (i. e. etch pit density EPD), no microdefects, and homogeneous electrical properties. Available 2" undoped, semi-insulating LEC crystals still exhibit EPDs in the range of 10^4 cm^{-2}. Dislocations, however, have a great influence on the homogeneity of integrated devices (Miyazawa et al 1986, Packeiser et al 1986). Also As precipitates are found in crystals grown from stoichiometric melts (Cullis et al 1980, Rüfer et al 1990).

The origin of dislocations building up the cellular structure is not fully understood. Alexander (1989) called for a strict separation between dislocation nucleation and multiplication and discussed in detail possible dislocation sources acting in a dislocation free matrix. As a result only two mechanisms remained: surface damage of any kind and point defect aggregates. To mention thermal stresses in this context is not correct (Alexander 1989). In order to generate dislocations alone thermal stresses have to be in the order of the theoretical shear strength $G/2\pi$ which is evaluated to about 6 GPa for GaAs and temperatures near the melting point using elastic constants provided by Jordan (1980a). Even the highest calculated stresses (Jordan et al 1980b, Duseaux 1983) during crystal growth remain two orders of magnitude smaller.

Starting from the assumption that dislocation sources exist at or near the crystal surface we calculated the Schmid factor distribution for all twelve glide systems at the periphery of a [001]-oriented GaAs wafer. The procedure is the same as employed by Jordan et al (1980b), but we used stress components as described in a theoretical paper by Szabó

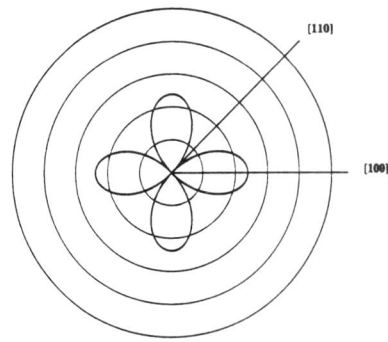

$[1\bar{1}0](\bar{1}\bar{1}1); [1\bar{1}0](111); [\bar{1}\bar{1}0](\bar{1}1\bar{1}); [\bar{1}\bar{1}0](1\bar{1}\bar{1})$
AB(c); AB(d); CD(a); CD(b)

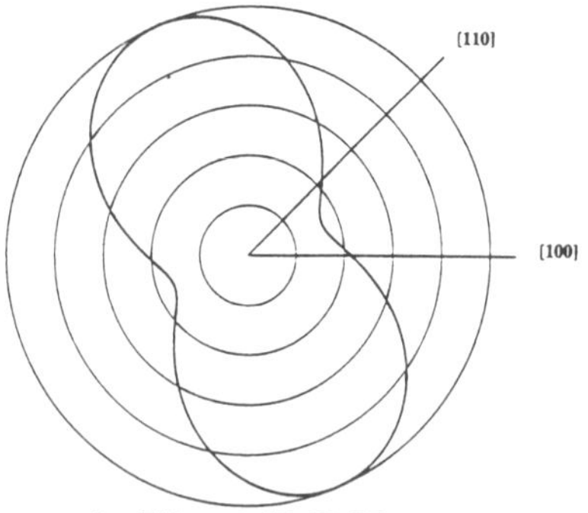

$[10\bar{1}](111); [\bar{1}0\bar{1}](\bar{1}\bar{1}1)$
BC(d); AD(c)

Figure 1: Schmid factor distribution over a [001] wafer for all four glide systems with $\vec{b}$ in plane of the wafer.

Figure 2: Schmid factor distribution over a [001] wafer for two glide systems with $\vec{b}$ at 45° to the wafer normal.

(1985) (r=R, z=R, σ_r=0, σ_θ=1, σ_z=-0.9, σ_{rz}=0). As an example the Schmid factor distribution for the given stresses and all glide systems with $\vec{b}$ in the plane of the wafer are shown in Fig. 1 and for two selected glide systems with $\vec{b}$ at 45° to the wafer normal in Fig. 2. The distribution for the other six glide systems with $\vec{b}$ not in plane is obtained by rotation of Fig. 2 by one, two and three times 45°, respectively. The maximum Schmid factors always appear midway between ⟨100⟩- and ⟨110⟩-directions. In both figures the same scaling is used, and both cubic and Thompson notation (see e.g. Hirth and Lothe 1968) are given simultaneously. Such a model with a dominating tangential stress (σ_θ) was previously applied by Kitano et al (1986a) for the explanation of the dislocation structure in In-doped GaAs crystals which exhibit no cellular structure but slip lines with an eightfold symmetry at the outer periphery of [001] wafers. To our knowledge an interpretation of so called grown-in dislocations building up the cell walls by application of this model has never been tried.

2. EXPERIMENTAL

Wafers were taken from three commercially grown, undoped semi-insulating 2" GaAs single crystals. Two crystals were post-growth annealed at 1,100°C for 10 h (wafer #2 and #3); one crystal remained untreated ("as-grown", wafer #1). KOH etching revealed the well known W-profile and fourfold symmetry of the dislocation density with average values in the range of 1.6–3.8×10⁴ cm⁻². All wafers exhibited the cellular structure with some lineage, but no slip lines were visible. Care was taken to prepare only specimens from regions of the cellular structure in order to investigate mainly dislocations building up the cell walls.

Dislocations and precipitates were investigated by analytical transmission electron microscopy (TEM) using a Philips EM430T equipped with a double-tilt specimen holder and an energy-dispersive X-ray system (EDAX). In a complete dislocation analysis Burgers vector $\vec{b}$, line direction $\vec{u}$ and habit plane normal $\vec{n}$ were determined for about 800 dislocations. The nature of precipitates was analysed by both microdiffraction and

X-ray analysis whenever possible. Moreover, their shape, size and volume density was determined.

3. RESULTS

The overall dislocation patterns of etched wafers from ingots with or without post-growth annealing showed no marked differences. Dislocation analysis revealed no other $\vec{b}$ than the expected type $a/2\langle 110\rangle$. In Table 1 the results of dislocation type analysis are summarized. The two columns "sessile" and "?" denote dislocations on a $\{111\}$ plane with $\vec{b}$ lying on another $\{111\}$ plane and dislocations that belong to no $\{111\}$ plane at all, respectively. The line directions are mainly parallel $\langle 110\rangle$ or $\langle 112\rangle$. By far the major part of dislocations (70–80%) were lying on $\{111\}$ glide planes in a glissile configuration and only 20–30% were sessile or had climbed out of their glide planes.

Table 1: Analysis of dislocation type in three different wafers.

wafer	number	screw	30°	60°	edge	sessile	?
#1	284	20	105	52	41	7	60
	(100%)	(7%)	(37%)	(18%)	(15%)	(2%)	(21%)
#2	269	22	83	82	32	24	26
	(100%)	(8%)	(31%)	(30%)	(12%)	(9%)	(10%)
#3	242	26	69	53	30	11	53
	(100%)	(11%)	(28%)	(22%)	(12%)	(5%)	(22%)

The distributions of Burgers vectors are presented in Table 2. Only a distinction was made between $\vec{b}$ laying in plane (AB or CD) or not (AC, AD, BC or BD). A considerable number of in plane Burgers vectors was found in all three wafers at the $\langle 110\rangle$-periphery. This is not consistent with our model which predicts a disappearing Schmid factor for glide systems with Burgers vectors AB or CD at the $\langle 110\rangle$-periphery as can be seen in Fig. 1. Therefore detailed investigations of Burgers vector distributions in single specimens were performed taking into account possible dislocation reactions. Because of limited space we present only one analysis of a sample taken from the $\langle 110\rangle$-periphery of wafer #3 in Table 3. At this peripheral site the Schmid factor of glide systems AD(b), BC(a), BD(a), AC(b) reaches nearly 90% of its maximum value. The maximum lies at the

Table 2: Results of Burgers vector analysis in three different wafers.

wafer	$\vec{b}$	$\langle 100\rangle$-periphery	$\langle 110\rangle$-periphery	center of the wafer
#1	in plane	39 (31%)	17 (24%)	13 (18%)
	out of plane	85 (69%)	53 (76%)	58 (82%)
#2	in plane	44 (60%)	14 (24%)	33 (29%)
	out of plane	29 (40%)	44 (76%)	82 (71%)
#3	in plane	45 (34%)	24 (34%)	5 (23%)
	out of plane	86 (66%)	47 (66%)	17 (77%)

Table 3: Burgers vectors of a sample taken from the ⟨110⟩-periphery of wafer #3 (Thompson notation).

Burgers vector	BC	AD	BD	AC	AB	CD
occurrences	10	2	4	2	7	10

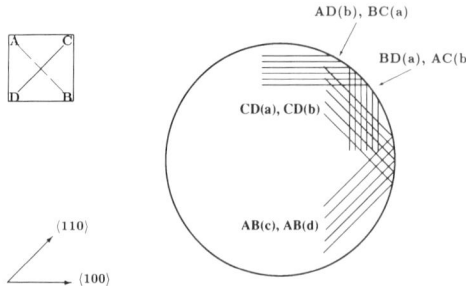

Figure 3: Projection of propagation directions (i. e. $\vec{b}$) on the [001] wafer plane for glide systems expected at a selected ⟨110⟩-periphery. Arrows mark the sites were the Schmid factors for the glide systems concerned assume their maximum values. A Thompson Tetrahedron is also included.

arrows in Fig. 3. Therefore, dislocations in these glide systems can be generated directly. The straight lines in Fig. 3 denote the propagation directions of the dislocation in the glide systems concerned i.e. the projection of $\vec{b}$ on the wafer plane. Also the propagation directions of the four glide systems with $\vec{b}$ in plane (having their maximum Schmid factor in ⟨100⟩-directions) are drawn at one of the two neighbouring ⟨100⟩-directions. As can be seen from their propagation directions in Fig. 3 these dislocations cannot reach the periphery of the wafer in ⟨110⟩-direction by glide processes so that their appearance must be due to dislocation reactions. The reactions between the involved glide systems are the following ones:

$$
\begin{aligned}
BC(a) + AD(b) &\longrightarrow \quad no\,reaction & (a) \\
BC(a) + CA(b) &\longrightarrow \quad AB(along\,CD) \quad [Lomer] & (b) \\
BC(a) + DB(a) &\longrightarrow \quad CD(a) & (c) \\
AD(b) + CA(b) &\longrightarrow \quad DC(a) & (d) \\
AD(b) + DB(a) &\longrightarrow \quad BA(along\,CD) \quad [Lomer] & (e) \\
AC(b) + BD(a) &\longrightarrow \quad no\,reaction & (f)
\end{aligned}
\tag{1}
$$

There are only three different results: no dislocation reaction at all because of perpendicular Burgers vectors (the resulting one would be of type a⟨100⟩), a Lomer dislocation and glissile one with $\vec{b}$ =CD on plane a. Since seven of the ten $\vec{b}$ =CD in Table 3 belong the glide systems CD(a), the most probable reactions produce exactly the most frequently appearing glide system.

All investigated specimens contained precipitates which were identified as hexagonal As. They were of tetrahedral shape with rounded apexes and bounding GaAs {111} matrix planes. In "as-grown" material 80% of the precipitates were found along a dislocation line whereas in post-growth annealed wafers this amount is nearly 95%. The most prominent effect of the annealing treatment can be seen in the distribution of sizes of the precipitates in Fig. 4. The comparison between "as-grown" (Fig. 4, left side) and

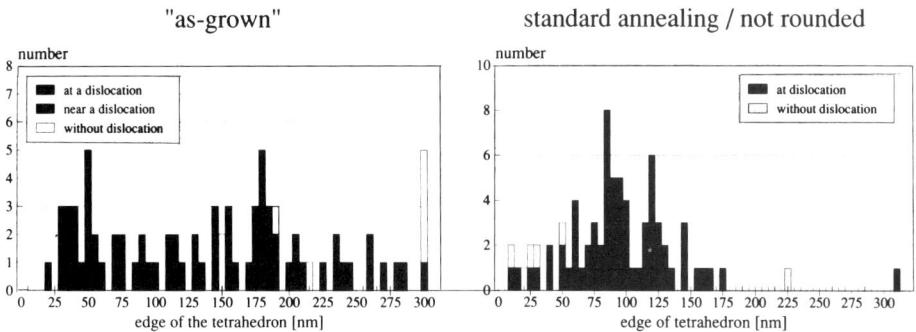

Figure 4: Size distribution of As precipitates in "as-grown" (wafer #1, left side) and post-growth annealed material (wafer #3, right side).

post-growth annealed (Fig. 4, right side) material directly shows that in "as-grown" GaAs there is no distribution around an average value but a statistical scattering of dimensions. Table 4 lists the evaluated values for the total number of observed precipitates, their average edge dimensions, volume density, and calculated excess amount of As atoms precipitated.

Table 4: Analysis of As precipitates in three different wafers.

wafer	number	$\bar{l}$ [nm]	volume density	excess of As
#1	80	146	1.0×10^{8} cm^{-3}	3.6×10^{15} cm^{-3}
#2	78	85	2.4×10^{8} cm^{-3}	2.9×10^{15} cm^{-3}
#3	74	94	1.9×10^{8} cm^{-3}	1.3×10^{15} cm^{-3}

4. DISCUSSION

The assumption of dislocation sources at or near the periphery of the crystals is obvious because of the following experimental observations. Firstly, a GaAs specimen with rough surfaces is deformed much easier than one with polished surfaces (Laister and Jenkins 1969). Secondly, in-situ observations of deformation experiments with a synchrotron beam showed the first dislocations penetrating the crystals from the outer surfaces (Tohno et al. 1986). Neither the nature of these sources nor their formation is known yet. Probably the surface degradation by evaporation of As observed already at 730°C (Siethoff et al 1990) plays an important role. In addition the boric oxide in the LEC process should also not be neglected, since deformation under boric oxide is much easier than in an Argon atmosphere or in air (Djemel and Castaing 1986).

Furthermore, we assumed that Schmid factors alone determine which glide systems are activated by dislocation sources. This assumption is justified by results of Kitano et al (1986a, 1986b). These authors were able to explain slip line patterns at the periphery of wafers in as well [001]- and ⟨111⟩-GaAs or -InP solely by Schmid factor distributions. In different deformation experiments always dislocations in glide systems with maximum Schmid factors were generated (e.g. Di Persio and Abbas 1989, Ono and Matsui 1987). In a sophisticated experiment Ono (1988) simulated stress conditions at special peripheral sites on a wafer by selected deformation geometries. He also found only glide systems and dislocation reactions predicted by Schmid factors.

On the size distribution of As precipitates there are no results in the literature up to now. As annealing at temperatures above 1,000°C dissolves all As precipitates back into the GaAs matrix (Clark et al 1988), precipitation has to take place somewhere between 1,000°C and 700°C in both cases. Since thermal rates during crystal growth and after annealing are the same ($\approx 100°C/h$) the effects must be due to the annealing treatment itself. Via "pipe diffusion" of As interstitials, As_i, along the dislocation lines the locally different excess As in "as-grown" GaAs is distributed more homogeneously at least in a cylindrical volume around the dislocation. For subsequent precipitation the maximum diffusion distance amounts to half the distance between two neighbouring precipitates along a dislocation line which is about 3 μm. This results in a diffusion coefficient for "pipe diffusion" of As_i of $D=4 \times 10^{-12}$ cm^2/s between 1,000°C and 700°C which is several orders of magnitudes higher than reported values for As diffusion in bulk GaAs (see e.g. Willoughby 1983).

REFERENCES

Alexander H 1989 Rad. Effects and Defects in Solids 111/112(1/2) 1
Clark S, Stirland D J, Brozel M R, Smith M and Warwick C A 1988 Semi-insulating
    III-V Materials, eds G Grossmann and L Lebedo (Bristol: Adam Hilger) p 31
Cullis A G, Augustus P G and Stirland D J 1980 J. Appl. Phys. 51 2556
Di Persio J and Abbas M 1989 Inst. Phys. Conf. Ser. No. 100 391
Djemel A and Castaing J 1986 europhys. lett. 2 611
Duseaux M 1983 J. Cryst. Growth 61 5761
Hirth J P and Lothe J 1965 Theory of Dislocations (New York: McGraw-Hill) pp 300-3
Jordan A S 1980a J. Cryst. Growth 49 631
Jordan A S, Caruso R and von Neida A R 1980b Bell System Techn. J. 59 593
Kitano T, Ishigawa T, Ono H and Matsui J 1986a Japan. J. Appl. Phys. 25 L530
Kitano T, Ono H and Matsui J 1986b Japan. J. Appl. Phys. 25 L761
Laister D and Jenkins G M 1973 J. Mater. Sci. 8 1218
Miyazawa S 1986 Defects in Semiconductors, ed H von Bardeleben,
    Mat. Sci. Forum Vol. 10–12 pp 3-10
Ono H 1988 J. Cryst. Growth 89 209
Ono H and Matsui J 1987 Appl. Phys. Lett. 51 801
Packeiser G, Schink H and Kniepkamp H 1986 Semi-insulating III-V Materials,
    eds H Kukimoto and S Miyazawa (Tokyo: Ohmsha) pp 561-6
Rüfer H, Uelhoff W, Schloßmacher P and Urban K 1990, Proc. Int. Conf. on the
    Science and Technology of Defect Control in Semiconductors, ed K Sumino
    (North-Holland: Amsterdam) pp 691-4
Siethoff H, Behrensmeier R, Ahlborn K and Völkl J 1990 Phil. Mag. A 61 233
Szabó G 1985 J. Cryst. Growth 73 131
Tohno S, Shinoyama, Katsui A and Takaoka H 1986 Appl. Phys. Lett. 49 1204
Willoughby A F W 1983 MRS Symp. Proc. Vol. 14 237

Inst. Phys. Conf. Ser. No 117: Section 6
Paper presented at Microsc. Semicond. Mater. Conf., Oxford, 25–28 March 1991

343

The evolution of prismatic dislocation loops in heavily Si-doped GaAs

E Dobročka and R Gleichmann*

Institute of Electrical Engineering, Slovak Academy of Sciences, 842 39 Bratislava, Czechoslovakia
*Institute of Solid State Physics and Electron Microscopy,Weinberg 2, O-4050 Halle(Saale), Germany

ABSTRACT: Si-doped horizontal gradient-freeze GaAs crystals have been investigated by HVEM. Triangular extrinsic faulted loops in the as-grown samples have been found to transform to prismatic ones in {110} planes as a result of an annealing at 1050° C. A possible mechanism for the observed transformation has been proposed. The analysis of the forces acting between the loop segments and the loop energy calculation have been performed for various types of loops. The role of the extrinsic character of the loops in the process of prismatic loop formation is discussed.

1. INTRODUCTION

Annealing induced changes of defects in heavily Si-doped horizontal gradient- freeze GaAs crystals have been studied by HVEM. Large density of extrinsic faulted dislocation loops in the {111} planes has been found in the as-grown samples with free carrier concentration (N_d-N_a) exceeding the value 4.10^{18} cm^{-3}. The annealing at 1050° C for 5 and 20 hours in an As overpressure resulted in either growth of larger circular stacking faults (SFs) or complete dissolution of the loops depending on their size. In addition, in the lower parts of the crystals where large (up to 0.3 μm) triangular SFs have been found in the as- grown samples, prismatic loops (PLs) in the {110} planes have developed. All the results obtained along with the details of the annealing and the sample preparation are presented elsewhere (Gleichmann, Dobročka and Tomek 1991). The aim of this work is to describe a possible mechanism of the PLs formation and to provide a theoretical analysis supporting the proposed mechanism.

2. TRANSFORMATION OF THE STACKING FAULTS TO PRISMATIC LOOPS

The heat treatment has not been preceded by a preannealing at a temperature >1100° C that would result in the dissolution of all defects in the as-grown samples as   was done in the work of Chen and Spitzer (1981). Therefore it is reasonable to suppose that the PLs have developed during annealing directly from the SFs (Fig.1). Both the SFs and the PLs are approximately of the same size suggesting that the transformation proceeded via rearrangement of the interstitials in the loops. The process of PL formation can be described in terms of a sequence of three reactions schematically shown in Fig.2: 1 - annihilation of the SF in the (111) plane leaving behind the triangular perfect loop ABC with b=1/2[$\bar{1}$10] (Fig.2a). 2 - diffusion of the interstitials from the corners B, C to the central part of the segment BC. This diffusion is caused by an additional climb force arising after the SF annihilation and it occurs always along the segment which is in edge orientation for the given **b**. This process results in a bending of the segment BC and it continues until the rhombic loop AB'DC' is achieved (Fig.2b). 3 - rotation of the loop plane from (111) to (110) around the [$\bar{1}$10] direction, the loop segments glide in the appropriate {111} planes (Fig.2c). The second and the third steps of the transformation can proceed simultaneously, i.e. the segments AB' and AC' can start to glide even before the part B'DC' is fully developed. However, to perform the theoretical analysis it is more convenient to consider all three steps separately.

3. FORCES ON THE LOOP SEGMENTS

Calculation of the forces acting between the loop segments has been performed following the method given by Hirth and Lothe (1968). The results are shown in Fig.3, where the variation of the components

© 1991 IOP Publishing Ltd

 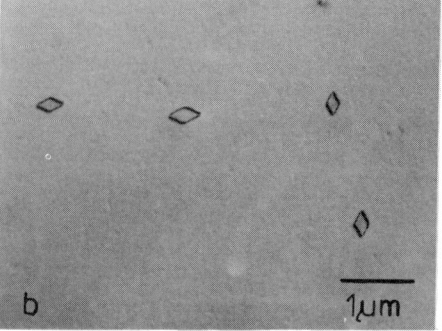

Fig.1 TEM micrographs of the as-grown (a) and the annealed (5h 1050° C) samples (b)

of force $\bar{F}$ along the loop segments are plotted for different types of loops. The coordinate $\bar{x}$ is normalized to 1 and gives the position along the loop segments. The actual values of the force can be obtained after multiplication of $\bar{F}$ by $\mu b^2/(2\pi(1-\nu)L)$, where μ, b, ν (=0.3) and L are the shear modulus, the magnitude of the Burgers vector $1/2 <110>$, the Poisson ratio and the length of the loop segment, respectively.

The curve 1 in Fig.3a shows the climb force (the component lying in the loop plane and oriented in the sense of inward normal of the dislocation line) exerted by the segments AB and AC on BC in the triangular faulted loop with $\mathbf{b}=1/3[\bar{1}1\bar{1}]$. The traction of the SF and the osmotic force exerted by the surrounding interstitials are not considered. The symmetry of the curve is apparent and the force has a pronounced minimum at the centre of the segment. Obviously the climb force distribution is the same for the other two segments due to the loop symmetry. The curves 2 and 3 give the climb force on the segments BC and AB in the triangular perfect loop with $\mathbf{b}=1/2[\bar{1}\bar{1}0]$. The curve 3 involves also the self force of the segment AB. Evidently, there is a difference between the forces acting on the segments BC and AB. This asymmetry is more pronounced in Fig.3b, where the differences between the curves 2 and 1 (curve 2') and 3 and 1 (curve 3') are shown. These curves clearly suggest that the increase of the climb force around the corners B and C is asymmetrical resulting in the core diffusion towards the centre of the segment BC.

In Fig.3c the climb force (curve 1) and the force perpendicular to the loop plane (curve 2) are plotted for the segment B'D of the loop AB'DC'. The climb force reaches the highest value at the acute corner D as expected. In the calculation of the curve 2 only the effect of the segments AB', AC' and the self force of the segment B'D have been considered, because the interaction between B'D and C'D does not

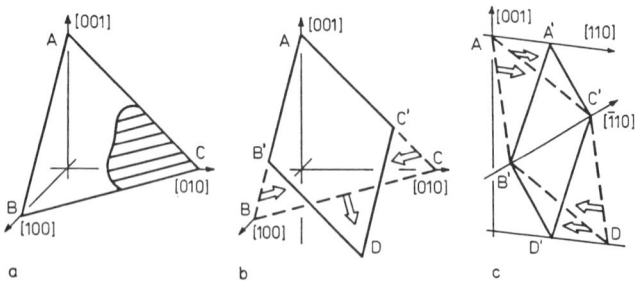

Fig.2 Schematic diagram of three steps of the loop transformation

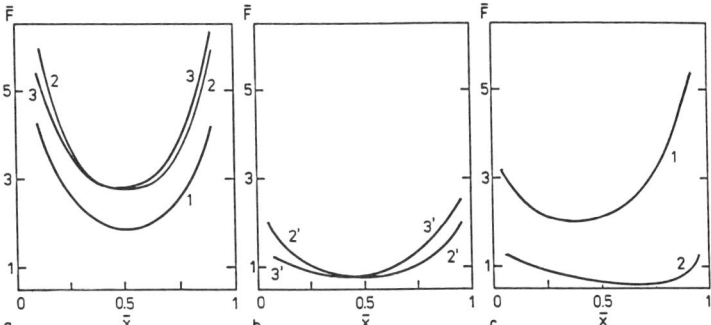

Fig.3 Forces acting on the loop segments. See text for details

contribute to the loop rotation. The force given by the curve 2 is oriented in the direction [111], i.e. the interaction between the segments of the rhombic loop AB'DC' is acting against the expected rotation of the loop plane towards the (110) orientation.

4. CALCULATION OF THE LOOP ENERGIES

Energies of four types of loops have been calculated using the method given by Hirth and Lothe (1968). W_1 - energy of the triangular faulted loop (including the SF energy) in the (111) plane with $b=1/3[\bar{1}\bar{1}\bar{1}]$. W_2 - triangular perfect loop ABC in the (111) plane with $b=1/2[\bar{1}\bar{1}0]$. W_3 - rhombic loop AB'DC' in the (111) plane with $b=1/2[\bar{1}\bar{1}0]$. W_4 - prismatic loop A'B'D'C' in the (110) plane with $b=1/2[\bar{1}\bar{1}0]$. The calculation has been carried out using the cut off parameter $\rho=b/8$ as proposed by Hirth and Lothe (1968) and the SF energy of 55 mJ/m^2. In order to assess the reactions involved in the loop transformation the relative differences $\Delta W_1/W_1=(W_2-W_1)/W_1$, $\Delta W_2/W_2=(W_3-W_2)/W_2$ and $\Delta W_3/W_3=(W_4-W_3)/W_3$ have been evaluated. The compared energies correspond to loops containing the same number of interstitials. The curves 1,2 and 3 in Fig.4a show the dependence of these parameters on the side length $\bar{L}$ of the primary triangular loop. $\bar{L}$ is given in the units of b - the magnitude of the Burgers vector $1/2\langle110\rangle$. It is seen that the annihilation of SFs starts to proceed above the size $\bar{L}=450(=0.18$ μm). The transformation of the triangular perfect loops to rhombic ones beginsimmediately after SF annihilation, this process lowers the loop energy for any value of $\bar{L}$. The loop rotation further lowers the loop energy, however, only for the values of $\bar{L}>15$. For smallerloops the energy minimum does not occur at the (110) orientation but it is shifted towards the (111) plane. This is shown in Fig.4b where the normalized rhombic loop energy $\Delta W/W=(W(\varphi)-W(111))/W(111)$ is plotted against φ - the angle of deviation of the loop plane from the (110) orientation - for three lengths of the diagonal B'C' - 5b(curve 1), 50b(2) and 500b(3). W(111) is the loop energy in (111) orientation. Similar curves were obtained by Bullough and Foreman (1964) using another method of loop energy calculation.

The loop rotation can be decomposed into two processes: a rigid rotation preserving the shape the loop has in the (111) orientation and an appropriate shortening of the loop segments at a given orientation of the loop plane. The corresponding normalized energy differences $(W_r(\varphi)-W(111))/W(111)$ - curve 2, and $(W(\varphi)-W_r(\varphi))/W(111)$ - curve 3, where $W_r(\varphi)$ is the loop energy after rigid rotation, are given in Fig.4c for B'C'=500b. It is clearly shown that the rigid rotation towards the (110) orientation increases the loop energy, while the segment shortening results in an energy decrease. Curve 1 is the sum of the curves 2 and 3. The contribution of both processes to the total energy varies with the loop size that explains the obtained shift of the loop energy minimum in Fig.4b.

5. DISCUSSION

The results of the theoretical analysis support the proposed mechanism of the loop transformation and are in agreement with the TEM observations. The triangular perfect loops are really not stable after SF

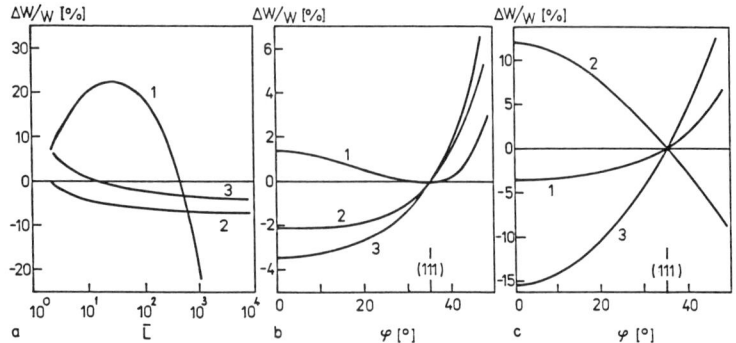

Fig.4 Normalized differences of the loop energies. See text for details

annihilation and start to transform to rhombic loops as   is seen in Fig.1a. However, the transformation is not finished during the crystal growth and an additional annealing at relatively high temperature is necessary to complete the reaction.

The apparent disagreement between the sense of the force perpendicular to the loop plane (curve 2 in Fig.3c) and the observed rotation of the loops can be explained on the basis of the loop energy calculation. The sense of the force that results from the interaction between the loop segments agrees with the sense of rigid rotation leading to the decrease of the loop energy. The segment shortening does not contribute to this force and this process is responsible for the observed loop rotation towards the (110) orientation.

The extrinsic character of the loops seems to be substantial for the process of prismatic loop formation. The intrinsic triangular loops are known to convert to SF tetrahedra via dissociation of the Frank partial dislocations (Silcox and Hirsch 1959). This conversion proceeds at the loop size well below the critical size for the SF annihilation (Morita, Kasahara and Kawado 1985, De Cooman, McKernan and Carter 1987) that may prevent the formation of the intrinsic PLs. The dissociation of the Frank partial (if occurs) around the extrinsic SF can be expected to form an obtuse bend between the fault planes and this configuration does not enable the SF tetrahedra formation. Therefore the extrinsic triangular SFs can grow until reaching the critical size for the SF annihilation and the subsequent PL formation.

REFERENCES

Bullough R and Foreman A E J 1964 Phil. Mag. 9 315
Chen R T and Spitzer W G 1981 J. Electronic Mater. 10 1085
De Cooman B C, McKernan S and Carter C B 1987 Phil. Mag. Lett. 56 85
Gleichmann R, Dobročka E and Tomek K 1991 phys. stat. sol.(a) to be published
Hirth J P and Lothe J 1968 Theory of Dislocations (New York: McGraw - Hill)
Morita E, Kasahara J and Kawado S 1985 Jpn. J. Appl. Phys. 24 1274
Silcox J and Hirsch P B 1959 Phil. Mag. 4 72

Inst. Phys. Conf. Ser. No 117: Section 6
Paper presented at Microsc. Semicond. Mater. Conf., Oxford, 25–28 March 1991

347

Observation of point defects and dislocations on GaAs{110} by scanning tunneling microscopy

G Cox, Ph Ebert and K Urban

Institut für Festkörperforschung, Forschungszentrum (KFA) Jülich, D-5170 Jülich, Germany

ABSTRACT : Scanning Tunneling Microscopy has been used to study the penetration of dislocations on the {110} surface of plastically deformed GaAs single crystals (p-type). Only perfect dislocations producing a double atomic step at the surface were observed. The dislocation cores were localized within two lattice constant at the surface. This led to a considerable strain field around the dislocation core, which could be directly seen in the bending of the As rows in front of the dislocation core. As in n-type material no band bending was observed around the dislocation cores. From this it can be concluded that the dislocations do not carry a strong electrical charge. In contrast to this, positively charged point defects on the As sublattice were observed around the dislocation cores, which can be interpreted as As vacancies.

1. INTRODUCTION

Scanning Tunneling Microscopy (STM) has been widely used to study the geometric and electronic structure of semiconductor surfaces on the atomic scale. The cleaved {110} surface of GaAs has been extensively studied by Feenstra and coworkers (1985, 1987). This surface exhibits a simple (1x1) reconstruction with zig-zag rows of Ga- and As atoms. Since the empty surface electron states are preferentially localized above the Ga atoms and the filled ones above the As atoms, both sublattices can be imaged separately by switching the tunneling voltage from positive to negative values (Feenstra et al 1987). Cox et al (1990 a) showed that the STM can be used to study not only ideal surfaces but also point defects and dislocations penetrating the surface of plastically deformed crystals. On n-type material they were able to image perfect and partial dislocations with atomic resolution. Due to the absence of band bending around the dislocation cores they concluded that the dislocations did not carry a strong electrical charge at the surface of n-doped GaAs. In this paper the authors report on the extension of their studies onto p-type material.

2. EXPERIMENTAL

The dislocations were induced by plastic deformation under a constant load of 40 MPa acting along the [213] direction of p-doped (Zn: $10^{18} cm^{-3}$) GaAs. The samples were deformed by compressions between 3% and 4% at 400 C° under an argon atmosphere. The primary slip system to be activated is the $(\bar{1}1\bar{1})$ slip plane with the $[0\bar{1}1]$ slip direction (Schmid factor 0.47). Transmission Electron Microscopy (TEM) analysis showed the dislocation density to be about $10^{8} cm^{-2}$. From the deformed samples rectangular pieces were cut with the long axes along the primary $[0\bar{1}1]$ slip direction. The samples were cleaved under ultra high vacuum conditions ($p < 10^{-8}$ Pa) in an analyzing chamber equipped with Auger and Leed facilities. The cleavage was performed perpendicular to the primary slip direction, exposing a $(0\bar{1}1)$ cleavage plane. The samples were transferred with the help of a magnetic transfer system to the STM chamber and investigated with a STM which was designed in our institute. More experimental details about the microscope and the sample geometry are given in Cox et al (1990 a).

3. RESULTS AND DISCUSSION

Fig.1 shows an STM image with atomic resolution on the As sublattice. The As rows run along the $[01\bar{1}]$ direction with a corrugation amplitude of about 0.03 nm along the [100] direction.

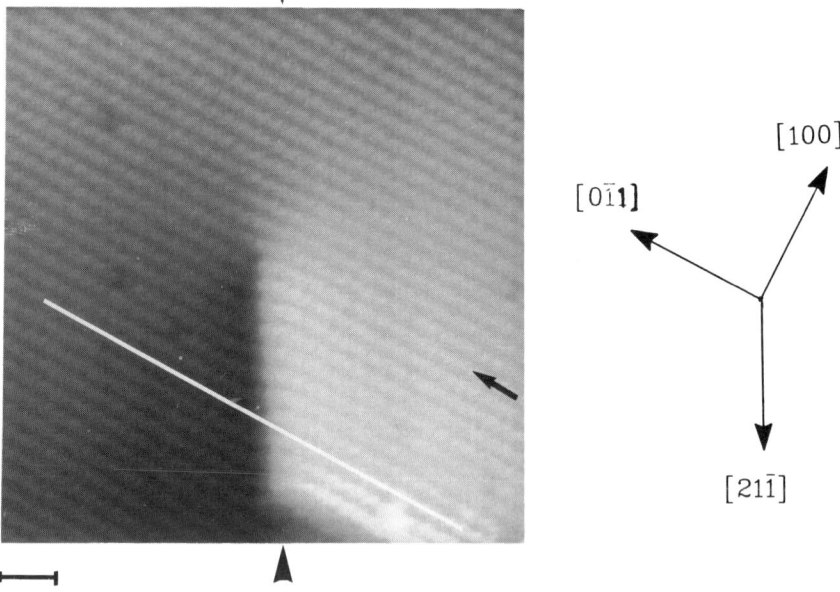

2.5 nm

Fig.1: Grey-scale image of a perfect dislocation producing a double atomic step at the surface. The white line shows that the As rows on the lower and the upper terrace are in register. The extra inserted As row is marked by an arrow on the right side. The corrugation profile along the line, marked by the arrows at the top and at the bottom of fig.1, is shown in fig.2.

In the middle of the image the core of a perfect dislocation can be seen from which a double atomic step extends along the [21$\bar{1}$] direction, which is the common direction between the primary slip plane and the surface. The step height can be determined during the experiment by measuring the elongation of the calibrated z-piezo-stack onto which the tip is mounted or by comparing the registration of the As-rows on the lower and the upper terrace of the dislocation induced step. Since these rows are in register (see white line in fig.1) the step height is two atomic layers. This would correspond to a Burgers vector $\vec{b}$ = a/2[0$\bar{1}\bar{1}$]. However, performing a Burgers circuit around the dislocation core, one can see an inserted As row on the upper terrace ending at the dislocation core (see arrow on the right hand side in fig.1). This extra row produces an additional shift of a/2[21$\bar{1}$] along the [21$\bar{1}$] direction, so that the total Burgers vector is $\vec{b}$ = a[10$\bar{1}$] which is under 30° to the surface plane. Such a Burgers vector could result from a reaction of a dislocation with Burgers vector $\vec{b}$ = a/2[0$\bar{1}\bar{1}$] and an in plane dislocation with Burgers vector $\vec{b}$ = a/2[21$\bar{1}$]. The latter could be produced at the cleavage front during the cleavage process. This is supported by the fact that all dislocations observed with the STM at the surface of the p-type material were not dissociated into partial dislocations, whereas TEM analysis of the samples showed all dislocations in the bulk to be dissociated.

The dislocation core is highly localized, which can be seen in the formation of the double atomic step height. Fig.2 shows the corrugation profile along the [21$\bar{1}$] direction between the arrows shown in fig.1. The location of the dislocation core along the corrugation is marked by an arrow in fig.2. The full step height is formed within two lattice spacings along the [21$\bar{1}$] direction.

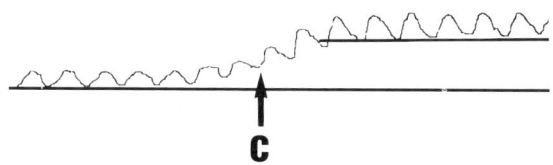

Fig.2: Corrugation profile along the line, marked by the arrows shown in fig.1. The point 'C' marks the location of the dislocation core along the corrugation profile.

This highly localized core leads to a substantial stress around the dislocation core, which can be seen in the bending of the As rows in front of the dislocation core. Near the dislocation core the As rows are shifted by about one lattice constant along the [100] direction. This shift decreases slowly within about 10 nm in front of the dislocation core.

Around the dislocation cores point defects have been observed on the As sublatice with a concentration of about 10^{12}cm^{-2} (see fig.1). Fig.3 shows that such a point defect is located at one As-atom site. By taking images simultaneously at positive and negative sample polarities, the defect could only be observed on the As sublattice. The same type of defect has previously been found on surfaces of n-type material, too, and has been interpreted as an As vacancy (Cox, et al 1990 b).

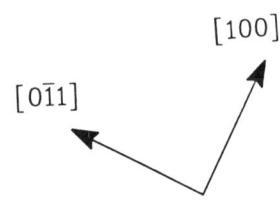

0.56 nm

Fig.3: Point defect on the As sublattice, taken at a negative sample voltage of -2.5 V.

In contrast to the observed dislocation cores, the size of the point defects varied considerably when changing the tunneling voltage to lower values. Fig.4 shows a sequence of four images where the tunneling voltage has been decreased from -2V (upper left image) to -0.9V (lower right image). When the tunneling voltage is decreased to lower values, the localized point defect appears as an approximately 5 nm wide depression on the surface. The voltage dependent size of the defect can be explained by an downwards band bending of the valence band due to a positive electrical charge associated with the point defect. This results in a decrease of the number of tunneling states and thereby in a decrease of the tunneling current around the defect, to which the feedback regulation circuit of the STM corresponds by moving the tip closer to the surface. The lower the tunneling voltage the more the tunneling current is affected by the decreased number of available tunneling states. A similar effect has been found for Ga vacancies for n-type material (Cox 1990 b). There the Ga vacancies carried a negative charge which led to the same effect for the sample conduction band.

4. SUMMARY

Using a scanning tunneling microscope point defects and dislocations were studied on the (0$\bar{1}\bar{1}$) surface of p-doped GaAs single crystals. In contrast to TEM studies

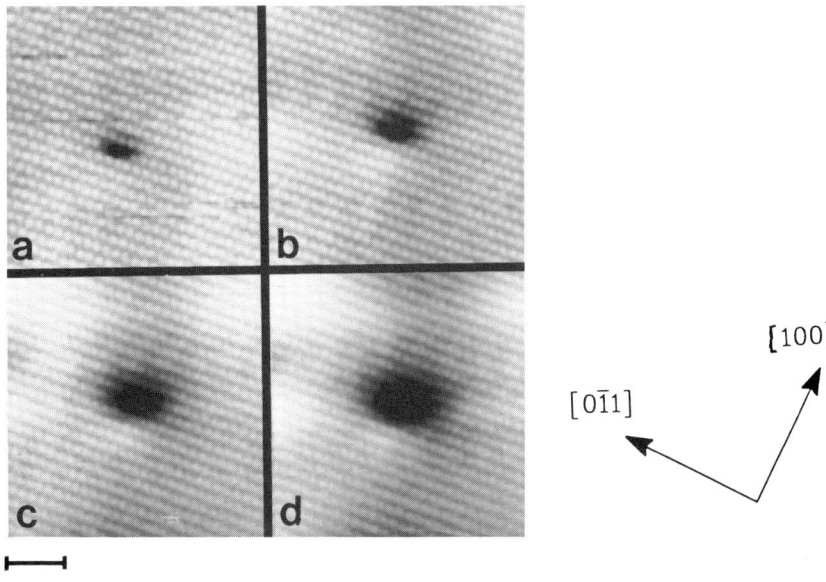

[100]

[0$\bar{1}$1]

3.0 nm

Fig.4. Four STM images of the As point defect shown in fig.3. The images were taken at sample voltages of -2V (a), -1.5V (b), -1.2V (c), and -0.9V (d).

(Gottschalk et al), all dislocations were found not to be dissociated into partial dislocations at the surface. Around the dislocation cores point defects on the As sublattice could be observed. In contrast to the dislocation cores, the point defects had a voltage dependent size, which can be explained by a downwards band bending due to a positive charge associated with the point defects. Since no such band bending was observed around the dislocation cores for both n-type and p-type material, it can be concluded that the dislocation cores do not carry a strong electrical charge.

The authors would like to thank J. Krüger and H. Alexander for the sample preparation and valuable discussions.

REFERENCES

Cox G, Szynka D, Graf KH, Poppe U, Urban K, Kisielowski-Kemmerich C, Krüger J, Alexander H, 1990 a, Phys. Rev. Lett. 14, 2402
Cox G, Graf KH, Szynka D, Urban K, 1990 b, Vacuum 41, 591
Feenstra R, Fein AP, 1985, Phys Rev B 6, 1394
Feenstra R, Stroscio JA, Tersoff J, Fein AP, 1987, Phys. Rev. Lett. 58, 1192
Gottschalk H, Patzer P, Alexander H, 1978, Phys stat sol (a), 45, 207

Inst. Phys. Conf. Ser. No 117: Section 6
Paper presented at Microsc. Semicond. Mater. Conf., Oxford, 25–28 March 1991

Microdefects in Si-doped HB GaAs crystals investigated by TEM, DSL photoetching and laser scattering tomography

C Frigeri°, J L Weyher° and P Gall*

° CNR-MASPEC Institute, via Chiavari 18/A - 43100 PARMA (Italy)
* Centre d'Electronique de Montpellier, USTL, F-34060 MONTPELLIER (France)

ABSTRACT : Microdefects giving rise to high microroughness after DSL photoetching in Si-doped GaAs crystals have been studied by TEM. They turned out to be small dislocation loops which sometimes were also decorated by particles. The smallest loops were faulted and of the 1/3<111> type, whereas the largest ones were of the 1/2<110> type. Such microloops are at the origin of the surface microroughness after DSL as they produce small etch hillocks. The origin of the microloops is also discussed.

1. INTRODUCTION

A problem still being studied in the field of the characterization of bulk GaAs, both as-grown and annealed, is the origin of the microstructure roughness detected in such material after wet chemical etching, such as A/B and DSL, which is widely used for quick examination of the GaAs wafers (for a review on the argument see Stirland 1990). The most common explanation for such an etch microroughness is the presence of microprecipitates, usually related to the excess of As, either decorating dislocations or dispersed in the matrix (Stirland 1990). On the other hand, as far as DSL photoetching is concerned, surface microroughness would also form if a very high density of dislocations were present because this type of etching produces hillocks at the dislocation sites (Weyher and van de Ven 1986). Recently it has been shown that dislocation loops were the defects responsible for the formation of the surface microroughness after DSL in Si-implanted and then annealed undoped GaAs wafers (Frigeri et al 1990). This paper deals with a TEM study aimed at establishing the cause of the microroughness produced by DSL etching in as-grown GaAs crystals.

2. EXPERIMENTAL DETAILS

The specimens investigated were wafers cut from (001) GaAs crystals grown by the horizontal Bridgman (HB) method. The crystals were Si-doped to a free electron concentration of $(1\text{-}2) \cdot 10^{18}$ cm^{-3}. Observations were performed with diluted Sirtl-like photoetching (DSL) (Weyher and van de Ven 1983) in conjunction with interference contrast optical microscopy, laser scattering tomography (LST) (Gall et al 1988) and TEM. TEM observations were carried out with either a 2000 FX (point-to-point resolution 0.31 nm) or a 2010 (p.-to-p. resolution 0.23 nm) JEOL microscope at 200 kV. The plan-view specimens were either chemically thinned (chlorine in methanol) or ion milled. <110> oriented specimens for HREM observations were also prepared by using the standard procedures for the preparation of cross-sectional specimens (Bravman and Sinclair 1984) starting from slabs cut parallel to a [110] direction from the bulk wafers in the area of interest.

3. RESULTS AND DISCUSSION

Two types of etch features have been observed in the HB wafers after DSL photoetching, namely, a) isolated etch hillocks due to grown-in dislocations (Weyher and van de Ven 1986) and b) surface microroughness whose origin was not clear. An image of such microroughness is shown in

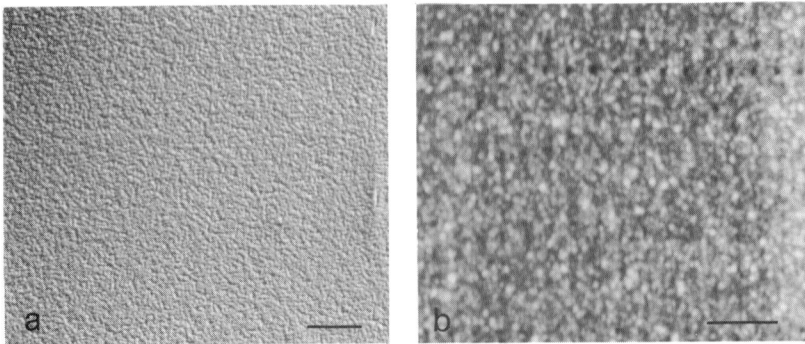

Fig. 1 - a) Interference contrast optical microscope image of surface microroughness after DSL etching in an HB wafer. The bar is 200 μm. b) LST image of the same area. The bar is 20 μm.

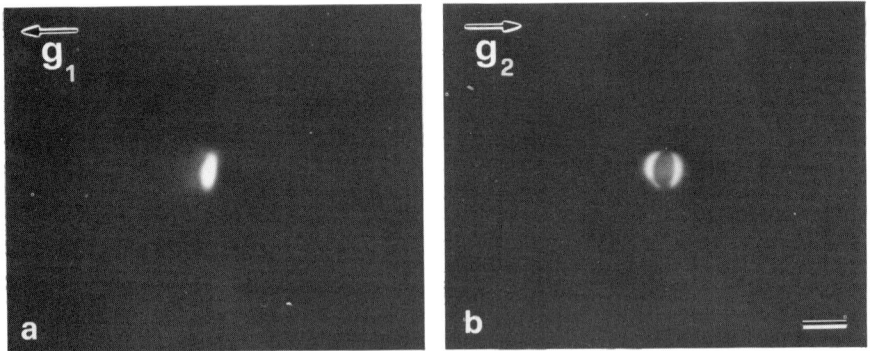

Fig. 2 - Weak beam dark field TEM images of the same dislocation loop. a) $g_1 = [2\bar{2}0]$, b) $g_2 = [\bar{2}20]$. The bar is 20 nm.

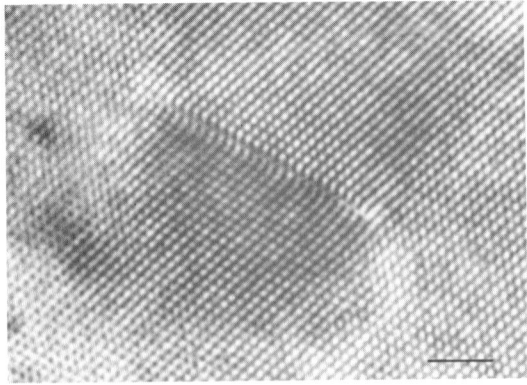

Fig. 3 - <011> bright field HREM image of a dislocation loop. The bar is 2 nm.

Fig. 1a. The microroughness was detected only in some regions of the wafer that covered a surface ranging from 30% to 80% of the total wafer area. Any estimation of the density of the etch features was impossible because they could not be well resolved even after shallow (~1 μm etch depth) photoetching. The grown-in dislocation density ranged from ~1 to ~$10 \cdot 10^3$ cm^{-2}, depending on the type of crystal and position in the wafer. A typical LST map from an area exhibiting a high degree of microroughness is given in Fig. 1b. It shows that in such areas a very high density (~$2 \cdot 10^9$ cm^{-3} in this sample) of light scatterers exists in the bulk. Before performing the LST measurements the DSL etched wafers were repolished.

TEM observations have shown the presence of small strain centres in those wafer regions exhibiting surface microroughness after DSL and strong light scattering. No other defects were detected. TED did not reveal any extra spot. Typical strain centre density and size range were ~$8 \cdot 10^8$ cm^{-2} and 10-30 nm, respectively. The largest strain centres showed clear double arc contrast in both the bright field and (g, 3g) weak beam dark field mode and turned out to be dislocation loops as they had outside-inside contrast behaviour on reversing the diffraction vector (Fig. 2). The loops were of the intrinsic (vacancy type) nature as established by the method suggested by Dahmen (1989). The diffraction vectors $g_1 = [2\bar{2}0]$ and $g_2 = [\bar{2}20]$ of the [001] pole were used, g_1 being arbitrary defined as + g as it pointed toward the [1$\bar{1}$2] pole where the habit plane of the loop was edge-on with respect to the [001] orientation. The identification of the strain centres as dislocation loops was also confirmed by HREM observations (Fig. 3). The presence of dislocation loops in HB and LEC crystals Si-doped to nearly the same level as ours has been previously reported (e.g., Darby et al 1980, Chu et al 1981, Fornari et al 1989). The largest loops had a Burgers vector of the type **b** = 1/2a<110> and exhibited strong contrast for both g = [220] and g = [2$\bar{2}$0]. On the other hand, the smallest loops, in addition to some large ones, always gave strong contrast for g = [220] and very weak contrast for g = [2$\bar{2}$0] or vice versa and all gave contrast for both g = [040] and [400]. They also looked faulted (Fig. 2b). Such a behaviour suggests that they are of the Frank type with **b** = 1/3a<111> (Meekison et al 1989). Some loops have been found to be decorated by a precipitate as shown in the (g, 3g) weak beam dark field image of Fig. 4 where the 'parallel' type Moirè fringes have a spacing of ~2.4 nm. No lattice plane of hexagonal arsenic, that has been reported to be the origin of precipitates in undoped LEC GaAs (Lodge et al 1985), has been found to fit such a spacing.

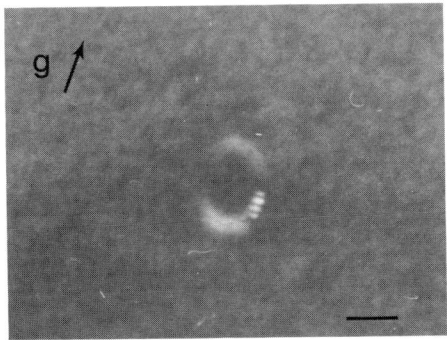

Fig. 4 - (g, 3g) WBDF TEM image of a dislocation loop decorated with a particle. g = [220]. The bar is 20 nm.

The microloops detected by TEM are at the origin of the surface microroughness after photoetching. At such microdefects, in fact, the etching rate is lower than in the defect-free matrix as they are effective recombination sites for the holes which are generated by light during DSL etching and used to dissolve the GaAs molecule (van de Ven et al 1986, Frigeri and Weyher 1989). A similar correlation between dislocation loops and DSL microroughness was also found in Si-implanted and annealed semi-insulating LEC GaAs (Frigeri et al 1990). The high degree of microroughness and the difficulty in resolving the small etch hillocks which formed at the microloops is due to the high dislocation loop density. Arsenic precipitates are known to produce pits after DSL, whereas other kinds of precipitates may be resistant to the DS etchant

(Weyher et al 1990). It is expected, therefore, that the microroughness in our samples could also be due to the precipitates attached to some dislocation loops (Fig. 4). However, as the density of the decorated loops is smaller than that of the undecorated ones by at least two orders of magnitude, the precipitates should contribute to a much lesser extent to the formation of the microroughness. It is well established that light scattering in the LST method occurs at precipitates either decorating the dislocation lines or dispersed in the matrix (Gall et al 1988 and references therein). The low density of decorated loops in our samples might suggest that the undecorated loops may also scatter light. To reliably establish this, however, further experimental work is required.

In order to explain the formation of the microloops in GaAs crystals Si-doped to $(1-2)\cdot10^{18}$ cm^{-3} the dopant incorporating reactions should be considered after the model by Giling et al (1986). In n-type GaAs silicon is preferentially incorporated on Ga sites. If it is assumed that the formation of Ga interstitials is unlikely this would mean that As vacancies are formed (Giling et al 1986). The density of these excess As vacancies is larger than the intrinsic As vacancy concentration if the dopant density exceeds a certain value which should be somewhat around $1\cdot10^{18}$ cm^{-3} (Giling et al 1986). At the same time a reduction of the As interstitial density also occurs. At the high doping levels typical of the crystals investigated here local increase of the silicon concentration might thus produce non-equilibrium As vacancies that can then collapse into vacancy-type loops. The habit plane of such loops is {111} at the early beginning (Holt 1966), i. e. when they are small, in agreement with our results. As the loops increase their size they also become unstable and reduce their energy by removing the stacking fault by a shearing process involving a Shockley partial, according to, e. g., (Mazey et al 1962)

$$a/3\,[111] + a/6\,[11\bar{2}] = a/2\,[110]$$

and dislocation loops with a Burgers vector of the type 1/2a <110> form as found here.

ACKNOWLEDGEMENTS

Part of the work is financially supported by the EEC Contract # SC 1/0247 (J. L. W.). J. L. W. also acknowledges NATO financial support through grant # 0358/88. Thanks are due to S Bertoni of ENICHEM for permission to use the JEOL 2010 microscope.

REFERENCES

Bravman J C and Sinclair R 1984 J. Electr. Microsc. Tech. 1 53
Chu Y M, Darby D B and Booker G R 1981 Inst. Phys. Conf. Ser. 60 331
Dahmen U 1989 Ultramicroscopy 30 102
Darby D B, Augustus P D, Booker G R and Stirland D J 1980 J. Microsc. 118 343
Fornari R, Frigeri C and Gleichmann R 1989 J. Electron. Mater. 18 185
Frigeri C and Weyher J L 1989 J. Appl. Phys. 65 4646
Frigeri C, Weyher J L and de Potter M 1990 Proc. XIIth Congress for El. Microsc. (San Francisco: San Francisco Press) vol 4 684
Gall P, Fillard J P, Castagné M, Weyher J L, and Bonnafé J 1988 J. Appl. Phys. 64 5161
Giling L J, Weyher J L, Montree A, Fornari R and Zanotti L J. Crystal Growth 1986 79 271
Holt D B 1966 J. Mater. Sci. 1 280
Lodge E A, Booker G R, Warwick C A and Brown G T 1985 Inst. Phys. Conf. Ser. 76 217
Mazey D J, Barnes R S and Howie A 1962 Phil. Mag. 7 1861
Meekison C D, Gold D P, Booker G R, Hill and Boys D R 1989 Inst. Phys. Conf. Ser. 100 507
Stirland D J 1990 International Conf. on the Science and Technology of Defect Control in Semiconductors, Ed K Sumino (Amsterdam: North Holland) pp. 783-794
van de Ven J, Weyher J L, van der Meerakker J E A M and Kelly J J 1986 J. Electrochem. Soc. 133 799
Weyher J L, Gall P, Frigerio G and Zanotti L 1990 J. Crystal Growth 106 175
Weyher J L and van de Ven J 1983 J. Crystal Growth 63 285
Weyher J L and van de Ven J 1986 J. Crystal Growth 78 191

Inst. Phys. Conf. Ser. No 117: Section 6
Paper presented at Microsc. Semicond. Mater. Conf., Oxford, 25–28 March 1991

Laser scanning microscope investigation of individual dislocations in as-grown LEC GaAs: geometry, glide, climb and particle decoration

P Kidd[*1], D J Stirland[2] and G R Booker[1]

1 Department of Metallurgy and Science of Materials, University of Oxford, Parks Road, Oxford OX1 3PH, UK.
2 Plessey Research Caswell Ltd, Caswell, Towcester, Northants. NN12 8EQ, UK.
* Present Address: Dept M.S.E., University of Surrey, Guildford GU2 5XH, UK.

ABSTRACT: We have used the LSM to make a comprehensive study of dislocations present in as-grown In-doped and undoped LEC SI GaAs. These dislocations had mainly arisen by glide from the surface to the interior of the ingot, followed by climb and particle decoration, during the growth and subsequent cooling of the ingot. The observations that were made included large abrupt changes in particle decoration as individual dislocations changed direction, dislocation segments with complex cross-slip geometries, dislocation segments in the form of large helices, and differences in the dislocations and particle decoration in In-doped and undoped ingots. Mechanisms to explain these behaviours are described.

1. INTRODUCTION

We have shown during the last few years how the laser scanning microscope (LSM), operating in transmission with infrared light, can be used as a non-destructive method to image the As or As-rich precipitate particles which decorate the dislocations present in as-grown bulk GaAs (Kidd 1987a,b,1990). The particle sizes range from 30 to 300nm, and are spaced 3 to 10µm apart along the dislocations. Individual particles are imaged as dark spots 2µm across, this being the spatial resolution of the method. The spot contrast is a measure of the particle size, particles down to 30nm across being detected. Observation of the particles reveals the geometry of the individual dislocations. Differences in particle size and distribution associated with individual segments of dislocations can be directly seen. Focussing down through the specimen enables individual dislocations to be followed over distances of several millimetres.

2. EXPERIMENTAL

The specimens examined were blocks cut from the shoulder regions of as-grown In-doped and undoped liquid encapsulated Czochralski (LEC) semi-insulating (SI) GaAs ingots (Fig 1). The micrographs shown are LSM images corresponding to different regions within the slabs. Figs 2, 3 and 4 refer to the In-doped specimen, and Fig 7 to the undoped specimen.

3. In-DOPED GaAs

Figs 2a and 2b show (001) views of decorated dislocation lines in the In-doped specimen. The sense of the image is that of looking down from the seed where the left-hand side of each image is closer to the surface of the

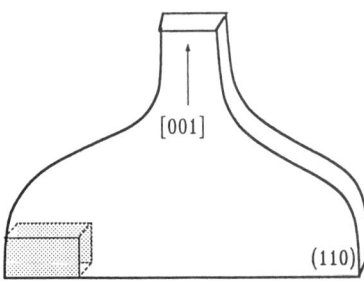

Fig 1. The location of specimens in In-doped and undoped ingots.

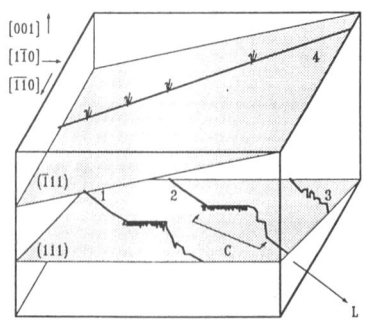

Fig. 2c. Schematic diagram showing the orientations of the dislocations 1-4.

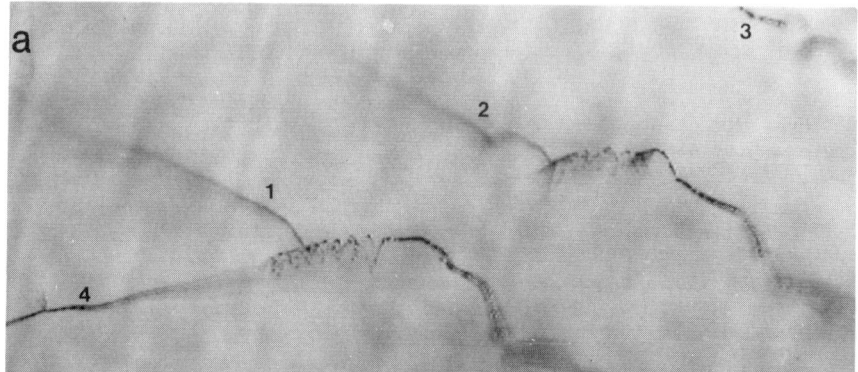

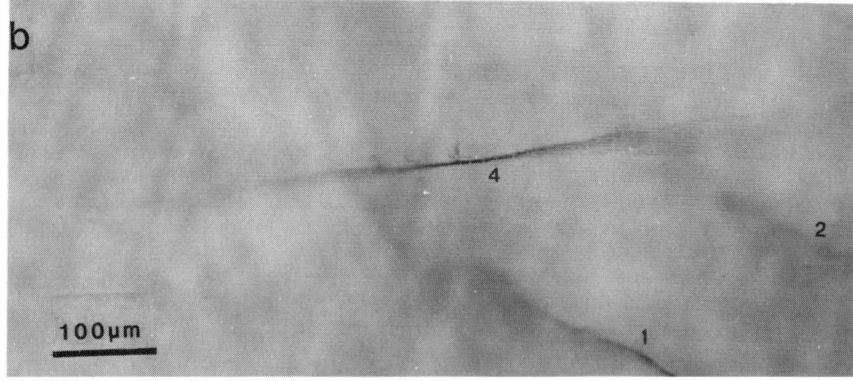

Figs.2a and 2b. (001) LSM views of decorated dislocation lines in the In-doped GaAs specimen.b) shows the region 500μm directly above a)

ingot. The focal depth of each image is 30-40μm. Those dislocation segments exhibiting sharp black dots or lines are in focus. As the dislocation line comes up above, or down below, the focal plane, the dislocation line image broadens and exhibits less contrast. Fig 2b shows the region 500μm directly above Fig 2a. Four dislocation lines, numbered 1 to 4, are present in the pair of images. Their line directions were determined by analysing a focal series through the specimen (not shown here) and are indicated diagrammatically in Fig 2c. Such dislocations were observed throughout the block with a density of 10^3cm^{-2}. Most of the dislocation lines were parallel to 1, 2, and 3, only two were observed parallel to 4. All four dislocations shown in Figs 2a and 2b lie in {111} slip planes.

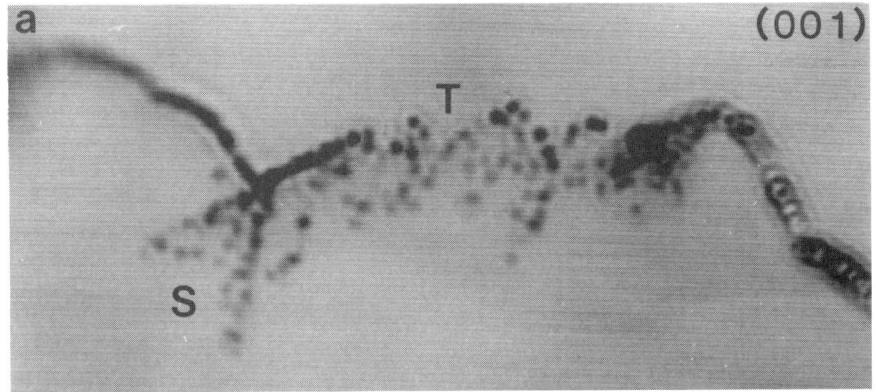

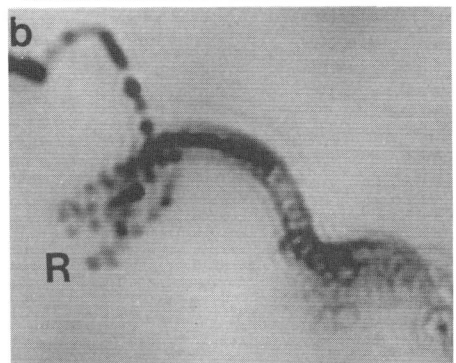

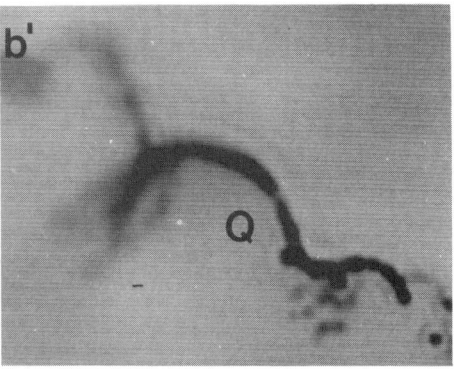

Figs. 3a–c. (001) high magnification images of segments of dislocations 2,3 and 4, in Fig 2, respectively. Fig 3b' is at focal position 70μm above 3b.

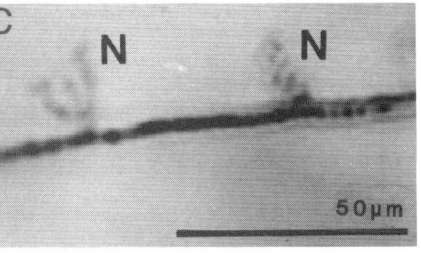

The arrow labelled L in Fig 2c shows the overall line direction for dislocations 1, 2, and 3. However, sections of these otherwise straight dislocations bow out between apparent pinning points, P, to form large cusps, such as C in Fig 2c. The striking observation is the dramatic change in configuration of the dislocation as its line direction changes along such a cusp. Fig 3a shows a segment of cusp C at higher magnification. Two distinct features are observed. S is an 'arrowhead' with a sharp geometrical shape, which originates from a single point on the dislocation line, and T is an example of an irregular helix along the dislocation segment which lies parallel to [1$\bar{1}$0]. Both these features were observed only on dislocation segments which lay parallel to [1$\bar{1}$0]. Part of dislocation line 3 is shown in Fig 3b. The two images are taken with the focal position of the right-hand image 70μm above that on the left. In this case the line direction is close to [0$\bar{1}$1] and consists of R, a similar but less angular form of S, where bunches of dislocation loops appear to have originated from a single point, and Q, a large-period helix with its axis along [0$\bar{1}$1]. Dislocation line 4 is entirely straight, and is decorated at 50–100μm intervals by 20μm sized features, two of which, labelled N, are shown at higher magnification in Fig 3c.

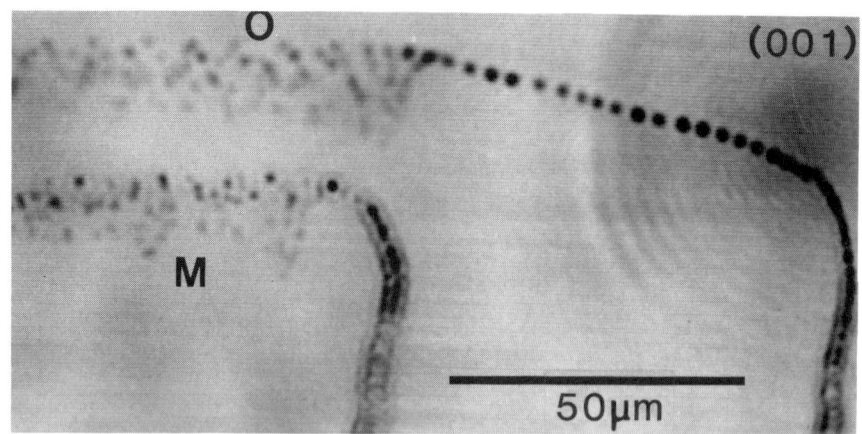

Fig. 4a. (001) LSM image of particles decorating dislocations in the In-doped specimen (above).
Fig. 4b. Schematic diagram showing the orientations of the dislocations (right).

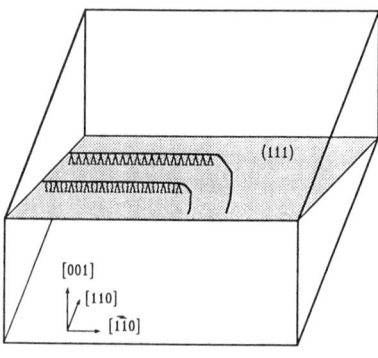

Fig 4a shows particles decorating two dislocation segments elsewhere in the In-doped specimen. The segments which show complex decoration, often in the form of overlapping triangles, O, or irregular helices, M, lie parallel to [1$\bar{1}$0]. They start at the surface of the ingot and extend for about 3mm to the region shown here. The complex decoration stops abruptly as the dislocation changes direction and curves up out of the field of view. These dislocations lie in the (111) plane as shown diagrammatically in Fig 4b.

The analysis of the effects shown here will be reported in greater detail elsewhere. As shown in Fig 5a, with reference to Thompson's tetrahedron, the dislocations in Fig 4a correspond to segments of the dislocation half loops described by Matsui (1987) and Ono et al (1987) which propagate along ⟨110⟩ in a tangential stress field which may dominate under a radial thermal gradient. The tips of these loops are screw in character (Matsui 1987) and are likely therefore to have Burgers vectors a/2[0$\bar{1}$1] or a/2[$\bar{1}$01]. The segments with the more complex decoration would therefore be 60° dislocations. Fig 5b and 5c show possible origins of the segments shown in Figs 2 and 3 These dislocations correspond to the glide dislocations reported by Ogawa(1989), Matsui (1987) and Ono et al (1987), and which are considered to have propagated in radial stress fields which dominate under axial thermal gradient conditions.

Our interpretation of the decoration observed in Figs 2,3 and 4 is as follows: The dislocations belonging to the group 1,2,3, have crossed over with those parallel to 4 as they glided in from their sources on the surface. This created jogs in the dislocation lines. Some jogs have acted as pinning points so that the dislocation lines are forced to bow out between

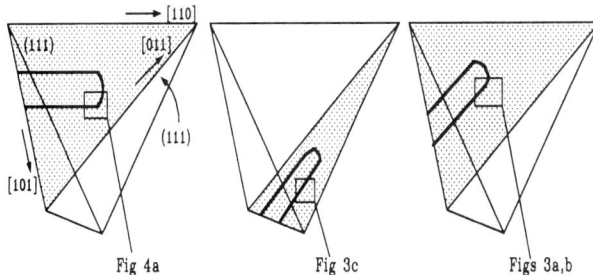

Fig 4a   Fig 3c   Figs 3a,b

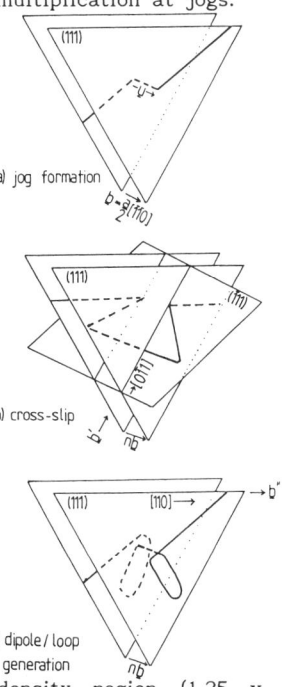

Fig.5.   (left)Schematic
diagrams showing the
dislocation half loops
containing the segments
shown in Figs. 2-4.

Fig.6 . (right) Schematic
diagrams of dislocation
multiplication at jogs.

(a) jog formation

(b) cross-slip

(c) dipole/loop
generation

them. Thus, segments of dislocations with
different line directions are created. The helix
shown in Fig 3b shows climb of a screw
dislocation, with Burgers vector a/2[0$\bar{1}$1], meaning
that the complex decoration, T, observed in the
[1$\bar{1}$0] segments in Fig 3a is due to climb of a 60°
dislocation. The similarities between features T,
in Fig 3a and M in Fig 4a support this
interpretation. The features on the dislocation
lines labelled S, R, N are examples of dislocation
multiplication at jogs where the dislocations have
formed dipole loops or have cross-slipped at the
jogs. Mechanisms for the formation of these
features are shown schematically in Fig 6. The
differences in the configurations of S, R and N
will depend on the type of jog formed. Cross-slip
occurs when both the jog and the Burgers vector
share a common slip plane (Fig 6b) and dipoles or
loops are formed when the jog is sessile.

4. UNDOPED GaAs

Fig 7 shows a (110) view of a low dislocation density region (1.25 x
10^{4}cm^{-2}) in the undoped ingot. In regions where the dislocation density is
higher (2 x 10^{4}cm^{-2}), the particles occur in clouds and are associated with
tangles of dislocations in cell walls (Kidd 1987). Here the particles
clearly reveal individual dislocation lines and the lines are straight and
inclined relative to [001], consistent with them lying on {111} slip planes.
For undoped GaAs, the particle contrast is about one tenth of that in the
In-doped GaAs, meaning that the particles decorating dislocations in the
undoped material are smaller. In general, the decoration of dislocations in
the undoped specimen shows less variation than that seen in the In-doped
specimen. However, there are some changes in particle spacing along
dislocation lines. Different particle spacings are interpreted as being
indicative of different dislocation core stuctures. The dislocation segments
labelled X and Y both show looped segments x and y. The average particle
spacing on X is 10μm, while the average spacing on Y is 4μm, and these two
spacings are maintained even when the dislocations change line direction.
The formation of the looped segments x and y provide examples of dislocation
multiplication inside the ingot. The faint line labelled L is in focus and
is due to decoration by smaller and more closely spaced particles. Smaller
particles, s, are also present between the particles decorating x. They
appear to decorate bowed-out segments of x. Dislocation line Z is wavy and
appears not to lie in a simple {111} glide plane.

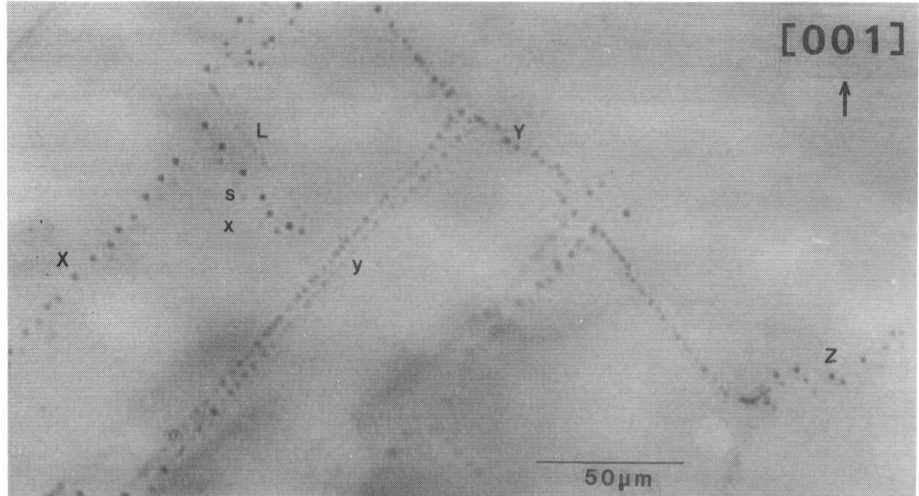

Fig.7. (110) LSM image of particles decorating dislocations in undoped GaAs.

5. SUMMARY

In summary, this technique is uniquely able to image individual particles decorating dislocation lines in bulk specimens, with a resolution of $2\mu m$, whilst at the same time non-destructively sampling large volumes of material and revealing dislocation line directions over several millimetres. For the glide dislocations in undoped as-grown GaAs, different dislocation lines exhibit differences in particle spacing. For In-doped as-grown ingots, dramatic differences in their behaviours are observed as their line directions change. In both cases the particle decoration reveals the operations of dislocation multiplication mechanisms.

The precipitation of particles on dislocation lines will be controlled by the local concentrations and mobilities of point defects such as As interstitials and Ga vacancies, the density of dislocations, their core structures, and the temperatures at which they are formed. The introduction of In into the melt increases the hardness of the material, thus changing the temperature and stress at which dislocations are formed. These factors contribute to the observed differences between In-doped and undoped material for dislocation interaction and particle precipitation.

ACKNOWLEDGEMENTS
We are pleased to acknowledge support by the Science and Engineering Research Council, UK, and the Procurement Executive, Ministry of Defence.

REFERENCES
Kidd P, Booker G R, Stirland D J, 1987a, in: Microscopy of Semiconducting Materials, Oxford, Inst. Phys. Conf. Ser. No. 87, eds. A G Cullis and P D Augustus (I.O.P. Bristol) pp 275-280.
Kidd P, Booker G R, Stirland D J, 1987b, Appl. Phys. Lett. **51** p1331.
Kidd P, Stirland D J, Booker G R, 1990, Materials Letters, **9** pp521-525.
Ogawa T, J. Crystal Growth, 1989, **96** pp777-784.
Matsui J, 1987, in: Microscopy of Semiconducting Materials, Oxford, Inst. Phys. Conf. Ser. No. 87, eds. A G Cullis and P D Augustus (I.O.P. Bristol) pp 275-280.
Ono H, Kitano T, Matsui J, 1987, Appl. Phys. Lett. **51** p238

Inst. Phys. Conf. Ser. No 117: Section 6
Paper presented at Microsc. Semicond. Mater. Conf., Oxford, 25–28 March 1991

363

Electron beam radiation effects on the properties of unpassivated GaAs-MESFET

K Kaufmann, P Koschinski and L J Balk

Universität Duisburg   Fachgebiet Werkstoffe der Elektrotechnik
Leiter: Univ—Prof. Dr.—Ing. E. Kubalek
Sonderforschungsbereich 254, Kommandantenstr.60, D—4100 Duisburg 1, F.R.G.

ABSTRACT: The influence of electron radiation as used in scanning electron microscopy on the properties of GaAs—MESFETs has been investigated. For this purpose GaAs—MESFETs fabricated on MOVPE—grown layers have been exposed selectively to electron beams of different energies and currents. Whereas a significant change of the output characteristics of the MESFETs has not been observed, the characteristics of the gate—diodes are strongly changed due to electron impact. Both improvement at very low doses and deterioration at high doses of the gate—diode has been found, the effects associated with microscopic effects as observed in EBIV—micrographs.

1. INTRODUCTION

Scanning electron microscopy (SEM) is widely used for investigating semiconducting materials and devices. Furthermore, electron beam lithography for submicrometer device fabrication has become very important. But specimens are mechanically, thermally and electrically loaded by application of electron radiation. Hitherto, an alteration of semiconducting specimens based on III—V—compounds has not been investigated in detail. An alteration of the electron mobility in High—Electron—Mobility—Transistors (HEMT) caused by electron irradiation during processing has been reported by Fink et. al. (1990), the change of existing crystal defects with electron radiation thought to be important. Other investigations by Nel and Auret (1989) conclude that the crystal structure of GaAs may be changed by electrons of energies ranging from 400eV to 3keV already. In this paper it is investigated how GaAs—MESFETs are influenced by an electron radiation as commonly used in SEM. A knowledge of this topic is necessary to avoid an alteration of MESFET parameters, e.g. during a check of their structure by SE—images. Furthermore, a variation of microscopic properties may lead to a misinterpretation of quantitative techniques, such as for EBIC/V—(Electron—Beam—Induced—Current/Voltage)—micrographs.

2. EXPERIMENTAL PROCEDURE

The structure of the investigated GaAs—MESFETs are shown in Fig.1, the layers being grown by MOVPE (Metal—Organic—Vapour—Phase—Epitaxy). The conducting layer of 170nm thickness has been n—doped ($n=3\times10^{17}cm^{-3}$) with silicon and a gate recess has been made. The gate metallization consisted of a gold layer (thickness 200nm) on a thin chromium layer (10nm). The gate and the channel area of six unpassivated GaAs—MESFETs have been scanned several times by the electron beam, however, while adjusting the specimens the MESFETs have been entirely irradiated for a short time only. The scanned areas were $6\times16\mu m^2$ and are drawn in Fig.5a. Electron energies of $W_{PE}=4keV$ and $W_{PE}=20keV$ and beam currents of $I_{PE}=$ 10pA, 100pA, 10nA, and 100nA in the combinations as indicated in Fig.3 have been used. The radiation dose D there is defined as:

$$D = I_{PE} \cdot t \,/\, A$$

I_{PE} is the electron beam current, t the irradiation time and A the irradiated area.

The irradiation of the specimens has been interrupted several times in order to measure the output—characteristics of the MESFETs, gate currents, gate—capacities and the characteristics of the

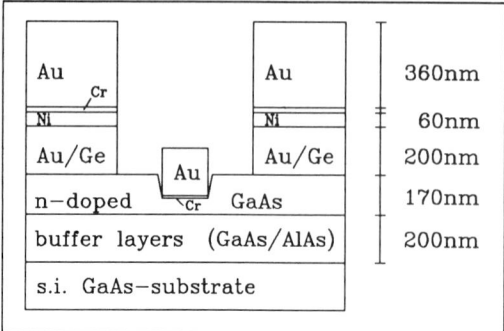

gate—(Schottky)—diodes within the SEM vacuum chamber. All measurements except for the gate—capacities have been carried out at very low frequencies. EBIV—micrographs have been taken to gain information about the electrical fields inside the MESFETs. For these measurements beam currents of $I_{PE}=1pA$ and lock—in—amplification have been used to minimize the influence on the specimen.

Fig.1:.Cross—section of the irradiated GaAs—MESFETs

3. EXPERIMENTAL RESULTS

A significant change of the MESFET output characteristics has not been observed. Both a slight improvement and a slight deterioration occurred. On the other hand the properties of the gate—(Schottky)—diodes have been strongly changed. For example, in Fig.2 the current—voltage characteristics of an irradiated gate—diode are shown for $W_{PE}=20keV$ and $I_{PE}=10nA$. After an electron irradiation with a small dose the properties are improved. The reverse current becomes smaller, the breakdown voltage more negative. After a further irradiation the change is opposite and the reverse current becomes higher until at a dose of 73Ascm^{-2} ohmic behavior is reached. Fig.3 gives an overview of the change of all tested MESFETs, by plotting the reverse voltage being necessary for a reverse current of 1μA, versus the dose. At low doses up to 1Ascm^{-2} a slight increase of the reverse voltage can be observed, above this value the reverse currents decrease. This could only be measured for higher beam currents, as for the measurements at low beam currents very high irradiation times would have been necessary.

Similar to the current—voltage characteristics of the gate—diodes the gate currents, the input resistances of the MESFETs and the gate—capacities have been changed. The input resistance of the MESFETs could be dropped from some MΩ to 100kΩ by electron beam irradiation, see Fig.4. After irradiation with high doses the gate—capacity decreases faster than with the reverse voltage proportional to 1/sqrt(U_r) as expected due to the theory for the metal—semiconductor diode. Parallel to these changes the contrasts of the irradiated specimen areas in EBIV—micrographs are altered. The MESFET shown in the micrographs of Fig.5 has been irradiated selectively by an electron beam with $W_{PE}=20keV$ and $I_{PE}=10nA$. After a short irradiation time the small bright contrast on the right side of the gate in Fig.5a vanishes (see Fig.5b). When the MESFETs are irradiated further, a bright contrast spreads from both edges of the gate (Fig.5c) and finally extends over the whole irradiated channel area (Fig.5d). The contrast in the gate area itself has not been changed.

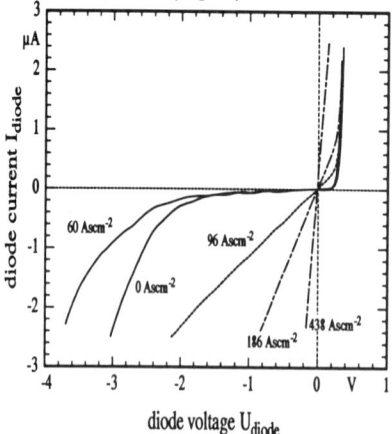

Fig.2: Gate diode characteristic for different doses ($W_{PE}=20keV$, $I_{PE}=10nA$)

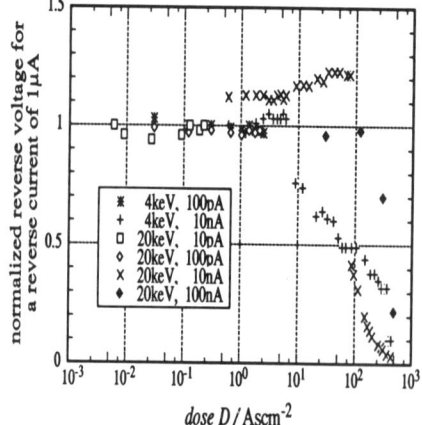

Fig 3: Normalized reverse voltage for 1μA reverse current depending on beam parameters and dose

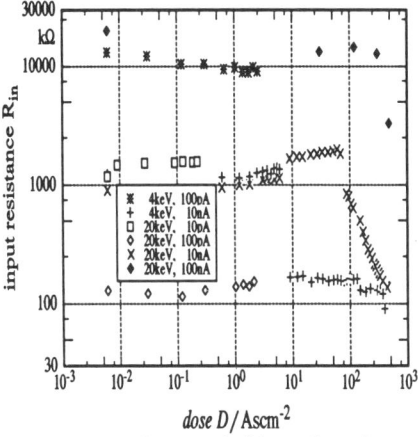

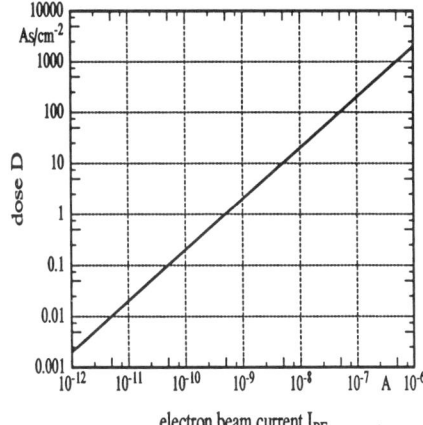

Fig.4: Input resistance of all investigated MESFETs in dependence on dose D

Fig.6: Radiation dose D vs. electron beam current for an irradiation of a $30\mu m^2$ area for 10 minutes

4. DISCUSSION

The improvement of the characteristics of the gate—diode may be explained by an electron beam induced desorption of surface adsorbates, e.g. solvents. The surface adsorbates may cause a slight surface conductivity and depletion layers, which can be detected by EBIV. The effect of the deterioration of the gate—diode can be simply modelled by an equivalent circuit consisting of the best gate—(Schottky)—diode measured after improvement by electron irradiation and a parallel resistance. Using this equivalent circuit it is possible to calculate the forward—current—voltage characteristics of the damaged diodes by using the best idealty factor n, the reverse current I_S of the improved diode, and by estimating a fixed parallel resistance R. The descriptive formula is

$$I = I_S \left(\exp(qU/nkT) - 1\right) + U/R$$

The origin of the parallel resistence has to be explained. The deterioration of the characteristics of the gate—diode occurs at $W_{PE}=4keV$ as well as at $W_{PE}=20keV$. In the first case the primary electrons do not penetrate the gate metal, so they can alter neither the gate—metal interface nor the n—doped channel below the gate by kinetic processes. An alteration of the interface by thermal annealing due to the electron beam induced temperature rise can be excluded, too. On the other hand, the parallel resistance may be caused by a conducting surface layer between the gate and the ohmic contacts. But this conducting surface layer cannot be caused by contamination, as the resistance of contamination on metal structures on SiO_2/Si have been measured to be greater than $20M\Omega$. These metal lines had the same spacing as the contacts of the MESFETs and were irradiated in the same manner. A possible mechanism may be diffusion of gold atoms into the bulk or along the surface of the n—doped channel. It is well known that gold diffuses easily in GaAs. For that reason, during device processing, a thin layer (10nm) of chromium is evaporated on the GaAs as diffusion barrier before gold deposition. At the edges, which in reality are not sharp, the diffusion may be stimulated by the electron beam as an edge effect. But a mechanical alteration of the edges could not be detected in SE—(secondary electron)— and optical micrographs (maybe due to a too poor resolution). Both surface and bulk resistance would be changed by additional gold atoms this being a probable reason for the changed diode characteristic and the reduced input resistance. Taking into consideration the EBIV—micrographs one learns that alteration of the gate—diode begins actually at the edges of the gate—metal. On the other hand the bright contrasts in the EBIV—micrographs in the irradiated channel area cannot be explained by a conducting surface layer. Even if a conducting surface layer with a depletion layer inside the channel is formed, the level of the EBIV—signal should not be that high. The signal is higher than at the edges of the gate in the unirradiated areas of the MESFETs and as high as at the mesa—edges, where the gate metal lies directly on the buffer layer. In the area of the mesa—edges the minority carriers generated in the buffer layer can be collected directly by the Schottky—diode. On the other hand, the minority carriers generated in the middle of the MESFETs have to surmount the potential barrier between the semiinsulating buffer layer and the n—doped channel. Because of this reason the collection efficiency should be lower, as is discussed by Flesner et al. (1982). Perhaps the origin of the bright contrasts can be explained by an electron beam induced lowering of the barrier height. This may be caused by diffusion of gold or doping atoms or by generation or alteration of crystal defects as reported by Leamy and Kimerling (1977).

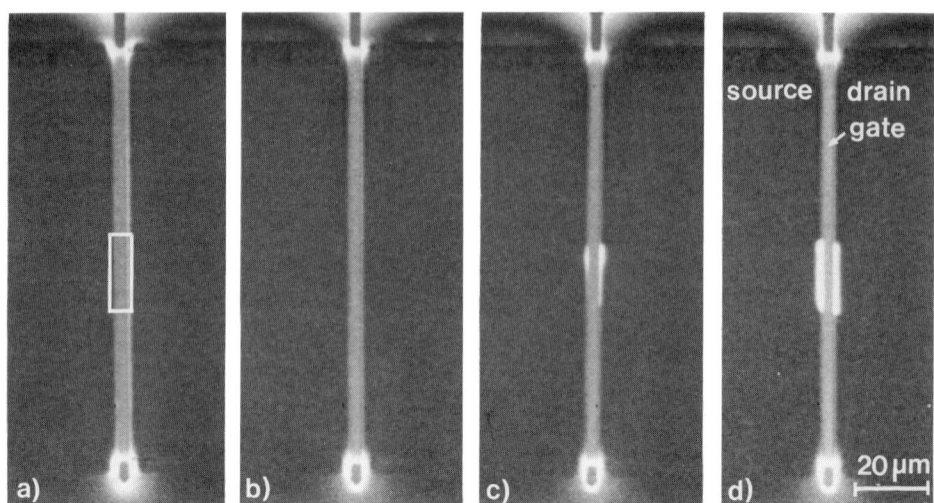

Fig.5: EBIV—micrographs of an irradiated MESFET. Beam parameters are $W_{PE}=20keV$ and $I_{PE}=10nA$ and the doses are a) $0Ascm^{-2}$, b) $0.6Ascm^{-2}$, c) $60Ascm^{-2}$, and d) $276Ascm^{-2}$.

5. CONCLUSIONS

It has been shown that GaAs—MESFETs may be damaged during investigation with SEM. Fig.6 shows the dose dependence on the beam current for the case that a micrograph is taken within 10 min of a specimen area of $6x5\mu m^2$. The critical dose of $1Ascm^{-2}$ is already reached for $I_{PE}=5nA$. Although it may be clear ad hoc, with this work it is shown that for investigating such semiconductor devices only very small beam currents should be used. This means that modes incorporating high doses due to a principly low signal—to—noise situation, like cathodoluminescence or scanning electron acoustic microscopy may cause damage in this kind of specimen, as these techniques usually require electron beam currents from at least 1nA up to $1\mu A$. Depletion layers induced by surface adsorbates may lead to a misinterpretation of EBIV—micrographs. These layers of adsorbates may be changed by electron irradiation and seem to vanish even after an irradition with small doses and low beam currents.

ACKNOWLEGDEMENTS

We would like to thank the departement for solid state electronics of Duisburg University for supplying the MESFETs, especially U. Doerk, K. Ntikbasanis, and M. Joseph. The work was financially supported by the Deutsche Forschungsgemeinschaft.

REFERENCES

Fink T, Smith D D and Braddock W D 1990 IEEE Trans. Electron Devices 37 1422
Flesner L D, Davies N M and Wieder H H 1982 J. Appl. Phys. 53 3873
Leamy H J and Kimerling L C 1977 J. Appl. Phys. 48 2795—2803
Nel H J and Auret F D 1989 Jap. J. Appl. Phys. 28 2430—2435

Inst. Phys. Conf. Ser. No 117: Section 6
Paper presented at Microsc. Semicond. Mater. Conf., Oxford, 25–28 March 1991

Defect agglomeration and annealing in ZnTe during Ar$^+$-ion and electron irradiation

G Lu and F Phillipp

Max-Planck-Institut für Metallforschung, Institut für Physik,
Heisenbergstr. 1, 7000 Stuttgart 80, FRG

ABSTRACT: Agglomerates of atomic defects in ZnTe single crystals were produced by 3 keV - 9 keV Ar$^+$-ion irradiation. They were identified as interstitial-type dislocation loops. Their size is about 20nm except in the case of 9 keV irradiation at elevated temperature where they grow considerably larger. Annealing of the loops occurs during electron irradiation in an electron microscope (20kV - 400kV). Loop growth and annealing are discussed in terms of mechanisms for enhanced diffusion and defect formation induced by electron - hole recombination.

1. INTRODUCTION

During preparation of transmission electron microscopy (TEM) specimens of semiconducting materials by ion milling it was found that the Ar$^+$ ions produce considerable damage. This damage in II-VI semiconductors appears in TEM as densely distributed small defects which obscure the true structure of the specimen and can be diminished when I^{2+} ions are used for thinning. (Cullis et al 1985). In this paper, we report on the formation of defects in ZnTe during Ar$^+$ irradiation, their identification and the annealing during electron irradiation in an electron microscope.

2. EXPERIMENT

ZnTe specimens were ground mechanically to about 30 μm and then ion beam thinned by Ar$^+$-ions under selected conditions, with a Gatan DIM machine, operated with Ar$^+$ ions of energies between 3 keV to 9 keV. Thinning and formation of the radiation damage took place simultaneously. During irradiation the specimen is either with or without liquid nitrogen (LN$_2$) cooling. The specimen temperature during the irradiation was estimated to be higher than 420 K in the case without LN$_2$ cooling. When cooled with LN$_2$, the specimen temperature is then considered higher than 230 K. The angle between the ion beam and the specimen surface is normally from 10° to 15°. The specimens were examined with a Siemens Elmiskop 102, which allows operation at 20kV to 100kV for electron irradiation studies at room temperature. An AEI-EM7 high voltage electron microscope was used for irradiation with electrons of higher energies up to 400kV and for *in situ* temperature studies. A JEM-4000FX was used for high resolution electron microscopy (HREM) investigation.

3. RESULTS

Fig. 1 shows the radiation damage produced by 7 keV and 9 keV Ar$^+$ ions without cooling during ion milling. It can just be recognized in the specimen thinned by 7 keV Ar$^+$ ions that the defects are dislocation loops sized below 20nm (Fig. 1a). The defects produced by

9 keV Ar$^+$ ions are much larger. In many cases they reach sizes of several hundred nanometers and grow through the specimen (Fig. 1b). Specimens thinned by Ar$^+$ ions of 3 keV to 7 keV show more or less similar defect feature as in Fig. 1a, regardless of the specimen temperature during the irradiation. When the ion energy is 9 keV and the operation is cooled by LN$_2$ the defects produced are still small loops as those in Fig. 1a. HREM shows that loops are of interstitial type. One example is given in Fig. 1c. Hundreds of loops in over 15 specimens were examined and not a single vacancy loop was encountered.

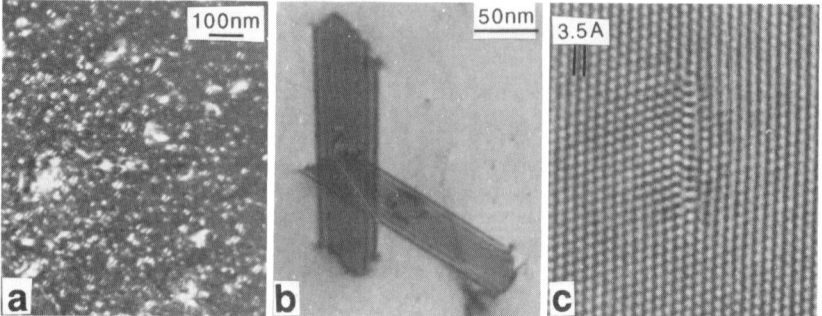

Fig.1 Small dislocation loops produced by radiation of 7 keV Ar$^+$ (a), and the grown up loops by 9 keV Ar$^+$ irradiation (b) at room temperature. A (110) HREM image of a loop (c) showing that it is of interstitial type.

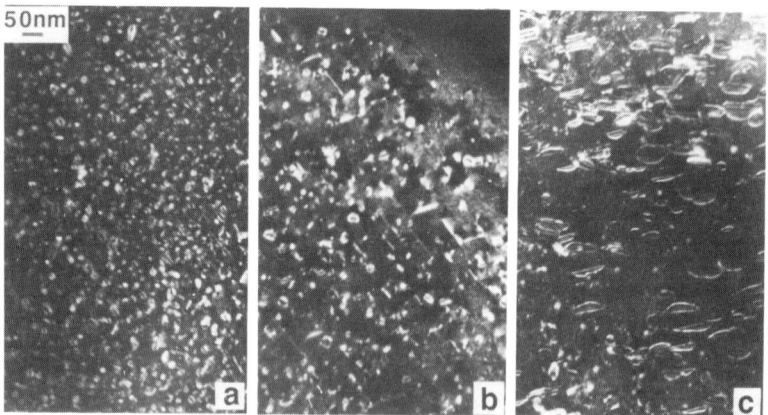

Fig.2 Weak beam images showing dislocation loops at different growth stages when irradiated with 9 keV Ar$^+$ without cooling for (a) 10 minutes, (b) 10 + 10 minutes interrupted and (c) continuous 20 minutes.

Fig. 2 shows the temporal development of the loops during 9 keV irradiation without cooling. After the first 10 minutes small loops of high density are observed, similar to those produced by lower energy ions (Fig. 2a). Another 10 minute irradiation of the same specimen at the same conditions causes the loops to grow up to larger sizes (Fig. 2b). If a specimen is irradiated for 20 minutes without interruption the loops grow even larger (Fig. 2c), which may be attributed to a temperature dependence of the loop growth since during the interruption of the irradiation the specimen cools down to room temperature.

The dislocation loops produced by ion irradiation are found to be annealed out during intense electron irradiation in an electron microscope. As an example, Fig. 3 shows a specimen irradiated with 5 keV ions, before (Fig. 3a) and after (Fig. 3b) irradiation with 100kV electrons at room temperature for 80 minutes. The density of the loops in the area illuminated by the condensed electron beam is clearly seen to be reduced with respect to the surrounding area, where the loop density essentially remained unchanged. Similar annealing was observed for electron energies between 20kV to 400kV and illumination times of 20 to 120 minutes. Although no quantitative studies were performed it can be stated that the annealing is almost independent of the electron energy but is strongly enhanced by increasing the electron current density.

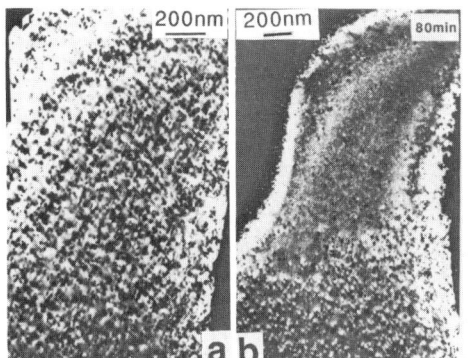

Fig.3 Annealing of the loops produced by 5 keV Ar$^+$ under irradiation of 100kV electrons, (a) before and (b) after the irradiation.

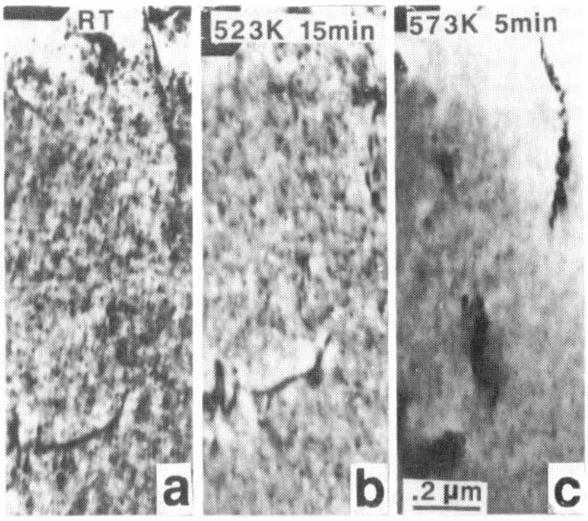

Fig. 4 Observation of *in situ* heating in a TEM, (a) room temperature, (b) 473 K and (c) 523 K.

The annealing of the loops is insensitive to the specimen temperature during electron irradiation in the range between 173 K to 523 K. It is, however, much more effective when the temperature is higher than 523 K. Fig. 4 shows a series of images taken during an *in situ* heating experiment at an electron energy of 200kV. After 15 minutes at 523 K (Fig. 4b) the defect structure is not greatly different from that at room temperature (Fig. 4a), only a small fraction of the loops is annealed out. However, after only 5 minutes at 573 K (Fig. 4c), the loops are annealed out almost completely. It should be noticed that even at 573 K the electron beam still plays a important role in the defect annealing process, since in the unilluminated area outside the beam, many loops remained.

4. DISCUSSION

During ion irradiation of solids the primary defects, i.e., vacancies and interstitial atoms, are produced in a shallow layer close to the surface in collision cascades. For ZnTe the mean range of the Ar$^+$-ions is ca. 10 nm at 9 keV (Wilkens 1991), thus at an angle of 15° the cascades are located ca. 2.5 nm underneath the surface. In a cascade damage process a separation of vacancies and interstitials takes place. Whereas the vacancies are concentrated in the depth corresponding to the mean ion range, interstitials will be displaced deeper into the crystal via replacement collision sequences (Hertel et al 1980). Keeping this in mind it is easy to understand why only interstitial-type dislocation loops are observed: The vacancies are produced very close to the surface and can easily anneal out at the surface. Furthermore, this shallow layer is sputtered away during the ion irradiation. Thus, only the interstitials survive and form clusters. During the course of sputtering the layer containing interstitial loops produced previously is etched away as well and new loops are formed, thus a balance between etching and loop production is established. This idea accounts well for the observations except for the results of the 9 keV irradiation without cooling. Here, further mechanisms have to be operative which allow the loops to expand deep into the bulk, much deeper than the range of the replacement collision sequences. Two mechanisms will be considered to account for the large penetration depth of the interstitial loops: Firstly, the loops agglomerated at a depth position corresponding to the range of the replacement collision sequences may collect interstitials by drift diffusion in the strain field of the loop, thus climbing mainly in a direction parallel to the surface. This results in elongated elliptical loops. However, this loop shape is energetically unfavourable from line tension considerations. A re-arrangement of the interstitials within the loops may take place yielding loops of more circular, or, if the near-surface part is etched away, semi-circular shape. It is indeed observed that many loops have shapes of semi-circles. Secondly, a high growth speed of the loops into the bulk can further be supported by pipe diffusion (e.g. Shaw 1973). At present it is not clear yet why only under specific conditions, i.e. 9 keV ion energy and high temperature, the large loops are observed. Certainly temperature plays an important role, which favours both re-arrangement and pipe diffusion. This is supported by the observation that the loops grow large at 9 keV irradiation without cooling but remain small when specimen cooling is employed. Furthermore, continuous irradiation causes the loops to grow larger than intermittent irradiation where the specimen cools down during the interruptions. Whether the strong influence of the ion energy may be explained by beam heating is not yet clear and further experiments are planned to elucidate the reason of the apparent threshold between 7 keV and 9 keV.

The annealing of the loops when subjected to electron irradiation is certainly not due to the absorption of vacancies produced by knock-on damage. Among others the most conclusive argument is that the annealing occurs at electron energy far below the threshold for atom displacement. The electron energy necessary to displace Zn atoms from their lattice site was determined to be 110 keV (Bryant and Baker 1973) and it should be even higher for the displacement of the heavier Te atoms. Therefore we suggest the annealing to be driven by non-radiative recombination processes (Kimerling 1978), i.e. valence electrons of ZnTe are excited by the incident electrons to the conducting band and recombine with holes subsequently. If either electrons or holes are trapped at levels inside the band gap introduced by the dislocation loops, this recombination is localized. The energy released by the recombination is supposed to be transferred to the atoms at the dislocation core, cause interstitials to leave the loop, to enhance their diffusion (Frank et al 1980), and thus result in the annealing (Lu et al 1991). The energy released during a recombination event cannot exceed the gap energy which is 2.66 eV in ZnTe. As estimated by Bailly (1968) the formation energy of the Zn vacancy in ZnTe is 1.66 eV. If the formation energies of other types of point defects are assumed to be of similar values, the non-radiative recombination processes are indeed expected to be effective in the production of point defects and to cause the annealing of the loop. This mechanism would also explain well the observation that the annealing does not strongly depend on the energy of the incident electrons, since the cross section for electron excitation in the keV range is almost constant.

REFERENCES

Bailly F 1968 Lattice Defects in Semiconductors ed Ryukiti R. Hasiguti (University of Tokyo Press and the Pennsylvania State University Press) p 231

Bryant F J and Baker 1973 Rad. Effect and Defects in Semicond. (Inst. Phys. Conf. Ser. 16) p 42

Cullis A G, Chew N G and Hutchison J L 1985 Ultramicroscopy **17** 203

Frank W, Gösele U and Seeger A 1980 Radiation Physics of Semiconductors and Related Material, eds G P Kekelidze and V I Shakhovtsov (Tbilsi State University Press) p 110

Hertel B, Kammholz K and Diehl J 1980 Radiation Effect **48** 45

Kimerling L C 1978 Solid State Electronics **21** 1391

Lu G, Niu F and Wang R, Rad. Effects and Defects in Crystals, in the press

Shaw D 1973 Atomic Diffusion in Semiconductors, ed D Shaw (London and New York: Plenum Press) p 46

Wilkens M private communication

REM observations of fracture behaviour and crystal polarity of ZnO single crystals

L -M Peng and J T Czernuszka
Department of Materials, University of Oxford
Parks Road, Oxford OX1 3PH, UK

ABSTRACT: Low index surfaces of ZnO single crystals have been examined by using reflection electron microscopy (REM). It is found that the fracture behaviour and detailed atomic structures of the $(1\bar{1}00)$ and $(11\bar{2}0)$ faces of ZnO crystals depend sensitively on the cleavage direction and applied stress. The polar (0001) surface is observed to consist of domains of Zn-terminated and O-terminated faces.

1. INTRODUCTION

In recent years the technique of reflection electron microscopy (REM) has been shown to have considerable potential as means for studying the structures of surfaces and crystal defects within the bulk crystal (Cowley, 1986; Yagi, 1987). In particular, REM requires minimal effort of specimen preparation and can be performed in a conventional transmission electron microscope without any modification to either the microscope or the specimen stage. In this paper we will report some of our recent observations of the anisotropic fracture behaviour of ZnO single crystals on the $(1\bar{1}00)$ and $(11\bar{2}0)$ faces, and the crystal polarity of the polar (0001) surface.

2. EXPERIMENTAL PROCEDURE

Single crystals of hydrothermal grown ZnO single crystals were obtained from R.S.R.E., Malvern. The REM observations were performed in a commercial Philips CM12 transmission electron microscope, fitted with a LaB_6 filament. To obtain a low index fracture surface, the crystals were aligned, sawn and polished parallel to a face, which is perpendicular to the required fracture surface, to a thickness of approximately $300\mu m$. The fracture surfaces were then obtained by cleaving the wafers along certain directions. All REM images were obtained with a 120keV accelerating voltage, and have been so arranged that the incident beam direction is vertically downward in all figures.

3. RESULTS

Shown in figure 1 are four REM images of the $ZnO(1\bar{1}00)$ surface. The incident glancing angle is about 38mrad, giving a foreshortening factor of 26. In figures 1a and 1b, the initial (0001) wafer was cleaved along $[11\bar{2}0]$ direction, and the incident beam azimuth is near the $[0001]$ zone axis. Figure 1a shows a network of surface steps being parallel

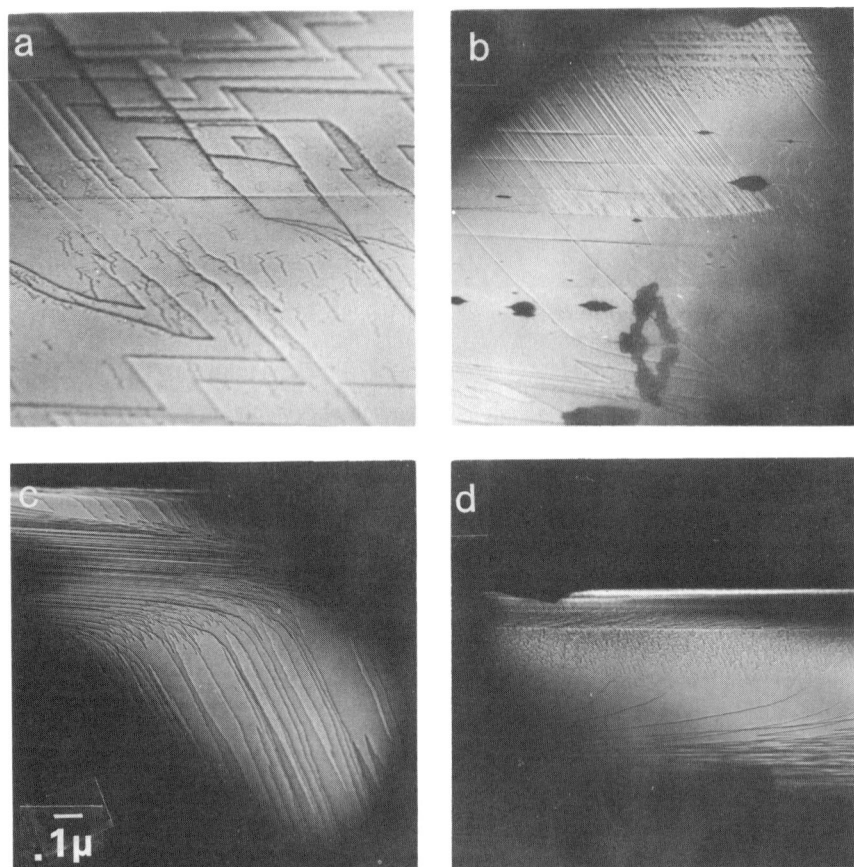

Figure 1: REM images of the ZnO($1\bar{1}00$) surface.

predominantly along [0001] and [$11\bar{2}0$] directions. Figure 1b shows a slip trace left by one basal dislocation. This slip trace is seen also to serve as a boundary between two regions of different step density, suggesting a rather important role of the interaction between the surface steps and dislocations for the fracture process.

The specimen used for figures 1c and 1d was obtained by cleaving a ($11\bar{2}0$) wafer along the [0001] direction. The incident beam azimuth is near [$11\bar{2}0$]. Figure 1c shows a rather nice water fall pattern, and a fracture centre around the upper left corner. Figure 1d shows a region somewhere away from such a fracture centre. It is seen that the fracture speed was apparently slowed down in the lower part of the figure, by the creation of many fine surface steps. The configuration of these fine surface steps also agrees with the previous observation of figure 1a, i.e. on the ($1\bar{1}00$) surface the steps show no preference between [0001] and [$11\bar{2}0$] directions.

Shown in figure 2 are REM images from the ZnO($11\bar{2}0$) surface. In figures 2a and 2b the incident beam azimuth is near [0001], and the surface was obtained by cleaving a (0001) wafer along [$1\bar{1}00$] direction. In contrast with the ($1\bar{1}00$) surface, in both figures 2a and 2b the surface steps show apparent preference about [0001] direction.

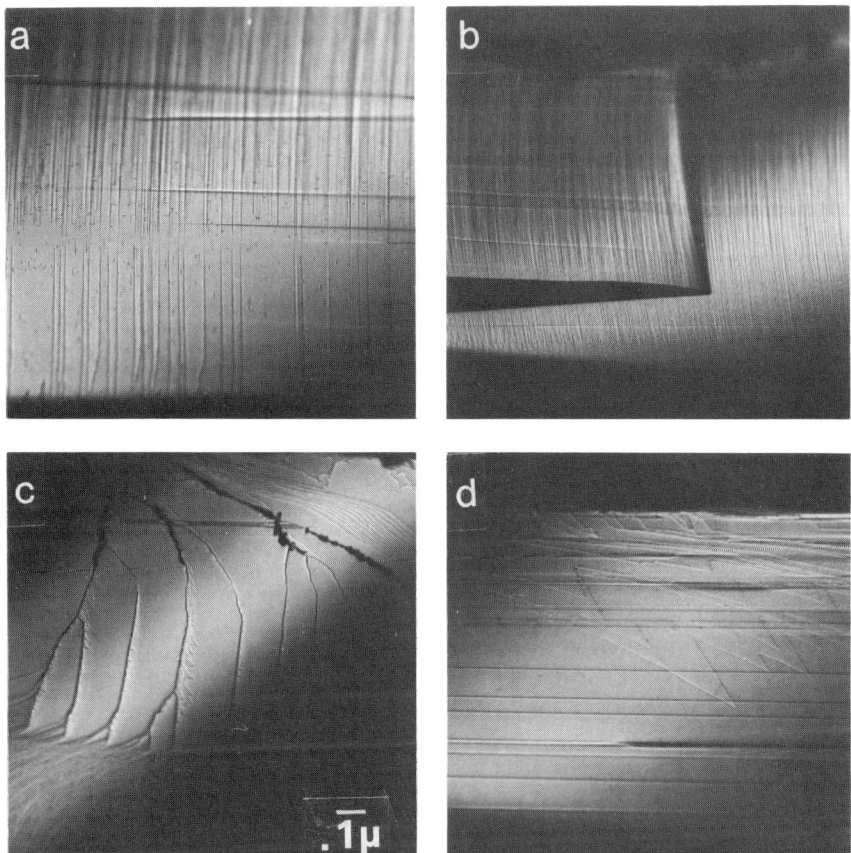

Figure 2: REM images of the ZnO(11$\bar{2}$0) surface.

Also shown in these figures are some inclined screw dislocations (Peng and Czernuszka, 1991), and in figure 2b a rather peculiar wedge shaped step starting from the centre of the image, with increasing step height to the left of the figure. In figures 2c and 2d the beam azimuth is near [1$\bar{1}$20] direction, and the surface was obtained by cleaving a (1$\bar{1}$00) wafer along the [0001] direction. Figure 2c shows a typical river fracture pattern, which differs from that of the (1$\bar{1}$00) surface, as shown in figure 1c, in many detailed aspects. Away from the fracture centre, figure 2d shows that the steps preferentially lie along [0001] direction, in agreement with the observations of figures 2a and 2b. The directional preference of surface steps on the (1$\bar{1}$00) and (11$\bar{2}$0) surfaces can be easily explained in terms of the atomic structures of the surfaces. Also, it is found that though all < 11$\bar{2}$0 > directions are possible directions for the dislocation Burgers vectors, in figure 2a and 2b only those inclined dislocations are observed, in contrast with figure 2d in which only normal emerging screw dislocation are seen. This interaction of the crack stress field with pre-existing dislocations is shown to have a strong effect on the final dislocation configuration. These results will be dealt with in more details in a separate paper.

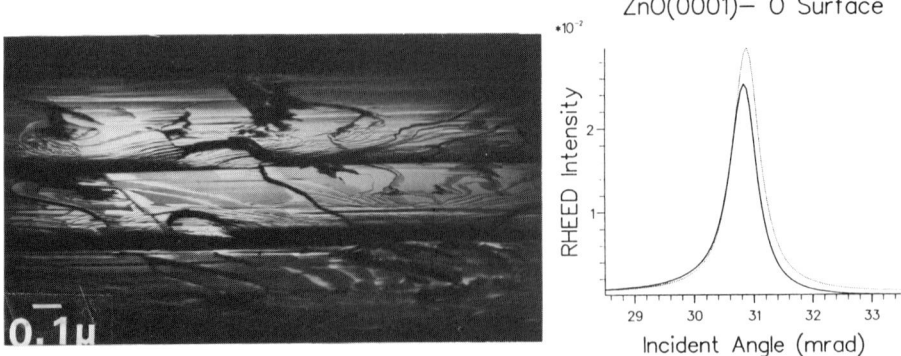

Figure 3: (a)REM image of the ZnO(0001) surface; (b)corresponding RHEED rocking curves. In (b) the solid line is for Zn-terminated face and dotted line is for O-terminated face.

The crystallographic polarity of ZnO crystals can also be imaged easily using REM. Shown in figure 3a is a REM image, taken from the polar (0001) surface. This image shows domains of bright and dark contrast separated by atomic steps. Dynamical RHEED calculations have been made under the same incident conditions as that of figure 3a. The reflection rocking curves are given in figure 3b, showing that within the angular range of incident illumination the Zn-terminated face always gives lower reflectivity than that of the O-terminated face on the O surface. It is thus concluded that the domain with bright contrast in figure 3a is O-terminated face, and the dark domain is Zn terminated face. This is the first REM observation of crystallographic polarity in non-centrosymmetric crystals. REM has the advantage over CBED and etching techniques in that it is non destructive while still giving spatial information.

4. CONCLUSIONS

The nature of the steps which appear on the fracture surfaces have been shown to be dependent on the cleavage direction. Only some of these steps are associated with intersecting dislocations. The final dislocation configuration is also strongly dependent on cleavage direction. The crystallographic polarity of the (0001) surface has been determined in a non-destructive manner. It has been shown that both O-terminating and Zn-terminating domains exist on the same fracture surface.

REFERENCES:

Cowley, J.M. 1986 Progr. Surface Sci., 21, 209
Peng, L.-M. and Czernuszka, J.T. 1991 Phil. Mag. A (in press)
Yagi, K. 1987 J. Appl. Cryst. 20, 147

Convergent beam and selected area electron diffraction analysis of the structure of In$_2$Se$_3$ crystals

C De Blasi, D Manno, G Micocci and A Tepore

Dipartimento di Scienza dei Materiali - Universita' di Lecce
G.N.S.M./C.I.S.M. Lecce - Italy

ABSTRACT: Structural analysis of In$_2$Se$_3$ crystals, grown from the melt by the Bridgman-Stockbarger method, has been performed at room temperature by electron diffraction techniques. Convergent beam diffraction patterns have been obtained from good quality regions of crystals, while selected area diffraction patterns have been obtained from regions affected by defects. These diffraction experiments have shown the structure to be prevailingly rhombohedric, while only a small fraction of the crystal volume is hexagonal. The stacking sequence of such crystallographic modifications has been observed.

1. INTRODUCTION

In recent years particular attention has been turned to the study of In$_2$Se$_3$ single crystals, because the electrical and optical properties of this material are of considerable interest for possible applications in technology. In$_2$Se$_3$ is a semiconductor of the A$_2^{III}$B$_3^{VI}$ family, whose compounds have structures which are defective with respect to their metal atoms: one third of the sites in the cation sublattice is vacant (Newman, 1961). These materials are able to trap ions or molecules selectively at vacancies, therefore they are suitable for sensors of small particles and could be used as detectors for ionizing radiations (Koshkin *et al* 1973, Koshkin *et al* 1977). In addition In$_2$Se$_3$ has shown attractive characteristics for possible applications in electrochemical devices, such as solid solution electrodes (Julien *et al* 1986).

Several authors have investigated the In$_2$Se$_3$ structure (Semiletov 1961, Popovic *et al* 1971, Van Landuyt *et al* 1975, Likforman *et al* 1980, Julien *et al* 1985, Kambas and Julien 1982) and they have reported that various crystallographic modifications and phase transitions occur in the material. In particular, the results report the structure to be dependent both on the method of preparation and on the experiment temperature.

In this paper we report the structural analysis of In$_2$Se$_3$ crystals grown in our laboratory, the analysis being performed at room temperature by electron diffraction.

2. EXPERIMENTAL

In$_2$Se$_3$ ingots were grown from the melt by using the Bridgman-Stockbarger method. The starting charge was obtained by synthesis of appropriate quantities of the elements according to the Spandau and Klanberg procedure. Details both of the experimental apparatus and of growth conditions have been previously reported (De Blasi *et al* 1989). Several growth runs supplied polycrystalline ingots with a lot of large layered regions, from which we have obtained specimens for the analysis.

Rutherford backscattering spectroscopy has given the chemical composition of the material as being close to the stoichiometric one. Measurements of electrical conductivity of the samples, performed by the Van der Pauw method, have shown the crystals to be n-type.

A systematic investigation of the crystal structure has been performed by using a Philips EM 400 T electron microscope operating at nominally 120 kV. Repeated cleavages of the ingots have supplied samples with very thin regions transparent to the electron beam. The zones of good quality and free of defects have been analyzed by the Convergent Beam Electron Diffraction (CBED) technique, focusing the beam onto areas of about 40 nm in diameter. The transmitted disk, recorded according to the Tanaka technique (Tanaka *et al* 1980), and the Whole Pattern (WP), recorded with a low diffraction camera length, have been observed by locating the [00.1] high symmetry zone axis. Their symmetries have been correlated to the point and space group of the structure by using the tables of Buxton *et al* (1976). The lattice parameters have been evaluated both using the zero layer disk and Higher Order Laue Zone (HOLZ) rings. The regions of poor quality, containing lattice defects, have been analyzed by the Selected Area Diffraction (SAD) technique, performed in zones of controlled area with diameters of about 5 μm.

3. RESULT AND DISCUSSION

Figure 1 shows a CBED pattern observed in a thin region: the symmetry both of Tanaka diffraction (a) and of the WP (b) is 3m, which can be attributed either to the 3m diffraction group (point group 3m) or to the $6_R mm_R$ diffraction group (point group $\bar{3}$m), according to Buxton *et al* (1976). Mirror reflection has been observed in any pair of $\pm \vec{g}$ vectors, excited in the Bragg condition: therefore we can affirm the point group to be 3m and the structure to be rhombohedral with space group R3m. By referring the rhombohedral lattice to an hexagonal unit cell, the *a*-axis has been evaluated by the position D of the (hk.0) reflections in the zero order [00.1] Laue zone, and the *c*-axis by using the radius R of the [00.1] n order Laue zone ring, according to the equations:

$$a = \frac{2\lambda L}{D^2} \sqrt{(h^2 + k^2 + hk)/3} \qquad\qquad c = \frac{2\lambda L^2 n}{R^2} \qquad (1)$$

Computations have given the values *a*=0.403 nm and *c*=2.89 nm.

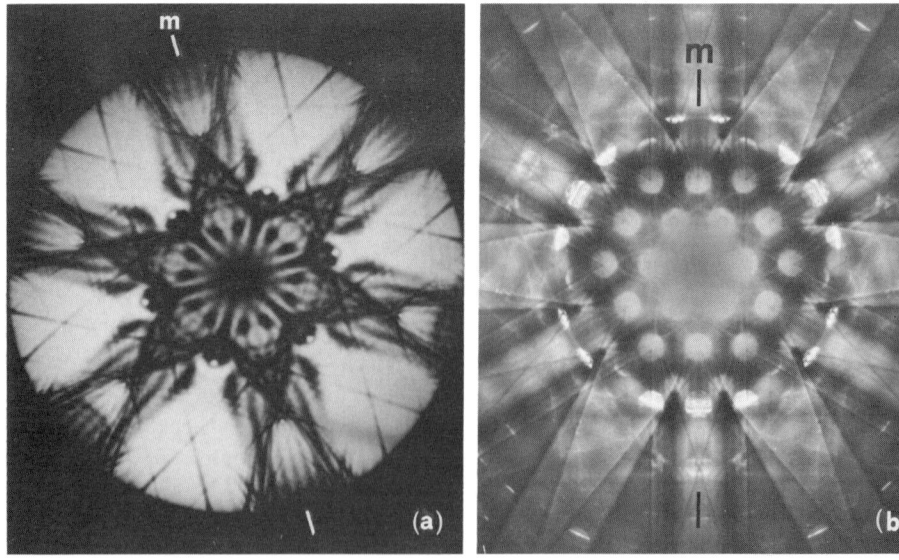

Figure 1 *Tanaka (a) and WP (b) recorded from a thin region of In$_2$Se$_3$ sample*

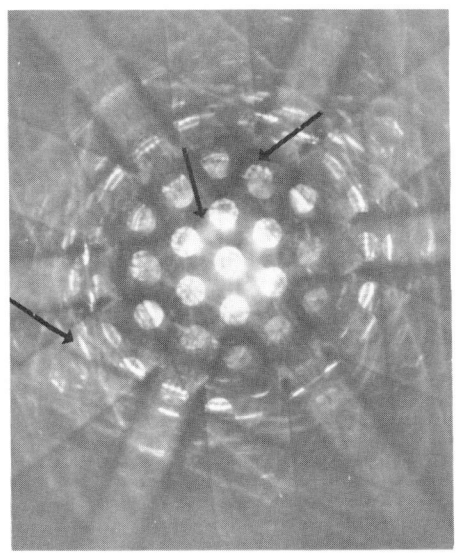

Figure 2 *CBED showing the faint [00.1] hexagonal pattern (arroved) and the strong [00.1] rhombohedral one.*

Figure 3 *SAD pattern showing the [00.1] diffractions of the hexagonal (H) and of the rhombohedral (R) structures.*

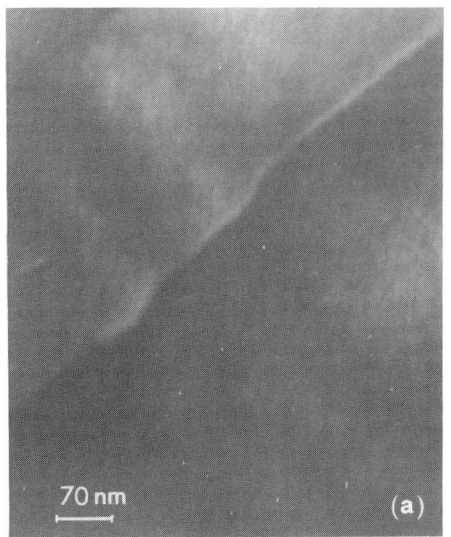

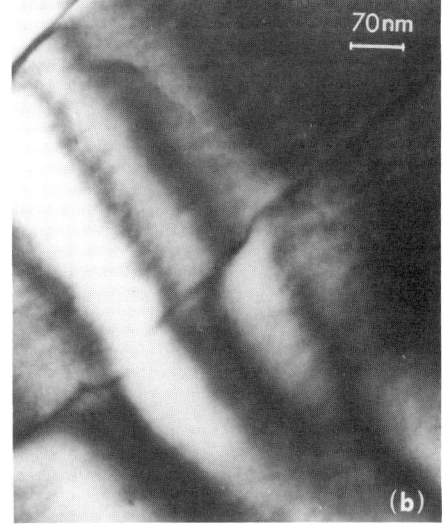

Figure 4 *Dark field images obtained (a) by (10.0)ₕ and (b) by (2̄1.0)ᵣ beams of Figure 3.*

A typical WP observed in a slightly thicker region is given in Figure 2: the picture shows two different kinds of patterns which differ in intensity. Their analysis has shown that the strong reflections belong to the [00.1] zone axis pattern of the rhombohedral phase and that the faint ones are due to a hexagonal structure with lattice parameters given by a=0.403 nm and c=1.91 nm. Therefore, the CBED observations suggest that In_2Se_3 is made of hexagonal and rhombohedral structural modifications.

Similar results have been obtained by the SAD technique, as presented in Figure 3. The diffraction pattern shows a net of faint reflections belonging to a hexagonal lattice and a net of strong reflections whose indexes comply with the rule 2h + k + l = 3n and therefore belonging to a rhombohedral lattice.

In order to gain information about the stacking sequence of the two phases, the (10.0)$_H$ beam and the ($\overline{2}\overline{1}$.0)$_R$ beam have been used to obtain the dark-field images shown in Figure 4a and in Figure 4b, respectively. By comparing the two images it is evident that the two phases are inserted one upon the other along the c-axis of the crystal.

By appropriately tilting the samples, different SAD zone axes of each structure have been excited and the lattice parameters have been computed. Calculations have given values in very good agreement with the ones obtained by the CBED technique.

4. CONCLUSIONS

An accurate crystallographic analysis of In_2Se_3 crystals, grown from the melt by the Bridgman-Stockbarger method, was performed at room temperature both by selected area and by convergent beam electron diffraction techniques and has shown the structure to be prevailingly rhombohedral with equivalent hexagonal unit cell parameters a=0.403 nm and c=2.89 nm. These results are in reasonable agreement with those quoted by Popovic *et al* (1971), although the sign of the space group is different.

In addition, we have found a fraction of the volume in the samples to crystallize in the hexagonal structure with crystallographic constants a=0.403 nm and c=1.91 nm. The two types of structure have been observed to be placed one upon the other along the c-axis of the lattice.

REFERENCES

Buxton B F, Eades J A, Steeds J W and Rackham G M 1976 *Philos. Trans. Roy. Soc. London* **A281** 171.
De Blasi C, Drigo A V, Micocci G, Tepore A and Mancini A M 1989 *J. Cryst. Growth* **94** 455
Julien C, Eddrief M, Balkanski M, Hatzikranitis E and Kambas K 1985 *Phys. Stat. Sol. (a)* **88** 687
Julien C, Eddrief M, Kambas K and Balkanski M 1986 *Thin Solid Films* **137** 27
Kambas K and Julien C 1982 *Mater. Res. Bull.* **17** 1583
Koshkin V M, Gal'chinetskii L P, Kulik V M, and Minkov B I 1973 *Solid. Stat. Comm.* **13** 1
Koshkin V M, Gal'chinetskii L P, Kulik V M, Gusev G K and Ulmanis V A 1977 *Sov. At. Energy,* **42** 321
Likforman A, Fourcray P, Guittard M, Flahaut J, Poirier F and Szydlo N 1980 *J. Solid State Chem.* **33** 91
Newman P C 1961 *J. Phys. Chem. Solids,* **23** 19
Popovic S, Celustka B and Bidjin D 1971 *Phys. Stat. Sol. (a)* **6** 301
Semiletov S A 1961 *Soviet Phys. Cryst.* **5** 673
Tanaka M, Saito R, Ueno K and Harada Y 1980 *J. Electron Microsc.* **29** 408
Van Landuyt J, Van Tendeloo G and Amelinckx S 1975 *Phys. Stat. Sol. (a)* **30** 299

Inst. Phys. Conf. Ser. No 117: Section 6
Paper presented at Microsc. Semicond. Mater. Conf., Oxford, 25–28 March 1991

A TEM study of inclusions and domains in CuInSe$_2$ crystals grown by the vertical Bridgman technique

S M Casey, C A Mullan, C J Kiely, R C Pond and R D Tomlinson*

Department of Materials Science and Engineering, The University of Liverpool, P.O.Box 147, Liverpool, England.

*Department of Electrical and Electronic Engineering, The University of Salford, Salford M54WT, England.

ABSTRACT: A study of the composition along a boule of directionally solidified CuInSe$_2$ was carried out using microprobe analysis. The stoichiometry was not found to vary a great deal along the length of the boule but a number of inclusions were found to exist at either end of the ingot. These inclusions were examined using SEM, TEM and EDX analysis and were found to be indium rich at one end of the boule and copper rich at the other end. A crystallographic analysis of the inclusions showed a definite relationship with the matrix structure. Finally evidence is presented of a domain structure existing within the CuInSe$_2$ matrix.

1. INTRODUCTION

CuInSe$_2$ is a direct band gap ternary semiconductor which is now used in polycrystalline thin film form as an absorber layer in efficient solar cell devices. Recently it has been demonstrated by Tomlinson (1989) that directional solidification methods can be used to grow polycrystalline ingots of CuInSe$_2$ which contain crystals of centimetre dimensions. TEM studies of polycrystalline thin films have been reported by Tseng at al.(1990) but to date very little microscopy on the directionally solidified material has been carried out. Kiely at al.(1991) have recently published the first in-depth TEM study of CuInSe$_2$ crystals grown by the vertical Bridgman technique which concentrated on (i) the basic crystallography of CuInSe$_2$ (ii) the structure of dislocations in CuInSe$_2$ and (iii) the qualitative analysis of Cu^{In} and In^{Cu} antisite defect populations using diffuse diffraction effects. In this paper we report on the chemistry and crystallography of inclusions and domains that are frequently found within the Bridgman boule.

2. BRIDGMAN GROWTH AND THE CuInSe$_2$ PHASE DIAGRAM

The details of the precise growth conditions and the problems associated with the vertical Bridgman growth technique have been reported by Tomlinson(1989). A pre-reacted charge of stoichiometric CuInSe$_2$ powder was encapsulated in a silica ampoule, thoroughly mixed and superheated for two days at 1150°C. It was then traversed through a two zone vertical furnace in which the hot zone temperature was set at 1050°C whilst the cold zone was kept at 900°C with a temperature gradient between the two of 20°C /cm. The whole charge was then traversed through the furnace at a rate of 1mm/hour until the whole of the resolidified charge was relocated in the cold zone. The ingot was then cooled at 1°C/hour through the sphalerite(δ)/chalcopyrite(γ) solid state transformation temperature at 810°C.

A recent study of the pseudobinary phase system by Fearheiley (1986) indicates that the compound melts incongruently at the stoichiometric composition since the maximum on the liquidus occurs at 1005°C with a composition of 55% In_2Se_3. This suggests that a non-homogeneous elemental distribution and crystalline inclusions may exist.

3. EXPERIMENTAL

Slices were cut from specific points along the length of the Bridgman ingot for microstructural and compositional analysis. Samples for SEM and microprobe analysis were prepared by mechanical polishing followed by a one minute etch in 0.1% Br_2/methanol solution. TEM samples were made by mechanically polishing down to 80 µm followed by 4kV argon ion milling (employing nitrogen cooling and a 15° incidence angle). TEM samples were examined in a Philips EM400T electron microscope and microprobe analysis was performed in a JEOL JXA 50A electron probe microanalyser using virtual standards.

4. RESULTS AND DISCUSSION

(a) Microprobe Analysis

Second phase formation was observed in the first to freeze and last to freeze zones of the Bridgman ingot. Only approximately 32mm of a 70mm length boule was found to be inclusion free. Table 1 shows a summary of our composition analysis of the $CuInSe_2$ matrix at various positions along the ingot. The composition is shown to be relatively uniform (although slightly copper deficient) along most of the ingot length. This correlates with recent STM images of this material obtained by Abou-Elfotouh et al. (1991) which clearly show copper vacancies. At the 60/70 position the ingot is seen to be slightly indium rich and selenium deficient. Such deviations from stoichiometry are important in this system since they control the conductivity type and level in this crystal.

Microprobe analysis of inclusions showed their compositions to be quite variable but they all contain significant quantities of copper, indium and selenium. Analysis of the phase diagram suggests that In_2Se_3 and Cu_2Se may be expected to form at the first and last to freeze zones respectively, however these selenide phases have not been experimentally observed.

(b) Inclusions

Figure 1 shows an SEM image close to the last to freeze zone in the $CuInSe_2$ crystal where copper rich inclusions are to be expected. Facetted inclusions, approximately 15µm in diameter were observed in clusters within the $CuInSe_2$ matrix. These inclusions are distributed in a cellular fashion across the complete diameter of the boule at this position . TEM analysis of these facetted particles shows them to have a definite orientation relationship with the $CuInSe_2$ matrix. Figure 2(a) shows the [021] chalcopyrite diffraction pattern obtained from pure $CuInSe_2$ whereas figure 2(b) shows an [021] SADP from a region containing inclusions. Extra reflections labeled X and Y are evident and dark field imaging using one of these reflections causes one of the two sets of epitaxial inclusions to light up. Such a dark field image of one of these particles is shown in figure 2(c).

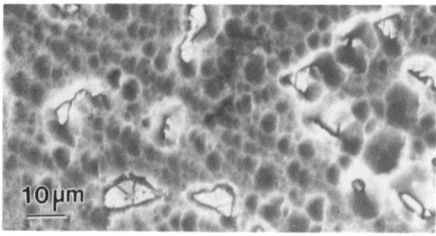

Position	% Cu	% In	% Se	Inclusion
14/70	23.7	26.6	49.7	Yes
28/70	23.7	26.2	50.1	Yes
44/70	23.3	26.6	50.1	No
52/70	23.6	26.8	49.6	No
60/70	24.1	26.0	47.9	Yes

Figure 1 SEM image of inclusions at   Table 1
copper rich end of boule

Characteristic deformation twinning faults are often seen within these inclusions. It seems reasonable to speculate that these inclusions precipitate out from the δ-sphalerite phase during cooling since they have a definite orientation relationship with the $CuInSe_2$ matrix. We have ruled out the possibility of the extra reflections X and Y arising from superlattice ordering effects similar to those observed in other ternary semiconductors such as InGaP, by obtaining SADPs from a number of other zone axes. A more complete analysis of the crystallography and microchemistry of these inclusions will be presented elsewhere.

 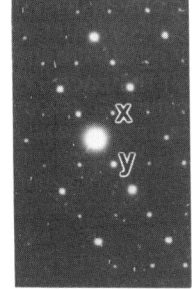 

Figure 2(a) [021] SADP of $CuInSe_2$

Figure 2(b) [021] SADP of inclusion at copper rich end of ingot

Fig 2(c) Dark Field Image of inclusion taken in reflection X.

Inclusions from the first to freeze zone have also been examined by TEM. The 1μm diameter particle in figure 3(a) was shown by EDX analysis to be indium rich with an approximate weight percentage composition of 14% copper, 50% indium and 36% selenium. Selected area diffraction analysis has shown the inclusion to have a defect chalcopyrite structure which is misorientated by about 2° from the $CuInSe_2$ matrix. Figure 3(b) shows the SADP [021] pattern for this inclusion which exhibits a definite 5-fold superlattice periodicity along the [11$\bar{2}$] direction. Frangis et al. (1986) have previously reported a five layer stacking periodicity for a $CuIn_5Se_8$ phase. Furthermore, Tseng and Wert (1989) have reported the existence of 7-layer and 14-layer polytype structures in indium rich phases. The (11$\bar{2}$) stacking periodicity within the indium rich inclusion shown in 3(a) is seen to vary considerably from point to point and this polytype structure is thought to arise from a mixture of cubic and hexagonal stacking of (11$\bar{2}$) type planes.

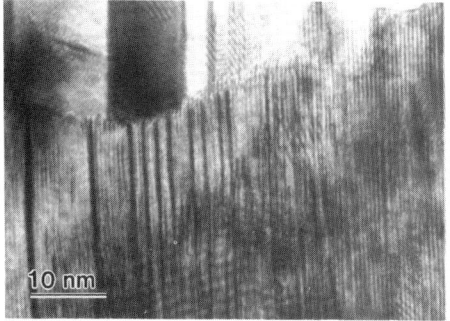 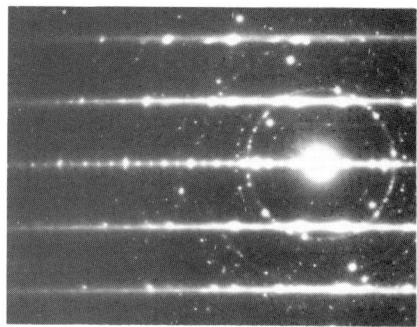

Figure 3(a) Bright Field Image of Indium rich inclusion showing irregular (11$\bar{2}$) stacking

Figure 3(b) SADP from region of inclusion which exhibits 5 fold periodicity along indium rich end of ingot

(c) Domains

On cooling through 810°C, the CuInSe$_2$ undergoes a solid state phase transformation from a cubic sphalerite structure (δ) to a tetragonal (γ) structure. The sphalerite crystal exhibits the space group F$\bar{4}$3m which has twenty four symmetry operations, whereas the chalcopyrite structure has a space group I$\bar{4}$2d with eight point symmetry operations. Consequently, three crystallographically equivalent variants of the lower symmetry structure should coexist, which should be separated by domain boundaries. These correspond to having the tetragonal c axis orientated in three orthogonal directions. These variations have been experimentally observed as demonstrated by the [001] and [100] SADP's shown in figure 4 which were obtained from adjacent regions on the same CuInSe$_2$ sample without any specimen tilting. In addition, since we have considered a non-centrosymmetric parent structure, three further domains, related to the three described above by the inversion operation, can in principle coexist. These latter domains have not been experimentally observed to date in the Bridgman grown CuInSe$_2$ samples.

Figure 4(a)   SADP from an [001] oriented CuInSe$_2$ domain

Figure 4(b)   SADP from an [100] oriented CuInSe$_2$ domain

5.CONCLUSIONS

Microprobe analysis has been used to analyse the composition of a directionally solidified CuInSe$_2$ boule. The matrix stoichiometry was not found to vary a great deal along the length of the boule but indium rich and copper rich inclusions were found at the first and last to freeze ends of the ingot respectively. TEM analysis of the inclusions showed them to have specific orientation relationships with the CuInSe$_2$ matrix. The sphalerite/chalcopyrite solid state phase transformation at 810°C has been seen to result in a domain structure occurring within the crystalline form of CuInSe$_2$.

REFERENCES

Abou-Elfotouh F, Moutinho H, Bakry A, Coutts T J and Kazmerski L L 1991 Solar Cells (in press)
Kiely C J, Pond R C, Kenshole G and Rockett A 1991 Phil. Mag. A (in press)
Fearheiley M L 1986 Solar Cells **16**, 96
Frangis N, Van Tendeloo G, Manolikas C, Van Landuyt J and Amelinckx S 1986 Status Solidi **A96**, 53
Tomlinson R D 1989 PVSEC Sydney Australia p. 467
Tseng B H, Lommasson T C, Yang L C, Wert C A, Rockett A and Thornton J A 1990 J. Appl. Physics, **67**, 2673
Tseng B H and Wert C A 1989 J. Appl. Physics **65**, 2255

Inst. Phys. Conf. Ser. No 117: Section 6
Paper presented at Microsc. Semicond. Mater. Conf., Oxford, 25–28 March 1991

383

Structural characterisation of Nb-doped lead zirconate titanate ferro-electric thin films

Louise Weaver, Lynnette D Madsen[1], and Ellen M Griswold[2]

Northern Telecom Electronics Limited (NTEL), P.O.Box 3511, Station C,
Ottawa, Ontario, Canada K1Y 4H7;
[1]Department of Materials Science, McMaster University,
Hamilton, Ontario, Canada L8S 4M1;
[2]Department of Physics, Queen's University,
Kingston, Ontario, Canada K7L 3N6.

ABSTRACT: Ferroelectric thin films of Nb-doped rhombohedral Lead Zirconate Titanate (PNZT) sputter deposited and annealed on Pt coated Si substrates were studied. The composition was determined by wavelength dispersive x-ray analysis (WDX). Glancing-angle XRD was used to identify the crystal systems and determine the lattice parameters. The preferred orientation of (101) was observed by both XRD and electron diffraction. Scanning and transmission electron microscopy revealed cracking of the films and grains of 2-5nm in diameter. Mono-domains are suggested as the mechanism for ferroelectricity.

1. INTRODUCTION

Ferroelectric thin films of lead zirconate titanate (PZT) have found application in the semiconductor industry in memory devices. In bulk ceramics, doping with niobium (Nb), was found to increase the electrical resistivity and give square hysteresis loops with low coercive fields and high remanent polarizations (Jaffe et al 1971). Since these characteristics are favourable for memory applications, thin films of Nb-doped PZT (PNZT) were examined for similar properties.

PZT, $PbZr_{1-x}Ti_xO_3$, has the general formula ABO_3. The perfect perovskite structure, space group Pm3m, is cubic with large (A) cations at the cube corners, small (B) cations at the body centres, and the oxygen ions at the face centres (Lines and Glass 1977). Most perovskite compounds are slightly distorted at room temperature and hence have lower symmetry (Roth 1957). These departures from the ideal structure result in dielectric and magnetic properties including ferroelectricity. For PZT, the large cations are lead (Pb) and the small cations are either zirconium (Zr) or titanium (Ti). The stoichiometry of PZT is achieved by the volatilisation of PbO during annealing, and for this reason the material initially deposited contains excess Pb. PZT crystallises from the high temperature cubic form into either a tetragonal or rhombohedral phase at room temperature according to the Ti/Zr ratio (Fig 1a). For the rhombohedral phase (Ti/Zr < 52/48), the cell is distorted along the body diagonal (Fig. 1b).

In PNZT, the dopant Nb will replace either Ti or Zr in the body centre position and will cause a decrease in the number of Pb atoms to maintain charge neutrality (Dih and Fulrath 1978). For small additions of Nb (< 1wt%) both increased resistivity and reduction in grain size have been observed (Dih and Fulrath 1978). In this work, the effect of Nb doping has been investigated in terms of crystal structure and lattice parameters. A discussion of the electrical properties of Nb doped films has been presented elsewhere (Griswold et al 1991).

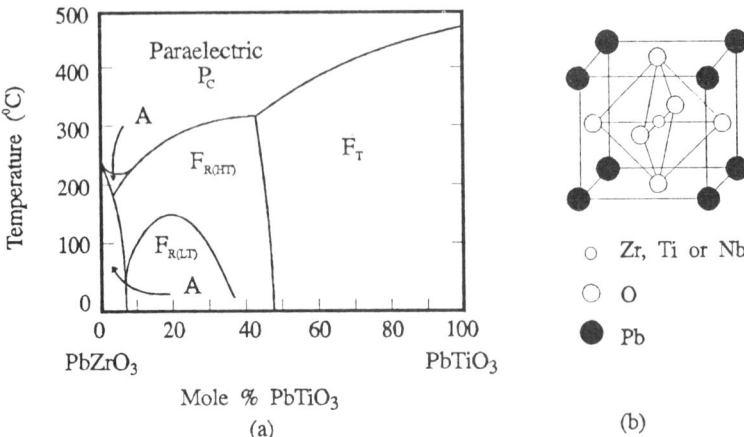

Figure 1: a) Phase diagram for PZT with F (ferroelectric) and A (antiferroelectric) phases (after Jaffe, Cook and Jaffe 1971) and b) PNZT shown schematically for the perovskite cubic phase. The centre atom can be Ti, Zr or Nb.

2. EXPERIMENTAL

PNZT films of $\sim 0.8\,\mu$m were deposited at 200 °C using reactive dc magnetron sputtering (Griswold et al 1991). The films were made on 2" x 2" square substrates of sputter deposited platinum (Pt) on <100> silicon (Si). After deposition, these films were annealed in air ambient at atmospheric pressure in a Lindberg furnace for 8 hr at 450 and 550 °C using ramp rates of 5 °C/min.

A Rigaku DMAX glancing angle x-ray diffractometer was used to obtain x-ray spectra of the as-deposited and annealed films and for lattice parameter determination. For compositional analysis WDX was used. SEM was carried out on a Cambridge S360 and TEM on a JEOL 2000FX STEM. The TEM samples were prepared in cross-section and plan-view by mechanical thinning and ion milling.

3. RESULTS

3.1 As-deposited Sample

As-deposited, the film composition, determined by WDX, was found to be non-uniform and in SEM back-scatter mode it was possible to discern areas rich in Pb (Fig. 2). The presence of excess Pb is desirable for obtaining stoichiometric PZT after annealing (Takayama and Tomita 1989). The X-ray spectrum (Fig. 3a) indicated some crystallisation had also occurred during the deposition at 200 °C. The strong peaks at 2Θ values of 23° and 28° are characteristic of PbO-type compounds and the double peak at 51° and single peaks at 31°, 36° and 48° correspond to complex oxides, such as pyrochlore, $A_2B_2O_7$. The peaks in these spectra are relatively broad indicating high stress. A severe adhesion problem between the Pt and Si substrate caused delamination in some instances. Although SEM showed a smooth and continuous film, TEM images at higher magnification of the PNZT near the Pt interface revealed the presence of some cracks (Fig. 4a). Plan-views at different depths showed cracking of both the Pt and PNZT had occurred, but that it decreased significantly away from the interface. Since the thermal expansion coefficient for Pt is significantly different to $Pb_2(Zr,Ti)_2O_6$, cracking could be expected to occur.

3.2 450 °C Sample

The composition of the film annealed at 450 °C did not show the expected reduction in the quantity of Pb from the as-deposited film. The x-ray spectrum, although changed from that of the as-deposited film (Fig. 3b), is still indicative of a pyrochloric type compound, rather than a perovskite structure and still shows the peak at 28 °. The adhesion between the Pt and Si substrate had improved slightly. Cracking of the PNZT

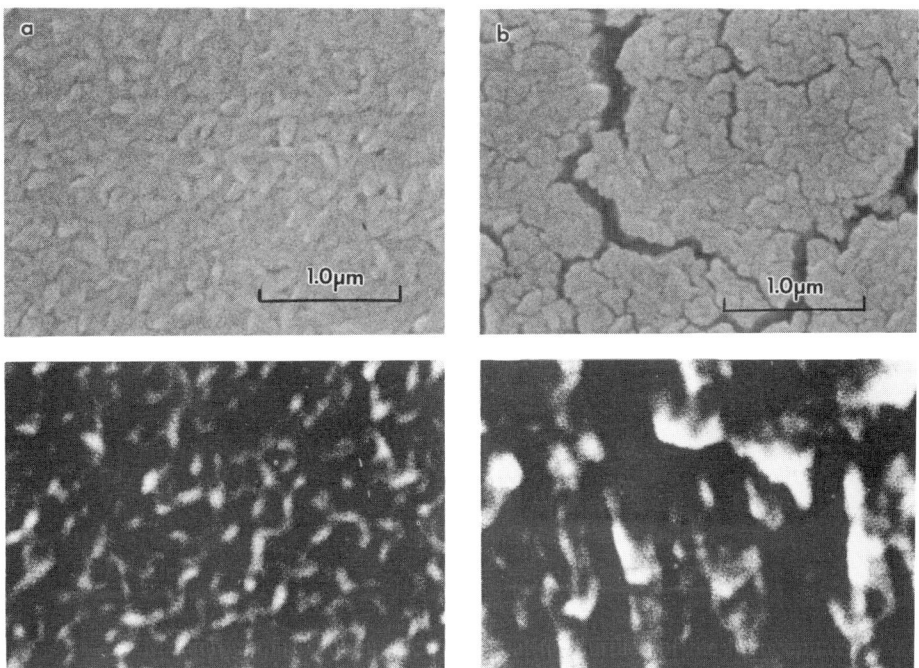

Figure 2: Backscattered and SEM images of the same areas in PNZT films (a) as deposited and (b) annealed at 550°C.

film, apparent only in TEM in the as-deposited sample, had progressed and was observed even at low magnifications at the surface (Fig. 4b).

3.3 550 °C Sample

Annealing at 550 °C reduced the Pb content by ~10-20%. The non-uniformity and Pb rich areas observed in the as-deposited films were retained. However, there was sufficient Pb loss to allow the film to transform to a perovskite structure with a Zr:Ti ratio of 58:42 at%. According to the phase diagram (Fig 1a) this ratio corresponds to the HT rhombohedral phase, specifically the R3m space group (Michel et al 1969). The Nb concentration of 0.5at% is much lower than that reported in the literature

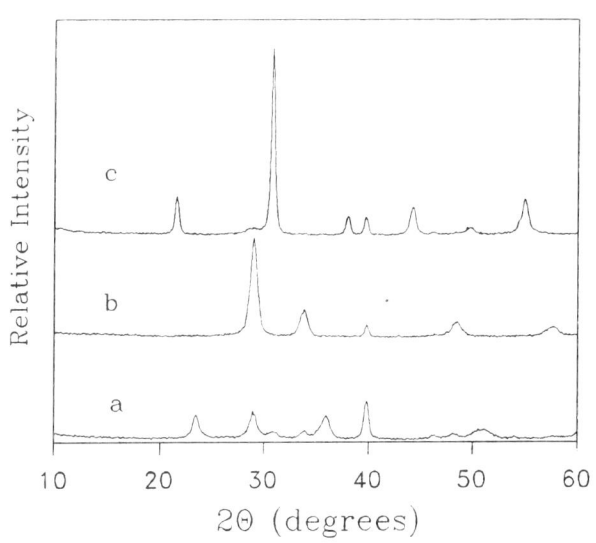

Figure 3: Glancing-angle XRD spectra of (a) as-deposited and annealed PNZT films for (b) 450 °C and (c) 550 °C.

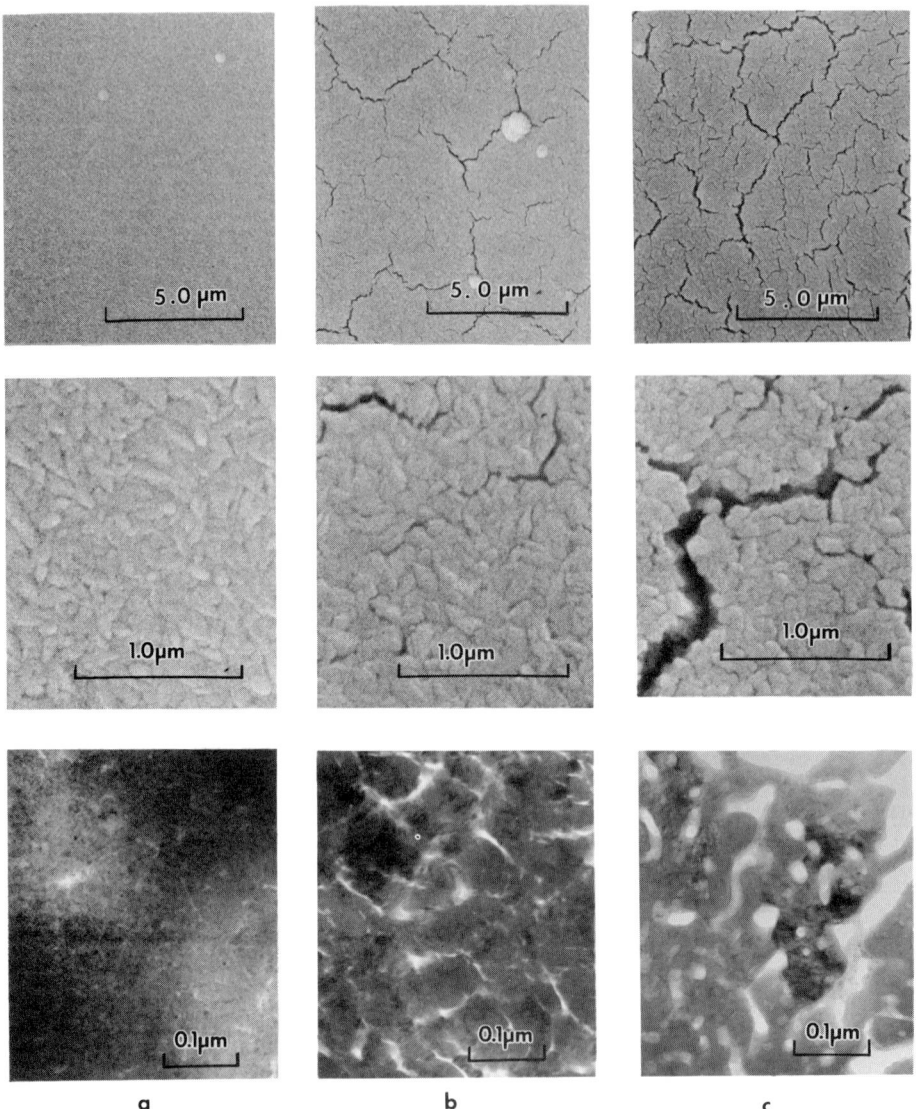

Figure 4: SEM images at 5K and 30K mag. and TEM plan-view micrographs at 100K (top to bottom) for (a) as grown and annealed PNZT films for (b) 450 °C and (c) 550 °C.

for bulk PNZT ceramics (Gerson 1960, Dih and Fulrath 1978) and other PbTiO$_3$ thin films containing Nb (Chapman 1969, Francis and Payne 1990).

X-ray spectrum of the films annealed at 550 °C confirmed that crystallisation into the rhombohedral structure had occurred and indicated the presence of the following crystallographic planes: (100), (101), (111), (200), (102), and (212) at 22, 31, 38, 44.5, 49, and 55° respectively. The strong peak at 31° indicates a preferred orientation in the [101] direction. The lattice parameters were determined as follows: a= 4.1Å and α = 89.9°.

These parameters aided in the identification of the dominant orientations observed in electron diffraction patterns.

The preferred orientation in the [101] direction found in the XRD spectrum was reflected in diffraction patterns from the sample in general (Fig. 5a). When SAD was carried out on regions of similar diffraction contrast, spot patterns were obtained (Fig. 5b). Although the 101 zone dominated, other zones, namely, the 100, 111, 113, 331, 531, 123, 201 were also present elsewhere on the sample. These regions were composed of very small individual grains, ~2-5nm, separated by low angle grain boundaries (Fig. 5c).

The cracks in this sample were larger and denser than for the 450 °C sample and again were carried through into the sub-micron level (Fig. 4c). In this case, there was no delamination of the Pt film from the Si substrate due to the formation of PtSi at 550 °C (Segmüller and Murakami 1988) evident in cross-sectional TEM micrographs.

4. DISCUSSION

Electrical testing performed on PNZT films of the same composition indicated good ferroelectric properties. Previous studies of larger sized grains showed twinning occurred as a strain relief during the phase transformation from the cubic to the rhombohedral or tetragonal phase (Rice and Pohanka 1979) with each twin then acting as a ferroelectric domain. TEM studies of lanthanum-modified $PbTiO_3$ have indicated a minimum grain size of $0.3\,\mu m$ is necessary for twinning to occur (Demczyk et al 1987). In the PNZT films studied here, the low angle grain boundaries between small grains are believed to relieve strain and allow the ferroelectric effect to take place with each grain or several grains acting as a mono-domain. This has been suggested for bulk Nb doped PZT (Martirena and Burfoot 1974) but has not been reported previously as occurring in thin films.

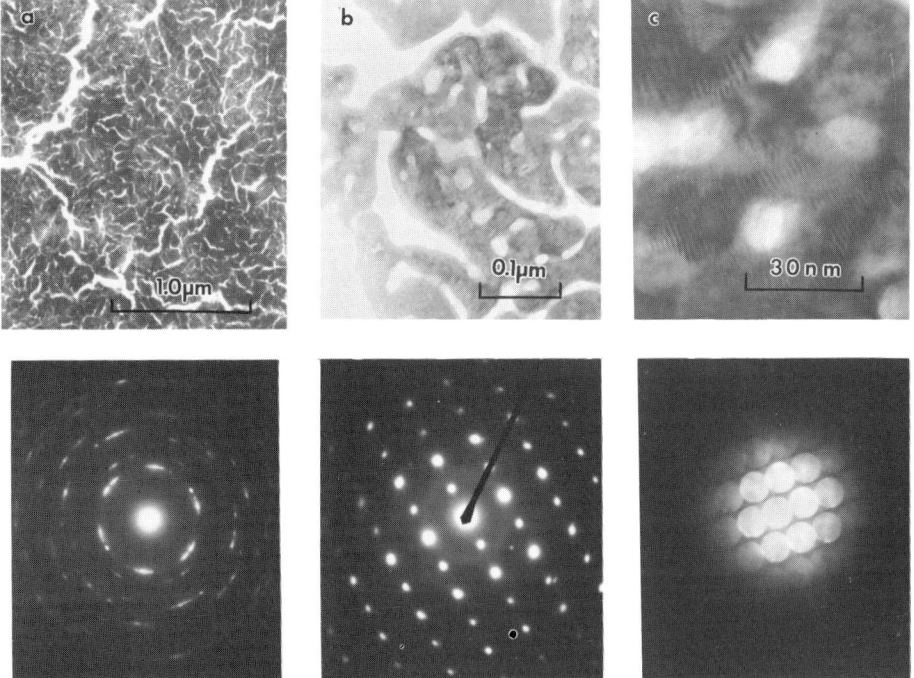

Figure 5: The preferred orientation seen in the XRD is reflected in the electron diffraction patterns obtained for (a) a large area, (b) the area between the cracks and (c) individual grains.

In comparison to undoped PZT films, slight alterations in the lattice parameters and increased resistivity were found, but no noticeable improvement in terms of electrical properties was noted. This could be due to the very low Nb content or alternatively, increased domain-wall mobility with Nb-doping linked to electrical improvements (Gerson 1960) may not apply to the thin films studied here.

Cracking appeared to be associated with the high angle grain boundaries that separated regions of similar crystallographic orientations. Separation of grains has been observed previously in $PbTiO_3$ by Matsuo and Sasaki (1966). This has been attributed to stress at grain boundaries induced by the thermal expansion of individual grains with differing crystallographic orientations. This effect is exacerbated by the thermal mismatch between the Pt and PZT films.

5. CONCLUSIONS

R3m rhombohedral phase PZT thin films with 0.5 at% Nb doping were investigated. The formation of very small grains and the absence of twins in these films suggests that the ferroelectric effect is occurring by a mono-domain mechanism rather than by the usually observed twining mechanism. The effect of Nb, which provides easier domain wall motion may be insignificant with a small grain structure and could account for the lack of enhanced electrical properties observed. PNZT thin films suffered from severe cracking due to either stress relief or mismatch with the substrate. These cracks tended to occur at high angle boundaries and separated regions of grains of similar crystallographic orientation. Within these regions were small grains separated by low angle boundaries.

ACKNOWLEDGEMENTS

Useful discussions with Drs. I. Calder of NTEL, G. Carpenter of CANMET, Y-P. Lin of Bell-Northern Research, M. Sayer of Queen's University and G. Weatherly of McMaster University were most appreciated. The glancing-angle XRD analysis was carried out by Dr. R. Pascual of Queen's University. WDX was provided by Surface Science Western. Thanks are also extended to Mr. D. Mayer of NTEL for his assistance with the TEM sample preparation. This work is funded in part by scholarships from NTEL, the Natural Sciences and Engineering Research Council of Canada and the Ontario Centre for Materials Research. The associated presentation expenses were aided by an EMAG bursary.

REFERENCES

Chapman D 1969 J. Appl. Phys. 40 2381
Demczyk B, Khachaturyan A and Thomas G 1987 Scripta Metallurgica 21 967
Dih J and Fulrath R 1978 J. Amer. Ceram. Soc. 61 448
Francis F and Payne D 1990 Mat. Res. Soc. Symp. Proc. 200 173
Griswold E, Sayer M, Amm D and Calder I 1991 Can. J. Phys. 69 (in print)
Jaffe B, Cook W and Jaffe R 1970 Piezoelectric Ceramics (Academic: London) pp 136, 152-6
Lines M and Glass A 1977 Principles and Applications of Ferroelectrics and Related Materials (Clarenden:Oxford) p 241
Martirena H and Burfoot J 1974 J. Phys. C: Solid State Phys. 7 3182
Matsuo Y and Sasaki H 1966 J. Amer. Cer. Soc. 49(4) 229
Michel C, Moreau J-M, Achenbach G, Gerson R and James W 1969 Solid State Commun. 7 865
Gerson R 1960 J. Appl. Phys. 31 188
Rice R and Pohanka R 1979 J. Amer. Cer. Soc. 62(11-12) 559
Roth R 1957 J. Res. Nat. Bur. Stds. 58(2) 75
Segmüller A and Murakami M 1988 in Analytical Techniques for Thin Films, eds. Tu K and Rosenberg R (Academic:London) pp 143-200
Takayama R and Tomita Y 1989 J Appl. Phys. 65(4) 1666

Inst. Phys. Conf. Ser. No 117: Section 7
Paper presented at Microsc. Semicond. Mater. Conf., Oxford, 25–28 March 1991

389

In situ electron microscopy of GaAs MBE monolayer growth

N Inoue

NTT LSI Laboratories, Morinosato-Wakamiya, Atsugi 243-01 Japan

ABSTRACT: A molecular beam epitaxy(MBE) - scanning electron microscopy(SEM) - scanning reflection electron microscopy(SREM) hybrid system is described. The resolution and observation rate are discussed. Three kinds of applications are demonstrated; observation of quick transient growth processes, measurement of material parameters under actual growth conditions, and in-situ monolayer growth control. Unique application of SEM to growth study are the observation of Ga droplet formation and Ga monolayer lateral growth. Results from a scanning tunneling microscope(STM) mounted on the system are shown.

1. INTRODUCTION

Recently, quantum effect devices have attracted a great deal of attention due to their potential for ultra-fast operation. However, partial monolayers on the quantum well interface make the well thickness nonuniform. In addition electron scattering from this rough interface slows down the electrons. So, precise growth control down to the monolayer level is essential. Quantum well structures are usually grown by molecular beam epitaxy. Molecular beams are cut abruptly by shutters, forming ultra-thin layers. However, perfect monolayer growth technology still has to be developed. This is primarily due to the fact that we really don't know much about the growth mechanism or the problems involved in growth control.

Previous growth models have been based on ex-situ observation. This leaves a lot of ambiguity. Another analytical method is computer simulation. Unfortunately, the use of slightly inadequate parameters results in a far from realistic picture of the growth mechanism. We might be able to get some new insights if we could actually see what was going on by in-situ observation during growth. One "in-situ method" that is commonly used in conventional MBE is reflection high energy electron diffraction (RHEED). This is actually in-situ measurement, not observation. It provides only ambiguous information averaged over a wide area of the sample surface. In-situ electron microscopy would provide much more detailed information about small features on the surface.

Surface reflection electron microscopy (REM) using diffracted electrons has been developed by Halliday and Newman (1960). Then scanning reflection electron microscopy (SREM) was developed by Cowley (1975). But application to in-situ observation of MBE growth was not provided when we started this project eight years ago, though there was a "microprobe-RHEED" for poly-silicon (Ichikawa 1982). And thereafter a trial of "SEM-STEM-MBE" was made for GaAs (Petroff 1986). We have successfully analyzed Si terrace growth processes during annealing using in-situ REM (Inoue 1987). However, for the in-situ observation of growth, we need a high resolution and   quick response   electron microscope, and precise growth control in a growth system, and the latter may not be satisfied in the small REM specimen room. So we have developed an MBE-SEM-SREM hybrid system for GaAs (Yamada et al, 1988). The first successful application to in-situ observation of MBE was made by Ichikawa on Si(1987). The present study is an application to GaAs   and its differences from the work with Si are that this is the first real-time, continuous observation and that this evolved a new application of SEM, not SREM, to monolayer growth study. Step-by-step observation of GaAs MBE has been reported recently (Isu 1988). To understand growth in more detail, we have recently mounted a scanning tunneling microscope (STM) in this system.

2.   MBE-SEM-SREM HYBRID SYSTEM

2.1 Outline

The schematic illustration of the hybrid system is shown in Fig. 1. We attached the ultra-high-vacuum optical column of the SEM horizontally, not vertically, to the small MBE chamber. This solved the most serious problem with GaAs, which was the clouding of the lenses by the upward As flux. The electron beam is focused on the surface of the sample so as to produce Bragg reflection. Then the intensity of the diffracted electrons is measured with a photomultiplier (PM) placed in front of a phosphor screen. The image of the surface is displayed on a CRT as the electron beam scans the sample. This is called the SREM mode. We can also obtain conventional SEM images by using a secondary electron detector (SED). The images are foreshortened in the vertical direction by about 20 times due to the grazing incidence of the electron beam. The changes are continuously recorded with a video tape recorder (VTR). We use a LaB_6 filament because it is more resistant than a field emission filament to a low vacuum environment with a high vapour pressure As flux. The whole system is mounted on a vibration isolation stage on a concrete platform.

The difficulty in constructing such a hybrid system is that it is very delicate, and both components have to operate under optimum conditions at the same time. The resolution and observation rate have been improved recently (Inoue 1991a,b).

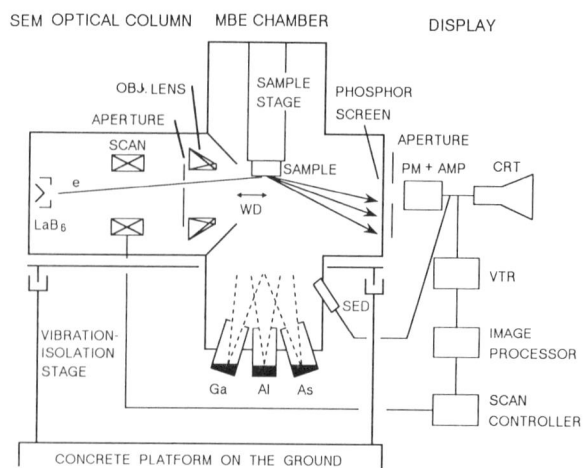

Fig.1 An MBE-SEM
-SREM hybrid system

2.2 Resolution

The resolution is limited by many factors. Among them, the beam size is the most important. We developed a high gain amplifier(AMP) and use a high-brightness filament, which
made it possible to reduce the beam current. The relationship between the electron beam
size and the beam current is shown in Fig. 2. The parameters are the working distance
(WD) and the aperture size of the objective lens (OL). We use a JSM-840 optic column
from the JEOL Company. In our system, the working distance (WD) is 39 mm. The beam
current is a low 10^{-10} A for both the SEM and SREM modes. As a result, the beam size is
expected to be 20 nm. Beam divergence is an important consideration in making high contrast SREM images. In this system it is 3 mrad. with a 50-micron OL aperture.

We tried to eliminate sample vibrations in two ways. One was to mount the whole system
on a vibration isolation stage. The other was to take advantage of the horizontal optic
column and develop a vertical sample stage. It is stiffer than the horizontal sample stage
usually used in SEM.

To estimate the effect of stray magnetic fields, we intentionally created an AC magnetic
field. The electron beam is deflected by about 3 microns by an 8 mG magnetic field. This
agrees well with our calculations, so we must keep the magnetic field below 0.1 mG to
ensure that its effect is less than the beam size. Stray magnetic fields were reduced as much
as possible by using DC current for the evacuation and sample heating devices, and by using
magnetic shields.

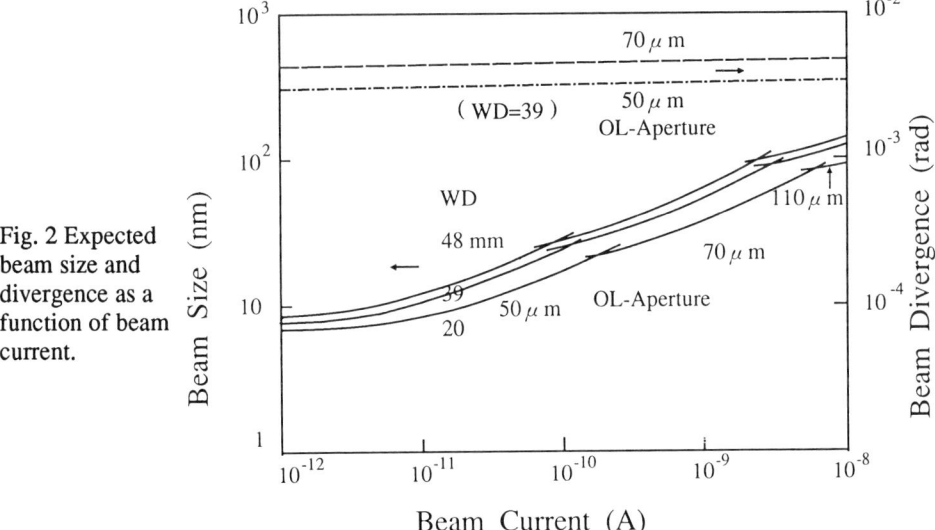

Fig. 2 Expected
beam size and
divergence as a
function of beam
current.

Figure 3 shows the resolution of SEM images obtained with this system, 3(a) shows the
surface roughness on the sample holder and (b) shows the multiple steps on the cleaved
surface. Features 30 nm in size can be resolved. The resolution is fairly close to the designed beam size of 20 nm, though more degradation was expected due to the hybridization.
Figure 4 shows the pictures for the SREM mode: (a) shows surface undulations. For crystalline features, the apparent resolution is about 200 nm. The dark spots in (b) are Ga droplets. For noncrystalline features, the apparent resolution is better, about 50 nm. The slightly
worse resolution of SREM compared to SEM is due to poor S/N.

2.3 Observation Rate

The observation rate is nearly as important as the resolution. At the start of our research, it took 10 seconds to complete 1 frame. But real-time observation is possible, because the raster is scanned continuously and the surface is uniform; so the variation in the vertical direction, or time axis, shows the change with time. A high observation rate either requires a high beam current, which increases the beam size, or degrades the contrast. In either case, the resolution gets worse. We have developed an image processor and reduced the background signals as much as possible. For SREM, we developed a high-gain amplifier(AMP) and improved the S/N by reducing the beam divergence to 3 mrad and using a small aperture on the phosphor screen. As a result, TV rate (60 frames/sec) observation is possible in the SEM mode and 1 frame/sec is possible in the SREM mode using an external scan controller.

3. WHAT CAN BE SEEN?

There have been a few reports on in-situ observation. Unfortunately, due to difficulty in growth control, the results obtained were not always surprising but confirmed models based on other techniques. We want to see things that are not preservable outside the growth chamber including otherwise inaccessible transient processes. It is very exciting to explore the unknown and encounter the unexpected.

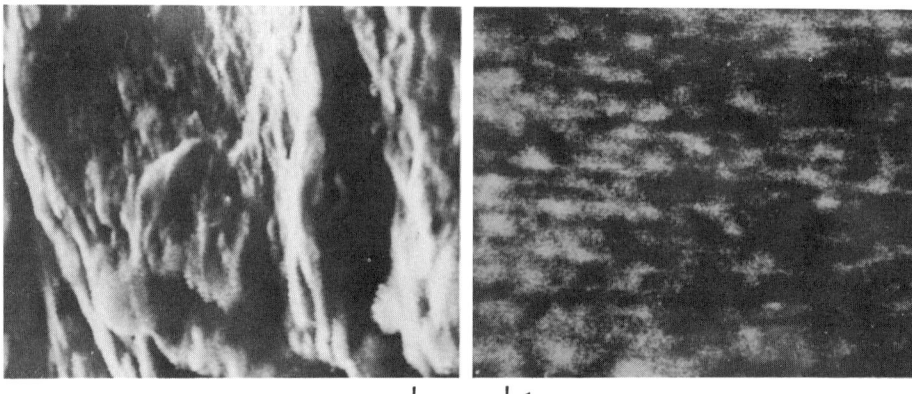

|—————| 1μm

(a) Roughness on the sample holder   (a) Undulation

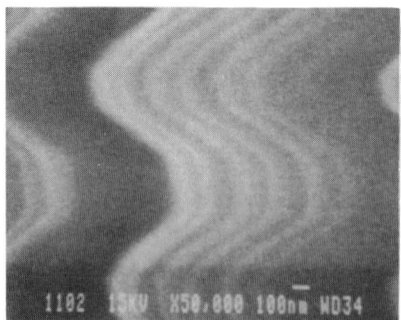   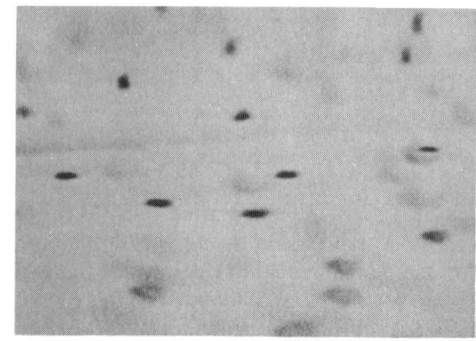

(b) Multiple steps on cleavage   (b) Ga droplets and mounds from droplets

Fig.3 SEM resolution   Fig.4 SREM reslution

What can we see with this instrument? Surface undulations and multiple steps are easily observable as shown in Fig.4(a) and 3(b). One of the things we did not expect was the Ga droplets shown in Fig.4(b). The other was the coexistence of Ga- and As- top layers shown in Fig.8(b). Such a situation was never even dreamed of.

We found a difference in the intensity of secondary electrons from an As-top layer and a Ga-top layer. As shown in Fig. 5, when the As shutter is closed and then Ga is supplied to an As-top layer, the secondary electron intensity increases with the Ga coverage (the surface becomes bright), until the surface is entirely covered with Ga, and droplets appear. By using this result, the locally separated but coexisting Ga- and As-top layers can be distinguished by SEM observations as shown in Fig.8(b) (as well as by SREM observations). This is useful because SEM has a better resolution than SREM has. Moreover, it does not require a grazing incident angle, so we might be able to view the surface without any foreshortening. The observation of monolayer growth is a completely new application of SEM.

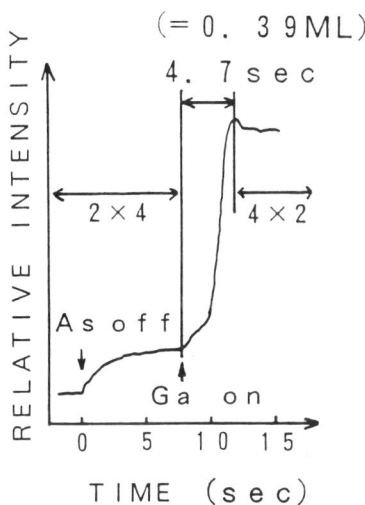

Fig.5 Secondary electron intensity change with Ga surface coverage. Reconstruction changes are shown.

4. APPLICATIONS

We developed three kinds of applications. One is the observation of transient growth processes. Another is the measurement of material properties that control the growth mechanism but cannot be obtained in any other way. And the third purpose is, of course, using this technique for in-situ control of monolayer growth. Examples are shown below.

4.1. Transient Growth Processes

The first application is the real-time observation of quick, transient growth processes related to surface roughness, such as roughening and smoothing, Ga droplet formation and GaAs mound formation on droplets.

The example shown here is the observation of the surface roughening and smoothing processes during buffer layer growth (Inoue 1991a). Preparation of an atomically smooth substrate surface is essential for monolayer growth. Previously it has been accomplished by growing a buffer layer to a thickness of at least a few hundred nanometers while monitoring the growth by in-situ RHEED measurements. In contrast, we watched the actual roughening processes occur under unsuitable growth conditions and the smoothing processes under suitable conditions.

Figure 6 shows single frames from the video tape, along with the corresponding RHEED patterns. After removal of the oxide film, even though we can't see it by SREM, the surface is a little rough. The corresponding RHEED pattern consists of three-dimensional spots (a). After roughening by inadequate growth, the RHEED pattern also has three-dimensional spots. The undulations are as large as half micron as shown in (b). During the smoothing process, the RHEED pattern changed to a 2x4 reconstruction before the surface roughness disappeared (c). In this case, the growth of only 100 monolayers in 4 minutes eliminated the roughness (d).

The SREM observations provide quantitative information about the roughness and how we can prepare a smooth surface. Step-by-step observation of roughening processes are provided by Isu (1988).

Ga droplet formation and surface roughening due to the droplets were also investigated. When the As shutter was closed and then 1 monolayer of Ga was supplied, dark ovals appeared on the surface (Fig.4(b)). They are droplets of liquid Ga. Next, the Ga shutter was closed and then the As shutter was opened. The dark ovals are quickly removed leaving small spots behind. The dark dots still remained after supplying 200 layers of As. We confirmed by ex-situ SEM that these dots are mounds. The gray, large ovals in Fig.4(b) are mounds grown by repeated growth. So, this is the first hard evidence that droplets cause roughness (Osaka 1990a).

SREM   RHEED Pattern

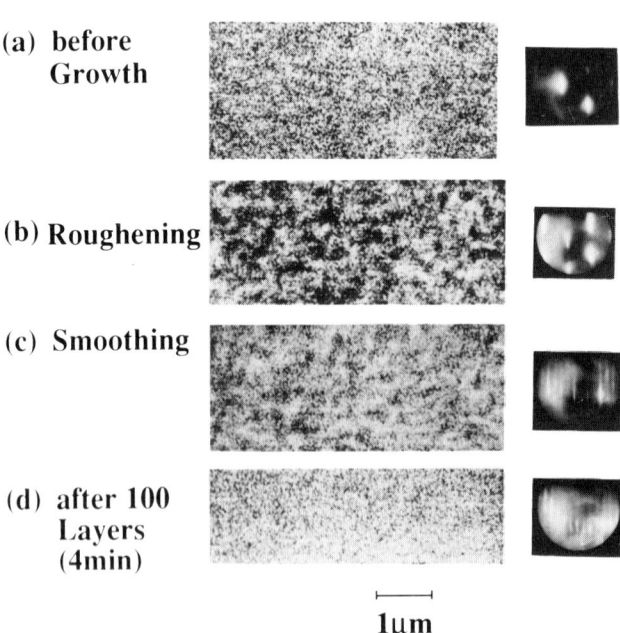

(a) **before Growth**

(b) **Roughening**

(c) **Smoothing**

(d) **after 100 Layers (4min)**

1μm

Fig.6 Surface roughening and smoothing by buffer layer growth observed by SREM (left) and RHEED (right).

4.2. Material Property Measurement under Actual Growth Conditions

In order to understand and control the growth, we have to know the material properties that determine the growth, under actual growth conditions. We'll discuss 3 examples with one in detail.

First, it was established that the Ga-top layer has a pseudo-self-limiting nature. As shown in Figs. 5 and 4(b), as soon as the surface is covered completely by a Ga-top layer and Ga atoms become supersaturated, they form droplets. In other words, it cannot form 2 continuous layers (Osaka 1990a). Droplet formation suggests that the surface tension of liquid Ga is large and that the Ga-Ga binding energy is small.

Second, as the Ga-top layer has a pseudo-self-limiting nature, the Ga atoms supplied before the formation of droplets bond to the As-top atoms. Then the supplied amount of Ga equals the As surface coverage. In-situ observation provides us with a new method of determining surface coverage or stoichiometry (Inoue 1990, Kanisawa 1991a). The maximum As coverage on a 2x4 reconstructed surface was found to be 75%, which agrees with the missing dimer model. This suggests that this method is fairly accurate.

The third example is the surface migration distance. The number and location of Ga droplets do not change after a few monolayers of Ga are supplied. So, Ga atoms arriving on the surface between the droplets are caught by the droplets as shown in Fig.7(a). Ga atoms migrate half the distance between the Ga droplets (Yamada 1988,1989). Thus, observation of the droplets gives us a new way of estimating the lower limit on the Ga migration distance on a Ga top layer. Droplet formation was found to be due to diffusion limited nucleation of supersaturated Ga atoms (Osaka 1991).

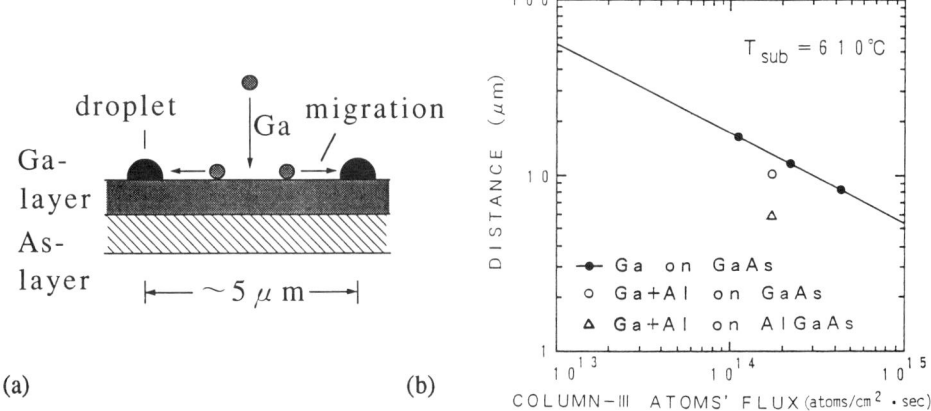

(a)   (b)

Fig.7 Ga and Ga/Al migration distance on Ga and Al/Ga top layers
(a) Measurement method, (b) column-III flux dependence.

Figure 7(b) shows how the Ga and Al/Ga alloy migration distance changes with Ga/Al flux on Ga and Al/Ga top layers (Kanisawa 1991b). The distance is as large as a few microns and this is more than 2 orders of magnitude larger than the Ga migration distance on the As-top layer obtained by RHEED oscillation measurement on vicinal surface (Neave 1975).
It should be noted that these data represent the migration of Ga atoms over many terraces, while the RHEED oscillation corresponds to migration within a terrace which is around 10 nm in size. Interlayer migration plays an important role in obtaining a smooth surface.

4.3.   In-situ Monolayer Growth Control

The last application of in-situ observation is growth control by in-situ monitoring.   We wanted to take advantage of the above two characteristics of Ga, the self-limiting nature and the extremely large migration distance of Ga on a Ga layer for the growth of atomic layers, but the growth requires an As supply.   These advantageous characteristics may be interfered with by the As, but successfully utilized as follows.

When As was supplied continuously to a Ga-top layer with Ga droplets, faint bright areas appeared and soon disappeared. We figured out that Ga atoms were being emitted from the droplets and bonding with As atoms. In this way a Ga monolayer grows laterally over the As-top layer by the alternate supply of Ga and As. But a Ga-top layer thus formed is covered again by the continuously supplied As, so the growth of the monolayer is incomplete.

To ensure complete growth we examined the growth sequence, quantity of atoms supplied, temperature and background As pressure by in-situ observations.   Figure 8 summarizes the results and illustrates what happens (Osaka 1990b). (a) First, the surface is covered with Ga top layer with droplets on. (b) Next, an As-top layer is formed by a 1 monolayer supply of As in a pulse. (b1) It is then possible for Ga atoms in the droplets to find stable As sites. In this way Ga atoms move out from the droplets and attach themselves to the nearest As atoms. (b2) The Ga atoms that follow migrate over the new Ga overlayer until they reach the As layer. Coexisting As- (dark) and Ga- (bright around droplets) top layers are clearly observed by SEM. Since the migration takes place on a Ga layer, the migration distance is very large. (c) This process continues until the surface is completely covered with Ga.

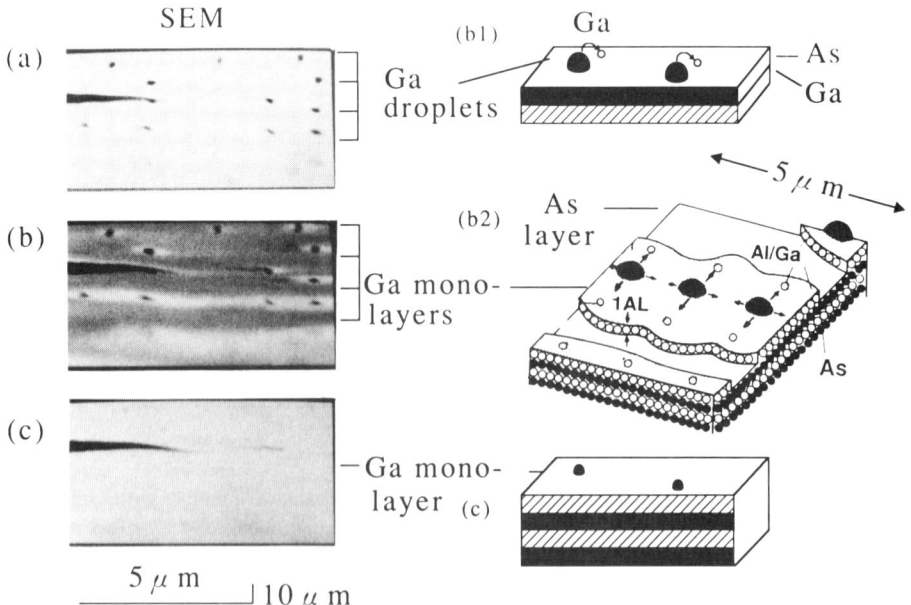

Fig.8 Ga Monolayer lateral growth on As top layer observed by SEM

Figure 9 compares the conventional method and our new method. In conventional growth with a co-supply of Ga and As, Ga migration is interfered with by the As-top layer. The resulting domain size is then not large enough compared to the electron wavelength in quantum effect devices. This induces electron scattering (Sakaki 1987). On the other hand, in our new method, Ga and As are supplied alternately with an interval around 10 seconds. Ga migrates over the Ga-top layer over 1 micron. If we make quantum devices by this method, we can expect a much higher performance.

At a high observation rate of 1 second/frame, we determined the time necessary for complete growth very precisely (Inoue 1991b). This gives new and important information to material scientists about diffusion kinetics in the time-domain. As previous studies have been carried out on steady-state systems, such information has been unobtainable up to now.

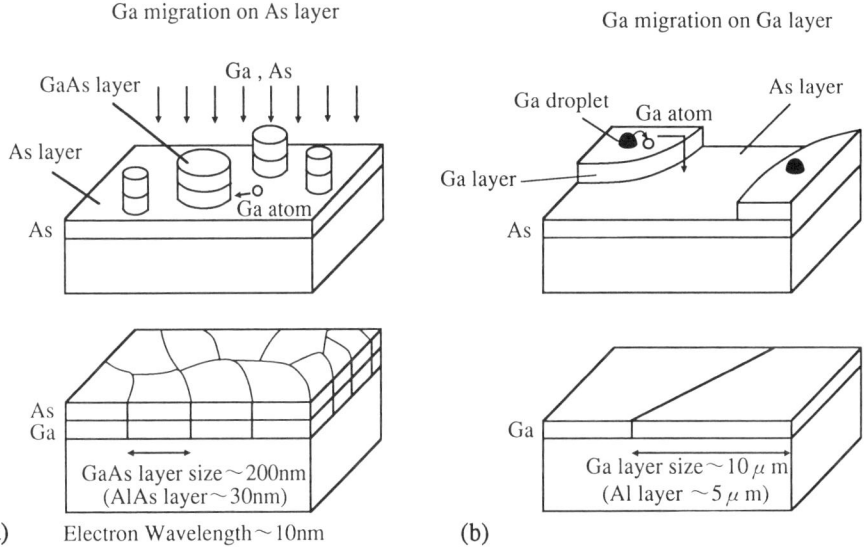

Fig.9 Comparison of the (a) conventional and (b) new growth methods.

5. MBE-EM-STM HYBRID SYSTEM

We have recently installed an STM on this MBE-electron microscope hybrid system. The sample is prepared in the growth chamber and quenched without excess As flux. Then it is quickly transferred to the STM chamber.

Figure 10 is an example of our observations. Terraces as large as 1500A were realized by high temperature, slow growth and directly observed (Tanimoto 1991). We hope that the STM will be a powerful tool to improve the growth control.

6. SUMMARY

We have developed a high-resolution system for real-time, in-situ observation of GaAs MBE growth. New phenomena were observed. New measurement methods were established and a new growth procedure was developed; impact of in-situ observation lies in its ability to show us what may at first seem unbelievable, but which is, in the end, undeniable.

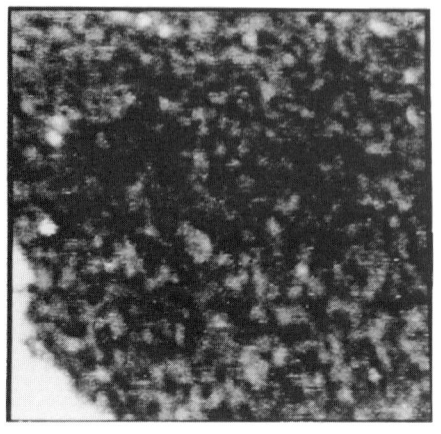

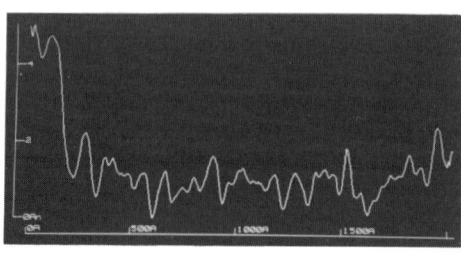

(a)   (b)

Fig.10 STM Image of as-grown GaAs. (a) Large terrace covers most of the 1600A square area with monolayer step in the left bottom corner. (b) Height profile on the diagonal line. The vertical bar corresponds to the monolayer step height.

ACKNOWLEDGEMENT

I am indebted to my colleagues Jiro Osaka, Kiyoshi Kanisawa, Shigeru Hirono, Masafumi Tanimoto, Takako Takigami and Koji Yamada for their contributions to the work reported here. I appreciate very much the continuous help of Dr. Seiichi Nakagawa of JEOL Company. Discussions with Professors Katsumichi Yagi and Bruce Joyce are gratefully acknowledged. I am also grateful to Professor Ron Newman, Drs. Anthony Cullis, Akira Ishida and Kazuo Hirata for encouragement.

REFERENCES

Cowley L M et al 1975 Rev. Sci. Instr. 46 826
Halliday J S and Newman R C 1960 Brit. J. Appl. Phys. 11 158
Ichikawa M and Hayakawa K 1982 Japan. J. Appl. Phys. 21 145
Ichikawa M and Doi T 1987 Appl. Phys. Lett. 50 1141
Inoue N, Tanishiro M and Yagi K 1987 Japan. J. Appl. Phys. 26 L293
Inoue N 1990 Defect Control in Semiconductors, ed K Sumino (Elsevier) pp 129-140
Inoue N 1991a Proc. VI-th MBE Conf.(1990) to be published in J. Crystal Growth
Inoue N et al 1991b submitted to J. Appl. Phys.
Isu T et al 1988 Japan. J. Appl. Phys. 27 L2259
Kanisawa K et al 1991a to be published in Proc. ICVGE7
Kanisawa K et al 1991b to be published in Appl. Phys. Lett.
Osaka J et al 1990a J. Crystal Growth 99 120
Osaka J and Inoue N 1990b Mater. Res. Soc. Symp. Proc. 159 33
Osaka J 1991 submitted to J. Crystal Growth.
Petroff P M 1986 Proc. XI-th Int. Cong. Electron Microscopy 137
Sakaki H et al 1987 Appl. Phys. Lett. 51 1934
Tanimoto M et al 1991 submitted to STM '91
Yamada K et al 1988 MBE-V LN-9; Appl. Phys. Lett. 55(1989) 622

Inst. Phys. Conf. Ser. No 117: Section 7
Paper presented at Microsc. Semicond. Mater. Conf., Oxford, 25–28 March 1991

399

TEM investigations of LP-MOVPE grown GaAs/Ge heterostructures

P Franzosi, L Lazzarini, G Salviati, M Scaffardi and G Timò*

Maspec-CNR Institute, via Chiavari 18/A, 43100-I Parma
*CISE spa, via Reggio Emilia 39, 20134-I Segrate (MI)

ABSTRACT: TEM investigations have been carried out on GaAs/Ge single heterostructures grown by Low Pressure Metalorganic Vapour Phase Epitaxy at different growth rate deposition and with different thicknesses of the epitaxial layer. Misfit dislocation networks have been observed in specimens with the epitaxial layer thicker than 0.45 μm: they were confined within 50 nm of the interface and were mainly of 60° type. The samples grown at lower deposition rate exhibit twins at the heterointerface on the {111} composition planes.

1. INTRODUCTION

The GaAs/Ge heterostructure (HS) is the basic system for fabricating high conversion efficiency ($\eta > 20$ %) solar cells of the GaAlAs/GaAs/Ge type to be employed in space applications. The cell performance is influenced by the quality of the GaAs/Ge junction also; in fact, the occurrence of recombination centres at the junctions, which are related to the presence of defects, reduces the ideal conversion efficiency of the cell. Since the electrical junction occurs approximately at the "metallographic" GaAs/Ge interface region, it becomes very important to investigate the nature and the precise localization of the crystal defects.

The structural characteristics of the GaAs/Ge HS have not yet been carefully investigated; indeed, misfit dislocations (MDs) due to the lattice mismatch between GaAs and Ge and other kinds of defects originating at the interface may be present in these HSs. Therefore, the dependence of the defect nature, concentration and distribution on the growth conditions merits further investigation.

The aim of this paper is to report the main results of a careful investigation of the crystal quality of GaAs/Ge single HSs by conventional transmission electron microscopy (CTEM). Moreover high resolution electron microscopy (HREM) has been employed to characterize the interface quality and the nature of defects.

2. EXPERIMENTAL

The GaAs/Ge HSs have been grown by the Low Pressure Metalorganic Vapour Phase Epitaxy (LP-MOVPE) technique at the CISE laboratories. High quality Ge crystals 6° off (001) towards [110] were used as substrates; such a substrate miscutting was indeed reported to be effective in reducing the formation of antiphase domains (APDs), which may generate at the interface between a polar semiconductor and a non polar one (Cho et al. 1985, Mizuguchi et al. 1986, Pukite and Cohen 1987, Strite et al. 1990). At the growth temperature of 700 °C, the trimethylgallium and arsine fluxes were adjusted in such a way that a growth rate ranging from 1 to 10 μm/h was obtained. The thickness of the epitaxial layers investigated ranged from about 0.24 to 5.50 μm.

TEM investigations have been performed by using a JEOL 2000FX instrument. Both (001) planar sections and (110) cross sections have been prepared by mechano-chemical and argon ion-beam thinning of the samples at liquid nitrogen temperature in order to reduce the sample damage. All the HREM pictures were taken under [110] axial bright field (BF) conditions.

3.   MISFIT DISLOCATIONS AND PLANAR DEFECTS

The lattice parameter of GaAs is slightly smaller than that of Ge; as a consequence, the GaAs/Ge is a low lattice-mismatched system ($\Delta a/a = -7.42 \times 10^{-4}$), similar to the well-known $Ga_{1-x}Al_xAs/GaAs$ and $In_{1-x}Ga_xAs/InP$. This mismatch causes a lattice strain of the epilayer and it is accomodated by an elastic deformation if the epilayer thickness is lower than a crytical value t_c. In contrast, when the epilayer is thicker than t_c, the associated strain energy is larger than the MD nucleation energy; therefore MDs are generated to release the strain. It must be stressed that for solar cell applications the GaAs epilayers is usually a few µm thick.

The efficiency of a dislocation to release the strain is proportional to the edge component of its Burgers vector b_e projected on the interface; when the strain is completely released, the MD density is given by the ratio between the lattice mismatch and b_e. Two types of MDs, the pure edge 90° and the mixed 60° type, are expected to be present in this kind of HS; the 90° type dislocations are the most efficient ones for releasing the strain, but the 60° type dislocations are dominant in systems with small lattice mismatch (Matthews et al.1974, Mader et al. 1974, Choi et al. 1988) because of their lower nucleation energy and multiplication barrier.

It has been found that MDs are present in samples with a layer thickness t>0.45 µm and are missing in samples with t<0.29 µm; therefore, the value of the critical thickness t_c falls in this interval. The dislocations are parallel to three directions: one coincident with the [1$\bar{1}$0] direction; the other two occurring at an angle of about 4° either side of the [110] direction, as expected for a substrate miscutting of 6º (Fig. 1).

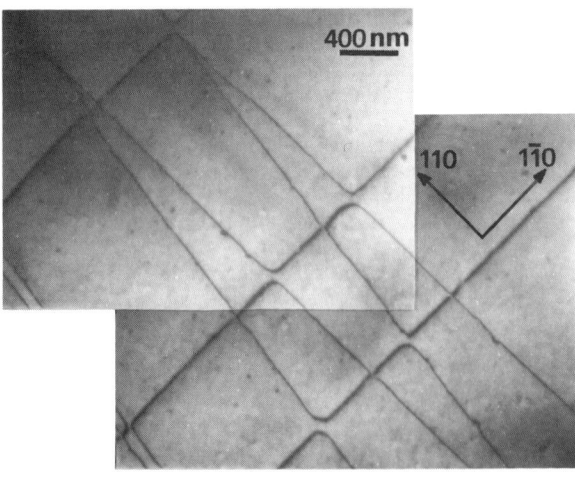

Fig. 1: BF-CTEM (001) plan view of the GaAs/Ge HS with t_{GaAs} =0.8 µm showing interacting MD network.

It has been found that the MDs are confined within 50 nm of the interface and do not thread further into the epitaxial layer (Fig. 2a). HREM investigations of the interface allowed us to carry out the analysis of the Burgers vector of some MDs. In all cases the dislocations have been found to be of 60° type (Fig. 2b); the Burgers vector **b** is indeed inclined at an angle of 45° to the interface and is of the a/2 <011> type.

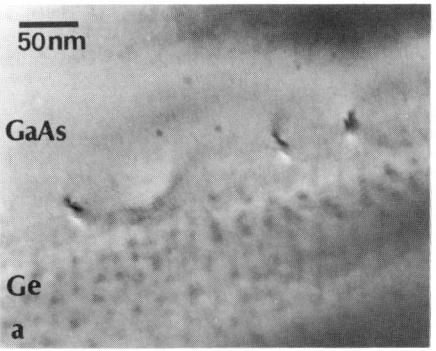
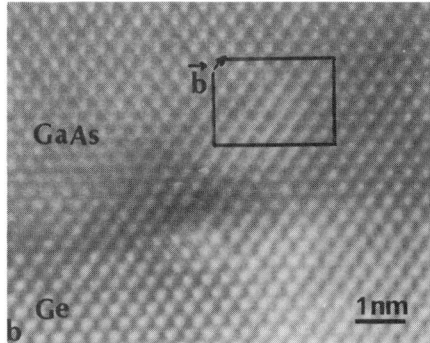

Fig. 2: a) (110) cross section BF-CTEM image of the GaAs/Ge interface of a sample with tGaAs =5.4 µm showing some dislocations normal to the foil plane.
b) (110) cross section HREM micrograph of a 60° misfit dislocation at the GaAs/Ge interface. The Burgers vector lies inclined towards the interface on a {111} plane.

The epilayer crystal quality is strongly affected by the growth deposition rate: layers grown at low growth rate (1 µm/h) exhibit planar defects which are not found to be present in samples grown at high deposition rate (10 µm/h). It has been recently argued that the lattice mismatch plays a minor role in the formation of planar defects like twins or stacking faults (Pirouz et al. 1987, Ernst and Pirouz 1987, Lee and Tsoi 1987). They would probably be originated during the early stages of the epitaxial growth as a consequence of the formation of supercritical nuclei faceted along low-energy planes such as the {111} crystal planes; the growth of stable nuclei occurs by the deposition of atoms on the facets and errors in stacking on the facets during deposition give rise to twins and stacking faults. In Fig. 3a planar defects are observed in a 240 nm thick GaAs layer. The defects have been seen to propagate from the interface to the top of the layer. HREM investigations and the diffraction pattern obtained in the nanodiffraction mode revealed that these defects exhibit a twin character (Fig. 3b); the twinning process takes place on the {111} composition planes and the <112> directions are the traces of these planes.

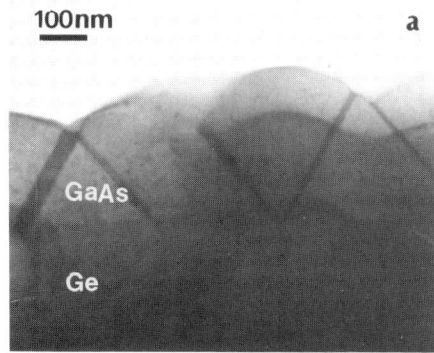
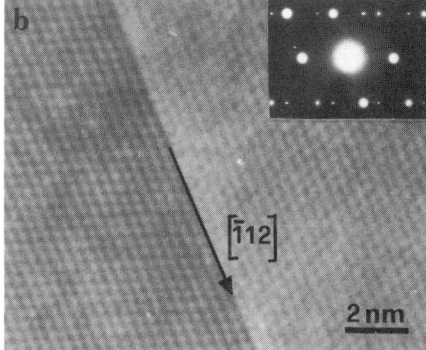

Fig. 3: a) (110) cross section BF-CTEM image of planar defects in a sample with tGaAs = 300 nm. b) (110) cross section HREM image of one of the twins of fig. 3a. In the inset the diffraction pattern of the twinned region obtained in nano-diffraction conditions is shown.

Finally, another kind of defect has been observed in all the investigated specimens; such defects were localized at the interface between GaAs and Ge and extended into the bulk epilayer (Fig. 4). Their lateral dimensions never exceeded 600-700 Å and their average distance was about 1500 Å. Strite et al. (1990) have identified features very similar to these ones as APDs: we also suggest that they could be APDs despite the fact that they would not be present in the investigated samples due to the miscutting of the substrate.

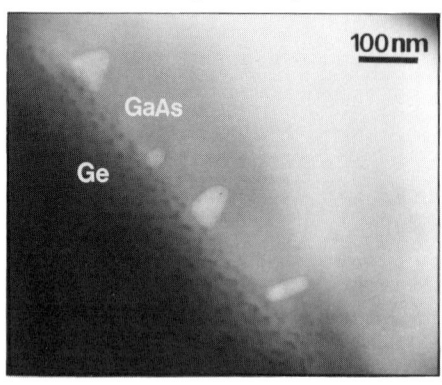

Fig. 4:   (110) cross section BF-CTEM picture of the GaAs/Ge interface region showing the presumed APDs.

4.   CONCLUSIONS

GaAs/Ge HSs have been grown by the LP-MOVPE technique and investigated by CTEM and HREM.

It has been found that MDs networks are present in the GaAs/Ge HSs when the layer thickness is larger than 0.45 µm. The MDs are confined within 50 nm of the interface and do not thread further into the bulk epilayer. The analysis of the Burgers vector carried out in the HREM mode has shown that the investigated MDs were mixed 60° type. In contrast, no MDs have been observed in HSs with a layer thickness t<0.29 µm.

The growth rate has been found to affect the layer crystal quality: the samples with t<t_c are nearly perfect if grown at a high deposition rate, while exhibit twins if grown at a low rate. It has been ascertained that the twinning takes place on the {111} composition planes. Finally, other defects originating at the interface have been found in all the samples investigated independently from the growth conditions: we suggest they could be APDs and, in this respect, further investigations are in progress.

REFERENCES

Cho N-H, De Cooman B C, Carter C B, Fletcher R and Wagner D K 1985 Appl. Phys. Lett. 47 879
Choi C, Otsuka N, Kolodziejski L A, Melloch M R and Gunshor R L 1988 Proc. of Symp. "Dislocations and Interfaces in Semiconductors", Ed. by Rajan K, Narayan J and Ast D, pp 141
Ernst F and Pirouz P 1987 J. Appl. Phys. 64 4526
Lee J W and Tsoi H L 1987 J. Vac. Sci. Technol. B 5 819
Mader S, Blakeslee A E and Angilello J 1974 J. Appl. Phys. 45 4730
Matthews J W and Blakeslee A E 1974 J. Cryst. Growth 27 118
Mizuguchi K, Hayafuji N, Ochi S, Murotani T and Fujikawa K 1986 J. Cryst. Growth 77 509
Pirouz P, Chorey C M, Cheng T T and Powell J A 1987 Inst. Phys. Conf.Ser. 87 175
Pukite P R and Cohen P I 1987 J. Cryst. Growth 81 214
Strite S, Unlu M S, Adomi K, Gao G-B, Agarwal A, Rockett A, Morkoc H, Li D, Nakamura Y and Otsuka N 1990 J. Vac. Sci. Technol. B 8 1131

Inst. Phys. Conf. Ser. No 117: Section 7
Paper presented at Microsc. Semicond. Mater. Conf., Oxford, 25–28 March 1991

Reduction of dislocation density in GaAs on Si substrates by use of Si interlayers and initial Si buffer layer

Akihiro Hashimoto, Naoharu Sugiyama and Masao Tamura

Optoelectronics Technology Research Laboratory
5-5 Tohkodai, Tsukuba, Ibaraki 300-26, Japan

ABSTRACT: Reduction of the dislocation density in GaAs on Si by the use of a Si interlayer and an initial Si buffer layer is reported. The Si interlayer was inserted into the GaAs epitaxial layer on a Si substrate by low-temperature Si deposition. Molten KOH etching and cross-sectional transmission electron microscope (XTEM) observations showed a reduction of dislocation density by 10^{-2}~10^{-3} times at the top surface as a result of the Si interlayer insertion compared to that of conventional growth. An additional reduction of the dislocation density by the initial Si buffer layer grown at 680 °C coupled with the interlayer was observed.

1. INTRODUCTION

Reduction of threading dislocations in the GaAs on Si system is an important subject in view of the actual device applications such as optoelectronic integrated circuits (OEIC's). It has been reported that it is possible to control the threading dislocations in epitaxial layers using an $In_xGa_{1-x}As$/GaAs strained layer superlattice (SLS) (Fischer et al 1986a), impurity doping into the epitaxial layers (Ohbu et al 1989), and post-growth annealing at high temperatures, such as 800 °C (Yamaguchi et al 1988). Although these techniques are very effective for reduction of the threading dislocation density, it seems to be difficult to reduce the dislocation density to less than 10^6 cm^{-2}.

In general, the dislocation propagation velocity in a Si layer is much slower than that in a GaAs layer (Imai et al 1983), and the critical stress for dislocation formation in Si is higher than that in GaAs because of the strong covalent Si-Si bond (Labusch et al 1965). These facts imply that the Si layer is "hard" for both dislocation formation and propagation. In this paper, the reduction of dislocation density obtained by the use of thin Si interlayers is reported, and the improvement of the crystal quality produced by a Si interlayer combined with an initial Si buffer layer to obtain a clean and smooth surface is also described.

2. EXPERIMENTAL

Si (001) wafers inclined toward the [001] direction by 2° were used as substrates. A typical growth sequence together with the surface-cleaning procedure is shown in Fig.1. After surface-cleaning of Si substrates, 10 ML AlAs layers were grown first at 500 °C by an alternating source supply method. GaAs layer growth at 600 °C by a conventional MBE method followed. Then, thin Si layers were deposited at 250 °C and, finally, 2 μm-thick GaAs layers were grown at 600 °C.

When required, initial Si layers were grown at 680 °C for 1 hour just after Si surface cleaning. The same growth sequence as shown in Fig.1 was then used. All growth of GaAs and Si were made using a multichamber MBE system. The Si was deposited by electron beam evaporation in ultra high vacuum (UHV). The background pressure of the exchange chamber, which was used for the sample transfer between the III-V MBE chamber and the Si MBE chamber, was less than 3×10^{-10} torr. The RHEED pattern of the Si (100) surface did not change after the sample transfer in the UHV.

The etch pit density (EPD) of the top surface of GaAs on Si was judged by molten KOH etching for 4 minutes at 300 °C. Threading dislocation behavior was studied by cross-sectional transmission electron microscope (XTEM) observations.

3. RESULTS

Figure 2 shows the surface morphology of the sample with the thin Si interlayer etched by molten KOH, and an XTEM image of the same sample. The EPD was about 1.2×10^6 cm^{-2}. The XTEM image clearly shows the reduction of dislocation propagation into the upper grown layer for the region 1 μm from the interface between the GaAs layer and the Si substrate. Some dislocations at the Si interlayer are bent, and some escape from the sample. The dislocation density estimated from this micrograph is also ~10^6 cm^{-2}, which is coincident with the etching results. In the case of the combination of the initial Si layer grown at 680 °C for 1 hour with the Si interlayer, the EPD was lowest

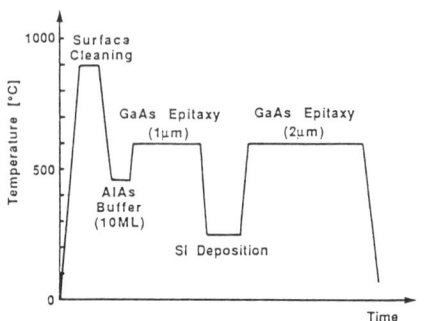

Fig.1   Growth sequence of GaAs on Si with a Si inter-layer.

EPD at the top surface

XTEM Image

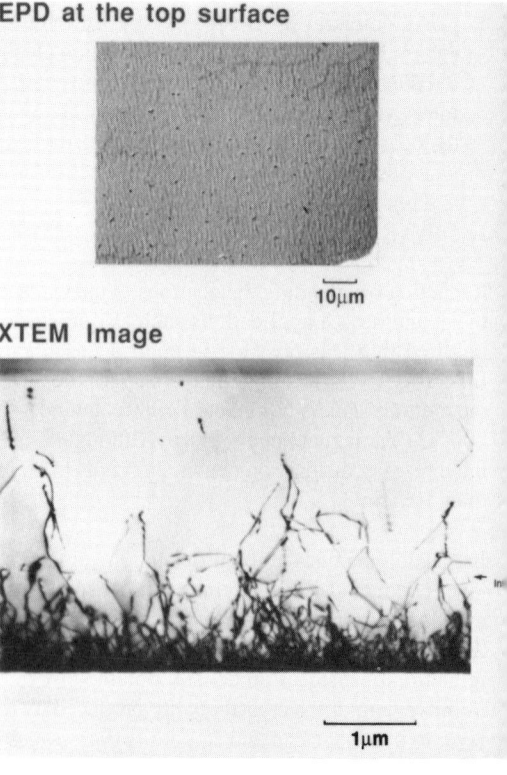

Fig.2   Surface morphology of a molten KOH-etched sample with a Si interlayer and an XTEM image of the sample.

$(8\times10^5 \text{ cm}^{-2})$ as shown in Fig. 3, although the initial Si growth condition was not optimized.

The multi-interlayer effect on the dislocation propagation was also examined as shown in Fig. 4. It is clearly seen that the dislocations are bent by the first Si interlayer and the propagation of some dislocations is also blocked by the second Si interlayer. Therefore, multi-interlayer insertion has the potential to reduce the dislocation density at the top surface of the GaAs epitaxial layer.

Optimization of the Si interlayer thickness was made by the use of the sample structure shown in Fig. 5. The XTEM image of the sample showed the strong effect of the Si interlayer thickness on the upper epitaxial layer. In the case of a Si interlayer thickness above 13 Å, anti-phase domain (APD) structures and stacking faults emerged as shown in Fig. 6. Therefore, the optimum Si interlayer thickness may be about 10 Å.

4. DISCUSSION

As stated above, it was clarified that the Si interlayer has an excellent possibility of being a dislocation blocking layer. In the case of a $In_xGa_{1-x}As$/GaAs SLS, the reduction of dislocation density in the GaAs was more effective with an underlying SLS layer than without it and, conventionally, 5 to 10 period SLS's were used for such a reduction (Fischer et al 1986b). In the case of the Si interlayer, however, the blocking effect appeared at the Si interlayer. The thin Si layer insertion resulted in almost the same effect as the conventional multi-period SLS layers regarding dislocation blocking. This may be due to an interfacial strain field induced by the 4 % lattice mismatch between the GaAs and Si as well as the "hardness" of the Si

EPD at the top surface

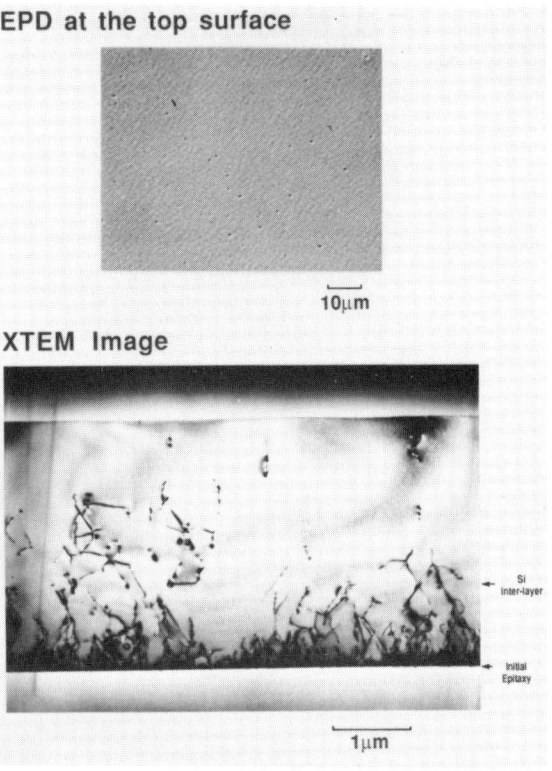

XTEM Image

Fig.3 Surface morphology of a molten KOH-etched sample with both a Si initial layer and an interlayer, and an XTEM image of the same sample.

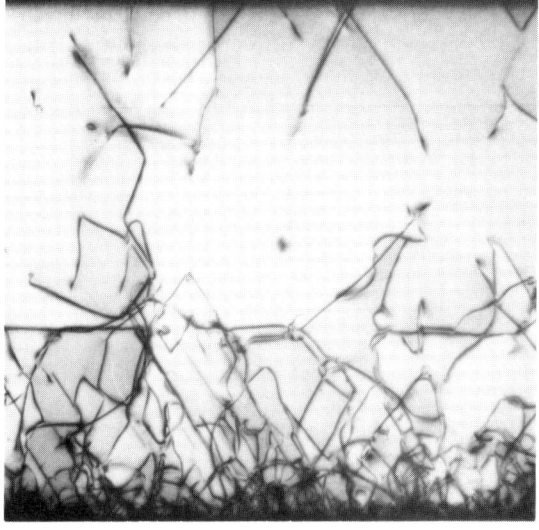

Fig.4 Multi-interlayer effect in GaAs on Si.

interlayer caused by the strong covalent Si-Si bond. As stated in the introduction, the dislocation velocity in Si is much slower than that in GaAs and $In_xGa_{1-x}As$ layers (Imai et al 1983). Therefore, the critical stress to form new dislocations in a Si layer is considered to be higher than that in the case of GaAs. The harder Si layer is thought to be the cause of the excellent dislocation blocking ability which exceeds that of $In_xGa_{1-x}As$ or other III-V compound materials.

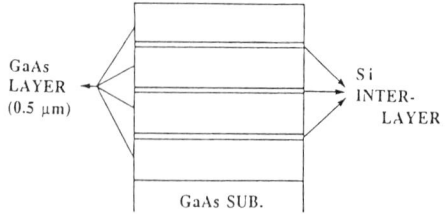

Fig.5 Sample structure

Improvement of the crystal quality by both the Si interlayer and the initial Si buffer-layer indicates that the initial Si buffer layer gives the possibility of reducing dislocation propagation into the top layer by suppressing the "seeds" of threading dislocations such as surface roughness and contamination, as well as antiphase boundaries and so on.

5. SUMMARY

A reduction of the dislocation density in GaAs on Si was achieved by the insertion of a thin Si interlayer.   Threading dislocations were well blocked by this Si interlayer. The dislocation blocking by

Fig.6 An XTEM image of the sample shown in Fig.5.

the Si interlayer was considered to be due to the difference in material properties between GaAs and Si, such as the difference in the covalent bond strength between the two materials, regarding dislocation formation and propagation. The initial Si epitaxy just after thermal surface cleaning coupled with the insertion of the Si interlayer also provided an additional reduction of the dislocation density. These results strongly suggest that optimized conditions of both the Si interlayer insertion and initial epitaxy will result in a more marked reduction of the dislocation density beyond the present state of the art.

REFERENCES

Fischer R, Morkoç H, Neumann D.A, Zabel H, Choi C, Otsuka N, Longerbone M, and Erickson L.P, 1986a J. Appl. Phys. 60, 1640.
Fischer R, Neuman D, Zabel H, Choi C and Otsuka N, 1986b Appl. Phys. Lett. 48, 1223.
Imai M and Sumino K, 1983 Phil. Mag. A47, 599.
Labusch R, 1965 Phys. Stat. Sol. 10, 645.
Ohbu I, Ishino M, and Mozume T, 1989 Appl. Phys. Lett. 54, 396.
Yamaguchi M, Yamamoto A, Tachikawa T, Itoh Y, and Sugo M, 1988 Appl. Phys. Lett. 53, 2293.

Inst. Phys. Conf. Ser. No 117: Section 7
Paper presented at Microsc. Semicond. Mater. Conf., Oxford, 25–28 March 1991

407

Defects in ALMBE grown GaAs on Si

A Vilà, A Cornet, A Herms, J R Morante, Y González*, L González* and F Briones*

L.C.M.M., Dept. Física Aplicada i Electrònica. Univ. de Barcelona, Diagonal 645-647, 08028 Barcelona, Spain.
* Centro Nacional de Microelectrónica CSIC, Serrano 144, 28006 Madrid, Spain.

ABSTRACT: ALMBE (Atomic Layer Molecular Beam Epitaxy) has been reported as an interesting technique in order to reduce the problems related with 3D growth. Of special interest is the case of the growth of epilayers on misoriented substrates. Samples of GaAs grown on Si by the ALMBE technique have been studied by plan view and cross section TEM and HREM. The influence of the substrate misorientation on the dislocation density is also presented.

1. INTRODUCTION

In spite of the high lattice mismatch between GaAs and Si and of the differences in their coefficient of thermal expansion, heteroepitaxial growth of GaAs on Si has practical interest for the production of integrated Si and GaAs circuits for electronic and optoelectronic devices. The main problem in this procedure is to obtain good crystallinity that gives good electric characteristics. Several methods have been tested in order to improve the crystallinity, such as the use of an initial Ge, Ga or As layer on Si, the growth of GaAs layers directly on slightly tilted (001) Si, annealing cycles, the introduction of strained multilayers that bend the dislocations into the bulk, but none of them has proved to be definitive (see Fang et al. 1990 for a review). The ultimate goal of research is to decrease the total dislocation density. One of the proposed alternatives consists in the use of approximately twodimensional growth methods, such as ALMBE (Watanabe H. et al (1987), M.A. Tischler et al., (1990)) to obtain layers with a low dislocation density.

In this work we present a TEM analysis of the defects arising in heteroepitaxial ALMBE layers of GaAs grown on Si (001) substrates. The influence of substrate tilt on dislocation density is also presented.

2. EXPERIMENTAL DETAILS

The substrate was (001) Si with the surface tilted towards [110]. The tilt was varied between 0 and 4°. In all samples, growth of the GaAs layer started by using the ALMBE technique at low temperature (300-400°C) (see Briones et al. (1989) for more details). The layers were grown at a 0.8 μm/h rate until a layer of thickness 2μm was formed. Plan-view and cross sectional samples were observed with a Hitachi 800 microscope operated at 200 keV and a Phillips 430 ST operated at 300 keV for the high resolution images.

3. RESULTS AND DISCUSSION

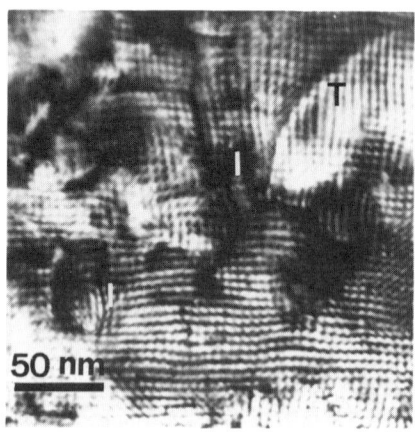
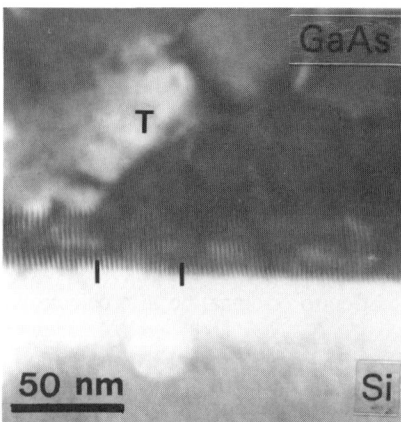

Figure 1. *Bright field images of a plan-view (a) and a cross sectional specimen (b) of the sample B, showing moiré patterns due to interference from the GaAs and Si lattices.*

Bright field images of plan-view and cross-sectional specimens (Figures 1.a and 1.b respectively) show moiré patterns due to the interference from the GaAs and Si lattices. The crystalline quality of the materials can be seen from the uniformity and parallel nature of these patterns, because they increase the visibility of nonuniformities in either of the two lattices. Thus, irregularities in the moiré pattern are due to interface dislocations (marked with I), and the dark lines which run irregularly on the interface are due to threading dislocations in the layer (marked with T).

When the sample (prepared in cross section) is oriented with the electron beam parallel to the [110] direction the moiré pattern cannot be seen (Figure 2). This Figure shows a bright field image of a whole layer, the threading dislocation

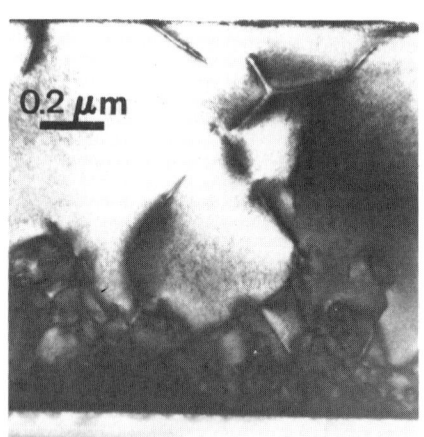

Figure 2. *TEM micrograph showing the presence of threading dislocations propagating from the interface.*

density, calculated to be $10^9 \, \text{cm}^{-2}$ near the interface, is reduced in the uppermost part of the film, resulting in a very smooth surface with a dislocation density of $10^7 \, \text{cm}^{-2}$ in the final 0.5 μm. There is a large decrease in twodimensional defects, the density of stacking faults being of the order of detectibility of the TEM ($1 \times 10^4 \, \text{cm}^{-2}$). As we can see in table I there is little influence from the misorientation on the threading dislocation density near the interface.

In order to identify the dislocations present at the interface, we have recorded dark field images in plan-view (figure 3). When the pictures are obtained in weak beam conditions with $\mathbf{g}=(220)$ only one set of dislocations is visible (Figure 3.a), while when $\mathbf{g}=(2\bar{2}0)$ is used, one can see another set of dislocations running in an orthogonal direction (figure 3.b). Most misfit dislocations have a $<110>$ dislocation line and are Lomer type with $\mathbf{b}=a/2<110>$. The average spacing between misfit dislocations along the interface is measured in both directions, parallel and perpendicular to the tilt axis (Table I).

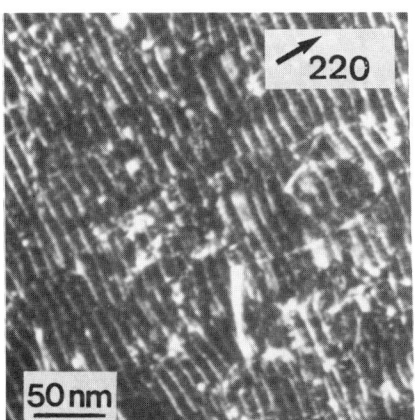 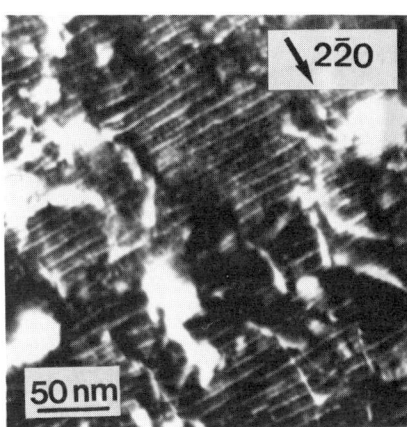

Figure 3. Plan-view weak beam dark field images. The parallel lines in the images are misfit dislocations and the irregularly arranged lines are the threading dislocations.

SAMPLES	TILT	DISL.DENS.(cm^{-2})	[110] SPACING (nm)	[$\bar{1}$10] SPACING (nm)
A	0°	$3\ 10^9$	9.4	9.5
B	2°	$1.2\ 10^9$	8.5	9.4
C	4°	$1.1\ 10^9$	9.5	9.5

Table I

Not all the dislocations in the interface are pure edge type. Otsuka et al. (1986) used HREM to show two types of misfit dislocation: one with its Burgers vector parallel to the interface (Lomer dislocations) and the other with its Burgers vectors inclined from the interface by 45° (60° dislocations). High resolution images of our samples verify the presence of these two types (figure 4). The average spacing found between Lomer dislocations agrees with the one obtained by plan-view observations. The same results have also been obtained by measuring the spacing between satellite diffraction spots formed by diffraction in the relaxed lattices (figure 5)

As we can observe in table I, no significant influence of the misorientation on the spacing between dislocations has been found. The role of the 60° misfit dislocations in the strain relief mechanism is, except in the sample B, of the order of 10%. The great reduction of the twodimensional defect density associated generally with a 3D growth mechanism, as well as the perfection of the edge dislocation grid, confirms that the growth of epilayers by ALMBE tends to be twodimensional.

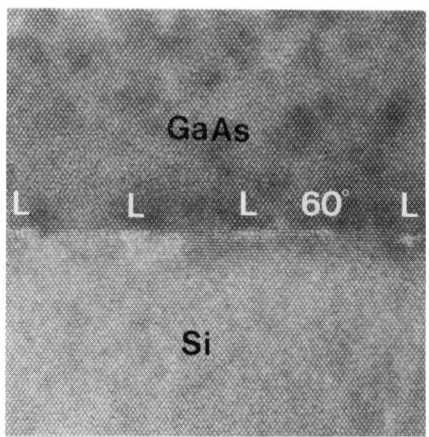

Figure 4. *High-resolution electron microscopy image of the GaAs-Si interface showing the presence of Lomer and 60° dislocations.*

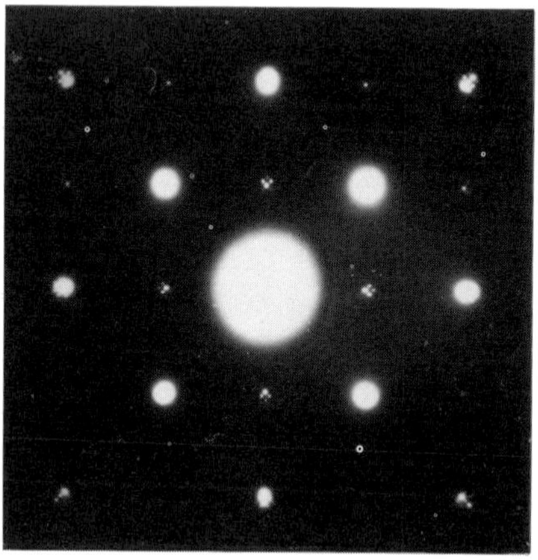

Figure 5. *Diffraction pattern showing the satellite spots.*

REFERENCES

Briones F, González L and Ruiz A 1989 Appl. Phys.A, **49**, 729

Fang S F, Adomi K, Iyer S, Morkoç H and Zabel H 1990 J.Appl.Phys., **68**, R31

Otsuka N, Choi C, Kolodziejski L A, Gonshar R L, Fischer R, Pengi C K,Morkoç H, Nakamura Y and Nagakura S 1986 J.Vac. Sci. Technol., **B4**, 785.

Tischler M A and Bedair S M 1990 Atomic Later Epitaxy, ed. by T Suntola and M Simpson, Chap. 4, Blackie and son.

Watanabe H and Usui A 1987 Inst. Phys. Conf. Ser. No. 83, pp 1.

Inst. Phys. Conf. Ser. No 117: Section 7
Paper presented at Microsc. Semicond. Mater. Conf., Oxford, 25–28 March 1991

411

Overgrowth of GaAs/AlGaAs layers on patterned surfaces

Y X Guo, S Andersen[1], R Høier and K Johannesen[2]

Department of Physics, University of Trondheim-NTH, N-7034 Trondheim,
[1]Division of Applied Physics, SINTEF, N-7034 Trondheim,
[2]ELAB, RUNIT, N-7034 Trondheim, Norway

ABSTRACT: Threading dislocations, stacking faults and microtwins have been observed and characterized in GaAs/AlGaAs epilayers grown by MBE on patterned substrates by means of transmission electron microscopy. These defects are usually found to originate at the sidewalls of the substrate grooves. The favoured plane for the epilayer growth near the sidewall is found to be (411).

1. INTRODUCTION

Epitaxial growth on patterned substrates in the GaAs/AlGaAs system is of potential interest in the purpose of building advanced electronic and optoelectronic devices. This is because the resulting products are of 2-dimensional lateral confinements to the photons and electrons. An example of application is the phased array laser (Johannessen and Stanley 1990). Applications may however be limited by the presence of the defects, such as threading dislocations, stacking faults and microtwins. In order to get rid of such defects, necessary characterization of the microstructure should be carried out to understand their formation and propagation. Cross-sectional observation of the epitaxial structures by transmission electron microscopy (TEM) has become an important method for examining the quality of the epilayers providing information on the materials at atomic level. The present work is a cross-sectional TEM study of the molecular beam epitaxy (MBE) grown GaAs/Al$_x$Ga$_{1-x}$As (x=0.4) superstructure on the patterned GaAs substrates. Possible defects as well as facet to facet variations due to a sudden change in direction of the substrate surface have been investigated.

2. EXPERIMENTS

Specimens were grown by MBE on patterned GaAs substrates at two different temperatures (620°C and 720°C) with a total of 15 periods of superlattices consisting of 50nm Al$_{0.4}$Ga$_{0.6}$As and 50nm GaAs layers. The plain GaAs substrates with surface normal parallel to [100] were patterned by standard photolithographic methods with alternating 5µm openings and 5µm grooves in sets of ten. 0.75µm deep grooves were etched by using a 1:8:160 H$_2$SO$_4$:H$_2$O$_2$:H$_2$O solution. The grooves lined up along the [01$\bar{1}$] direction of the substrate have trapezoidal cross-sections with the side walls lying on {111}. Thin [01$\bar{1}$] foils for TEM observation were made by ion-beam milling and the examinations were performed in a Philips CM30 instrument.

3. RESULTS AND DISCUSSION

The growth of high quality epitaxial layers of AlGaAs on GaAs planar surfaces can be achieved by the MBE technique, because there is only a small mismatch between the two compounds. On a regular basis, the heterointerface can be made atomically flat and free

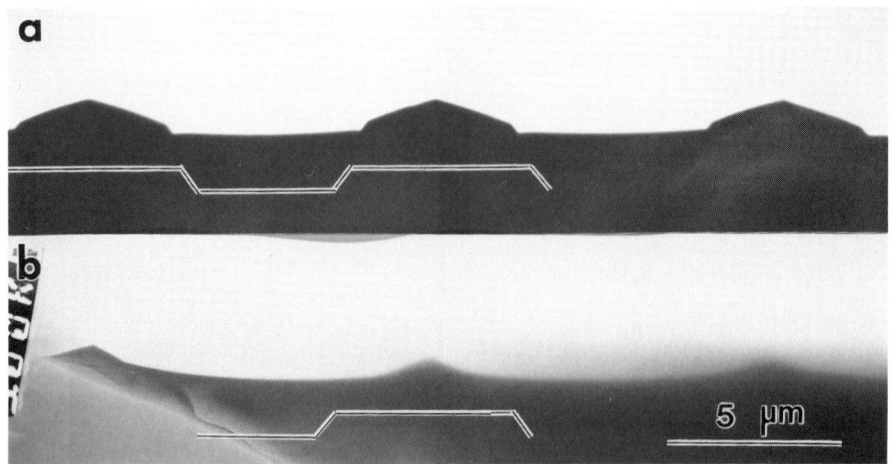

Fig.1 Surface topography of specimens overgrown at (a) 620°C and (b) 720°C

from lattice defects. Defect-free structures may also be made in the case of growth of the $Al_{0.5}Ga_{0.5}As$ interlaced with ultrathin (10 monolayers) of GaAs marker layers or vice versa (Guha et al, 1989). Fig.1a and 1b show the topography of the MBE epilayer   superlattices grown at 620°C and 720°C on the patterned substrate respectively. It has in the present studies been focused on the orientation of the superlattices near the sidewalls of the groove and the defects in the same areas. The origin of these defects have been investigated as a function of the substrate temperature.

3.1. Epilayers Grown on 620°C Patterned Substrate

Fig.1a is an overview of the epilayers at this temperature. Comparing with that grown at a higher temperature as shown in Fig.1b, the overgrown layers are seen to be more convex on top of the mesas, although the flattening of the growth-front in both cases has the same tendency. This indicates that the migration of material at the lower temperature from mesa via sidewall into the recess is less than that at the higher temperature. Attention has been payed to the epilayers near the sidewall where the orientation of the substrate surface has a sudden change. Fig.2 is a bright field image of this area. It is found that at the sidewalls of the substrate, threading dislocations and stacking faults can be formed in the epilayers. The dislocations observed have been characterized by standard diffraction contrast methods to be of the screw type with a Burgers vector $\mathbf{b}=a[110]/2$. By high resolution electron microscopy (HREM), the stacking faults are determined to be intrinsic (insert in Fig.2). This is found by comparing atomic planes related to the ABC stacking sequence along <111> directions shown by arrows. The defects seem to be caused by the small irregular atomic steps on the sidewall and/or at positions where a sudden change in direction of the substrate surface occurs. They could also be due to a relaxation of the concentration of internal stress built up at the sidewall because of the lattice mismatch of the heterostructures. From the bottom of the groove and upwards along the sidewall of the groove, the layers change their directions with a sharp corner between neighbours. Along such a route, layer normals make sudden changes according to the sequence: [100]→[111]→[411]→[100]. [111] is the normal of the groove sidewall. [411] seems to be rather stable. This result agrees with a theoretical simulation of the growth over patterned substrates (Ohtsuka and Suzuki 1989). As a matter of fact, the layers which have a mean normal of [411] are slightly bent. Their normal starts along [311] and continuously changes to[811]. Electron diffraction and HREM micrographs show that the *crystal orientation* of the epilayers remains unchanged although the layers themselves obviously change their directions.

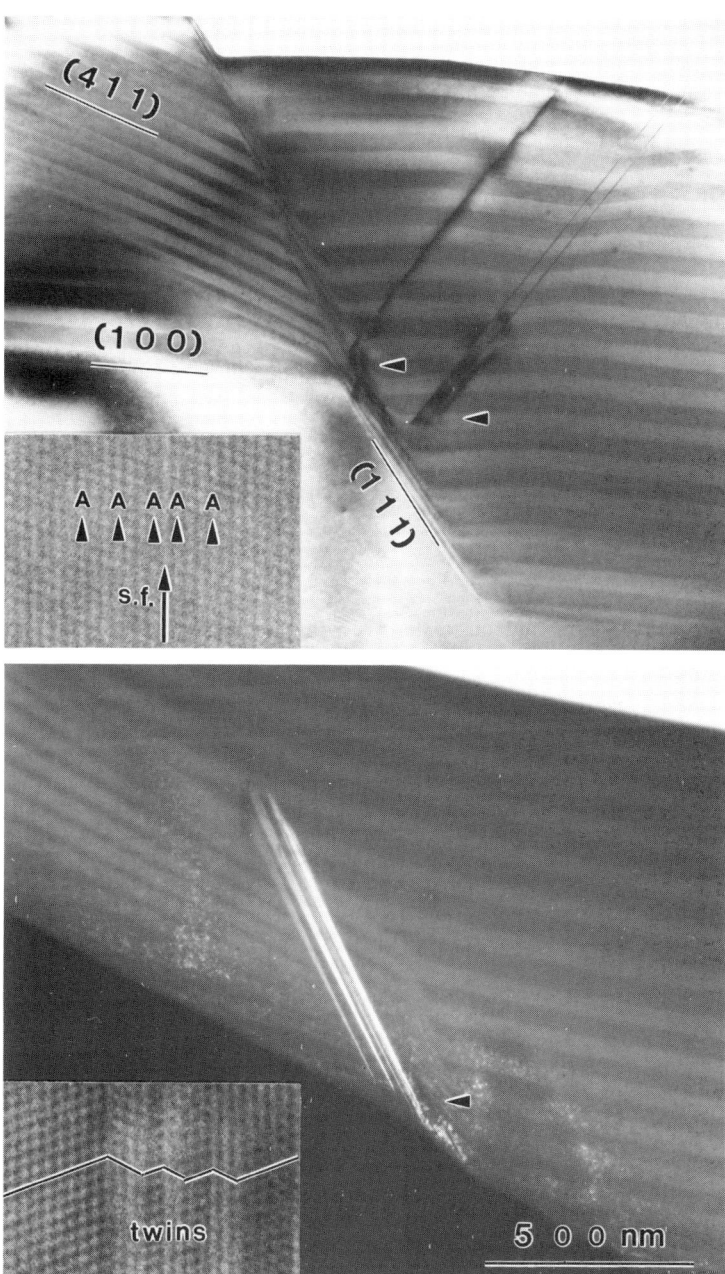

Fig.2 (above) Specimen grown at 620°C. Threading dislocation (**b**=a[110]/2) and intrinsic stacking faults (s.f.) oringinate at sidewall (arrow-heads). The slightly wavy epilayers are shown above the sidewall. The favoured growth plane is seen to be (411) in the epilayers above the mesas.

Fig.3 (below) Specimen grown at 720°C. Beam-induced defects grow near the sidewall.

3.2. Epilayers Grown on 720°C Patterned Substrate

Following Fig.1b, a higher magnified dark field image is shown in Fig.3. No defects were found in the epilayers at the beginning. This means that the growth on the patterned substrate with higher temperature can provide materials free of defects. Comparing with the epilayer grown at 620°C, the whole layers are shown to be smooth rather than wavy near the sidewalls. This point is in good agreement with the assumption that a degradation in quality of the crystal growth near the sidewalls occurs for substrate temperature lower than 700°C (Meier et al 1989 and Guha et al 1989). It is obvious from this that the quality of the epilayers grown on a patterned substrate depends strongly on the temperature of the substrate.

It is found that by keeping a high intensity electron beam focused on the epilayers for some time, threading defects form at and/or near the sidewalls. Fig.3 shows this case. The defects formed in this way are not only dislocations and stacking faults, but also microtwins (insert in Fig.3). Since the beam induced defects were only formed in the vicinity of the sidewalls, it implies that an internal stress was introduced in the epilayers near the sidewalls during the overgrowth, even at the higher temperature. The build-up of the internal stress is probably due to the fact that the surface normal of the substrate is not parallel to the growth direction of the epilayers. Still, at the higher substrate temperature, the favoured direction of the growth is [411] with a slight bend on the top of the sidewalls.

4. CONCLUSION

Threading dislocations, stacking faults and microtwins have been observed in GaAs/AlGaAs heterostructures grown on patterned GaAs substrates. For low temperature growth (620°C), the dislocations are of a general screw type with $b=a[110]/2$, and the stacking faults are intrinsic. Beam induced defects, including microtwins, are found in the specimen made at the higher substrate temperature (720°C). The common feature for the defects is that they all originate at the sidewall of the grooves. This indicates that in the case of formation of such defects, care should be taken to avoid internal stress near the sidewall during the process of MBE growth. The favoured epilayer plane of overgrowth near the sidewalls of the grooves is (411) whether the growth is performed at a lower (620°C) or higher (720°C)⁻ substrate temperature.

ACKNOWLEDGEMENT

Y.X.Guo is grateful for a grant from NTNF, Norway.

REFERENCES

Guha S, Madhukar A, Kaviani K, Chen L, Kuchibhotla R, Kapre R, Hyugachi M and Xie Z 1989 Mat. Res. Soc. Symp. Proc. 145 27
Johannessen K and Stanley C R (1990) (to be published)
Meier H P, Van Gieson E, Walter W, Harder C, Krahl M and Bimberg D 1989 Appl. Phys. Letters 54 1347
Ohtsuka M and Suzuki A 1989 J. Crystal Growth 95 55

Inst. Phys. Conf. Ser. No 117: Section 7
Paper presented at Microsc. Semicond. Mater. Conf., Oxford, 25–28 March 1991

415

TEM study of GaAs films grown on GaAs substrates by the close-spaced vapor transport technique

N Guelton, RG Saint-Jacques, D Cossement, G Lalande, JP Dodelet
INRS-Energie, C.P. 1020, Varennes, Québec, Canada, J3X 1S2

ABSTRACT: Epitaxial GaAs films are grown by CSVT on (100)GaAs substrates. The effect of various CSVT conditions on the sample morphology and on the defect density is examined. SEM and TEM are used to characterize the layers. As also observed in other processes, it is found that the defect density decreases with increasing epilayer thickness. When the spacing between the source and substrate is increased, in the CSVT process, the defect density decreases but the GaAs surface morphology degrades. If the transporting agent (H_2O) is injected when the system temperature reaches 700°C it is found that the defect density decreases to 10^4 - 10^5 cm^{-2}.

1. INTRODUCTION

Close-spaced vapor transport (CSVT) is an efficient and cost effective technique that allows the growth of epitaxial layers of semiconductors (Perrier and al. 1988). In this technique a temperature gradient is maintained between the closely spaced (~ 1 mm) solid source and substrate. A transporting agent (water vapor) is used to react with the source and form volatile compounds which are subsequently decomposed at the surface of the substrate to form a thin film of the source material. The reaction chamber consists of a quartz tube containing the source and the substrate separated by a spacer and held between two graphite susceptors. The deposition rate is strongly dependent on the kinetics of transport of the gaseous species (Ga_2O, As_2, As_4), the source and substrate temperatures, the water vapor pressure, the distance between the source and substrate and the thermodynamics of the reactions involved (Côté and al. 1986).

The aim of our research program is to produce by this low cost deposition technique the high power conversion efficiency solar cell n-Ge/n$^+$-GaAs/n-GaAs/p-GaAs/ZnSe (Lombos 1983). Heteroepitaxy of GaAs on (100)Ge substrates and homoepitaxy of GaAs on (100)GaAs have been undertaken by CSVT. The present work reports the contribution of TEM and SEM analysis to obtain homoepitaxial GaAs layers of good quality (defect free layers with specular surfaces). It describes the influence on the density of the different types of defects of (i) the film thickness, (ii) the spacing between the source and substrate and (iii) the temperature of the system when water vapor is injected.

2. EXPERIMENTAL

After introduction of the source and the substrate into the reactor and purge with dry nitrogen, the heating system is turned on. The source temperature is maintained at 800°C and the substrate temperature at 760°C. The transport reagent (water vapor at 4.58 Torr) is carried into the reactor by bubbling H_2 through water either when the system is still at room temperature or when it reaches 700°C. Various spacer thicknesses are used (0.3; 1.5; 2 mm). Under these conditions, GaAs films are grown to thicknesses ranging from 0.5 to 7 μm. When the spacer thickness is increased from 0.3 to 2 mm the growth rate decreases from 0.5 to 0.15 μm min^{-1}. At the end of the deposition, the heating system is turned off and a dry H_2 flow is maintained until the graphite susceptors are cooled down to room temperature.

An ion miller is used to back thin planar TEM samples. Since back thinned planar views show the defect density of the top of the film, it is possible to study the evolution of the defect densities with thickness, from samples grown to various thicknesses under similar growth conditions.

3. RESULTS AND DISCUSSION

3.1 Effect of film thickness

Table 1 shows the dislocation densities, N_d, as a function of thickness, t, of films grown with a 0.3 mm spacer and with water vapor injected at room temperature.

Table 1:   Dislocation density as a function of the thickness of the homoepitaxial GaAs layer measured from TEM planar view micrographs.

t (μm)	0.48	0.93	3.2	7.3
N_d(cm^{-2})	7.3×10^8	3.5×10^8	1.6×10^8	7.0×10^7

All dislocations have their origin at the GaAs/GaAs interface where their density is as high as 10^9 cm^{-2}. The dislocation density quickly decreases with increase of thickness of the film. Fig. 1 shows the defect densities for both the thinnest and thickest GaAs films investigated.

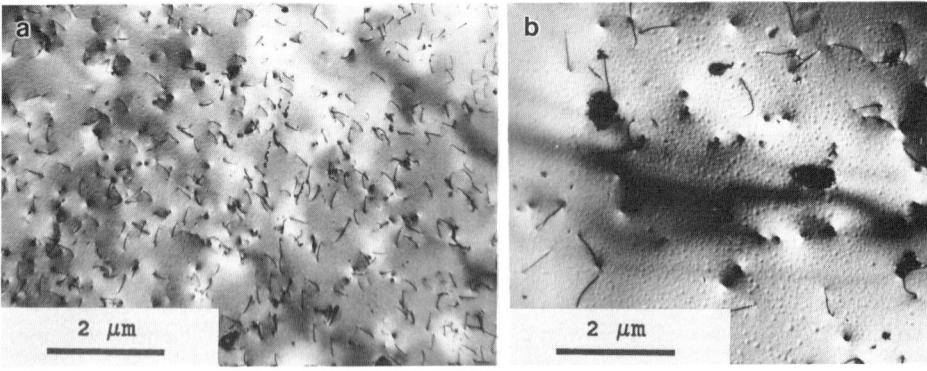

Fig. 1:   TEM planar view micrographs of the homoepitaxial GaAs layer at a) 0.48 μm and
b) 7.3 μm from the interface.

This decrease can be explained by an annihilation process between dislocations (Sheldon and al. 1985) or by a deviation process of the dislocation line (Fang and al. 1988). In the first case, a dislocation line can bend toward another line so that they annihilate one another. In the second case, a dislocation line can bend away from the surface and run parallel to the growth plane so that the line does not reach the surface at all.

The stacking fault density is much smaller than the dislocation density (Fig. 2). Errors in stacking on the {111} facets of the nuclei during deposition give rise to stacking faults. The propagation along these planes leads to the mutual annihilation of stacking faults which are then confined to a short distance from the GaAs/GaAs interface (Ernst and Pirouz 1989).

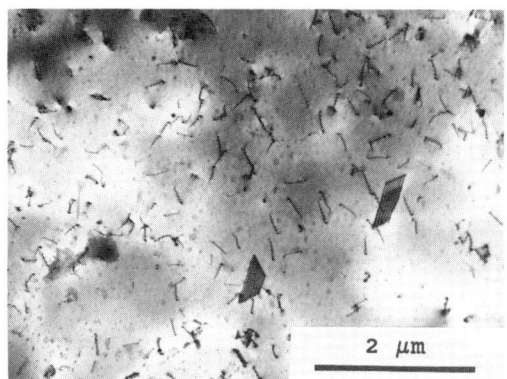

Fig. 2: TEM of a planar view micrograph of the 0.48 μm thick film. It shows a much smaller stacking fault density than the dislocation density.

3.2 Effect of the spacer thickness

As seen above, at 7 μm from the interface the dislocation density is reduced but remains high. It is possible to reduce it significantly further by choosing a thicker spacer. A film grown to 7 μm with a spacer of 2 mm reveals a dislocation density of $5.4.10^6$ cm^{-2} which is one order of magnitude lower than the dislocation density of the layer grown to 7.3 μm with a spacer of 0.3 mm, all the other growth conditions being identical. However, in the case of a 2 mm spacer, the GaAs morphology degrades; its surface contains a high density of pits (Fig. 3) and appears to be milky to the naked eye. This reduction in dislocation density and the degradation of the surface morphology could be related to the decrease of the growth rate when a thicker spacer is used. The net result is that less defects cross the film but their emergence is more apparent.

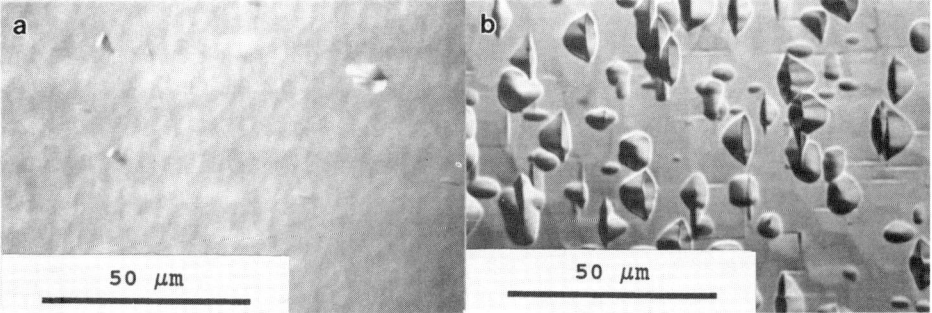

Fig. 3: SEM micrographs showing defect pits at the surface of the homoepitaxial GaAs layer a) for a spacer of 0.3 mm and b) for a spacer of 2 mm.

3.3 Effect of the temperature of the system when the water vapor is injected

Recent experiments involving only a change in the operating conditions allowed a further drastic reduction of the defect density. The transporting reagent was previously injected when the system was at the room temperature. If the transporting reagent is injected when the source temperature reaches 700°C, higher quality GaAs films with dislocation densities less than 10^5 cm^{-2} are obtained for thicknesses greater than 0.54 µm. It should be possible to reduce even further this density since the deposition was carried out with a spacer of only 0.3 µm. This defect density, which is too small to be estimated from TEM planar views has been estimated by preferential etching.

Another experiment, performed under the usual growth condition but with no injection of water vapor in the reactor, revealed that a very thin film of GaAs (less than 0.1 µm) has however been deposited on the substrate. Thus some transport reaction took place. It means that some residual water vapor was present in the reactor. This water comes mainly from the graphite susceptors where it adsorbs during the reactor loading. Its partial pressure, which has been estimated at 10^{-2} Torr, is sufficient for the nucleation and coalescence of GaAs to be carried out. This value shows that the water pressure injected via H_2 (4.58 Torr) is far in excess to that required for nucleation.

If no water vapor is injected into the reactor until the source temperature reaches 700°C, the nucleation can be carried out properly with the appropriate quantity of water adsorbed on the graphite susceptors. The exact quantity of injected water at 700°C is no longer critical for the growth stage since nucleation has already taken place. But if water vapor is injected when the system is still at room temperature, the excess of transporting agent seems to perturb the nucleation or degrade the surface of the substrate by oxidation and leads to a high density of dislocations.

4. CONCLUSION

Variation in the deposition parameters are correlated with defect densities in order to optimize the growth conditions. It is found that (i) increasing the GaAs layer thickness from 0.5 to 7 µm, (ii) using a spacer of at least 1.5 mm, and (iii) injecting water when the system has reached a temperature high enough for the nucleation to be carried out, cause a decrease in the defect density. The main conclusion of this study is that optimization of the CSVT conditions leads to an improvement in the quality of epitaxial GaAs layers. This optimization reduces dislocation density to less than 10^5 cm^{-2}. The use of these experimental conditions for the system GaAs/(100)Ge may however not produce such an important increase of the defect density since the lattice mismatch of this system is smaller than 0.13%.

Côté D., Dodelet J.P., Lombos B.A. and Dickson J.I. 1986, J. Electrochem. Soc. 132, 1925.
Ernst F., and Pirouz P. 1989, J. Mater. Res. Vol. 4, Jul./Aug. 1989.
Fang S.F., Adomi K., Lyer S., Morkoç H., Zabel H., Choi C. and Otsuka N. 1990, J. Appl. Phys. 68(7), R31.
Lombos B.A., Bartkowski M., Buchbinder M., Côté D. and Dodelet J.P. 1983, 5th Photovoltaic Solar Energy Conference, Proceedings, Athens, Greece, 17-21 october 1983.
Perrier R., Phillipe R. and Dodelet J.P. 1988, J. Mater. Res. 3(5), 1031.
Sheldon P., Yacobi B.G., Jones K.M. and Dunlavy D.J. 1985, J. Appl. Phys. 58(11) 4186.

Inst. Phys. Conf. Ser. No 117: Section 7
Paper presented at Microsc. Semicond. Mater. Conf., Oxford, 25–28 March 1991

419

The growth of GaAs on $Sc_{0.3}Er_{0.7}As$ epilayers

Jane G Zhu, Chris J Palmstrøm* and C Barry Carter

Department of Materials Science and Engineering, Cornell University, Ithaca, New York 14853, USA
*Bellcore, 331 Newman Springs Road, Red Bank, New Jersey 07701, USA

ABSTRACT: The formation of $GaAs/Sc_{0.3}Er_{0.7}As/GaAs$ heterostructures by molecular beam epitactic growth is investigated. Twinned grains are often observed in the top GaAs layer at the $GaAs/Sc_{0.3}Er_{0.7}As$ heterojunction region for the samples grown on (100) GaAs substrates. A comparison is made of the island growth of GaAs on $Sc_{0.3}Er_{0.7}As$ for (100) and (311) orientations. The crystalline quality of the GaAs on top of $Sc_{0.3}Er_{0.7}As$ is significantly improved for the samples grown on (311)A GaAs substrates.

1. INTRODUCTION

Several epitactic compound metallic materials grown on III-V semiconductors have been studied in recent years, including rare-earth (RE) arsenides (e.g., Palmstrøm et al 1990, Sands et al 1990). Most of the rare-earth monoarsenides, e.g., $Sc_xEr_{1-x}As$, have the rocksalt structure with lattice constants very close to the lattice constant of GaAs; the value of x can be adjusted so that the rare-earth monoarsenides and the GaAs are exactly lattice-matched at x=0.32. This close match of the lattice parameters makes it possible to grow epilayers of the semimetal on, and in, the compound semiconductor which has exciting potential applications in novel device fabrication. However, the difference in crystal structures between that of (RE)As and GaAs substantially affects the growth of GaAs/(RE)As/GaAs heterostructures. It has been demonstrated that the rare-earth arsenides can be grown on GaAs in a layer-by-layer mode with very good crystal quality, but the GaAs grown on the rare-earth arsenides is too low in quality to meet the requirement for novel devices (Zhu et al 1990). Therefore, the investigation on the growth of GaAs on (RE)As is essential for GaAs/(RE)As/GaAs heterostructures towards device applications. Until the present time, there are only some preliminary results reported on the island growth of GaAs on (RE)As/GaAs (Zhu 1990a et al and Finstad et al 1991). In the present paper, we report the study on the growth of GaAs on $Sc_{0.3}Er_{0.7}As$, which had, in turn, been grown on GaAs substrates by molecular beam epitaxy (MBE). The microstructure of $GaAs/Sc_{0.3}Er_{0.7}As/GaAs$ layers has been characterized extensively using transmission electron microscopy (TEM).

2. EXPERIMENTAL

Samples were grown on GaAs substrates in a VG V80H MBE system. Er and Sc were evaporated from effusion cells. A 500-nm-thick, undoped, GaAs buffer layer was grown on GaAs substrates at 600°C. After the substrate was cooled to 400°C, a $Sc_{0.3}Er_{0.7}As$ layer was grown at a growth rate of 0.37 μm/hr. Then, a thin GaAs layer, with various thicknesses, was deposited at a rate of ~ 0.1 μm/hr on top of the $Sc_xEr_{1-x}As$ at different growth temperatures for the study on the nucleation of GaAs on $Sc_{0.3}Er_{0.7}As$. The As_4 shutter was open all the time during the growth of $GaAs/Sc_{0.3}Er_{0.7}As/GaAs$

heterostructures. Additional details concerning the sample growth can be found in the papers by Palmstrøm et al (1990) and Zhu et al (1990). The growth rates were calibrated from reflection high-energy electron diffraction (RHEED) oscillations. A thin amorphous-Si cap layer was deposited *in situ* to prevent the oxidation of $Sc_{0.3}Er_{0.7}As$ upon exposure to air. The samples were characterized by TEM in both plan and cross-section views using JEOL 1200EX and JEOL 4000EX electron microscopes.

3. GROWTH ON (100) GaAs SUBSTRATES

In contrast to the layer-by-layer growth of $Sc_{0.3}Er_{0.7}As$ on GaAs, island growth is observed for GaAs on top of $Sc_{0.3}Er_{0.7}As$. Figure 1 shows a pair of plan-view bright-field images of GaAs islands grown on $Sc_{0.3}Er_{0.7}As(2.8 nm)/GaAs$ at growth temperatures of 330°C for sample (a) and 480°C for sample (b). The average thickness of the top GaAs layer for both samples is 5.6 nm. The coverage of GaAs on $Sc_{0.3}Er_{0.7}As$ is better when the GaAs is grown at relatively low temperatures. The GaAs islands are elongated along <011> directions. Cross-section TEM has revealed that only a small number of GaAs islands consist entirely of single-crystal material corresponding to the standard epitactic relationship. The majority of GaAs islands comprise a mixture of epitactically aligned material and both singly and doubly twinned material with the twin boundaries parallel to {111} planes (Zhu et al 1990a). A cross-section high-resolution image (Fig. 2) shows part of a relatively large GaAs island on $Sc_{0.3}Er_{0.7}As/GaAs$ from the sample (b). At the interface between the twinned GaAs and $Sc_{0.3}Er_{0.7}As$, a different phase boundary has been found, namely {122} GaAs on (100) $Sc_{0.3}Er_{0.7}As$ with $<01\bar{1}>_{GaAs}$ // $<01\bar{1}>_{ScErAs}$ and $<\bar{4}11>_{GaAs}$ // $<011>_{ScErAs}$. The periodicity at the {122}GaAs/(100)$Sc_{0.3}Er_{0.7}As$ interface is increased by a factor of three as indicated by the arrows, which corresponding to the (3×1) (referred to the $Sc_{0.3}Er_{0.7}As$ surface unit cell) coincidence lattice at this interface. Large areas of {122} GaAs have been observed grown on (100) $Sc_{0.3}Er_{0.7}As$.

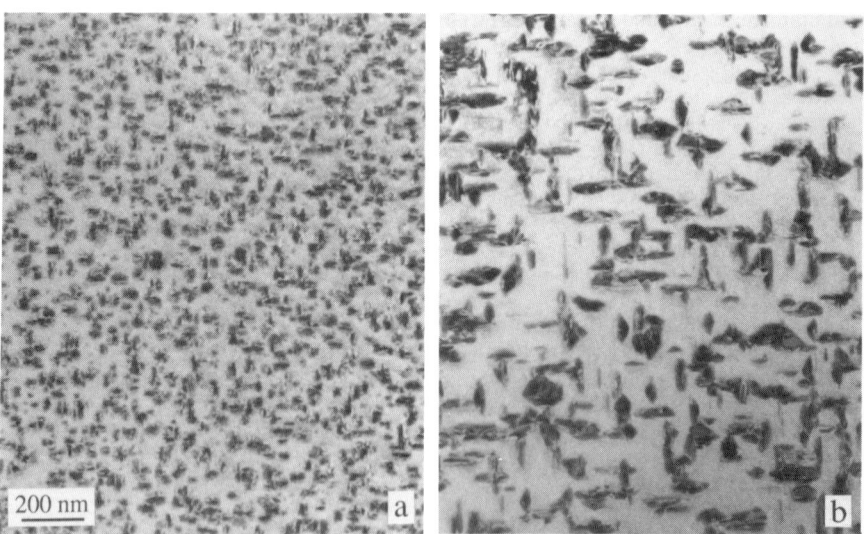

Fig. 1. Plan-view bright-field images (same magnification) of sample (a) and (b) showing GaAs islands, grown at 330°C and 480°C respectively, on $Sc_{0.3}Er_{0.7}As(2.8 nm)/GaAs$.

4. GROWTH ON (311)A GaAs SUBSTRATES

Wetting of GaAs on $Sc_{0.3}Er_{0.7}As$ has been significantly improved for the samples grown on (311)A GaAs substrates. Figure 3 shows a plan-view bright-field image from a sample with 5.6 nm (average thickness) of GaAs deposited on $Sc_{0.3}Er_{0.7}As(5.6 nm)/GaAs$ at

Fig. 2. High-resolution image showing {122} GaAs on (100) $Sc_{0.3}Er_{0.7}As$.

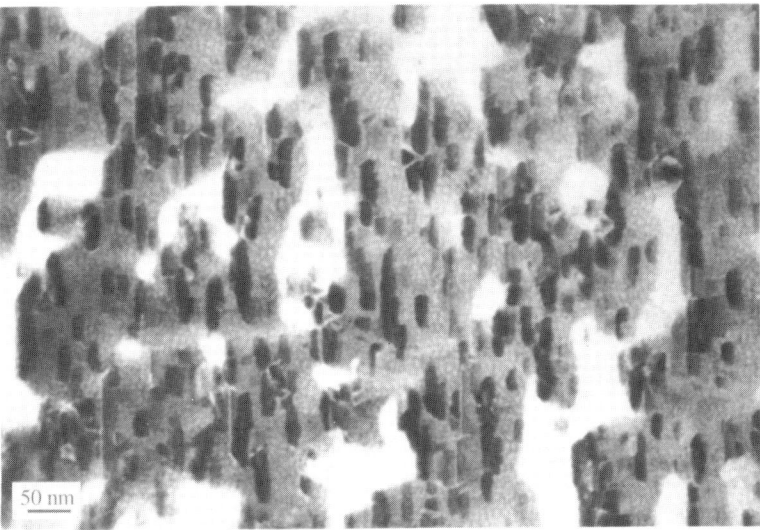

Fig. 3. Plan-view image of GaAs on $Sc_{0.3}Er_{0.7}As$(5.6 nm)/GaAs grown on (311)A GaAs substrate.

about 490°C. The areas with very bright contrast are the open areas on $Sc_{0.3}Er_{0.7}As$. The close-to-rectangular areas with dark contrast, which are parallel to <01$\bar{1}$> direction, are the projections of inclined {111} microtwins (Zhu et al 1991). Comparing with the samples shown in Fig. 1, which were grown on (100) GaAs substrates, the coverage of GaAs on

$Sc_{0.3}Er_{0.7}As$ is much better for the sample grown on (311)A GaAs substrate. Figure 4 shows a cross-section high-resolution image from another sample grown on (311)A GaAs substrate. The crystalline quality of the top GaAs layer has been significantly improved although there are some stacking faults and microtwins on {111} planes. The $GaAs/Sc_{0.3}Er_{0.7}As$ interfaces are not as sharp as those in the samples grown on (100) substrates. The surface facets on $Sc_{0.3}Er_{0.7}As$ may help the nucleation of GaAs on $Sc_{0.3}Er_{0.7}As$ (Zhu et al 1991).

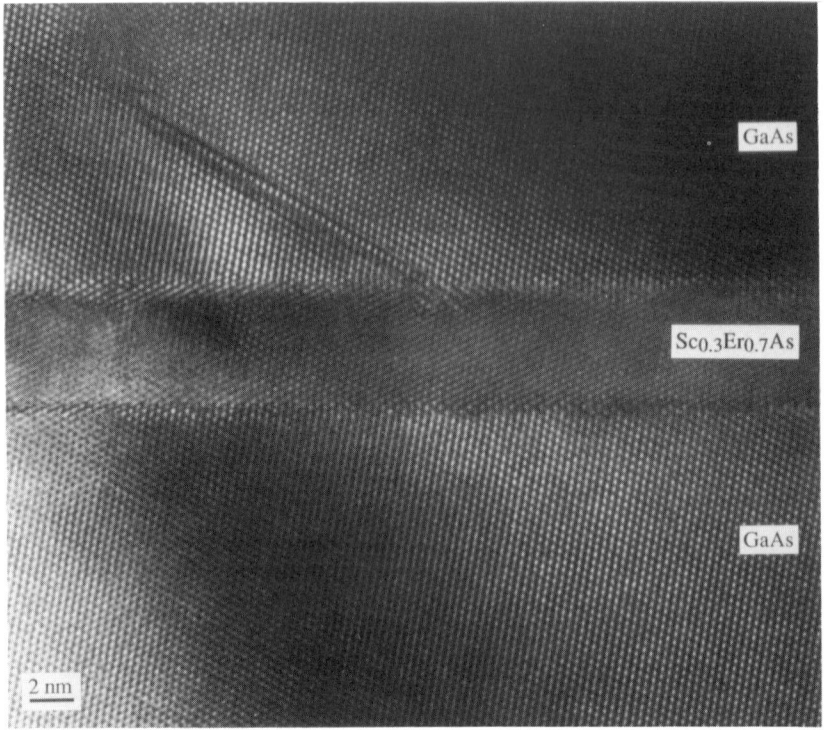

Fig. 4. <01$\bar{1}$> cross-section high-resolution image of $GaAs/Sc_{0.3}Er_{0.7}As/GaAs$ grown on (311)A GaAs substrate.

ACKOWLEDGMENTS

JGZ acknowledges support by the Semiconductor Research Corporation Center for Microscience and Technology under grant No. 90-SC-069. The electron microscopes are part of a central facility provided by the Materials Science Center at Cornell and are supported, in part, by the National Science Foundation.

REFERENCES

Finstad T G, Palmstrøm C J, Mounier S, Keramidas V G, Zhu J G and Carter C B 1991 to be published in Mat. Res. Soc. Symp. Proc.
Palmstrøm C J, Mounier S, Finstad T G and Miceli P F 1990 Appl. Phys. Lett. 56 382
Sands T, Palmstrøm C J, Harbison J P, Keramidas V G, Tabatabaie N, Cheeks T L, Ramesh R and Silberberg Y 1990 Mat. Sci. Reports 5 99
Zhu J G, Carter C B, Palmstrøm C J and Mounier S 1990 Appl. Phys. Lett. 56 1323
Zhu J G, Palmstrøm C J and Carter C B 1990a Mat. Res. Soc. Symp. Proc. 198, 177
Zhu J G, Palmstrøm C J and Carter C B 1991 to be published

Inst. Phys. Conf. Ser. No 117: Section 7
Paper presented at Microsc. Semicond. Mater. Conf., Oxford, 25–28 March 1991

The reaction at the InAs-GaAs(001) interface during MBE growth

X Zhang, D W Pashley[‡], I T Ferguson, P N Fawcett, A E Staton-Bevan[‡] and B A Joyce.

The London Interdisciplinary Research Centre for Semiconductor Materials, Imperial College, London SW7 2AZ.
[‡]Department of Materials, Imperial College, London SW7 2BP.

ABSTRACT: A series of layers of InAs has been grown on GaAs (001) by MBE with a wide range of growth conditions. Cross-sectional TEM studies showed the presence of (InGa)As protrusions into the GaAs at the interface in all samples. The formation of these protrusions is thought to be due to the presence of In droplets which react with the GaAs at the early stage of InAs deposition.

1. INTRODUCTION

Interface interactions and the resulting structures formed during growth of lattice mismatched systems by MBE are of crucial importance in determining the mechanism of strain relief as well as having a direct impact on the electrical and optical properties of the films.

Atomically abrupt InAs-GaAs (001) interfaces grown by MBE even under In stabilised conditions have been claimed (e.g. Schaffer et al 1983). However, it is generally believed that during the growth of In based III-V films, In arriving at the substrate surface or growth front can aggregate or form clusters under certain conditions (Foxon & Joyce 1978, Zhang et al 1991). This would result in a modification to the structure of the interface. In this paper we present a careful study of the InAs-GaAs (001) interfaces carried out using cross-sectional TEM.

2. EXPERIMENTAL PROCEDURES

The samples were grown in either a VG Semicon V80H or a purpose built MBE machine having a specially designed RHEED facility. More than fifteen samples have been studied. The growth parameters were varied within the following ranges: substrate temperature: 440 to 540^{0}C; As/In flux ratio: 2:1 to 20:1; As source type: As$_2$ or As$_4$; growth rate: 0.1 to 1 monolayer/s; InAs layer thickness: 0.3 to 5.1um. A 0.5um GaAs buffer layer was grown before the deposition of InAs and all samples were un-doped.

TEM {110} cross-sections were prepared using a standard Ar⁺
beam thinning method. Two JEOL microscopes, a 2000FX and a
2010, were used in the TEM studies.

3.   RESULTS AND DISCUSSION

TEM cross-sections of randomly chosen GaAs buffer layers
without the InAs showed very smooth and pit-free surfaces.
The cross-sections of all InAs-GaAs samples revealed the
presence of protrusions into the GaAs substrate at the InAs-
GaAs interfaces, but their lateral sizes and depths of
penetration varied over a large range.

Fig.1 A (110)
cross-section
shows the
presence of
protrusions,
indicated by the
arrows at the
interface and in
the insert.

Fig.2 EDX spectra
taken from the
area below,
within and above
the protrusion
shown.

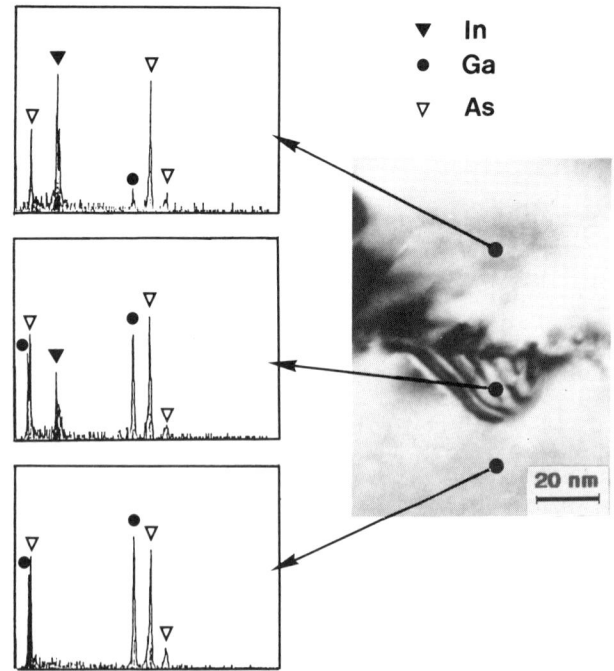

▼ In
● Ga
▽ As

A majority of the samples showed small protrusions up to 50nm
in lateral size and 20nm in depth of penetration. A
typical example is shown in Fig.1. The interfaces of this
group of samples are usually sharp and straight except where
the protrusions, indicated by the arrows in Fig.1, are
present. Using the EDX technique with 3nm probes these

protrusions are found to contain In. This is shown in Fig.2 where direct comparisons can be made between the spectra taken from the areas below, within and above the protrusion shown. Fig.3 is a high resolution lattice image of a similar protrusion. Misfit dislocations can be clearly seen along the interface of the protrusion and the GaAs substrate. They are generally 60^0 type with b=a/2<110> as shown in the insert.

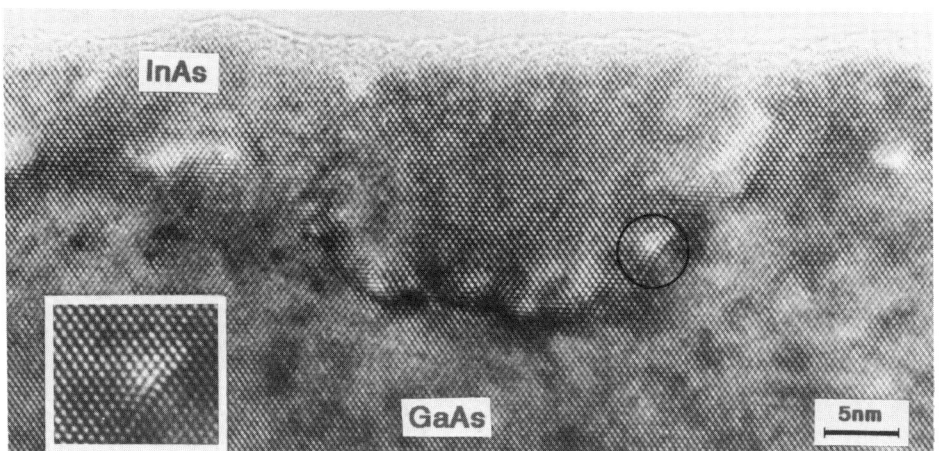

Fig.3 A (110) high resolution micrograph of a protrusion in to the GaAs. Most dislocations at the protrusion and GaAs interface are 60^0 type, an example is marked by the circle and shown enlarged in the insert.

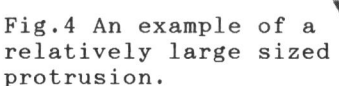

Fig.4 An example of a relatively large sized protrusion.

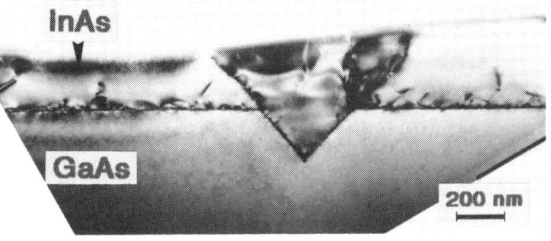

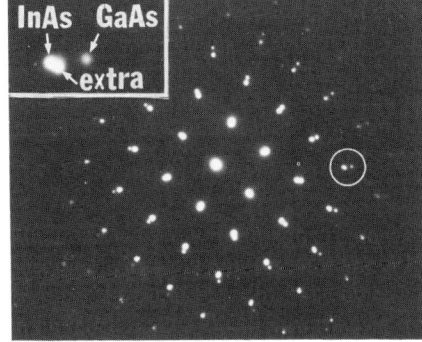

Fig.5 An electron diffraction pattern from an area containing GaAs, a large protrusion and InAs. The circled group of reflections are shown enlarged in the insert where the extra reflection corresponding to the presence of protrusion is indicated.

A few samples have protrusions much larger in size. The corresponding microstructure is more complex as shown in Fig.4. They reveal a more developed crystallographic shape and the depth of penetration can be as deep as 400nm. EDX analysis from the protrusion area revealed the presence of In as in the case of smaller protrusions. Fig.5 is an electron diffraction pattern taken from an area containing a protrusion. Additional reflections between those of GaAs and InAs, due to the presence of the protrusion, can be clearly seen. The lattice spacing calculated from these extra reflections is 0.596+/-0.001nm which corresponds to $In_{0.76}Ga_{0.24}As$ assuming validity of Vegard's law.

Electron diffraction measurements, mainly on large protrusions, indicate that the protrusions are $In_xGa_{1-x}As$ compounds where 0.75<x<0.9. The protrusions shown in Fig.1 are too small to be able to give clear indications of extra reflections by selective area diffraction. However, from the EDX analysis data including some evidence for the presence of Ga in the InAs layer adjacent to the protrusion, see Fig.2, we believe that these protrusions are also $In_xGa_{1-x}As$ compounds where 0<x<0.9.

It has been deduced from desorption measurements that In has the tendency to aggregate (Foxon &. Joyce 1978, Zhang et al 1991). At a typical growth temperature, such accumulation will result in In liquid droplets. Similar observations of Ga droplet formation have been made during MBE growth of GaAs (Osaka et al 1990) and recently large clusters of In were observed directly on the surface of InAs films on GaAs (001) (Suzuki et al 1990). It is believed that in the present case, such In droplets form during the early stages of InAs deposition and dissolve the GaAs causing the characteristic protrusions into the substrate as seen in Fig.1. This dissolution and growth of the InAs layer would continue simultaneously until the change in composition of the protrusions causes them to solidify as (InGa)As compounds.

4.   CONCLUSIONS

Evidence has been shown which indicates that during the growth of InAs on GaAs by MBE, In forms droplets which dissolve GaAs and causes the formation of (InGa)As protrusions at the interface.

REFERENCES

Foxon C T and Joyce B A 1978 J. Crystal Growth **44** 75
Osaka J, Inoue N, Mada Y, Yamada K and Wada K 1990 J. Crystal Growth **99** 120
Schaffer W J, Lind M D, Kowalczyk S P and Grant R W 1983 J. Vac. Sci. Technol. **B1** 688
Suzuki Y, Chikawa Y and Akazaki T 1990 Appl. Phys. Lett. **56** 1856
Zhang J, Gibson E M, Foxon C T and Joyce B A 1991 J. Crystal Growth (in the press)

ACKNOWLEDGEMENT

The authors are grateful to JEOL (UK) Ltd. and John Critchell for the assistance with the HREM work.

Inst. Phys. Conf. Ser. No 117: Section 7
Paper presented at Microsc. Semicond. Mater. Conf., Oxford, 25–28 March 1991

In situ UHV REM study of the structure of silicon surfaces

A L Aseev, A V Latyshev, A B Krasilnikov and L V Litvin

Institute of Semiconductor Physics, Academy of Sciences of the
USSR, Siberian Branch, 630090 Novosibirsk, USSR

ABSTRACT: Reflection electron microscopy in ultra high
vacuum (UHV REM) developed on the basis of transmission
electron microscopy with top entry stage was applied for
the *in situ* study of clean silicon surfaces during sublimation, surface reconstruction and initial stages of epitaxial growth.

The Si(111) surface contains steps which are seen on the REM
images due to the superposition of the diffraction contrast
because of surface deformations near the steps and the phase
contrast because of the phase shift upon defocusing (see e.g.
Yagi 1987). The formation of a step at the emergence points of

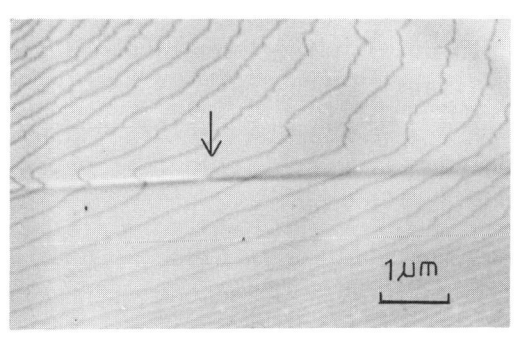

Fig. 1. REM images of the
Si(111) surface containing
monoatomic steps and emergence points of disloca-
tions (marked by arrows).

dislocations having a screw component of Burgers vector is
seen in fig. 1. The sublimation process leads to step movement

in step-up direction. During annealing the crystal by direct
electric current a reversible rearrangement of a regular monoatomic step system to step bunching takes place (fig.2).   The

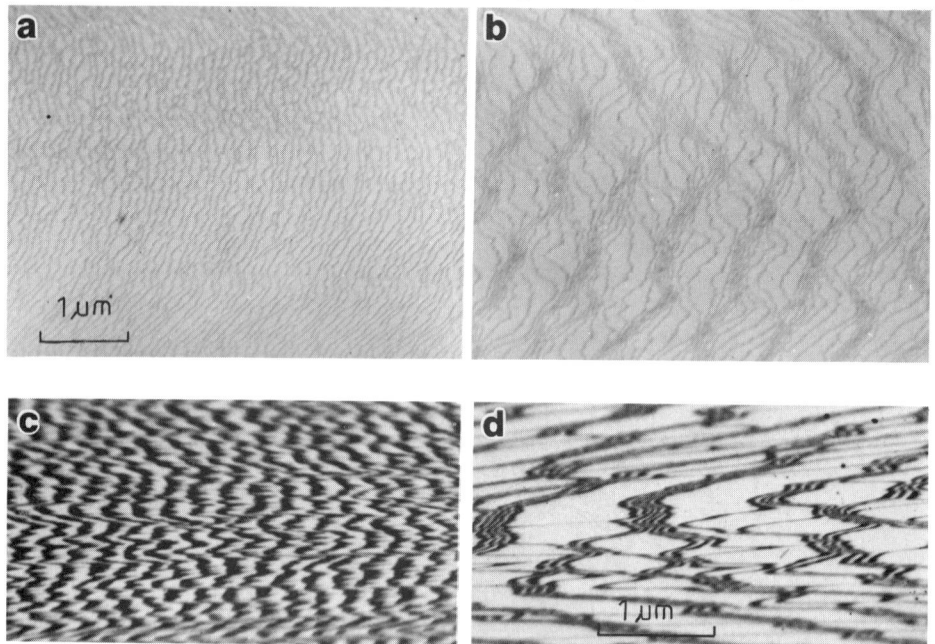

Fig. 2. REM images of the same   areas   of Si(111) (a,b)   and
Si(100) (c,d) surface before (a,c) and after (b,d)   step
bunching. For the observation of   Si(100)   surface   the
superstructural reflection (1/2   order)   was   used.   The
directions of the dimerization on the neighbouring terraces of
reconstructed Si(100) surface differ by 90°, so the terraces
separated by monoatomic steps have various surface structure
(1x2) and (2x1) and give alternatively bright and dark
contrast.

step bunching depends on the temperature and the direction of
the heating electric current. The qualitative model of step
bunching by Latyshev et al (1989) and the theoretical consideration of the electromigration induced step bunching on Si
surfaces by Stoyanov (1991) include the adatom drift which
change the fluxes of the adatoms on the steps.

It should be pointed out that on the Si(100) surface the additional factor connected with the anisotropy of adatom diffusion a reconstructed surface takes place and so the step bunching depends on the initial monoatomic step density (Litvin

et al 1991). More over, on the Si(100) surface the formation of pairs of monoatomic steps takes place which leads to a change of relative areas with (1x2) and (2x1) structures which depends on the direction of the heating electric current (Latyshev et al 1990).

It is essential to take into account step bunching on Si(111) and Si(100) surfaces during thermal treatment of the crystal through annealing by a direct electric current. We use this effect to obtain surface regions with required distances between the steps.

The monoatomic step rearrangement takes place also during the various superstructural transitions on the Si(111) surface. The movement of segments of monoatomic steps which are the boundaries between reconstructed and unreconstructed areas of the surface is observed during the formation of various structures: Si(7x7)Ge and Si(5x5)Ge for epitaxy of Ge, Si(5x1)Au for epitaxy of Au and , also, Si(7x7) on the clean Si(111) surface (Latyshev et al 1991). The probable reason of such movement is

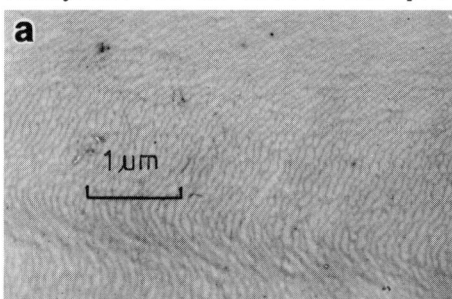

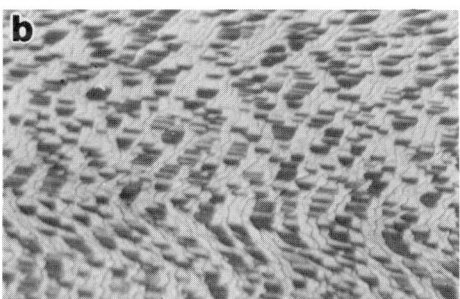

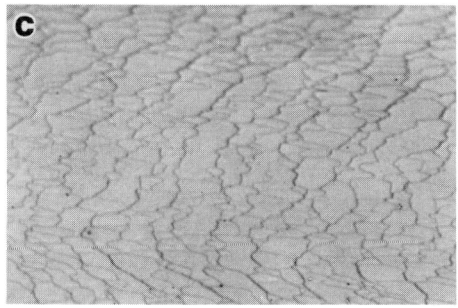

Fig. 3. REM images of the same areas of Si(111) surface before (a), during (b) and after (c) the formation of Si(5x1)Au structure.

the difference in free energies between the reconstructed and unrecon-structed surfaces which leads to the violation of the balance between the atom incorporation into and generation from steps separating the areas with various surface structures. Such a movement leads to reversible monoatomic step clustering (fig. 3) when the polycentric nucleation of superstructural domains (see fig. 3b) and the diffusion exchange between the steps take place. For example, the investigation of the (1x1)⇒(7x7) transition on the clean Si(111) surface which contained a step band system revealed the shift of the steps on 0.2-0.3 width of the terrace in step-down direction on the flat regions between step bands (Latyshev et al 1990) and step clustering in the bands where the distances between the steps are much lower (fig. 4). There are kinetic limitations on the

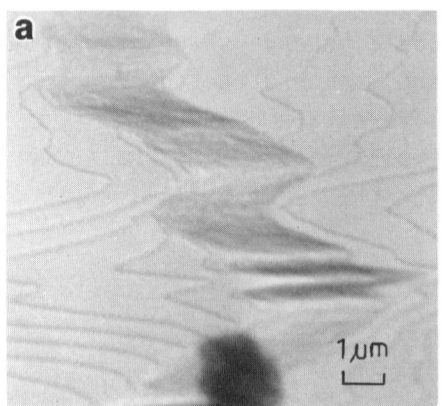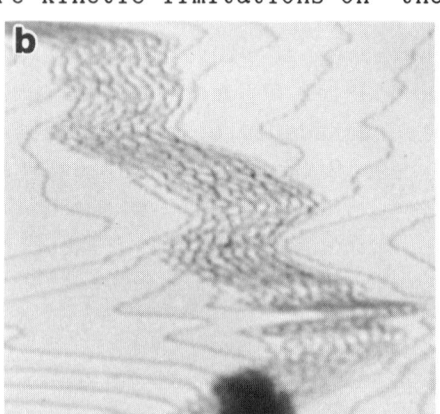

Fig. 4. REM images of the same areas of Si(111) surface before (a) and after (b) the (1x1)⇒(7x7) reconstruction.

process of step clustering because of the necessity of significant mass transfer. This allows the required surface micromorphology to be obtained during the initial stages of epitaxy by the choice of growth conditions.

REFERENCES

Latyshev A V, Aseev A L, Krasilnikov A B and Stenin S I 1989 Surf.Sci. 213 157
Latyshev A V, Aseev A L, Krasilnikov A B and Stenin S I 1990 Surf.Sci. 227 24
Latyshev A V, Krasilnikov A B and Aseev A L 1991 J.Electron Microscopy Technique (in press)
Litvin L V, Krasilnikov A B and Latyshev A V 1991 Surf.Sci.Let. (in press)
Stoyanov S 1991 Jpn.J.Appl.Phys. 30 1
Yagi K 1987 J.Appl.Cryst. 20 147

Inst. Phys. Conf. Ser. No 117: Section 7

Paper presented at Microsc. Semicond. Mater. Conf., Oxford, 25–28 March 1991

431

Limited-thickness epitaxy of semiconductors down to room temperature

DJ Eaglesham, H-J Gossmann, M Cerullo, LN Pfeiffer and D Windt

AT&T Bell Laboratories, 600 Mountain Avenue,Murray Hill,NJ07974,USA

ABSTRACT: We use TEM of thin layers to study the low-temperature limit to semiconductor molecular beam epitaxy (MBE). In Si, Ge/Si and GaAs MBE we show that at low temperature limited-thickness epitaxy occurs: layers initially grow epitaxially, and the amorphous phase forms beyond some epitaxial thickness, h_{epi}. TEM and doping experiments show that the defect density always remains low in these films, and growth on H-terminated Si suggests that in Si, at least, segregation of residual H in the MBE chamber is not responsible: this limited-thickness epitaxy is probably linked to surface increasing roughness. Ge marker layers are used to define the surface morphology in Si(100) growth at low temperature, and TEM measurements confirm a significant increase in surface roughness just before the breakdown of epitaxy. We conclude that increasing surface roughness arising from growth at temperatures where surface diffusion processes are very slow changes the kinetics for nucleation of amorphous zones on the crystalline surface, and is thus responsible for the final breakdown of epitaxy.

1. INTRODUCTION

The low-temperature limit to homoepitaxial growth has always seemed fairly clear: in growth of any material where there is a stable amorphous phase we expect to see epitaxy unless surface diffusion processes are significantly slower than the deposition rate, R. Since surface diffusion is thermally activated, there should be a (rate-dependent) minimum temperature for epitaxy (Venables, 1975). However, relatively few experiments have attempted to determine this minimum temperature and its rate-dependence in semiconductor epitaxy, and considerable uncertainty remains.

2. LIMITED-THICKNESS EPITAXY

We have recently studied the low-temperature growth of Si (Eaglesham et al. 1990a) Ge/Si (Eaglesham et al. 1991a), and GaAs (Eaglesham et al. 1991b) Surprisingly, we observe a breakdown in epitaxy during growth at a fixed R and T. Fig. 1 shows a Si(100) layer grown near room temperature following growth of a high-temperature (600°C) buffer layer (to establish a clean growth surface) and deposition of a ≈3 monolayer Ge marker. (The Ge grows coherently and monatomically flat to this thickness (Eaglesham et al. 1990b), and control experiments confirm that the subsequent Si growth is not affected by the Ge marker). Nucleation of the amorphous phase clearly occurs only beyond an epitaxial thickness, h_{epi}. Limited-thickness epitaxy occurs in this system over a wide range of deposition rates and temperatures, with the thickness increasing with increasing temperature on an Arrhenius curve with a rate-dependent "activation energy", as shown in Fig. 2. A similar Arrhenius behaviour is seen in Ge/Si , but in GaAs the effect is complicated by an apparent discontinuity in the curve This may be linked to earlier measurements of a low-temperature limit in GaAs MBE which was closely associated with the limit for As_4 dissociation on the GaAs surface (Neave and Joyce 1978).

Fig. 1 Breakdown in epitaxy during deposition at fixed temperature and rate. [110] HREM cross-section showing low-temperature growth after deposition of a thin Ge marker layer.

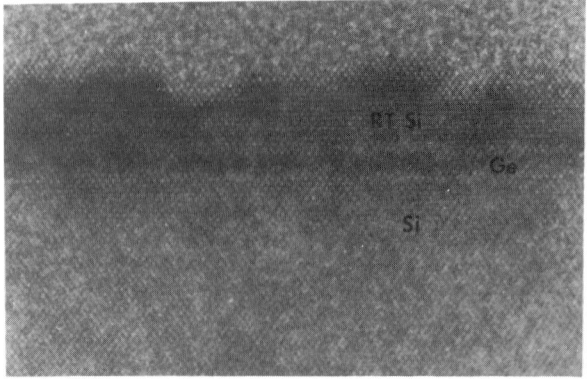

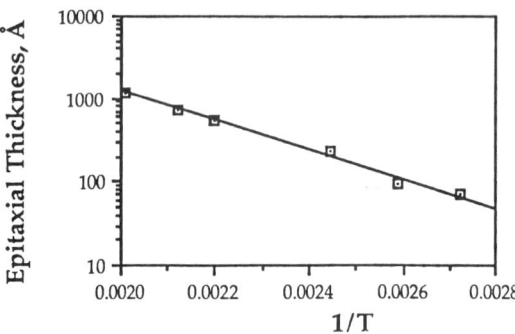

Fig. 2 Temperature dependence of the epitaxial thickness in Si(100) homoepitaxy, from a sequence of experiments similar to Fig.1. The activation energies corresponding to growth rates of 0.7 and 11 Å/s are 0.4 and 1.5 eV respectively.

3. POSSIBLE EXPLANATIONS FOR THE BREAKDOWN IN EPITAXY

It should be stressed that the observation of formation of the amorphous phase after low-temperature growth of some thickness requires a radical rethinking of the nature of epitaxial vs. amorphous deposition. Measurement of the density of amorphous zones in these films (using, for convenience, higher-temperature layers where h_{epi} is ≈ 2000Å) show that the formation of the amorphous phase is an abrupt process: we are not concerned with either continuous deterioration of crystalline quality or a constant probability of nucleation of the amorphous phase (see e.g. Jorke et al. 1990). The growing surface must therefore change either structurally or chemically between the initial high-temperature buffer surface (where we see good epitaxy) and the surface which finally nucleates the amorphous phase. Structural defects seem unlikely: we do not in general see extended defects prior to formation of the amorphous phase (although in layers grown at high rates and high temperatures we do observe the void chains reported by Perovic et al. (1990)). We have shown that even at moderate doping levels electrical dopants such as Sb are 100% active (Gossmann et al. 1990), so that point defect concentrations must also stay low. Chemical modification of the surface, particularly by gas-phase contaminants seems more probable, In particular, very high pressures of H_2 (10^{-5}Torr) have a record of causing H segregation under usual Si MBE conditions (550°C) which ultimately causes the breakdown of epitaxy even at these high temperatures.   Like most UHV chambers, the three Si MBE systems used in these studies are all dominated by H_2 in the 10^{-10} Torr range, so it is difficult to eliminate H segregation as a possible cause. However, several results suggest that H does not play a role in limited-thickness epitaxy. First, increases in the H_2 partial pressure by nearly an order of magnitude do not affect the observed h_{epi}. Second, SIMS (secondary ion mass spectrometry) does not show detectable H segregation during low-temperature growth (i.e. $< 10^{18}$cm^{-3} peak concentrations). In addition, we have recently been studying growth on H-terminated Si as an alternative to the Shiraki cleaning technique. Dipping wafers in either dilute HF or buffered oxide etchant (7:1 NH_4OH:HF,

BOE) leads to a perfectly H-terminated surface which resists oxidation even on air-exposure (Higashi et al. 1990). Even with a complete monolayer of H we appear to be able to carry out Si epitaxy at temperatures down to 370°C (Eaglesham et al 1991c), *below* the H desorption temperature (Schulx and Henzler 1983). This tends to reinforce our conclusion that H segregation at the low concentrations observed in SIMS is not sufficient to suppress epitaxy.

4. SURFACE ROUGHNESS IN LOW-TEMPERATURE Si(100) EPITAXY

We conclude that the evolution of the surface which is responsible for the breakdown in epitaxy at h_{epi} is not bulk structural defects (either point or extended) or surface chemistry. By a process of elimination, this would suggest that surface defects (such as steps, double steps, facets onto (111) or 2x1 domain boundaries) are responsible. This implies a direct link between increasing surface roughness during growth at temperatures where surface diffusion is extremely slow and the nucleation of the amorphous phase. Making this link explicit, however, is proving a considerable challenge. Indirect evidence for increased surface roughness during low-temperature growth comes from RHEED oscillation studies (Aarts and Larsen 1988): the heavy damping of RHEED oscillations demonstrates that the first few monolayers have rapidly increasing step densities. However, RHEED cannot provide evidence that surface roughness is still increasing after 100-1000Å (where breakdown typically occurs but RHEED oscillations have long since damped out). To study roughness at these much larger thicknesses we have recently grown epilayers containing a sequence of Ge monolayers up to well beyond h_{epi}. The Ge marker should approximately follow the surface morphology at that time, so that the width of a monolayer imaged in TEM should represent an average through the sample of the surface morphology at a given point in time (this width is obviously convoluted with an instrument response: for typical large-defocus bright-field images this image resolution will only be 10-15Å). Fig. 3 shows a (400) BF image of a typical sample. In this case after an initial clean (leaving a visible line of residual carbides) we grew a high-temperature Si buffer (900°C), a 900°C Ge marker, a Si layer during cooling to 350°C, and a series of Ge markers through a thick Si layer at 350°C. The diffusional broadening of the 900°C Ge demonstrates that these images are extremely sensitive to broadening of the Ge profile. In subsequent Ge markers we expect no diffusion, and the ≈15Å width of the first Ge line is probably limited by the image resolution at large defocus. The width of subsequent lines should be a measure of the mean height deviation of the surface morphology. The substantial increase in width of the last two markers before formation of the amorphous phase demonstrates for the first time that massive amounts of roughening occurred before the breakdown in epitaxy. In Fig. 4 we show quantitative measurements from Fig. 3: the image was digitised and integrated parallel to the Ge lines, and a series of Gaussian fits made to the intensity profile of each line: the width of the last two lines (35Å) suggests that the surface roughens to have a standard deviation of 5-10 monolayers before the amorphous phase forms.

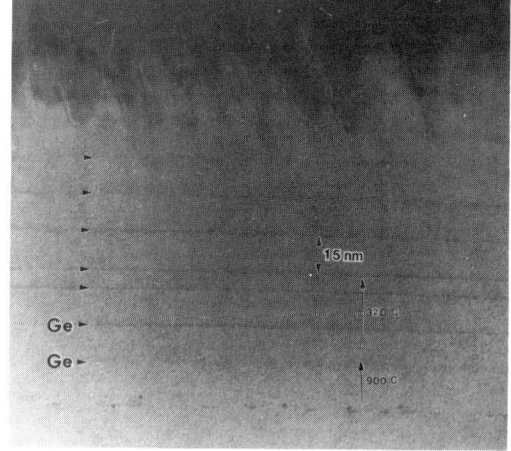

Fig. 3 Sequence of Ge markers through a high-temperature buffer and a low-temperature layer. Shortly before the amorphous phase forms the surface roughness increases to give a Ge marker width nearly equivalent to the diffusion in the high-temperature buffer. Bright-field image near the (400) 2-beam condition, large defocus.

While the observation of massive roughness near the epitaxial thickness does not conclusively demonstrate that roughness is linked to the nucleation of the amorphous phase, the results at least show that the mechanism we have proposed is reasonable. We are currently exploring a wide variety of experimental and modelling techniques in an attempt to demonstrate the link more conclusively, and to try to establish an atomic mechanism whereby roughness influences the structure of subsequent layers. Specific surface structures which may be linked to formation of the amorphous phase are {111} facets (which could definitely be linked to breakdown for Si(100), but are more problematical for GaAs and Si(111)), 2x1 domain boundaries or a step-induced change in the local reconstruction (again, likely on (100), but rather difficult to generalise into a more universal form of behaviour), and rare step configurations associated with a catastrophic degree of roughness (such as groups of double-height steps or, in the limit, void formation of the kind seen on a large scale for higher-temperature growth).

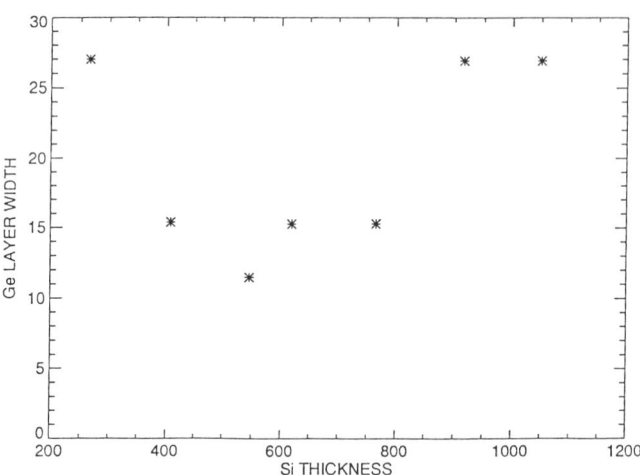

Fig. 4 Fitted Gaussian widths of the Ge markers imaged in Fig. 3, suggesting an effective resolution of $\approx 15\text{Å}$ in the large-defocus image and a $\approx 35\text{Å}$ broadening prior to breakdown.

In conclusion, we have shown that at low temperature most semiconductors undergo limited-thickness epitaxy, with growth being initially epitaxial on a flat, clean surface but breaking down into amorphous deposition later. We believe that the later formation of the amorphous phase is linked to surface roughness, and we have shown that low-temperature growth can lead to massive surface roughness prior to the breakdown of epitaxy.

REFERENCES

Aarts J and Larsen PK, 1988 in *RHEED and Reflection Electron Imaging of Surfaces*, Ed. PK Larsen and PJ Dobson (Plenum, New York), p. 449.
Eaglesham DJ, Gossmann H-J, and Cerullo, M 1990a *Phys. Rev. Lett.*, **65**, 1227.
Eaglesham DJ and Cerullo M 1990b Phys. Rev. Lett. **64**, 1943.
Eaglesham DJ, and Cerullo M 1991a *Appl. Phys. Lett.* to be published.
Eaglesham DJ, Pfeiffer LN, West KW, and Dykaar DR 1991b Appl. Phys. Lett. **58**, 65.
Eaglesham DJ, Higashi GS and Cerullo M 1991c *Appl. Phys. Lett.* to be published.
Gossmann H-J, Schubert EF, Eaglesham DJ, and Cerullo M, *Appl. Phys. Lett.* **57**, 2440.
Higashi GS, Chabal YJ, Trucks GW, and Raghavachari K, *Appl. Phys. Lett.* **56** 656 (1990).
Jorke H, Herzog H-J and Kibbel H, 1989 *Phys. Rev.* B **40**, 2005.
Perovic DD, Noel J-P Houghton DC and Weatherly GC 1990, *Proc. 1st Topical Symp. on Si based heterostructures*, AVS Toronto.
Neave JH and Joyce BA, 1978, J. Cryst. Growth **43**, 204
Schulx G and Henzler M 1983, *Surf. Sci.* **124**, 336.
Venables JA 1975 in *Epitaxial Growth, Part B*, Ed. JW Matthews, Academic Press, New York, p. 389.
Wolff SH, Wagner S, Bean JC, Hull R, and Gibson JM, *Appl. Phys. Lett.* **55**, 2017 (1989).

Inst. Phys. Conf. Ser. No 117: Section 7
Paper presented at Microsc. Semicond. Mater. Conf., Oxford, 25–28 March 1991

435

Microvoid formation in low temperature molecular beam epitaxy-grown silicon

G C Weatherly[+], D D Perovic[*], J -P Noël[#] and D C Houghton[#]

+ Department of Materials Science and Engineering, McMaster University, Hamilton, Canada L8S 4L7.
* Cavendish Laboratory, University of Cambridge, Cambridge, U.K.,CB3 0HE.
# Institute for Microstructural Sciences, National Research Council, Ottawa, Canada K1A 0R6.

ABSTRACT: When $(100)_{Si}$ crystals are grown by molecular beam epitaxy at low temperatures ($< \sim 400°C$), the growing interface no longer remains planar, but develops a series of cusps. At the base of each cusp, a small ($\sim 3nm$) cylindrically-shaped defect region is formed, but this is morphologically unstable and breaks down to form a linear array of spherical defects trailing behind the interface. Detailed electron microscopy studies have shown that these defects are voids. The high vacancy concentration introduced by the molecular beam epitaxy process is occluded with a characteristic motif which parallels that found in many phase transformations in binary alloys.

1. INTRODUCTION

The goal of precise control of dopant distributions at the atomic level during growth of Si by molecular beam epitaxy (MBE) has led many groups to consider just how far the growth temperature may be lowered in practice. In their study of low temperature Si-M B E growth, Jorke et al (1989 a, 1989 b) observed the breakdown of epitaxial growth as a function of Si deposition rate for both constant and variable temperature growths. More recently Eaglesham et al (1990) confirmed the existence of a limiting epitaxial thickness which exhibited an exponential temperature dependence as theoretically predicted by Jorke et al (1989 a). A single crystal to polycrystalline transition has also been observed in low temperature (200°C) growth of GaAs (Liliental - Weber 1990). In this contribution we report on the generation and characterization of microvoid regions within low temperature M B E -Si which are formed prior to the breakdown of epitaxial growth.

2. EXPERIMENTAL PROCEDURE

The Si-based structures of this study were grown in a VG Semicon V80 M B E system using deposition rates in the range of 0.1-1 nm/s on (100) Czochralski Si substrates rotated at 30 rpm. The substrate surface prior to growth was cleaned in-situ by heating to above 850°C under a 0.01 nm/s Si flux to remove a UV-grown oxide. The substrate temperature was calibrated above 350°C with an optical pyrometer (bandwidth: 0.9-1.08 μm). The heater power during growth was regulated using a thermocouple in contact with the Ta heater assembly and the substrate temperature for growth runs below 350°C was estimated from an extrapolation of the pyrometer calibration. Samples grown both at constant temperature and with a decreasing temperature ramp were examined. Samples were studied using transmission (TEM) and field-emission scanning-transmission electron microscopy (STEM) in cross-sectional and plan-view geometries; thin foils were prepared using argon atom milling.

3. RESULTS

a) Temperature Ramp Experiments

The result of a typical temperature ramp experiment is illustrated in Fig. 1. As the growth temperature is lowered, the transition from a perfect single crystal through a polycrystalline film to an amorphous phase is found. For the particular growth conditions used in this experiment, the transition occurred at approximately 200°C, although as both Jorke et al (1989 a) and Eaglesham et al (1990) have shown, not only the growth temperature but also the growth conditions and epitaxial layer thickness are important variables in determining this transition. A close examination of Fig. 1 shows that in a critical temperature range above that required for polycrystalline film growth, a series of aligned cylindrical or spherical defects, a few nm in diameter, are formed. These cylindrical channels are often observed to continue through to the amorphous phase.

b) Constant Temperature Growth

All samples grown in the temperature range from 300 - 400°C contain cylindrical or spherical defects aligned in the growth direction (Fig. 2). The defects are associated with cusps at the growth front, each cusp leaving cylindrical or spherical defects in its wake as growth proceeds. The cylindrical defects are morphologically unstable and undergo a classical Rayleigh instability, forming a series of aligned spherical defects by a process which is probably surface diffusion controlled.

c) Defect Identification

Two techniques were used to identify the nature of the defects; contrast experiments in the electron microscope and variable-energy positron annihilation spectroscopy (Perovic et al, 1991 a). Both demonstrated unequivocally that the defects are voids. The defects show strong structure-factor but no strain field contrast (Fig. 3). By comparing the contrast behaviour with the predictions from dynamical theory (van Landuyt et al 1965; Howie and Hutchison 1986), the effective shift in extinction distance was shown to be consistent with voids. The possibility

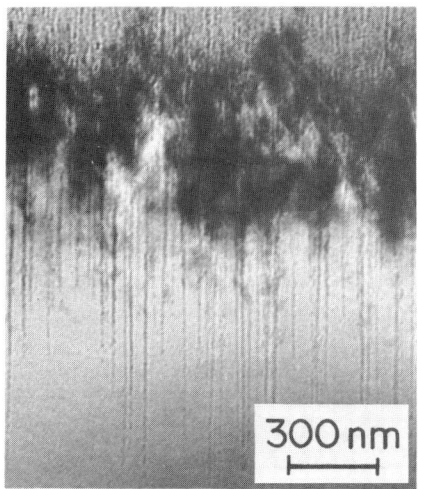

Figure 1 Temperature ramp experiment showing single crystal to amorphous transition. Note void regions close to polycrystalline film.

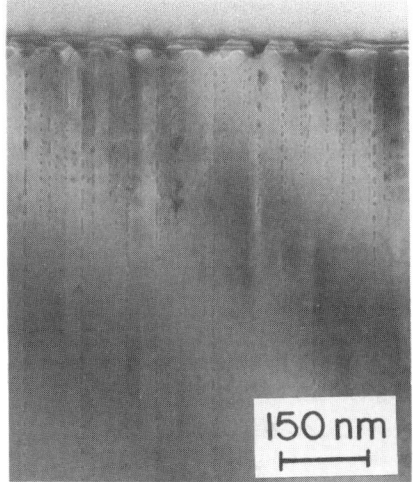

Figure 2 Spherical or cylindrical voids forming at cusps on growth front. Growth temperature =400°C.

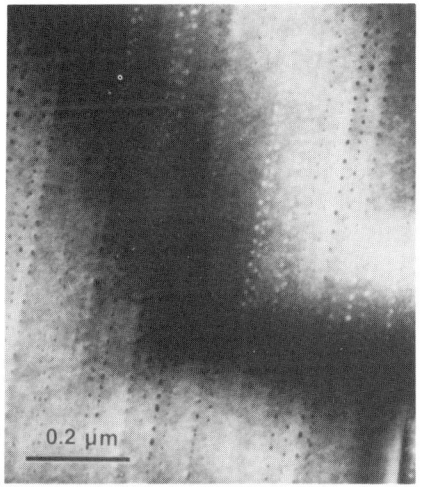

Figure 3 Structure factor contrast from rows of spherical voids.

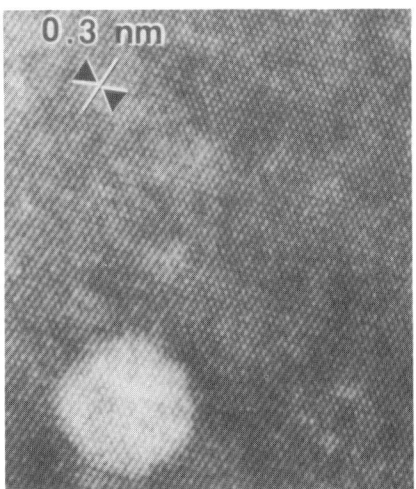

Figure 4 High resolution ([011], 7 beam) image of voids after a post-growth anneal of 30s at 700°C.

that the defects are regions of an amorphous phase can be discounted. Annealing of samples for 30s at 700°C (well above the temperature at which solid phase epitaxial regrowth of amorphous Si should occur) produced some faceting on {111} (Fig. 4), but the structures were otherwise identical to those found in the as-grown samples.

4. DISCUSSION

At the relatively fast growth rates and low growth temperatures typical of Si-MBE a supersaturation of point defects is introduced during growth. In our experiments, it appears that in a critical temperature range (300 - 400°C) this excess vacancy concentration is occluded during growth, in association with the development of a non-planar growth interface (see Fig.2). Voids which form at each cusp on the interface act as efficient vacancy sinks. As growth proceeds, a regular void structure develops, forming a hexagonal array with an average interpore spacing of about 50nm. (This corresponds to the annihilation of an excess vacancy concentration of about 0.5%)

The behaviour is very reminiscent of that found in cellular solidification, in the solidification of rod eutectics and monotectics and in discontinuous precipitation (Porter and Easterling 1981). In each of these transformations, solute is redistributed at (or slightly ahead of) the growth interface to produce an aligned structure, parallel to the growth direction. A very rough measure of the effective diffusivity (D) governing the redistribution of solute (or vacancies in our case) is given by $D \sim xv$, where x is the pore spacing and v is the growth velocity. Substituting in x = 50 nm and v = 0.5 nm/s, D is approximately 2.5×10^{-11} mm^2/s. This is very much larger than the rate of volume diffusion of Si in the temperature range of 300 - 400 °C, even allowing for a high non-equilibrium vacancy concentration. In turn this suggests that the transport probably occurs through a narrow region of relatively high disorder at the growth front (Perovic et al 1991 b).

REFERENCES

Eaglesham D.J., Gossman H.J. and Cerullo M. 1990 Phys. Rev. lett. 65 1227
Howie A. and Hutchison J.L. 1986 J. Microsc. 142 131
Jorke H., Herzog H.J. and Kibbel H. 1989a Phys. Rev. B. 40 2005
Jorke H., Kibbel H., Schäffler F. and Herzog H.J. 1989b Thin Solid Films 183 307
Liliental-Weber Z. 1990 Mat. Res. Soc. Symp. Proc. 198 371
Perovic D.D., Weatherly G.C., Simpson P.J., Schultz P.J., Jackman T.E., Aers G.C., Noël J.P. and Houghton D.C. 1991a Submitted to Phys. Rev. B. Communications
Perovic D.D. , Weatherly G.C. and Kirkaldy J.S. 1991b to be published
Porter D.A. and Easterling K.E. 1981 Phase Transformations in Metals and Alloys (Von Nostrand Reinhold)

Inst. Phys. Conf. Ser. No 117: Section 7
Paper presented at Microsc. Semicond. Mater. Conf., Oxford, 25–28 March 1991

The structural morphology of SiGe alloy layers grown on Si by low pressure chemical vapour deposition

A G Cullis, D J Robbins, A J Pidduck and P W Smith

DRA Microelectronics Division, Royal Signals & Radar Establishment, St Andrews Road, Malvern, Worcs WR14 3PS

ABSTRACT: The structural morphology of SiGe alloy layers grown on Si by chemical vapour deposition is examined by transmission electron microscopy and atomic force microscopy. It is shown that, for high growth temperatures and high Ge concentrations in the alloy, layers generally form with an undulating upper surface. Indeed, well-defined ripple arrays are produced which exhibit periodic strain fluctuations. This observation is correlated with theory which indicates that the formation of an undulating morphology lowers the energy of the system. It is also found that the upper surface of a Si cap deposited on such a layer rapidly becomes planar with increasing thickness, this being likely to result from the reduction in surface energy so achieved.

1. INTRODUCTION

The growth of SiGe alloy layers on Si substrates is becoming increasingly important as new device applications are foreseen. However, in order to optimise device performance, it is vital to understand and control layer growth characteristics. In the present paper, we focus on the study of the origin and nature of layer topographical distortions which can occur under specific ranges of growth conditions.

The variation in surface roughness with changes in SiGe layer composition, growth temperature and Si cap thickness is examined. It is demonstrated that strain modulations accompany SiGe layer thickness undulations and a theoretical correlation is made with the free energy of the strained layer system.

2. EXPERIMENTAL DETAILS

Layer growth in the present work was carried out by low pressure chemical vapour deposition in a reactor with ultrahigh vacuum background pressures (Robbins and Young 1987). Wafers of 76mm-diameter (001) Si were used as substrates after precleaning by 3min exposure to UV-ozone. Within the reactor, immediately after thermal desorption of the passivating oxide at $900^{\circ}C$ in 130Pa hydrogen, epitaxy was performed by thermal decomposition of silane- and germane-hydrogen mixtures at 13Pa total pressure and a substrate temperature of $750^{\circ}C$. After growth of a nominally 120nm Si buffer layer the SiGe alloy was grown, sometimes followed by deposition of a thin Si cap and further alloy layers. The evolution of surface topographical features was monitored in real time by *in situ* laser light scattering (Robbins et al 1987). Post growth characterisation primarily relied upon plan-view and cross-sectional transmission electron microscopy (TEM) carried out using a JEM 4000EX instrument. Observations were also made by atomic force microscopy (AFM) using a Digital Instruments Nanoscope II system equipped with a microfabricated silicon nitride cantilever stylus.

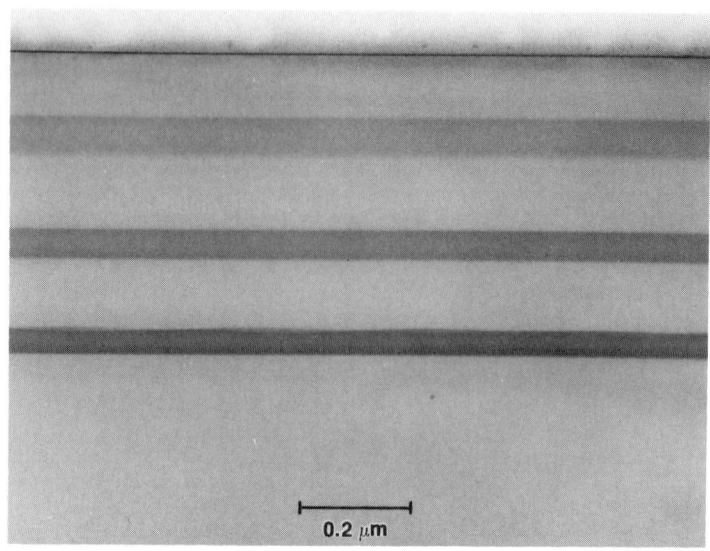

Fig. 1. Cross–sectional TEM image (bright field, **g**=004) showing sample with three SiGe alloy layers (dark contrast) grown at 750°C. The lowest layer has an undulating upper surface and contains the highest Ge concentration.

3. RESULTS

It was found that, under certain GeSi layer growth conditions, topographical undulations were formed at alloy layer surfaces. In Fig. 1, a cross–sectional TEM image shows three SiGe alloy layers with Si spacers and a Si cap, the whole structure having been grown at a temperature of 750°C. The Ge concentration increases in steps from ~10% in the top layer to ~20% in the bottom layer. Note that the top two alloy layers are relatively uniform and planar. However, the bottom layer exhibits ripples on its upper surface with a periodicity of a few hundred nanometers. Therefore, it is concluded that, at a given growth temperature, high concentrations of Ge in an alloy predispose layers to the occurrence of surface undulations. Also, it is clear that the growth surface is flattened by subsequent Si overgrowth.

Figure 2 presents plan–view and cross–sectional TEM images showing a distribution of ripples across a single sample with SiGe alloy (50nm) and Si capping layers grown at 750°C. In (a) is shown a bright–field [001]–axial multibeam image in which the SiGe alloy ripple array is clearly visible. The ripples lie along orthogonal <100> directions and form an interlocking network. Strain variation is produced by the presence of ripples and this is most clearly shown in two–beam 400 and 040 images (b) and (c) in which only ripples normal to the **g**–vector of the operating reflection give linear dark/bright contrast in each case. The periodic lattice strain is, thus, directed normal to the ripple rows. The actual ripple structure is shown in cross–section in (d) with a [100] foil normal. However, when the 040 reflection with **g**–vector parallel to the surface is strongly excited, once again periodic strain contrast (dark/bright vertical stripes) is given by the ripple array (e) due to the distortions of the vertically aligned lattice planes.

Ripples on the surfaces of samples such as those just described are very clearly revealed by use of AFM imaging. Figure 3 presents an AFM image showing ripples on the surface of an uncapped sample with a SiGe layer (Ge concentration ~20%) grown at 750°C. In agreement with the TEM observations, the ripples lie along orthogonal

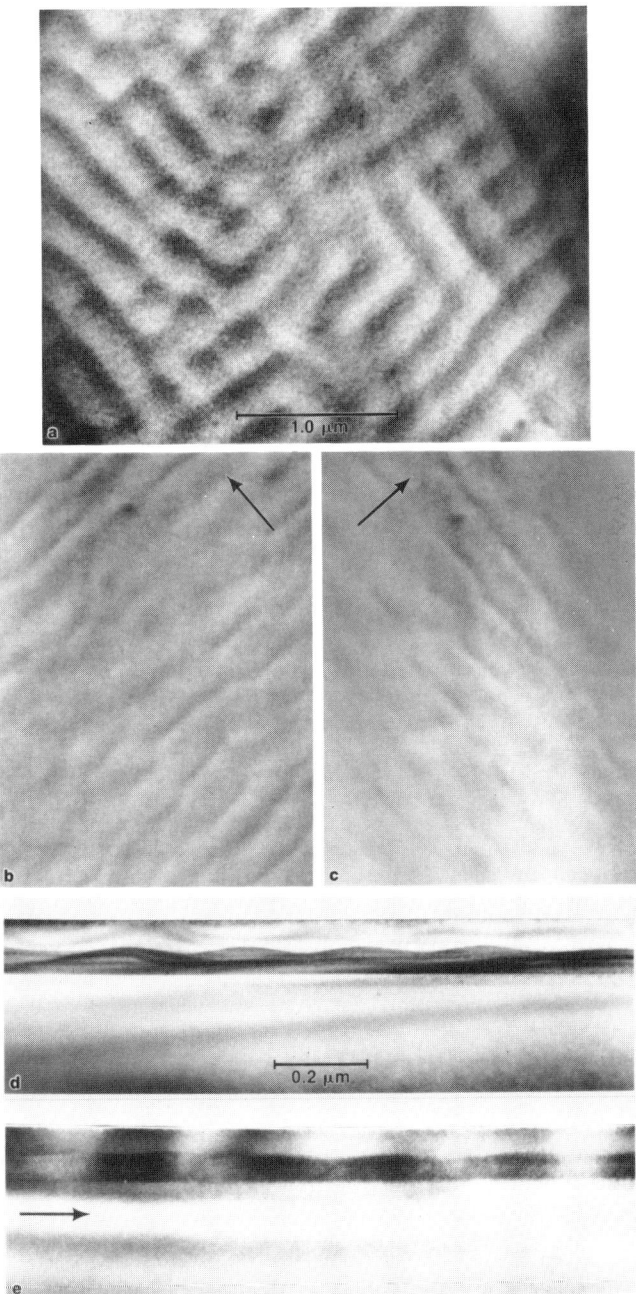

Fig. 2. TEM images demonstrating distribution of ripples across a single sample with SiGe alloy (50nm) and Si capping layers grown at 750°C: (a) plan–view, bright field, [001] multibeam image showing SiGe alloy ripple array; (b) and (c) plan–view, bright–field images obtained using orthogonal 400 and 040 reflections (g–vectors indicated) showing variation in ripple contrast; (d) and (e) cross–sectional, bright–field images showing ripple configuration, (e) using in–plane 400–type reflection.

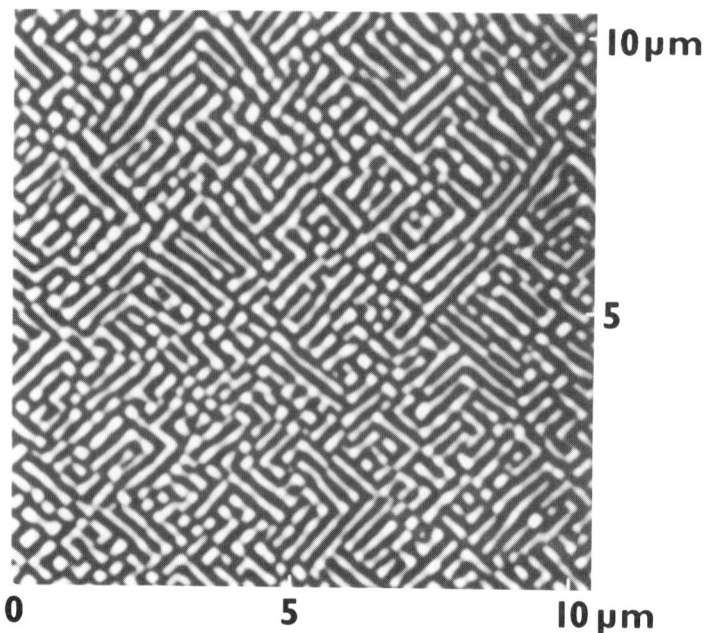

10 µm

5

0   5   10 µm

Fig. 3. AFM image of uncapped SiGe alloy layer surface showing intersecting <100>-aligned ripples. The peak–to–valley heights (white–to–black) are in the range 20–40nm.

<100> directions and have a periodicity of ~300nm. Indeed, the ripple pattern results in the maxima forming discrete SiGe alloy islands. Furthermore, the ripple amplitude is measured by AFM to be 20–40nm, in agreement with that observed by TEM. It should be noted that the relatively pointed form of the ripple minima is probably an artefact depending upon the stylus diameter, which was shown by scanning electron microscopy to be ~50nm.

The cross–sectional TEM images of Fig. 4 show four SiGe alloy layers, each with a different thickness of Si cap overgrowth. All layers were deposited at 750°C and the Ge concentration in each alloy layer was in a range from ~20% to ~25%, thus ensuring that ripple formation took place. In (a) there is no Si cap and the alloy surface terminates in a regular array of ripples. In (b), (c) and (d) there is a gradually increasing Si cap thickness and it is immediately evident that the first Si to be deposited predominantly fills in the ripple minima (eg at X in (b)) while little Si covers the ripple maxima. This process continues until, at a Si thickness of ~30nm, the alloy ripples are completely covered and little upper surface roughening remains.

In complete agreement with the TEM work just presented, the series of AFM plots shown in Fig. 5 demonstrates the rapid smoothing which occurs as an undulating SiGe layer is capped with Si. The plot in (a) shows the deep undulations present in the initial uncapped alloy layer. After sufficient Si deposition almost to fill the SiGe layer depressions (b), the peak–to–valley amplitude has decreased by a factor of four to 5–10nm. At this stage, the periodicity is smeared to about 500nm, indicating a significant surface Si atom diffusion length. After ~60nm of Si deposition (c), the roughness amplitude is reduced to less than 2nm: a rippled texture is still faintly evident with a periodicity of ~1 µm and this is believed to be characteristic of a step flow growth mechanism (Pidduck et al 1989). (The cylindrical form of plot (c) is an artefact of the

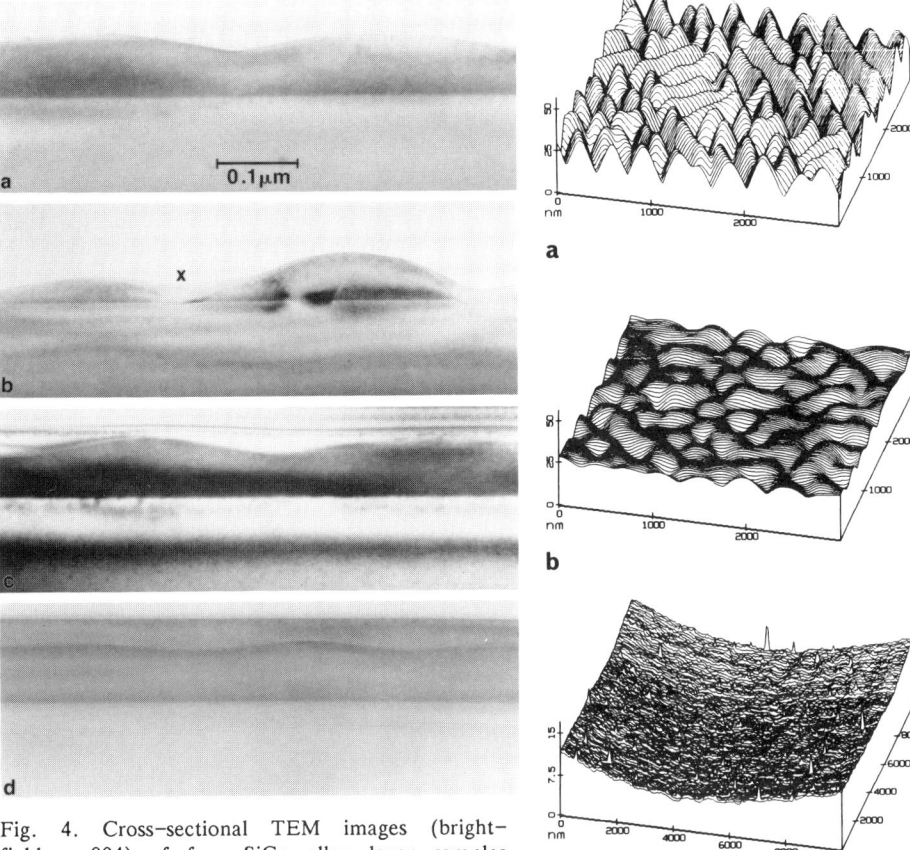

Fig. 4. Cross–sectional TEM images (bright–field, g=004) of four SiGe alloy layer samples with increasing Si cap thickness from (a) to (d). Note that deposited Si preferentially fills in depressions, as at X in (b).

Fig. 5. AFM plots showing surface smoothing as SiGe layer is capped with Si: (a) uncapped alloy surface; (b) and (c) increasing Si cap thickness.

AFM scanner.) The results just outlined also correlate well with *in situ* laser light scattering measurements during growth (Robbins et al 1991).

4. DISCUSSION AND CONCLUSIONS

It is clear from the TEM and AFM work presented here that pseudomorphic (strained) SiGe alloy layers grown on (001) Si can, depending upon Ge concentration and growth temperature, show surface undulations which are correlated with layer strain fluctuations. For growth of a 20% Ge alloy layer at lower temperatures (~600°C) the fluctuations appear random in shape and distribution although, for growth at 750°C, they form periodic ripples which lie along [100] and [010] directions. However, no undulations or strain fluctuations were observed, using TEM, in 10% Ge alloy layers grown at this latter temperature.

The elastic strain energy in an alloy layer of constant thickness has been compared with that in a similar layer having periodic thickness fluctuations. The strain energy in the latter is reduced when the lattice planes are expanded in thicker regions and contracted in thinner regions, measured in relation to the unstrained Si substrate, ie when the periodic shear strain correlates with the thickness fluctuations.   The driving force inducing thickness fluctuations in GeSi alloy layers increases with the strain in the layer, ie with the Ge concentration.   However, growth in this low energy form requires that surface diffusion distances are large enough to allow the thickness variations to build up. Therefore, the tendency to form such surface distortions increases with Ge concentration and with growth temperature.   The reduction in strain energy due to the presence of the fluctuations is offset by an increase in surface energy arising from the higher density of surface steps.   The reduction in this surface energy is the driving force for the rapid planarization of an epitaxial Si capping layer grown on an undulating alloy surface.

ACKNOWLEDGEMENT

We would like to thank Dennis Haslop (Environmental Monitoring Systems) for making the AFM measurements.

REFERENCES

Pidduck A J, Robbins D J, Young I M and Patel G 1989 Thin Solid Films $\underline{183}$ 255
Robbins D J, Cullis A G and Pidduck A J 1991 J Vac Sci Technol B $\underline{9}$
Robbins D J, Pidduck A J, Cullis A G, Chew N G, Hardeman R W, Gasson D B, Pickering C, Daw A C, Johnson M and Jones R 1987 J. Crystal Growth $\underline{81}$ 421
Robbins D J and Young I M 1987 Appl. Phys. Lett. $\underline{50}$ 1575

Inst. Phys. Conf. Ser. No 117: Section 7
Paper presented at Microsc. Semicond. Mater. Conf., Oxford, 25–28 March 1991

445

TEM study of strain relaxation processes in metastable Si/SiGe/Si structures for heterojunction bipolar transistors (HBTs)

M Hockly, CG Tuppen, CJ Gibbings, ASR Martin, ZA Shafi* and P Ashburn*

British Telecom Research Laboratories, Martlesham Heath, Ipswich IP5 7RE, UK.
*Department of Electronics and Computer Science, University of Southampton, Southampton SO9 5NH, UK.

ABSTRACT: Control of strain relaxation in SiGe HBT structures is desirable because of the potentially adverse effect of mismatch dislocations on device performance. Various test devices were formed based on an MBE-grown planar heterostructure which is metastable with respect to strain relaxation. TEM studies showed the predominant mismatch dislocation geometry to be the three-segment (dipole) configuration. Ion implantation and annealing greatly enhanced dislocation densities and produced cross slip. A particular geometry was shown to produce asymmetric strain relaxation.

1. INTRODUCTION

The SiGe HBT has received considerable attention in recent years as a way of achieving very high speed operation in a Si-based structure. However, the strain in the SiGe base layer makes such HBT structures liable to mismatch dislocation formation, either during growth or in the course of subsequent device processing. Even in cases where the level of relaxation produced by such dislocations is extremely small, the dislocation spacing can be less than typical lateral device dimensions, in which case most devices are likely to contain one or more dislocations, with potentially adverse effects on device performance. Results are presented of a TEM study of relaxation behaviour in various test device regions formed in a metastable Si/SiGe/Si heterostructure (Fig. 1). The effects of differences in mesa geometry, thermal treatment and ion implantation are described and interpreted.

Si	$0.3\,\mu m$	n	emitter
$Si_{0.85}Ge_{0.15}$	$0.1\,\mu m$	p	base
Si		n	collector

Fig. 1. HBT planar structure

2. THEORY

The approach generally used to determine the stability at equilibrium of strained layer structures is described by Matthews and Blakeslee (1974). For the simple case of a single strained layer only (Fig. 2a) they consider the driving stress, σ_ε, arising from mismatch strain, and the thickness-dependent retarding stress, σ_d, arising from dislocation line tension, acting on the threading segment of a glide dislocation in a {111} plane. If the effective stress, $\sigma_{eff} = \sigma_\varepsilon - \sigma_d$, is positive then the layer is unstable with respect to relaxation by interfacial dislocation formation. The rate of relaxation of an unstable layer is limited by factors such as the rates of dislocation formation and multiplication, and the glide velocity; these factors are dependent on parameters including temperature and σ_{eff}. As a result, in many cases including the present, metastable structures can be expected, i.e. ones which are liable to significant relaxation in the equilibrium state but in which little or no relaxation has occurred in practice because of kinetic limitations (Dodson and Tsao 1987). Thus, for example, Bean et al (1984) showed that for SiGe growth at a typical temperature of 550°C, metastable layers many times the equilibrium critical thickness can be obtained.

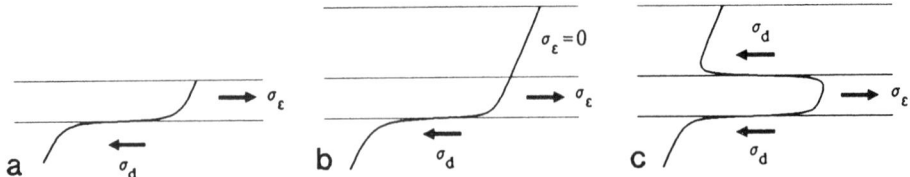

Fig. 2. Mismatch relaxation mechanisms. a) Two-segment mechanism, single strained layer; b) Two-segment mechanism, buried strained layer; c) Three-segment or dipole mechanism, buried strained layer.

The general approach outlined above can be applied to typical HBT heterostructures, as described in Tuppen and Gibbings (1990). Two basic dislocation configurations are possible, the two-segment version (Fig. 2b) and the three-segment or dipole version (Fig. 2c). For the present structure (Fig. 1), calculations show that the complete structure is metastable with respect to relaxation via the dipole mechanism. During growth, the SiGe layer becomes metastable when the thickness exceeds the critical value of ~20 nm (mechanism of Fig. 2a), and in the early stages of growth of the capping layer the structure is metastable with respect to the two-segment mechanism, before the dipole mechanism becomes dominant.

3. EXPERIMENTAL

Growth of the Si/SiGe/Si structure was carried out on a Si (001) substrate at 535°C and at a growth rate of 0.1 nm s^{-1} in a VG Semicon V80 MBE growth system fitted with an Inficon Sentinel III flux controller. The wafer was processed using a test pattern layout, to produce various HBT test devices and other test areas. Device and test areas were defined by surrounding trenches plasma etched to below the level of the SiGe layer, and are henceforth referred to as mesa areas. A limited number of test areas were studied by TEM and the relevant process treatments are referred to in the description of results. After stripping off metallisations and oxides, plan view TEM specimens were prepared in a conventional manner using mechanical dimple polishing and ion beam milling. TEM examination was carried out in a JEOL 200CX electron microscope operated at 200kV.

4. RESULTS

The extent of relaxation was generally very small but varied between areas that had been subjected to different processing. The overwhelmingly dominant configuration of mismatch dislocation corresponded to the dipole form (Fig. 2c), as was confirmed by tilting experiments and other observations. As seen in [001] plan view (Fig. 3), the dipole pairs had a characteristic fixed spacing determined by their location on {111} planes and the thickness of the SiGe layer. The interfacial segments are 60° in character. The dipoles extended over considerable distances; they could generally be tracked in the TEM for distances of 100 μm and more, and few examples of turnover points with threading segments in the SiGe layer (A in Fig. 3) were observed. A limited number of the dislocations were not in the form of individual dipoles, but occurred in groups of 5 - 15 dislocations. These groups usually contained several dipoles, but also a number of other less regular dislocations, not all interfacial, some of which were observed to bend and thread up to the surface of the capping layer.

Most of the abovementioned features were found to be generally typical of the mismatch dislocations observed, both within mesa areas and in the larger areas outside (which still comprise the complete grown structure). In the latter areas typical spacings between adjacent dislocation dipoles were ~15 μm; this corresponds to relaxation of only ~0.2%. Within test mesa areas, higher linear densities and additional phenomena were often observed, some of which will now be described.

One area examined was a square mesa of dimension ~140 μm, which had been subjected to a dual p^{+} implant (B^{+}, 120kV, 2x10^{15} cm^{-2}; BF^{+}, 35kV, 2x10^{15} cm^{-2}) over virtually all of its area. This produces a p^{+} doping profile extending from the surface down to the SiGe (base) region and is used to form the base contact in the HBT devices. The implant was subjected to a rapid thermal anneal at 825°C for 10s. The defect structure in this type of implanted mesa was markedly different from that in unimplanted regions. There was a very high density (~10^{11} cm^{-2}) of small loops (up to 50 nm in diameter) and other fine scale defects. Mismatch dislocation dipoles were present in much higher densities (typical spacing ~1 μm) than in unimplanted regions (spacing ~15 μm), and a high incidence of cross slip was observed (Fig. 4). The vast majority of the dipoles extended, often via one or more cross slips, to termination points at the mesa edges (trench walls). The observed dislocation spacing of ~1 μm corresponds to relaxation of ~3%.

Another of the areas studied demonstrated how the dislocation configuration can be manipulated by appropriate choice of geometry and processing. A long, thin rectangular mesa, ~825 μm x 45 μm, was implanted in a limited area at each end, with the deep p^+ implant described above. The dislocation network in the main (unimplanted) area of this mesa, between the implanted end regions, is extremely asymmetric, comprising closely spaced dislocations (~1 μm apart) lying along the <110> direction parallel to the long axis of the rectangular mesa (i.e. extending between the two implanted end regions, Fig. 5), and infrequent, widely spaced dislocations (~20 μm apart) along the perpendicular <110> direction parallel to the short axis of the mesa. The asymmetry ratio is therefore ~20:1.

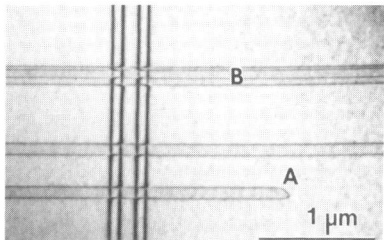

Fig. 3. TEM plan view. Typical dislocation dipoles along orthogonal <110> directions. One dislocation from each pair is in upper, one in lower interface of buried SiGe layer. A - turnover point with segment threading from lower to upper interface. B - two overlapping dipoles. BF 220.

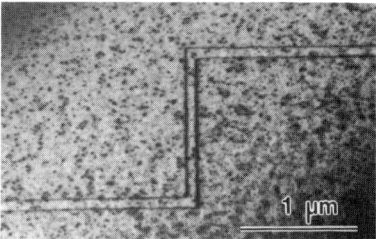

Fig. 4. TEM plan view. Ion-implanted region, showing implantation damage (loops and fine-scale defects) and dipole which has cross slipped twice. BF 220.

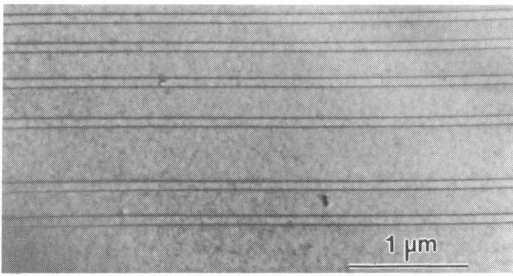

Fig. 5. TEM plan view. Highly asymmetric dislocation network (and consequent asymmetric strain relaxation) produced by enhanced dislocation nucleation in selectively implanted areas. Implanted areas to left and right of area shown. BF220.

The continuity of dislocation lines across trenches was also investigated. Because the trenches are etched to a depth below the strained layer, interfacial dislocations cannot exist in the trenches themselves. Dislocations generated during layer deposition and therefore present *before* the trench formation (the first stage of processing) will continue along the same line on either side of a trench with a short break at the trench itself. Dislocations formed by glide *after* the formation of the trenches will terminate at the wall of any trench they encounter. It was found that some of the low density of dislocations that are present outside mesa areas, continued across trenches. In particular, almost all dislocations in the form of irregular groups did so, while the behaviour of dislocation dipoles was mixed. In regions such as implanted mesas, where the density of dislocations was very much higher, the additional dislocations terminated at the trench walls. There was evidence that some dislocations may be generated at irregularities at trench walls.

5. DISCUSSION

The structure was found to be metastable, with kinetically limited relaxation occurring via the dipole mechanism and strongly affected by nucleation and glide velocity considerations. The observations are in general accord with the theoretical predictions described in section 2. The dislocations formed during growth and prior to the trench formation (i.e. those which continue across trenches and are present outside mesa areas), were relatively low in density and corresponded to a very small proportion of relaxation. A significant proportion of these were in the form of groups of 5-15 which typically included several dipoles and other individual dislocations. These match with the initial configuration often produced by heterogeneous nucleation sources during growth (Tuppen et al 1989). Other dislocations which were outside mesas and which terminated at trenches, were mostly dipoles; these are consistent with dislocation glide activity, possibly emanating from the same sources but occurring after trench formation, e.g. during the implantation anneal, at which stage the dipole mechanism is dominant.

In the ion-implanted areas, it is clear that, during annealing, the implantation damage in the strained layer region can provide nucleation sources for mismatch dislocations and that these can glide relatively rapidly over considerable distances. The density of loops is far greater than the density of dislocations. However, the critical diameter required for a loop of suitable character to nucleate a mismatch-relieving dipole is effectively the same as the critical thickness for the dipole mechanism, i.e. ~50 nm (Tuppen and Gibbings 1990). This corresponds to the maximum loop diameter observed, possibly indicating that, out of the high density of loops initially present, only the limited number larger than 50 nm were able to act as nucleation sources. According to Tuppen and Gibbings (1990), the anneal temperature and time are sufficient to produce glide over distances of 2 - 3 mm, i.e. much greater than the mesa dimension of 140 μm, even after allowance is made for some reduction in glide velocity as a result of the high defect density. This is in conformity with the observation that nearly all the dipoles generated in the mesas terminated at the trench walls. The high density of loops and similar defects with associated strain fields can clearly interfere with the glide of mismatch dislocations, and evidence of this was observed. In this context the high incidence of cross-slip is not surprising and is in agreement with other findings (Hull et al 1990).

The striking asymmetry of the dislocation network in the long thin rectangular mesa is quite consistent with the observation that suitably ion-implanted and annealed regions were efficient in nucleating dislocations. The spacing of the dislocations running parallel to the long axis of the rectangle (~1 μm) was very similar to that in the implanted square mesa, and this is consistent as the relevant dimension of the implanted end regions was similar (~140 μm parallel to the long axis of the rectangle). The spacing for dislocations running parallel to the short axis was similar to that in the regions outside the mesas. This rectangular mesa provided a clear example of how appropriate choice of geometry and processing can be used to manipulate dislocation configurations, e.g. to produce asymmetric strain, which may have potential device applications.

6. CONCLUSIONS

In a typical, metastable SiGe HBT structure, mismatch dislocation formation was found to conform well to theoretical predictions. The dipole configuration, with very few threading segments, predominated. Dislocation densities increased in ion implanted regions, but corresponded in all areas to a low percentage of strain relaxation. The production of asymmetric relaxation by particular choice of geometry was demonstrated.

The authors thank Andrew Webster for skilful TEM specimen preparation. This work was partly funded under a Department of Trade and Industry Information Engineering Directorate collaborative award.

REFERENCES

Bean JC, Feldman LC, Fiory AT, Nakahara S and Robinson IK 1984 J. Vac. Sci. Technol. A2 436
Dodson B W and Tsao J Y 1987 Appl. Phys. Letters 51 1325
Hull R, Bean J C, Bonar J M, Higashi G S, Short K T, Temkin H and White A E 1990 Appl. Phys. Letters 56 2445
Matthews J W and Blakeslee A E 1974 J. Crystal Growth 27 118
Tuppen C G and Gibbings C J 1990 J. Appl. Phys. 68 1526
Tuppen C G, Gibbings C J and Hockly M 1989 J. Crystal Growth 94 392

Inst. Phys. Conf. Ser. No 117: Section 7
Paper presented at Microsc. Semicond. Mater. Conf., Oxford, 25–28 March 1991

449

TEM characterisation of the influence of B-concentration on recrystallisation of polycrystalline silicon

J R A Carlsson, S F Gong, X -H Li, and H T G Hentzell

Department of Physics and Measurement Technology
Linköping University
S-581 83 Linköping, Sweden

ABSTRACT: B-doped polycrystalline Si films were prepared by electron beam evaporation and heat treatment in order to study the influence of B-concentration on the recrystallization of polycrystalline Si. It is shown that: (1) at a B concentration between 1 and 10 at. %, post annealing at 1100 °C results in an enhanced solid phase epitaxial growth; (2) at a B concentration higher than 10 at. %, post annealing results in a retarded recrystallization of polycrystalline Si; (3) a relatively stable alloy can form at a B-concentration of ~35 at. % up to 1100 °C.

I. INTRODUCTION

Mei. et. al. (1982) have previously shown that P could enhance the recrystallization of polycrystalline Si (polysilicon). Zheng et. al. (1987) and Ghannam et. al. (1987) have shown that this is also the case for As and B, respectively. Furthermore, our previous work [Gong. et. al. (1988)] showed that, when annealed at high temperatures, a heavily Sb-doped poly-Si film could transform to an epitaxial Si film on a (100) Si substrate, and that the SPE growth rate depended on the Sb-concentration. Campisano et. al. (1983) have shown that the maximum SPE growth rate occurs for Sb concentrations above the solid solubility. However, up to now, there are no detailed studies of the functional dependence of SPE growth versus B-concentration, especially not for concentrations above the B solid solubility in Si. In this work we have made a detailed study of polysilicon recrystallization with respect to B concentration and annealing temperature.

II. EXPERIMENTAL DETAILS

Three groups of samples were prepared as shown in Table. 1. All samples were deposited on (100) Si substrates which were cut from a Si wafer with a resistivity of ~30 Ωcm. The depostition was carried out in an electron beam evaporation system with two 20 kV electron guns.The vacuum pressure was 5×10^{-7} Torr before deposition. A quartz tube furnace was used to anneal the samples in 99.9995 % pure Ar gas. The samples were prepared for cross sectional transmission electron microscopy (XTEM) analysis by mechanical thinning and ion-milling using Ar^+-ions with an incident energy of 5 kV as described by Gong et. al. (1990). The TEM, a Phillips EM400T with a LaB_6 filament, was operated at 120 kV. The Auger electron spectroscopy (AES) equipment is a Varian 981-2001 scanning spectrometer with a base pressure of approximate 6×10^{-10} Torr. The AES was operated at 3 keV with an electron beam current of 1.0 mA and a beam size of < 10 µm. The maximum energy resolution in the single pass cylindrical mirror analyzer is 0.25 %. The samples were sputter-etched with a 3 keV Ar ion-beam for depth profiles.

III. EXPERIMENTAL RESULTS

A. Temperature dependence

Fig.1 shows the microstructure for two samples with 5 at. % B annealed at 600 and 1000 °C for 30 min, respectively. Fig. 1a shows the micrograph annealed at 600 °C. The bright field image shows that the film is amorphous, which is also the case for the as-deposited film. The arrrows in all figures indicates the interface between the films and the Si substrates. The film crystallized when it was annealed at higher temperatures than 700 °C. It is noted that the average grain size increases with increasing annealing temperature. Fig. 1b shows the micrograph of the sample annealed at 1000 °C. The bright field image shows that some alignment close to the interface has occured, indicating that the epitaxial regrowth has started.

Table. 1. Parameters of sample preparation and reaction products. Substrate: p-type (100) Si; pressure: 5×10^{-7} Torr; depostition rate: ~3-4 nm/s for Si, ~.01-.2 nm/s for B; annealing time: 30 min; film thickness: 200- 300 nm.

Molar ratio (at. % B)	Annealing temp. (°C)	Reaction products
5	600	Amorphous
5	1000	Polycrystalline, with alignment
1	1100	Epitaxial film
10	1100	Half regrown, half polycryst.
2 / 35 / 6	1100	SPE /amorphous /polycrystalline

B. Concentration dependence

Fig. 2 shows micrographs of the samples with the B concentrations of 1 and ~10 at. %, annealed at 1100 °C for 30 min. Fig. 2a shows the micrograph of the sample with 1 at. % B, indicating that the film has totally regrown. Twinning defects in the (111) direction can be seen from the bright field image. The formation of twins during SPE has been reported and discussed previously by Narayan (1982). It is noticed that the number of defects in the film

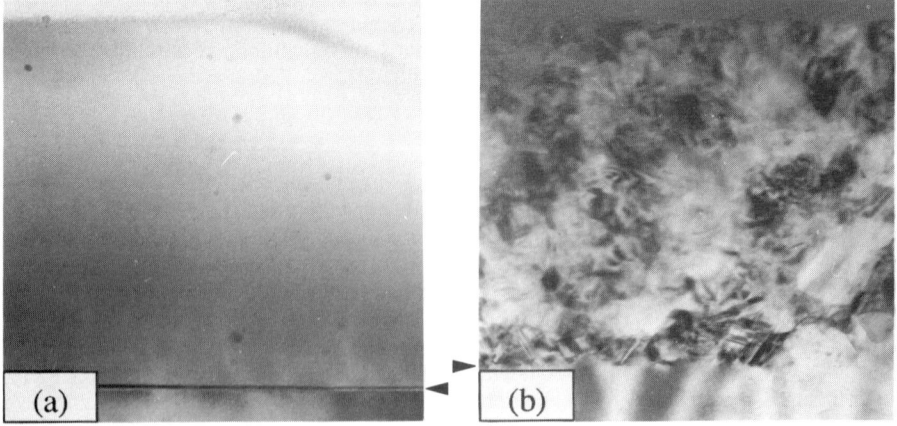

Fig. 1 (a). XTEM micrograph of the sample with 5 at.% B annealed at 600 °C. The structure is amorphous. (b). XTEM micrograph of the sample with 5 at.% B annealed at 1000 °C. The structure is polycrystalline and some alignment close to the interface can be seen.

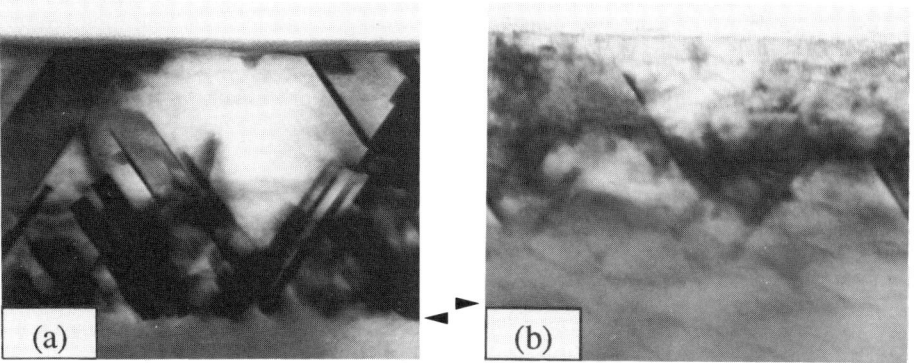

Fig. 2 (a). XTEM micrograph of the sample with 1 at.% B annealed at 1100 °C. The film is totally regrown and contains few twins. (b). XTEM micrograph of the sample with 10 at.% B annealed at 1100 °C. Half of the film is epitaxially regrown and the other half is still polycrystalline.

increases with increasing B-concentration [Carlsson et. al. (1991)]. Fig. 2b shows the micrograph of the sample with ~10 at. % B. Half of the film is polycrystalline and the other half is epitaxially regrown. It is likely that a B-pile up at the epitaxial regrowth front has supressed the SPE as discussed by Gong et. al. (1988). Fig. 3 shows the micrograph of the sample with a varied B-concentration of 2 at. % / 35 at. % / 6 at. % annealed at 1100 °C. The bright field image shows that the layer close to the interface, with 2 at. % B, is totally regrown. In the middle of the film, where the concentration is 35 at. % B, there is an amorphous Si-B alloy with some crystallites. The top layer of the film, with 6 at. % B, is polycrystalline with relatively large grains.

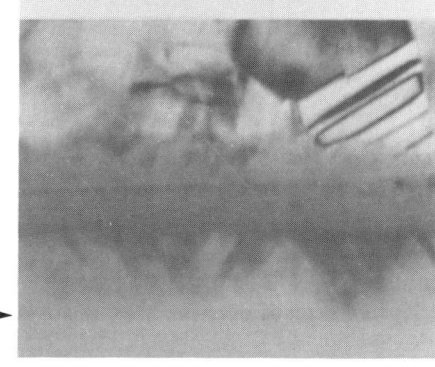

Fig.3. XTEM micrograph of the sample with a varied B-concentration of 2 at. % / 35 at. % / 6 at. % . The layer at the interface is totally regrown. The middle region, corresponding to the 35 at. % B, is amorphous. The top layer, with 6 at. % B is polycrystalline with relatively large grains.

IV. DISCUSSION

In this work we have shown that polysilicon doped with 1 - 10 at. % B regrew epitaxially on a (100) Si substrate when heat treated at 1100 °C, indicating a strong enhancement of recrystallization. We have also shown that at a B concentration above 10 at. %, the epitaxial regrowth is supressed. At B concentrations lower than 1 at. %, Kim et. al. (1990) have reported that the grain size in lightly B-doped polysilicon appeared to be slightly larger than in undoped films, indicating a slight enhancement of recrystallization of polysilicon. According to both our and Kim´s results, the influence of B concentration on recrystallization of polysilicon can mainly be divided into three regions. The first region is for B concentrations below 1 at. % (from Kim´s results). In this region the grain growth is slightly enhanced. For concentrations lower than the solid solubility of B in Si enhancement of SPE can be explained by either Fermi-level shifting or electronic excitation as shown by e. g. Adekoya et. al. (1988) and Wautelet et. al. (1988), respectively. The second region is for B concentrations between 1 and 10 at. %. In this region the SPE growth rate, i. e., the recrystallization rate, is enhanced considerably (see Fig. 2). In this concentration region Fermi-level shifting and electronic excitation is no longer valid, so the enhancement must be explained in some other way. One explanation might be that when the B-concentration exceeds the solid solubility in Si, interstitial B atoms and B precipitates decrease the Si-Si bonding energy probably by strain and atomic interaction. However the effect of strain has been ruled out by Pai and Lau (1985). Thus it seems that the atomic interaction between B and Si atoms, which decreases the Si-Si bonding strength, may be responsible for the enhancement. Our previous work has shown that increasing a metal content (up to ~ 20 at. %), e. g., Al or Ti, resulted in a decrease of the crystallization temperature of amorphous Si, indicating that incorporation of a small amount of metal spices weakens the Si-Si bonding strength. If one assumes that B plays a similar role in Si as Al (both are p-type dopants in Si), the Si-B interaction resulting in a decrease of Si-Si bonding strength is understandable. The third region is above 10 at. % B. In this region the recrystallization is supressed (see Fig. 2e and Fig. 3). This might be due to the fact that B segregates during regrowth, and a B-rich layer at the growth front prevents further growth. Moreover, when the B-concentration exceeds a certain value, e. g., ~ 35 at. % the Si-B bonding may predominate, and hence an amorphous Si-B alloy forms (see Fig. 3).The essential factor which permits the formation of an amorphous alloy is the existence of a negative heat of mixing between the two elements. The amorphous Si-B alloy observed in this study is stable up to 1100 °C, indicating a strong short range order which is difficult to break.

V. SUMMARY

The influence of B concentration on recrystallization of polysilicon has been studied in a composition range of 1 - ~ 35 at. % B, the results show:
(1) Codeposited Si samples with 5 at. % B annealed at 500 - 600 °C for 30 min are amorphous. After annealing at a temperature between 700 and 900 °C the film is polycrystalline. At a temperature above 1000 °C solid phase epitaxial growth starts.
(2) Incorporation of B in Si with a concentration between 1 and 10 at. % enhances the recrystallization of polysilicon and results in an enhanced epitaxial growth on a (100) Si substrate.
(3) Incorporation of B in Si with concentrations higher than 10 at. % supresses or hindersthe recrystallization of polysilicon. A relatively stable amorphous Si-B alloy can form at a B concentration of ~ 35 at. % up to 1100 °C.
(4) With a B concentration of 1 - 2 at. % the epitaxially regrown film has fewer defects than that with a higher B concentration.

VI. ACKNOWLEDGEMENT

We wish to express our gratitude to Ricardo Garcia- Padron and Anders Robertsson for XTEM- photos and experimental guidance. The Swedish National Board for Thechnical Development is acknowledged for their financial support.

REFERENCES

Adekoya W O, Hage-Ali M, Muller J C, and Siffert P, J. Appl. Phys. 64 (2), 666 (1988).
Campisano S U, and Chang C T, Appl. Phys. A 31, 157 (1983).
Carlsson J R A, Gong S F, Li X.-H., Hentzell H T G, to be published in J. Appl. Phys.
Ghannam M Y, and Dutton R W, Appl. Phys. Lett. 51 (8), 611 (1987).
Gong S F, Hentzell H T G, and Radnoczi G, Appl. Phys. Lett. 53 (10), 902 (1988).
Gong S F, Hentzell H T G, Robertsson A, and Radnoczi G, IEE-G 137 (1), 53 (1990).
Kim H-J, and Thompson C V, J. Appl. Phys. 67 (2), 757 (1990).
Mei L, Rivier M, Kwark Y, and Dutton R W, J. Electrochem. Soc. 129 (8), 1791 (1982).
Narayan J, J. Appl. Phys. 53 (12), 8607 (1982).
Pai C S, Lau S S, Appl. Phys. Lett. 47 (11), 1214 (1985).
Wautelet M, Gehain E D, and Quenon P, Phys. Stat. Solidi. (b) 149, 465 (1988).
Zheng L R, Hung L S, Philips J R, and Mayer J W, J. Appl. Phys. 62 (11), 4426 (1987).

Inst. Phys. Conf. Ser. No 117: Section 7
Paper presented at Microsc. Semicond. Mater. Conf., Oxford, 25–28 March 1991

Substrate effects on the ordering and microstructure of epitaxial layers

Jean-Pierre Chevalier and Richard Portier

C.E.C.M.- C.N.R.S., 15, rue Georges Urbain,
F 94407 Vitry Cédex

ABSTRACT: The recent results concerning ordering in III-V and GeSi epilayers are reviewed, paying particular attention to the number of ordered variants observed. Simple symmetry considerations are proposed to account for these observations and these are based on the effect of the substrate symmetry on the ordering dictating the number of variants. This is extended to the case of polarity selection in the epitaxy of GaAs/Si.

1. INTRODUCTION

Recent results on ordering in semiconductor systems suggest that the substrate and its misorientation play a role in the selection of the number of ordered variants occurring. A parallel with the epitaxy of GaAs on Si, where the substrate misorientation enables monodomain GaAs epilayers to be obtained, can also be made. It is thus tempting to introduce symmetry considerations, frequently used to determine the number of variants resulting from ordering.

An initial approach to this would be simply to take into consideration the symmetry of the two crystallographic systems taken separately. However, deviations from this can be diverse. At an atomic scale, for some of the ordered epilayers, notably for ordering on {111} planes in III-V compounds, there is a clear consensus showing that of the 4 ordered orientation variants expected, at most only 2 are present.

On a microstructural scale, differences in the nature of defects on apparently symmetry related planes betray this symmetry breaking. This has been observed for GaAs/(001)Si, where notable differences in the nature and density of defects are seen for $[1\bar{1}0]$ and $[110]$ cross-sections. The origin of this asymmetry lies at the atomic level, and is due to the selection of a given polarity of the GaAs epilayer by appropriate substrate conditions. One out of the two possible polarities, or variants can be obtained, and the formal link with the ordering is apparent.

The aim of this paper is to examine these results in the light of symmetry, including the symmetry effects of surface misorientation. In choosing to deal essentially with symmetry properties, considerations concerning the atomic scale mechanisms are avoided, and we simply rely on the persistence of the substrate surface symmetry.

2.   ORDERING IN SEMICONDUCTOR EPILAYERS

2.1   General Comments

Kuan et al (1985) were the first to report ordering in the $Al_xGa_{1-x}As/(110)$ and (001) GaAs system. Ordering occurred on {100} planes, and is equivalent to that of the classical CuAu I type (layered tetragonal, space group $P\overline{4}2m$). Soon after, ordering on the {111} planes (CuPt type order, trigonal R3m space group in III-V and $R\overline{3}m$ in IV compounds) was observed in a number of different systems, e.g.: GeSi/(100)Si by Ourmazd and Bean (1985), $Ga_xIn_{1-x}P/(100)GaAs$ by Ueda et al (1987), $Ga_{0.47}In_{0.53}As/(001)InP$ and $Ga_xIn_{1-x}As_yP_{1-y}/(001)InP$ by Shahid et al (1987), $Al_xIn_{1-x}As/(001)InP$ by Norman et al (1987) and $GaAs_{1-x}Sb_x$ (Imh et al 1987). More complex ordering (such as chalcopyrite, $I\overline{4}2d$ or famatinite, $I\overline{4}2m$, e.g. Wyckoff, 1964,1965, Miller et al 1981) has also been observed by e.g. Nakayama and Fujita (1986) in $In_{1-x}Ga_xAs/(001)InP$ and Jen et al (1986) in $GaAs_{0.5}Sb_{0.5}/(001)InP$.

Since then a substantial number of authors have reported ordering in a wide range of alloyed III-V semiconductors, notably for the $Ga_xIn_{1-x}P/(001)GaAs$ system, and also for GeSi alloys, for different substrates, different substrate orientations, exact or misoriented substrates, different epitaxial techniques and conditions. The ordering observed is thus a general phenomenon and not an experimental accident. The situation is not, unfortunately as simple as this: other structural effects occur and the ordering is often accompanied by modulations, whose origins are not perfectly established (i.e. compositional and/or strain fluctuations) but which may be linked to the growth process. These modulations appear in the electron diffraction patterns as diffuse scattering which are more or less well localised and correspond to modulated or mottled contrast in images (e.g. see Bellon et al 1988, Gomyo et al 1988a, Yasuami et al 1988, Murgatroyd et al 1990, Hsieh et al 1990). Diffuse scattering due to other structural causes, such as antiphase boundaries (e.g. Kuan et al 1985) or shape effects of small ordered regions (e.g. Baxter et al 1990) is also observed. None of these effects will be mentioned further.

Some comment also has to be made concerning the degree of long range order, conventionally defined by the Bragg-Williams parameter S, and directly related to the superlattice intensity, which varies as S^2. The maximum value of S (S=1) can only occur if the alloy composition corresponds to the appropriate stoichiometry for the superstructure considered, e.g. 1:1 or 1:3 on the ordering sublattice. It should then be obvious that if the composition is different from this then the ordering will necessarily be imperfect, and the superlattice reflections weaker.

For more complicated structures this reasoning does not apply, since ordering may occur on crystallographically different sites. For example, there are a number of other possible ordered superstructures on the basis of the zinc blende lattice, some based on variations of the stannite/famatinite type of ordering. For the 1:1 ratio of the ordering species, this is the simple tetragonal lattice ($P\overline{4}2m$), and for a 1:3 ratio, the famatinite (Cu_3SbS_4-type, $I\overline{4}2m$), closely related to the stannite (Cu_2FeSnS_4-type, $I\overline{4}2m$) structure (e.g. Wyckoff 1965, Miller et al 1981). In the case of the stannite/famatinite like structure, the ordering can occur on the 2a, 2b and 4d sites of the $I\overline{4}2m$ space group and it is possible to evolve in a continuous manner from the $I\overline{4}2m$ to $P\overline{4}2m$ via imperfect ordering on the 2b sites. The intensity of superlattice reflections need not vary in a simple manner, as a function of composition or of the degree of order.

We will examine the results obtained, especially for ordering on {100} and {111} planes. Particular attention will be paid to the number of ordered variants observed as a function of substrate orientation, and misorientation.

2.2 Ordering on {100} planes.

For $Al_xGa_{1-x}As/(001)GaAs$ Kuan et al (1985) observed only 2 orientation variants with their c-axes normal to each other and to the growth direction. Superlattice reflections had weak intensities, and appear only for an Al content, x=0.75 and x=0.25, i.e. far from the stoichiometry for the $L1_0$ superstructure, and for MOCVD growth at 800°C and 700°C respectively. For growth on (110) substrates, ordering is stronger, and only 1 variant, with anti-phase boundaries, is present. Here the c-axis lies in the substrate plane, and is parallel to the [001] direction. Ordering is observed over a range of compositions (from x=0.50 to 0.75) and MOCVD growth temperatures (from 600°C to 800°C).

MBE $In_{0.5}Ga_{0.5}As/(110)InP$ has been studied by both Kuan et al (1987) and Ueda et al (1991). Here different substrates were used : (110) tilted 4° towards either the {111}A or {111}B pole (Kuan et al 1987) and exact (110) and (110) tilted 3° or 5° towards the '<00$\bar{1}$>' direction (Ueda et al 1991). There is agreement on the main tendencies: ordering is always observed, only 1 variant is observed, and the intensity of the superlattice reflections increases for tilts away from the exact orientation.

Finally, apparently the only other case where ordering on {100} planes is reported, concerns the OMVPE $GaAs_{0.5}Sb_{0.5}/(001)InP$ (Jen et al, 1986, 1987). The diffraction patterns are more complex were interpreted in terms of a mixture of $L1_0$ and chalcopyrite superlattices. For the $L1_0$ superstructure, two variants are reported with their c-axes lying in the substrate plane and perpendicular to the growth direction, as observed by Kuan et al (1985) for $Al_xGa_{1-x}As$.

2.3 Ordering on {111} planes

The very large majority of recent papers concerns ordering on {111} planes. The most studied system is $Ga_xIn_{1-x}P/(001)GaAs$, and ordering has been observed over a range of experimental conditions. MOVPE is generally used, but recently McDermott et al (1990) have reported ordering in GaInP films grown by ALE. For MOVPE growth, on (001) substrates, either misoriented, 2.5° off towards (110) (Ueda et al 1987), 6° off towards [110] (Bellon et al 1988), 2° off towards [011] (Gomyo et al 1988a), or at exact orientation (Kondow et al 1988a, Morita et al 1988), there is now a clear consensus that only 2 ordered {111} variants are observed *at most*. In all cases, after a few problems in indexing, these have been identified as the 2 {111}B variants. For (001) substrates misoriented towards [1$\bar{1}$0], Augarde et al (1989) and Bellon et al (1989) report that only 1 variant is observed. In the similar $Ga_{0.47}In_{0.53}As/(001)InP$, Shahid et al (1987) also observe only one {111} variant; however, they make no mention of the substrate misorientation. As far as the effect of temperature is concerned, there are suggestions that a higher degree of ordering is obtained for temperatures ranging from 650° C to 700° C, with respect to both lower and higher temperatures (Gomyo et al 1988a, Suzuki et al 1988a and Kondow et al 1988b, Nozaki et al 1988). Whether the layers are lattice matched to the substrate or not (i.e. strained or not), does not appear to play a major role (e.g. Ihm et al 1987). Furthermore, {111} ordering appears to be 'robust' with respect to composition and is still observed for compositions, e.g. $Ga_{0.7}In_{0.3}P$ (Kondow et al 1989), where other ordered structures, such as famatinite, might be expected.

Similar results are obtained for other alloys where ordering occurs on the III sublattice, such as

$Al_xIn_{1-x}As/(001)InP$ (Norman et al 1987) and also in quaternary systems such as MOVPE $AlGaInP/(001)GaAs$ (Suzuki et al 1988b, Yasuami et al 1988, Kondow et al 1988c and Chen et al 1990). Finally, ordering has also been observed on the V sublattice, as in the case of $GaAs_{1-x}Sb_x$ grown by MBE (Imh et al 1987, Otsuka et al 1989, Murgatroyd et al 1990), InAsSb (Jen et al 1989), GaAsP (Plano et al 1988) and for mixed quaternaries like InGaAsP (Shahid et al 1987, Shahid and Mahajan 1988, Plano et al 1988). In all these cases only two variants are observed, and it seems clear that these are the same as those observed for ordering on the III sublattice.

What can be said about {111} ordering in III-V compounds ? It occurs in a wide variety of systems and for a large range of experimental conditions. It does not appear to be an equilibrium phase since when epitaxy is carried out on other substrate orientations, e.g. on (111)B and (110) GaAs (Gomyo et al 1988b, Augarde et al 1989, Ueda et al 1989) no ordering is observed. Furthermore thermal annealing at temperatures below the growth temperature leads to disordering and Zn diffusion destroys the order on the III sublattice (Dabkowski et al 1988, Gavrilovic et al 1988, Plano et al 1988). This has led a number of authors to suggest that the ordering is driven by the substrate and frozen in by the growth of the layer (e.g. Norman et al 1987, Gomyo et al 1988, Augarde et al 1989, Bellon et al 1989, Murgatroyd et al 1990). This idea is consistent with the studies of Kurtz et al (1990) who showed that there existed a critical regime of temperature and growth rate that permitted ordering. Two simple models have been proposed to account for substrate driven ordering, one is a bond length or size effect model (Norman et al 1987, Gomyo et al 1988, Augarde et al 1989, Bellon et al 1989) and the other is based on surface reconstructions which leads to the initial ordering (Murgatroyd et al 1990). Recent theoretical considerations by Froyen and Zunger (1990) also suggest that the ordering is driven by a surface reconstruction and frozen in during growth.

It is very interesting to compare these results with those observed for GeSi/(001)Si. Ourmazd and Bean (1985) first reported the occurrence of {111} order in this system. Subsequently, LeGoues et al (1990a, 1990b) also observed long range ordering in thick GeSi films, and concluded that this ordering was substrate driven and then frozen in. Interestingly, they found that no ordering occurred for growth on (111) substrates (LeGoues et al 1990b). Muller et al (1989) investigated the role of strain on the observed ordering and concluded that it was independent of the choice of substrate lattice parameter (i.e. not strain driven). Of great importance here, all the authors observed the 4 possible ordered variants.

To conclude this section on {111} ordering, it suffices, for the purpose of this paper, to say that this kind of ordering appears to be substrate driven, but only on (001) substrates, that for III-V alloys, at most 2 of the 4 possible ordering variants are observed and that only 1 is observed for certain substrate misorientations. In contrast, for GeSi/(001)Si, the 4 possible variants are always observed.

2.4   Other forms of ordering

Other ordered structures have been reported in $GaAs_{0.5}Sb_{0.5}/(001)InP$ (Jen et al 1986,1987) and in $In_{1-x}Ga_xAs/(001)InP$ (Nakayama and Fujita 1986). The former report a mixture of {100} type and chalcopyrite type ordering, whilst the latter report the famatinite structure. In the case of partial long range order it is extremely difficult to distinguish between these, since they are all variations around a tetragonal ordered cell of the zinc blende structure. The differences reside in the composition and in the occupation of the different possible sublattices, leading to different diffracted intensities in the superlattice reflections. These reports may well correspond to some kind of partially ordered superstructure related to the famatinite/stannite structure.

3.   EPITAXY OF GaAs ON (001) Si

It is tempting, in the approach chosen here, to include the epitaxy of GaAs on Si (although the GaAs structure is obviously not induced by the substrate !), since epitaxy aims at selecting one of the two polar variants of the GaAs structure. It is beyond the scope of this paper to review the field of GaAs/Si (see e.g. Fang et al 1990), and we will consider here a few results which illustrate asymmetry in the defect microstructure in both the first and second stages of GaAs growth. The first stage corresponds to heteroepitaxy (the growth of a thin layer at relatively low temperatures) and this is followed by growth of a thick GaAs layer with the usual conditions for homoepitaxy.

At the end of the first stage, for growth on slightly misorientedsubstrates, i.e. 3.5° from (001) towards [1$\bar{1}$0], electron microscopy images of the [110] and [1$\bar{1}$0] cross-sections show very different defect structures (Vannuffel et al 1989, Tsai and Matyi 1989, Vannuffel 1990, Xie et al 1990). In the [110] section numerous planar defects, lying on {111} planes are observed, whilst in the [1$\bar{1}$0] section hardly any are visible. Of the 4 possible sets of {111} planar defects, only two are observed, which is somewhat reminiscent of ordering in GaInP. In all these cases the GaAs layer was single domain. After the second stage (during which annealing has occurred) the asymmetry is not so blatant, but persists and lies in differences in the density of 60° and pure edge dislocations along [110] and [1$\bar{1}$0], as described by Zhu and Carter (1990) and Vannuffel (1990).

Vannuffel (1990) (see also Vannuffel et al 1990) has proposed that it is the polarity (determined by the misoriented substrate) that leads to the observed asymmetry. Dislocations propagating from the free surface to the interface will be of different type depending on whether they are along [110] (β type) or [1$\bar{1}$0] (α type). Due to differences in mobility of the partial dislocations, stacking faults will preferentially appear on the (1$\bar{1}$1) and ($\bar{1}$11) planes for the dissociation of the 60° β dislocations, whilst the 60° α dislocations will not dissociate, leading to the observed asymmetry of planar defects. After annealing, recombinations occur producing 60° and pure edge dislocations along [110] and [1$\bar{1}$0].

The origin of this asymmetry lies in the selection of one of the polar variants of GaAs by the Si substrate. This particular point has been widely discussed (e.g. Fang et al 1990) in terms of double steps on the (001) Si surface, induced by either substrate misorientation (e.g. Sakamoto et al 1989), high temperature treatment, As stabilisation, or a combination of these.

4   SYMMETRY CONSIDERATIONS

From the above review we have seen that the number of possible variants in ordering or in polarity can be determined by the substrate, i.e. its crystallography, its orientation and misorientation. Whenever variants are considered symmetry comes to mind as a tool for predicting the number of possible variants. In this context, it also has the advantage of plausibly taking into account surface reconstructions since their symmetry will be related to that of the substrate. In choosing to ignore explicitly surface reconstructions, we will thus obtain the maximum number of equivalent variants. Symmetry considerations, however, will not give any information on the mechanisms involved, nor on which of the ordered structures or variants will occur.

We first make the distinction between what can be called spontaneous volume ordering and substrate driven ordering. For the former, it is necessary to consider the intersection of the disordered and ordered space groups and then see what is the order of this intersection group in the disordered space group. For the latter we will present an approach which takes into account the symmetry of the substrate surface. How to do this is not necessarily obvious and we suggest a simple method to do this.

4.1 Symmetry in the case of spontaneous volume ordering

For what can be called 'spontaneous volume ordering' the number of orientation variants can be readily obtained as follows. Let G_D be the space group of the 'disordered phase' with point group g_D (since we are looking for the orientation variants, we need only consider the point groups). Let g_O be the point group of the ordered phase. The intersection group, g_i will be:

$$g_i = g_O \cap g_D \tag{1}$$

g_i is equal to g_O for a group-subgroup transition.

The order (i.e. the number of symmetry elements) of g_D divided by that of g_i will give the number of orientation variants. These are tabulated below for the structures considered :

Structure (Space group)	Point group	N° of Variants
For III-V ordering ($g_D = \overline{4}3m$) :		
Cubic Au_3Cu type ($P\overline{4}3m$)	$\overline{4}3m$	1
Trigonal CuPt type (R3m)	3m	4
Chalcopyrite ($I\overline{4}2d$)	$\overline{4}2m$	3
Stannite/famatinite ($I\overline{4}2m$)	$\overline{4}2m$	3
Layered tetragonal ($P\overline{4}2m$)	$\overline{4}2m$	3
For GeSi ordering ($g_D = m\overline{3}m$) :		
Trigonal CuPt type (R$\overline{3}$m)	$\overline{3}$m	4
For GaAs on Si ($g_D = m\overline{3}m$) :		
Zinc Blende (F$\overline{4}3$m)	$\overline{4}3m$	2

The number of variants given here corresponds only to the orientation variants. If we take translations into account then, more variants, related by anti-phase vectors will be introduced.

4.2 Symmetry in the case of substrate driven ordering

What has to be considered to assign the appropriate symmetry for the substrate surface ? This is not an obvious question and we propose to consider the overall symmetry. The substrate surface will be defined by its normal. This defines the growth direction as well as planes which cut the (substrate) crystal. The corresponding symmetry can be defined by a new group whose operations leave the surface normal unchanged. This will already reduce the symmetry and the number of variants. A similar approach has been previously proposed by Portier and Gratias (1982) and Pond (1986, 1989). Substrate misorientation can be accounted for by defining a new normal inclined in the appropriate direction. Stepped surfaces are thus not explicitly considered. It is, of course, necessary to include the relevant epitaxial relationships between the ordered epilayer and the substrate (e.g. for {111} ordering the 3-fold axis will be parallel to one of the substrate <111> directions). The formalism is straightforward:

Let G_D be the space group of the ('disordered') substrate, with g_D being the associated point group. Since we are only concerned with orientation variants, for the time being, we need only deal with the point groups. Let $\mathbf{n}$ be the substrate normal, and we will define the 'surface' group $g_s^{[\mathbf{n}]}$ such that :

$$g_s^{[\mathbf{n}]} = \{ \, g \in g_D : g[\mathbf{n}] = [\mathbf{n}] \, \} \qquad (2)$$

We next consider the intersection of $g_s^{[\mathbf{n}]}$ with the point group g_O of the ordered epilayer and for a given epitaxial relationship. this gives the intersection group g_i^e :

$$g_i^e = g_O \cap g_s^{[\mathbf{n}]} \qquad (3)$$

The number of equivalent orientation variants is given by the order of $g_s^{[\mathbf{n}]}$ divided by the order of g_i^e.

Substrate misorientation can now be introduced into the surface group $g_s^{[\mathbf{n}]}$, yielding a new group, $g_s^{[\mathbf{n}']}$, with reduced symmetry. The number of orientation variants will be obtained from the order of $g_s^{[\mathbf{n}']}$ divided by the order of the $g_i'^e$, defined as :

$$g_i'^e = g_O \cap g_s^{[\mathbf{n}']} \qquad (4)$$

4.3 Consequences with respect to ordering

For 'spontaneous volume ordering', the major consequence is that the substrate orientation should have no effect on the ordering. The absence of ordering on {111} oriented substrates for both III-V (Gomyo et al 1988b, Augarde et al 1989, Ueda et al 1989) and GeSi (LeGoues et al 1990b) epilayers implies that the ordering must be substrate driven. That fewer than the volume symmetry dictated number of variants occur, for III-V compounds, corroborates this. We will now only consider the symmetry effects on substrate driven ordered epilayers.

We will first consider the {111} ordering of GeSi/(001)Si. The Si point group is m$\overline{3}$m, which is lowered to the 4mm 'surface group' on applying equation (2) for $\mathbf{n}$=[001]. The ordered phase has $\overline{3}$m point group, and the intersection group g_i^e (3), for the ordered phase ternary axis parallel to, say, [111], will be m, of order 2. 4mm being of order 8, there are thus 4 orientation variants. This is in agreement with the observations of Muller et al (1989) and LeGoues et al (1990a,b).

For the same type of ordering in III-V compounds, things are a little different. Following the same procedure as above, the substrate point group is $\overline{4}$3m and the ordered trigonal phase has the 3m point group. We consider a [001] surface normal, and applying (2), this gives :

$$g_s^{[n]} = mm2 \ (\text{of order 4}) \tag{5}$$

The intersection of this with 3m, with the trigonal axis parallel to <111>B, gives m, which is of order 2. There are therefore 2 equivalent <111>B orientation variants, and if we consider the trigonal axis parallel to <111>A, we will have a further 2 equivalent <111>A variants. The A and B orientation variants are, of course, not equivalent. The experimental results clearly show that 2, most probably the {111}B variants occur, for the exact substrates (e.g. Kondow et al 1988a, Morita et al 1988). This agrees with the symmetry considerations, and it can only be for kinetic or free energy reasons that the {111}A variants do not occur.

We can now consider the effect of misorientation. The mm2 point group (5) consists of 2 mirrors on (110) and (1$\overline{1}$0), which can be labelled m_1 and m_2 respectively, with a 2-fold axis parallel to [001]. We will consider two misorientations, one towards [110] and the other towards [1$\overline{1}$0]. For the sake of simplicity we will take the trigonal axis as being parallel to [1$\overline{1}$1]B. The disorientation will reduce the 'surface' group to the m point group (of order 2), and these will be m_1 and m_2 depending on the tilt direction. This will affect the intersection group g'_i^e, since this will depend on the orientation of the trigonal axis. For the case chosen here, tilt towards [110], i.e. lying in the m_2 plane, will yield an intersection group unity (of order one), thus still giving 2 equivalent variants. For the converse, the intersection group will be m_1, of order 2, and so there will only be one variant. This is completely consistent with what Augarde et al (1989) and Bellon et al (1989) report for epitaxy on different misoriented substrates.

We can now deal with all the tetragonal based epitaxial ordered structures, since these all have the same, $\overline{4}$2m, point group. It suffices to say that for the c-axis // [001], there is only one variant and that for the c-axis // [100] there will be two equivalent variants. Kuan et al (1985) for CuAu I type ordering in $Al_xGa_{1-x}As$/(001)GaAs observe the two variants with their c-axes lying in the substrate plane, but not the third; this is not surprising since the former are not equivalent to the latter. Why the latter is not observed can only be accounted for by energetics. For ordering in the famatinite/stannite tetragonal structure ($\overline{14}$m2), the experimental observations (Jen et al 1986,1987, Nakayama and Fujita 1986) are not sufficiently clear to be compared here.

Finally for the case of the epitaxy of GaAs/(001)Si, concepts of substrate driven ordering are clearly not relevant. However, it is interesting to see whether the substrate misorientation can reduce the number of variants from 2 to 1. The demonstration is trivial. For substrates misoriented towards [110] or [1$\overline{1}$0], the 'surface' group drops to m. The intersection with $\overline{4}$3m is also m, and they both have the same order. There is thus only 1 variant. For misorientation towards [100], the surface group is still m, but the mirror lies on the (010) plane. The intersection group is thus unity, and we now have 2 variants. Experimentally the substrates are nisoriented towards <110> to obtain single domain GaAs (e.g. Fang et al 1990).

5   DISCUSSION AND CONCLUSIONS

The first point to be discussed concerns the physical meaning of the 'surface' symmetry defined by the group $g_s^{[n]}$. This corresponds to that of a set of *equivalent* planes with normal n, implicitly accounting for steps between *equivalent* planes. If we consider a single surface plane, then it will have lower symmetry, and the symmetry relations between steps have to be introduced. The second point concerns the symmetry relation between macroscopically nisoriented surfaces and steps; further work is in progress concerning this point.

To conclude, it is reasonable, in an initial approach, to propose an overall symmetry treatment as we have attempted to do. These very simple symmetry considerations are consistent with all the observations and provide a unifying framework for these very diverse observations on rather different systems. It should be emphasised that this formalism can only lead to necessary, but not sufficient, conditions in order to obtain a given number of variants in heteroepitaxy.

ACKNOWLEDGMENTS

It is a pleasure to thank P. Bellon and C. Vannuffel for many enlightening discussions and for access to their results.

REFERENCES

Augarde E, Mpaskoutas M, Bellon P, Chevalier J-P and Martin G P 1989 Inst. Phys. Conf. Ser. No 100 155

Baxter C S, Broom R F and Stobbs W M 1990 Surface Science 228 102

Bellon P, Chevalier J-P, Martin G P, Dupont-Nivet E, Thiebaut C and Andre J-P 1988 Appl. Phys. Lett. 52 567

Bellon P, Chevalier J-P, Augarde E, Andre J-P and Martin G P 1989 J. Appl. Phys. 66 2388

Chen G S, Wang T Y and Stringfellow G B 1990 Appl. Phys. Lett. 56 1463

Dabkowski F P, Gavrilovic P, Meehan K, Stutius W, Williams J E, Shahid and Mahajan S 1988 Appl. Phys. Lett. 52 2142

Fang S F, Adomi K, Iyer S, Morkoç H, Zabel H, Choi C and Otsuka N 1990 J. Appl. Phys. 68 R31

Froyen S and Zunger A 1990 preprint

Gavrilovic P, Dabkowski F P, Meehan K, Williams J E, Stutius W, Hsieh K C, Holonyak Jr N, Shahid M A and Mahajan S 1988 J. Crystal Growth 93 426

Gomyo A, Suzuki T and Iijima S 1988a Phys. Rev. Lett. 60 2645

Gomyo A, Suzuki T, Iijima S, Hotta H, Fujii H, Kawata S, Kobayashi K, Ueno Y and Hino I 1988b Jap. J. Appl. Phys. 27 L2370

Hsieh K C, Baillargeon J N and Cheng K Y 1990 Appl. Phys. Lett. 57 2244

Ihm Y E, Otsuka N, Klem J and Morkoç H 1987 Appl. Phys. Lett. 51 2013

International Tables for Crystallography 1983 Vol A ed. T Hahn pub International Union of Crystallography

Jen H R, Cherng M J and Stringfellow G B 1986 Appl. Phys. Lett. 48 1603

Jen H R, Jou Y T, Cherng M J and Stringfellow G B 1987 J. of Crystal Growth 85 175

Jen H R, Ma K Y and Stringfellow G B 1989 Appl. Phys. Lett. 54 1154

Kondow M, Kakibayashi H and Minagawa S 1988a J. of Crystal Growth 88 291 and 89 614

Kondow M, Kakibayashi H, Minagawa S, Inoue Y, Nishino T and Hamakawa Y 1988b Appl. Phys. Lett. 53 2053

Kondow M, Kakibayashi H, Minagawa S, Inoue Y, Nishino T and Hamakawa Y 1988c J. of Crystal Growth 93 412

Kondow M, Kakibayashi H, Tanaka T and Minagawa S 1989 Phys. Rev. Lett. 63 884

Kuan T S, Kuech T F, Wang W I and Wilkie E L 1985 Phys. Rev. Lett. 54 201

Kuan T S, Wang W I and Wilkie E L 1987 Appl. Phys. Lett. 51 51

Kurtz S R, Olson J M and Kibbler A 1990 Appl. Phys. Lett. 57 1922

LeGoues F K, Kesan V P and Iyer S S 1990a Phys. Rev. Lett. 64 40

LeGoues F K, Kesan V P, Iyer S S, Tersoff J and Tromp R 1990b Phys. Rev. Lett. 64 2038

McDermott B T, Reid K G, El-Masry N A, Bedair S M, Duncan W M, Yin X and Pollak F H 1990 Appl. Phys. Lett. 56 1172

Miller A, MacKinnon A and Weare D 1981 Solid State Physics 36 119

Morita E, Ikeda M, Kumagai O and Kaneko K 1988 Appl. Phys. Lett. 53 2164

Müller E, Nissen H-U, Ospelt M and von Känel H 1989 Phys. Rev. Lett. 63 1819

Murgatroyd I J, Norman A G and Booker G R 1990 J. Appl. Phys. 67 2310

Nakayama N and Fujita H 1986 Inst. Phys. Conf. Ser. 79 289

Norman A G, Mallard R E, Murgatroyd I J, Booker G R, Moore A H and Scott M D 1987 Inst. Phys. Conf Ser. 87 77

Nozaki C, Ohba Y, Sugarawa H, Yasuami S and Nakanisi T 1988 J. Crystal Growth 93 406

Otsuka N, Ihm Y E, Hirotsu Y, Klem J and Morkoç H 1989 J. Crystal Growth 95 43

Ourmazd A and Bean J C 1985 Phys. Rev. Lett. 55 765

Plano W E, Nam D W, Major Jr J S, Hsieh K C and Holonyak Jr N 1988 Appl. Phys. Lett. 53 2537

Pond R C 1986 J. Crystal Growth 79 946

Pond R C 1989 Critical Reviews in Solid State and Materials Science 15 441

Portier R and Gratias D 1982 J. de Physique C4 17

Sakamoto K, Sakamoto T, Miki K and Nagao S 1989 J. Electrochem. Soc. 136 2705

Shahid M A, Mahajan S, Laughlin D E and Cox H M 1987 Phys. Rev. Lett. 58 2567

Shahid M A and Mahajan S 1988 Phys. Rev. B 38 1344

Srivastava G P, Martins J L and Zunger A 1985 Phys. Rev. B 31 2561

Suzuki T, Gomyo A, Iijima S, Kobayashi K, Kawata S, Hino I and Yuasa T 1988a Jap. J. Appl. Phys. 27 2098

Suzuki T, Gomyo A and Iijima S 1988b J. Crystal Growth 93 396

Tsai H L and Matyi R J 1989 Appl. Phys. Lett. 55 265

Ueda O, Takikawa M, Komeno J and Umebu I 1987 Jpn. J. Appl. Phys. 26 L1824

Ueda O, Takechi M and Komeno J 1989 Appl. Phys. Lett. 54 2312

Ueda O, Nakata Y and Fujii T 1991 Appl. Phys. Lett. 58 705

Vannuffel C, Beaucour J, Andre J-P and Chevalier J-P 1989 Inst. Phys. Conf. Ser. 100 115

Vannuffel C 1990, Thesis, Université de Paris VI

Vannuffel C, Schiller C and Chevalier J-P 1990 'Proceedings of the XIIth. International Congress for Electron Microscopy 4 592

Wyckoff R W G 1964 Crystal Structures Vol 2 pub Interscience Publishers

Wyckoff R W G 1965 Crystal Structures Vol 3 pub Interscience Publishers

Xie Q H, Fung K K, Ding A J, Cai L H, Huang Y and Zhou J M 1990 Appl. Phys. Lett. 57 2803

Yasuami S, Nozaki C and Ohba Y 1988 Appl. Phys. Lett. 52 2031

Inst. Phys. Conf. Ser. No 117: Section 7
Paper presented at Microsc. Semicond. Mater. Conf., Oxford, 25–28 March 1991

TEM, TED and HREM studies of atomic ordering and associated domains in MOCVD In_xGa_{1-x} As layers: Dependence on growth temperature and growth rate

T -Y Seong, A G Norman[+], J L Hutchison, G R Booker, A G Cullis[*], S J Bass[*] and L L Taylor[*]

Department of Materials, University of Oxford, Parks Road, Oxford OX1 3PH, UK
[*]Royal Signals and Radar Establishment, St. Andrews Road, Great Malvern, Worcs. WR14 3PS, UK

ABSTRACT: We report the first detailed TEM, TED and HREM results for atomic ordering in (001) MOCVD $In_{0.53}Ga_{0.47}As$ layers. Ordering of the CuPt-type was observed in layers grown below 650°C, with a maximum degree of ordering occurring at 550°C. At 600°C the degree of ordering was lower for a slower growth rate. In the layers grown at 550°C, separate columnar-like ordered regions of the two observed {111} variants were present and these contained antiphase boundaries (APBs) whose density and orientation strongly depended on the growth rate. Mechanisms are suggested to explain the observations.

1. INTRODUCTION

Hetero-epitaxial layers of III-V compound semiconductors such as $In_xGa_{1-x}As$ are of considerable importance for optoelectronic and high speed microwave devices. $In_xGa_{1-x}As$ layers have been grown using a variety of techniques including liquid phase epitaxy (LPE), vapour phase epitaxy (VPE), molecular beam epitaxy (MBE) and metal organic chemical vapour deposition (MOCVD). Structural studies of some of these $In_xGa_{1-x}As$ layers grown by LPE (Nakayama and Fujita 1986), MBE (Kuan et al 1987), MOCVD (Norman 1987) and VLE (Shahid et al 1987) revealed the existence of a variety of atomically ordered structures such as famatinite, CuAu-I and CuPt-type. The CuPt-type is the most frequently reported ordered structure for ternary and quaternary III-V alloy semiconductor (001) epitaxial layers. It arises from a growth-surface-induced atomic ordering mechanism (Murgatroyd 1987, Norman 1987, Suzuki et al 1988, Bellon et al 1989, Bernard et al 1990 and Murgatroyd et al 1990). Transmission electron microscopy (TEM) and photoluminescence (PL) examinations showed that the presence of CuPt-type ordering causes a reduction in the band gap of MOCVD $In_xGa_{1-x}P$ layers (Gomyo et al 1987). Bass et al (1986) reported the growth condition dependence of the optical and electrical properties of MOCVD $In_xGa_{1-x}As$ layers and showed that the low temperature PL deteriorates at growth temperatures < 636°C, the spectra in general consisting of several relatively broad peaks lying 20-30meV below the expected position of the near-band-edge exciton line. This shift of the PL peaks motivated us to investigate in detail the structural properties of MOCVD $In_xGa_{1-x}As$ layers grown in the temperature range 500 to 650°C. In this paper we report the first detailed study of CuPt-type atomic ordering in MOCVD $In_xGa_{1-x}As$ layers. The effects of growth temperature and growth rate on the ordering and the associated domains are described, and the mechanisms responsible are discussed.

2. EXPERIMENTAL

$In_{0.53}Ga_{0.47}As$ epitaxial layers were grown on (001) InP and GaAs substrates by atmospheric pressure MOCVD using trimethylgallium (TMG), trimethylindium (TMI), arsine and phosphine as sources (Bass et al 1986). The layer thickness in this study was approximately $1\mu m$. The growth conditions investigated were growth temperatures of 500, 550, 600 and 650°C using a growth rate of 0.2nm/s and 550, 600 and 650°C using a growth rate of 2.0nm/s. Plan-view and two orthogonal <110> cross-sectional specimens were prepared for TEM observations using standard techniques. The thinned materials were then examined using a Philips CM 20 operated at 200kV and a JEOL 4000EX operated at 400kV.

[+] Present address: IRC for Semiconductor Materials, Blackett Laboratory, Imperial College, Prince Consort Road, London SW7 2BZ, UK

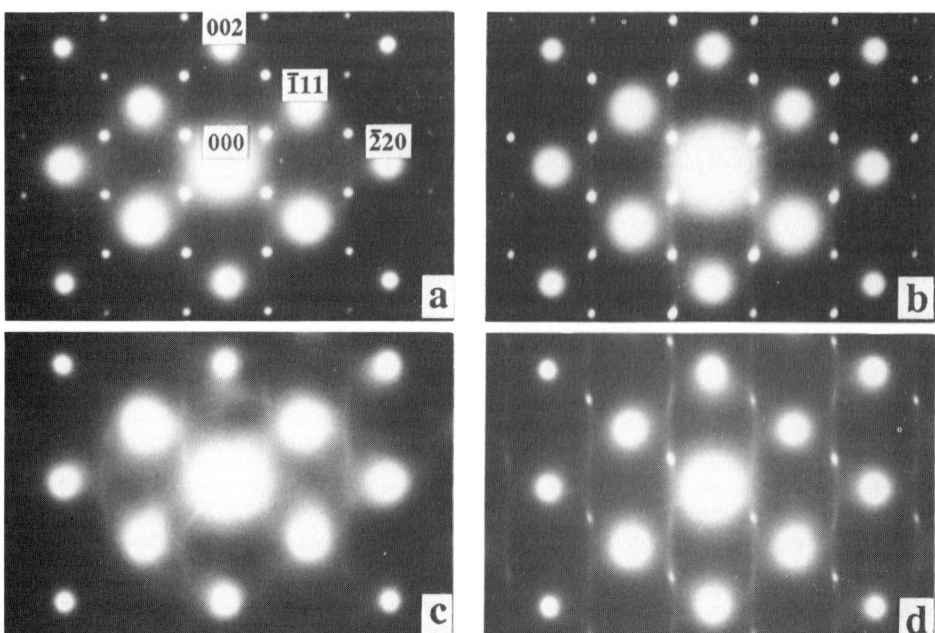

Figure 1. [110] TED patterns of (001) MOCVD $In_{0.53}Ga_{0.47}As$ layers showing evidence of CuPt-type atomic ordering on {111} planes. (a) Growth temperature T_g = 550°C (growth rate 0.2nm/s), (b) T_g = 550°C (2.0nm/s), (c) T_g = 600°C (0.2nm/s), (d) T_g = 600°C (2.0nm/s).

3. RESULTS AND DISCUSSION

Detailed transmission electron diffraction (TED) studies were performed on orthogonal <110> TEM cross-section samples of the layers to investigate the structure and degree of any atomic ordering introduced during growth. The [110] and [$\bar{1}$10] cross-sections of the samples were distinguished by reference to the results of a polarity determination performed using chemical etching (Murgatroyd et al 1990) and convergent beam electron diffraction techniques (Taftø and Spence 1982) on the GaAs substrate of a sample grown at 550°C (0.2nm/s). This sample exhibited strong CuPt-type ordering on only the ($\bar{1}$11) and (1$\bar{1}$1) planes which give rise to $\frac{1}{2}$ {111} superlattice spots visible in the [110] but not the [$\bar{1}$10] cross-section. The [$\bar{1}$10] TED patterns of all the layers showed only the fundamental zinc-blende diffraction spots and no variation in these patterns was observed with change in growth conditions. The [110] TED patterns of some of the layers however showed several different features which were strongly dependent on the growth conditions used. For the layers grown at 550°C using growth rates of both 0.2 and 2.0nm/s, very strong superlattice spots were present, lying half way between the rows of the fundamental zinc-blende spots parallel to the <111> directions, see Figs. 1(a) and (b). This is consistent with the existence of a superlattice structure in the layers along the <111> directions with a period twice that of the zinc-blende structure. The mixed group III atom sublattice of the InGaAs layer consists of alternating In-rich and Ga-rich {111} planes. The TED results indicate that only the ($\bar{1}$11) and (1$\bar{1}$1) planes are ordered. Similar $\frac{1}{2}$ {111} superlattice spots were present in the [110] diffraction patterns of several of the other layers and their intensity was strongly dependent on layer growth conditions. The TED results are summarised in table I. It can be seen that for a growth rate of 0.2nm/s, the $\frac{1}{2}$ {111} superlattice spots were of strong intensity for the layer grown at 500°C, reached a maximum intensity at 550°C, became weak at 600°C and were absent at 650°C. For the layers grown at 2.0nm/s the superlattice spots again exhibited a maximum intensity at 550°C, decreased to a medium intensity at 600°C and were absent at 650°C. At a growth temperature of 550°C, changing the growth rate from 0.2 to 2.0nm/s had a negligible effect on the intensity of the superlattice spots, e.g. see Figs. 1(a) and (b). For a growth temperature of 600°C, however, a large increase in intensity of the superlattice spots occurred on increasing the growth rate from 0.2 to 2.0nm/s, see Figs. 1(c) and (d).

For both growth rates the intensity of the superlattice spots, a measure of the volume fraction and degree of atomic ordering, decreases with increasing growth temperature above 550°C. This effect may be due to a reduction in the degree of surface atomic ordering as the critical temperature for the ordering is approached. However, Bernard et al (1990) have calculated that the CuPt-type ordered structure is unstable in the bulk of the layer and it has been suggested (Bellon et al 1989, Kurtz et al 1990) that competing processes may be associated with the atomic ordering. The ordering is proposed to arise at the surface of the layer during growth, by a process of rapid surface diffusion of the Group III atoms, forming ordered surface monolayers which are then overgrown and frozen into the bulk of the layer. However, since the CuPt-type ordered structure is predicted to be unstable in the bulk, the ordered structure could subsequently disorder in the bulk of the layer during further growth by the slower process of bulk diffusion to lower the free energy state in the bulk. As the growth temperature increases the rate of bulk diffusion and hence disordering increases. Thus above 550°C, although the surface ordering of atoms may still be occurring, the observed degree of ordering in the layers may decrease due to an increased rate of disordering by bulk diffusion. A similar argument may explain the observation that at 600°C the intensity of the superlattice spots decreases with decreasing growth rate. The slower growth rate and hence longer time could enable more disordering of the ordered structure to occur beneath the growing surface by bulk diffusion leading to the reduced degree of atomic ordering observed in comparison to the high growth rate sample. For the layers grown at 550°C the rate of disordering by bulk diffusion may be much lower, hence resulting in no appreciable difference in the degree of atomic ordering observed for the two different growth rates. The bulk interdiffusion coefficients of III-V alloy semiconductors are generally thought to be extremely low in the absence of significant levels of impurities. Post-growth annealing of a CuPt-type ordered $In_{0.53}Ga_{0.47}As$ layer at its growth temperature of 550°C for 91 hours performed in the present investigation resulted in only a small reduction in the degree of ordering. However enhanced bulk interdiffusion of superlattice structures is observed to occur close to the free surface of samples (Kim et al 1990) due to an injection of point defects from the sample surface. Thus it has been suggested (Kurtz et al 1990) that enhanced bulk interdiffusion may occur in a transition region just beneath the growing layer surface, leading to disordering of the CuPt-type ordering during growth at high temperatures and slow growth rates in $In_{0.50}Ga_{0.50}P$ layers, i.e. precisely as observed above for our $In_{0.53}Ga_{0.47}As$ layers. The intensity of the superlattice spots decreased with growth temperature below 550°C for the layers grown at 0.2nm/s. This effect is probably a consequence of a kinetic limit for the degree of surface ordering as the mobility of the Group III atoms on the growing surface is reduced at low temperatures. No layers were grown at 500°C (2.0nm/s) and so the effect of growth rate at this temperature was not determined.

Table I Effects of growth temperature and growth rate on atomic ordering.

Growth temperature (°C)	Growth rate (nm/s)	Relative intensity of superlattice spots	Shape of superlattice spots	Relative intensity of wavy [001] lines of diffuse intensity
500	0.2	strong	circular	weak
550	0.2	v. strong	circular	weak
600	0.2	weak	elongated	weak
650	0.2	none	—————	none
550	2.0	v. strong	elongated	weak
600	2.0	medium	elongated	strong
650	2.0	none	—————	none

The $\frac{1}{2}\{111\}$ superlattice spots present in the TED patterns of some of the samples were connected by weak wavy lines of diffuse diffracted intensity running along the [001] growth direction. The fundamental diffraction spots in these patterns were free of any such diffuse intensity indicating that the diffuse intensity was associated with imperfections in the superlattice structure, e.g. antiphase boundaries (APBs) or order twin boundaries (Baxter et al 1991), rather than any modulation of the fundamental lattice. The $\frac{1}{2}\{111\}$ superlattice spots in the samples grown at 500°C and 550°C (0.2nm/s) were circular, see Fig. 1(a). However, the superlattice spots in the samples grown at 550°C (2.0nm/s) were elongated in a direction tilted ~20° from the [001] growth direction with the superlattice spots from the two different variants being tilted in opposite directions, Fig. 1(b). This suggests the existence of planar imperfections, e.g. APBs, in the superlattice structure lying tilted ~20° from the (001) plane. For the layer grown at 600°C (0.2nm/s), Fig. 1(c), only very weak $\frac{1}{2}\{111\}$ superlattice spots connected with weak wavy [001] lines of diffuse intensity occurred, indicating a very low degree of atomic ordering with many imperfections such as APBs. In the sample grown at 600°C (2.0nm/s), Fig. 1(d), one set of tilted $\frac{1}{2}\{111\}$ superlattice spots of medium intensity was much stronger than the other set and was connected by

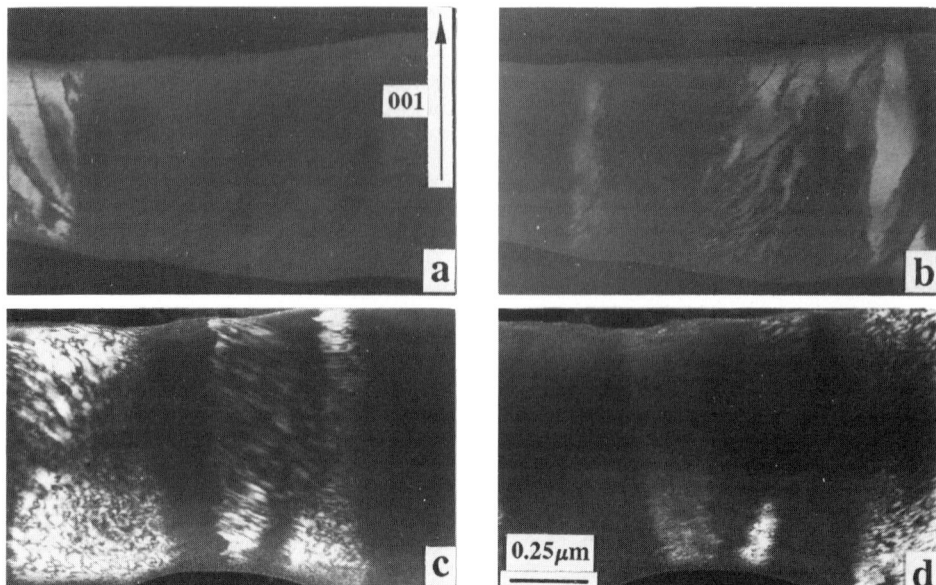

Figure 2. (a) and (b) $\mathbf{g}\frac{1}{2}\{331\}$ DF cross-section micrographs of same area of $In_{0.53}Ga_{0.47}As$ sample grown at 550°C (0.2nm/s) showing regions ordered on ($\bar{1}11$) and ($1\bar{1}1$) planes. (c) and (d) $\mathbf{g}\frac{1}{2}\{331\}$ DF cross-section micrographs of same area of $In_{0.53}Ga_{0.47}As$ sample grown at 550°C (2.0nm/s) showing regions ordered on ($\bar{1}11$) and ($1\bar{1}1$) planes. Note higher density of APBs in (c) and (d).

relatively strong wavy [001] lines of diffuse intensity. This indicates the presence of a relatively high density of imperfections tilted from the (001) plane, e.g. APBs, in the ordered regions of this sample. TED patterns similar to Figs. 1(c) and (d) have recently been reported by Cao et al (1991) for $Ga_{0.5}In_{0.5}P$ layers grown at a high growth rate of 3.3nm/s. This elongation and tilting of the superlattice spots is related to a change in the morphology of the ordered domains, as revealed by the TEM studies described below.

The microstructure of the more highly ordered layers was investigated by obtaining TEM DF images using superlattice spots. Figs. 2(a) and (b) show $\frac{1}{2}\{331\}$ DF micrographs taken of the same area of a cross-sectional sample of the layer grown at 550°C (0.2nm/s) and these reveal the regions ordered on ($\bar{1}11$) and ($1\bar{1}1$) planes, respectively, which appear bright. Columnar-like regions of the two different ordered variants are present. These ordered regions originate at the InGaAs layer/buffer layer interface and propagate to the layer surface. They range from ~0.2 - 0.75μm in width. Dark contrast features are visible in the ordered regions, which appear to be random in orientation, and it is concluded from HREM results, described below, that these features are inclined antiphase boundaries (APBs). Not all of the layer is ordered and fairly sharp boundaries exist between regions of the two different variants and between ordered and unordered material. These boundaries run roughly parallel to the [001] growth direction. Figs. 2(c) and (d) show $\frac{1}{2}\{331\}$ DF micrographs taken from a [110] TEM cross-section of the sample grown at 550°C (2.0nm/s), and these reveal regions ordered on the ($\bar{1}11$) and ($1\bar{1}1$) planes. Again columnar shaped ordered regions of the two variants are present. These range in width from ~0.1 - 0.6μm. A high density of dark ribbon-like contrast features is visible in the ordered regions, considered to be APBs and described below. A much higher density of APBs is present in this sample than in the layer grown at 550°C (0.2nm/s), i.e compare Figs. 2(a) and (b) and Figs. 2(c) and (d). This much higher density of APBs in the fast growth rate sample may be a result of the group III atoms having less time at the growing surface to diffuse to their lowest energy positions before being overgrown by the next monolayer. In some areas of the sample grown at the high growth rate, the APBs are aligned and tilted with respect to the (001) plane by ~20°, and the APBs in the two different variants are tilted in opposite directions. These tilted arrays of APBs are responsible for the elongation and tilting of the superlattice spots visible in the [110] TED pattern of this sample (Fig.1(b)).

Figs. 3(a) and (b) show $\frac{1}{2}\{331\}$ DF micrographs of the same area of the layer grown at 550°C (2.0nm/s) obtained by tilting a (001) plan-view sample ~18° about the orthogonal <100> tilt axes to the <103> poles, and reveal the areas ordered on ($\bar{1}11$) and ($1\bar{1}1$) planes, respectively, as bright regions. The complementary nature of

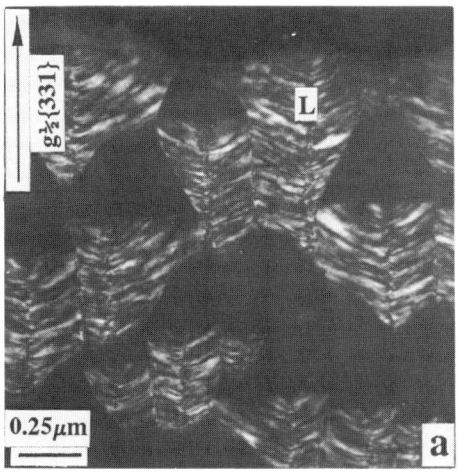
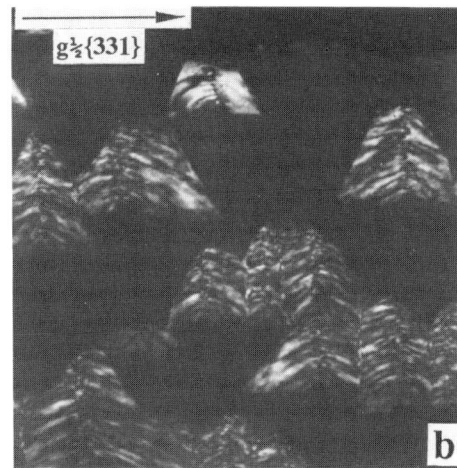

Figure 3. $\mathbf{g}\frac{1}{2}\{331\}$ DF plan-view micrographs taken close to [103] and [013] poles of same area of layer grown at 550°C (2.0nm/s). (a) shows regions ordered on $(\bar{1}11)$ planes, (b) shows regions ordered on $(1\bar{1}1)$ planes. Note characteristic shapes and contrast of ordered regions and linear features, L.

these micrographs shows that almost the whole layer is ordered. The ordered regions of the two variants occupy separate areas when viewed in this orientation. The observed ordered regions have characteristic shapes and contrast features and range in size from ~0.2 - 0.7μm across. The origin of some of the contrast features is not yet clear although they are probably related to the high density of APBs present in this sample, Figs. 2(c) and (d). The regions of the different variants have sharp boundaries which show some preferred alignment. Boundaries between different variants in these micrographs taken close to the [103] and [013] poles lie along <311>, <131> and <331> directions. Boundaries lying along the <331> directions, which appear to be less sharp, could be lying on the orthogonal [110] and [$\bar{1}$10] planes as suggested from the [110] cross-sectional TEM pictures. The sharper boundaries along the <311> directions probably correspond to boundaries on {311} planes. Linear contrast features, e.g. marked L, running through approximately the middle of the domains, and forming boundaries between separate regions of the same variant, are also visible in the micrographs running along one of the <331> directions, which could correspond to features lying on one of the {110} planes.

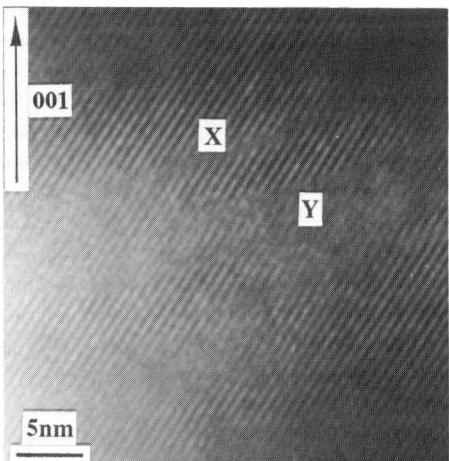

Figure 4. [110] cross-section dark field HREM image, formed using small objective aperture which just included $(\bar{1}1\bar{1})$ and $\frac{1}{2}(\bar{3}3\bar{3})$ spots, of sample grown at 550°C (2.0nm/s). {111} superstructure fringes, spacing 0.678nm visible in ordered regions. X marks sharp APB lying parallel to electron beam, Y marks broad APB which may be lying inclined to electron beam.

In Fig. 4 is shown a [110] cross-section dark field HREM image obtained from the sample grown at 550°C (2.0nm/s). This image was obtained with a JEOL 4000EX operated at 400kV using a small objective aperture which just included the $(\bar{1}1\bar{1})$ fundamental diffraction spot and a $\frac{1}{2}(\bar{3}3\bar{3})$ superlattice spot. The ordered regions in this image show {111} superstructure fringe contrast with a spacing of ~0.678nm, which is double the {111} plane spacing of the $In_{1-x}Ga_xAs$ epitaxial layer. {111} superstructure fringes should shift by half a fringe spacing when crossing an APB (Bellon et al 1989). APBs are clearly visible in Fig. 4, e.g. at points marked X and Y. Some of the APBs are sharp, e.g. X, and lie parallel to the electron beam, whereas others are broad, e.g. Y, and may be lying inclined to the electron beam. This shows that the dark ribbon-like contrast features observed in the [110] cross-section $\frac{1}{2}\{331\}$ DF micrographs of Figs. 2(a) and (b) described earlier are APBs. The presence of tilted arrays of APBs in the ordered regions is responsible for the elongation and tilting of superlattice spots observed in the TED patterns of some samples and may also be associated with the origin of the wavy [001] lines of diffuse intensity running through the superlattice spots in the [110] TED patterns.

4. CONCLUSIONS

We have shown that atmospheric pressure MOCVD (001) $In_{0.53}Ga_{0.47}As$ epitaxial layers grown at temperatures below 650°C exhibit CuPt-type ordering on ($\bar{1}11$) and ($1\bar{1}1$) planes. The degree of ordering is strongly dependent on both growth temperature and growth rate. The maximum degree of atomic ordering occurred for a growth temperature of 550°C. The results may be explained if competing processes of growth surface-induced ordering and bulk disordering are associated with the observed CuPt-type atomic ordering. Large, separate columnar-like ordered regions of the two variants were observed in the layers grown at 550°C. In the high growth rate samples arrays of APBs, tilted ~20° from the (001) plane, were present in the ordered regions which gave rise to elongated tilted superlattice spots present in the [110] TED patterns obtained from these layers.

ACKNOWLEDGEMENTS

The authors would like to thank Prof. Sir Peter Hirsch for laboratory space and the British Council (TYS) for financial support.

REFERENCES

Bass S J, Barnett S J, Brown G T, Chew N G, Cullis A G, Pitt A D and Skolnick M S 1986 J. Crystal Growth 79 378

Baxter C S, Stobbs W M and Wilkie J H 1991 J. Crystal Growth in press

Bellon P, Chevalier J P, Augarde E, André J P and Martin G P 1989 J. Appl. Phys. 66 2388

Bernard J E, Dandrea R G, Ferreira L G, Froyen S, Wei S-H and Zunger A 1990 Appl. Phys. Lett. 56 731

Cao D S, Reihlen E H, Chen G S, Kimball A W and Stringfellow G B 1991 J. Crystal Growth 109 279

Gomyo A, Suzuki T, Kobayashi K, Kawata S, Hino I and Yuasa T 1987 Appl. Phys. Lett. 50 673

Kuan T S, Wang W I and Wilkie E L 1987 Appl. Phys. Lett. 51 51

Kurtz S R, Olson J M and Kibbler A 1990 Appl. Phys. Lett. 57 1922

Murgatroyd I J 1987 D. Phil. Thesis University of Oxford

Murgatroyd I J, Norman A G and Booker G R 1990 J. Appl. Phys. 67 2310

Nakayama H and Fujita H 1986 Proc. GaAs and Related Compounds Conf. Karuizawa Japan 1985, Inst. Phys. Conf. Ser. 79 289

Norman A G 1987 D. Phil. Thesis University of Oxford

Shahid M A, Mahajan S, Laughlin D E and Cox H M 1987 Phys. Rev. Lett. 58 2567

Suzuki T, Gomyo A and Iijima S 1988 J. Crystal Growth 93 396

Taftø J and Spence J H C 1982 J. Appl. Crystallogr. 15 60

Inst. Phys. Conf. Ser. No 117: Section 7
Paper presented at Microsc. Semicond. Mater. Conf., Oxford, 25–28 March 1991   469

A TEM study of the influence of growth conditions on the ordering of $Ga_xIn_{1-x}P$

C S Baxter, W M Stobbs and J H Wilkie[1]

Department of Materials Science, University of Cambridge, Pembroke Street, Cambridge, CB2 3QZ
[1]British Telecom Research Laboratories, Martlesham Heath, Ipswich, IP5 7RE

ABSTRACT: In this paper we report the initial results of a systematic study of the effect of growth conditions on the morphology of {111} ordered structures in $Ga_xIn_{1-x}P$. The results are interpreted as variations of a basic structural model of fine-scale interlocking laminae of the $(1\bar{1}1)$ and $(\bar{1}11)$ variants grouped into large-scale antiphase domains. Our results cast doubt on existing approaches to modelling the growth and we suggest an alternative viewpoint which could be explored further as a possible route to understanding the ordering behaviour.

1. INTRODUCTION

The observation of a reduction in the band gap energy of $Ga_xIn_{1-x}P$ coupled with the occurrence of 'Cu-Pt' type ordering on the $(1\bar{1}1)$ and $(\bar{1}11)$ planes (Suzuki et al 1988a) has led to a plethora of work on the ordering of $Ga_xIn_{1-x}P$ and related compounds, see eg. references in Stringfellow 1989. From the assortment of reported results it is clear that variations in the growth conditions of these materials can lead to a variety of images and diffraction patterns. The structural model (Baxter et al 1991) summarised here can account consistently, by variations on a single theme, for all published TEM data of {1$\bar{1}$1} ordering in III-V compounds. The details of our model, as well as simple structural considerations, have caused us to query the existing 'step-growth' models for explaining the development of these ordered structures. We are therefore investigating the influence of various growth parameters in an attempt to arrive at a better understanding of the growth process.

2. MODEL OF THE MORPHOLOGY OF ORDERED STRUCTURES IN $Ga_xIn_{1-x}P$

In a previous paper (Baxter et al 1991) we described and justified experimentally a structural model for the morphology of {111} ordered $Ga_xIn_{1-x}P$ when grown on substrates with slight (<0.5°) or no misorientation from (001). A characteristic of the model, summarised schematically in figure 1, is the way it describes structural features of the ordering of different size scales. At the fine scale the material consists of alternating laminae of the $(1\bar{1}1)$ and $(\bar{1}11)$ variants. These laminae can extend up to ~100nm laterally in the (001) plane but are frequently no more than a single $Ga_xIn_{1-x}P$ unit cell in thickness (ie. 5.7Å). Evidence for these fine laminae comes from lattice images showing the $\frac{1}{2}${$1\bar{1}1$} planes and also from relatively high magnification dark field images such as those in figure 2 obtained with the electron beam close to perpendicular to [001]. Figures 2a and 2b were formed using ordering reflections from each variant and the complementarity of the two variants can be seen most clearly in the upper part of these images where some of the laminae are slightly thicker than usual.

On a coarser scale the laminae of each variant (considered separately) are grouped into relatively large domains within which all the laminae of the particular variant are in phase with each other but out of phase with the laminae of the same variant in the adjacent domain. We have observed domain widths in the range 2-400nm, depending on growth conditions, and the wider domains can extend throughout the entire thickness of the $Ga_xIn_{1-x}P$ layers. The domains are generally in the form of plates which are inclined at angles from ~10-40° (higher angles have been seen by others for $Ga_xIn_{1-x}P$ on misoriented substrates, eg. figure 5 in Bellon et al 1989). Dark field images taken at lower magnifications than figure 2 show these plate-like domains, as in figure 3, and characteristically the domains relating to each variant are inclined by equal but opposite amounts. When the fine-scale laminar structure is not resolved images of this type have the perplexing appearance that both the variants apparently co-exist in all parts of the material. The appearance of the domain boundaries in strong and weak beam images is in accordance with their being a phase shift of π across the boundaries, and the effect of the presence of antiphase domains on the diffraction pattern is interesting. For both wide and narrow domain material the intensity in the $^1/_2\{\bar{1}11\}$ diffraction spots is concentrated in positions displaced away from the central $^1/_2\{\bar{1}11\}$ position in directions normal to the plates, see figure 4. This is exactly the result expected from an array of antiphase domains, where for an ideally regular array there would be no intensity at the exact $^1/_2\{\bar{1}11\}$ positions. The fine laminae already described generate diffracted intensity streaked perpendicular to the laminae, ie. parallel to [001]. This [001] streaking is displaced from the exact $^1/_2\{\bar{1}11\}$ positions by the intensity distribution generated by the antiphase domains, and the overlap of the [001] streaking of adjacent $^1/_2(1\bar{1}1)$ and $^1/_2(\bar{1}11)$ spots leads to the familiar 'wavy' diffraction patterns of fine domain material. As we discuss elsewhere (Baxter et al 1991) the grouping of the laminae into antiphase domains necessarily implies that <u>all</u> the laminae are an integral multiple of the 'a' lattice parameter in thickness everywhere except at the domain boundaries.

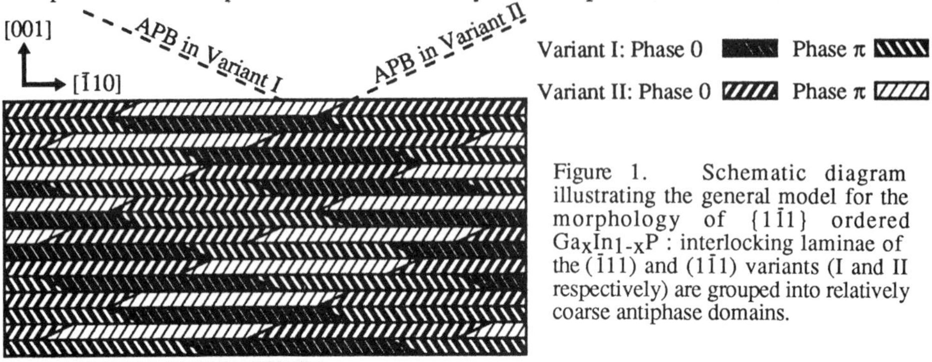
[001]
[$\bar{1}$10]
APB in Variant I
APB in Variant II
Variant I: Phase 0
Phase π
Variant II: Phase 0
Phase π

Figure 1. Schematic diagram illustrating the general model for the morphology of $\{1\bar{1}1\}$ ordered $Ga_xIn_{1-x}P$: interlocking laminae of the $(\bar{1}11)$ and $(1\bar{1}1)$ variants (I and II respectively) are grouped into relatively coarse antiphase domains.

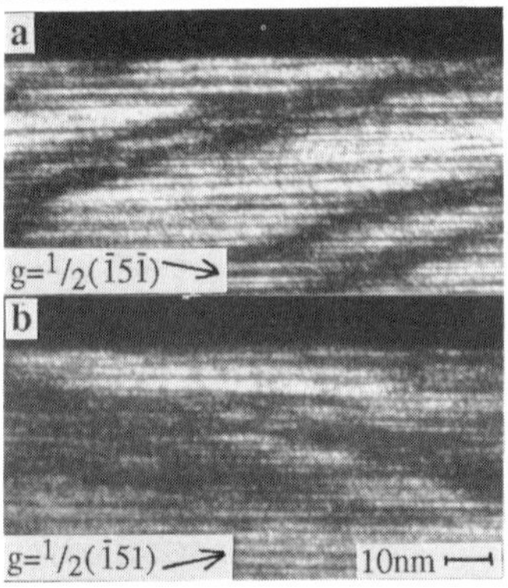
a
g=$^1/_2(\bar{1}5\bar{1})$
b
g=$^1/_2(\bar{1}51)$
10nm

Figure 2. Dark field images of the $(1\bar{1}1)$ [a] and $(\bar{1}11)$ [b] variants. The complementarity of the variants is most easily seen for the coarse laminae at the top of the $Ga_xIn_{1-x}P$ layer.
⟵

⟶
Figure 3. Dark field image of the $(\bar{1}11)$ variant showing the coarse plate-like antiphase domains. The laminae in the domains are not resolved in this image.

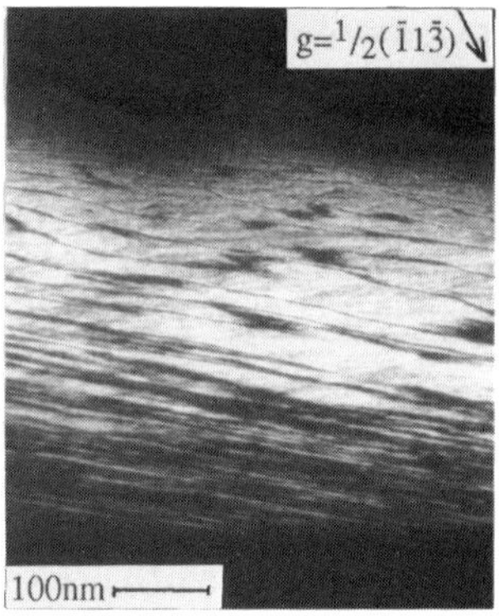
g=$^1/_2(\bar{1}1\bar{3})$
100nm

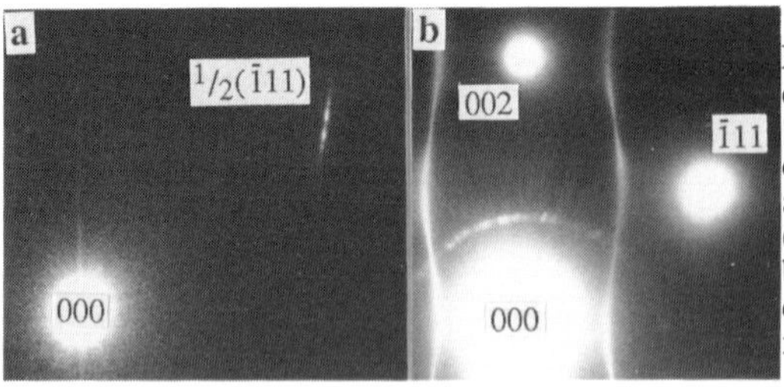
a
$^1/_2(\bar{1}11)$
000
b
002
$\bar{1}$11
000

Figure 4. Enlarged portions of [110] diffraction patterns. For both relatively wide [a] and narrow [b] domain material the diffracted intensity tends to be concentrated away from the exact $^1/_2\{\bar{1}11\}$ positions. (The [001] streaking does not appear in [a] because of the short exposure needed to enable the structure of the spot to be resolved.)

3. EXPERIMENTAL

All our results are for $Ga_xIn_{1-x}P$ with nominal composition x=0.52 grown by atmospheric pressure MOVPE on GaAs substrates. Since the substrate orientation is itself an important variable, wherever possible we have grown several layers on the same substrate during one growth run, with for example just the growth temperature varying between the layers. The $Ga_xIn_{1-x}P$ layers are separated by thick GaAs layers - this approach also saves on TEM specimen preparation time! This main set of results are occasionally supplemented by results from other $Ga_xIn_{1-x}P$ layers grown separately to illustrate particular points.

4. RESULTS AND DISCUSSION

4.1 Effect of Growth Rate. The results of changing the growth rate (by changing the V/III ratio but keeping the overall flow constant) are reasonably straightforward. While keeping the other growth conditions constant (using a single (001) substrate at 670°C), a faster growth rate is associated with finer scale domains which are inclined at a lower angle. This result is most easily illustrated by comparing the $^1/_2\{\bar{1}11\}$ reflections shown in figure 5 - the overall morphology of all the layers is described well by the model discussed above. Changing the rate from 0.15 nms^{-1} to 0.72 nms^{-1} results in a change in domain width from ~10nm to ~3nm. Finer-scale features associated with faster growth are typical of many types of growth experiment. The changes of scale, if not the changes in angle, thus appear to be reasonably comprehensible and by comparison with the results given below it will be seen that the effects of changing the growth rate are small compared with the effects of changing other growth parameters.

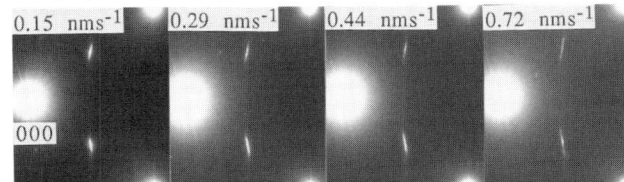

Figure 5. Portions of [110] diffraction patterns from $Ga_xIn_{1-x}P$ layers grown at different rates (as marked). Changes in the shape and inclination of the $^1/_2\{\bar{1}11\}$ diffraction maxima can be seen.

4.2 Effect of Growth Temperature. $Ga_xIn_{1-x}P$ layers were grown at four temperatures on a nominally (001) substrate (actually misoriented $\gtrsim 0.5°$ towards $[1\bar{1}0]$), keeping the growth rate constant at 0.29nm/sec and the V/III ratio at 100. The morphologies which developed show considerable differences as can be seen in the diffraction patterns and image shown in figure 6. The general trend is that the size and angle of inclination of the domains increases with increasing temperature, and that, at least in the lower part of the range investigated, the degree of ordering increases with increasing temperature. The alternating laminae themselves remain on the same fine scale, regardless of the growth conditions. From other samples we have found that qualitatively similar degrees of ordering (though not morphologies) can be obtained in a fairly broad temperature range of ~650-700°C, and higher temperature results can be complicated by intrinsic problems with the growth. Thus the domain structures show trends in size with growth temperature which are in accordance with typical observations of growth processes but the thicknesses of the interlocking laminae apparently do not. This suggests that these laminae may be an intrinsic feature of this ordered state rather than a microstructure which arises purely from the details of growth.

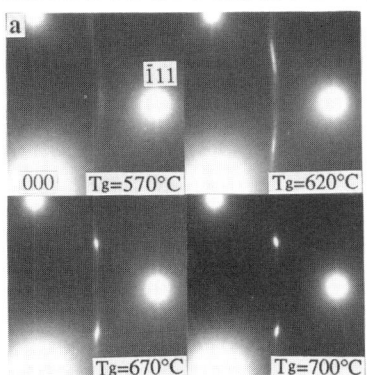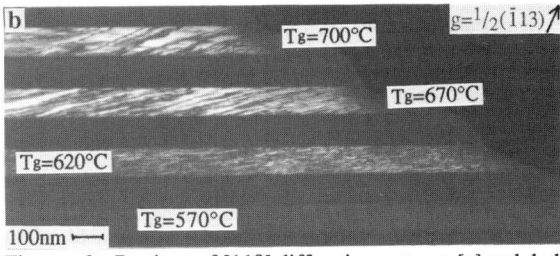

Figure 6. Portions of [110] diffraction patterns [a] and dark field image [b] of $Ga_xIn_{1-x}P$ layers grown on the same substrate but at different temperatures. Both the image and diffraction patterns illustrate the changes in morphology associated with changing the growth temperature.

4.3 Effect of Substrate Orientation. The striking result that the effect of using a substrate surface oriented a few degrees from (001) towards the normal of one of the variants is to inhibit the occurrence of the other variant is well established (eg. Bellon et al 1989). Here we report some additional results obtained using such a substrate oriented 4° towards $[1\bar{1}0]$. $Ga_xIn_{1-x}P$ layers were grown at the same temperatures as those just described with this misoriented substrate placed adjacent to the (001) substrate in the reactor. The expected asymmetry of the variants appeared, but interesting differences between the domain morphologies on the two substrates are also apparent. The general effect is that the domain structure which appears at a particular temperature on the misoriented substrate tends to be similar to one which would be expected at a higher temperature on the (001) substrate - at a particular temperature the misoriented specimen has larger domains inclined at higher angles. The other point to be considered is the inter-relationship of the two variants - with a 4° misorientation both variants do still occur albeit that ($\bar{1}11$) is much "weaker" than ($1\bar{1}1$). The basic results are illustrated in figure 7 (from a specimen grown separately): a) the diffraction pattern contains weak streaking parallel to [001]; b) a dark field from the strong ($1\bar{1}1$) variant shows weak (and variable) modulations; c) both variants still apparently coexist 'everywhere' when considered at low magnifications. These results suggest that the structural pattern is still essentially in accordance with our model, but it is not clear how the different "strengths" of the variants are accommodated structurally. The main possibilities are that the laminae of the "weaker" variant are of reduced size laterally and relatively widely spaced, that the ($\bar{1}11$) variant has a lower order parameter, or a combination of these. Given the influence of the substrate orientation on both the domain morphology and the relative strengths of the variants it is clear that if a flat substrate develops undulations then the growth morphology intrinsic to the orientation may in fact be entirely disguised by that caused by the varying misorientation (see Baxter et al 1990 for an example). One effect of this is that $Ga_xIn_{1-x}P$ grown at high temperatures (700-720°C) produces material for which images of the two variants are to some extent complementary on a coarse scale and have a less modulated structure.

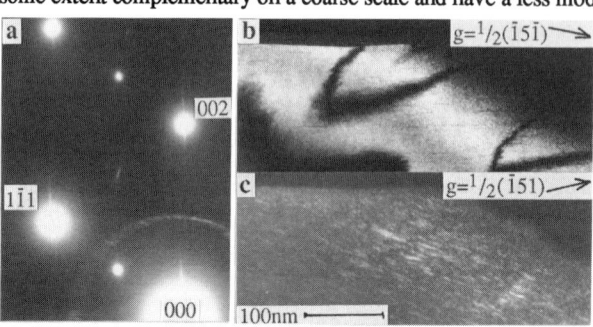

Figure 7. a) Portion of [110] diffraction pattern, b) ($1\bar{1}1$) variant dark field image and c) ($\bar{1}11$) variant dark field image from $Ga_xIn_{1-x}P$ grown on a substrate oriented 4° from (001) towards [$1\bar{1}0$]. It is not clear how the obvious difference in "strengths" of the variants is accommodated structurally.

5. IMPLICATIONS FOR GROWTH MODELS

Given both our interpretation of the interlocking structure and the results reported here we find the approaches to modelling the growth which consider the local effects of a step (eg. Suzuki et al 1988b) unsatisfactory. It seems unlikely that a structure so dependent on the presence of steps for its development would contain domains of widths seemingly independent of the step spacing to the extent that a single coarse domain can contain tens of steps while in fine domain material there can be ~100 domains per step. It is also very difficult to see how a localllised step-growth model could produce a structure of alternating laminae of the two variants given that growth at the surface has to be correlated with several previously grown atomic layers to produce the observed coarse domains. The inclination of the domains remains puzzling - since there is a considerable range of inclinations, the inclination is presumably not linked to the preferred adoption of a crystallographic habit plane. The sense of the inclination is apparently always the same, and it should be noted that where surface steps have generated one stronger variant, the sense of the boundary inclination in the strong variant is such that the boundaries lean 'backwards' up the steps rather than projecting out from them.

We are not as yet in a position to put forward an alternative growth model. However we feel it may be useful to consider the problem of the growth of ordered structures from the viewpoint that a structure of interlocking fine laminae is the intrinsically "desired" (low energy?) structure. We can then speculate that a high step density might inhibit the growth of one variant (rather than actively promoting the other) and thus the development of the "preferred" structure could be prevented.

ACKNOWLEDGEMENTS. We thank Professor C.J. Humphreys for the provision of laboratory facilities and the SERC and BTRL for financial support (CSB).

REFERENCES

C.S. Baxter, R.F. Broom and W.M. Stobbs 1990 Surface Science **228** 102
C.S. Baxter, W.M. Stobbs and J.H. Wilkie 1991   J. Crystal Growth - in press.
P. Bellon, J.P. Chevalier, E.Augarde, J.P. André and G.P. Martin 1989 J.Appl.Phys. **66** 2388
G.B. Stringfellow 1989 J. Crystal Growth **98** 108
T. Suzuki, A. Gomyo, S. Iijima, K. Kobayashi, S. Kawata, I. Hino and T. Yuasa 1988a
    Jap.J.Appl.Phys. **27** 2098
T. Suzuki, A. Gomyo and S. Iijima 1988b J. Crystal Growth **93** 396

Inst. Phys. Conf. Ser. No 117: Section 7
Paper presented at Microsc. Semicond. Mater. Conf., Oxford, 25–28 March 1991

473

Formation of disordered regions in atomically ordered GaInP/AlGaInP heterostructures

R F Broom, W Heuberger, P Unger and P Roentgen

IBM Research Division, Zurich Research Laboratory, 8803 Rüschlikon, Switzerland

ABSTRACT: We describe studies on modifying the group III sub-lattice ordering of GaInP and AlGaInP alloys. Annealing at temperatures of 900-935 °C for 30-60 sec leads to selective area disordering in uncapped areas. This technique is found to be an effective and reproducible means of achieving local disorder, having submicron area resolution, minimal influence on electrical parameters, and uncritical processing conditions. Additionally, it allows a local change in band gap of up to 60 meV without physically destroying the AlGaInP/GaInP heterostructure.

1. INTRODUCTION

The III-V compound alloys GaInP and AlGaInP are important for optoelectronic devices (e.g. lasers) which operate in the visible range of the spectrum, 600-700 nm wavelength. Under a variety of fabrication conditions these materials exhibit atomic ordering of the group III elements on $(\bar{1}11)$ and $(1\bar{1}1)$ planes. The ordering is observed to be associated with a reduction in the band gap of the material, compared to that in which the group III elements are randomly distributed within their respective atomic planes (Gomyo et al 1988a, Suzuki et al 1988a, Kondow et al 1988). Doping the material with a high impurity concentration, ca. 10^{18} cm^{-3}, introduced during growth (Suzuki et al 1988b), or by diffusion, (Morita et al 1988, Gavrilovic et al 1988, Goral et al 1990) disorders the group III elements and raises the band gap. Fully disordered material can be grown by a suitable choice of substrate orientation (Gomyo et al 1988b, Minagawa et al 1989), however, it is useful to be able to vary the band gap locally, which can only be done by disordering of an ordered material. Control of the order/disorder is therefore of importance in determining critical device parameters such as laser wavelength, carrier localization and optical absorption, for example the realization of non-absorbing laser mirrors by Ueno et al (1990). Here we report the results of our studies on the selective removal of the ordering after the initial epitaxial growth, with emphasis on capless annealing, which has been found to have significant advantages.

2. EXPERIMENTAL

The ternary and quaternary $Ga_{0.52}In_{0.48}P$ and $(Al_xGa_{1-x})_{0.52}In_{0.48}P$ epitaxial films, including superlattices (SL's) and quantum wells, (QW's) were grown by low pressure MOVPE, lattice matched, on (001) 2° off-oriented GaAs substrates. The growth temperatures were 680-700 °C with group V/III ratios of 60-200, resulting in atomically ordered layers. Two sample types were studied, having the following principal features:

Structure "A" consisted of 200 nm AlGaInP, followed by 100 nm GaInP, a further 100 nm AlGaInP, a 3.0 nm thick QW of GaInP and a top cladding layer of AlGaInP. Structure "B", for MQW GRINSCH lasers, (Roentgen et al 1990) had AlGaInP cladding layers about 1.0 μm thick, enclosing a SL composed of 5 or 6 layers of 10 nm thick GaInP, separated by 4 nm of AlGaInP. The lower part of the cladding was Si-doped while the region above the SL was Zn-doped. A Zn-doped GaAs layer completed the structure.

Disordering by diffusion was based on rapid thermal annealing (RTA) with elemental Si or a heavily Zn-doped GaAs layer, capped with 150 nm of Si_3N_4. A second technique, selective area disordering, through openings in the Si_3N_4 mask (no additional doping impurity), had advantages in being simple, reproducible and minimizing change in the intrinsic properties of the material. Characterization was done by photoluminescence (PL) and transmission electron microscopy (TEM).

3. RESULTS

The annealing results are summarized in Table I, where the first three columns list the structure, RTA conditions and surface treatment respectively. The last two columns show qualitatively the degree of ordering observed in the GaInP and AlGaInP by dark field TEM, using mainly the $\underline{g} = 1/2(1\bar{1}3)$ diffraction vector and a specimen orientation close to [110].

TABLE I. Annealing results. PL energy-gap shift in mV, ordering in GaInP and AlGaInP observed by TEM. "N.O." = not observable.

Struct.	RTA °C, sec.	Surface coverage	PL ΔE_g (mV)	TEM GaInP	TEM AlGaInP
A.1	925, 60	Si_3N_4 only	14	strong	strong
A.2	925, 60	$Si + Si_3N_4$	58	N.O.	moderate
A.3	925, 60	$GaAsZn + Si_3N_4$	65	N.O.	N.O.
A.4	935, 30	Si_3N_4*	60	See text	
B.1	900, 30	Si_3N_4*	≥ 45		moderate
B.2	935, 30	Si_3N_4*	-		weak
B.3	935, 30	Si_3N_4**	-		v. weak

* Patterned, 5 μm wide, Si_3N_4 stripes, 1.6 μm apart
** 1.5 μm wide, Si_3N_4 stripes.

Figures 1-4 illustrate some features of the results. The TEM micrographs in Figs. 1a and b show the ordering after the treatments A.1 and A.2 of Table I. Arrows indicate the QW and the quaternary and ternary layers are marked by Q and T. In Fig. 1b the ternary is totally disordered while the quaternary is little affected by the Si diffusion. To disorder the quaternary required either Zn as diffusion species as in sample A.3, or, higher temperatures.

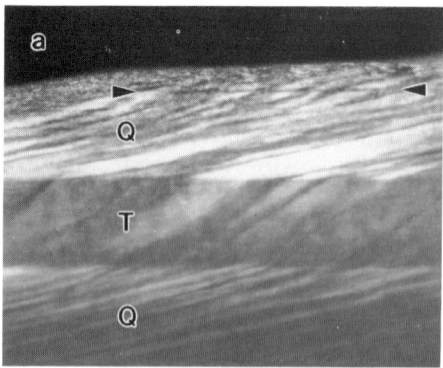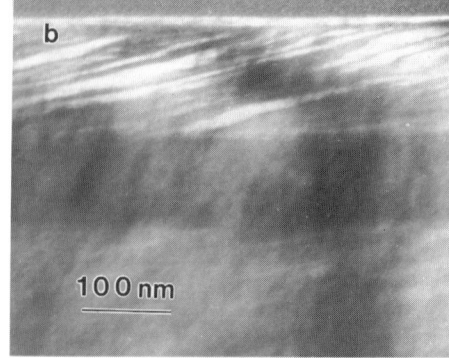

Figure 1. Dark field micrograph of structure A. (a) RTA 925 °C, 60", Si_3N_4 cap. (b) as (a) but $Si + Si_3N_4$ cap.

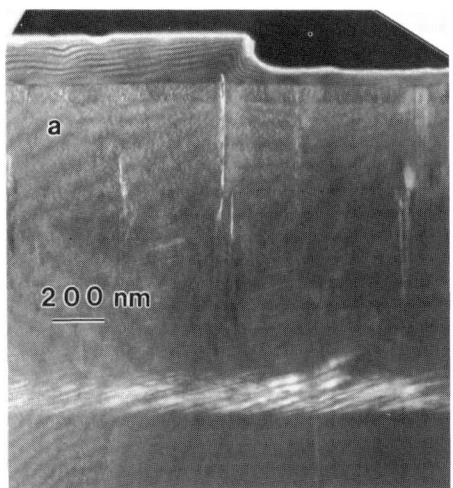

Figure 2. Dark field micrograph of structure B.
(a) RTA 900 °C, 30″,
(b) RTA 935 °C, 30″, (SL region only).

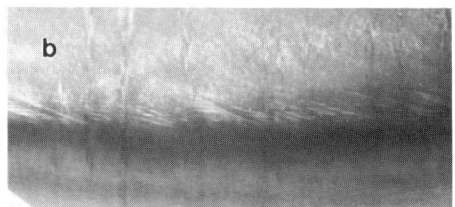

Figure 2a is a TEM micrograph of the upper region of sample B.1. The step in the surface corresponds to the edge of the 1.6 μm wide opening in the Si₃N₄, extending to the right of the figure. Ordering is visible in the region of the SL (fabricated with a higher V/III ratio than the surrounding AlGaInP), but there is no visible lateral variation, probably because of the large distance from the top. At higher temperature, Fig. 2b of sample B.2, most of the order at the SL is removed. A semiconductor laser after capless RTA at 900 °C exhibited an emission energy increase by 45 mV, with only minimal change in electrical parameters.

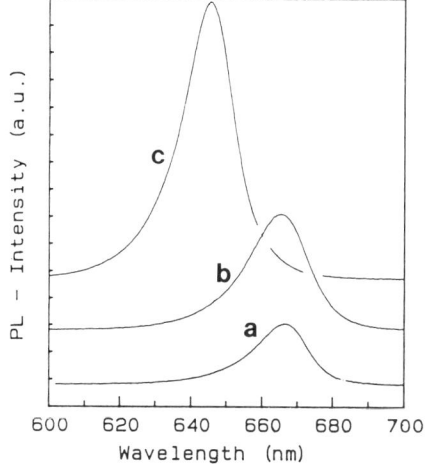

Figure 3. 300 K PL spectra of sample A.4.
(a) unannealed, (b) annealed with Si₃N₄ cap,
(c) annealed without cap.

The three PL spectra of Fig. 3, corresponding to luminescence of the GaInP layer of sample A.4, are from (a) the unannealed material, (b) annealed with Si₃N₄ cap and (c) annealed without cap. Clearly, there is only a small wavelength shift with the capped compared to the uncapped surface. Note the increase in intensity of the spectrum of (c). In all cases the linewidth is constant at FWHM = 52 ± 1 mV, indicating that the disordered GaInP is quite uniform. Figure 4 is a TEM micrograph of the boundary region between coated and uncoated surfaces, (Sample A.4). Some 100 nm of the top surface, including the QW, has been etched off, however it can be seen that the GaInP along the centre of the picture is ordered beneath the coating whereas it is becomes progressively disordered to the right. The transition region is quite sharp, being within 200-300 nm.

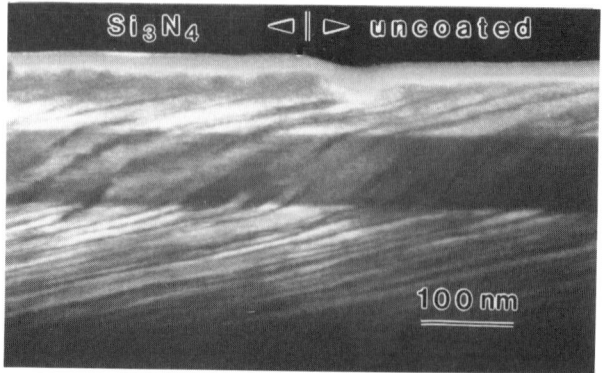

Figure 4. Dark-field micrograph of sample A.4 at the boundary between capped and uncapped regions, showing diminishing order in the GaInP beneath the uncapped area.

4. CONCLUSIONS

Rapid thermal annealing, combined with a patterned Si_3N_4 capping layer, is shown to be an effective means of forming disordered layers in GaInP, having sub-micron resolution. The method has several advantages. For example, it is simple and fast, it is reproducible and it is relatively insensitive to the conditions. Capless annealing has a minimum influence on intrinsic device properties, but the ability to form lateral variations in the band gap, without disturbance of the heterostructure, offers the flexibility of designing improved devices. Experimentally, we find that the ordered AlGaInP is more stable than GaInP in the sense that it is harder to disorder.

REFERENCES

Gavrilovic P, Dabkowski F P, Meehan K, Williams J E, Stutius W, Hsieh K C, Holonyak N Jr., Shahid M A and Mahajan S 1988 J. Crystal Growth **93** 426
Gomyo A, Suzuki T and Iijima S 1988a Phys. Rev. Lett. **60**(25) 2645
Gomyo A, Suzuki T, Iijima S, Hotta H, Fujii H, Kawata S, Kobayashi K, Ueno Y and Hino I 1988b Jpn. J. Appl. Phys. **27**(12) L2370
Goral J P, Kurtz S R, Olson J M and Kibbler A 1990 J. Electron. Mat. **19**(1) 95
Kondow M, Kakibayashi H, Minagawa S, Inoue Y, Nishino T, and Hamakawa Y 1988 Appl. Phys. Lett. **53**(21) 2053
Minagawa S, Kondow M, Kakibayashi H 1989 Electron. Lett. **25**(21) 1440
Morita E, Ikeda M, Kumagai O, aand Kaneko K 1988 Appl. Phys. Lett. **53**(22) 2164
Roentgen P, Bona G L, Heuberger W, Unger P, and Jaeckel H, Proc. 12th IEEE Int. Semiconductor Conf. Sept. 9-14, 1990 Davos, Switzerland PD-11 23 (IEEE Catalog No. 90CH2918-1)
Suzuki T, Gomyo A, Iijima S, Kobayashi K, Kawata S, Hino I, and Yuasa T 1988a Jpn. J. Appl. Phys. **27**(11) 2098
Suzuki T, Gomyo A, Hino I, Kobayashi K, Kawata S and Iijima S 1988b Jpn. J. Appl. Phys. **27**(8) L1549
Ueno Y, Fujii H, Kobayashi K, Endo K, Gomyo A, Hara K, Kawata S, Yuasa T and Suzuki T 1990 Jpn. J. Appl. Phys. **29**(9) L1666

Effects of annealing on phase separated microstructures in InGaAsP epitaxial layers

TL McDevitt*, S Mahajan, DE Laughlin, WA Bonner** and VG Keramidas**

Department of Materials Science, Carnegie Mellon University, Pittsburgh, PA 15213, USA

**Bellcore, Red Bank, NJ 07701, USA

ABSTRACT :   The influence of annealing on phase separated microstructures in lattice-matched InGaAsP epitaxial layers, grown on (001) InP substrates by liquid phase epitaxy, has been investigated. It is shown that the fine and coarse contrast modulations commonly observed in these layers are coupled. This observation is consistent with the suggestion that the fine scale structure causes the coarse modulations.   Further, the carrier mobility and the carrier concentration increase on annealing.

1. INTRODUCTION

Following the theoretical work of de Cremoux et al. (1981), Onabe (1982) and Stringfellow (1982) on the occurrence of phase separation in ternary and quaternary III-V compound semiconductors, Henoc et al. (1982), Launois et al. (1982), Mahajan et al. (1984), Norman and Booker (1985), Chu et al. (1985), Treacy et al. (1985), Mahajan and Shahid (1989), Mahajan et al. (1989) and McDevitt (1990) have investigated the microstructural characteristics of InGaAsP epitaxial layers. Two types of contrast modulations are observed in lattice-matched layers grown on (001) InP substrates by liquid phase epitaxy (LPE) (Henoc et al. 1982, Mahajan et al. 1984, Norman and Booker 1985, Chu et al. 1985, Treacy et al. 1985, Mahajan and Shahid 1989, Mahajan et al. 1989 and McDevitt 1990).   A fine-scale speckle structure is seen that is aligned along the <100> directions lying in the (001) growth plane (Mahajan et al. 1984 and Norman and Booker 1985). The wavelength of these modulations is ~12-15nm (Henoc et al. 1982, Mahajan et al. 1984, Norman and Booker 1985 and Chu et al. 1985), and depends on the growth temperature as well as on the orientation of the underlying substrate (McDevitt 1990). This structure is associated with two-dimensional strains along the directions of the modulations (McDevitt 1990). In addition, the layers show coarse contrast modulations whose period is ~125-150nm; these modulations are also oriented along the [100] and [010] directions (Henoc et al. 1982, Mahajan et al. 1984, Norman and Booker 1985, Chu et al. 1985, Treacy et al. 1985, Mahajan and Shahid 1989, Mahajan et al. 1989 and McDevitt 1990).

Several investigators have attempted to rationalize the formation of the two types of contrast modulations (Henoc et al. 1982, Launois et al. 1982, Mahajan et al. 1984, Norman and Booker 1985, Chu et al. 1985, Mahajan and Shahid 1989 and Mahajan et al. 1989).   Two divergent viewpoints have emerged.   Since the wavelength of the coarse modulations is very large, it is impossible to argue that these modulations could develop by phase separation in the bulk because bulk diffusion is extremely slow in these materials. To obviate this difficulty, Launois et al. (1982) have suggested that the modulations evolve by phase separation occurring at the surface while the layer is growing.   An interesting question pertains to the formation of the fine scale speckle structure?   Norman and Booker (1985) envisage that this is caused by phase separation in the bulk.   Mahajan et al. (1984), Mahajan and Shahid (1989) and Mahajan et al. (1989) agree that the speckle structure is due to phase separation, but have argued that the

---

*Present address: IBM Corporation, Essex Junction, VT 05452-4299

coarse contrast modulations may form to accommodate strains associated with the fine scale structure. Their argument is that the wavelength of the coarse modulations is very large and cannot evolve by surface diffusion during the time available between the deposition of two successive monolayers.

It is apparent from the above discussion that the fine and the coarse modulations are decoupled from each other according to Launois et al. (1982) and Norman and Booker (1985), whereas Mahajan et al. (1984), Mahajan and Shahid (1989) and Mahajan et al. (1989) envisage them to be coupled. Since the wavelength of the coarse modulations is about ten times longer than that of the speckle structure, the reversion of coarse modulations on annealing via bulk diffusion should take hundred times longer than that for the fine structure. On the other hand, the two features in the coupled case should revert simultaneously on annealing. We have carried out such experiments on InGaAsP epitaxial layers, and these results constitute the present paper.

2. EXPERIMENTAL DETAILS

For this study, InGaAsP epitaxial layers emitting at 1.3μm were grown at 600° C by LPE on (001) InP substrates. Some of the layers were annealed in the temperature range of 750-900° C. In order to prevent the thermal decomposition of the material during annealing, a dielectric capping layer was deposited on the surface of the epitaxial layers. The encapsulant consisted of 1-2nm of SiO followed by ~100nm of Si_3N_4. In addition, the encapsulated specimens were surrounded by pieces of InP to enhance thermal stability. Following the anneal, the dielectric layers were removed by etching in HF. The as-grown and annealed specimens were prepared for transmission electron microscopy by chemical etching as described by Chu and Sheng (1984). Thinned samples were examined in a Philips 420 electron microscopy operating at 120 keV.

3. RESULTS

Shown in Fig. 1 are a series of dark-field electron micrographs obtained from the as-grown and the annealed layers; these images were taken with 220 reflections under dynamical conditions. Several important microstructural characteristics are evident in the micrographs. First, the as-grown layer, shows the two types of contrast modulations. The contrast experiments have shown that the strains associated with the modulations are along the [100] and [010] directions (McDevitt 1990). Second, when the layer is annealed at 750° C for 30 min., the contrast from both types of modulations is weak and becomes progressively weaker as the annealing temperature is raised. Third, the layer annealed at 900° C, Fig. 1(f), shows speckle structure that is much finer than that observed in the as-grown layer, Fig. 1(a).

Concomitantly, the changes in microstructures on annealing have been followed using selected area electron diffraction, and these results are reproduced as Fig. 2. The satellite spots which are seen clearly in the pattern obtained from the as-grown layer are barely discernible in the layer annealed at 800° C for 30 min., Fig. 2(c). Furthermore, the patterns from the specimens annealed at 850 and 900° C show very weak diffuse intensity.

Figure 3 shows the contrast behavior of the fine scale structure observed in the sample annealed at 900° C. Comparing these results with those of the as-grown sample (McDevitt 1990) it is inferred that the contrast behavior of the speckle structure is identical to that of the fine scale structure observed in as-grown layers. The speckle structure is aligned along the [100] and [010] directions and has principal strains parallel to the directions of the modulations. The major difference between the annealed and as-grown microstructures is that the periodicity of the contrast in the annealed specimen is much smaller. The measurements of the period in Fig. 3 give a value of ~3-4nm, whereas the periodicity of the fine scale modulations in the as-grown samples is 8-10nm.

The effects of a 850° C anneal on carrier mobility and carrier concentration are depicted in Fig. 4. All layers show n-type conductivity. The mobility increases sharply after a 30 min. anneal. However, additional annealing does not produce a further increase in mobility. Surprisingly, the carrier concentration also shows a marked increase on annealing.

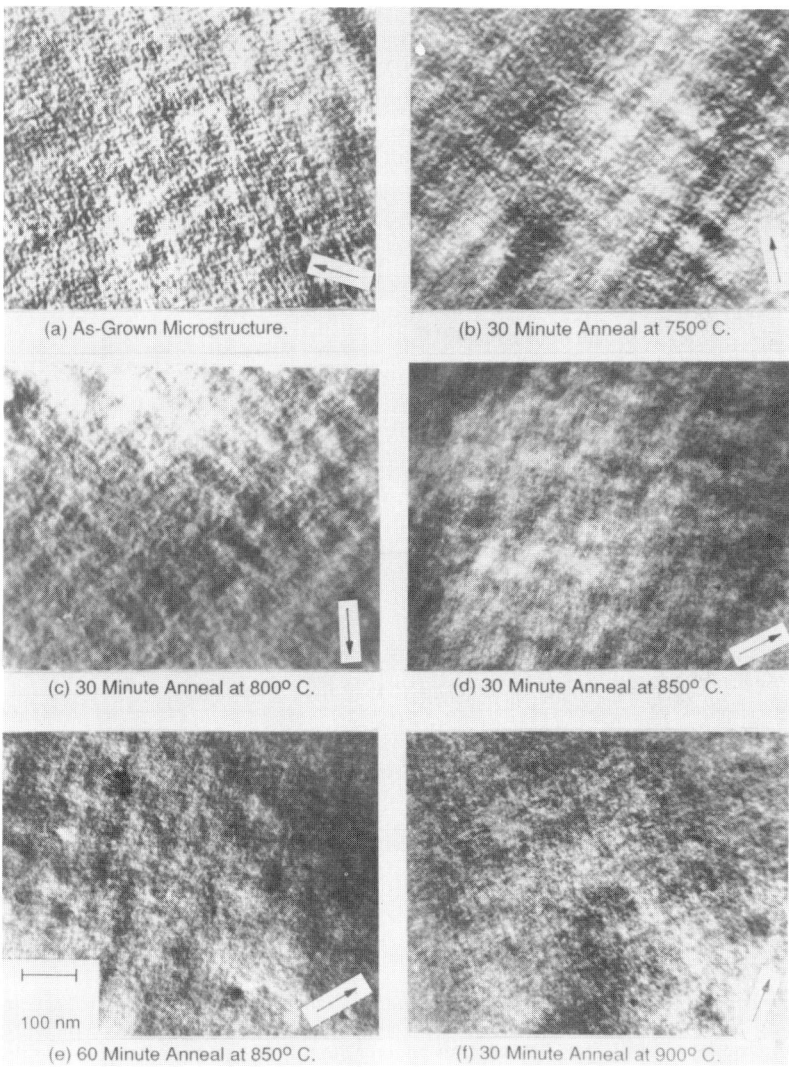

(a) As-Grown Microstructure.

(b) 30 Minute Anneal at 750° C.

(c) 30 Minute Anneal at 800° C.

(d) 30 Minute Anneal at 850° C.

100 nm

(e) 60 Minute Anneal at 850° C.

(f) 30 Minute Anneal at 900° C.

Figure 1. Dark-field electron micrographs showing the behavior of the fine and coarse contrast modulations during the annealing of (001) InGaAsP epitaxial layers at different temperatures. All micrographs were taken under dynamical diffraction conduction with the 220 reflection satisfied.

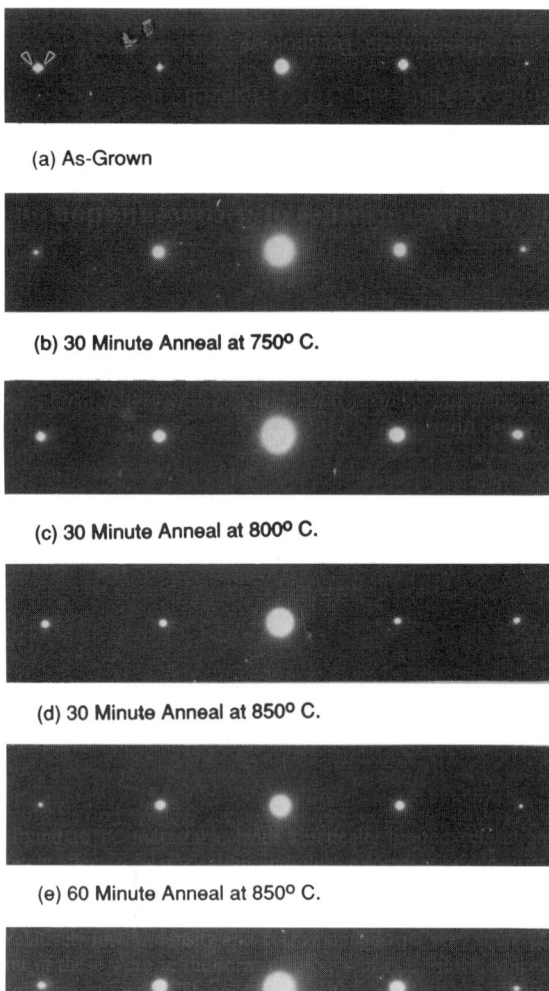

(a) As-Grown

(b) 30 Minute Anneal at 750° C.

(c) 30 Minute Anneal at 800° C.

(d) 30 Minute Anneal at 850° C.

(e) 60 Minute Anneal at 850° C.

(f) 30 Minute Anneal at 900° C.

Figure 2. Sections of the selected area diffraction patterns, showing the $\bar{4}00$ to 400 reflections, obtained from the specimens shown in Fig. 1. The pattern from the as-grown film shows satellites indicated by arrows in (a), whereas the patterns from the specimens annealed at 850 and 900° C show very weak diffuse intensity.

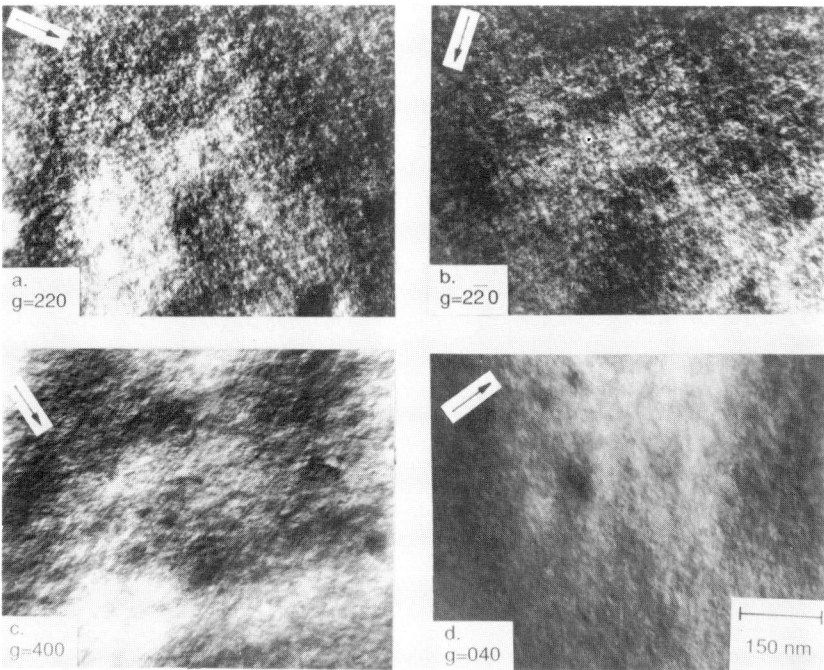

Figure 3. Dark-field electron micrographs showing the contrast behavior of very fine speckle microstructure observed in an InGaAsP layer after annealing at 900° C for 30 mins. The operating reflection in each micrograph is delineated by an arrow.

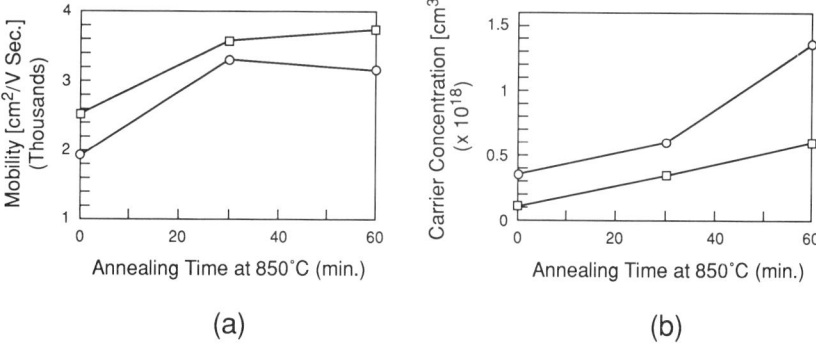

(a)   (b)

Figure 4. Plots showing the dependence of (a) carrier mobility and (b) carrier concentration in an InGaAsP layer on annealing time at 850° C. Both the mobility and the carrier concentration increase on annealing.

4. DISCUSSION

Several interesting observations emerge from the preceding investigation. First, the fine and the coarse contrast modulations in InGaAsP epitaxial layers are coupled. Second, an extremely fine speckle structure is observed even after annealing at 900° C for 30 min., but the coarse modulations are not visible.

The above results clearly demonstrate that the two types of modulations are coupled. Therefore, as argued in the Introduction Section, they are inconsistent with the explanations of Launois et al. (1982) and Norman and Booker (1985); i.e., the two features could not have formed independently.

The origin of the coarse modulations can be rationalized as follows. By studying the dependence of the speckle structure in InGaAsP layers on the orientation of the underlying substrate, McDevitt et al. (1991) have demonstrated unequivocally that the fine scale structure results from two-dimensional phase separation occurring at the layer surface during growth. In addition, phase separation is not observed along the growth direction for each of the four orientations examined. If the speckle structure were to evolve in the bulk as suggested by Norman and Booker (1985), decomposition would have dominated along the growth direction because of the thinness of the layer the transformation-induced strains could be accommodated more easily. Coupling the above result with the presence of strains along the [100] and [010] directions due to phase separation, it is inferred that the layer is under a biaxial stress state.

Now imagine a situation where the underlying substrate has been removed and the layer is thinned so that it is transparent to 120 keV electrons. As suggested by Alerhand et al. (1988) the ground state for such a system is surface domains. The break up of surface into domains would cause periodic, localized bending of planes in the surface regions. We envisage that this effect is responsible for the coarse contrast modulations.

Treacy et al. (1985) have observed that the contrast from the coarse modulations depends on the thickness of the region under examination. They do not see the modulations in the thicker and thinner regions. This observation can be rationalized as follows. Since the two-dimensional surface stresses are responsible for the periodic buckling discussed above, an optimal balance between the thickness of the region and the magnitude of the biaxial stresses is required. In the thinner regions, the stresses are not high enough for buckling to occur, whereas the thicker regions require much higher buckling stresses.

A plausible explanation for the observed speckle structure in the layer annealed at 900° C, Figs. 1(f) and 3, is that the fine scale structure contains a range of wavelengths. Thus, the structure observed on annealing represents remanence of the coarser component of the microstructure.

The observed increase in carrier mobility may be rationalized in terms of the Blood-Grassie model (1984). Their approach assumes that phase separated microstructures can be modeled as a series of heterojunctions between materials differing slightly in compositions. They have proposed that the mobility of carriers in such a system would be lower than in a random alloy because the carriers may be confined to troughs in the conduction band. The carriers would then migrate by a percolation process occurring along these minima in the band structure. Further, we at present do not understand the concomitant increase in carrier concentration on annealing. It could be that there are carrier traps in the phase separated material that are eliminated on annealing.

The microstructure of InGaAsP epitaxial layers appears to be simpler than that envisaged earlier (Henoc et al. 1982, Launois et al. 1982, Norman and Booker 1985 and Treacy 1985). In the (001) layers, the two-dimensional composition modulations occur along the [100] and [010] directions. Their wavelength and amplitude can be tailored by changing the growth temperature, the growth technique and the orientation of the underlying substrate McDevitt et al. (1991). Since these microstructural features appear to affect the electronic properties (Launois et al. 1982 and McDevitt 1990), it should be possible to manipulate microstructures to produce epitaxial layers having unique electronic behavior.

In summary, it has been demonstrated by annealing experiments that the fine and coarse contrast modulations observed in InGaAsP layers, grown by LPE, are coupled. The fine scale speckle evolves by two-dimensional phase separation occurring at the surface while the layer is growing, whereas the coarse contrast modulations are an artifact of thin foils. Furthermore, the carrier mobility is increased on annealing.

ACKNOWLEDGEMENTS

The work at Carnegie Mellon was carried out under the auspices of the Department of Energy and TLM, SM and DEL gratefully acknowledge the award of the grant DE-FG02-ER45329.

REFERENCES

Alerhand OL, Vanderbilt D, Meade RD and Joannopoulas, JD 1988 Phys. Rev. Lett. 61 1973
Blood P and Grassie ADC 1984 J. Appl. Phys. 56 1866
Chu SNG, Nakahara S, Strege KE and Johnston, Jr WD 1985 J. Appl. Phys. 57 4610
Chu SNG and Sheng TT 1984 J. Electrochem. Soc. 131 2663
de Cremoux, Hirtz P and Ricciardi J 1981 Inst. Phys. Conf. Ser. No. 56 115
Henoc P, Izrael A, Quillec M and Launois H 1982 Appl. Phys. Lett. 40 963
Launois H, Quillec M, Glas F and Treacy MMJ 1982 Inst. Phys. Conf. Ser. No. 65 537
Mahajan S, Dutt BV, Temkin H, Cava RJ and Bonner WA 1984 J. Cryst. Growth 68 589
Mahajan S and Shahid MA 1989 MRS Proceeding Vol. No. 144 169
Mahajan S, Shahid MA and Laughlin DE 1989 Inst. Phys. Conf. Ser. No 100 143
McDevitt TL, 1990 Ph.D. Dissertation, Carnegie Mellon University
McDevitt TL, Mahajan S, Laughlin DE, Bonner WA and Keramidas GB 1991 submitted for publication to Phys. Rev. B
Norman AG and Booker GR 1985 J. Appl. Phys. 57 4715)
Onabe K 1982 Jpn. J. Appl. Phys. 21 1323
Stringfellow GB 1982 J. Cryst. Growth 58 194
Treacy MMJ, Gibson JM and Howie A 1985 Phil. Mag. A 51 389

Inst. Phys. Conf. Ser. No 117: Section 7
Paper presented at Microsc. Semicond. Mater. Conf., Oxford, 25–28 March 1991

485

Phase separation and associated defects in MBE $InAs_ySb_{1-y}$ epitaxial layers

T -Y Seong[1], A G Norman[2], J L Hutchison[1], I T Ferguson[2], G R Booker[1], R A Stradling[2,3] and B A Joyce[2]

[1]Dept. of Materials, University of Oxford, Parks Road, Oxford OX1 3PH, UK
[2]IRC for Semiconductor Materials, Blackett Laboratory, Imperial College, Prince Consort Road, London SW7 2BZ ,UK
[3]Dept. of Physics, Blackett Laboratory, Imperial College, Prince Consort Road, London SW7 2BZ ,UK

ABSTRACT: TEM, TED and HREM examinations have been performed on (001) MBE layers of nominal composition $InAs_{0.5}Sb_{0.5}$ grown using constant ratios of group V fluxes. For growth at 430°C, single composition material occurred. For growth at ≤ 400°C, alternating, tetragonally distorted, elongated platelets of two cubic phases of compositions $\sim InAs_{0.40}Sb_{0.60}$ and $\sim InAs_{0.78}Sb_{0.22}$ occurred, corresponding to a strained layer superlattice structure. These two-phase structures formed spontaneously at the growing layer surface and were stable during subsequent annealing at 370°C. It is suggested that they arise due to the presence of a miscibility gap at these low temperatures.

1. INTRODUCTION

The ternary alloy $InAs_ySb_{1-y}$ is of increasing interest for infrared detectors in the 3-12μm wavelength region (Osbourn 1990). Growth of $InAs_ySb_{1-y}$ alloys (0.5<y<0.7) has proved difficult by equilibrium growth techniques such as liquid phase epitaxy (LPE) due to the presence of a thermodynamically predicted miscibility gap (de Cremoux 1982, Onabe 1982, Stringfellow 1983). It is expected that alloys are thermodynamically unstable when grown within the miscibility gap and phase separation may occur. Nevertheless $InAs_ySb_{1-y}$ has been successfully grown over the complete range of compositions by non-equilibrium techniques such as metal organic chemical vapour deposition (MOCVD) (Chiang and Bedair 1984) and molecular beam epitaxy (MBE) (Lee et al 1985, Yen et al 1987), but structural characterisation of the layers has not been extensively performed.

In our previous work (Seong et al 1990), single thick MBE $InAs_ySb_{1-y}$ layers were grown on InAs buffer layers on (001) GaAs substrates at 370°C, using a constant ratio of Group V fluxes, with nominal values of y = 0, 0.2, 0.4, 0.6, 0.8 and 1.0. This temperature is lower than those generally used to grow such $InAs_ySb_{1-y}$ layers. Transmission electron microscopy (TEM) structural studies showed that single-phase material grew for y = 0, 0.2 and 1.0, but two-phase material grew for y = 0.4, 0.6 and 0.8. The two-phase material consisted of alternating plates of composition $\sim InAs_{0.3}Sb_{0.7}$ and $\sim InAs_{0.7}Sb_{0.3}$, the plates lying approximately on the (001) plane and having thicknesses in the range 20 to 200nm. The crystal lattice of the $InAs_{0.3}Sb_{0.7}$ phase was tetragonally distorted with c/a > 1, while the $InAs_{0.7}Sb_{0.3}$ phase was tetragonally distorted with c/a < 1. This phase separation was attributed to the presence of a miscibility gap at 370°C. Atomic ordering of the CuPt-type occurred on two sets of {111} planes in these layers. In order to investigate this behaviour further, TEM and transmission electron diffraction (TED) studies have now been performed on MBE layers of nominal composition $InAs_{0.5}Sb_{0.5}$ grown at temperatures in the range 340 to 430°C. The results obtained strongly suggest that phase separation occurs at the growing surface in these layers. High resolution electron microscopy (HREM) studies have been performed on the crystallographic defects present in these phase separated layers.

2. EXPERIMENTAL DETAILS

The $InAs_{0.5}Sb_{0.5}$ samples were grown by MBE in a VG Semicon V80H system: the details of growth will be given elsewhere (Ferguson et al). The substrates were (001) Cr-doped semi-insulating GaAs on which a fully relaxed InAs buffer layer was grown. $InAs_{0.5}Sb_{0.5}$ layers were then grown in the temperature range 340 to

430°C using a constant ratio of Group V fluxes. Two orthogonal [110] and [$\bar{1}$10] cross-section thin foil specimens were prepared by mechanical polishing and Ar$^+$ ion-beam milling using a liquid N$_2$ cold stage. The specimens were then examined by TEM and TED, using standard two-beam diffraction conditions and selected area electron diffraction techniques in a Philips CM 20 operated at 200kV and a Philips CM 12 operated at 120kV, and by HREM using a JEOL 4000 EX operated at 400kV (point resolution 0.16nm). [110] and [$\bar{1}$10] cross-sections were distinguished using chemical etching (Murgatroyd et al 1990) and convergent beam electron diffraction techniques (Taftø and Spence 1982).

3. RESULTS AND DISCUSSION

3.1 TEM and TED Results

The InAs$_{0.5}$Sb$_{0.5}$ samples were examined in cross-section by TEM dark field (DF) using the compositional dependent 002 reflection. In Fig. 1a and b are shown **g**[002] DF micrographs taken from the [110] and [$\bar{1}$10] cross-sections of the layer grown at 340°C. In the [110] cross-section , Fig. 1a, a strong plate-like contrast is visible with the platelets lying approximately parallel to the layer surface. The platelets alternate in contrast from dark to light. These correspond to material with two different alloy compositions indicating that the layer has undergone phase separation. Consideration of the structure factors for InAsSb shows that the darker contrast corresponds to an Sb-rich phase and the lighter contrast to an As-rich phase. The abrupt change in contrast between the platelets indicates that there are relatively sharp boundaries between the two phases. The morphology of the platelet structure when viewed in the [110] direction is reasonably regular.   TED measurements, described below, indicated that a strained layer superlattice structure had formed. Comparison of this micrograph with that of Fig. 1b taken from the [$\bar{1}$10] cross-section shows a marked anisotropy of the platelet geometry along the orthogonal <110> directions parallel to the layer surface. The platelets appear much less regular and planar when viewed along the [$\bar{1}$10] direction. In this orientation lenticular shaped platelets of the light contrast, As-rich phase, appear to be embedded in a continuous matrix of the dark contrast, Sb-rich phase. The length of the As-rich platelets along the [110] direction as measured from [$\bar{1}$10] cross-sections varied from 0.02-0.5μm. Conversely, along the [$\bar{1}$10] direction the corresponding length was often greater than 1μm. The As-rich phase thus occurs as elongated, lenticular cross-section platelets oriented with their long axis along the [$\bar{1}$10] direction, surrounded by a matrix of the Sb-rich phase. As the substrate temperature increased from 340 to 400°C, the platelets progressively became more planar and uniform, and increased in length and thickness, as shown in the [110] cross-section micrographs of the layers grown at 370 and 400°C, Fig. 1c and d. Layers grown at 430°C and above were essentially homogeneous, showing no phase separation, Fig. 3, although atomic ordering on two sets of {111} planes did occur.   A high density of crystallographic microdefects was present in the layers, Fig. 1a-d. These included threading dislocations together with stacking faults and microtwins lying on inclined {111} planes. Many of the microtwins originated and then terminated at interfaces between plates, Fig. 1a. Some dislocations were also present running parallel to the platelet interfaces and these are associated with the lattice-misfit between the two phases, see below.

Selected area TED patterns were taken from the InAs$_{0.5}$Sb$_{0.5}$ layers with the incident electron beam parallel to the <110> directions. For the layers that had undergone phase separation the higher order "spots" were split into two spots in the [001] direction, but not in the <110> directions, Fig. 2, indicating the existence of two different lattice parameters perpendicular to the layer surface. From this splitting the relaxed cubic lattice parameters of each phase were deduced using the procedure described by Seong et al (1990) and the composition of the two phases derived using Vegard's law. The results are summarised in Table I. For example, for the layer grown at 340°C, the lattice parameters perpendicular to the layer surface were 6.40Å and 6.04Å, whilst that parallel to the surface was 6.25Å for both phases. Thus the lattices of the two phases were tetragonally distorted, one with c/a = 1.024 and other with c/a = 0.966. The relaxed cubic lattice parameters correspond to alloy compositions of InAs$_{0.38}$Sb$_{0.62}$ and InAs$_{0.80}$Sb$_{0.20}$ respectively. Previously we found, by TEM and TED, that varying the nominal alloy composition markedly changed the relative volume of these phases but not significantly their compositions (Seong et al 1990).   Energy dispersive X-ray (EDX) microanalysis using a VG HB 501 scanning transmission electron microscope (STEM) gave compositions for the two different phases which confirmed those deduced from TED. Table I shows that although there is not a convincing progressive change in either y$_1$ or y$_2$ as the growth temperature increases, there is a progressive decrease in δy (0.42 to 0.38 to 0.34), which is the behaviour that would be expected if the effect were due to a miscibility gap. Calculations of the miscibility gap for InAsSb (de Cremoux 1982, Onabe 1982, Stringfellow 1983) show that the difference in composition (δy) corresponding to the two ends of the gap decreases as the temperature increases in this range.

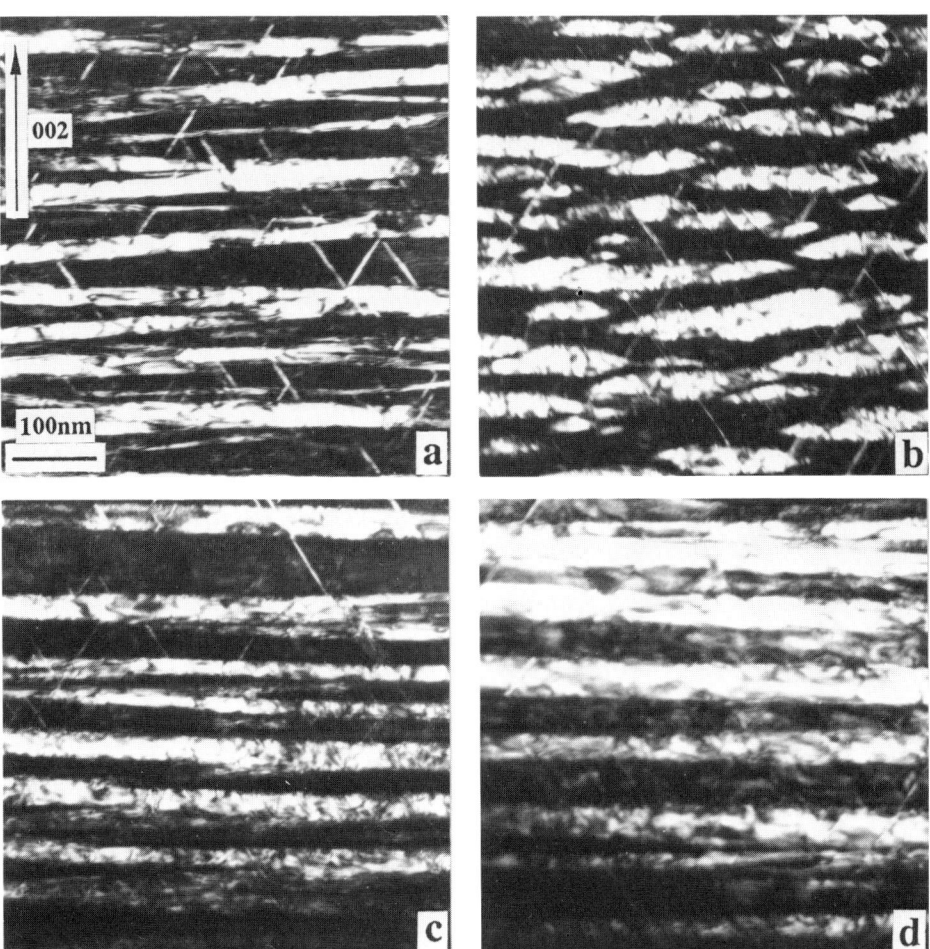

Figure 1. **g**[002] DF cross-section micrographs of MBE layers of nominal composition InAs$_{0.5}$Sb$_{0.5}$ showing plate-like contrast due to phase separation. a. [110] cross-section, growth temperature (T$_g$) 340°C, b. [$\bar{1}$10] cross-section, T$_g$ = 340°C, c. [110] cross-section, T$_g$ = 370°C, d. [110] cross-section, T$_g$ = 400°C.

Figure 3 shows a **g**[002] DF TEM micrograph taken from a [110] cross-section sample of an InAs$_{0.5}$Sb$_{0.5}$ layer grown at 430°C for 1 hour to give a thickness of 0.85μm, at which point the substrate temperature was decreased to 370°C and growth continued for a further 1 hour to give an additional layer of thickness 0.85μm. The sample was then annealed at 370°C for a further 2 hours under an arsenic flux. The layer grown at 430°C is reasonably homogenous with no significant phase separation even after being at 370°C for 3 hours. This shows that there is a critical growth temperature between 400 and 430°C above which phase separation does not occur and that the phase separation observed in layers grown below 400°C does not occur by a process of diffusion in the bulk. Strong phase separation did occur in the layer grown at 370°C and this was thermally stable with regard to annealing for 2 hours at the growth temperature. The 370°C layer in this sample was effectively grown onto a lattice-matched substrate and so the platelet structure observed in the layers of the previous specimens is not a consequence of the initial nucleation of the InAsSb on the lattice-mismatched InAs buffer layers normally used. We conclude from these observations and the anisotropy of the platelet geometry that the platelet structure can only be generated at the growing surface.

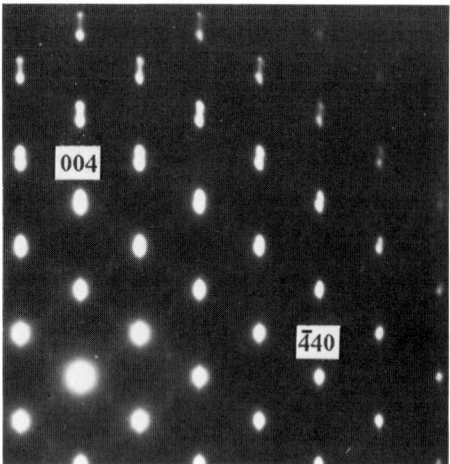

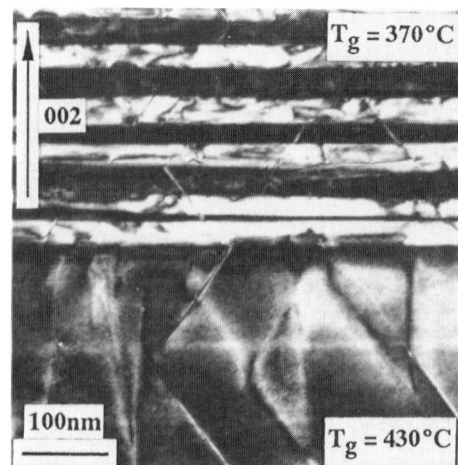

Figure 2. [110] pole TED pattern of phase separated MBE InAs$_{0.5}$Sb$_{0.5}$ layer (T$_g$ = 340°C) showing splitting of diffraction spots in [001] direction but not in [$\bar{1}$10] direction.

Figure 3. g[002] DF cross-section micrograph of MBE InAs$_{0.5}$Sb$_{0.5}$ layer: upper region (T$_g$ = 370°C) is phase separated, lower region (T$_g$ = 430°C) is homogeneous although annealed at 370°C /3 hours.

Table I. Transmission electron diffraction results for phase-separated InAs$_y$Sb$_{1-y}$ layers

T$_g$	a$_p$	c$_1$	c$_2$	a$_{r1}$	a$_{r2}$	y$_1$	y$_2$	δy
340	6.25	6.40	6.04	6.32	6.14	0.38	0.80	0.42
370	6.25	6.35	6.03	6.30	6.14	0.42	0.80	0.38
400	6.24	6.39	6.10	6.31	6.17	0.40	0.74	0.34

T$_g$ = growth temperature (°C); a$_p$ = measured lattice parameter parallel to (001) layer surface of phases 1 and 2 (Å); c$_1$ = measured lattice parameter perpendicular to layer surface of phase 1 (Å); c$_2$ = measured lattice parameter perpendicular to layer surface of phase 2 (Å); a$_{r1}$ = relaxed lattice parameter of phase 1 (Å); a$_{r2}$ = relaxed lattice parameter of phase 2 (Å); y$_1$ = composition of phase 1; y$_2$ = composition of phase 2; δy = y$_2$-y$_1$

Several authors (e.g. Hénoc et al 1982, Mahajan 1983, Norman and Booker 1985) have reported a periodic coarse scale (100-200nm) "tweed-like" strain contrast along the [100] and [010] directions of (001) LPE Ga$_x$In$_{1-x}$As$_y$P$_{1-y}$ layers grown inside the predicted miscibility gap (de Cremoux et al 1981). STEM EDX microanalysis (Hénoc et al 1982) revealed this contrast to be associated with a periodic modulation in composition along the [100] and [010] directions of the layers with variations in the composition parameters x and y as high as 0.1 being measured. No such similar composition variations were present in the [001] growth direction of the layers. These composition variations were suggested to arise by a process of surface spinodal decomposition. Petroff et al (1982) reported a plate-like structure of different compositions in (110) MBE Ga$_{1-x}$Al$_x$As layers but it was not observed in (001) layers. A miscibility gap in this alloy system is not thermodynamically predicted for the bulk alloy. It was thus proposed by Petroff et al (1982) that the platelet structure arose as a consequence of phase separation occurring at the growing surface due to the existence of an exchange-reaction-induced miscibility gap. Madhukar (1983) and Van Vechten (1985) considered this segregation and proposed that it resulted from the kinetics of the MBE growth process, i.e. was a "kinetic segregation". Maksimov (1981, 1984, 1991) has reported a composition modulation along the growth direction in ternary III-V alloy (001) epitaxial layers such as GaAsP and GaInP grown by LPE and vapour phase epitaxy (VPE). The amplitude of composition variations observed was typically very small, e.g. 0.3-1.0 atomic % P for VPE GaAsP with a modulation period of 0.04-0.4μm A model was proposed for this "automodulation" based on a non-linear dependence of the sticking coefficient for different atoms on their surface concentration.

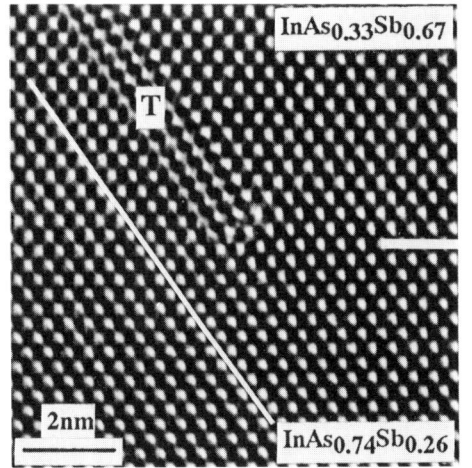

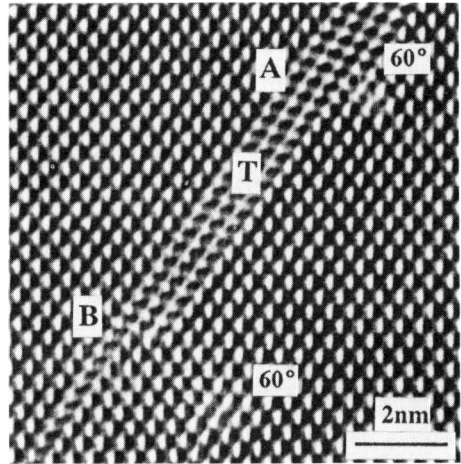

Figure 4. [110] HREM image of microtwin, T, originating at boundary between platelets in phase separated MBE InAs$_{0.4}$Sb$_{0.6}$ layer (T$_g$ = 370°C). Deviation of {111} planes due to the two tetragonal distortions is visible on crossing boundary.

Figure 5. [110] HREM image of phase separated MBE InAs$_{0.4}$Sb$_{0.6}$ layer (T$_g$ = 370°C) showing a 4 plane wide twin lamella, T, that decreases in width to 3 planes and then a single plane at points A and B respectively. 60° dislocations also visible.

The characteristic features of the platelet structure observed in our (001) InAs$_y$Sb$_{1-y}$ alloy layers are different to the cases described above. In this work a large composition variation is observed along the [001] growth direction of the layers, with tetragonally distorted, anisotropic geometry platelets, of two very different composition cubic phases, possessing abrupt interfaces, being present. The results obtained on the annealed layer grown at 430°C strongly suggests that the phase separation occurs at the surface of the InAsSb layers during growth. The anisotropy observed for the platelet structure in layers grown at and below 400°C is consistent with phase separation occurring at the growing surface of the layer with islands of the As-rich phase continuously forming on plates of the Sb-rich phase and then being subsequently overgrown by further plates of the Sb-rich phase. The long axis of the As-rich phase platelets lies along the [$\bar{1}$10] direction in which a much higher growth rate of GaAs by MBE has been reported in comparison with the [110] direction (Kawabe and Sugaya 1989). This anisotropy of the MBE growth rate along the orthogonal <110> directions may be a result of anisotropic cation diffusion rates (Ohta et al 1989) or more likely due to the different atomic structure of steps along the <110> directions (Horikoshi et al 1990, Joyce et al 1990). Islands with edges elongated parallel to [$\bar{1}$10] have been directly revealed on GaAs (001)-2x4 surfaces by scanning tunnelling microscopy (Pashley 1989). The increase in platelet size with increase in growth temperature is also indicative of a surface diffusion process. Therefore, we suggest that the phenomena observed in the InAs$_y$Sb$_{1-y}$ alloy layers results from phase separation occurring at the growing surface, associated with the presence of a miscibility gap, leading to the continuous formation of islands of the different phases which subsequently laterally overgrow each other producing the characteristic platelet structure observed in the layers.

3.2 HREM Results

In this section we present some HREM results obtained on the defects present in an InAs$_{0.4}$Sb$_{0.6}$ layer grown at 370°C which exhibited phase separation (Seong et al 1990). This layer contains a plate-like structure consisting of tetragonally distorted elongated platelets of two cubic phases of composition InAs$_{0.33}$Sb$_{0.67}$ and InAs$_{0.74}$Sb$_{0.26}$. The TEM studies reported earlier indicated that the phase separated InAsSb layers contained a high density of microtwins, many of which originated and terminated at boundaries between the phase platelets. Fig. 4 shows a [110] orientation HREM image of a microtwin, T, originating at the boundary between platelets of the different compositions. The presence of such microtwins may lead to some strain relaxation occurring between the tetragonally distorted platelets of the two phases. Similar microtwins have been reported to relax strain in strained Si/Ge superlattice structures grown on Ge (001) by Wegscheider et al (1990) who proposed that they were formed by successive glide of 90° (a/6)<211> Shockley partial dislocations on adjacent {111} planes. As the platelets of the two phases are tetragonally distorted, one with c/a > 1 and the other with c/a < 1, it is expected that {111} lattice planes crossing the platelet interfaces should show an angular deviation calculated to be ~ 1.55° for the phase compositions in this sample. This deviation is clearly illustrated in Fig. 4 by the extrapolation of a line drawn parallel to the {111} lattice fringes in one phase across the interface into the other phase. In Fig. 5 is shown a four-plane-wide twin lamella, T, that decreases in width to first three planes and then to a single-plane stacking fault at the marked points A and B respectively. Undissociated 60° dislocations are present at the positions where the narrowing of the twin lamella occurs, i.e. at steps A and B. Deviation of the {111} lattice fringes due to the tetragonal distortion of the platelets of the two phases is again present. The HREM results indicated that the interfaces between platelets were abrupt to 1-2 monolayers.

4. CONCLUSIONS

In conclusion, we have shown that phase separation spontaneously occurs in (001) MBE $InAs_{0.5}Sb_{0.5}$ layers grown at temperatures $\leq 400°C$ producing tetragonally distorted, elongated platelets of two cubic phases, of compositions $~InAs_{0.40}Sb_{0.60}$ and $~InAs_{0.78}Sb_{0.22}$, corresponding to a naturally occurring strained layer superlattice structure. Our results strongly suggest that this phase separation occurs at the surface of the growing layers. Some of the misfit strain between the phase platelets is relieved by the formation of microtwins.

ACKNOWLEDGEMENTS

The authors would like to thank Prof. Sir Peter Hirsch FRS for the provision of laboratory space and the SERC and British Council (TYS) for financial support.

REFERENCES

Chiang P K and Bedair S M 1984 J. Electrochem. Soc. 131 2422
de Cremoux B, Hirtz P and Ricciardi J C 1981 Proc. GaAs and Related Compounds Conf. Vienna 1980 Inst. Phys. Conf. Ser. No. 56 115
de Cremoux B 1982 J. de Physique 43 no. 12 suppl. C5 19
Ferguson I T, Norman A G, Joyce B A and Stradling R A to be published
Hénoc P, Izrael A, Quillec M and Launois H 1982 Appl. Phys. Lett. 40 963
Horikoshi Y, Yamaguchi H, Briones F and Kawashima M 1990 J. Crystal Growth 105 326
Joyce B A, Neave J H, Zhang J, Vvedensky D D, Clarke S, Hugill K J, Shitara T and Myers-Beaghton A K 1990 Semicond. Sci. Technol. 5 1147
Kawabe M and Sugaya T 1989 Jpn. J. Appl. Phys. 28 L1077
Lee G S, Lo Y, Lin Y F, Bedair S M and Laidig W D 1985 Appl. Phys. Lett. 50 927
Madhukar A 1983 Surface Sci. 132 344
Mahajan S 1983 Proc. Microsc. Semicond. Mater. Conf. Oxford 1983 Inst. Phys. Conf. Ser. No. 67 259
Maksimov S K and Nagdaev E N 1981 Phys. Stat. Sol. (a) 68 645
Maksimov S K 1984 Phys. Stat. Sol. (a) 83 685
Maksimov S K 1991 This proceedings
Murgatroyd I J, Norman A G and Booker G R 1990 J. Appl. Phys. 67 2310
Norman A G and Booker G R 1985 Proc. Microsc. Semicond. Mater. Conf. Oxford 1985 Inst. Phys. Conf. Ser. No. 76 257
Ohta K, Kojima T and Nakagawa T 1989 J. Crystal Growth 95 71
Onabe K 1982 Jpn. J. Appl. Phys. 21 964
Osbourn G C 1990 Semicond. Sci. Technol. 5 S5
Pashley M D 1989 Phys. Rev. B 40 10481
Petroff P M, Cho A Y, Reinhart F K, Gossard A C and Wiegmann W 1982 Phys. Rev. Lett. 48 170
Seong T Y, Norman A G, Booker G R, Droopad R, Williams R L, Parker S D, Wang P D and Stradling R A 1990 Proc. MRS Fall Meeting USA 1989, Mat. Res. Soc. Symp. Proc. 163 907
Stringfellow G B 1983 J. Appl. Phys. 54 404
Taftø J and Spence J H C 1982 J. Appl. Crystallogr. 15 60
Van Vechten J A 1985 J. Crystal Growth 71 326
Wegscheider W, Eberl K, Abstreiter G, Cerva H and Oppolzer H 1990 Appl. Phys. Lett. 57 1496
Yen M Y, Levine B F, Bethea C G, Choi K K and Cho A Y 1987 Appl. Phys. Lett. 50 927

Inst. Phys. Conf. Ser. No 117: Section 7
Paper presented at Microsc. Semicond. Mater. Conf., Oxford, 25−28 March 1991

491

Composition automodulation of semiconductor epitaxial films on the basis of A_3B_5 solutions

S K Maksimov

CFPM, Moscow Institute of Electronic Technology, Moscow, 103498, USSR

ABSTRACT: Composition automodulation along the growth direction was observed in three component epitaxial A_3B_5 films grown by VPE or LPE. This effect was explained as a result of a non−equilibrium phase transition. Typical parameters of automodulation for the VPE and LPE films are reported. The effects of automodulation on parameters and degradation of devices produced on the basis of these films are discussed.

1. INTRODUCTION

This paper represents a short review of our previous publications on the topics of i) composition automodulation in epitaxial films of triple solutions of A_3B_5 compounds, ii) electron microscopy techniques of investigations of these films, iii) explanation of the phenomenon of automodulation, iv) the effect of automodulation on the parameters of light−emitting diodes and degradation of devices produced on the basis of composition−modulated films. The automodulation is observed in all investigated epitaxial films of GaAsP and GaInP obtained from different commercial sources (USSR, Japan, USA) or grown specially (Maksimov and Nagdaev 1979, Maksimov et al 1982). The composition non−uniformity was found in GaAlAs films produced by MOS−Hydride technology, but with different properties (Ilichev et al 1986). Automodulation is a new type of irregularity in compound epitaxial films arising during the epitaxial crystallisation and differing in origin from composition inhomogeneities due to, for example, a eutectic transformation or spinodal decomposition. Automodulation arises due to an essentially thermodynamic non−equilibrium of conditions in which epitaxial growth takes place and it results in unstable composition variations at least at the growth temperature.

2. SPECIFIC FEATURES OF ELECTRON MICROSCOPY INVESTIGATIONS

From the standpoint of electron microscopy, composition inhomogeneity results in variations of 1) absorption parameters, 2) average structure factors, 3) deformation (displacement) fields. Intensity variations on images arising due to the 1st and 2nd parameters can be investigated simultaneously. The further effect of these variations will be described as "structural contrast". Peculiarities of images due to the 3rd parameter will be called "deformation contrast" (Maksimov and Nagdaev 1981, 1982a, b).

If in the electron micrographs points differing in intensity by 5% can be resolved and if intensity variations arise due to structure contrast, then for $GaAs_{1-x}P_x$, composition modulation with amplitude $2\Delta x \geqslant 0.12$ and for $Ga_{1-x}Al_xP$, modulation with $2\Delta x \geqslant 0.07$ can be seen in images in the 220 reflection and can be distinguished in the 002 image for $2\Delta x > 0.01$ (Maksimov and Nagdaev 1981). In the analysis of deformation contrast (Maksimov and Nagdaev 1981, 1982a, b) the following model with two assumptions was examined: 1) the modulation direction coincides with [001] and 2) the modulation is macroscopic. The composition $x(z)$ is a periodic function along z. The one-dimensional structure consists of layers of two different compositions x_2 and x_1; $<x>$ is the average composition. For modulated crystal, the variation of the interference error Δs_g for a reflection g can be written in the form:

$$\Delta s_g = \delta_g \sin\theta \ \Delta x^{[001]} + \delta_g \theta_B \cos\theta \ \cos(g \ z_1')\Delta x^{[001]} \qquad (1)$$

where $\theta_g = -(1/<a>)\{(C_{11} + C_{12})/C_{11}\}u_0$; $<a>$ is the lattice parameter of the cubic crystal of the composition $<x>$; C_{11}, C_{12} are the elastic constants; u_0 is the coefficient of the concentration expansion; θ is the angle between the axis of projection and the conjugation plane of different composition layers; $\Delta x^{[001]} = |x_1 - x_2|/2$; θ_B is the Bragg angle and z_1' is the vector along the intensity oscillation direction in the image.

The first term of (1) describes variations of Δs_g arising due to the displacement field gradient along the scattering column. The second term characterises the variations of Δs_g under the transition from one scattering ring column to another, arising due to variations in g (dilatational contrast). Since the second term of (1) includes θ_B as a factor, the contribution of this term dominates only for $\theta\sim0$. Such rigid conditions for the observation of dilatational contrast make its observation practically impossible for any reflection, except reflections g (in our case, 004, 008 reflections, etc) having a small angle with I (the interface vector). The sensitivity of electron micrographs to composition variations with small amplitudes increases sharply if the deformation contrast is used. For GaAsP utilisation of the deformation contrast allows one to observe modulation with an amplitude of 0.1 mole % of GaP (relative deformations of 2×10^{-4}).

For micrographs with dilatational contrast the bright–field images exhibit higher sensitivity to small composition variations than the dark–field images. The sensitivity of bright–field micrographs to the composition modulation at negative interference errors is higher than that at positive errors, the maximum sensitivity being observed at $w_0 < -1.0$. At thicknesses $\tau = \tau/\xi_g'$ (ξ_g' is the anomalous absorption length), the absolute contrast is independent of the foil thickness and determined by the modulation amplitude, the interference error and the image type. The high–τ images are most sensitive to composition variations. For studies of the composition modulation, the optimum thickness of the foil is $\tau = (1.0-1.5)$. At these thicknesses and $w = 0$, the amplitude of the intensity oscillations can be used for estimations of the magnitude of the composition variations. Under these conditions, the lower the P concentration the brighter are the corresponding images on the positive micrographs.

For superposition of images of regions of different compositions the contrast on the micrographs depends on the combined effects of the factors described by both the terms of (1). In addition, the phases of waves propagating through the different composition regions differ by an angle α:

$$\alpha = 2\pi(\Delta g \ r) \qquad (2)$$

where $\Delta g = g_1 - g_2$, g_i are the diffraction vectors corresponding to the layers of the 1st and 2nd compositions. At stepwise composition changes, the image of an inclined interface between layers of different composition can be regarded to a first approximation as the superposition of two images: a moiré fringe image and a displacement fringe image. The relative contribution of contrast of a particular type is determined by the relation between the rates of variations of α and the wave amplitude, ie by the relation between the tilt angle θ and Δx. The images are characterised by two systems of the intensity oscillations, one of which has a periodicity of $1/|\Delta g|$ (the moiré oscillation) and the second of ξ_g (the displacement fringe oscillation). As a rule, the main oscillations are due to the moiré effect. Displacement oscillations have a substantially smaller amplitude resulting in the appearance of "ripples", which distort the shape of the moiré maxima. The moiré oscillations allow us to estimate the amplitude of the composition variations and the displacement fringe contrast permits us to determine the sequence of alternation of the layers.

In the micrographs of stepwise interfaces, moiré oscillations are observed at any tilt angle θ but, for the extended interfaces, they arise only in certain cases; if the thickness of the transition layer along the scattering column: i) $\Delta t \ll 0.25\xi_g$ or ii) $\Delta t = n\xi_g$. Investigations of the appearance of moiré fringes make it possible to identify the type of interface and to estimate the boundary thickness.

3. COMPOSITION AUTOMODULATION IN EPITAXIAL FILMS OF THE A_3B_5 COMPOUNDS

Epitaxial films of A_3B_5 were studied by Maksimov and Nagdaev (1979), Maksimov et al (1982). Films of GaAsP on GaAs and GaAsP on GaP were produced by VPE technology in industrial and experimental reactors, those of InGaP on GaAs were obtained with the help of the experimental technology described by Batov et al (1978). All $GaAs_{1-x}P_x/GaAs$ films had a graded layer (in which x changed from x = 0 up to x = 0.4) of $30\mu m$ thickness and "the constant composition layer" with x = 0.4. The $GaAs_{1-x}P_x/GaP$ had x = 0.9. The $GaAs_{0.6}P_{0.4}$ films can be divided into groups: those produced i) in the experimental reactor, ii) in the industrial reactor by the traditional technology used in the production of light-emitting diodes, iii) by the sandwich technology, and iv) obtained from different commercial sources (as standard commercial light-emitting diodes). Films from groups i) and ii) were grown by the chloride-hydride technology in a hydrogen stream at a temperature of 1025K for 2.5 hours. The growth direction coincided with [001]. The InGaAs/GaAs films were grown in the temperature field under isothermic conditions at the crystallisation front. The crystallisation temperature changed from 1023K (beginning) to 958K (end). The composition of the films changed from $In_{0.52}Ga_{0.48}P$ near the interface up to $In_{0.65}Ga_{0.35}P$ with a gradient of 0.6 mole % of InP. The lateral component of the temperature gradient (the motivation force for crystallisation) was equal to 20x. The growth direction coincided with [111]. Obviously, from the thermodynamical standpoint, the VPE and LPE growth conditions were essentially non-equilibrium.

Specimens for electron microscopy investigations were prepared by the cross-section technique, so the growth direction lay in the foil plane. Electron microscopy investigations were carried out on the basis of the principles set forth in the 2nd section. Misfit dislocations were observed at the GaAsP/GaAs interface. The interface remained planar. Under conditions resulting in the maximum sensitivity to the composition variations, the micrographs of all specimens (irrespective of their origin) had specific features testifying to the presence of composition automodulation in the epitaxial films, but peculiarities of the modulation were different for the VPE and LPE processes. The direction of the composition modulation coincides with the growth direction ([001] for the GaAsP films and [111] for the GaInP films). In GaAsP films the automodulation starts immediately from the GaAsP/GaAs interface (Fig. 1). The first layer near this interface is always As-rich. Interfaces between layers of different compositions were strictly parallel to the GaAsP/GaAs boundary. The intensity oscillations due to the modulation are visible in micrographs corresponding to 004, 008 reflections, but absent in those in the 220 reflections ($g_{220} \perp I$) (Fig. 1b). Analysis of the specific features of micrographs demonstrated that lattices of the conjugation layers are distorted tetragonally. For layers of the 1st type c/a > 1, for the 2nd c/a < 1 (a is the lattice parameter lying in the conjugation plane, c is normal to this plane), the layers of both types have the same a parameter. Such a situation arises due to the conjugation of layers having cubic lattices in the free state. The specimen which, according to the microscopy data, has d ~ 40nm and $2\Delta x$ = 8.0 mole % of GaP was studied also by the Auger technique. According to the Auger measurements it has d = 30nm and $2\Delta x$ = 6 mole %.

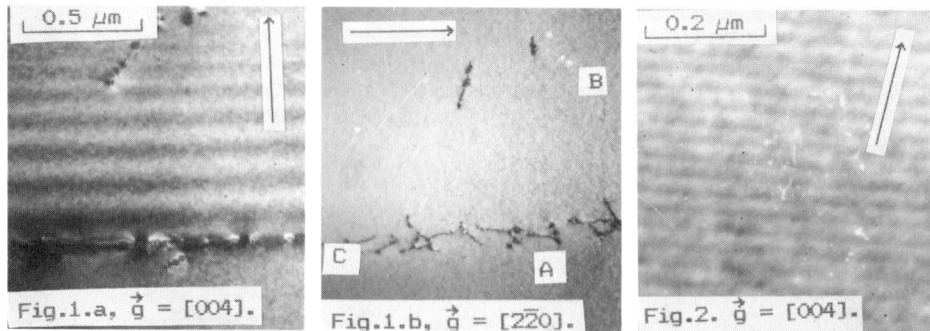

Fig. 1.   GaAsP/GaAs, 'A'–GaAs, 'B'–GaAsP, 'C'–misfit dislocations at the hetero–interface.   Intensity oscillations are visible in images in the 004 reflection and absent in 220.   Modulation starts from the interface and the period of modulation changes.

Fig. 2.   Region of GaAs$_{0.6}$P$_{0.4}$: the period of modulation is strictly constant.

The modulation parameters (d and 2Δx) are changing in the graded region and varying at the transition from one film to another, but strictly constant for the GaAs$_{0.6}$P$_{0.4}$ area in one film (Figs. 1,2).   Usually the amplitude of composition variations is less than 1.0 mole % of GaP but in some films it can increase up to 5.0–7.0 mole %.   Interfaces between layers of different composition belong to the stepwise type, their extension along the [001] direction Δd ≤ 2.0nm (moiré pattern is seen at changing θ continuously).   Typical modulation periods vary from approximately 30nm up to 300nm.   In general, a decreasing amplitude is accompanied by a reducing period.   Usually, an interaction between the modulated structure and isolated dislocations is not observed, but if boundaries with high dislocation density arise during the crystallisation or cooling, twisting of layers can take place.   The modulation parameters are sensitive to the presence of impurity, but this effect requires more investigation.   For increasing VPE temperature d and 2Δx decrease.   Increasing the rate of the flow of the initial components in the reactor results in a decrease of the modulation period and amplitude.

Other regularities are observed for the automodulation arising in LPE.   The misfit dislocations at the GaInP/GaAs interface are absent (at LPE temperature the lattice constants of Ga$_{0.48}$In$_{0.52}$P and GaAs are equal), but the boundary has pits due to etching of the GaAs surface at the beginning of the LPE process.   Near the interface, composition inhomogeneities are present which are probably explained by the action of the concentration overcooling–oversaturation mechanism.   They have a length in the [111] direction (the growth direction) of ~100nm and in the [1$\bar{1}$0] direction of ~10nm (Fig. 3a).   If one moves from the GaInP/GaAs interface this concentration non–uniformity is gradually suppressed and the modulation along the growth direction appears.   After a distance of ~10μm this modulation is dominant (Fig 3b).   It has the period d of ~200nm and an amplitude 2Δx of ~2.5 mole % InP.

4. POSSIBLE MECHANISM OF THE COMPOSITION AUTOMODULATION

Cyclic periodical processes are well known in chemical reactions.   They are known also for crystallisation processes but have damped oscillatory amplitude.   Several effects are described which can explain the formation of composition–modulated films: the N–form dependence of diffusion coefficients on the temperature; a phase transformation in surface–absorbed phases; an autocatalytic growth mechanism.   The latter corresponds best to our results.   This mechanism of automodulation of epitaxial layers was researched most carefully by Suris et al (1982) which suggested the following model: 1. The automodulation is a metastable state arising due to kinetic phenomena.   2. A buffer layer

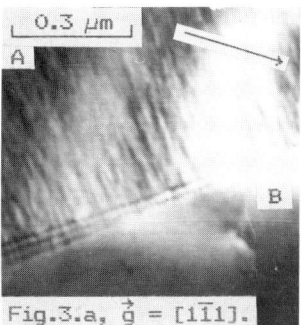
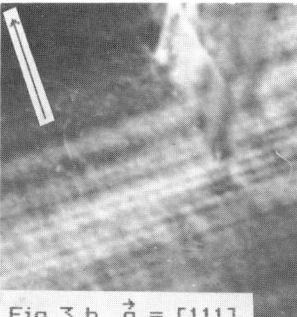

Fig.3.a, $\vec{g} = [1\bar{1}1]$.

Fig.3.b, $\vec{g} = [111]$.

Fig. 3. GaInP/GaAs a) The GaInP/GaAs interface; A – GaInP, B – GaAs. b) Region at distance ~10 μm from boundary, showing change in the type of concentration variation.

(for example, of Karman type) exists at the crystal vapour (liquid) phase boundary. 3. The crystal surface has steps; the step region is considered as a transition crystal layer determining an interaction between the film and the buffer layer. The types of process on the surface depend on the thermodynamic conditions. Under highly non-equilibrium conditions even a single atom can be a nucleation centre. 4. Components move through the buffer layer by diffusion and enter the crystal by adsorption through the step surface. 5. Affinity of the same atoms in the anion sublattice is more than the interaction between different atoms resulting in the selfcatalysis of the adsorption, the selfcatalysis is quadratic (accommodation of some atom into an angle formed by the same atoms has a high probability). 6. The characteristic time of reception of components in the buffer layer T_r is less than that of concentration changes on the step boundary T^*. In the framework of this model, the automodulation arises if $T_r \ll T^*$, which leads to the stepwise regularity of the composition variations.

5. EFFECT OF AUTOMODULATION ON PARAMETERS AND DEGRADATION OF DEVICES

Composition modulation results in an electroluminescent level for light-emitting diodes made on the basis of the automodulated GaAsP films. A satisfactory level (the quantum efficiency $\Omega > 5\%$) is achieved by diodes made of films with a modulation amplitude of less than 2.0 mole % GaP. An increase of the modulation amplitude in the original films to 5.0 mole % GaP leads to a drastic drop of luminous emittance ($\Omega < 2\%$).

In studying degradation processes three groups of light-emitting diodes with Ω of ~5% were subjected to different treatments (Maksimov 1982). The 1st group was operated for 5000h with a current density of 5Acm^{-2}. The 2nd was irradiated by neutrons with a dose of 10^{12}cm^{-2} and then kept at room temperature for 5000 h. The 3rd was subjected to the combined effect of irradiation and operation. As a result, the Ω level of the groups fell by a factor of 3 (the 1st and 2nd groups) to 5 (the 3rd group). The changes of defect structure produced by operation were of the same type for all groups. Dislocations with screw components underwent helical twisting. In regions of the graded layers with $x < 0.2$ the above treatments result in a reduction of the modulation amplitude $2\Delta x$. In the region $x = 0.4$, the period d remained constant and $2\Delta x$ increased. In regions of the graded layer with $0.2 < 2\Delta x < 0.4$, $2\Delta x$ increased but a single composition oscillation could split into oscillations with shorter periods (Volkov and Nagdaev 1986). In samples of the 1st and 2nd groups $2\Delta x$ increased up to 10 mole % GaP, while in those of 3rd group the increase was up to 15 mole % GaP. The stepwise boundaries between layers were replaced by extended boundaries. Their extension was estimated as ~5.0nm. The composition changed only within the boundaries and remained constant inside layers.

In composition modulated crystals (superlattices), boundaries between layers of different compositions have effective surface and elastic energies. The surface and the elastic energies increase with $(2\Delta x)$ and with area of boundaries per mole (nS), which results in an increased full energy of the system (F). F of modulated $AB_{1-x}C_x$ compounds is described by the expression:

$$F=(1-x)F_{AC}+xF_{AC}+(1/2)RT[(x+\Delta x)\ln(x+\Delta x)+(1-x-\Delta x)\ln(1-x-\Delta x)+(x-\Delta x)\ln(x-\Delta x)+$$
$$(1-x+\Delta x)\ln(1-x+\Delta x)]+\Omega x(1-x)-\Omega(\Delta x)+(B/<a>^2)(\Delta a)^2(\Delta x)+4nS\,|\Delta\sigma|\Delta x........(3)$$

Ω is the mixing factor; B depends on the boundary characteristics; $\Delta a = a_{AC}-a_{AB}$; a_{AC}, a_{AB} and $<a>$ are parameters of the elementary cells of AC, AB and $AB_{1-x}C_x$; $\Delta\sigma = \sigma_{AB}-\sigma_{AC}$; the σ_{AB}, σ_{AC}, are the surface tension coefficients of AC, AB. Several conclusions can be obtained by analysis of (3). The main conclusions are: i) the value of the "thermodynamic force" ($Z = \delta F/\delta(\Delta x)$) which is responsible for the diffusional transfer in composition modulated crystals can be changed by a selection of x, nS and Δx; ii) each set of x and nS is associated with a single equilibrium modulation amplitude corresponding to the F minimum. For decreasing Z, a reduction of device degradation rates can be expected, as is confirmed in some experimental papers.

REFERENCES

Batov I N et al 1978, Izvestiya LETI (in Russian) (Leningrad: LETI) 220 71
Ilichev E A et al 1986 J Techn Fiz (in Russian) 56 2245
Maksimov S K 1984 Phys Stat Sol (a) 83 685
Maksimov S K et al 1982 FTT (in Russian) 24 628
Maksimov S K and Nagdaev E N 1979 DAN SSSR (in Russian) 245 1369
Maksimov S K and Nagdaev E N 1981 Phys Stat Sol (a) 68 645
Maksimov S K and Nagdaev E N 1982a Phys Stat Sol (a) 69 505
Maksimov S K and Nagdaev E N 1982b Phys Stat Sol (a) 72 135
Suris R A et al 1982 in Structure Defects: Methods of their Investigations, ed Maksimov S K (in Russian) (Moscow: MIET) pp 3–16
Volkov S I and Nagdaev E N 1986 Pis'ma J Techn Fiz (in Russian) 12 106

Inst. Phys. Conf. Ser. No 117: Section 7
Paper presented at Microsc. Semicond. Mater. Conf., Oxford, 25–28 March 1991

497

Dynamic observations of misfit dislocations in strained layer heterostructures

R Hull, J C Bean, D Bahnck, J M Bonar* and L J Peticolas

AT&T Bell Laboratories, Murray Hill, NJ 07974, USA
*University of Southampton, Southampton

ABSTRACT: By annealing metastable strained layer semiconductor heterostructures in-situ in the transmission electron microscope, we are able to directly observe dynamic misfit dislocation processes in real-time. We present results from such studies of $Ge_x Si_{1-x}$/Si (100), (110) and (111) heterostructures. Misfit dislocation nucleation, propagation and interaction mechanisms are discussed, and the effects of surface orientation, layer geometry, stress and strain analyzed.

1. INTRODUCTION

Heteroepitaxial integration of semiconductor layers with different lattice parameters offers exciting new opportunities in fundamental and device physics. There have been several studies of the fundamental mechanisms of strain relaxation via misfit dislocation introduction in lattice-mismatched heterostructures, and a coherent picture of these mechanisms is starting to emerge. In this paper we review some of our results relating to misfit dislocation kinetics obtained from transmission electron microscope (TEM) studies, including in-situ strain relaxation observations. We present results from the $Ge_x Si_{1-x}$/Si system, comparing relaxation processes for different substrate surfaces and different epilayer geometries, stresses and strains. The outline of this paper will be as follows: First we will outline some important existing models of strained layer epitaxy; secondly we will extend the concepts of strained layer epitaxy on the conventional (100) surface to (110) and (111) surfaces, thirdly we shall discuss our in-situ experimental techniques and finally we shall discuss separately misfit dislocation nucleation, propagation and interaction mechanism. Examples will be drawn from the $Ge_x Si_{1-x}$/Si (100), (110) and (111) systems.

2. EXISTING MODELS OF STRAINED LAYER EPITAXY

The classic model for predicting the critical layer thickness, h_c, at which misfit dislocations are first energetically favored in strained semiconductor heterostructures is the Matthews-Blakeslee "Mechanical Equilibrium" model (Matthews and Blakeslee, 1974). By reference to Figure 1, the forces are analyzed upon a dislocation threading through the substrate and epilayer. A lattice mismatch force, F_a, exists due to the strain, ε, between the substrate and epilayer lattices. This will act so as to drive the epilayer threading segment to the right of Figure 1, creating interfacial misfit dislocation length. Balancing this applied force is the line tension force due to the self-energy of the interfacial misfit dislocation created, F_T. If $F_a > F_T$, then interfacial misfit dislocation length is created. The definition of critical thickness is $F_a = F_T$, and substituting standard expressions for these forces in an uncapped heterolayer on a (100) surface yields:

$$2 \cos \lambda \ G\varepsilon \frac{(1+v)}{(1-v)} bh_c = \frac{Gb^2(1-v\cos^2\theta)}{(1-v)} \ln\left[\frac{\alpha h_c}{b}\right] \qquad (1)$$

Here the applied force is on the left hand side of the equation and the line tension term on the right. G is the epilayer shear modulus (taken here to be 64 GPa), v the Poisson ratio (0.28), b is the magnitude of the Burgers vector, $\underline{b}$, of the misfit dislocation ($3.9\,\text{Å}$ for $a/2 <011>$ glide dislocations), λ the angle between $\underline{b}$ and the interfacial normal to the dislocation line direction ($60°$), θ is the angle between $\underline{b}$ and the line direction $\underline{u}$ ($60°$) and α is a factor describing the dislocation core energy (taken here to be 4 as suggested in Hirth and Lothe, 1968). ε is the lattice strain between $Ge_x Si_{1-x}$ and Si ($=0.041$ x). The results of this model for the $Ge_x Si_{1-x}/Si$ (100) surface are plotted in Figure 2. Also shown in this figure are the experimental measurements of h_c in this system after Bean et al., 1984 at a growth temperature, $T_g = 550°C$, and Kasper et al., 1975 at $T_g = 750°C$. These experimental curves represent the epilayer thickness at which misfit dislocations are observed by the relevant detection technique employed (Fritz, 1987). They thus represent iso-relaxation contours rather than the true thickness at which misfit dislocations first exist. Nevertheless, it is clear that at low Ge concentrations and low growth temperatures, epilayers may be grown substantially beyond the equilibrium critical thickness before significant strain relaxation occurs. This is due, of course, to the presence of kinetic barriers to nucleating and propagating misfit dislocations. These kinetic effects were initially analyzed by Matthews, et al. (1970) but later and more comprehensively by Dodson and Tsao (1987).

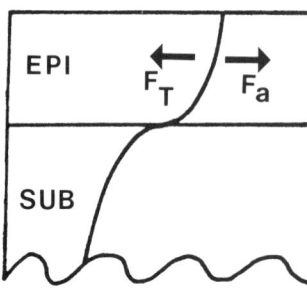

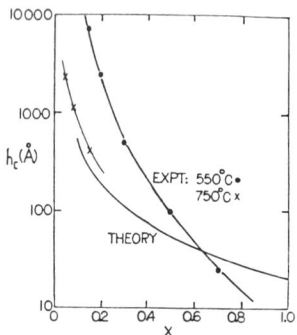

Figure 1: Schematic illustration of the Matthews-Blakeslee Mechanical Equilibrium model for the forces acting on a dislocation threading through an epitaxial layer.

Figure 2: Predictions of the Matthews-Blakeslee theory for the critical thickness h_c, for 60-degree $a/2<011>$ misfit dislocations at a $Ge_x Si_{1-x}/Si(100)$ interface. Also shown are experimental data for MBE-grown structures at $550°C$ (Bean et al, 1984) and $750°C$ (Kasper et al, 1975).

In the Dodson-Tsao formulation, it is assumed that misfit dislocations, once nucleated, propagate with a velocity, v, given by:

$$v = v_0 \sigma_e^m \, e^{\dfrac{-E_v(\sigma)}{kT}} \qquad (2)$$

Here σ_e is the excess stress, obtained by converting the Matthews-Blakeslee net force $F_a - F_T$ to a stress by dividing over the characteristic area upon which the force acts, hb sec ϕ, where ϕ is the angle between the {111} glide plane and the interface normal (sec $\phi = \sqrt{3/2}$ for (100) epitaxy). This yields for σ_e:

$$\sigma_e = \frac{2G\varepsilon(1+\nu)}{(1-\nu)} \cos\lambda\cos\phi - \frac{Gb\cos\phi(1-\nu\cos^2\theta)}{4\pi(1-\nu)h} \ln\left[\frac{\alpha h}{b}\right] \qquad (3)$$

The quantity $E_v(\sigma)$ represents the thermal activation energy for dislocation glide. Typical values obtained from deformation experiments in bulk Si and Ge are 2.2 eV and 1.6 eV respectively (Alexander and Haasen, 1968; Imai and Sumino, 1983; Patel and Chaudhuri, 1966). Dodson (1988) considered the possibility that $E_v(\sigma)$ is stress dependent. The exponent m in σ_e is typically of the order 1 in bulk experiments, particularly for very pure crystals (Imai and Sumino, 1983). Equation (3) must be modified for capped $Ge_x Si_{1-x}$ layers, where for cap thicknesses $> \sim$ the epilayer thickness, dislocations propagate by simultaneous loop relaxtion of both interfaces, with a threading segment traversing the epitaxial layer and joining the two interfacial misfit segments. The line tension in this geometry may be reasonably approximated by $2\sigma_T$, i.e. twice the line tension term in equation (3), for cap thicknesses greater than the $Ge_x Si_{1-x}$ epilayer thickness.

Dodson and Tsao also modelled certain assumptions about dislocation nucleation, favoring a combination of pre-existing sources and multiplication processes. Although we disagree in detail on the nature of this nucleation source (see later section), their model is important because it sets up the framework for a predictive kinetic equation of strain relaxation, as a function of time t at temperature, T. We build upon many elements of the Matthews-Blakeslee and Dodson-Tsao models in our own subsequent analysis.

3. EXTENSION TO DIFFERENT SURFACES

If we assume that misfit dislocation motion is by glide of expected a/2 <110> Burgers vectors on inclined {111} slip planes, then the interfacial dislocation geometry for the (100), (110) and (111) interfaces is as schematically illustrated in Figure 3. On the (100) surface the four inclined slip planes intersect the interface along orthogonal in-plane <011> directions producing the observed orthogonal misfit dislocation grid. On the (110) surface, however, there are only two inclined {111} slip planes as the other two {111} planes are normal to the interface and from equation (3) experience no resolved mismatch stress, σ_a. The inclined slip planes intersect along only one in-plane [1$\bar{1}$0] direction. On the (111) surface there are three inclined {111} slip planes intersecting the interface along the three in-plane <011> directions with hexagonal symmetry.

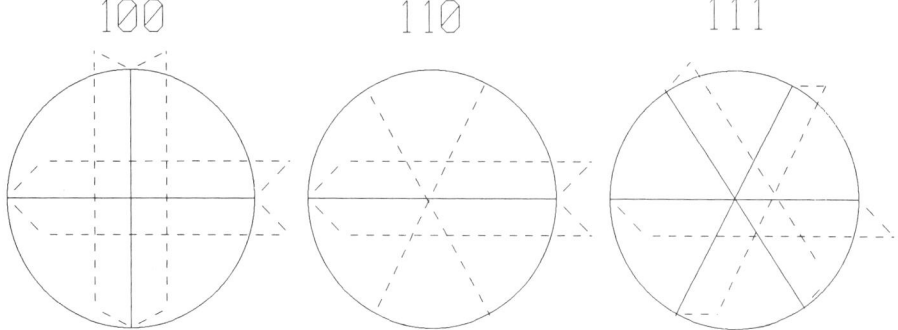

Figure 3: Schematic illustration of the symmetries of interfacial misfit dislocations at (100), (110) and (111) interfaces. Solid straight lines show the interfacial misfit dislocations; dashed lines outline intersecting {111} glide planes.

	100	110	111
$\cos \lambda$	0.50	0.71	0.29
$\cos \phi$	0.82	0.58	0.94
$\cos \lambda \cos \phi$	0.41	0.41	0.27
σ_a (MPa)	760	760	540
σ_e (MPa)	610	660	390

Table 1: Angular factors, $\cos\lambda$, $\cos\phi$, $\cos\lambda\cos\phi$ for 60-degree a/2<110> dislocations in different surfaces. Also shown are the magnitudes of σ_a and σ_e for a coherent 1000Å Ge$_{0.2}$Si$_{0.8}$ heterostructure.

The resolved stresses on the misfit dislocations on these three different surfaces are very different as summarized for a/2<110> glide dislocations in Table 1. The variation of $\cos\lambda$, $\cos\phi$, $\cos\lambda\cos\phi$ are shown, together with the variation of σ_a and σ_e for a 1000 Å $Ge_{0.2}Si_{0.8}$ epilayer coherently (i.e. essentially free of misfit dislocations) grown onto the relevant Si surface. It is apparent that for equivalent Burgers vector (which as we shall see later may not be the case) the applied stress, σ_a, is equivalent in (100) and (110) and substantially less in (111). The angular term, $\cos\lambda$, which is the primary angular term in determining the critical thickness as may seen from equation (1), increases in the order (111) < (100) < (110). The critical thickness, h_c, should thus decrease in the order (111) > (100) > (110) for equivalent Burgers vectors.

4. EXPERIMENTAL

We use the difference between equilibrium critical thickness and the experimentally determined critical thickness in Figure 2 to our advantage in the experiments described here. Samples are grown by MBE at 550°C such that their thicknesses lies between the equilibrium and experimental values of h_c. Thus, the as-grown structures are essentially unrelaxed but would be expected to relax via nucleation and propagation of misfit dislocations if annealed to suffieciently high temperatures. We supply the neccessary thermal energy for these processes by annealing these structures in-situ in a JEOL 2000FX TEM, using a Gatan single-tilt heating goniometer. This enables us to directly observe and quantify dynamic misfit dislocation phenomena, most notably propagation velocities from video recordings. For such in-situ experiments, samples are annealed in the plan-view geometry, because of concerns regarding surface diffusion effects across exposed thin-foil surfaces in the cross-sectional geometry.

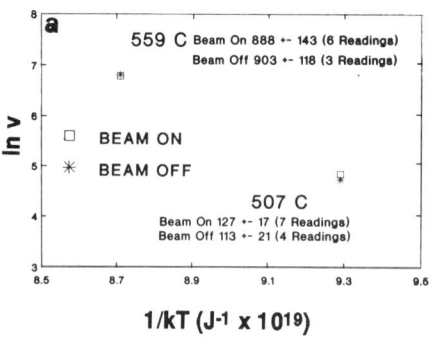

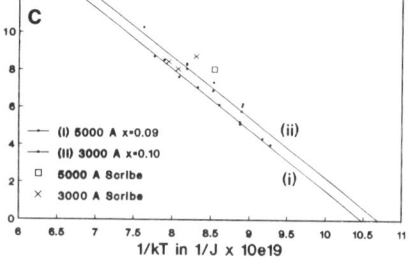

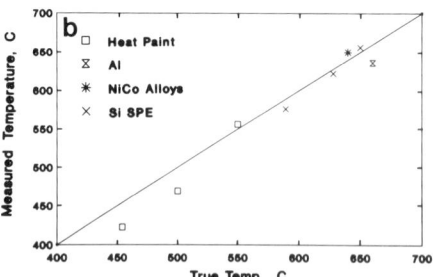

Figure 4: (a) Misfit dislocation velocities measured with and without 200 kV electron irradiation in a 3000 Å Si/1400 Å $Ge_{0.14}Si_{0.86}$/Si(100) structure. The numbers given after "Beam On" and "Beam Off" correspond to the mean and standard deviation of velocities in $Å.s^{-1}$. (b) Calibrations of the goniometer furnace thermocouple by various techniques. (c) Comparison of in-situ TEM and scribe line-emiitted velocities for (i) 3000 Å $Ge_{0.10}Si_{0.09}$/Si(100) and (ii) 5000 Å $Ge_{0.09}Si_{0.91}$/Si(100) structures.

We have taken considerable care to quantify the effects of electron beam irradiation, local sample temperature and thin foil relaxation. In Figure 4(a) we compare misfit dislocation velocities within a given structure for the electron beam on and off at 200 kV. The irradiation intensity on the sample for the "beam on" case is of the order 1 mA.cm^{-2}. No significant variation in dislocation velocity is observed. This observation is consistent with the results of Maeda et al (1991) who observed radiation-enhanced dislocation glide in Si only at substantially higher irradiation intensities. In Figure 4(b), we plot calibrations of the goniometer thermocouple (which is welded to the outside of the goniometer furnace) via various techniques. Most notably, we have measured local sample temperature by the solid phase epitaxial regrowth velocity of an amorphous/crystalline

Si interface in the plan-view geometry. This calibration uses the same sample material and geometry as our in-situ relaxation observations and thus should be very accurate. In Figure 4(c), we attempt to estimate the effects of thin foil relaxation upon our experiments. In general, as described in Hull et al, 1988, we make measurements from relatively thin epilayers and from relatively thick areas of the TEM sample to minimize abrogation of the rigid substrate approximation. Thus for 200 kV electrons, we can image through a maximum of 1-2 μm of Si and if we restrict the epilayer thicknesses to of the order $1000\,\text{Å}$ or less, then the Si substrate thickness in imaged areas is more than one order of magnitude greater than the epilayer thickness. Relaxation effects in such a regime should not generally be substantial. We have attempted to experimentally justify this statement by measuring dislocation velocities in relatively dilute $Ge_x Si_{1-x}$ ($x\sim 0.10$) films which are $3000\,\text{Å}$ and $5000\,\text{Å}$ thick. We then compare these velocities to dislocation velocities inferred from the time taken for dislocations to propagate from a scribe line (which is known to efficiently generate misfit dislocations in these structures (Tuppen and Gibbings, 1990)) in an unthinned region of the TEM sample to the etched region. In this case, dislocation propagation is almost entirely through a region where the substrate is 100 μm thick and thin foil relaxation effects will be absent. As can be seen from Figure 4(c), good agreement is obtained between dislocation velocities obtained from the two techniques for the $3000\,\text{Å}$ film, but a discrepancy of ~ 2x is observed for the $5000\,\text{Å}$ film, where the epilayer thickness approaches the substrate thickness in the sample area imaged in our in-situ experiments.

$Ge_x Si_{1-x}$ alloy layer thicknesses and compositions are accurately measured for each sample studied using a combination of cross-sectional TEM and Rutherford Backscattering Spectroscopy (RBS), as described in Hull et al, 1991. Structures are grown by Molecular Beam Epitaxy (MBE) as described elsewhere (Bean et al, 1984). TEM samples are made by conventional back-side etching (plan-view) and argon-beam sputtering (cross-section) techniques.

5. MISFIT DISLOCATION NUCLEATION

The mechanisms and arguments of misfit dislocation nucleation have been treated in many other papers (e.g. Matthews et al, 1976; Eaglesham et al, 1989; Fitzgerald, 1989; Hull and Bean, 1989a), and in this section we shall simply summarize existing arguments and then report two new series of results.

The primary problem in explaining misfit dislocation nucleation is in invoking a source which is sufficiently plentiful and of sufficiently low activation energy. One attractive mechanism for producing high densities of misfit dislocations is a multiplicative source such as the Hagen-Strunk source (Hagen and Strunk, 1978) which acts regeneratively between intersecting dislocation segments. However, in common with many other authors, we have never observed this source in the GeSi/Si structures we have studied. Neither have we observed any other regenerative source arising from misfit dislocation interactions. Another possibility is homogeneous nucleation of either complete dislocation loops within the epitaxial layer, or half-loops at a free epilayer surface. The energetics of this process was first described by Matthews et al (1976) and since by many other authors (e.g. Eaglesham et al, 1989; Fitzgerald, 1989; Hull and Bean, 1989a). The ubiquitous approach consists of balancing the strain energy released within the loop with the loop self-energy (plus the energies of any steps created or removed, stacking fault energies etc.). This generally produces a nucleation barrier, ΔE at a critical loop radius, R_c. The magnitude of this activation barrier depends upon the exact form of the expression used for the dislocation self-energy (and in particular the value assumed for the dislocation core energy parameter, α), whether or not step and/or stacking fault energies are included in the calculations, and, most critically, upon the epilayer strain, ε. Although published calculations differ in detail for the above reasons, all authors conclude that for the elastic parameters of the diamond cubic and zincblende semiconductors under consideration, homogeneous nucleation only becomes favourable at strains > 2-4%, and activation barriers at strains < 1% are too high for source operation. This leads to the concept of "nucleation limited relaxation" at low strains. If multiplicative mechanisms are ruled out, required source densities to completely relax a 1% strain are of the order $10^5\,\text{cm}^{-2}$. In the absence of homogeneous nucleation, it is difficult to imagine any sources which will be present in sufficiently high densities. Possible candidates include growth-related defects such as particulates, stacking faults, interfacial oxide particles etc. Under typical MBE growth conditions, the density of these sources should be of the order $10^3 - 10^4\,\text{cm}^{-2}$, which is too low a density to allow efficient strain relaxation. This paucity of dislocation sources is believed to be the primary reason for the large metastable regime at low Ge concentrations in Figure 2.

An illustration of the effects of nucleation limited relaxation is shown in Figures 5 and 6. A buried layer heterostructure is synthesized consisting of $3000\,\text{Å}$ Si/$1400\,\text{Å}$ $Ge_{0.14} Si_{0.86}$/Si(100). The as-grown structure is relatively stable to thermal relaxations, and does not substantially relax even on annealing to temperatures >

800°C, as shown in Figure 5. If the same structure is implanted with 1×10^{15} cm^{-2} B$^+$ at an energy of 100 kV, which puts the implant peak in the base region, the thermal stability of the structure is greatly degraded as illustrated in Figure 5. Note that the implant does not amorphize the structure, but clearly introduces high densities of point defects. By analogy to implantation in bulk semiconductors, post-implant annealing causes the point defects to aggregate into dislocation loops. In bulk semiconductors prolonged annealing causes these loops to slowly shrink and disappear. In the present strained layer structures, however, these loops act as ideal incipient misfit dislocations and provided that they are larger than the approriate critical loop radius, R$_c$, for misfit dislocation nucleation then they will expand to form misfit dislocations. This greatly increased source density causes the much poorer thermal stability exhibited in Figure 5. Note that our in-situ observations show that the dislocation velocity is actually retarded with respect to unimplanted structures, due to interactions with the point defect atmosphere (Hull et al, 1990). This causes uneven motion and cross-slip. In Figure 6, incipient misfit dislocation loops and an example of cross-slip in an implanted structure are illustrated. This dependence of the strained layer stability upon implantation has important technological ramifications for device contacting (Hull et al, 1990).

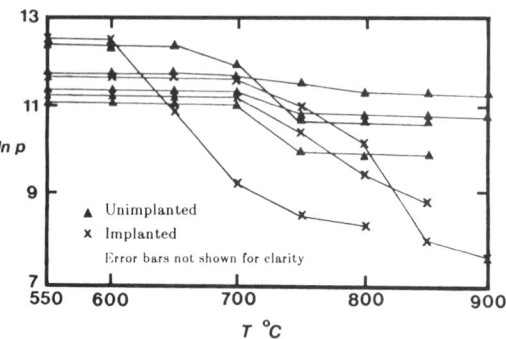

Figure 5: Relaxation of 3000Å Si/1400Å Ge$_{0.14}$Si$_{0.86}$/Si(100) heterostructures. The quantity p is the average distance between misfit dislocations in Å and T is the in-situ TEM annealing temperature, with successive 3 minute anneals at each temperature for which a datum point is given. Triangles correspond to the unimplanted structure and crosses to a 1×10^{15} cm^{-2} 100 kV B-implanted structure.

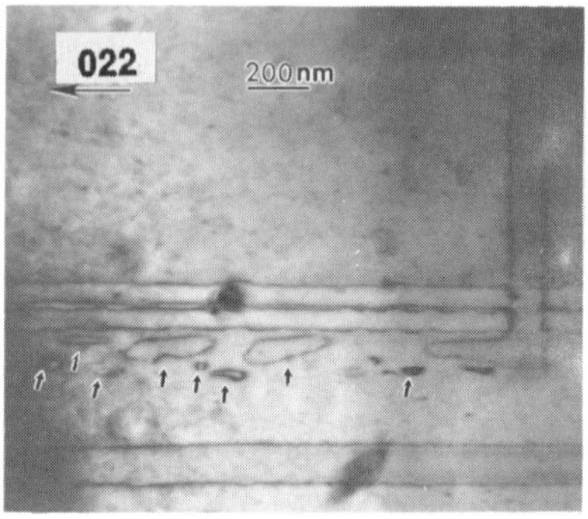

Figure 6: Plan view TEM image of incipient misfit dislocation loops formed during in-situ annealing at ~800°C of an implanted buried layer structure.

All of the above discussion, and most previous work, has ignored the effects of dislocation dissociation. It has long been recognized that total or perfect dislocations (i.e. dislocations whose Burgers vectors are equal to lattice translation vectors) may dissociate into two partial dislocations whose Burgers vectors are not equal to lattice translation vectors and which commonly introduce stacking faults into the host lattice. In fcc, dc and zb lattices, for example, the common dissociation reaction is:

$$a/2<110> = a/6<211> + a/6<12\bar{1}> \qquad - (4)$$

The partial dislocations on the right side of the equation are Shockley partials. This reaction is favoured because the total dislocation energy, which is proportional to b^2, is less after dissociation. The partials move apart under the influence of mutual repulsion due to the interaction of their stress fields, producing a stacking fault between them. The equilibrium configuration is attained when the stacking fault energy created balances the interaction energy between the partials. In bulk unstressed Si and Ge, the equilibrium partial separations are of the order $30-50$Å, yielding stacking fault energies of the order 70 mJm^{-2} (George and Rabier, 1987).

For the general 60-degree (angle between dislocation Burgers vector and line direction) a/2<110> interfacial glide dislocation in (100) epitaxy, the two dissociated Shockley partials make angles of 30 degrees and 90 degrees with respect to the line direction. In Figure 7, we plot for (100) epitaxy the equilbrium critical thickness from equation (1) for undissociated a/2<110> 60-degree dislocations and a/6<112> 30- and 90-degree dislocations. The latter partial dislocations include an extra stress term due to their stacking fault energy, taken here to be 65 mJ.m^{-2} for all Ge$_x$Si$_{1-x}$ compositions. It can be seen that for higher Ge compositions, the 90-degree partial has a lower critical thickness than the 60-degree total dislocation, indicating a greater excess stress acting upon it. (The excess stress acting upon the 30-degree partial is substantially less than that acting upon either of the other two dislocation structures considered.) However, the 90-degree dislocation is not free to move alone for (100) epitaxy, because as pointed out by Viegers et al (1985) and Maree et al (1987), consideration of the Thompson tetrahedron (Thompson, 1953) construction shows that the 90-degree partial in this orientation will yield a high-energy stacking fault with nearest-neighbour stacking violation, i.e. of the type ABC/CAB. Such a fault is extremely unlikely. For a low energy fault of the type ABC/BCA, the 30-degree partial would have to move first. The 90-degree partial following it would now have a very large excess stress acting upon it as it would now effectively be removing a stacking fault as it moves. Thus dislocations in (100) epitaxy are expected to be very closely dissociated (or even undissociated) total dislocations, as is almost invariably observed.

For compressive strain in (110) and (111) epitaxy, however, consideration of the Thompson tetrahedron shows that on these surfaces leading 90-degree partials can produce the low energy ABC/BCA type fault. In Figure 8, we plot the critical thicknesses for the 3 types of dislocation on the (110) surface, again with a stacking fault energy of 65 mJ.m^{-2} for the partials. For higher Ge concentrations, the 90-degree partial again appears to be favoured over the 60-degree total dislocation, thus we might expect to observe isolated partial dislocations on this surface. This is indeed observed, as illustrated experimentally in Figures 9 and 10. Figure 9 shows plan view TEM images of a 250 Å Ge$_{0.33}$Si$_{0.67}$ epilayer grown on Si(110). The expected uniaxial interfacial dislocation symmetry of Figure 3 is observed. This yields an orthorhombic unit cell following relaxation. Note that if the diffraction vector, $\underline{g}$, is placed parallel to the dislocation line direction, $\underline{u}$, far fewer dislocations are visible than for $\underline{b}$ perpendicular to $\underline{u}$. The "missing" dislocations with $\underline{g}$ parallel to $\underline{u}$ correspond to $\underline{g}.\underline{b}=0$ and must therefore have $\underline{b}$ perpendicular to $\underline{u}$; as $\underline{g}.\underline{b}x\underline{u}$ is also zero, invisibility is also very strong. These dislocations are therefore edge and most probably the 90-degree a/6<112> partials. This is confirmed by direct imaging of the stacking faults in a similar structure in cross-section in Figure 10.

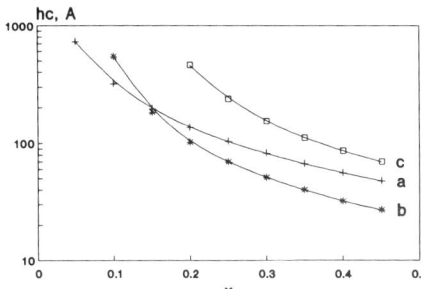

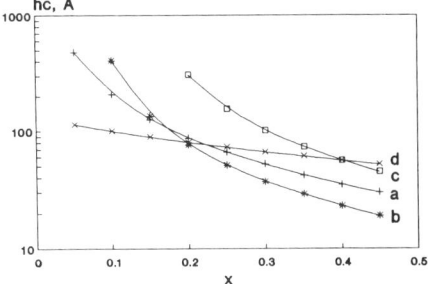

Figure 7: Equilibrium critical thickness for misfit dislocations at the Ge$_x$Si$_{1-x}$/Si(100) interface for (a) undissociated 60-degree a/2<110>, (b) a/6<112> 90-degree, (c) a/6<112> 30-degree. (b) and (c) include a stacking fault energy of 65 mJ.m^{-2}.

Figure 8: Equilibrium critical thickness for misfit dislocations at the Ge$_x$Si$_{1-x}$/Si(110) interface for (a) undissociated 60-degree a/2<110>, (b) a/6<112> 90-degree, (c) a/6<112> 30-degree ((b) and (c) include a stacking fault energy of 65 mJ.m^{-2}) and (d) a/6 <112> 30-degree with removal of 65 mJ.m^{-2} stacking fault energy, but with a reduced applied stress corresponding to a local reduced strain from a pre-existing 90-degree a/6<112> dislocation.

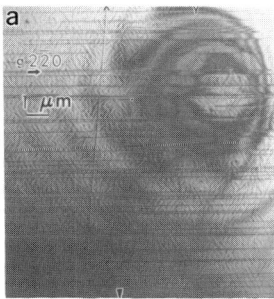

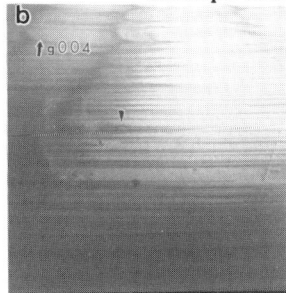

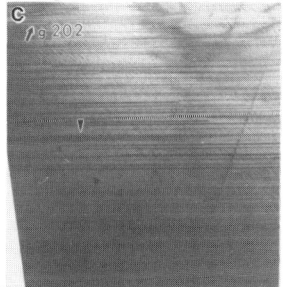

Figure 9: Bright field plan view images of a 250 Å Ge$_{0.33}$Si$_{0.67}$/Si(110) structure with (a) $\underline{g}$ = [2$\overline{2}$0] in the [110] pole, (b) $\underline{g}$ = [004] in the [110] pole and (c) $\underline{g}$ = [202] in the [11$\overline{1}$] pole.

Although the excess stress acting on a 90-degree partial is much greater than that acting on the 30-degree partial if the contribution from the stacking fault energy is included, the 30-degree partial *following* the 90-degree partial will effectively be repairing a stacking fault. In this configuration the stacking fault energy stress should be deducted from the excess stress rather than added on, and calculations suggest that the 30-degree partial should then follow very closely on the heels of the 90-degree partial, making it surprising that large densities of isolated 90-degree partials are observed. The solution to this apparent dilemma probably lies in the fact that the presence of the pre-existing 90-degree partial lowers the local strain and thus the applied stress acting on the following 30-degree partial. It is also possible that the 90-degree partial is more mobile than the 30-degree partial; calculations (Heggie and Jones, 1987) suggest that it has a lower Peierls barrier to kink nucleation and/or motion.

Similar arguments and experimental observations apply to the existence of isolated 90-degree partials on the (111) surface. The existence of separate partials on (110) and (111) surfaces is unfortunately likely to render them inapplicable for technological purposes because of the concomittant stacking faults extending through the epitaxial layer.

6. MISFIT DISLOCATION PROPAGATION

There have been extensive studies (e.g. Alexander and Haasen, 1968; Imai and Sumino, 1983; Patel and Chaudhuri, 1966; George and Rabier, 1987) of $a/2<110>$ dislocation glide propagation velocities in bulk Si and Ge in the stress range $\sim 1 - 100\,MPa$. It is generally found that equation (2) applies to these bulk studies in the higher stress regime, with activation energies of about 2.2 eV in intrinsic Si and 1.6 eV in intrinsic Ge, and exponential factors m ~ 1 in pure materials. Slight dependences of velocity upon dislocation structure (screw vs. 60-degree) and stronger dependences upon doping levels are often reported.

The generally accepted model for dislocation propagation in bulk crystals is the Hirth-Lothe diffusive kink pair model (Hirth and Lothe, 1968). In this model, dislocations propagate by nucleation of atomic-length segments, known as kinks, transverse to the dislocation line direction; these kinks then propagate along the dislocation line direction to move the dislocation one atomic spacing. In bulk materials, kinks neccessarily nucleate in pairs, and this configuration must also apply to buried $Ge_x Si_{1-x}$ layers studied here, as shown in Figure 11(a). For thin uncapped $Ge_x Si_{1-x}$ layers, single kinks may conceivably nucleate at the free surface, Figure 11(b), and we consider this possibilty in subsequent analysis.

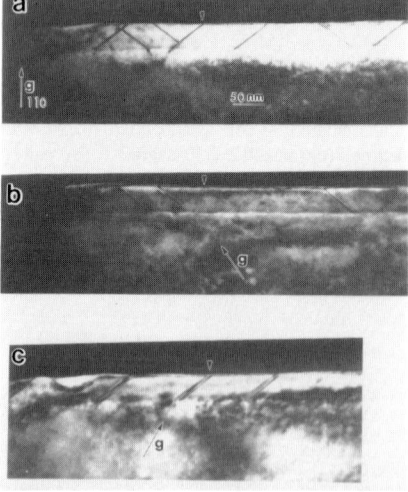

Figure 10: Cross-sectional images in the $[1\bar{1}0]$ pole of a $410\,\text{Å}$ $Ge_{0.33} Si_{0.67}/Si(110)$ structure with g = (a) [220], (b) [111] and (c) $[11\bar{1}]$.

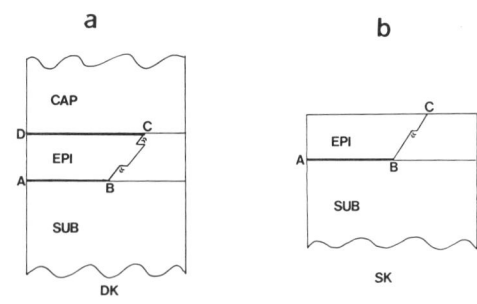

Figure 11: Kink nucleation modes in strained epilayers: in (a), buried layers, the kink pair mecahnism must operate; in (b), uncapped epilayers, single kink nucleation at the surface may be possible.

The mathematics of kink pair nucleation and propagation have been described in detail by Hirth and Lothe (1968), and we simply reproduce their salient results here. The rate of kink pair nucleation per unit dislocation length of the threading arm traversing the epitaxial layer is:

$$J = \frac{v_D bq\sigma_{excess}}{kT} e^{-\frac{E_m + 2F_k}{kT}} \qquad - (5)$$

Here v_D is an attempt frequency, which we take in our modelling to be the Debye frequency, b is the magnitude of the dislocation Burgers vector, q is the kink jump distance transverse to the dislocation line in the kink pair nucleation stage, $2F_k$ is the activation energy for kink pair nucleation and E_m is the kink migration activation energy.

The kink propagation velocity is given by:

$$v_k = \frac{v_D a^2 bq\sigma_{excess}}{kT} e^{-\frac{E_m}{kT}} \qquad - (6)$$

Where a is the kink jump distance along the dislocation line. Equations (5) and (6) may be combined to give a dislocation propagation velocity:

$$v = \frac{2v_D abq^2 \sigma}{kT} \frac{L}{L+X} e^{-\frac{F_k' + E_m}{kT}} \qquad - (7)$$

Here L is the length of the propagating dislocation (threading) arm, and X is the average distance between kinks, given by $2ae^{\frac{F_k}{kT}}$. Two limits to this general equation exist: (a) when the length of the propagating dislocation length is very large and L >> X. In this configuration, kinks typically collide with each other before reaching the end of the propagating dislocation arm. This is the condition generally assumed in studies of bulk semiconductors, and in this case equation (7) simplifies to:

$$v = \frac{2v_D abq^2 \sigma}{kT} e^{-\frac{F_k' + E_m}{kT}} \qquad - (8)$$

Note that v in this equation is independent of L, and that the exponential numerator is $(F_k + E_m)$. This numerator thus corresponds to the glide activation energies measured in bulk experiments, i.e. $(F_k + E_m) = 1.6$ eV in Ge and 2.2 eV in Si. The relative contributions of F_k and E_m to this total are not well known. Recent electron microscopy experiments (e.g. Hirsch et al, 1981; Gottschalk et al, 1987) have tended towards values of $F_k \sim 1.0$ eV in Si, whereas pulse loading techniques have yielded values of $F_k \sim 0.6$ eV (Farber et al, 1986).

The limit (b) of equation (7) corresponds to the case when the propagating dislocation length is very small, L << X and kinks reach the end of the propagating arm before colliding with each other. In this configuration:

$$v = \frac{v_D bq^2 \sigma L}{kT} e^{-\frac{2F_k' + E_m}{kT}} \qquad - (9)$$

The exponential numerator is now $(2F_k + E_m)$ and v is proportional to L. The cross-over between the collision and non-collision regimes, equations (8) and (9) will clearly depend upon the magnitude of X which is in turn given by F_k. As F_k is not accurately known, the value of X is uncertain. Two signatures of the non-collision regime, however, are that the dislocation velocity depends upon the propagating dislocation length and the activation energy increases from $(F_k + E_m)$ to $(2F_k + E_m)$. A value of $F_k = 0.6$ eV gives $X \sim 1\mu m$.

Another factor which should be considered is the work done against the applied stress by the nucleating kink pair, which will lower the kink pair nucleation energy. This process was analysed by Seeger and Schiller (1962) who considered the self energy of the created kinks, their interaction energy and work done against the stress field as a function of the separation, s, of the kinks within one pair. By minimizing, the critical kink separation, s_0, is found and substituted back into the total energy expression. This yields a corrected value for F_k of:

$$F_k' = 2F_k - (\frac{G(1+v)\sigma_a b^3 q^3}{2(1-v)\pi})^{\frac{1}{2}} \qquad - (10)$$

The stress used in this expression is the applied stress, σ_a, as the nucleating kink arms are generally far from the interface compared to their separation at the critical configuration, and thus experience negligible interactions with the interfacial misfit segments. This correction is relatively insignficant for typical applied stresses of the order a hundred MPa or less accessed in bulk experiments, but for the present system where the applied stress in the $Ge_x Si_{1-x}$ layer is ~ 4x GPa, it can significantly lower the kink nucleation energy. Note that the Seeger-Schiller correction is only likely to be accurate where the kink arms do not overlap in their critical configuration: analysis by Hirth and Lothe (1968) suggests that kink overlap will occur at stresses $\geq \dfrac{\sigma_p}{6}$, where σ_p is the Peierl's stress of the epitaxial layer. This quantity is not well known for Si or Ge, but is probably in the range 5-10 GPa. Thus the Seeger-Schiller correction will probably break down at higher Ge concentrations in the present system. For a fuller discussion see Hull et al, 1991.

We have applied equation (7) to our measured dislocation velocities in $Si/Ge_x Si_{1-x}/Si(100)$ heterostructures. All parameters in equation (7) are known, except for the individual magnitudes of F_k and E_m. We treat F_k as a fittable parameter subject to the constraint that $F_k + E_m$ = 2.2 eV in Si and 1.6 eV in Ge, with linear interpolation for intermediate compositions. The Seeger-Schiller correction is also incorporated. A detailed analysis of this work is given in Hull et al, 1991.

In Figure 12, we show experimental results and modelling from four buried $Ge_x Si_{1-x}$ structures. Good agreement is obtained for a value of $F_k = 1.0 \pm 0.1$ eV in Si. This is the zero-stress value, i.e. without the Seeger-Schiller correction. At the applied stresses in these heterostructures, this correction lowers F'_k to of the order 0.5 eV. Thus although we are in fact studying structures in the range L << X, the sum $2F'_k + E_m$ is still of the order 2.0 - 2.2 eV. These measured activation energies, and absolute velocities, are in reasonable agreement with those measured by Tuppen and Gibbings (1990) and Houghton et al, obtained by large-area optical microscopy following ex-situ annealing, although our interpretation of trends in the data differ somewhat (Hull et al, 1991).

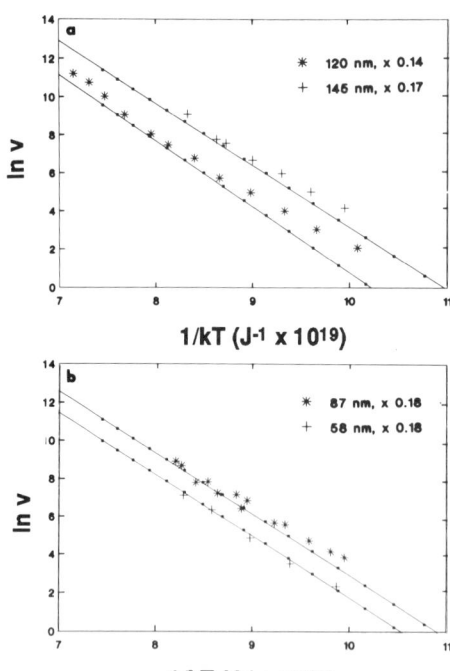

Figure 12: Measured dislocation velocities vs. predictions of the double kink theory for $Ge_x Si_{1-x}$ epilayers buried beneath a 3000 Å Si cap, with F_k = 1.0 eV at zero stress in Si. Ge compositions, x, and epilayer thicknesses, h, are given in the legends. v is in Å/s.

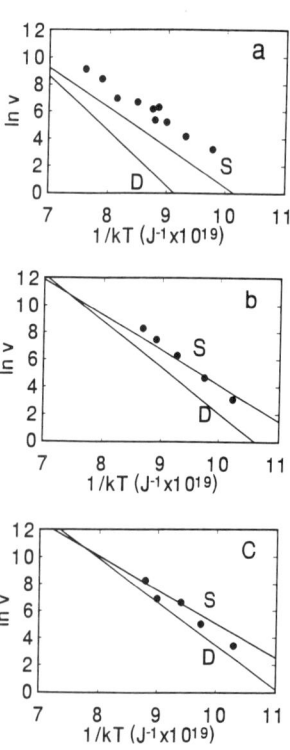

Figure 13: Comparison of single (S) and double (D) kink theories with measured dislocation velocities (solid circles) for uncapped structures: (a)   4000 Å $Ge_{0.06} Si_{0.94}$,   (b) 1000 Å $Ge_{0.15} Si_{0.85}$,   (c) 685 Å $Ge_{0.18} Si_{0.82}$. v is in Å/s

It should be noted that the Hirth-Lothe model for L >> X, equation (8), does not agree well with bulk experimental data in that the prefactor is a factor $\sim 10^3 x$ too low. This may be due to the presence of obstacles along the propagating dislocation line, charge on the kink sites, or an attempt frequency, v_D, substantially different from the Debye frequency. At the very high stresses in the present experiments, the effect of obstacles may be relatively insignificant rendering the prefactor in the Hirth-Lothe expression more accurate. Nevertheless, given the problems in extrapolating the Hirth-Lothe model quantitatively to bulk materials, we do not regard our present modelling as absolutely predictive; rather as an attempt to analyze our data and evaluate consistency in the framework of one particular model.

For very thin uncapped epilayers, we have considered the possibility of single kink nucleation at the free surface (Hull et al, 1991). By considering the interface between the epilayer surface and vacuum, the activation energy for single kink nucleation, including the Seeger-Schiller correction, is half that of the nucleation energy for kink pairs. The prefactor in the nucleation rate expression is also reduced relative to the double kink nucleation rate, however, by a factor $\dfrac{s_o}{4L}$ where s_o is the critical kink separation in a kink pair or between a single kink and its image in vacuo. This difference in prefactor arises primarily from the fact that double kink nucleation may occur anywhere along the propagating dislocation line of length L, whereas single kink nucleation may only occur within a distance $\frac{1}{2} s_o$ of the free surface, where s_o is typically only of the order 1 nm. The above analysis must be regarded as an approximation in that the actual configuration in the present experiments is not a crystal-vacuum interface, but a crystal-oxide-vacuum interface. This analysis predicts a transition thickness between predominant single kink and predominant double kink nucleation which increases with decreasing Ge concentration and decreasing temperature. In Figure 13, we show predictions of single vs. double kink theory vs. experimental data for several uncapped structures where we expect single kink nucleation to dominate. Further analysis and results are given in Hull et al, 1991.

A final fascinating possibility is that by analyzing misfit dislocation motion on (110) and (111) surfaces, we will be able to separately analyze and model velocities of partial and total dislocations. This should give insight, for example, into the motion of dissociated dislocations and into the magnitude of the Peierl's barrier for total vs. partial dislocations.

7. DISLOCATION INTERACTIONS

The interactions between misfit and threading dislocations via their stress fields can be of dominant importance in strained layer relaxation, particularly in the later stages of relaxation where dislocation densities are high. The primary interaction mechanism is the blocking of intersecting misfit dislocations with parallel Burgers vectors, as the back-stress between dislocations cancels out the Dodson-Tsao excess stress (Hull et al, 1989b; Freund 1990). This mechanism is particularly effective in thinner epilayers and lower residual strains. An example of such blocking mechanisms is shown in Figure 14. These blocking events substantially limit the average misfit dislocation length attained, and thus for a given amount of strain relaxation, threading dislocation densities are substantially increased. This blocking mechanism may thus be the limiting factor in producing low threading dislocation density relaxed epilayers, which are desirable for many device applications.

Dislocation interactions also have more subtle effects in modifying effective activation energies vs. non-interacting regimes (Hull et al, 1991). Interaction and intersection events will produce jogs on propagating dislocation arms, hindering motion. With increasing dislocation density in the strain relaxation process, such effects may start to dominate dislocation nucleation and propagation.

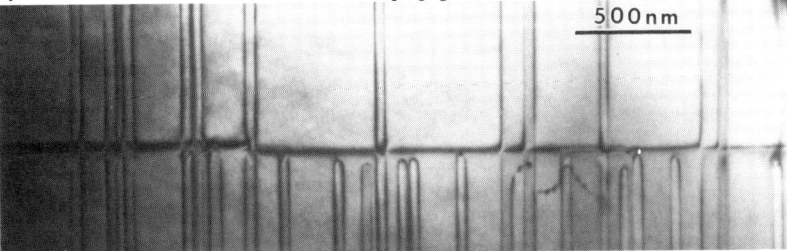

Figure 14: Dislocation pinning events in a 3000Å Si/585Å $Ge_{0.18}Si_{0.82}/Si(100)$ structure.

8. CONCLUSIONS

We have summarized misfit dislocation nucleation, propagation and interaction mechanisms for $(Si)/Ge_x Si_{1-x}/Si(100)$, (110) and (111) surfaces. Misfit dislocation symmetries, energetics and microstructures were observed to be markedly different on the three surfaces, with partial dislocation motion dominating on the (110) and (111) surfaces. Misfit dislocation velocities were measured by the in-situ TEM technique and modelled by the Hirth-Lothe diffusive double kink model. A kink nucleation activation energy of $F_k = 1.0 \pm 0.1$ eV was inferred, and evidence for single kink nucleation at the surface of thin uncapped epilayers presented. The crucial role of misfit dislocation interactions was also summarized. Studies of these thin strained layers are of crucial importance both for studying relaxation mechanisms in strained layer epitaxy, and for the insight they provide into the fundamental physics of dislocation motion, where they provide the opportunity to attain well-characterized stresses in the GPa regime.

ACKNOWLEDGEMENTS

We would like to acknowledge expert assistance in RBS measurements by F. Unterwald, growth of $Ge_x Si_{1-x}/Si(111)$ crystals by Y.H. Xie, and invaluable discussions with B. Dodson, D.J. Eaglesham, J.P. Hirth, E.P Kvam, D.M. Maher and P. Pirouz.

REFERENCES

Alexander H and Haasen P 1968, Solid State Physics 22, 27
Bean J C, Feldman L C, Fiory A T, Nakahara S and Robinson I K 1984, J. Vac. Sci. Technol. A2, 436
Dodson B W 1988, Phys. Rev. B38, 12383
Dodson B W and Tsao J Y 1987, Appl. Phys. Lett. 51, 1325
Eaglesham D J, Kvam E P, Maher D M, Humphreys C J and Bean J C 1989, Phil. Mag. A59, 1059
Farber B Y, Iunin Y L and Nititenko V I 1986, Phys. Stat. Sol. 97, 469
Fitzgerald E A 1989, J. Vac. Sci. B7, 782.
Freund L B 1990, J. Appl. Phys. 68, 2073
Fritz I J 1987, Appl. Phys. Lett. 51, 1080
George A and Rabier J 1987, Revue Phys. Appl. 22, 941
Gottschalk H, Alexander H and Dietz V 1987, Inst. Phys. Conf. Ser. 87 (Institute of Physics, Bristol, England), 339
Hagen W and Strunk H 1978, Appl. Phys. 17, 85
Heggie M and Jones R 1987, Inst. Phys. Ser. Conf. 87 (Institute of Physics, Bristol, England), 367
Hirsch P B, Ourmazd A and Pirouz P 1981, Inst. Phys. Conf. Ser. 60 (Institute of Physics, Bristol, England), 29
Hirth J P and Lothe J 1968, "Theory of Dislocations" (McGraw-Hill, New York)
Houghton D C 1990, Appl. Phys. Lett. 57, 1434 and 2124
Hull R, Bean J C, Werder D J and Leibenguth R E 1988, Appl. Phys. Lett. 52, 1605
Hull R and Bean J C 1989a, J. Vac. Sci. A7, 2580
Hull R and Bean J C 1989b, Appl. Phys. Lett. 54, 925
Hull R, Bean J C, Bonar J M, Higashi G S, Short K T, Temkin H and White A E 1990, Appl. Phys. Lett. 56, 2445.
Hull R, Bean J C, Bahnck D, Peticolas L J, Short K T and Unterwald F C 1991, J. Appl. Phys., in press.
Imai M and Sumino K 1983, Phil. Mag. A47, 599
Kasper E, Herzog H-J and Kibbel H 1975, Appl. Phys. 8, 199
Maree P M J, Barbour J C, van der Veen J F, Kavanagh K L, Bulle-Lieuwma C W T and Viegers M P A 1987, J. Appl. Phys. 62, 4413.
Matthews J W, Mader S and Light T B 1970, J. Appl. Phys. 41, 3800
Matthews J W, Blakeslee A E and Mader S 1976, Thin Solid Films, 33 253
Maeda K, Yamashita Y, Maeda N and Takeuchim S 1991, Proc. Mat. Res. Soc. 184, in press; and references therein
Patel J R and Chaudhuri A R 1966, Phys. Rev. 143, 601
Seeger A and Schiller P 1962, Acta. Met. 10, 348
Thompson N 1953, Proc. Phys. Soc. 66B, 481
Tuppen C G and Gibbings C J 1990, J. Appl. Phys. 68, 1526
Viegers M P A, Bulle-Lieuwma C W T, Zalm P C and Maree P M J 1985, Proc. Mat. Res. Soc. 37, 331

Inst. Phys. Conf. Ser. No 117: Section 7
Paper presented at Microsc. Semicond. Mater. Conf., Oxford, 25–28 March 1991

509

Misfit dislocations in GaSb/GaAs (001) heterostructures

A Rocher, J M Kang, H Atmani, J Crestou, G Vanderschaeve[1], L Lassabatère[2] and R Bonnet[3]

CEMES/LOE, 29, rue Jeanne Marvig, F-31400 Toulouse, France.
[1] Laboratoire de Physique des Solides, INSA, F- 31077 Toulouse, France.
[2] LESIC, USTL, Place E. Bataillon, F-34095 Montpellier, France.
[3] LTPCM, ENSEEG- Domaine universitaire- BP 75, F-38042 St Martin d'Hères, France.

ABSTRACT : The GaSb/GaAs heterostructure, with 8% lattice mismatch, reveals a complete relaxation of the misfit strain. This relaxation is connected to the formation of a highly regular network of misfit dislocations which is a consequence of the growth conditions: an ideal GaAs surface due to a GaAs buffer layer, and an adequate temperature allowing the direct relaxation of GaSb by both an island growth and a periodic creation mechanism of Lomer dislocations which is briefly discussed. The very low density of threading dislocations is also due to constructive interactions between interfacial dislocations especially during island coalescence.

1. INTRODUCTION

Threading dislocations, propagating parallel to the growth direction into the epilayers, are acknowledged to impair both the electrical and the optical properties of semiconductor epilayers. The reduction of their density constitutes an important objective for epitaxy growth techniques. A detailed understanding of the nature of misfit dislocations and of their mutual interactions is needed in order to control and reduce the density of these unwanted defects.

In the case of a mismatched heterostructure, the interface is ideally constitued by a periodic array of dislocations, accommodating the lattice mismatch. Rocher and Raisin (1991) have shown that the (001) GaSb/GaAs system, with its 8% lattice mismatch, exhibits a quasi-perfect array of misfit dislocations which are all Lomer type with a <110> direction and a <110> Burgers vector contained in the (001) interface. This system exhibits a threading dislocation density, measured at the level of the interface, at least two orders of magnitude smaller than in the 4% mismatched GaAs/Si system. The object of this article is a brief discussion concerning some interactions between misfit dislocations which preserve the perfection of the interface and introduce a low defect density in the display.

2. EXPERIMENTS

The growth of GaSb is processed by MBE in two steps as described by Raisin et al. (1986) : a 1 μm thick buffer layer of GaAs is first grown at 580°C. Its role is to make the best possible surface quality of the (001) GaAs substrate. GaSb is then deposited at its conventional temperature of homoepitaxy, 470°C. The deposited thickness is always larger than the theoretical critical thickness (< 1nm) which is suitable to give significant information on the misfit dislocation network of the GaSb/GaAs system.

Specimens with GaSb layers thinner than 80 nm are suitable for plan view observations by TEM. The sample preparation needs only the chemical etching of the rear side of the substrate because the GaSb film is already thin enough to give GaSb/GaAs samples transparent to the electron beam accelerated at 200kV. Such observations, with the electron beam parallel to the growth direction [001], give a two dimensional image which allows a complete investigation of the interface and its related defects.

2.1 Misfit dislocations

Dislocations organized as a grid with a mesh smaller than 100 Å are visible individually only by using the weak beam technique. The sharpness of the dislocation images is improved by increasing the deviation from the Bragg position of the operative reflection g, the 3 g or 4 g being in Bragg position.

Observation performed on a thick layer of GaSb using a {220} weak beam shows only one set of dislocations (fig. 1); dislocations are very straight with a mean length of about 300nm. The spacing between misfit dislocations, D_d, is equal to 54 ± 5 Å which is in good agreement with the theoretical spacing calculated with Lomer dislocations in a stress free system.

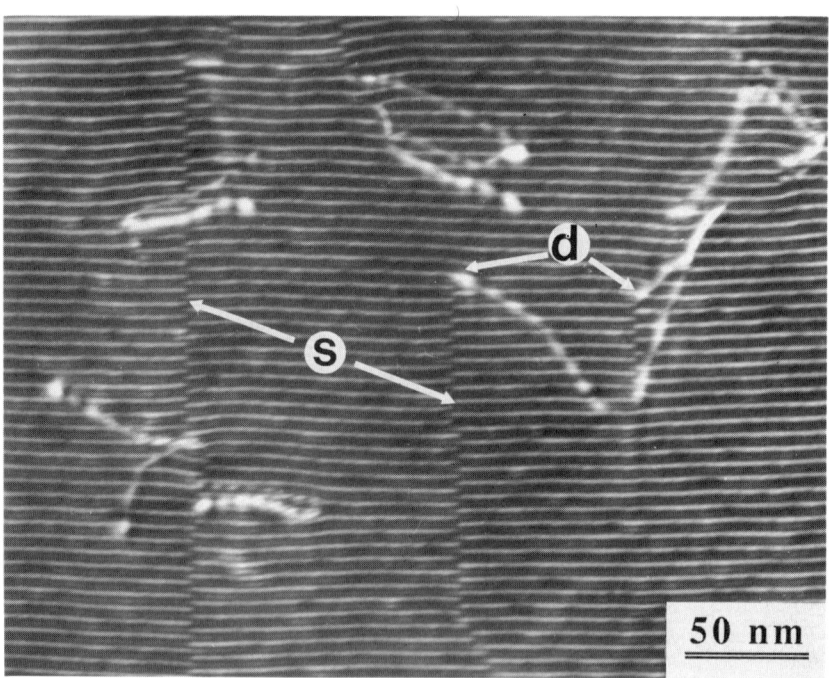

Fig. 1 : Weak beam image of the GaSb/GaAs interface for a 77 nm thick film of GaSb. Note the perfect alignement of the dislocations and the $D_d/2$ shifts between dislocation subarrays, marked S. Threading dislocations, related to the end of the shifts, are marked **d**.

The (400) reflection is able to image simultaneously the two sets of interfacial dislocations not as a grid but as an array of black and white triangles. This contrast is related to the algebraic sum (R1+R2) of the displacement fields of the two sets of misfit dislocations. Nevertheless, misfit dislocations can be observed individually with a non-conventional weak-beam condition. Such an observation is obtained using the $(2\bar{2}0)$ reflection in (- g, 3g) WB condition and the (400) reflection in Bragg position as drawn in the insert of Fig. 2. In this case, the two families of dislocations are imaged independently. i) the [110] Lomer dislocations are observed with the conventional weak beam contrast {(220)xR1 $\neq$ 0 and (220)xR2 = 0}. ii) the (400) reflection is taken as the incident beam for the (220) reflection. The interaction potential between them is related to the $(2\bar{2}0)$ reflection which images only the $[1\bar{1}0]$ Lomer dislocations {$(2\bar{2}0)$xR1 = 0 and $(2\bar{2}0)xR2 \neq 0$}. Under these conditions, the variations of intensity related to two set of dislocations do not interfer and each set is seen individually.

When the deposited thickness is smaller than 30 nm, the GaSb film is not continuous. As revealed on Fig. 2, the 3 nm thick layer of GaSb consists of isolated islands. The elongated shape of the islands is related to an anisotropy of the growth rates of the island facets. Similar results have been obtained on InSb/GaAs by Zhang et al. (1990). Fig. 2 shows the dislocation network associated with the GaSb islands: it is organized as a periodic array very close to the ideal case on the whole interface. Nevertheless, each island seems to contain at least one defect which is probably at the origin of the nucleation of the first GaSb cluster. After complete coalescence, appearing at about 500Å, the crystal growth becomes a 2-D process.

2.2 Dislocation interactions

The dislocation network is very close to the ideal grid on a very large scale as shown on fig. 1. The defect free area is quite large and can be evaluated on fig. 1 as being larger than 100x100 nm^2. The mean length of misfit dislocations is of the order of 300 nm, much larger than the distances between GaSb island.

The interfacial density of misfit dislocations is of the order of $4 \times 10^6 / cm^2$ to obtain an unstrained structure; it corresponds to $4 \times 10^{12} / cm^2$ dislocation intersections which are subject to the reaction and the creation of threading dislocations. In the case of Lomer type the two sets of misfit dislocations are perpendicular and they do not interact elastically. Their intersections are mainly exempt of threading dislocation as shown on fig. 2.

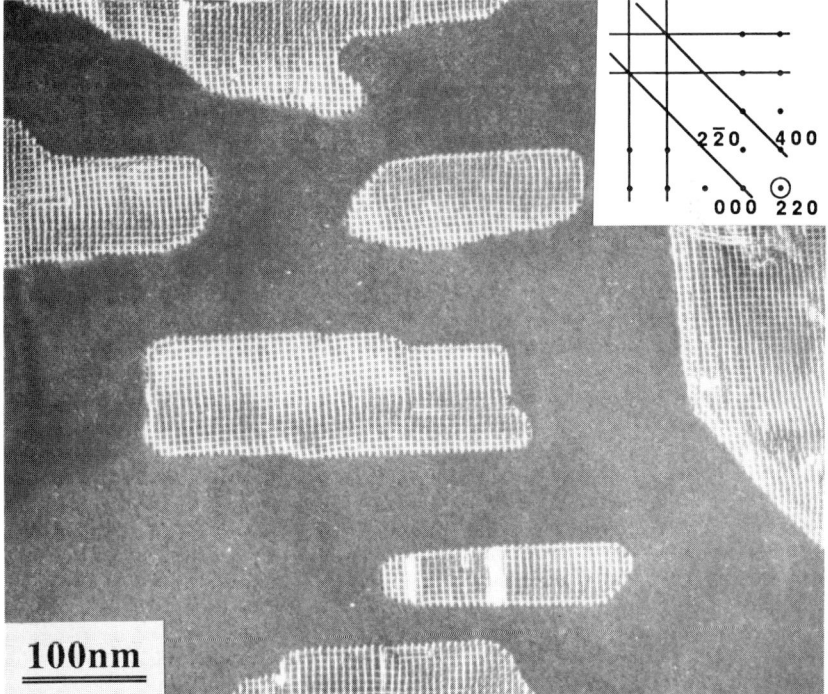

Fig. 2 : 2-D weak beam image of the GaSb/GaAs interface for a 3nm film of GaSb. Note the perfect array of square misfit dislocations.

The configuration of the most observed dislocation reaction is imaged on fig.1. It is characterized by a $D_d/2$ shift between Lomer dislocation sub-arrays. This shift is always observed to be equal to $D_d/2$. There are no intermediate values. Vannuffel (1990), Zhu and

Carter (1990) explain this shift by the interaction of Lomer dislocations with a 60° dislocation. The 60° dislocations have been identified both as perfect and dissociated dislocations. This reaction is not yet completely understood.

The importance of misfit dislocation length, greater than 250 nm as observed on fig. 1, shows that the joining of GaSb islands is not a simple phenomenon of coalescence. When the distance between island becomes small enough, a constructive reaction between dislocation sub-arrays leads to the alignment of Lomer dislocations perpendicular to the coalescence line of the adjacent islands as shown on fig. 3. The coalescence effect can be then obtained without emission of defects into the GaSb epilayer. Calculations indicate that the driving force for this arrangement could be the minimization of the elastic energy related to the interaction between the corresponding dislocation sub-arrays. This result needs further investigation.

Individual threading dislocations are seen on Fig.1. They are related to imperfections of the misfit dislocation network: The threading dislocation, marked **d** on Fig.1, is bent from the interface into the growth direction. It is probably a supernumerary misfit dislocation which bent during coalescence.

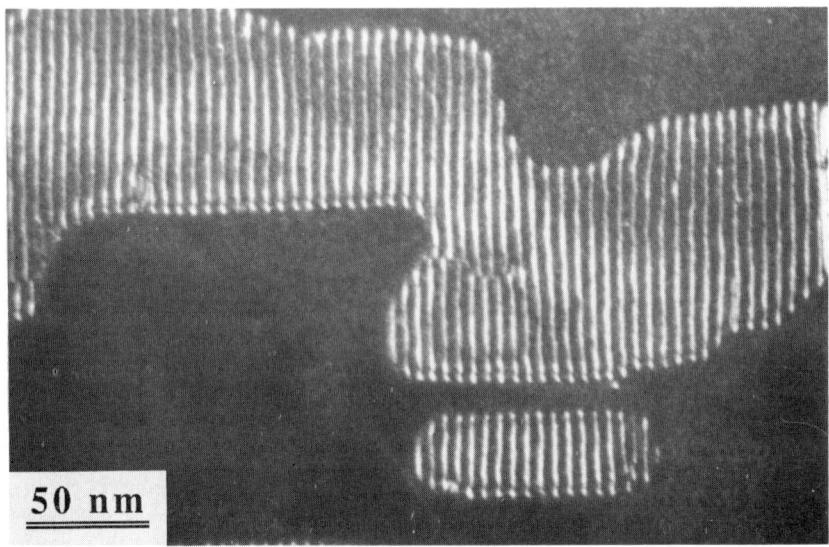

Fig. 3 : Weak beam image of interface GaSb/GaAs for a 3 nm film of GaSb. Note the perfect alignement of the dislocations belonging to different islands.

3. DISCUSSION

TEM , Raman spectroscopy analysis performed by Raisin, Landa et al.(1991) and X-Ray diffraction analysis made by Bourret et al. (1990) show that the GaSb films are stress-free. This result is directly related to the high quality of the interfacial dislocation network due to both a specific creation mechanism of misfit dislocations and their constructive interactions.

In the GaSb/GaAs system, the creation of Lomer dislocations appears to be a mechanism directly related to the 3-D growth. Bourret et al. (1990) have shown that the islands are faceted with (111) planes. These (111) facets play a major role in this creation mechanism: in fact the GaSb grows epitaxially not on the (001) GaAs plane but on (111) GaSb planes with an absence of misfit strain except at the atomic level of the interface. Each (111) facet makes with the (001) interface an edge parallel to the <110> directions.

The basic idea concerning the creation mechanism of the Lomer dislocation could be described using its atomic structure calculated by Bonnet et al. (1991) shown on fig.4: i) the (111) GaSb planes are developed epitaxially from the position 1 to 12 with a good continuity between the

(111) planes of both GaSb and GaAs. ii) the next plane grows also epitaxially except at the atomic level of the interface where the GaSb doublet, referenced 1, does not bond the GaAs doublets 13 and 1 but jumps to the next GaAs doublets referenced 1 and 2. This jump creates directly the   5 and 7 ring core of a Lomer dislocation which is then formed at the periphery of the island. This process is very likely since, as discussed by Kiely et al. (1989), it can occur without the coordinated motion of a large number of atoms and it is not subject to an energy formation barrier. Subsequent dislocations form at approximately even intervals as the strain in the island periodically accumulates. This speculative model needs confirmation.

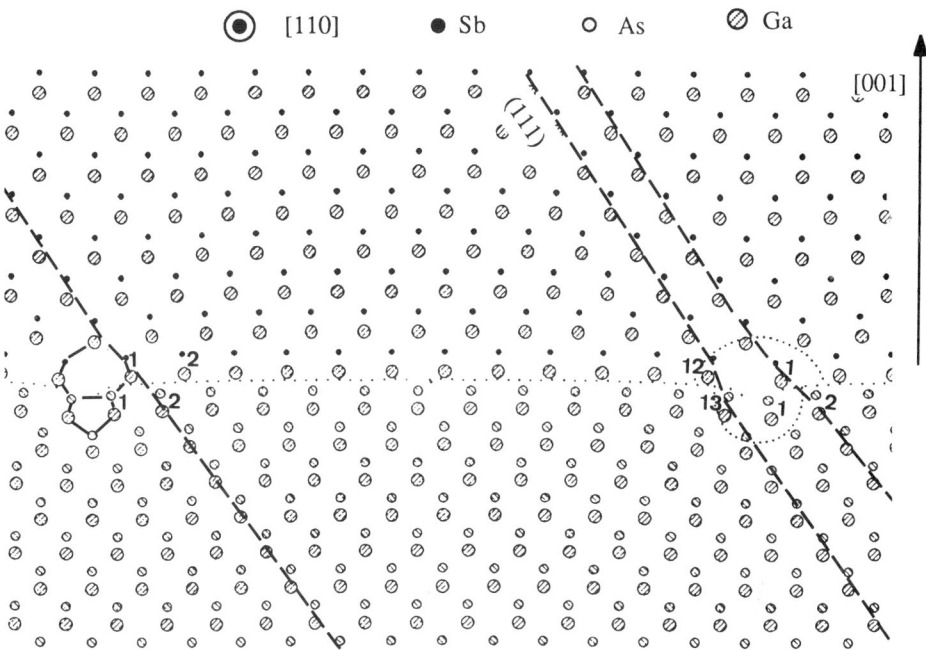

Fig. 4 : Creation mechanism of Lomer dislocations discussed on its <110> atomic scheme. Note the sketch of the GaSb epitaxial growth on the (111) planes. GaAs and GaSb are perfectly bonded except at the level of the core of the dislocation where the GaSb doublet of atoms 1 which should be bonded with their corresponding 13th and 1st GaAs doublets, jumps to bond the 1st and 2nd ones. The core of the Lomer dislocation, constitued by 5 and 7 rings, is then directly obtained.

This perfect relaxation of the misfit stress is also directly related to the surface quality furnished by the GaAs buffer layer. The second important parameter in the growth process is the temperature, which is not taken into account in the models of critical thickness. In this case, the growth temperature, 470°C, is large enough to enable a direct plastic relaxation of the deposited GaSb.

4. CONCLUSION

We have seen that the threading dislocations are directly related to the imperfections of the interfacial misfit dislocation network. Their density is much lower in GaSb/GaAs than in GaAs/Si as discussed by Fang et al. (1990). These results are explained by the difference of quality of their misfit dislocation networks. Contrary to a common idea, there is then no direct correlation between the value of the mismatch and the density of threading dislocations.

The Lomer dislocations, especially when they are developed with a well defined mechanism as observed in GaSb/GaAs, are useful to relax completely the stress due to the lattice mismatch and there is no other fundamental reason for creating threading dislocations in the epilayer. In addition, the perfection of the GaAs surface, due to a buffer layer, allows a reorganization of the misfit dislocation network during growth. The coalescence does not constitute an important source of defects propagating in the GaSb epilayer.

These results furnish some indications concerning the origin of both the misfit dislocations and the threading dislocations. The GaSb/GaAs system permits us to conclude that it is possible to define experimental conditions that considerably reduce the number of dislocations threading into the epilayer. When the misfit dislocation network is perfectly organized at the interface, the stress due to the lattice mismatch is well relaxed and the density of threading dislocations can be very low.

REFERENCES

Bonnet R, Loubradou M, Catana A and Stadelmann P 1991, to be published in Met. Trans A.

Bourret A, Fuoss P H, Rocher A and Raisin C 1990, in MRS Symposium "Advances in surface and thin film diffraction", Boston

Fang S F, Adomi K, Iyer S, Morkoç H, Zabel H, Choi C, Otsuka N, 1990, J. Appl. Phys., 68, (7), R31

Kiely C J, Chyi J I, Rockett A and Morkoç H 1989, Phil. Mag. A, 60, n°3, pp 321-37

Raisin C, Saguintaah B, Tetegmousse H, Lassabatere L, Girault B & Allibert C 1986, Ann. Telecommun. 41, 50

Raisin C, Rocher A, Landa G, Carles R and Lassabatere L 1991, E-MRS Fall Meeting Strasbourg, to be publihed in Appl. Surf. Science

Rocher A, Da Silva F W and Raisin C, 1990, Rev. Phys. Appl.25, 957.

Rocher A and Raisin C 1991 Series Springer Proceedings in Physics: Polycrystalline Semiconductors II , eds. J.H. Werner and H.P. Strunk.

Vannuffel C, Thèse de Doctorat de l'Université de Paris VI, July 1990.

Zhang X, Staton-Bevan A E, Pashley D W, Parker S D, Droopad R, Williams R L and Newman R C 1990, J. Appl. Phys., 67, pp 800-806.

Zhu J G and Carter C B 1990, Phil. Mag. A, 62, N°3, pp 319-28.

Inst. Phys. Conf. Ser. No 117: Section 7
Paper presented at Microsc. Semicond. Mater. Conf., Oxford, 25–28 March 1991

Interfacial dislocation arrays for on- and off-axis epitaxy of InGaAs on GaAs

Philip Kightley and Peter J Goodhew

Department of Materials Science and Engineering, University of Liverpool, PO Box 147, Liverpool, L69 3BX, England.

ABSTRACT: The structure of the misfit dislocation array as a function of the tilt of the substrate has been studied. It is shown that for vicinal substrates misfit dislocations are easily capable of adopting line directions other than <110> to remain within the interfacial plane. For wafers tilted toward (010) from (001) then four separate line directions are adopted, two in each approximate <110> direction. A simple geometric formula is presented that describes the expected angle of separation of these dislocations from <110>. In addition to the ~orthogonal dislocation reactions that are expected, a profuse generation of edge dislocations is observed at the low angle intersections between the nearly parallel dislocations. The formation of these edge segments is studied by diffraction contrast analysis including stereomicroscopy. The occurrence of polygonal dislocation features is explained.

1 INTRODUCTION.

Misfit Dislocations (MD) formed for moderately small mismatched InGaAs grown on (001) GaAs substrates generally adopt <110> line directions which are assumed to be in the interface plane. If an exactly oriented wafer is used for epitaxy then indeed dislocations are observed with <110> line directions (Chang et al (1989), Goodhew and Kightley (1991)). If a vicinal substrate is used then the misfit dislocation line direction changes to remain within the interfacial plane (Kightley et al (1991)). Figure 1 illustrates the intersection of the {111} slip planes with the (001) plane. The square base represents the <110> sides of a portion of (001). It is the base of a pyramid of {111} planes and for the (001) plane the {111} intersect with **u** parallel to <110>. If the wafer is grown off-axis, in the figure the bold plane represents the interface of a wafer tilted toward (010) from (001) by ß degrees, then the {111} planes, which are the planes of motion of dislocations as they create misfit dislocations, no longer intersect the interface with <110> directions. To enable optimum misfit relief a threading dislocation must form its misfit segments following the lines of intersection of the glide planes with the interface plane and thus **u** will no longer be parallel to <110>. When the interface plane is tilted toward (010) from (001) then an offset is generated in both the [110] and [-110] directions. For this offset, and assuming that the dislocations when forming all lie upon {111} planes, then the offset angle, θ, viewed in a (001) projection is given as;

$$\theta = \tan^{-1}\left(\frac{N}{R}\right) \quad \text{for} \quad N = \frac{\sqrt{2}\,\tan(\beta°)}{2\,\tan(54.7°)} \quad \text{and} \quad R = 1 - \left(\frac{\sqrt{2}\tan(\beta°)}{2\,\sin(54.7°)\tan(60°)}\right)$$

so that for a wafer tilt of ß≈2° toward (010) the offset angle θ≈1°. The term N describes the lateral displacement of the dislocation line from a <110> if projected into the (001) plane and the term R describes the distance from the intersection of the vicinal substrate plane with the (001) plane which is normal to the measurement N.

2 OBSERVATIONS

2.1 Line Directions of Misfit Dislocations at the Interface.

It is shown in the plan view TEM micrograph of figure 2 that MD which have formed at the interface between an $In_{0.12}Ga_{0.88}As/GaAs$ multilayer (100Å pairs) and its GaAs substrate, which has a tilt $\approx2°$ to (010) from (001), adopt line directions that are $\approx1°$ away from the <110> directions in agreement with our simple prediction shown above. This is confirmed by measurements of both diffraction contrast micrographs and superimposed diffraction patterns and dislocation images which show θ to be $1.25°±0.5°$. The micrograph of figure 2 is one of a montage that extends for $\approx90\mu m$ x 10 μm and in this area all dislocations form long straight lines and are generated from single slip operations. No proof for multiplication of misfit dislocations has been found in any of the samples studied here and it is very likely that the MD are formed from substrate dislocations in the manner proposed by Matthews et al (1970) and proven for the InGaAs/GaAs system in a series of elegant x-ray topographs by Green et al (1990). To confirm the observation of the variation of **u** with substrate tilt two 0.5 μm layers of $In_{0.12}Ga_{0.88}As$ were simultaneously grown on both an on-axis substrate and on one that was 2° off-axis in the direction shown above. The micrographs are shown in figures 3a and 3b. Clearly when a substrate tilt is present the dislocations are easily capable of forming a line direction that enables them to maintain their presence at the strained layer/substrate interface for these low tilts.

Figure 1   The intersection of the {111} glide planes with (001) and vicinal surfaces tilted toward (010) from (001).

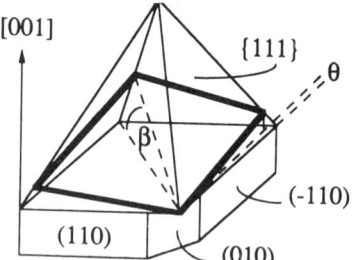

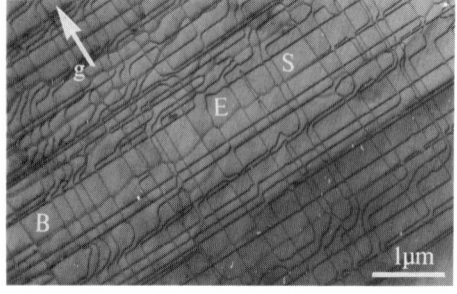

Figure 2 (001) plan view TEM micrograph illustrating a 1° offset from <110> for the dislocation lines. Where the dislocations react to form edge segments (E) one side of the interaction is sharp (S) whereas the other side is blunt (B).

a)   b)

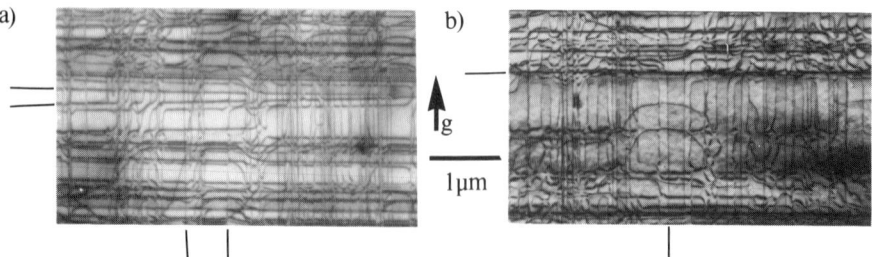

Figure 3   Single 5000Å $In_{0.12}Ga_{0.88}As$ layer grown on a) Off-axis 2°toward (010) from (001) substrate. b) On-axis substrate. Note Four line directions for the ~60° dislocations in sample a) and two for sample b).

2.2 Dislocation Reactions at Low-angle Intersections.

Four different line directions are observed that enable the dislocations to remain at the strained layer to substrate interface and from the micrographs it is apparent that as well as the 'usual' approximately orthogonal intersections normally associated with such an array there are 'low-angle' intersections also between dislocations separated by an angle of 2θ. The dislocations cross many others and often react to form segments of edge dislocation. Indeed there is a profuse generation of edge dislocation segments which push into the buffer layer in a similar manner to that described by Fitzgerald et al (1988). An example of the formation of an edge dislocation segment at a low angle intersection is shown in figures 4a and 4b. The two 60° misfit segments combine to form a single (a/2)<110> dislocation with its Burgers vector in the (001) plane. Note that the dislocation segments here come to a point. The dislocations in the sample from figure 2 can 'zip-up' to form segments (which are further segmented by orthogonal intersections) upto 30 microns long. In the low magnification micrograph of figure 2 one end of the reacting segments is pointed and the other is blunt. Stereomicroscopy shows that the pointed intersection is within the plane of the sample while the edge dislocation at the blunt end is deep in the buffer layer, joined to the misfit segment in the original interface by two threading portions. If the dislocations are constrained only to glide into the buffer layer to react to form the edge segment then they may only 'zip-up' in one direction from the point of intersection as they should remain upon their glide planes. This is depicted in the schematic of figure 5.

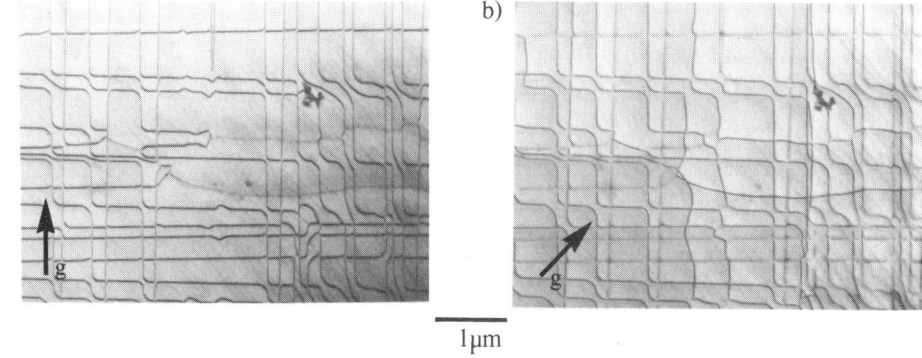

Figure 4 Formation of an edge dislocation segment at a low angle intersection of 60° dislocations. a) 220 reflection and b) 040 reflection.

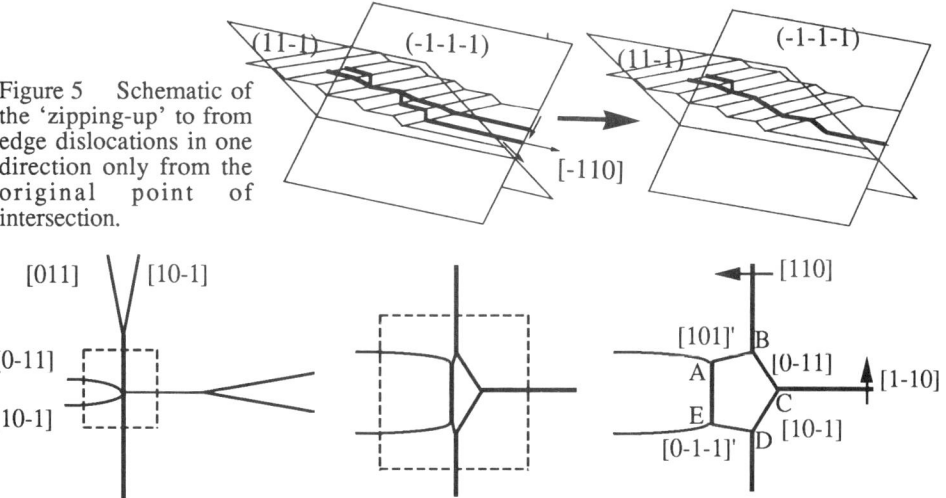

Figure 5 Schematic of the 'zipping-up' to from edge dislocations in one direction only from the original point of intersection.

Figure 6 Schematic of the formation of the polygon of dislocation indicated in figure 4. These polygons occur with a density of ≈1x10⁶ cm⁻² in this material.

Often polygons of dislocation are formed after reaction of the dislocation array. Figures 4a and b show one such polygon. Simple diffraction contrast analysis shows the segments of the polygon to have the Burgers vectors indicated in figure 6. The (a/2) factor is omitted for brevity. The polygon occurs by reaction of the intersecting dislocations. The [011] and [10-1] dislocations react to form an edge segment [110]. As the approximately orthogonal dislocations [0-11] and [10-1] 'zip-up' to form the edge segment [1-10] they cross the dislocation [110] and react forming extra short segments of dislocation. The reaction centres are labelled in figure 6 and are A:[0-11]+[110]=[101], B:[101]-[110]=[0-11], C:[0-11]+[10-1]=[1-10], D:[0-1-1]+[110]=[10-1] and E:[10-1]-[110]=[0-1-1]. The [110] segment contained in the polygon is laterally displaced from its original line. This movement is likely to be (001) glide.

3 DISCUSSION and CONCLUSIONS.

The simplest argument that explains the accommodation of the offset in the dislocation line is by the moving threading dislocation leaving portions of itself when it is incident at surface step risers. The threading segment is likely to align itself to an inclined <110> direction so the action of leaving a section of itself at the step will incur the constant offset necessary for the dislocation to remain in the interface plane. The exact atomic scale processes that enable the offset to occur are difficult to establish from diffraction contrast studies. Several simplifying assumptions may be made that enable a reasonable hypothesis to be formed upon the nature of the offset on the dislocation line. Firstly vicinal surfaces are expected to consist of (001) plateaus separated by steps which will be expected to have <110> ledge directions. Secondly misfit dislocations extend their line by the repeated generation of kinks on the dislocation which move to the interface. The kinks may be sourced singly at the surface or by the dissociation of thermally generated double kinks on the threading dislocation segment at the interface. When these diffusing kinks are arrested on the plateaus at both sides of a surface step an offset is incurred in the overall dislocation line enabling it to stay in the interfacial plane.

Many edge dislocation segments are formed by the low angle intersections of these nearly parallel misfit dislocations. The action of 'zipping-up' is shown to occur in one direction from the original point of intersection generating many of the product edge dislocations in the buffer layer of these samples. An example of a common polygonal dislocation feature is shown to be formed from simple dislocation reactions.

ACKNOWLEDGEMENTS.

Richard Beanland of Ohio State University for his help in interpreting the polygon reaction, Peter D Augustus and Robert R Bradley from Plessey Research, and R C Pond from the University of Liverpool are all acknowledged for their useful discussions.

REFERENCES.

Chang K H, Bhattacharya P K and Gibala R. (1989) J.Appl.Phys. **66** 2993
Fitzgerald E A, Ast D G, Kirchner P D, Pettitt G D, and Woodhall J M. (1988)
    J.Appl.Phys. **63** 693
Goodhew P J and Kightley P. MRS Fall meeting, Boston, USA. In press.
Green G S, Tanner B K, Barnett S J, Emeny M, Pitt A D, Whitehouse C R and Clark G F.
    (1990) Phil.Mag.Lett. **62** 131
Kightley P, Goodhew P J, Augustus P D and Bradley R R. J.Cryst.Growth. In press.
Matthews J W (1976) J.Vac.Sci.Technol. **12** 126

Inst. Phys. Conf. Ser. No 117: Section 7
Paper presented at Microsc. Semicond. Mater. Conf., Oxford, 25–28 March 1991

519

Characterisation of strained In$_{0.54}$Ga$_{0.46}$As layers grown on InP substrates

F Peiró, A Vilà, A Cornet, A Herms, J R Morante, S Clark*, R H Williams*

L.C.M.M. Dept. Física Aplicada i Electrònica. Universitat de Barcelona.
Diagonal,645. Barcelona 08028.
* Department of Physics. University of Wales. PO Box 913. Cardiff. UK.

ABSTRACT: TEM observations of In$_{.54}$Ga$_{.46}$As strained layers grown on InP substrates have been performed. A coarse quasiperiodic structure has been found in all the samples. The period of this structure decreases when layer thickness increases. The strain relaxation has been correlated with both the change in the period and the presence of defects.

1. INTRODUCTION

It is now well established that the lattice mismatch between a heteroepitaxial film and a substrate can be accomodated by elastic strain and/or by misfit dislocations. Van der Merwe (1963) argued that beyond a critical thickness, h$_C$, a strained layer will no longer grow coherently and misfit dislocations will be incorporated at the epilayer/substrate interface to relax the lattice mismatch strain. This topic is of considerable technological importance because of the effect of the misfit and threading dislocations upon the electronic properties of devices. Force and energy balance models (Matthews et al. (1976), People and Bean (1985), Miles and McGill (1989)) have previously reported explanations of the strain relaxation process. However, there is no clear consensus as to which is the actual mechanism involved in specific material systems.

In this work we examine the strain relaxation in a system where a composition modulation is present parallel to the interface. The parallel strain evolution with epilayer thickness has been correlated with both the change in the modulation period and the presence of defects.

2. EXPERIMENTAL PROCEDURES

All epilayers were grown using a VG Semicon V8OH Molecular Beam Epitaxy (MBE) system. Growth was carried out at 515° C on InP (001) semi-insulating Fe-doped substrates. The substrates were etched prior to growth, at 50°C in H$_2$SO$_4$: H$_2$O : H$_2$O$_2$ prepared in the ratio 7:1:1. Five epilayers were grown at a fixed alloy composition of x=54.3% ± 0.2% with different thicknesses as shown in Table I. On initiation of growth a step increase in temperature of the group III sources was employed to limit In and Ga flux transients.

TEM studies were performed on plan view, cleaved wedge and cross section samples. The cross-sectional specimens were thinned by I$^+$ ion bombardment. Plan-view [100] TEM specimens were prepared by mechanical polishing of the substrate side to a thickness of 30μm. A final thinning was carried out by ion milling in a cooled stage using low Ar$^+$ intensities. Thicker layers had to be etched to make samples thin enough for TEM studies. The observations have been performed using Hitachi H-800 NA microscope operating at 200 keV. Elastic strain was measured by Double Crystal X-Ray Diffraction (DCXRD).

2. RESULTS AND DISCUSSION

Micrograph of a cleaved wedge sample, imaged in the <010> direction, is shown in Fig.1. Interference fringes seem as perfectly straight in the InP. However, the undefined contrast upwards from the interface, points out a variation in composition through the layer and parallel to the interface. The most important feature that has been found in all the samples, is the existence of a tweed structure, with strong dark contrast roughly along both the [001] and [010] directions. For two beam conditions with g=<0$\bar{2}$2> two sets are in contrast (fig. 2), but when the sample is imaged in g=<004>, only the set of bands perpendicular to g remain visible. The diffraction contrast characteristics of these coarse modulations have been described in detail and reviewed by Treacy et al (1985).

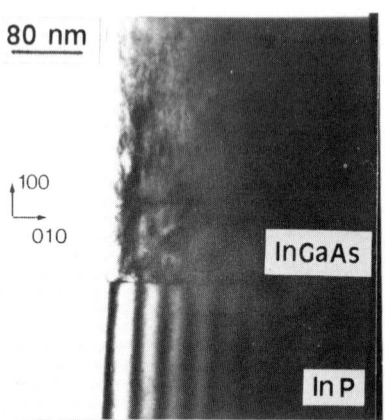

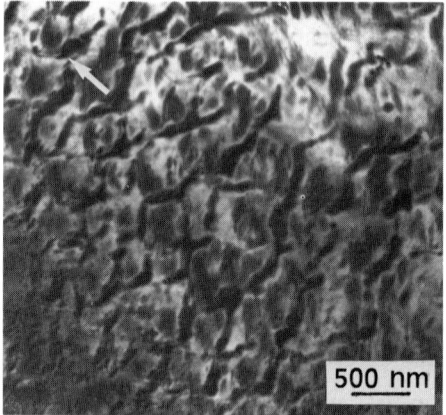

Figure 1. Micrograph of a cleaved wedge sample showing interference fringes on the InP substrat. The indefined contrast on the layer suggests a variation in composition.

Figure 2.   Plan   view   [100] micrograph of the sample A. The arrow   marks   g=0$\bar{2}$2.   The coarse modulation lies along both [010] and [001] directions.

From Energy Dispersive X-Ray Analysis (EDX), it is currently assumed (Henoc et al. 1982), that the presence of a composition modulation at the layer-substrate interface is responsible for the coarse pattern observed. Although the mechanism of this modulation is not well established (Glas 1989, Mahajan et al. 1989), the role played by the associated elastic energy must be taken into account to explain the stability of III-V alloys grown on InP (Glas 1987). In the same way the presence of this elastic energy must be included when an energy balance calculation for lattice relaxation in strained layers is performed.

In order to study the influence of the composition modulation on strain relaxation, we have studied epilayers with thickness beyond the critical value, h_c. In Table 1, we summarize the thicknesses as well as the wavelength Λ of the coarse, quasiperiodic, tweedlike pattern present in each sample. The wavelength Λ of the modulation has been found to decrease as layer thickness increases. It should be noted that samples D and E have similar values of Λ, despite large difference on thickness values.

Plan view micrographs of the thicker samples (Fig.3), show a large density of defects in comparison with the thinner samples, (Fig.2). The main type of defects observed are stacking faults bordered by partial dislocations, V-shaped defects and threading dislocations. Defects start to appear in sample C and their densities rise as layer thickness increases, reaching values of 10^7 cm^{-2} for stacking faults and 10^6 for threading dislocations, for sample E.

It must be noted that a network of misfit dislocations at the interface has not been observed in any of the samples. HREM images (Fig.4) showing an interface free of Lomer and 60° dislocations corroborate this feature. The existence of a composition modulation at the interface suggests that regions of localized strain will be present, limiting the propagation and interaction of dislocations to form misfit segments generally observed in mismatched layers. On the contrary, when a region above the interface is imaged (Fig.5) a set of misfit dislocations lying on <022> can be observed. In this case the structure is not affected by the modulation induced strain because, as suggested by Strunk (1989), the propagation mechanism of such dislocations is a climbing process.

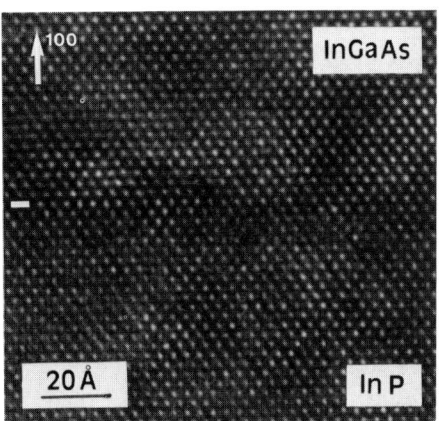

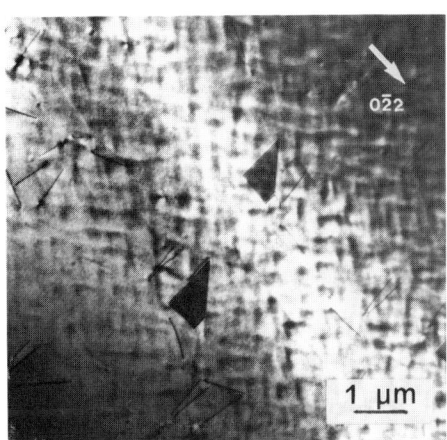

Figure 3. *Cross-section HREM image of the sample E. No Lomer and 60° dislocations along the interface are observed.*

Figure 4. *Plan view image of the thickest sample showing stacking faults, V-shape defects and threading dislocations.*

Measurements of remaining strain parallel to the growth plane have been performed by DCXRD (Clark et al. 1991). Figure 6 shows the evolution of the stress with the layer thickness. A and B samples, even with different thicknesses, present similar values of strain but different period of the modulation (table I). From this result, we suggest (Peiró et al.) that there is an adaptation of the period in order to absorb the increment of elastic energy introduced by the greater thicknesses. When the layer thickness exceeds a critical value, the relaxation of energy by means of defect nucleation is more favourable than a further decrease of the period. So, despite the large difference in thickness, samples E and D present similar values of the period but large difference in the density of defects.

SAMPLE	THICKNESS (μm)	WAVELENGTH (nm)	$2\pi t/\Lambda$
A	0.29	405	4.4
B	0.49	365	8.6
C	0.74	320	14.5
D	0.98	240	26.7
E	1.96	235	49.3

Table I: *Sample characteristics. The measurement error of the coarse structure period (Λ) has been estimated to be ±10%.*

In the thinnest samples, the energetic balance of the system leads to an average strain value corresponding to a totally strained layer matched to the InP substrate. The relaxation of the energy by means of defect nucleation commences at 0.5 μm, so it is concluded than the apparent critical thickness corresponds to this value. The remaining strain (from 0.96 μm) is linked to the fact that the modulation is still present.

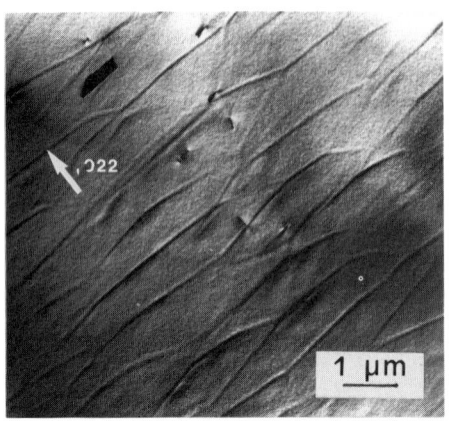

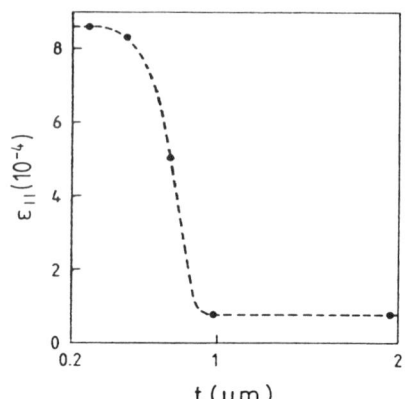

Figure 5. *Dislocation lines and stacking faults upward from the interface in sample E.*

Figure 6. *Evolution of the stress with the layer thickness.*

3. CONCLUSIONS

The introduction of energy relaxation induced by the presence of a modulation of composition leads to a critical thickness value greather than the one predicted by the Matthews model (Matthews et al. 1976), where only the energy relaxation due to the presence of misfit dislocations is taken into account.

REFERENCES

Clark S A, Macdonald J E, Westwood D I and Williams R H, in press 1991.
Glas F 1987 J. Appl. Phys. **62**, 3201.
Glas F 1989 NATO ASI series B203 (Plenum Press, New York) pp. 217-233.
Henoc P, zrael A, Quillec M and Launois H 1982 Appl. Phys. Let. **40**, 963
Mahajan S, Shadid M A and Laughlin D E 1989 Inst. Phys. Conf. Ser. **100**, 143.
Matthews J H, Blakleslee A E and Mader S 1976 Thin Sol. Films **33**, 253
Miles R H and McGuill T C 1989 J. Vac. Sci. Technol. **B7** 753.
Peiró F, Cornet A, Morante J R, Clark S and Williams R H to be published.
People R and Bean J C 1985 Appl. Phys. Lett **47** 327.
Treacy M M J, Gibson J M and Howie A 1985 Philos. Mag. A **51**, 389.
Van de Merwe J H 1963 J. Appl. Phys. **34** 123

Inst. Phys. Conf. Ser. No 117: Section 7
Paper presented at Microsc. Semicond. Mater. Conf., Oxford, 25–28 March 1991

Equilibrium defect structure of annealed II–VI/GaAs interfaces

A Schwartzman and R Sinclair*

Division of Engineering, Brown University, Providence, Rhode Island 02912, U. S. A.
*Department of Materials Science and Engineering, Stanford University, Stanford,
   California 94305, U. S. A.

ABSTRACT: High-resolution electron microscopy is used to characterize the as-deposited and annealed defect structure for (001) epitaxy of CdTe and ZnTe buffer layers on GaAs. The annealed interface is an equilibrium microstructure, consisting of a periodic array of Lomer misfit dislocations and zero residual strain in the thin film. The as-deposited microstructure corresponds to a metastable distribution of defects, in that there is a small amount of residual strain in the thin films and the variety of neighboring defects can react to form Lomer misfit dislocations. This paper investigates the means by which complete relaxation can occur through dislocation reactions during annealing of thick deposited films.

1. INTRODUCTION

Difficulty in growing defect free, II-VI compound semiconductor, bulk crystals results in their practical use, almost exclusively, in the form of thin films. Typically, good quality epitaxial layers are grown on single crystal GaAs substrates. They are used as buffer layers before processing the active device layers. The original interface between the GaAs substrate and the II-VI buffer layer is examined here with high-resolution electron microscopy (HREM) for (001) epitaxy of CdTe or ZnTe thin films. Owing to their large lattice mismatches, strain relaxation by formation of misfit dislocations becomes energetically favorable within one monolayer. This paper utilizes HREM as a means of imaging the interfacial defect arrangement, which is an indication of the local residual strain in the thin film. In turn, mechanical equilibrium analysis indicates if the microstructure represents an equilibrium configuration. With a knowledge of the Burgers vectors before and after annealing (based on Burgers circuit analysis not discussed here), energetically favorable dislocation reactions that transform the as-deposited interface to the annealed one are considered.

2. ANNEALED INTERFACES

A 9 μm CdTe film, grown by congruent evaporation in an ultra-high vacuum chamber (Bean et al 1986), was vacuum annealed for 100 hours at 600 °C and 10^{-5} Torr. The resulting defect structure is a periodic array of perfect edge, Lomer misfit dislocations with a spacing corresponding to a strain-free thin film, as depicted in Fig. 1. Since this is the most efficient manner to obtain a completely relaxed thin film, it represents the equilibrium microstructure for this epitaxial geometry. The Lomer misfit dislocation is characterized by two extra half-planes along each of the two {111} slip planes that intersect the (110) image projection plane and are arranged symmetrically about the dislocation core. The measured spacing between Lomer misfits is 3.1 nm for the annealed CdTe film. A similar equilibrium microstructure resulted for an annealed ZnTe film on GaAs, where now the periodic array of Lomers has an increased spacing of 5.4 nm, corresponding to the smaller lattice mismatch.

Once misfit dislocations appear, beyond the critical growth thickness, the remaining residual strain in the thin film is $\varepsilon = f - \delta$, where ε is the residual strain, f is the lattice mismatch and δ

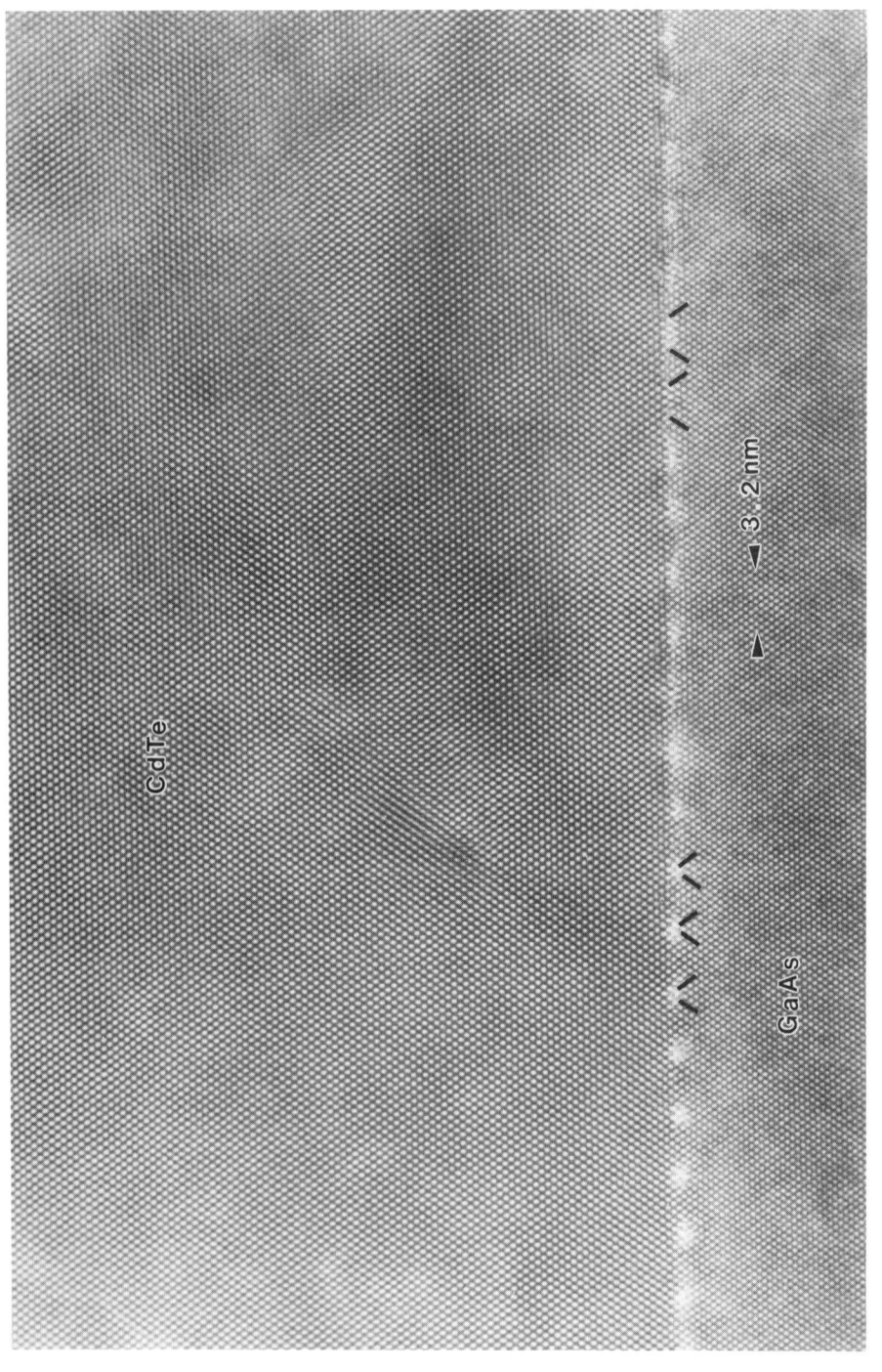

Fig. 1. HREM image of an annealed CdTe/GaAs interface.

is the strain due to the misfit dislocations. The amount of strain relaxation by misfit dislocation formation is $\delta = b_{eff}/s$, where b_{eff} is the effective component of the Burgers vector of magnitude b that accommodates the mismatch and s is the spacing between dislocations. The effective Burgers vector component is the projection of the Burgers vector edge component onto the interface. Lomers are the most efficient strain relieving dislocations since $b_{eff} = b$ (60° misfit dislocations are half as efficient). As the film grows much thicker than its critical thickness, the residual strain decreases to zero and the density of misfit dislocations increases to some fixed amount determined by the type of dislocation and the lattice mismatch. For complete relaxation by Lomer formation, we have $s_{eq} = b/f$. Substitution of the appropriate values yields the calculated equilibrium Lomer spacings of 3.1 and 5.4 nm for the CdTe/GaAs and ZnTe/GaAs heterojunctions respectively, in agreement with the measured values.

3. AS-DEPOSITED INTERFACES

The thick as-deposited films are mostly relaxed, with a typical defect arrangement illustrated in Fig. 2 for an MBE grown ZnTe film (Wagner *et al* 1988). In general, there are interfacial misfit dislocations (mostly 60°'s and some Lomers) and extended defects penetrating into the thin film (stacking faults bounded by either Shockley or Frank partials, and more complicated defect structures due to interacting perfect and/or partial dislocations on opposite slip planes). Consecutive interfacial 60° dislocations (delineated by just one extra half-plane corresponding to the edge component) lie on alternating slip planes. If they have opposite screw components, not projected onto the image plane, then these alternating 60° dislocations can climb along the interface during the annealing process to produce a periodic array of Lomers. If they have similar screw components, then the neighboring 60° dislocations will repel each other during climb, as marked in the right-hand part of Fig. 1 for the annealed CdTe film.

Only the simplest type of extended defect is seen in Fig. 2, a single extrinsic stacking fault bounded by Shockley partials. Since the lattice mismatch is so large, these films in compression evolve through island growth. Thus, there is the question as to whether the stacking faults form during growth or dissociation of 60° dislocations gliding towards the interface. In the latter case, the order of Shockley partials is not arbitrary. For extrinsic faults in compressive films, the leading partial at the interface is the 90° Shockley (and 30° partial for intrinsic stacking faults). Such is the case shown in Fig. 2 (and other stacking faults not shown here), where the short line corresponds to the extra half-plane associated with the pure edge 90° partial. Note that if the 30° partial combines with the 90° during annealing, a 60° perfect dislocation is formed, which can then climb to combine with the other 60° just to the right on the opposite slip plane to create a Lomer (assuming opposite screw components). In turn, this newly formed Lomer during the annealing process is close to the equilibrium separation from the as-deposited Lomer to the left, marked by the double arrows.

This paper demonstrates that as-deposited interfaces for thick CdTe and ZnTe buffer layers on GaAs correspond to a metastable defect arrangement. That is, there is some residual elastic strain in the thin film (even though equilibrium mechanical analysis predicts that there should be none) and neighboring defects can react to form Lomer dislocations. Annealing produces the equilibrium microstructure, a periodic array of Lomer misfits with spacing corresponding to a completely relaxed thin film.

Funding from the Materials Science Program, Division of Basic Energy Sciences Division, Department of Energy is gratefully acknowledged. Samples were provided by Ken Zanio of Ford Aerospace (CdTe) and Brent Wagner of Georgia Tech Research Institute (ZnTe).

REFERENCES

Bean R C, Zanio K R, Hay K A, Wright J M, Saller E J, Fischer R and Morkoc H 1986 J. Vac. Sci. Technol. A4 2153
Wagner B K, Oakes J D and Summers C J 1988 J. Cryst. Growth 86 296

Fig. 2. HREM image of an as-deposited ZnTe/GaAs interface.

Inst. Phys. Conf. Ser. No 117: Section 7
Paper presented at Microsc. Semicond. Mater. Conf., Oxford, 25–28 March 1991

Defect analyses of MBE-grown ZnSe films

S B Sant*, G C Weatherly** and R W Smith*

*Department of Metallurgical Engineering, Queen's University, Kingston, Ontario, Canada K7L 3N6
**Department of Materials Science and Engineering, McMaster University, Hamilton, Ontario, Canada L8S 4L7

ABSTRACT: Molecular beam epitaxy (MBE) grown ZnSe films studied by X- ray pole figures and transmission electron microscopy (TEM) clearly show epitaxial and twinned regions as well as the presence of sub-boundaries. Planar specimens of ZnSe/Ge were in-situ TEM annealed, for 5.5 hours at 873K. The twins and sub-boundaries are thermally very stable which would indicate that they arise during the growth process. The occurrence of these twins in the ZnSe film is explained by nucleation and growth of normal and twinned nuclei.

1. INTRODUCTION

A promising candidate for blue light emitting devices is ZnSe, a II-VI compound semiconductor with the zinc-blende structure with a direct energy band-gap of 2.7 eV at room temperature. Suitable substrates for the MBE growth of ZnSe are (100)GaAs and (100)Ge because of their relatively small lattice mismatch with ZnSe, 0.25% and 0.17% respectively while (100)Si has a 4.4% mismatch.

The mechanism of twin formation in epitaxial ZnSe films has not been clearly understood even though TEM has been extensively used to study the defects present in ZnSe films [Stutius and Ponce 1985, Williams and Wright 1987, Sant et al 1987, Sant et al 1990, Sant 1989]. *In-situ* TEM annealing could provide some clues to a possible mechanism of twin formation since stresses may have developed due to the lattice mismatch and thermal expansion coefficient difference between the film and the substrate. A comparison of in-situ TEM annealing of planar specimens and bulk vacuum annealing experiments is reported elsewhere [Sant et al 1990].

Minima in the interfacial energy curves occur for twin growth. This would lead us to believe that two types of nuclei are able to form, those in the epitaxial orientation as well as those in a twin relationship. In the light of this, a growth model has been formulated that is based on the free energy of formation of the different nuclei [Sant 1989].

2. EXPERIMENTAL PROCEDURE

The growth of ZnSe films by MBE has been described earlier [Park et al 1985, Park and Mar 1986] as have the TEM specimen preparation and the *in-situ* TEM annealing experiments [Sant 1989]. Both cross-sectional and planar specimens were studied using diffraction contrast in an Hitachi H-800 TEM operated at 200 kV. The as-grown specimens were typically 100 mm² with one of the {011} edges longer than the other. For cross-sectional specimens it is common practice to need to preserve knowledge of different <011>; this procedure however does not give us the exact <011>. We will see that this is important due to the anisotropy of the twin distribution. The misorientation of the low-angle boundaries was characterized using a 1 nm probe in a HB5 dedicated STEM.

In order to study the bulk crystallography of twins in the as-grown specimens, X-ray pole figures were obtained. In these experiments, the Schulz reflection method was employed with the 2θ angle set for the ZnSe {111} reflection. The pole figures were obtained by changing the angles of rotation and inclination continuously for 6 spiral scans.

3. RESULTS

3.1 X-Ray Pole Figures

Fig. 1 is a {111} pole figure of an as-grown ZnSe/(100)GaAs specimen. We notice that there are 4 intense {111} poles that arise from the film having (100) at the centre of the pole figure. There are other {111} poles that are located at 17°, 55° and 80° from the centre of the pole figure. We know that in the zinc-blende crystal, as in face-centered cubic lattices, twins are formed by a 180° rotation about a <111> axis. We perform this manipulation, on a stereographic projection, by rotating the (100) pole about the [111] axis and find that the pole (hkl), the resultant of the above rotation, now lies 56° from the (111) pole and 18° from the (111) pole and is in fact the {122} pole. We find that twin oriented regions have the ($\overline{1}$22) planes parallel to the (100) planes of the substrate. We shall see how this information is used to develop a model for the formation of epitaxial and twin oriented nuclei based on free energy calculations. The orientation relationships of different regions in the film can be given as [Sant 1989]:

Epitaxial Orientation:

$(100)_f$ // $(100)_s$; $[0\overline{1}1]_f$ // $[0\overline{1}1]_s$ and $[011]_f$ // $[011]_s$

Twin Orientation:

$(\overline{1}22)_f^t$ // $(100)_s$; $[0\overline{1}1]_f^t$ // $[0\overline{1}1]_s$ and $[411]_f^t$ // $[011]_s$

where f, s and t refer to the film, substrate and twin respectively.

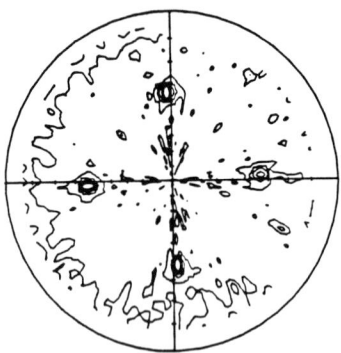

Fig. 1: X-Ray pole figure showing location of the {111} poles.

3.2 *In-situ* TEM annealing

Fig. 2 shows two-beam TEM bright field images of the specimen that underwent the *in-situ* TEM annealing experiments. Planar defects exhibiting stacking fault-like fringe contrast have been positively identified as twins [Sant 1989]. It can be observed that there is no appreciable change in the microstructure of the defects upon *in-situ* annealing other than some dislocation line reorientation. Contrast changes expected if the low-angle boundaries try to rotate or move into lower energy configurations were not observed. The thermal stability of the twins leads us to believe that their origin is likely to be associated with a nucleation and growth induced process. Micro-diffraction patterns were obtained across the boundary using the 1 nanometer probe of the HB5 dedicated STEM. The misorientation of the sub-boundary was calculated by measuring the relative displacements of certain pairs of Kikuchi lines. On average, the misorientation was determined to be $0.45°$.

3.3 Twins

Fig. 3 is a cross-sectional micrograph obtained from a sample that consisted of alternate layers of ZnSe and Si, each 30 nanometers thick, grown on a (100) Si substrate with a final ZnSe film grown on the multilayer structure. In the centred dark field image using a twin reflection, the dark layers are those of Si while the intermediate layers are ZnSe. It can be clearly seen that a number of twins are present in the ZnSe layers with some of the twins appearing to cross the ZnSe/Si interface. The twins appear edge-on and we see micro-twins having dimensions of approximately 3 nanometers in width and 12 nanometers in length.

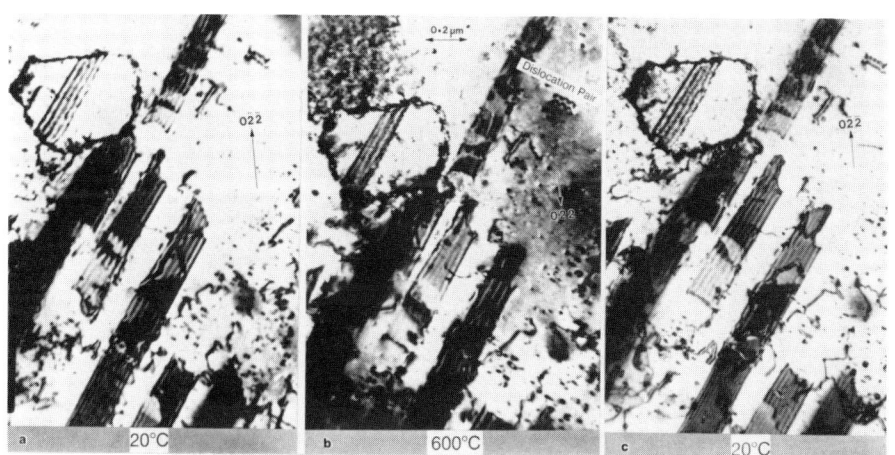

Fig. 2: *In-situ* TEM annealing of planar ZnSe/(100)Ge (a) 293K (b) After 5.5 hours at 873K (c) After cooling to 293K.

Fig. 4 is a set of bright field image, dark field image and selected area diffraction pattern obtained from a planar specimen of ZnSe/(100)Ge. We see that defects exhibiting typical stacking fault-like fringe contrast can be clearly seen in fig. 4(a). However when the specimen is tilted in the microscope we obtain the bright field image, with no fringe contrast, in fig. 4(b) which gives rise to the diffraction pattern in fig. 4(c). Intutively, we might deduce that there are no spots due to twins. However, we do notice one interesting feature. The diffraction spots have two different intensities. Fig. 4(d) is a centred dark field image obtained using the circled weak diffraction spot in fig. 4(c) and we see that the twins are in very good contrast. This then demonstrates that the weak spots arise from the twins and upon indexing the pattern we find that the strong diffraction spots are from the superimposition of the matrix and the twin reflections. Upon indexing the pattern we find that the foil normal is [011] for the matrix and [411] for twins.

We can now correlate the above diffraction pattern analysis with the stereographic projection. Upon twinning, which is a 180° rotation about the [111] axis, the [011] pole transforms to the [411] pole and likewise the [411] pole to the [011] pole. This is what we find from the orientation relationship

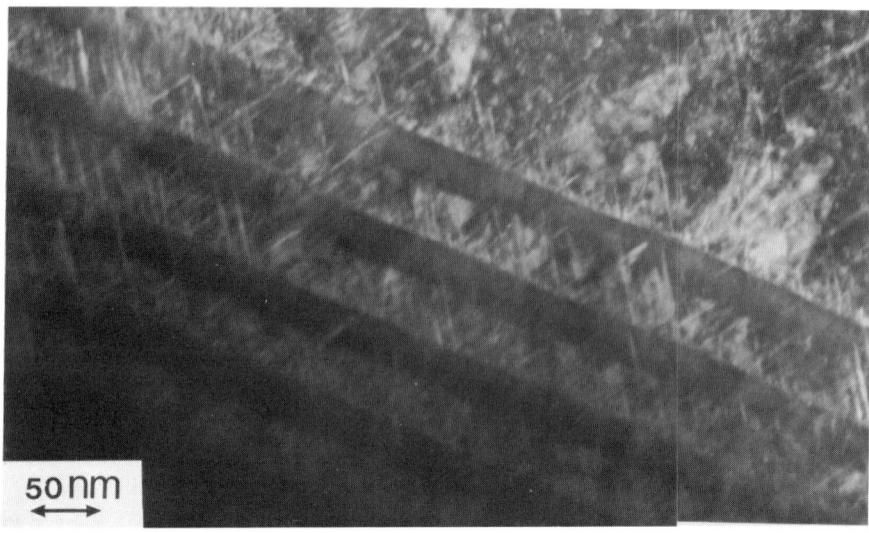

Fig. 3: Centred dark field cross-sectional TEM image from ZnSe/(100)Si showing twins and alternate ZnSe and Si(dark) layers.

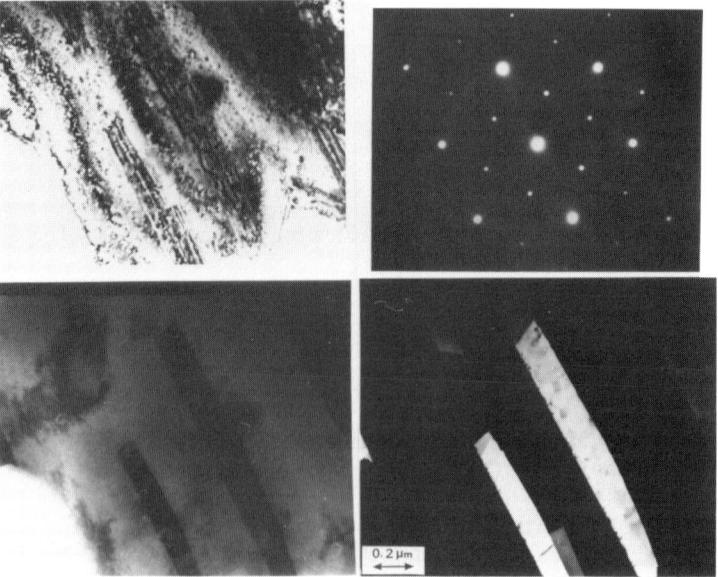

Fig. 4: TEM micrographs of planar ZnSe/(100)Ge (a) bright field showing fringe contrast
(b) diffraction pattern of (c), (c) bright field image and (d) centred dark field image showing twins.

of the diffraction pattern in fig. 4(c). The twin rotation on the stereographic projection also shows that the $[0\bar{1}1]$ pole becomes the $[0\bar{1}\bar{1}]$ pole and all the $\{111\}$ poles become the $\{115\}$ poles.

3.4 Model

The presence of twins is not unique to ZnSe, but, has been reported for a large number of materials with the zinc-blende structure. Studies of MBE-grown GaAs on Si has shown that microtwins form during the growth of the initial islands [Hull and Fischer-Colbrie 1987, Tsai and Matyi 1989, George at al 1990]. Given this evidence and the results presented here we can develop a model that will explain the occurrence of twins in MBE-grown ZnSe films. Assuming no lattice misfit, we can see that the epitaxial nuclei form coherent interfaces along the two orthogonal $<011>$ directions. On the other hand, the twin nuclei and the substrate form a coherent interface along the $[0\bar{1}1]$ direction while in the orthogonal direction the interface is semi-coherent because $d_{011} = 3d_{411}$.

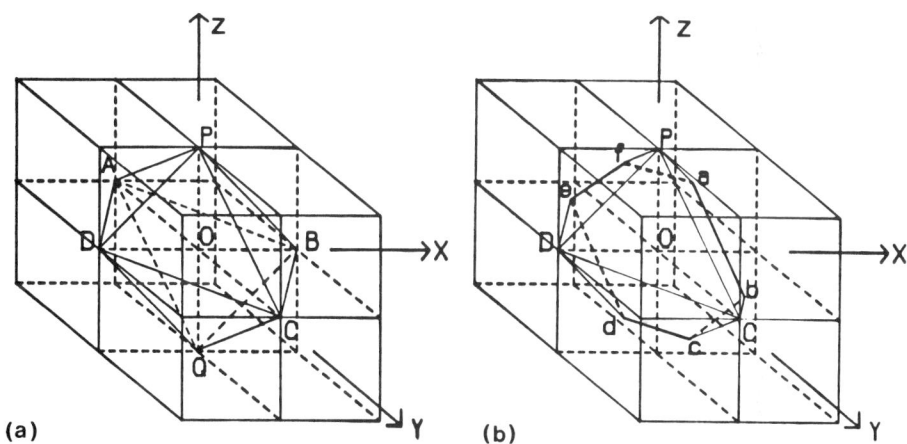

(a)　　　　　　　　　　　(b)

Fig. 5: Shape of nuclei having $\{111\}$ facets (a) epitaxial nucleus; ABCD is the (100) base in contact with the (100) substrate surface, APD, DPC, CPB and APB are the $\{111\}$ pyramidal facets. (b) twin oriented nucleus; abcdef is the $(\bar{1}22)$ base in contact with the (100) substrate surface, Paf, Pabc, Cbc, DCdc, Ded, DefP and DPC are the $\{111\}$ facets.

Our results and those of several other researchers suggest that the initial film forms by the nucleation of three dimensional islands. For the epitaxial orientation the nuclei that form can be imagined to be made up of discrete pyramids that have the (100) face in contact with the (100) substrate surface and the pyramidal faces are the $\{111\}$ faces, fig. 5(a). This is not unreasonable since the $\{111\}$ planes have the lowest surface energy. Let us imagine that one of these pyramids is in the twin orientation, fig. 5(b). In the case of zinc-blende materials this means a 180° rotation about a $<111>$ axis. The shape of the twinned nucleus is obtained by starting with the dipyamid having $\{111\}$ faces and then looking at the intersection of the $(\bar{1}22)$ plane with this dipyramid. The base is the $(\bar{1}22)$ plane, abcdef, which is in contact with the (100) substrate surface as expected from the pole figure experiments. The $\{111\}$ faces are the exposed faces since they are still the planes with the lowest surface energy. The growth of the twin nuclei will be much faster along the coherent [0$\bar{1}$1] direction. As a result, upon coalescence of the islands, the thickness of the twin will be much smaller than its length. The propensity of twins to be present in greater number along one direction can also be explained. The four {111} planes intersect the (100) surface along the two orthogonal <011> directions and since the growth of the twin nuclei is faster along the [0$\bar{1}$1] direction in the {111} planes this would lead to a larger number of twins in one direction [Pirouz et al 1987]. Details of the calculation for the free energy of formation of the nuclei as a function of their orientation will be described elsewhere [Sant et al 1991].

4. CONCLUSION

MBE-grown ZnSe films on (100)GaAs and (100)Ge substrates have been characterized using X-ray pole figure and TEM. It was found that the primary defects were twins that are thermally very stable. A model developed explains the occurrence of twins due to the formation of initial nuclei that have the epitaxial and twin orientations.

ACKNOWLEDGEMENTS

The authors would like to acknowledge the support of the Natural Sciences and Engineering Research Council of Canada and 3M Canada Inc for providing the MBE grown ZnSe samples. One of the authors (SBS) acknowledges useful discussions with Dr. Uwe Erb, Dr. Roberto Pascual and Professor F.R.N. Nabarro.

REFERENCES

George, T., Weber, E.R., Nozaki, S., Wu, A.T., Noto N., and Umeno, M., 1990, J. Appl. Phys. 67, 2441.
Hull, R., and Fischer-Colbrie, A., 1987, Appl. Phys. Lett., 50, 851.
Park, R.M., Mar, H., and Salansky, N.M., 1985, J. Vac. Sci. Technol., B3, 676.
Park, R.M., and Mar, H., 1986, J. Materials Research, 1, 543.
Pirouz, P., Chorey, C.M., Cheng, T.T., and Powell, J.A., 1987, Inst. Phys. Conf. Ser. No. 87: Sec. 2, 175.
Sant, S.B., Kleiman, J., Melech, M., Park, R.M., Weatherly, G.C., Smith, R.W. and Rajan, K., 1987, Inst. Phys. Conf. Ser. No. 87: Sec. 2, 129.
Sant, S.B., Smith R.W., and Weatherly, G.C., 1990, Phil. Mag. Letters 61, 273.
Sant, S.B., 1989, Ph.D. Thesis, Queen's University, Canada.
Sant, S.B., Weatherly G.C., and Smith, R.W., 1991, to be published.
Stutius, W. and Ponce F.A., 1985, J. Appl. Phys., 58, 1548.
Tsai H.L., and Matyi, R.J., 1989, Appl. Phys. Lett., 55, 265.
Williams, J.O. and Wright A.C., 1987, Phil. Mag. A, 55, 99.

Inst. Phys. Conf. Ser. No 117: Section 7
Paper presented at Microsc. Semicond. Mater. Conf., Oxford, 25–28 March 1991

533

Microstructural and electrical properties of single crystal ZnTe/CdTe/CdS p-i-n solar cells and related materials

K Durose, P D Brown, M Y Simmons, H M Al-Allak and A W Brinkman

Applied Physics Group, SEAS, University of Durham, South Road, Durham, DH1 3LE, U.K.

ABSTRACT: The structural quality of epilayers grown during the development of a ZnTe/CdTe/CdS p-i-n solar cell structure is reported upon, together with preliminary device measurements. CdTe has been grown epitaxially on $\{0001\}$ and $\{10\bar{1}6\}$ CdS substrates, the best morphologies being obtained for the S variants of these planes. However, CdTe epilayers grown on both of these planes contained a high density of lamellar twins arising from bulk growth and surface nucleated mechanisms respectively. (CdZn)Te alloys grown on (100) GaAs contained distributions of dislocations resembling those observed in similarly mismatched systems. EBIC studies of a trial solar cell structure have shown that there is enhanced recombination at the wurtzite-sphalerite interface.

1. INTRODUCTION

The ZnTe/CdTe/CdS p-i-n solar cell structure reported by Meyers (1988) avoids the requirements of making conductive p-type CdTe and contacting it with a metal. Meyers has been able to produce polycrystalline cells of this type having efficiencies of about 10%. In this laboratory a programme to produce single crystal ZnTe/CdTe/CdS solar cells by MOVPE on CdS substrates is underway. The single crystal devices should offer high efficiencies and allow the limiting aspects of device performance to be investigated. Also the high degree of materials control available with MOVPE will enable graded or multi-composition (CdZn)Te layers to be inserted into the device structure making better use of the solar spectrum.

In this work the development of the epitaxial growth processes necessary to fabricate the single crystal solar cell structure is recorded with a special emphasis on the structural quality of the materials grown. The systems addressed here are CdTe/CdS and (CdZn)Te/GaAs (Δa_0 7.9-14.6%), and the latter is also of interest as a lattice matched hybrid substrate for (CdHg)Te device fabrication.

2. EXPERIMENTAL

The first objective was to obtain good CdTe epitaxy on CdS substrates. CdS adopts the wurtzite lattice (a=4.137Å, c=6.716Å) whereas CdTe has the sphalerite lattice (a_0=6.477Å), the bond length misfit between the two being about 9.7%. Growth trials on the polar basal plane surfaces of CdS, ie (0001)Cd and (000$\bar{1}$)S were therefore conducted. Awan et al (1987) showed that evaporated CdS films having superior morphologies could be deposited on $\{221\}$ CdTe. Since these films had the $\{10\bar{1}6\}$ orientation growth trials have also been conducted on (10$\bar{1}$6)Cd and (10$\bar{1}$6)S surfaces. These planes are tilted from the respective basal planes by about 16°. Basal plane oriented samples purchased from Eagle Pitcher were n-type and had resistivities in the range 5-8 Ωcm. $\{10\bar{1}6\}$ oriented slices were cut from boules grown from the vapour phase in this laboratory. Polar variants of the $\{10\bar{1}6\}$ plane can be located by first identifying the polarity of the basal plane using the procedure described by Warekois et al(1962,1966) and then tilting the boule accurately on the X-ray goniometer before cutting. Surfaces were prepared for epitaxy using the method described by Simmons et al (1991) prior to growth at 325°C. (100) GaAs substrates purchased from ICI Wafer Technology were used for the (CdZn)Te growth trials and these were degreased and cleaned with a 4:1:1 mixture of H_2SO_4, H_2O_2 and H_2O at 40°C prior to growth. A variety of growth temperatures ranging from 325°C to 400°C and a constant II–VI ratio of about 15:1 were used. The Zn:Cd ratio was varied over the full range. A trial device structure comprising 0.1μm ZnTe and ~1μm CdTe (both undoped) on (000$\bar{1}$)S oriented CdS was grown at 325°C. All epitaxy was carried out at atmospheric pressure in hydrogen carrier gas using the precursors diisopropyl telluride, dimethyl zinc and dimethyl cadmium. XTEM samples were prepared by Ar$^+$ ion beam thinning in a liquid nitrogen cooled kit and finished using I$^+$ at room temperature.

3. RESULTS AND DISCUSSION

Growth of CdTe on {0001} CdS produced epilayers with smoother surfaces on the S–face than on the Cd–face (Simmons et al 1991). XTEM studies of the samples grown on the {000$\bar{1}$}S face confirmed that the epitaxial relationship was {111}<110>CdTe//(000$\bar{1}$)B<1$\bar{2}$10>CdS. These films also contained a high density of thin lamellar twins (typically 10-20nm thick) which had their {111} composition planes parallel to the heterointrface. The twin lamellae were terminated by a variety of lateral Σ=3 boundaries.

Growth on the {10$\bar{1}$6}S surface of CdS yielded CdTe with the best morphology of all. However XTEM showed that these layers still contain a high density of lamellar twins and that these are inclined to the heterointerface. Fig.1 shows such an epilayer taken with **B** close to <110> in the CdTe and to <1$\bar{2}$10> in the CdS. The twin lamellae are almost parallel to the basal plane of the CdS. Fig.2 shows the selected area diffraction pattern corresponding to the area shown in fig.1 and it is evident that there is a misorientation between the close packed planes in the epilayer and substrate of 2°. This could be caused by a slight deviation of the substrate orientation from the exact {10$\bar{1}$6}S plane.

Twinning in {111} oriented CdTe films probably originates from growth accidents which act so as to nucleate these low energy defects. This is confirmed by the fact that unrelated parallel twin boundaries are present throughout the entire thickness of the {111} layers and this is true for CdTe on any substrate (see Brown et al 1987 and Aindow et al 1989 for example). However, for other orientations, eg (100) twins and stacking faults (when present) invariably extend from the heterointerface indicating that they are surface nucleated, probably at substrate steps. Twins in CdTe/(10$\bar{1}$6)S CdS are probably also surface nucleated, the (10$\bar{1}$6) surface being almost certainly made up of basal and prismatic plane segments. Clearly the use of this unusual orientation does not eliminate twinning in CdTe and further efforts to do so will be made using CdS oriented on its prismatic planes.

(CdZn)Te epilayers of all compositions (measured by EDAX) could be grown on (100) GaAs substrates at 400°C although excess DMZn was always required to achieve Zn incorporation, even at low concentrations. Growth at temperatures less than 400°C precluded Zn incorporation. These findings are contrary to those of Ahlgren et al (1989) who report that Zn is preferentially incorporated and that excess dimethyl cadmium is required to control the composition of the films. However, in that work diethyl telluride was used, and, more significantly a II-VI ratio of 1:1 was employed. Work is currently in progress in this laboratory to assess the influence of the II-VI ratio on the structure and composition of (CdZn)Te/GaAs.

Fig.3 shows fringe contrast with a periodicity of ~30Å from a sample having the composition $Cd_{0.06}Zn_{0.94}Te$ arising from the interfacial dislocation array. For this alloy on GaAs the dislocation spacing expected for complete stress relief by pure edge $a_o/2$<110> dislocations is ~52Å and is less if relief is effected by 60° dislocations having **b** inclined to the interface.

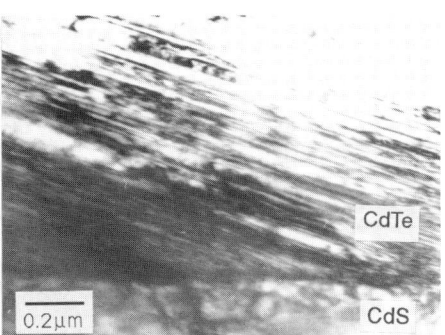

Figure 1. XTEM micrograph of epitaxial CdTe/(10$\bar{1}$6)S CdS showing lamellar twins parallel to the(0001)S plane in CdS.

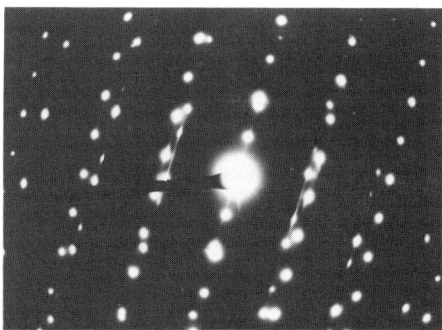

Figure 2. Selected area diffraction pattern taken from the area shown in figure 1 showing the 2° misorientation between the close packed planes

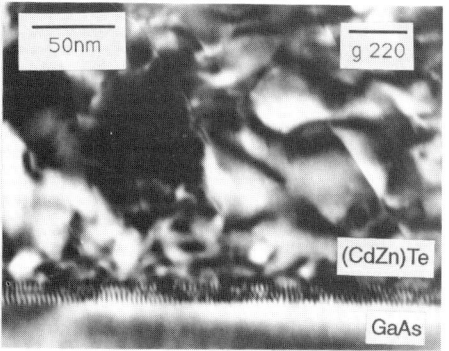

Figure 3. XTEM micrograph showing the Cd$_{0.06}$Zn$_{0.94}$Te/(100)GaAs interface.

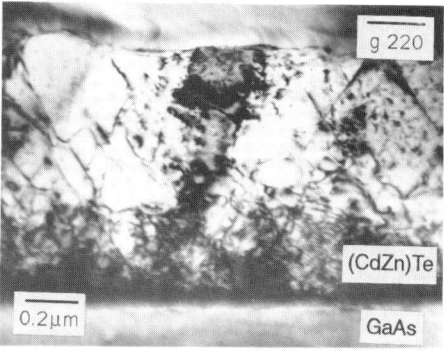

Figure 4. XTEM micrograph of a 0.9µm thick Cd$_{0.06}$Zn$_{0.94}$Te/(100)GaAs epilayer. The surface dislocation density isincreased by ~100 by growing at 400°C rather than the more usual 325°C

ZnTe CdTe CdS substrate

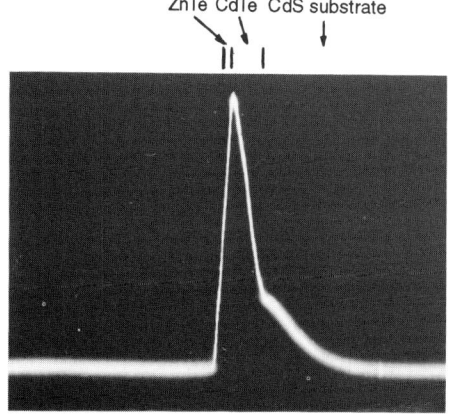

Figure 5. SEM/EBIC profile of a trial ZnTe/CdTe/CdS diode. The highest external current is genrated by excitation close to theZnTe/CdTe interface.

While the experimental spacing would appear to indicate that a variety of dislocation types are active in stress relief a conclusive assessment cannot easily be made from the diffraction contrast method used here. In fig.4 the entirety of the 0.9μm thick layer is visible. The distribution of dislocations is similar to that observed in CdTe/GaAs by Petruzzello et al (1987). Immediately above the interfacial aray there is a dense tangle of dislocations about 0.2 μm deep. A lower dislocation density of $\sim 10^{10} cm^{-2}$ is present at the surface. Similar dislocation distributions have been seen in CdTe and ZnTe epilayers on (100) GaAs grown in our laboratory at 325°C and this is typical for systems with large lattice mismatch. However, the surface dislocation density in the alloy is two orders of magnitude higher than is seen in binaries of comparable thickness and is attributed to the relatively high growth temperature used in order to incorperate Zn.

I-V measurements on the trial ZnTe/CdTe/CdS structure (grown with undoped epilayers) yielded an open circuit voltage of up to 0.6V and a short circuit current of 5 mA cm^{-2}. Temperature dependant I-V measurements yielded constant slopes on semilogarithmic plots which is typical of the current transport mechanism being a multi-step tunneling process. An EBIC profile of a cleaved device is shown in fig.5. This shows that the highest external currents are generated by excitation from close to the ZnTe/CdTe interface. Enhanced minority carrier recombination at the highly mismatched wurtzite/sphalerite interface and at the twin boundaries in CdTe is probably responsible for this behaviour.

4. CONCLUSIONS

The work has demonstrated the feasibility of growing a single crystal ZnTe/CdTe/CdS p-i-n solar cells on a CdS substrates. The non-metal variants of the {0001} and {10$\bar{1}$6} CdS surfaces have been identified as the best for epitaxy. However, EBIC indicates that enhanced recombination at the wurtzite/sphalerite interface will act so as to reduce device efficiency. Furthermore twinning in CdTe/CdS cannot be eliminated by growth on {1016} substrates. Growth trials of (CdZn)Te on (100)GaAs substrates indicates that all compositions may be achieved, the dislocation distribution in the film examined being similar to that normally encountered in ZnTe/GaAs and CdTe/GaAs. It is likely that better quality ternary material can be grown directly onto CdTe/CdS and its incorporation into a p–i–n solar cell will be attempted.

REFERENCES

Ahglren E P, 1989 J.Vac. Sci. Technol. A7(2) p331
Aindow M et al 1989 Inst. Phys Conf. Ser. No 100: Section 3 p223
Awan G R, Brinkman A W, Russell G J and Woods J 1987 J.Crystal Growth 85 p447
Brown P D, Hails J E, Russell G J, Brinkman A W and Woods J 1987 Appl. Phys Lett 50 p1144
Meyers P V 1988 Solar Cells 23 p59
Petruzzello J, Olego D, Ghandi S K, Taskar N R and Bhat I 1987 Applied Physics Letters 50 (20) p1423
Simmons M Y, Brown P D and Durose K 1991 J.Crystal Growth 107 p447
Warekois E P, Lavine M C, Mariano A N and Gatos H C 1962 J. Appl. Phys 33 (2) p690
Warekois E P, Lavine M C, Mariano A N and Gatos H C 1966 J. Appl. Phys 37 p2203

Inst. Phys. Conf. Ser. No 117: Section 7
Paper presented at Microsc. Semicond. Mater. Conf., Oxford, 25–28 March 1991

HREM investigation of epitaxial layer and interface structure in the CaF$_2$/Si heterosystem

V Yu Karasev,[*] A N Kiselev, E V Orlova, S M Pintus[*], A A Velitchko[*] and O A Zalabasov

Institute of Crystallography, USSR Academy of Sciences
Leninsky pr.59, 117333 Moscow, USSR
[*]Institute of Semiconductor Physics, USSR Academy of Sciences
pr.acad.Lavrentyeva 13, 630090 Novosibirsk, USSR

ABSTRACT: The structure of 150-800 nm thick CaF$_2$/Si films obtained by MBE combined with SPE on Si(111) and (001) was investigated. CaF$_2$/-/Si(111) films contain microtwins and grain boundaries, normal to the growth surface. An abnormal contrast intermediate layer was observed at the CaF$_2$/Si(001) interface. (111) and (1̄1̄1) Si extra half-planes were observed. Increasing of the CaF$_2$ amorphous layer thickness to 50-100 nm in the SPE process results in formation of polycrystalline films.

1. INTRODUCTION

Epitaxially grown CaF$_2$ on Si is a promising insulating material for fabricating CMOS structures and 3-D ICs. This heterostructure allows one to investigate the interface structure between materials with different lattice and chemical bond types (ion bond in CaF$_2$ and covalent bond in Si).

As a rule, the CaF$_2$ film surface, grown on Si(001) is strongly faceted (Philips 1986) (Fathoner and Schowalter 1984). The faceted growth surface reduces the advantages of MBE in obtaining multilayered heterosystems containing epitaxial dielectric layers. At present, rapid annealing (RTA) of CaF$_2$ films in the growth chamber is employed (Preffer 1986), allowing one to radically improve the growth surface morphology and to use it consequently for further operations of deposition of Si or other materials. Nevertheless, RTA temperature (~1100°C) considerably exceeds the growth temperature (~600°C) and may result in structure defect formation due to thermal strain, caused by different thermal expanding coefficients of Si and CaF$_2$.

Solid state epitaxy (SPE), when used on the initial stage of film formation, also enables one to smooth the growth surface. The method was successfully employed for obtaining ~10 nm thick CaF$_2$/Si films with consequent deposition of GaAs using the common MBE technique (Fathoner et al 1986).

2. EXPERIMENT

Prior to SPE the Si substrates were annealed at 850°C during 30 min in 10^{-8} Pa vacuum. A (2x1) superstructure was registered by RHEED for Si(001) and a (7x7) superstructure - for Si(111). The thermally cleaned Si substrate was cooled to room temperature (RT) and an ~10 nm thick CaF$_2$ layer was deposited onto it at RT. The structure was then gradually heated to 500-600°C, until strands corresponding to monocrystalline epitaxial CaF$_2$

were observed on the monitor. Further growth of CaF_2 film was carried out at the same temperature, followed by RTA at 900°C.

In the MBE process calcium fluoride was sublimed from a molecular source, heated to 1200°C onto a Si substrate, and heated to 550°C. The growth speed was ~0.4 mkm/hour and film thickness - 150-800 nm. Prior to RTA the structure was cooled to RT in 15-20 minutes time and then removed from the furnace. After heating the latter to 900°C the structure was returned to the furnace for 30 sec. Heating of the structure and its subsequent cooling took only a few seconds.

[100] cross-sections, mechanically ground and Ar^+ ion thinned were investigated in a Philips EM 430 ST electron microscope at 200 and 300 kV at Sherzer defocus. Some areas of the micrographs (512x512 pixel) were digitized with a Perkin Elmer densitometer. Fourier band filtering was used to improve the images.

3. RESULTS AND DISCUSSION

SPE on the initial stage of obtaining CaF_2/Si(111) films results in formation of microtwins in the film thickness, nucleating on the interface (Fig.1). The twin planes are parallel to the interface. The twin boundaries contain dislocations in planes, normal to the growth surface, and resemble large-angle grain boundaries. Formation of the twins in the interface indicates that CaF_2 nucleation in twinned and non-twinned positions is equiprobable on Si($\bar{1}$11).

Fig.2 illustrates a <110> cross-section of a CaF_2/Si(001) structure. ($1\bar{1}1$) and (111)Si half-planes are seen in the interface. SPE-growth of CaF_2 on Si(001) prevents twinning and grain boundary formation. In addition, it enables one to exclude faceting and thus improve the CaF_2 film growth surface morphology.

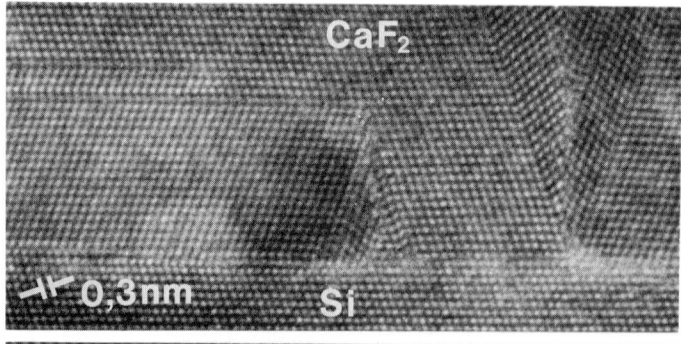

Fig.1. <110> cross-section of a CaF_2/Si(111) structure, imaged at 300 kV.

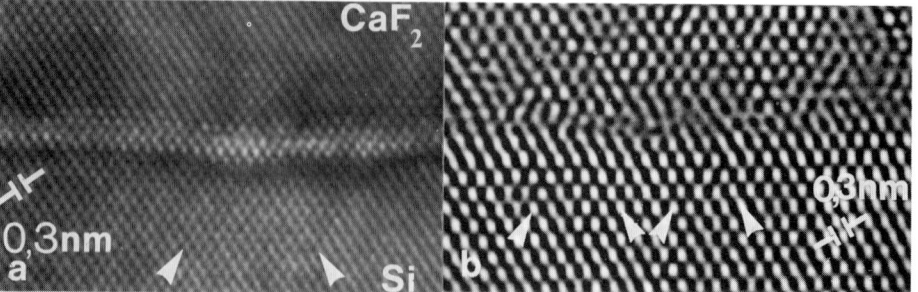

Fig.2.   <110> cross-sectional image (300 kV) of a CaF_2/Si(001) structure (a) and filtered image (b). Extra half-planes are arrowed.

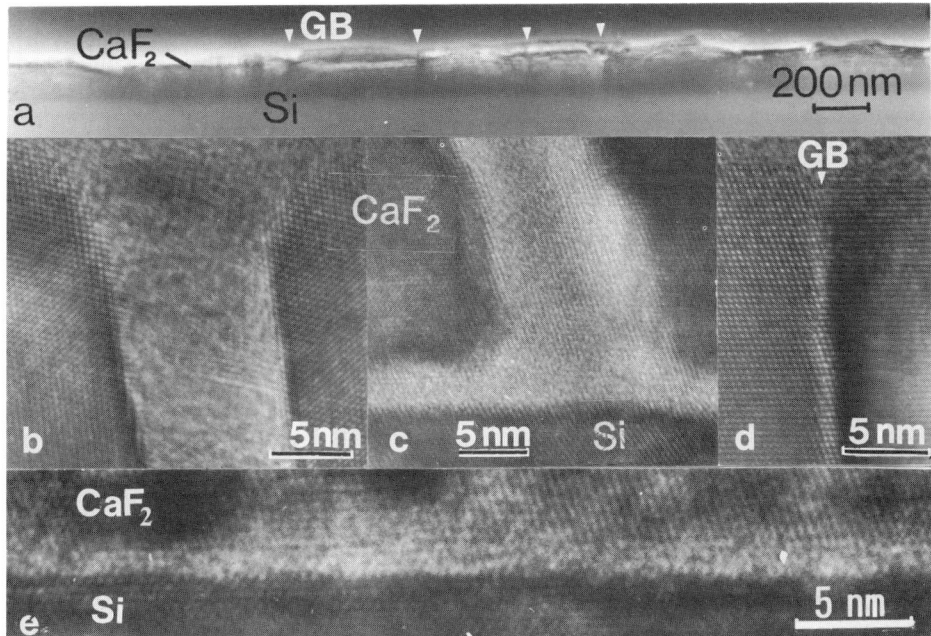

Fig.3. <110> cross-section of a (111)CaF$_2$ /Si(001) structure (amorphous layer thickness during SPE was more than 50 nm). GB - grain boundary. a) CaF$_2$ layer formed by platelet-like grains; b) upper part of an open GB; c) bottom part of (b); d) a GB without a gap; e) (111)CaF$_2$ /Si(001) interface.

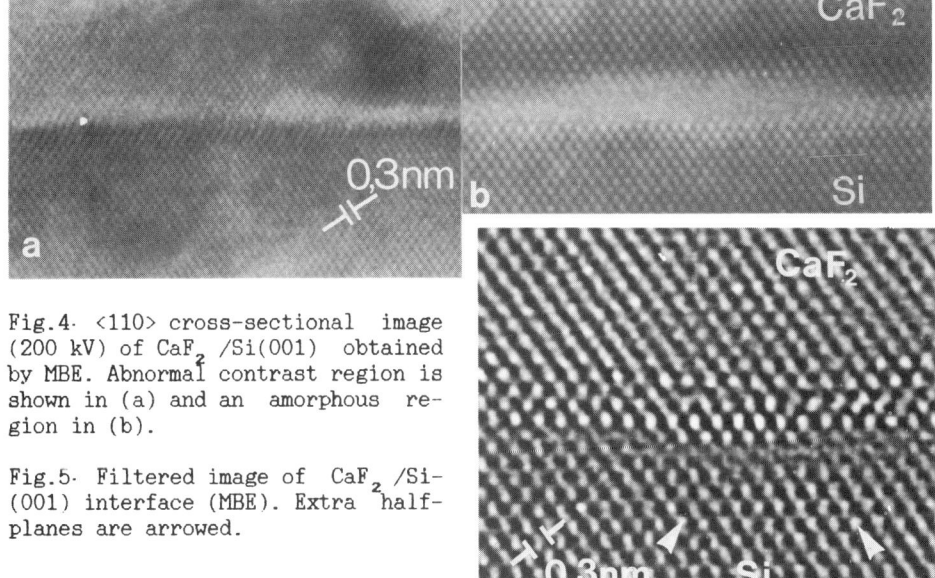

Fig.4. <110> cross-sectional image (200 kV) of CaF$_2$ /Si(001) obtained by MBE. Abnormal contrast region is shown in (a) and an amorphous region in (b).

Fig.5. Filtered image of CaF$_2$ /Si-(001) interface (MBE). Extra half-planes are arrowed.

One may assume that the structure of films grown by SPE depends critically on the CaF$_2$ amorphous layer thickness. It seems that some critical thickness of the amorphous layer exists, above which nucleation centres may form not only in the interface, but also in the film thickness, leading to polycrystalline film formation. Fig.3 shows an example of such a structure.One can see at low magnification (Fig.3a) that the film thickness is ~180 nm. The film consists of plate-like grains, 600-900 nm in size along the interface. In one case the grains are divided by a considerable gap (Fig.3b,c), while in the other the gap is absent (Fig.3d). In both cases the grains are somewhat misoriented relative to each other. The CaF$_2$ grain orientation differs from the substrate orientation (Fig.3c) and the (111)CaF$_2$/Si(001) system is obtained.

The MBE-grown CaF$_2$/Si(001) structure after RTA at 900°C has some specific features. (110) cross-section HREM images (Fig.4a) usually show an intermediate layer ~1 nm thick with abnormal contrast. The image motif on some areas of this layer indicates partial amorphization of the material forming it (Fig.4b). Analysis of filtered interface images (Fig.5) allows one to assume that the image motif variation derives from the shift of [110] atomic columns relative to their normal positions in CaF$_2$ and Si. This shift is irregular along the interface. Two extra ($1\bar{1}1$) and ($1\bar{1}1$) half-planes in CaF$_2$ are sometimes observed, which may derive from the presence of edge-type misfit dislocations (MD) with the Burgers vector a/2 <110> in the interface. However, the exact position of MD nuclei is difficult to define due to the complicated structure of the intermediate layer.

The intermediate layer atomic structure irregularity may derive from partial escape of fluorine atoms from the interface during RTA (Batstone and Philips 1989). Local amorphization of this intermediate layer may result from: a). radiation damage of the structure in the microscope; b). crystalline structure damage by Ar$^+$ ions in the thinning process.

4. CONCLUSION

HREM of SPE- and MBE-grown CaF$_2$ layers on Si(111) and (001) has revealed a 1 nm wide region near the interface with a rather irregular crystalline lattice image motif. The CaF$_2$ film on Si(111) contains twins and grain boundaries, perpendicular to the interface. Using Si(001) substrates eliminates formation of twins and dislocation boundaries. In the latter case Si($1\bar{1}1$) and ($\bar{1}11$) extra half-planes are seen in the interface, evidently linked with presence of edge-type MD with the Burgers vector of a/2 <110> in the interface. A similar interface structure is observed when standard MBE technique with subsequent RTA is employed.Deviation of SPE modes from the optimal ones (in particular, increasing of the amorphous layer thickness to 50-100 nm) results in polycrystalline film formation after annealing.

REFERENCES

Batstone J L and Julia M Philips 1989 Mat.Res.Soc.Symp.Proc. **139** 351-6
Fathoner F W, Schowalter L I 1984 App.Phys.Lett. **45** No 5 519-21
Fathoner F W, Lewis N, Hall E L and Schowalter L I 1986 App.Phys.Lett.
    60 No 11 3886-94
Philips J M 1986 J.of Electrochem.Soc.: Solid State Sci. and Technol.
    133 No 1 224-7
Preffer H 1986 App.Phys.Lett. **48** No 9 596-8

Inst. Phys. Conf. Ser. No 117: Section 7
Paper presented at Microsc. Semicond. Mater. Conf., Oxford, 25–28 March 1991

541

Theory of RHEED intensity oscillations in MBE

L -M Peng and M J Whelan
Department of Materials, University of Oxford, Parks Road, Oxford OX1 3PH, UK

ABSTRACT: A practical computing procedure has been developed for calculating RHEED from MBE growing surfaces. A generalized birth-death growth model has been used to describe MBE growth, and the diffraction from MBE growing surfaces is treated dynamically. Our method is shown to be able to reproduce almost all features of measured RHEED intensity oscillations from low-index surfaces.

1. INTRODUCTION

Reflection high energy electron diffraction (RHEED) has became well established as one of the most powerful and versatile technique for growth and surface studies of thin films prepared by molecular beam epitaxy (MBE). In particular the technique of RHEED intensity oscillations has been used routinely to calibrate beam fluxes and alloy compositions and to control the thicknesses of quantum wells and superlattice layers (see articles in Larsen and Dobson, 1988). In this paper we report a simple theory which will be shown to be able to explain almost all features of measured RHEED intensity oscillations from low-index surfaces.

2. GROWTH MODELS

Following Cohen and coworkers (1989), we use a birth-death growth model and consider epitaxial growth on a low-index substrate surface. By assuming that the net diffusion of the ith species from an upper layer to the layer immediately below is proportional to a product of the mobile i^{th} (i.e. uncovered) species on the upper layer and the number of sites which are uncovered in the second layer below, the layer coverage $\theta_n^{(i)}$ of the ith species on the nth layer then satisfies a set of non-linear differential equations

$$
\begin{aligned}
\frac{d\theta_n^{(i)}}{dt} = {} & d^{(i)}(t) \sum_j [\theta_{n-1}^{(j)} - \theta_n^{(j)}] + k^{(i)}[\theta_{n+1}^{(i)} / \sum_j \theta_{n+1}^{(j)}] \sum_j [\theta_{n+1}^{(j)} - \theta_{n+2}^{(j)}] \sum_j [\theta_{n-1}^{(j)} - \theta_n^{(j)}] \\
& - k^{(i)}[\theta_n^{(i)} / \sum_j \theta_n^{(j)}] \sum_j [\theta_n^{(j)} - \theta_{n+1}^{(j)}] \sum_j [\theta_{n-2}^{(j)} - \theta_{n-1}^{(j)}].
\end{aligned}
\tag{1}
$$

This set of equations gives the diffusive growth model of epitaxial growth, in which $d^{(i)}$ is the deposition rate for the i_{th} species, and $k^{(i)}$ is a diffusion parameter.

An alternative to the diffusive growth model is to treat the transfer of atoms between layers in accordance with the lateral structure of the layers, and this is the distributed growth model as discussed by Cohen et al. (1989). In this model we assume that among all the i_{th} species arriving on top of the n_{th} layer per unit time, only a fraction $\alpha_n^{(i)}$ will transfer to the n_{th} layer below and a fraction of $1 - \alpha_n^{(i)}$ will remain on top of the n_{th} layer. The set of non- linear rate equations then takes the form

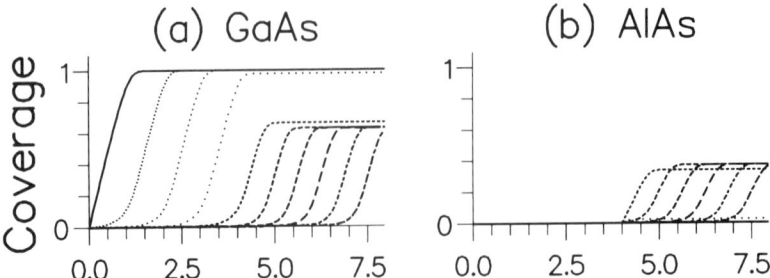

Figure 1: Layer coverages of (a)GaAs and (b)AlAs. A diffusive growth model is used, with $k^{Al} = 30$ and $k^{Ga} = 40$.

$$\frac{d\theta_n^{(i)}}{dt} = d^{(i)}(t) \sum_j [\theta_{n-1}^{(j)} - \theta_n^{(j)}]$$

$$+ \alpha_n^{(i)} d^{(i)}(t) \sum_j [\theta_n^{(j)} - \theta_{n+1}^{(j)}] - \alpha_{n-1}^{(i)} d^{(i)}(t) \sum_j [\theta_{n-1}^{(j)} - \theta_n^{(j)}]. \tag{2}$$

in which

$$\alpha_n^{(i)} = A^{(i)} \frac{p_n}{p_n + p_{n+1}}, \tag{3}$$

and p_n is the total perimeter of the nth layer. The main advantage of this model is that there is considerable flexibility in defining the dependence of the perimeter on the coverage, and a simple approximation is to define

$$p_n \propto \sum_j \theta_n^{(j)} [1 - \sum_j \theta_n^{(j)}]^{1/2}. \tag{4}$$

Shown in figure 1 are ten surface layer coverages. Four monolayers of GaAs were initially deposited on a GaAs(001) substrate, with $d^{Ga} = 1.0$ (monolayer per second). The deposition rate for Al was the increased from zero to 0.59, while that for Ga was maintained at $d^{Ga} = 1.0$. This corresponds to the growth of $Al_{0.37}Ga_{0.63}As$.

3. DYNAMICAL RHEED FROM GROWING SURFACES

We now consider the dynamical diffraction processes. To avoid the complexity of treating the non-periodically distributed surface disorders on the growing layers and the often unknown atomic details of the reconstructed surface, in what follows we make a so-called systematic reflection approximation, considering only the reflection by planes parallel to the substrate surface so that only a potential averaged over the plane need be considered. To simulate the structural variation of growing surface layers along the surface normal direction, we divide the surface layers into an assembly of thin slices parallel to the surface, and assume that each slice at coordinate z_i normal to the surface has a constant potential $U_0(z_i)$. Within each thin slice the z-dependent part of the electron wave function may be written as

$$\psi_n(z) = T_n \exp(ik_z z) + R_n \exp(-ik_z z),$$

in which $k_z(z) = \sqrt{\chi_z^2 + U_0(z)}$, and χ is the electron wave vector in the vacuum. The scattering matrix **M** for relating the electron wave function and its surface normal derivative on the upper and lower surfaces of an assembly of thin slices, each having thickness t_i, is given by (Peng and Whelan, 1991a)

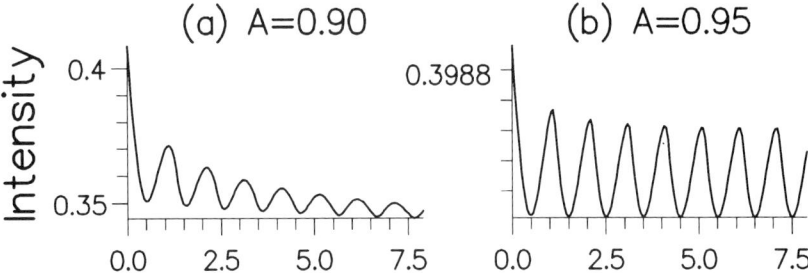

Figure 2: RHEED intensity oscillations during distributed homoepitaxial growth on a GaAs(001) substrate. in (a) A=0.90 and in (b)A=0.95.

$$\mathbf{M} = \prod_i \begin{pmatrix} \cos[k_z(z_i)t_i] & \sin[k_z(z_i)t_i]/k_z(z_i) \\ -k_z(z_i)\sin[k_z(z_i)t_i] & \cos[k_z(z_i)t_i] \end{pmatrix}, \tag{5}$$

and the specular reflected beam amplitude R_0 is given by

$$R_0 = -\frac{[M_{11} - M_{12}/i\chi_z] - [M_{22} - i\chi_z M_{12}]}{[M_{11} - M_{12}/i\chi_z] + [M_{22} - i\chi_z M_{12}]}. \tag{6}$$

The averaged scattering potential $U_0(z)$ can be related to the surface layer coverage,

$$U_0(z) = \sum_n \sum_i \theta_n^{(i)} U_0^{(n,i)}(z), \tag{7}$$

where

$$U_0^{(n,i)}(z) = \frac{16\pi}{a_0^2} \frac{m}{M_0} \sum_{k=1}^4 \sqrt{\frac{\pi}{b_k^{(i)}}} a_k^{(i)} \exp[-4\pi^2(z - z_n^{(i)})^2]/b_k^{(i)},$$

a_k and b_k are the Gaussian parameters of Doyle and Turner (1968), a_0 is the lattice constant, and m and m_0 are the relativistic and rest electron mass respectively.

4. RHEED INTENSITY EVOLUTION DURING GROWTH

The RHEED intensity evolution curves shown in figure 2 are calculated for the homoepitaxial growth of GaAs and for specular reflected beam, using the distributed growth model with A=0.90 for figure 2a and A=0.95 for figure 2b. The primary beam of energy 20keV is incident on the growing surface at 50mrad. Almost all features of measured intensity oscillations from a low-index surface are reproduced here. Among other features there is an initial transient, after which the intensity damps slowly to a nearly constant but non-zero value. At moderate temperatures or diffusion rates the oscillations are sinusoidal (figure 2a), while at higher diffusion rates they are more cusp-like (figure 2b)

In figure 2 it is seen that the RHEED intensity maxima are always at complete layer coverage and the intensity minima are at about half layer coverage. Detailed calculations show, however, that this is not a general rule. Both the phase and the amplitude of the intensity oscillations are sensitive to not only the diffraction conditions but also the growth conditions. Figure 3 shows RHEED intensity oscillations calculated for a different diffraction condition and varying diffusion parameters. These curves show that, in contrast to figure 2, the reflected beam intensity increases immediately after the growth is started and the intensity minima are at about complete layer coverage for large values of k, but for small k shift occur

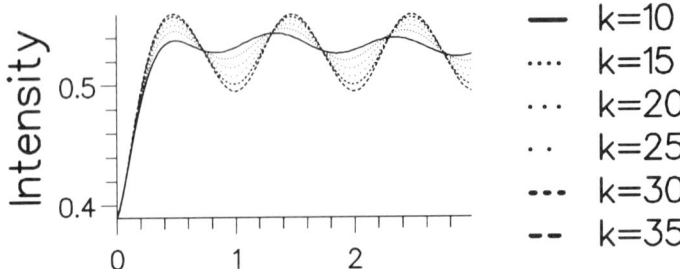

Figure 3: RHEED intensity oscillations for the growth of GaAs on a GaAs(001) substrate. The primary beam energy is 10keV, and the incident glancing angle is 30mrad.

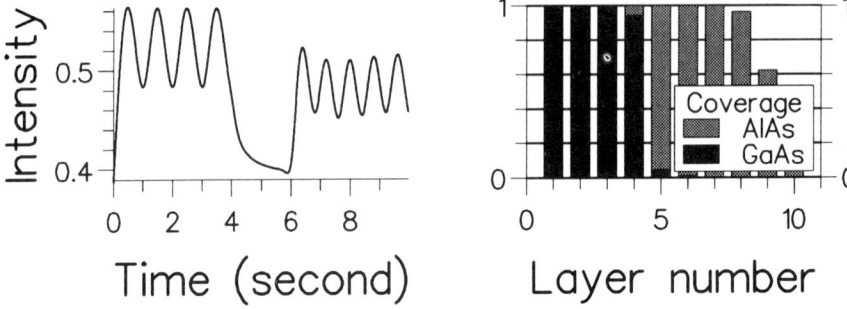

Figure 4: (a)RHEED intensity oscillations during diffusive epitaxial growth of a AlAs/GaAs heterostructure and (b) the resulting histograms of layer coverages.

in the maxima and minima. It is also clear that the first few periods of the oscillation are decreases for small values of k.

Shown in figure 4 are RHEED intensity oscillations for the growth of a AlAs/GaAs heterostructure on a GaAs(001) substrate, and the resulting histogram of layer coverages. The growth is interrupted for two seconds prior to switching on the Al molecular beam, and the improvement in the interface abruptness using growth interruption is very striking. It should be pointed out that the improvement under other conditions may not be as great as that shown in figure 4. For details see Peng and Whelan (1991b).

5. CONCLUSIONS

In summary, a practical computing procedure has been developed for calculating RHEED from MBE growing surfaces, and our calculations have reproduced almost all features of measured RHEED intensity oscillations from low-index surfaces.

REFERENCES

Cohen, P.I. et al. 1989 Surface Sci. 216, 222
Doyle, D.A. and Turner, P.S. 1968 Acta Cryst. A24, 390
Larsen, P.K. and Dobson, P.J. 1988 Eds. RHEED and Reflection Electron Imaging of Surfaces (Plenum, New York)
Peng, L.-M. and Whelan, M.J. 1991a Proc. R. Soc. London A432, 195
Peng, L.-M. and Whelan, M.J. 1991b Proc. R. Soc. London A (in press)

Inst. Phys. Conf. Ser. No 117: Section 7
Paper presented at Microsc. Semicond. Mater. Conf., Oxford, 25–28 March 1991

Liquefaction and solidification of laser pulsed alloy films

O Bostanjoglo and P Thomsen-Schmidt

Optisches Institut TU Berlin, Strasse des 17. Juni 135, W 1000 Berlin 12

ABSTRACT: Laser-induced liquefaction and freezing were visualized by high-speed transmission electron microscopy (TEM) on the nanoseconds time scale in Ge-Te films. Collision-limited melting, macroscopic thermo-capillary convection and crystallization with velocities $\approx$ 0.5 m/s were observed. The composite nature of the films complicated each of these processes by adding diffusion limited liquid etching, microscopic thermoca-pillary flow caused by segregation and amorphization due to recalescence. Time resolved EM is shown to be a powerful technique to study micro-patterning by laser pulses in real time.

1. INTRODUCTION

Alloy films are important in different fields, such as high-T_c superconductivity, optical coating, integrated circuits, optical storage. Their application is even extended by micropatterning. Short laser pulses are very effective for this purpose as large power densities are deposited in a well controlled way and treatment can be confined to small areas. However, as heating/cooling rates are high, the involved phase transitions are not properly known. Atomic mixing and segregation add to their complexity in alloys. This paper describes liquefaction and solidification induced by a laser pulse in free-standing Ge-Te films (60-80 nm). These films were selected as an example of alloys with transformations differing from those in elemental films. The phase transitions were induced by a pulse from a Q-switched frequency doubled Nd: YAG laser (gaussian, pulse/spot width 3 and 25 ns/17 μm FWHM), and investigated by time-resolved TEM (Bostanjoglo and Thomsen-Schmidt 1989, 1990).

2. RESULTS AND DISCUSSION

2.1 Liquefaction

As the solid/vapor interface energy exceeds the sum of solid/liquid and liquid/vapor interface energies there is no nucleation barrier for surface melting. It is collision-limited in a crystal and proceeds within the Debye period $\approx$ 1 ps. Liquefaction of a solid composed of several materials with different melting points is more complicated. If the laser pulse heats the solid above the highest melting point the mixture melts like a simple crystal. If, however, the temperature remains below the maximum melting point, liquefaction starts as fast melting but continues as a dissolving process, which is diffusion-limited and is expected to proceed in a Ge-Te liquid on a time scale $D^2/ 8D_{diff}$ $\approx$ 100 ns, given by the film thickness $D \approx$ 60-80 nm and the atomic diffusion constant $D_{diff} \approx$ 5·10^{-9} m²/s. These two effects are shown in Fig. 1.

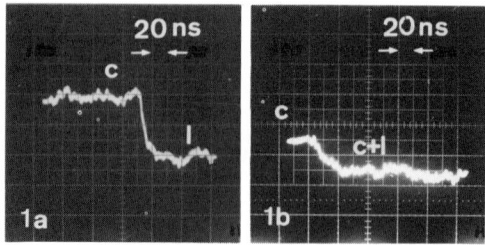

Fig. 1 Liquefaction of $Ge_{0.4} Te_{0.6}$ films by a 3 ns laser pulse. The oscillograms give the bright-field image intensity within an area of 2.6 μm ∅.
a) High-energy laser pulse (700 nJ) melting all components. Liquefaction is completed within the resolution limit of 3 ns (c crystal, l liquid).
b) Low-energy laser pulse (130 nJ) inducing fast partial melting (not resolved) and a slow liquid etching process.

After liquefaction several flow processes can occur depending on the laser pulse energy. A pulse of intermediate energy generates a liquid/solid mixture and excites violent fluctuations, which frequently are frozen-in (Fig. 2). These short-wave flow processes are driven by inhomogeneous contractions probably due to mismatch in the density of the coexisting liquid and solid. For $Ge_{0.4} Te_{0.6}$ the difference is expected to be 2%.

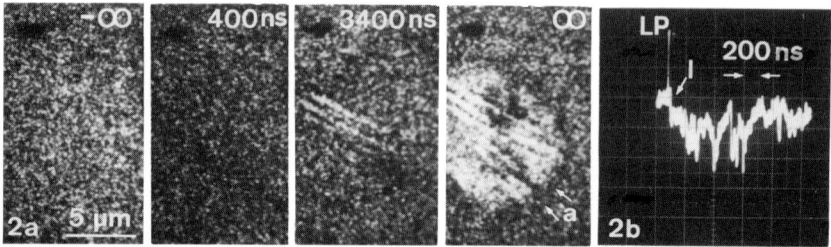

Fig. 2 Localized flow in a $Ge_{0.4} Te_{0.6}$ film partially liquified by a 25 ns laser pulse. a) Short exposure images before (-∞), 400 ns, 3400 ns, ∞ after the maximum of the laser pulse. Exposure time 40 ns. The final image (∞) shows frozen-in folds in the recrystallized area. Material amorphized by recalescence "a". b) Typical bright-field image intensity within an area of 2.6 μm ∅, showing liquefaction "l" and violent fluctuations caused by to localized motion (LP laser pulse).

At higher laser energies another flow process, the thermocapillary or Marangoni convection (Schwabe and Scharmann 1981) is triggered by a local variation of the surface tension σ , which is caused by thermal gradients. These can be macroscopic due to the laser pulse and microscopic due to liberated crystallization heat of precipitates. Elemental and eutectic films exhibit only macroscopic thermal gradients produced by the laser across the liquid. They drive material from the hot center to the colder periphery. This flow may open a hole after a nucleation time

(1) $$t_n \approx D/[(\Delta\sigma)^2/\rho\eta r]^{1/3}$$

were D, ρ, η, r denote thickness, density, viscosity, radius of the liquid area, and $\Delta\sigma$ is the difference of surface tensions between periphery and center of the liquid. The denominator is the velocity of Marangoni flow. Once the hole is opened it expands, now due to the local surface tension, with a velocity

(2) $$v \approx \sigma D/(\eta r_h)$$ r_h radius of hole

Different stages of this single-hole perforation mechanism in eutectic Ge-Te films are shown in Fig. 3. Assuming a starting temperature difference of 100 K across the liquid area (r = 5 μm), eq. (1) gives $t_n \approx$ 170 ns. This, and v $\approx$ 9 m/s agree roughly with measured values e.g. from Fig. 3.

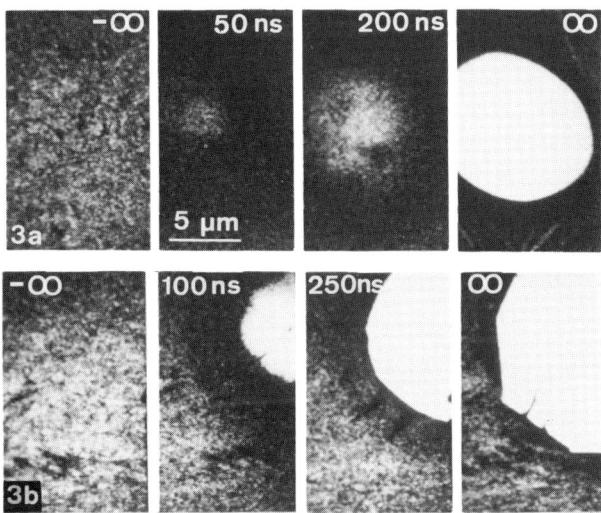

Fig. 3 Perforation of a liquid eutectic $Ge_{0.15}Te_{0.85}$ film by the single-hole mechanism due to macroscopic capillary flow. Exposures are before and at the indicated times after the maximum of a 25 ns laser pulse. Exposure time 50 ns. a) Nucleation of a hole due to Marangoni flow. b) Growth of a hole driven by the surface tension ($\approx$ 10 m/s). Mark the low image intensity of the melt.

An additional mechanism of thermocapillary flow operates in segregating alloy films. The latent heat of the crystallizing precipitates generates local thermal gradients. These drive Marangoni flows on a microscopic scale creating crowds of holes which break the film into closely spaced solidifying islands (Fig. 4).

2.2 Solidification

Solidification is triggered at a certain supercooling. If the latter is not too low and cooling not too fast the liquid crystallizes, otherwise a glass is formed. Fig. 5 shows crystallization of a $Ge_{0.4}Te_{0.6}$ film, starting at the periphery of the liquid and proceeding with $\approx$ 0.5 m/s towards the center. In addition, material at the periphery is amorphized ("a" in Fig. 2a and 5), probably due to heating by liberated enthalpy of crystallization.

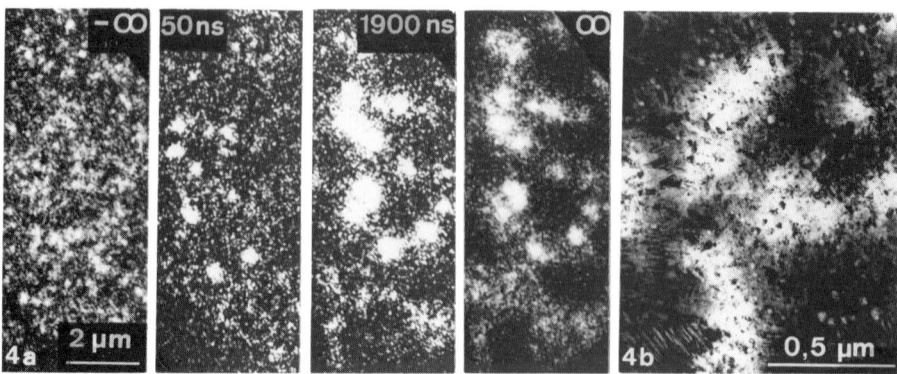

Fig. 4 Perforation of a segregating liquid $Ge_{0.8}$ $Te_{0.2}$ film by microscopic Marangoni flow due to crystallization heat from Ge precipitates.
a) Short exposure images before, 50 ns, 1900 ns, ∞ after the maximum of a 25 ns laser pulse. Exposure time 50 ns. b) Typical final structure showing the Ge precipitates.

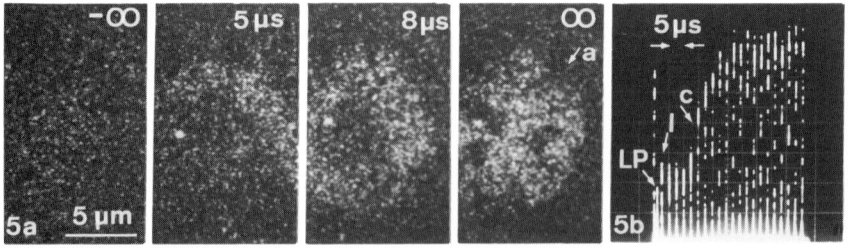

Fig. 5 Crystallization of a $Ge_{0.4}$ $Te_{0.6}$ film after liquefaction by a 25 ns laser pulse. a) High-speed frames before, 5 µs, 8 µs, ∞ after laser pulse. Exposure time 40 ns. Material amorphized by recalescence "a". b) Typical bright-field image intensity within an area of 1.3 µm $\emptyset$, showing liquefaction "l" and crystallization "c". A train of electron pulses was used to avoid electron beam heating .

A decision between the existing diffusion- and entropy-limited models of crystallization (Stolk et al 1989) cannot be made yet. For this purpose the crystallization velocity must be determined at different temperatures, which remains to be done.

This work was supported by the Deutsche Forschungsgemeinschaft.

REFERENCES

Bostanjoglo O and Thomsen-Schmidt P 1989 Appl. Surf. Sci. **43**, 136
Bostanjoglo O and Thomsen-Schmidt P 1990 Appl. Surf. Sci. **46**, 392
Schwabe D and Scharmann A 1981 J. Cryst. Growth **52**, 435
Stolk PA, Polman A and Sinke W C 1990 Mat. Res. Soc. Symp. Proc. **157**, 363

Inst. Phys. Conf. Ser. No 117: Section 8
Paper presented at Microsc. Semicond. Mater. Conf., Oxford, 25–28 March 1991

549

Convergent beam diffraction and microscopy of single and multiple quantum well structures

D Cherns

H H Wills Physics Laboratory, University of Bristol, Tyndall Avenue, Bristol BS8 1TL

ABSTRACT: An electron rocking curve method of analysing the composition and strain profiles in plan-view semiconductor multi-layers is described. It is shown that rocking curves for periodic multiple quantum wells, observed both by large angle convergent beam electron diffraction and imaging techniques, display superlattice satellite reflections which are sensitive to near monolayer changes in the period and individual layer thicknesses and to layer strains greater than about 2×10^{-4}. For single quantum well structures, wells down to ~8Å thick have been characterised and individual steps in well thickness imaged in some cases. The techniques used are explained and illustrated with recent results for near lattice matched ($InP/In_{0.53}Ga_{0.47}As$) and strained layer ($Si/SiGe$ and $CdTe/CdMnTe$) systems.

1. INTRODUCTION

The materials characterisation of quantum well structures is important both for the fundamental understanding of their electrical and optical properties, and for improving growth procedures. Factors which can influence carrier energy levels and recombination routes include layer thicknesses, interface atomic roughness and sharpness and any layer strains present. Much information can be obtained by high resolution electron microscopy or diffraction contrast techniques applied to cross-sectional specimens. However, although individual layers or interfaces can be examined at the near-atomic level, structure is only seen in projection. Thus it is difficult to extract a meaningful two-dimensional model of the growth. Moreover, characterisation of strain is difficult owing to surface relaxations which cause plane bending near the specimen surfaces. In the plan-view geometry, useful information on the composition and strain profiles in quite complex multilayer structures has been obtained using X-ray rocking curves despite the fact that scattered contributions from different layers superimpose (eg see Fewster (1988)). In this paper a similar approach is described for electron diffraction, henceforth termed the electron rocking curve method. Electron rocking curves are generated either in diffraction using convergent beam electron diffraction (CBED) or large angle CBED (LACBED), or over a restricted angular range by imaging a slightly bent sample. The method in using plan-view samples enables growth topology to be observed directly and over much larger areas (eg up to 1mm^2) than are possible in cross-sections. In the plan-view geometry layer strains also remain essentially unrelaxed. It is shown that layer thicknesses in both periodic multiple quantum well (MQW) and single quantum well (SQW) structures can be measured to near monolayer sensitivity and layer strains down to $\sim 2 \times 10^{-4}$ quantified. Unlike the X-ray method, local variations can be observed in layer thicknesses with individual interface atomic steps visible in some cases.

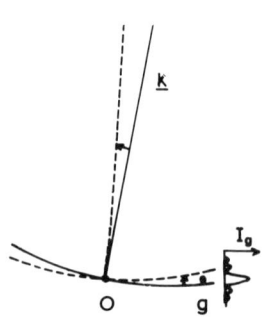

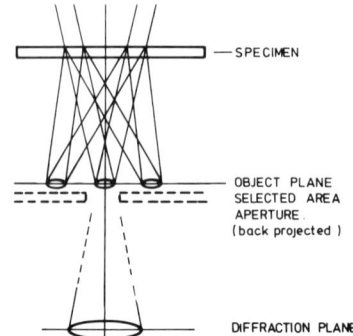

Fig 1. The diffraction function for g is scanned by a converging incident beam.

Fig 2. A schematic illustration of LACBED.

2.   METHODS

Electron rocking curves can be conveniently observed in CBED or LACBED patterns.   As illustrated in Fig 1, changing the angle of the incoming beam causes the Ewald sphere to be swept through the diffraction function for a reflection g which is thus mapped out as the diffraction disc for g is traversed. It is important to note that diffraction angles are small for the transmission electron microscope ($\sim 1°$) such that the diffraction function approximately parallel to the incident electron beam is mapped.   For a horizontal plan-view multilayer this is just the direction in which structure in the diffraction function depends on the layer thicknesses and composition.

In our work electron rocking curves are mostly derived using the LACBED method.   In LACBED (eg see Vincent (1989) for a review) the sample is not at the focus of the incoming beam (Fig 2).   Since different parts of the specimen see different directions of the incident beam, the illuminated area is mapped onto the diffraction disc, producing a shadow image whose spatial resolution is given by the minimum probe size.   Moreover, diffracted beams are separated in the focal plane of the objective lens such that a selected area aperture can be used to select a single diffracted beam.   As well as removing the problem of beam overlap encountered in conventional CBED, allowing a convergence or rocking angle of up to 6° in our case, the aperture filters out inelastic background, improving pattern contrast.   Thus the selected area aperture behaves like the objective aperture in imaging mode.   Studies by Jordan et al (1991a) have shown that filtering can be achieved down to angles $\sim 5 \times 10^{-4}$ rad around a diffracted beam ($\sim 1/20$th of a Bragg angle) giving exceptionally high quality diffraction detail at large deviation parameters.   For most studies LACBED patterns are recorded under two-beam conditions for simplicity, ie with non-systematic row reflections reduced to a minimum. Fig 3(a) shows an example of a 200 dark field LACBED pattern recorded under these conditions from a 1050Å InP/20Å $In_{0.53}Ga_{0.47}As$/1150Å InP SQW sample. The sample was grown by MOCVD on an $In_{0.53}Ga_{0.47}As$/InP substrate which was removed using selective etches. The pattern shows useful rocking curve detail over the entire 6° angular range of the pattern. In fact, useful rocking curve detail was observed up to ~10-12° tilt from the Bragg position (obtained in further exposures after tilting the sample) corresponding to a very large deviation parameter, $s \sim 7 \times 10^{-2}$ Å^{-1}.

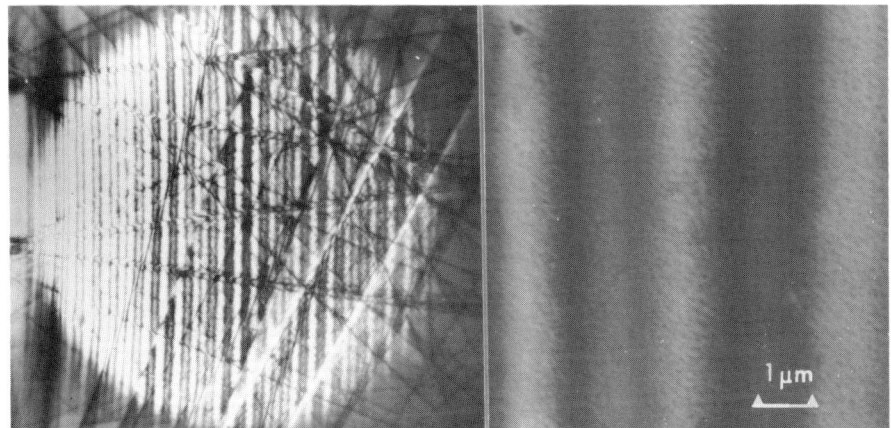

Fig 3. (a) A 200 dark field LACBED pattern at 250kV from a SQW sample comprising 1050Å InP/20Å $In_{0.53}Ga_{0.47}As$/1150Å InP, (b) a 200 dark field image from a bent region of the same sample (see also Jordan et al (1989)).

In Fig 3(a) different fringes and different parts of the same fringe derive from different areas. Although spatial variations can be clearly observed in some LACBED patterns (see Fig 5), structural changes in the SQW sample in Fig 3(a) occur on a scale less than the spatial resolution, ie the minimum probe size, such that spatial variations are averaged. The probe size in this case was about 500Å, a practical limit set by the need for sufficient beam intensity to record very weak diffraction detail. However, in order to observe spatial variations we can use the more limited range of the rocking curve seen in imaging, either of a bent crystal or of a flat crystal observed by the convergent beam imaging (CBIM) technique. Fig 3(b) illustrates a dark field 200 image taken with a parallel incident beam of a bent region of the sample in Fig 3(a). A small range of the rocking curve ($\Delta s \sim 3 \times 10^{-3}$Å^{-1}) is observed and oscillations in the fringe intensity in Fig 3(a) are clearly reproduced in the image. Moreover, structural features less than 500Å in scale are observed which are found to be due to a variation in the quantum well thickness (see later). Experimentally we have found that useful images are limited to values of deviation parameter about a factor of 3 smaller than accessible in

LACBED, emphasising the efficient background inelastic filtering achieved in the LACBED method.

3.   THEORETICAL BACKGROUND

In order to interpret results such as those in Fig 3 it is useful to outline briefly some background theory of electron rocking curves.   It should first be noted that in the majority of experiments two-beam diffraction conditions are used. Where non-systematic row interactions occur these are clearly observed in LACBED patterns associated with diffraction contours which cross at angles to the main diffraction contours (eg Fig 3(a));   thus such interactions can be avoided in interpreting results.   Systematic row interactions which occur at constant deviation parameter (parallel to the main contour) are much less obvious.   However, ignoring such interactions for the moment, since most of the detail in LACBED patterns occurs at relatively large deviation parameters s, a qualitative understanding can be achieved using a kinematical approach. In this approximation the scattered amplitude is given by

$$A_g = \int_0^t F_g(z)\exp(-2\pi i(sz + \underline{g}.\underline{R}))dz$$

(1)

where t is the foil thickness, $\underline{R}$ is any displacement field and the structure factor F_g is a function of depth z.

The form of A_g for MQW and SQW structures has been discussed by previous authors.   Many of the salient points can be summarised by considering the MQW case;   discussion of the SQW case is deferred until section 5.   Vincent et al (1987b) examined the case of periodic lattice-matched MQWs with no layer intermixing and showed that superlattice peaks are expected at $\underline{g} + n\underline{q}$ where q is parallel to the growth direction with $|\underline{q}| = \lambda^{-1}$ where λ is the period and n is an integer. In addition the superlattice peaks should be modulated in amplitude by an envelope function given by

$$A_n = \lambda \frac{(\sin \pi n d_1 / \lambda)\Delta f}{\pi n}$$

(2)

where d_1 is the thickness of one of the constituent layers and Δf is the difference in structure factor between the layers.   Cherns et al (1991) extended the treatment to include layer strains.   Such strains will not to first order affect reflections from planes perpendicular or parallel to the growth direction.   However, planes inclined at an angle θ to the growth direction are rotated in opposite directions in the two layers, the relative angle of rotation $\delta\theta$ being given by

$$\delta\theta = \frac{\varepsilon (1+\upsilon)}{2 (1-\upsilon)}\sin 2\theta$$

(3)

where ε is the relative strain and υ is Poisson's ratio.   The rotation is equivalent in equation (1) to a depth dependence of deviation parameter s. When strain is included, the result is again that regularly spaced superlattice peaks appear but these are now modulated by a function which is asymmetrical.   As strain increases the identity of the main reflection becomes less clear as the superlattice peaks separate into two groups (Fig 4).   This behaviour can be readily understood by noting that the modulation function is in fact the shape transform of a single period of the structure.   Thus for large angular rotations the modulation function essentially consists of two peaks centred on the Bragg angle for each of the layers taken separately (Cherns et al 1988a).

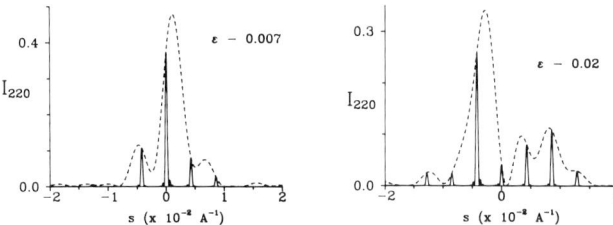

Fig 4. Kinematical simulation of rocking curves for a 10 period 109Å Si/73Å Si$_{0.75}$Ge$_{0.25}$ MQW for an inclined 220 reflection ($\theta = 45°$).

Although kinematical theory explains most features of MQW rocking curves, dynamical effects have been explored.   Jordan et al (1989) treated periodic MQWs using a Bloch wave approach in which the crystal unit cell included a period λ of the superlattice.   Thus superlattice reflections become primary reflections of the superlattice unit cell.   Three-beam interactions involving the straight-through beam, a superlattice reflection and a systematic row reflection of the basic lattice were shown to cause splitting of the superlattice reflection into two peaks which swap intensity as the exact three-beam condition is traversed.   An example of this splitting is seen in the ±2q reflections in Fig 7.   Since the three-beam condition depends on the projection of q in the beam direction and on the accelerating voltage, it is possible to tune the interaction either by tilting the crystal or by changing the accelerating voltage.   Such interactions are very sensitive to q and, in principle, can be used for structural analysis.

To avoid the limitation to exactly periodic structures, an alternative approach has been developed by Rossouw et al (1991).   This method is also a Bloch wave treatment but uses the conventional unit cell and solves the boundary conditions for transmission of Bloch waves from one layer to the next.   The method allows an arbitrary number of interfaces at which changes in structure factor, crystal orientation and any rigid body displacement may occur.   Results have been found to give good agreement with those obtained by the approach of Jordan et al (1989) and give a detailed match to experimental results (eg see Cherns et al (1991)).

4.   APPLICATIONS TO PERIODIC MQWs

The electron rocking curve method has been applied to periodic MQWs of AlGaAs/GaAs, InP/In$_{0.53}$Ga$_{0.47}$As, Si/SiGe and CdTe/CdMnTe. Studies of AlGaAs/GaAs (Vincent et al (1987a)) concentrated on reflections from planes perpendicular to the growth plane and hence only examined the compositional modulation since $\theta = 0$ (cf Equation (3)). Studies of LACBED patterns and images in the 200 reflection indicated that the period λ could be determined to near monolayer sensitivity and that there was some evidence of spatial variations in λ. Similar studies of InP/In$_{0.53}$Ga$_{0.47}$As MQWs, (Vincent et al (1987b)) showed up to 20 superlattice sidebands in $g = 200$, benefitting from a much greater structure factor difference than in AlGaAs/GaAs. Jordan (private communication) has examined spatial variations in an MOCVD-grown 60Å InP/60Å In$_{0.53}$Ga$_{0.47}$As MQW. Examples of a LACBED pattern and a complementary dark field image are shown in Fig 5. With $d_1 = ½$ we expect the superlattice peaks at $n = 2, 4, 6$ etc to disappear (Equation (2)). In fact it can be seen that the $n = 2$ peak is present but is not continuous, indicating that some regions deviate fom the ideal structure. The spatial variation is seen more clearly in the image taken with the crystal oriented at the Bragg condition for the $n = 2$ superlattice peak. A quantitative analysis suggested that a variation in the average layer thicknesses to a 70Å/50Å ratio was sufficient to account for the results and subsequent cross-sectioning suggested that local fluctuations of about this magnitude were retained throughout growth.

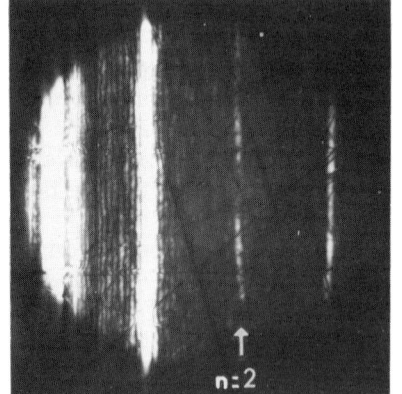 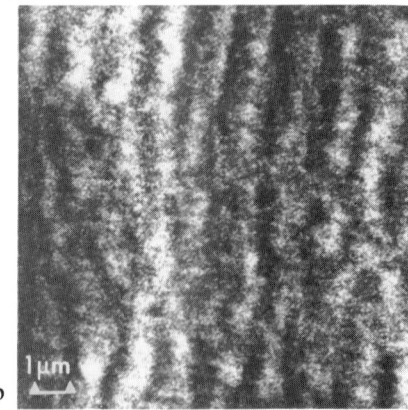

Fig 5. (a)A 200 dark field LACBED pattern from a 60Å InP/60Å In$_{0.53}$Ga$_{0.47}$As MQW sample at 250kV. (b) 200 dark field image exciting the $n = 2$ sideband (courtesy of I K Jordan, unpublished).

Strains in MQWs can be investigated using diffracting planes inclined to the growth direction. Figs 6 and 7 illustrate some recent results for Si/SiGe and CdTe/CdMnTe MQWs, in both cases using the 202 reflection from planes inclined at 45° to the growth direction. In CdTe/Cd$_{0.8}$Mn$_{0.2}$Te (Fig 6) there is little difference in the 202 structure factors between the CdTe and

$Cd_{0.8}Mn_{0.2}Te$ layers such that the presence of the superlattice satellites is almost wholly due to strain. The experimental rocking curve is in good agreement with the dynamical prediction for the expected structure. Since the layers are of equal thickness and structure factors are close, the resulting rocking curve is nearly symmetrical. In general this is not the case where strain is involved and Fig 7 shows an example of an asymmetric rocking curve for a Si/SiGe MQW. A good match to these results, including the dynamical splitting of the superlattice reflections $\pm 2q$, has also been achieved (Cherns et al (1991)). More detailed studies of Si/SiGe MQWs by Cherns et al (1991) have shown that layer strains ($\sim 10^{-2}$) can be determined to about $\pm 20\%$ by the electron rocking curve method. However, much lower strains can be observed and the graph in Fig 7 shows, using a kinematical model, how the intensity of superlattice peaks for the example given would be expected to vary with strain. It can thus be seen that an asymmetry due to strain is probably detectable for strains as low as 2×10^{-4}, and that strains greater than this magnitude can be expected to be the dominant factor in determining superlattice intensities.

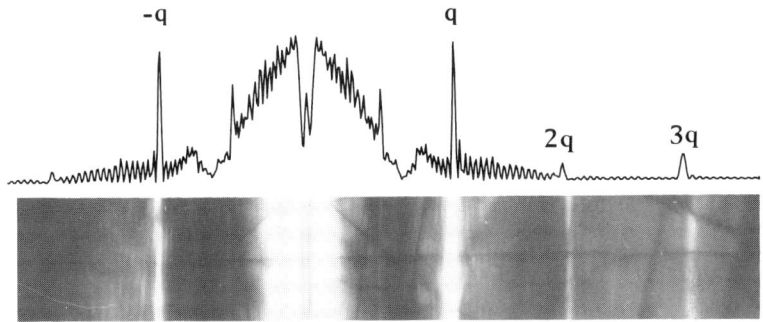

Fig 6. A dark field LACBED pattern at 250kV in g = 202 ($\theta = 45°$) from a 100Å CdTe/100Å $Cd_{0.8}Mn_{0.2}Te$ MQW showing superlattice reflections almost wholly due to the misfit strain ~0.004. The simulation is dynamical, with absorption (courtesy of S Diamond and Dr C J Rossouw, unpublished).

5.   APPLICATIONS TO SQWs

In work at Bristol a range of SQW structures in the AlGaAs/GaAs and InP/$In_{0.53}Ga_{0.47}As$ systems has been studied by the electron rocking curve method. Even very thin quantum wells (down to ~8Å) in much thicker barrier layers (typically 1000-5000Å in total thickness) are found to have a marked effect on the rocking curves. This was first reported by Cherns et al (1988b) for a 30Å GaAs quantum well in nearly 5000Å $Al_{0.3}Ga_{0.7}As$ and subsequently by Jordan and others for thin $In_{0.53}Ga_{0.47}As$ quantum wells in InP (eg Jordan et al (1989b, 1991b)). The influence of a buried quantum well on the rocking curve can be explained by reference to Fig 8 which shows the type of amplitude-phase diagram expected, ie the graphical representation of Equation (1). A change of radius in the amplitude-phase diagram can occur at the quantum well, either through a difference in structure factor or orientation (due to strain), since the radius is proportional to F_g/s. Thus, if a significant phase change occurs across the well, it can be seen that the resultant amplitude can differ markedly from the single crystal result. For a purely compositional modulation (no strain) the amplitude-phase diagram predicts that the rocking curve of a SQW structure should contain "single crystal" fringes spaced at $\Delta s = t^{-1}$ which are modulated in intensity at intervals of $\Delta s = t_x{}^{-1}$ where x is the thickness of any of the constituent layers. This can be seen by reference to Fig 3, which shows that fringe intensities oscillate every 2 fringes approximately corresponding to the ratio $t:t_1 \sim 2$ where t_1 is the thickness of one of the InP barrier layers.

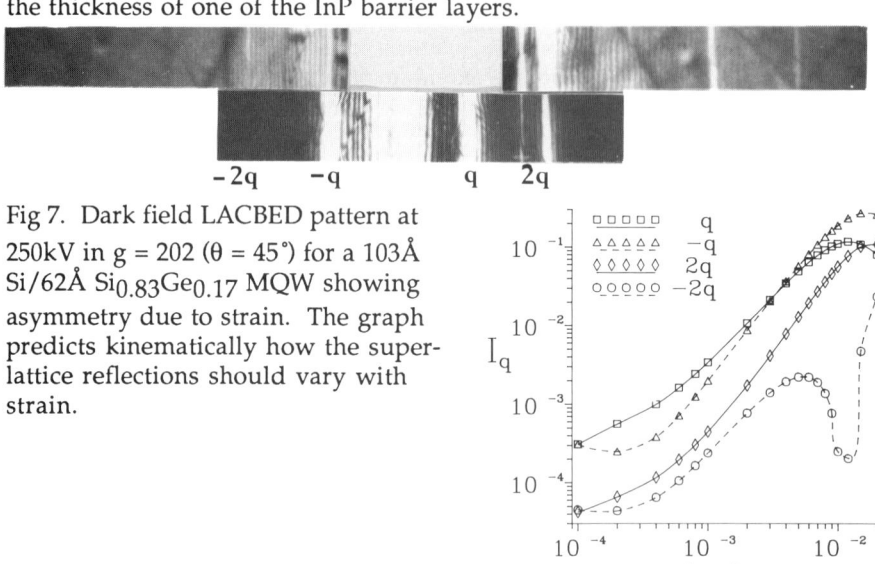

Fig 7. Dark field LACBED pattern at 250kV in g = 202 (θ = 45°) for a 103Å Si/62Å $Si_{0.83}Ge_{0.17}$ MQW showing asymmetry due to strain. The graph predicts kinematically how the super-lattice reflections should vary with strain.

Jordan and others have shown that the thickness of $In_{0.53}Ga_{0.47}As$ quantum wells in the range 8-50Å in InP can be obtained to near monolayer precision using a kinematical analysis of LACBED patterns. For quantitative analysis, rocking curves are first digitised and then autocorrelated with theoretical curves for trial structures using a standard computing routine. A detailed match to the experimental results is found to require the thickness of each of the constituent layers to be known to monolayer precision (eg see Jordan et al (1989b), (1991b) for further details). Spatial variations in the well thickness in SQW samples have been observed using the imaging method illustrated in Fig 3(b). The explanation for the dark and light features on the adjacent strong and weak fringes is that these represent a reduction in the amplitude of the fringe modulation consistent with a reduction of quantum well thickness. These features are therefore pits in the quantum well which are believed to be present at the $In_{0.53}Ga_{0.47}As/InP$ interface and of depth in the range 5-10Å. Experiments on a range of samples have shown that rough and smooth wells can be distinguished with smooth wells showing linear features which are believed to be individual interface atomic steps (Fig 9). As in the case of MQWs, the presence of strain in SQWs leads to rocking curve asymmetries with even comparatively low strains being important. It is now clear that small strains in nominally lattice-matched InP/In$_{0.53}$Ga$_{0.47}$As SQWs are important in allowing us to study very thin wells, both in LACBED patterns and in images. The results in Fig 9 are an example of this, showing steps in an 8Å well; further details are to be found in the paper by Jordan et al (1991b) in these proceedings.

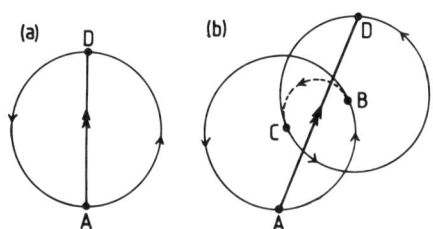

Fig 8. Amplitude-phase diagram for (a) a single crystal and (b) a SQW.

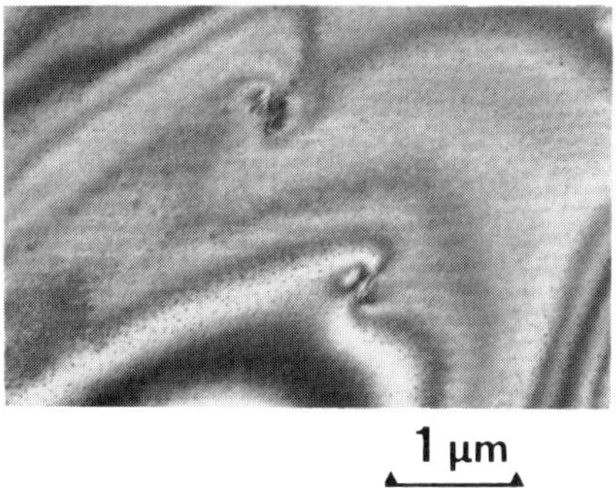

Fig 9. Dark field image in g = 202 (θ = 45°) from a 820ÅInP/8ÅInGaAs/ 700ÅInP SQW showing well steps.

6. DISCUSSION

It has been shown that the electron rocking curve method can give information on the composition and strain profiles in MQW and SQW samples. Both for MQW and SQW structures, layer thicknesses can be determined to near monolayer sensitivity. Individual interface atomic steps have been observed in InP/In$_{0.53}$Ga$_{0.47}$As SQWs. Due to the existence of selective etches both for InP and In$_{0.53}$Ga$_{0.47}$As it has proved possible to isolate single interfaces by preparing bicrystals rather than tricrystals enabling us to characterise step densities for individual interfaces, and to correlate structure with optical properties (Jordan (1991)). The results for MOCVD grown structures suggest that atomic terraces are present, supporting the idea that growth of a new layer proceeds by addition to pre-existing steps. The method is currently being extended to MBE grown SQWs. For MQW structures the observation of superlattice satellites by electron and X-ray rocking curves can be compared. The X-ray method provides a spatial average such that it is difficult to distinguish between interface roughness and sharpness, or between local and whole sample bending. The situation is much improved in the electron rocking curve method which has a spatial resolution of 500Å or better using LACBED, and near-atomic resolution in imaging mode. The effect of averaging interface roughness or inter- diffusion is to attenuate higher order satellites (Fig 10). Studies are currently in progress to use electron rocking curves to examine interdiffusion in CdTe/CdMnTe layers (S Diamond: private communication).

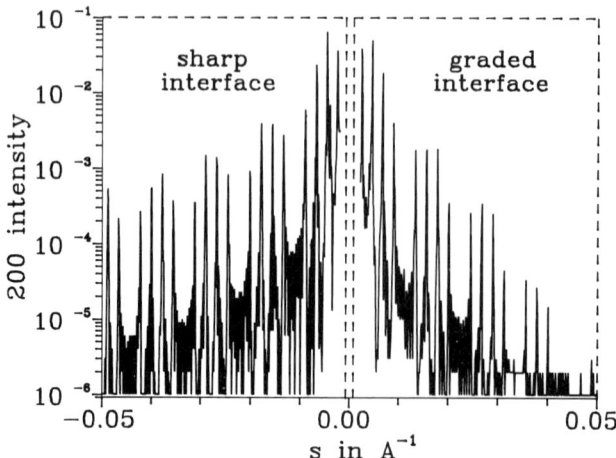

Fig 10. Dynamical simulation of the symmetrical 200 rocking curve for a 400Å InP/100Å In$_{0.53}$Ga$_{0.47}$As MQW assuming abrupt interfaces (LHS) and interfaces linearly graded over 20Å (RHS) (courtesy of Dr C J Rossouw).

ACKNOWLEDGEMENTS

The author is grateful to I K Jordan, Dr C J Rossouw, R Touaitia & S Diamond for the provision of illustrations for this review.

REFERENCES

Cherns D, Kiely C J and Preston A R 1988a Ultramicroscopy 24 355.
Cherns D, Jordan I K and Vincent R   1988b Phil. Mag. Lett. 58 45.
Cherns D, Touaitia R, Preston A R, Rossouw C J and Houghton D C
    1991 Phil. Mag. : in press.
Fewster P F 1988 J. Appl. Cryst. 21 524.
Jordan I K, Preston A R, Qin L C and Steeds J W 1989 Inst. Phys. Conf.
    Ser. 98 131.
Jordan I K, Cherns D, Hockly M and Spurdens P 1989b Inst. Phys. Conf.
    Ser. 100 293.
Jordan I K 1991 PhD thesis, University of Bristol.
Jordan I K, Rossouw C J and Vincent R 1991a Ultramicroscopy: in press.
Jordan I K, Grigorieff N, Cherns D, Hockly M and Spurdens P   1991b:
    these proceedings.
Rossouw C J, Al-Khafaji M, Cherns D, Steeds J W and Touaitia R
    1991 Ultramicroscopy: in press.
Vincent R, Cherns D, Bailey S J and Morkoc H 1987a Phil. Mag. Lett. 56, 1.
Vincent R, Wang J, Cherns D, Bailey S J, Preston A R and Steeds J W
    1987b Inst. Phys. Conf. Ser. 90 233.
Vincent R 1989 J. Electron Microscopy Tech. 13 40.

Accurate determination of local multilayer period from zone-axis conventional CBED patterns

H Gong and F W Schapink

National Centre for HREM, Laboratory of Materials Science, Delft University of Technology, Rotterdamseweg 137, 2628 AL Delft, the Netherlands

ABSTRACT: As observed previously, a number of fine lines are contained in satellite HOLZ (sHOLZ) reflection discs of CBED from GaAs/AlAs multilayers. Employing computer simulation based on dynamical diffraction theory, the dependence of the location of these lines on multilayer period and various other parameters is investigated. It is found that by choosing an adequate line in a sHOLZ reflection disc from a multilayer with suitable specimen thickness, the multilayer period can be determined almost without error.

1. INTRODUCTION

Accurate determination of semiconductor-multilayer periods is important for research and industry. In electron microscopy, a multilayer period is generally determined from cross-section specimens. However, apart from problems of specimen preparation, this inevitably involves averaging along the interfaces, prohibiting the accurate determination of multilayer periods in many cases. Recently, average multilayer periods could be determined in plan view by using convergent-beam electron diffraction (CBED) techniques. The first experiment was carried out by Gat and Schapink (1987). These authors could relate the multilayer period to the diameter of satellite HOLZ (sHOLZ) rings in zone-axis conventional CBED patterns geometrically. However, this geometrical relation did not lead to very accurate results. Shortly after that, Vincent, Cherns, Bailey and Morkoc (1987) showed that another CBED method, the large-angle CBED (LACBED), could also be used to determine multilayer periods accurately in plan view. However, a large flat thin area of the specimen is required for LACBED observation and the spatial resolution is much lower than that of the conventional CBED method. In this paper, the accurate determination of local multilayer (GaAs/AlAs) period from some lines appearing in sHOLZ reflection discs of zone-axis conventional CBED patterns will be investigated by employing computer simulation based on dynamical diffraction theory. It will be shown that by choosing an adequate line in a sHOLZ reflection disc and by taking into account the effect of some parameters, multilayer periods can be determined very accurately.

2. COMPUTER SIMULATION OF sHOLZ DISCS

To determine the period in a GaAs/AlAs multilayer from fine lines in sHOLZ reflection discs, computer simulation based on a three-dimensional version of the eigenvalue equations (Jones, Rackham and Steeds 1977, Blom and Schapink 1985) is employed. This version allows HOLZ diffraction effects to be taken into account by directly including HOLZ beams in the calculation. To model the multilayer, a large unit cell composed of n_1 GaAs and n_2 AlAs zinc-blende-structure cells along [001] is adopted, and na ($n=n_1+n_2$) is the period of the multilayer. The very small difference of lattice parameters between GaAs (a=0.5653 nm) and AlAs (a=0.5660 nm) has been neglected, and a=0.5658 nm is employed unless indicated otherwise. It should be noted that n may be an integer or an integer plus 1/2, which can easily be understood from the formation of a GaAs/AlAs multilayer. When n is an integer, the third index of rings of HOLZ reflections is an integer. When n is not an integer, the third index of the fundamental FOLZ (fFOLZ) ring of reflections becomes a fractional index, which is different from the traditional convention of an integer index; nevertheless, this does not invalidate our results. Atomic scattering factors for electrons are obtained from the International Tables for X-Ray Crystallography, Vol. IV (1974). Debye-Waller factors for GaAs at different temperatures are obtained from Reid (1983). However, no Debye-Waller factors for AlAs are available; as a first approximation we assume that the Debye-Waller factors for AlAs are the same as those of GaAs. Absorption is treated by using the perturbation approach (Humphreys 1979), which has been proved adequate for simulating HOLZ reflection discs (although it may be invalid for simulating the transmitted disc sometimes) for several systems by Tanaka et al (1988).

3. RESULTS AND DISCUSSION

As has been observed previouly, rings of sHOLZ reflections appear in zone-axis CBED patterns from [001] GaAs/AlAs superlattices. These rings approximately correspond to the intersections of the Ewald sphere and the sHOLZ layers originating from period modulation. The radius R of the m-th sHOLZ ring is related geometrically to the wave vector K and the distance H between this sHOLZ layer and the zero layer by (following Steeds 1979)

$$R=(2KH)^{1/2} \tag{1}$$

where H=m/P. It can be seen that when R is measured, multilayer period P can be determined (Gat and Schapink 1987). However, in a sHOLZ reflection disc two bright lines exist (Gat and Schapink 1987, Gong, de Haan and Schapink 1989), which makes the determination of R difficult. The two lines in a sHOLZ reflection correspond (Gong and Schapink 1989a) to the intersections of the relevant sHOLZ dispersion sphere and branches 2 and 5 of the zero-layer dispersion surface according to the dynamical diffraction theory. In the following, we will employ computer simulation based on the dynamical diffraction theory to investigate period determination from lines in a sHOLZ reflection disc.

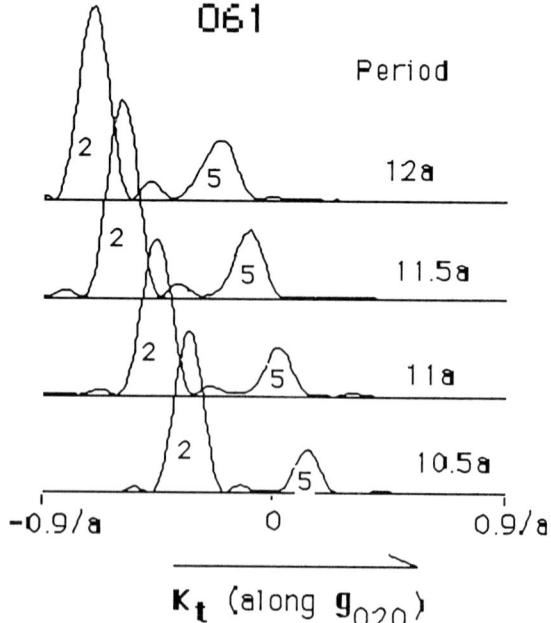

Fig. 1 The dependence of sHOLZ lines position on multilayer period. Computer simulated intensity curves for a (061) sHOLZ reflection disc from a [001] GaAs/AlAs superlattice. The specimen thickness is 120 nm. 120 kV. 70 beams.

Fig. 1 shows the computer simulated intensity curves of the 061 sHOLZ reflection disc for small multilayer-period changes for a [001] GaAs/AlAs superlattice. In figure 1, it can be seen that the location of the peaks (i.e. lines in the disc) is very sensitive to even the smallest period change (0.5a), indicating that a multilayer period can be determined by a comparison of the computer simulated and experimental results. However, to determine multilayer periods accurately, the dependence of sHOLZ lines on various other parameters should be established. It has been known that the location of the zero-layer branch 2 of the dispersion surface is very sensitive to Al concentration, whereas the location of branch 5 is almost independent of composition (Eaglesham et al 1986, Gong and Schapink 1989b). Since Al content in a GaAs/AlAs superlattice is often not exactly known, line 5 (rather than line 2) in a sHOLZ reflection disc should be employed for period determination. Now let us consider the choice of a suitable sHOLZ reflection for period determination. We will concentrate on a 6.7 nm GaAs/AlAs superlattice. Ideally, any sHOLZ reflection disc can be employed for accurate period determination. However, the line position in some reflection discs (eg. {441}) can be affected by sHOLZ beam coupling (Gong and Schapink 1989a). Although we can still use such reflections for period determination, a lot of computation time may be required because many sHOLZ beams should be incorporated in matrix diagonalization generally. It has been found that the location of the peaks in 061 intensity curves is not affected by HOLZ beam coupling; therefore, it is sufficient to include only one sHOLZ beam (061) in the calculation.

Table 1. Sensitivity of the location of line 5 in 061 sHOLZ disc to different parameters. The period is 12a and the specimen thickness is 120 nm unless stated otherwise. 120kV. 70 beams.

Parameter	Parameter variation	The shift of line 5 (units of a^{-1})
Composition (Al)	50 at%	0.00
Accelerating voltage	1 kV	0.01
Lattice parameter (a)	0.002 nm	0.01
Specimen thickness	from 30 to 70 nm	0.11
	from 70 to 100 nm	0.02
	from 100 to 320 nm	0.00
Single GaAs layer n_1a	a	0.00
Period non-uniformity	1.5a	0.00
Cell tetragonal distortion	4%	0.01
Temperature	200 $^{\circ}$C	0.03
Absorption (V'/V)	0.1	0.00
Multilayer period	0.5a	0.13

Note: 'Period non-uniformity' is investigated by choosing a very large unit cell in the simulation.

Now let us investigate the dependence of line 5 in the 061 disc on various other parameters. To discuss the effect of various other parameters, intensity curves of 061 sHOLZ disc are generated by computer simulation and the position of line 5 is determined from the curves. Small changes of composition, lattice parameter and accelerating voltage have been known to have a detectable effect on the location of lines in fundamental HOLZ reflection discs (eg. Eaglesham et al 1986, Jones et al 1977); however, the effect of these parameters on the location of lines in sHOLZ discs is not clear yet. Except for these three parameters, it is also necessary to discuss the effect of specimen thickness, single GaAs (or AlAs) layer thickness n_1a (or n_2a) while keeping $(n_1+n_2)a$ constant, period non-uniformity, lattice tetragonal distortion, temperature and absorption. Table 1 shows the effect of a variation of these parameters, and as a reference, also the effect of the smallest period change of two-atom layers on the location of line 5 in 061 reflection disc. From table 1 the following results are obtained. A large composition change does not show a detectable effect on the position of line 5 (although it shows a large effect on the position of line 2), just as expected. The effect of a change in accelerating voltage or lattice parameter on the location of a line in the 061 disc can be neglected. This appears different from the case for the lines in a fFOLZ disc, which can be explained approximately by using eq. (1). Next let us discuss the effect of the other parameters. It can be seen that except for specimen thickness, the effect of the remaining parameters can all be neglected. When the specimen is very thin (<70 nm), the line location depends heavily on specimen thickness. This is due to the overlapping of the lines in this thickness range. When specimen thickness changes from 70 nm to 100 nm, the line location is also changed, still due to the overlapping effect. In the range of 100-320 nm, the lines are distinct and no overlap happens, and therefore, the line location does not show dependence on specimen thickness. Since we generally locate the convergent probe on the area from which distinct HOLZ lines appear in CBED experiments, specimen thickness appears not to be a problem for accurate period determination. All the above results suggest that the effect of these various parameters on the location of a distinct line 5 can be neglected, therefore, multilayer period can be determined very accurately. By comparing the computer simulated and experimental results, the average double-layer of the 6.7 nm GaAs/AlAs multilayer (Gat and Schapink 1987) is determined to be exactly composed of 48 single atom layers. The multilayer period is then 6.788 nm, with relative inaccuracy less than 0.06%. Here, the inaccuracy of the period is due to the maximum uncertainty of the average atom-layer spacing of the GaAs/AlAs system. If the average spacing between atom layers are known exactly, multilayer periods can be determined almost without error.

4. CONCLUSION

By employing computer simulation based on the dynamical diffraction theory, the location of bright lines in sHOLZ reflection CBED discs from a GaAs/AlAs superlattice is found to be extremely sensitive to multilayer period. To assess the accuracy of determining multilayer period from the location of sHOLZ lines, the dependence of sHOLZ lines on various other parameters (such as accelerating voltage, composition, single GaAs (or AlAs) layer thickness, period non-uniformity, lattice tetragonal distortion, lattice parameter, temperature and absorption) is investigated. It is found that except for specimen thickness, the effect of the other parameters on sHOLZ line 5 can all be neglected. From a proper thickness range, the multilayer period can be determined very accurately from line 5 in a chosen sHOLZ reflection disc.

REFERENCES

Blom N and Schapink F W 1985 J. Appl. Cryst. <u>18</u> 126
Eaglesham D J and Humphreys C J 1986 XIth Int. Congress on Electron Microsc., Kyoto, 209
Gat R and Schapink F W 1987 Ultramicroscopy <u>21</u> 389
Gong H, de Haan C D and Schapink F W 1989 in: Evaluation of Advanced Semiconductor Materials by Electron Microscopy, ed. Cherns D (Plenum Press, New York) p 75
Gong H and Schapink F W 1989a Inst. Phys. Conf. Ser. <u>98</u> 91
Gong H and Schapink F W 1989b Inst. Phys. Conf. Ser. <u>98</u> 263
Humphreys C J 1979 Rep. Prog. Phys. <u>42</u> 1825
Jones P M, Rackham G M and Steeds J W 1977 Proc. R. Soc. Lond. A. <u>354</u> 197
Reid J S 1983 Acta Cryst. A <u>39</u> 1
Steeds J W 1979 in: Introduction to Analytical Electron Microscopy, eds. Hren J J, Goldstein J I and Joy D C (Plenum Press, New York) p 387
Tanaka M, Terauchi M and Kaneyama T 1988 Convergent Beam Electron Diffraction II (Tokyo: JEOL) p 108
Vincent R, Cherns D, Bailey S J and Morkoc H 1987 Phil. Mag. Lett. <u>56</u> 1

Inst. Phys. Conf. Ser. No 117: Section 8
Paper presented at Microsc. Semicond. Mater. Conf., Oxford, 25–28 March 1991

563

The detection of strain within InP-InGaAs single quantum well structures using large angle convergent beam electron diffraction

IK Jordan,[1] N Grigorieff,[1] D Cherns[1], M Hockly[2], PC Spurdens[2], MR Aylett[2] and EG Scott[2]

[1]H H Wills Physics Laboratory, University of Bristol, Tyndall Avenue, Bristol BS8 1TL

[2]British Telecom Research Laboratories, Martlesham Heath, Ipswich IP5 7RE

ABSTRACT: The large angle convergent beam electron diffraction technique has been used to obtain rocking curves from MOVPE and MBE grown InP-In$_{.53}$Ga$_{.47}$As single quantum well samples. Reflections sensitive to strain within the well layer were studied. From analysis of the curves an estimate of the mean strain present within the well in the samples has been made. Strains as low as $1.2 \cdot 10^{-3}$ have been measured. Plan-view specimens were used, reducing the problem of surface relaxation effects often encountered in thin cross-sectioned foils.

1. INTRODUCTION

The Bristol group has used the method of large angle convergent beam electron diffraction (LACBED) to study composition profiles in a number of single and multiple quantum well systems (Vincent et al (1987), Cherns et al (1988), Jordan et al (1989a) and Cherns et al (1991)). Here we extend the method to detect the low level of strain present in nominally lattice matched InP-InGaAs SQW (Single Quantum Well) samples. Structures grown using both atmospheric pressure MOVPE and gas source MBE were studied, (001) oriented substrates were used in all cases.

The LACBED method enables the selection of a single convergent beam disc. The maximum convergence angle is limited only by the electron optics (typically 6° in our case) (Vincent 1989). A single disc is selected using the selected area aperture, this aperture also acts as a transverse momentum filter and significantly reduces the level of thermal diffuse scattering within the pattern (Jordan et al 1991). The filtering enables the detection of detail within a LACBED disc at higher values of deviation parameter than are possible in comparable conventional CBED discs, or dark field (DF) images.

The technique differs from conventional CBED in that an image of the illuminated area (typically 2 to 3 μm in diameter) maps onto the diffraction disc, with spatial resolution usually limited by the probe size. For the information within a LACBED disc to be readily interpretable the specimen must be reasonably flat over the illuminated region of sample. This is easily achieved when preparing plan-view specimens of InP-InGaAs samples as excellent selective etches exist. Specimens with up to 2mm^2 of flat electron transparent areas have been produced. Electron transparent areas comprise a tricrystal formed from a thin InGaAs layer sandwiched between InP layers of suitable thickness (500 to 3000Å). To allow the use of selective etches a thin InGaAs buffer layer must be grown to isolate the tricrystal from the substrate.

2. LACBED FROM InP-InGaAs SQWs

To successfully model all the features within a rocking curve from a multilayer structure dynamical diffraction theory must be used (Jordan et al 1989b, Rossouw et al 1991). However, at large values of deviation parameter, s, typically greater than 0.003Å^{-1}, the main features in the scattered amplitude in a rocking curve can be calculated using a kinematic approach. The amplitude A is given by:

$$A \propto \int_{o}^{t} F_{hkl}(z)\exp\{-2\pi i(sz+\mathbf{g}.\mathbf{R}(z))\}dz \qquad (1)$$

F_{hkl} is determined by the phased sum of the scattering from the two fcc sublattices, these add in phase for reflections such as 400 and 220 but in antiphase for 200. $\mathbf{R}(z)$ is a measure of the displacement within the crystal due to strain. In the structures studied here the layers are constrained to the same lattice parameter in the plane of the foil. Any lattice mismatch in the SQW results in a tetragonal distortion involving elongation or compression parallel to the growth direction.

In previous studies we concentrated on reflections with $\mathbf{g}$ in the plane of the foil. The term in $\mathbf{g}.\mathbf{R}$ is then zero, and eqn. 1 can be considered as the Fourier transform of the structure factor. Analysis of rocking curves from compositionally sensitive reflections enables the measurement of mean well thickness to an accuracy of ±5%. For well layers greater than approximately 25Å the thickness may be obtained by simply measuring the value of deviation parameter, s_{min}, at which the modulation present in the rocking curve falls to a minimum (see, for example, Fig. 1, Jordan et al (1989)). The thickness, t, of the well layer is given by $t=s_{min}^{-1}$. In thin wells the position of modulation loss cannot be detected directly, as it occurs at high values of deviation parameter where the diffracted intensity is very low. Computer matching of the intensity profiles of the experimental and theoretical rocking curves must then be used (Jordan 1991).

Similar modelling can be performed on rocking curves formed in reflections with a component of $\mathbf{g}$ out of the plane of the foil. An estimate of the mean strain level within the well layer of the sample can then be made.

3. MEASUREMENT OF THE MEAN STRAINS WITHIN SQW GROWN USING MOVPE AND MBE.

3.1. 32Å well layer

Fig 1 shows a montage of 2 202 DF LACBED images obtained from a sample of 1550Å total thickness, with a well layer of nominally 32Å. The specimen was grown using MBE. Due to the low intensity of the LACBED discs only a limited range of s may be obtained (exposures of 2 to 3 minutes were used to capture the discs shown in fig. 1). The deviation parameter ranges from -0.025Å^{-1} at the left of the montage to $+0.025\text{Å}^{-1}$ at the right. A similar montage obtained from a sample with the same structure, but grown using MOVPE, is shown in fig. 2. All patterns were obtained with a specimen tilt, θ_t, of approximately 55°. The rocking curves show fringes with a spacing of s_f equal to t_{proj}^{-1}, where t_{proj} is the projected thickness of the sample (i.e. 1550Å/cos 55° in this case).

A strong modulation in the fringe intensity may also be seen. This effect is predominantly due to strain within the InGaAs well layer, the 202 reflection being relatively insensitive to variation in composition. An asymmetry in the modulation about s=0 may also be observed. The modulation decays to a minimum in the two areas of the pattern arrowed. As for the case of rocking curves formed in unstrained reflections the mean well thickness, t_{mw}, may be calculated from the spacing between consecutive losses of modulation, and is given by:

$$t_w = 2 \cdot (s_r - s_l)^{-1} \cdot \cos\theta_t \qquad (2)$$

where s_r and s_l are the positions of the first modulation loss to the right and left respectively of the Bragg peak. The substitution into eqn. 2 of values obtained from fig. 1 give a well thickness of 35Å, in good agreement with that specified.

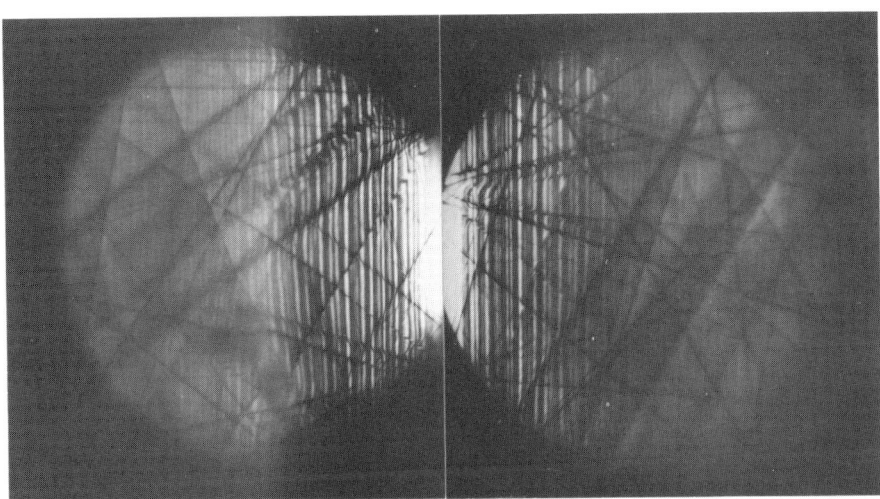

Fig. 1. A montage of 2 202 DF LACBED discs obtained from SQW samples containing a nominal 32Å well grown using MBE.

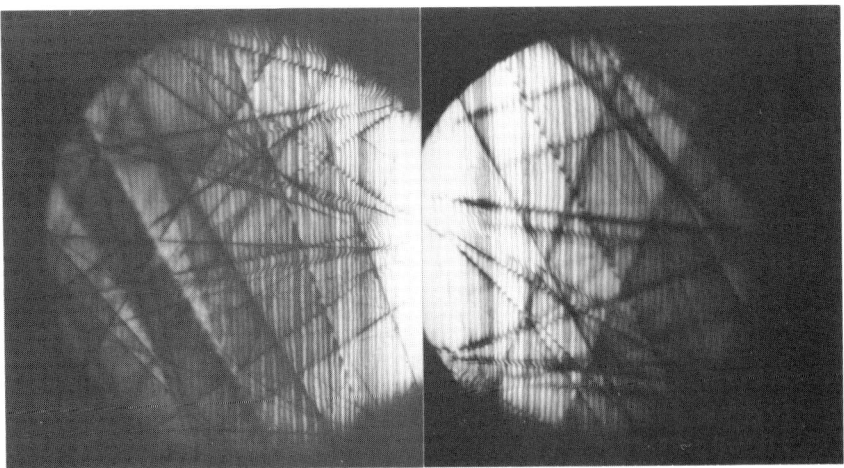

Fig. 2. A montage of 2 202 DF LACBED discs obtained from SQW samples containing a nominal 32Å well grown using MOVPE.

An estimate of the degree of strain within the well layer may be obtained from the level of asymmetry in the positions of s_r and s_l. To ensure consistency all values of deviation parameter must be measured relative to the Bragg position of the unstrained InP layers. This may be found by study of the position of high order deficiency lines, which cross the 202 LACBED disc.

A montage of 202 type rocking curves, calculated using eqn. 1, for a structure similar to that which generated figs. 1 and 2 is shown in fig. 3. The level of strain included in the calculation varies from 0 at the top to $1.0 \cdot 10^{-2}$ at the bottom, a 55° tilt angle was assumed. The curves best matching the patterns in fig. 1 and 2 are arrowed; they indicate a mean strain within the well layer of $2.4 \cdot 10^{-3}$ (MBE) and $1.2 \cdot 10^{-3}$ (MOVPE), with an accuracy of $\approx \pm 10\%$. (All strains are quoted relative to the relaxed cubic unit cell).

Fig. 3. A montage of simulated 202 type rocking curves, The level of strain varies from 0 (top) to $1.0 \cdot 10^{-2}$ (bottom). A 55° tilt angle was assumed.

3.2. 9Å well layer

Fig. 4 shows 2 202 DF LACBED discs obtained from SQW samples containing a nominal 9Å well layer. One sample was grown using MBE, the other using MOVPE. Modulation due to strain can be clearly seen in both discs. The well layer was too thin in the samples for a loss of modulation to be visible within the disc; therefore in this case it was not possible to obtain an estimate of the well thickness and strain without computer simulation.

Initially the thickness of the constituent layers of the sample are measured by comparison of rocking curves from 200 images (Jordan 1991). As the structure is then known the strain can be found by comparison of the 202 curve with theoretical rocking curves calculated with varying strain. At present these comparisons are performed by eye, limiting the level of accuracy possible, but in the future it is intended that the process will be computerised, using a routine similar to that developed to analyse 200 LACBED discs.

The mean strains present in fig. 4 were measured at $5 \cdot 10^{-3}$ (MBE grown) and $4 \cdot 10^{-3}$ (MOVPE grown). The accuracy of the measurements is at present only estimated at $\approx \pm 30\%$.

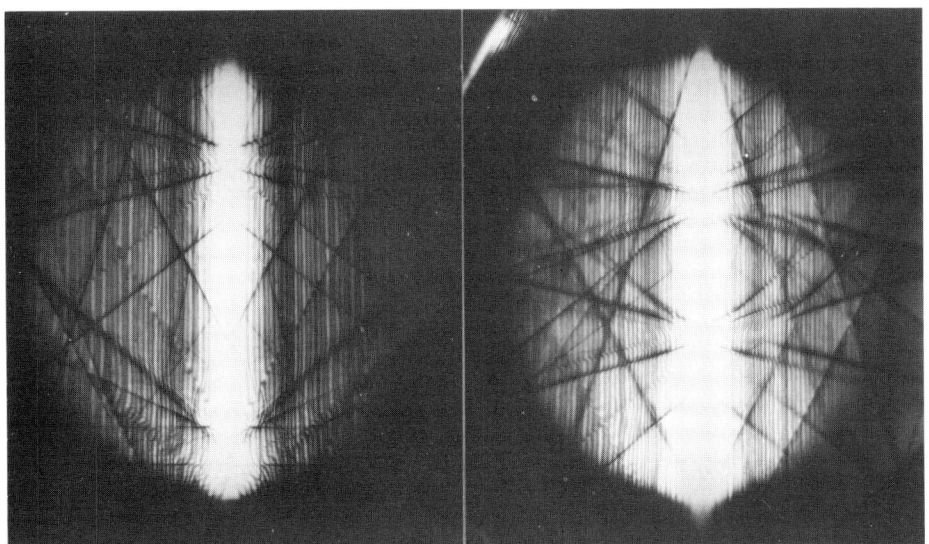

Fig. 4. 2 202 DF LACBED discs obtained from SQW samples grown using MBE (left) and MOVPE (right). The samples contained a nominal 9Å well layer.

4. DISCUSSION

The results indicate that small strains may be detected in SQW samples using this method. At present conventional x-ray double crystal diffractometry from SQW samples is typically limited to wells thicker than 100Å due to the low diffracted intensity levels involved. As the samples are viewed in plan-view the problem of further strain relaxation during specimen preparation does not arise.

In matching the experimental rocking curves with those predicted from kinematic theory it was assumed that the strain was equally distributed throughout the InGaAs well layer, with no strain present in the bulk InP. Clearly this is an over-simplification. Lyons (1989) has shown that thin layers of relatively high strain may occur due to gas flow switching at interfaces. Interfacial strain cannot be eliminated entirely even for ideal growth conditions as monolayers of InGaP and InAs will form at the InP to InGaAs and InGaAs to InP interfaces respectively.

Strain localisation at interfaces will, however, mainly effect the low frequency Fourier components of the rocking curve. In the study of SQW samples we are necessarily limited to analysis of the high order components (i.e. relatively low deviation parameter), It is therefore valid to use a model incorporating a uniform strain field to match theoretical and experimental rocking curves. Further experiments are to be performed to study this. It is also possible to determine the sign of the strain in the well layer from the direction of the asymmetry within rocking curve, but this has not been performed on the samples described here.

In the future it should be possible to extend the technique to the study of more highly strained systems. Analysis of rocking curves formed in out of plane reflections, incorporating strain effects, may also produce a way of measuring mean well thicknesses in the AlGaAs-GaAs system. Here study of in plane 200 type rocking curves has so far not proved satisfactory, due to the small structure factor fluctuation present.

5. CONCLUSIONS

The preliminary results presented here indicate that it is possible to detect an average level of strain as low as $4 \cdot 10^{-3}$ within a well layer only 9Å thick, using a structure of appropriate geometry. In the future it is hoped that the accuracy of the technique could be improved by automating the matching of experimental and theoretical rocking curves. The use of plan-view specimens reduces the problems of surface strain relaxation often encountered when preparing cross-sectioned TEM foils. Samples grown using MBE and MOVPE were found to incorporate similar levels of strain, but detailed comparisons cannot be made as the samples used do not form a systematic set.

ACKNOWLEDGEMENTS

IKJ and NG would like to thank BTRL for financial support. IKJ also acknowledges funding from the Wolfson foundation.

REFERENCES

Cherns D, Jordan I K and Vincent R (1988) *Phil. Mag. Lett.* **58** 45
Cherns D, Touaitia R, Preston A R and Rossouw C J (1991) *Phil. Mag.* In Press.
Jordan I K (1991) PhD thesis, University of Bristol
Jordan I K, Cherns D, Hockly M and Spurdens P C (1989) *Inst. Phys. Conf. Ser.* **100** 293
Jordan I K, Rossouw C R and Vincent R (1991) *Ultramicroscopy* In Press.
Lyons M H (1989) *J. Cryst. Growth* (1989) **96** 339
Rossouw C R, Al-Khafaji M, Cherns D, Steeds J W and Touaitia R (1991) *Ultramicroscopy* In Press.
Vincent R (1989) *J. Electron microscopy Tech.* **13** 40
Vincent R, Wang J N, Cherns D, Bailey S J, Preston A R and Steeds J W (1988) *Inst. Phys. Conf. Ser.* **90** 233

Inst. Phys. Conf. Ser. No 117: Section 8
Paper presented at Microsc. Semicond. Mater. Conf., Oxford, 25–28 March 1991

569

Scanning tunneling optical spectroscopy of InAsP/InP quantum well structures

L Q Qian and B W Wessels

Department of Materials Science and Engineering and Materials Research Center, Northwestern University, Evanston, Illinois 60208, USA

ABSTRACT: A scanning tunneling microscope combined with an optical probe was used to determine the electronic and optical properties of $InAs_xP_{1-x}/InP$ quantum well structures. The measured optical spectra at 295 K exhibit well resolved transitions attributable to interband transitions involving the heavy hole valence band and the n = 1 conduction sub-band. The observed transition energies are in good agreement with values measured using conventional photoconductivity and photoluminescence spectroscopy on the same samples. The potential of this technique for the local determination of the electronic and optical properties of semiconductor nanostructures is considered.

1. INTRODUCTION

The direct measurement of the electronic and optical properties of nanoscale semiconductor structures is fundamental to future understanding of the physics of low-dimensional structures. Recently we have reported that the scanning tunneling microscope (STM) combined with an optical probe, can be used to determine the electronic and optical properties of bulk semiconductors (Qian et al, 1990 a,b). The technique which we call scanning tunneling optical spectroscopy (STOS) can potentially be used to locally determine the electrical properties of a wide range of structures including homojunctions, heterojunctions and quantum well structures. In this paper we report on our recent application of STOS to the study of InAsP/InP quantum well structures. From the measurements the optical transitions for the buried quantum wells were directly determined. The measurements indicate that scanning tunneling optical spectroscopy is sensitive to sub-surface structure. The technique potentially enables imaging of internal interfaces on the nanometer scale.

2. EXPERIMENTAL

All the experiments were conducted at 295 K in air using a fully automatic Nanoscope I STM (Digital Instruments Inc.). Electrochemically etched tungsten tips were used for the tunneling junction. Fig. 1 shows a schematic of the experimental set up. The optical excitation source was a 250 W tungsten-halogen lamp. The light was dispersed through an ISA 0.25 m grating spectrometer equipped with appropriate filters to eliminate interference from visible wave-lengths. The monochromatic light was then focussed on the STM tip-sample junction with a short focal length lens. The light was modulated at the frequency of 1 kHz with a mechanical chopper. This frequency was beyond the response of STM feedback. The photoexcited tunneling current was measured using phase sensitive detection and the amplifier output was recorded on an x-y recorder. The optical spectra were taken using the constant current STM mode. The measured photocurrent was in the range of 0 - 10 pA which was more than two orders of magnitude smaller than the STM DC tunneling current (1 - 7 nA).

The quantum well samples studied were $InAs_xP_{1-x}/InP$ grown on an InP(100) semi-

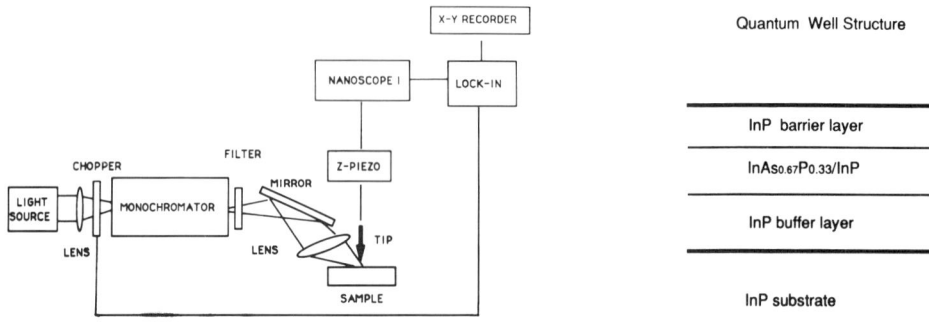

Fig. 1. Schematic of STOS apparatus

Fig. 2. Schematic of the well structure

insulating substrate by atmospheric pressure organometallic vapor phase epitaxy. Fig. 2 shows a schematic of the well structures. Details of the growth conditions and apparatus are presented elsewhere (Schneider *et al*, 1989). The structures consist of InAs$_{0.67}$P$_{0.33}$/InP with wells of a width 2.1 – 5.1 nm and an InP barrier layer of 100 nanometers. The quantum well was grown on a 150 nm InP buffer layer.

Fig. 3 shows the STOS spectrum (curve a) as well as photoluminescence spectrum (curve b) for a structure of InAs$_{0.67}$P$_{0.33}$/InP with well of width 5.1 nm. Several transitions are observed and are labelled A - D. The predominant transition corresponds to the E1H1 transition between the n=1 heavy-hole, valence band and the n= 1 conduction sub-band. In addition to the E1H1 transition, higher energy transitions are observed. The transition labelled B is tentatively attributed to a transition involving the light hole band and the n = 1 conduction sub-band (E1L1). The transition labelled D corresponds to the band-gap transition involving the InP barrier layer.

Table I lists the E1H1 transition energies for four quantum well samples. For comparison, the transition energies observed by independent photoconductivity (PC) measurements as well as photoluminescence (PL) measurements on the same samples are also listed. Excellent agreement between the E1H1 transition energies measured by the various techniques is found.

Table I. Measured and Calculated Transition energies for InAs$_{0.67}$P$_{0.33}$/InP at 295 K

Sample No.	Well thickness (nm)	STOS (eV)	PC (eV)	PL (eV)	Calculated (eV)
302	4.5	0.86	0.86	0.86	0.86
322	3.9	0.89	0.89	0.89	0.89
325	5.1	0.84	0.84	0.84	0.84
372	2.1	1.03	1.01	1.03	1.03

As the well becomes narrower, a systematic shift of the E1H1 transition energy to higher energies is observed which is consistent with increased quantum confinement. For comparison, the transition energies were modelled using a standard finite square well, taking into account strain and using well width L as a given variable. As listed in table I, the agreement between experiment and theory is quite good.

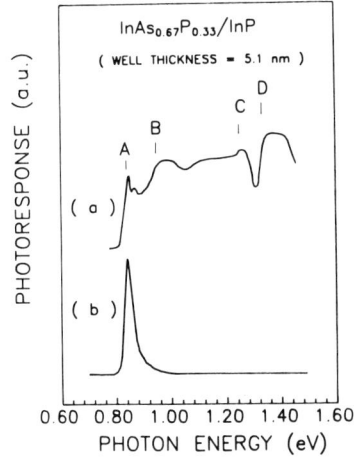

PHOTORESPONSE (a.u.)

InAs$_{0.67}$P$_{0.33}$/InP

(WELL THICKNESS = 5.1 nm)

D
C
A B
(a)

(b)

0.60 0.80 1.00 1.20 1.40 1.60
PHOTON ENERGY (eV)

Light
Ef
Tip
Ef
Quantum well

Fig. 3. STOS spectrum (curve a) and photoluminescence spectrum (curve b) for an InAsP/InP quantum well (well thickness = 5.1 nm)

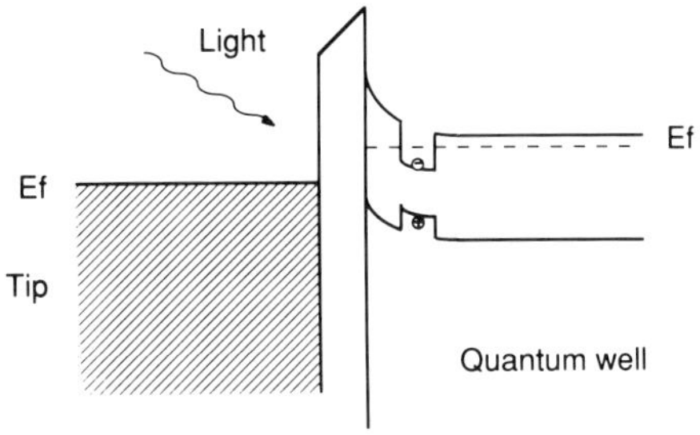

Fig. 4. Energy level diagram for sample and tip under negative bias

3. DISCUSSION

The measurements on the quantum well structures indicate that scanning tunneling optical spectroscopy is sensitive to the presence of buried layers. In this case, the quantum well is 100 nm below the surface. As to the mechanism, for the most part, only the carriers that are photoexcited and drift or diffuse to the surface will influence the tunneling current as indicated in Fig. 4.

Thus the depth at which influence of a buried layer can be observed depends on several factors including optical absorption length, carrier diffusion length and depletion layer width. For lightly doped material this can extend more than several microns into the material. Thus presumably imaging of deep buried structures should be realizable depending on experimental conditions and the material properties.

As to the lateral resolution of the technique, the same parameters are important. Thus the lateral resolution will depend on the specific electronic properties of the material. Fig. 5 shows an STOS image for an Fe-doped semi-insulating InP sample patterned with 5 μm gold lines. The gold lines were grounded during the measurements. For these measurements a 10 mw He-Ne laser was used to illuminate the STM junction. These preliminary experiments on InP structures indicate that submicron resolution is realizable. The limit of resolution needs to be determined and is currently under study.

While the carrier diffusion length plays a key role in determining the resolution of STOS in planar structures, in non-planar structures the geometrical boundaries limit carrier transport. Thus in low dimensional structures such as quantum wires and quantum dots that are non-planar, significant improvement in the resolution should be obtainable. Fundamental information on the quantum nature on these low dimensional structures should be realizable.

In conclusion, scanning tunneling optical spectroscopy of InAsP/InP quantum wells have been demonstrated. The relevant optical transitions have

2 μm

Fig. 5. STOS image of InP

been measured and are in good agreement with values measured using photoluminescence and photoconductivity as well as theoretical calculations. This technique should enable the detailed study of the quantum properties of low dimensional semiconductor structures. It should be applicable to a wide range of semiconductor materials.

ACKNOWLEDGEMENT

The authors thank R P Schneider Jr for preparing the quantum well structures. Helpful discussions with Professor Y W Chung regarding STM technique are gratefully acknowledged. This work was supported by the NSF through a grant from the materials Research Center at Northwestern University (DMR 8821571). Extensive use of the Surface Science Facility at Northwestern is acknowledged.

References

Qian L Q and Wessels B W 1991a Appl. Phys. Lett. <u>58</u>, 1295
Qian L Q and Wessels B W 1991b Appl. Phys. Lett. in press
Schneider R P, Li D X and Wessels B W 1989 J. of Electrochem Soc. <u>136</u>, 3490

Inst. Phys. Conf. Ser. No 117: Section 8
Paper presented at Microsc. Semicond. Mater. Conf., Oxford, 25–28 March 1991

Structural analysis of III−V superlattice structures by combined TEM and XRD observations

M J Yates, M A G Halliwell, S J Amin, M R Taylor, M R Aylett and S D Perrin

British Telecom Research Laboratories, Martlesham Heath, Ipswich, IP5 7RE

ABSTRACT: Transmission electron microscopy (TEM) can give sensitive dimensional information from cleaved edge sections of GaInAsP multilayers on InP. X-ray diffraction and simulation can give more precise compositional information but when many layers are present there can be too many unknown variables to find a unique solution. It is demonstrated that in complicated structures, the use of TEM data directly in XRD simulations can improve both the quality and confidence of deduced compositional information, beyond the scope of either individual technique.

1.   INTRODUCTION

In this paper the combined approach of x-ray diffraction (XRD) and transmission electron microscopy (TEM) is discussed for the routine quality assessment of $Ga_xIn_{1-x}As_yP_{1-y}$ grown lattice matched to InP for opto-electronic telecommunication devices. XRD is widely used to measure the period and overall mismatch of superlattices non-destructively. If the superlattice period is regular and a large number of satellite reflections can be recorded it is possible to determine the compositional variation as a function of position within the superlattice unit cell. This is achieved by matching the measured satellite reflections to calculated values using kinematic diffraction theory (Fewster 1986; Lyons et al, 1989). In many cases, when the periodicity is not perfect or the number of periods is too small this procedure is not applicable. In these circumstances simulation of diffraction data using dynamical theory (Halliwell, Lyons and Hill, 1984) is a possibility but the lattice parameter variation as a function of depth must be known in detail. For each individual layer and interface region a thickness and lattice mismatch are required. Consequently the number of unknowns is usually too large for the problem to be solved by XRD alone.

Dimensional information using conventional TEM is excellent. High resolution electron microscopy (HREM) on cross-sections being widely used to quantify interface abruptness (eg Petford-Long et al 1989; Ourmazd 1989). Recently, 90° cleaved edges have become the favoured method for rapid assessment of interface abruptness and layer widths by examination of extinction fringes in B = [010] images (Kakibayashi & Nagata 1985; Spycher et al 1989). However, even using cleaves the measurement of mismatch of < 10 000ppm (parts per million) which is required by crystal growers is very difficult. This is because the effect on extinction fringes of either slight compositional changes or elastic relaxation in the specimen is small and, in any case, simulation of fringe positions is subject to several uncertainties.

In this context the information from TEM and XRD is complementary. We use examples to show the advantages and limitations of incorporating TEM measured thicknesses directly into model structures for XRD rocking curve simulation. By eliminating uncertainties such as layer and interface thickness from simulated trial structures, information on mismatch composition can be more accurately inferred.

2.   EXPERIMENTAL

Samples were grown on [001]±0.5° InP substrates by atmospheric pressure MOVPE using the method described by Nelson et al (1988). The MOVPE kit routinely produces high quality material with low compensation, low mismatch < 500ppm and specular surface morphology; this makes excellent devices of the type examined here (Bagley 1990). The structures grown contained superlattices in the form of a multiple quantum well in sample A and a longer period test structure in sample B. All layers were nominally lattice matched to InP and quaternary compositions corresponded to optical emission at 1μm, 1.1μm, 1.2μm and 1.5μm. TEM was conducted on a JEOL 200CX operated at 200kV. Observations on cleaved-edge samples were made in bright-field with $\underline{B}$ = [010], using an objective aperture to select only the 000 beam, giving a spatial resolution of ~0.8nm. The alignment of $\underline{B}$ with the [010] pole was always better that 6x10^{-4}radians.   X-ray diffraction was performed with a Philips High Resolution Diffractometer using Cu Kα radiation and a four reflection monochromator (Bartels 1983) which gave a signal to noise ratio of >10^5. All experimental rocking curves were recorded about the 004 reflection, simulations were performed using dynamical theory based on that described by Halliwell et al (1984).

3.   RESULTS

3.1   GRIN-SCH-MQW laser

Sample A was a 9-well Graded Refractive INdex - Separate Confinement Heterostructure - Multiple Quantum Well (GRIN-SCH-MQW) laser with the nominal structure given in Fig 1c, from which only a small number of superlattice satellites were observed. The experimental rocking curve Fig 1a, showed strong first order satellites originating from the MQW region; the fine subsidiary structure between the satellites originated from interference amongst all the other layers. Careful TEM examination of cleaved edges revealed that the layers were of high quality (Fig 2) with interfaces of $\leq$10Å width.

The XRD simulation Fig 1b, was produced using dimensional data from TEM and average MQW mismatch from the experimental rocking curve. A close fit of satellite peak heights was found only when a large mismatch difference between the Q1.5 wells (+790ppm) and Q1.2 barriers (-2000ppm) was assumed. No interfacial layers were included in the trial structure, in view of the TEM result and because significantly mismatched interfaces were not anticipated between Q1.2 and Q1.5 layers. The central peak was not reproduced exactly because there were two other quaternary compositions present of unknown mismatch.

(c)	Nominal	Simulated	
	(nm)	(nm)	(ppm)
InPcap	250.0	250.0	0
Q1.0	10.0	8.7	-350
Q1.1	10.0	8.7	-400
Q1.2	10.0	9.9	-2000
9xQ1.55 ⎤ MQW	8.0	6.6	+790
8xQ1.2 ⎦	6.0	6.0	-2000
Q1.2	10.0	9.9	-2000
Q1.1	10.0	8.7	-400
Q1.0	10.0	8.7	-350
InP buffer			

Fig 1. Sample A. Experimental (a) and simulated (b) x-ray rocking curves. Box (c) shows the nominal structure and the simulated structure.

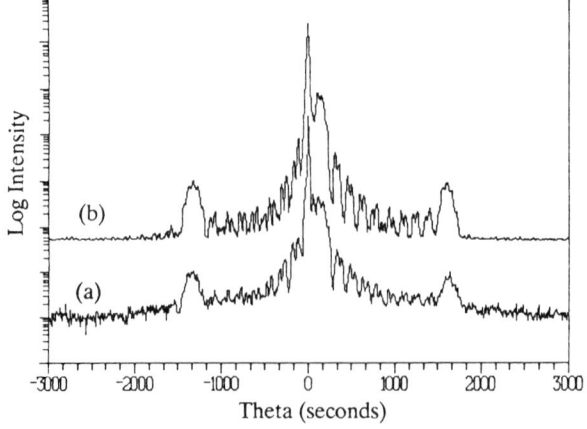

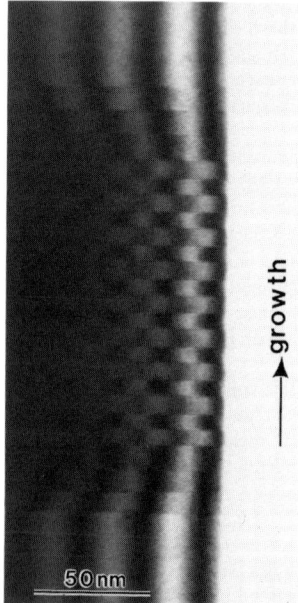

Fig 2. Sample A, cleaved edge bright-field B = [010] micrograph.

3.2 20 Period Superlattice

A complicated XRD result Fig 3a was obtained from sample B, an InP/GaInAs superlattice with the nominal structure shown in Fig 3d but which lacked perfect periodicity. This was a test structure in which every alternate InGaAs→InP interface was paused for 5s under a phosphine flush, with no group III's present. The most interesting feature of the rocking curve is the presence of double-period satellite peaks (first order near ±230secs) some of which showed slight splitting behaviour, characteristic of a structural change at a reasonably discrete point in the stack. The double-period satellites were a consequence of alternate, paused InGaAs→InP interfaces being fundamentally different from the non-paused interfaces, thus creating a double-sized period. The TEM micrograph Fig 4, established that although the measured period appeared constant, the paused InGaAs→InP interfaces were different from the non-paused ones and, futhermore, became sharper in the top half of the stack. The different interface characters were also confirmed by the fringe structures observed at projected interfaces examined using 202 type reflections.

Simple XRD simulations confirmed that no close similarity to experiment could be achieved without including significant interfacial width and a difference between paused and unpaused interfaces. Using interfacial widths based on the TEM results, improved fitting was obtained Fig 3b although simulated peaks were still sharper than experiment. A closer resemblance was achieved by including a random fluctuation of layer thickness and mismatch with standard deviations of 0.3nm and 200ppm respectively, shown in Fig 3c. Although the fit could not be regarded as accurate in detail, some general features such as the subsidiary peaks and main satellite positions were reproduced, thus allowing reasonable estimates of layer and interface mismatch to be deduced. In such a complicated structure the uncertainties of interfacial mismatch made further elaboration of the simulation unjustifiable.

Fig 3. Sample B.
X-ray rocking curves: experimental (a), simulated with interfaces (b), and including a random width and mismatch fluctuation (c). Box (d) shows nominal and simulated structures - interfaces are denoted by Δ.

(d)		Nominal (nm)	Simulated (nm)	(ppm)
InP cap		500	500.0	0
	InP	20	18.0	0
	Δ GaInAsP		1.9	+6000
	GaInAs	20	18.9	-1200
	Δ GaInAsP		1.9	+6000
10x	InP	20	18.0	0
	Δ GaInAsP		1.1	+2500
	GaInAs	20	19.7	-1200
	Δ GaInAsP		1.9	+6000
InP buffer				

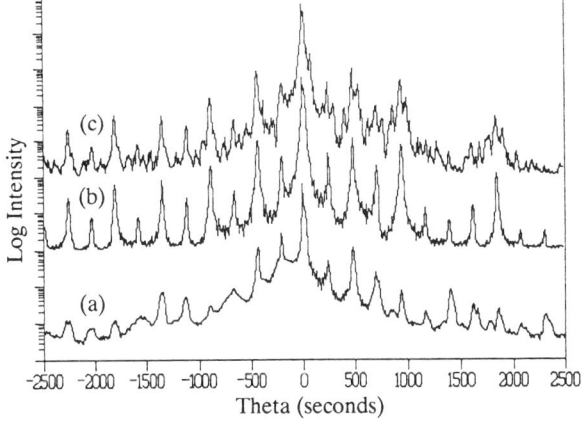

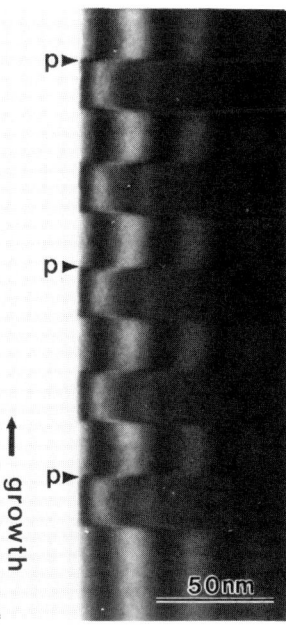

Fig 4. Sample B, cleaved edge bright-field B = [010] micrograph of first 5 periods showing paused interfaces P.

4. DISCUSSION

In the examples given the TEM data and rocking curves originated from areas of the wafer in reasonable proximity. Experience has shown that where this is the case, excellent agreement is achieved between the superlattice period measured by TEM and XRD. MOVPE growth usually shows minor variations in growth-rate and mismatch across a 50mm wafer; non-uniformities are composition dependent but typical variations might be as little as ±8% in thickness and ±100ppm in mismatch (Nelson et al. 1988). Within this context, for a superlattice there is little difficulty in sampling a different part of the wafer by XRD and for the simulation, scaling all the TEM dimensions to fit the period measured by XRD.

There is an obvious difference between the area of interface sampled by each technique. Long scale undulations of lateral wavelength, say >500nm are unlikely to be seen by TEM on cleaves, whilst the 1mm^2 XRD sampling area would interpret the undulations as a graded interface. If present, such undulations might be inferred by a close comparison of the techniques. Graded interfaces observed by TEM will certainly by sampled by XRD. However the question of whether such grading is associated with significant departures from lattice match (Lyons 1989) must be resolved by XRD alone, unless the mismatch is gross enough to influence the position of extinction fringes in cleaved edge TEM. Grading in the InP/GaInAsP system grown by MOVPE is well reported (eg Spurdens et al 1991) and is usually attributed to persistent arsenic incorporation. In this study the usefulness of diffraction as a very sensitive indicator of differences between two types of gas switching in MOVPE has been exemplified in sample B.

Since the primary purpose of combining TEM and XRD data is to derive more certain knowledge of mismatch, the question of accuracy must be addressed. Unfortunately this is somewhat intractable because it depends foremost on the quality of correlation between experimental and theoretical rocking curves. The best fits are achieved with the most perfect structures. In periodic structures the average mismatch of the period is directly calculable with an uncertainty based simply on the error in locating the substrate and satellite peak centres. For the samples discussed here the average mismatch is accurate to about ±100ppm. Satellite intensities are dependent on mismatch in the periodic region from which they originate, and not primarily from surrounding non-periodic layers which tend to add broad background structure. Thus estimation of mismatch from non-periodic layers must receive separate consideration. In cases such as sample A where the satellite heights are well matched the uncertainty in mismatches for the periodic layers is about ±250ppm. In sample B the figure is probably slightly worse. However, to keep these figures in perspective they should be related to their corresponding change in alloy composition; a ~700ppm change in mismatch corresponds to only a 0.01 change in x fraction.

5.   CONCLUSIONS

The scope for using x-ray rocking curve simulations to calculate layer mismatches in two different types of periodic InP|GaInAsP structures has been shown. For these structures the use of cleaved edge TEM in specifying dimensions of trial structures used for the simulations is shown to offer improved confidence in the deduction of layer mismatch. Combined XRD simulation and TEM have been demonstrated to be a powerful method for the characterisation of different gas switching effects at MOVPE hetero-interfaces.

The authors would like to thank P C Spurdens for constructive discussions. Gratitude is extended to A Webster for diligent TEM specimen preparation.

REFERENCES

Bagley M, Sherlock G, Cooper D M, Westbrook, Elton D J, Wickes H J, Spurdens P C, Devlin W J 1990 Electronic Lets 28 8 512
Bartels W J 1983 J. Vac. Sci. & Tech. B1 338
Fewster P F 1986 Philips J. Res. 41 268
Halliwell M A G, Lyons M H and Hill M J 1984 J. Crystal Growth 68 523
Kakibayashi H and Nagata F 1985 Jpn. J. Appl. Phys. 24 L905
Lyons M H, Scott E G and Halliwell M A G 1989, Inst. Phys. Conf. Ser. No 100: section 6
Nelson A W, Spurdens P C, Cole S, Walling R H, Moss R H, Wong S, Harding M J, Cooper D M, Devlin W J and Robertson M J 1988 J. Crystal Growth 93 792
Ourmazd A 1989 J. Crystal Growth 98 72
Petford-Long A K, Booker G R, Hockly M and Taylor M R 1989 Inst. Phys. Conf. Ser. No 100, sect. 4
Spurdens P C, Taylor M J, Hockly M and Yates M J 1991 J. Crystal Growth 107 215
Spycher R, Buffat P A, Stadelmann P A, Roentgen P, Heuberger W and Graf V 1989 Inst. Phys. Conf. Ser. No 100 section 4 299

Inst. Phys. Conf. Ser. No 117: Section 8
Paper presented at Microsc. Semicond. Mater. Conf., Oxford, 25–28 March 1991

Wafer uniformity of an MBE grown $Al_xGa_{1-x}As$/GaAs tunnelling structure

N Rimmer[1,2], R T Syme[1,3], J E F Frost[1], D A Ritchie[1], G A C Jones[1], M J Kelly[1,3] and W M Stobbs[2]

[1]Semiconductor Physics Group, Cavendish Laboratories, Cambridge CB3 0HE
[2]Department of Materials Science and Metallurgy, Pembroke street, Cambridge CB2 3QZ
[3]GEC Marconi Ltd, Hirst Research Centre, East Lane, Wembley HA9 7PP

ABSTRACT: The form and properties of a triple barrier $Al_xGa_{1-x}As$/GaAs tunnelling structure were evaluated as a function of position on an MBE grown wafer using SIMS, PL, TEM, and its I-V characteristics. The TEM data on layer uniformity, thicknesses and compositions were correlated with evaluations of the forms of the barriers as deduced from the PL. SIMS provided the doping levels. Using all these data as input parameters in standard theoretical models for the I-V characteristics, the doping asymmetries thence deduced were compared with those measured. While the radial changes in the structure were thus determined, we were also able to assess the relative reliability of the different characterisation methods and the efficacy of the different modelling approaches.

1. INTRODUCTION

The common aims of the device designer and the MBE grower are to improve the properties and reliability of a given electronic device structure. The achievement of such goals requires that the crystal grower has access to a detailed and accurate description of the structure produced for a particular set of growth conditions. The effects of changing a given structural parameter on the properties used to evaluate the structure thus also have to be understood. Ideally we might hope to find sets of unique relationships between the value of a specific property and a specific characteristic of the structure, but we can see from fig.1 that the relationships between the properties which can be measured and the structural parameters are generally convoluted. In this schematic diagram a heavy line represents a strong dependence and interconnections indicate the lack of a unique relationship. As we can see, most techniques necessarily probe several structural parameters simultaneously so that it is only on comparing the data obtained in several different ways that we can gain confidence in the values inferred for any one parameter. The approach leads to a 'self consistent loop' type of solution allowing the data obtained from the combination of the results from TEM, PL and SIMS to be used to assess the modelling of the I-V curves. It is the tunnelling behaviour which is the most difficult to model since it depends not only on all the parameters which can be determined from the other measurements but also on other variables, the importance of which can be considered only when experimental I-V characteristics can be compared with simulations with no free fitting parameters. Our aim here, while presenting the results of a characterisation of a triple barrier tunnelling structure (the design specification for which is shown in fig. 2), has been to determine the viability of the approach we describe above.

2. RESULTS

Figure 3 shows how a quarter wafer of the structure shown in fig.2, as grown by MBE on a (001) substrate, was divided up so that related parts of it could be assessed as a function of radial position using SIMS, PL, TEM and the measurement of its I-V characteristics.

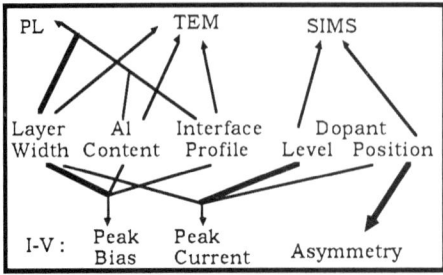

Fig 1 : Structure - Property Relationships

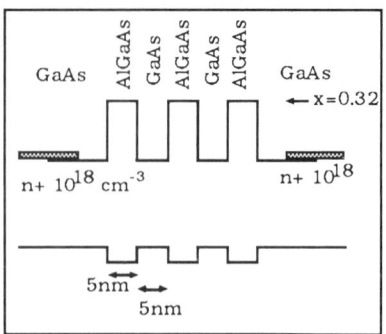

Fig 2 : Design Specification

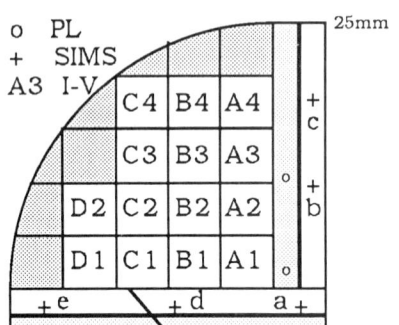

Fig 3 : Positions for PL, SIMS, I-V

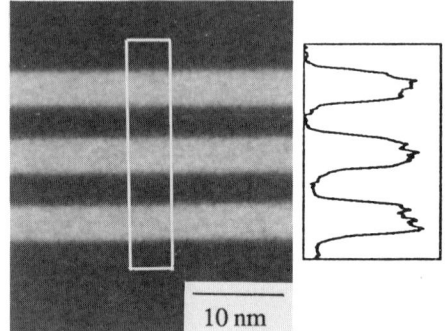

Fig 4 : 002 dark field image and profile

We will consider firstly the data we obtained on the structure using TEM. Images such as that shown in fig.4, taken in the compositionally sensitive 002 reflection, were digitised to yield profiles such as that also shown here. These were used to measure layer thicknesses and the values of x for the $Al_xGa_{1-x}As$ barriers by the methods described by Bithell et al (1989). Table 1 shows the layer and interface widths together with the aluminium compositions for the indicated positions on the wafer. The error in the layer widths was ±0.2nm while the values of x had an error of ±0.03. These accuracies are sufficient that we can be confident that the MBE growth rate was about 13% higher for the given wafer at its centre than at a radius of 21 mm; 4% values tend to be predicted. The same Ga source was used at the same temperature for the growth of both the GaAs and the $Al_xGa_{1-x}As$. Thus taking the well thicknesses as a function of position to indicate associated Ga flux changes, values of x can be inferred from the barrier thicknesses. On this basis, and taking the measured value of x at 2mm to be correct, then the values of x at 10mm and 21mm would be expected to be 0.36 and 0.34 respectively. Considering the accuracies of both approaches it would appear that the value of x can be taken to be about 0.32±0.03 and, relative to the changes in layer thickness, constant to within this error with position. The interface widths were inferred from the intensity profiles after subtracting a figure of 0.5nm (as indicated by dark field multislice simulations of abrupt barriers) from the separation of the 10% and 90% intensity values. This technique is not as accurate as the Fresnel Method but the values obtained are consistent with data obtained using this latter method on similar structures (Ross et al 1991) and we can be confident that the interfaces are spread compositionally by between 0.5 and 1.0nm. On a coarser scale, the structure's depth exhibited 10nm undulations at a period of about 250nm.

Table 1.TEM values for layer widths and compositions and SIMS data for dopant asymmetry.

mm to centre	Layer Thickness (nm)					interface (nm)	Al (x) content	Dopant Asymmetry
	barrier 1	well 1	barrier 2	well 2	barrier 3			
2	6.00	5.86	5.89	5.80	5.93	0.87	0.34	12 nm
10	5.85	5.55	5.79	5.51	5.77	0.76	0.32	17 nm
21	5.01	4.97	5.12	5.08	5.20	0.91	0.31	13 nm

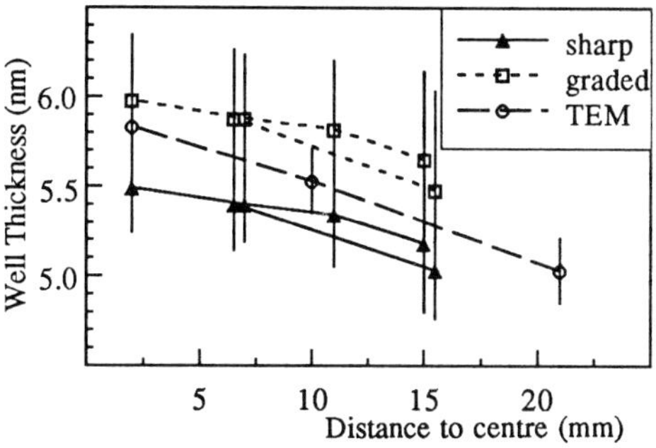

Fig 5 : PL and TEM measured well widths

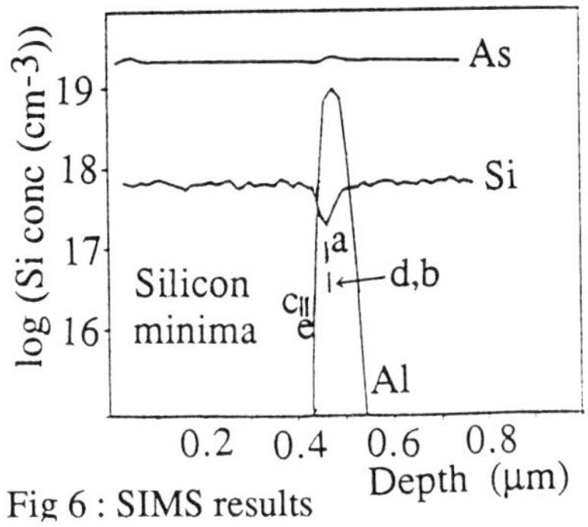

Fig 6 : SIMS results

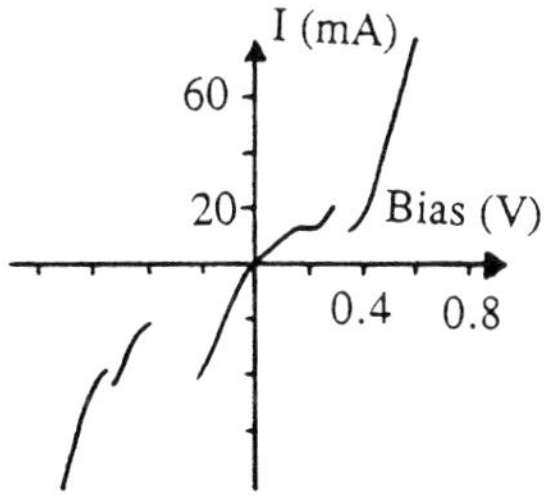

Fig 7 : I-V characteristic for chip A1

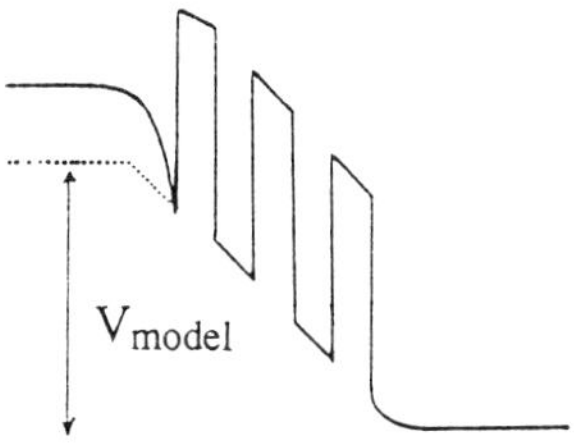

Fig 8 : Biased conduction band profile

PL spectra (courtesy of K L Schumacher) showed two peaks, the smaller, higher energy, peak at about 1.60 eV being due to transitions from the $n=1$ electron to heavy or light hole confined levels in the quantum wells. Six spectra were recorded in two perpendicular radial directions. No PL peak due to the wells was detected further than 16mm from the centre of the wafer, suggesting that the wells further out than this (which have been shown by SIMS and TEM to be closer to the surface) were removed by the $H_2O_2:H_2SO_4:H_2O$ etch used to remove the absorptive n^+ capping layer. The position of the quantum well peaks shifted to higher energies with increasing distance from the centre of the wafer indicating a reduction in well width. The experimental results were modelled using calculations of the energy levels for both abrupt and diffuse wells using the model described by Davies (1987). Davies noted that this analysis may contain systematic errors in that the energy of the confined electron state above the conduction band is calculated, whilst the measured energy is the difference between the electron and hole confined states. Thus, the band gap must be subtracted from the measured energy and a correction made to allow for the hole confinement energy. The well widths as a function of position, as calculated from the PL data, are shown in fig.5; the solid lines are for wells taken to have sharp interfaces and the dotted lines assume 1nm grading. The layer widths measured by TEM are plotted for comparison (dashed lines) and, given that the roughness measured by TEM was less than was assumed in the calculations, we gain increased confidence in the use of a figure of 1.52 eV for the band gap of GaAs as used here.

An example of SIMS data (courtesy of A Chew, Loughborough) obtained using a 10keV Cameca Cs^+ ion source is presented in figure 6 for position a at the centre of wafer. The mean doping level away from the barriers was measured to be 6×10^{17} cm^{-3}. We were also particularly interested in any asymmetry of the dopant profile relative to the barrier structure in relation to the interpretation of the I-V characteristics described below; this is why Si sensitive

Cs+ ions were used. The distances between the minima in the Si signal and the maximum Al signal are shown in fig.6 and table 1.

A typical 77K I-V curve is shown in figure 7. Two point probe current-voltage measurements were made on six 120µm square mesas per chip (as shown in fig. 3) and the resistance between two substrate contacts was measured in order to estimate the resistance in the system which was not due to the barriers, the value obtained being 2 Ω. There is asymmetry in the bias and current at the first resonant tunnelling peak as well as in the low bias resistance. The sense of this asymmetry did not change as a function of radial distance as is consistent with the SIMS data. The positive and negative bias voltages for the first peak, as measured for positions corresponding to those at which the structural characterisations were made, are given in table 2 together with values, (V), corrected for the series resistance. These latter values can be compared with those, also given here, as calculated using simulations based on Davies' (1987) model. The structural parameters obtained using TEM (in which we have reasonable confidence given their consistent trends relative to the PL data) were used in these calculations together with the doping level from the SIMS data and an assumed undoped layer thickness. The doping asymmetry was however treated as a free fitting parameter and the best fit values, as used for the simulated peak positions, are shown in table 2 and can be compared with the approximate measurements made using SIMS, as given in table 1. While the values obtained for A1 and C1 are smaller than those measured, they are still surprisingly large. The data for Chip D1 could be modelled only if a low value for the layer widths of about 3.2nm was assumed and this would require the TEM data to be much more inaccurate than it is.

Table 2. Measured and simulated bias voltages for first I-V peaks with inferred Si asymmetry

Chip	Bias at First Peak		Simulated Peaks		Fitted
	substrate +ve	substrate -ve	sub +ve	sub -ve	Asymmetry
A1	0.28 (0.25)	0.22 (0.14)	0.25	0.14	3.8 nm
C1	0.37 (0.29)	0.30 (0.18)	0.26	0.16	3.0 nm
D1	0.45 (0.36)	0.41 (0.29)	0.30	0.16	?

There are, however, several effects which the model does not include fully. The most important of these is the space-charge effect due to the way electrons will tend to accumulate at the first barrier, thus locally lowering the conduction band to form an additional well (fig.8). The accumulation region will result in an increase in the bias at the resonant tunnelling condition. In order to correct for this in the model it would be necessary to solve the Poisson equation at each point on the band profile. If the undoped region before the barrier is large then the well will be deep enough to have a confined level and this is the cause of the extra structure observed in the I-V characteristic before the resonant peak for a positive bias. The effects are less pronounced with a negative substrate bias because the band bending occurs over a smaller region, hence the well is less deep. Nonetheless, the differences in the degree to which fits can be obtained for the A1 and C1 curves relative to those for D1, using simulations unmodified as described above, suggest that still further effects must be included.

4. CONCLUSIONS

We may conclude that fairly accurate data for barrier widths, roughnesses and Al contents can be obtained using TEM, and that PL data can be fitted to the values obtained for the widths and roughnesses, lending greater confidence to both approaches. It is SIMS which must be used at present to provide the doping levels but, if I-V characteristics are to be modelled uniquely and to the accuracies which are now physically interesting, better methods are needed for the measurement of doping asymmetries. We are grateful to SERC and GEC for financial support.

REFERENCES

Bithell E G and Stobbs W M 1989 Phil. Mag. A60 39
Davies R A 1987 GEC J. Research 5 65-75
Ross F M and Stobbs W M 1991 Ultramicrosc. in press

Inst. Phys. Conf. Ser. No 117: Section 8
Paper presented at Microsc. Semicond. Mater. Conf., Oxford, 25–28 March 1991

581

TEM studies of buried heterostructure laser diodes before and after overstress testing

U Bangert, A T R Briggs*, A R Goodwin* and P Charsley

Physics Department, University of Surrey, Guildford
*STC Technology Ltd, London Road, Harlow, Essex

ABSTRACT: Transmission electron microscopy studies have been carried out on degraded buried heterostructure lasers and unaged controls. Degradation was produced by deliberately overstressing the lasers at high currents and high temperatures.

1. INTRODUCTION

Early InGaAsP/InP buried heterostructure (BH) lasers exhibited variable reliability performance. Potential generic mechanisms of degradation are parasitic electrical leakage currents due to non-radiative recombination in or close to the active region and current leakage remote from the active region. This paper deals with some aspects of early results of the microstructural degradation in InGaAsP/InP BH lasers of varying Zn doping levels, which might cause parasitic current leakage.

2. EXPERIMENTAL

The specimens were aged BH lasers and unaged controls of the design shown schematically in fig.1. Prior to the TEM investigation, the lasers underwent accelerated lifetests at 125°C and 100 mA to 200 mA for durations of up to 600 hours during which electrooptic performance was regularly monitored. TEM investigations were undertaken on lasers of low and high Zn doping of the p-infill layer close to the regrown mesa-sidewall. The increased Zn levels in the high-Zn doped lasers in the shaded areas in fig.1, arise due to higher Zn incorporation during regrowth at lower temperatures for an initial growth period.

Cross-sectional TEM specimens were obtained by surrounding a pair of lasers, standing on their facets and facing each other with their top contact sides, with GaAs in a close-packed fashion. This 'stack' was glued together with M-Bond 600 and 3 mm disks containing the lasers in the centre were extracted with a high speed drill. These disks were mechanically polished from both sides down to approximately 50 μm thickness and then Ar-beam thinned at LN temperature. After perforation the electron transparent region was extended to contain the active lasing region by milling at room temperature using an Ar/I-beam. TEM studies were undertaken with a 2000FX EM.

3. RESULTS

The unaged lasers showed no defects in the active stripe sidewalls and very few elsewhere in the structure. Sometimes, long straight dislocations or stacking faults descend on the inclined 111-planes from the top contact region and intersect the epitaxial layers. Fig.2 shows a segment of a dislocation loop, probably originating in the region of the oxide mask, used in processing, which has reached the top interface of the active region. The reason for these dislocations could be strains due to lattice mismatch of the GaInAsP, the oxide used to define the mesas (removed before overgrowth) or the top-oxide window with respect to the InP.

Fig.3 shows the active stripe of a typical low-Zn doped laser, life-tested for 4 hrs at 125°C. Dislocation loops (arrowed) arise on ABC and ABD planes in the inserted thompson tetrahedron near to, but not on, the sidewall planes (ACD and BCD) and extend into the p-InP confining layer. This suggests that they may originate from pre-existing dislocations rather than from sidewall defects. They appear to be perfect loops with 110-type Burgers' vector.

Fig.4 (left hand side and bottom axis) shows the threshold current characteristics for low-Zn doped lasers. Note the initial low values and the rapid degradation.

Fig.5 shows the active region of a typical high-Zn doped laser diode, tested at 125°C for 600hrs. The micrograph shows severe microstructural degradation of the sidewalls and the interface with the p-InP top confining layer. Note extended faulted loops on planes ABC or ABD. Fig 4 (left hand side and top axis) shows the corresponding current threshold curves. The current degradation rate is 100 times lower in the high-Zn doped lasers. However, the initial threshold current values are three times as high in this structure.

Fig.6 shows a close-up of the defect structure near the sidewall of a degraded high-Zn doped laser diode. It contains large straight loops, possibly decorated, on ACD and BCD planes (marked L); large loop tangles on ABC and ABD planes, some of them containing stacking faults (marked F) and small precipitates within the loop tangles (marked P). In contrast to the low-Zn doped lasers they all appear to be partial dislocation loops.

4. DISCUSSION

Both sets of lasers had similar electrooptic properties when removed from test and sectioned (fig.4). It is surprising therefore that the rapidly degraded low-Zn doped lasers show relatively few defects whilst the high-Zn doped lasers have extensive defect formation yet were still able to function. Clearly the low-Zn doped lasers are much more sensitive to the defects which form.

There are two mechanisms, by which defect formation may give rise to degradation: (i) defects can act as centres for non-radiative recombination of electrons and holes, or (ii) defects can act as conducting channels for electrons, connecting the active layer to the n-type InP layer of the blocking structure (the location of the defects in the low-Zn doped devices suggests this). This channel then results in a greatly increased leakage current through the blocking layer.

In both cases the vicinity of a forward biased junction allows for recombination enhanced defect reactions (Lang and Kimerling, 1974) to further promote defect growth and degradation during device operation. Both mechanisms, to be effective, rely on the ability of the defects to capture electrons. The crucial difference between the high-Zn and low-Zn doped devices is the location of the p-n junction in the active layer. During growth of the blocking layer, zinc will diffuse into the active region from the p-InP adjacent to the sidewalls and above the active layer (Van Gurp et al, 1989). Staining techniques which locate the p-n junction show that, in the case of the high-Zn doped lasers, the junction is markedly displaced from its as-grown position. As the position of the p-n junction is moved into the active layer, fewer electrons are able to reach the defects located above the active layer and thus their effectiveness is reduced. This explains the greater sensitivity of the low-Zn doped lasers to defects, as in this case they are closer to the junction.

The fact that the voltage across the device at high current (100 mA) does not change significantly as the device degrades makes hypothesis (ii) of increased current leakage via the blocking layers as the sole contributor to the degradation unfavourable. On the other hand, mechanism (i) acting on its own also seems unlikely, because the small number of defects in the low-Zn doped lasers can be hardly sufficient to give rise to such rapid and large degradation simply by acting as non-radiative recombination centres. The situation might, however, be rather complex in that the defects in the low-Zn doped lasers act electrooptically differently (e.g. show higher electron conductivity) from the defects in the high-Zn doped lasers (e.g. capture more electrons non-radiatively). As our observations indicate, the latter defects are of a different nature and most likely are influenced by Zn diffusion (Ball et al, 1981). It is hoped that further investigations will reveal which of the mechanisms dominates in which type of laser.

REFERENCES
Ball R K, Hutchinson P W and Dobson P S 1981 Philosoph. Mag. A 43 1299
Van Gurp G J, Van Dongen T, Fontijn G M, Jacobs J M and Tjaden D L A 1989 J. Appl. Phys. 65 553
Lang D V and Kimerling 1974 Phys. Rev. Lett. 33 489

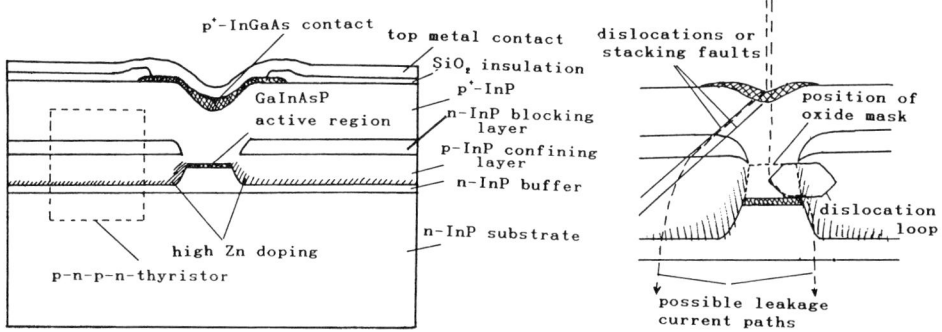

Fig.1 Schematic diagram of BH laser cross section. Right hand diagram shows region around active stripe with possible dislocation arrangements.

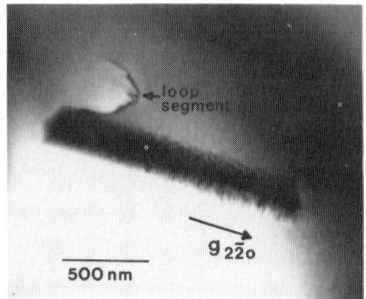

Fig.2 Cross section of active stripe in unaged laser.

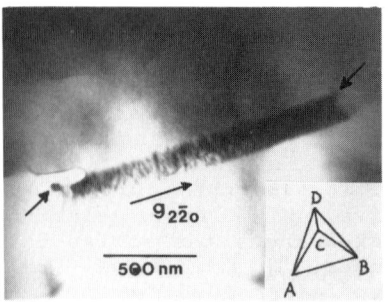

Fig.3 Active region in degraded low-Zn doped laser.

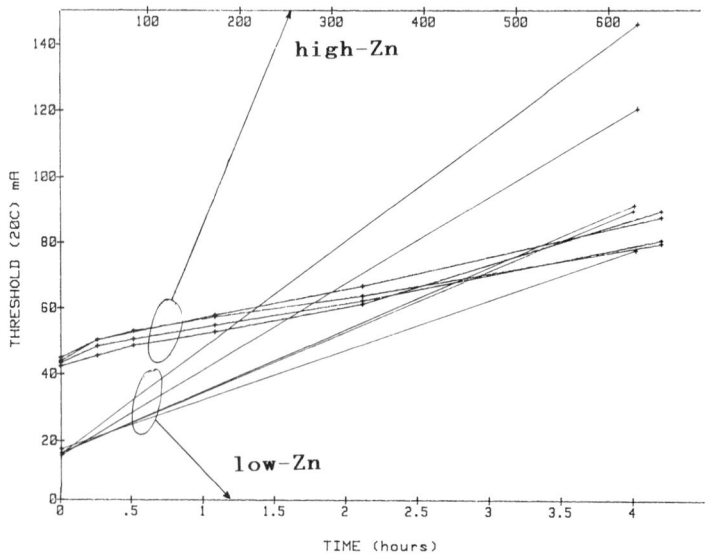

Fig.4 Threshold current versus aging time for low- and high-Zn doped laser diodes

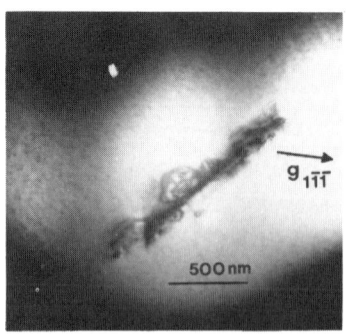

Fig.5 Cross section of active region in degraded high-Zn doped laser

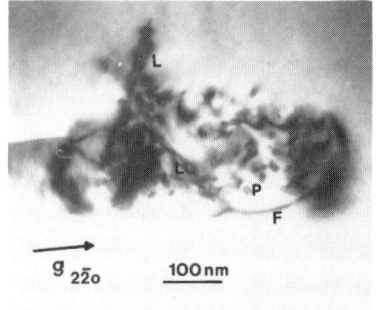

Fig.6 Enlarged defect structure in side-wall of a high-Zn doped laser

Electron microscopy investigations of stress-related degradation of AlGaAs/GaAs quantum well laser diodes

A Fried, A Jakubowicz, S B Newcomb* and W M Stobbs*

IBM Research Division, Zurich Research Laboratory, 8803 Rüschlikon, Switzerland

*Department of Materials Science and Metallurgy, Cambridge University, Cambridge, UK

ABSTRACT: The effects of process-induced stress on the formation of defects in AlGaAs/GaAs quantum well laser diodes with a ridge-type waveguide are examined using EBIC, CL and TEM, and the degradation patterns at, and adjacent to, mirror facets are correlated with stress induced by the laser's waveguide overlay. Lasers were separately degraded by operation and by electron irradiation in an SEM. These experiments showed that regions near the lower ridge edges exhibit enhanced susceptibility to degradation. The effect of the ridge on degradation is similar to that of a stripe window.

1. INTRODUCTION

We report here an EM investigation of mechanical stress/strain-related degradation of single quantum well graded-index separate-confinement heterostructure (GRIN-SCH) GaAs/AlGaAs laser diodes with a ridge-type waveguide. The effect of mechanical stress on degradation has been extensively studied, in particular for stripe lasers (Robertson et al 1981 and Kirkby et al 1979), and is still of interest (van der Ziel and Chand 1990). The non-homogeneous distribution of defects in stripe lasers was found to result from strain-enhanced defect migration (Wakefield 1979). These earlier studies concentrated on defect analysis in operated devices. Since degradation is a dynamic process, the result of any defect analysis depends on the device history. Moreover, since degradation depends on various parameters it is often difficult to determine their respective roles. We have concentrated on the degradation of that part of the material which is closest to the laser mirror, since it is often here that degradation is initiated, and find that the operation-induced degradation/defect pattern near the mirror, in planes parallel to it, is a combined effect of stress-related "weak spots" within the laser and the spatial distribution of power in an operating device. Our experimental results, supported by theoretical calculations, show that the stress-related effect of the ridge on degradation is similar to that of a stripe window.

2. EXPERIMENTS AND DISCUSSION

The "reconstruction" of the original strain field from an operation-induced degradation map faces the serious problem that such a map reflects the distribution of power in an operating device as integrated over the entire time of operation. This implies that there is generally no one-to-one correlation between such a map and the strain field, as has been recently demonstrated by Jakubowicz (1991) who, in parallel with laser operation, used electron irradiation in an SEM to degrade the laser mirror facets, the resultant defects being revealed by using EBIC. The approach with other SEM modes was shown to provide a convenient microscopic way of revealing regions exhibiting enhanced susceptibility to degradation. The scanning process in an SEM allows a precisely controlled deposition of equal amounts of energy over any selected area of the structure under investigation. This is crucial to producing a legible degradation/defect pattern, where the local defect state correlates with the original strain field responsible for generating this pattern. The SEM electron energy is well below the threshold energy for single-atom displacement, and therefore other mechanisms, such as electron beam generated x-ray irradiation, or non-radiative recombination of electron-hole pairs are responsible for degradation of the semiconductor. The latter mechanism was observed in the electron irradiation-induced degradation of a p-n junction in a laser (Jakubowicz 1991). Since non-radiative recombination is known to be one of the mechanisms of laser degradation, electron irradiation provides a relevant technique for laser degradation studies. Here we compare devices degraded during their operation with those degraded only by electron irradiation in an SEM.

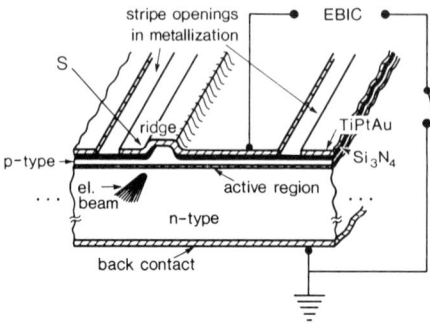

Fig. 1: The experimental setup.

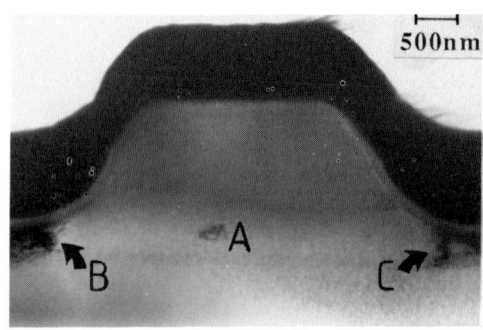

Fig. 2: TEM image of mirror after 100 hrs.

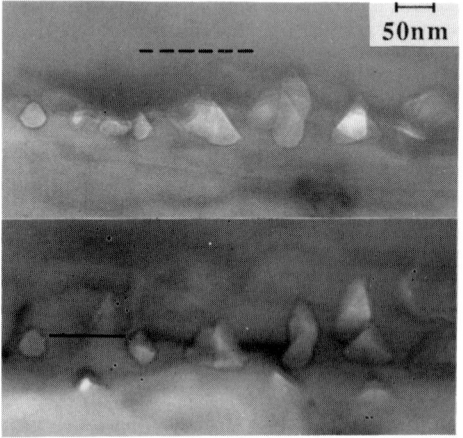

Fig. 3: TEM stereo pair of central "pores".
Active layer: —— Tilt axis: - - - -

The laser diodes investigated, as shown schematically in Fig. 1, have been described by Harder et al (1986). After 100 hours of operation (10 mW optical power) of a laser with a cleaved unprotected mirror, it was examined by TEM. The low magnification image in Fig. 2 shows the mirror facet as retained in a back-thinned foil. There is damage, not seen in unused lasers, in the central part of the active region as marked at A, and defect structures have developed around the lower corners of the waveguide (at B and C). Images of the retained facet region of a failed laser are shown in Figs. 3-5 demonstrating the defect development. Figure 4 shows how the defects at the ridge corners are small loops and how not only has a dislocation structure developed at the center of the mirror, but how there are now also pores, or inclusions, formed in the center as a result of a chemical reaction. The relationship of these "pores" to the active layer is shown in Fig. 5 while the way in which their depth is related to the outer mirror surface is demonstrated by the stereo-pair in Fig. 3.

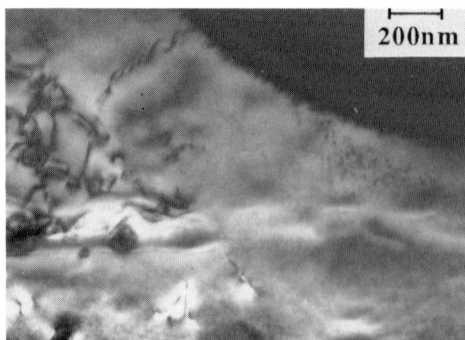

Fig. 4: Center to edge of mirror for failed laser.

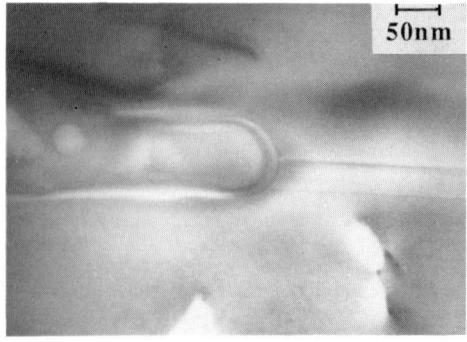

Fig. 5: "Pores" at active layer at center.

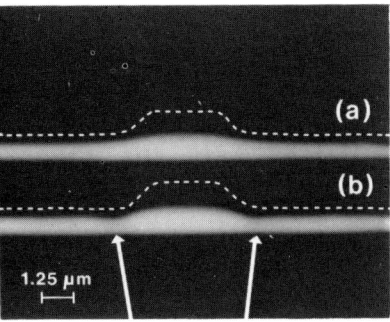

Fig. 6: EBIC images of lasers: a) undegraded and b) as 15 kV, 8×10^{-10} A electron scanned. Dashed line indicates ridge structure. Arrows indicate local changes at the ridge corners.

Figure 6 compares EBIC images of two identical lasers: (a) a non-degraded one and (b) as exposed only to electron irradiation along the active region. The image of the irradiated laser has two symmetric features, not seen for (a) close to the lower corners of the ridge and extending outside of it. We emphasize that these features are the first degradation symptoms during electron irradiation so it may be inferred that these regions have an enhanced susceptibility to degradation. There may be several possible explanations both for the way the laser operation-induced damage (Figs. 2-5) develops and, in particular, for its distribution in planes parallel to the mirror. The most probable of these is that the operation-induced degradation pattern reflects two contributions: the presence of "weak spots" in the laser structure, (regions of enhanced susceptibility to degradation), and the distribution of power in the operated device, integrated over the entire time of operation.

Jakubowicz (1991) demonstrated both that the weak spots at the lower ridge corners, detected by electron irradiation, probably originate from the strain field around the ridge caused by its overlay and that the effect of the ridge on degradation is similar to that of a window. In stripe lasers, it was shown that the edges of the stripe window exert a force on the semiconductor. This force results in a strain field which in turn promotes the migration of defects during laser operation (Robertson et al 1981). That the irradiation of the cleaved facet below the opening in the metallization (i.e. a window) and below the ridge (see Fig. 1) resulted in very similar degradation patterns strongly suggests that a similar mechanism is involved in degradation during operation of a stripe- and a ridge-type laser. It was also concluded from the similarity of the two degradation patterns that in both cases the strain fields are similar in the electron irradiated part of the material. Since the insulating nitride layer underneath the window in the metal is continuous in our samples (Fig. 1), it must be the discontinuity of the metal that produces the strain field around the window. Hence, the strain field around the ridge most probably results from the geometric perturbation of the metallization in the ridge region, in particular at the ridge edges where the force exerted by the metallic overlay changes its direction sharply (arrow S in Fig. 1).

We have calculated the stress distribution in a plane parallel to the mirror facet in an approximate analytical approach, and compared the pattern with the results of spatially and spectrally resolved CL measurements. Figure 7 shows the calculated distribution of stress below a ridge and a window of the same dimension, evaluated along the p-n junction, and Fig. 8a shows the respective CL wavelength shift dependence on position, measured at 5 K for the quantum well (QW) emission. By monitoring the QW line we ensured the best possible spatial resolution perpendicular to the active region; the lateral resolution can also be expected to be improved by the p-n junction electric field which reduces the lateral spread of carriers. The wavelength shift points to tensile stress below the ridge of about 1 kbar. In many cases the shift was much smaller, and in some devices it reached the limits of our method. One of the measurement problems is the limited spatial resolution of CL, which is comparable with the size of our structures. Another difficulty is due to the effect of electron irradiation on the material. CL requires rather high excitation, and therefore

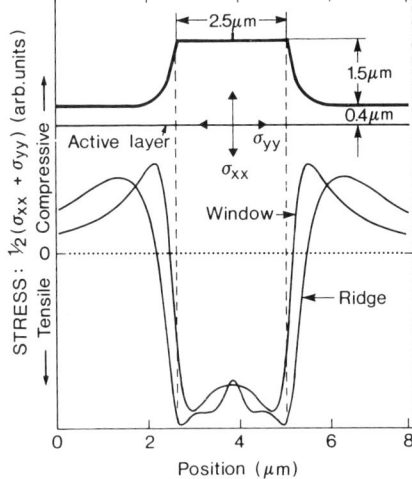

Fig. 7: Linear combination of normal, σ_{xx}, and parallel, σ_{yy}, stress components along the active layer for a ridge and a window.

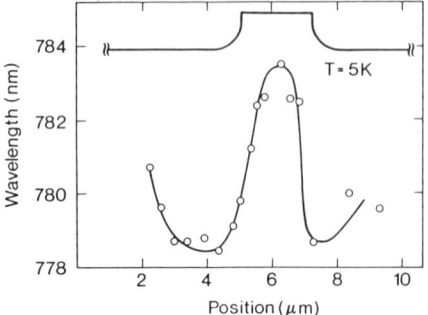

Fig. 8a: QW line λ along active layer. Higher λ at ridge center indicates tension there.

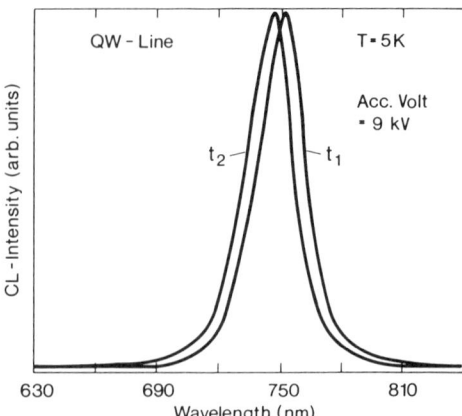

Fig. 8b: CL spectra of QW line at same point on active layer. t_1: before, and t_2: after electron irradiation showing blue shift.

defects may form, which can modify the QW emission. Figure 8b is an example of an electron irradiation-induced CL shift of the QW line. One can expect such effects as intermixing to be dependent on stress. This can be the contrast source for CL imaging, similar to the case of EBIC measurements.

From Figs. 6, 7 and 8a it can be concluded that the "weakest" spots of the structure correlate with high stress gradients, i.e. where two micro-regions, one under compressive the other under tensile stress, are close to each other. This is consistent with the strain-enhanced defect migration model. From our TEM observations (consistent with the different characters of the corner and central damage) we conclude that the central damage originates in the maximum light intensity in the ridge center and is due to a temperature-driven process, while the corner defects are due to local stress. Obviously, these two mechanisms are interdependent in an operating laser. In addition, high temperature-induced stress can also be an important factor.

In summary our TEM, EBIC and CL investigations, supported by theoretical calculations, indicate that regions near the lower corners of a ridge waveguide show an enhanced susceptibility to degradation. This is due to enhanced defect migration/reactions in a strain field, the latter originating from the ridge overlay. The stress-related effect of the ridge is similar to that of a stripe window.

REFERENCES

Harder C, Buchmann P and Meier H 1986 Electron. Lett. 22 1081
Jakubowicz A 1991 to be published in J. Appl. Phys.
Kirkby P, Selway P R and Westbrook L D 1979 J. Appl. Phys. 50 4567
Robertson M J, Wakefield B and Hutchinson P 1981 J. Appl. Phys. 52 4462
van der Ziel J P and Chand N 1990 J. Appl. Phys. 68 2731
Wakefield B 1979 J. Appl. Phys. 50 7914

Inst. Phys. Conf. Ser. No 117: Section 8
Paper presented at Microsc. Semicond. Mater. Conf., Oxford, 25–28 March 1991

589

Strain release in GaAs/Ga$_{1-x}$In$_x$As strained layer superlattices grown on (112) substrates

T E Mitchell and O Unal

Center for Materials Science, Los Alamos National Laboratory, Los Alamos, NM 87545

ABSTRACT: GaAs/Ga$_{1-x}$In$_x$As strained layer superlattices with well-widths of 7nm, barrier widths of 14nm and periods of 10 to 30 have been examined by TEM for (112) substrates. Individual layers are below the critical thickness while the overall SLS is above its critical thickness. Two sets of primary 79° dislocations are observed lying along <$\bar{1}\bar{3}$2> directions, while for larger periods two additional sets of secondary 60^o dislocations are observed lying along [1$\bar{1}$0]. This is discussed in terms of the resolved shear stresses resulting from coherency strains.

1. INTRODUCTION

Strained layer superlattices (SLS) are of interest for the tailoring of materials for electronic applications. For both the individual layers and the total SLS, there are critical thicknesses above which dislocations are generated to relieve the coherency strains. This can be treated in terms of a strain energy criterion or a stress criterion (see, for example, Matthews and Blakeslee 1974, Matthews 1975, Jesser and Van der Merwe 1989, Nix 1989, Hirth and Feng 1990, Jesser and Fox 1990). For (001) substrates, 60° dislocations with 1/2<110> Burgers vectors are observed at the interface. These are produced by glide of dislocations on all four {111} slip planes to give a square network (Kavanagh et al 1988, Unal et al 1990). For (111) substrates, a triangular network is produced by glide on the three other {111} planes (Unal 1991). In the present study, the dislocation arrangement on the less symmetrical (112) substrate has been examined. The observations are discussed in terms of the resolved shear stress acting on the various slip systems.

2. EXPERIMENTAL

Superlattices were grown on (112) GaAs substrates at 540°C by molecular beam epitaxy. Each SLS period consisted of 7nm of the GaAs/Ga$_{1-x}$In$_x$As well (x = 0.15) and 14nm of the GaAs barrier. The misfit is 1.05% and the corresponding critical layer thickness is ~30nm (see, for example, Fig. 4) so that the layers should be coherent. The SLS was the intrinsic component of a p-i-n junction where the p layer is a 250nm Be-doped GaAs top cap and the n layer is a 50nm Te-doped GaAs buffer layer. Both 10 period and 30 period SLS's were examined by TEM. The misfit between SLS and substrate is 0.35% and the corresponding critical thickness is ~120nm (6 periods) so that dislocations should be observed in both specimens. Cross-sectional and plan view thin foils were prepared by standard techniques and examined in a Philips CM30 electron microscope operating at 300 kV.

3.RESULTS

Plan view images of the 10 period and 30 period specimens are shown in Figs. 1(a) and (b) respectively. The dislocation density is seen to be higher in the 30 period specimen. Stacking fault tetrahedra are also observed, presumably formed during growth. The dislocations lie mostly along the intersection of the (1$\bar{1}$1) and ($\bar{1}$11) planes with the (112) substrate; g.b analysis showed that these have Burgers vectors of b = 1/2[011] and 1/2[101] respectively. Short reaction segments between these two sets of dislocations are frequently observed. The reaction produces dislocations with a Burgers vector of 1/2[$\bar{1}$10]; they appear to lie in the substrate along the [11$\bar{1}$] direction rather than along [110], which would correspond to a Lomer-Cottrell dislocation. A lower density of secondary dislocations lying along [110] is also seen in Fig. 1; g.b analysis showed that these have b = 1/2[0$\bar{1}$1] and 1/2[$\bar{1}$01].

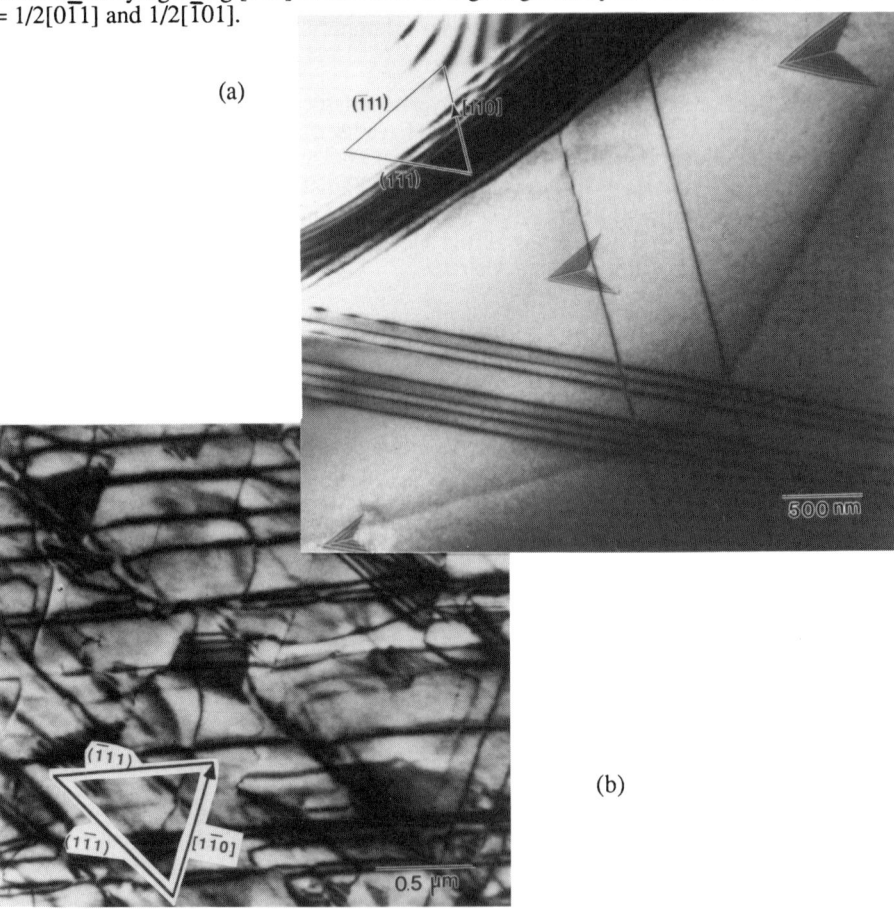

(a)

(b)

Fig. 1. Bright-field TEM images of (a) 10 period and (b) 30 period SLS in plan view, g = $\bar{2}$20. The primary dislocations lie along the (1$\bar{1}$1) and ($\bar{1}$11) traces and the secondary dislocations lie along [1$\bar{1}$0]. Stacking fault tetrahedra are also seen, as well as the intersection of the SLS with the foil at the top left corner in (a).

Corresponding cross-sectional images are shown in Fig. 2. It is seen that the dislocations are concentrated at the buffer/SLS and cap/SLS interfaces, as expected. For the 30 period sample, threading dislocations are observed and some penetration of dislocations into the substrate, although these may have been pre-exisiting. The above observations are all for (112) substrates with an As surface. Defects in SLS s grown on (112) substrates with a Ga surface have also been studied (Unal 1991). These will not be shown but, to summarize, the SLS itself is non-uniform, the dislocations are less regular and there is a high density of microtwins.

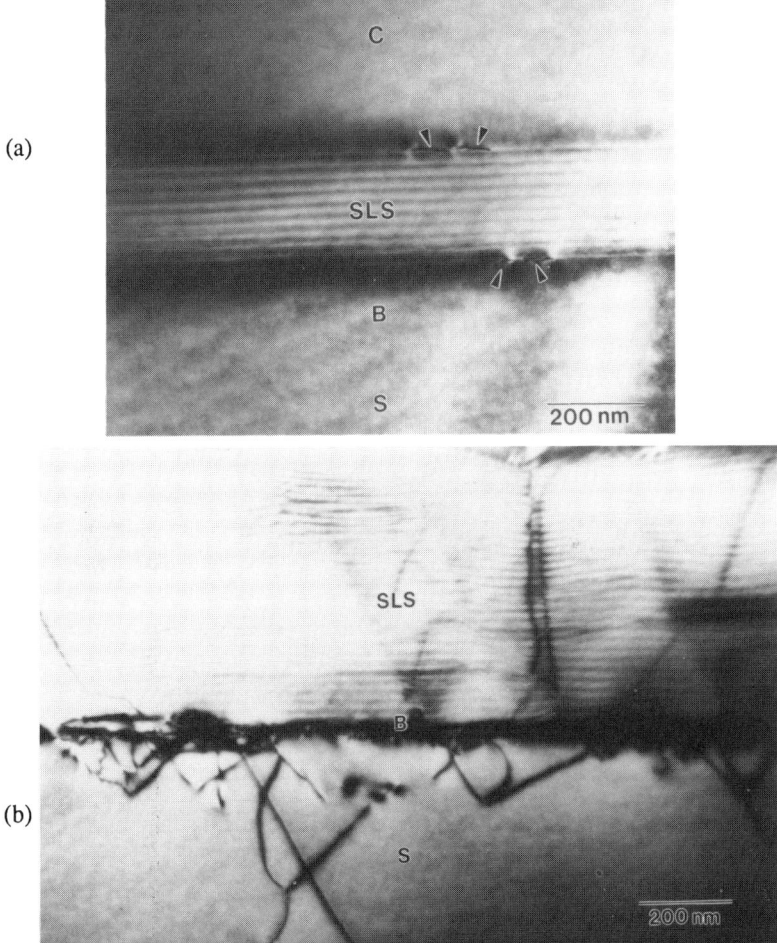

Fig. 2. Bright-field TEM images of (a) 10 period and (b) 30 period SLS in cross-section. C = cap, B = Buffer, S = substrate. A few dislocations are seen at the C/SLS and B/SLS substrates in (a) while many more are evident in (b), including threading dislocations and some penetration of dislocations into the substrate.

4. DISCUSSION

Due to coherency strains the SLS is under a biaxial compressive stress, -σ. This can be replaced by a uniform pressure, -σ, and a tensile stress, σ, along [112]. Dislocation motion can be considered to be in response to the latter stress. Accordingly a [112] sterographic projection is shown in Fig. 3. The primary slip systems are ($\bar{1}$11)[011] and ($1\bar{1}$1)[101]. These have a Schmid factor of 0.41 and correspond to the predominant dislocations in Fig. 1. Note that these are 79° dislocations lying along <$\bar{1}$32> rather than the traditional 60° dislocations; clearly strain relief is more important than Peierls energy considerations in this case. The secondary slip systems with a Schmid factor of 0.27 are ($1\bar{1}$1)[110], ($\bar{1}$11)[110], (111)[0$\bar{1}$1] and (111)[$\bar{1}$0$\bar{1}$]. The latter two are observed in Fig. 1 because they are needed to relieve strain along [1$\bar{1}$0] while the primary dislocations only relieve strain along [111].

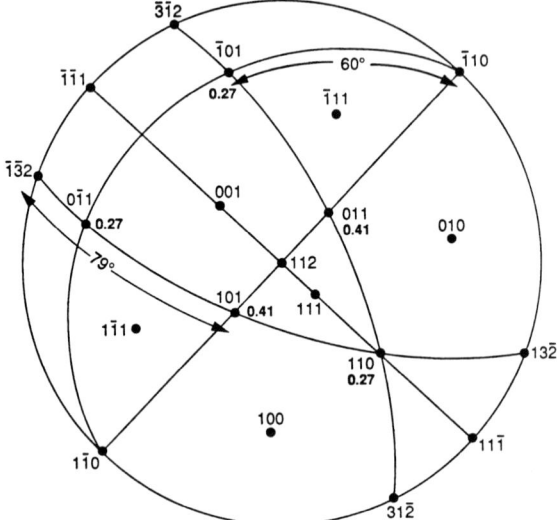

Fig. 3. [112] sterographic projection showing the 79° angle between the 1/2<101> Burgers vectors and <$\bar{1}$32> directions of the primary dislocations and the 60° angle between the 1/2 <$\bar{1}$01> Burgers vectors and [$\bar{1}$10] direction of the secondary dislocations. Schmid factors are indicated next to the corresponding slip directions.

The resolved shear stress criterion can be used to analyze misfit strain relaxation. If we assume that there is no dislocation source problem, the resolved shear stress to bow out a dislocation loop is (Hirth and Lothe 1982):

$$\tau = \frac{\mu b}{4\pi h(1-v)}[\ln(\frac{2h}{\rho})-1+\frac{1}{2}v]+\tau_f = \sigma\cos\phi\cos\theta$$

$$= 2\mu\left(\frac{1+v}{1-v}\right)\varepsilon\cos\phi\cos\theta = 2\mu\left(\frac{1+v}{1-v}\right)\left(\varepsilon_0 - \frac{b}{\lambda}\cot\phi\cos\theta\right)\cos\phi\cos\theta \tag{1}$$

where μ is the shear modulus, v is Poisson's ratio, h is the thickness, ρ is the core radius (b/2), τ_f is a friction stress, $\cos\phi\cos\theta$ is the Schmid factor, ε is the residual strain, ε_0 is the misfit strain and λ is the dislocation spacing. It is assumed that the critical radius of curvature is h (allowing for image forces). The term containing λ is the strain release due to dislocation motion. From equation (1), assuming $\tau_f = 0$, we can derive the critical thickness,

$$h_c/b = [\ln(h_c/b) + 0.5]/k_1\varepsilon_0 \ , \tag{2}$$

the dislocation density,

$$(\lambda/b)^{-1} = [\varepsilon_0 - (\ln(h/b) + 0.5)/k_1(h/b)]/k_2 \ , \tag{3}$$

and the residual strain

$$\varepsilon = [\ln(h/b) + 0.5]/k_1(h/b) \tag{4}$$

where k_1 and k_2 are constants which depend on orientation and slip system. For (112) substrates, the primary system has $k_1 = 12.62$, $k_2 = 0.463$ while the secondary system has $k_1 = 8.41$, $k_2 = 0.817$.

Critical thickness is plotted against ε_0 in Fig. 4. It is seen that the critical thickness is more than 60% higher for the secondary system than the primary system. The curve for the primary system is identical for (001) substrates and very close to that given by the

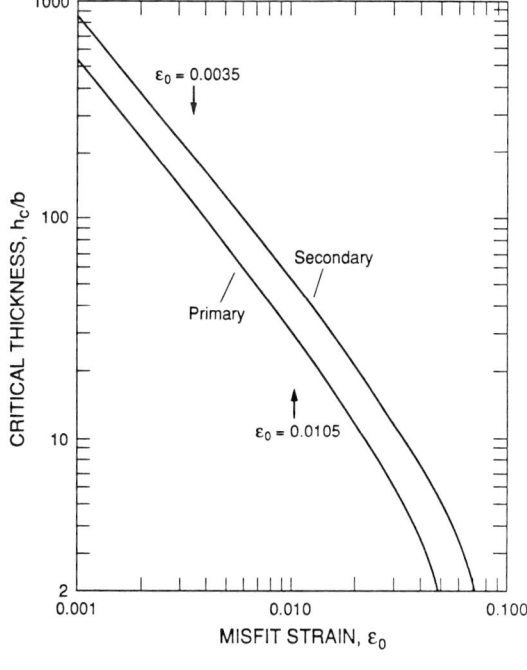

Fig. 4. Log plot of critical thickness versus misfit strain calculated from Equation (2) for glide of primary and secondary dislocations. Values of misfit for the overall SLS (0.0035) and the individual layers (0.0105) are arrowed.

Matthews-Blakeslee equation (see also Hirth and Feng 1990). Dislocation density (Equation 3) is plotted against thickness in Fig. 5. The dislocation density is predicted to increase rapidly at first and then more slowly as thickness is increased beyond the respective critical values for the primary and secondary systems. This is generally

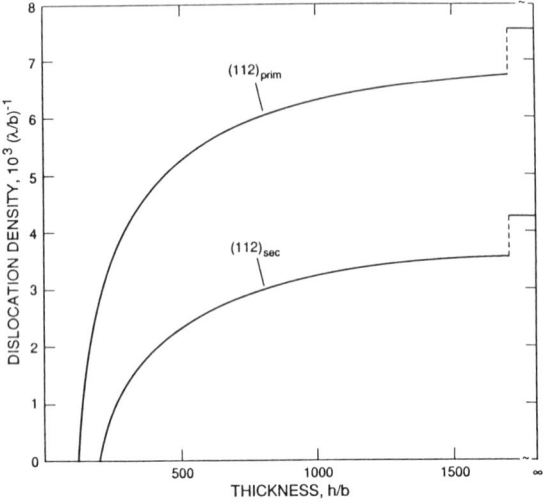

Fig. 5. Plot of dislocation density (reciprocal spacing) versus thickness calculated from Equation (3) for glide on the primary and secondary systems. The 10 and 30 periods correspond to h/b = 525 and 1575 repectively.

observed and it is certainly true that the primary dislocation density is higher than the secondary. However, there is poor numerical agreement. For example, the primary dislocation spacing in Fig. 1(b) is ~160nm whereas the expected spacing from Fig. 5 is ~60nm. Apparently strain release is even slower than shown in Fig. 5 (or by Equation 4 - not plotted here). A likely reason for this is the frictional stress which was taken to be zero in evaluating Equation (1) but can easily be given a finite value, as discussed by Jesser and Fox (1990). We find that frictional stresses $~10^{-3}\mu$ are needed to explain the slow strain relief (Mitchell and Unal 1991). Such stresses could arise from a combination of the Peierls stress and a solution hardening stress due to the In substitution.

This research was supported by the U.S. Department of Energy-Office of Basic Energy Sciences. B. K. Laurich and D. L. Smith kindly supplied the materials and participated in useful discussions.

REFERENCES

Hirth J P and Lothe J 1982 Theory of Dislocations (New York: Wiley) p 752
Hirth J P and Feng X 1990 J. Appl. Phys. 67 3343
Jesser W A and Van der Merwe J H 1989 Dislocations in Solids, ed F R N Nabarro
   (Elsevier) pp 422-460
Jesser W A and Fox B A 1990 J. Electron. Mater. 19 1289
Kavanagh K L et al. 1988 J. Appl. Phys 64 4843
Matthews J W and Blakeslee A E 1974 J. Cryst. Growth 27 118
Matthews J W 1975 J. Vac. Sci. Technol. 12 126
Mitchell T E and Unal O 1991 J. Electron. Mater. to be published
Nix W D 1989 Met. Trans. 20A 2217
Unal O, Laurich B K and Mitchell T E 1990 Mater. Res. Soc. Symp. Proc. Vol. 183, eds
   R. Sinclair, D. J. Smith and U. Dahmen (Pittsburgh:MRS) pp 193-196
Unal O 1991 Ph.D. Thesis, Case Western Reserve University

Inst. Phys. Conf. Ser. No 117: Section 8
Paper presented at Microsc. Semicond. Mater. Conf., Oxford, 25–28 March 1991

Contrast effects in strained layer InGaAs/GaAs superlattices

Philip Kightley, Chris J Kiely and Peter J Goodhew

Department of Materials Science and Engineering, University of Liverpool, Liverpool, England, L69 3BX.

ABSTRACT: The contrast behaviour of $In_xGa_{1-x}As$/GaAs strained layers for imaging under systematic row conditions where the composition sensitive $g=200$ reflection is precisely at the Bragg condition has been studied. The contrast behaviour of the layers as a function of x is kinematically modelled. Because of the quadratic expression in x for $In_xGa_{1-x}As$ contrast reversal occurs at some x. The point of reversal is shown to be sensitive to the atomic scattering factors used in the calculation. The relative contributions of other factors to the layer contrast such as 1) multiple scattering, 2) inelastic scattering, 3) thin film surface relaxation effects, 4) systematic row effects, 5) thickness/absorption effects and 6) tetragonal distortion are considered.

1 INTRODUCTION

Dark field 002 imaging is used extensively to obtain high contrast cross-sectional micrographs of $In_xGa_{1-x}As$ alloys. The 002 diffraction intensity is described by a quadratic function of x and thus at some point contrast reversal with the GaAs matrix should occur as the relative intensity of the InGaAs rises to be greater than that of the GaAs. Here we consider two points. We first present a simple kinematical calculation of the intensity profile for which the effects of different electron scattering factors can be deduced. We then consider the relative contributions of other possible factors to the InGaAs/GaAs layer contrast.

2 CALCULATIONS and DISCUSSION

2.1 Kinematical $In_xGa_{1-x}As$ Intensity Profile

For the 002 reflection in GaAs the sine terms of the imaginary part of the structure factor cancel to zero and the As beams are π out of phase with the Ga beams. The structure factor becomes $F_{002}=4(f_{Ga}-f_{As})$. The structure factor for $In_xGa_{1-x}As$ can similarly be derived so that $F_{002}=4(f_{Ga}-f_{As}+(f_{In}-f_{Ga})x)$. The kinematical intensity of the $In_xGa_{1-x}As$ can thus be shown to be;

$$I_{002}=\frac{h^2\lambda^2\left(4\left(f_{Ga}-f_{As}+(f_{In}-f_{Ga})x\right)\right)^2 m^2}{\left(a_{GaAs}+x(a_{InAs}-a_{GaAs})\right)^6 m_0^2} \qquad (1)$$

Where h is the thickness of the TEM foil, λ is the wavelength of the electron and m/m_0 is the relativistic mass correction factor. Three sets of electron scattering factors have been used in our treatment. Baxter et al (1987) have previously calculated the kinematical $In_xGa_{1-x}As$ intensity profile using atomic scattering factors derived from the Thomas-Fermi-Dirac (TFD) statistical model and they identified an apparent anomaly in the experimentally determined and theoretically predicted contrast cross-over point. We have performed a similar calculation (shown in figure 1) for $\sin\theta/\lambda=0.15$, of the intensity of the 002 reflection against the concentration of x using TFD data. If the Ga scattering factor is altered by 4%, from 5.2 to 5.0Å, whilst the remaining factors are held constant then the contrast cross-over shifts from x=0.25 to x=0.425. The point of contrast reversal in the kinematic formulation is not altered significantly by changing accelerating voltage or foil thickness. The TFD data are quoted at best to only ±10% thus it is not safe to predict the exact intensity profile for $In_xGa_{1-x}As$ from this data. Using the computations of Coulthard (1967) for a relativistic Hartree-Fock wave function calculation, Doyle and Turner (1967) extracted the kinematical electron scattering factors for many elements. A general equation is given where a_i, b_i and c are parameters determined by curve fitting procedures. The term $v=\sin\theta/\lambda$.

$$f_s = \sum_{i=1\rightarrow4}^{N} a_i e^{(-b_i v^2)} + c$$

(2)

More recently Sheng and Fang-Hua (1984) have recalculated these values. We have calculated the contrast profile for both the Doyle and Turner (DT) and the Sheng and Fang-Hua (SF) data sets. Note that no values are given for c by Doyle and Turner so the calculations with these factors contain only the 8 parameters of a and b. $\sin\theta/\lambda$ depends on the lattice parameter of the $In_xGa_{1-x}As$ alloy and hence is dependant on x. It varies between 0.1768 and 0.1651 as the alloy fraction of InAs increases.   The term v in equation (2) is expressed as a function of alloy fraction so that $v = \sin\theta/\lambda = [0.1768-0.0117x]$. Figure 2 shows the curves calculated using SF, DT and TFD data for an accelerating voltage of 200kV. Clearly the calculations involving the Hartree-Fock scattering factors have correctly shifted the intensity curve toward a higher value of x for the contrast cross-over point. The cross-over has shifted from x~0.25 for the TFD data to x~0.47 for either the DT or SF data. It is interesting to note that the SF and DT data give almost identical curves.

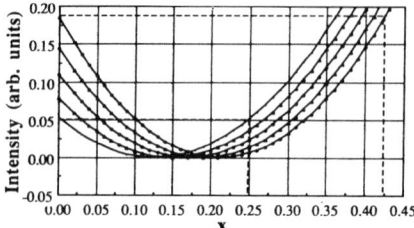

Figure 1   002 intensity profile for TFD data varying f_{Ga} from 5.2 to 5.0.

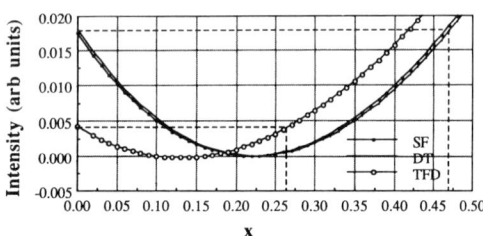

Figure 2   002 intensity profiles for TFD, SF and DT data.

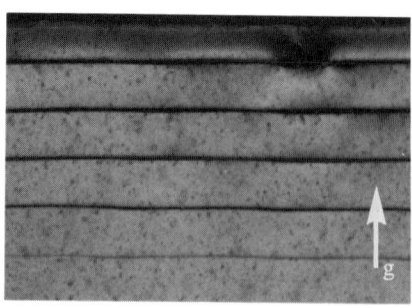

Figure 3   [110] cross section of InGaAs layers of x=0.10, 0.19, 0.28, 0.37 and 0.45

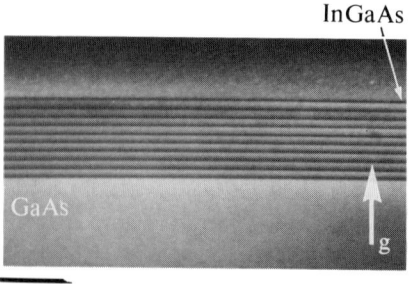

0·1μm

Figure 4   [110] cross-section of a 10 period 30Å InAs/100Å GaAs multilayer.

2.2 TEM observations

In order to locate the exact contrast cross-over, multilayer GaAs/In$_x$Ga$_{1-x}$As samples were grown with x increasing up the multilayer stack. Figure 3 is a [110] cross-section TEM sample where there are single In$_x$Ga$_{1-x}$As layers of x=0.10, 0.19, 0.28, 0.37 and 0.45. The layers in the dark field 002 image all appear dark with respect to the GaAs. Only the SF and DT curves can both qualitatively predict the contrast for these layers. Baxter et al showed that even for layers of x~0.6 the correct 'kinematical' contrast is still not obeyed. Figure 4 shows a 200 DF micrograph from one of a series of InAs/GaAs multilayers examined. In all cases the InAs layers display the 'wrong' contrast, ie the 200 reflection is less intense for InAs then GaAs. The contrast profile as a function of x is thus more complex than may be described by simple kinematical theory.

2.3 Other Factors

We have shown that errors can arise in calculating relative layer intensities because of inaccuracies in the values used for the atomic scattering amplitudes of electrons. In this section we discuss other possible contributing factors to layer intensities in 002 DF images.

a) *Multiple scattering*. Multiple scattering through strong reflections can in some circumstances make a significant contribution to the 002 intensity. For instance at, or near, <110> zone axes strong 111 reflections may become multiply scattered (eg -111 + 1-11 → 002) and thus make the intensity for both the GaAs and InGaAs stronger than expected. This effect can be eliminated by careful choice of two beam (or more realistically systematic row) reflection conditions in which 111 beams are not excited. We did this for figure 4.

b) *Inelastic scattering*. Because the 002 reflection is weak then the contribution of inelastically scattered electrons, as pointed out by Baxter et al (1987), may be significant. When the 002 reflection is at the exact Bragg condition (deviation prameter s=0) then the weak 002 excess Kikuchi lines will also be included in the objective aperture and add to the intensity of both the InGaAs and GaAs layers. We have noted in agreement with Baxter et al (1987) that imaging from the 002 Kikuchi line gives similar contrast to the 002 elastic reflection and imaging of the background intensity of inelastically scattered electrons gives little detectable difference in intensity.

c) *Thin film surface relaxation effects*. Gibson et al (1985) showed that surface relaxation effects can be important in determining the relative contrasts of SiGe/Si superlattices. They show that if the superlattice period, Λ, is comparable to the thickness of the TEM specimen, h, (ie Λ/h→1) then a significant proportion of the superlattice strain will be relieved at the foil surfaces leading to anomalous layer contrast. Since the same argument applies to InGaAs/GaAs superlattices then 004 SADPs should be acquired. The superlattice spots shown in figure 5 are formed at the 004 reflection of the InAs/GaAs sample shown in figure 4. The sample is sufficiently thick that the periodicity of the superlattice reflections chosen corresponds to the periodicities contained in the superlattice. This is a good indicator that the surface relaxation effects are having a minimal effect upon contrast.

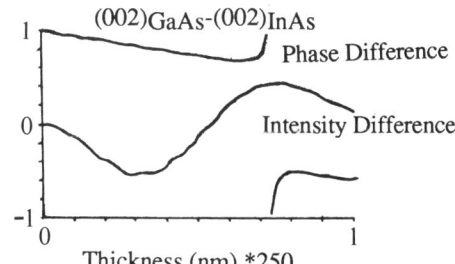

Figure 5   004 Superlattice spots from the sample shown in figure 4.

Figure 6   Intensity difference between InAs and GaAs as a function of TEM sample thickness.

d) *Excitation of systematic row and dynamical diffraction effects.*   In reality a perfect two beam reflection is never achieved. The experimentally obtained pattern is best described as a systematic row condition with **g**=002 at the exact Bragg condition. Some preliminary systematic row Bloch wave calculations using the Stadelmann EMS package for a pure InAs/GaAs imaged under 002 coditions have been done. In the calculation the appropriate Debye-Waller and absorption parameters for In, Ga and As have been included and the intensity as a function of TEM sample thickness has been modelled. Figure 6 is the difference between the InAs and GaAs intensity profiles as a function of thickness. The first dynamical contrast reversal is expected to occur at a sample thickness of 1250Å. We are currently extending this analysis to the $In_xGa_{1-x}As/GaAs$ system for various values of x.

e) *Tetragonal distortion.*   In the superlattices studied here the In(Ga)As layers are tetragonally distorted in order to match the the GaAs substrate. Formally we can say that the InGaAs is no longer cubic (with space group F-43m) but rather needs to be represented by a body-centered tetragonal cell with space group I-4m2. For a pure InAs layer the tetragonal distortion causes an expansion in the 'c' axis from 6.0583Å to ~6.95Å; increasing the effective lattice mismatch from ~7% to ~23%. This significantly increases the separation of the 002 reflections in reciprocal space and means that when the GaAs is at the exact Bragg condition the InAs reflection is displaced from the Ewald sphere; for example at 100keV $s=0.002Å^{-1}$. It should be noted that the change to a tetragonal unit cell does not greatly alter F_{002} for InGaAs, as can be demonstrated by analysing the diffraction geometry from a planes of constant phase viewpoint.

3 CONCLUSIONS

The intensity profile of $In_xGa_{1-x}As/GaAs$ multilayers (x=0 to 1) cannot be accurately described by the kinematically derived structure factor contrast using current values of electron scattering factors. If the electron scattering factors are accurate then other factors must be influencing the overall contrast. In this treatment we have shown that multiple scattering events, inelastic scattering and thin film surface relaxation effects have a negligible effect on the contrast. It is hoped that dynamical calculations of the excitation of systematic rows and of the effects of tetragonal distortion will enable the real profile to be determined.

REFERENCES

Baxter C S, Stobbs W M, Monserrat K J and Tothill J N (1988) EMAG 209
Coulthard M A (1967) Proc.Phys.Soc. **91** 44
Doyle P A and Turner P S (1967) Acta. Cryst. **A24** 390
Gibson J M, Treacy M M J, Bean J C and Hull R. (1985) Inst.Phys.Conf.Ser.No.**76** 277
Sheng J J and Fang-Hua L (1984) Chin.J.Phys.(USA) **33** 845

ADDENUM   DF 200 Cross-sectional imaging for AlGaAs.

An obvious extension of the DF imaging work for the ternary InGaAs is a similar treatment for the ternary AlGaAs. Calculating the structure factors in a similar manner using both Doyle-Turner and Sheng-Fang-Hu data and then evaluating intensity gives the contrast profile opposite. Because the Al structure factor is considerably smaller than the Ga the cross-over is moved to a low aluminum concentration (x→0) and becomes effectively invisible. The structure factor contrast thus increases approximately linearly with Al concentration and does not obviously display the contrast cross-over behaviour expoected for InGaAs. This is qualitatively correct.

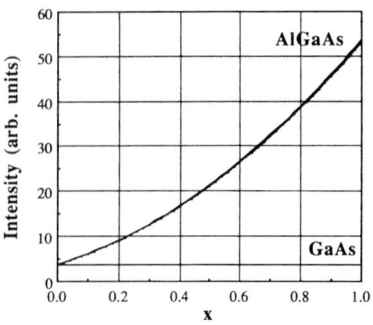

Inst. Phys. Conf. Ser. No 117: Section 8
Paper presented at Microsc. Semicond. Mater. Conf., Oxford, 25–28 March 1991

Strain analysis of GaInAs/GaAs structures using TEM image simulation of 90°-wedges

A J Harvey, D A Faux, U Bangert and P Charsley

Physics Department, University of Surrey, Guildford

ABSTRACT: A new characterisation technique for the analysis of strain is presented. The method matches TEM images of thickness fringes from 90°-cleaved wedge samples with simulated TEM images obtained from theoretical calculations of the strain relaxation at the edge of the wedge. The technique is applied here to a GaInAs/GaAs strained single quantum well structure. The remarkable sensitivity of the thickness fringe contrast to the misfit strain, f, and the deviation parameter, s, is demonstrated.

1. INTRODUCTION

The use of 90° wedges cleaved on {110} planes for TEM has two significant advantages over the use of conventionally thinned specimens: the samples are simple to prepare and the specimen geometry is known precisely. TEM observations from wedge samples with layers of some degree of mismatch, which intersect the wedge edge in a perpendicular direction, reveal a shift in the thickness fringes due to strain relaxation at this edge (Bangert and Charsley, 1989). As the geometry is known it is possible to calculate the strain relaxation. Simulated TEM images can be obtained by implementing the calculated strain distributions in the equations of the dynamical theory of diffraction contrast. It will be demonstrated here that with this new technique it is possible to obtain very close agreement between simulated and experimental images and that high sensitivity for variations in the strain determination can be achieved, whilst retaining high spatial resolution. Since the misfit strain in the bulk must be assumed in order to calculate the strain distribution it is possible from these matches to deduce the bulk strain value.

2. EXPERIMENTAL

The samples were $Ga_{1-x}In_xAs$ layers with x of the order of 0.2, sandwiched between a GaAs substrate of {100} orientation and a GaAs capping layer. 90° wedges, cleaved on {110} planes were viewed at an angle of 45° in a 2000FX JEOL electron microscope (see fig.1). Theoretical strain distributions for wedge specimens containing a single mismatched layer were obtained using finite-element calculations. These strain values were used to derive values for the displacement along columns parallel to the electron beam so that the diffraction contrast images could be computed with the two-beam dynamical equations. The program requires the input of several parameters: the anomalous absorption, the extinction distance, the g-vector, the deviation parameter s and the misfit strain f. Knowing the first three parameters, only the latter two parameters needed to be varied in order to obtain the best match between simulated and experimental TEM images. The comparison between experimental and simulated images was done visually.

3. RESULTS AND DISCUSSION

Fig.2 shows a series of simulated images for a 040 reflection according to the geometry of fig.1. Within the columns the misfit strain value changes while s remains constant; whithin the rows the image development with changes in s at constant f is shown. It can be seen that the contrast is clearest for s-values approximating to zero and that for relatively small changes in both parameters distinctive changes in the images take place. It is hence possible to obtain a match within small error limits. The last image in fig.2 is an experimental TEM image of a GaInAs layer with 18% In using the 040 reflection and s close to zero. The best match is the simulated image in fig.2b with $s=4x10^{-4}\text{Å}^{-1}$ and f=1.26% which corresponds exactly to the strain expected from the above In-content.

Fig.3 shows a series of simulated images for the 400 reflection. Again the misfit strain value changes along the vertical side and the s-value along the horizontal side of the figure. Again strong dependence of the contrast on both parameters can be observed. The overall contrast features are very different in the 400 reflection. Contrast changes close to the interface are pronounced, whereas in the 040 reflection contrast changes further away from the interfaces as well as within the strained layer dominate the images. This is due to the different relaxation behaviour of the two sets of planes, arising from different boundary restrictions in the 100 and the 010 directions. The last image in fig.3 is an experimental image of a single quantum well containing 20% In. The best match is the simulation in fig.3f with $s=4x10^{-4}\text{Å}^{-1}$ and f=1.4%. The deduced misfit strain is in excellent agreement with the strain value expected from the above In-content.

As shown in fig.4 the 400 reflection is extremely sensitive to changes in f. In fact changes as small as 0.01% can still be detected in the simulated images.

4. CONCLUSION

Simulated TEM images of 90° wedges containing mismatched layers are found to be extremely sensitive to changes in the misfit strain. By varying the misfit strain value and the deviation parameter very close matches with experimental images can be obtained. It is with this method therefore possible to deduce bulk misfit strain values at individual interfaces with high sensitivity, which are in very good agreement with the expected values.

REFERENCE
Bangert U and Charsley P 1989 Phil. Mag. A 59 629

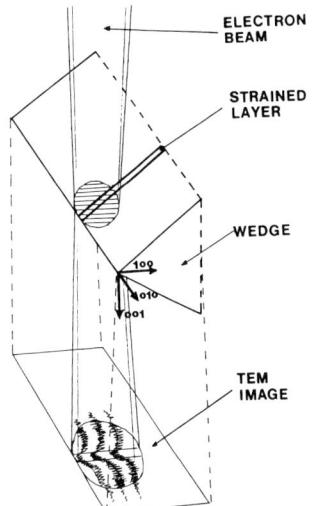

Fig.1 Wedge geometry and
crystal orientation
in the TEM

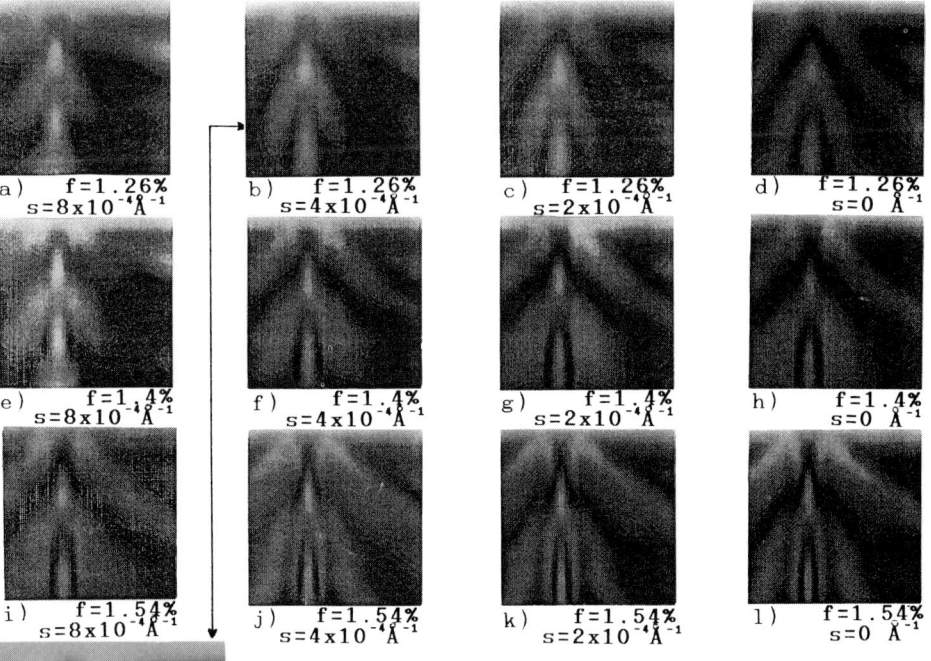

a) $f=1.26\%$
    $s=8\times10^{-4}\text{Å}^{-1}$

b) $f=1.26\%$
    $s=4\times10^{-4}\text{Å}^{-1}$

c) $f=1.26\%$
    $s=2\times10^{-4}\text{Å}^{-1}$

d) $f=1.26\%$
    $s=0$ Å^{-1}

e) $f=1.4\%$
    $s=8\times10^{-4}\text{Å}^{-1}$

f) $f=1.4\%$
    $s=4\times10^{-4}\text{Å}^{-1}$

g) $f=1.4\%$
    $s=2\times10^{-4}\text{Å}^{-1}$

h) $f=1.4\%$
    $s=0$ Å^{-1}

i) $f=1.54\%$
    $s=8\times10^{-4}\text{Å}^{-1}$

j) $f=1.54\%$
    $s=4\times10^{-4}\text{Å}^{-1}$

k) $f=1.54\%$
    $s=2\times10^{-4}\text{Å}^{-1}$

l) $f=1.54\%$
    $s=0$ Å^{-1}

m)

Fig.2 Simulated images for $g=040$ with
variations in s and f. Fig.2m shows the
experimental image

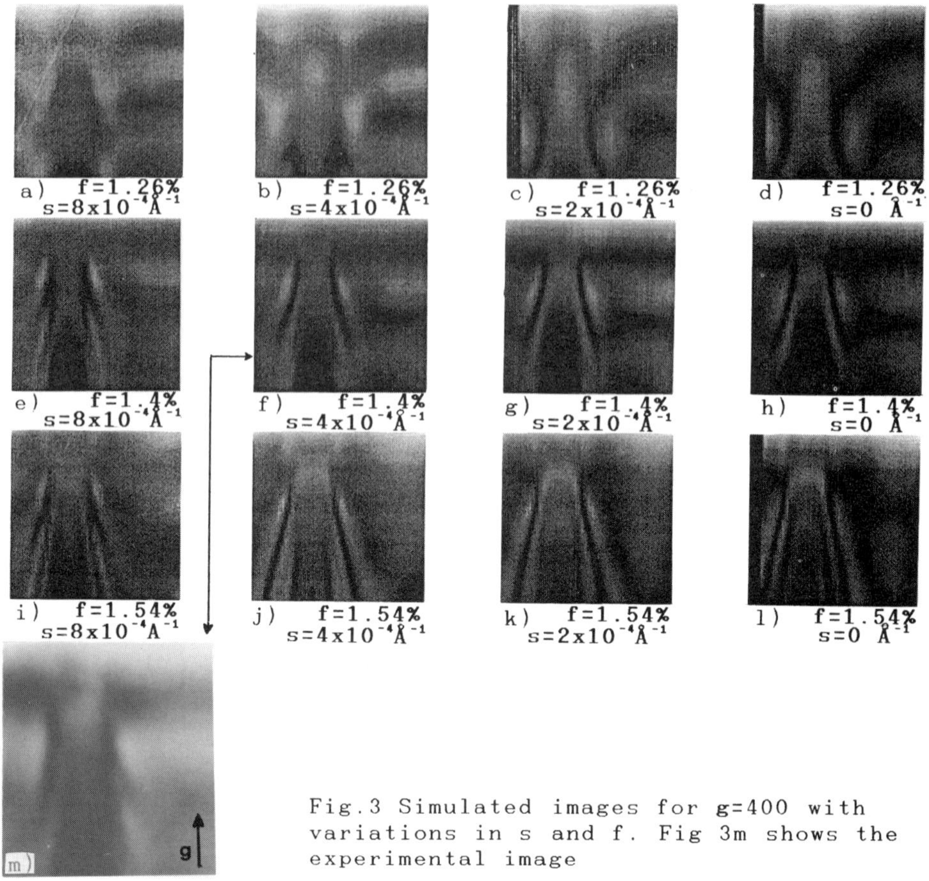

a) f=1.26%
s=8x10⁻⁴Å⁻¹

b) f=1.26%
s=4x10⁻⁴Å⁻¹

c) f=1.26%
s=2x10⁻⁴Å⁻¹

d) f=1.26%
s=0 Å⁻¹

e) f=1.4%
s=8x10⁻⁴Å⁻¹

f) f=1.4%
s=4x10⁻⁴Å⁻¹

g) f=1.4%
s=2x10⁻⁴Å⁻¹

h) f=1.4%
s=0 Å⁻¹

i) f=1.54%
s=8x10⁻⁴Å⁻¹

j) f=1.54%
s=4x10⁻⁴Å⁻¹

k) f=1.54%
s=2x10⁻⁴Å⁻¹

l) f=1.54%
s=0 Å⁻¹

m)   g

Fig.3 Simulated images for g=400 with variations in s and f. Fig 3m shows the experimental image

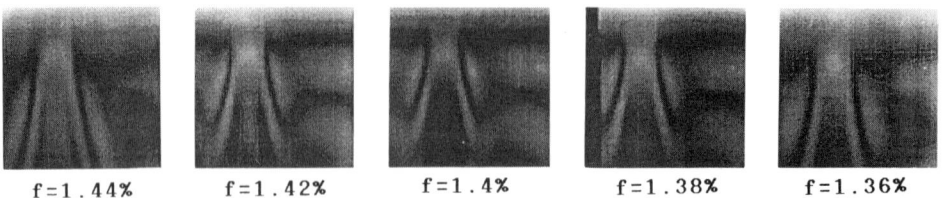

f=1.44%   f=1.42%   f=1.4%   f=1.38%   f=1.36%

Fig.4 Simulated images for g=400 with small variations in f and s=0 Å⁻¹.

Dislocation dipoles in strained InGaAs layers grown on GaAs

P Kightley, G Aragon-Herranz*, P J Goodhew and R C Pond

Department of Materials Science and Engineering, University of Liverpool, Liverpool, England, L69 3BX.
*Departamento de Quimica Inorganica, Universidad de Cadiz, Apdo 40, Cadiz, Spain.

ABSTRACT: Dipole formation at the interfaces of InGaAs/GaAs multilayers has been studied. Slip dipoles with sides along <110> are demonstrated and their use to reduce residual threading dislocation density for strained layer materials is shown. A detailed diffraction contrast analysis has been performed upon dipoles that have <010> line directions. These dipoles also relieve misfit and probably arise from a climb process.

1 INTRODUCTION.

Dislocation dipoles contain parallel segments that have opposite line directions. For InGaAs epitaxially grown on (001) GaAs, misfit dislocations (MD) first form from the threading dislocations that exist in the substrate (Green et al (1990)). For a single (001) InGaAs layer, MD segments form with **u** parallel to <110> as these are the lines of intersection of the glide planes with the interface plane and enable the dislocations to relieve misfit. If the single layer is capped with GaAs and MD are formed in the substrate/strained layer interface, the capping layer is also strained. When the initial (a/2)<110> misfit dislocations are formed in the substrate/strained layer interface, the two possible <110> line directions along which misfit segments may be formed can be predicted by calculation of the dislocation glide plane **bxu**. Similarly when the second misfit segments are formed in the strained layer/capping layer interface the two opposite line directions will be chosen. In at least half of these cases disocation dipoles will be formed; the remaining events will appear as right angled segments in an (001) image. In this paper two dipole phenomena are described. The first is the dipole generated by slip movements at InGaAs strained layers. The parallel <110> sides of the dipoles are formed at different times during the growth of the multilayer. The second, and more complex defect, is the generation of dipoles which have **u**=<010> which probably occur through climb. The dipoles reported by Matthews and Blakeslee (1974) for GaAsP/GaAs were generated in a single slip movement. Each segment of dislocation in the alternating interfaces of a multilayer sample had an opposite displacement, so by gliding together to form long MD segments bounding the strained layer the extra half plane associated with the 60° MD was added only to the region between the dislocations ie the GaAsP strained layer. Similar observations were made by Hull and Bean (1989) and Houghton (1990) in multilayers of SiGe/Si that were thermally processed after growth. In essence the slip dipole shown here is similar to these structures.

2 OBSERVATIONS.

2.1 <110> Dipole Generation from Slip.

The greater the number of misfit dislocation arrays, generated by $In_{0.2}Ga_{0.8}As$ strained multilayers formed as dislocation filters in GaAs/Si, the lower the residual threading dislocation density (RTDD). Figures 1a and b show how dipoles are generated between layers of alternating stress increasing the numbers of MD arrays . In figure 1a a strained layer superlattice (SLS) dislocation filter (which acts similarily to a single strained layer) in a GaAs/Si layer is covered by a thin (0.15 μm) cap layer (see g200 inset). A misfit dislocation array is formed at the GaAs/SLS interface but no such array is formed at the SLS/cap interface. An identical sample is shown in figure 1b, except that the cap layer is now 0.24μm thick. A misfit dislocation array is present at each interface; the second array is formed because the critical layer thickness for MD formation has been exceeded by thickening the cap. Stereomicroscopy has confirmed the presence of dipoles bounding each interface of the SLS. Indeed the spacer layer thickness between the strained layers, which determines the dipole density and length, becomes important in determining the RTDD. This is demonstrated in figure 1c which is a graph of RTDD against the number of SLS and thickness of spacer layer.

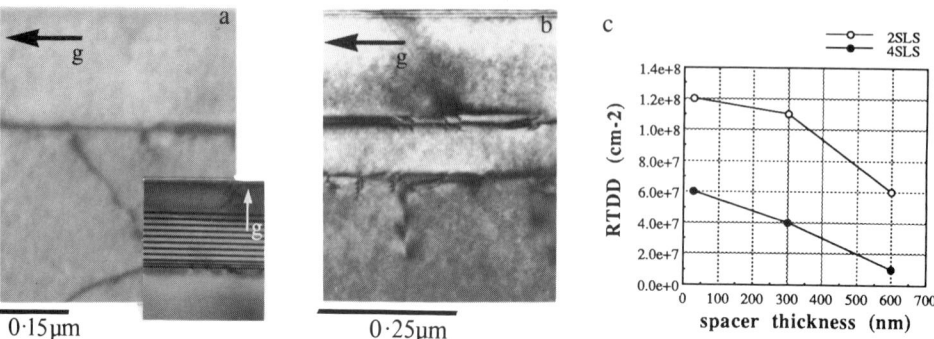

0·15μm   0·25μm

Figure 1   a) An InGaAs/GaAs ten period (100Å/100Å) SLS with 0.15μm cap layer. b) A similar sample with the cap layer thickened to 0.24μm. Note the extra MD array. c) The RTDD as a function of number of SLS and their separation.

2.2   <010> Dipole Generation from Climb.

A multilayer sample of ten periods of $In_{0.1}Ga_{0.9}As$/GaAs was analysed by (001) plan view TEM. The sample contained GaAs spacer layers fixed at 500Å thickness whereas the InGaAs layers ramped in increments of thickness starting at 50Å and finishing at 220Å. Stereomicroscopy showed the MD had formed at a variety of interfaces throughout the multilayer stack, providing some dipole structures similar to those shown above with segments of **u** parallel to ±<110> bounding the strained layer. In addition to these, more unusual dipoles with **u**=<010> were seen. Two examples of these are shown in figure 2 labelled D.

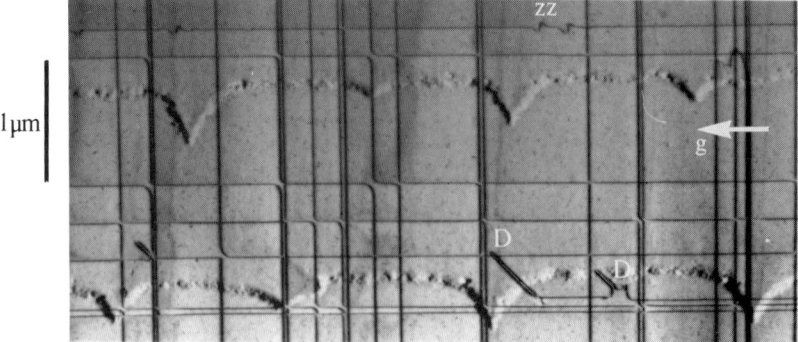

Figure 2   (001) plan view of the dislocation dipoles, labelled D, with **u** parallel to <010>.

Stereomicroscopy, figure 3a, shows that these dipoles lie with their <010> segments bounding a strained layer. The **u**=±<010> parallel segments are laterally separated by a short distance. Dipoles that are located at the same position in the multilayer stack have their segments separated by a fixed distance. This suggests that the parallel dipole segments are both on the same inclined plane. The geometry of the dipole is shown in figure 3b. The 60° MD segments of figure 3b are always in different interfaces and can either be parallel or at right angles in the (001) projection. Diffraction contrast experiments show that the dipoles have the same Burgers vector as the 60° misfit dislocation segments. During analysis the contrast of the defects cancels to zero for the 400 or 040 diffraction vectors because **g.bxu** and **g.b** are simultaneously zero.

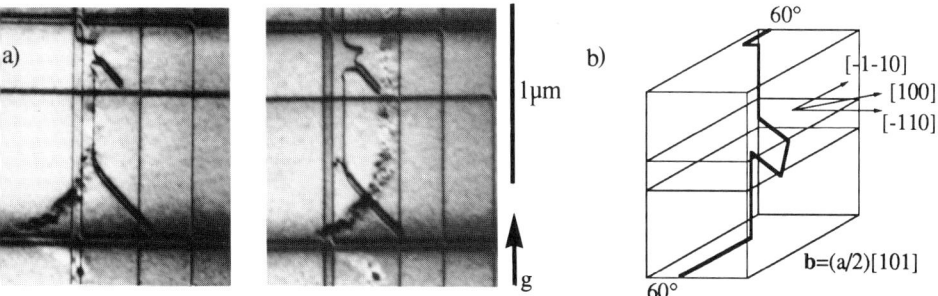

Figure 3   a) Stereomicrographs of the dislocation dipoles. b) Schematic of the geometry of the dipole extracted from the stereo micrographs

MD segments formed in this array are seen to be linked between different interfaces by threading segments. The link segments are steeply inclined to the (001) surface. It is from these segments that the dipoles form. The growing dipole has a choice of four **u**. For example, if a dislocation **b**=(a/2)[101] forms a dipole and the bottom segment has **u** parallel to [010] then **b^u** is 90° and the dipole segments are pure edge and relieve misfit. Note that for the bottom segment, if **u** is parallel to [0-10] then the sign of the misfit relief is reversed. If the dislocation forms with **u** parallel to [100] then **b^u** is 45° and the dislocation has its edge component normal to the interface, thus relieving no misfit. It is most likely that the dipole forms with edge segments of dislocation. Figure 4 represents the (a/2)[101] dipole discussed above. The parallel segments of the dipole could either bound a single InGaAs layer or a composite of several layers. The principal stresses are illustrated on the figure. For **b**=(a/2)[101] and **u** parallel to <010>, <0-10> or <10-1> the segments are pure edge and the climb (and glide) forces can be evaluated using the generalised form of the Peach-Koehler formula given by Hirth and Lothe (1968). For the ±<010> segments both glide and climb forces per unit length of dislocation are equivalent, $F_c/L=F_g/L=|\sigma_{xx}|/2\Diamond v\sqrt{2}$. Because the dipole plane is not an easy glide plane the climb force is more important. When these segments of dislocation reach the two strained layer interfaces the forces disappear, unlike those on the inclined segment between the two parallel dipole segments. For **u** parallel to [10-1] $F_g/L=0$ and F_c/L will always be equal to $|\sigma_{xx}|/2\sqrt{2}$. The parallel dipole segments lie on the same plane which, so far, remains unidentified. It is possible to specify that the joining segment has a general line direction of either **u** parallel to [n0-1] or [10-n] where $1\le n\le\infty$. It is interesting to note that in the limiting cases of n=∞ then Fc/L=0 in the former case and $F_c/L=|\sigma_{xx}|/2$ in the latter case: the steeper the inclined portion of the dipole, the bigger the climb force. F_g/L is zero for both cases.

2.2.1   DIPOLE FAMILIES.

Dipoles can be generated either when the two 60° misfit segments joined by the inclined section from which the dipole is formed are parallel or at right angles. The direction in which the dipole forms determines the misfit relief; thus for different Burgers vectors different line directions will be adopted. This enables four arrangements of dipole line direction for each arrangement of misfit dislocations. We also observe that the dipoles often form in pairs, as shown by the pair in figure 2.

When categorising these reactions it was observed that some dislocation lines contain a high density of 'zig-zag' defects. These defects are shown in figure 2 as zz. The geometry of these centres, derived in a similar manner to the dipole defect, is shown in figure 5.

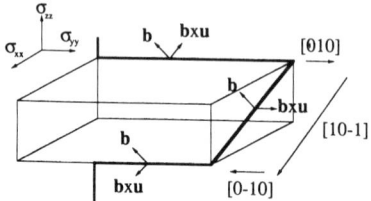

Figure 4   Schematic of the (a/2)<101> dislocation dipole

Figure 5   a)The geometry of the 'zig-zag'defect shown in figure 2 as zz. b) Interaction between orthogonal MD to form a small edge segment.

3 DISCUSSION.

A possible explaination of the 'zig-zag' defect could be through the loss of intersecting misfit dislocations. If a MD crosses an existing MD to form a small segment of edge dislocation then at this intersection small segments of the MDs move up and down on their glide planes and the small edge segment is inclined between them. This is illustrated in Figure 5b). If one of the orthogonal MD then glides out of the foil the edge segment is reformed as a segment of an original MD but the dislocation now contains a zig-zag defect.

In the absence of point defect interactions it would normally be expected that if a misfit relieving dipole were to form, then it should form in the manner described by Matthews et al (1974) with <110> line directions. It has been shown by Kimerling et al (1976), and later by Huang et al (1988), that 60° MD can interact with point defects. In the former case the interactions result in climb to change the line directions of segments of the dislocations. Our diffraction contrast analysis of the <010> dipoles, and the analysis of the forces acting upon the various dipole segments, strongly suggests that these defects are formed by climb requiring a high density of point defects. It is not clear whether the point defects introduced during growth are sufficient to account for the climb, or whether more have been introduced from the surface of the specimen. As pointed out by Kimerling et al (1976) this effect could be enhanced by ionization produced during electron beam analysis, leading to recombination enhanced defect motion.

REFERENCES

Green G S, Tanner B K, Barnett S J, Emeny M, Pitt A D, Whitehouse C R and Clark G F (1990) Phil.Mag.Lett. **62** 131.
Hirth J P and Lothe J, *Theory of Dislocations* (1968), McGraw-Hill, USA. p78.
Houghton D C (1990) Appl.Phys.Lett. **57** 1434
Huang Y J, Ioannou D E and Iliadis A   (1988) Scanning Microscopy **2** 129
Hull R and Bean J C (1989) Appl.Phys.Lett. (1989) **55** 1900
Kimerling L C, Petroff P and Leamy H J   (1976) Appl.Phys.Lett. **28** 297
Matthews J W and Blakeslee A E (1974) J.Cryst.Growth **27** 118

The University of Surrey/RSRE strained layer project is thanked for supplying the sample.

Inst. Phys. Conf. Ser. No 117: Section 8
Paper presented at Microsc. Semicond. Mater. Conf., Oxford, 25–28 March 1991

An analysis of a misfit dislocation array

R Beanland and P Kightley

Dept. of Materials Science and Engineering, Liverpool University, Liverpool, L69 3BX, UK

ABSTRACT: An analysis of a misfit dislocation array in a $Ga_{0.85}In_{0.15}As/GaAs$ superlattice on GaAs has been performed on a microscopic scale. The Burgers vectors of the misfit dislocations present were determined using plan-view transmission electron microscopy (TEM). This data has been used in a recent formulation of the Frank-Bilby equation to obtain the deformation of the layer. The state of strain is compared with the Matthews model of strain relief, and the layer is found to be metastable; only 45% of the strain energy of the layer has been relieved in comparison with an equilibrium value of 81%.

1. PLAN VIEW TEM

The sample was a ten period $Ga_{0.85}In_{0.15}As$:GaAs superlattice layer grown on (001) GaAs of-fcut by 2° towards [100]. Each superlattice layer was 10nm thick. The sample was back-thinned and chemically etched to perforation using bromine in methanol. A region of the dislo-cation array in the interface (3.4 μm x 4.9 μm) was analysed using bright field images using 220, 400 and 422 type reflections. All images were obtained at 200kV in a JEOL 2000cx.

The dislocation array consisted of a/2<101> 60° misfit dislocations which lay between the bottom layer of the superlattice and the substrate. The Burgers vectors of the dislocations were found using the invisibility criterion (e.g. Hirsch et al. 1967). In order to render a dislocation segment invisible, both of the conditions **g.b**=0 and **g.(bxu)**=0 must be satisfied, where **g** is the diffraction vector, **b** is the Burgers vector, and **u** is the dislocation line direction. In the pre-sent case, the dislocations mainly had line directions close to <110>, with occasional segments close to <100> at reactions. For an a/2[101] dislocation, **g.b**=0 is satisfied when **g**=040, $\bar{2}42$ and $\bar{2}4\bar{2}$; these reflections give **g.(bxu)**=0 for line directions of [0$\bar{1}$0], [$\bar{1}\bar{1}$0] and [1$\bar{1}$0] respec-tively. An image obtained with **g**=220 will give strong contrast for all line directions. Micrographs of the dislocation array are shown in Figure 1a, b, and c with **g**=04̄0, $\bar{2}42$ and 2$\bar{2}$0. No contrast is present for a/2[101] dislocations with line directions close to [0$\bar{1}$0] in Figure 1a. In comparison, there is residual contrast for a/2[101] dislocations with line direc-tions close to [1$\bar{1}$0] in Figure 1b. This may be due to the difficulty of obtaining a 422 type two-beam condition; often other reflections were lightly excited. Also, bands of contrast can be seen, running parallel to [100]. This contrast probably arises from relaxation at the surface of the sample, which cuts through layers of alternating strain. These bands cause periodic changes of contrast along the dislocation line. Other a/2<101> type dislocations may be identi-fied using images taken using further 422 type reflections. Edge dislocations with **b**=a/2[110] and a/2[1$\bar{1}$0] have zero contrast in images with **g**=2$\bar{2}$0 and 220 respectively - for example the dislocation marked E in Figure 1c. The Burgers vectors of all the dislocations present in the area of Figure 1 were ascertained using 400, 220 and 422 type images.

2. ANALYSIS OF THE MICROSTRUCTURE

The spacing of the dislocations is irregular, consistent with a/2<101> dislocations arriving at the interface by glide on {111} planes. Four distinct dislocation line directions close to <110> can be seen; this is also consistent with glide into the interface, as the line direction in this case

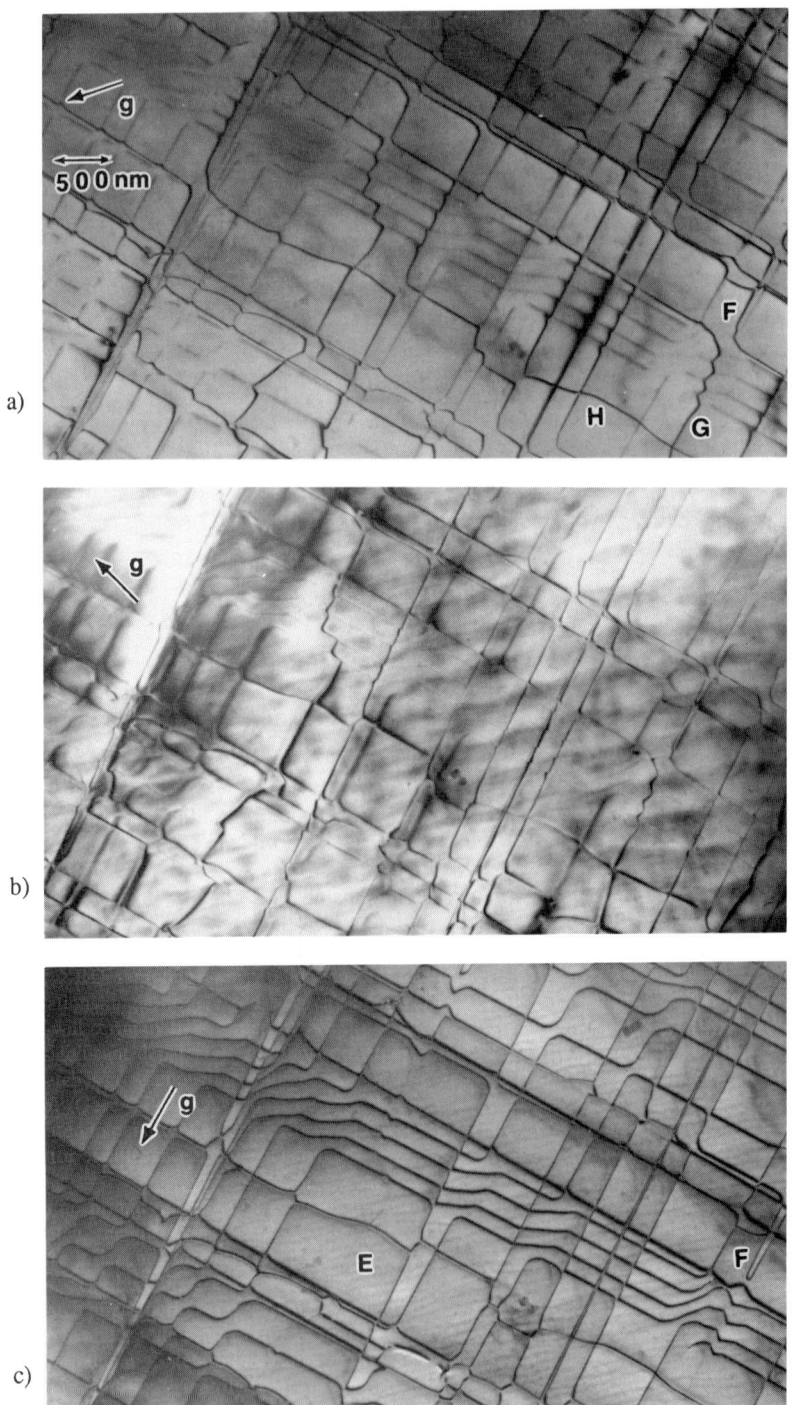

Figure 1. Bright field micrographs of the dislocation array. a) **g**=0$\bar{4}$0, b) **g**=$\bar{2}\bar{4}$0, c) **g**=2$\bar{2}$0.

is given by the intersection of the interface and the glide plane. All the dislocation lines are straight and extend across the entire region examined (except at reactions), i.e. cross slip has not occured. Few dislocations of the same Burgers vector are closely spaced, which indicates that multiplication or nucleation of dislocations on closely spaced glide planes is not a primary mechanism of dislocation introduction. There is a noticeable difference in the density of dislocations along the two <110> directions, consistent with the α/β nature of dislocations in III-V semiconductors (e.g Abrahams et al. 1972). By convention, those with the highest density are α dislocations, i.e. those which glide on $(\bar{1}11)$ or $(1\bar{1}1)$ and have line directions close to $[1\bar{1}0]$.

Many dislocation reactions are present in the array. All reactions that could be identified corresponded to configurations in which all segments relieve misfit. However, a few reactions could not be identified, especially where several parallel dislocations were closely spaced. Reactions between two almost parallel a/2<011> dislocations which produce an a/2<110> edge dislocation can be seen in several places (e.g. reaction F and dislocation G). The edge dislocation begins in the interface and extends into the substrate before splitting back into the two a/2<011> dislocations. Long edge dislocation segments are not usually straight (e.g. dislocation H). This indicates that either a) some climb has occurred, despite the relatively low growth temperature (670°C), or b) glide on the (001) plane has occurred.

3. ANALYSIS OF THE DEFORMATION PRODUCED BY THE ARRAY

It is assumed that the multilayer behaves in a similar manner to a single strained layer of average composition - i.e. the ten 20nm periods of GaAs:Ga$_{0.85}$In$_{0.15}$As are equivalent to a single 200nm layer of Ga$_{0.925}$In$_{0.075}$As. Also, the elastic constants of the layer are assumed to vary linearly with composition. The misfit is -0.542%; the layer is in compressive strain. The equilibrium critical thickness of a layer of average composition, calculated from the Matthews and Blakeslee model (1975), is 28.7nm; the layer is expected to be semicoherent. In practice, the critical thickness will be exceeded by the first Ga$_{0.85}$In$_{0.15}$As layer. Using the same model, the equilibrium elastic strain is -0.234%, i.e. 57% of the misfit is relieved at equilibrium. The density of each dislocation type is shown in Table 1. As all edge dislocations resulted from the combination of a/2<011> dislocations, only a/2<011> dislocations are considered. The α/β nature of each dislocation is also shown.

Glide system	b	glide plane	$\rho(10^6 m^{-1})$	Glide system	b	glide plane	$\rho(10^6 m^{-1})$
1 (β)	$a/_2[011]$	$(\bar{1}\bar{1}1)$	0	5 (α)	$a/_2[\bar{1}0\bar{1}]$	$(\bar{1}11)$	4.11
2 (β)	$a/_2[0\bar{1}1]$	(111)	2.05	6 (α)	$a/_2[10\bar{1}]$	$(1\bar{1}1)$	0.293
3 (β)	$a/_2[101]$	$(\bar{1}\bar{1}1)$	3.07	7 (α)	$a/_2[01\bar{1}]$	$(\bar{1}11)$	0.587
4 (β)	$a/_2[\bar{1}01]$	(111)	0.819	8 (α)	$a/_2[0\bar{1}\bar{1}]$	$(1\bar{1}1)$	2.93

Table 2. Dislocation densities of the a/2<011>:{111} glide systems present in Figure 1.

In order to calculate the deformation produced by the array, it is assumed that the characterised region of the array is characteristic of the array everywhere. The deformation produced by the array can be obtained from the Frank-Bilby equation (Beanland 1990, 1991). In the present case, the deformation produced by the array is given by

$$D_{(array)} = 10^{-3} \begin{bmatrix} 1.672 & -0.089 & -0.428 \\ -0.294 & 1.135 & 0.252 \\ 0.428 & -0.252 & -1.283 \end{bmatrix}$$

when written in an orthogonal coordinate frame with z normal to the interface and y parallel to [010]. In this reference frame, the deformation of a coherent layer of In$_{0.075}$Ga$_{0.925}$As is

$$D^* = \varepsilon_{//} \begin{bmatrix} 1 & 0 & 0 \\ 0 & 1 & 0 \\ 0 & 0 & -2c_{12}/c_{11} \end{bmatrix} = 10^{-3} \begin{bmatrix} -5.424 & 0 & 0 \\ 0 & -5.424 & 0 \\ 0 & 0 & 4.959 \end{bmatrix}$$

where $\varepsilon_{//}$ is the misfit strain. The actual deformation of the layer $D=D^*+D_{(array)}$ is

$$\mathbf{D} = \begin{bmatrix} D_{11} & D_{12} & D_{13} \\ D_{21} & D_{22} & D_{23} \\ D_{31} & D_{32} & D_{33} \end{bmatrix} = 10^{-3} \begin{bmatrix} -3.752 & -0.089 & -0.428 \\ -0.294 & -4.289 & 0.252 \\ 0.428 & -0.252 & 3.676 \end{bmatrix}$$

4. DISCUSSION

The dislocation array has relieved a substantial amount of the misfit; ~31% in the x direction, and ~21% in the y direction. However, this is much less than the 57% misfit relieved at equilibrium - the layer is in a metastable state. Also, there is a significant difference in the misfit relieved in the x and y directions. This is not due to the α/β nature of the dislocations, but results from the greater density of dislocations on the (111) glide plane. The other components of the deformation introduced by the misfit dislocation array produce rigid body rotations and shears of the layer. Components D_{13} and D_{31} represent a rigid-body rotation of 88" about the [010] direction. The sense of this rotation is to rotate the [001] direction away from the surface normal. Components D_{23} and D_{32} describe a rotation of 52" about the x direction, acting to rotate the [001] direction towards [0$\bar{1}$0]. Components D_{21} and D_{12} describe a complex shear of the interfacial plane, comprised of a simple shear of $\gamma=1.17\times10^{-2}$, and a rotation about the surface normal of 18". The layer is clearly in a metastable state; however, there is no accepted measure of metastability to date. Here we use the elastic strain energy of the layer as a measure of metastability. The strain energy per unit volume is given by

$$W = \frac{c_{11}}{2}(\varepsilon_1^2+\varepsilon_2^2+\varepsilon_3^2)+c_{12}(\varepsilon_1\varepsilon_2+\varepsilon_2\varepsilon_3+\varepsilon_3\varepsilon_1)+c_{44}(\gamma_4^2+\gamma_5^2+\gamma_6^2)$$

where ε_i and γ_i are the components of the engineering strain matrix;

$$\varepsilon = \begin{bmatrix} \varepsilon_1 & \gamma_6 & \gamma_5 \\ \gamma_6 & \varepsilon_2 & \gamma_4 \\ \gamma_5 & \gamma_4 & \varepsilon_3 \end{bmatrix} = \begin{bmatrix} D_{11} & D_{12}+D_{21} & D_{13}+D_{31} \\ D_{21}+D_{12} & D_{22} & D_{23}+D_{32} \\ D_{31}+D_{13} & D_{32}+D_{23} & D_{33} \end{bmatrix} = 10^{-3} \begin{bmatrix} -3.752 & -0.192 & 0 \\ -0.192 & -4.289 & 0 \\ 0 & 0 & 3.676 \end{bmatrix}$$

The strain energy per unit volume of a coherent layer of $In_{0.075}Ga_{0.925}As$ on GaAs is 3.550×10^6 J m^{-3}. A semicoherent layer in equilibrium has a strain energy of 0.660×10^6 J m^{-3} i.e. ~81% of the strain energy has been relieved. The actual strain in the layer gives an elastic strain energy of 1.957×10^6 J m^{-3}. Thus, only 45% of the strain energy has been relieved, i.e. there is about three times as much strain energy in the film than at equilibrium.

It should be noted that the above analysis is based on several assumptions, some of which may reduce the validity of the results. Probably the most serious assumption is that the region of the array examined is characteristic of the array everywhere. However, the sides of the area surveyed are about twenty times the thickness of the layer, and the influence of adjacent areas on the state of strain of the region examined is small.

5. SUMMARY

A misfit dislocation array present in an interface between an $In_{0.15}Ga_{0.85}As$:GaAs strained multilayer and a GaAs substrate has been analysed. The Burgers vectors of all the dislocations were identified using bright field plan view TEM. The structure of the array is consistent with glide of a/2<101> dislocations on {111} planes. No evidence for cross slip was observed, and all dislocations extended across the whole region examined. The deformation produced by the array has been obtained through use of the Frank-Bilby equation. The array has introduced rigid body rotations of the layer about three axes and a shear of the interfacial plane. The layer is metastable, having about three times the equilibrium strain energy.

REFERENCES

Abrahams M S Blanc J and Buiocchi C J 1972 Appl. Phys. Lett. **21**, 185
Beanland R and Pond R C 1990 Mat Res Soc Conf Proc **198** 111
Beanland R 1991 PhD thesis, Liverpool University
Matthews J W and Blakeslee A E 1975 J. Cryst. Growth **29**, 273

Inst. Phys. Conf. Ser. No 117: Section 8
Paper presented at Microsc. Semicond. Mater. Conf., Oxford, 25–28 March 1991

611

HREM investigation of MBE-grown single and multilayer heterostructures on (100)-GaAs substrates

M Hohenstein, F Phillipp, O Brandt[+], L Tapfer[+], and K Ploog[+]

Max-Planck-Institut für Metallforschung, Institut für Physik
[+]Max-Planck-Institut für Festkörperforschung
Heisenbergstr.1, D-7000 Stuttgart 80, Germany

ABSTRACT: MBE-grown InAs/GaAs heterostructures of various types were investigated by HREM applying imaging parameters that allow a reliable direct interpretation of the micrographs. Single InAs layers appear atomically flat and grow coherently up to a thickness of two monolayers. In thicker layers, however, local relaxation is observed and the lattice mismatch between InAs and GaAs results in the incorporation of misfit dislocations. In the multilayer systems intermixing may occur depending on the thickness ratio of InAs:GaAs.

1. INTRODUCTION

Thin layers of InAs in a GaAs matrix constitute interesting materials systems from various aspects. Due to the large lattice mismatch (7.16%) they are highly strained and thus show unique optical and electronic properties. Molecular beam epitaxy (MBE) allows the growth conditions to be controlled sufficiently well for producing such heterostructures in good structural quality down to the atomic scale (Ploog 1988). This is most evidently proved by high-resolution electron microscopy (HREM). For a quantitative determination of the thickness and abruptness of thin layers in heterostructures artefacts arising from the complex imaging process in HREM have carefully to be considered (see, e.g. Petford-Long 1989). Ourmazd et al. (1987,1990) developed a detailed procedure for chemical mapping. In a different approach a reconstruction scheme for the complex exit surface wave function given by Hohenstein (1991a) can be employed.
In the present work the structural characteristics of single and multilayer systems of InAs in GaAs matrices are investigated using optimized imaging parameters for a reliable interpretation of the micrographs.

2. EXPERIMENTAL

The heterostructures were grown on undoped (100) GaAs substrates (misorientation < 1°). A 1 μm thick GaAs buffer layer was deposited prior to the growth of the first InAs layer. In the single layer specimens the InAs layer was capped with 200 nm GaAs. The multilayer structures were grown with sequences of 1:8, 2:8 (10 periods), and 2:16 (20 periods) monolayers (ML) of InAs:GaAs. A special temperature modulation technique was employed and the growth process was monitored in situ by high-energy electron diffraction (Brandt et al. 1991 b). High-resolution double-crystal x-ray diffraction and double-crystal x-ray reflection topography were used for the determination of mean layer thicknesses, lattice strains, and large scale perfection (Brandt et al. 1990, 1991 a, b). HREM was carried out using JEOL JEM-4000EX and JEM-4000FX microscopes. <110> cross-sectional samples were prepared by the conventional sandwich technique and Ar[+]-ion milling.

3. OPTIMIZED IMAGING PARAMETERS FOR GaAs/InAs STRUCTURES

In general, the most severe artefacts in HREM imaging arise from the mixing of amplitude and phase contrast and the spherical aberration of the objective lens. In the particular materials system under investigation, however, defocus values and foil thickness ranges minimizing these artefacts can be found. A direct interpretation of the micrographs thus becomes possible.

In order to deduce those parameters we regard the complex exit surface wave function (ESWF) $O(x)$. After renormalization, so that the phase of the transmitted beam is zero (for a detailed discussion see Hohenstein (1991 b)), the complex ESWF can be split into a real $(O_r(x))$ and an imaginary part $(O_i(x))$, corresponding to amplitude and phase contrast. The image intensity $I(x)$ in real space x resulting from interference between the transmitted beam and the diffracted beams is given by

$$I(x) = \mathcal{F}^{-1}\left[2\ U(0)\ F(u,0)\left((\mathcal{F}O_r(x) \cdot \cos(2\pi X(u))) - \left(\mathcal{F}O_i(x) \cdot \sin(2\pi X(u))\right)\right)\right], \quad (1)$$

where $\mathcal{F}$ and $\mathcal{F}^{-1}$ are the operators for the direct and inverse Fourier transformation, $U(0)$ denotes the amplitude of the transmitted beam, u is the frequency in reciprocal space, $F(u,0)$ denotes the damping function depending on focus spread and beam convergence, and $X(u) = \lambda\ C_s\ u^4\ /4 + \lambda\ \Delta f\ u^2\ /2$ denotes the phase shift of the electron waves due to the spherical aberration C_s, defocus Δf and the electron wavelength λ. In the case of GaAs the weak phase object approximation holds only below a specimen thickness of 3 nm, so that the imaginary part dominates, whereas above this value the real part of the ESWF becomes increasingly important. Particularly in the range of experimentally most relevant values of specimen thicknesses between 7 and 15 nm multislice calculations show, that it is advantagous to use amplitude contrast. Therefore, imaging parameters are selected which set the phase shift $X(u)$ close to zero, so that the cosine contrast-transfer-function becomes

$$\cos(2\pi X(u)) \approx 1. \qquad (2)$$

Using a JEOL-4000 FX microscope $(C_s = 1.8\,mm)$ eqn. (2) is very well fulfilled for defocus values Δf between -20 and -50nm (Fig. 1) and spatial frequencies u up 4 nm^{-1}. This frequency range includes the GaAs (111) and (200) beams and is therefore sufficient to resolve monolayers of InAs- or GaAs-dumbbells. Furthermore, up to 4 nm^{-1} practically no damping occurs and the damping function $F(u,0)$ is constant within 10%. Thus we restricted ourselves to this frequency range using corresponding objective aperture and image processing to exclude non-linear contributions to

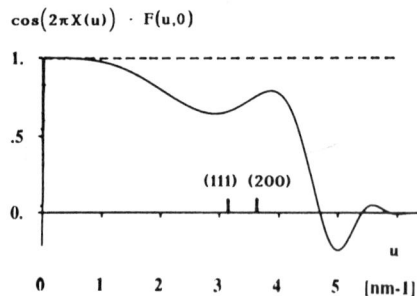

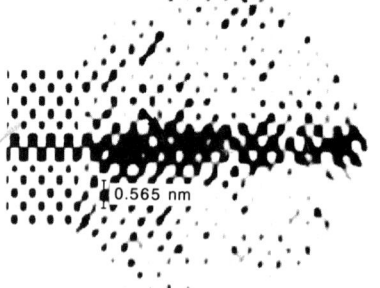

Fig. 1: Contrast transfer functions for amplitude contrast
The broken line corresponds to the contrast transfer function (CTF) of an ideal microscope ($C_s = 0$, $\Delta f = 0$). The solid line represents the CTF of the JEOL 4000 FX microscope with $C_s = 1.8$mm, $\Delta f = -40$nm, focus spread df=8nm, beam convergence dα=0.8mrad. For frequencies up to 4nm^{-1} the CTF is close to that of the ideal microscope, so that aberration free imaging is possible.

Fig. 2: Experimental and simulated image of 2 monolayers InAs in GaAs matrix.
Enlarged portion of 2:16 InAs:GaAs superlattice from Fig. 3 a. The experimental image (right) was recorded with $\cos2\pi X(u) \approx 1$ (Fig. 1) and Fourier filtered so that only contributions from frequencies up to 3.4nm^{-1} are present. The corresponding simulated image (left) for an exit surface wave function with exactly two monolayers InAs proves that with these imaging parameters the Fourier filtered experimental image may be directly interpreted. Therefore an InAs layer thickness between two and three monolayers can be deduced.

higher spatial frequencies. Under these conditions the image contrast is very well approximated by

$$I(x) \sim O_r(x). \tag{3}$$

Eqn. (3) and image simulations prove the resulting simple relationship between image contrast and object function. An example is given in Fig. 2 showing the nominal 2:16 multilayer structure in comparison with a simulated image. The ESWF for the simulation is calculated for exactly two monolayers InAs, which can be easily recognized in the image. Therefore the corresponding experimental micrograph may be directly interpreted.

4. RESULTS AND DISCUSSION

Adopting the imaging parameters derived above, single-layer structures with InAs layers of nominally 1 – 5 ML thickness were studied. Specimens containing up to 2 ML are found to grow coherently and no dislocations are observed in the HREM images (Fig. 4). This is in accordance with x-ray topography revealing these specimens to be dislocation free on a larger scale as well as with x-ray diffraction studies that show the lattice mismatch to be entirely accomodated by the elastic distortion of the InAs unit cell (Brandt et al. 1991a). The distortion along the growth direction measured directly in the HREM micrographs is 10%. From the contrast change in the HREM micrographs one can conclude that interface fluctuation is restricted to one ML on either side of the InAs.

Beyond a critical thickness of 2 ML, however, strong fluctuations of the InAs layer thickness are observed (Fig. 5), reflecting the onset of three-dimensional island growth. Furthermore, at the upper interface of the islands, which are found to reach thicknesses up to 17 ML, strong intermixing of InAs with the adjacent GaAs is observed which is attributed to strain enhanced In diffusion and segregation. This growth mode in our particular case is accompanied by the nucleation of 60°-dislocations at the island edges resulting from in plane relaxation (Fig. 6). However, they accomodate only a minor fraction of the misfit strain. By counting the lattice planes within the islands and the surrounding GaAs matrix we determined the lattice constant of the InAs island parallel to the interface to be the same as in the matrix within an accuracy of 0.5%.

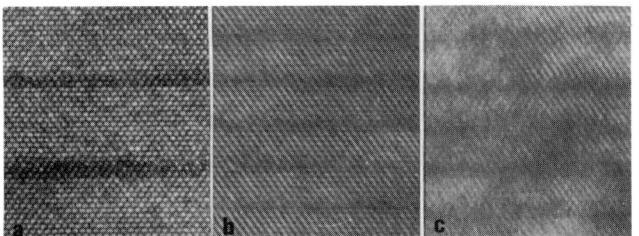

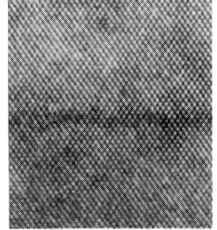

Fig. 3: InAs:GaAs superlattices
a) 2:16 InAs:GaAs superlattice. The interfaces appear abrupt on an atomic scale. b) 1:8 InAs:GaAs superlattice. Due to intermixing the sharpness of the interfaces is reduced. c) 2:8 InAs:GaAs superlattice. The GaAs and InAs layers are strongly intermixed. With increasing distance from the substrate the layer structure is nearly lost.

Fig. 4: Lattice image of a thin InAs-layer in a GaAs matrix.
<110>-sample containing nominal 0.8 monolayers InAs in a GaAs matrix. Both interfaces are of equivalent quality and no dislocations are observed.

An example of the structural quality of the multilayer systems is presented in Fig. 3. The thickness ratio InAs:GaAs determines the net strain stored in the multilayer and thus its critical thickness. The total thickness of all investigated structures exceeds the critical one. However, though the ratio of total and critical thickness is equal for the 1:8 and 2:16 structures, their interface properties are quite different. Whereas the interfaces of the 2:16 superlattice appear abrupt on an atomic scale (Fig. 3 a), they are less well defined in the 1:8 structure (Fig. 3 b), suggesting a starting interdiffusion of the layers. This process is more evident in the 2:8 (Fig. 3 c) superlattice, where the higher net strain induces strong intermixing and degraded interface quality.

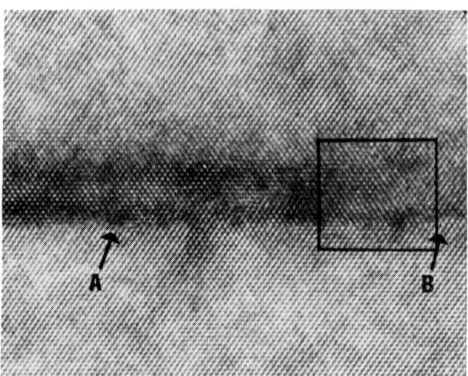

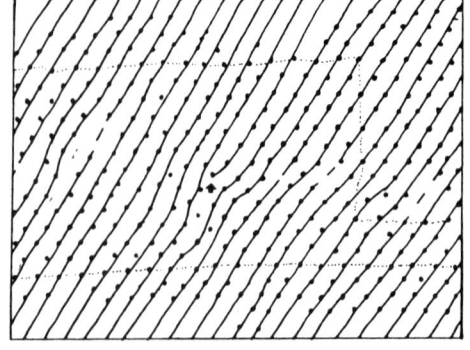

Fig. 5: Lattice image of an InAs island in a GaAs matrix.
HREM lattice image of a sample containing nominal five monolayers InAs in a GaAs matrix. The image shows InAs islands (A) with a thickness of approximately 17 monolayers included in regions with only three (B) monolayers. Fig. 6 shows an enlarged area of the edge of the island.

Fig. 6: Schematic representation of the edge of the InAs island
The edge of the InAs island from Fig. 5 shows a 60°-dislocation. Growth direction is from the bottom to the top of the image. The shape of the island is indicated by the broken line. The measured average lattice constant of the InAs-island is the same as that of GaAs. The resulting stress limits the lateral size of the InAs islands and leads to dislocations at the edge of the island.

5 . CONCLUSION

The scope of the present work was the investigation of thin InAs layers embedded in a GaAs matrix. We first determined the appropriate imaging parameters for this structure, which allow a direct interpretation of the HREM micrographs. Because a range of defocus values Δf from -20 to -50nm and specimen thicknesses from 8 to 15 nm fulfil the required conditions, the presented concept may be adopted to common experimental conditions.

Using this concept the layer thickness in a nominal 2:16 InAs:GaAs superlattice was determined to be between two and three monolayers, emphasizing the superior quality of MBE samples grown by the temperature modulation technique. Investigations of InAs islands with a thickness beyond the critical layer thickness showed, that the relaxation of the InAs island is restricted to local areas and does not result in a difference between the lattice constant in the InAs island and the GaAs matrix.

REFERENCES

Petford-Long A 1989 Ultramicroscopy 31 385
Ourmazd A, Tsang W T, Rentschler A, Taylor D W, 1987 Appl. Phys. Lett. 50 (20)
Ourmazd A, Baumann F H, Bode M, Kim Y, 1990 Ultramicroscopy 34 237
Hohenstein M 1991a Ultramicroscopy , in press
Hohenstein M 1991b Ph.D. thesis, University of Stuttgart, to be published
Brandt O, Tapfer L, Cingolani R, Ploog K, Hohenstein M and Phillipp F 1990 Phys. Rev. B 41 (18) 12599
Brandt O, Tapfer L, Ploog K, Hohenstein M and Phillipp F 1991a J. of Crystal Growth, in press
Brandt O, Ploog K, Tapfer L, Hohenstein M and Phillipp F 1991b submitted to Phys. Rev. B
Ploog K 1988 Angew. Chem. Int. Ed. Engl. 27 593

Acknowledgement
Thanks are due to M. Kelsch for the preparation of the HREM samples.

Structure and composition of GaAs$_{1-y}$Sb$_y$ quantum wells and superlattices determined by cross-sectional transmission electron microscopy

S Haq[1], A C Wright and J O Williams

Department of Chemistry, UMIST, P O Box 88, Manchester, M60 1QD, UK

[1] Now at: Sowerby Research Centre, British Aerospace, FPC 266, P O Box 5, Filton, Bristol, BS12 7QW, UK

ABSTRACT: Superlattice structures of GaAs$_{1-y}$Sb$_y$ - GaAs have been grown by MBE with up to 40% Sb content. The microstructures of the Sb wells have been investigated using cross-sectional TEM. It was found that there was a threshold level of the Sb content above which non-uniform wells occur. This threshold is consistent with the low Sb end of the miscibility gap of the GaAsSb system and above this spinodal decomposition occurs.

1. INTRODUCTION

The new generation of electronic devices consisting of superlattice structures in theory have the necessary prerequisite of sharp interfaces and uniformity of well material for optimum device operation. GaAsSb layered structures have considerable potential for use as infra-red and high mobility devices. However there exists for GaAsSb, in common with a variety of III-V ternary and quaternary alloy systems, a miscibility gap giving rise to spinodal decomposition. This results in the material being unstable causing alloy clustering and an inhomogeneity of chemical composition leading to deleterious device performance.

We have deposited GaAs$_{1-y}$Sb$_y$-GaAs superlattices by molecular beam epitaxy (MBE) over a range of antimony concentrations and we have investigated the effect of the apparent miscibility gap on compositional uniformity and interface quality of the grown structures by transmission electron microscopy (TEM).

2. EXPERIMENTAL

Growth was carried out in a Riber 2300 MBE system under group V stabilised conditions using elemental arsenic and antimony sources at a substrate temperature of 550° C. The GaAs$_{1-y}$Sb$_y$ - GaAs superlattices were deposited on (100)-oriented GaAs substrates. The substrates were etched in 6:1:1 sulphuric acid, peroxide, water after an initial solvent degrease prior to loading into the growth apparatus. A GaAs buffer layer of approximatly 0.3 microns thickness was deposited prior to the growth of the superlattice. The superlattice structures have a relative GaAs to GaAs$_{1-y}$Sb$_y$ thickness ratio of approximately 11:1. The composition of the layers was determined from double crystal X-ray diffraction measurements and comparison of the (400) rocking curves with computer generated data, and has been explained in greater detail elsewhere (Hobsen et al. 1988).

The TEM was performed with a Philips EM430 operating at 300kV. The cross-sectional specimens were prepared by sequential mechanical polishing of the glued sandwich structure to a thickness of 30 microns. The specimens were then thinned using reactive iodine beams to electron transparency in an Ion-Tech 700 series ion beam thinner. The incident gun angle was 14 degrees at 5kV, 20 microamps current with LN_2 cooling and sample rotation.

3. RESULTS

Figure 1 shows a high resolution electron micrograph of one of the $GaAs_{0.73}Sb_{0.27}$ wells of a 20 period structure with a nominal well width of 5 nm. This micrograph is viewed along a [110] direction and shows clearly the two sets of {111} fringes. It can be seen that the $GaAs_{0.73}Sb_{0.27}$ well is undulating with the modulation more prominent on the GaAsSb-GaAs well interface compared to the GaAs-GaAsSb interface. This gives rise to a variation of the well thickness on average of between 3 and 5 nm.

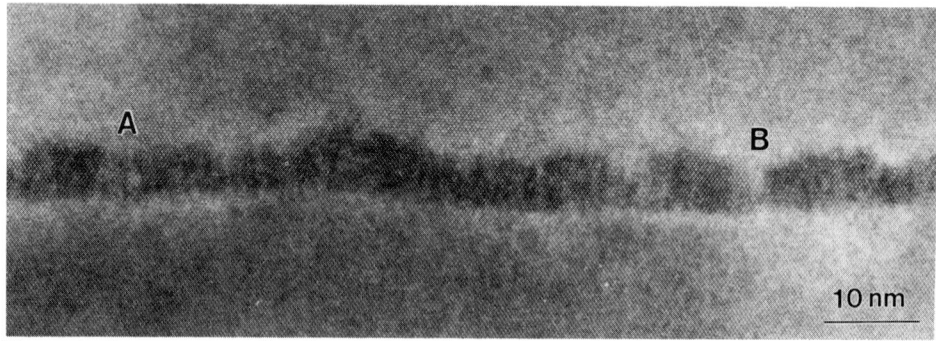

Figure 1. Cross-section of well with 27% Sb showing undulating upper interface.

It is also apparent that at various points throughout the GaAsSb well, eg at A, there are variations in the contrast. In fact at some points the contrast resembles that of the surrounding barrier material. This then suggests that within the well material there are several regions which are GaAs rich. It can be seen at several places, e.g. B, that there is continuity of the GaAs from the lower barrier through the lower interface of the well and up to the top GaAsSb-GaAs interface.

Figure 2 shows a high resolution micrograph of two $GaAs_{0.6}Sb_{0.4}$ of 3 nm thickness. Again, similar to the previous specimen, the non-uniform nature of the wells is clear in terms of the degree of abruptness of the interfaces and the general undulating nature of the layer. In fact, the micrograph shows that both interfaces are a great deal poorer at this Sb concentration of 40% compared to the lower antimony content specimen of figure 1. The poorer quality of the wells of this specimen was further illustrated by the RHEED patern observations which indicated 3-D growth was occurring from the start of the $GaAs_{0.6}Sb_{0.4}$ well deposition. This contrasts with the earlier specimen which although exhibiting an intensity modulated RHEED pattern did show a streaky (2 by 4) pattern at the start of the growth. It can also be seen from figure 2 that the well has broken down more frequently in this case and also over longer lengths, e.g. there are areas in the GaAsSb well over 4nm in length which appear to have little or no contrast pertaining to the GaAsSb but rather resemble the GaAs barrier material.

The micrograph of figure 3 is obtained from a specimen with a much lower antimony concentration (viz. 20%). This shows that the $GaAs_{0.8}Sb_{0.2}$ well is much improved as compared with those of a higher antimony concentration. The $GaAs_{0.8}Sb_{0.2}$ well thickness in this case is 10nm approximatly. The well thickness remains uniform of this value throughout the entire length of the specimen viewed in the TEM (several microns). In addition to the greater uniformity of well thickness it can be seen that neither the $GaAs_{0.8}Sb_{0.2}$ nor the interfaces associated with it are undulating. The upper and lower interfaces of the well are also more abrupt compared to the higher antimony concentration layers.

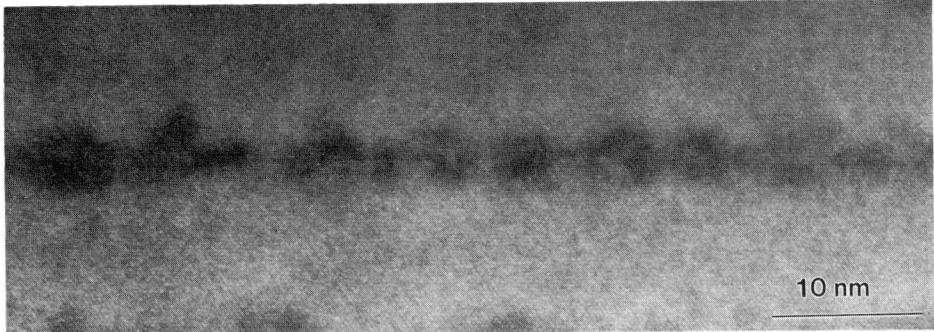

Figure 2. Cross-section of well with 40% Sb showing poorly defined interfaces.

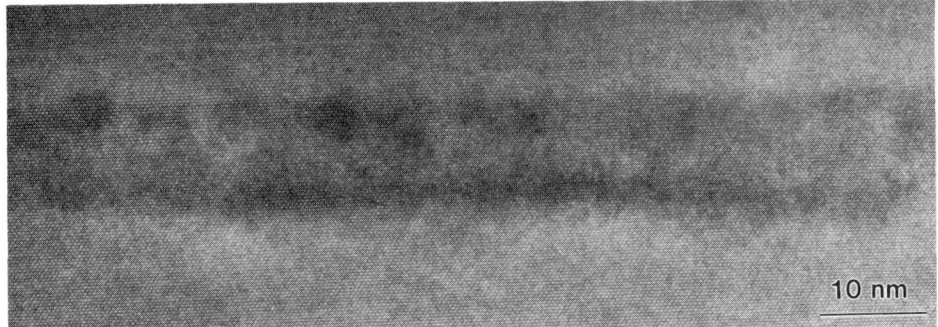

Figure 3. Cross-section of layer with 20% Sb showing well defined interfaces.

Figure 4. Cross-section of layer with 15% Sb showing well defined interfaces.

Figure 4 shows a high resolution micrograph of an 8nm $GaAs_{1-y}Sb_y$ well with an even lower Sb concentration (y=15%). This micrograph shows that at this value of Sb content the well has relatively sharp upper and lower interfaces. The $GaAs_{0.85}Sb_{0.15}$ well is also still uniform and shows no sign of the undulations observed with the wells grown at the higher Sb levels.

4. DISCUSSION

It is clear that there is a threshold for the antimony concentration above which the GaAsSb wells of the superlattice are of poor quality, being wavy, of non-uniform composition and having undulating interfaces. Below this concentration the wells are more uniform and have higher quality interfaces. It would appear that this threshold lies between 20 and 25 %. The threshold concentration appears to be consistent with the lower concentration end of the miscibilty gap for GaAsSb which appears to lie between $0.2<y<0.8$ at 600° C for near-equilibrium crystal growth. It has been reported that MBE, a non-equilibrium growth technique also gives rise to spinodal decomposition (Ihm et al. 1987). In addition, the well thickness as measured by TEM can vary considerably from that calculated from the growth conditions. Below 20% Sb, the error between measured and expected well thickness becomes negligible.

Spinodal decomposition is often identified by the characteristic tweed or granular structure observed under certain diffraction conditions. However, as the periodicity of this structure is 10nm it is not suprising that we have not observed this in the extremely thin GaAsSb well material we have used for this study.

5. CONCLUSION

We have shown that the MBE deposition of GaAsSb-GaAs structures with Sb concentrations > 25% results in a deterioration of interface quality and a breakdown in the GaAsSb well material. It is only possible to deposit uniform thickness wells with abrupt interfaces and without undulations with Sb concentrations less than 20%. This appears to agree with the lower concentration end of the miscibilty gap which gives rise to composition modulation along <100> direction at greater Sb levels. Furthermore it is difficult to predict the GaAsSb well thickness with any certainty above this threshold level.

ACKNOWLEDGEMENT

We thank Prof. K.E. Singer of the Solid State Electronics Group for supplying the MBE grown material and also the SERC for support.

REFERENCES

Hobsen G I, Khamsehpour B, Singer K E and Truscott W S 1988
  Fifth International Conference on Molecular Beam Epitaxy,
  Sapporo, Japan, p 483.

Ihm Y E, Otsuka N, Lem K and Morkoc H 1987 Appl. Phys. Letters, 51, 24

Inst. Phys. Conf. Ser. No 117: Section 8
Paper presented at Microsc. Semicond. Mater. Conf., Oxford, 25–28 March 1991

619

A TEM study of CdTe/ZnTe single quantum well structures grown on GaAs substrates by hot wall epitaxy

A Hobbs, O Ueda, I Sugiyama and K Shinohara

Fujitsu Laboratories Ltd, 10-1 Morinosato-Wakamiya, Atsugi 243-01, Japan

ABSTRACT: A TEM study has been made on CdTe/ZnTe Single Quantum Well (SQW) structures to directly determine the critical thickness for the generation of new defects. Both CdTe wells in ZnTe epilayers and ZnTe wells in CdTe epilayers, grown on GaAs substrates have been investigated for growth in the (001) and (111) orientations. It is found that critical thickness and defect structure depend strongly on growth orientation but not on the well material.

1. INTRODUCTION

The use of Strained Layer Superlattices (SLSs) as "dislocation filters" is one technique which can be applied in hetero-epitaxial growth to reduce defect densities to desired levels for device applications. To date, however, there are few reports of the successful reduction of dislocation densities in II-VI epilayers using these methods. To create an effective dislocation block, it is necessary that both the individual layer thickness and the total number of periods be optimised to enhance dislocation bending but avoid the production of new defects. In the CdTe/ZnTe system, with a lattice mismatch of 6.4%, the optimum SLS parameters are difficult to predict since such predictions rely on extrapolations of models designed for systems with lower mismatch. In addition, judging from the spread in experimentally observed values for the critical layer thickness and critical SLS thickness for the generation of new defects in the CdTe/ZnTe system (Clifton et al, 1988, Kisker et al, 1988, Miles et al, 1987), there would appear to be an uncertain dependence on both the growth technique and the growth temperature.

In an attempt to directly determine the critical layer thickness for defect generation in MBE growth, Cibert et al (1990) examined a test structure consisting of CdTe single quantum wells (SQWs) of varying thickness using cross-sectional TEM. Their results indicated that for (001) growth, CdTe wells thicker than 17Å give rise to the generation of both misfit dislocations and stacking faults. We have extended this study to both CdTe and ZnTe SQWs grown by Hot Wall Epitaxy (HWE) for both the (001) and (111) growth orientations.

2. EXPERIMENTAL

The epilayers used in this study were grown by HWE on GaAs (001) and (111)B substrates, each with a misorientation angle of 2°. Separate chambers were used for CdTe and ZnTe growth, the substrate being rotated between them to grow the SQW structure. Growth of both materials was carried out at 320°C at a chamber pressure of 5×10^{-8}Torr. The ZnTe and CdTe source temperatures were 570°C and 520°C, respectively.

Cross-sectional TEM specimens were prepared using Ar^{+} ion-milling and observations were carried out in an Akashi EM002B operated at 200keV.

Fig 1 shows a schematic of the SQW structure for the cases of ZnTe and CdTe wells. The initial buffer layer thickness was about 1μm in both cases and the SQWs were separated by 500Å thick

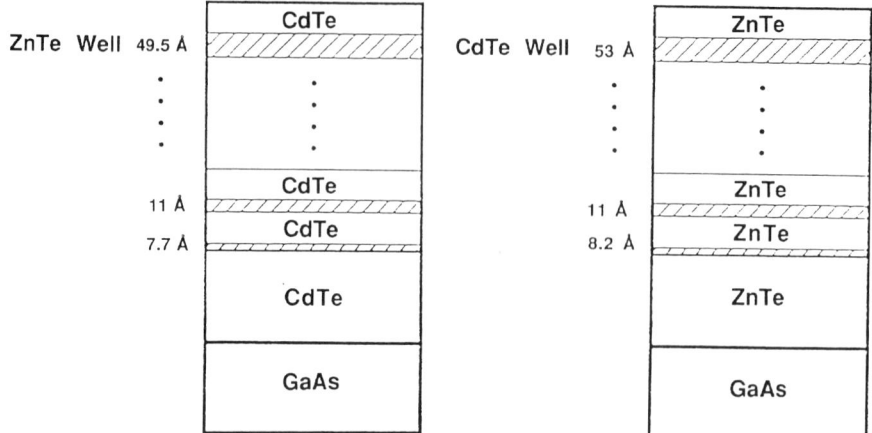

Fig. 1   Schematic diagram of epilayer structure including CdTe and ZnTe SQW's

barrier layers. This arrangement is similar to that used by Cibert et al (1990) and so allows a direct comparison between MBE and HWE for the case of (001) growth of CdTe SQW's. For both materials, the well thicknesses were varied from approximately 8Å to 50Å.

3.   RESULTS AND DISCUSSION

3.1.   (001)   Epilayers

Fig. 2 shows a cross - sectional TEM image of the (001) epilayer with CdTe wells. CdTe layers 1 - 3 are seen to be free from new defects, whereas from layer 4 upwards a dense tangle of defects exists. These have been identified as misfit dislocations and stacking faults; the latter are often seen in high - resolution images to terminate at the SQW interface to form partial misfit dislocations. Isolated misfit dislocations are sometimes found in SQW 3 (19Å). This result agrees with the findings of Cibert et al. (1990) and is probably due to monolayer fluctuations in the well thickness. The critical thickness for misfit dislocation generation in the CdTe wells is therefore probably just above 19Å, which is close to the result of 17Å obtained for MBE growth. This value is supported by PL data which indicates that the first three SQW's are unrelaxed.

Fig. 2   Bright - field cross - sectional image of (001) epilayer with CdTe SQW's

Fig. 3 shows the situation for the epilayer with ZnTe SQW's. The results for the generation of misfit dislocations are found to be very similar to the previous case. SQW 3 (18Å) again generates occasional misfit dislocations so the critical thickness is expected to be just above 18Å. Therefore, misfit strain alone and not well material determines the critical thickness in this system, implying that the energy required to generate misfit dislocations is similar in both CdTe and ZnTe. One

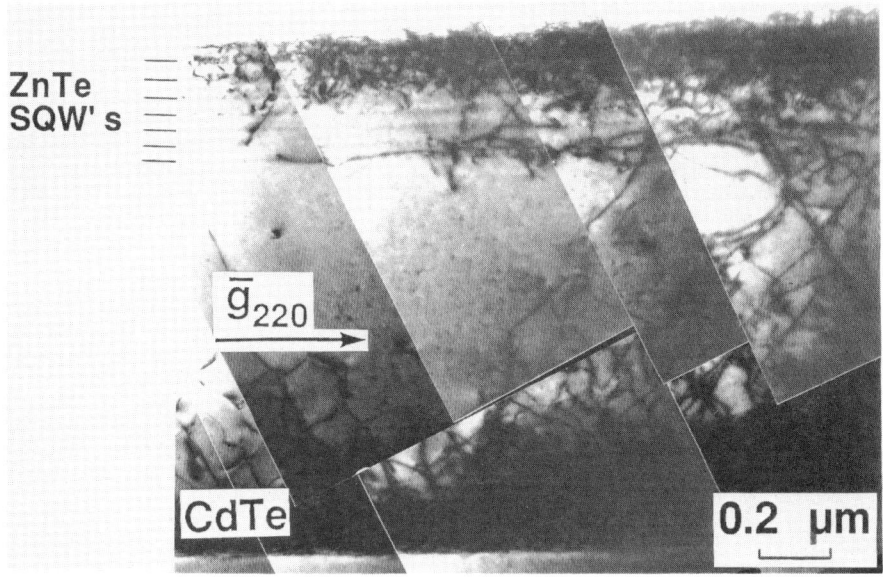

Fig. 3   Bright - field cross - sectional image of (001) epilayer with ZnTe SQW's

difference between the two systems is that stacking faults are sometimes found to originate at the 18Å ZnTe well and propagate through the CdTe barrier layer whereas for CdTe wells, stacking faults are generated only at wells of 28Å or thicker. This probably reflects the difference in stacking fault energy between CdTe and ZnTe.

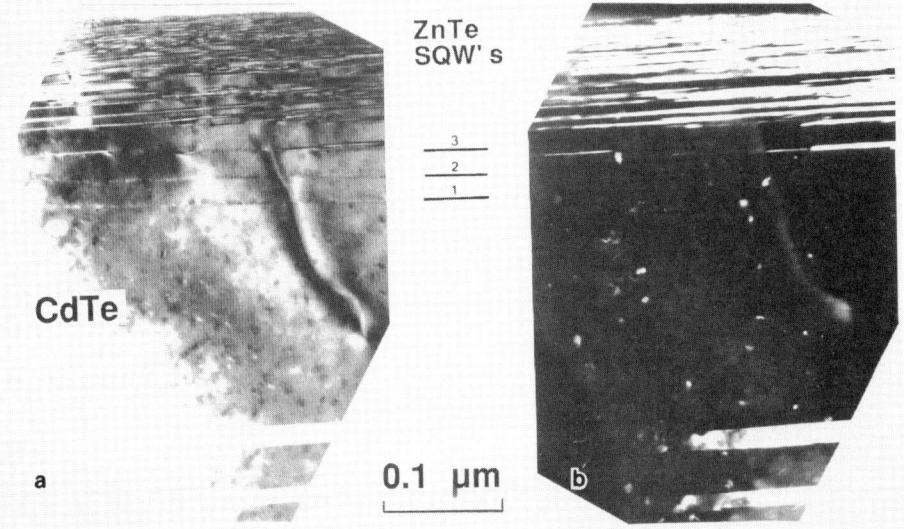

Fig. 4 Bright and dark - field images of (111) epilayer with ZnTe SQW's; a) Bright - field in 111 reflection, b) dark - field from (111) twin - spot.

3.2.   (111)   Epilayers

A similar analysis of (111) oriented epilayers with CdTe or ZnTe SQW's has shown the the critical thickness for the production of misfit dislocations is smaller than in the (001) case. For both CdTe and ZnTe SQW's large numbers of dislocations originate from well 3 (18 - 19Å) implying that the value of critical thickness is in the range 12 - 19Å in this case. Stacking faults are not found to occur for the (111) case. Instead lamellar twinning begins at the third SQW (18 / 19Å) and continues to the top surface. Fig. 4 shows a bright and dark field image of the (111) ZnTe SQW structure; the dark field image is taken from one of the twin spots in the diffraction pattern and so twinned regions appear bright. The first bright band is seen to correspond to the third ZnTe layer. A similar result is found for the case of CdTe SQW's. Thick twin lamellae are also visible in the CdTe buffer layer in these images. These twins are produced at the CdTe / GaAs interface but terminate before reaching the series of SQW's. The slight tilting of the twin lamellae is a result of the offset angle of the GaAs substrate.

4.   CONCLUSIONS

Both CdTe / ZnTe and ZnTe / CdTe SQW test structures have been examined by cross - sectional TEM for growth in the (001) and (111) orientations. This has allowed the critical layer thickness for the generation of defects to be determined as a function of well material and growth orientation.

It is found that critical thickness and defect structure depend strongly on the growth orientation. In (001) epilayers with either CdTe or ZnTe wells both misfit dislocations and stacking faults are introduced when the well thickness exceeds the critical value. For both materials the critical thickness is about 19Å for dislocation formation. However, a slight difference between the critical thickness values for stacking faults exists for the case of CdTe and ZnTe wells. In (111) epilayers lamellar twinning occurs above the critical thickness; no dependence on well material is found.

REFERENCES

Cibert J, Gobil Y, Le Si Dang, Tatarenko S, Feuillet G, Jouneau P H and Saminadayer K 1990
    Appl. Phys. Lett. 56 (3) 292
Clifton P A, Mullins J T, Brown P D, Russell G J, Brinkman A W and Woods J 1988 J. Cryst.
    Growth 93 726
Kisker D W, Fuoss P H , Krajewski J J, Amirtaraj P M, Nakahara S and Menendez J 1988
    J. Cryst. Growth 86 210
Miles R H, McGill T C, Sivanathan S, Chu X and Faurie J P 1987 J. Vac. Sci. Technol.
    B 5 (4) 1263

Inst. Phys. Conf. Ser. No 117: Section 8
Paper presented at Microsc. Semicond. Mater. Conf., Oxford, 25–28 March 1991

623

Strain relaxation in (001) and (111) CdTe/CdZnTe heterostructures

P H Jouneau, J Cibert, G Feuillet, R E Mallard, K Saminadayar, S Tatarenko and Le Si Dang

Equipe CNRS-CEA "Microstructures à Semiconducteur II-VI"
Centre d'Etudes Nucléaires de Grenoble, SP2M/PSC, BP 85 X, 38041 Grenoble Cedex, France
and Laboratoire de Spectrométrie Physique BP 87 38402 St Martin d'Hères Cedex, France

ABSTRACT: A combination of transmission electron microscopy, reflection high-energy electron diffraction and photoluminescence was used to investigate strain relaxation in CdTe/CdZnTe heterostructures grown by molecular beam epitaxy in both (001) and (111) orientations. The critical thicknesses were determined and the type of the interfacial defects was identified.

1. INTRODUCTION

The epitaxial growth of strained heterostructures with good optoelectronic properties requires a knowledge of the critical thickness above which misfit dislocations are formed to accommodate the lattice mismatch. We report here on a combination of reflection high-energy electron diffraction (RHEED), photoluminescence (PL) and transmission electron microscopy (TEM) measurements on the critical thickness for both (001) CdTe grown on (001) $Cd_{1-x}Zn_xTe$ (with $0.03 \leq x \leq 1$) and (111) CdTe grown on (111) ZnTe. Moreover, high resolution TEM allows us to assess the nature of the different types of misfit dislocations and their efficiencies for mismatch accommodation.

2. EXPERIMENTAL

The epilayers were grown by molecular beam epitaxy on semi-insulating (001) GaAs substrates (Cibert et al 1990a). In the case of (111)-oriented systems, the substrates were slightly misoriented from 2° to 6° around the [110] axis to avoid the formation of twins (Tatarenko et al 1990). RHEED oscillations at 40 keV were monitored in-situ during the growth. TEM analysis of the cross-sectional samples was performed at 200 keV using a JEOL 200 CX microscope and at 400 keV using a JEOL 4000 EX. The specimens were prepared in <110> cross-sections by mechanical polishing and subsequent Ar^+ ion milling.

3. RESULTS AND DISCUSSION

3.1 (001) CdTe/Cd$_{1-x}$Zn$_x$Te

Fig. 1 represents the RHEED intensity oscillations of the specular spot during the growth of CdTe on ZnTe and of ZnTe on CdTe. These oscillations result from a 2D growth with one period corresponding to the completion of one monolayer (ML). In both cases, five strong oscillations are observed during the growth of the first five monolayers and then the intensity decreases rapidly with further oscillations of weaker intensity. This decrease is interpreted as due to the onset of elastic strain relaxation at $n_C = 5$ ML. Hence the critical thickness for CdTe/ZnTe is the same as for ZnTe/CdTe, and the growth mode is still 2D even above the critical thickness.

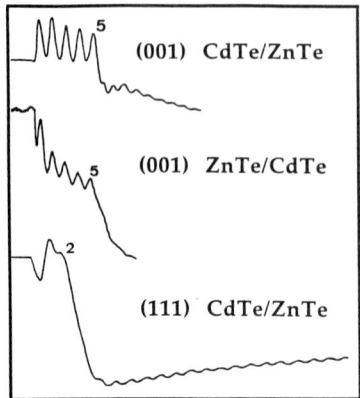

Fig. 1 RHEED specular intensity variation during the growth of (001) CdTe on ZnTe, (001) ZnTe on CdTe and (111) CdTe on ZnTe

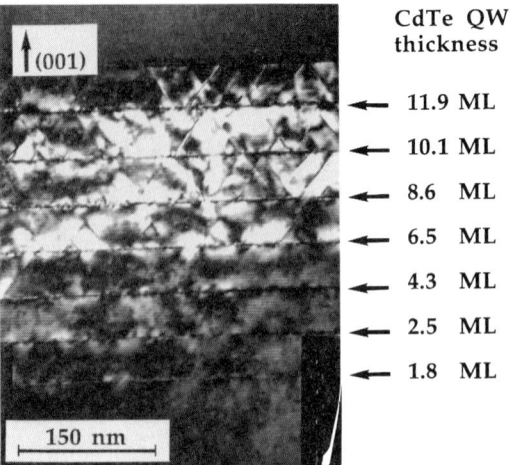

Fig. 2 TEM micrograph of a (001)-multiple QW sample with 7 CdTe QWs between thick ZnTe barriers. Bright field image with g=002

TEM observations are in support of these measurements. Fig. 2 is a conventional TEM micrograph of a CdTe multiple quantum well (QW) sample with progressive well thickness from 1.8 ML to 12 ML between thick(60nm) ZnTe barriers. It clearly shows that many defects originate from the fourth QW (6.5 ML thick) and from the QWs with larger thicknesses, and propagate within the barriers and/or within the QW. No such defects are generated from the QWs with lower thickness. These defects were identified by high resolution TEM (Fig. 3) as being stacking faults bounded by two partial dislocations, at least one of them lying at one interface. Other types of dislocations are found at these interfaces, namely 90° Lomer, 60° undissociated, and Frank or Stair-Rod dislocations resulting from the interaction of these different sorts of dislocations. Each type contributes differently to misfit accommodation, thus the strain relaxation at each interface of the QWs can be deduced from the relative population of these defects. For the last QW (12 ML), the relaxation is found to be only 20%, indicating a rather slow relaxation process.

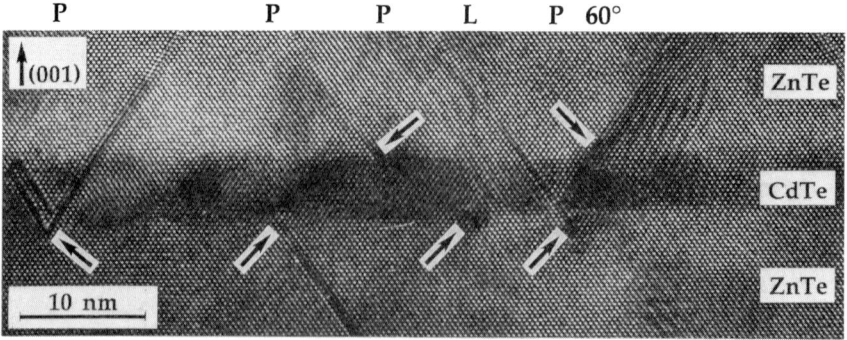

Fig.3 High Resolution TEM micrograph on QW No 7 (n=11.9 ML) showing the different types of dislocations [Lomer (L), 60°, partial (P)] and the generation of stacking faults at the interfaces.

Low temperature PL measurements on the same QWs structure confirm these results. The first three QWs with thicknesses below 5 ML present intense excitonic lines and the next four QWs beyond the critical thickness exhibit PL lines about 50 times weaker due to the relaxation defects (Cibert et al 1990a).

These different techniques are used to determine the critical thickness for various lattice mismatches. Final results, complemented with data for low mismatch quoted from Fontaine et al (1987) and Chami et al (1988) are given in Fig. 4 (Cibert et al 1990b). These results can be compared with theoretical values deduced from the various available models (e.g. Maree et al 1987): our experimental points lie between the theoretical curves. However the calculations are all based on perfect dislocations of the 1/2<110> type (Lomer or 60°) while our TEM observations revealed that strain-relaxation involves other defects like partial dislocations.

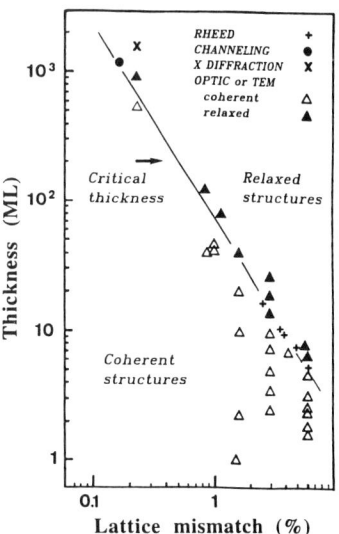

Fig. 4 Critical thickness *vs* mismatch for (001) CdTe on $Cd_{1-x}Zn_xTe$ as deduced from different techniques. The same critical thickness was found for CdTe on ZnTe and for ZnTe on CdTe.

3.2. (111) CdTe/ZnTe

Fig. 5 is a TEM cross section of a multiple QWs sample with five QWs of varying thicknesses ranging from 2 ML to 6 ML in between thick (40 nm) ZnTe barriers grown on a slightly misoriented GaAs substrate. The first QW is free from dislocations; as for the following quantum wells, relaxation has obviously occurred through formation of stacking faults and microtwins. This indicates a critical thickness for plastic deformation for the (111) orientation of the order of 2 ML. The same critical thickness was also observed for pure (111) growth. This was checked on the intensity of the specular RHEED spot which reveals only 2 rather intense oscillations when the (111) growth of CdTe on ZnTe is initiated (Fig. 1). An abrupt   decrease of the PL intensities is also observed for QWs thicker than 2 ML.

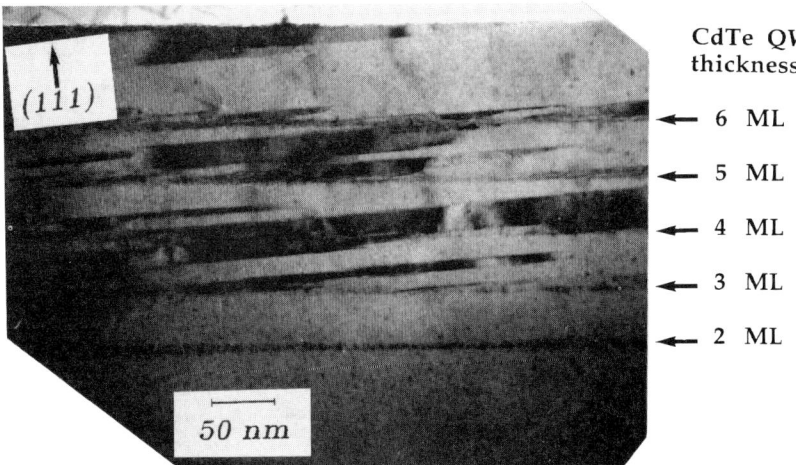

Fig. 5 TEM bright field image, with g=220, of a (111)-multiple QW sample with 5 CdTe Qws between thick ZnTe barriers grown on a slightly misoriented substrate.

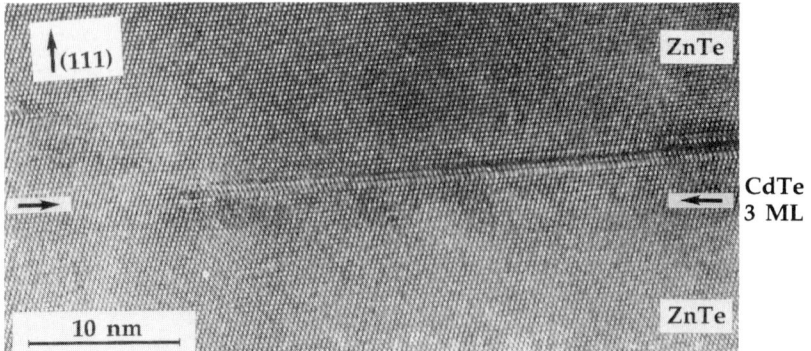

Fig. 6 <110> high resolution image of the 2nd QW (3 ML), same sample as in Fig. 5.

The nature of the partial dislocations bounding the stacking faults was identified by high resolution TEM. For instance in the case of Fig. 6, the stacking fault is of extrinsic nature and the partial dislocation lying at the top surface of the QW is a 30° partial. This is in agreement with the relaxation of a tensile stress as experienced by a ZnTe layer growing on a partially relaxed CdTe quantum well .

Twinning is known to be a very efficient high-stress low temperature deformation process in semiconductors with the diamond-like structure (Androussi 1989). In this process, twins and/or widely dissociated dislocations occur because of the very different friction forces that partial dislocations constituting perfect dislocations experience upon deformation. For epitaxial growth of the (111) type, the growth plane is the primary glide plane for the misfit dislocations whereas, for (001) growth, the primary glide planes are inclined to the growth plane; when created the misfit dislocations will thus experience a larger stress in the (111) case which could explain the larger extension of the stacking faults in this case as compared to the case of epitaxial growth of the (001) type. This difference in stresses may also be partly responsible for the different critical thicknesses in the (001) and (111) cases. For (111) growth, the two {111} planes are not equivalent as far as the stresses on the dislocations are concerned. We would thus expect much less dislocation interaction such as the formation of Frank or Stair-Rod dislocations as observed in the (001) case. This in turn can also influence the width of the stacking faults.

REFERENCES

Androussi Y, Vandershaeve G, Lefebvre A 1989 Phil. Mag. A 59 1189
Chami A C, Ligeon E, Danielou R, Fontenille J, Lentz G, Magnea N and Mariette H 1988 J. Appl. Phys. 52 1874
Cibert J, Gobil Y, Le Si Dang, Tatarenko S, Feuillet G, Jouneau P H and Saminadayar K 1990a Appl. Phys. Lett. 56 292
Cibert J, André R, Deshayes C, Feuillet G, Jouneau P H, Le Si Dang, Mallard R, Nahmani A, Saminadayar K and Tatarenko S 1990b Proc. 5th Int. Conf. on Superlattices and Microstructures, Berlin, J. Superlattices and Microstructures (to be published)
Fontaine C, Gaillard J P, Magli S, Million A and Piaguet J 1987 Appl. Phys. Lett. 50 903
Maree P M J, Barbour J C, Van der Veen J F, Kavanagh K L, Bulle-Liewma C W T and Viegers P A 1987 J. Appl. Phys. 62 4413
Tatarenko S, Cibert J, Gobil Y, Feuillet G, Ligeon E, Le Si Dang and Saminadayar K 1990 J. Cryst. Growth 101 126

Inst. Phys. Conf. Ser. No 117: Section 8
Paper presented at Microsc. Semicond. Mater. Conf., Oxford, 25–28 March 1991

627

TEM studies of Cd:Zn:S-based superlattices and epitaxial layers

P D Brown, Y Y Loginov[†], J T Mullins[*], T Taguchi[*] and K Durose

Applied Physics Group, School of Engineering and Applied Science, Durham Univeristy, Science Site, South Road, Durham, DH1 3LE.
† Krasnoyarsk State University, 79 Svobodnii Pr., Krasnoyarsk, 660062, USSR.
* Dept. of Electrical Engineering, Osaka University, Yamada Oka 2-1, Suita, Osaka 565, Japan.

ABSTRACT: Epitaxial CdS and (Cd,Zn)S, and (Cd,Zn)S‖ZnS and CdS‖(Cd,Zn)S superlattices have been grown by low pressure MOVPE. Cubic and hexagonal layers were formed on (001) and ($\bar{1}\bar{1}\bar{1}$)B GaAs substrates respectively. All layers exhibited complex stacking disorders and this was attributed to the tendency of CdS to adopt the wurtzite phase. Strong anisotropy in the microstructural defect content of CdS‖(Cd,Zn)S superlattices, which were almost lattice matched to (001) GaAs, was observed, whereas (Cd,Zn)S‖ZnS superlattices, which have a larger mismatch, exhibited inclined planar defects on all four {111} planes.

1. INTRODUCTION

Wide band-gap II-VI compounds are of interest for their luminescence and non-linear optical properties. The ternary alloy (Cd,Zn)S has a band-gap variable between 2.5eV and 3.7eV, and (Cd,Zn)S‖ZnS superlattices have potential for applications in the blue region of the spectrum. Indeed, such superlattices have shown stronger, non-linear, excitonic emission than the corresponding ternary alloys, large exciton binding energies and a quantum confined Stark effect (Endoh and Taguchi, 1990; Taguchi et al, 1990). CdS‖(Cd,Zn)S superlattices have potential at green wavelengths and may be lattice matched to GaAs. The microstructure of these superlattices is likely to be influenced by the dimorphism of CdS (it usually has the wurtzite lattice but can also adopt the sphalerite lattice) and the strong propensity for polytypism in ZnS. Indeed, Cullis et al (1989a,b) report the formation of cubic and hexagonal CdS grown by atmospheric MOVPE at 450°C on (001) and (111)A GaAs respectively. Also, the structural quality of CdS‖ZnS superlattices on (001) GaAs, as determined by PL (Endoh et al, 1990) and TEM studies (Parbrook et al, 1990), is found to be much poorer than ZnS‖ZnSe superlattices even though these systems have comparable (large) lattice mismatches with respect to GaAs.

2. EXPERIMENTAL

Epitaxial CdS and (Cd,Zn)S, and 120 period (Cd,Zn)S‖ZnS and CdS‖(Cd,Zn)S superlattices were grown on (001) and ($\bar{1}\bar{1}\bar{1}$)B GaAs substrates using a low pressure

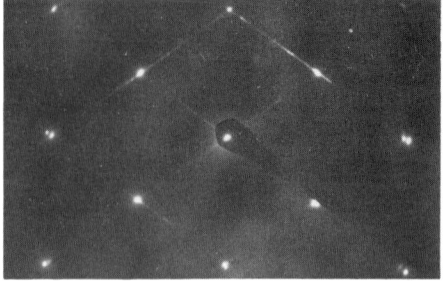

Fig. 1 CdS/(001) GaAs (g=220)

Fig. 2 [1$\bar{1}$0] diffraction pattern showing microtwin and polytype spots

MOVPE reactor as previously described
(Kawakami et al, 1987). The low pressure technique eliminates the gas phase
reactions associated with atmospheric
MOVPE. The reactor pressure was $\approx$0.3
torr and all layers examined in this study
were grown at a temperature of 350°C
unless otherwise stated, using H_2S (10%
in H_2, 20sccm), $(CH_3)_2Cd$ (0.13% in He_2,
15-30sccm) and $(CH_3)_2Zn$ (1.06% in He_2,
5-10sccm), with a hydrogen dilution flow
of 70sccm. The GaAs substrates were

Fig. 3 (Cd,Zn)S/(001) GaAs (g=220)

degreased in trichloroethylene vapour, etched in 4:1:1 H_2SO_4:H_2O_2:H_2O and then
boiling HCl, washed in 18MΩ deionised water, dried under N_2 and then heat treated
at 550°C for 10min in the growth reactor under the H_2 dilution flow prior to growth.
Samples were prepared for observation in the TEM using conventional techniques
with iodine reactive ion sputtering being used for the final stage of sample preparation.

3. RESULTS

Epitaxial CdS on (001) -oriented GaAs was found to be cubic and the defect microstructure comprised a high density of inclined planar defects (Fig. 1). The associated [1$\bar{1}$0] diffraction pattern from the epilayer contained an array of extra spots
with maxima at $\frac{1}{3} < 111 >$ positions (Fig. 2), and this indicates the presence of
microtwins and polytypes within the material. The extra spots showed no particular
periodicity indicating that no type of polytype was dominant. The ternary alloy
$Cd_{0.65}Zn_{0.35}S$/(001) GaAs (Fig. 3) was also found to be cubic. In this instance,
strong anisotropy in the microstuctural defect distribution was found with a high
density of inclined stacking faults or microtwins being observed only for the [1$\bar{1}$0]
substrate projection as determined by microdiffraction (Taftø and Spence, 1982).
The orthogonal [110] epilayer orientation contained a dense tangle of threading dislocations. In contrast, the ternary alloy $Cd_{0.4}Zn_{0.6}S$/($\bar{1}\bar{1}1$)B GaAs grown at 375°C
contained a complex array of planar defects lying parallel to the epilayer/substrate
interface (Fig. 4). A similar defect microstructure was also found in epitaxial

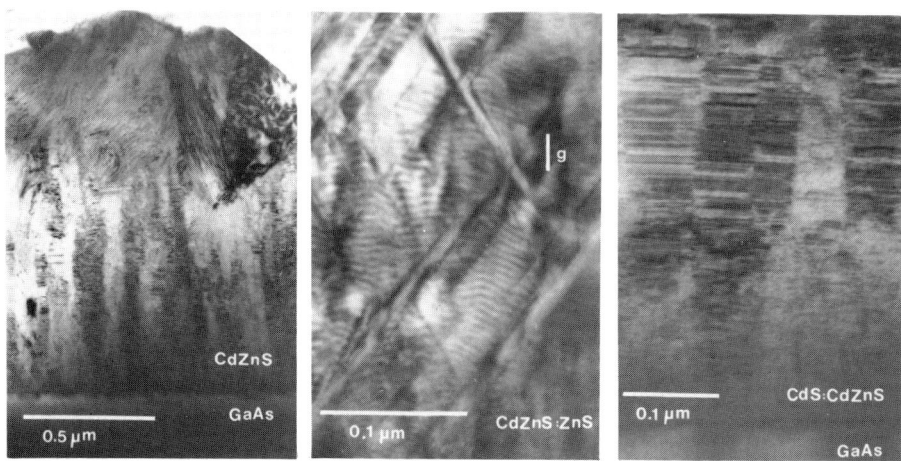

Fig. 4 (Cd,Zn)S/($\bar{1}\bar{1}\bar{1}$)B GaAs
Fig. 5 (Cd,Zn)S$\|$ZnS/(001) GaAs (g=004)
Fig. 6 CdS$\|$(Cd,Zn)S/($\bar{1}\bar{1}\bar{1}$)B GaAs

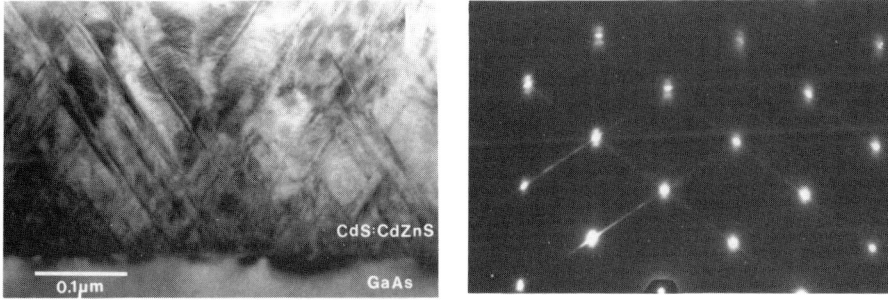

Fig. 7 CdS$\|$(Cd,Zn)S/(001) GaAs (g=004)
Fig. 8 [1$\bar{1}$0] diffraction pattern showing superlattice and polytype spots

CdS/($\bar{1}\bar{1}\bar{1}$)B GaAs. Diffraction patterns demonstrated that both of these epilayers were hexagonal with $< 11\bar{2}0 >_{epilayer}\|< 110 >_{GaAs}$. Streaking within the diffraction patterns was indicative of the presence of thin planar defects with an irregular spacing, while polytype spots were also present. In addition, grains near to the surface of the (Cd,Zn)S/($\bar{1}\bar{1}\bar{1}$)B GaAs sample (Fig. 4) were observed and these were found to be cubic.

Fig. 5 shows a section from a cubic $Cd_{0.3}Zn_{0.7}S\|ZnS$/(001) GaAs superlattice with a $\approx$45Å period. Microtwins were found to be present on all four inclined {111} planes, but no evidence of a polytype structure was found. The CdS$\|$(Cd,Zn)S/($\bar{1}\bar{1}\bar{1}$)B GaAs superlattice structure shown in Fig. 6 was found to be hexagonal and contained a similar distribution of irregular planar faults as found in CdS and (Cd,Zn)S on ($\bar{1}\bar{1}\bar{1}$)B GaAs. An estimate of the fringe spacing near the surface, however, indicted a period of $\approx$ 31Å which corresponded to that of the target superlattice structure in this instance. The CdS$\|$(Cd,Zn)S/(001) GaAs superlattice shown in Fig. 7 has a $\approx$ 33Å period and was found to exhibit a similar anisotropic distribution of microtwins as epitaxial (001)(Cd,Zn)S. A [1$\bar{1}$0] diffraction pattern (Fig. 8) taken from a 45Å period CdS‖(Cd,Zn)S superlattice confirmed that this (001)-oriented superlattices was cubic. Pairs of superlattice spots, present for all of the superlattice structures investigated, were found to be broadened or split. This was attributed to morphological roughness within the superlattice structure. Streaked extra spots and polytype spots were also present.

4. DISCUSSION

All (001)-oriented samples were cubic even though bulk-grown CdS tends naturally to form the wurtzite phase and ZnS tends to twin and form polytypes. Also, it is interesting to note that (001) (Cd,Zn)S‖ZnS superlattices contained only microtwins, while (001) -oriented CdS‖(Cd,Zn)S superlattices, and epitaxial CdS and (Cd,Zn)S additionally exhibited polytypism, more so within samples containing higher proportions of Cd, as indicated by the presence of arrays of additional spots within the diffraction patterns. CdS, (Cd,Zn)S and CdS‖(Cd,Zn)S superlattices grown on ($\bar{1}\bar{1}\bar{1}$)B -oriented GaAs substrates were all found to be hexagonal and contained a complex irregular array of thin planar defects and polytypes parallel to the epilayer/substrate interface. As a general trend it is considered that Zn acts to stabilise cubic (Cd,Zn)S and draws the system away from the unstable CdS hexagonal-cubic transition region. Potential for the development of (Cd,Zn)S‖ZnS superlattices remains provided the microtwin and threading dislocation contents can be reduced by refined growth strategies and post growth processing procedures. Exploitation of CdS‖(Cd,Zn)S superlattices will probably be severely hindered by the tendency for high Cd containing layers to form polytypes.

ACKNOWLEDGEMENT

PDB would like to acknowledge SERC for financial support under grant No. F05855.

REFERENCES

Cullis A G, Smith P W, Parbrook P J, Cockayne B, Wright P J and Williams G M, 1989a Appl. Phys. Lett. 55 2081

Cullis A G, Williams G M, Cockayne B, Wright P J, Smith P W, Parbrook P J and Halsall M P, 1989b Inst Phys Conf Ser 100 217

Endoh Y, Kawakami Y, Taguchi T and Hiraki A, 1990 Jpn J. Appl. Phys. 16 L2199

Endoh Y and Taguchi T, 1990 Mat. Res. Soc. Symp. 161 211

Kawakami Y, Taguchi T and Hiraki A, 1987 J. Vacuum Sci. Technol. B5 1171

Parbrook P J, Wright P J, Cockayne B, Cullis A G, Henderson B and O'Donnell K P, 1990 J. Crystal Growth 106 503

Taftø J and Spence J C H, J Appl. Cryst. 15 60

Taguchi T, Endoh Y and Nozue Y, 1990 Proc. 20th Int. Conf. on Physics of Semiconductors 1159

Inst. Phys. Conf. Ser. No 117: Section 8
Paper presented at Microsc. Semicond. Mater. Conf., Oxford, 25–28 March 1991

Electron scattering contrast mechanisms of doped semiconductor layers

D D Perovic[†], G C Weatherly[*], J H Paterson[†], C J Rossouw[§] and T E Jackman[+]

† Cavendish Lab., University of Cambridge, Madingley Rd., Cambridge, CB3 0HE U.K.
* Dept. of Materials Science and Eng., McMaster University, Hamilton L8S 4L7 Canada
§ H.H. Wills Physics Lab., University of Bristol , Tyndall Ave., Bristol BS8 1TL U.K.
+ Inst. for Microstructural Sciences, National Research Council, Ottawa, K1A 0R6 Canada

ABSTRACT: The origin of the electron microscope contrast of B- and As-doped Si multi-layers viewed in cross-section has been investigated. It will be demonstrated that the built-in elastic displacements associated with misfitting dopant atoms is a principal source of electron scattering contrast when imaging layers of relatively low dopant concentration. In the absence of atomic misfit, any observed contrast must be associated with the difference in atomic scattering factors of the dopant and matrix atoms. Experimental two-beam and high-angle dark field image contrast results have been quantitatively predicted using Bloch-wave scattering theory.

1. INTRODUCTION

With the development of crystal growth techniques such as molecular beam epitaxy (MBE), it is possible to fabricate modulation-doped superlattices consisting of alternating, ultra-thin layers of *n*- and/or *p*-type material of relatively high dopant concentration. Thus electron microscopy is ideally suited for characterizing doping layers provided the observed contrast can be correctly interpreted. Surprisingly, only a few workers to date have studied the influence of doping on the electron microscope contrast of semiconductor crystals. The first study was carried out by Hall *et al.* (1966) who used conventional TEM (CTEM) to measure increases in the absorption parameters of Si doped with As, Cu and B atoms. They suggested that two possible mechanisms were responsible for enhanced absorption:

      (i) additional scattering associated with the dopant atoms and
      (ii) elastic displacement of the matrix atoms surrounding misfitting dopant atoms.

More recent work has focussed on the use of a high-angle annular detector (HAAD) on a dedicated scanning transmission electron microscope (STEM) to image low levels of dopants (~ 1 *at.%*) (Pennycook *et al.* 1986, Treacy *et al.* 1988). Unlike Pennycook *et al.*, Treacy *et al.* concluded that the enhanced contrast from Sb-doped Si was associated with impurity atom-matrix misfit strains and was estimated to contribute about 6 times more to the contrast than the true atomic number (Z)-contrast effect associated with the additional scattering from Sb atoms (ie. mechanism (i)). However, in light of x-ray diffraction data, it appears that Treacy *et al.* have overestimated the misfit strain contribution by one order of magnitude (see §5).

In this work, we have been interested in generally understanding the true electron scattering mechanisms in doped semiconductors using B- and As-doped Si multi-layers as model systems. Accordingly, it is then possible to establish the optimum conditions for imaging doped semiconductor layers at very low concentrations ($<< 1at.\%$).

2. EXPERIMENTAL DETAILS

The <100>-oriented B- and As-doped Si multi-layer samples used for electron microscopy in this study were grown by MBE using coevaporation and low-energy ion implantation respectively; further details can be been found elsewhere (Perovic *et al.* 1991). The substrate temperature was varied (350-800°C) in several samples in order to study the effects of different dopant atom configurations (ie. substitution vs. precipitation) on the observed contrast. Secondary-ion mass spectroscopy and Rutherford backscattering/channelling have been used to profile dopant-atom concentrations and configurations.

Cross-sectional samples with <110> surface normals were prepared for TEM observation in the standard way by mechanical thinning followed by Ar atom milling. CTEM examination was carried out on several conventional electron microscopes including: a Philips EM430T (250 kV), an Hitachi H-800 (200 kV) and a JEOL JEM2000EX (200 kV). In addition, several samples were studied by STEM using a VG HB501UX (100 kV).

3. CTEM/STEM RESULTS

The initial observations of a B-doped (4×10^{20} B/cm^3) multi-layer which piqued our interest in this subject is shown in Fig. 1. The relatively strong *dark* contrast resulting from an increase in both the normal and anomalous absorption coefficients, and the significant increase in extinction distance from the B-Si regions were unexplainable using a simple Z-contrast description (Perovic *et al.* 1991). From two-beam experiments, it was found that a reduction in the elastic (Bragg) scattering from the B-doped regions results due to the presence of atomic displacements of the surrounding Si atoms which effectively behave as "frozen-in" (static) phonons. In this way, the enhanced "absorption" within the B-doped layers should quasi-elastically scatter electrons, away from the Bragg positions, to relatively high angles which can be imaged in STEM with an annular detector.

Figure 2 shows a <110> zone-axis HAAD-STEM image of the same structure shown in Fig. 1 where the B:Si layers now appear bright relative to the Si. Furthermore, as in Fig. 1, the magnitude of the contrast is observed to decrease as the B-atoms precipitate from the fully substitutional condition at higher growth temperatures. Therefore, there exists a direct correspondence between elastic strain and the resultant image contrast (Perovic *et al.* 1991).

Figure 3a is a digitally recorded HAAD image of an As-B:Si *nipi* doping superlattice. In order to enhance low contrast levels, the raw image was processed using SEMPER (Saxton *et al.* 1979). Specifically, the grey-levels were remapped using a square-root look-up table following background levelling. The B-doped regions (3.5×10^{20} B/cm^3) again appear bright as in Fig. 2. Furthermore, weak contrast from the As-doped (8×10^{18} As/cm^3) regions can be seen between the B-Si layers; this is clearly shown in the 1-D intensity profile of Fig. 3b taken along a line normal to the interfaces and then averaged over the complete image.

It is known from several x-ray diffraction studies that substitutionally-doped As induces negligible strain in Si for concentrations below 1×10^{19}As/cm^3. (The As-doped layers of Fig. 3 are known to be fully substitutional from ion-channelling studies). Accordingly, the observed HAAD contrast from the As:Si regions must be associated with simple large-angle Rutherford scattering (ie. Z-contrast) owing to the significantly larger atomic number of As relative to Si.

4. BLOCH-WAVE CALCULATIONS

In order to verify the relative contribution of either scattering mechanism described in §1, theoretical calculations based on Bloch-wave theory were performed. The strain-dependent scattering mechanism has been treated based on thermal diffuse scattering (TDS) theory using an Einstein model. Accordingly, the increase in dopant-layer scattering can be represented by an effective Debye-Waller factor which is proportional to a static mean-square displacement $<v_g>^2$ parallel to a given set of reflecting planes. An Eshelby-type model can be used to theoretically estimate $<v_g>$ from a measure of the "macroscopic" misfit strain, e_o. Using e_o values determined from x-ray diffraction, it was found (Perovic *et al.* 1991) that the increased magnitude of the two-beam absorption coefficients and the increase in extinction distance from B-doped regions (cf. Fig. 1) can be successfully accounted for using TDS theory. Alternatively, the contrast from As-doped Si regions can be interpreted simply based on differences in scattering cross-section (Lenz model) between As and Si atoms.

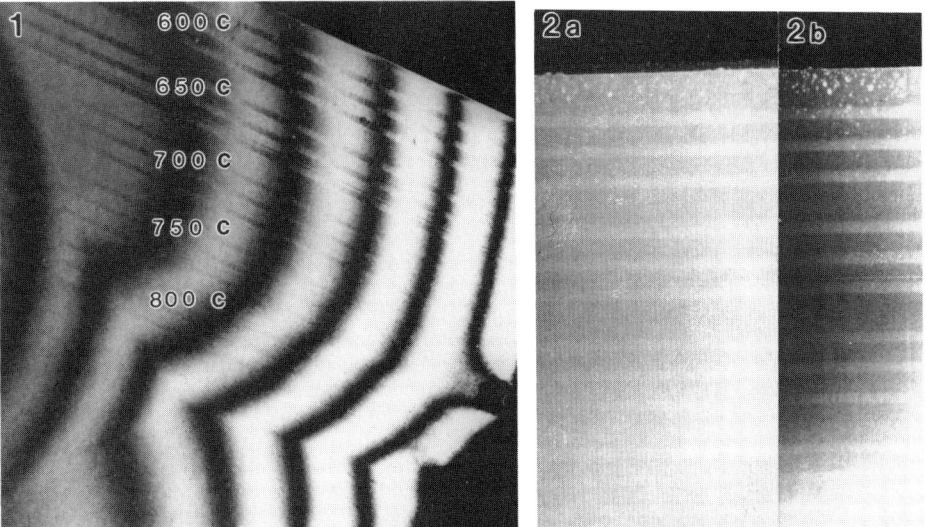

Fig. 1: CTEM two-beam (g= 022) dark field image of a B-Si multi-layer taken from a wedge profile specimen. The B-Si layers have been grown at various substrate temperatures as indicated. At higher growth temperatures the B-atoms change from a fully substitutional configuration (ie. 600 °C) to a precipitated state (ie. 800 °C).

Fig. 2: HAAD-STEM image of the structure shown in Fig. 1 taken along the <110> zone-axis with a 100 μm objective aperture. Note the reversal in B:Si contrast relative to Fig.1.
(a) raw intensity image (ie. black level= 0); (b) high contrast setting.

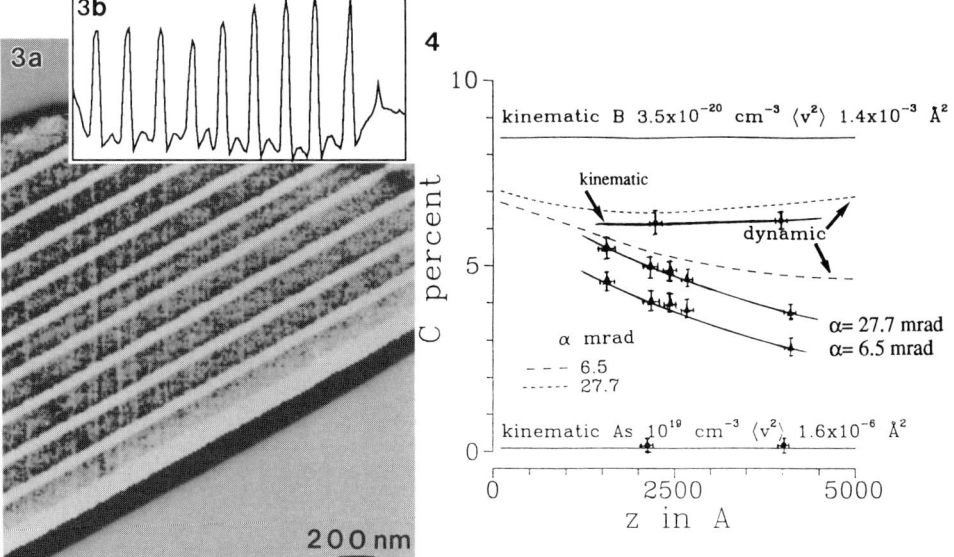

Fig. 3:(a) Digitally recorded HAAD-STEM image, using a 100 μm objective aperture (α= 27.7 mrad), of an As-B:Si multilayer structure grown at 700 °C. (b) 1-D intensity profile of (a).

Fig. 4: Multi-beam Bloch-wave calculation for the B- and As-doped layers imaged in Fig. 3. Experimental measurements (with error bars) are added for kinematic and dynamic conditions.

More recently, we have carried out many-beam, dynamical calculations to compare with the HAAD-STEM results. A Bloch-wave treatment was used wherein eigenanalysis routines enable calculation of diffracted beam intensities with TDS absorption, based on an Einstein model, included via perturbative methods (Rossouw *et al.* 1990). The total dynamical intensity as a function of foil thickness (z) is obtained by summing all diffracted beam intensities and associating this with an effective mean free path ($\lambda(z)$) for TDS where $\lambda(z)$ may be for total TDS or for any limited angular scattering range, as defined by the HAAD for example. For comparison with images taken away from a zone-axis orientation, the diffracted intensities were also calculated for kinematical scattering conditions under otherwise identical conditions.

Fig. 4 shows the result of a <110> zone-axis calculation using 51 beams. The contrast, defined as [I(doped)/I(undoped)-1], is plotted for two incident beam convergence angles for the B- and As-doped layers shown in Fig. 3. Experimental data for both dynamical and kinematical conditions have been added for comparison. The measured values were taken from relatively thick foil regions to avoid surface stress relaxation contributions. The contrast under exact zone-axis diffraction conditions is reduced because "absorption" or scattering to high-angles due to TDS is enhanced more in the unstrained Si layer than in the strained layer since the TDS is more localized in the unstrained Si. As the degree of "peaking" is decreased, either by increasing the convergence angle or by tilting to kinematical diffracting conditions, the dynamically enhanced intensity from Si is reduced more severely from Si than from the less localized (strained layer) source thus increasing the doped-layer contrast.

5. DISCUSSION

The results of Fig. 4 indicate that relatively good agreement between experiment and theory has been obtained for both B- and As-doping contrast. In particular, the dependence of HAAD contrast on foil thickness, incident beam convergence, crystal orientation and $<v_g>^2$ has been well-predicted. It should be noted that the semi-empirical $<v_g>^2$ values employed in the calculations make use of "macroscopic" misfit strain data as obtained from x-ray diffraction studies. This will avoid any possible inaccuracies associated with deformation potential effects which are not accounted for in strain calculations based solely on differences in atomic radii.

It is evident that significant contrast enhancement results for doped semiconductor layers which induce a significant static Debye-Waller factor effect. Further evidence supporting a TDS description for doping contrast comes from small aperture, two-beam experiments in the CTEM. Images were taken with a displaced objective aperture positioned equal distances parallel and perpendicular to the g= 022 systematic row. The dopant contrast was observed to be significantly higher perpendicular to the systematic row where intraband scattering is dominant (ie. contrast is preserved). This contrast behaviour follows the theoretical phonon scattering predictions of Rez *et al.* (1977) and may explain the anomalous contrast effect observed by Baxter *et al.* (1988) in g= 002 dark field images of InGaAs/GaAs multi-layers.

ACKNOWLEDGEMENTS

The authors would like to thank Dr. W.O. Saxton for his help with image processing.

REFERENCES

Baxter, C.S., Stobbs, W.M., Monserrat, K.J. and Tothill, J.N., 1988, *Proc. Analytical Electron Microscopy, EMAG '87*, ed. G.W. Lorimer, (London, Inst. of Metals), p. 209.
Hall, C.R., Hirsch, P.B. and Booker, G.R., 1966, *Phil. Mag.*, **14**, 979.
Pennycook, S.J., Berger, S.D. and Culbertson, R.J., 1986, *J. Microsc.*, **144**, 229.
Perovic, D.D., Weatherly, G.C., Egerton, R.F., Houghton, D.C. and Jackman, T.E., 1991, *Phil. Mag.*, **63**, (4).
Rez, P., Humphreys, C.J. and Whelan, M.J., 1977, *Phil. Mag.*, **35**, 81.
Rossouw, C.J., Miller, P.R., Drennan, J. and Allen, L.J., 1990, *Ultramicrosc.*, **34**, 149.
Saxton, W.O., Pitt, T.J. and Horner, M., 1979, *Ultramicrosc.*, **4**, 343.
Treacy, M.M.J, Gibson, J.M., Short, K.T. and Rice, S.B., 1988, *Ultramicrosc.*, **26**, 133.

Inst. Phys. Conf. Ser. No 117: Section 8
Paper presented at Microsc. Semicond. Mater. Conf., Oxford, 25–28 March 1991

635

Strain characterisation in Si/SiGe superlattices by convergent beam electron diffraction

R Touaitia[1], D Cherns[2], C J Rossouw[2,4] and D C Houghton[3]

[1] Interface Analysis Centre, University of Bristol, Oldbury House, 121 St Michael's Hill, Bristol BS2 8BS, UK
[2] H H Wills Physics Laboratory, University of Bristol, Tyndall Avenue, Bristol BS8 1TL, UK
[3] Division of Physics, National Research Council of Canada, Ottawa, Canada K1A OR6
[4] On leave from CSIRO Division of Materials Science and Technology, Clayton, Australia 3168

ABSTRACT: A method is described whereby strain in $Si_{1-x}Ge_x/Si(001)$ superlattices can be derived using convergent beam electron diffraction (CBED) and large angle CBED. The method uses plan-view samples, thus avoiding strain relaxation in cross-sectional samples. Rocking curves for reflections from planes inclined to the interface were asymmetric and showed superlattice peaks whose intensities were sensitive to strain. A quantitative analysis of the experimental results using kinematical and dynamical simulations showed good agreement for the expected strains to within an accuracy of $\pm 20\%$.

1. INTRODUCTION

Strained layers systems are very promising for novel optoelectronic devices. In the Si/SiGe system, by varying the Ge concentration, ie strain, the band gap can be varied over the 1.3-1.55μm wavelength range used in optical communication (Osbourn (1982) and People and Bean (1986)). It is therefore very important to be able to characterize strain in the strained layer system. Previous work on Si/SiGe used convergent beam imaging (CBIM) technique on cross-sectional samples (Humphreys et al (1988), Eaglesham et al (1989)). In this technique, the shift of HOLZ lines across an interface, which is due to strain, is measured. However, discrepancies have been found between experiment and the expected result from tetragonal distortion. This is believed to be due to surface relaxation effect in cross-sectional samples (Treacy et al (1985), Perovic et al (1990)).

In previous work, the LACBED technique has been used to measure strain in bicrystals (Cherns (1989), Cherns et al (1989)). The technique uses plan-view samples, thus avoiding the problem of surface relaxation. For a strained bicrystal, inclined planes undergo a small rotation so that reflections from these planes are split along the foil normal (fig. 1a). As the rocking curve in LACBED technique is scanned along the beam direction (fig. 1b), strain can be observed as a split peak.

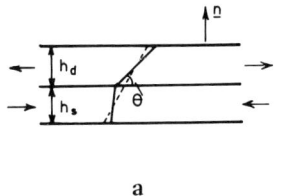

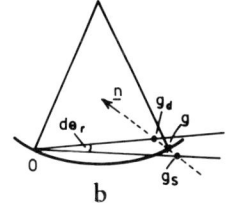

Fig. 1 (a) Rotation of inclined planes in strained bicrystal. (b) Reflections for planes common to both crystals are split along the specimen normal $\underline{n}$.

a

b

In this study, LACBED and CBED are used to generate rocking curves from plan-view Si/SiGe samples from which strain can be characterized. The rocking curves, which showed superlattice peaks, are asymmetric for reflections from planes inclined to the 001 growth direction and are sensitive to strain. Superlattice peaks are separated in reciprocal space by $q = 1/\lambda$, where λ is the fundamental repeat distance of the superlattice cell. Experimental rocking curves were compared to simulated rocking curves using kinematical and dynamical theory from which a measure of strains was possible to within an accuracy of $\pm 20\%$.

2. EXPERIMENTAL METHODS AND RESULTS

Three different Si/SiGe samples, grown by MBE on (001)Si in a VG Semicon V80 system, were used in this study. Their specifications are shown in Table 1. Transmission electron microscopy (TEM) samples were prepared by mechanical thinning and Ar ion milling. TEM has been performed in a Philips EM430 microscope. CBED pattern were recorded at 100 kv and LACBED patterns at 250 kv with a convergence angle of 6° and a minimum probe size of 150 Å.

Sample number	Si thikness(Å)	$Si_{1-x}Ge_x$ thickness (Å)	x
26	477	162	0.33
346	103	62	0.17
349	109	73	0.25

Table 1 Specifications for the Si/SiGe superlattices examined. Layer thicknesses were determined from cross-sectional TEM.

The samples showed only few dislocations suggesting that almost all the misfit between the multilayers is accommodated by strain. Fig. 2 shows a dark field LACBED patterns from sample 346. In fig. 2a, the reflection used is 220 which is from planes perpendicular to the interface. The pattern was recorded near the [001] zone axis. The sample was tilted between exposures to pick up a wide range of deviation parameter. It can be seen that the rocking curve is symmetrical. The rocking curve in fig. 2b is from 022 inclined planes. This reflection is found near a [111] pole after a tilt of about 55° from the horizontal. It can be seen that the rocking curve is now asymmetrical, with, for example, the peak at 2q being stronger than the corresponding peak at -2q. The main features are explained using theoretical simulations as discussed later.

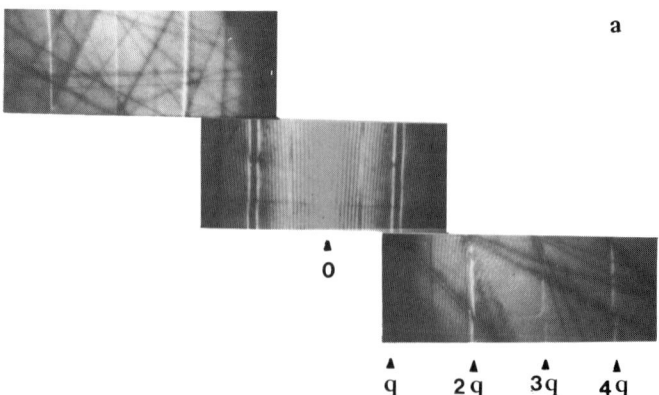

a

0   q   2q   3q   4q

Fig. 2a **220** Dark field LACBED pattern from sample 346. The sample was oriented about 10° off the horizontal and tilted by few degrees approximately about [110] to record the three patterns.

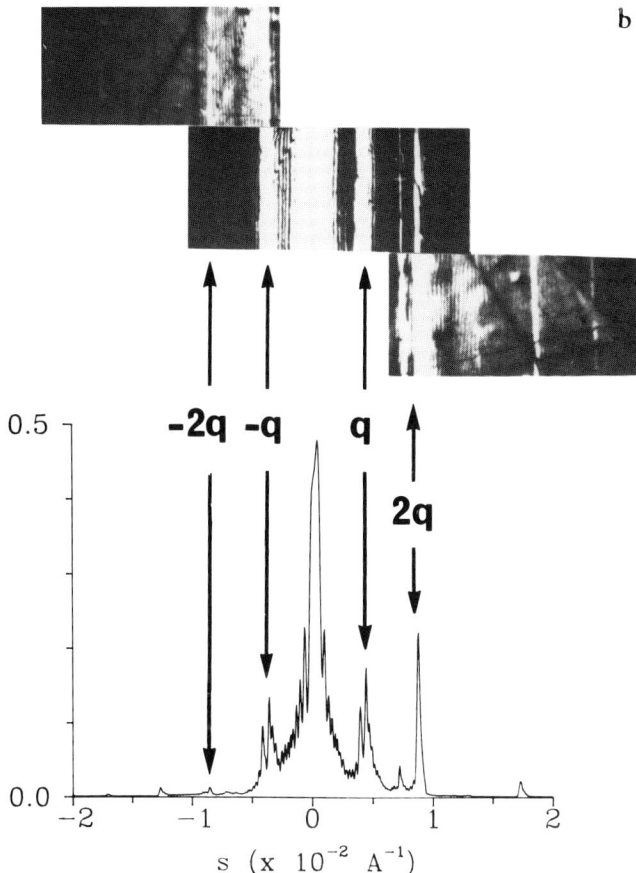

Fig. 2b   **022** dark field LACBED pattern. The diffracting planes are inclined at 45° to the growth plane.   The dynamical simulation shows a good agreement with experiment.

Fig. 3a shows a LACBED bright field pattern taken along the [102] zone axis from sample 26. It can be seen that the 040 rocking curve from planes perpendicular to the interface is symmetrical while the rocking curve from the {422} inclined planes is more complex and shows a series of deficiency lines. HOLZ reflections in a conventional [102] CBED pattern were also split into series of branches as shown in fig. 4. It can be seen that the branches for refections g perpendicular to the tilt axis, such as ($\overline{1}4$ 4 6) and (9 5 $\overline{5}$) are divided into two groups.

3. DISCUSSION

Kinematically, the diffracted amplitude for a reflection g is given by

$$A = \frac{i\pi}{\xi_g} \int_0^t F_{hkl}(z) \exp\{-2\pi i s_g z\}\, dz \qquad (1)$$

where $F_{hkl}(z)$ is the structure factor at a depth z in the foil, ξ_g being the extinction distance associated with g. A superlattice structure can be seen as a convolution of one period λ of the superlattice with a finite regular one dimensional lattice of spacing . In this case, A can be written in the following form using the convolution theorem

$$A = A_1 \ X \ A_2 \qquad (2)$$

where A_1 is the diffraction function for one period, ie a Si/SiGe bicrystal. A detailed description of the kinematical calculation is given elsewhere (Cherns et al (1991)).

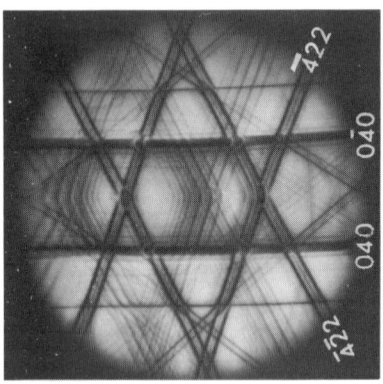

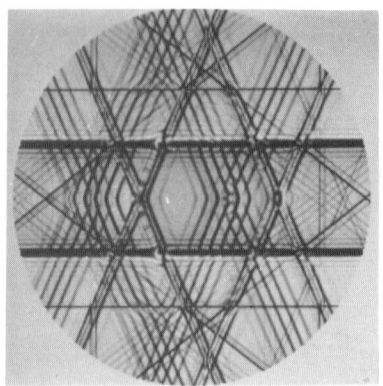

Fig. 3a Bright field LACBED pattern taken along [102] zone axis from sample 26.

Fig. 3b Dynamical simulation of the LACBED pattern in fig. 3a.

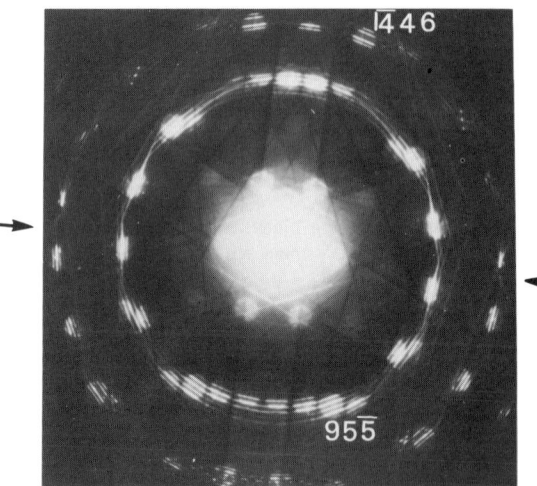

Fig. 4 CBED pattern taken along [102] zone axis from sample 349 showing complex branching in the HOLZ rings. The branching is most extensive for reflections perpendicular to the tilt axis (see arrows).

Fig. 5 shows computed kinematical rocking curves, to be correlated with experiments in fig. 2b where g=022. It is assumed that all the mismatch is accommodated by elastic strain. Fig. 5a shows the rocking curves for the two layers in the bicrystal ie Si and SiGe. In fig. 5b, the rocking curve for one period is plotted as a dashed curve, while the solid line shows the diffracted intensity for the whole crystal. It can be seen that the bicrystal rocking curve acts as an envelope for the superlattice peaks. The position of the two peaks in the intensity envelope layers (fig. 5a) depend on the value of strain. This shift, coupled with the difference in the structure factor and thicknesses of the two layers (which control the peaks heights and widths) results in an asymmetry in the bicrystal rocking curve, hence an the asymmetry of the superlattice reflection intensities. Our previous results (Cherns et al (1991)) show that for small strains, $I_q = I_{-q}$ and $I_{2q} = I_{-2q}$ in fig. 2b confirming that

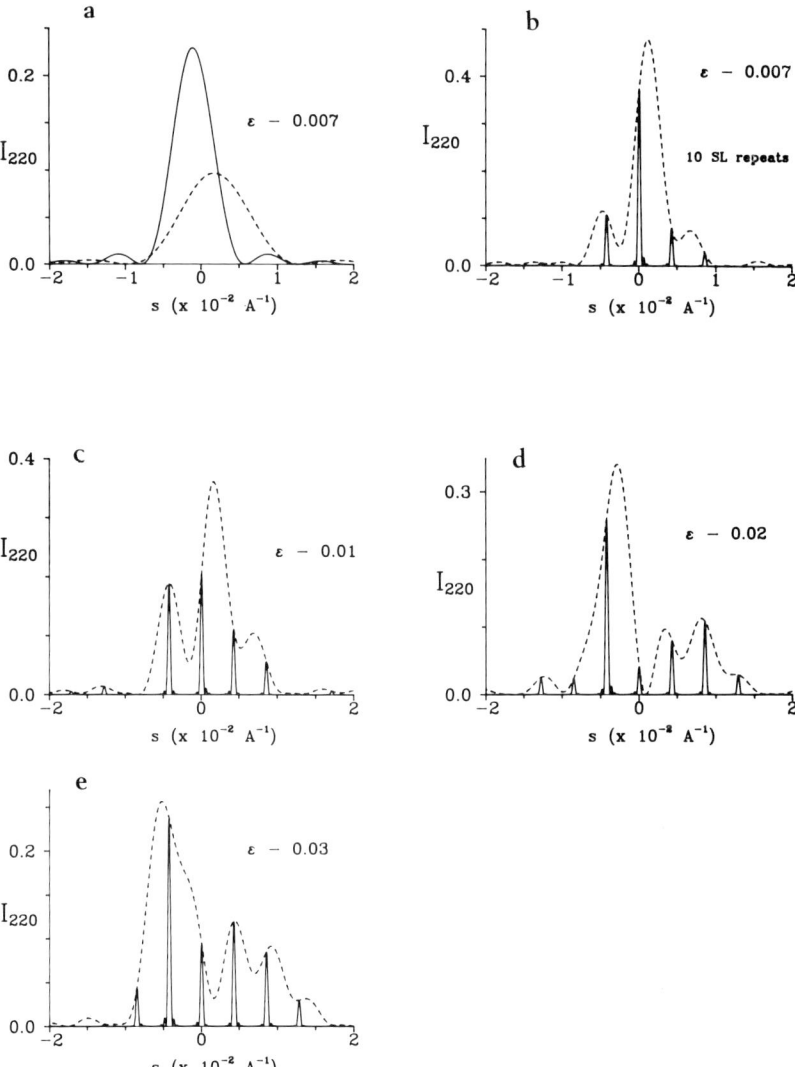

Fig. 5 Kinematical rocking curves (| A | ^{2}, equation 1) for g = 022, sample 346. (a) shows rocking curves for the Si and SiGe layers separately and (b) shows the rocking curve for one period of the multilayers (dashed) which acts as an envelope for the superlattice peaks. As the strain increases, the crystal rocking curve shows a transition from a single to a double set of superlattice reflections. Intensities are given in arbitrary units $(c - e)$.

the rocking curve is symmetrical. As the value of strain increases, the intensities of the satellite peaks increase dramatically and the q and -q or 2q and -2q peaks have no longer the same intensity. Asymmetry between superlattice reflections of oposite sign may be used to detect and measure strain. A rough estimate is obtained via the simple kinematical approach (c.f fig. 5b). However, due to the presence of dynamical effects, such as the splitting of the satellite 2q, the rocking curve in fig. 2b is compared with a dynamical simulation (fig. 2c) which shows a quantitative agreement with experiment. A detailed description of the dynamical computation is given elsewhere (Rossouw et al (1990)). By estimating the areas under the peak at $+q$ and $-q$, we can derive the ratio $I_q : I_{-q} = 2.3 \pm 0.5$. Experimental intensities were obtained by digitising the rocking curve in fig. 2b using an Optronics P-1000 Photoscan. The ratio $I_q : I_{-q}$ was found to be about 2.2 ± 0.5, in good agreement with the theory. We believe that strains as low as 2×10^{-4} should be measurable.

The effect of strain on the rocking curve is shown in fig. 5. For relatively small strain (fig. 5b), rocking curves show a strong peak flanked by superlattice reflections separated by q. As the strain increases (fig. 5(c-e)), the rocking curves have no obvious central peak. For high strains, the two peaks in the bicrystal rocking curve move further apart, and the crystal rocking curve consists of two separate sets of superlattice peaks as observed for {422} reflections in fig. 3. We also expect this behaviour from HOLZ reflections, such as (14 4 6) and (9 5 5) in fig. 4, as the splitting of the diffraction envelope function is greater than for low order reflections. The dynamical simulation of the [102] LACBED pattern in fig. 3b is in good agreement with the experimental pattern in fig. 3a. The series of deficiency lines associated with the {422} reflections are superlattice reflections separated by q. It can be deduced that the two main groups of deficiency lines are associated with the two main peaks of the diffraction envelope for one Si/SiGe period as discussed earlier. Thus by fitting the maxima of the intensity envelope between theory and experiment, with strain as parameter, we find that value of strain is correct within $\pm 20\%$ or better.

4. CONCLUSIONS

We have shown that strain in multilayered structures can be characterized by convergent beam electron diffraction techniques. The rocking curves show superlattice reflections whose intensities fall within an envelope of the rocking curve for one structure period. The strain which induces an asymmetry into the rocking curve, can be measured either by measuring the intensity for opposite superlattice satellites or by fitting the intensity envelope of one structure period.

ACKNOWLEDGEMENTS

Financial support is gratefully acknowledged from the Algerian Government (RT) and from the Wolfson Foundation (CJR).

REFERENCES

Cherns D (1989) *"The Evaluation of Advanced Semiconductor Materials by Electron Microscopy"*, NATO ASI Series B (Physics), ed. D Cherns, 203, 59
Cherns D and Preston A R (1989) *J. Electron Microscope Technique* 13, 111
Cherns D, Touaitia R, Preston A R, Rossouw C J and Houghton D C (1991) *Phil. Mag.* (in press)
Eaglesham D J, Maher D M, Fraser H L, Humphreys C J and Bean J C (1989) *Appl. Phys. Lett.* 54, 222
Humphreys C J, Maher D M, Fraser H L and Eaglesham D J (1988) *Phil. Mag.* 58, 787
Osbourn G C (1982) *J. Appl. Phys.* 53, 1586
People R and Bean J C (1986) *Appl. Phys. Lett.* 48, 538
Rossouw C J, Al-Khafaji M, Cherns D, Steeds J W and Touaitia R (1991) *Ultramicroscopy* (in press)

Inst. Phys. Conf. Ser. No 117: Section 8
Paper presented at Microsc. Semicond. Mater. Conf., Oxford, 25–28 March 1991

Misfit dislocation injection in strained layer epitaxy

D D Perovic[†] and D C Houghton[*]

† Cavendish Lab., University of Cambridge, Madingley Rd., Cambridge, CB3 0HE U.K.
* Inst. for Microstructural Sciences, National Research Council, Ottawa, K1A 0R6 Canada

ABSTRACT: The nucleation of misfit dislocations in Ge_xSi_{1-x}/Si ($x < 0.3$) heterostructures has been investigated microscopically using diffraction contrast in the TEM and macroscopically using large-area imaging techniques such as Nomarski interference microscopy of chemically etched surfaces. Heterogeneous nucleation at strained layer interfaces and at free surfaces via half-loop injection is illustrated. A series of structures has been studied which possesses a wide density range of internal heterogeneities ($N_0 = 10^2$-10^6 cm^{-2}) for which a linear increase in misfit dislocation density with increasing N_0 was found. Bulk nucleation measurements from post-growth annealed specimens indicate the existence of a low activation energy ($Q_n = 2.5 \pm 0.2$ eV) for misfit dislocation nucleation independent of heterostructure geometry for misfit strains $< \sim 1\%$.

1. INTRODUCTION

The unique opto-electronic and transport properties of Ge_xSi_{1-x}/Si strained layer devices depend critically on the retention of structural perfection throughout epitaxial growth and subsequent thermal processing. Accordingly, several research groups have studied the coherency breakdown phenomenon inherent in these metastable structures in order to understand the mechanisms of elastic strain relaxation.

Surprisingly, the nucleation of misfit dislocations in strained layer epitaxy is still the least well understood process in misfit strain relaxation. Many workers have considered the nucleation stage from a theoretical basis, in particular the surface nucleation of half loops (see Jesser and van der Merwe (1989) for a recent review). Recently, these models have been revised to include more realistic dislocation core energies and heterostructure geometries (Hull *et al.* 1989, Kamat and Hirth 1990). However, these theoretical studies have been unable to predict nucleation behaviour and have been hampered by a paucity of experimental data.

On the other hand, very few workers have experimentally explored microscopic nucleation mechanisms in the Ge_xSi_{1-x}/Si system. Humphreys *et al.* (1989) re-evaluated the so-called "diamond defect" which acts as a multiply regenerative, heterogeneous source of 60° misfit dislocations. This source was suggested to originate from interstitial precipitation based on bond-breaking arguments. Alternatively, in our previous work (Perovic *et al.* 1989) it was found that intentionally introduced, coherent SiC precipitates located at the substrate-epitaxial layer interface can act as efficient sources of 60° misfit dislocations. In addition, we also illustrated the existence of surface half-loop nucleation in a $Ge_{0.25}Si_{0.75}$/Si epitaxial layer (Perovic *et al.* 1990). The one earlier report of surface nucleation by Kvam *et al.* (1988) has been re-interpreted by Humphreys *et al.* (1989) in terms of the "diamond defect" source. However, we have never observed the "diamond defect" in our TEM experiments.

In this study we present new TEM evidence for the internal nucleation of 60° misfit segments at Ge_xSi_{1-x}/Si ($x < \sim 0.3$) interfaces. Furthermore, we employ large-area imaging techniques, which are much more statistically significant as compared to TEM analysis, to make bulk measurements of misfit dislocation nucleation rates and velocities for Ge$_x$Si$_{1-x}$/Si structures of various geometries (ie capped and uncapped individual alloy layers and superlattices) which possess a wide range effective stress to drive misfit dislocation formation.

2. EXPERIMENTAL DETAILS

Ge$_x$Si$_{1-x}$/Si heterostructures were grown by MBE on <100>-oriented Si substrates using procedures described elsewhere (Houghton 1990). The composition (x) and strained layer thicknesses were chosen to provide a wide range of effective stress (0-750 MPa) and were determined by double-crystal x-ray diffraction and TEM.

As-grown specimens were rapid thermal annealed in a forming gas ambient for times and temperatures ranging from 5-200 sec and 450-1000 °C respectively. Plan-view electron microscopy was used to study the dislocation nucleation process in detail. In addition, Nomarski interference microscopy of chemically etched specimens (Houghton 1990,1991) was employed to measure misfit dislocation nucleation rates and velocities from bulk samples.

3. MICROSCOPIC ANALYSIS (ELECTRON DIFFRACTION CONTRAST)

Fig. 1 is a weak-beam image ($g= \overline{6}60$) from a plan view sample with an [001] foil normal showing surface-nucleated half-loops. The sample has been tilted ~45° such that the electron beam direction is near the [011] zone-axis. One can see several 60° dislocation half loops lying on different {111} variants. The straight interfacial misfit segments are joined to the free surface by two threading segments which are pure screw character and exhibit the oscillatory contrast characteristic of steeply inclined dislocations. The sense of the inclined segments were determined from several stereo pairs which confirmed that the half-loops terminate at the free surface above the strained interface. Specifically, the half-loops shown in Fig. 1 have piled up on separate {111} planes. One set of half-loops lie on a (1$\overline{1}$1) variant with Burgers vector $b= \pm a/2[10\overline{1}]$ whereas the other set lie on a (1$\overline{1}$1) variant with $b= \pm a/2[01\overline{1}]$. Thus, from the above example it is clear that surface nucleation is a viable mechanism for misfit dislocation nucleation at misfit strains as low as ~1%. However, it must be noted that these sources have rarely been observed in our TEM studies, and only in relatively thin, uncapped single strained layer structures in the as-grown state.

A much more common dislocation source is shown in Fig. 2a. The image is obtained from a region of a plan-view sample where the thinned foil surface intersects the multi-layer heterostructure; the strained Ge$_x$Si$_{1-x}$ layers are clearly delineated due to surface stress relaxation contrast effects (Perovic and Weatherly 1991). In this way one can accurately locate defect positions without the need for cross-sectional analysis. The 4-segment threading dislocation shown in Fig. 2a is observed to nucleate at a strained interface near the bottom of the superlattice. From the oscillatory contrast of the steeply inclined segments it can be seen that the 4-segment group consists of two half-loops lying on different planes steeply inclined to one another. These sources have frequently been observed in several heterostructures and most often are nucleated at the first strained multi-layer interface . It was then of interest to establish whether these threading arrays can nucleate misfit relieving segments given sufficient thermal activation to overcome the metastability imposed during epitaxial growth. Fig. 2b shows an example where one of the threading segments lying on a slip plane has generated a 60° misfit segment. Of the several examples observed, it is was clear that each 4-segment array generated only one misfit dislocation. Work is currently under way in attempt to elucidate the nature of this interfacial source since no clear evidence has been observed for the existence of a *microscopic* heterogeneity at the origin of the threading arrays.

4. MACROSCOPIC ANALYSIS (CHEMICAL ETCHING/NOMARSKI INTERFERENCE)

Using Nomarski interference microscopy, it is possible to quantitatively measure the misfit dislocation nucleation rate following post-growth annealing of metastable heterostructures by scanning representative areas (several cm^2) of the etched surfaces far from scratches and edge

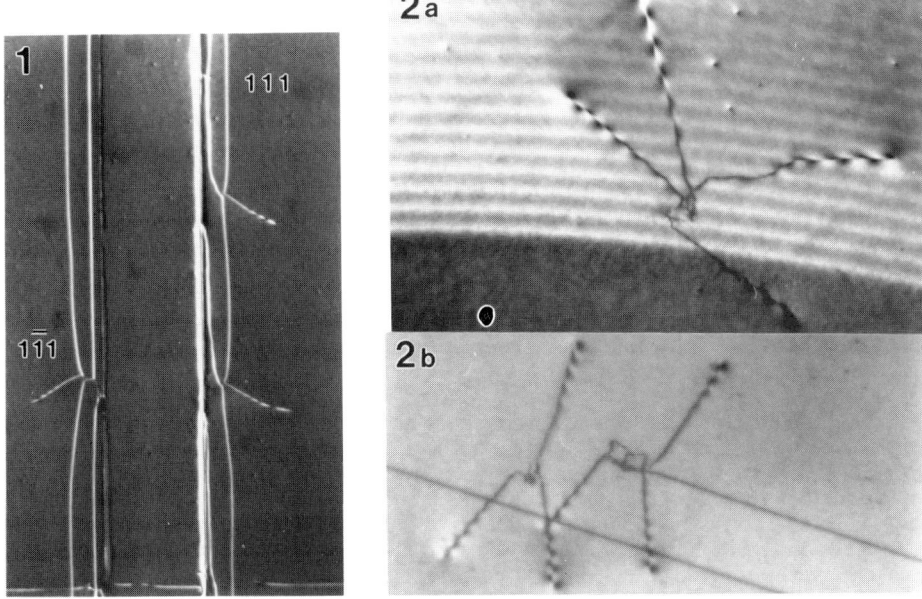

Fig. 1: Weak-beam image showing two sets of surface nucleated half-loop pile-ups.
Fig. 2:(a) Plan-view image showing the nucleation of a 4-segment threading array at a Ge$_x$Si$_{1-x}$ multi-layer interface; (b) generation of a misfit segment following thermal activation

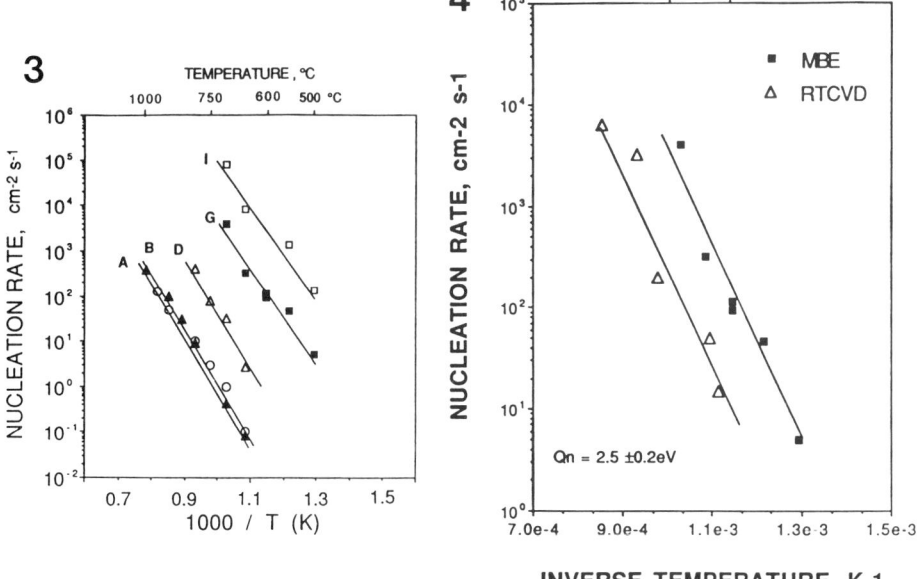

Fig. 3: Misfit dislocation nucleation rate vs. anneal temperature for selected Ge$_x$Si$_{1-x}$ hetero-structures with varying effective stress: A= 73 MPa, B= 100 MPa, D= 140 MPa, G= 580 MPa and I= 730 MPa.
Fig. 4: Comparison of misfit dislocation nucleation rate for structures possessing identical effective stresses grown by MBE (N_o= 6x10^3 cm^{-2}) and RTCVD (N_o~ 10^3 cm^{-2}).

effects. Thus, the number of new misfit segments formed per unit area, per unit time can be determined.

Fig. 3 compares the nucleation rates of several Ge_xSi_{1-x} heterostructure types possessing a large range of effective stresses (73-730 MPa). Although the nucleation rate is observed to increase markedly with effective stress as expected, the activation energy for nucleation (2.5 ±0.2 eV) is effectively constant and essentially the same as the activation energy (2.25 eV) for misfit dislocation glide observed in the same samples (Houghton 1990, 1991).

Fig. 4 compares the measured nucleation rates for two Ge_xSi_{1-x} heteroepitaxial layers of identical effective stress as grown by MBE and rapid thermal CVD. The lower N_o characteristic of RTCVD material results in a lower misfit dislocation nucleation rate under otherwise identical conditions. This trend has been confirmed by studying a series of MBE-grown layers possessing a wide range of intentionally introduced heterogeneous sources ($N_o \sim 10^2$-10^6 cm^{-2}) under otherwise identical conditions. The misfit dislocation nucleation rate was found to increase linearly with increasing N_o.

More importantly, the increase in nucleation rate with thermal activation persists to densities well-above the source density (N_o) present in the as-grown material. Accordingly, there is strong evidence suggesting that nucleation may occur at sites in addition to the obvious heterogeneities revealed by chemical etching. It should be noted that the measurements described above correspond to the initial stages of misfit strain relaxation where dislocation interactions and multiplication are not significant. In fact, the non-existence of multiplication events is evident from the observed linear increase of the misfit dislocation nucleation rate as plastic relaxation occurs.

5. SUMMARY

From a quantitative study of misfit dislocation nucleation rate measurements in the Ge_xSi_{1-x} system, it was found that a sufficient density of misfit dislocation sources exist to initiate misfit strain relaxation. Furthermore, the observed nucleation rate data taken from many heterostructures indicated that the nucleation of misfit segments always takes place via a low activation energy event which efficiently generates 60° misfit segments at the required interfaces. This is consistent with our TEM observations which illustrate surface and interfacial nucleation mechanisms which may be due to localized perturbations associated with compositional (ie. Ge fluctuations) and/or structural (ie. interfacial ledges) heterogeneities.

ACKNOWLEDGEMENTS

The authors would like to thank Prof. G.C. Weatherly and Prof. L.M. Brown for their contributions to this work and J.C. Sturm for the provision of the RTCVD sample.

REFERENCES

Houghton, D.C., 1990, *Appl. Phys. Lett.*, **57**, 1434, 2124; 1991, *J. Appl. Phys.*, (in press).
Hull, R., Bean, J.C. and Buescher, J., 1989, *J. Appl. Phys.*, **66**, 5837.
Humphreys, C.J., Maher, D.M., Eaglesham, D.J. and Salisbury, I.G., 1989, *Inst. Phys. Conf. Ser.*, **100**, 241.
Jesser, W.A. and van der Merwe, J.H., 1989, *Dislocations in Solids*, Vol. 8, (Amsterdam, North Holland), pp. 421-460.
Kamat, S.V. and Hirth, J.P., 1990, *J. Appl. Phys.*, **67**, 6844.
Kvam, E.P., Eaglesham, D.J., Maher, D.M., Humphreys, C.J., Bean, J.C., Green, G.S. and Tanner, B.K., 1988, *Mat. Res. Soc. Symp. Proc.*, **104**, 623.
Perovic, D.D., Weatherly, G.C., Baribeau, J.-M. and Houghton, D.C., 1989, *Thin Solid Films*, **183**, 141.
Perovic, D.D., Weatherly, G.C. and Houghton, D.C., 1990, *Mat. Res. Soc. Symp. Proc.*, **160**, 65.
Perovic, D.D. and Weatherly, G.C., 1991, *Ultramicroscopy*, (in press).

Grazing incidence X-ray diffraction studies of strain relaxation in monolayer-thick films

J E Macdonald[1], A A Williams[1], J M C Thornton[1], R G van Silfhout[2], J F van der Veen[2], M S Finney[3] and C Norris[3]

[1] Physics Department, University of Wales College of Cardiff, P.O. Box 913, Cardiff CF1 3TH, U.K.
[2] FOM Institute for Atomic and Molecular Physics, Kruislaan 407, 1098 SJ Amsterdam, The Netherlands.
[3] Physics Department, University of Leicester, Leicester LE1 7RH, U.K.

ABSTRACT: Grazing incidence x-ray diffraction provides an extremely sensitive probe of strain relaxation in ultrathin layers. Results for Ge/Si(001) demonstrate that the critical thickness for strain relaxation is 3-4 monolayers (ML). Comparison with STM, TEM and molecular dynamics results demonstrate that relaxation is driven by island formation rather than dislocation generation. The use of a sub-monolayer Sb surfactant layer during Ge deposition supresses the islanding process and delays the onset of strain relaxation.

1. INTRODUCTION

X-ray diffraction has been used extremely successfully for characterising heteroepitaxial semiconductor layers during the last decade or so. Rocking curves through Bragg peaks for the substrate and epilayer may be interpreted simply to deduce strain perpendicular to and parallel to the interface, compositions, crystalline quality and other parameters from the angular separation and widths of the respective peaks. It is known that care must be taken with such an approach when treating thin epilayers as the peak position may be shifted from that predicted from straightforward consideration of the respective lattice spacings (Fewster and Curling 1987). These effects may be numerically simulated using an iterative solution of the Takagi-Taupin equations within a dynamical scattering framework (Halliwell et al 1984). The kinematic scattering approximation also provides a straightforward, transparent framework for simulation, which also simulates accurately such peak shifts (Clark et al 1991). However the kinematic approximation cannot account for the diffracted profile at the peak of the Darwin curve for perfect crystal substrates. The intense current interest in heavily strained layers requires the extension of such techniques to characterise ultrathin layers a few atomic layers thick, since the calculated critical thickness for a lattice mismatch of 4% is about 10Å (Matthews and Blakeslee 1975). X-ray diffraction has been used successfully to characterise strain in superlattices of such materials, but no results have been obtained for single epilayers of a few monolayers (ML) thickness. This is partly due to the limited flux of conventional laboratory x-ray generators. However synchrotron-based work has not yielded much success since the substrate scattering dominates the epilayer scattering in the conventional diffraction geometry.

While x-ray diffraction has enjoyed considerable success in characterisation of strained layers, it has proved to be a rather insensitive probe of the critical thickness for the onset of strain relaxation, even for relatively thick films. The critical thickness for strain relaxation as observed using x-ray diffraction is consistently higher than the calculated values (People and Bean 1985). This is partly the result of the metastable nature of the epilayers after growth with molecular beam epitaxy (MBE) at normal temperatures: annealing at higher termperatures results in substantial relaxation which is thermally-activated (Houghton et al 1989). However, Fritz (1987) and others have pointed out that x-ray diffraction and other experimental techniques lack sensitivity to the initial generation and migration of dislocations, since they detect the resulting change of lattice spacings, which occurs only after dislocations have permeated considerably throughout the epilayer. Consequently, the importance of techniques which image the dislocations directly has been emphasised. Such defect-revealing studies on samples which have been annealed at a sufficiently high temperature to overcome the kinetic barrier to relaxation show extremely good agreement with theory (Houghton et al 1989).

Detecting the onset of strain relaxation in ultrathin layers is further hampered by the need for techniques which have sufficient sensitivity to the onset of dislocations or of slight changes in lattice parameters in films of a few monlayers thickness. To date, these have been Medium Energy Ion Scattering (Bevk et al 1986), Reflection High Energy Electron Diffraction (RHEED) (Sakamoto 1990), and Low Energy Electron Diffraction (LEED) (Eberl 1989). In the case of Ge/Si(001), all these techniques yield a critical thickness of 6 ML, at which the Ge epilayer begins to relax (1 ML = 1.41 Å). In this paper we describe the use of Grazing Incidence X-ray Diffraction (GIXRD) to monitor the onset of strain relaxation in this system. The results show that the actual onset of relaxation occurs at 3-4 ML, but that this occurs in a small fraction of the Ge epilayer. A large increase in the amount of relaxed material occurs at 6 ML, which was detected with the other techniques. The work thus illustrates the sensitivity of GIXRD as compared with these surface science techniques. The results have important implications for the mechanism of strain relaxation in ultrathin layers.

2. GRAZING INCIDENCE DIFFRACTION AS A PROBE OF RELAXATION

The sensitivity of the grazing incidence scattering geometry to strain relaxation can be understood from consideration of Fig. 1. For a strained, unrelaxed layer, the lattice parameter of the epilayer and the substrate are identical in the plane of the interface thus causing a tetragonal distortion of the epilayer unit cell. As the layer relaxes, its in-plane lattice parameter increases towards its bulk value and the tetragonal distortion is reduced thus changing $a_{\parallel}$, the lattice parameter parallel to the interface. The conventional diffraction geometry, as employed in Double Crystal X-ray Diffraction (DCXRD) studies of strained layers, involve scans along the scattering vector $Q_{\perp}$ perpendicular to the interface, with $Q_{\parallel}$ being small or zero. Consequently the onset of strain relaxation is detected as a shift in the epilayer peak as the tetragonal distortion is relieved. If the epilayer peak is broad, as is the case for ultrathin layers, then slight changes in lattice spacing in limited regions of the epilayer are obscured until a substantial fraction of the epilayer has relaxed. In the grazing incidence geometry both incident and diffracted beams are close to the plane of the interface and thus $Q_{\perp} \approx 0$. The strain distribution is probed by scanning radially outwards in reciprocal space (the so-called $\vartheta/2\vartheta$ scan) through the region around a substrate Bragg peak. In this case, the peaks from both the substrate and the unrelaxed ultrathin epilayer coincide and are both narrow in $Q_{//}$ (whereas the epilayer peak is broad in $Q_{\perp}$). Thus weak side-peaks arising from slight relaxation in small localised regions of the sample may be detected far more easily in this diffraction geometry, even when most of the layer is fully strained.

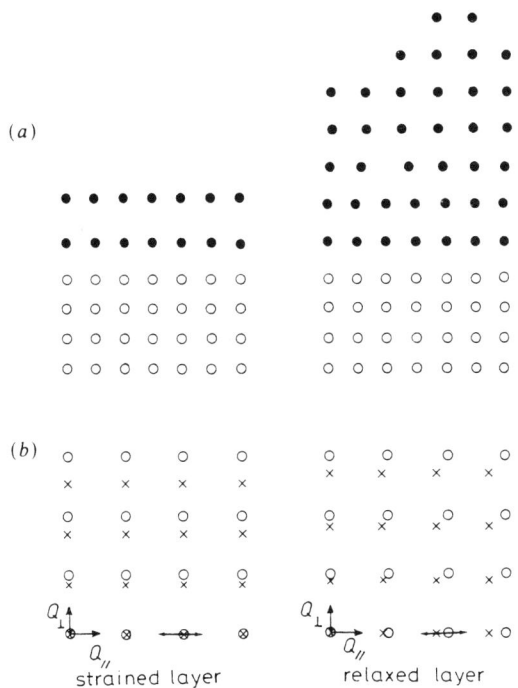

Fig. 1 A schematic side-view of (a) direct space (b) reciprocal space for a fully strained and partially relaxed layer. In (a) the open and filled circles denote substrate and epilayer atoms respectively. In (b) the open circles and crosses denote the peak position for the substrate and a fully relaxed epilayer respectively. A radial scan in the grazing incidence geometry is shown as an arrow.

The grazing incidence geometry offers other benefits for studies of strain relaxation in ultrathin films. The intrinsic peak width is much narrower for the epilayer due to the much larger extent of the film parallel to than normal to the interface. This factor, combined with the absence of interference effects giving rise to peak shifts for thin films (Curling and Fewster 1987), yields a direct measure of the in-plane strain distribution. The scattering from the substrate crystal truncation rod, which can obscure the epilayer peak in the conventional geometry, may also be resolved entirely and consequently does not obscure the epilayer scattering. The limited beam penetration of the grazing incidence geometry also suppresses the background due to thermal diffuse scattering from the bulk. These factors combine to yield a sensitivity to the onset of relaxation in ultrathin layers which is superior to that obtained using the other available techniques. This strength complements the conventional diffraction geometry which is ideally suited to characterising strain in coherent epilayers. Both diffraction geometries may be combined on a suitably modified four-circle diffractometer.

3. STRAIN RELAXATION IN Ge/Si(001)

Such grazing incidence scans for Ge/Si(001) are displayed in Fig. 2. The scans were taken in-situ during growth on the surface diffraction facility at the Daresbury synchrotron. Experimental details and detailed discussion are given by Williams et al (1991). The peak profile remains unchanged for a coverage $\theta \leq 3\text{ML}$ due to the coherent epitaxial nature of the Ge layer. The wings of the peak are not substantially broadened indicating the high crystalline quality of the overlayer. At 4ML, a weak shoulder appears on the Bragg peak due to the onset of strain relaxation. At $\theta \simeq 6$ ML, a substantial increase in the amount of relaxed material is observed, as seen by electron diffraction and ion scattering. The epilayer contains almost fully-relaxed Ge at $\theta = 11$ ML, but lattice constants intermediate between those of Ge and Si are also observed.

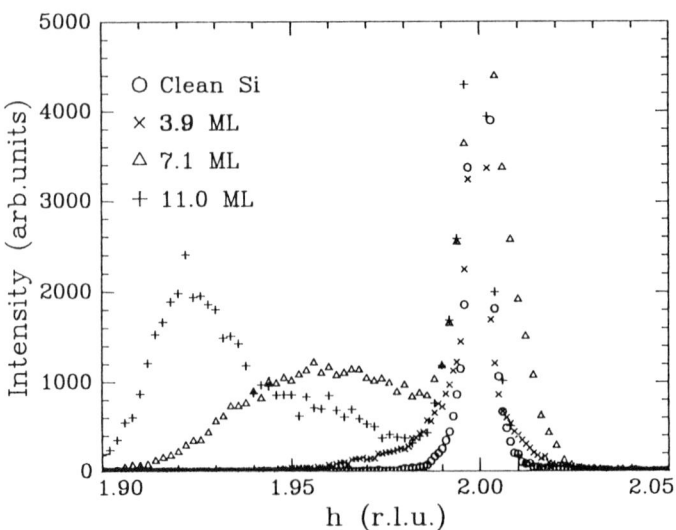

Fig. 2   Radial scan at grazing incidence through the (2,2,0) Bragg peak for Ge epilayers on Si(001). The onset of relaxation is observed at 4ML, coinciding with the onset of islanding. Fully relaxed Ge at h=1.92 is observed at a coverage of 11ML. At this coverage, in-plane lattice parameters intermediate between those of bulk Ge and Si are seen.

Insight into the mechanism of strain relaxation may be obtained from comparison with electron microscopy (TEM) and scanning tunnelling microscopy (STM) results. The onset of relaxation at 3-4 ML coincides exactly with the coverage at which islanding of the Ge overlayer occurs (Zinke-Allmang et al 1989, Mo et al 1990). Furthermore, the STM images show that, in the range 3-6 ML, the islands are small (few hundred Ångstroms in length and about 30Å high), rectangular clusters having well-formed (105) facets. At 6 ML, these small islands coalesce to form large islands having (113) facets (Mo et al 1990). This correspondence between the islanding behaviour and the strain relaxation strongly suggests that the strain relaxation relates to islanding of the surface. The question arises whether the islanding is the result of dislocation formation or whether it occurs without dislocations. This may be answered by considering three investigations.

TEM images indicate that the macroscopic islands occuring at about 8 ML coverage are dislocation-free, even though the island height is $\simeq$ 500 Å, well in excess of the calculated critical thickness of $\simeq$ 15 Å (Matthews and Blakeslee 1975). Secondly, comparison of the equilibrium energies of a relaxed, dislocated interface and that for a Ge island with the corresponding fully-strained layer, using empirical potentials in a molecular dynamics calculation, indicates that islanding of the surface is favorouble after 2-3 ML whereas $60°$ dislocations would be expected at $\theta = 10$ ML (Matthai and Ashu 1990, Ashu and Matthai 1991). A third important piece of evidence arises from our recent experiment that strain relaxation, as well as islanding, may be supressed by employing a surfactant layer during growth, as described below.

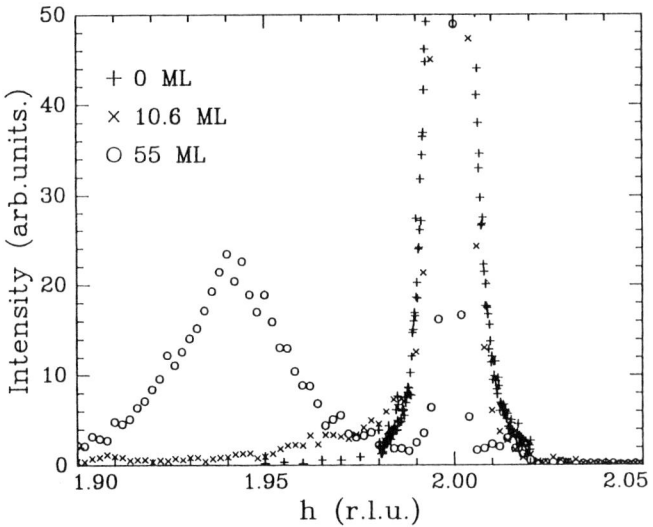

Fig. 3 Radial scans at grazing incidence for Ge/Si(001) grown with a Sb surfactant layer. Note that relaxation is delayed until 10ML which coincides with the formation of V-shaped defects. Fully-relaxed Ge is not observed even after 55ML.

Le Goues and coworkers have shown that deposition of about a monolayer of As or Sb onto the Si surface prior to deposition of Ge supresses the growth of islands (Le Goues et al 1989, Copel et al 1990). The group V element continuously segregates to the surface, modifying the surface energetics, and thus prevents the islanding process. We have repeated the above GIXRD study on a Ge film grown with a 0.7 ML Sb surfactant layer. The islanding was supressed, as seen from reflectivity measurements, and the onset of strain relaxation was delayed until $\theta = 10$ ML (Fig.3). At this coverage, V-shaped defects involving grain boundaries are known to occur (Le Goues et al 1989), which also coincides with the Matthews-Blakeslee critical thickness for dislocations. If the relaxation at 3 ML for pure Ge/Si(001) is caused by dislocations, then these would not be inhibited by the presence of the surfactant layer.

4. CONCLUSIONS

Grazing incidence x-ray diffraction has been demonstrated to be an extremely sensitive probe of the onset of relaxation in ultrathin films. Strain relaxation in Ge/Si(001) sets in at a coverage of 3-4 ML as a result of islanding of the Ge surface, the relaxation being caused by elastic deformation around the island edges. The islanding process may be supressed by use of a As or Sb surfactant layer during growth, which results in the delay of strain relaxation up to a coverage of 10 ML. This may have useful applications for modifying growth modes in MBE. Further work is required to investigate whether this is a general phenomenon in III-V semiconductors.

REFERENCES

Ashu and Matthai 1991 Appl. Surface Science, in press
Bevk J, Mannaerts J P, Feldman L C, Davidson B A and Ourmazd A 1986 Appl. Phys. Letters 49 286
Clark S A, Macdonald J E, Williams R H and Barnett S J 1991 to be submitted, Appl. Phys. Lett.
Copel M, Reuter M C, von Hoegen M H and Tromp R M 1990 Phys. Rev. B42 11682
Eaglesham D J and Cerullo M 1990 Phys. Rev. Letters 64 1943
Eberl K, Friess E, Wegscheider W, Menczigar U and Abstreiter G 1989 Thin Solid Films 183 95
Fewster P F and Curling C J 1987 J. Appl. Phys. 62 4154
Fritz I J 1987 Appl. Phys. Letters 51 1080
Halliwell M A G, Lyons M H and Hill M J 1984 J. Cryst. Growth 68 523
Houghton D C, Gibbings C J, Tuppen C G, Lyons M H and Halliwell M A G 1989 Thin Solid Films 183 171
Le Goues F K, Copel M and Tromp R 1989 Phys. Rev. Lett. 63 1826
Matthai C C and Ashu P 1990 Coll. de Phys. 51 C1-873
Matthews J W and Blakeslee A E 1975 J. Vac. Sci. 12 126
Mo Y W, Savage D E, Swartzentruber B S, Lagally M G 1990 Phys. Rev. Lett. 65 1020
People R and Bean J C 1985 Appl. Phys. Letters 47 322
Sakamoto T, Sakamoto S, Miki K, Okumura H, Yoshida S and Tokumoto H 1990 Kinetics of Ordering and Growth at Surfaces, ed M G Lagally (New York: Plenum)
Williams A A, Thornton J M C, Macdonald J E, van Silfhout R G, van der Veen J F, Finney M S, Johnson A D and Norris C 1991 Phys. Rev. B43 5001
Zinke-Allmang M, Feldman L C, Nakahara S and Davidson B A 1989 Phys. Rev. B39 7848

The influence of interfaces in X-ray thickness measurements of hetero-epitaxial layers

Shara Amin and Mary Halliwell

British Telecom Research Laboratories, Martlesham Heath, Ipswich IP5 7RE, UK

ABSTRACT: X-ray rocking curves from heteroepitaxial layers contain thickness fringes. It is possible to extract values for the thicknesses of heteroepitaxial layers by Fourier transforming these curves. This paper describes a study of the effect of interfaces on the relative intensities of peaks in the Fourier transformed data. Applications of this study include the measurement of homoepitaxial layer thicknesses by introducing a thin mismatched interfacial layer, and the assessment of interface abruptness in heteroepitaxial structures.

1. INTRODUCTION

Using x-ray scattering experiments and theoretical simulations, detailed investigations of highly mismatched interfacial layers were performed. It is shown that these highly strained interfacial regions strongly influence the intensity of the thickness fringes from the layer directly above. By introducing a highly mismatched layer between two semiconductor layers of similar lattice parameter, the thickness of the upper layer can be measured. The experimental studies used epitaxial layers of indium phosphide on indium phosphide substrates. The mismatched layer was created either by exposing the substrate to a beam of arsine or by growing 6 Å of indium arsenide. This enabled the thicknesses of homoepitaxial layers of indium phosphide grown by chemical beam epitaxy (CBE) and molecular beam epitaxy (MBE) to be measured.

Previous published work has concentrated on the effect of interfaces on the diffraction data recorded from MQW structures (Lyons 1989). Here for the first time we consider how highly mismatched interfacial layers influence the scattering from simple one- and two-layer structures.

A simple model is proposed to describe the compositional variation through the interface. We report on results of simulations of theoretical rocking curves based on the above model and also compare these results with experimental measurements. Applications to the characterisation of homo and heteroepitaxial layers are described.

2. THE INTERFACE MODEL

We assume that there is a thin interfacial layer between two semiconductor layers and that this thin layer is only a few monolayers thick and has a lattice parameter which differs from the adjacent layers by M, which we will express in parts per million (ppm). We assume that the mismatch (M) within the interfacial layer is very high, and constant and that it decreases sharply to zero on both sides. The model is shown in Fig. 1. We will assume the interfacial layer thickness (dt) is equal to a few monolayers.

Theoretical rocking curves are calculated by solving the Takagi-Taupin equations (1969, 1964), which give the rate of change of the ratio of diffracted to incident beam amplitude as a function of depth below the surface; for more detail see Halliwell et al (1984). The model described in Fig. 1 is introduced into the simulation to represent highly mismatched interfacial layers.

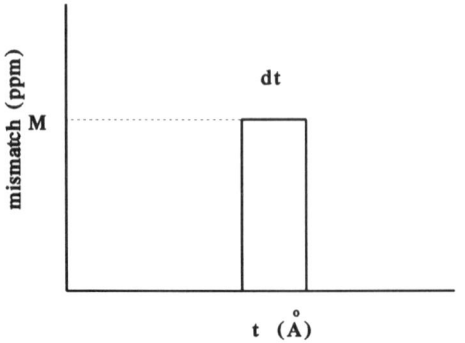

Fig. 1 - Model for highly mismatched interfacial layer

3. THE EFFECT OF INTERFACIAL LAYERS ON ROCKING CURVES

3.1 Single homoepitaxial layer structure

Firstly we consider a single layer of indium phosphide on an indium phosphide substrate. Since the layer and the substrate have the same lattice parameter the compositional changes will be zero. We can then simulate the rocking curve for a variety of thicknesses and obtain rocking curves identical to the x-ray scattering from an indium phosphide substrate.

Consider a 0.25 μm layer of indium phosphide on an indium phosphide substrate. If we introduce a thin and highly mismatched (dt = 6Å, M = +12000 ppm) layer at the interface, see structure (a) in Fig. 2, as proposed in section 2, then the interfacial layer will introduce a phase change between the incident and reflected beams. The phase change leads to the formation of a set of interference fringes, the amplitude of which depends on the mismatch of the interface. Fig. 2(a) shows the calculated rocking curve for the 004 CuKα reflection from this structure. The effect of reducing the mismatch of the interfacial layer to 6000 ppm (see structure (b) in Fig. 2) is shown in Fig. 2(b).

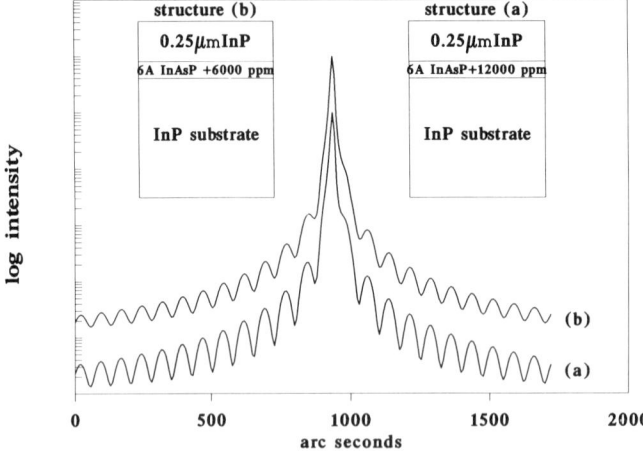

Fig. 2 - Calculated rocking curves for a 0.25 μm layer of InP on InP. (a) with 6 Å interfacial layer mismatched by 12000 ppm and (b) with 6 Å interfacial layer mismatched by 6000 ppm

The interference fringes are thickness fringes from the top layer. The layer thickness (t_l) is related to the fringe separation (dθ) by the general expression (Batterman, Hildebrandt, 1968, Bartels and Nijman 1978):

$$t_l = \lambda \sin(e)/d\theta \sin(2\theta) \qquad (1)$$

where λ is the x-ray wavelength, e is the angle between the diffracted beam and the sample surface, θ is the Bragg angle and $d\theta$ the fringe spacing in radians. The relationship between fringe spacing and layer thickness for 004 CuKα rocking curves from indium phosphide is shown in Fig. 3.

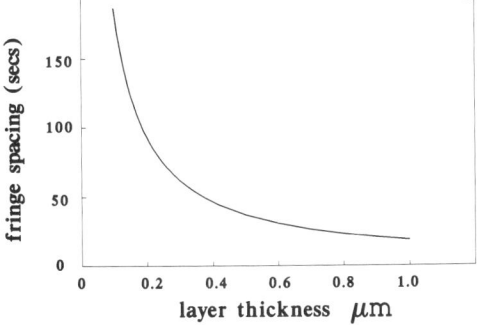

Fig. 3 - Relationship between fringe spacing and layer thickness for 004 CuKα rocking curves for InP

In this paper the Fourier transform technique (Macrander et al 1988) is used to analyse the frequency components present in rocking curves, so that the thickness of individual layers can be calculated. Equation (1) is used to convert the fringe frequency to a layer thickness. It was first reported by Amin and Halliwell (1990,1991) that the Fourier components in the Fourier transform data can be greatly enhanced by taking the first differential and then performing one or more successive self-convolution processes, and it is this approach which we will be using throughout the current paper.

The Fourier transform of the rocking curve in Fig. 2(a) is shown in Fig. 4, confirming that the thickness fringes seen correspond to the layer thickness of structure (a).

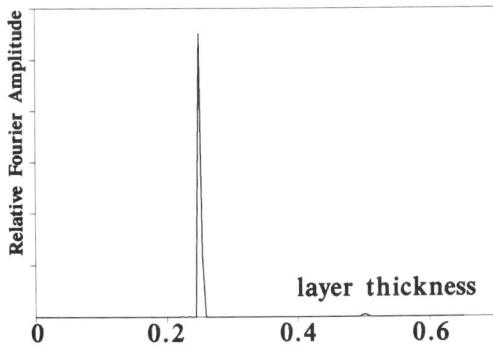

Fig. 4 - Fourier transform of rocking curve (a) in Fig. 2.

From the results shown in Fig. 2 we conclude that the highly mismatched layer amplifies the thickness fringes from the layer immediately above, and the degree of amplification depends on the value of the mismatch at the interface.

3.2 Two layer heteroepitaxial structure

Secondly we consider a two layer structure. We will consider an indium phosphide substrate with a 0.2 μm layer of gallium indium arsenide with a lattice parameter 500ppm smaller than the substrate (layer A) and an indium phosphide cap 0.3 μm thick (layer B) as shown in structure (a) of Fig. 5. Now we introduce an artificial thin layer which has a high mismatch (dt = 6 Å M = 12000 ppm) between the substrate and layer A, as shown in structure (b) of Fig. 5. This thin highly mismatched layer will mostly affect the thickness fringes of the layers A + B directly above. This can clearly be seen by comparing the Fourier transform of the rocking curve from structure (a) and (b). In Fig. 5(b) the peak at 0.5 μm is the dominant feature, whereas in Fig. 5(a) the peaks at 0.3 and 0.5 μm are more equal in intensity.

If we add a similar mismatched interface between layers A and B rather than between layer B and the substrate as shown in structure (c) of Fig. 5 then the Fourier transform is dominated by a strong sharp peak from layer B, see Fig. 5(c). Note that the only difference between structures (b) and (c) of Fig. 5, is the position of the highly mismatched thin layer.

Now consider the situation where we have two highly mismatched interfacial layers as shown in structure (d) of Fig. 5. In this example we assume both thin layers to have similar sign, then we would expect that thickness fringes from both layer B and A+B will be amplified. The result of the Fourier transform is presented in Fig. 5(d) which has clear peaks at the positions corresponding to the thicknesses of layers, B and A+B.

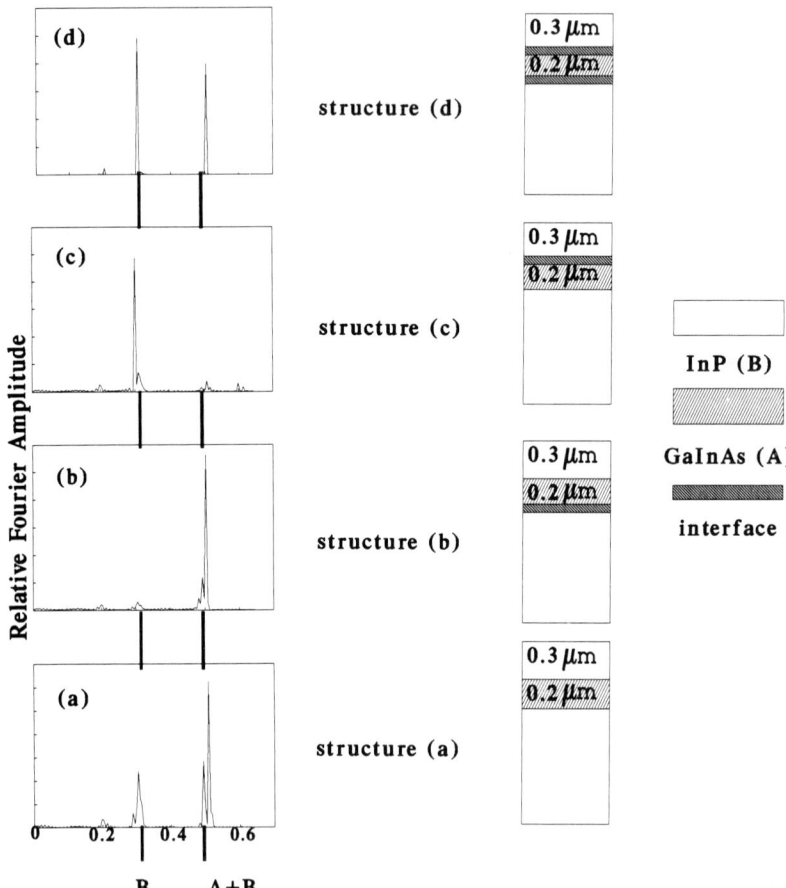

Fig. 5 - Fourier transforms of rocking curves from two layer heteroepitaxial structures.

4. APPLICATIONS

4.1 Thickness of homoepitaxial layers

The samples studied were single indium phosphide layers grown on indium phosphide using Chemical Beam Epitaxy (CBE) and Molecular Beam Epitaxy (MBE). The introduction of an artificial high mismatch thin layer at the interface for the CBE sample was achieved by allowing an arsine flux to impinge on the substrate for about 10 seconds, for the MBE sample the artificial high mismatch thin layer was introduced by growing about 6 A layer of indium arsenide between the indium phosphide substrate and the indium phosphide layer. X-ray scattering data were recorded using a Philips High Resolution Diffractometer. The experimental rocking curve for the first sample, is presented in Fig. 6(a) (nominal layer thickness 0.26 μm) and for the second sample is shown in Fig. 6(b) (nominal layer thickness 0.20 μm). Clearly the amplitude of the thickness fringes of the MBE sample is very much higher than that of the CBE sample, this is because the mismatch of the thin layer in the CBE sample is very much smaller than that of the MBE sample. From the Fourier transforms (Figs 6(c) and (d)) of these experimental curves, the layer thicknesses were measured to be 0.268 $\pm$ 0.005 μm and 0.218 $\pm$ 0.005 μm. Thickness measurements made in this way are now in routine use for determining growth rates in this laboratory.

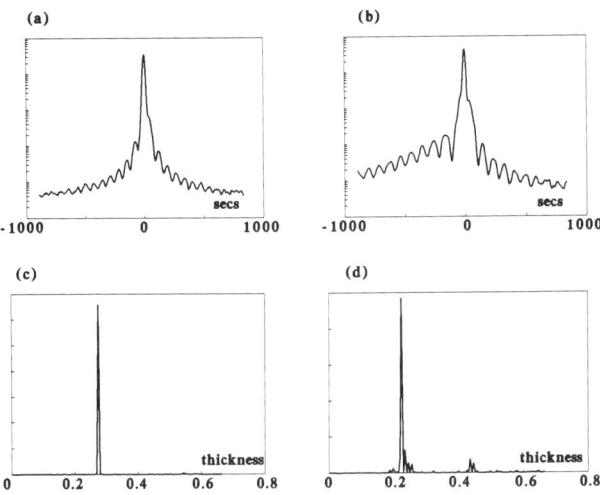

Fig. 6 - Experimental rocking curve from (a) CBE layer (b) MBE layer. Fourier transforms of rocking curve data for (c) the CBE layer and (d) the MBE layer.

4.2 Abruptness of heteroepitaxial interfaces

The amplitude of the thickness fringes depends on the magnitude of the mismatch of the layer. However in real experimental rocking curve there are many examples in which the layer is almost lattice matched to the substrate, but the x-ray scattering curves contain very strong thickness fringes. This indicates that a significantly mismatched layer is present at the interface, that is the lattice parameter change across the interface is not a step function as intended, but has some interfacial spike. Such interfaces are described as graded or non-abrupt. They can be modelled in a similar manner to the approach used in section 2, by introducing a thin highly mismatched layer at the interface.

Fig. 7(a) shows the experimental rocking curve for an indium phosphide substrate with a 0.17 μm layer of $Ga_{(1-x)}In_xAs_{(1-y)}P_y$ capped by 0.23 μm of indium phosphide. The simulated rocking curve assuming an abrupt interface (Fig. 7(b)) has much less intense fringes. The presence of an non-abrupt interface can be readily confirmed by comparing the Fourier transform of the experimental (Fig. 7(c) and the simulated data (Fig. 7(d)). The relative strengths of the Fourier peaks in Fig. 7(d) suggests both interfaces are non-abrupt.

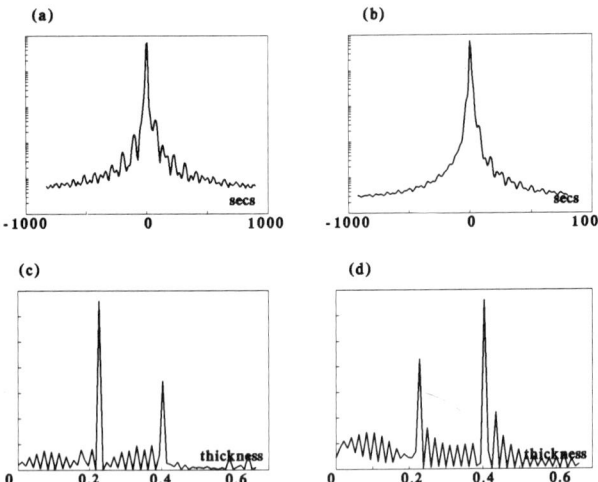

Fig. 7(a) - Experimental rocking curve from an InP substrate plus a layer of $Ga_{(1-x)}In_xAs_{(1-y)}P_y$ capped by InP (b) simulated rocking curve for structure with abrupt interfaces. Fourier transforms of diffraction data for (c) the experimental curve and (d) the simulated curve.

5. CONCLUSION

A simple model is proposed for highly strained interfacial layers between two similar or two different crystalline layers. These interfacial layers were found to change the phase difference between the incident and the reflected x-ray beams in such a way as to allow constructive interference. As a result this will drastically influence the amplitude of the thickness fringes. It was also demonstrated that such an interface will have the greatest effect on the thickness fringes from the layers directly above. These phenomena have been seen experimentally by studying a number of CBE and MBE grown epitaxial layers. The most striking result from this work is that it is now possible to measure thickness of semiconductor layers which have a similar lattice parameter to the layer directly beneath, provided a highly mismatched interfacial layer is present between the two layers. Using the combination of experiment and simulation and by comparing the Fourier components in the Fourier transform data it is possible to determine the position of a highly strained interface in a multilayer sample. This new method has already found useful applications in the characterisation of epitaxial layers.

ACKNOWLEDGEMENTS

We thank our colleagues Peter Skevington, Tony Dann and Paul Spurdens for providing the samples.

REFERENCES

Amin S and Halliwell M A G 1990 British Crystallographic Association Spring Meeting, Exeter University
Amin S and Halliwell M A G 1991 To be published in J Appl Cryst
Bartels W J and Nijman W 1978 J. Cryst Growth 44 518
Batterman B W and Hildebrandt G 1968 Acta Cryst A24 150
Halliwell M A G, Lyons M H and Hill M J 1984 J. Cryst. Growth 68 523
Lyons M H (1989) J. Crystal Growth 96, 339
Macrander A T, Lau S, Strege K, and Chu S N G 1988 Appl. Phys. Lett. 52, 1985
Takagi S 1962 Acta Cryt. 15, 1331
Takagi S 1969 J. Phys. Soc. Japn. 26, 1239
Taupin D 1964 Bull. Soc. Franc. Miner. Crist. 87, 469

Inst. Phys. Conf. Ser. No 117: Section 9
Paper presented at Microsc. Semicond. Mater. Conf., Oxford, 25–28 March 1991

657

X-ray diffraction and SEM investigation of the crystal quality of GaAs/Ge heterostructures

C Bocchi, B Bollani*, P Franzosi, L Lazzarini, D Passoni* and G Timò*

MASPEC-C.N.R., Via Chiavari 18/A, 43100 Parma, Italy
*CISE S.p.A., Via Reggio Emilia 39, 20134 Segrate, Italy

ABSTRACT: GaAs/Ge heterostructures have been grown by metal organic vapour phase epitaxy and investigated by X-ray and SEM techniques. X-ray double crystal diffractometry has been used to study the elastic strain release, while the generation of misfit dislocations has been investigated by X-ray topography and SEM integral cathodoluminescence. The results have not been found to agree with the theoretical predictions. From the shape of the Bragg peaks, some information about the interdiffusion processes and the crystal quality of the layers have also been obtained.

1.   INTRODUCTION

GaAs/Ge heterostructures (HSs) are attractive for making high conversion efficiency solar cells (Tobin et al. 1985, Chang et al. 1987, Partrain et al. 1987, Tobin et al. 1988). Since GaAs and Ge have closely matched lattice parameters and thermal expansion coefficients, the GaAs/Ge HSs are considered of non-strained type; however, since for solar cell applications the GaAs layers may be a few μm thick, the study of the strain relaxation through the formation of misfit dislocations (MDs) deserves attention. Moreover, it must be considered that the electrical performance of the solar cells is strongly influenced by the interdiffusion processes at the interface, that can produce autodoping effects and modify the electrical characteristics of both the substrate and the layer.
The present paper reports the growth of GaAs/Ge HSs by metal organic vapour phase epitaxy (MOVPE) and their investigation by X-ray double crystal diffractometry (DCD), X-ray topography (XRT) and SEM integral cathodoluminescence (ICL).

2.   EXPERIMENTAL

GaAs/Ge HSs have been grown by MOVPE at 700 °C, using trimethylgallium and pure arsine; as the substrates, high quality Ge crystals 6° off (001) toward [110] have been employed. The V/III flux ratio ranged from 200 to 20 and the growth rate from 0.1 to 1.0 μm/h; therefore, the final layer thickness was in the 0.25 to 6.0 μm range.
The DCD rocking curves have been recorded by a computer controlled double crystal goniometer having an angular resolution of 0.1 arcsec; the Cu Kα₁ radiation, the 004 symmetric reflection and a high quality Ge monochromator have been used. The samples were mounted on the goniometer with the [110] axis normal to the diffraction plane and care was taken to minimize all the spurious effects which can produce a peak broadening. The XRT images have been obtained by a conventional Lang camera in the back reflection geometry, using the Cu Kα₁ radiation and the 115 asymmetric reflection. Finally, SEM-ICL investigations have been performed using a Si photovoltaic detector parallel to the electron beam and about 1 cm from the beam axis; in this way the radiation emitted upwards by the GaAs epilayers was collected.

3.   RESULTS AND DISCUSSION

The generation of the MDS has been investigated by XRT and SEM-ICL. As an example, Fig. 1 shows the XRT micrographs of a few typical GaAs epilayers whose thickness ranges from 0.29 to 5.50 μm. As seen, for $t_1 \leqslant 0.29$ μm, a low dislocation density is observed; since the same defect configuration characterizes the substrates, it is not clear if either dislocations located within the substrate or dislocations threading into the layer are imaged. In any case no MDs are revealed. In contrast, when $t_1 \geqslant 0.45$ μm MDs parallel to the interface are observed; for t_1 in the 0.45 to 0.80 μm range the MDs appear as narrow and faint lines, therefore they are single dislocations. Finally, for $t_1 > 0.80$ μm MDs arranged in bands are clearly observed. It can be seen in Fig. 1 that the MDs are parallel to

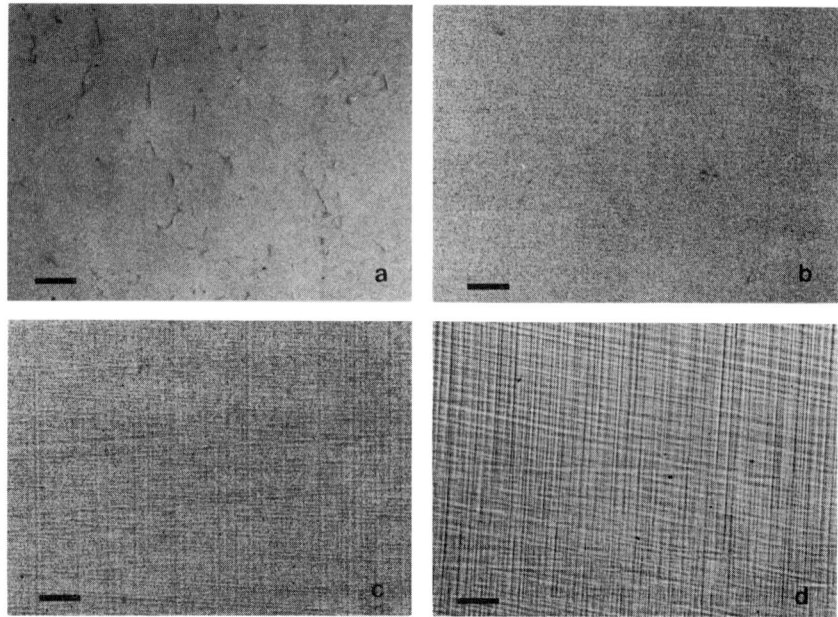

Fig. 1 XRT images showing the defect distribution as a function of the layer thickness. a: 0.29 μm; b: 0.45 μm; c: 0.85 μm; d: 5.50 μm. The lenght of the markers is 200 μm.

three different directions; the first one coincides with the [110] direction, the others occur at 4° either side of the [1$\bar{1}$0] direction, as it is expected for a sample surface 6° off (001).
On the basis of the present results, it may be concluded that the critical thickness t_c for the MDs generation is in the 0.29 to 0.45 μm range. The value predicted by the elastic equilibrium theory of Matthews and Blakeslee (1974) is 0.22 μm; much larger disagreements are obtained with the non-equilibrium theories (People and Bean 1985, Marée et al. 1987).
By the DCD experiments, the elastic tetragonal distortion of the epilayer unit cell and the strain release as a function of t_1 have been investigated. Fig. 2 reports the parallel residual strain $e_{//}$ as a function of t_1. For $t_1 < t_c$, a perfect lattice accomodation on the interface plane is observed, while for $t_1 > t_c$ a progressive strain release is seen; however, the release is not complete even for a layer as thick as 6.0 μm. Fig. 2 also shows for comparison the theoretical curve calculated on the basis of the equilibrium theory of Matthews and Blakeslee (1985). The experimental slope is smaller than the theoretical one; however, it must be stressed that a larger disagreement has been found with the results of the non-equilibrium theories.

The shape of the DCD Bragg peak and its full width at half maximum give information on the crystal quality and homogeneity. Fig. 3 shows the width of the Bragg peaks of both the epilayer (triangles) and the substrate (dots) as a function of t_l; for comparison the theoretical

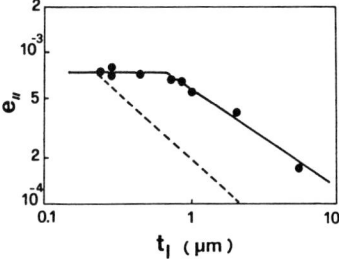

Fig. 2 Parallel residual strain as a function of the layer thickness. Dots: experimental points; broken line: prediction of the equilibrium theory.

predictions of the dynamical diffraction theory (Halliwell et al. 1984) calculated for the layer and the substrate are reported as broken and dotted lines respectively. It may be observed that the width of the layer peak is almost coincident with the theoretical one for $t_l < t_c$; for larger t_l values the peak width increases rapidly by increasing t_l up to about 1.5 μm, then it decreases again. In contrast the width of the substrate peak is found to increase continuously by increasing t_l. These results can be explained considering the effect of the MDs; their deformation field is in fact able to produce a peak broadening in both the layer and the substrate and the broadening increases by increasing the MDs density. The decreasing of the layer Bragg peak width for $t_l > 1.5$ μm suggests that the MDs are located in a relatively thin layer close to the interface and do not propagate into the bulk layer.

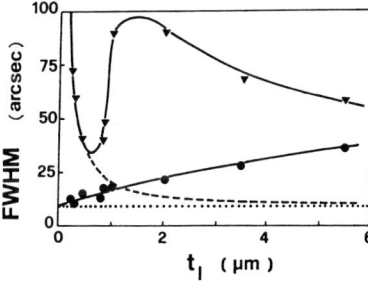

Fig. 3 Width of the Bragg peaks as a function of the layer thickness. Triangles: experimental data for the layer; broken line: theoretical predictions for the layer; dots: experimental data for the substrate; dotted line: theoretical prediction for the substrate.

The peak shape analysis is also able to give information about the interdiffusion processes. Fig.s 4a and 4b show two HSs grown at a high and a low growth rate respectively; the growth rate ratio is about 10:1, but the final layer thickness is almost identical (0.29 μm). As seen, in Fig. 4a the agreement between the experimental curve (continuous line) and the simulated one (broken curve) is excellent; this means that the layer is a nearly perfect crystal with a homogeneous lattice parameter distribution. In contrast, Fig. 4b shows a severe disagreement between the experimental and the simulated curves. More in detail, the substrate peak exhibits a diffracted intensity tail at the low angle side; this is probably due to the indiffusion of As and/or Ga into the Ge substrate. It must be stressed that no differences

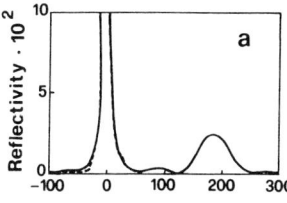

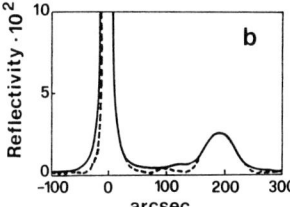

Fig. 4 DCD rocking curves of two SHs grown at different growth rates. a: high rate; b: low rate. In both cases the layer thickness is 0.29 μm.

have been found in samples submitted to different pre-growth thermal treatments; more specifically, layers treated in H_2 and AsH_3 atmospheres respectively exhibited almost identical rocking curves. Finally, the experimental DCD curve in Fig. 4b shows a broadening of the layer peak and does not exhibit well defined thickness interference fringes. However, as seen in the XRT and SEM-ICL images, the layers grown at a low

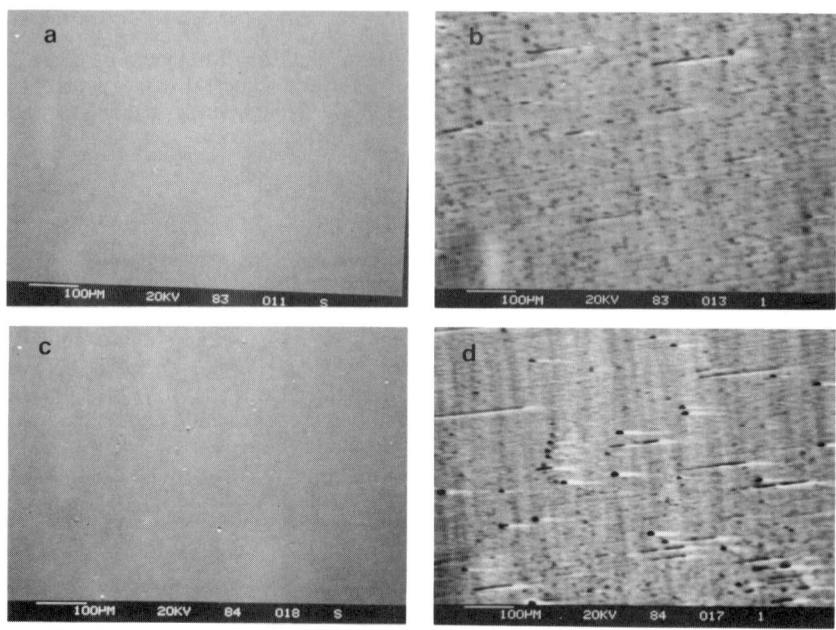

Fig. 5 SEM results obtained in two HSs grown at different growth rates; in both cases the layer thickness is 2.0μm. a and b: high rate; c and d: low rate. a and c: SEM images of the surface morphology; b and d: SEM-ICL images of the defect distribution.

growth rate do not exhibit a larger concentration of defects than those grown at a high rate. Typical results of the SEM-ICL investigation are shown in Fig.5; as seen, the layer grown at a low rate exhibit a slightly larger concentration of morphological imperfections that behave as non-radiative recombination regions. In conclusion, the broadening of the layer Bragg peak observed in the layers grown at a low rate may be related either to a Ge diffusion into the layer or to the presence of submicron defects within the layer.

REFERENCES

Chang K I, Yeh Y C M , Iles P A, Tracy J and Morris R K 1987 Conf. Rec. 19th IEEE
    Photovoltaic Specialists Conf. pp 273-279
Halliwell M A G, Lyons M H and Hill M J 1984 J. Crystal Growth 68 523
Marée P M J, Barbour J C, van der Veen J F, Kavanagh K L , Bullè Lieuwma C W T and
    Vierges M P A 1987 J Appl. Phys. 62 4413
Matthews J W and Blakeslee A E 1974 J. Crystal Grwoth 27 118
Partrain L D, Kuryla M S, Weiss R E, Ransom R A, McLeod P S, Fraas L M and Cape J
    A 1987 J. Appl. Phys. 62 3010
People R and Bean J C 1985 Appl. Phys. Lett. 47 322
Tobin S P, Vernon S M, Bajgar C, Haven V E and Davis S E 1985 Conf. Rec. 18th IEEE
    Photovoltaic Specialists Conf. pp 134-139
Tobin S P, Vernon S M, Bajgar C, Haven V E, Geoffroy L M and Lillington D R 1988
    IEEE Electron. Dev. Lett. 9 256

Inst. Phys. Conf. Ser. No 117: Section 9
Paper presented at Microsc. Semicond. Mater. Conf., Oxford, 25–28 March 1991

661

Double crystal X-ray diffraction and TEM of partially relaxed strained-layer structures

P Kidd and R Dixon

Strained Layer Structures Research Group, c/o Materials Science and Engineering, University of Surrey, Guildford, SURREY, GU2 5XH, UK.

ABSTRACT: This paper reports the analysis of partially relaxed single surface layers and buried multilayer structures of $In_xGa_{(1-x)}As$ on GaAs. Using Double Crystal X-ray Diffraction (DCXD) real and simulated rocking curves are compared. TEM in plan-view and cross section reveals the misfit dislocation density and distribution of dislocations within layers. The effect of misfit dislocations on relaxation and DCXD peak broadening is discussed.

1. INTRODUCTION

It has now been well established that the rocking curves obtained by double crystal x-ray diffraction of high quality, defect free, strained layer structures can give measurements of the strain, composition and thickness of consituent layers (Halliwell 1984). However, the presence of dislocations in any structure causes broadening and lowering of the diffraction peaks leading to loss of vital fine detail in the rocking curve. The extent of peak broadening is dependent not only on the total dislocation density but also on the distribution of the dislocations throughout the layers. This paper reports analysis of rocking curves obtained from a series of specimens containing strained layer structures of $In_xGa_{(1-x)}As$ on GaAs, with x in the range 0.08 to 0.20, which are designed to deliberately include partially relaxed layers. The layers were grown by MBE at growth temperature 509°C, on (001) GaAs substrates.

For both DCXD and TEM all specimens were taken from the annular regions of the wafers along the [100] diameters in order to eliminate systematic variations in the layer which arise from inhomogenies across the substrate wafer. Two types of structure were examined. The first is a series of eight specimens with single surface layers of nominally x = 0.2 $In_xGa(1-x)As$ with thicknesses of 25, 50, 70, 100, 200, 400, 800, 3000nm respectively. The second, ME580, is a partially relaxed multilayer structure, with three $In_{(0.19)}Ga_{(0.81)}As$ wells, 10nm, 30nm, 40nm thick, separated by 100nm GaAs barriers and with a 100nm GaAs capping layer.

The dislocation densities and their distributions between various layers in the specimens were examined by TEM in plan view (001) and cross-section (110), using a JEOL 2000FX microscope. DCXD rocking curves were obtained for each of the specimens, for surface symmetrical (004) and asymmetrical high and low angle (115) reflections, using a Bede Scientific 150 diffractometer, CuKα radiation and (001) GaAs as the first crystal in the (+ -) non-dispersive mode. Rocking curve simulations were computed using Bede Scientific Rocking curve Analysis by Dynamical Simulation (RADS) software.

2. RESULTS

2.1. Single surface layer

For each of the single surface layer specimens four (004) rocking curves were recorded where the projections in (001) of the incident and diffracted planes lie along the four ⟨110⟩ directions. From these four readings tilt between the layer and substrate is measured and is corrected for in measurements of peak splitting on the corresponding (115) rocking curves. The layer composition and extent of relaxation for each specimen was then derived from the rocking curves by comparing the measurement of layer and peak splitting with values generated by RADS.

Graph 1a shows the average relaxation versus layer thickness. Some asymmetry in relaxation was observed for the thickest layers, but for the purposes of this study, and since the asymmetry was small, an average relaxation was used. Graph 1b shows the interfacial dislocation density versus layer thickness, measured using TEM of plan-view specimens. Direct correlation of the DCXD relaxation data with the TEM dislocation density data requires knowledge of the distribution of Burgers vectors between the dislocations for each of the specimens. This will be discussed in more detail elsewhere.

Graph 2a shows the average (004) full width at half maximum (FWHM) of the layer peak for each of the specimens. For a study of the extent to which dislocations affect the FWHM of a layer peak it is neccessary to account for peak broadening by other factors. For each layer there is a theoretical FWHM which is generally greater than the substrate peak due to the finite thickness of the layer and the fact that the diffracting conditions for the InGaAs layer are slightly dispersive. These two

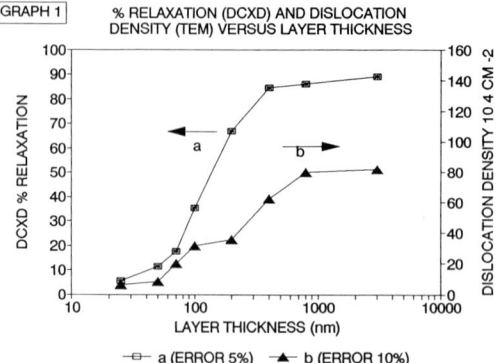

GRAPH 1 — % RELAXATION (DCXD) AND DISLOCATION DENSITY (TEM) VERSUS LAYER THICKNESS

— a (ERROR 5%)   — b (ERROR 10%)

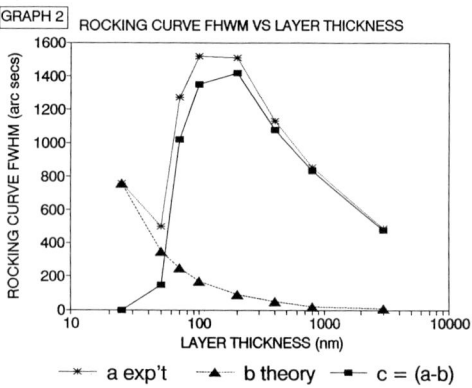

GRAPH 2 — ROCKING CURVE FHWM VS LAYER THICKNESS

— a exp't   — b theory   — c = (a-b)

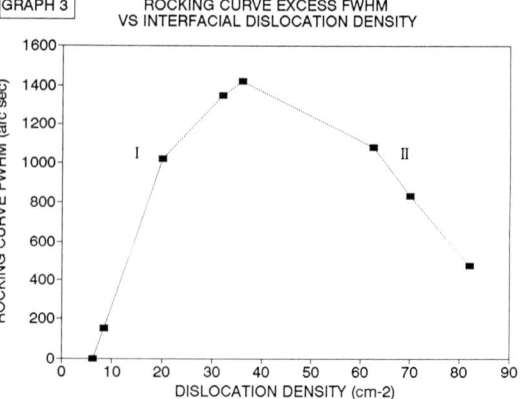

GRAPH 3 — ROCKING CURVE EXCESS FWHM VS INTERFACIAL DISLOCATION DENSITY

effects are predicted in the RADS simulation, and the resulting theoretical values for the layer FWHM are shown in Graph 2b. Graph 2c is obtained by subtracting 2b from 2a and shows the excess FWHM versus layer thickness. This broadening of the excess FWHM is attributed to misfit dislocations in the interface. Other effects such as the apparent change in layer thickness for tilted layers and curvature of the wafer, which gives rise to peak broadening in both substrate and layer, are found to be negligible for these specimens. Grading of the In composition can also give rise to asymmetry of the layer peak (Halliwell 1984) but was not observed here.

For direct comparison of dislocation density with DCXD layer peak FWHM Graph 3 shows the excess FWHM versus interfacial dislocation density [D]. The graph is split into two regimes, I, steadily increasing FWHM with [D] and II, reducing FWHM with increasing [D]. The second part of the graph is explained by the fact that although the dislocation density is increasing with thickness in the layers with thicknesses above 100nm, the dislocations are still confined to the interfacial region. Thus as the layer thickness increases, an increasing percentage of almost totally relaxed good quality crystal has grown above the dislocated regions.

2.2 Buried multilayer structure

Fig 2a is a cross-sectional TEM image of ME580 where the specimen is tilted away from the [110] pole such that the interfaces of the wells and barriers are inclined with respect to the image plane. The dislocation lines such as P run from top to bottom in the foil and lie close to [110] in the (001) interfacial plane. No dislocations were observed in the top 40nm thick InGaAs well. A few dislocations were observed in the second 30nm well giving an upper limit on the dislocation density in this well of $1 \times 10^4 \mathrm{cm}^{-2}$, which is known from measurements of $\mathrm{In(0.2)Ga(0.8)As}$ single layers to have negligible effect on relaxation. Fig 2b shows a plan view TEM micrograph of the specimen showing misfit dislocations with a density of $15 \times 10^4 \mathrm{cm}^{-2}$.

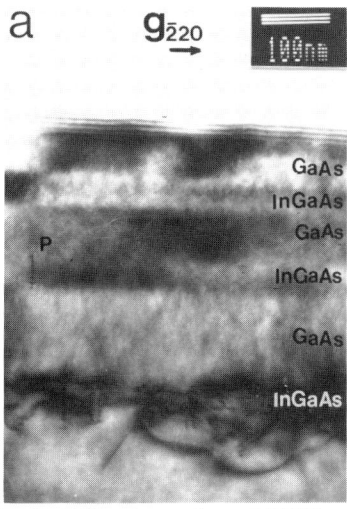

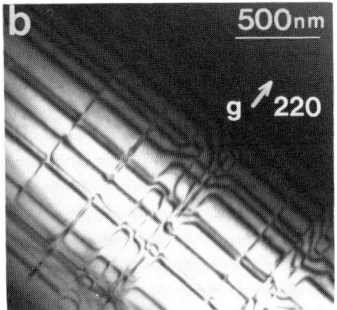

Fig 2, TEM of specimen ME580
a) cross-section, b) plan view.

Fig 3a shows the experimental (004) rocking curve for this specimen. Three diffraction peaks are generated corresponding to diffraction from the GaAs substrate, X, InGaAs wells, Y, and GaAs barriers, Z. Virtually all the relaxation of the InGaAs occurs in the first 10nm thick well. The subsequent wells and barriers are each fully strained with respect to the layer below. That is, they share a common lattice parameter in the (001) plane which is between the relaxed values for the GaAs and InGaAs. For the three peaks on the rocking curve, the peak splitting X-Y corresponds to the average residual strain in the InGaAs wells, (expansion of the InGaAs az parameter) and Z–X to the residual strain in the GaAs barriers (contraction of the GaAs a_Z parameter). The splitting Z–Y is dependent on the mismatch between the InGaAs and GaAs and is thus a direct measure of the InGaAs composition. Fig 3b shows the best fit simulated rocking curve. The composition In(0.19)Ga(0.81)As was found to give the correct Z–Y value. X–Y and Z–X were found to correspond to 70% relaxation of the first InGaAs well, and all subsequent layers fully strained with respect to their neighbours. The FWHM of the InGaAs wells the GaAs barriers and the substrate are experimentally broader than the simulated rocking curve, consistent with the TEM observation that the dislocations extend into the GaAs on either side of the first InGaAs well.   The loss of fine structure in the experimental rocking curve means that it is not possible by this method to verify the widths of wells and barriers in the structure.

The InGaAs wells in this structure have a total thickness of 80nm and are relaxed by 70% with respect to the substrate for a misfit dislocation density of 15 x $10^4 cm^{-2}$. For the 70nm single layer a dislocation density of 20 x $10^4 cm^{-2}$ corresponded to 18% relaxation. It would appear that the information from the single surface layers directly predicts the dislocation density in the multilayer structure but not the degree of relaxation of the InGaAs wells. We have found that a number of criteria are necessary when considering the relaxation of buried multilayer structures, these include the thickness of individual wells with regard to their critical thicknesses, the isolation of wells from each other and the total strain energy of the multilayer stack. Further studies of these effects will be reported elsewhere.

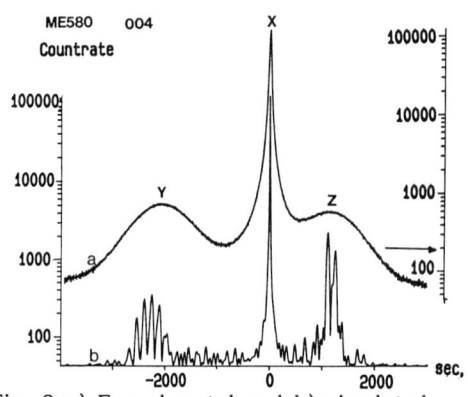

Fig. 3 a) Experimental and b) simulated DCXD (004) rocking curves for ME580.

3. SUMMARY

We have systematically investigated the change in Rocking Curve FWHM with extent of relaxation and dislocation density for a series of single In(0.2)Ga(0.8)As surface-layers. For the layers with thicknesses below 100nm the excess FWHM was proportional to the misfit dislocation density. For layers with thicknesses above 100nm the FWHM was dependent also on the proportion of dislocation-free material present above the dislocated interface. For partially relaxed buried multilayer structures we have shown that even though fine structure is lost from the rocking curve, due to peak broadening by dislocations, the extent of relaxation may still be measured from the peak splitting. The extent of relaxation of buried multilayer structures can not be predicted simply from the relaxation of single surface-layers.

ACKNOWLEDGEMENTS
The strained layers were grown at RSRE, Malvern by M. T. Emeny and L. K. Howard in support of the Surrey Strained Layer Structures Research project. The work is funded by SERC.

REFERENCE
Halliwell M A G, (1984) J. Crystal Growth, Vol 68, pp523-531.

Inst. Phys. Conf. Ser. No 117: Section 9
Paper presented at Microsc. Semicond. Mater. Conf., Oxford, 25–28 March 1991

665

X-ray diffraction studies of periodic and quasi-periodic SiGe/Si superlattice structures

S J Barnett, D C Houghton[*], A D Pitt and J-M Baribeau[*]

DRA Electronics Division, RSRE, St Andrews Road, Malvern, Worcs WR14 3PS
[*]Institute for Microstructural Sciences, National Research Council of Canada, Ottawa, Canada K1A 0R6

ABSTRACT: Double crystal X-ray diffraction coupled with rocking curve simulations is widely used to assess the structural quality of epitaxial semiconductor structures. Interface grading/roughening and alloy composition/layer thickness variations can all be quantified. Here we use the technique to study quasi-periodic $Si_{1-x}Ge_x$/Si superlattices. We examine the effects of structural imperfections on the rocking curves of such structures. This data we compare and contrast with similar data from regular periodic superlattice structures.

1. INTRODUCTION

High resolution X-ray rocking curves from semiconductor superlattice structures reveal a wealth of structural information. Often determination of the key structural parameters (period and average alloy composition) can be obtained by direct measurement from the rocking curve. Detailed information about the structure is obtained by fitting simulated rocking curves to the experimental data. In this way the thickness of individual superlattice layers, interface grades (Fewster et al 1991; Barnett et al 1988), random (Lyons et al 1987) and systematic (Barnett et al 1989) variations in layer thickness/composition can be determined.

In recent years several papers have been published which demonstrate and explain X-ray diffraction from quasi-periodic Fibonacci superlattice structures (Merlin et al 1985; Clemens et al 1987 , Terauchi et al 1988, Chen et al 1990).   Most authors used a kinematical diffraction model to calculate the expected rocking curves from such structures. Clemens et al (1987) have shown, by use of an approximation, that rocking curves taken from Fibonacci structures with small random fluctuations in layer thickness exhibit similar trends to those of superlattices with the same fluctuations; ie the effects on peak intensities and widths are larger for diffraction peaks further from the zeroth order reflections.

In this paper we use dynamical diffraction theory to simulate the effects of random thickness variations on the rocking curves of quasi-periodic structures. We compare these results with equivalent data from periodic superlattice structures.   We also demonstrate the sensitivity of the rocking curve from Fibonacci sttructures to a missing layer in the structure. The samples used consisted of $Si_{1-x}Ge_x$ and Si layers grown in sequence to form a structure with Fibonacci repeat characteristics. Two varieties of Fibonacci sequence are used;

$$
\begin{array}{lll}
\text{Bronze sequence} & S_{j+1} = S_{j-1} + 3S_j \\
\text{Silver sequence} & S_{j+1} = S_{j-1} + 2S_j
\end{array}
$$

where: $S_1 = A \equiv Si$ and $S_2 = AB \equiv Si/Si_{1-x}Ge_x$

2.   EXPERIMENTAL

The $Si_{1-x}Ge_x/Si$ samples were grown by MBE at 500–550°C on 100mm diameter Czochralski Si substrates at a base pressure of 5×10^{-10} mbar. The grown structures showed typical etch pit densities of $<10^3$ cm^{-2}. The fluxes from the electron beam evaporated Si and Ge sources were controlled by feedback from Sentinal III sensors and the shutters were operated under computer control.

X-ray rocking curves were measured on a computer controlled Apex goniometer equipped with a four reflection Ge beam conditioner. This system is extremely versatile; the narrow spectral width of the conditioned output beam allows rocking curves to be recorded form different materials and reflections almost free of wavelength dispersion effects. For the measurements presented here the 220 reflections were used from the Ge <110> monochromator crystals ($\frac{\Delta\gamma}{\lambda} = 1 \times 10^{-4}$, $\Delta\theta = 12$ arc sec). The 004 sample reflection was used with Cu $K\alpha_1$ radiation. Rocking curves were simulated using dynamical diffraction theory in the form of the Takagi–Taupin equations simplified to assume a one–dimensional variation in lattice parameter and scattering factor. An analytical solution to the above equation for a thin uniform layer was used to calculate the rocking curve from an arbitrary structure in the manner analogous to that employed by Halliwell et al (1983). This approach is extremely useful and very flexible, enabling structures of arbitrary layer repeat sequence to be simulated relatively easily.

3.   RESULTS AND DISCUSSION

Fig. 1 shows experimental and simulated rocking curves for a Bronze series, S_5 Fibonacci structure. The calculated dimensions of the individual layers within the structure are shown in the figure caption. This particular thickness/composition combination was chosen by fitting the intensity and position of the more significant peaks in the simulated rocking curve to those in the experimental data. The fit is extremely good. The intensity of most of the significant peaks agree to within a few percent.

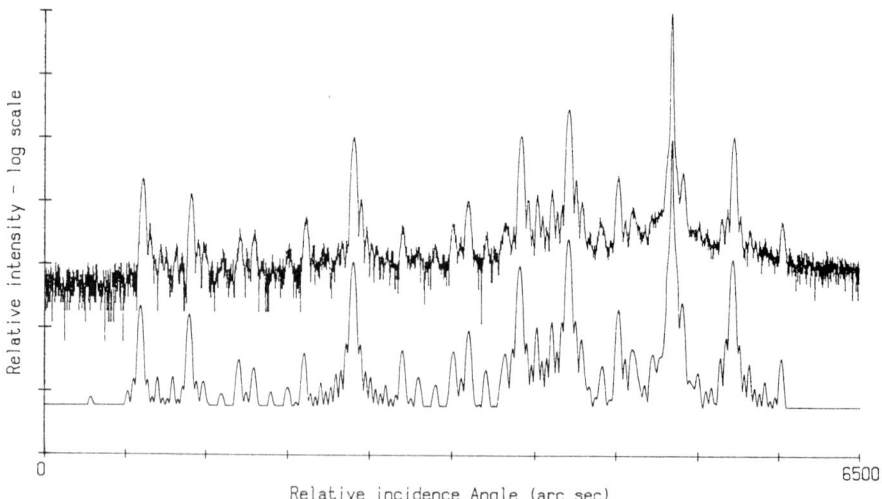

Figure 1.   Experimental (upper line) and simulated (lower line) rocking curves of a Fibonacci structure with Bronze repeat (S_5) sequence. Layer dimensions: A $\equiv$ Si (7.7nm), B $\equiv$ $Si_{0.75}Ge_{0.25}$ (4.7nm)

The sensitivity of this fitting procedure can be gauged from Fig. 2 which shows a comparison of two theoretical plots for a Silver series structure. The two simulations are

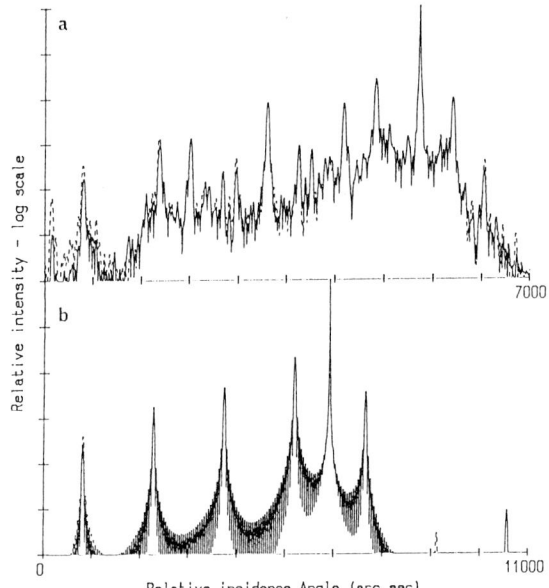

Figure 2. Simulated curves of a) Fibonacci superlattice, b) regular superlattice with x changing by 0.02 between solid and dotted lines.

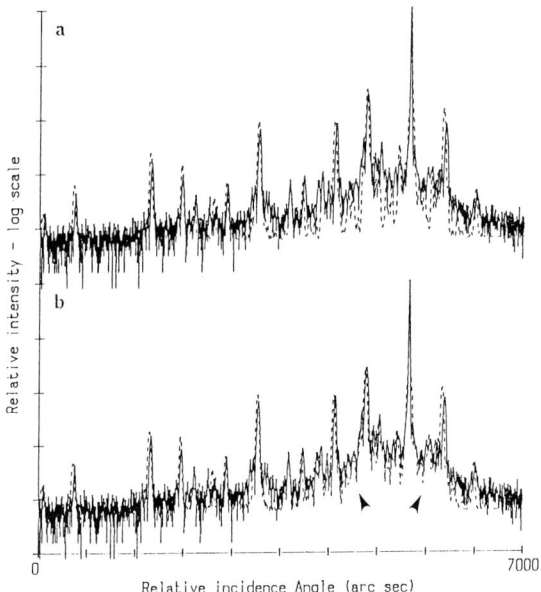

Figure 3. Experimental (solid) and simulated curves for a Silver series Fibonacci sample a) perfect structure and b) missing layer six.

for alloy compositions differing by x = 0.02, but the average alloy composition in both structures is equal. For comparison, a similar plot (Fig. 2b) is shown for a regular superlattice of equal thickness, containing the same number of layers and having the same average alloy composition as the Fibonacci structure shown in Fig. 2a. For a given departure from the 'average lattice' peak, the Fibonacci curves differ by a greater percentage than those of the regular superlattice.

A demonstration of the sensitivity of the rocking curve to a disruption in the Fibonacci structure is shown in Fig. 3. This silver series Fibonacci structure is missing layer number 6. The ideal structure should be BAABABAAB ..... but transmission electron microscopy showed the actual structure to be BAABAAAB.... Fig. 3a shows the experimental data and a simulation of the perfect structure while Fig. 3b shows the simulation modified by the missing layer. Although the missing layer changes the intensity of the more significant peaks very little, the intensity of the more minor peaks, in the region marked by the arrows, (Fig. 3b) increases and more closely matches the experimental data.

The effects of random layer thickness variations on the rocking curves of a Fibonacci repeat structure can be seen in Fig. 4a for a 10% variation in layer thicknesses. The data can be compared with a similar plot for a regular superlattice (Fig. 4b). As might be expected, both curves exhibit similar trends.

The deviation of the rocking curve from that of the perfect structure is a minimum in the region of the average lattice peak (zeroth order superlattice reflection). However, for a given departure from this average lattice peak, the Fibonacci simulation exhibits a greater percentage deviation form the perfect structure simulation than does the corresponding superlattice curve.

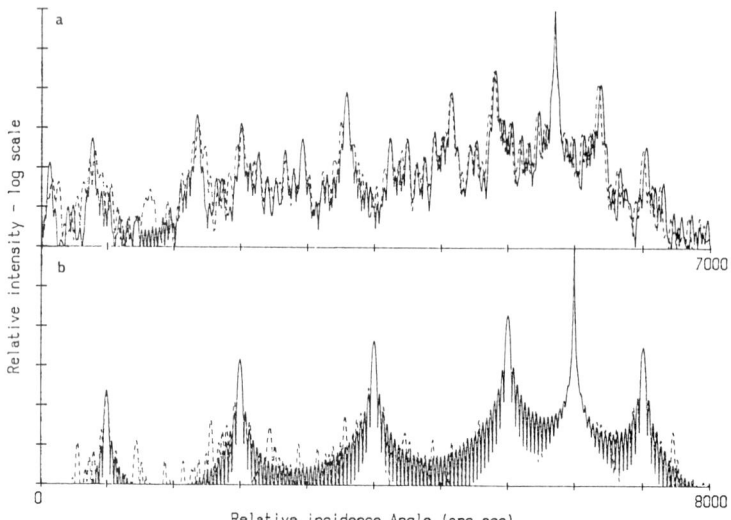

Figure 4: Simulated rocking curves of a) Fibonacci structure and b) regular superlattice, solid line – perfect structure and dotted line – 10% layer thickness variations.

4.   CONCLUSIONS

The experimental and simulated data here demonstrate the application of X–ray diffraction for assessing quasi–periodic structures. Rocking curves from such structures are compared with those from similar regular periodic superlattices and it is shown that, in some cases, the Fibonacci rocking curve is more sensitive to structural imperfections.

REFERENCES

Barnett S J, Brown G T, Courtney S J, Bass S J and Taylor L L 1988 J. Appl. Phys. **64** 1185

Barnett S J, Brown G T, Houghton D C and Baribeau J–M 1989 Appl. Phys. Lett. **54** 1781

Chen G and Zuqin W 1990 Sol. State. Comm. **76** 269

Clemens B M and Gay J G 1987 Phys. Rev. Lett. **B35** 9337

Fewster P F, Norman A L and Curling C J 1991 Semicond. Sci. Technol. **6** 5

Halliwell M A G, Juler J and Norman A G 1983 Inst. Phys. Conf. Ser. **67** 365

Lyons M H, Halliwell M A G, Tuppen C G and Gibbings C J 1987 Inst. Phys. Conf. Ser. **87** 609

Merlin R, Bajema K, Clarke R, Juang F–Y and Bhattachrga P K 1985 Phys. Rev. Lett. **55** 1768

Terauchi H, Noda Y, Kamigaki K, Matsunaka S, Nakayama M, Kato H, Sano N and Yamada Y 1988 J. Phys. Soc. Jap. **57** 2416

Inst. Phys. Conf. Ser. No 117: Section 9
Paper presented at Microsc. Semicond. Mater. Conf., Oxford, 25–28 March 1991

TEM and triple-crystal diffractometry investigation of the distribution of dislocations across the depth of epitaxial structures

T S Argunova, R N Kyutt, S S Ruvimov and M P Scheglov

Ioffe Physical-Technical Institute of the USSR Academy of Sciencies, Leningrad 194021, USSR.

ABSTRACT: New nondestructive methods to reveal misfit dislocation (MD) networks and to determine the depth of their locacation are suggested. This problem is solved either by X-ray penetration depth variation or with the aid of triple-crystal diffractometry (TCD). In the last case the distribution of $(\Delta d/d)_{\perp}$ and $(\Delta d/d)_{\shortparallel}$ across the depth of epitaxial structures (ES) is constructed from the two-dimensional intensity distribution. X-ray diffractometry results are compared with those obtained by TEM.

1. INTRODUCTION

The depth distribution of the dislocations in ES severely affects device operation. On the basis of results obtained during a study of various ES via TEM, X-ray topography and TCD we have suggested a few new nondestructive methods to determine the depth of MD networks. This problem can be solved in two different ways. The first one is X-ray penetration depth variation caused by diffraction geometry. The second method uses a certain feature of the sublayer in the ES, namely the interplanar spacing is selected with the aid of TCD. By use of a differential triple-crystal scheme, the diffracted intensity is registrated versus the specimen α and analyzer $\Delta\vartheta$ deviation angles from the Bragg condition. When the specimen is rotated and the analyzer is set, ω-scanning curves are registrated and the end point of the wave vector of the diffracted beam (observation point) moves along directions normal to the reciprocal lattice vector H (q_α axis). During this procedure X-rays reflect from planes with equal spacing. When the specimen and the analyzer rotate relative to each other with velocity 1:2 the observation point moves along directions parallel to H (q_H axis).
To avoid the influence of bending and misorientation and to obtain the correct distribution of intensity along H we propose the construction of the relationship $J(q_H)$ in integral form. In this construction the intensity corresponding to each point $q_H = -k\,\Delta\vartheta\cos\vartheta_B$ at the q_H axis is the sum of intensities diffracted from planes with equal lattice spacing $\int I(\alpha)d\alpha$ (1). The construction $J(q_H)$ characterizes the variation of interplanar spacing. In practice ω-scanning curves

are measured at different angular positions of the analyzer
around Bragg angles of the layer and the substrate. Then the
integral intensity J(qH) calculated according to (1) is plotted as a function of qH. The width of each ω-scanning curve
is caused by the structural perfection of layers or sublayers
with certain lattice parameters (Kyutt et al 1989).

2. RESULTS AND DISCUSSION

2.1 MD networks in thin near-surface layers

A considerable decrease of the X-ray penetration depth is
known to be obtained by the use of grazing incidence geometry. The double-crystal diffraction scheme previously used by
Marra et al (1979) has been applied in our experiment. The
reflection curves in grazing incidence geometry are very sensitive to MD networks localised in a near surface layer up to
100 nm. An essential broadening of the reflection peaks, the
half-width of which reached up to 10 angular minutes was observed for specimens $GaAs_{1-x}P_x/GaAs$ containing dislocation
networks at a depth up to 100 nm (Fig.1) (Scheglov et al
1988). The location depth, the density of the networks and
the type of MD´s have been determined by TEM. It is worth no-
ticing that the grazing incidence angle variation permits the
penetration depth to be changed and the depth of MD´s location to be estimated.

2.2 The defects distribution in ES on low absorption subs- trates

In studied ES of $Si_{1-x}Ge_x/Si$ (100) the concentration of Ge
varied from X=0 at the interface to X=10-15% at the surface.
The distribution of the strain in thick layers on the basis of
low absorption substrates can be estimated from Laue-geometry
diffraction. The oscillations of intensity are distributed
along the vector H (Fig.2) while those caused by interference
of diffracted waves are distributed along the direction normal
to the interface. Thus there is a correspondence between intensity oscillations in Fig.2 and variation of $(\Delta d/d)_{\parallel}$ across
the depth. The peaks are well separated from each other and
this is evidence of an abrupt variation of $(\Delta d/d)_{\parallel}$ between
sublayers. The last result may be interpreted as due to the
presence of MD networks at interfaces between sublayers. One
can easily determine the integral intensity of each peak and
calculate the thickness of the sublayer by use of kinematical
theory. The steps in $(\Delta d/d)_{\parallel}$ distribution across the depth of
the layer are calculated from distances between peaks in the
curve J(qH). The MD density can be evaluated from these steps.
Fig.2 shows that steps in $(\Delta d/d)_{\parallel}$ distribution correspond to
MD networks. There is a good agreement between thicknesses of
sublayers and MD densities obtained by means of TCD and TEM.

2.3 TCD asymmetrical Bragg reflection measurements

Asymmetrical Bragg reflection ω-scanning measurements provide
two values: the interplanar spacing $(\Delta d/d)$ and the angle
between scattering planes of the layer and the substrate.

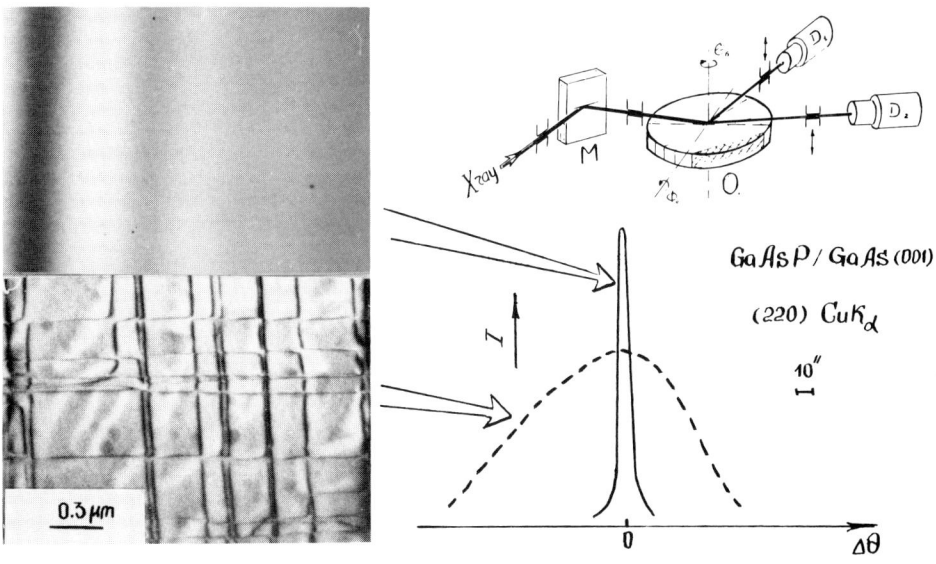

Fig.1. The influence of MD networks in near surface layer on the grazing incidence geometry rocking curves.

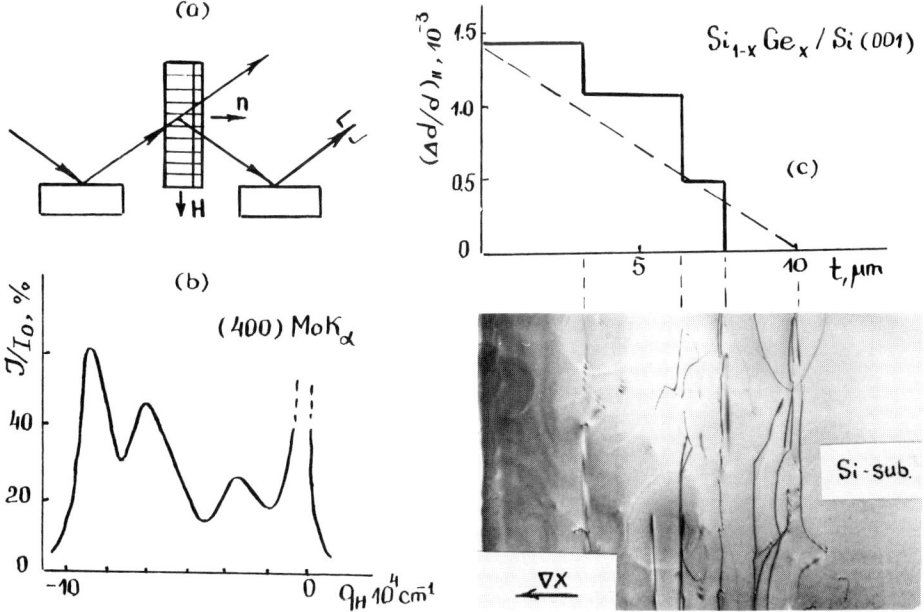

Fig.2. Depth distribution of $(\Delta d/d)_{\shortparallel}$ (c) obtained from Laue diffraction measurements (a,b) and TEM cross section image.

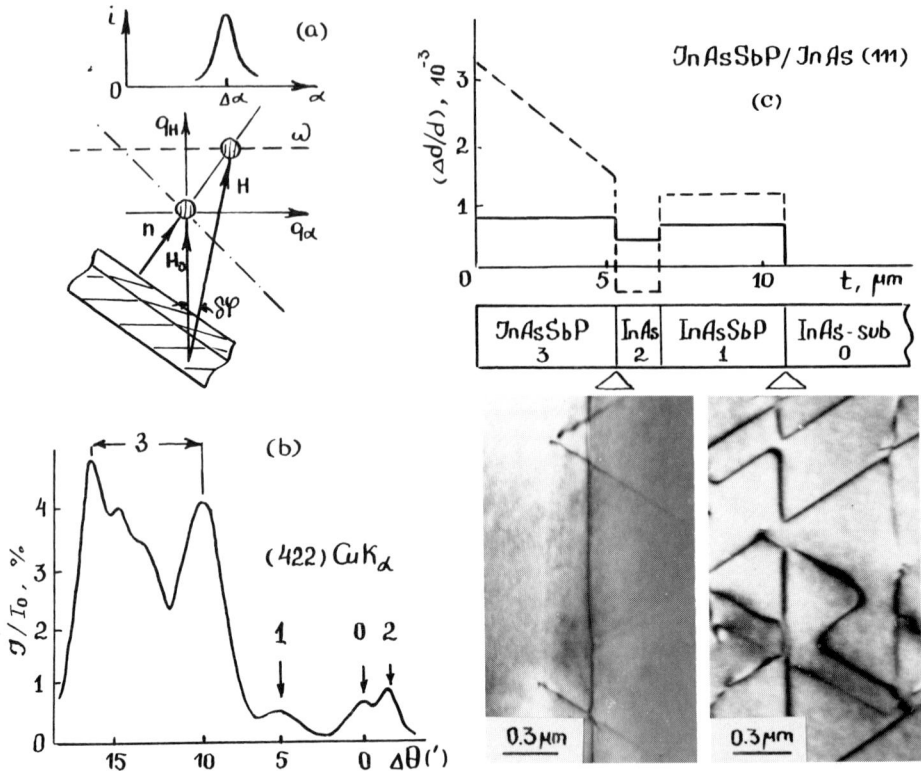

Fig. 3. The picture of asymmetrical Bragg diffraction in reciprocal space (a), intensity distribution along H (b),
obtained depth distribution of ($\Delta d/d$)$_\text{II}$ (solid line)
and ($\Delta d/d$)$_\perp$(dotted line) and TEM images of interfaces.

The first value is obtained from the distribution $J(q_H)$. The
second one is an additional displacement of the intensity
maximum on the ω-scanning curve. Values of ($\Delta d/d$) and $\delta\varphi$ are
connected with the variation of ($\Delta d/d$)$_\text{II}$ and ($\Delta d/d$)$_\perp$. If the
epitaxial layer sequence is known, values obtained allow the
distribution of ($\Delta d/d$)$_\text{II}$ and ($\Delta d/d$)$_\perp$ across the depth of ES to
be constructed. Analysis of $J(q_H)$ and $\delta\varphi(q_H)$ constructions or
the distribution of ($\Delta d/d$)$_\text{II}$ across the depth of ES shows the
presence of MD networks and provides their location depth
values. For example (Fig.3) use is made of this method and the
presence, the density and the location depth of MD networks in DH InAsSbP-InAs are determined.

REFERENCES

Kyutt R N and Argunova T S 1989 Fizika tverdogo tela 31 40
Marra W C, Eisenberger P and Cho A I 1979 J.Appl.Phys. 50 6927
Scheglov M P, Ruvimov S S, Kyutt R N and Sorokin L M 1988 Jur-
nal Tehnicheskoy Fiziki 58 583

Inst. Phys. Conf. Ser. No 117: Section 9
Paper presented at Microsc. Semicond. Mater. Conf., Oxford, 25–28 March 1991

X-ray double crystal topography of semiinsulating GaAs crystals

C Ferrari and P Franzosi

MASPEC-C.N.R., Via Chiavari 18/A, 43100 Parma, Italy

ABSTRACT: Semiinsulating GaAs crystals grown by the liquid encapsulated Czochralski technique, (001) oriented and two inches in diameter, have been investigated by X-ray double crystal topography in a nearly non-dispersive configuration. This made it possible to study both the long range elastic strains and the dislocation distribution; in particular, dislocations arranged along cell boundaries, lineage and slip bands have been observed. Using a simple grain boundary model, it has been found that all these features are produced by $1/2<110>$ dislocations originating at the crystal edges, gliding on the $\{111\}$ planes and interacting with each other.

1. INTRODUCTION

GaAs semiinsulating crystals grown by the liquid encapsulated Czochralski (LEC) technique are known to exhibit a relatively high dislocation density; moreover the dislocations tend to be arranged in typical features that are slip bands, mainly present close the crystal edges, and cell structures, prevalently located in the central crystal zone. In some cases dislocation lineage, a few cm in length, is also observed. Up to date no clear and simple mechanisms for the generation of these features have been reported in the literature.

The aim of this paper is to report a careful investigation of the crystal quality of LEC GaAs semiinsulating crystals by the X-ray double crystal topography (DCT) method; the results may give some contribution to the understanding of the defect formation mechanisms.

2. EXPERIMENTAL

Several undoped semiinsulating GaAs crystals grown by the LEC technique have been investigated; all the samples were (001) oriented and two inches in diameter. The DCT investigation has been performed using a double crystal camera equipped with a perfect Si monochromator set at the Bragg angle for the asymmetric 224 reflection; the CuKα radiation and a conventional fine focus X-Ray tube have been used. The monochromator could be elastically bent in order to compensate the sample bending (Jenichen et al 1985). Since the GaAs 115 reflection used for recording the DCT images and the Si 224 reflection have nearly identical Bragg angles the chromatic dispersion was negligible. All the samples were mecano-chemically polished on both surfaces and 400 μm thick; since no sample fixing on the horizontal holder was required any unintentional sample bending was avoided. A map of the elastic strain in the samples could be obtained by keeping the monochromator flat and overlapping several exposures, corresponding to different angular positions of the sample, on the same photoplate; in contrast, the defect distribution in large sample areas could be investigated by partly compensating the sample curvature. At a tube power of 1200 watts a single exposure could be taken in 3 hours on Orwo K6 nuclear plates.

3.   RESULTS

Fig. 1 shows a typical DCT image, obtained with the flat monochromator; four exposures, corresponding to angular positions of the sample differing from each other by 15 sec. of arc, are superimposed in the image. Since each black stripe in the picture is a locus of constant incidence angle, the local curvatures of the sample could be easily calculated. Curvature radii ranging from 62 m in the center of the wafer down to 11 m close to the (100) sample edges, which exhibit a higher dislocation density, have been measured in this way. From the sample curvature a lattice displacement map similar to those found by other authors (Barnett et al 1987, Tanner et al 1985) can be deduced. In principle it is not possible to distinguish the effect of the lattice tilt and the lattice parameter variation by mean of a single topograph; nevertheless it was shown by Barnett et al (1987) that very low lattice parameter variations, probably due to stoichiometry changes, characterize undoped semiinsulating GaAs crystals. It must be stressed that the dislocation distribution is in agreement with the model of Jordan et al (1980) for the induced thermal stress during the growth.

Typical dislocation arrangements, already observed by other authors (see for example Brown et al 1984), are shown in the Fig. 1: slip bands, lineage and cell structures especially in the centre of the wafer; in particular the cell structures near the sample edges (Fig. 2) appear elongated in the <110> directions. It can be observed that the single dislocations can be in general resolved in the cell walls and in the slip bands but the same is not possible in the lineage features owing to their density. A second topograph is shown in Fig. 3 in which the tilt of the lattice planes near the lineage is clearly seen; from Fig. 3 the tilt angle could be estimated to be 9 sec. of arc. The perpendicular component of the lattice rotation was measured by taking a topograph of the sample rotated by 90°. Since no contrast was seen in this case it was possible to deduce that the lattice rotation associated with lineage features lying on the diametrical axis has essentially a zero twist component as observed by Brown et al (1984). In contrast, for the lineage b in Fig. 3 a strong contrast is observed even if the scattering plane is parallel to the lineage direction.

The dislocation density in the lineage b of Fig. 3 could not be measured by DCT; therefore it was determined by the SEM catodoluminescence technique (see for example Yacobi et al 1990). From Fig. 4, a linear dislocation density of 1.1×10^3 cm^{-1} in the lineage of fig. 3a has been evaluated.

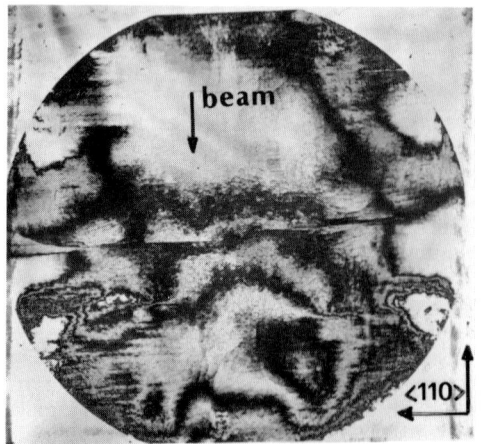

Fig. 1: double crystal topograph of a 2" SI GaAs (001) oriented wafer grown by the LEC technique, 511 CuKα reflection. Four exposures separated by 15 sec of arc are overlapped. The sharp straight contrast is associated with the lineage features.

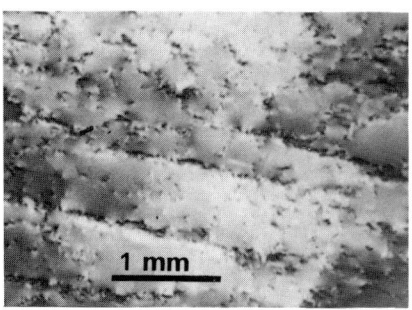

Fig. 2: magnified area from topograph of Fig 1 showing cell structures elongated in a <110> direction.

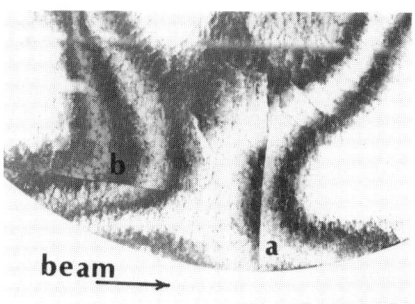

Fig. 3: detail of a DC topograph of a SI GaAs showing lineage at 90°; the sharp contrast of the vertical lineage evidences a crystal lattice rotation perpendicular to the scattering plane.

4.   DISCUSSION

The rotation angle θ induced by a set of parallel dislocation lines arranged in a grain boundary is given by the Frank (1950) formula:

$$\Sigma_i \, N_i b_i = (r \times l)\theta$$

in which b_i and N_i are the Burgers vector and the number of the dislocation of the kind i crossed by the path **r** on the boundary plane and **l** is the rotation axis. It was shown by Brown (1981) that in LEC grown GaAs crystals the majority of the dislocations are of the 1/2<110> type and are generated at the sample edges. The thermal stresses then cause the dislocation movement mainly by slip on the easy glide system <110>{111}. An array of such dislocation lines is able to produce a grain boundary in which a rotation of the crystal lattice has a strong twist component i.e. the rotation axis is not parallel to the boundary plane. This fact can explain the behaviour of the lineage in Fig. 3b; similar observations about arrays of

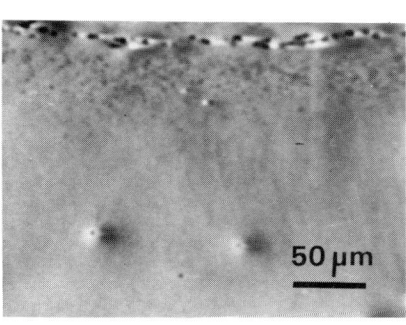

Fig. 4: SEM trasmission cathodoluminescence micrograph of the lineage in Fig. 3a obtained using a Si detector; a dislocation density of 1.1×10^3 cm^{-1} has been measured.

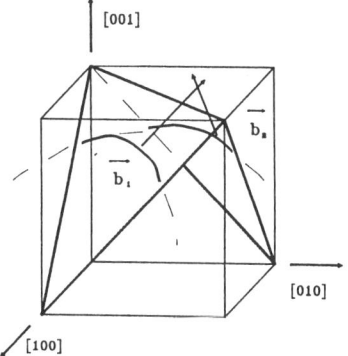

Fig. 5: simple model showing the lineage formation from 1/2<110> dislocations loops moving by glide on {111} planes.

screw dislocations in GaAs ingots were made by Matsui (1987). We suppose that the pure tilt character of the grain boundary corresponding to the lineage in Fig 3a is given by a superposition of two arrays of dislocation lines with Burgers vectors a/2[011] and a/2[101] and lying on a (0-11) plane as sketched in Fig. 5. Assuming that the densities in the two dislocation arrays are equal, the Frank formula can be simplified as follows:

$$D^{-1} = b \cdot n / \theta$$

n being the normal to the boundary plane. The calculated dislocation density $D = 1.1 \times 10^3$ cm^{-1} is in excellent agreement with the value measured by the cathodoluminescence micrograph of Fig. 4. The rotation axis coincides in this case with the <11-1> direction.

All the results reported can be easily interpreted by supposing that the open dislocation loops generated at the (100) sample edges move by glide on {111} planes along <110> directions. In the crystal centre loops gliding on all the four {111} panes may interact together and produce cell structures; in the vicinity of the crystal edges, i.e. where the dislocations originate, they are essentially of a single type and produce slip bands and lineage with a strong twist rotation of the crystal lattice; the lineage on the {110} diametrical planes is obtained by the superposition of two kinds of dislocations and induce a lattice rotation around a (111) axis.

5.   CONCLUSIONS

The DCT technique has been used for investigating both the long range elastic strains and the dislocation nature and distribution in undoped semiinsulating GaAs crystals grown by the LEC method. By carefully analysing the cell structures, the slip bands and the lineage features it has been deduced that they are essentially of the same nature and are produced by the interaction of the a/2<110> type dislocations gliding on the four {111} planes. This mechanism is in agreement with the thermal induced stress model of Jordan et al (1980).

ACKNOWLEDGEMENTS

The authors are grateful to dr L Lazzarini for the cathodoluminescence investigation and to Mr. M Scaffardi for his technical assistance.

REFERENCES

Barnett S J , Brown G T and Tanner B K 1987   Inst Phys Conf Ser No 87 pp 615
Brown G T , Skolnick M S , Jones G R , Tanner B K and Barnett S J 1984 Conference on III V Semiinsulating Materials Kah-nee-ta pp 76
Brown G T , Warwick C A , Young Y M and Booker G R 1981 Inst Phys Conf Ser No 67 pp 371
Frank F C 1950 Conf on Plastic Deformation of Crystalline Solids Carnegie Inst. of Techn and Office of Naval Research pp 150
Jenichen B, Kohler R and Mohling W 1985 Phys Stat Sol (a) 89 pp 79
Jordan A S, Caruso R and von Neida A R 1980 Bell Syst Tech 59 pp 593
Matsui J 1987 Inst Phys Conf Ser No 87 pp 249
Tanner B K, Barnett S J and Hill M J 1985 Inst Phys Conf Ser No 76 pp 429
Yacobi B G and Holt D B 1990 Cathodoluminescence Microscopy of Inorganic Solids Plenum Press New York

X-ray transmission topographic observations of nickel silicide precipitation along dislocations in silicon

B Pichaud and F Minari

Laboratoire de Physique Cristalline, URA 797 CNRS, Faculté des Sciences de St Jérôme,
Av. escadrille Normandie-Niemen, 13397 Marseille Cedex 13, FRANCE

ABSTRACT: $NiSi_2$ precipitates (5 to 50μm in diameter) were observed by X ray transmission topography in Si saturated with Ni. The number and the size of the precipitates is a function of the dislocation density. Whatever the reflection the well-defined images of the precipitates are always homogeneous and circular: these particles are probably coherent with the Si matrix. Dislocations act as very efficient gettering centres especially at junctions or along non-crystallographic (kinked) segments.

1. INTRODUCTION

One of the crucial problems in device manufacturing is the contamination of Si wafers with metallic impurities. Furthermore as dislocation generation can hardly be avoided, at the same time, their interactions with metallic impurities must be taken into account. The study of these interactions could lead to an estimation of the gettering efficiency of process-induced dislocations.

Precipitation of iron, nickel, cobalt, copper etc... was already observed along dislocations in silicon by transmission electron microscopy (Nes et al. 1973, Vanderwalker 1984, Honda et al. 1987, Ryoo et al. 1988). However this type of observation is very localized and it is not so easy to extend its conclusions to large volumes of crystal. X ray transmission topography (XRT) owing to its capability of imaging large crystals can be used to study this particular dislocation-precipitate interaction at a different scale; the drawback of this method is its poor resolution which needs low dislocation densities and large precipitates.

In this paper we report XRT examination of nickel silicide precipitates along dislocations. We developed well-controlled dislocations in FZ silicon crystals then we introduced nickel atoms by diffusion at high temperature. After cooling XRT showed precipitates, the contrast, size and location of which were studied in connection with the presence of the stress-induced dislocations.

2. EXPERIMENTAL PROCEDURE

Rectangular samples 1x3x0.1 cm^3 were cut from a dislocation-free FZ (111) phosphorus doped ($\rho=10\Omega.cm$) silicon wafer. These rectangular plates were optically polished on both faces and chemically etched with CP4 in order to smooth the surface and to blunt the edges. Dislocations were introduced from scratches made on one face of the sample, then they were developed in elongated half-loops by a two step procedure: the sample was elastically deformed by cantilever bending at room temperature, then heated under stress at 700°C during 3h in high vacuum (Pichaud et al. 1984). A nickel layer (100nm thick) was deposited by evaporation in high vacuum (10^{-6} Torr) on the unscratched surface, a further heat treatment performed at 1000°C for 20min allowed nickel diffusion into the whole sample: diffusion coefficient D= $3.10^{-5}cm^2$ s^{-1} (Weber 1983). The sample cooling, in the stressed state, was obtained by moving the furnace away and by ventilation of the quartz tube containing the sample: this allows a cooling rate of approximately 50° s^{-1} between 1000°C and 700°C.

After unloading at room temperature, the sample was examined by XRT using AgKα radiation and different reflecting planes. After XRT observations the samples were analysed by Inductively Coupled Plasma -Atomic Emission Spectometry in order to verify the total Ni concentration.

3.   RESULTS AND DISCUSSION

The sample thickness corresponds to kinematic conditions for the propagation of X-rays (μt=0.8). After deformation, residual stresses modify the diffraction conditions so that the observed contrasts are of rather dynamical type i. e. white dislocation on the background (Yang et al. 1989).

Accounting for the thermal conditions of the experiment we can assume that we obtained $NiSi_2$ precipitates (Seibt et al. 1989). We observe in figure 1 different zones of a sample containing different dislocation densities and different distributions of precipitates the size of which is a function of the dislocation density (Table 1). Concerning these precipitates two main features may be distinguished: the characteristics of their images (contrast, shape, size) and the nature of their interactions with dislocations.

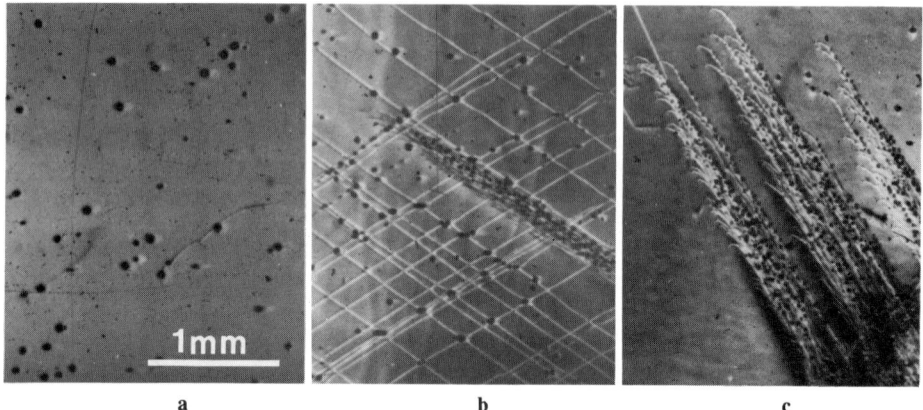

a   b   c

Figure 1: X-ray transmission topograph (reflection 111) showing $NiSi_2$ precipitates in different regions of a Si crystal,(a) dislocation-free zone,(b) homogeneous distribution of dislocations (ρ=7.2 10^2 cm.cm^{-3}), (c) dislocation bundles. Same black-white contrast as the original plate.

3.1 Precipitate image

In fig.1a the image of each precipitate may be examined without any perturbational effect of stress field associated with dislocations. Using different reflecting planes, the precipitates appear as well defined homogeneous black spots associated with white shadows. The distance between spot and shadow varies with the depth of the precipitate in the plate: thus figure 2 shows two precipitates, one without white shadow and the other for which the distance spot-shadow is maximum. This figure corresponds to the extreme case of two precipitates respectively situated near the exit surface and near the entrance surface of X rays. These characteristics are probably a consequence of both the existence of a dynamical image accompanying the kinematical one and of the small mismatch (0.4%) between Si and $NiSi_2$ lattices.

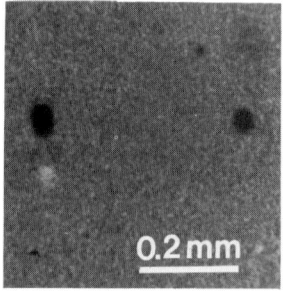

mean image radius R(μm)	32	23	14	11	11	9
precipitates density N_v (cm^{-3})	3.8 10^3	1.1 10^4	5.5 10^4	1.1 10^5	6.7 10^4	1.7 10^5
dislocation density (cm.cm^{-3})	0	7.2 10^2	10^3	3.2 10^3	2.5 10^3	7.5 10^3

Figure 2: black-white contrast
of the precipitates, reflection 111

Table 1:mean radius of the precipitate image , number of precipitates
per unit volume and dislocation density in different zones of a crystal

Irrespective of the reflecting plane the image of a precipitate, or at least the black spot, is always nearly circular indicating that the precipitate, or at least the deformation field around it, is nearly spherical. Table 1 gives the mean image radius R in regions containing different densities of precipitates N_v: it is easy to see that the product R^3N_v is constant. The amount of Ni dissolved is the same anywhere within the sample and corresponds to the solubility limit of Ni in silicon (the analysis gives $3 \ 10^{17}$at.cm^{-3}). From the values of N_v and assuming that the composition of the precipitates is $NiSi_2$, one can derive the mean theoretical radius of the precipitates r in the corresponding regions. Figure 3 shows that r and R are proportional in agreement with the model by Mott et al. (1940) which established that the deformation at a distance ρ from a precipitate is proportional to $(r/\rho)^3$. In our case, R may be considered to be a particular value of ρ: the one for which the elastic deformation of the lattice allows the Bragg law to be satisfied. Moreover, R and r are not very different $(R \approx 3.5r)$, accounting for the extreme sensitivity to misorientation of Bragg conditions in nearly perfect crystals, the deformation field around the precipitate is very localized. This is consistent with the fact that the precipitates are probably coherent with the silicon matrix (Seibt et al. 1989).

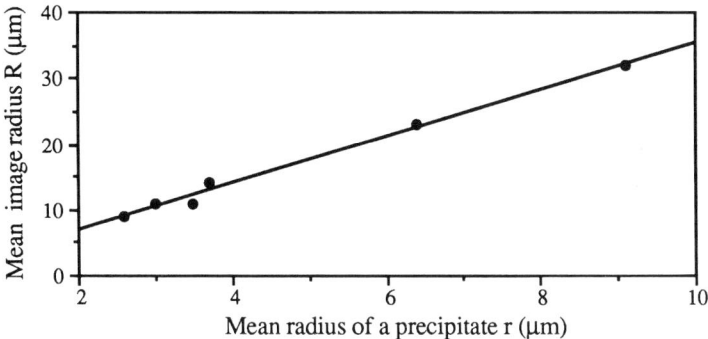

Figure 3: linear relationship between the mean image radius and the mean radius of the precipitate corresponding to the measured quantity of Ni

3.2 Dislocation-precipitate interaction

It is obvious from figure 1b and c that the presence of dislocations favoured precipitates nucleation as compared to the perfect crystal. Figure 4 gives the dependence of the precipitate density on the dislocation density. It shows a linear relationship between these two quantities which means that, from a global point of view, the nucleation sites are homogeneously distributed along dislocation lines.

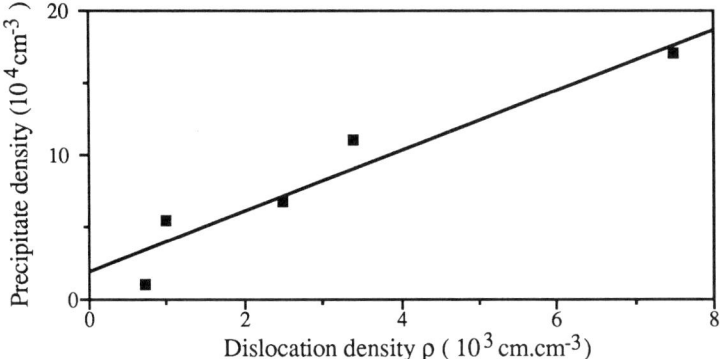

Fig 4: relation between the precipitate density andthe dislocation density ρ in the case of precipitates homogeneously distributed along dislocations.

In the neighbourhood of dislocation bundles, precipitates are more rarely observed than in zero dislocation zones. The corresponding denuded zones extend for distances of the order of 1mm (fig.1c). It may be concluded that the dislocations promote a very efficient gettering effect as their presence suppresses precipitation within wide regions.

However, precipitates are not always homogeneously distributed along dislocation lines and two distinct types of preferential sites are observed: dislocation junctions (figure 1b) and non-crystallographic parts of lines (curved segment AB in figure 5). In the last case, the long half loops were developed at 700°C but, due to their further annealing at 1000°C during Ni diffusion, they exhibit smooth shapes (this temperature is situated in the athermal regime of glide), and not the hexagonal shape characteristic of lower temperatures (Pichaud et al. 1986). Nevertheless, along these half loops, well developed straight parts often remain nearly parallel to <110> directions. There are few precipitates along these crystallographic parts while precipitates are numerous elsewhere.

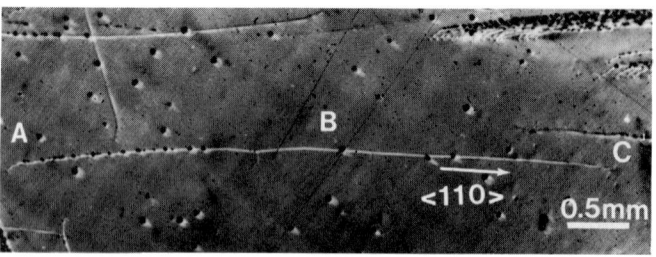

Figure 5: preferential precipitation along parts of a dislocation half-loop deviating from a crystallographic direction <110>.

The characteristic of both types of location is that they contain geometrical defects such as jogs due to dislocation crossing or kinks allowing a dislocation line to curve, using, on a microscopic scale, Peierls valleys parallel to <110> . These geometrical defects are likely to constitute favourable sites for nickel silicide precipitation as they correspond to wider cells as compared to perfect lattice or straight dislocation line (Jones 1983), and contain dangling bonds while straight dislocations are reconstructed (Jones 1983). These sites may act as very efficient sources or sinks of point defects which favour the precipitation kinetics.

4.   CONCLUSION

We analysed by XRT the contrast, the size and the distribution of $NiSi_2$ precipitates in deformed silicon. We showed that:
- the precipitates exhibit a very well defined image probably due to their coherent character with the silicon matrix. The size of the image is not very far from the size of the precipitate.
- the precipitates are preferentially nucleated along dislocations and they are more numerous along lines of non-crystallographic orientation and at dislocation junctions.
   The larger efficiency of these sites along dislocations may allow one to promote a way of improving gettering of metallic impurities by process-induced dislocations using heavily kinked or jogged dislocations.

REFERENCES

Honda K, Nakanishi T, Ohsawa A and Toyokura N 1987 Inst. Phys. Conf. Ser. 87 469
Jones R 1983 J. Phys. C4 61
Mott N F and Nabarro F R N 1940 Proc. Phys. Soc. 52 86
Nes E and Washburn J 1973 J. Appl. Phys. 44 3682
Pichaud B, Jean P and minari F 1986 Phil. Mag. A 54 479
Pichaud B and Minari F 1984 Application of X ray topographic methods to Materials Science,
      eds S Weissmann, F Balibar and J F Petroff, (N Y : Plenum Press) pp 385-92
Ryoo K, Drosd R and Wood W 1988 J. Appl. Phys. 63 4440
Seibt M and Schröter W 1989 Phil. Mag. A 59 337
Vanderwalker D M 1984 Phys. Stat. Sol. (a) 86 507
Weber E R 1983 Appl. Phys. A 30 1
Yang P, Green G S and Tanner B K 1989 Inst. Phys. Conf. Ser. 100 467

Inst. Phys. Conf. Ser. No 117: Section 10
Paper presented at Microsc. Semicond. Mater. Conf., Oxford, 25–28 March 1991

Developments in high resolution scanning cathodoluminescence microscopy

C A Warwick

AT&T Bell Laboratories, Room 4C-410, Crawfords Corner Road, Holmdel, NJ, USA, 07733-1988.

ABSTRACT: The advent of quantum well structures has given rise to new physics and new technology. Quantum well properties are determined by interfaces between chemically distinct, heterostructural compounds. The study of these interfaces is of vital importance to the progress of this field, and many studies using a variety of techniques have been made. The scanning cathodoluminescence microscope has made major contributions to the understanding of interfaces in quantum well devices but has previously been limited to a spatial resolution of $\approx 1\,\mu m$. We have achieved $60\,nm$ spatial resolution in this mode by: 1) using a small probe generated by an intense field emission source; 2) using a $1 \approx 2\,kV$ beam to inject both near the surface and with minimal lateral scattering ($10 \approx 30\,nm$); and 3) by detecting photons from pre-diffusion radiative recombination events. Carriers living longer than $\approx 10\,ps$ recombine non-radiatively at the surface; the surface acts as a picosecond "shutter". We force surface recombination to dominate the lifetime by injecting very close to the surface.

1. INTRODUCTION: Why use photonics?

The ease of manufacture of silicon solid state electronics has fuelled the information age. However, in some applications, specifically those involving the transmission of data by guided electromagnetic waves, electronics has a severe handicap. Electronic devices can presently generate electromagnetic waves up to frequencies of many GHz, but these waves suffer strong absorption in their guiding media (eg copper wave-guides, co-axial cable). The metallic conduction band electrons in the guiding media respond to the wave, and undergo a dissipative motion. As a result the signal is lost to heat. In contrast, for waves at much higher frequencies, in the $\approx 200\,THz$ (infra-red) range, the interaction with the atomic shell electrons in certain glasses is almost lossless. Indeed, with high purity silica fibers, and appropriate selection of photon energy, (eg $1.55\,\mu m$), losses can be reduced to as low as $0.154\,dB\,km^{-1}$, ie the absorption is 2 or 3 orders lower than copper wave-guides at microwave frequencies. Low dispersion in this band leads to a low "blurring" of the signal pulses and hence an enormous signal bandwidth. These considerations have been the driving forces to build *photonic* (also called *lightwave*) transmission systems, using electronic modulation of infra-red semiconductor lasers. The loss to convert from electrical to photonic signals and back gives a minimum distance over which photonic transmission gives lower net loss. The numerical value of this distance depends on data rates, signal-to-noise ratios and technological implementation, but typically it is $\approx 1 km$. Thus building sized computer networks are implemented in $10 Mbs^{-1}$ co-axial systems such as *Ethernet*®, whereas campus sized networks may use $150 Mbs^{-1}$ *Fiber Distributed Data Interface (FDDI)*® technology. This limit has restricted the use of photonics to *km* range transmission.

But recently it has been realized that photonics has another advantage, one which could broaden its use in the future. An infra-red photon at $0.8 eV$ can generate an electron-hole pair in a $0.8 eV$ band-gap semiconductor, which has only a few thermally generated carriers, from the room temperature thermal background, $kT = 0.026 eV$. On the other hand, an electronically generated GHz photon from a co-axial line has a photon energy in the $\approx \mu eV$ range, and thus behaves classically, and requires many photons and much more energy to generate a detectable voltage in a classical detector (*eg* an antenna, a tuned circuit and rectifier). This quantum behavior of infra-red photon detectors can be used to effect what has been termed a "quantum impedance transformation" by Miller (1989) Briefly, he shows that information transmitted using photonic energies ($\approx 0.8 eV$) can use less energy than that transmitted by low photon energy, classical *EM* waves available from electronic devices. This fact takes on practical significance when it is realized that the performance (*ie* speed) of digital processors is ultimately limited by the energy dissipated per clock cycle: the processor must not generate more heat than its heat sink can remove at a safe processor operating temperature. A commonly used figure of merit is speed-power product, *ie* energy, which should be as low as possible for a given operation.

Thus the goal of digital processor designers is to use less energy per clock cycle, so that more cycles per second can be executed at constant power dissipation. (The parallel development of lower thermal impedance heat sinks is close to the ultimate limit set by nucleate boiling of refrigerant fluids such as Freon. The cost of these is prohibitive in many cases.) Present electronic devices use $\approx 10^6$ more energy than the limit set by quantum thermodynamics, for *switching*. Nevertheless, most of the heat generated by an electron digital processor is in *transmitting* the signals from one chip of the processor to another: Every clock cycle, cables, of length $\approx l$, where l is of order of the processor size, have to be fully charged and discharged, generating heat energy of lCV^2 where C is the capacitance per unit length and V is the voltage swing (typically $5V$, but always $>> 0.026V$). Interestingly, this energy must be dissipated (in the load devices), even if the interconnections themselves are lossless (*eg* superconducting).

The quantum nature of infra-red photons enables photonic interconnection at much lower energy cost. The obstacle which prevents the use of this considerable advantage is a technological one related to the difficulty of manufacturing dense photonic circuits with a low *switching* energy. In terms of *switching* energy the very advantage that made photonic signals attractive in the first place, *ie* low interaction with matter, makes it rather difficult to devise a satisfactory photonic "transistor"; *ie* one with a high gain and a low switching energy. Such a device is necessary to obviate the need for wasteful conversion back and forth between electronic and photonic energy at each point in the processor.

One successful approach is the *S-SEED* device, a three terminal, all-photonic differential switch, whose operation relies on the quantum confined Stark effect (*QCSE*), in which a room-temperature exciton, confined in a thin quantum well structure, has its absorption peak shifted by application of a *DC* bias, which can be optically controlled by a control

Figure 1. Scanning Cathodoluminescence Microscope in the author's laboratory

beam. The shift in absorption peak modulates a power beam, to give the output beam. Efficient $QCSE$ depends on a lack of absorption linewidth broadening since this controls the gain of the device. In present day S-$SEED$s, even at room temperature, the primary source of broadening is interfacial roughness.

This paper describes the efforts in hand to understand the nature of real quantum wells, whose properties are defined and determined by the interfaces between chemically different compounds, such as heterostructural $GaAs$ and $Al_xGa_{1-x}As$.

There are two difficulties in this work: the conceptual and practical difficulties of dealing with roughness on length scales which cover 8 orders of magnitude, and the fact that the interface is only defined between volumes of material sufficiently large to potentially contain significant numbers of Al cations. The former problem is solved by considering the interface to be a 2-D waveform in Fourier space, having a amplitude *vs* spatial frequency spectrum spanning the huge range from $(cm)^{-1}$ to $(0.3\,nm)^{-1}$, and by compiling data from a variety of complimentary techniques, each covering a "window" of the frequency domain bounded in each case by a spatial resolution and a field of view. The latter problem is due to the fact that a Ga cation in $GaAs$ is indistinguishable from one in $Al_xGa_{1-x}As$, and is considered by assigning a component of the roughness termed "random alloy roughness" to this effect.

2. INTERFACIAL ROUGHNESS STUDIES

2.1 Transmission Electron Microscopy

High quality interfaces are almost universally between compounds of the same *structure* but different *chemical composition*, with one exception, that of the Si/SiO_2 interface.

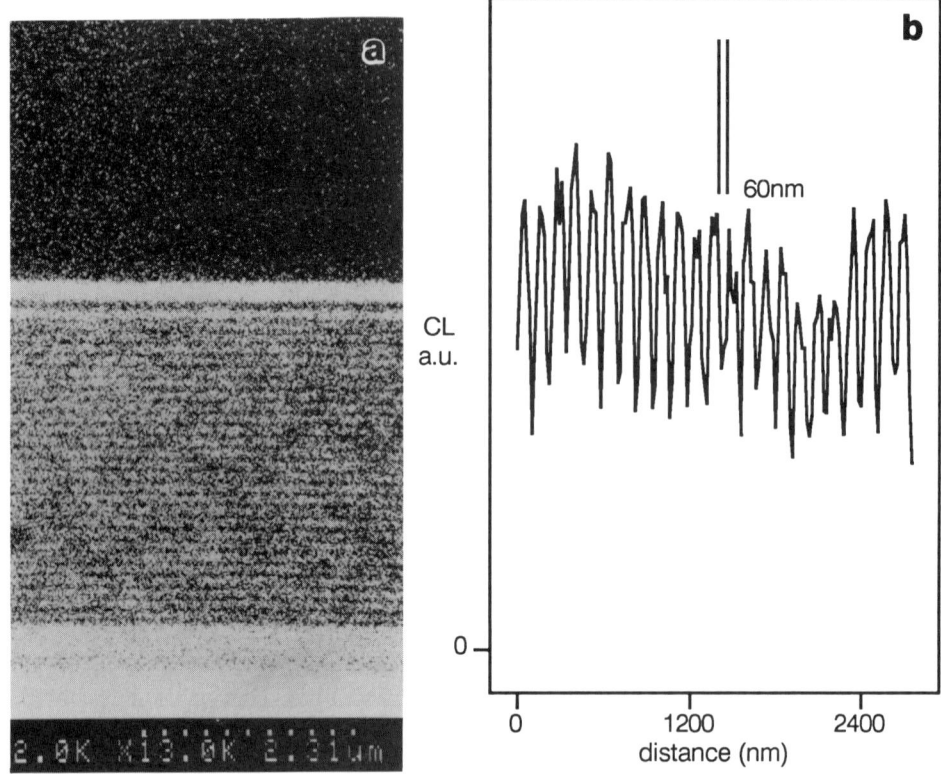

Figure 2. a) Cross-section of a dielectric stack mirror b) line trace from a). The interfaces are 60 nm apart

Conventional diffraction contrast *TEM* is therefore unsuited to determining the interfacial structure. Chemical contrast in the HREM gives information on a length scale from near atomic dimension to 30 to 100 nm, the field of view being restricted by the stringent sample preparation requirements. Ourmazd *et al* (1989) have demonstrated that the highest quality single quantum wells have interfacial roughness that can be characterized over the range of spatial frequencies mentioned above as having an Fourier amplitude of 0.3-0.6 nm and a peak spatial frequency of $\approx (2\text{-}3 \, nm)^{-1}$

2.1 Photoluminescence

Weisbuch *et al.* (1981) were one of the first groups to use photoluminescence (PL) to infer an interface shape. However, as noted above it is difficult to form a complete picture of the interface by using only one technique. PL gives indirect information about how an ensemble of objects the size of the exciton ($\approx 15 \, nm$ diameter) experience a local average quantum well thickness. Goldstein *et al.* (1983) observed multiple peaks in PL spectra from single quantum wells, indicating a multi-valued preferred thickness existed, and the difference in preferred thicknesses was close to one atomic layer or monolayer (*ML*). [1 $ML \approx 0.28 \, nm$ for *GaAs*]. This result gave indirect evidence to the idea that monolayer flat islands existed but when a more complete picture was formed from spatially resolved experiments, it became generally agreed that this picture was naive.

2.2 Conventional Scanning Photoluminescence & Cathodoluminescence

Scanning luminescence can be conveniently achieved by using a focussed beam of photons or electrons (cathodoluminescence, *CL*). Bimberg *et al* (1987) have published detailed scanning CL studies of high quality quantum wells. They emphasis the fractal nature of the interfaces, showing successively finer island-like features as the magnification is increased. Islands previously thought to be flat, proved to have sub-structure at higher resolution, which shows they were not flat at all.

Further evidence that the mono-layer flat island picture was naive came from Warwick *et al* (1990) who showed the peak splitting varied by up to 38% for a given sample, which could not be accounted for by this picture. They proposed an interface shape which was characterized by a bimodal roughness spectrum.

PL and CL work motivated in part the desire for higher spatial resolution by several groups, including Wada *et al.* (1988) and the present author.

2.3 Pre-diffusion Scanning Cathodoluminescence

In the present work we have achieved $60\,nm$ spatial resolution.

This is accomplished by:

a. using a low accelerating voltage to inject both near the surface and with minimal lateral scattering ($\approx 10\,nm$)

b. using a small probe generated by a bright, field-emission source

c. detecting photons from pre-diffusion radiative recombination events. Carriers living longer than $\approx 10\,ps$ recombine non-radiatively at the surface; the surface acts as a picosecond "shutter". We force surface recombination to dominate by injecting very close to the surface ($1kV$ electrons penetrate $\approx 10\,nm$).

We also are aided in these efforts by efficient collection optics and a novel roof-mounted low-drift ($\approx 0.02\,nm\,s^{-1}$) liquid helium sample holder (*Oxford Instruments CF302-SP*). Sample surface temperature under excitation was determined to be 5K, from comparisons of PL and CL spectra from a high purity *GaAs* sample. Also vital are the Special Evacuation System on the specimen chamber of our Hitachi S-800-SES SEM which is oil-free and metal-sealed, and can withstand baking at 100°C. As a result it has a base pressure (sample at room or helium temperature) below $5\times10^{-7}Pa(4\times10^{-8}Torr)$. The low water vapor and hydrocarbon ambient practically eliminates the surface contamination that prevents low-temperature low-voltage observation in conventional *SEMs*.

In addition, the *SEM* is mounted in a cut-out in an air-suspended optical table, thus eliminating vibration and aiding optical alignment. The detector is a thermo-electrically cooled *GaAs* photo-multiplier tube used in photon counting mode. Counts are stored in direct digital form in a $4MB$ frame buffer, memory-mapped into an *AT&T 6386E* personal computer, which also controls the 0.75m *SPEX* 1702 spectrometer. The frame buffer is part of an Hitachi Digital Image Processing System, that can also control the scan amplifiers. The microscope is shown in figure 1. Micrographs and spectra were recorded

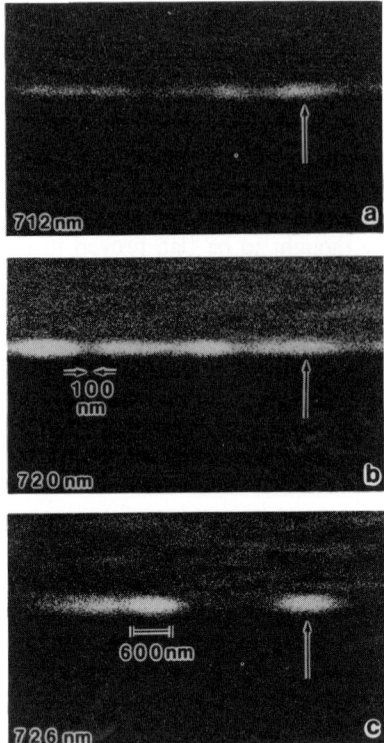

Figure 3. Cross-section of a single quantum well a) 712 *nm* CL b) 720 *nm* CL c) 726 *nm* CL

using $1 \approx 2kV$ electron beam with a probe current of $\approx 3pA$.

3. EXPERIMENTAL RESULTS

Figure 2a shows a cross-sectional Scanning Cathodoluminescence Micrograph of a dielectric stack mirror, used in *S*-*SEED*s and, as in this case, in Surface Emitting Lasers. The interfaces are 60 *nm* apart. Figure 2b shows the line trace from this structure. The contrast is $\approx 50\%$, easily satisfying the Rayleigh Criterion.

An important part of the *S*-*SEED* structure is its quantum wells. Figure 3 shows three monochromatic CL images of the same area of the same single quantum well, again viewed in cross-section. This type of quantum well is typical of one of the ≈ 30 quantum wells used in *S*-*SEED* starting material. This sample has been studied previously using scanning PL and scanning tunneling microscopy by Kopf *et al.* (1991). In short, that paper showed that this sample has a remarkable macroscopic average thickness uniformity as evinced by the constant splitting of the PL peaks. However, the splitting observed corresponded to a change in quantum well width significantly less than 1 *ML*, $\approx 0.9ML$, which could be caused by fine scale roughness, superimposed on a low spatial frequency change in thickness. A detailed interpretation of the growth mechanism and resulting

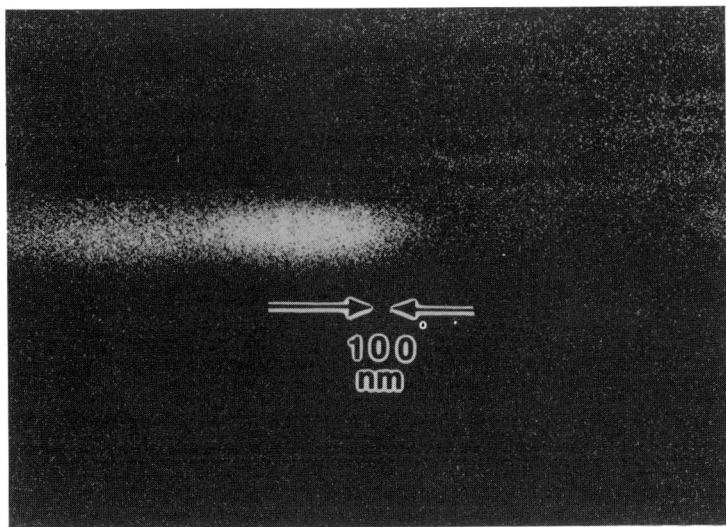

Figure 4. Cross-section of a single quantum well 726 *nm* CL: higher magnification

structure is in preparation (Warwick and Kopf, 1991) but, referring to figure 3, the most important thing to note is that the images at different wavelengths are partly complementary and partly not. In other words, parts of the well emits light mainly at one of the three wavelength peaks but other sections emit at more than one wavelength, such as the region arrowed in figure 3. This is categorical proof that areas as small as 100 *nm* contain smaller areas of quantum well whose thicknesses are different by $\approx 1 ML$. The finer the resolution, the more detail is resolved. Thus the interfaces have a fractal like quality. Techniques with low resolution reveal only the low spatial frequency variation and give the misleading impression that large, monolayer-flat islands exist. Figure 4 shows the same sample at higher magnification, illustrating the high resolution that can be achieved by the method.

4. CONCLUSIONS

High resolution Scanning Cathodoluminescence is giving us direct insight into the shape of interfaces which determine the properties of the quantum wells used in novel, experimental devices. As this work progresses and an understanding of the growth process is achieved, this will lead to better control and ultimately better devices.

ACKNOWLEDGEMENTS

Rose Kopf and Benjamin Tell kindly provided the single quantum well sample and surface emitting laser sample respectively. Thanks are due to Abbas Ourmazd for many stimulating discussions. Trademarks are the property of their respective owners.

REFERENCES

Bimberg D, Christen J, Fukunaga T, Nakashima H, Mars D E, and Miller J N 1987 *J. Vac. Sci. & Technol.* **B5** 1191
Goldstein L, Horokoshi Y, Tarucha S and Okamoto H 1983 *Japan. J. Appl. Phys.* **22**, 1489
Miller D A B 1989 *Optics Letters* **14** 146
Ourmazd A, Taylor D W, Cunningham J, and Tu C W 1989 *Phys. Rev. Lett.* **62** 933
Wada H 1988 *Japan J. Appl. Phys. Lett.* **27** L1952
Warwick C A, Jan W Y, Ourmazd A and Harris T D 1990 *Appl. Phys. Lett* **56** 2666
Warwick C A and Kopf R F 1991 *Appl. Phys. Lett* to be published
Weisbuch C, Dingle R, Gossard A C and Wiegman W 1981 *Solid State Commun.* **38** 709

Inst. Phys. Conf. Ser. No 117: Section 10
Paper presented at Microsc. Semicond. Mater. Conf., Oxford, 25–28 March 1991

Characterisation of GaAs/AlAs quantum well structures by HREM and LTSCL

D B Holt, C E Norman, G Salviati*, S Franchi* and A Bosacchi*

Department of Materials, Imperial College of Science, Technology and Medicine, London   SW7 2BP, UK
*C.N.R. MASPEC Institute, Via Chiavari 18/A, 43100 Parma, Italy

ABSTRACT: At 1 keV the CL emission from a nominally 3 nm type II single quantum well (SQW) is attributed to indirect transitions from states in the AlAs to the Γ_h states in the GaAs SQW.   The width of this peak agrees well with HREM observations of the SQW heterointerface roughness, on the basis of a particle–in–a–box model.   At 3 keV, the SQW CL peak broadens and shifts towards higher photon energies, attributed to additional higher energy, direct transitions arising due to saturation of the AlAs states.   Growth interruption at the heterointerfaces broadens the SQW CL band.   Photon excitation occurs unexpectedly deep in the specimens and the CL intensity increases with irradiation time.

1.   INTRODUCTION

Quantum well (QW) structures are interesting because of e.g. the quantum size effect i.e. the dependence of energy levels on well width when this is less than the de Broglie wavelength.   Recently, interest in QW structures has spread from the much studied GaAs/AlGaAs system to the more difficult to grow GaAs/AlAs system studied here. Depending on the layer thickness involved, the GaAs QW may be either of type I ("straddled") or of type II ("staggered") as shown in Figures 1(a) and (b).   Type II structures occur in GaAs wells < ~ 3 nm thick, between confining AlAs layers. Considering both the Γ and X conduction band edges, it is found that there are electron energy levels at both Γ_e and X_e.   The energy of the electron in the lowest level in the well (Γ_e) depends on the well width, L, increasing as L decreases below ~ 30 nm (above which there is no quantum size effect).   The energy level of electrons in the AlAs confining layers (X_e) is however virtually constant so there exists a well width for which the Γ_e and X_e electron levels are equal.   In such a well both direct ($\Gamma_e \rightarrow \Gamma_h$) and indirect ($X_e \rightarrow \Gamma_h$) recombination are possible (Figure 1(c)).   Theory (Wilson 1988) indicates that, in the GaAs/AlAs system, this width is near 3 nm, the width of the QWs studied here.

High resolution (transmission) electron microscopy (HREM) is widely used for characterizing QW structures, especially their heterointerfaces.   Preparing HREM samples is however slow and difficult and the volume sampled is small so many specimens must be prepared to ensure that the results are representative.   Low temperature spectroscopic cathodoluminescence (LTSCL) requires minimal specimen preparation and has the advantages that either relatively large areas can be investigated or individual features may be studied with a spatial resolution of the order of a μm (for reviews see e.g. Yacobi and Holt 1990).   Two groups applied LTSCL microscopy to type I QWs of unusually high interfacial quality and observed splitting of the exciton emission into discrete lines corresponding to different well widths (Petroff *et al* 1987, Bimberg *et al* 1987).   Few such CL spectra have been published and the accompanying monochromatic CL images recorded using different excitonic lines displayed a "columnar" structure of allegedly atomically flat areas (of the order of μm across) in the SQW interfaces (Bimberg *et al*) or less sharply defined island areas of similar size (Petroff *et al*).   This conflicts with the HREM evidence of much more closely spaced monotomic steps in QW interfaces (Ourmazd 1989).

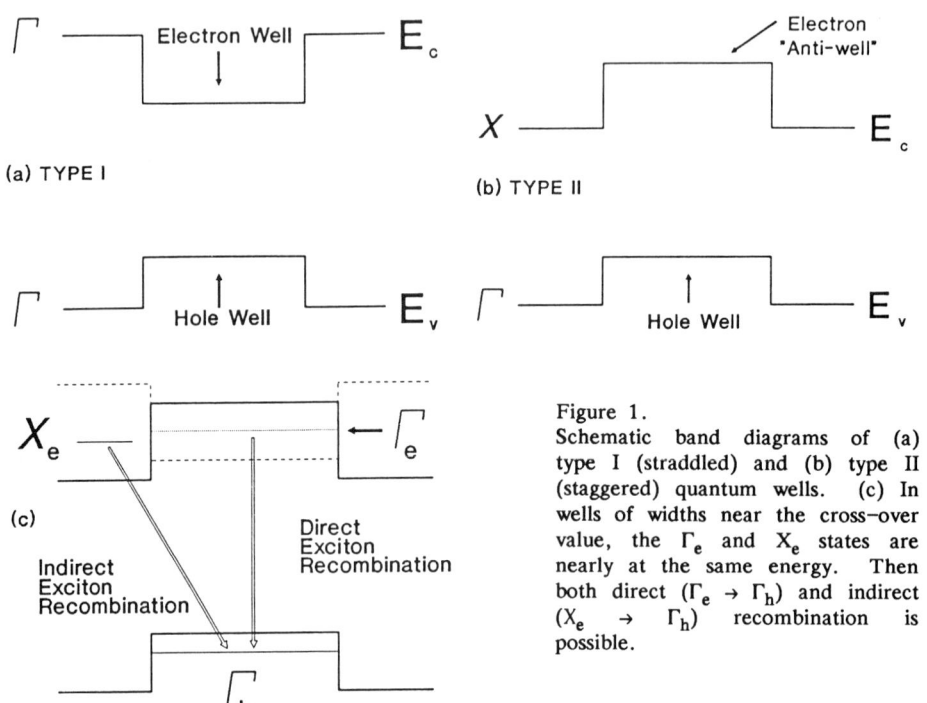

Figure 1.
Schematic band diagrams of (a) type I (straddled) and (b) type II (staggered) quantum wells.   (c) In wells of widths near the cross-over value, the Γ_e and X_e states are nearly at the same energy.   Then both direct ($\Gamma_e \rightarrow \Gamma_h$) and indirect ($X_e \rightarrow \Gamma_h$) recombination is possible.

This work reports further results from a systematic study (Holt *et al* 1991) by LTSCL and HREM of a series of   GaAs/AlAs single and multi-QW structures purpose-grown by MBE to provide a comparison of the results of the two techniques in assessing the effect of growth interrupts on interface smoothness.

2.   EXPERIMENTAL

The spectroscopic CL observations were carried out using an upgraded version of the monochromator system described previously (Saba and Holt 1983) and a JEOL JSM–840A SEM.   Liquid helium temperatures were obtained and measured using an Oxford Instruments CL301H continuous flow cryostat CL stage.   High resolution (transmission) electron microscopy (HREM) was performed on a JEOL 2000FX TEM.   The specimens were thinned in (110) cross–section in a Gatan 600 dual ion mill at liquid nitrogen temperature using first Ar ions and then I ions, in order to minimize possible ion bombardment–induced artifacts that were sometimes found in the AlAs layers.

Six wafers with the structure in Figure 2 were grown for this work with the listed values of growth interrupts at the interfaces marked A and B.   Interruption is supposed to smooth the interfaces but at the possible expense of increased contamination. Smoothing is needed as AlAs growth surfaces tend to be uneven due to the low surface mobility of the Al.

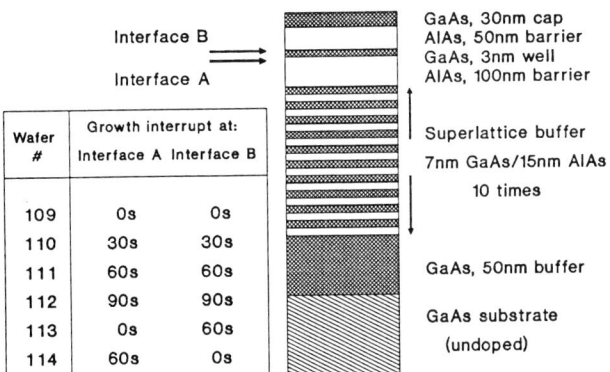

Wafer #	Growth interrupt at:	
	Interface A	Interface B
109	0s	0s
110	30s	30s
111	60s	60s
112	90s	90s
113	0s	60s
114	60s	0s

GaAs, 30nm cap
AlAs, 50nm barrier
GaAs, 3nm well
AlAs, 100nm barrier

Superlattice buffer
7nm GaAs/15nm AlAs
10 times

GaAs, 50nm buffer

GaAs substrate
(undoped)

Figure 2.
The stucture of the wafers grown for this work with the growth interruption times applied at the A and B heterointerfaces in the six wafers.

3. RESULTS

Figure 3(a) and (b) show the LTSCL spectra of specimen #111 (60 sec growth interrupts) at 8K and beam energies 1 and 3 keV, with the origins of the emission bands marked. From 1 to 3 keV only the 3 nm SQW band changes, broadening and shifting the peak to a shorter wavelength i.e. a higher photon energy. The full width at half maximum (FWHM) rises with E_b from 1 to 6 keV as shown in Figure 4 for #109 (zero growth interrupts). High beam currents ($\sim$ 1 μA) were used to obtain adequate CL intensity. Varying the the sample temperature at a single beam energy confirmed, however, that no significant thermal broadening effects occurred below 40K so the observed increase in FWHM cannot be due to local beam heating.

HREM (Figure 5) revealed that the SQWs had relatively rough interfaces so the typical well width was 3 ± 1 nm i.e. 10 ± 3 monolayers. Steps occurred in the lower interface (A in Figure 2) at intervals of a few nm and the upper interface (B) was noticeably smoother (Figure 5). Figure 6 plots the effect of growth interrupt times on the 1 keV FWHM of the SQW band.

4. DISCUSSION

The appearance of the bands in the LTSCL spectra at low E_b are surprising if the depth of penetration of the incident electron beam is assessed using widely accepted semi-empirical approximations. Both the Kanaya and Okayama and the Gruen (Everhart and Hoff) expressions (see e.g. pp. 57–58 in Yacobi and Holt 1990) predict penetration ranges for a 1 keV electron beam < 0.05 μm in these samples and yet the MQW is excited although it lies deeper than 0.18 μm in these wafers. At 3 keV the predicted ranges are still < 0.16 μm but the bulk GaAs buffer layer is strongly excited although it lies deeper than 0.4 μm. The observation of CL excitation at depths ❭ predicted beam penetration ranges may be due to one or more of these effects: (i) the penetration range may be greater than the calculated values at low E_b and low temperature, (ii) beam–induced carrier diffusion, which has a negligible effect in degrading resolution at ambient temperatures, may become important in extending activation at low temperatures and (iii) higher energy photons generated by recombination in the 3 nm SQW that travel deeper into the wafer can excite lower energy transitions in the MQW and GaAs buffer layer (the "readsorbed recombination radiation" (RRR) effect).

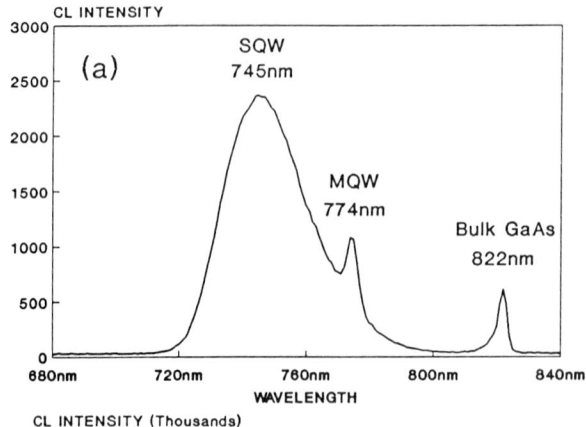

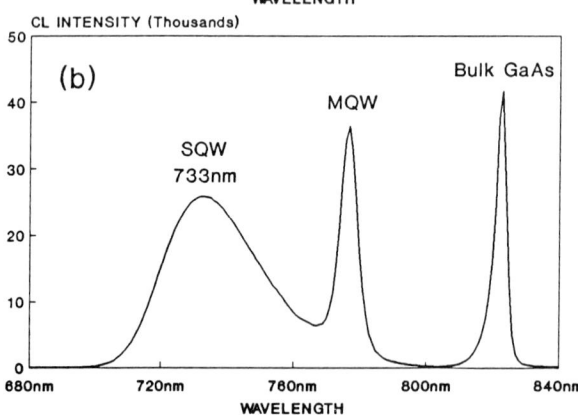

Figure 3.
CL spectra of a specimen from wafer #111 at 8K and (a) 1keV and (b) 3keV.

The form of the SQW emission spectra changes significantly with beam energy (Figure 3). At E_b = 1 keV, the shortest wavelength, highest photon energy SQW CL bands have small FWHM values (Figure 6) in close agreement with theoretical prediction by a simple particle-in-a-box model for the <u>indirect</u> recombination band from a QW of width 3 nm and roughness ± 1 nm. The model predicts that direct recombination in such wells would produce much broader emission bands. Thus the CL and HREM assessments of interface roughness agree.

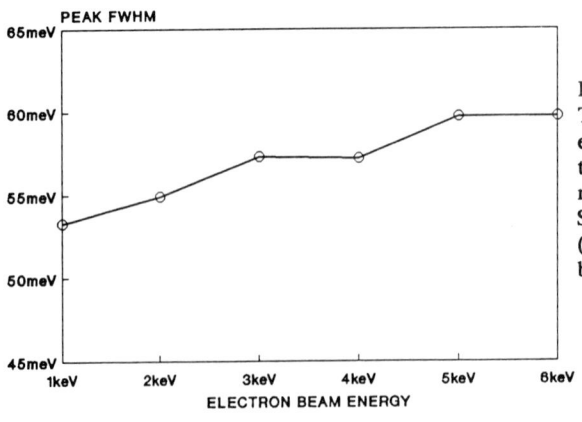

Figure 4.
The effect of increasing incident electron energy on the low temperature full width at half maximum (FWHM) of the 3nm SQW CL band of specimen #109 (zero interrupts) at a constant beam power of 3.7mW.

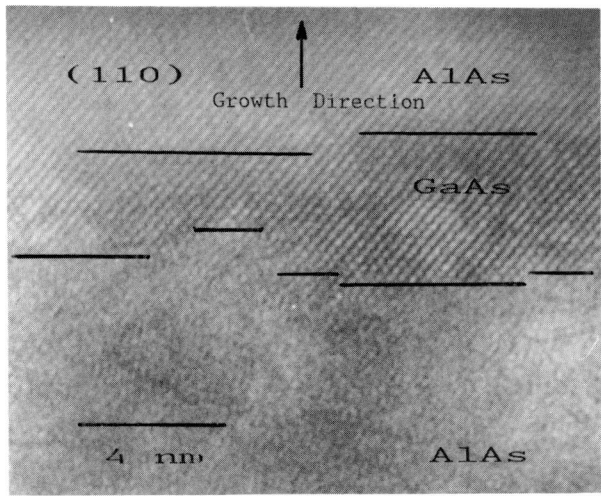

Figure 5. High resolution cross–sectional TEM image of a specimen of wafer #111 ([110] axial bright field). Lattice fringes are visible in the upper AlAs layer only. Both heterointerfaces are rough but the lower (A) interface is markedly more so than the upper (B) interface.

The observed SQW 1 keV CL band is also in good agreement with the corresponding PL band obtained from the specimen with 60 sec growth interrupts. At 1 keV all the trapped electrons can be accommodated in the lowest energy X_e states so only indirect (type II) excitonic recombination occurs. For higher beam energies the SQW CL band becomes much broader and the peak shifts to a higher photon energy. At these higher beam energies, electrons are also confined in the higher energy Γ_e states, direct excitons form and higher photon energy, direct recombination radiation appears. The superimposition of the direct and indirect radiation bands gives rise to the SQW band peak shift and broadening.

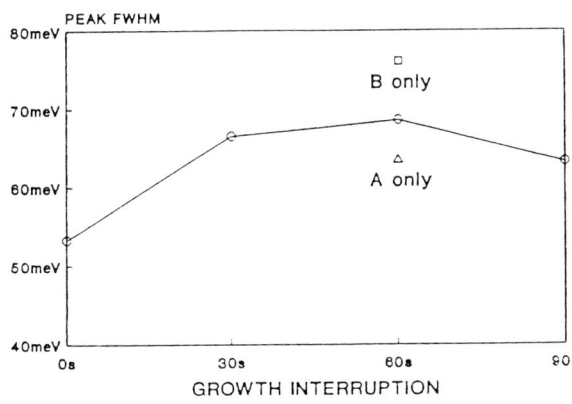

Figure 6.
The low temperature FWHM of the SQW CL band as a function of growth interrupt time (equal on both heterointerfaces of the SQW) for $E_b = 1$ keV. The points for #113 and #114 are marked as having 60s interrupts on interface B only and A only, respectively.

The width (FWHM) of the SQW band increases with growth interrupt time. This may be related to increasing interface contamination or, more probably, to the increasing size of the flat interface terraces between monolayer steps relative to the exciton size. The effect of a growth interrupt on only one of the interfaces of the SQW on the CL band width is shown in Figures 6 and 7. Specimen #113 had a 60 sec interrupt on face B only and #114 on face A only. There is a peak shift between the 1 keV SQW bands from these specimens (Fig. 7) and the sum of the two is similar to the SQW band of #111 which had 60 sec interrupts on both the A and B interfaces.

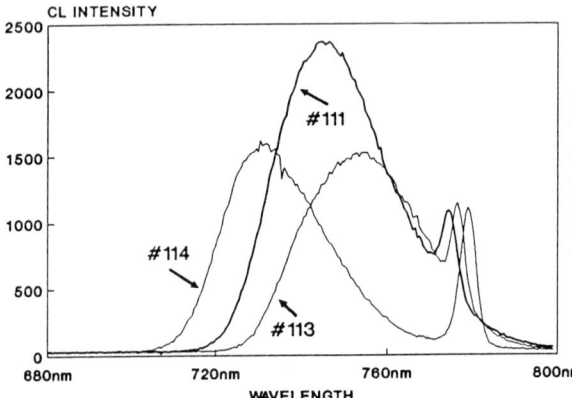

Figure 7.
The 1keV CL spectra of specimens #111, #113 and #114. The sum of the #113 and #114 SQW bands is similar to that for #111 which had the same growth interrupts as the other two together.

The integral CL intensity of all these specimens increases with time scanned. As an example, an increase of more than a factor of six was observed from a specimen of wafer #109 over a period of 20 seconds using a stationary 3keV electron beam with a current of $1.8\,\mu A$. There are indications that other characteristics of the CL may also be affected, most notably that the higher energy, direct recombination radiation appears to increase in intensity more markedly with time than the indirect recombination radiation. Work on this phenomenon is continuing.

REFERENCES

Bimberg, D., Christen, J., Fukunaga, T., Nakashima, H., Mars, D.E. and Miller, J.N. 1987 J. Vac. Sci. Technol. B5 1191

Holt, D.B., Norman, C.E., Salviati, G., Franchi, S. and Bosacchi, A. 1991 Mat. Sci. Eng. B to be published.

Ourmazd, A. 1989 in Point and Extended Defects in Semiconductors (G. Benedek, A. Cavallini and W. Schroter, Editors) Plenum Press: New York) pp.135

Petroff, P.M., Cibert, J., Gossard, A.C., Dolan, G.J. and Tu, G.W. 1987 J. Vac. Sci. Technol. B5 1204

Saba, F.M. and Holt, D.B. 1983 in Microscopy of Semiconducting Materials 1983. Conf. Series No. 67 (Inst. Phys.: Bristol) pp. 333

Wilson B.A. 1988 IEEE J. Quantum Electron. QE-22 1763

Yacobi, B.G. and Holt, D.B. 1990 Cathodoluminescence Microscopy of Inorganic Solids (Plenum Press: New York)

Inst. Phys. Conf. Ser. No 117: Section 10
Paper presented at Microsc. Semicond. Mater. Conf., Oxford, 25–28 March 1991

695

Low temperature cathodoluminescence studies of GaAs/AlGaAs quantum dot structures

G M Williams, A G Cullis, C M Sotomayor Torres[1], S Thoms[1], S P Beaumont[1], C R Stanley[1], D Lootens[2] and P Van Daele[2]

DRA Electronics Division, RSRE, St Andrews Road, Malvern, Worcs WR14 3PS, UK
[1]Nanoelect Res Cent, Dept of Electronics & Elect Eng, Glasgow Univ, Glasgow G12 8QQ
[2]Lab of Electromagnetism & Acoustics, Gent Univ., B–9000 Gent, Belgium

ABSTRACT: Low temperature (15K) scanning cathodoluminescence (CL) has been employed to study in detail MBE grown GaAs/AlGaAs quantum dot prototype structures fabricated by reactive ion etching. The CL images demonstrate the excellent uniformity of luminescence that can be achieved from these dot arrays. Regions of non–uniformity have also been observed and by comparing total light CL images with secondary electron images problems associated with the dry etching stage of dot fabrication have been identified. Free standing dot arrays have been overgrown with AlGaAs using MOCVD and preliminary results of CL investigations of these structures are also presented. Spectral information from all the quantum wells is observed and has been studied as a function of dot size. Data is also compared with 5K photoluminescence spectra obtained from the same structures.

1. INTRODUCTION

Two–dimensional quantum well (QW) structures have produced many novel optical and transport properties in a range of semiconductor materials. By further modifying the physical dimensions of the quantum well device (Thoms et al 1986) it is also possible to generate structures with one and zero degrees of carrier freedom, these structures being known as quantum wires and quantum dots respectively. Evidence for one (Cibert et al 1987) and zero (Reed et al 1986) degrees of carrier freedom have been reported and novel electronic and optical effects are predicted.

Low temperature scanning CL has previously been used to study such structures (Cibert et al 1986; Wang et al 1990). In this paper, however, we present the first report of a 15K scanning CL study of both free–standing AlGaAs/GaAs dot arrays and dots subsequently overgrown with AlGaAs.

2. EXPERIMENTAL DETAILS

The GaAs/AlGaAs (30% Al) structure under investigation was grown at $645\,^{\circ}C$ using molecular beam epitaxy (MBE). Following the growth of a GaAs buffer layer and 80nm AlGaAs layer three GaAs QWs 8nm, 6nm and 4nm thick were grown, being spaced by 20nm AlGaAs barrier layers; the surface was capped with 10nm of GaAs. Dot mesa structures 250nm high were fabricated in the MBE layers using a $SiCl_4$ reactive ion etching (RIE) process (Thoms et al 1986). Arrays containing free standing dots of one particular width in the range 60nm to 1000nm were produced. Following the dot fabrication one set of arrays was overgrown with 200nm of AlGaAs using MOCVD at a substrate temperature of $750\,^{\circ}C$.

The scanning CL system used to study these structures is based on a Cambridge Stereoscan 150 Mk2 SEM. The instrument was operated at 10kV with an LaB$_6$ electron source giving a beam current of <45nA and beam diameter of ~200nm. A high efficiency parabolloidal mirror with reflecting light guide collected the photons emitted from the specimen, which was cooled to ~15K using liquid helium. The RSRE designed collection optics employed both a standard photomultiplier tube for total light imaging and a monochromator/GaAs S1 detector system for spectrum acquisition and monochromatic imaging, both coupled to a Link QX2000 computer-controlled multi-channel analyser. Secondary electron images of the structures were obtained using a high resolution JEOL JSM 6400F field emission microscope. Photoluminescence (PL) spectra were acquired at 5K using sample excitation by the 633nm line of a 2.1mW helium neon laser focussed to a 50μm spot.

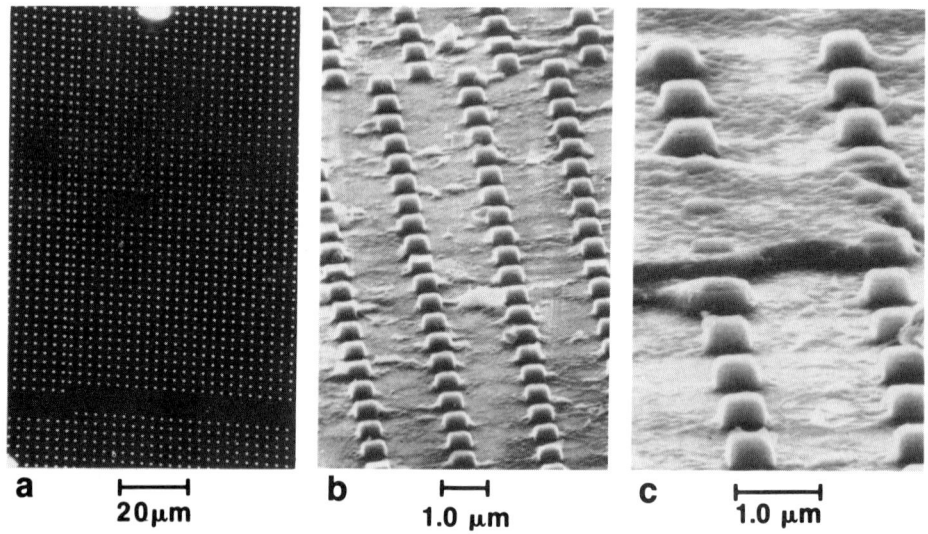

a ⊢——⊣
   20μm

b ⊢—⊣
   1.0 μm

c ⊢——⊣
   1.0 μm

Fig. 1. An array of 500nm dots is imaged in (a) using total light 15K CL, and in (b) and (c) using secondary electron contrast. Note the unetched region in (c).

3. RESULTS AND DISCUSSION

3.1   Free-Standing Dot Structures

Total light CL images revealed large areas of uniformly luminescent structures such as the array of 500nm dots shown in Fig. 1a. A small number of failed areas of dots were present, possibly due to their coincidence with extended defects. Bright irregular shaped regions, several microns in size, were also observed occasionally (see Fig. 1a). High resolution secondary electron images of this array confirmed that large areas of well-defined 500nm dots had been correctly fabricated (see Fig. 1b). However, as seen in Fig. 1c, areas of unetched quantum well material were also present. The unetched material illustrated corresponded to a bright region in the array shown in Fig. 1a and is thought to be the result of ineffective lift-off of the SrF$_2$/AlF$_3$ mask caused by surface contamination prior to the e-beam lithography stage of fabrication.

Figure 2a shows the spectrum obtained from a region of 500nm dots specifically selected to exclude artefacts such as the bright regions discussed above. Comparing this with the spectrum shown in Fig. 2b taken from a 200μm square control area several differences are

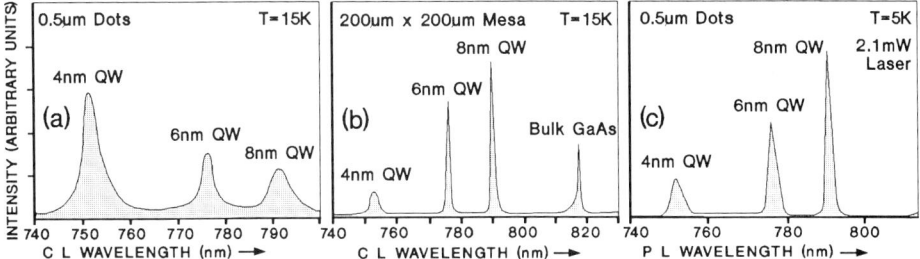

Fig. 2. The CL spectrum from the 500nm dots is shown in (a). The peak ratios and widths are both different from the CL spectrum taken from the 200 μm square control region (b). A 5K PL spectrum from the same array of 500nm dots is shown in (c).

apparent. Firstly, it is important to note that the integrated intensity of all the peaks is reduced. In particular, the 4nm quantum well peak is attenuated by a factor of approximately 100. Secondly, peak broadening has occurred, with the FWHM value for the 6nm QW peak increased from 1.86meV to 5.8meV and the 8nm QW peak FWHM increased from 1.7meV to 9.4meV. These effects are undoubtedly due to side-wall damage produced during the RIE process which is likely to introduce both defects and changes in the stoichiometry of the AlGaAs and GaAs. A 5K PL spectrum taken from the same array of 500nm dots is shown in Fig. 2c. Although broadening of the peaks has occurred the PL peak ratios resemble more closely the control area CL spectrum (Fig. 2b) than the CL spectrum taken from the 500nm dots (Fig. 2a). The area of collection for the PL spectrum is the 50 μm laser spot which may have randomly included bright unetched regions of material similar to that illustrated in Fig. 1.

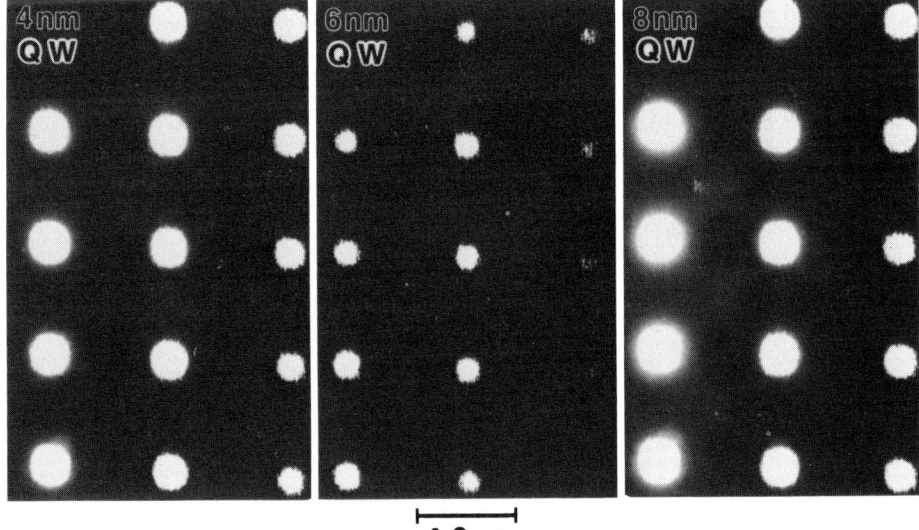

Fig.3. The same fifteen 1000nm dots are shown in each of the three monochromatic CL images above. Each image is formed using light from one particular GaAs QW. Note that the top left hand dot was physically intact but did not luminesce.

Cathodoluminescence is also able to probe the various QW emissions for any one particular dot using the monochromatic imaging facility. Figure 3 shows three different images of fifteen 1000nm dots, each image being generated by emission from a particular quantum well. The various QW intensities within each dot and from dot-to-dot reveal obvious non-uniformities. The top left-hand dot, although physically intact, was not luminescent. It is important to note that, overall, the intensity of the 6nm quantum well emission is most attenuated, in agreement with the relative integrated intensities of the peaks in the spectrum obtained from these dots.

The overall trend in the intensity variation of the QWs was similar for all the areas which were examined and may be associated with an inhomogeneous distribution of point defects that are generated during the RIE process. The SrF_2/AlF_3- GaAs interface and the GaAs buffer layer may act as a sink, leaving the maximum point defect density in the region of the 6nm QW, hence reducing its luminescence.

3.2 Overgrown Dot Structures

Both CL and PL spectra indicate a decrease in the luminescence efficiency as the dot size is reduced and, below 500nm, no CL emission could be detected from free-standing structures. However, in the overgrown structures increased luminescence is detected by both CL and PL (Arnot et al 1989) and dots as small as 250nm have been imaged using total light CL. Cathodoluminescence spectra obtained from all the overgrown structures showed further broadening of the quantum well peaks relative to the free-standing dots. A red shift in all peaks of 1 to 5nm relative to the $200\mu m$ square mesa was also observed and this shift increased with decreasing dot size. The peak broadening and red shift are both thought to be associated with strain induced by the AlGaAs overlayer.

4. CONCLUSIONS

Low temperature scanning CL is uniquely able to probe quantum dot prototype structures on a dot-to-dot basis. The MOCVD AlGaAs overgrown dots have been shown to exhibit improved luminescence efficiency, although strain in the final structure appears to affect both the quantum well peak widths and positions. This effect may be reduced by modification of the MOCVD growth procedures. The structures studied are at this time too large to produce the zero degree of carrier freedom predicted for true quantum dots. However, luminescence investigations of the type reported here are proving essential to the development of optimum fabrication processes.

ACKNOWLEDGEMENTS

The authors would like to thank Mrs O D Dosser for her technical assistance.

REFERENCES

Thoms S, Beaumont S P, Wilkinson C D W, Frost J and Stanley C R 1986 Microelectronic Eng. 5 249
Cibert J, Petroff P M, Dolan G J, Werder D J, Pearton S J, Gossard A C and English J H 1987 Superlattices and Microstructures 3 35
Reed M A, Bate R T, Bradshaw K, Duncan W M, Frensley W R, Lee J W and Shih H D 1986 J. Vac. Sci. Technol. B4 358
Cibert J, Petroff P M, Dolan G J, Pearton S J, Gossard A C and English J H 1986 Appl. Phys. Lett. 49 1275
Wang J, Steeds J W and Arnot H 1990 Microsc. Microanal. Microstruct. 1 241
Arnot H E G, Watt M, Sotomayor Torres C M, Glew R, Cusco R, Bates J and Beaumont S P 1989 Superlattices and Microstructures 3 459

Inst. Phys. Conf. Ser. No 117: Section 10
Paper presented at Microsc. Semicond. Mater. Conf., Oxford, 25–28 March 1991

699

Cathodoluminescence characterisation of GaInAs/InP quantum-well wire structures

S Nilsson, A Gustafsson, L Montelius, A Semu, K Georgsson and L Samuelson

Department of Solid State Physics, Lund University, Box 118 S-221 00 Lund, Sweden.

ABSTRACT: We have fabricated quantum-well wires of GaInAs/InP by electron-beam lithography and subsequent metalorganic reactive ion etching, having the patterned polymethylmethacrylate resist as an etching mask. The wires were characterized by low-temperature cathodoluminescence, taking spectra as well as monochromatic images. The images exhibit homogeneous cathodoluminescence arising from the wires and the spatial resolution is ≤ 2000 Å, to our knowledge one of the highest ever observed. A clear blue shift of the spectral distribution was found and correlated to the increased exposure dose i.e. with the decreased linewidth of the wires.

1. INTRODUCTION

Because of the expected changes of the physical properties in low-dimensional systems, for example density of states, the low-dimensional structures having quantum confinement in two or three dimensions hold great promise for novel semiconductor physics and for device applications (e.g. Sakaki 1980; Petroff et al 1982; Suemune et al 1988). Among the different approaches carried out to fabricate these structures, lithographical methods are the most widely used and have been shown to yield some lateral control of the crystalline properties. Various cathodoluminescence (CL) techniques have been demonstrated to be suitable tools for the characterization of low-dimensional structures, qualitatively as well as quantitatively (Christen 1990). In particular, the easily obtained spatial resolution in CL comes to the fore in the characterization of low dimensional structures.

In this work we report on GaInAs/InP quantum-well wire (QWW) structures made by electron-beam lithography and reactive ion etching. The structures were characterized by low-temperature CL techniques, taking spectra as well as monochromatic images. A remarkable high spatial resolution, ≤ 2000 Å, is observed in the CL images as well as a correspondence between an observed blue shift of the CL spectral distribution and a reduction of the lateral size of the QWWs.

2. EXPERIMENTAL

Details about the epitaxial growth are reported by Seifert et al (1990). Briefly, the sample used for processing was grown by metalorganic vapour phase epitaxy on an InP (100) substrate. The epitaxial layers consisted of ≈ 270 nm InP buffer layer, followed by ≈ 34 nm GaInAs, lattice-matched to InP as a reference layer, followed by a single quantum well,

≈ 30 Å thick. The quantum well is embedded between ≈ 25 nm InP spacer layer and ≈ 35 nm InP capping layer.

For the definition of the etching mask, the sample was covered by polymethylmethacrylate (PMMA) with a molecular weight of 950 k to a thickness of ≈ 300 nm. The resist was exposed by a scanning electron microscope (SEM) equipped with an LaB$_6$ filament. Five different fields each with an extension of $\approx 50 \times 50$ μm^2 were exposed. Each field contains a repeated grating-structure. Each grating-structure was nominally designed to have a period of 450 nm and a 1: 2 ratio of exposed to unexposed lines. The distance between adjacent grating-structures was ≈ 10 μm. Different exposure doses were applied for each field and consequently, different linewidths were obtained between exposed and unexposed lines. The subsequent transfer of the pattern onto the sample was made utilizing a radio-frequency biased metalorganic reactive ion etching (MORIE)-process having the patterned PMMA-layer as an etching mask. The etched sample was controlled by SEM measurements whereby an etch depth of ≈ 100 - 200 nm was determined. Furthermore, SEM investigation clearly reveals that the linewidths in the gratings get smaller with increased exposure doses. Fig. 1 displays schematically a cross section of the fabricated QWW structures. To our knowledge this is the first time a PMMA pattern is used as an etching mask in the MORIE-process, a comprehensive report on the fabrication of the wire structures is in preparation for publication.

Fig. 1. Schematic cross section of the GaInAs/InP QWW structures. The striped areas are GaInAs layers which prior to the processing corresponded to the single quantum well and the reference layer, respectively.

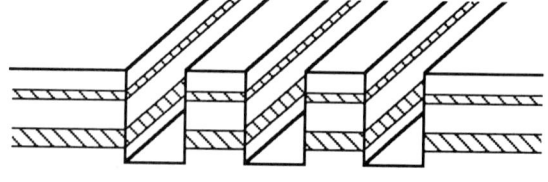

The CL technique used in this study is previously described (Gustafsson et al 1989). The experiments were performed using a SEM operated at 6 keV, modified with a CL collector and a continuous helium flow cold stage cooled to 25 K. The CL was dispersed by a 0.50 m monochromator (bandpass 30 Å) and focussed onto a cooled Ge detector. Spectrally resolved (monochromatic) images of the CL were obtained by scanning the electron beam digitally (64 $\times$ 64 pixels) over quadratic areas down to $\approx 2 \times 2$ μm^2.

3. RESULTS AND DISCUSSION

Fig. 2 shows CL spectra recorded at 25 K within a grating structure $\approx 50 \times 50$ μm^2. Three emission bands are clearly resolved; emission A originates from the ≈ 34 nm thick GaInAs reference layer, emission B is the QWW emission and emission C is attributed to the InP near band edge emission.

CL imaging was performed in order to probe the optical quality of the individual QWWs. The monochromator was set at a fixed wavelength given by the QWW emission peak, ≈ 0.92 eV according to Fig. 2, and the CL intensity variation was measured over quadratic areas down to $\approx 2 \times 2$ μm^2. A monochromatic image is shown in Fig. 3 where the brighter regions represent a higher luminescence yield. The image exhibits homogeneous CL arising from the wire structures and the lateral size (wire width) is determined to be ≈ 2000 Å. Although not displayed, it is worth pointing out that the CL image correlates perfectly with an SEM micrograph recorded in the same area.

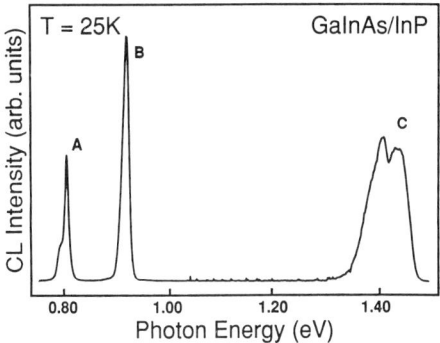

Fig. 2 CL spectra at 25 K from a grating structure ≈ 50 × 50 μm². The three emission bands originate from the ≈ 34 nm thick GaInAs reference layer (A), the GaInAs QWW (B) and the InP near band edge (C).

The CL images of the QWW structures exhibit a remarkable high spatial resolution, ≤ 2000 Å, which is in the same range as (or less than!) the generation volume of electron-hole pairs created by the electron beam (Yacobi and Holt 1990). In principle the diffusion of minority carriers from the generation volume can increase the excitation volume and hence reduce the spatial resolution. In this case however, the diffusing carriers are probably efficiently captured in the GaInAs reference layer, thereby not influencing the CL images of the GaInAs QWWs.

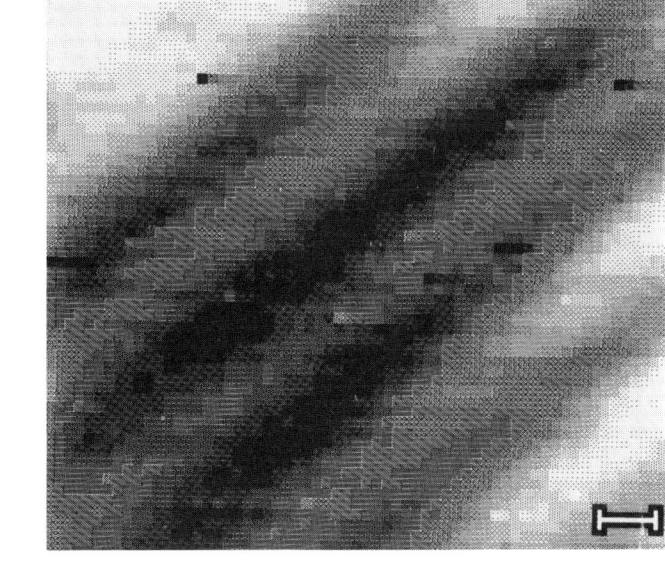

Fig. 3 Monochromatic CL image recorded at 25 K. The energy position of the monochromator was set at the QWW emission peak, ≈ 0.92 eV according to Fig. 2. Note the remarkable spatial resolution of ≤ 2000 Å. The marker scale is 200 nm.

Fig.4 displays three CL spectra of the QWW emission recorded in three different fields (grating-structures), b), c) and d), with different exposure doses, increased from b) to d). For comparison, an additional spectrum was recorded in the unmasked region, a), far outside the grating structures ≈ 100 μm. As can be seen, a clear blue shift of the spectral distribution is observed when the exposure dose is increased i.e. when the linewidth of the QWWs is decreased. It is also noted that the luminescence efficiency decreases with decreasing line width. The energy difference of the spectra corresponding to the unmasked region, a), and the most exposed grating structure, d), was determined to be ≈ 5 meV. Furthermore, the lateral sizes of the QWWs were investigated by SEM and the wire widths in the field b) and d) were found to be ≈ 2500 Å and ≈ 1500 Å, respectively. Further investigations are required to be able to account for the observed blue shift.

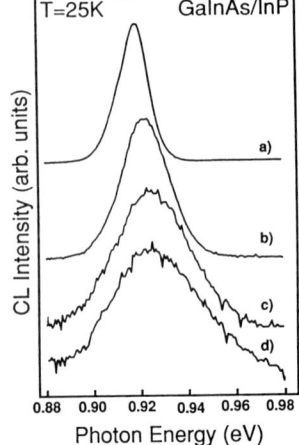

Fig. 4 CL spectra at 25 K displaying the QWW emission recorded in three different fields (grating-structures), b), c) and d), with different exposure doses, increased from b) to d). For comparison, an additional spectrum was recorded in the unmasked region, a). A clear blue shift of the spectral distribution is observed when the linewidth of the QWW structures is decreased.

To conclude, we have successfully fabricated QWWs of GaInAs/InP by electron-beam lithography and subsequent MORIE, having the patterned PMMA resist as an etching mask. CL images exhibit homogeneous CL arising from the QWWs. It is noteworthy that the CL spatial resolution obtained is ≤ 2000 Å, to our knowledge one of the highest ever observed. Furthermore, a clear blue shift was found in the CL spectra and correlated to the increased exposure dose, i.e. with the decreased linewidth of the QWWs.

ACKNOWLEDGEMENTS

R. Riemand is acknowledged for assistance with the electron-beam exposure and Epiquip AB for providing the high-quality epitaxial layers. This work was performed partly within nm-structure consortium in Lund and was supported by the Swedish Natural Science Research Council, by the Swedish Board for Technical Development and by Defence Materiel Administration.

REFERENCES

Christen J 1990 in Festkörperprobleme/Advances in Solid State Physics XXX, ed Rössler U (Braunschweig: Vieweg)

Gustafsson A, Gerling M, Jönsson J, Leys M R, Pistol M-E, Samuelson L and Titze H 1989 Microscopy of Semiconducting Materials 1989, eds Cullis A G and Hutchison J L (Bristol: IOP Publishing Ltd) pp 771-776

Petroff M, Gossard A C, Logan R A and Wiegman W 1982 Appl. Phys. Lett. 41 635

Sakaki H 1980 Jpn. J. Appl. Phys. 19 94

Seifert W, Fornell J-O, Ledebo L, Pistol M-E and Samuelson L 1990 Appl. Phys. Lett. 56 1128

Suemune I and Coldren L A 1988 IEEE J. Quantum Electron. 24 1178

Yacobi B G and Holt D B 1990 Cathodoluminescence Microscopy of Inorganic Solids (New York: Plenum Press)

Inst. Phys. Conf. Ser. No 117: Section 10
Paper presented at Microsc. Semicond. Mater. Conf., Oxford, 25–28 March 1991

Cathodoluminescence study of oval defects on QW structures

D Araújo, J -D Ganière, R Houdré and F K Reinhart

Institut de Micro-et Optoélectronique, Ecole Polytechnique Federale de Lausanne, 1015 Lausanne, Switzerland

ABSTRACT: We report a cathodoluminescence (CL) study at helium temperature of oval defects on quantum well (QW) structures grown by molecular beam epitaxy (MBE). We collected the luminescence emitted both from the facet shaped defect and from the defect free growth region. By comparing the experimental CL peak position with the numerically estimated values, we determined the Al concentration and the thicknesses of the layers in the defect area.

1. INTRODUCTION

The formation of defects during epitaxial growth is frequently observed in GaAs layers. Their presence is a serious problem in the fabrication of integrated circuits from MBE grown crystals. Surface defects are classified into several types, i.e., whiskers, polycrystalline circular areas, and spindle-shaped hollows. Their origins are assumed to be related to the conditions of the substrate surface or spitting from the evaporating source [1]. The defect density varies greatly with growth conditions such as substrate temperature, Ga/As flux ratio and growth rate, and therefore it is not clear which factor is dominant. The oval defects belong to a general class of defects crystallographically oriented along the (110) direction and their size ranges from a fraction to several hundreds of microns [1-5]. As they swell during the sample growth, it is important to study the smallest specimen in order to understand their origins. As most of the structural studies were made by morphological observations using electron microscopy, the chemical composition and growth rate fluctuations have not been analyzed yet.

We present, in the following, a cathodoluminescence (CL) study, at helium temperature and spectrally resolved, of oval defects present on a quantum well (QW) structure to determine the chemical composition and the growth rate variations.

A QW is formed by several monolayers of a low gap material, such as GaAs, embedded in a matrix of a higher bandgap material, such as AlGaAs. The chemical composition variation in the real space along the growth axis induces a square well. In CL or photoluminescence (PL) experimentsone has access to the optical transition between the first quantum energy level of the conduction band (CB) and the first energy level in the valence band (VB), which we labelled $E_{1,h}$. If the thickness of the low gap layer is sufficiently small, quantum size effects invoke discrete levels in the valence and conduction bands. Hence, the QW is an ideal probe to determine locally the growth conditions as the energy levels are sensitive to the barrier height and the well width. The growth conditions are governed by the surface mobility and the sticking coefficient of the different atomic species for each crystallographic orientation.

The facet shape of the oval defects affects the QW width by changing locally the growth condition in the vicinity of the defect. On a QW structure, due to these different orientations present on the defect, the wavelength of the emitted luminescence changes from one point to another. Therefore, CL is a unique and useful tool to study the structural and chemical composition at a microscopic level of such local entities.

The shift of the $E_{1,h}$ peak on two QW of different width is sufficient to determine the well width and the barrier height variation in the defect. The growth rate and the Al concentration of the barriers are constant as the orientation does not change. To increase the precision and to probe the stucture at different depths, we performed the CL experiment on a QW structure with four different well widths and measured four different peak shifts. The sample structure chosen consists of four QW of 155, 90, 56, 25Å grown on an undoped GaAs (001) substrate separated by 100Å of $Al_{0.3}Ga_{0.7}As$ barriers. The last 25Å layer of GaAs covered with 0.1μm of $Al_{0.3}Ga_{0.7}As$.

2. EXPERIMENTAL RESULTS

In the defect free regions, we measured $E_{1,h}$ transitions at 1.538, 1.562, 1.602, 1.699eV (Fig.1a) corresponding to the four QW respectively. The sharp half width (FWHM) of the peaks (2, 5, 7, and 10meV respectively) indicates very good interfaces between the barriers and the wells. In the oval defect region, we observed a shift of the $E_{1,h}$ peak position on the four quantum wells. We measured additional peaks at 1.549, 1.583, 1.674eV (Fig.1b) corresponding to a change of the recombination energy in the three upper top wells. The oval defects contain a GaAs matrix that luminesces at 1.511eV. The peak shift of the first well cannot be estimated accurately due to the overlap with this 1.511eV peak. The broadening of all the shifted peaks reflects the roughness of interfaces within the defect.

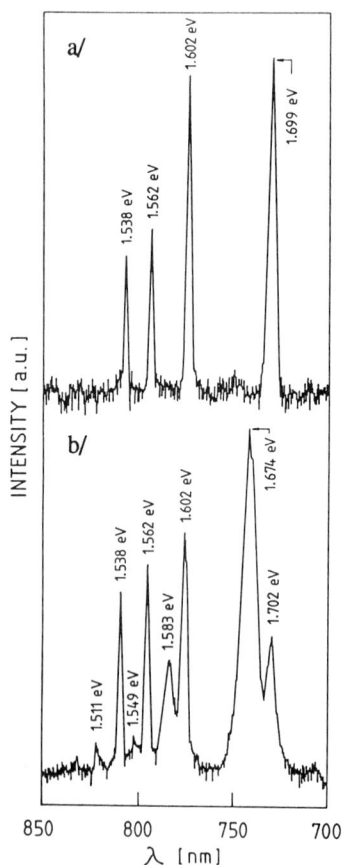

Fig.1:Cathodoluminescence spectra with a beam voltage of 20kV and a beam current of 30nA on:
a/ The defect free region
b/ The oval defect

We present in Fig.2 CL pictures at wavelengths corresponding to energies of 1.699, 1.674, 1.583eV. The minority carrier diffusion lengths deduced from the 1.699, 1.674eV pictures are less than 0.1μm which is 30 times too low for this material. We explain this short diffusion length by the difficulty the carriers encounter when crossing the interface between two twinned microcrystals of the defect. Using both secondary electron (SE) and CL modes, we can find the spatial origin of the luminescence. We observe a higher apparent diffusion length at 1.583eV which is not yet well understood. The deeper is the well, the higher the apparent diffusion length.

a/

c/

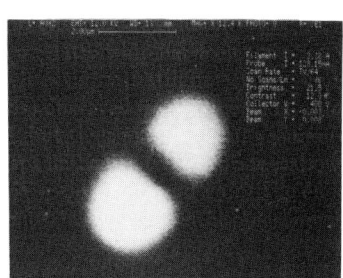

b/

Fig.2: Cathodoluminescence micrographs
of the oval defect at: a/ $E_{1,h}$ = 1.699eV
                                  b/ $E_{1,h}$ = 1.674eV
                                  c/ $E_{1,h}$ = 1.583eV

3. DISCUSSION

As seen by secondary electron (SE) mode of the scanning electron microscope (SEM), the oval defect is symmetric and contains four different crystal orientations. L.Däweritz *et al* [2] studied the morphology of oval defects. They observed by RHEED pattern the growth of the defects and proposed a model with multiply twinned regions to explain the presence of (111) facets. This was confirmed by other authors [3-5].

As different orientations are present within the oval defect, the sticking Ga/Al coefficients and the Al mobility may be affected. However, for these four differently oriented twinned microcrystals, the growth rate and the Al concentration of the barriers are constant. Hence, all the barriers have the same height and all the QW have the same width in one microcrystal part. From the symmetry of monochromatic CL pictures, where local spectra are all the same, we deduce that the Al incorporation and the growth rate are the same in the four twinned microcrystals.

Numerical calculations permit us to interpret the origin of these peak shifts. As the structure consists of four different QW, the peak shift, due to a variation of the barrier height and/or the layer thickness, is different for each one. By varying the Al concentration of the barriers and the width of the wells identically for all the layers we found the growth condition on the defect. We deduce a relative decrease of 18% of the Al barrier concentration and an increase of 15% of the quantum well width in the defect region. The recombination energies in the oval defect yield then QW of 105, 65, 28Å separated by $Al_{0.24}Ga_{0.76}As$ barriers. They correspond to the defect free well thicknesses of 90, 56, and 25Å. This suggests, as the group III elements usually have a sticking coefficient of 1 (maximum value) on a (001) surface, that the stronger effect is an enrichment in Ga on the facets due to a diminution of the Ga surface mobility in this growth direction.

4. CONCLUSION

We were able to determine the layer thickness and the composition of oval defect present on QW structure grown by MBE. Monochromatic CL pictures confirm that the four microcrystals of the oval defect have the same structure. The comparison of the spectra from the defect free region and the oval defect allowed us to quantify the Al concentration of the barriers and the width of the QW in the oval defect. CL is an unique tool for the characterization of local particularities on QW structures. Further reflection electron microscopy (REM) investigations are in progress to assesthe facet orientation of the oval defect.

ACKNOWLEDGEMENTS

This work was made possible through grants from the Swiss National Science Foundation (n° 2.979-0.88). The authors want to thank P.Buffat for helpful discussions, G.Peter and B.Garoni for the mechanical support.

REFERENCES

[1]   Fujiwara F, Kanamoto K, Ohta Y N, Tokuda Y and Nakayama T 1987 J.Crystal Growth 80 104
[2]   Däweritz L, Hey R and Berger H 1984 Thin Solid Films 116 165
[3]   Kakibayashi H, Nagata F, Katayama Y and Shiraki Y 1984 Jpn.J.Appl.Phys. 23 L846
[4]   Suzuki Y, Seki M, Horikoshi Y and Okamoto H 1984 Jpn.J.Appl.Phys. 23 164
[5]   Zandbergen H W, Weyher J and Van Landuyt J 1987 J.Crystal Growth 84 476
[6]   Hollan l and Schiller C 1972 J.Crystal Growth 13 14 319
[7]   Kamon K, Shimazu M, Kimura K, Mihara M and Ishii M 1987 84 126
[8]   Papadopoulos A C, Alexandre F and Bresse J F 1988 Appl.Phys.Lett. 52 224

Inst. Phys. Conf. Ser. No 117: Section 10
Paper presented at Microsc. Semicond. Mater. Conf., Oxford, 25–28 March 1991

Depth resolved luminescence characterisation of chemically treated GaAs surfaces

S Myhajlenko, RA Puechner, JL Edwards and DB Davito[+]

CSSER, Arizona State University, Tempe, AZ 85287-6206 USA.
+ Epitronics Corporation, 21002 N. 19th. Avenue, AZ 85027 USA.

ABSTRACT: We report on depth resolved CL profiling experiments performed on sulphide treated GaAs surfaces. Results are presented which highlight the effect of surface recombination velocity changes on CL properties. Specifically, we report on luminescence improvements and spectral changes, some of which correlate with PL observations. Notably, 1000-fold improvements in room temperature PL are not observed in CL, where typical gains are less than 10. We discuss the low temperature PL and CL data with respect to surface dead layer properties.

1. INTRODUCTION

Inorganic sulphide solution treatment of GaAs has received considerable attention in recent years as a method for passivating the electronic properties of the surface. The deleterious effects associated with the high surface state density of GaAs has hampered progress in specific technologies, for example, MIS-devices. Recently, Yablonovitch et al. (1987) have reported on the beneficial reduction in surface recombination velocity (at least 50-fold) with different sulphide treatments. In addition, photoluminescence (PL) has been successfully applied to monitor this passivation process (Skromme et al. 1987). In this work, we have optimised the ammonium sulphide [$(NH_4)_2S$] treatment from the PL perspective. We have extended measurements to evaluate the changes in the surface recombination properties by performing cathodoluminescence (CL) depth profiling experiments.

2. EXPERIMENTAL

We have used nominally undoped p- and n-type (100) oriented GaAs samples grown by MOCVD on SI GaAs substrates. Layer specifics were as follows: p-type, 8.0 μm thick with $p=4x10^{14}$ cm^{-3}, and for n-type, 9.1 μm thick with $n=1.5x10^{14}$ cm^{-3}. In summary, the samples were treated as follows: (i) NH_4OH oxide strip, (ii) 14 hours soak in buffered $(NH_4)_2S$, (iii) de-ionised water rinse, and (iv) dried with N_2. This treatment was found to be relatively stable (slow degradation in passivation properties over a period of one week in air), and produced a specular surface finish (unlike Na_2S which leaves a salt-like film). Control samples (untreated) were simply the same material which were air-aged over a period of months in the laboratory. This oxide coated surface was chosen as the reference because of its better stability compared with a freshly etched surface which exhibited transient luminescence properties. Prior to CL evaluation, room temperature PL measurements were performed in order to verify luminescence improvement with treatment. Therefore, freshly treated samples were exposed to air for approximately 1 hour prior to loading into the SEM vacuum. The low temperature, single photon counting CL system used in this work was implemented on a JEOL 840 SEM.

3. RESULTS

We found that the improvement in the room temperature peak PL intensity with treatment was typically x300 - x1000. These dramatic changes were not observed with CL, where the improvement was usually <10. In Fig.1 we show room temperature CL depth profiles, from p- and n-type material, obtained by scanning an area (5000 μm²) at 33.3 ms per frame (30 Hz reduced TV-scan rates) under constant incident beam power (200 μW). The TV-rate scanning was done to reduce contamination effects. The data has been normalised to the respective treated sample maximum intensity. The intensity-axis represents CL signal integration in a 50 meV window on the low energy side of band edge. This was done to reduce self-absorption effects. Two observations to note: (i) the intensity profile shape and offset were different for the p- and n-type layers when CL generation was close to the epi-substrate interface - see region beyond marker A, and (ii) the CL signal cut-off (detectivity limit) was similar for both treated and untreated material - see marker B. This signal cut-off region is commonly attributed to the 'dead-layer'.

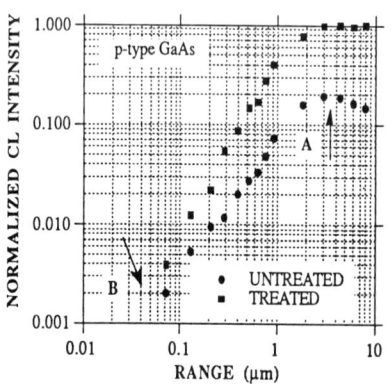
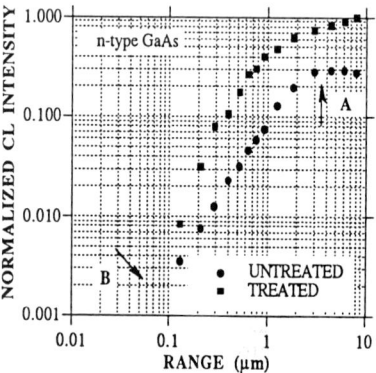

Fig.1 Integrated CL intensity versus range

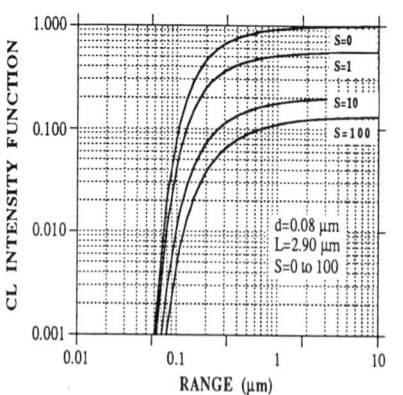

Fig. 2 Theoretical CL Depth Profile

The general shape of the CL depth profile can be understood by using the phenomenological model developed by Wittry and Kyser (1967). In Fig.2 we show CL plots derived from this model, where we have used the Gaussian approximation suggested by Orton and Blood (1990), and the range calculation recommended by Joy and Luo (1989). The following parameters are defined; S is the reduced surface recombination velocity (surface recombination velocity/ diffusion velocity), d is the luminescence dead layer term, and L is the ambi-polar diffusion length (given the excitation conditions used). Also, we have assumed that the recombination is surface and interface dominated, and $L = 2.9$ μm (n-type layer) was calculated from Hall data. Note, the epi-substrate interface was not taken into account in the generation of Fig.2.

Low temperature measurements (T < 25K) revealed further interesting differences in the luminescence behaviour of treated versus untreated GaAs. We present results obtained from the n-type layer to highlight these property changes - please note that similar trends in behaviour were observed with p-type material. The results can be summarised as follows: (i) the low temperature improvement in luminescence with treatment (PL and CL) was much reduced compared with that observed at room temperature, and was typically < x1.5 with CL, (ii) untreated (air-aged) samples were observed to exhibit an absorption 'notch' in the free exciton (FE) line, treated samples did not - see arrow marker in Fig.3 (CL at 20K), (iii) the same phenomenon was also observed with PL at 2K- see Fig.4, but high excitation power was necessary (> 5 W/cm^2) to generate the notch effect (the magnitude of the dip was also found to be temperature dependent), and (iv) treated samples generally exhibited spectral luminescence characteristics consistent with a condition of higher excess carrier injection relative to untreated material.

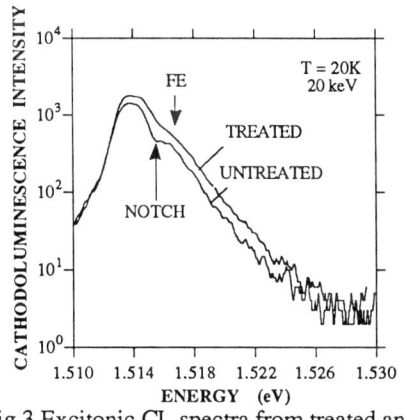

Fig.3 Excitonic CL spectra from treated and untreated n-type GaAs

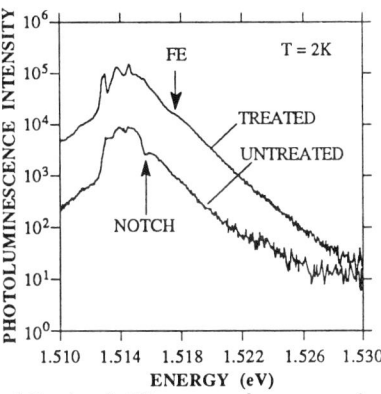

Fig.4 Excitonic PL spectra from treated and untreated n-type GaAs

This injection level difference in spectral character was determined from extensive PL and CL excitation dependence studies performed at various temperatures. This effect was primarily manifested by a difference in the exciton/free-to-bound transition ratio at a given temperature and excitation power. This ratio is a useful measure of the recombination traffic branching between competing radiative routes - namely, single particle (free-to-bound) versus coupled particle (exciton) transitions. In short, treated samples had higher exciton luminescence efficiency - see Fig.5, which shows a CL ratio depth profile acquired at 15K. The CL data presented in Figs.3 and 5 were obtained by scanning a 3000 μm^2 area of the sample at reduced TV-rates (1 μW beam power). These experimental conditions ensured linear CL response with excitation.

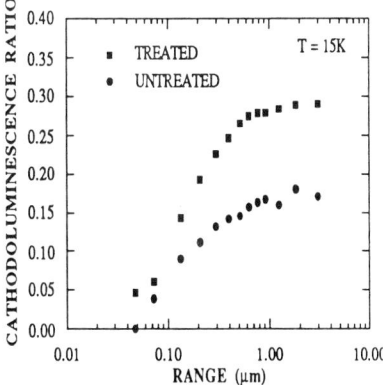

Fig.5 CL (exciton/free-to-bound) transition ratio versus electron range

4. DISCUSSION

The discussion is divided into two sections. Firstly we consider the implications of the room temperature CL observations. Secondly, we speculate on the nature of the dead layer as inferred from the room and low temperature data.

The Wittry model predicts that for a given dead layer thickness (d), L determines the degree of CL intensity modulation for changes in surface recombination velocity. The larger L, the greater the change in CL for large changes in S (> 50). Our measurements have thus far been limited to epi-layers, where the relatively small improvements in CL with treatment were adequately accounted for by Wittry's model, assuming surface-interface dominated recombination with a large reduction in S (compare general form of Fig.1 with Fig.2). The low temperature data of Fig.5 also lends support to the reduction of S. Of additional interest was the different CL profile behaviour for electron penetration close to the epi-substrate interface. A peaking of the CL intensity with penetration would be expected from generation volume spreading effects. This would reduce the excess carrier density, and subsequently the CL signal. Such an effect was seen to a small extent with untreated material. Bear in mind that we had incorporated some self-absorption correction to the data of Fig.1. However, sulphide treated GaAs (both p- and n-type) did not exhibit this behaviour. At present we do not understand the mechanisms involved in this discrepancy.

The concept of a dead layer is important to the interpretation of these results. The luminescence dead layer has been found to be dependent on the depletion width (Hollingsworth and Sites, 1982). In our experiments (room and low temperature) it is reasonable to assume that the injection of excess carriers would essentially flatten the bands, thereby reducing the surface and interface electric fields. The 'dark' surface depletion widths for the present samples are estimated to be 2 - 3 μm (with comparable interface depletion widths!). The experimentally extrapolated CL cut off values (dead layer thickness) were typically in the range 300 - 700 Å - this would indicate the presence of some residual electric fields. Skromme et al. (1987) have reported observing the FE notch from bare surface samples at 1.8 K with PL, whereas the same samples treated with Na_2S did not exhibit the notch effect. This was similar to our observations with $(NH_4)_2S$. Schultheis et al. (1987) have attributed the FE notch to rapid field ionisation of excitons near the surface of air-exposed GaAs material. Our results do not completely tally with this interpretation. Although the FE notch was observed with air-exposed (untreated) samples (p- and n-type), the dead layer thickness for treated and untreated material was nominally the same (within 100 Å). This would imply similar residual fields. An alternative speculation may involve anomalous absorption / interference effects associated with the dead layer. These various factors are being investigated further. In conclusion, our results confirm that the CL properties of GaAs are also sensitive to surface conditions, but to a lesser degree than PL properties.

REFERENCES

Joy D C and Luo S 1989 Scanning 11 176.
Hollingsworth R E and Sites J R 1982 J. Appl. Phys. 53, 5357.
Orton J W and Blood P 1990 The Electrical Characterisation of Semiconductors:
  Measurement of Minority Carrier Properties (Academic Press) pp 38-9.
Schultheis L, Köhler K and Tu C W 1987 Excitons in Confined Systems - Springer
  Proceeding in Physics No.25 pp 110-18.
Skromme B J, Sandroff C J, Yablonovitch E and Gmitter T 1987 Appl. Phys. Lett. 51
  2022.
Wittry D B and Kyser D F 1967 J. Appl. Phys. 38 375.
Yablonovitch E, Sandroff C J, Bhat R and Gmitter T 1987 Appl. Phys. Lett. 51 439.
This work was funded by ASU-Research Incentive Award No.021882.

Inst. Phys. Conf. Ser. No 117: Section 10
Paper presented at Microsc. Semicond. Mater. Conf., Oxford, 25–28 March 1991

711

Investigations of hillocks in strained GaAs$_x$P$_{1-x}$

A Gustafsson and L Samuelson

Department of Solid State Physics, Lund University, Box 118 S-221 00 Lund, Sweden.

ABSTRACT: A sample consisting of one single thin strained layer of GaAs$_x$P$_{1-x}$ in between GaP barriers was found to contain several hexagonal hillocks. The hillocks are the result of a higher growth rate locally. Cathodoluminescence imaging reveals that the emission from the the hillocks is shifted towards lower energy, indicating a higher arsenic content compared to the perfectly strained layer. The arsenic content of the perfectly strained layer was 25%, whereas the individual hillocks were found to contain up to 32% arsenic.

1. INTRODUCTION

In epitaxial growth the ideal surface is perfectly smooth without any irregularities in the morphology. To achieve the ideal surface the growth conditions must be optimised. However, growth conditions more or less off from the ideal growth conditions can result in a rough surface containing a high density of growth formations, so called hillocks. Typical for GaP and related systems is the hexagonal hillock. These hillocks can be observed in samples grown by metalorganic vapour phase epitaxy (MOVPE) on (001) oriented substrates under certain growth conditions (Baliga and Gandhi 1974). It has previously been established that the hillocks are a result of a too high growth rate in relation to the temperature of the substrate during growth (Biefeld 1982). The rate at which atoms impinge on the substrate from the gas phase is much higher than the rate at which atoms are added to the substrate surface. If the surface mobility of the ad-atoms is low the growth will not take place at the the steps and kinks on the surface, as in the usual MOVPE (Stringfellow 1982), but rather via nucleation of two-dimensional islands. This can give rise to imperfections in the crystal structure, like stacking faults and dislocations, around which the growth rate is enhanced. Besides, impurity atoms also have a tendency to be adsorbed preferentially at these imperfections. The areas where the enhanced growth rate has taken place can eventually be observed macroscopically as hillocks on the growing surface.

An important factor when studying luminescence from hillocks containing strained layers is the strain induced effect on the energy gap. The shifts of the energy bands can be calculated by the theory of deformation potentials (Mathieu et al 1979). Calculations utilizing this technique have previously been reported for the system under study, i.e thin strained layers of GaAs$_X$P$_{1-X}$ grown in compression in between GaP barriers on (001) oriented substrates (Pistol et al 1988). Furthermore it was shown that the peak energy position of the photoluminescence (PL) emission, at 2K, was in good agreement with the values predicted by the theoretical calculations. Fig. 1 shows the energy-band structure of the ternary alloy of GaAs$_X$P$_{1-X}$. The solid lines represent the unstrained Γ and X energy bands and the dashed line represents the calculated peak energy position of the PL from a thin strained layer, corrected for the binding energy of the M$_O{}^X$. M$_O{}^X$ originates from the decay of either excitons or separate charge carriers localized by fluctuations in the alloy. The solid dots represents the PL measurements of layers thicker than 100 Å, i.e. layers exhibiting no shift due to quantum confinement. This additional shift is observed for layers thinner than 100 Å (Pistol et al 1988).

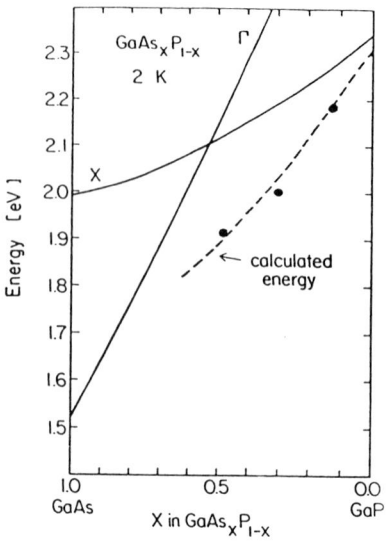

Figure 1. Photoluminescence transition energies for different compositions. The solid lines represents the X and Γ bands of the unstrained alloy and the dashed line is the calculated band gap shift due to strain, corrected for the binding energy of the M$_O{}^X$

2. EXPERIMENTAL

The sample studied was grown by MOVPE in a horizontal reactor operated at atmospheric pressure and with radio frequency heating. The gases used in the growth were trimethylgallium (TMG) arsine (AsH$_3$) and phosphine (PH$_3$) with hydrogen as a carrier gas. First a buffer layer was grown using a growth rate of 5 monolayers (ML) per second to a thickness of about 5 μm. Then the strained layer of 160 Å was grown followed by a top layer of 250 Å. For both layers a lower growth rate of 1 ML per second was used. The sample contains 25% arsenic, which means it is well below the critical thickness for strain relaxation. The critical thickness was found to be 200 Å for a layer containing 30% arsenic (Pistol et al 1991).

The cathodoluminescence (CL) study was carried out with a scanning electron microscope (SEM) modified with a continuous helium flow cold stage. The SEM was operated at 20 keV and the sample temperature was 25K. The CL was collected by a parabolic mirror and dispersed by a 0.50 m single monochromator. The band width chosen for recording   the spectra was 15 Å and for the imaging of the hillocks a 30 Å band width was chosen. A cooled GaAs photomultiplier was used to detect the luminescence. Further details about the CL system can be found elsewhere (Gustafsson et al 1989).

3. RESULTS AND DISCUSSION

Monochromatic images of the CL originating from the strained layers from areas free from hillocks are smooth with some spots, corresponding to the exit points of threading dislocations. This is an indication that the layer is perfectly strained contrary to a partially relaxed layer, which exhibits dark lines in the <110> directions. These dark lines are associated with the non-radiative recombination of misfit dislocations (Gustafsson et al 1989). Concerning the hillocks, the hillocks themselves appear dark in the images monitoring the emission from the strained layer, which is shown in Fig. 3a. However when the detection window is displaced to lower energy the hillocks appear bright except for the centre, Fig 3b. The darker centre part of the hillocks supports the assumption that the hillocks are formed at dislocations.

Figure 2. An optical micrograph of a hexagonal hillock. The typical extension of the hillocks is 10 - 20 μm.

In Fig. 4 the luminescence spectra from several of the hillocks are shown. It can be noted that the shift in energy differs from one hillock to the other, however all the peaks are shifted to lower energy compared to the CL emission originating from the strained layer outside the hillocks. Based on previous studies, the shift can be attributed to a higher arsenic content in the thin layers in the hillocks.

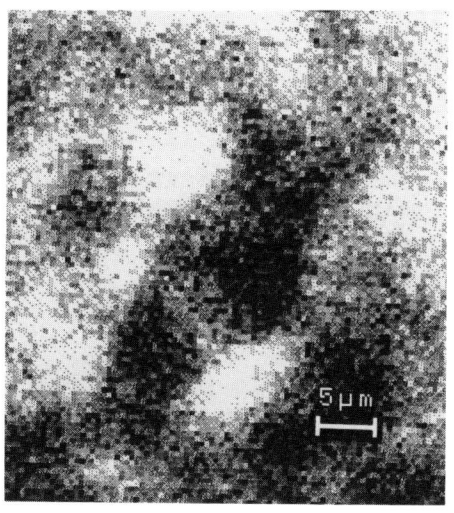
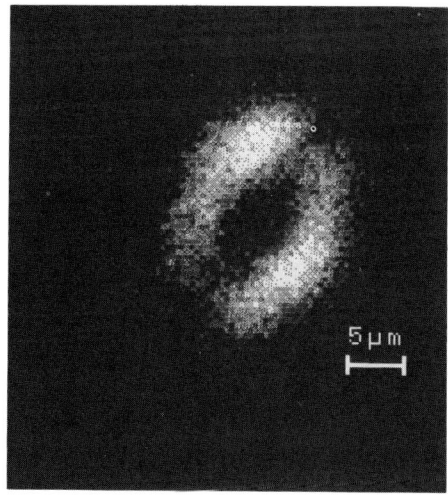

a)   b)

Figure 3. Monochromatic CL images of an area containing a hillock. a) when the detection window is set at the peak energy position corresponding to the arsenic content of the strained layers outside the hillocks, the hillock appear dark on a bright background. b) when the detection window is moved to lower energy the hillock appear bright on a dark background.

For the same MOVPE reactor used in this study it was shown that the arsenic content of the solid phase of $GaAs_xP_{1-x}$ increases with increased growth rate (Leys et al 1988). This provides a possible explanation for the higher arsenic content in the faster growing hillocks than in the surrounding layer, thus shifting the peak energy position to lower energy. The shift in energy corresponds to a change in the arsenic content of the individual hillocks in a range from nominally 25% up to 32%. It is worth pointing out that, in comparison with a pure relaxation of the strained $GaAs_xP_{1-x}$ layer, the CL energy position at the hillocks is shifted in the reverse direction, as can be seen from Fig. 1. This rules out the possibility that relaxation of the strain is responsible for the shift of the peak energy position at the hillocks.

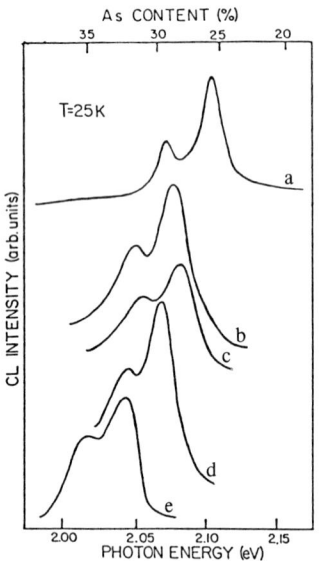

Figure 4. A series of CL spectra. a) recorded from the layer surrounding the hillocks and b) - e) taken at several different hillocks, showing a shift to lower energy, which is attributed to a higher arsenic content in the hillocks. The lower abscissa shows the photon energy of the emission and the upper abscissa shows the arsenic content derived from the peak energy position of the emission (zero phonon line).

ACKNOWLEDGEMENTS

The Authors are grateful to Dr. M-E Pistol and Dr. S Nilsson for helpful discussions. This work was supported by the Swedish Natural Council and by the Swedish Board for Technical Development.

REFERENCES

Baliga B J and Gandhi S K 1974 J. Crystal Growth 26, 314

Biefeld R M 1982 J. Crystal Growth 56, 382

Gustafsson A, Gerling M, Jönsson J, Leys M R, Pistol M-E, Samuelson and L Titze H 1989 Microscopy of Semiconducting Materials 1989, eds A G Cullis and J L Hutchison (IOP Publishing Ltd., Bristol) p. 771-776

Leys M R, Titze H, Samuelson L and Petruzzello 1988 J. Crystal Growth 93, 504

Mathieu H, Merle P, Ameziane E L, Archilla B and Camassel J 1979 Phys. Rev. B 57, 5428

Pistol M-E, Leys M R and Samuelson L 1988 Phys. Rev. B 37, 4664

Pistol M-E, Gustafsson A, Gerling M, Samuelson L, and Titze H 1991 J. Crystal Growth 107, 458

Inst. Phys. Conf. Ser. No 117: Section 10
Paper presented at Microsc. Semicond. Mater. Conf., Oxford, 25–28 March 1991

715

Electron beam excitation and profiling of CdSe-ZnSe multiple quantum well and strained layer superlattice structures

C Trager–Cowan, P J Parbrook, D Clark, B Henderson and K P O'Donnell
Department of Physics and Applied Physics, University of Strathclyde
John Anderson Building, 107 Rottenrow, Glasgow G4 0NG, Scotland, UK

B Cockayne and P J Wright
Royal Signals and Radar Establishment, St Andrews Road, Malvern, Worcs WR14 3PS

ABSTRACT: The variation of penetration depth of an electron beam with energy is used to profile the luminescence properties of CdSe–ZnSe multiple quantum well and strained layer superlattice structures. Luminescence from different layers within the multilayered structures, gives rise to multicoloured emission.

1. INTRODUCTION

Electron beam pumped wide bandgap multiple quantum wells (MQW) and strained layer superlattices (SLS) are promising candidates for visible optoelectronic devices. ZnSe–ZnS, CdSe–ZnSe and CdS–ZnS SLS have been grown at the Royal Signals and Radar Establishment (RSRE, Malvern), using atmospheric metalorganic vapour phase epitaxy (MOVPE) at low temperatures, Wright et al (1989), Parbrook et al (1990). The development of optical devices utilising the entire wavelength spectrum from the UV to the near infra–red is possible using such SLS. The electron beam excitation of a number of these SLS has been reported previously, Trager–Cowan et al (1991). In this paper we report further work on CdSe–ZnSe multilayered structures.

2. ELECTRON BEAM EXCITATION

An electron beam has the useful property that it deposits energy at a depth controllable via the electron beam energy. By changing the energy of the beam it is possible to probe different depths in the sample under investigation. This allows variations in sample structure to be identified, and different layers within a heterostructure to be investigated.

Figure 1 shows a typical SLS structure and sketches the energy loss of various electron beams in this structure. Figure 1(c) shows the depth of maximum energy deposition as a function of the electron beam energy calculated by the empirical formula, Chin (1983),

$$X_E = 2.19 \times 10^{-8} AE^2 / \rho Z \ln(0.108 E Z^{-2/3}) \qquad (1)$$

where X_E is in μm, A, Z are the atomic mass and number, and ρ is the density of the target in g/cm^2. E is the energy of the electrons in eV.

Figure 1 illustrates that it is possible to excite all the layers in a typical SLS with the electron energies available (<30kV), i.e., the dimensions of the SLS structures are 'well matched' by the penetration depths of the available electron beams.

3.   EXPERIMENT

At Strathclyde an electron beam of up to 30keV in energy and of moderate current density ($\leqslant$ 3mA/cm^2 in a spot size of 300μm) is used to obtain cathodoluminescence spectra from samples cooled to around 100K. A schematic diagram of the apparatus is given in Figure 2. Note that the electron beam is incident on the front face of the sample and the cathodoluminescence from the edge of the sample is detected, i.e., the cathodoluminescence emitted at right angles to the exciting beam. This excitation/detection geometry reduces potential problems due to re-absorption of the cathodoluminescence by un-excited regions of the sample.

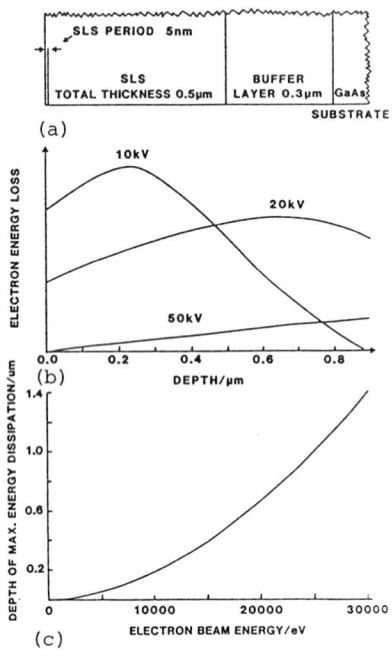

Figure 1

4.   RESULTS AND DISCUSSION

(i) The cathodoluminescence properties of a sample comprising a 100 period SLS of CdSe(1nm)–ZnSe(4nm); a buffer layer of ZnSe (0.3μm); and a GaAs substrate (300μm), were investigated, as the electron beam energy was increased from 3 to 30kV (a schematic diagram of this sample is drawn in Figure 1). Figure 3 shows the cathodoluminescence spectra obtained at 5kV and 15kV respectively. Figure 4 shows the variation of the peak intensity and of peak position of the SLS luminescence as a function of electron beam energy (constant beam current). The peak of the cathodoluminescence spectrum shifts by $\approx$27meV to higher energy as the electron beam energy is increased, and the linewidth is also observed to increase by $\approx$13meV. The peak shift and the peak luminescence intensity were observed to 'saturate' at an electron beam energy of $\approx$15keV.

The dependence of energy deposition on electron beam energy explains these effects. As seen from Figures 1 and 4, at electron beam energies between 3 and $\approx$15keV, the electron beam deposits most of its energy at depths increasing from 0.1μm to 0.4μm, i.e., in the SLS. At energies greater than $\approx$15keV the electron beam deposits most of its energy in the buffer layer and substrate. The peak shift observed below 15keV may therefore be attributed to some systematic change in the SLS. Transmission electron micrographs, Parbrook (1990), revealed a continous variation of the SLS well thickness: the nearer to the buffer layer a well lies the narrower it is. Quantum confinement increases with decreasing well thickness, therefore the observation of a "blue shift" of the emission with increasing electron beam energy is consistent with the observed variation in well thickness.

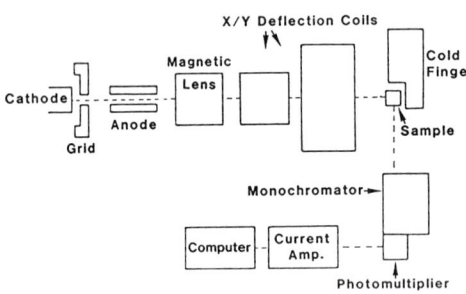

Figure 2

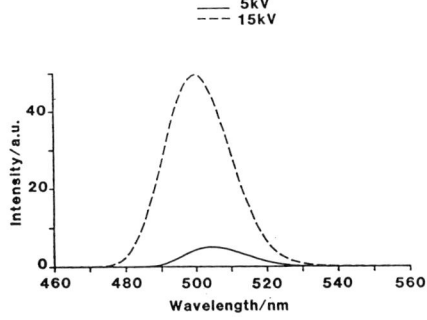

Figure 3

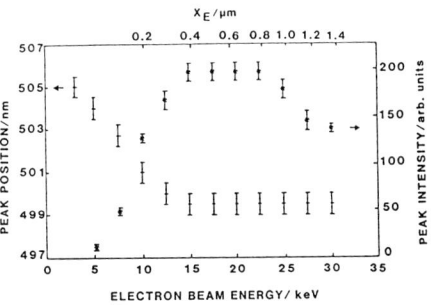

Figure 4

The average peak position of the emission is ≈2.47eV and the average linewidth of the emission is ≈110meV. Using a modified Kronig–Penney model, Bastard (1981), this peak position and linewidth implies that the electron beam is "sampling" a series of quantum wells whose widths vary between 2 and 3 monolayers (0.605 and 0.9075nm). This is consistent with the variation in well width deduced from the transmission electron micrographs. The observed increase in linewidth with increasing electron beam energy is attributed to the fact that the greater the energy of the beam, the wider its distribution, (see figure 1 (b)), i.e., the higher the energy of the beam, the greater the number of wells it samples.

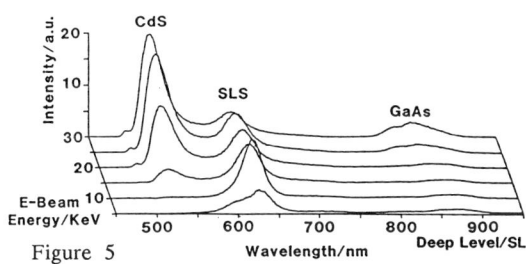

Figure 5

(ii) The cathodoluminescence properties of a sample comprising a 100 period SLS of CdSe(2.5nm)–ZnSe(2.5nm); a buffer layer of CdS (0.3μm); and a GaAs substrate (300μm), were investigated as the electron beam energy was varied between 5 and 30kV. The sample emits red, 'yellow' and green light in turn as the electron beam energy increases (figure 5). The red luminescence originates from the SLS, the green luminescence from the CdS buffer layer, and the 'yellow' light is the physiological combination of the red and green luminescence.

Deep level emission from the SLS was also observed at low electron beam energies. At high electron beam energies emission from the GaAs substrate was also observed.

The above illustrates that it is possible to obtain emission from different layers within a structure by changing the electron beam energy and in principle, this could be exploited to produce multicoloured light emitting devices.

(iii) Finally, the cathodoluminescence properties of a sample comprising 20 periods of a multiple quantum well with (1) 5A CdSe well, (2) 15A CdSe well and (3) 25A CdSe well, sandwiched between 30A ZnSe barrier layers (see figure 6); a buffer layer of ZnSe (0.3μm); and a GaAs substrate (300μm), were investigated as the electron beam was increased from 7 to 20kV (figure 6).

The emission at ≈478nm and ≈610nm was deduced from calculations (using the modified Kronig–Penney model of Bastard (1981) to originate from the 5 and 25A quantum wells respectively. The emission at ≈820nm originates from the GaAs substrate. The emission at ≈444nm was originally thought to originate from the ZnSe buffer layer, however the variation in intensity of this peak as a function of exciting electron beam energy was found to be the same as that of the luminescence from the quantum wells (figure 7). If the emission at 445nm originated from the buffer layer, the variation in its intensity would be different from that of the quantum wells, as is the case for the luminescence originating from the GaAs substrate (figure 7). This implies that the luminescence at 445nm must originate from within the multiple quantum well structure. It was finally concluded that this luminescence must originate from the ZnSe barrier layers.

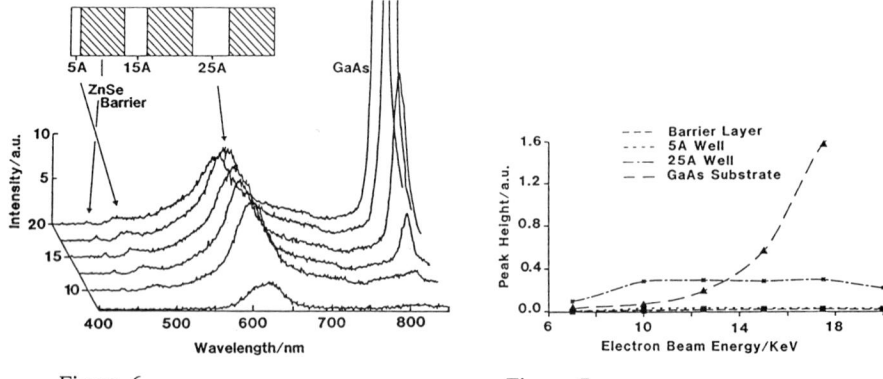

Figure 6   Figure 7

5.   SUMMARY

Cathodoluminescence has been observed from a number of multilayered structures and it has been shown that an electron beam can be used to resolve from which layer/set of layers luminescence of a given wavelength is being emitted, using the ability to vary the energy deposition of an electron beam to probe into these structures.

A systematic decrease in the well width of a 100 period CdSe(1nm)–ZnSe(4nm) SLS was identified. Multicoloured emission was obtained from a sample comprising a 100 period CdSe(2.5nm)–ZnSe(2.5nm) SLS and a CdS buffer layer. Emission was observed from the 5 and 30A CdSe wells of a 20 period multiple quantum well structure, emission at 444nm was identified to come from the ZnSe barrier layers as opposed to the ZnSe buffer layer.

6.   ACKNOWLEDGEMENTS

We would like to thank Professor E. C. Lightowlers and Dr A. T. Collins of King's College, London for the loan of the cathodoluminescence system used in this work.
This work has been supported by the University of Strathclyde Research and Development Fund. The samples investigated in this work were grown with support from the SERC/MoD under grant number GR/E/26853. C. Trager–Cowan is presently supported by a SERC Research Fellowship.

REFERENCES

Bastard G 1981 Phys. Rev. B 24 5693
Chin T N 1983 Microscopy of Semiconducting Materials, Inst. Phys. Conf. Ser. 67 343
Parbrook P J, Wright P J, Cockayne B, Cullis A G, Henderson B and
    O'Donnell K P 1990 J. Cryst. Growth 106 503
Trager–Cowan C, Parbrook P J, Clark D, Green G, Wiseman A B, Henderson B
    O'Donnell K P, Cockayne B and Wright 1991 P J J. Lumin. 48 & 49 773
Wright P J, Parbrook P J, Cockayne B, Jones A C, Orrell E D,
    O'Donnell K P and Henderson B 1989 J. Cryst. Growth 94 441

Inst. Phys. Conf. Ser. No 117: Section 10
Paper presented at Microsc. Semicond. Mater. Conf., Oxford, 25–28 March 1991

Low temperature injection luminescence using a scanning tunneling microscope

Lars Montelius, Fredrik Owman[1] , Mats-Erik Pistol and Lars Samuelson

Department of Solid State Physics, University of Lund, Box 118, S-221 00 Lund, Sweden

ABSTRACT: Besides its conventional use in the current imaging mode, the scanning tunneling microscope (STM) offers the possibility of studying minority carrier injection in semiconductors. In this paper we present the first observation of low-temperature spectrally resolved radiative recombination resulting from the minority carrier injection from the STM metal tip. Electrons are in this case injected across the vacuum gap into the conduction band of p-type InP material. We denote this process as scanning tunneling luminescence (STL). We will report on the voltage dependence of the luminescence and compare STL with photoluminescence.

1. INTRODUCTION

Modern electronic and opto-electronic devices such as semiconductor lasers, resonant-tunneling devices etc are dependent on the possibility of growing epitaxial films of very high quality and of patterning these films in all three directions. Devices, whose performance is governed by concepts from quantum physics are becoming very interesting for the semiconductor industry. For characterization of semiconductor materials that are structured on the nanometer scale a very high spatially resolving tool is needed. For this "microcharacterization" of materials the cathodoluminescence method has attracted great interest (Christen 1990; Nilsson et al 1991) but the resolution is limited by principally the electron penetration depths and carrier diffusion, which are of the order of 1-2μm. However, during the past 10 years the use of Scanning Tunneling Microscopy (STM) for various types of investigations of electronic properties of surfaces of conductors and semiconductors has been exploited. The STM monitors essentially the electronic states close to the Fermi Level with atomic resolution (Binnig et al 1982; Stroscio et al 1988) and thereby it has the potential for making detailed investigations on the atomic scale. In recent years different methods have been proposed for investigations of subsurface properties such as "ballistic electron emission microscopy" (Kaiser and Bell 1988) and for investigations of light emission based upon inverse photoemission and fluorescence processes due to the decay of plasmons (Gimzevski et al 1988; Coombs et al 1988) resulting from electron injection using the STM. Abraham et al (1990) presented luminescence microscopy with nanometer resolution of a GaAs/AlGaAs heterostructure, in which work they could resolve individual quantum wells. In the present work, we demonstrate for the first time spectrally resolved injection luminescence at low temperatures with the use of an STM as the minority carrier injector.

2. EXPERIMENTAL DETAILS

The studies presented here were performed in a compact temperature compensated STM utilizing a piezo-tube for the X,Y and Z-movements of the tip. The microscope has

---

[1] Present Adress: Department of Physics and Measurement Technology, University of Linköping, S- 581 83 Linköping, Sweden

demonstrated atomic resolution on highly oriented pyrolitic graphite at room temperature, at which temperature the lateral scan range is approximately 10 nm/V. At lower temperatures the scan range is reduced to around 10Å/V. For our studies, the microscope is suspended on springs and vibrationally damped inside a liquid Helium cryostat equipped with optical windows, thus allowing the extraction of photons as well as making possible optical microscope observation of the sample-tip region. In order to efficiently collect the light from the tunneling region a parabolic mirror is mounted close to the tunneling tip. Furthermore, the design of the microscope allows samples to be cleaved *in situ* in either protective atmosphere (He-gas) or in liquid Helium in order to minimize contamination of the semiconductor surface. Besides the conventional constant current mode, with a feedback loop for the movement, the microscope may be used to study the dependence of the tunneling injection on the voltage between tip and sample, as well as on the tip-sample separation. For recording of the STL spectra the tip sample region is magnified and imaged onto the entrance slit of a 0,5 m monochromator equipped with a photo-multiplier tube for photon detection. The sample that was studied here was a p-type (Zn-doped) InP wafer grown in the (100) direction, and consequently the cleavage surface that was studied was the (110) orthogonal surface. The bias was delivered with a high precision voltage supply unit and the current was measured with a current amplifier allowing current levels up to several μA while operating the microscope in the constant current mode. The tips were made out of a PtRh-wire that was cut with scissors into small pieces resulting in enough sharp tips for tunneling.

3.   RESULTS AND DISCUSSION

In Fig. 1 the STL spectrum shown was obtained by injection of minority carriers from the STM metal tip into the semiconductor. The STL-spectrum are typical of near band-edge recombination of donor excitons.The dotted line in Fig. 1 displays the photoluminescence (PL) spectrum from the sample aquired after the STL spectrum was taken. As can be seen the agreement between ordinary PL and STL measurements is very good. The spectra are typical of near band-edge recombination of donor excitons. No STL was observed at higher energies, in particular no emission was seen at an energy corresponding to the bias voltage. This study is, to our knowledge, the first observation of low-temperature spectrally resolved radiative recombination resulting from the minority carrier injection from the STM metal tip.

Fig. 1 STL spectrum obtained during scanning due to minority carrier injection from the STM tip into the semiconductor. The spectrum are typical for near-band edge donor exciton recombination. The tunneling current was 60μA and the temperature around 150-200K. The bias on the sample was 1.6V. The dotted spectrum is obtained under ordinary photoluminescence conditions.

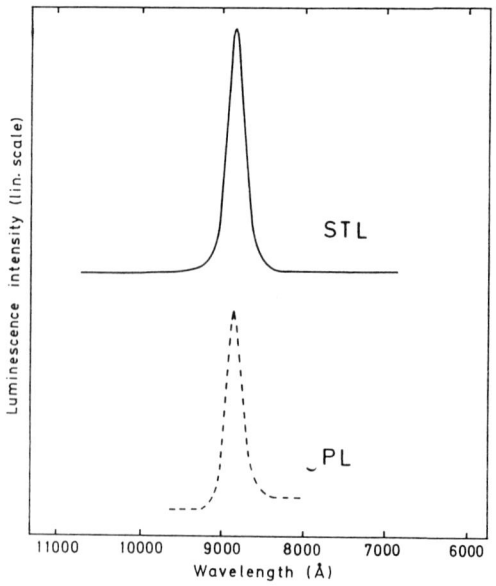

Electrons are in this case injected across the vacuum barrier region into the conduction band of p-type InP (Fig. 2) resulting in a radiative recombination process. This mode of recombination is in direct analogy to minority carrier injection luminescence previously employed using MIS contacts. The STL spectrum was obtained after cleaving the sample and during scanning of the STM tip, with the feedback system activated. The STL spectrum was taken for tunneling conditions and it should be noted that a crashed tip also causes an emission which should not be confused with emission caused by tunneling. The temperature was around 150-200 K and the bias was 1.6V with a positive polarity on the sample and the

Fig. 2 Idealized band diagram for the STM metal tip, insulating vacuum region and a p-type semiconductor with a positive bias on the semi-conductor with respect to the tip of a) 0V and b) V>V_threshold.

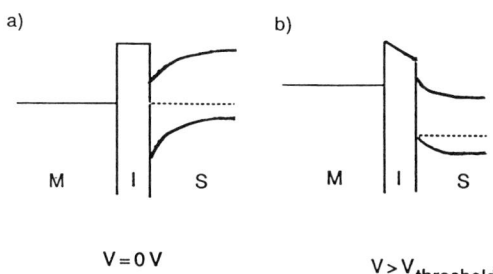

tip at virtual ground potential. The current was 60 µA, indicating that the tip was close to the surface. We expect the tip to be rather blunt since it was made by cutting a wire, and this in turn means that no observation having atomic resolution of the surface was possible during scanning. It should also be mentioned that at some positions on the sample a strong band-to-acceptor emission was observed, thus opening up new exciting possibilities for mapping local variations of donor-acceptor concentrations on the surface. The PL spectrum gives direct evidence for an integrated strong acceptor-bound transition (Fig 1). This will also be reported elsewhere.

We have also investigated the voltage dependence of the luminescence which is shown in Fig. 3 revealing that the intensity of the luminescence is exponentially increased with increasing bias level. The treshold voltage was deduced to be close to 1.3 V, a value which should correspond to the metal-semiconductor barrier energy. From Fig. 2 it is clear that the threshold voltage for minority carrier injection into the conduction band and subsequent radiative recombination is close to the corresponding bandgap of the material.

Fig. 3 The dependence of bias on the luminescence intensity. A threshold value of approximately 1.3 eV can be deduced from the figure. The constant tunnel current was 60 µA.

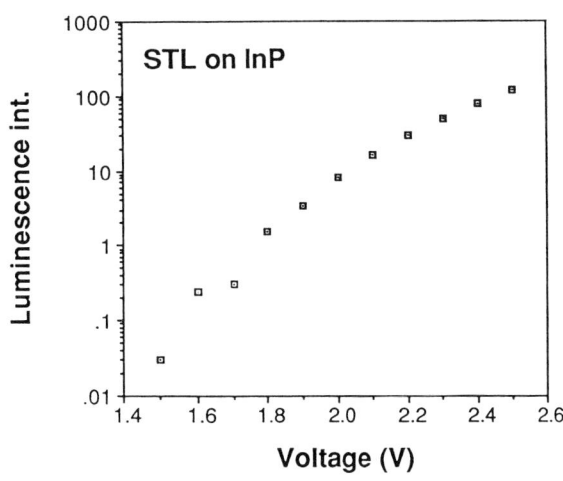

ACKNOWLEDGEMENT

Special thanks to Jeper Mygind and Ole Albrektsen at the Danish Technical Highschool (DTH), Lyngby for help with the initial version of the microscope. Per Mårtensson is acknowledged for help with the computer programme that controls the scanning tunneling microscope and for valuable discussions. This work was performed partly within the nm-structure consortium in Lund, and was supported by the Swedish Natural Science Research Council and by the Swedish Board for Technical Development.

REFERENCES

Abraham D L, Veider A, Schönenberger Ch, Meier H P, Arent D J and Alvarado S F 1990 Appl. Phys. Lett. 56 1564
Binnig G, Gerber Ch, Rohrer H and Weibel E 1982 Phys. Rev. Lett. 49 57
Christen J 1990 in Festkörperprobleme/Advances in Solid State Physics XXX, ed Rössler U (Braunschweig: Vieweg)
Coombs J H, Gimzevski J K, Reihl B, Sass J K and Schlittler R 1988 Z. Phys. B 72 497
Gimzevski J K, Reihl B, Coombs J H, and Schlittler R 1988 J. Microscopy 152 325
Kaiser W J and Bell L D 1988 Phys. Rev. Lett. 60 1546
Nilsson S, Gustafsson A, Montelius L, Semu A, Georgsson K and Samuelson L 1991 Microscopy of Semiconducting Materials 1991, conference proceeding
Stroscio J A, Feenstra R, Newns D M and Fein A P 1988 J. Vac. Sci. Technol. A6 499

Inst. Phys. Conf. Ser. No 117: Section 10
Paper presented at Microsc. Semicond. Mater. Conf., Oxford, 25–28 March 1991

EBIC diffusion length evaluation by the moment method

D Cavalcoli, A Cavallini and A Castaldini.
Dept. of Physics, via Irnerio 46, University of Bologna, I-40126 Bologna, Italy

ABSTRACT: The diffusion length of minority carriers in n-type Fz Si is obtained with the electron beam induced current (EBIC) technique in planar configuration. The charge collection current as a function of the beam-junction distance is analysed by the moment method, which is based on the calculation of the variance of the derivative of the current profile. This method requires no assumptions on the surface recombination velocity and, thus, provides a diffusion length evaluation free from its influence. The data are also analysed with the asymptotic method, which requires conventional assumptions on the surface recombination velocity.

1. INTRODUCTION

For many years the electron beam induced current (EBIC) technique of scanning electron microscopy has been widely used to determine electronic properties in semiconductors (wide literature in Holt and Joy, 1989), in particular the minority carrier diffusion length L. However, the evaluation of the actual value of this parameter has always been confined to the limiting cases corresponding to a surface recombination velocity $s = 0$ and $s = \infty$ (Ioannou and Davidson, 1980; Davidson and Dimitriadis,1978). The charge collection problem with arbitrary surface recombination velocity was first analysed by von Roos(1978) but the application of his results appears rather difficult because of the extensive numerical calculation. Kuiken and van Opdorp (1985) analysed asymptotic expansions of the fraction Q of generated minority carriers that flows into the collecting junction, but their method has, in practice, a limited applicability since it requires data acquisitions at large scan distances ($x' >> L$), which are not always available in electronics circuits. Donolato (1985) yielded a method for determining L based on the calculation of the moments of the induced current profile derivative. Since this method does not require information on s, it provides a determination of L free from its influence. This paper deals with the application of this last method.

2. EXPERIMENTAL

The planar collector geometry, shown in the inset of Fig.1, has been used to evaluate the diffusion length L by the moment method . The value of L is obtained by recording the decay of the charge collection current I as a function of the beam-junction edge distance x' (Fig.1), and analysing the data on the basis of analytical expressions provided by the theory. The electron beam, impinging normally to the sample surface (x, y), gives rise to an ideal steady state point source at (x', z') and correspondingly, to a collected current $I(x')$. Low injection dose is used.

Under these conditions, Donolato (1985) recognizes that for any s the current $I(x')$ collected by the diode can be expressed in the form of a convolution product of the function $J(x)$ which represents the current density of minority carriers from the source to the surface when $s = \infty$, independent of the actual surface conditions, times the probability $c(x)$ that a minority carrier reaching the surface flows into the junction, only determined by the surface conditions. Defining the normalized current profile $i(x') = I(x')/I(-\infty)$, and calculating the variance σ^2

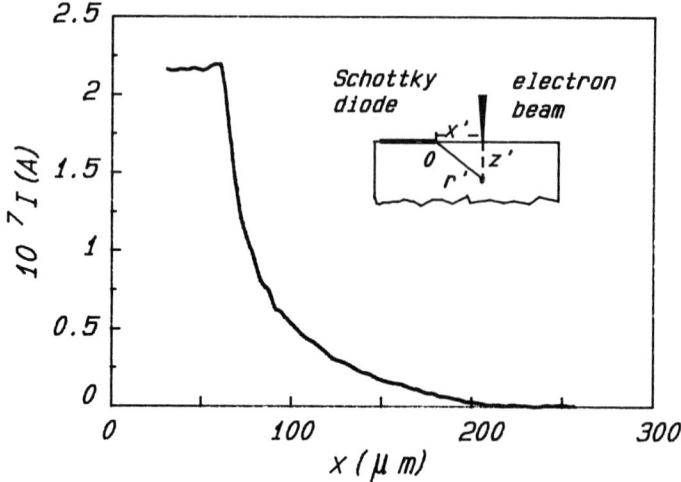

Fig1.   EBIC profile. Microscope conditions: beam current $I_b = 5 \cdot 10^{-9} A$, electron energy $E_b = 30 keV$. In the inset the experimental arrangement used.

of the derivative $i'(x')$, that is its moment of second order about the mean m

$$\mu_2[|i'|] = \int_{-\infty}^{+\infty} (x - m)^2 |i'(x)| dx, \tag{1}$$

one finds that:

$$\sigma^2 = \sigma_s^2 + Lz' \tag{2}$$

where the term σ_s^2 is a function of s, but not of z'. Thus, L can be evaluated from the slope of σ^2 vs z' (Eq.2), without knowing the value of s. Since σ_s^2 has the dimension of a squared length, it can be written as

$$\sigma_s^2 = wL^2/2 \tag{3}$$

where w is a dimensionless function of s and L which ranges from 0 for $s = \infty$ to 1 for $s = 0$.

This method has been applied to measure the minority carrier diffusion length in P doped, $< 111 >$ Fz Si, with majority carrier concentration equal to $5 \cdot 10^{15} cm^{-3}$. A thin layer of gold has been evaporated on the sample surface to realize a Schottky junction.

The normalized EBIC profiles have been analysed to evaluate m and the related second order moment σ^2; besides, they have been elaborated following the asymptotic method (Davidson et al 1980; Ioannou et al 1982).

3. RESULTS AND DISCUSSION

The σ^2 vs z' diagrams for two values of the beam current are shown in Fig.2. The z' values have been evaluated according to the expression $z' = 0.41 \cdot R$ with R the primary electron range (in μm). The diagrams $\sigma^2(z')$ shown in Fig.2 bring into evidence that the slope does not depend on the beam current I_b. To check the results obtained by the moment method, the EBIC profiles have been analysed also with the asymptotic method. In this theory the general expression

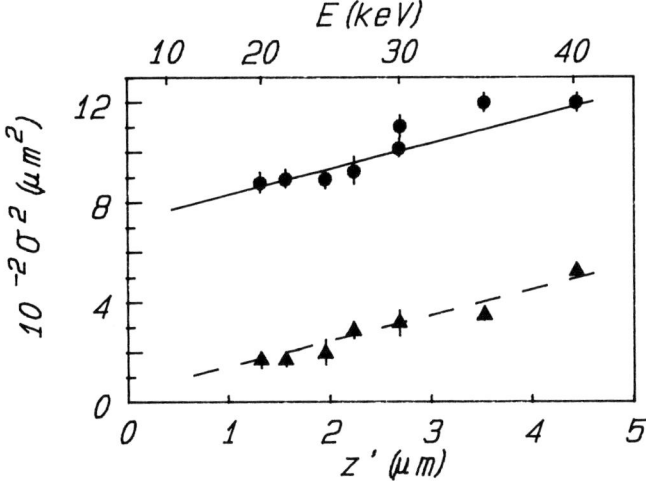

Fig2. Variance σ^2 of the derivative of the current profile as a function of the penetration depth z' for $I_b = 5 \cdot 10^{-9} A$ (•) and $I_b = 5 \cdot 10^{-10} A$ (▲). The lines represent the least-squares fit to the experimental data (see table 1).

of the EBIC profiles, valid for large x'/L values, is:

$$I(x') \propto \frac{\exp(-x'/L)}{x'^p} \tag{4}$$

where $p = 1/2$ for $s = 0$ and $p = 3/2$ for $s = \infty$. Thus, the semilogarithmic plot of $I(x') \cdot x'^p$ vs x' shows a linear trend, whose slope gives the diffusion length. The application of the model is shown in Fig.3, where the two calculated diagrams refer to the same experimental profile.

As can be observed, the choice of the p value strongly influences the evaluation of L and it is sometimes difficult to decide which of the two hypotheses on s is the correct one.

From the results obtained by the moment method it's possible to have some information on the surface recombination velocity value. In fact the adimensional parameter w, which is a function of s, has been evaluated from the values of σ_s^2 by means of eq.(3) (the values are reported in table I). Since the w values are very close to zero, the hypothesis of large surface recombination velocity seems to be the most reliable. This result is corroborated by the good agreement with the diffusion length value obtained with the asymptotic method for $s = \infty$.

The dependence of w on I_b can be interpreted as a consequence of the variation of the surface recombination velocity s with the injection dose δp, that is the density of the injected carriers. As a matter of fact, s depends on both the surface state density and the surface charge, and the variation of the beam current I_b results in a δp change. Therefore, an effect of δp on s, and thus on σ_s^2, will affect the value of w as well.

Table 1. Mean values of diffusion length obtained by applying the models described in the text. The mean value of L over a large number of measurements is reported for the asymptotic method.

Model	Hypothesis on s	$< L > (\mu m)$	$I_b(A)$	w
moment	not required	110 ± 20	$5 \cdot 10^{-10}$	$\simeq 0$
	"	104 ± 7	$5 \cdot 10^{-9}$	0.14
asymptotic	$s = 0$	53 ± 8		
	$s = \infty$	106 ± 19		

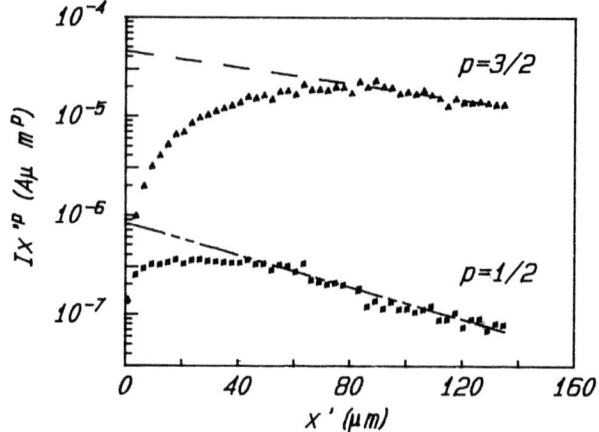

Fig3.   Diagram of $I x^p$ vs beam-junction distance x. The data have been obtained at $E_b = 30 keV$ and $I_b = 5 \cdot 10^{-9} A$ (see Table 1)

4. CONCLUSIONS

Evaluating intrinsic properties, such as integral quantities, of the EBIC profiles presents a double advantage: first of all, such quantities are evaluated from profile data at low scan distances, where the current is high and known with good accuracy; furthermore, these quantities are independent of the choice of the pure diffusion process origtrein, that is from the actual position of the electrical junction. Besides, the moment method does not require assumptions about the s value, and L does not depend on s. The moment method evaluations and the comparison with the usual asymptotic technique results have allowed us to advance an hypothesis on the value of s: in the present experimental condition the most reliable value is $s \approx \infty$.

Another relevant finding of the present contribution is the following. The second order moment σ^2 is the sum of two terms: the first, σ_s^2, depends on s and L, while the second, Lz', only depends on L but not on s. In other words, the contribution of the surface is only present in the first term of the expression of σ^2. This has made it possible to deduce that, for the injection conditions used, s varies with the injection dose, while L does not depend on this parameter, that is the injection conditions do not affect the bulk properties, while they influence the surface.

ACKNOWLEDGEMENTS

The authors gratefully acknowledge Dr. C. Donolato for helpful discussion. This work was supported by MPI grants of the Italian Ministry of Education.

REFERENCES

Davidson S M and Dimitriadis C A (1980) Journal of Microscopy, **118**, 275.
Donolato C (1985) Solid State Electron. **28**, 1143.
Holt D B and Joy D C (1989) SEM Microcharacterization of Semiconductors
   (Academic Press, London 1989).
Ioannou D E and Dimitriadis C A (1982) IEEE Trans. Electron Devices **ED-29**, 445.
Kuiken H K and van Opdorp C (1985), J.Appl.Phys. **57**, 2077.
von Roos O (1978) Solid State Electron. **21** 1069.

Inst. Phys. Conf. Ser. No 117: Section 10
Paper presented at Microsc. Semicond. Mater. Conf., Oxford, 25–28 March 1991

Can stereo-EBIC images provide useful information?

A Bergonzoni and U Valdrè

Electron Microscopy Centre, Department of Physics of the University,
Via Irnerio 46, I-40126 Bologna, Italy.

ABSTRACT: It is shown that pairs of EBIC images, recorded at two different specimen tilts with respect to the electron beam, may provide, when examined in a stereoviewer, additional qualitative information not obtainable from flat EBIC images, e.g., on the 3D-distribution of the defective structure present in the sample and on the geometry of single defects. Furthermore it is possible to recognize the component of the EBIC contrast due to crystal defects from that due to spurious surface features, and to measure the depth of single defects.

1. INTRODUCTION

EBIC (Electron Beam Induced Current) imaging is a well established technique and its basic contrast mechanism is rather well understood, so that numerically computed images of defects of various type compare well, at least qualitatively, with experimental images. Various parameters affect the image contrast, especially the angle between the electron beam and the normal to the specimen surface, which alters the shape, size and orientation of the generation volume of the electron-hole (e-h) pairs created by the beam (Donolato 1978). The fairly large dependence of image contrast on non-geometrical parameters has probably been the reason why the Stereo-EBIC image technique has not, to our knowledge, received attention. However there seem to be good reasons for attempting the recording and analysis of Stereo-EBIC images: in fact, the geometrical contribution to the contrast may be important (particularly for defects lying normal or parallel to the specimen surface), although EBIC images are not just the result of the effect of geometrical factors. Hence the Stereo-EBIC technique should provide additional information on defect position and type, which is not available in conventional EBIC images. We therefore decided to record and analyse Stereo-EBIC pictures to verify if 3D-information is obtainable and, this being the case, to assess the usefulness of the method and its limits.

2. MATERIALS AND METHODS

Three different types of silicon specimens were used:
    (a). Si single-crystal (n-type), (100) oriented, resistivity 3.7 $\Omega\cdot$cm, deformed by bending in Argon atmosphere at 730 C, on which a Schottky barrier was prepared by deposition of an evaporated gold film about 25 nm thick.
    (b). Si single-crystal (n-type), (111) oriented, resistivity 100 $\Omega\cdot$cm; deformation and Schottky barrier preparation the same as for sample (a).
    (c). Si single-crystal (p-type), (100) oriented, resistivity 1 $\Omega\cdot$cm,

implanted with P⁺ to form a planar p-n junction, and containing oxidation induced stacking faults and other defects.

Observations were made in a Philips scanning electron microscope SEM 515, which is delivered with a standard stereo stage of eucentric type. The tilt axis is normal to the microscope front; x,y,z stage traverses and stub rotation (which is normal to the x,y plane and, approximately, parallel to the z-traverse) are all built inside the stereo-axis mechanism. Since z-traverse is obtained by means of a cantilever, the effective motion takes place along arcs of a circle. The angular range of the stereo stage is from +60° (anticlockwise) to -8° (clockwise).

Stereo-pair micrographs are taken by inclining the specimen, between successive exposures, around the tilt axis of equal angles (± θ) on both sides with respect to the electron beam axis; non-geometrical effects which may contribute to the contrast are in this way minimized. We have found, empirically, that the optimum tilt range is ± 10° to ± 15°. Parallax differences of about 1 mm can be obtained in stereo-images magnified 800-1000x, which is a convenient magnification considering the EBIC resolution is approximately 0.5 μm. Such a rather large parallax difference is necessary, even if it is measured by means of a micrometer, because EBIC images of defects are not sharp, hence measurements are affected by a large systematic error (see Sect.3.4). Angles higher than 15° have not been used to avoid excessive distortion of EBIC images due to loss of symmetry in the e-h pair generation region. Although the data given by Hudson and Makin (1970) for TEM stereo-imaging are not strictly applicable in the present case, our 10°-15° angular range is consistent with that corresponding to TEM images of equal magnification.

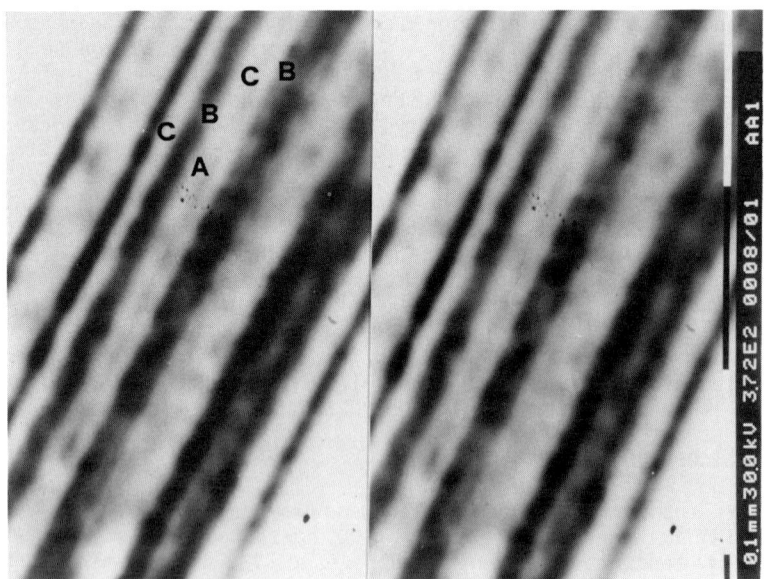

Fig. 1. Example of a pair of stereo-images of sample (a). Tilt axis, vertical; Tilt angle ± 15°; V = 30 KV. A, Upper surface details; B,C, high and low dislocation density bands respectively. Marker = 100 μm.

Specimens should be initially (i.e., before tilting) mounted horizontally in the eucentric stereo stage, so that the junction, or the barrier, plane will be perpendicular to the beam. When trying to satisfy this requirement three problems are encountered which are due to the special geometry of the SEM 515 stereo stage. First, the 8° maximum clockwise tilt angle of the stereo stage is too small with respect to the optimum tilt range. This problem has been overcome by mounting the specimen in a special tapered stub whose top surface normal is 7° off with respect to that of the standard stub and the z-level of the tapered stub centre is the same as that of the standard stub at $\theta = 0°$. The stub is placed in the stage with its slope parallel to and containing the stereo-tilt axis. At a nominal tilt $\theta = +7°$ the specimen is therefore horizontal and at eucentric level.

Second, the stage is no longer eucentric with respect to the y-axis when the tapered stub is used. Refocusing and magnification compensation are therefore necessary, as well as having to track the specimen area on tilting. Trying to restore the initial specimen level is not convenient.

Third, in order to reduce possible risks of artifacts, the plane formed by stereo- and z-axes should be horizontal, otherwise the plane slope shows up when the images are stereoviewed. Horizontality is satisfactorily checked by means of a water-level and corrected by the z-controller.

3. RESULTS

3.1 Defective Structures at Slip Planes

Figure 1 shows a stereo-pair which refers to specimen (a). It is an example of the possibility that stereo-EBIC imaging offers of evaluating the location of dislocation arrays on slip planes and of distinguishing between dust and/or scratches lying at the specimen top surface or at the interface between gold and silicon. Stereoviewing shows clearly and simultaneously the separation between details (A), which may be located at the surface (or at the Au/Si interface) and dislocation bands in the bulk aligned along slip planes, hence simplifying the correlations between various details. Although this information can be obtained from secondary electron images (SEI), SEI contrast might be lower than in EBIC images and either images cannot be obtained simultaneously, unless the microscope is equipped with two viewing screens (or a dual split screen); in each case a careful comparison between EBIC and SE images is required. Figure 1, in addition, shows that high dislocation density bands (B) are intercalated with bands (C) of low dislocation density with a 3D structure such as that sketched in fig.2, representing a cross-section of the bands. Such an undulated structure is produced by the motion of dislocations on (111) slip planes during the deformation process. The wavelike structure has been smoothed out at the surface by the polishing of the sample in a CP4-A bath, performed before depositing the gold layer. The reason(s) why it is not

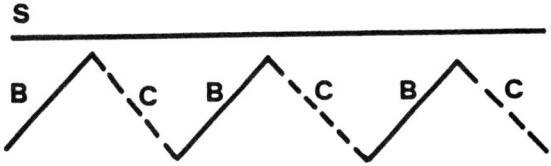

Fig. 2. Cross-section sketch of dislocation bands of Fig.1. S, Top surface; B,C, Bands with high and low dislocation density respectively.

visible up to the surface is not clear, since the space charge region is of the order of a few tenths of μm (it might be due to a high surface recombination, or to the value of the diffusion length; an effect connected to the sensitivity of parallax values can be excluded).

3.2   Detection of the Depletion Region

Stereo-pair of Fig.3 from specimen (b) shows again an (apparent) absence of defects in a layer between surface and slip planes. In this case, however, the junction is deeper than for specimen (a) of Fig.1 and the apparently depleted region is probably the space charge region of the Schottky barrier. Wu and Wittry (1978) have reported a barrier depth of about 2 μm for specimens of doping level similar to sample (b). The high collection current of this region quite likely obliterates the signal producing dislocation visibility. Defects are concentrated in narrower bands than in specimen (a) and their 3D shape is therefore less noticeable.

3.3   Structure of Single Defects

Figure 4 shows stereo-EBIC images of oxidation induced stacking faults in specimen (c). These defects lay on (111) planes and are bound by a 1/3 Frank partial dislocation of semicircular shape. The characteristic EBIC image and contrast of these defects, narrow and sharp at the upper limbs, large and smooth in the middle part, depends on their depth (Donolato 1979). Stereo-EBIC images clearly show this structure.

3.4   Depth Measurements and Assessment of the Type of Various Defects

Stero-pair of Fig. 5 has been used to measure the depth, z, of several details, numbered 1 to 8, in sample (c). The depth is referred to point P situated at the specimen top surface. In order to gather information on

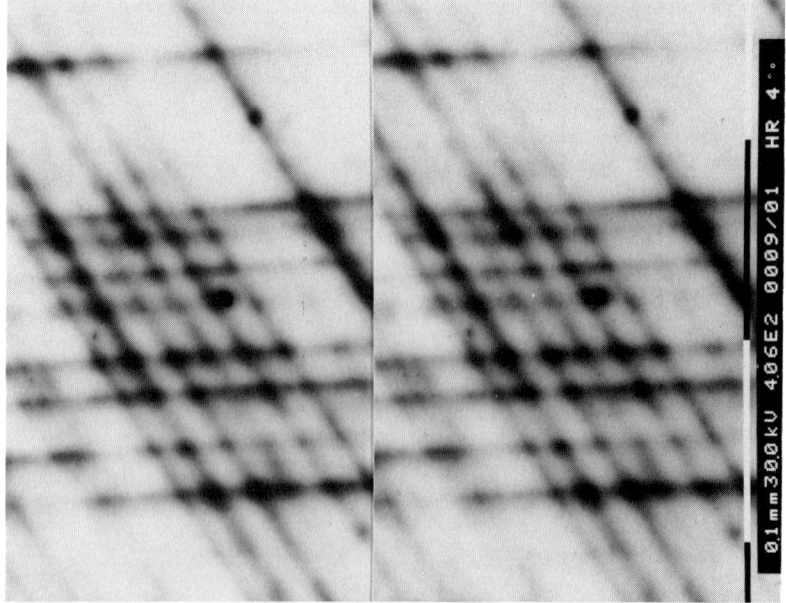

Fig. 3. Pair of stereo-images of sample (b). Tilt axis, vertical; Tilt angle, ± 10°; V = 30 kV; Marker = 100 μm.

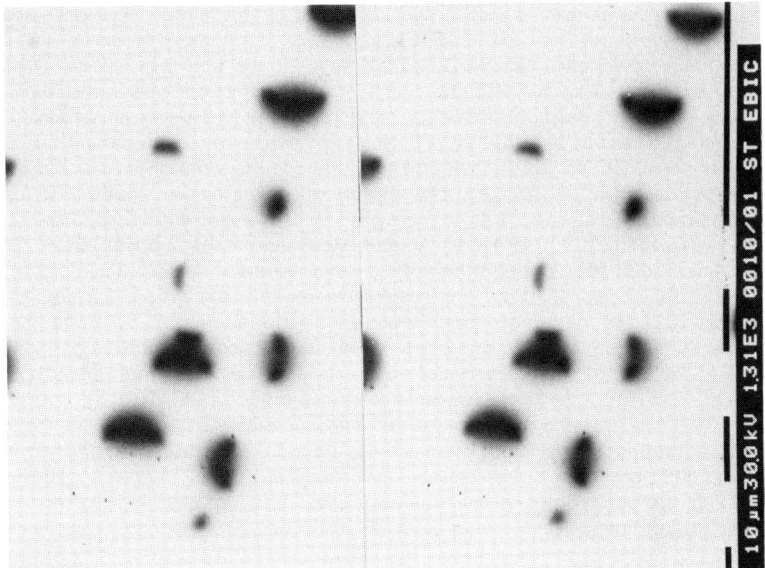

Fig. 4. Example of a pair of stereo-images of sample (c) containing stacking faults on (111) planes. Tilt axis, vertical; Tilt angle, ± 15°; V = 30 kV; Marker = 10 μm. (Specimen kindly supplied by Dr L. Pasemann).

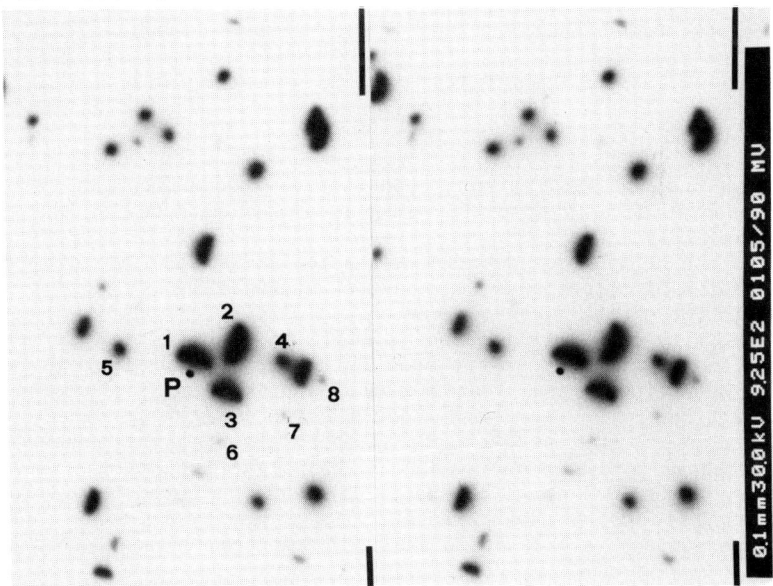

Fig. 5. Example of a pair of stereo-images of sample (c) showing stacking faults 1,2,3 on (111) planes, dislocations 4,5 running perpendicular to the specimen surface and point-like defects 6,7,8. Tilt axis, vertical; Tilt angle, ± 15°; V = 30 kV; Marker = 100 μm. (Specimen kindly supplied by Dr L. Pasemann).

each defect, stereo-pairs of the same region were taken and depth measurements repeated for different values of the accelerating voltage, V. The Table shows the results obtained at V = 20 kV and V = 30 kV for defects numbered 1 to 8 in Fig.5. Depths were measured from the specimen surface down to the defect region showing the sharpest image contrast. Details 1,2,3,4,5 (depth extended defects) show a systematic dependence of their depth on increasing V, as expected by the EBIC contrast mechanism which strictly depends on V, i.e., on the depth of the region where e-h pairs are mainly generated. A change of V results in an apparently different defect level, corresponding to the level where EBIC contrast is maximum. This fact is responsible for the presence of a systematic error in depth values, estimated to be about 0.5 μm for depth-extended defects. Systematic depth changes with voltage have not been found for defects 6,7,8. Combining depth information with the shape of most common defects (Donolato 1978, 1979, 1980) it can be concluded that defects 1,2,3 are stacking faults, whereas 4,5 are dislocations perpendicular to the surface and 6,7,8, having their length smaller than the sensitivity of the measurements, may be referred to as point-like defects. In addition it is found that defects are confined in a region extending down to a depth not greater than 2.7 μm; since the p-n junction lies about 2.8 μm from the surface, the defective region is within the ion implanted n-side.

TABLE. Measured depth z (μm) of defects 1-8 performed at 20 and 30 kV. All values are affected by the same statistical error ± 0.1 μm (standard deviation) over 6 repeated measurements for each defect.

Detail No	1	2	3	4	5	6	7	8
20 kV	1.9	1.1	1.6	1.6	1.4	0.8	1.0	1.4
30 kV	2.3	1.5	1.9	2.0	1.9	0.9	1.0	1.3

5. CONCLUSIONS

Qualitative and, to some extent, quantitative information can be obtained by the observation of stereo-EBIC images, as demonstrated by the examples given above. They refer to:
(i)   Location of spurious details at the top surface and at interfaces.
(ii)   Location of slip planes in arrays of dislocations.
(iii) Detection of the depletion region.
(iv)   Capability of distinguishing between point-like defects and dislocations normal to the specimen surface; capability of improving the detection of the structure of a single defect.
(v)   Detection of the layer within which defects may be confined.
(vi)   Depth measurements. For depth-extended defects, the measured values are affected by a systematic error of about 0.5 μm, due to the voltage dependence of the EBIC image contrast. It should be noted that in order to perform accurate depth measurements it is necessary to have a reference mark at the surface (e.g. surface dust or scratches) of negligible height.

REFERENCES

Donolato C 1978 Optik 52 19
Donolato C 1979 Scanning Electron Microscopy I 257
Donolato C 1980 J. Appl. Phys. 51 1624
Hudson B and Makin M J 1970 J. Phys. E.: Sci. Instrum. 3 311
Wu C J and Wittry D B 1978 J. Appl. Phys. 49 2827

Inst. Phys. Conf. Ser. No 117: Section 10
Paper presented at Microsc. Semicond. Mater. Conf., Oxford, 25–28 March 1991

733

Quantitative EBIC investigations of deformation-induced and copper decorated dislocations in silicon

T S Fell and P R Wilshaw

Department of Materials, Parks Road, Oxford OX1 3PH, United Kingdom

ABSTRACT: Deformation induced dislocations in silicon have been found to exhibit different degrees of electrical activity dependent upon whether they were deformed at 420°C or 650°C. Furthermore the effects of copper decoration are also found to vary between dislocations produced at these temperatures. Quantitative analysis of EBIC measurements in terms of the charge at dislocations reveals this behaviour to be due to different defect states.

1. INTRODUCTION

The charge controlled model of recombination at dislocations developed by Wilshaw (1984) enables information about fundamental dislocation parameters to be derived from quantitative EBIC contrast measurements. From these measurements, made as a function of both temperature and incident electron beam current, data such as the density of occupied dislocation states and their position in the band gap can be determined (Wilshaw and Fell 1989). Due to the EBIC technique's high spatial resolution ($\approx 1 \mu$m), these measurements provide a distinct advantage over bulk analysis techniques such as DLTS and EPR when attempting to identify the particular features of the structure of dislocations that cause them to be electrically active.

In this paper results are presented which reveal significant differences in dislocation recombination activity dependent on the temperature at which they were produced and also on the presence of copper impurities. These results are interpreted in terms of the recombination model and the likely causes of the electrical activity observed are discussed.

2. ANALYSIS OF RESULTS

For brevity, the results presented here are described only in terms of the general trends predicted by Wilshaw's recombination theory. For a more detailed description of the method used to quantitatively analyse the EBIC results see Wilshaw and Fell (1989).

The important features of the theory with regard to this work are summarised in the theoretical curves of EBIC dislocation contrast versus temperature shown in Fig. 1. In this diagram parameters such as the dislocation level position E_0, and the concentration of defect states N_D, have been chosen to illustrate the different regimes of behaviour on two curves. Wilshaw (1984) found that in regime 1 the dislocation charge (which is proportional to dislocation EBIC contrast), is mainly determined by the recombination process such that the dislocation states are not all occupied. In regime 2, for a small N_D, all the dislocation states can be occupied without the Coulombic potential which this produces bending the conduction and valence bands sufficiently to raise the defect states to the Fermi level. Thus once in regime 2 the charge on the dislocation and hence its EBIC contrast is approximately independent of temperature. Regime 3 occurs when enough states are occupied to raise the dislocation level to the Fermi level where it becomes pinned. Under these circumstances, as the Fermi level moves towards the centre of the band gap with increasing temperature, the occupancy of the pinned dislocation level adjusts accordingly to reduce the dislocation charge, hence reducing the dislocation EBIC contrast.

Therefore from dislocations demonstrating Regime 2 behaviour quantitative analysis of the dislocation charge can determine the concentration of states N_D, and establish that the position of the states E_0, must be below that of the Fermi level at the temperatures measured. Conversely, from dislocations demonstrating Regime 3 behaviour quantitative analysis reveals E_0 but can only establish a lower limit for N_D, the number of states available for occupation.

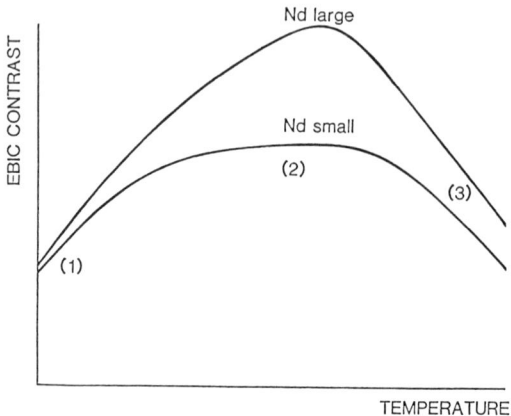

Figure 1. Theoretical Curves of EBIC dislocation contrast vs. temperature.

3. EXPERIMENTAL

EBIC measurements have been performed using a system similar to one described previously (Wilshaw et al 1983) which is now based around a Philips 505 SEM with a LaB_6 gun.

In a two stage deformation procedure, which has been described in detail elsewhere (Wilshaw and Fell 1989), dislocations are nucleated in float zone, swirl free, n-type $10^{15}cm^{-3}$ silicon ingots by a 900°C pre-deformation compression along a [213] axis. A main deformation is then performed at either 650°C under a resolved shear stress (RSS) of 30MPa or at 420°C under a RSS of 300MPa. Under these conditions the dislocation sources activated in the bulk of the ingots during the predeformation generate many concentric hexagonal loops on (111) planes. In order to investigate the effects of copper impurities dislocation recombination activity some ingots were given a single, gentle wipe with a copper wire immediately prior to the main deformation.

4. RESULTS

Previous quantitative analysis of EBIC measurements was performed by Wilshaw (1984) on dislocations deformed under high stress at 420°C by Alexander and co-workers in Cologne (Wessel and Alexander 1977). These dislocations demonstrated relatively high levels of recombination activity with EBIC contrast of typically 12% being measured. From contrast measurements as a function of temperature these dislocations were found to demonstrate Regime 1 and 2 behaviour. Quantitative analysis revealed this activity to be caused by a deep defect state >0.52 eV from the conduction band edge E_c with an $N_D \approx 2.9 \times 10^6 cm^{-1}$.

4.1 Results Comparing 420°C and 650°C Dislocations.

The dislocations deformed in Oxford under high stress at 420°C have been found in the present work to demonstrate behaviour identical to those originally investigated by Wilshaw, indicating identical states. As these two results were obtained from dislocations produced by different groups using different starting material this provides some evidence that impurity decoration is not a significant cause of the electrical activity of these dislocations since there would inevitably be differences in the purity of the starting materials and deformation apparatus used. In addition, intentional introduction of impurities through incomplete cleaning procedures also showed results inconsistent with the 420°C activity being due to impurities.

The dislocations deformed under low stress at 650° C demonstrated much lower levels of electrical activity than their 420° C counterparts, with EBIC contrast measurements typically ≤2%. It is also reported, for the first time, that contrast vs temperature measurements on these 650° C dislocations have shown them to demonstrate both Regime 1 and Regime 3 behaviour, i.e. their dislocation states become pinned to the Fermi level as the Fermi level moves towards the centre of the band gap, see Fig. 2. Quantitative analysis of these results found the dislocation level responsible for this activity to have an occupation limit 0.34-0.37 eV below E_c.

4.2 Results Comparing 420° C and 650° C Copper Decorated Dislocations.

The 420° C copper decorated dislocations demonstrated Regime 1 and 2 behaviour identical to the uncontaminated 420° C dislocations described above and consistent with $N_D \approx 2.9 \times 10^6 cm^{-1}$ and $E_o > 0.52$ eV, see Fig. 3. However, the 650° C copper decorated dislocations, whilst still exhibiting Regime 3 behaviour showed considerably higher activity than the uncontaminated 650° C dislocation with typical contrast levels of $\approx 8\%$. Quantitative analysis of these results found a level with occupation limit 0.48-0.51 eV below E_c.

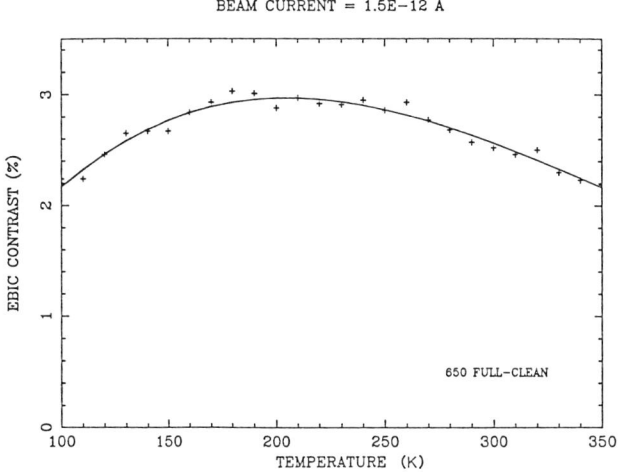

Figure 2. EBIC dislocation contrast vs. temperature curve for a 650° C uncontaminated dislocation.

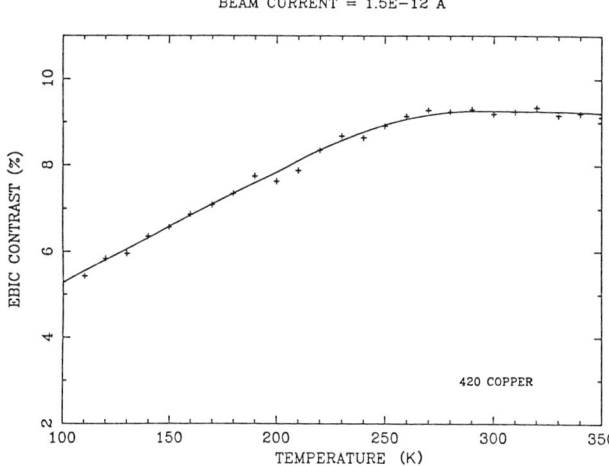

Figure 3. EBIC dislocation contrast vs. temperature curve for a 420° C copper decorated dislocation.

5. DISCUSSION

The dislocations showing lowest electrical activity investigated in this work were the uncontaminated, 650°C deformed dislocations. The states present at these dislocations were found to be sufficiently shallow that when charged they were readily raised by their coulombic potential to intersect the Fermi level. Decorating these 650°C dislocations with copper introduced a deeper impurity level in sufficient numbers so that it was pinned to the Fermi level. The effect of this deeper copper level becoming pinned is to raise the original shallow 650°C state completely above the Fermi level thus rendering it inactive as a recombination centre and also we assume, undetectable by techniques such as DLTS.

Deformation at 420°C produces a very deep state in the band gap but in insufficient numbers to raise it to the Fermi level. However, when all these 420°C states are occupied the Coulombic potential they produce is sufficient to raise any shallower states associated with the dislocations, including the copper impurity level if present, above the Fermi level thus rendering them inactive as recombination sites. Hence the identical nature of all the 420°C dislocations, irrespective of differences in impurity decoration, is fully consistent with the charge controlled model. This result shows that the commonly held view that the higher the concentration of impurities at dislocations the higher the recombination activity, is not always correct. This model however only remains valid for impurity decoration below the level required for precipitation at the dislocation in which case the activity may then be due to the precipitates themselves.

In conclusion the demonstration of Regime 1 and Regime 3 behaviour by the 650°C dislocations has, for the first time, experimentally confirmed all 3 regimes of the charge controlled model of recombination at dislocations.

With regard to identifying the cause of the electrical activity at dislocations three conclusions can be drawn:

1) High stress deformation at 420°C produces a very deep deformation induced level (> 0.53eV below E_c), in low concentrations, These states appear intrinsic to dislocations produced at these temperatures and are probably associated with some structural aspect of dislocations.

2) Copper decoration of dislocations induces a relatively deep (0.48-0.51eV) extrinsic impurity level at the dislocations. However for 420°C dislocations these states lie above the Fermi level and hence do not contribute to the recombination activity.

3) A shallow (0.34-0.37 eV) defect level of unknown origin is present at 650°C deformed dislocations. This may be due to residual impurities present and is lifted above the Fermi level when copper is added.

REFERENCES

Wessel K and Alexander H 1977 Phil. Mag. 35 1523
Wilshaw P 1984 D. Phil. thesis Oxford University
Wilshaw P and Fell T 1989 Inst. Phys. Conf. Ser. 104 85
Wilshaw P, Ourmazd A and Booker G 1983 J. Physique 44 C4-445

Inst. Phys. Conf. Ser. No 117: Section 10
Paper presented at Microsc. Semicond. Mater. Conf., Oxford, 25–28 March 1991

Characterisation of dislocations in the presence of transition metal contamination

V Higgs*, C E Norman **†, E C Lightowlers* and P Kightley***

*Department of Physics, King's College, Strand, London WC2R 2LS, UK
**Department of Materials, Imperial College, London SW7 2BP, UK
* * *Department of Materials, University of Liverpool, Liverpool L69 3BX, UK
†Present Address, MASPEC Institute, Via Chiavari 18/A, 43100, Parma, Italy

ABSTRACT: Dislocations generated in high purity FZ silicon and stacking faults (OISF and ESF) in CVD silicon have been characterised by EBIC, PL and TEM. The samples were intentionally contaminated with copper or nickel with surface contamination levels between 0.003-10 monolayers. EBIC contrast is observed from all the contaminated defects, whether decorated on the atomic scale or by precipitates. As the level of contamination is increased to a degree such that precipitation occurs the dislocation related luminescence is lost and non-radiative recombination dominates. The injection level dependence of the EBIC contrast from both lightly (i.e no precipitates) copper contaminated dislocations and from precipitate bearing dislocations bounding the stacking faults is observed to follow the form predicted by the Wilshaw-Booker theory of dislocation EBIC contrast.

1. INTRODUCTION

Dislocations produce pronounced effects on the electrical properties of silicon and are known to produce deep levels in the energy gap. A range of techniques has been applied to determine the electronic nature of extended defects and to establish a relationship between the observed electrical properties and the deep levels. The cause of their activity is still not clearly understood. Dislocation related photoluminescence bands (D-bands) observed from silicon containing dislocations have been attributed to a wide range of causes, but their precise origins remain unclear. The role that impurities play must be considered, especially the transition metals which are extremely fast diffusers and which produce deep levels within the band gap. The effect of the transition metal distribution around the dislocation must also be taken into account, whether present as an impurity cloud or as precipitates.

We have previously demonstrated (Higgs *et al* 1990) that dislocations free from contamination ($< 10^{11} cm^{-3}$) give no measurable EBIC contrast and no dislocation related luminescence, whilst dislocations in material contaminated with low levels of copper ($\approx 10^{13} cm^{-3}$) show both strongly. In the present study, dislocations produced in high purity FZ silicon and oxidation induced (OISF) and epitaxial (ESF) stacking faults in CVD grown silicon were intentionally contaminated with copper and nickel. The effect of contamination was investigated as a function of concentration over the range 0.003 to 10 monolayers.

2. EXPERIMENTAL

Deformation-induced dislocations were produced in samples cut from an ingot of ultrapure FZ silicon, with a very low level of transition metal contamination ($< 10^{11} cm^{-3}$), using a specially designed stress cell made entirely from quartz, in which the samples could be heated and stressed uniaxially. The density and distribution of dislocations was determined using preferential defect etching.

Both commercially grown CVD epitaxial (n/n^+ and n/p^-)layers with different thicknesses (3-15μm) and epitaxial layers grown using low temperature and low pressure (thickness$\approx$1-15 μm) have already been fully characterised using a wide range of techniques (Higgs *et al* 1991) This enabled a matrix of samples to be selected which contained varying densities of ESFs with different levels of transition metal contamination. In addition, epitaxial layers containing low levels of metal contamination and no defects (ESFs, dislocations or S-pits) and high purity FZ silicon wafers were oxidised under clean conditions to produce OISF's.

Copper and nickel were introduced onto the sample surface by backplating from ultra-pure hydrofluoric acid. The concentration of the metal ions in the plating solution was determined by atomic absorption spectroscopy, and the surface concentration of metal contamination has been checked by total X-ray reflection fluorescence. The samples were annealed in RCA cleaned quartz furnace tubes in flowing nitrogen gas.

Photoluminescence (PL) measurements were made at 4.2K with the luminescence excited by an Argon laser tuned to 514nm. The luminescence signal was analysed by a Nicolet 60 SX Fourier transform spectrometer and detected using a cooled North Coast germanium diode detector. After PL characterisation the same samples were further characterised by TEM.

EBIC investigations were performed in a JEOL JSM-840A SEM using a Matelect ISM5 EBIC amplifier. The EBIC signal was collected using electron transparent Schottky barriers or, where appropriate, the p-n junction already existing in the sample. The barriers were evaporated onto the polished (111) face of the plastically deformed silicon so that the dislocations intersected the sample surface beneath the barriers.

3. RESULTS AND DISCUSSION

No D-band luminescence was observed from the OISFs in the FZ silicon or from stacking faults in epilayers grown at low temperature and low pressure. All the commercially grown epilayers that contained ESFs showed weak D-bands (D1 and D2 only), whereas D-band luminescence was only observed from the layers grown by low temperature CVD at abnormally high growth pressures (>120 torr). A detailed analysis of the metal impurities contained in the epilayers revealed that the D-band luminescence was observed only when there was a relatively high level of transition metal contamination (Higgs *et al* 1991). TEM investigation of the epilayer samples revealed no variations in the structure of the ESFs. In addition, there was no evidence for metal precipitation at the ESFs.

Subsequent contamination of the epilayer samples with either copper or nickel at a low level (≈ 0.003 monolayers) produced a significant increase in the intensity of the D-bands (D1-D4). In addition, D-band features could now be observed from the epilayers that prior to intentional contamination showed none. The same trend was observed for the samples that contained OISFs; Following metal contamination the OISFs became luminescent. For all the defects examined it was found that as the level of contamination was further increased (> 0.1 monolayers) the D-band features decreased in intensity until they could no longer be observed.

These results are consistent with our previously reported work on dislocations (Higgs *et al*

1990) that transition metal contamination plays a crucial role in determining the luminescence properties of the defects. We observed that the as-deformed samples had no observable D-band luminescence and no detectable EBIC contrast (see Figure 1a). Following deliberate contamination ($\approx$ 0.003 monolayers) the D-bands were observed and the dislocations became visible in EBIC. The EBIC contrast of dislocations in a copper contaminated sample of FZ silicon plastically deformed at 750^0C (dislocation density $\approx 10^7 \text{cm}^{-2}$) is shown in Figure 1b. TEM measurements showed no evidence of precipitation on the dislocations or in the surrounding strain fields down to the 20Å level.

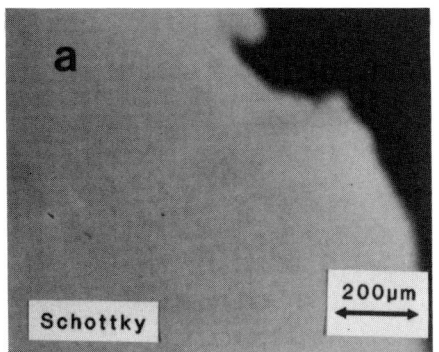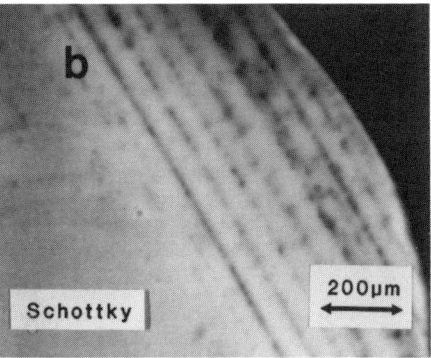

Figure 1. EBIC contrast from a) an uncontaminated and b) a Cu contaminated (0.003 monolayers) sample of FZ Si plastically deformed at 750^0C.

TEM investigations have revealed that there are two distinct regimes of decoration present at the defect. When the level of contamination is less than 1 monolayer there is no evidence of precipitation along the dislocation lines. At higher levels, TEM reveals the presence of metal related precipitates along the dislocation lines. The effect of the degree of decoration (ie. the physical nature of the contamination) of the defects on both the electrical and optical activity has been investigated further. An epilayer sample containing OISFs that had been contaminated with a high level of nickel ($\approx$ 1 monolayer), and which now showed no observable D-bands, was inspected by TEM. Figure 2a shows a TEM micrograph of an OISF in this sample. It is clear that there is precipitation on the partial dislocation bounding the stacking fault. On closer inspection of Figure 2a an hexagonal platelet is identifiable, which is indicative of nickel type precipitates. EDX was performed on these precipitates and confirmed the presence of nickel. Figure 2b shows an EBIC micrograph of an OISF in this nickel contaminated sample. The partial dislocations bounding the stacking fault are clearly visible. An epilayer specimen containing ESFs which had also been nickel contaminated ($\approx$ 1 monolayer) to a level at which it showed no D-band features was also analysed by TEM. Nickel precipitates were again detected on the stacking fault partials. An EBIC micrograph showing two such ESFs is presented in Figure 2c, and as in the case of the OISF, the partials can be clearly observed. Similar results were obtained for the case of copper contamination. Both results suggest that when there is precipitation on the bounding partials, dislocation related luminescence is lost.

Another epilayer sample containing ESFs contaminated with a low level of nickel ($\approx$ 0.003 monolayers) and which showed D-bands was also examined. TEM revealed no evidence of precipitation but the ESFs could still be detected in EBIC. Figure 3 shows an EBIC micrograph of such an ESF.

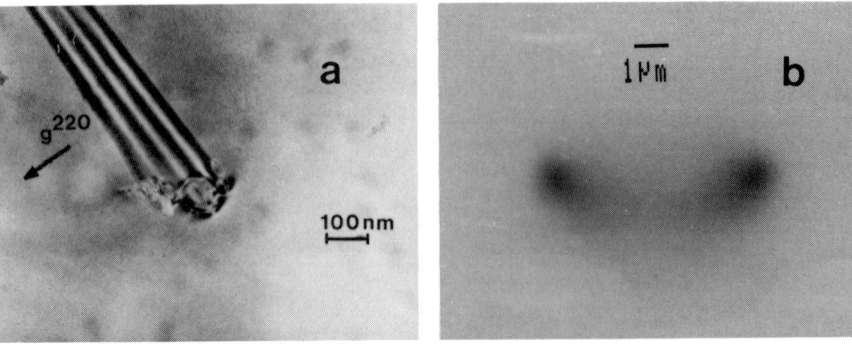

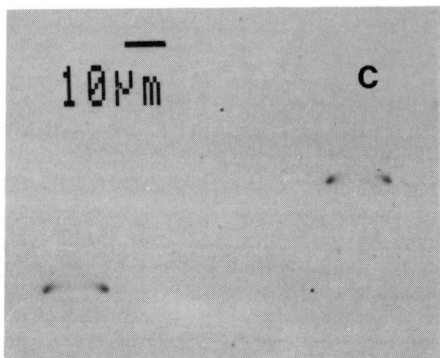

Figure 2. a) (001) Plan view TEM showing part of an OISF with Ni related precipitates on the partial bounding the fault; b) EBIC micrograph of an OISF in a heavily Ni-contaminated (1 monolayer) specimen; c) EBIC micrograph showing ESFs in a heavily Ni-contaminated (1 monolayer) specimen.

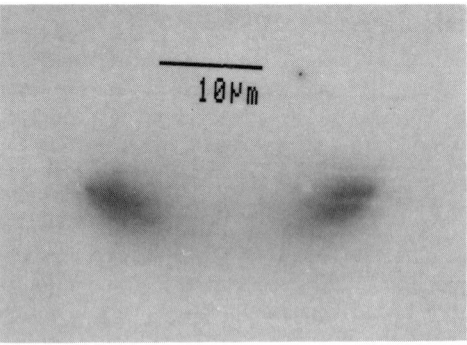

Figure 3. EBIC micrograph of an ESF in a lightly Ni-contaminated (0.003 monolayers) specimen.

A more detailed EBIC investigation was carried out to examine the effect of the two regimes of dislocation decoration. Figure 4 shows the variation of EBIC contrast with incident electron beam current (I_b) for a dislocation in the lightly copper contaminated (0.003 monolayers) FZ deformed sample. The incident beam electron energy used was 15keV. There are two distinct regimes of contrast versus $\log(I_b)$ behaviour. At low beam currents ($I_b < 10^{-9}$ A) the contrast is roughly independent of I_b. At higher beam currents, the contrast decreases proportional to $\log(I_b)$. Similar behaviour was also observed for the contrast from OISFs with copper precipitates on the bounding partial dislocations, an example of which is also shown in Figure 4. This form of behaviour can be described by the Wilshaw-Booker theory of dislocation EBIC contrast (Wilshaw and Booker 1985). This theory predicts that at high beam currents there will be a change in the rate of carrier recombination at the dislocation. This arises because at high injection levels the hole capture rate is so fast that the band bending around the dislocation line must decrease for electron capture to stay apace. The theory also predicts that the dislocation contrast, which is assumed to be proportional to the dislocation recombination strength and thence the degree of band-bending around the dislocation line, will fall linearly with $\log(I_b)$.

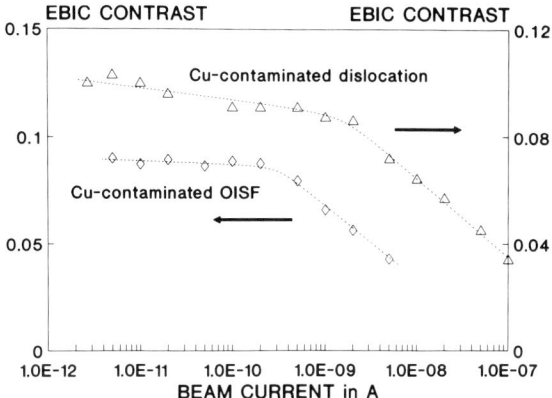

Figure 4. EBIC contrast versus $\log(I_b)$ for a lightly Cu-contaminated (0.003 monolayers) dislocation and a heavily Cu-contaminated (1 monolayer) OISF.

The Wilshaw-Booker theory was devised and tested on EBIC contrast data obtained from dislocations parallel to the sample surface, whereas in our specimens the dislocations are threading up to the sample surface. It is well known that geometrical considerations, such as defect depth and defect position relative to the edge of the depletion region, can greatly affect the magnitude of the EBIC contrast which will be measured from otherwise similar dislocations (Pasemann *et al* 1982). In order to test the validity for our specimen surface/dislocation orientation we examined a specimen kindly supplied by Alexander of the Universität zu Köln, which was produced using the same technique and apparatus as the specimens initially investigated by Wilshaw and Booker (1985) but with the dislocations in a similar orientation to those in our samples.

The variation of EBIC contrast with incident beam current for a dislocation in this specimen is shown in Figure 5 and can be seen to follow the same behaviour that has been observed both for inclined dislocations in the intentionally contaminated FZ silicon and for surface parallel dislocations investigated initially by Wilshaw and Booker in similar specimens. For comparison, data recorded from a screw dislocation by Wilshaw and Booker (1985) is also plotted in Figure 5. At low beam currents the contrast decreases very slowly with increasing beam current but at higher beam currents the contrast decreases more rapidly, proportional to $\log(I_b)$. We therefore conclude that our results on inclined dislocations are valid.

It is important to note that strong dislocation related luminescence is observed from the sample supplied by Alexander. Our work therefore suggests that this sample contains transition metal contamination.

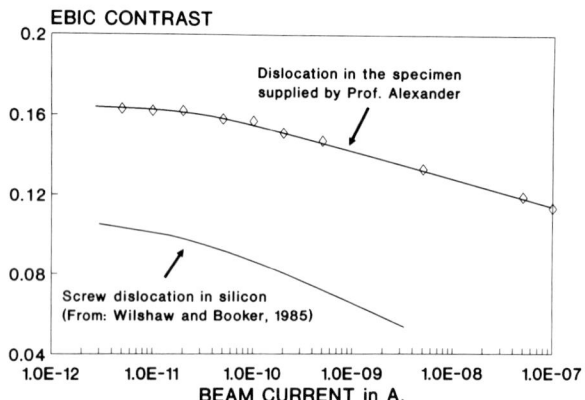

Figure 5. EBIC contrast versus $\log(I_b)$ for an inclined dislocation in the specimen deformed by Alexander. Data obtained by Wilshaw and Booker (1985) from a surface-parallel screw dislocation in a similarly deformed specimen are shown for comparison.

ACKNOWLEDGEMENTS

The authors would like to thank Prof. H. Alexander (Universität zu Köln) for supplying the deformed silicon specimen and extend their gratitude to Dr. D. B. Holt (Imperial College) for making the sample available for analysis. In addition, we (VH and ECL) would also like to thank the SERC for financial support.

REFERENCES

Higgs V, Norman CE, Lightowlers EC and Kightley P, Proc. 20th Int. Conf. on the Physics of Semiconductors, eds E M Anastassakis and J D Joannopoulos, Vol. 1, pp 706-709.
Higgs V, Goulding M and Kightley P, to be published 1991.
Pasemann L, Blumtritt H and Gleichmann R 1982 Phys. Stat. Sol. (A) **70** 197.
Wilshaw PR and Booker GR 1985 Inst. Phys. Conf. Ser. No.**76** 329-336.

Inst. Phys. Conf. Ser. No 117: Section 10
Paper presented at Microsc. Semicond. Mater. Conf., Oxford, 25–28 March 1991

743

EBIC investigation of charged dislocations in Si

I E Bondarenko and E B Yakimov

The Institute of Microelectronics Technology and High Purity Materials USSR
Academy of Sciences, Chernogolovka

ABSTRACT: The dependence of EBIC contrast of charged dislocations with
different impurity atmospheres on electron beam current has been studied. The
results obtained have been discussed taking into account the effect of centres
in impurity atmosphere on dislocation recombination properties.

1. INTRODUCTION

The EBIC contrast of one and two-dimensional defects strongly depends on electron
beam current. It was observed for dislocations by Wilshaw and Booker (1987) and
Bondarenko and Yakimov (1987), grain boundaries by Sundaresan et al (1984) and
dislocation slip planes by Bondarenko et al (1986) in Si. In all cases the dependencies observed were explained by the influence of a space charge region near the
charged defects on EBIC contrast and by the dependence of the stationary value
of defect charge on electron beam excitation level. Measurements of V-A characteristics of the microcontact with dislocation etch pits showed that a space charge
cylinder was formed near dislocations in n-Si introduced at temperatures $T_D < 700°$
C and after annealing at temperatures higher than $850°$ C these cylinders disappeared
(Eremenko et al 1975, 1978). So, dislocations introduced at rather low temperature
capture the charge carriers and are charged whereas those annealed at higher temperatures are uncharged. Comparison of these results with those obtained by EBIC
(Bondarenko and Yakimov 1987) shows that the dependence of dislocation EBIC
contrast on electron beam current can only be observed for charged dislocations and
may be considered as evidence of the proposed explanation of EBIC contrast dependence on electron beam current. But our investigations have shown (Bondarenko
and Yakimov 1988) that dislocation EBIC contrast depends also on the dislocation
impurity atmosphere state. In our opinion this result gives the possibility to develop
a more appropriate model to describe the EBIC contrast of charged dislocations. In
the present paper the results of EBIC investigations of charged dislocations in Si
are presented. Qualitative explanation of these results which takes into account the
dislocation impurity atmosphere is proposed.

2. EXPERIMENTAL

The investigations were carried out on two Cz-grown n-Si crystals (Cz-1 and Cz-2) with resistivity $\rho = 10$ and 7.5 Ohm · cm, respectively. The Cz-2 crystal had a higher oxygen concentration than the Cz-1 one. In addition, two Fz n-type Si crystals were used, one of which had $\rho = 10$ Ohm · cm. The other with $\rho = 15$ Ohm · cm was additionally doped with gold to a concentration of $8 \cdot 10^{13}$ cm $^{-3}$. All the crystals were doped with phosphorus. This set of samples was chosen for the purpose of varying the state of dislocation impurity atmosphere over a wide range because this state, as shown by Bondarenko et al (1980), is essentially dependent on the impurity content of the as-grown crystals as well as on the conditions of plastic deformation and subsequent thermal treatment.

Isolated semihexagonal loops were introduced into the samples by four-point bending around the $< 112 >$ axis at $600°$ C. $60°$ segments of these halfloops intersecting the $\{111\}$ surface at $57°$ were investigated. To control the state of impurity point defect atmosphere the value of starting stresses τ_{st} was measured on the same samples. EBIC investigations were carried out on Schottky barriers formed by gold evaporation on the chemically prepolished surface of the samples. The contrast value was calculated as $C = (I_o - I_d)/I_o$, where I_o and I_d are induced current values when the electron beam is far from the dislocation and on it, respectively.

3. EXPERIMENTAL RESULTS

The dependencies of dislocation contrast on electron beam current in the samples with different impurity content are shown in Fig.1. The contrast values for gold-free

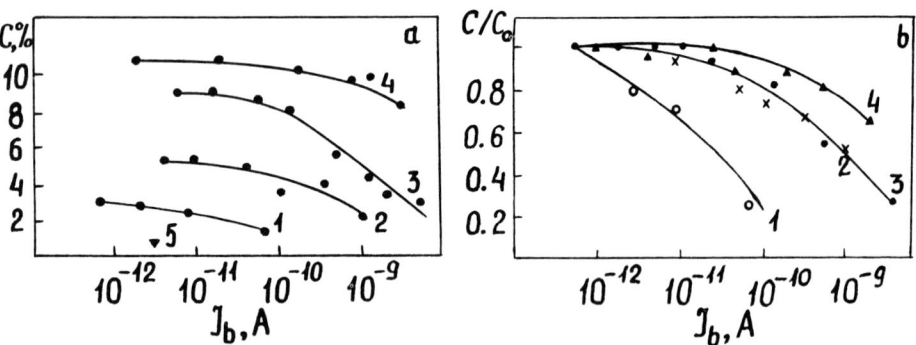

Fig.1. Dependence of EBIC contrast on beam current (a) and normalized dependencies for the same samples (b). Curve 1 is for Cz-1; 2 is for Fz-Si$< Au >$; 3,4 are for Cz-2 (curve 3 for quenched sample); 5 is for Fz-Si .

Fz-Si did not generally exceed 1% which prevented us from studying the EBIC contrast dependence on beam current with sufficient accuracy. The contrast for all other samples decreases with increasing beam current. It should be noted that for gold-free Cz crystals the contrast value increases with increasing τ_{st}. This shows that an increase of oxygen-related centre concentration in the dislocation impurity atmosphere leads to increase of its recombination activity. The difference between curves 3 and 4 obtained for the dislocations introduced under the same conditions but cooled under different conditions confirms this most distinctly. Indeed, quenching of deformed samples or cooling them under load leads to a decrease of the contrast value and changes its dependence on beam current. At the same time quenching results in a significant decrease of starting stress due to a decreasing number of point defect complexes near the dislocation (Bondarenko et al 1980). The changes observed show that dislocation contrast is associated with such complexes. In gold-doped Fz-Si τ_{st} is small but gettering of gold by dislocations may be revealed by DLTS (Aristov et al 1987) and this is the reason for the contrast increase observed.

4. DISCUSSION

The space charge cylinder (SCC) around the dislocation should result in increasing minority charge carrier capture on the dislocation under electron beam excitation. Therefore a question about the existence of this cylinder is very important. Analysis of the results of plastically deformed Si investigations by the Hall effect (Eremenko et al 1978) and the diode effect on individual dislocations (Eremenko et al 1975) shows that an electrostatic barrier with a height of about 0.3-0.4 eV is formed near dislocations introduced at temperatures below 700° C. The investigations of the temperature dependence of charge carrier concentration and dislocation density at which conductivity type inversion in n-Si was observed show (Eremenko et al 1978) that in n-Si the SCC radius is the same for both Cz-Si and Fz-Si with equal phosphorus concentration. Therefore the difference in EBIC contrast is associated with impurity atmospheres which, as it was shown by Bondarenko et al (1980), were formed near the dislocation during plastic deformation. These atmospheres contain oxygen atoms along with intrinsic point defects and fast diffusing impurities, the number of which depends on the crystal impurity content.

Let us consider the recombination processes on a charged dislocation. Under electron beam excitation the hole flow to the dislocation is

$$I_p = k_1 I_b \qquad (1)$$

The electron flow is

$$I_e = c_e N_D (1 - f) N_d \exp(-e\phi/kT) - c_e N_D f \, N_C \exp(-E_1/kT), \qquad (2)$$

where c_e is the probability of electron capture on the dislocation state, N_D is the density of centers along dislocations, N_d is the shallow impurity concentration, f is the filling factor of dislocation centers, ϕ is the barrier height, k_1 is a constant, N_C is the energy state density in conduction band, E_1 is the depth of the dislocation energy level. In the stationary state $J_e = J_p$, hence

$$\phi = \frac{kT}{e} \ln \frac{c_e N_D (1 - f) N_d}{c_e N_D f \, N_C \exp(-E_1/kT) + k_1 I_b} \qquad (3)$$

From (3) it is easily seen that at small beam currents I_b, ϕ maintains its equilibrium value and at higher I_b, $\phi \sim a - \ln I_b$. Thus, it is easy to explain the dependence $C(I_b)$ assuming that the contrast value is proportional to ϕ. The only question is the contrast dependence on impurity content. To explain this dependence Bondarenko and Yakimov (1990) assumed that some dislocation related centres can effectively take part in the recombination processes and do not practically change the radius of the space charge cylinder around the dislocation. In this case a change in the concentration of such centres will involve a change of part of the total flow of holes, recombining through N_D centres and, in turn, affect the contrast value as well as the dependence of the filling factor of N_D centres on I_b and, hence, the I_b range in which the barrier height is independent of I_b . Such properties may be associated with the centers situated near the edge of SCC. But if we take into account the probability for minority charge carrier to be captured inside space charge cylinder we should assume that the concentration of traps is $\sim 10^{17} - 10^{18} \mathrm{cm}^{-3}$. A recent DLTS study of dislocations shows (Koveshnikov et al 1991) that traps around dislocation are distributed as $N \sim N_o \exp(-r/a)$, where N_o is about $10^{15} \mathrm{cm}^{-3}$, r is the distance from the dislocation and a is about $1 \mu m$. Therefore such trap distribution can not explain the results in the frame of model by Bondarenko and Yakimov (1990) but gives us a possibility to propose another explanation. It should be taken into account that under the exitation level used the barrier can decrease and traps outside the barrier can be saturated. The lower the trap concentration, the smaller the value of I_b at which they are saturated. Therefore an increase of I_b leads to decrease of region in which traps are not saturated and effects the recombination processes. This determines the dependence of contrast on I_b . An increase of trap concentration increases the contrast C as well as I_b value at which C begins to decrease with I_b .

REFERENCES

Aristov V V, Bondarenko I E et al 1987 Phys. Stat. Sol.(a) 102 687

Bondarenko I E, Eremenko V G, Nikitenko V I and Yakimov E B 1980 Phys.Stat.Sol. (a) 60 341

Bondarenko I E and Yakimov E B 1987 Izv. AN SSSR, ser.Fiz. 51 703

Bondarenko I E, Blumtritt H et al 1986 Phys. Stat. Sol.(a) 95 173

Bondarenko I E and Yakimov E B 1988 Solid State Phenomena. 1-2, 59

Bondarenko I E and Yakimov E B 1990 Phys.Stat.Sol.(a) 122 121

Eremenko V G, Nikitenko V I and Yakimov E B 1975 Zh.Eksp.Teor.Fiz. 69 990 [Sov.Phys. JETP. 1976 42 503]

Eremenko V G, Nikitenko V I, Yakimov E B and Yarykin N A 1978 Fiz.tech. poluprov. 12 273 [Sov.Phys.Semicond. 12 157]

Koveshnikov S V, Feklisova O V et al 1991 Phys.Stat.Sol. to be published

Sundaresan R, Fossum J G and Burk D E 1984 J.Appl. Phys. 56 964

Wilshaw P R and Booker G R 1987 Izv. AN SSSR, ser.Fiz. 51 1582

Inst. Phys. Conf. Ser. No 117: Section 10
Paper presented at Microsc. Semicond. Mater. Conf., Oxford, 25–28 March 1991

Low-voltage EBIC imaging of doped microstructures

H Blumtritt and U Werner
Institute of Solid State Physics and Electron Microscopy, Weinberg 2
O-4050 Halle (Saale), Germany

ABSTRACT   The principles and potential applications of a combined Schottky barrier and p–n junction EBIC technique are presented, which allows imaging of submicron doped VLSI structures through application of low beam energies (2–5 keV). On the basis of a two-dimensional Monte-Carlo simulation of EBIC profiles the lateral dimensions of the imaged structures can be determined with accuracies of better than 0.05 μm, which is demonstrated for the measurement of the effective channel length of MOS transistors.

1. INTRODUCTION

The conventional EBIC technique (electron beam induced current) in scanning electron microscopy (SEM) is subjected to several, well-known restrictions if applied to the investigation of complex, highly integrated microcircuits. This results from the structured surface layers and from the fact that most of the active p–n structures are not directly connected with the contact pins so that it is not possible to pick up the generated EBIC signals. Furthermore, a sufficient lateral resolution for imaging the submicron structures would require the preparation of windows in the surface layers or of cross section samples to expose some suitable p–n structures to the surface and to be able to use the low beam energy necessary for high resolutions (Leamy 1982). An alternative imaging technique – the combined p–n junction and Schottky barrier EBIC technique, presented in this paper – may overcome most of the restrictions and extend the possibilities of the EBIC technique in the field of studying VLSI MOS devices (Blumtritt 1988). Furthermore, it uses a simple geometry of charge collection thus making it easier to theoretically model EBIC signals for quantitatively interpreting the measurements.

2. PRINCIPLES OF TECHNIQUE AND PREPARATION

The basic idea of the low-voltage EBIC technique is the use of a highly electron-transparent metal electrode on the naked silicon substrate which fulfils a double function: Together with the lower-doped areas of bulk conductivity type it forms a surface Schottky barrier – as usually used for EBIC investigations of crystal materials without p–n junctions – and it simultaneously makes parallel ohmic contacts to all highly doped active structures of opposite conductivity. In this way, a continuous charge-collecting space charge region (SCR) of locally varying depth position and thickness is obtained (Fig. 1), enabling all doped microstructures to be imaged independent of their dimensions and positions within the large metal-covered region of the microcircuit chip. As the first step of the quite simple preparation all surface layers of the circuit have to be removed – either by selective etching or even by only one lift-off etching with HF in an

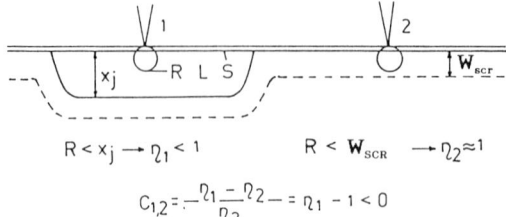

$$R < x_j \rightarrow \eta_1 < 1 \qquad R < W_{SCR} \rightarrow \eta_2 \approx 1$$

$$C_{1,2} = \frac{\eta_1 - \eta_2}{\eta_2} = \eta_1 - 1 < 0$$

ultrasonic bath. After rinsing the substrate in distilled water and methanol the Schottky barrier metal is evaporated up to a thickness of a few 10 nm; as usual, Al is used for p-type substrates and Au for n-type ones. As backside contact a Au-Ga compound is scratched in.

Fig. 1 Charge collection geometry of the technique and definition of the quantities used

3. IMAGING CONDITIONS AND CONTRAST CALCULATIONS

The original idea of improving the resolution consists in the reduction of the lateral extent of the generation volume by using very low beam energies. Simultaneously, optimum imaging conditions for the shallow implanted p-n structures are met, if the generation volume (or the primary electron range R) lies fully within the layer to be imaged (see Fig. 1). The low-voltage condition can be defined by $R < X_j$ and $R < W_{scr}$ and for today's VLSI circuits it is attained between 3 and 5 keV beam energy (X_j up to some 100 nm). The lowest limit is given by the thickness of the metal layer and the working stability of the SEM. In practice, with a conventional thermal gun SEM it was possible to carry out EBIC investigations of p-bulk (n-channel) MOS circuits at 1.5 to 2 keV, with the beam diameter already limiting the lateral resolution. If the condition $R < W_{scr}$ is fulfilled, then the Schottky barrier produces a maximum EBIC (no recombination assumed within SCR, collection efficiency nearly 1) with all the other structures appearing with   dark contrast. Still on the basis of the well-known expressions for the collection efficiency of extended planar p-n junctions (for example Kittler and Schröder 1983) the contrast dependence on the beam energy E_b, junction depth X_j, diffusion length L, and surface recombination velocity S can be calculated for microstructures to be as small as 1 µm. The plots in Fig. 2a reveal that except for very small L it is the surface recombination (at the metal-semiconductor contact in the p-n regions) that considerably or even mainly contributes to the contrast. This easily explains some experimental findings of irradiation-dependent contrast enhancements by the assumption that beam-induced interface charges shift S towards higher values. In practice, this effect can only be used if one remains well below the dose of total Schottky barrier damage, therefore the use of digital image storage and signal evaluation systems may become essential for such investigations.

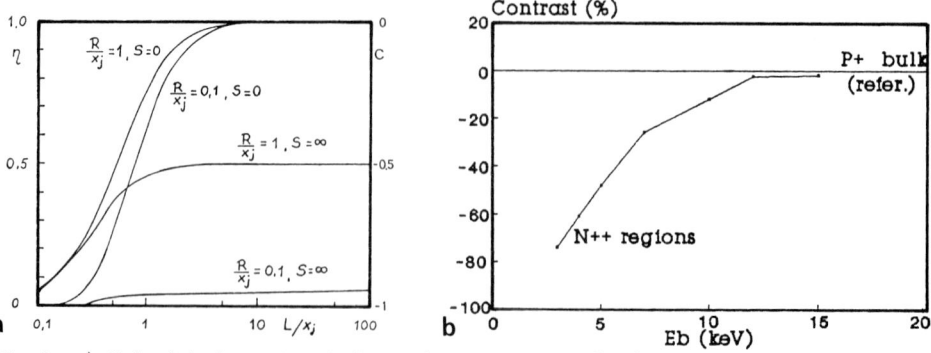

Fig 2. a) Calculated contrast dependence on sample data and imaging conditions; b) Measured for a n^+p structure with $X_j = 0.3$ µm

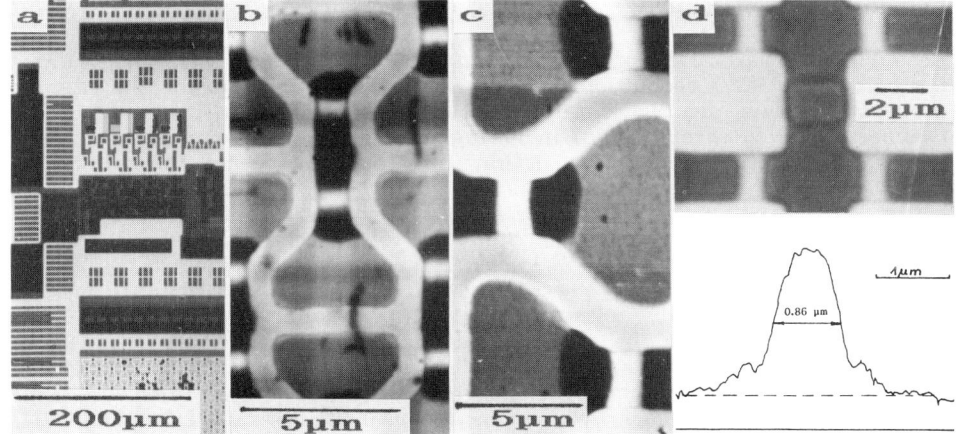

Fig. 3 Applications of the low-energy technique to VLSI microcircuits
   a) Doped structures in a CMOS device, b) Crystal defects in a DRAM
   c) Varying channel length, d) Submicron channels with EBIC profile

4. APPLICATION TO VLSI DEVICE INSPECTION AND FAILURE ANALYSIS

A few examples may illustrate the possibilities of the special EBIC technique
presented (Figs. 3, 4). It most easily allows the detection of crystal defects
and doping irregularities in any wrong-working   inner components of a
microcircuit, preselected, for example, by conventional function tests.
Structures of different doping concentrations clearly are in contrast to each
other (Fig. 3). The strongly doped source and drain regions are imaged with
contrasts of more than 50%, explained by the low L.
For fresh specimens of p-bulk circuits, weakly doped n-regions with large L
(such as the channels of depletion transistors) were found not to have
contrast. After some irradiation during scanning strongly reduced EBIC
occured (=enhanced dark contrast with the bright bulk regions, see Fig. 4).
In some cases, contrasts increased by more than 10 times from near or below
the detection limit up to about 50%, explainable with a shift of S values from
near zero up to quite large values caused by a beam-induced generation of
negatively charged interface states.
Fig. 3d clearly reveals that even submicron enhancement channels are imaged
with the same maximum signal as the longer ones. Up to now, the ultimate
limit has not been attained experimentally, but from contrast profile
simulations it can be concluded to lie between 0.1 and 0.2 μm.

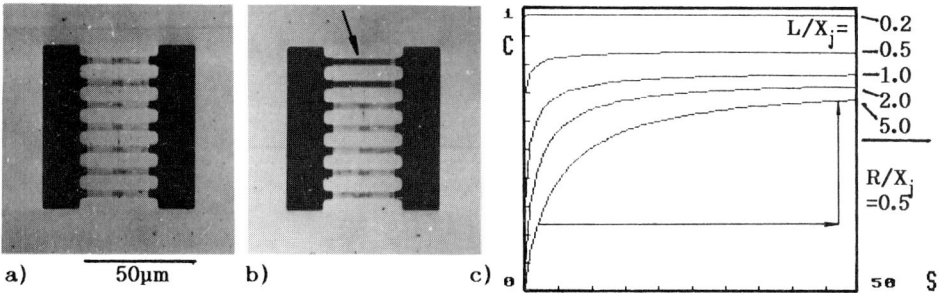

Fig. 4 Irradiation-induced contrast enhancement of slightly doped n-layers
   (depletion channels); a) fresh sample, b) irradiation dose 10^{-3} Ascm^{-2}
   c) explanation by increase of the surface recombination velocity S

5. RESULTS OF EBIC PROFILE SIMULATIONS

Measuring the lateral dimensions of doped microstructures with an accuracy far below the EBIC resolution requires a complete modelling of the imaging process. To this aim a Monte-Carlo simulation programme was developed (Werner and Heydenreich 1988) including a realistic expression of the spatial generation function and a free definition of the charge-collecting SCR; the basic ideas and first applications of measuring the effective channel length l_{eff} of MOS transistors are also presented by Werner and Blumtritt (1991). From sets of EBIC profiles, calculated for the special sample and imaging conditions, correction values W_c were determined, revealing the difference between the half-widths of the EBIC channel profiles and channel lengths looked for. It is possible either to roughly determine l_{eff} (if L and S are entirely unknown) from the calculated dependencies $W_c(L,S)$, (see Fig. 5a), or to accurately determine it on the basis of some additional knowledge of the above quantities. For high contrasts (80%) experimentally observed at the sample of Fig. 3d, for example, L can be determined from calculated C(L) plots to lie between 0.06 and 0.07 μm, yielding as the accurate correction value $W_c=0.175$ +/-0.02 μm (Fig. 5b). The most accurate length of the channel (EBIC profile width 0.86 μm, Fig.3d) is therefore 0.685 +/-0.02 μm.

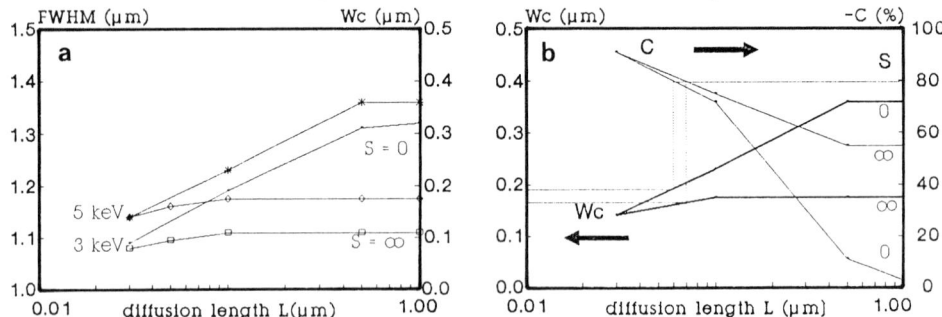

Fig. 5 a) Half-width of profiles simulated for $l_{eff}=1$μm; b) determination of correction values from calculated and measured contrast

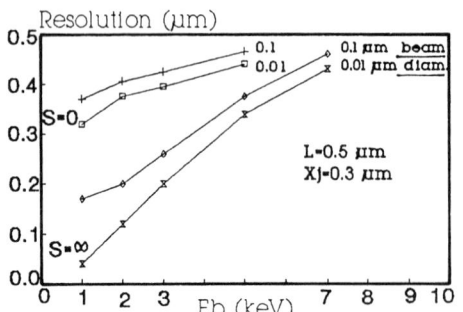

Fig. 6   Calculated influence of the beam diameter on resolution limit

Finally, the limits of resolution (defined as the 10%-90% slope at the profile tails) will be discussed (Fig.6). In the extreme low-voltage operation the beam diameter remarkably worsens the resolution, no longer being negligible compared to the effect of the generation volume. Thus, reducing the beam diameter down to 10 nm (for Epe < 3keV) by using a field-emission SEM should at best improve the resolution in the above given examples to values below 0.2 μm.

REFERENCES

Blumtritt H 1988 Veröff. 12. Tagung "Elektronenmikroskopie" Dresden, 307
Kittler M and Schröder K-W 1983 phys. stat. sol. (a) 77 139
Leamy H J 1982 J. Appl. Phys. 53 R51
Werner U and Heydenreich J 1988 Cryst. Res. Technol. 23 85
Werner U and Blumtritt H 1991 present conference

Inst. Phys. Conf. Ser. No 117: Section 10
Paper presented at Microsc. Semicond. Mater. Conf., Oxford, 25–28 March 1991

EBIC contrast simulation for characterising doped microstructures

U Werner and H Blumtritt

Institute of Solid State Physics and Electron Microscopy,
Weinberg 2, O-4050 Halle(Saale), Germany

ABSTRACT: To improve the information which the Electron Beam Induced Current (EBIC) yields about doped microstructures, a simulation was developed of the generation, propagation and collection of minority charge carriers in silicon targets. This simulation is mainly based on two Monte Carlo (MC) procedures, viz. on one for treating the electron scattering and one for treating the diffusion of minority charge carriers inside a target with a defined microstructure. By evaluating simulated EBIC profiles with respect to the defined microstructure rules can be given allowing a correction of experimental metrological data of doped microstructures.

1. INTRODUCTION

Using EBIC in scanning electron microscopy to get information about the geometry of Space Charge Regions (SCR) is an excellent method of inspecting integrated circuits. In many cases, the EBIC profile – resulting from the electron beam scanning of the SCR – yields only indirect information about the geometric SCR structure. This fact originates from convolutions, the root of which is based on the generation volume of the minority charge carriers and on their diffusion in the semiconducting target. If these processes of convolution are ignored no successful interpretation is possible, especially for relatively small SCR. Hence, on the basis of MC descriptions of these convolution processes a simulation was developed to calculate EBIC profiles of defined SCR structures in silicon. This simulation is applied in order to reveal the dependence of the behaviour of EBIC profiles on the SCR geometry.

2. SIMULATION PROCEDURES

The EBIC simulation is built up of three steps: As the first step the geometric structure of the SCR has to be defined by creating a two-dimensional array via a dialogue with the computer. The segments of this array get a collection probability of 1 or 0 if they are assigned to the SCR or to the region without charge, respectively. In Fig. 1, a defined SCR is depicted which represents a channel structure that is investigated by using a special Schottky barrier technique (Blumtritt and Werner 1991). As the second step of the simulation the computer calculates the distribution of the detection probability of minority charge carriers. The detection probability describes the probability of a minority carrier – which is emitted by a point source – to be detected by the defined SCR (possibly after the carrier has travelled a certain distance to the SCR). This probability is calculated by using a MC simulation of the diffusion process of the minority charge carriers (Werner and Heydenreich 1988): In order to save computing time the transport length of the diffusing minority carriers is increased impact by impact. By considering the basic relation

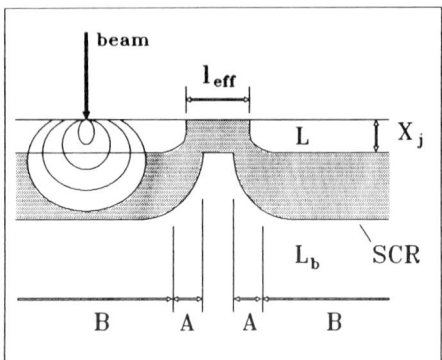

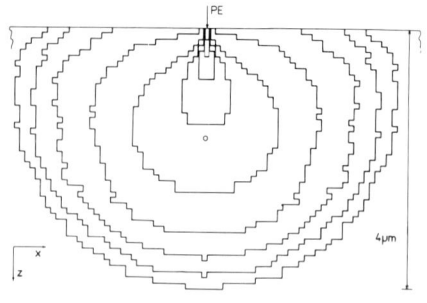

Fig. 1. Sketch of the defined channel structure (grey region, $l_{eff} = 2\mu m$, $X_j = 0.5\mu m$) and the generation volume of minority charge carriers caused by the incident electron beam.

Fig. 2. Generation volume of minority charge carriers at 20 keV calculated by using a Monte Carlo simulation of the interaction of the primary electron beam with a Si-target.

between transport, absorption and diffusion length along the path of the carriers, the phenomenology of the diffusion process is described correctly. Furthermore, the efficiency of the MC simulation is enhanced by simultaneously calculating the detection probability for all point sources lying at the same depth. If the distribution of the detection probability of the channel structure defined in Fig. 1 is simulated as stated above we get Figs. 3a and b. Both distributions are calculated for a relative surface recombination velocity of S = 0, enabling the view from the surface into the depth of the target to be seen. The distribution of Fig. 3a is governed by relatively large diffusion lengths, i.e. a diffusion length of L = 1µm for the region between the surface and the SCR, and of L_b = 3µm for the bulk region (see Fig. 1). In this figure the strong convolution owing to the large diffusion lengths and the disappearing surface recombination causes a slight local change of the detection probability. However, the distribution of Fig. 3b is due to relatively small diffusion lengths, i.e. L = 0.1µm and L_b = 0.3µm. Owing to these small diffusion lengths, contrary to Fig. 3a, here the distribution displays the SCR structure with a high contrast. This stage of the simulation takes a relatively long time (in the order of 1 hour). Therefore, a set of interesting distributions of the detection probability was calculated and stored. On the basis of these stored distributions the last step of the simulation can be accomplished relatively fast (in the order of some seconds) to yield EBIC profiles with dependence on the primary energy and the beam diameter. At this step the distribution of the detection probability has to be convoluted with the distribution of the generation of electron hole pairs, i.e. with the interaction volume of the electron beam incident into the Si-target. Since this volume calculated by a MC simulation of the electron target interaction (see Fig. 2) is not flexible enough for an energy- and a beam-dependent description, a suitable analytic formula was derived by fitting it to the numeric MC results (Werner et al 1988). This analytic formula takes into account the generation caused by the incident spreading electron beam as well as by the diffusion-like propagation of the primary electrons.

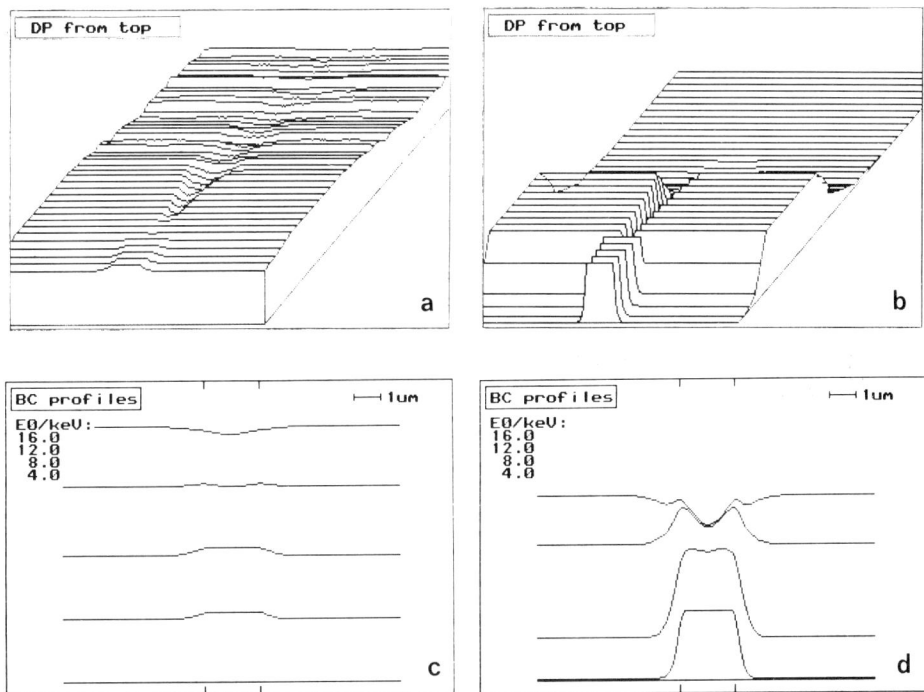

Fig. 3. Distributions of the detection probability (DP) for the SCR depicted in Fig. 1 (a and b) and profiles of the barrier current (BC) for different energies of the scanning electron beam (c and d). Target parameters: $L = 1\mu m$, $L_b = 3\mu m$, $S = 0$ for a and c. $L = 0.1\mu m$, $L_b = 0.3\mu m$, $S = 0$ for b and d.

3. RESULTS AND CONCLUSIONS

The results of the convolution between the distributions of the detection and generation probability, i.e. the EBIC profiles, are shown in Figs. 3c and d for the SCR sketched in Fig. 1, for a beam diameter of 0.1μm and for target parameters given above: These profiles are relatively smooth for large diffusion lengths (Fig. 3c) and relatively rich in contrast for small ones (Fig. 3d). With increasing incident energy the contrast changes in both cases. Figuratively speaking, for lower incident energies the EBIC "perceives" the SCR from the top, whereas for higher incident energies from the bottom. The symmetric maxima appearing in the 12keV-profile of Fig. 3c and in the 12 and 16keV-profiles of Fig. 3d are produced by the part originating from the incident spreading electron beam (see Fig. 2) rather than by the spherical part of the generation volume. This can be concluded from a detailed investigation of the 16keV-peaks: At 16keV the spherical part of the generation volume has a diameter of about 3μm, whereas the half-peak width is lower than 0.5μm (despite the diffusion of the carriers). Thus, the maxima are considered to be mainly formed by minority carriers generated by the incident spreading beam inside region **A** in Fig. 1. This fact revealing the influence of the structure of the generation volume on the EBIC profiles is more pronounced by the amplitudes of the EBIC signal in Fig. 4. Here, the distance between the amplitude maxima is a very rough measure of the length of the channel. To get a practical method of accurately determining the

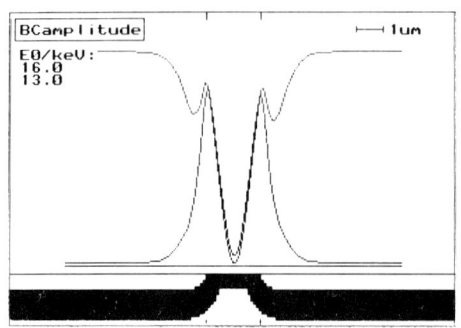

Fig. 4. Representation of stretched amplitudes of EBIC profiles (subtraction of background) shown in Fig. 3d. The defined SCR is outlined in the lower region of the figure.

channel length, the half widths W_h of simulated EBIC peaks were measured at lower primary energies with the complete generation volume lying entirely inside the channel region while the beam scans the channel. Comparing these half widths with the length of the defined channel structure (sketched in Fig. 1, but here l_{eff} = 1μm and X_j = 0.3μm) yields basic data for a correction formula. Since in practice diffusion length and surface recombination are usually unknown quantities this formula was derived to be valid under such conditions. Accordingly, EBIC profiles similar those shown in Figs. 3c and d were simulated for sets of target parameters which were taken from an extended range of diffusion lengths and relative surface recombination velocities ranging from L = 0.03 to 1μm and from S = 0 to ∞, respectively. Despite the supposed absence of L- and S-data the correction formula for the effective channel length derived for a primary energy of 5keV

$$l_{eff} = W_h - (0.25 \pm 0.11)\mu m$$

shows a relatively small uncertainty of ±0.11μm. This fact is not only due to the surface recombination but also (especially if S is small) to the detection of minority carriers by part **B** of the SCR (see Fig. 1). Undoubtedly, this detection causes an EBIC, which, however, does not contribute to the peak but only to the background of the profile. Thus, it is true that the channel structure with its flank sides (indicated by **B** in Fig. 1) diminishes the contrast of the peak but simultaneously also the half-peak width W_h. If a relatively high surface recombination (S near ∞) governs the carrier diffusion at low energies (5keV) the convolution is virtually entirely caused by the small generation volume and hence, the diffusion length has no remarkable influence on the half width. The correction formula attained

$$l_{eff} = W_h - (0.16 \pm 0.02)\mu m$$

consequently possesses a very small uncertainty of ±0.02μm. The range of validity of the above correction formulas can supposedly be extended to other channel lengths (than 1 μm) as long as the generation volume does not simultaneously interact with the two parts A of the SCR (see Fig. 1) during the scanning process. In other words, this only refers to the range of channel lengths with the EBIC peak showing a plateau. But with respect to the depth of the junction (X_j = 0.3μm) as well as to the shape of part **A** of the SCR the formulas are strictly bound to the defined values. The case presented is therefore an example of how the described simulation can be used to find correction formulas if there is first information about the probable structure of the SCR to be investigated. Furthermore, we have shown that a more advanced evaluation of the results of the simulation (considering the peak contrast) further improves the correction (Blumtritt and Werner 1991).

REFERENCES

Blumtritt H, Werner U 1991 present conference proceedings
Werner U, Heydenreich J 1988 Cryst. Res. Technol. <u>23</u> 85
Werner U, Koch F, Oelgart G 1988 J. Phys. D: Appl. Phys. <u>21</u> 116

Inst. Phys. Conf. Ser. No 117: Section 10
Paper presented at Microsc. Semicond. Mater. Conf., Oxford, 25–28 March 1991

EBIC contrast injection level dependence of dislocations in silicon

C E Norman[†] and D B Holt

Department of Materials, Imperial College, London SW7 2BP, UK
[†]Present address: MASPEC Institute, Via Chiavari 18/A, 43100 Parma, Italy

ABSTRACT: The recombination strength of decorated, process–induced dislocations above p–n junctions in silicon photodetectors is found to vary with the incident electron beam current, in the form predicted by the Wilshaw–Booker theory of dislocation EBIC contrast. An injection level dependence of the diffused p–n junction collection efficiency, detectable only under very low injection conditions, is reported for the first time. This is thought to arise due to the phosphorus doping concentration gradient above the p–n junction.

1. INTRODUCTION

The effect of varying the beam–induced carrier injection level on the EBIC contrast of dislocations in silicon photodetector devices has been investigated. Such measurements are important in testing the general applicability of the Wilshaw–Booker theory of dislocation EBIC contrast (Wilshaw and Booker 1987) to decorated, process–induced defects above p–n junctions in addition to cleaner, intentionally–induced defects below Schottky barriers. The dislocations in the present work and their host materials have all been well characterized previously. They include misfit dislocations in Avalanche Photodiodes [APDs] (Lesniak and Holt 1985, Norman and Holt 1989) and Quadrant Photodetectors [QPDs] (Norman 1990), plus dislocations associated with post–anneal damage in implanted QPDs (Holt *et al* 1989, Norman 1990). TEM investigations (Lesniak and Holt 1985) showed the APD misfit dislocations to be decorated with impurities, sometimes in the form of large precipitates which were thought to be phosphorus related. Evidence of dislocation–induced diffusion retardation has also been observed in all the QPD devices (Norman 1990) suggesting that the QPD dislocations are also heavily decorated with phosphorus. High resolution EBIC micrographs of the QPD misfit dislocations suggest that they too are decorated by precipitates. In addition, the dislocations in all the devices will be decorated unavoidably by impurities (eg. transition metals) present in the starting float–zone silicon.

2. EXPERIMENTAL

The EBIC measurements were made using a JEOL JSM–840A SEM equipped with a Matelect ISM–5 EBIC monitor. Early results were obtained with the ISM–5 under manual control, later results were recorded under computer control to avoid unecessary irradiation of the samples. The practical operating beam current (I_b) range of such a combination is approximately $10^{-12}A < I_b < 10^{-6}A$, the lower end of the range governed by the accuracy of the ISM–5 and the upper end limited by using a JSM–840A with tungsten filament. Electrical contacts to the p–n junction specimen devices were made using $25\,\mu m$

gold wire and silver-loaded epoxy. Where possible these contacts were formed on existing bond pad metallization in order to avoid high contact resistances or Schottky-type contact behaviour. Fortunately the specimens which had no such pre-formed bond pads had high surface dopant concentrations which facilitated ohmic contact formation. No Schottky nature was detected in any of the contacts.

3.   RESULTS AND DISCUSSION

Figure 1 shows the variation of EBIC contrast with $\log(I_b)$ for typical examples of the defects found in APDs and QPDs: (i) APD misfit dislocations, (ii) QPD misfit dislocations and (iii) QPD post-implantation dislocations. All of the dislocations show three distinct regimes of contrast versus $\log(I_b)$ behaviour. At low beam currents the contrast increases linearly with $\log(I_b)$, denoted Regime I. At middle-range beam currents the contrast is roughly independent of I_b (Regime II) whereas at very high beam currents the contrast decreases linearly with $\log(I_b)$ (Regime III).

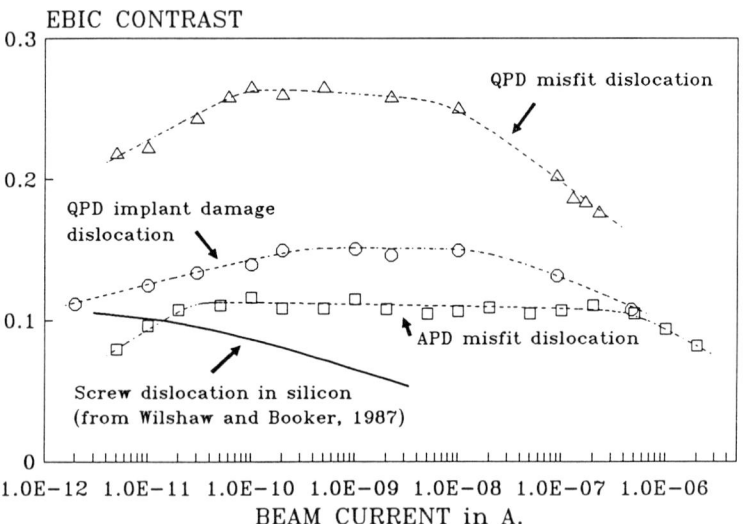

Figure 1.
EBIC contrast versus $\log(I_b)$ for dislocations in APDs and QPDs. Data from Wilshaw and Booker (1987) are also shown for comparison

In QPD devices the EBIC contrast of some dislocations was found to be more sensitive to changes in injection level than that of others, particularly in regime I. Figure 2(a) shows contrast data obtained from a dislocation forming part of the unremoved implantation damage in a QPD device. In regime I the contrast of this dislocation is much more injection-sensitive than the contrast from a similar dislocation represented by the circles in figure 1. The three regimes of contrast behaviour can be clearly seen as the contrast changes by more than a factor of two between its minimum and maximum values. The contrast, C, is given by:

$$C = I^*/I_0$$

where: I^* is the reduction of the EBIC current at the dislocation,
          I_0 is the EBIC signal level measured a large distance from the dislocation.

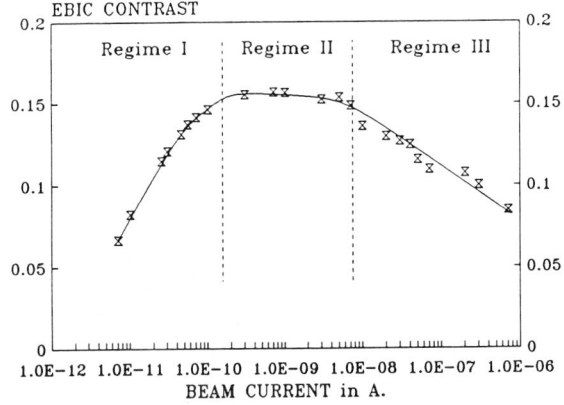

Figure 2(a).
EBIC contrast versus $\log(I_b)$ for a post-implant damage dislocation in a QPD.

The Wilshaw–Booker theory of dislocation EBIC contrast predicts only the behaviour observed in regimes II and III. To examine the contrast behaviour in regime I, the signal current, I_o, and the defect current, I^*, were both normalized to the beam current and plotted against $\log(I_b)$, as shown in figure 2(b).

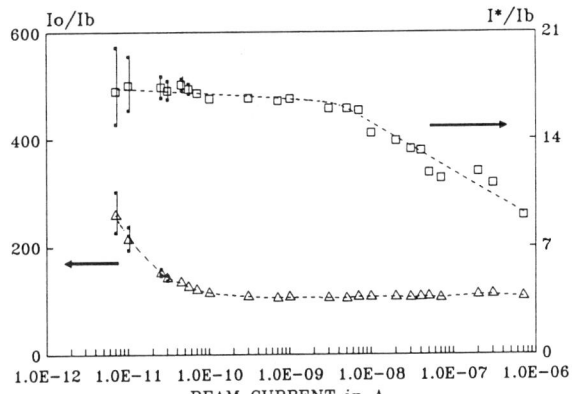

Figure 2(b).
Normalized EBIC signal (I_o/I_b) and defect (I^*/I_b) currents from which the contrast data shown in figure 2(a) was calculated. (NB. Error bars are shown where the error exceeds the size of the representing symbols.)

The rather large error bars arise at low beam currents because I_b becomes of the same order of magnitude as the specified accuracy of the Matelect ISM-5 (ie. $\simeq 10^{-12}$A). On inspection of figure 2(b) it becomes apparent that the contrast behaviour in regime I is wholly accounted for by a dramatic variation in the normalized EBIC signal level. This indicates that in regime I the EBIC contrast is not directly related to the defect recombination strength (Wilshaw 1989).

At low beam currents the collection efficiency of the p–n junction is greatly enhanced. This injection dependence of the normalized EBIC signal level does not appear to have been reported previously for the case of low injection into the surface of a diffused p–n junction. Injection dependences of the signal current for Schottky barrier configurations have been reported (eg. Leamy 1982), but for higher injection levels. They arise due to the screening effect of the neutral plasma of carriers that is formed when the injected carrier density is large. This screening effect reduces the electric field experienced by the carriers in the depletion region under the Schottky barrier (Tove and Seibt 1972).

The enhancement of the p–n junction collection efficiency is believed to be attributable to band bending which occurs due to differences between the doping level at the surface and in regions below the surface. Such band bending gives rise to the bulk electron voltaic effect which has often been observed on cross–sections of differentially–doped devices. The phosphorus concentration gradient obtained from a SIMS depth profile of a QPD device suggests that the electric field thereby produced will be around 100V/cm in the region where the beam–induced carriers are generated, ie. in a layer $\triangleq 1\,\mu m$ thick immediately below the device surface. This field strength is very modest compared to the much larger electric fields which can be developed across depletion regions, suggesting that only at very low carrier injection levels would the screening effects be sufficiently reduced that the minority carriers (holes) can experience the field which encourages them towards the collecting junction.

In regimes II and III the signal current is directly proportional to the beam current. The injection dependence behaviour of regime III therefore arises solely due to a change in the rate of recombination at the dislocation. This is evidenced by the variation of the normalized defect current (I^*/I_b) with increasing I_b. As I_b increases the dislocation can only act as a recombination centre for a progressively smaller proportion of the beam–induced electron–hole pairs. The Wilshaw–Booker theory predicts that at high injection levels the hole capture rate is so fast that the band bending around the dislocation line must decrease for electron capture to stay apace. It also predicts that the contrast (which is assumed to be proportional to the dislocation recombination strength and thence the band–bending around the dislocation line) will fall linearly with $\log(I_b)$ which is found to be true for the defects in the QPD devices.

The onset of regime III behaviour for dislocations in APD and QPD devices occurs at orders of magnitude higher beam currents than for dislocations in the specimen of Wilshaw and Booker (1987). The onset occurs at $I_b \triangleq 5 \times 10^{-7}$A in APDs and $I_b \triangleq 10^{-8}$A in QPDs, compared to $I_b \triangleq 10^{-11}$A in the data of Wilshaw and Booker. It was noted that the phosphorus doping concentrations in the vicinity of all the above defects also vary by similar orders of magnitude, ie. the APD doping is an order of magnitude higher than the QPD doping, which is itself four orders of magnitude higher than the doping in Wilshaw's specimen. This was initially thought to be significant, but recent experiments (Higgs *et al*, This conference) have determined the onset of regime III in very lightly–doped, intentionally copper–contaminated FZ silicon to occur at relatively large beam currents (typically $I_b \triangleq 10^{-9}$A). It therefore appears that the onset of regime III cannot be related in a simple way to the doping density alone. It is much more likely to be a strong function of the number and type of impurity centres and/or precipitates decorating the dislocations.

REFERENCES

Higgs V, Norman CE, Lightowlers EC and Kightley P. (This Conference).
Holt DB, Napchan E and Norman CE. (1989)   Inst. Phys. Conf. Ser. No.104,   Ch.2, 205–210.
Leamy HJ. (1982)   J. Appl. Phys. <u>53</u>, (6),   R51–R80.
Lesniak MP and Holt DB. (1985)   Inst. Phys. Conf. Ser. No.76,   337–342.
Norman CE. (1990) Ph.D. Thesis,   University of London.
Norman CE and Holt DB. (1989)   Inst. Phys. Conf. Ser. No.100, Sect.10,   731–736.
Tove PA and Seibt W. (1972)   Nucl. Instrum. Methods. <u>51</u>, 261–269.
Wilshaw PR. (1989)   Ultramicroscopy. <u>31</u>,   177–182.
Wilshaw PR and Booker GR. (1987)   Izv. Acad. Nauk. Ser. Fiz. <u>51</u>, (9), 1582–1586

Inst. Phys. Conf. Ser. No 117: Section 10
Paper presented at Microsc. Semicond. Mater. Conf., Oxford, 25–28 March 1991

759

Electron microscopical characterisation of the structural and electrical homogeneity of grain boundaries in direct-bonded silicon

H Blumtritt, R Gleichmann, A Höpner and T D Sullivan[+]
Institute of Solid State Physics and Electron Microscopy, Weinberg 2
O-4050 Halle (Saale), Germany
[+]Cornell University, Bard Hall, Ithaca, NY 14853, permanent address:
IBM/GTD, Dept. E63/967-2, Essex Jct., VT 05452, U.S.A.

ABSTRACT: The relation between the recombination activity and the local defect structure was studied by combined SEM/EBIC and TEM on several large – angle twist boundaries produced by hot-pressing. The coincidence boundaries Σ 5 and Σ 13 showed a homogeneously high activity owing to dislocation networks adapting small deviations from ideal coincidence. The 45^0 "non-coincidence" boundary had a low basic activity locally enhanced either by captured deformation-induced dislocations or by punching effects on inclusions in the boundary.

1. INTRODUCTION

Direct bonding (DB) of silicon crystals (Shimbo et al 1986) is a promising technique in future device technology. Up to now little attention has been paid to the structure and the electrical effects of the grain boundaries (GBs) unavoidably forming in the bonded interfaces. Thus it is still an open question, whether or not there exist special orientations with a minimized boundary activity. In the present work, large-angle model grain boundaries have been produced by DB using the earlier developed hot-pressing technique (Föll and Ast 1979). Electron beam induced current (EBIC) and transmission electron microscopy (TEM) were applied to relate the electrical activity of the boundaries to the local defect structure.

2. EXPERIMENTAL

Three types of large-angle twist GBs – the two coincidence boundaries Σ 13 (23^0) and Σ 5 (37^0) and the "non-coincidence" boundary with 45^0 twist angle – have been produced from (100) oriented n-type Czochralski-grown silicon slices of 5x5 mm². The slices had been cut by a wire saw with an accuracy of the orientation of better than 1^0. After careful cleaning they were bonded face-to-face in a high-density graphite die at 1275^0C (radio frequency heating) under a pressure of 6 MPa for 24 hours in Ar-H15% atmosphere. For the electron microscope investigations, from the bonded stacks slices were cut normal to the (100) GB planes and polished to a mirror finish by using diamond paste. After chemical etching in HF/HNO_3 and finally in HF the slices were rinsed in methanol and vacuum-coated with 30 nm of gold to form Schottky diodes. In a conventional scanning electron microscope (SEM) EBIC investigations were performed at a beam energy between 15 and 35 keV to record the local recombination contrasts of the GBs and to select interesting samples for a subsequent TEM analysis. For the chemical thinning of the narrow specimens (5x0.7 mm²) from their backside a special wax masking technique was applied, allowing typically several hundred micrometers of the GBs to be investigated. For this purpose, a JEOL high-voltage TEM was used operating at 1 MeV and equipped with a 45^0 double-tilt goniometer stage. After a first TEM inspection, in some cases the desirable direct EBIC/TEM correlations of identical defect regions could only be ascertained by successive thinning steps by ion milling.

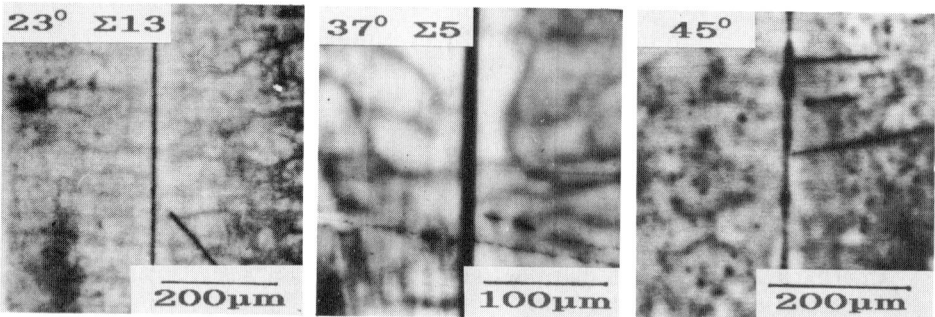

Fig. 1   EBIC micrographs (35 keV) of different twist grain boundaries

3. RESULTS AND DISCUSSION

First of all, various pieces of each GB sample were inspected by low-magnification EBIC in order to assess the homogeneity of the defect contrasts and to select representative GB segments for detailed studies. Fig.1 shows respective images for each type of GB.   While the 23^0 and the 37^0 GBs are characterized by a strong, quite homogeneous recombination contrast of about 40 % (at 35 keV), the 45^0 GB shows distinct fluctuations of the local strength by more than one order of magnitude with the maximum contrast being comparable to the other GBs. Besides this, in the bulk of all samples high densities of deformation-induced dislocations with more or less developed polygonization are observed. Remarkable is the existence of a zone of low dislocation density close to the GBs pointing to repulsive stress fields at the GBs acting on the glide dislocations having originated at the outer surfaces. A TEM micrograph of the 37^0 sample (Fig.2a) demonstrates this zone, too, and indicates captured dislocations as the source of repulsive forces.

It is commonly accepted today that dislocations in heat-treated silicon usually are   impurity-decorated   and   generate   well-detectable   EBIC   contrasts (Heydenreich et al 1981). For the 37^0 GB shown here the captured dislocations cannot be the main source of recombination activity, since their strong density fluctuations are not associated with corresponding fluctuations of the EBIC contrast. Therefore, the major part of the active centres   should be inherent in the boundary itself. Indeed, weak-beam investigations (on a sample

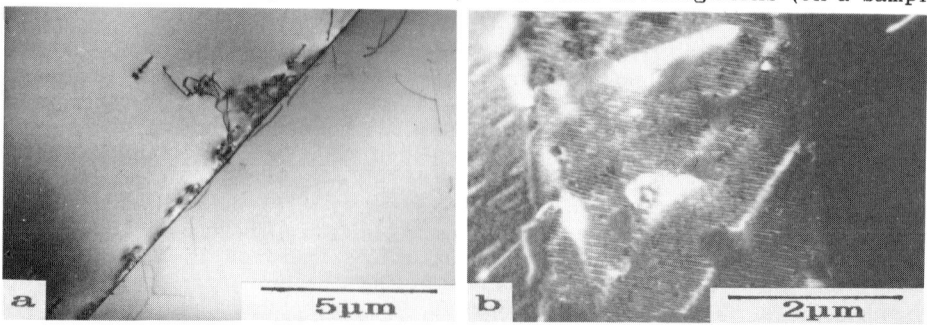

Fig. 2   Survey micrograph (a) and weak-beam image (b) of a 37^0 GB

cut 30^0 inclined to the GB) revealed dense dislocation networks (Fig.2b), resulting from unintentional tilt and twist components with respect to the perfect Σ 5 GB, which were measured by Kikuchi line shifts to have 0.25^0 tilt and 0.9^0 twist deviations. It should be pointed out here that there were almost similarly strong contrast levels of various GBs, which were produced by this technique both in n- and p-type silicon, which necessarily have different misorientations or even of small-angle GBs having a considerably lower network density (Gleichmann et al 1990). Thus, the GB activity cannot directly be correlated to the network densities but is most probably dominated by the point defect decoration in as far as a sufficiently high density of the dislocations is attained. The existence of the networks directly proves the absence of oxide interlayers, which pose a major problem to DB processes (Ahn et al 1987, Perreault et al 1990); some few amorphous inclusions (Fig. 2b) are the exception.

The "non-coincidence" 45^0 GB shows a different contrast behaviour (Fig. 1c). There are strong fluctuations between about 40% contrast and near the detection limit, indicating structural inhomogeneities and a surprisingly low basic activity of the GB. Directly correlated EBIC and TEM studies given in Figs. 3 and 4 showed several origins of local contrast enhancements. Again the main effects are related to dislocations the density fluctuations of which have a stronger impact owing to the low basic activity of the boundary itself. The lowest level – for example in the positions marked by arrows in Fig. 3 – corresponds to GB segments having distances in the order of 1 μm between the dislocations in and near the defect plane. Densities of the captured dislocations locally increased by up to one order of magnitude – as in detail A – cause a distinct contrast enhancement. Somewhat stronger changes, sometimes in a dot-like form, are detected at dislocation tangles originating from small particles in the GB (detail B).

The very strongest contrasts occur in places where TEM detects unusually thick amorphous interlayers (Fig.4, detail A), which have punched out a dense three-dimensional network of dislocations. The dislocation-rich zone exceeds the generation volume thus being a very effective recombination region compared to the other nearly two-dimensional dislocation arrangements at the GBs. A far more pronounced example of this seldom occurring effect was

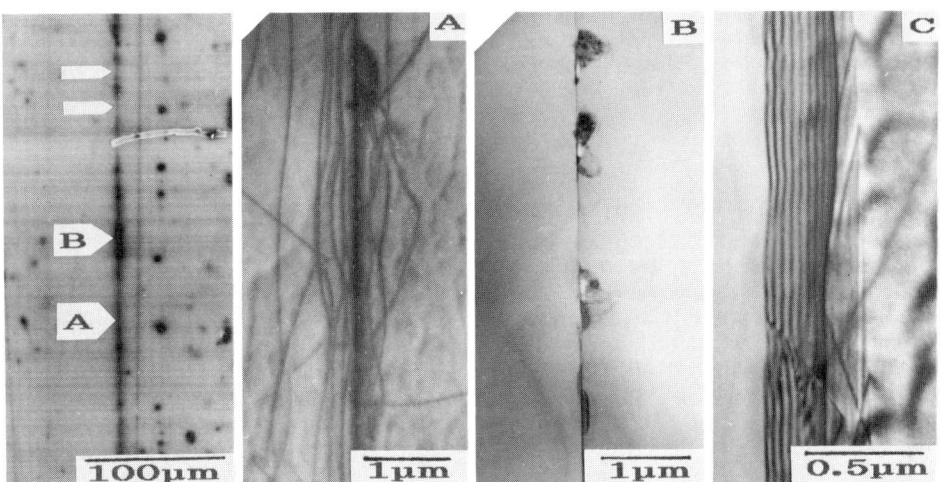

Fig. 3   EBIC and TEM (details A–C) images of medium-contrast GB segments (45^0), arrows mark nearly undisturbed GB like C ("basic" activity)

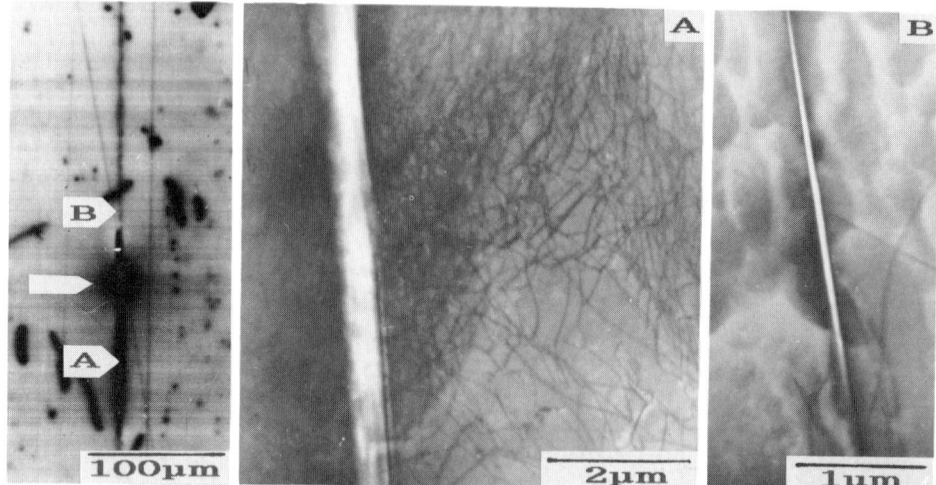

Fig. 4   EBIC and TEM images of a 45° GB with oxide interlayers;
A and arrow: thick patch with dislocation emission; B: thin layer

found in the position marked by an arrow with the patch thickness being
1.2 µm. Here, a TEM study failed owing to the large specimen thickness, but
a light-element energy-dispersive X-ray analysis revealed the amorphous
material in this position as silicon oxide.

For thin oxide layers (< 0.1 µm, see detail B) no dislocation emission was
observed. Owing to the simultaneous presence of captured dislocations such
layers could not be proved to significantly contribute to the recombination.

The present results indicate that probably the lowest boundary activity
should not be expected for good-coincidence GBs (low Σ) since, in practice,
minor deviations are not avoidable and cause dense dislocation networks
which easily can getter impurities. The very low basic activity of the 45°
interface, on the other hand, points to a weak decoration capability compared
to dislocation networks. According to high-resolution TEM (Furukawa et al
1986) the structure of such "non-coincidence" boundaries is an amorphous-
like layer (about 1 nm thick) which should not depend on the exact value of
the twist angle.

REFERENCES

Ahn K-Y, Gösele U and Smith P 1988 Mat. Res. Soc. Symp. Proc. vol. 107, 501
Föll H and Ast D G 1979 Philos. Mag. A <u>40</u> 589
Furukawa K, Shimbo M, Fukuda K and Tanzawa K 1986 Ext. Abstr. 18th Intern.
    Conf. on Solid State Devices and Materials, Tokyo, p. 533
Gleichmann R, Blumtritt H, Höpner A and Sullivan T D 1990 Polycrystalline
    Semiconductors II, eds J H Werner and H-P Strunk (in press)
Heydenreich J, Blumtritt H, Gleichmann R and Johansen H 1981 Scanning
    Electron Microscopy 1981 vol I ed. by SEM Inc., AMF O'Hare (Chicago) 351
Perreault G C, Hyland S L and Ast D G 1990 priv. comm., subm. to Philos.Mag.
Shimbo M, Furukawa K, Fukuda K and Tanzawa K 1986 J. Appl. Phys. <u>60</u> 2987

Inst. Phys. Conf. Ser. No 117: Section 10
Paper presented at Microsc. Semicond. Mater. Conf., Oxford, 25–28 March 1991

763

Temperature dependent EBIC contrast investigations of grain boundaries and precipitates in CdTe

G N Panin and E B Yakimov

The Institute of Microelectronics Technology and High Purity Materials
USSR Academy of Sciences, Chernogolovka

ABSTRACT: The temperature dependence of EBIC and REBIC contrast of some defects in CdTe crystals have been investigated. It is shown that all investigated defects are charged. The dependencies obtained have been discussed taking into account the influence of the depletion region formed near the defects as well as point defects in this region on EBIC and REBIC signal formation.

1. INTRODUCTION

The electron beam induced current (EBIC) mode of SEM has been widely used for defect characterisation in semiconductor crystals. As a rule this technique is used for revealing defects and their qualitative characterisation. To obtain the information about the nature of defects and to measure their properties it is necessary to carry out some additional experiments. For these purposes the measurements of EBIC contrast dependence on electron beam current and temperature are very promising. Unfortunately, up to now the theory of temperature dependence of EBIC contrast has not been available. Recently Wilshaw and Booker (1987) and Fell and Wilshaw (1989) proposed such a theory for charged dislocations which predicts a curve with a maximum depending on the electron beam excitation level. Curves with one or two maxima were observed by Bode et al (1988) who explained the results obtained in the frame of Shockley-Read statistics (Shockley and Read, 1952). In the present paper the results of EBIC contrast temperature dependence investigations of some defects in CdTe crystals are presented. It is shown that this dependence is described by a curve with one or two maxima. The existence of potential barriers near the defects under investigation is controlled on the same samples by remote contact EBIC (REBIC) measurement, a method which reveals local electrical fields (Russel et al, 1980, Bubulac and Tennance, 1988).

2. EXPERIMENTAL

Single crystals of p-CdTe with a shallow acceptor concentration of about $10^{15}cm^{-3}$ grown by the Bridgman technique were used for investigations. From these crystals the samples with dimensions $3 \times 3 \times 10mm^3$ were cut. Then the samples were polished mechanically and chemically in a solution of 8%Br in HBr. After ion milling by Ar^+ with energy about 600eV, barrier structures were formed on the top surface of the samples by evaporation of Al with a thickness of about 50nm. The ohmic contacts were formed on the opposite surface by Au chemical deposition. The EBIC contrast value was calculated as $C=(I_o-I_d)/I_o$ where I_o and I_d are the induced current values when the electron beam is far from the defect and on it, respectively. For REBIC measurments the ohmic contacts were deposited on opposite ends of the samples. In this case the signal is formed as a result of separation of electron-hole pairs created by the electron beam in the electric field of the charged defects. Preliminary analysis of the samples under investigation was carried out by selective chemical etching and X-ray microanalysis. Sub-boundaries, twins and some large (up to 10μm) Te-rich inclusions formed in the matrix, as well as near the extended defects, were revealed.

3. EXPERIMENTAL RESULTS

Our investigations have shown that almost all defects revealed by chemical etching, ie subboundaries, twins, Te-rich inclusions and some other unidentified defects, can be seen in EBIC mode. REBIC investigations have shown that all these defects at room temperature are charged and potential barriers exist near these defects in the experimental conditions used $(I_b=10^{-10}A)$ which lead to effective separation of e-h pairs (Fig 1). It is seen from this figure that the REBIC image contains black-white features which are usually observed in this mode for charged defects (Russel et al, 1980) due to opposing directions of the electric field on different sides of the defect. Therefore, the comparison of EBIC and REBIC images of the same defects gives one the possibility of studying the dependence of EBIC contrast on the space charge region (SCR) formed near charged defects.

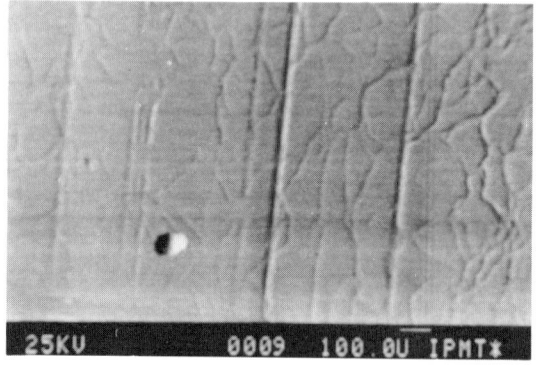

Fig. 1. REBIC image
of different electrically
active defects in CdTe

The temperature dependence of EBIC contrast and the REBIC signal for some defects is presented in Fig 2 and Fig 3 respectively. It should be noted that temperature dependencies of the REBIC signal are similar to those of EBIC contrast but, as a rule, with shifted maximum positions. Both the signals depend essentially on point defect atmosphere near the extended defect as well as on precipitate formation on it. For example, EBIC contrast of a twin is practically independent of temperature in the temperature range from 130 to 230K and decreases at T <130K. But electron beam irradiation $(I_b = 10^{-8}A)$ leads, in accordance with Yegorshev et al (1989), to an increase of contrast value as well as to a change of its temperature dependence (Fig 2, curves 2,3). After such treatment temperature dependence has a well pronounced maximum approximately at 200K. Te-rich inclusion increases the REBIC signal from twins (Fig 3, curve 2). As a rule, both EBIC and REBIC temperature dependencies for subboundaries and inclusions have two maxima (Fig 2,3). In some cases, however, our REBIC investigations have revealed the appearance once more of a potential barrier near such a defect and show that REBIC temperature dependence for these two barriers is different. Figure 4 shows the images of such a defect obtained at 300 and 170K.

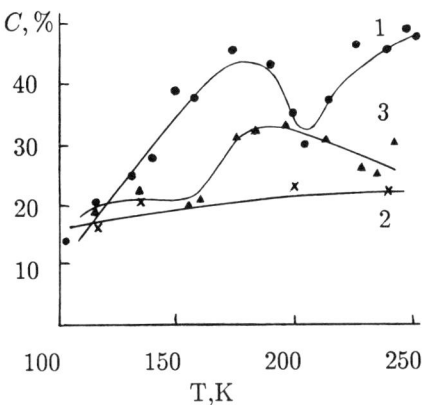

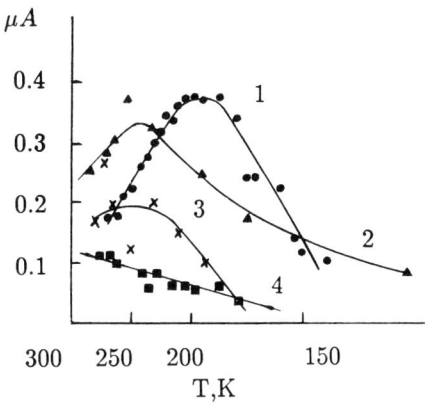

Fig. 2. Temperature dependence of EBIC contrast of inclusion (1) and twin in as-grown crystal (2) and after e-irradiation (3)

Fig. 3. Temperature dependence of REBIC signal for some inclusion (1), Te inclusion on twin (2), precipitates (3) and twin (4)

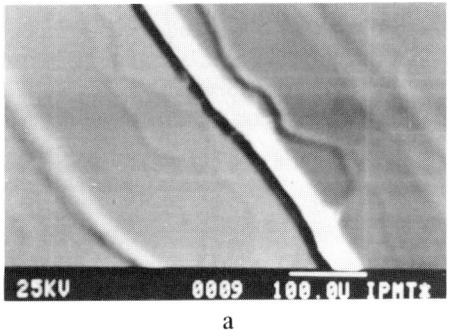

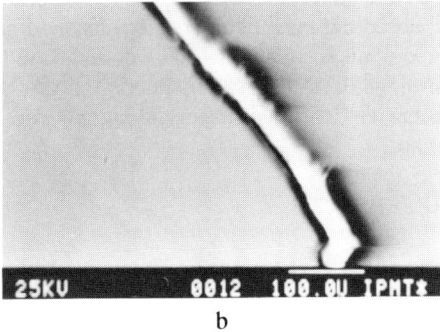

a   b

Fig. 4. REBIC images of an unidentified 2-dimensional defect at 300K (a) and 170K (b)

4. DISCUSSION

From our REBIC investigations some information about barrier height near the defect under investigation and its temperature dependence at the excitation conditions used can be obtained. In any case, we have seen a decrease of barrier height at low temperature which is easy to explain by a decrease of majority carrier capture for stationary minority carrier flow, dependent on the excitation level. Comparison of EBIC and REBIC images shows that an essential part of EBIC contrast is associated with SCR near the defect.   Therefore the results obtained can be qualitatively explained by the Wilshaw model (Wilshaw and Booker, 1987). The differences between these images can be explained by a dependence of the REBIC signal on the sample defect structure or by the influence of point defects and precipitates, situated near the defect, on its recombination properties. In our opinion the existence of such defects is the reason for EBIC temperature dependencies with two maxima.   In the case illustrated by Fig 4 these defects can be resolved but in many other cases they are distributed very close to the extended defect and change defect recombination properties, as well as SCR width. Some minority carriers can recombine through these defects and decrease the influence of electron beam on the barrier height (Bondarenko and Yakimov, 1990). The other possible way to maintain the barrier is carrier transport along the defect (Alexander et al, 1990).   The influence of point defects and precipitates surrounding an extended defect on its electrical properties may be a reason for differences of low temperature dependencies of REBIC signals for different types of extended defects. Indeed, if we take into account the similarity of EBIC and REBIC temperature dependencies and, for this reason, assume that EBIC contrast is proportional to barrier height $\varnothing$, we should obtain $C \sim T$ for all defects at low temperatures because, in spite of exponential dependence observed for some defects, in this case $\varnothing \sim cT(a-\ln I_b)$, where c and a are constants. Thus, it is at present very difficult to describe qualitatively the results obtained. Nevertheless, as shown in Figs 3 and 4, it is possible to use measurements of REBIC and EBIC temperature dependencies for defect identification.

5. CONCLUSION

Measurements of EBIC and REBIC temperature dependencies for the same defects in CdTe show that EBIC contrast is associated with a potential barrier formed near the charged defects. The results obtained can be qualitatively described by the Wilshaw model but for a quantitative description it is necessary to take into account the effect of a point defect atmosphere on the defect recombination properties, as well as minority carrier transport along a defect under local excitation by a focused electron beam.

REFERENCES

Alexander H, Dietrich S, Huhne M, Kolbe M and Weber G 1990 Phys Stat Sol(a) 117 417
Bode M, Jakubowicz A and Habermeier H-U 1988 Scanning 10 169
Bondarenko I E and Yakimov, E B 1990 Phys Stat Sol(a) 122, 121
Bubulac L O and Tennant W E 1988 Appl Phys Lett 52 1255
Fell T S and Wilshaw P R 1989 Inst Phys Conf Ser No 104 227
Russel G I, Robertson M J, Vincent B and Woods J 1980 J Mater Sci 15 939
Shockley W and Read, W T 1952 Phys Rev 87 835-842
Wilshaw P R and Booker, G R 1987 Izv AN SSSR, Ser Fiz 51 1582
Yegorshev V V, Panin G N and Yakimov E B 1989 Proc 12 Intern Congress on X-ray Optics and Microanalysis, ed S Jaslenska and L J Maksymowicz, Poland 904

EBĪC studies of WSe$_2$ "mixed" surface

L Margulis, D Mahalu, E Watkins and R Tenne

Department of Materials Research, the Weizmann Institute of Science, Rehovot 76100, Israel

ABSTRACT: A large number of well-defined hexagonal etch pits are produced on the WSe$_2$ surface after controlled anisotropic corrosion. As a result a **mixed** surface is created that exhibits better photovoltaic properties than the atomically smooth van der Waals surface. EBIC observations performed at low temperatures give direct evidence for enhanced photoactivity of ∥c facets.

1.   INTRODUCTION

WSe$_2$ belongs to the family of layered compounds extensively studied during the last ten years. This semiconductor, like some other transition metal dichalcogenides, is a promising material for both Schottky and liquid junction photovoltaic cells due to its high absorption coefficient for sunlight and its excellent chemical stability in air and liquid media (Tributsch 1977, Lewerenz et al 1980, Kautek et al 1980). The hexagonal crystal lattice of WSe$_2$ is built by periodic stacking of triple atomic layers Se-W-Se in the c-direction, with weak van der Waals (vdW) bonds between Se-layers. Consequently, WSe$_2$ is characterized by strong anisotropy of its electronic and mechanical properties (Wilson and Yoffe 1969) in mutually orthogonal (∥c and ⊥c) crystallographic directions. The WSe$_2$ crystal can be easily cleaved along the vdW (⊥c) planes exposing an atomically smooth Se surface. It was generally accepted that the collection efficiency of minority carriers, and hence light to electricity conversion ability of those crystals, depends on the quality of the exposed vdW surface. The charge carrier collection deficiency was usually associated with electrically active defects such as ∥c surface steps and near-surface dislocations (Canfield and Parkinson 1981, Lewerenz et al 1982).

Recently it was found (Mahalu et al 1990, Tenne and Wold 1985) that photo-voltaic properties of WSe$_2$ crystals can be considerably improved by **controlled anisotropic corrosion** (CAC). After this treatment, well-defined hexagonal etch pits are formed. Consequently, a **mixed** (i.e. including both ∥c and ⊥c facets) sur-face is created. Such a surface was found to have interesting optoelectronic properties together with better photoresponse than freshly cleaved smooth surfaces (Jakubowicz et al 1989). It is still not clear in detail what causes the improvement of the pho-toresponse. Obviously, the special geometry of the mixed surface should decrease the light reflectivity. The question is: does the geometry only lead to the enhance-ment of the photocurrent, or are other more complicated mechanisms involved in this process? Furthermore, what is the contribution of different components of the mixed surface, i.e. ∥c and ⊥c facets, to its photoactivity? To answer these questions a technique must be used that can measure the photoresponse of the mixed surface with submicron resolution. EBIC is such a technique.

2. EXPERIMENTAL

Single crystals of n-type WSe_2 were grown by chemical vapor transport with bromine as the transport agent. The crystals were approximately 0.3 mm thick with surface areas between 0.2 and 0.8 cm^2. CAC was used to obtain mixed surfaces on freshly cleaved WSe_2 crystals. Detailed description of the CAC technique can be found elsewhere (Jakubowicz et al 1989, Mahalu et al 1990).

EBIC experiments were performed using a "Philips 515" scanning electron microscope equipped with a "Hexland" cold stage, at temperatures 115 - 300 K. The accelerating voltage varied between 20 and 30 kV. Schottky barriers were obtained by evaporation of a 250Å thick gold layer. The gold was evaporated at an angle of 45° in order to provide equal thickness of the Schottky contacts on $\parallel$c and $\perp$c facets, at least for a certain part of the etch pit area. For increasing the signal/noise ratio, the Schottky contact areas were considerably restricted. The comparative EBIC measurements were performed at 45°-tilted position. Thus the identical orientation of $\parallel$c and $\perp$c facets relative to the electron beam was ensured.

3. RESULTS AND DISCUSSION

The morphology of the mixed surface created after CAC is shown in Fig. 1. The typical depth of the hexagonal etch pits is ca. 20 μm while the height of individual $\parallel$c steps varied between 0.01 and 0.5 μm.

Fig. 1 Mixed surface of WSe_2 single crystal. The regular shape of the hexagonal etch pits is slightly distorted because of the 45°-tilt. Bar is 20 μm

The main goal of EBIC experiments was the direct comparison of the charge collection efficiency of $\parallel$c and $\perp$c components of the mixed surface. Such a comparison was not made earlier (Mahalu et al 1988, Jakubowicz et al 1990) because of experimental difficulties. Since the EBIC signal from the Au/WSe_2 Schottky diodes was rather weak at room temperature, high beam currents (i.e. large spot sizes, of at least 0.2 μm), had to be used in order to obtain a good signal. Under these conditions, a good spatial resolution cannot be expected. Furthermore, severe electron damage is observed even after the first scan. Therefore observation with high magnifications is practically impossible at room temperature.

This problem can be avoided by lowering the temperature. It was found that cooling the WSe_2 crystal prevents almost totally the electron beam damage and simultaneously results in remarkable enhancement of the EBIC signal (Fig. 2). Using low temperatures enabled us to obtain EBIC images of a quality good enough for reliable resolving even very small $\parallel$c steps.

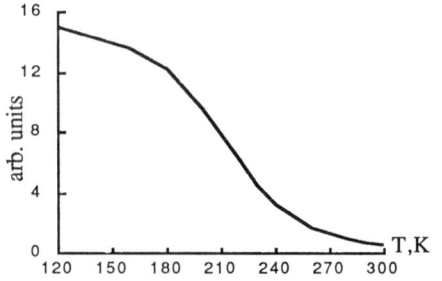

Fig. 2   EBIC dependence on temperature

An enhancement of the EBIC can be detected in the vicinity of ∥c facets even if the crystal surface is in a horizontal position, and the electron beam glides along ∥c facets (Fig. 3b). The difference in the EBIC signals for ∥c and ⊥c facets is revealed more distinctly when the crystal is tilted by 45° (Fig. 3c). In this latter case the enhanced EBIC signal for ∥c facets is clearly seen on the left side of the etch pit where ∥c and ⊥c facets form equal angles (45°) with the electron beam, while on the right side ∥c facets are shadowed and hence appear with dark contrast.

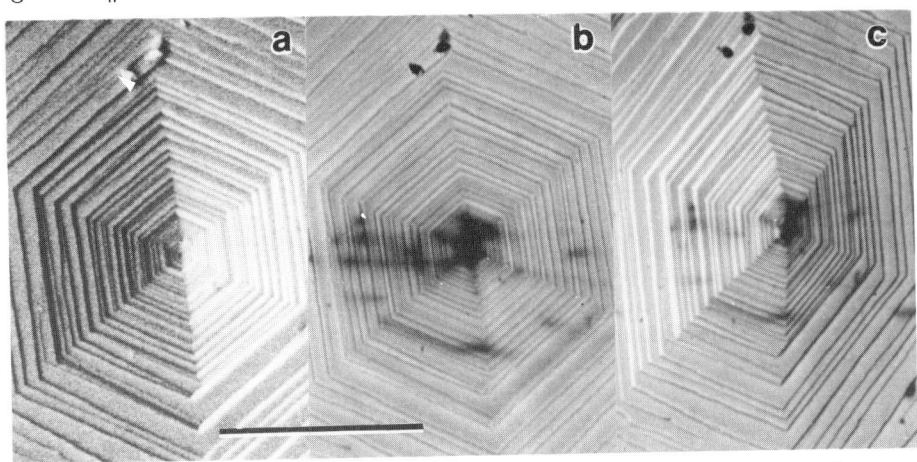

Fig. 3 Secondary electron (a) and EBIC (b, c) images taken in the horizontal and tilted positions (see schemes at the bottom). The near-surface dislocations (lines of diffuse dark contrast) can be seen in the EBIC pictures. The temperature is 115 K. Bar is 50 µm.

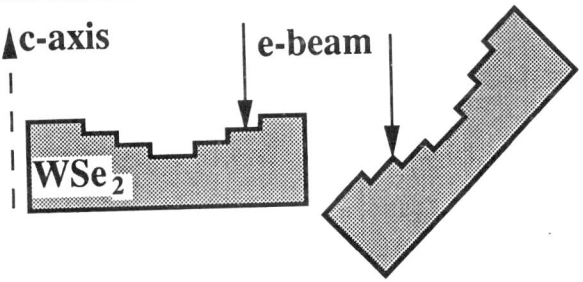

The comparative measurements of the EBIC signals for ∥c and ⊥c facets show that ∥c facets provide remarkably higher charge collection efficiency. The EBIC signal enhancement at these facets equals ca. 20% (at least for big enough facets) independent of the temperature and accelerating voltage (see, for example, Fig. 4).

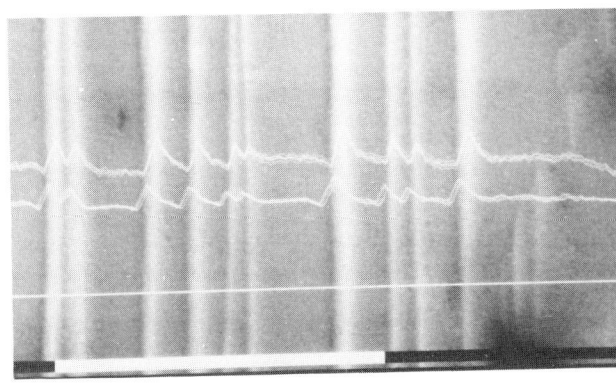

Fig. 4 EBIC image of ∥c facets and EBIC profiles for 150 and 180 K (upper and lower curves respectively). The crystal is tilted by 45°. EBIC zero line is also shown. It coincides with line scan. Bar is 10 µm.

The results obtained allow us to conclude that more than the morphology is responsible for the improvement of photovoltaic properties of the mixed surface. This surface created after CAC includes new elements - ∥c facets - which reveal enhanced photoactivity in comparison with vdW surface. This phenomenon may have special applications in optoelectronics.

ACKNOWLEDGEMENTS

We are very grateful to A Jakubowicz and R Matson for helpful discussions. L.M. thanks the Israel government for partial support. This research was supported by a grant from the Israel National Council for Research and Development and the KFA Jülich (Germany).

REFERENCES

Canfield D and Parkinson B A 1981 J. Am. Chem. Soc. **103** 1279

Jakubowicz A, Mahalu D, Wolf M, Wold A and Tenne R 1989 Phys. Rev. B **40**, 2992

Kautek W, Gerischer H and Tributsch H 1980 J. Electrochem. Soc. **127** 2472

Lewerenz H J, Heller A and Disalvo F J 1980 J. Am. Chem. Soc. **102** 1877

Lewerenz H J, Ferris S D, Doherty D J and Leamy H J 1982 J. Electrochem. Soc. **129** 418

Mahalu D, Jakubowicz A, Wold A and Tenne R 1988 Phys. Rev. B **38** 1533

Mahalu D, Peisach M, Jaegermann W, Wold A and Tenne R 1990 J. Phys. Chem. **94** 8012

Tenne R and Wold A 1985 Appl. Phys. Lett. **47** 707

Tributsch H 1977 Ber. Bunsenges. Phys. Chem. **81** 361

Wilson and Yoffe 1969 Adv. Phys. **18** 193

Inst. Phys. Conf. Ser. No 117: Section 10
Paper presented at Microsc. Semicond. Mater. Conf., Oxford, 25–28 March 1991

Analysis of defects in bulk semiconductors using electron channelling contrast imaging

JT Czernuszka, NJ Long and PB Hirsch
Department of Materials, University of Oxford, Parks Rd., Oxford OX1 3PH.

ABSTRACT: Electron channelling contrast imaging (ECCI) is a SEM based technique for imaging and characterising the Burgers vectors of dislocations in bulk samples. Thus, thin foil relaxation effects do not occur and specimen preparation is relatively straightforward. In this paper we present preliminary results of a study on dislocations and precipitates in bulk samples of deformed Si. It has been shown that increasing signal acquisition times improves the signal to noise ratio and the depth to which dislocations can be imaged. The precipitates have been imaged using strain contrast and atomic number contrast.

1. INTRODUCTION

It is now well established that near surface defects in bulk specimens can be imaged and characterised using back-scattered electrons in the scanning electron microscope (SEM) (Morin et al, 1979; Czernuszka et al, 1990a). The technique is known as Electron Channelling Contrast Imaging (ECCI). The general requirements are that a high brightness electron source (a field emission gun , FEG) is used in conjunction with an electron detector situated in the forward scattering position. In this way, electrons which have suffered few collisions, and therefore have lost little energy, are collected. Although Morin and co-workers (Morin et al, 1979; Fontaine et al, 1983) used an energy filter between their specimen and detector, to increase the contrast, we have shown that this energy filter can be dispensed with by using a more sensitive detector with image processing techniques (Czernuszka et al, 1990a, 1990b).

We have previously shown that ECCI images can be obtained by operating the SEM at 30 keV and 100 keV. It was shown that the image contrast could be improved by working at lower accelerating voltages; but the depth to which useful contrast is obtained is reduced from $\sim$ 200nm at 100keV to $\sim$ 95 nm at 30keV (Czernuszka et al, 1990b). These results imply that the ECCI technique can be relatively easily performed using commercial FEG SEMs which generally operate at 30keV. In this paper we present the results of a preliminary study to show how images of near surface dislocations and the depths to which they can be observed are improved by suitable image processing procedures. In addition, examples of precipitates in deformed silicon are presented to highlight the general versatility of the ECCI technique. The results presented in this paper were taken at an accelerating voltage of 100 keV. Especial emphasis is put on the role of improving the Signal to Noise Ratio (SNR) of the image.

2. EXPERIMENTAL PROCEDURE

A VG HB501 FEG STEM has been fitted with a highly efficient back-scattered electron detector in the secondary electron port. The geometry of the detector is such that it is in the forward scattering position. The signal was acquired using a Synoptics Synergy framestore and digital scan generator and subsequently processed using Semper 6. Frame averaging using a Kalman filter was the most successful image processing technique.

Single crystals of silicon were compressed for single slip at 450°C and 650°C after a high temperature pre-strain (Wilshaw and Fell, 1989). Sections were taken parallel to the (111) primary glide plane and polished flat. In addition, several sections were cut 6° off the primary glide plane for depth dependence of contrast measurements. There is only a single tilt holder available for specimens in the SEM position in the STEM, so the specimens were aligned with the tilt axis parallel to a $[\bar{1}10]$ direction and tilted approximately 60° towards the detector. The precise orientation of the sample was determined from electron channelling patterns (ECPs).

3. RESULTS AND DISCUSSION

The first set of examples was designed to show how the dislocation images become clearer with longer acquisition times. In effect, the signal-to-noise ratio (SNR) is being improved. The images of Figure 1 were taken using a 220 reflection and a beam divergence of 2.5 millirads. The number of frames in each case was 50 with the frame time for Fig. 1a being 0.5 seconds and that for Fig. 1b 1.5 seconds. The improvement in image quality is readily noticeable, although the actual contrast is maintained at approximately 10%. In fact, to improve SNR all that is needed is to increase the total scan time. For example, a frame time of 10 seconds averaged over 10 frames produces the same SNR as a frame time of 5 seconds averaged over 20 frames (after adjusting for bandwidth limitations).

These observations have been extended further to determine the depth to which dislocations could be imaged. The 6° sectioned sample was examined. Fig. 2 shows typical results using a 400 reflection. By scanning for 100 seconds (Fig. 2a) the depth resolvable was $\sim$ 173 nm, but by scanning for 180 seconds (Fig. 2b) this depth could be extended to $\sim$ 210nm. These measurements were taken by taking line scans across the dislocation and the depth measured until the desired signal became equal to the noise. This depth is greater than the extinction distance for the 400 reflection in Si at 100 keV.

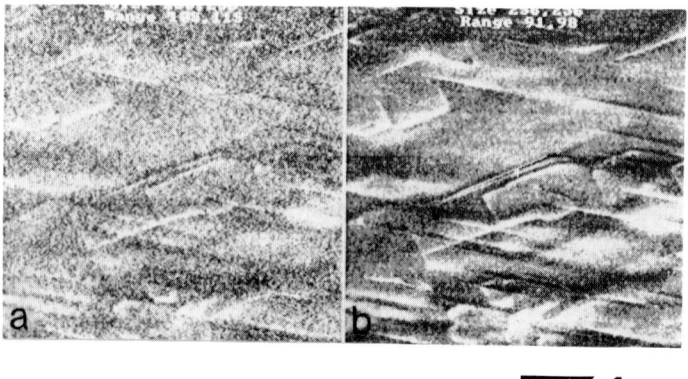

— **1 µm**

Fig 1. ECCI images of dislocations in Si deformed at 450°C. (a) for a scan time of 25 seconds and (b) for a scan time of 75 seconds.

Hence, because the contrast arising from near surface dislocations is low (of the order of a few %), it is important to improve the SNR by signal averaging over longer times. A reasonable compromise between SNR and scanning times is $\sim$ 200 seconds.

Of course, it is possible to image other defects such as stacking faults and precipitates. Figure 3 shows an example of what is probably a silicide precipitate in a silicon sample deformed at 650 °C. The thin disc is aligned along $[\bar{1}10]$. It is worth noting that the strain contrast is greater than any atomic number contrast, although further study is necessary to determine the magnitude of the strain. In fact there were only two of such precipitates found on a 4mm x 4mm sample. It is highly unlikely that these precipitates would have been observed using TEM and highlights the great potential of the ECCI technique.

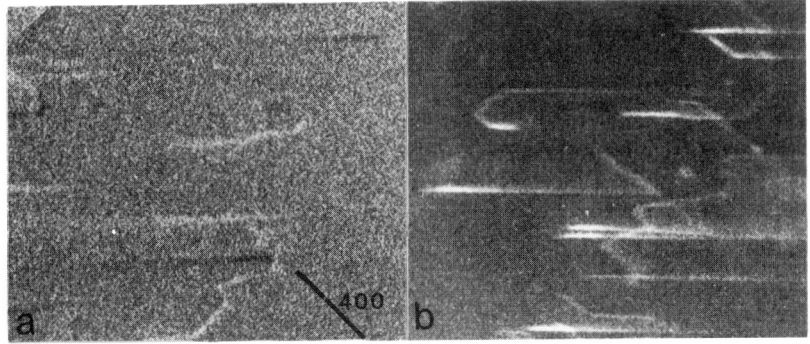

2μm

Fig.2 ECCI images of dislocations inclined 6° to the (111) surface using a 400 reflection. (a) was taken with a 400kHz bandwidth, with a 0.5 second frame time for 200 frames. (b) was taken with a 40 kHz bandwidth, with a 1.5 second frame time for 120 frames .

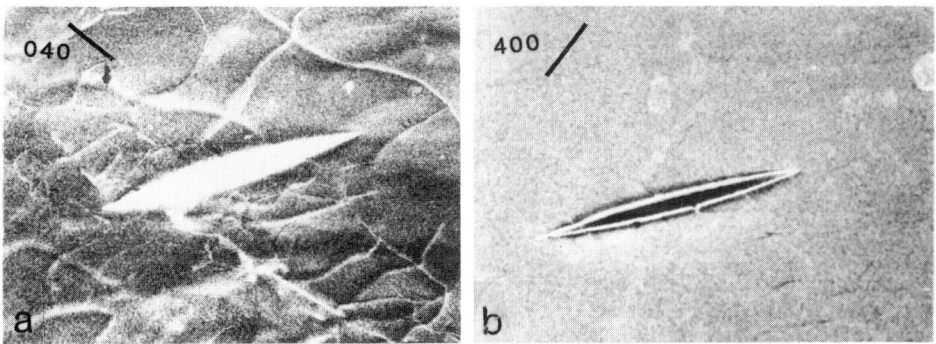

1μm

Fig. 3 ECCI images of precipitates in Si deformed at 650°C.

4. CONCLUSIONS

In conclusion, it has been shown that the signal to noise ratios of the ECCI images can be improved by increasing the signal acquisition times. By improving the SNR it has been possible to image dislocations to greater depths, up to over 200nm. It has also been possible to image precipitaes using strain contrast. Since only two precipitates were observed on the sample it is unlikely that they would have been imaged using TEM techniques.

ACKNOWLEDGEMENTS

Thanks are due to Dr TS Fell for loaning his samples, SERC for funding and the Royal Society for the provision of a Research Fellowship to one of us (JTC).

REFERENCES

Czernuszka JT, Long NJ, Boyes ED and Hirsch PB 1990a Phil. Mag. Letts. 62 227.
Czernuszka JT, Long NJ, Boyes ED and Hirsch PB 1990b MRS Fall Meeting, Boston USA.
Fontaine G, Morin P and Pitaval M 1983 Inst. of Physics Conf. Ser. No. 67, 213.
Morin P, Pitaval M, Besnard D and Fontaine G 1979 Phil. Mag. A 40 511.
Wilshaw PR and Fell TS 1989 Inst. of Physics Conf. Ser. No. 104, 85.

Inst. Phys. Conf. Ser. No 117: Section 10
Paper presented at Microsc. Semicond. Mater. Conf., Oxford, 25–28 March 1991

Direct electron beam-induced formation of nanometer carbon structures in STEM

V V Aristov, N A Kislov and I I Khodos

Institute of Microelectronics Technology and High Purity
Materials, USSR Academy of Sciences, 142432 Chernogolovka,USSR

ABSTRACT: Long-range growth of carbon-containing structures
beginning from the edge of the sample has been found to
occur under the action of the electron beam passing outside
the sample in the column of the microscope pumped down by an
oil diffusion pump. The mechanism of the phenomenon is
suggested.

1. INTRODUCTION.

It is well-known that a surface irradiated by electrons in
the vacuum chamber pumped down by an oil diffusion pump is
covered with a hydrocarbon layer (Stewart 1934). This layer
hinders conventional electron-microscopic studies. However,
this phenomenon may be used, for instance, in microelectronics
for fabrication of protective masks on the surface of objects,
since such layers formed by the electron beam prevent subsequent removal of the material upon chemical etching or ion
bombardment. Charalambous (1983) showed that a strip of hydrocarbon polymer grows following the beam if the electron beam
moves slowly. The strip is not interrupted even if the beam
keeps moving from the edge of the substrate into the hole area.
By this means, rods and bridges were produced in the gap between the two edges of a silicon dioxide film (Behringer 1986)

The aim of the present work is to study the mechanism of
growth of electron-induced carbon-containing structures in the
form of ledges and rods as well as the possibilities of the
formation of substrate-free nanometer carbon-containing structures in the scanning transmission electron microscope (STEM).

2. METHODS AND EXPERIMENTAL RESULTS

The formation of carbon-containing structures and the measurements were carried out in a VG HB501 STEM with an accelerating
voltage of 100 kV and a field emission cathode. The beam current was up to 0.5 nA, the beam diameter was $\approx$ 1 nm. The
accuracy of the determination of the beam position relative to
the characteristic features of the image was $\leq$3 nm. The
pressure in the microscope column pumped down by an oil
diffusion pump was $3 \cdot 10^{-7}$ Pa. The electron energy loss spectra

were recorded by a spectrometer with a resolution of 1.0 eV
and analyzed by a Link-860 system. The substrates were 40-80nm
thick films of amorphous carbon sputtered on NaCl and 40nm
thick Fe films sputtered on thin (≈60 nm) SiO_2 windows on
silicon (Enquist 1986).

The transmission images of the thin foils point to the growth
of a carbon-containing layer in the vicinity of the electron
probe on the specimen surface. Moreover, we found that the
passage of the electron beam outside the film at a distance of ≤35nm from its edge leads to the formation of a ledge
on the film edge in the region adjacent to the electron beam.
With further exposure the ledge increases reaching the beam
(Fig. 1). A rod protruding in the direction away from the substrate edge can be formed by successive movement of the electron beam at a certain distance from the ledge formed. This
technique enables formation of strap bridges in the breaks between two edges of the film (Fig. 2). The width of the second,
narrower, bridge did not exceed 10 nm. Initially the width of
the first bridge was the same, yet, it increased with the for-

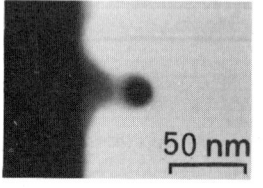

mation of the second one spaced at 70 nm
from the first. The widening of the first
bridge may result from the polymerization
of the hydrocarbon molecules on its surface. The polymerization is likely to be
induced by the secondary electrons emitted
during the formation of the second bridge.

50 nm

Fig .1. Image of the ledge formed by the electron beam passing
outside the Fe film at 35nm from the edge after 160s exposure.

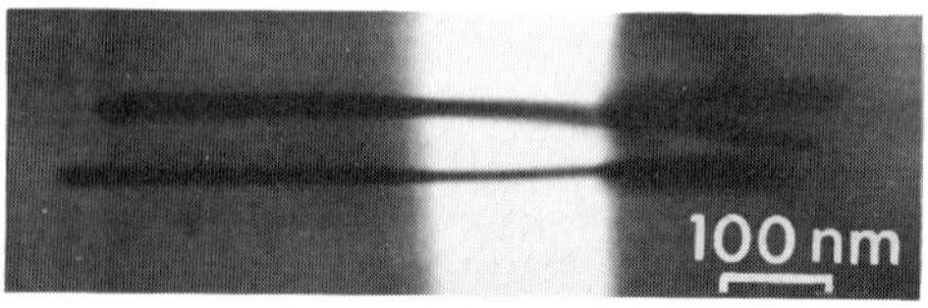

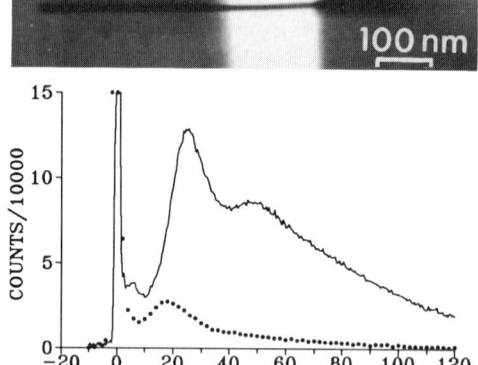

100 nm

Fig. 2. Bridges between two
edges of the carbon film.
The rate of electron beam
movement 6 nm/s.

Fig.3. EELS-spectra of primary beam electrons passing
through the ledge (solid
line) and outside the pre-
formed ledge at 10 nm from
its edge (dashed line).

The presence in EELS-spectrum of the 6 eV and 25 eV peaks
in the region of low losses
and a 284 eV absorption edge
(Fig.3, solid line) suggests
that the spectrum corres-
ponds to carbon (Egerton
1986). Fig. 3 also shows the EELS-spectrum (dashed line) recorded as the beam passed outside the preliminarily formed
ledge at a distance of 10 nm from its edge. (Throughout each
recording period (2 seconds) the growing ledge had no time to
touch the electron probe). It is seen that the peak has an energy
loss corresponding to 18eV, this value differing significantly

from the maximum of the plasmon loss energy for the beam
passing through the ledge (25 eV). In our case this fact
points to excitation of the surface plasmons whose maximum
energy for amorphous carbon is 17.7 eV (Egerton 1986).

Fig. 4a (open circles) shows time τ in which the growing carbon ledge reaches the electron beam as a function of probe
current i. It is seen that τ decreases with increasing probe
current i and at large currents it approaches asymptotically a
constant value. This suggests that the rate of formation of
carbon ledges is limited by the diffusion rate. The dependence
of time τ on distance x of the beam from the substrate edge
(Fig. 4b) is characterized by a nonlinear increase of τ with
increasing i.

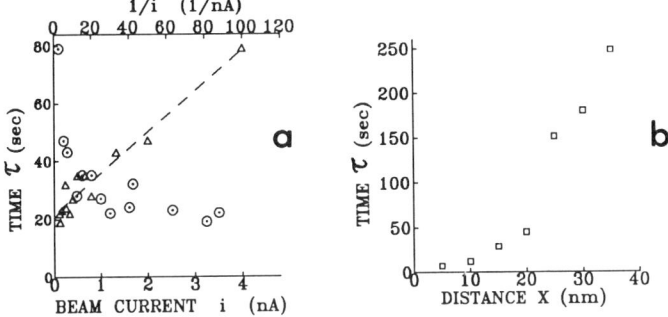

Fig. 4. Dependence of time τ on beam current i (open circles)
and inverse value of beam current $1/i$ (open triangle) (a), and
on the distance to the substrate edge x (b). In case (a) the
probe-substrate distance was 20 nm; in case (b) the probe
current $i = 1.6 \cdot 10^{-10}$ A.

3. DISCUSSION

3.1. It is possible to suggest that the hydrocarbon molecule
dissociates on the substrate surface due to excitation of its
valence electrons. Two channels of excitation are most
probable. One of them is transfer of some part of the energy
of the plasmon arising upon inelastic scattering of the beam
electrons to the hydrocarbon molecule (Marks 1982); the other
channel is direct resonance excitation of the molecule by the
passing electron beam (on the analogy of excitation of surface
plasmons on a small spherical particle by the passing electron beam (Acheche 1986)). Excitation of the molecule valence
shells may lead to the break of the chemical bonds up to
ionization of the molecule (Isaacson 1972). As is known,
dissociation products may polymerize losing their mobility and
forming a hydrocarbon layer. A specific concentration gradient
is created on the sample surface because of the mobile
hydrocarbon molecule depletion in the electron beam region.
The former determines the hydrocarbon diffusion flow into the
irradiated zone which is a sink region (Müller 1971).

To understand the experimental results, let us derive the
quantitative functional dependence between the main growth
characteristics of carbon-containing structures. Clearly, this

calls for simultaneous analysis of surface diffusion and the electron-induced process of molecule dissociation or polymerization. Let an excitation zone of width $2a$ be formed under the action of the electron beam passing at distance x from the edge of a flat substrate of thickness h (Fig. 5).

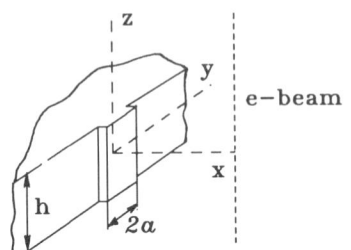

Fig. 5. Scheme of the formation of a carbon-containing ledge on the substrate edge.

Then using the solution of the continuity equation obtained in the work by Müller (1971) and assuming that all hydrocarbon molecules diffusing into the reaction zone take part in the formation of a hydrocarbon ledge, it is easy to find the relationship between the ledge growth rate v_x and the diffusion hydrocarbon flow J:

$$v_x = J \left(\frac{2\pi}{ah}\right)^{1/2} m_c \mu / \rho_c \tag{1}$$

where m_c is the carbon atomic mass, ρ_c the carbon (graphite) density, μ the number of carbon atoms going to form a ledge as a result of dissociation of a single hydrocarbon molecule. We assume that the microscope column contains polyphenyl ether $C_{30}H_{22}O_4$ with an intrinsic vapour pressure of 10^{-7} Pa. In this case $\mu \lesssim 30$. It may be expected that as the growing ledge approaches the electron beam, the probability of inelastic scattering of beam electrons will increase along with increasing effective cross-section of molecule dissociation. As a consequence of this, the dissociation rate and the linear rate of carbon ledge growth will increase. The exposure time required for the ledge to touch the beam may be determined from the relationship

$$\tau(x) = \int_0^x v_x^{-1}(\xi)\, d\xi \tag{2}$$

where ξ is the distance between the edge of the ledge and the center of the electron beam. Substituting (1) into (2) after simple transformations we obtain

$$\tau(x) = \frac{ahe\rho_c}{\nu\tau_a i m_c \mu} \int_0^x \frac{d\xi}{\Sigma(\xi)} + \frac{x}{v_d} \tag{3}$$

where $v_d = \pi\nu\rho^2 m_c\mu / \{ah\rho_c \ln[1.4\rho(ah)^{-1/2}]\}$ is the limiting rate in the case of the diffusion mechanism of growth, $\nu = \rho/(2\pi mkT)^{1/2}$ the effective flow of molecules coming from the gas phase with partial pressure ρ at temperature T, k is the Boltzmann constant, e electron charge, m the mass of a molecule, $\rho = (D\tau_a)^{1/2}$ the molecule diffusion length on the surface, D the surface diffusion coefficient, τ_a the lifetime of a molecule in the adsorbed state, Σ the full dissociation cross-section. The experimental results represented by open circles in Fig. 4a are well described by the linear dependence in the $\tau - 1/i$ coordinates (Fig 4a, open triangle). The intercept on the y-axis has the value $\tau = 21 \pm 7$ s and the tangent of the straight line slope is $tg\gamma = (0.59 \pm 0.05) \cdot 10^{-9}$ sA. Equating the second term of equation (3) to τ, we find $\rho = 7 \pm 2 \mu m$. The obtained value of ρ is typical of the hydrocarbon molecules migrating over the surface (Zdanov 1983).

3.2 Let us analyze mentioned elementary acts of scattering and dissociation. The full dissociation cross-section $\Sigma(\xi)$ in (3) may be represented as

$$\Sigma(\xi) = \sigma_1(\xi) + \sigma_2(\xi) = P_s(\xi)\sigma_s + P_m(\xi)\sigma_m$$

Here σ_1 and σ_2 are the cross-sections of dissociation for both main channels of energy transfer, P_s and P_m are the probabilities of electron beam excitation of the surface plasmon on the film face and the valency electrons of the hydrocarbon molecule, respectively; σ_s and σ_m are the corresponding dissociation cross-sections under the action of these excitation. Let us estimate the contribution of σ_1 to the total cross-section Σ. After differentiation of (3) with respect to x we obtain

$$\Sigma(x) = \frac{ahe\rho_c}{\nu \tau_a m_c \mu i (d\tau/dx - 1/\nu_d)} \tag{4}$$

Approximating the experimental dependence $\tau'(x)$ (Fig. 4c) by the third-degree polynomial and substituting $d\tau/dx$ into (4), we find

$$\Sigma(x) = (3.3 \pm 0.8) \cdot 10^{-22} x^{-2}$$

where x is the distance in nm.

The contributions of both dissociation channels may be separated by comparing the theoretical probability of plasmon excitation on the surfaces of the semicontinuous medium P_s (Acheche 1986) and of a spherical particle P_c (Ferrel 1985). Using the results of the work by Acheche (1986) and integrating with respect to all the possible frequencies the expression for the probability of excitation of a plasmon with energy $\hbar\omega$ on the surface of a semiinfinite medium by an electron passing at distance x from the surface, we find full excitation probability P_s

$$P_s(x) = \frac{he^2\Gamma\omega_s^2}{2\pi^2\hbar\varepsilon_0 v^2} \int_{\omega_0}^{\infty} K_0\left(\frac{2\omega x}{v}\right) \frac{\omega\, d\omega}{(\omega_s^2 - \omega^2)^2 + \Gamma^2\omega^2} \tag{5}$$

where $v = 1.644 \cdot 10^8$ m/s is the electron velocity at an accelerating voltage of 100 kV, $\varepsilon_0 = 8.85 \cdot 10^{-12}$ Fm^{-1} the dielectric constant, h the length of electron path at a distance x from the surface, $\hbar\omega_s = 17.7$ eV the energy of the surface plasmon loss peak for amorphous carbon, $\Gamma = 1.5 \cdot 10^{16}$ s^{-1} the damping factor determined experimentally from the spectrum in Fig. 3 as the halfwidth of the plasmon resonance peak. Since we are interested in plasmons with energy exceeding the chemical bond energy of organic molecules ($\approx$ 3-4 eV), we chose ω_0= 5.14·10^{15} s^{-1}as the lower limit of integration in (5).

The use of the theoretical dependence P (r) (r is the distance between the passing electron and the centre of the sphere of radius R_M<r at closest approach) is justified by the fact that the energy loss by valency electron excitation ($\approx$ 20 eV) (Isaacson 1972) is close to that of plasmon excitation on the surface of the molecule ($R_M \approx$ 0.5 nm). In this case, using exp. (12) from the work by Ferrel and Echenique (1985) in the approximation R_M<<r, and assuming that the hydrocarbon molecule may be in any point with coordinate y on the substrate face, the average excitation probability in the whole reaction zone of width $2a$ and height h may be estimated by means of the following expression

$$P_m = \frac{2\alpha c}{v} (\frac{\omega_1 R_m}{V})^3 \int_0^\infty \{K_0^2[\frac{\omega_1(x^2+y^2)^{1/2}}{V}] + K_1^2[\frac{\omega_1(x^2+y^2)^{1/2}}{V}]\}dy$$

Here α =1/137 is Sommerfeld's fine structure constant, c the velocity of light, K_0 and K_1 are the modified Bessel functions of orders 0 and 1, respectively, $\omega_1=\omega_p\cdot3^{-1/2}$.

Analysis of the correlation dependence between $\Sigma(x)$ and the calculated values of $P_m(x)$ and $P_s(x)$ showed that for $\Sigma(x)$ and $P_m(x)$ the correlation factor is close to unity at distance $x\lesssim$20nm. And a good correlation between $\Sigma(x)$ and $P_s(x)$ is observed for x >20nm. From the correlation dependences we find σ_s=(3.4±0.3)·10^{-24}m^2 and σ_m=(2.2±0.5)·10^{-18}m^2. The estimation shows that each event of direct resonance excitation of the hydrocarbon molecule results in its dissociation whilst the efficiency of the transfer of surface plasmon energy to the hydrocarbon molecule is fairly low.

It follows from the foregoing that to achieve the minimum size of structural elements it is important to take into account the effect of the "long-range"electron beam-substance interaction even in the absence of the substrate when the illumination by secondary electrons is lacking.

REFERENCES

Acheche M, Colliex C, Kohl H, Nourtier A and Trebbia P 1986 Ultramicroscopy 20 99
Behringer U W and Vettiger P 1986 J.Vac.Sci.Technol. 134 94
Charalambous P 1983 Ph.Dr.Thesis,Univercity of London
Egerton R F 1986 Electron Energy-Loss Spectroscopy in the Electron Microscopy (New York:Plenum Press)
Enquist F, Spetz A 1986 Thin Solid Films 145 99
Ferrel T L and Echenique R M Phys.Rev.Lett. 55 1526
M.Isaacson M 1972 J.Chem.Phys. 56 1803
Marks L D 1982 Solid State Commun. 43 727
K.H.Müller K H 1971 Optik 33 296
Stewart R L 1934 Phys.Rev. 45 488
Zdanov G S 1983 Poverkhnost. Fizika, Khimiya, Mekhanika 1 65

Backscattered electron compositional analysis of interfaces in bulk specimens using a deconvolution technique

P R Wilshaw, A Konkol* and G R Booker

Department of Materials, University of Oxford, Parks Road, Oxford, OX1 3PH, UK.
* Present address: MTA MFKI, Budapest, Hungary

ABSTRACT: An outline is given of a method for high resolution BSE analysis of chemical composition across interfaces in bulk semiconductor specimens. In this method the effects of electron beam diameter and beam spreading, which normally limit spatial resolution, are largely removed by a deconvolution technique in which these effects are measured experimentally. Line profiles giving 10-90% interface widths as low as 19nm have been obtained at 30kV from an abrupt interface.

1. INTRODUCTION

The basis of chemical composition analysis using back scattered electrons (BSE's) in an SEM is that the electron back scatter coefficient η increases monotonically with the mean atomic number of the material investigated. Thus, providing that η is accurately calibrated, chemical analysis can be performed on compounds where the composition can be described using only one variable parameter. Examples of such semiconductor compounds are $Cd_xHg_{1-x}Te$, $Ga_xAl_{1-x}As$ and $Ga_xIn_{1-x}As$ where x is the variable parameter. The change in η produced can be large and hence easily measured. For example the ratio of $\eta(HgTe)$ to $\eta(CdTe)$ is ~ 1.18 at 30kV and hence a BSE line profile across a specimen with such an interface perpendicular to the surface will give ~ 18% contrast. When these profiles are normalised to the back scatter coefficient of materials with known composition, they give what is effectively an atomic percent line profile across the interface (Hill 1985, Lyster and Booker 1987) and can, for example, be used to study the interdiffusion of the layers.

It is more common to use X-ray analysis in the SEM for composition investigations, however the BSE technique has several advantages which favour its use when only one compositional variable is present. These are: 1) The BSE signal is much larger because BSE yield and detection efficiency are both much larger than for X-rays. This allows much faster collection of data for a given signal to noise ratio in the composition profile. 2) X-rays are only weakly absorbed by the specimen and thus come from the entire interaction volume. BSE's are more strongly absorbed and most therefore come from the surface of the interaction volume where it has a smaller diameter. Thus better spatial resolution is obtained using BSE's for a given beam energy. 3) The BSE signal can be generated at lower accelerating voltages than for X-ray techniques for which the electron beam energy must be greater than the X-ray photon energy. Hence very low accelerating voltages are possible with BSE analysis which can minimise beam spreading effects.

2. FACTORS AFFECTING THE SPATIAL RESOLUTION OF BSE ANALYSIS OF INTERFACES

The BSE signal is very sensitive to surface topography and so careful preparation techniques are needed to produce a smooth flat surface with a mirror finish. If such a surface is not obtained surface topography features which affect the BSE signal can be confused with the changes in chemical composition which are to be measured. Once such a smooth flat specimen has been produced the measured composition profile across an interface depends not only on the actual composition but also on various features of the microscope analysis system. This is because the back scattered signal used for the analysis of the specimen does not originate from a probe of infinitely small size but rather from a region of finite size from which the composition signal obtained is an average. The finite size of the probe arises from two main reasons. One is the finite size of the electron beam incident on the specimen surface for which the electron intensity varies as a function of position in a way that is sometimes assumed to be Gaussian (Michael and Williams 1987). The other is that the beam incident on the specimen surface spreads out as it loses energy in the interaction volume from which the BSE signal originates. This beam spreading increases with increasing electron accelerating voltages as the electrons penetrate further into the specimen. In addition the effect that beam spreading has on the measured composition profile is further complicated by the characteristics and geometry of the electron detector. This is because many detectors have some threshold energy below which electrons are not detected and above which they produce a signal which depends on the electron energy. Since the energy spectrum of the electrons leaving the specimen depends on the region of the interaction volume in which they were generated, the characteristics of the detector alter the effective sensitivity of the system to the position in the interaction volume at which they were produced. Clearly to optimize the spatial resolution of the BSE (or X-ray) analysis the above factors should be minimised. This is often achieved in BSE analysis by reducing the accelerating voltage to reduce beam spreading effects until the electron beam diameter, which for given beam current increases with decreasing accelerating voltage, is of comparable size to the beam spreading in the specimen. Thus in this situation both factors contribute significantly to the spatial resolution achieved. Another approach, which is followed in this work, is to obtain a quantitative measure of the beam spreading and beam size effects and then use these to numerically calculate the actual chemical composition profile from the experimentally obtained profile. This approach uses deconvolution.

3. THE DECONVOLUTION METHOD

The deconvolution method (see for example Riley 1974) can be understood in terms of the actual composition distribution in a specimen, represented by $f(x)$ which is then measured by a system with a resolution function $g(y)$ to give an observed composition distribution $h(z)$. Here the symbols x, y, z all represent length but are denoted differently because the variable appears in different roles. Mathematically the process of obtaining the measured distribution from the true distribution can be written as:

$$h(z) = \int_{-\infty}^{\infty} dx f(x).g(z-x) \tag{1}$$

or $h(z) = f*g$, where $f*g$ is the convolution of f and g. It can also be shown that $H(\omega) = F(\omega).G(\omega)$ where H,F,G are the Fourier transforms of h,f,g respectively and w is the spatial frequency. Hence provided the resolution function of the system g is known, and its transform can be obtained, then the true composition distribution f can be found from the experimentally observed distribution h by deconvolution as follows:

$$f = \left(\frac{\hat{H}}{G}\right) \tag{2}$$

where $\hat{}$ denotes the inverse Fourier transform. Thus in the case of a real experimental situation the system resolution function g must first be found in order to obtain the true distribution function f.

4. MEASUREMENT OF THE SYSTEM RESOLUTION FUNCTION

In the case of BSE analysis of interfaces the system resolution function to be used in the deconvolution of the experimentally obtained data includes the effects of 1) the electron beam intensity/distance distribution, 2) spreading effects in the specimen, 3) the electron trajectories out of the specimen and 4) the detector geometry and characteristics. There are two possible approaches to finding this function. One is to perform BSE Monte Carlo calculations with the aim of simulating numerically the electron beam specimen interaction and this has been suggested by Newbury and Micklebust (1984). However it is not yet clear how accurately such simulations are able to describe the actual processes taking place and so far the effects of the detector geometry and characteristics have not been included in calculations. In addition such an approach cannot take account of the finite, and experimentally unknown, electron beam intensity / distance distribution, which will have a large effect on the system resolution function when a SEM has been optimised with respect to spatial resolution for BSE analysis. The approach used in this work is to measure the system resolution function experimentally by using a "calibration" specimen with an interface between materials of similar composition to the one to be analysed but with known composition profile. The "calibration" specimen is mounted immediately adjacent to the "unknown" specimen and both are polished together. Experimentally measured profiles are obtained from each so that all the factors contributing to the system resolution function are identical. Since the true composition distribution function, f is known for the "calibration" specimen, the method of deconvolution can be used to obtain the system resolution function g:

$$g = \left(\frac{\hat{H}}{F}\right) \tag{3}$$

where h is the measured profile. In this way all parts of the system resolution function are included in its measurement and excellent accuracy is achieved when this function is then used to obtain the actual composition profile from the "unknown" specimen.

In practise we chose a calibration specimen which had a very abrupt interface (<20nm) and then obtained g by taking the derivative of h(z). It can be shown that if f(x) = 0 for x < a and f(x) = 1 for x > a, then dh(z)/dz = g and this process is equivalent to the deconvolution shown by equation 3. It is, however, far less sensitive to noise than using a numerical deconvolution. It should be noted that the procedure outlined here produces an approximation to the true composition distribution, firstly because of the limited accuracy of numerical transform routines but more importantly because the system resolution function includes the effect of beam spreading in the specimen which is itself a function of the composition at that point and is thus not a system constant. However for the range of atomic numbers used here and by ensuring that the "calibration" specimen has a similar composition to the unknown, this was found to be an excellent approximation. We have also made the approximation that the "calibration" specimen with an interface width <20nm is ideally abrupt.

5. EXPERIMENTAL

Measurements were carried out on a JSM35X SEM using a tungsten filament and accelerating voltages in the range 15 - 35kV. All measurements were made with a beam current of 4x10-10A. An objective aperture of 200mm and a 15mm working distance were used. The back scattered electron detector was of the annular scintillator type. The BSE signal was digitised using a 14 bit analogue to digital convertor and each line scan contained 512 points. Data collection for rapid observation took ~3 seconds and low noise traces obtained by averaging many superimposed line scans which were suitable for numerical processing took 3 minutes. Numerical processing used the Fast Fourier Transform routine (Press et al. 1988).

The "calibration" specimen was HgTe grown on a CdTe substrate using a photo assisted MOVPE process to give an abrupt interface <20nm across. Other specimens examined were $Cd_xHg_{1-x}Te$ with x typically 0.2 grown on a CdTe substrate using LPE, these specimens possessing non-abrupt interfaces due to interdiffusion.

6. RESULTS

The effectiveness of the deconvolution technique was initially assessed by taking line scans at different positions on an abrupt interface and then using one such line scan to obtain the system resolution function and hence to deconvolute the others. This procedure was repeated for different accelerating voltages and the results are presented in Table 1. The unprocessed data shows 10-90% interface widths, mainly determined by the probe diameter and beam spreading effects, varying between 121 and 71nm. After deconvolution, widths as small as 27 and 19nm are obtained which are limited by the approximations in the technique. The higher values obtained at 20 and 15kV are limited by noise present in the unprocesed data. Figure 1 shows a typical processed line profile across an abrupt interface for 35kV.

Figures 2 and 3 show the usefulness of the technique for studying processes at interfaces such as diffusion. Figure 2 shows, superimposed, unprocessed line scans taken at two different accelerting voltages across a diffuse interface. The actual chemical composition across the interface is obscured by incident beam effects and also by the "dip and bump" effect (Lyster and Booker 1987) which is caused by the different trajectories of electrons across the interface whilst leaving the specimen. Figure 3 shows the deconvoluted data which is now independent of the accelerating voltage and reveals the compositional profile of the diffusion that has taken pace at the interface.

REFERENCES

Hill M 1985 Inst. Phys. Conf. Ser. 76 417
Lyster M and Booker GR 1987 Inst. Phys. Conf. Ser. 90 197
Michael J and Williams D 1987 J. Microscopy 147 289
Newbury D and Micklebust R 1984 in Electron Beam Interactions with Solids Kyser D, Newbury D,
    Niedrig H, Shimizu R (eds) SEM, Inc AMF O'Hare, IL60666 153
Press W, Flannery B, Teukolsky S, Vetterling W 1988 Numerical Recipes in C (Cambridge
    University Press)
Riley K 1974 Mathematical Methods for the Physical Sciences (Cambridge University Press)

ACKNOWLEDGEMENTS

The authors would like to thank M. Astles at RSRE Malvern, UK for the kind provision of samples.

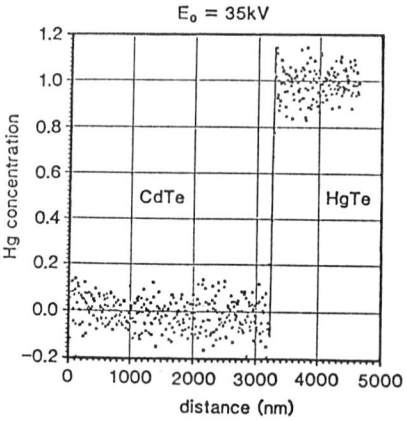

Figure 1. Deconvoluted profile for an abrupt interface showing a 10-90% width of ~30nm at 35kV.

E_0 (kV)	Unprocessed interface width	Deconvoluted interface width
35	121 nm	27 nm
30	113	19
25	109	19
20	83	49
15	71	45

Table 1. BSE analysis of an abrupt HgTe/CdTe interface. Measured width 10-90%

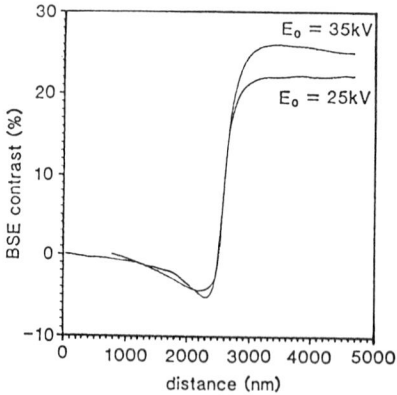

Figure 2. Line profiles across a diffuse CdTe/Cd.1Hg.9Te interface at 25 and 35kV.

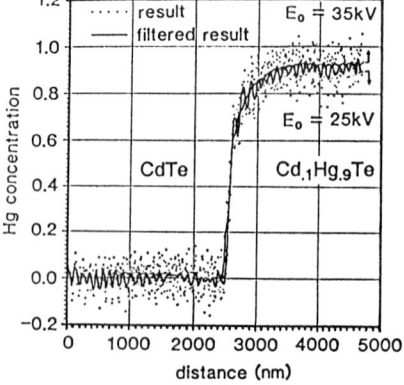

Figure 3. Deconvoluted profiles from fig. 2 Note that the profile is now independent of accelerating voltage and interdiffusion is revealed

Inst. Phys. Conf. Ser. No 117: Section 10
Paper presented at Microsc. Semicond. Mater. Conf., Oxford, 25−28 March 1991

785

New applications of the scanning infra-red microscope (SIRM) to inhomogeneities in bulk GaAs and Si

Z Laczik, P Török, G R Booker and R Falster[*]

University of Oxford, Department of Materials, Parks Road, Oxford OX1 3PH, UK
[*]MEMC Electronic Materials, Viale Gherzi 31, 28100 Novara, Italy

ABSTRACT: We have improved the performance and range of application of our scanning infra-red microscope (SIRM) for the investigation of inhomogeneities in bulk semiconductor materials, and two new applications are now described. First, the SIRM was used in the transmission polarized light mode to image strain fields associated with individual dislocations in LEC GaAs, and Burgers vectors were directly determined. Second, the SIRM was used in the transmission confocal mode so as to reduce the depth of field, and this enabled individual oxide particles of number density 5×10^9 cm^{-3} present in Cz Si to be directly observed without image overlap.

1. INTRODUCTION

We have used the scanning infra-red microscope (SIRM) during the past few years to image precipitate particles present in bulk semiconductor slabs and slices (Booker et al 1991). A new high performance SIRM and two new applications to semiconductors are now described.

In general, in a scanning optical microscope the light of a laser is brought to a focus by a probe-forming lens (usually a standard optical microscope objective lens) either at the surface of, or within, the specimen. Either the light beam or the specimen is then raster scanned and the transmitted or reflected/scattered light is detected by photodiodes. An image is built up pixel-by-pixel from the detector signal.

In our present SIRM the light of a high-brightness, low-noise semiconductor laser is focussed into a spot 1.5-2 μm across within the semiconductor specimen by a high numerical aperture (NA $\geq$ 0.85) lens. The wavelength of the laser is 1300 nm. The transmitted and/or scattered light is detected by Ge photodiodes or by a fibre-optic coupled low-noise/high-sensitivity Ge PIN photodiode cooled to liquid N_2 temperature. The specimen is mechanically scanned and a raster-image is built up in the control computer memory. Position sensors built into the scanning stages allow the use of feed-back electronics which enable a high scanning speed and a large maximum field of view. Slabs and slices up to 3 mm thick and wafers up to 150 mm in diameter can be examined.

2. POLARIZED MODE (PM) SIRM

Polarized light can be used with the SIRM in the transmission mode (by adding a polarizer/analyser to the system) so as to image local strain fields present within the specimens (Figure 1). Imaging the strain fields around individual dislocations is of particular interest because it enables the Burgers vectors of the dislocations to be determined. The mechanisms responsible for the contrast in this mode of the SIRM are similar to those in the polarizing mode of conventional transmission light microscopes, which have been described in detail (Booyens and Basson 1980a and b). The procedure using the SIRM is illustrated here for dislocations running in the [001] direction down the core of LEC In-doped GaAs ingot material.

By X-ray topography it was shown (Brozel et al 1986) that these core dislocations were of edge type with Burgers vectors perpendicular to the [001] growth direction and TEM evidence (Ono 1990) suggested that the

Burgers vectors were either a<100> or ½a<110>. Infra-red microscope (Clark et al 1985) and SIRM (Kidd et al 1987) investigations provided detailed information concerning the geometry and particle decoration of these dislocations. However, the Burgers vectors for these dislocations have not yet been conclusively determined.

Figure 2a shows (001) plan-view SIRM images of four end-on dislocations with 'parallel' polarizer and analyser set along the [110]

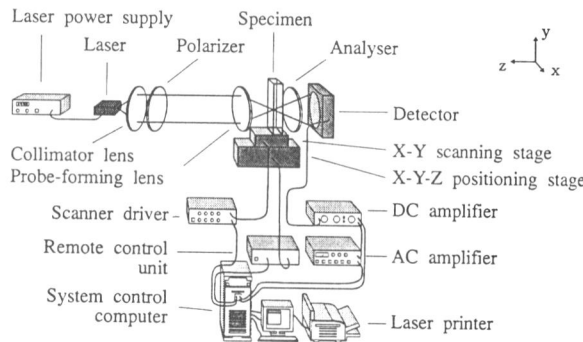

Figure 1 PM (Polarized Mode) SIRM system diagram.

direction. The individual dislocations appear as dark spots. The contrast in this case is due to light scattering by particles decorating the dislocations. Figures 2b to 2e show PM SIRM images corresponding to the same area, but with 'crossed' polarizer/analyser set at different orientations (illustrated by the column between Figures 2 and 3). The individual dislocations exhibit strong contrast, revealing the associated strain fields up to ≈50 μm from the dislocation. As the polarizer/analyser pair is rotated, the contrast changes in a consistent manner from the four-lobe to the two-lobe characteristic forms (see schematic diagrams on the left hand side

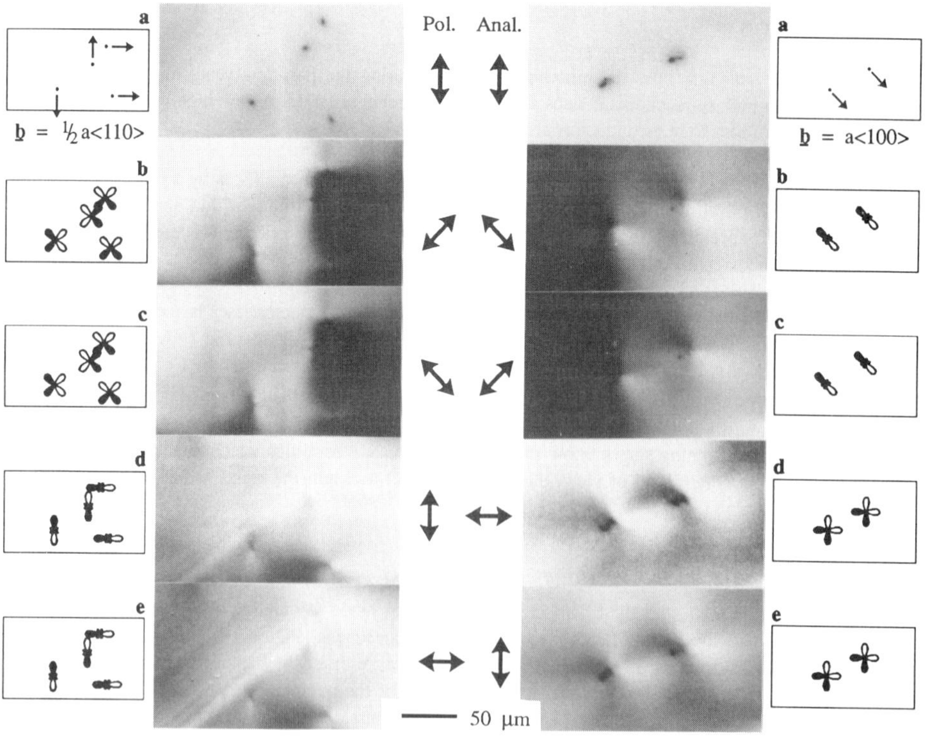

Figure 2 (001) plan-view PM SIRM images of four end-on dislocations in LEC GaAs with ½a<110> Burgers vectors of three different orientations.

Figure 3 (001) plan-view PM SIRM images of two end-on dislocations in LEC GaAs with a<100> Burgers vectors of one orientation. (Same specimen as for Figure 2, but different area.)

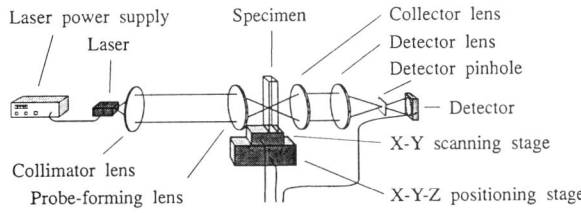

Figure 4 CM (Confocal Mode) SIRM system diagram.

of Figures 2b to 2e). Comparison with the calculated PM images for such dislocations (Booyens and Basson 1980a and b) shows that this type of contrast is characteristic of edge dislocations and that the Burgers vector is parallel to the long axis of the two-lobe contrast, which for these four dislocations is along the <110> directions, as illustrated by the left hand side of Figure 2a.

Figure 3 is similar to Figure 2 except that it shows two dislocations from a different area of the specimen. The Burgers vectors for these two dislocations lie along one of the <100> directions.

For the first time the Burgers vectors for these core dislocations were conclusively determined. Dislocations were identified with both a<100> and $\frac{1}{2}$a<110> type Burgers vectors, with two different directions and two different senses for each type.

3. CONFOCAL MODE (CM) SIRM

The depth of focus for the SIRM operating in the standard transmission mode is 15-50 μm, depending primarily on the probe-forming lens NA and the laser wavelength. For some applications, eg. when precipitate particles with very high number densities (>10^9 cm^{-3}) are to be imaged using scattering contrast, improved depth resolution is necessary to avoid overlapping particle images.

To decrease the depth of field, the SIRM was used in the confocal arrangement (Figure 4). In the confocal mode (CM), the transmitted light is focussed by collector and detector lenses onto a pin-hole in a metal disc, and only light passing through the pinhole is collected by the detector. The optics are aligned so that the image of the pin-hole and the image of the laser light source are confocal in the specimen. Under these conditions only light from the focal region of the probe-forming lens will pass through the pin-hole, and the depth resolution will be improved by the blocking of light from out-of-focus positions. This technique also improves the spatial resolution of the system (Wilson and Sheppard 1984). The procedure is illustrated here for oxide precipitate particles present in heat-treated Cz Si.

Figure 5 shows a focal series for a Cz Si specimen containing oxide particles with a number density of 5 x 10^7 cm^{-3}. The images were recorded with bright-field (BF) transmission CM SIRM using a 0.85 NA probe-forming lens and a 10 μm diameter pin-hole. A 'small' oxide particle (A) is in medium contrast in Figure 5b, in weak contrast in Figures 5c and is not observed in Figure 5a. Two closely spaced 'large'

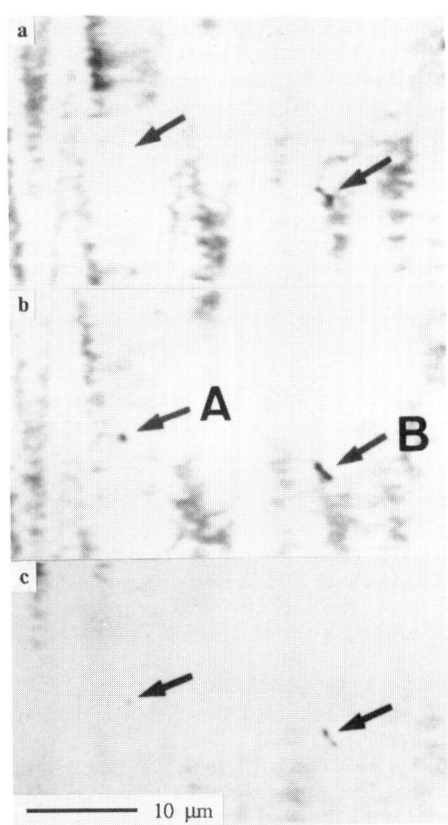

Figure 5 Focal series of CM SIRM images of Cz Si specimen containing oxide particles. The distance within the specimen between individual images was $\Delta z = 3.5$ μm.

oxide particles (B) are in strong contrast in Figure 5b. The upper of these particles is in medium contrast in both Figures 5a and c, while the lower particle is in strong contrast in Figure 5a and weak contrast in Figure 5c. On average, such particles were only visible in two consecutive images, which correspond to a depth of field dz of $7\,\mu m$ in the specimen. Experimental values for the depth of field, obtained by using two different probe-forming objective lenses and a range of pin-hole sizes, are shown in Figure 6. When using a 0.40 NA probe-forming lens, the introduction of a pin-hole in front of the detector reduced the depth of field from $50\,\mu m$ to $22\,\mu m$. In the case of the 0.85 NA probe-forming lens, the improvement was from $15\,\mu m$ to $7\,\mu m$.

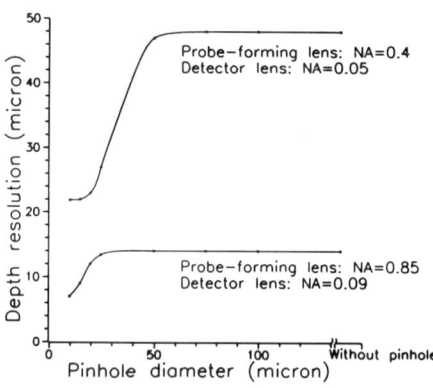

Figure 6 Experimental results for depth resolution vs. pinhole size for transmission CM SIRM.

Figure 7 shows an image of a Cz Si specimen containing oxide particles with a high number density (5×10^9 cm^{-3}). The imaging conditions were the same as for Figure 5. When in focus, the oxide particles appear as dark spots $1\,\mu m$ across (e.g. marked spot in Figure 5). Particles close to, but outside the in-focus region, are only visible with significantly reduced contrast, while particles still further away are not visible. This particle density would not be resolved in the standard mode SIRM because of extensive particle image overlap.

For the first time precipitate particles with very high number densities in semiconductor specimens were imaged by the SIRM in the BF transmission mode. Quantitative data concerning the depth of field and spatial resolution were obtained for the BF transmission confocal mode. A detailed comparison of these experimental data and imaging theory data will soon be published.

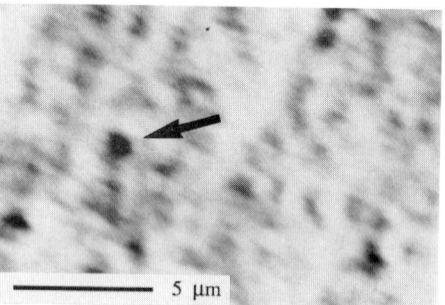

Figure 7 CM SIRM image of Cz Si specimen containing oxide particles. The number density of the oxide particles is 5×10^9 cm^{-3}.

ACKNOWLEDGEMENTS

We are grateful to Professor Sir Peter Hirsch for provision of laboratory facilities and to the SERC and MEMC Electronic Materials for financial support for ZL and PT respectively. We are also thankful to D J Stirland and P Kidd for helpful discussions.

REFERENCES

Booker G R, Laczik Z J, Kidd P, 1991, Proc. of Conference 'Defect Recognition and Imaging in Semiconductors Before and After Processing (DRIP4)' , Wilmslow, UK, Semiconductor Sci. and Technology (in press)

Booyens H and Basson J H, 1980a, J. Appl. Phys. 51 4368

Booyens H and Basson J H, 1980b, J. Appl. Phys. 51 4375

Brozel M R, Clark S and Stirland D J, 1986, 'SI III-V Materials', Nakone, Ohmsha, p133

Clark S, Brozel M R and Stirland D J, 1985, Proc. of Conference 'Defect Recognition and Image Processing in III-V Compounds (DRIP1)', Montpellier, France, p201

Kidd P, Booker G R and Stirland D J,1987, Appl. Phys. Lett. 51 1331

Ono H, 1990, J. Crystal Growth 102 949

Wilson T and Sheppard C, 1984, 'Theory and Practice of Scanning Optical Microscopy', Academic Press

Author Index

Subject Index*

*Page numbers refer to the first pages of the papers in which the citations appear